Handbook of NON-INVASIVE METHODS and the SKIN

Jorgen Serup
G.B.E. Jemec

CRC Press
Boca Raton   Ann Arbor   London   Tokyo

Library of Congress Cataloging-in-Publication Data

Catolog record is available from the Library of Congress.

This book contains information obtained from authentic and highly regarded sources. Reprinted material is quoted with permission, and sources are indicated. A wide variety of references are listed. Reasonable efforts have been made to publish reliable data and information, but the author and the publisher cannot assume responsibility for the validity of all materials or for the consequences of their use.

Neither this book nor any part may be reproduced or transmitted in any form or by any means, electronic or mechanical, including photocopying, microfilming, and recording, or by any information storage or retrieval system, without prior permission in writing from the publisher.

All rights reserved. Authorization to photocopy items for internal or personal use, or the personal or internal use of specific clients, may be granted by CRC Press, Inc., provided that $.50 per page photocopied is paid directly to Copyright Clearance Center, 27 Congress Street, Salem, MA 01970 USA. The fee code for users of the Transactional Reporting Service is ISBN 0-8493-4453-0/95/$0.00+$.50. The fee is subject to change without notice. For organizations that have been granted a photocopy license by the CCC, a separate system of payment has been arranged.

CRC Press, Inc.'s consent does not extend to copying for general distribution, for promotion, for creating new works, or for resale. Specific permission must be obtained in writing from CRC Press for such copying.

Direct all inquiries to CRC Press, Inc., 2000 Corporate Blvd., N.W., Boca Raton, Florida 33431.

© 1995 by CRC Press, Inc.

No claim to original U.S. Government works
International Standard Book Number 0-8493-4453-0
Printed in the United States of America 1 2 3 4 5 6 7 8 9 0
Printed on acid-free paper

Professor Albert M. Kligman, M.D., Ph.D.

Preface

This is the first complete handbook covering the field of noninvasive biophysical methods in clinical and experimental dermatology. It is the result of the coordinated efforts of the most prominent experts in many different fields. We have tried to cover all the major methods systematically and to present chapters of high educational value not only to young researchers just starting studies with noninvasive techniques but also to senior researchers with more experience.

Bioengineering and the skin and, more recently, also digital imaging techniques have been in a tremendous phase of expansion, improvement, and acceptance. It is no longer possible to be a master of every available technique.

There may exist a gap between the medical biologist and the medical physicist who as pure scientists are seekers of *know—why* on one side, and the clinical research workers concerned with the development of clinical medicine as seekers of *know—how* on the other side. A third but equally important group consists of medical practitioners and technicians who use the techniques in the daily clinical or experimental routine, and require the methods to be helpful in a cost efficient way. This book was edited to assist all three groups whether they are working with pure science, technology, or applied technique. It is our hope that a consensus of soundness and collaboration among users of noninvasive techniques at any level may be helped by this book.

The noninvasive methods are in no way competitors to clinical dermatology and dermatohistopathology as it is practiced today. The methods offer quantitative and supplementary information and an opportunity for *in vivo* assessment and in-depth exploitation. Usefulness to the patient or the user remains the ultimate endpoint.

The International Society for Bioengineering and the Skin was founded in 1979, the International Society for Skin Imaging (formerly the International Society for Ultrasound and the Skin) in 1991, and the International Society for Digital Imaging and the Skin in 1992. In 1995, a new journal, *Skin Research and Technology,* will be published as the formal organ of all three societies and, hopefully, the future forum for researchers in all continents (Editors: J. Serup, Denmark, and A. Zemtsov, U.S.).

The period of bad and good but inevitably cumbersome prototypes is over and we are privileged to have a great number of advanced and precise commercial equipment available. At the moment, we are in a phase of continuing validation, standardization, application, and formal acceptance by legal and medical authorities.

The demands for the near future are mainly educational. Formal teaching and training of young dermatologists and researchers in the noninvasive techniques and the understanding of *in vivo* skin structure and physiology is as limited as the subject is complex. This handbook on noninvasive methods was edited and directed to meet this challenge.

Tradition was always a heavy force in medicine, and courage, skills, and hard work were always needed to change the given situation. The field of noninvasive techniques and the skin owes everything to Professor Albert M. Kligman of the University of Pennsylvania for his contributions to modern dermatology, the technologies and the mind of dermatologists. He created the background philosophy needed for the progress of the methods and their critical evaluation, and made blind dermatologists see. Professor Ronald Marks of University of Cardiff, Wales, also did tremendous work with the development of prototypes and inspiration of young researchers during the fragile period of skepticism at the very beginning.

It is impossible to mention all the people who contributed to the development of the field, but you will find many of them among the authors of chapters in this book.

The editors wish to thank not only the authors for their generosity and support but also CRC Press for the work done to create this book.

The authors finally wish to express their gratitude for assistance and support from the Gerda and Aage Haensch Foundation, Denmark and the Leo Pharmaceutical Products, Ltd., Denmark.

Jørgen Serup, M.D., Ph.D.

Associate Professor of Dermatology
University of Copenhagen

Head, Department of Dermatological Research
Leo Pharmaceutical Products

Preface

Noninvasive biophysical methods for quantification of biological phenomena have come a long way from the early phase of one paper—one method. With the appearance of commercial and therefore industrially standardized biophysical tools, the methods have entered a phase of validation and increasing acceptance. But what is their practical role in research or clinical work? Medical knowledge in general is growing exponentially, and reliable information is now available on details of the many interdependent biochemical and physical events which combine into the living organisms we know. This knowledge is extensive, detailed, and most likely incomplete. In principle, biological knowledge is gathered in logical positivist experiments, in which the reactions of significant parameters are studied and related to the existing body of knowledge. Medicine is a physical science in all senses of the word. The biophysical methods offer the advantage of studying the effects of large numbers of biochemical, genetic, and other parameters simultaneously by quantification of biologically significant reactions, e.g., the abstract measurement of inflammation with laser-doppler (which measures blood flux) can give reliable information about magnitude and extent of inflammation without detailed knowledge of the many complex cellular and subcellular mechanisms involved. In a situation where detailed knowledge of all the involved mechanisms is not absolute or universal, such combined measurements offer great advantages for the study of practical problems. Although helpful in the study of biological reactions on the level of an individual or an organ, such an approach obviously does not explain detailed cellular mechanisms, and therefore cannot replace but only supplement other techniques in the quest for truth.

It is our hope that a similar synthesis of many different elements has been achieved in this book. Elements of pure technological science as well as pure clinical applications have been included, and while the former can inspire further development in this field of research, the latter is intended to help with the practical solution of practical problems. It is a handbook and should as such be of immediate use not only to those involved in clinical studies of pathophysiology but also to clinicians seeking new and reliable ways to monitor the progress of treatment.

Finally, we would like to thank the 110 experts who have contributed chapters, without whom this synthesis would not be possible, for their enthusiasm and efforts.

Gregor B.E. Jemec, M.D.

Assistant Professor of Dermatology
University of Copenhagen

The Editors

Jørgen Serup, M.D., Ph.D., is Associate Professor of Dermatology, Bispebjerg Hospital (Bioengineering and Skin Research Laboratory), University of Copenhagen, and the Head of the Department of Dermatological Research, Leo Pharmaceutical Products, Ballerup, Copenhagen, Denmark. Dr. Serup graduated from the University of Copenhagen in 1971 and received his research and clinical training at the Dermatological Departments of Rigshospital, Gentofte Hospital, and Bispebjerg Hospital. He is specialized in internal medicine as well as dermatology. In 1991 he became affiliated with Leo Pharmaceutical Products where he is responsible for dermatological development.

Dr. Serup is the founder and chairman of the Standardization Group on Noninvasive Methods of the European Society of Contact Dermatitis and a board member of the International Society for Bioengineering and the Skin, The International Society for Digital Imaging and the Skin, and the International Society for Skin Imaging. He is a member of the American Academy of Dermatology and the European Academy of Dermatology and Venereology (EADV). He served as an Associate Editor of *Skin Pharmacology* and is now Editor-in-Chief of *Skin Research and Technology,* official organ of the three societies.

Dr. Serup is among the pioneers in the development of ultrasound examination of skin, and the bioengineering and skin research laboratory established by him has contributed to the development, validation, and application of a number of other bioengineering techniques. The laboratory is a center of academic education and training within this field. He has published 170 papers. His medical dissertation from 1986 was on connective tissue and scleroderma. He has received honorary awards from the Universities of Bari, Italy, Seoul, Korea, and Sendai, Japan.

Gregor B.E. Jemec, M.D. is Senior Registrar and Assistant Professor of Dermatology at Bispebjerg Hospital, Department of Dermatology, University of Copenhagen, Denmark. Graduated in 1984, he worked at the Department of Dermatology, Oxford University, UK, from 1985 to 1986 as an assistant researcher. Dr. Jemec received his clinical and research training at the University of Copenhagen and was fully qualified as a Dermatologist in 1992. He was awarded the Samuel Friedman Scholarship by The Samuel Friedman Foundation in 1983, and the Florey Studentship by Queens College in Oxford, UK, in 1985.

Dr. Jemec was Honorary Secretary of the 3rd Congress of the European Academy for Dermatology and Venereology, and continues his involvement in the activities of the society. He is a member of the European Society of Dermatological Research (ESDR). Dr. Jemec's current major research interest is the study of the mechanical properties of skin, both measuring techniques and clinical applications. He has published more than 35 papers and book chapters.

List of Contributors

Professor P. Agache
Clinique Dermatologique
Centre Hospitalier et Universitaire
F-25030 Besancon, France

Dr. T. Agner
Gentofte Hospital
Department of Dermatology
Niels Andersensvej 65,
DK-2900 Hellerup, Denmark

Professor P. Altmeyer
Universitäts Hautklinik
St. Josef Hospital
Gudrundstrasse 56
D-44791 Bochum, Germany

Dr. T. Auer
Universitäts Hautklinik
St. Josef Hospital
Gudrundstrasse 56
D-44791 Bochum, Germany

S. Aygen, M.Sc.
Institut für Zentrale Analytik und Strukturanalyse
    der Universität
Witten/Herdecke, Germany

Professor J.C. Barbenel, Ph.D.
Bioengineering Unit
University of Strathclyde
106 Rottenrow
Glasgow G4 0NW U.K.

Professor A.O. Barel
Algemeine Biologische Scheikunde
Vrije Universiteit Brussel
Lokaal 6 G 304
Pleinlaan 2
B-1050 Brussel, Belgium

Dr. J.H. Barth
Department of Chemical Pathology and Immunology
General Infirmary at Leeds
Leeds LS1 3EX U.K.

Dr. T. Bauermann
Institut für Zentrale Analytik und Strukturanalyse
    der Universität
Witten/Herdecke, Germany

C. Bay, M.Sc.
Leo Pharmaceutical Products
Mathematical Statistical Department
Industriparken 55
DK-2750 Ballerup, Denmark

Dr. G. Belcaro
Via Vesjucci 65
I-65100 Pescare, Italy

Dr. E. Berardesca
Department of Dermatology
University of Pavia
IRCCS Policlinica S. Matteo
I-270199 Pavia, Italy

Dr. A. Bircher
Department of Dermatology
Kantonsspital
CH-4031 Basel, Switzerland

Dr. P. Bjerring
Department of Dermatology
Marselisborg Hospital
University Hospital of Aarhus
DK-8000 Aarhus, Denmark

Dr. U. Blume-Peytavi
Hautklinik und Poliklinik
Universitätsklinikum Steglitz
Freie Universität Berlin
Hindenburgdamm 30
D-12200 Berlin, Germany

Dr. C.H. Chang
Department of Dermatology
Kaohsiung Medical College Hospital
100 Shih Chuan 1st Road
Kaohsiung 807
Taiwan, R.O.C.

Dr. G.S. Chen
Department of Dermatology
Kaohsiung Medical College Hospital
100 Shih Chuan 1st Road
Kaohsiung 807
Taiwan, R.O.C.

S. Christiansen, M.Sc.
Technical University of Denmark
Institute of Manufacturing Engineering
Building 425
DK-2800 Lyngby, Denmark

Dr. D.C. Christopoulos
B2 Surgical Unit
Hippokrateion Hospital
University of Thessaloniki
Greece

Dr. P. Clarys
Algemeine Biologische Scheikunde
Vrije Universiteit Brussel
Lokaal 6 G 304
Pleinlaan 2
B-1050 Brussel, Belgium

P. Corcuff, M.Sc.
Laboratoires de Recherche Fondamentale L´Oreal
1 Av. Eugene Schueller
Boite Postale 22
F-93601 Aulnay-Sous-Boix Cedex, France

W. Courage, M.Sc.
Courage & Khazaka Electronic GmbH
Mathias-Brüggen-Strasse 91
D-50829 Köln, Germany

Professor W.J. Cunliffe
Department of Dermatology
Leeds General Infirmary
Great George Street
Leeds LS1 3EX U.K.

Dr. R.P.R. Dawber
Department of Dermatology
Churchill Hospital
Headington
Oxford OX3 7LJ U.K.

Professor B.L. Diffey
Medical Physics Department
Dryburn Hospital
Durham DH1 5TW U.K.

Professor S. Dikstein
The Hebrew University of Jerusalem
Unit of Cell Pharmacology
School of Pharmacy
PO Box 12065
Jerusalem 91120 Israel

Dr. K. Dirting
Universitäts Hautklinik
St. Josef Hospital
Gudrunstrasse 56
D-44791 Bochum 1, Germany

P. Dykes, Ph.D.
Department of Dermatology
University of Wales College of Medicine
Heath Park
Cardiff CF4 4XN U.K.

Dr. E. Anne Eady
Department of Microbiology
University of Leeds
Leeds LS2 9JT U.K.

C. Edwards, Ph.D.
Department of Dermatology
University of Wales College of Medicine
Heath Park
Cardiff CF4 4XN U.K.

J. Efsen, M.Sc.
Technical University of Denmark
Institute of Manufacturing Engineering
Building 425
DK-2800 Lyngby, Denmark

Dr. C. el-Gammal
Universitäts Hautklinik
St. Josef Hospital
Gudrundstrasse 56
D-44791 Bochum, Germany

Dr. S. el-Gammal
Universitäts Hautklinik
St. Josef Hospital
Gudrundstrasse 56
D-44791 Bochum, Germany

Professor H. Ermert
Institut für Hochfrequenztechnik
Ruhr-Universität Bochum
Universitätsstrasse 150
D-44780 Bochum, Germany

Dr. J. Faergemann
Department of Dermatology
University of Gotenburg
Sahlgren's Hospital
S-413 45 Gothenburg, Sweden

Dr. N. Farinelli
Department of Dermatology
University of Pavia
Policlinica S. Matteo
I-270199 Pavia, Italy

Professor T. Fischer
National Institute of Occupataional Health, IMD
S-171 84 Solna, Sweden

Professor B. Fornage
Department of Diagnostic Radiology
MD Anderson Hospital and Tumor Institute
1515 Holcombe Blvd
Houston TX 77030 U.S.

Professor B. Forslind
EDRG, Medical Biophysics
Department of Medical Biochemistry and Biophysics
Karolinska Institute
S-171 77 Stockholm, Sweden

Professor P. Frosch
Hautklinik der Städtischen Kliniken Dortmund
Beurhausstrasse 40
D-44137 Dortmund, Germany

A. Fullerton, Ph.D
Department of Dermatological Research
Leo Pharmaceutical Products
Industriparken 55
DK-2750 Ballerup, Denmark

Dr. J. Gassmueller
BioSkin Institut für Dermatologische Forschung
  und Entwicklung GmbH
Poppenbütteler Bogen 25
D-22399 Hamburg, Germany

Dr. M. Gniadecka
Department of Dermatology
Warsaw Medical School
ul. Koszykowq 82A
02786 Warszawa, Poland

Dr. R. Gniadecki
Dept. of Endocrinology
Warsaw Medical School
ul. Banacha 1A
02786 Warszawa, Poland

Dr. C.L. Goh
National Skin Centre
1 Mandalay Road
Singapore 1130 Singapore

H.N. Hansen, M.Sc.
Technical University of Denmark
Institute of Manufacturing Engineering
Building 425
DK-2800 Lyngby, Denmark

C.W. Hargens, M.Sc.
1006 Preston Road
Philadelphia, PA 19118 U.S.

Dr. R. Hartwig
Universitäts Hautklinik
St. Josef Hospital
Gudrunstrasse 56
D-44791 Bochum, Germany

Dr. A. Hoffmann
Universitäts Hautklinik
St. Josef Hospital
Gudrunstrasse 56
D-44791 Bochum, Germany

Dr. K. Hoffmann
Universitäts Hautklinik
St. Josef Hospital
Gudrundstrasse 56
D-44791 Bochum, Germany

Dr. P. Holstein
Department of Vascular Surgery
Bispebjerg Hospital
University of Copenhagen
Bispebjerg Bakke 23
2400 Kbh. NV., Denmark

Dr. T. Horio
Department of Dermatology
Kansai Medical University
Fumizono-cho 1
Moriguchi
Osaka 570, Japan

Professor E.F. Hvidberg
Himmeltoften 9A
DK-2830 Virum, Denmark

P. Jahn, M.Sc.
Schering AG
Diagnostika Koordination
Müllerstrasse 171-178
D-13353 Berlin, Germany

Dr. G.B.E. Jemec
Department of Dermatology
Bispebjerg Hospital
University of Copenhagen
Bispebjerg Bakke 23
DK-2400 Kbh. N, Denmark

Professor H. Irving Katz
Minnesota Clinical Study Center
7205 University Avenue N.E.
Fridley, MN 55432 U.S.

Dr. A. Kecskes
Schering AG
Dermatologie/Humanpharmakologie
Müllerstrasse 171-178,
D-13353 Berlin, Germany

J. Keiding, M.Sc.
Leo Pharmaceutical Products
Department of Dermatological Research
Industriparken 55
DK-2750 Ballerup, Denmark

Professor A.M. Kligman
Duhring Laboratories
Department of Dermatology
226 Clinical Research Building
422 Curie Boulevard
Philadelphia, PA 19104-6142 USA

Dr. N. Lange
National Institute of Neurological Disorders and Stroke
Federal Building, Room 7C04
Bethesda, MD 20892 USA

Professor N.A. Lassen
Department of Physiology and Nuclear Medicine
Bispebjerg Hospital
University of Copenhagen
Bispebjerg Bakke 23
2400 Kbh. NV. Denmark

J.L. Leveque, Ph.D.
Laboratoires de Recherche Fondamentale
L'Oreal
1 Av. Eugene Schueller
Boite Postale 22
F-93601 Aulnay-Sous-Boix, Cedex, France

Dr. J.S. Lindholm
Minnesota Clinical Study Center
7205 University Avenue N.E.
Fridley, MN 55432 U.S.

Dr. R.A. Logan
Department of Dermatology
Bridgend General Hospital
Quarella Road, Bridgend,
Mid Glamorgan, CF31 1JP U.K.

Professor H.I. Maibach
Department of Dermatology
University of California School of Medicine
Box 0989, Suite 110
San Francisco, CA 94143-0989 U.S.

V. Manny-Aframian M.Sc.
The Hebrew University of Jerusalem
Unit of Cell Pharmacology
School of Pharmacy
PO Box 12065
Jerusalem 91120, Israel

Professor R. Marks
Department of Dermatology
Uniersity of Wales College of Medicine
Heath Park
Cardiff CF4 4XN U.K.

Dr. A.D. Martin
School of Human Kinetics
University of British Columbia
6081 University Blvd.
Vacouver B.C. VGT 121, Canada

Professor J. Mignot
Laboratoire de Métrologie des Interfaces Techniques
Institut Universitaire de Technique
F-25009 Besancon Cedex, France

Dr. D. Miller
Bionet Inc.
P.O. Box 797686
Dallas, TX 75379 U.S.

Dr. O.H. Mills, Jr.
Hill Top Research, Inc.
223 Highway 18
East Brunswick, NJ 08816 U.S.

Dr. P.S. Mortimer
St. George's Hospital Medical School
Department of Dermatology
Blackshaw Rd.
London SW17 0QT U.K.

S. Møller, M.Sc.
Leo Pharmaceutical Products
Mathematical Statistical Department
Industriparken 55
DK-2750 Ballerup, Denmark

Dr. C. Nickelsen
Department of Obstetrics and Gynecology
Hvidovre Hospital
University of Copenhagen
Kettegårds Alle 30
DK-2650 Hvidovre, Denmark

Professor A.N. Nicolaides
Irvine Laboratory for Cardiovascular
Investigation and Research
St. Mary's Hospital Medical School
London W2 U.K.

Professor G.E. Nilsson
Department of Biomedical Engineering
Linköping University
S-581 85 Linköping, Sweden

Professor C.E. Orfanos
Hautklinik und Poliklinik
Universitätsklinikum Steglitz
Freie Universität Berlin
Hindenburgdamm 30
D-12200 Berlin, Germany

Dr. A. Pagnoni
S.K.I.N. Inc.
151 E. 10th Avenue
Conshohocken, PA 19428 U.S.

C. Passmann, M.Sc.
Institut für Hochfrequenztechnik
Ruhr-Universität Bochum
Universitätsstrasse 150
D-44780 Bochum, Germany

Dr. D.A. Perednia,
Department of Dermatology
Oregon Health Sciences University
3181 S.W. Sam Jackson Park Rd.
Portland, OR 97201-3098 U.S.

Dr. L.J. Petersen
Department of Dermatology
University of Copenhagen
Bispebjerg Hospital
Bispebjerg Bakke 23
DK-2400 Copenhagen, Denmark

Professor G. Pierard
Laboratory of Dermometrology
CHU du Sart Tilman
University of Liège
B-4000 Liège, Belgium

Dr. B. Pilz
Hautklinik der Städtischen Kliniken Dortmund
Beurhausstrasse 40
D-44137 Dortmund, Germany

Dr. J. Pinnagoda
Occupational Health and Environmental Safety Division
Southeast Asia Region
3M Singapore Pte Ltd.
9 Tajora Lane
Singapore 2678 Singapore

B. Querleux, Ph.D
Laboratoires de Recherche Fondamentale L'Oreal
1 Av. Eugene Schueller
Boite Postale 22
F-93601 Aulnay-Sous-Boix, Cedex, France

Professor E.F.J. Ring
Royal National Hospital for Rheumatic Diseases
Upper Borough Walls
Bath BA1 1RL U.K.

Dr. J.S. Roth
Department of Dermatology
College of Physicians and Surgeons of Columbia University
630 West 168th Street
New York, NY U.S.

Dr. D. Hugh Rushton
School of Pharmacy & Biomedical Sciences
University of Portsmouth
Portsmouth, U.K.

Dr. A. Röchling
Universitäts Hautklinik
St. Josef Hospital
Gudrunstrasse 56
D-44791 Bochum, Germany

Dr. S. Sarin
Department of Surgery
University College and Middlesex School of Medicine
The Middlesex Hospital
Mortimor Street
London, W1N 8AA U.K.

Dr. H. Schatz
Universitäts Hautklinik
St. Josef Hospital
Gudrundstrasse 56
D-44791 Bochum, Germany

Professor R.K. Scher
Department of Dermatology
College of Physicians and Surgeons of Columbia University
630 West 168th Street
New York, NY U.S.

Professor S. Seidenari
Universita' degli Studi di Modena
Istituto di Clinica Dermatologica
Via del Pozzo 71
I-41100 Modena, Italy

E. Seidenschnur, M.Sc.
Leo Pharmaceutical Products
Department of Regulatory Affairs
Industriparken 55
DK-2750 Ballerup, Denmark

Dr. P. Sejrsen
University of Copenhagen
Institute of Medical Physiology
Building 12.4.7
The Panum Institute
Blegdamsvej 3C
DK-2200 Copenhagen N, Denmark

Dr. J. Serup
Department of Dermatological Research
Leo Pharmaceutical Products
Industriparken 55
DK-2750 Ballerup, Copenhagen, Denmark

Dr. P. Sohl
Nyhavn 45
DK-1051 Copenhagen, Denmark

Dr. M. Stücker
Universitäts Hautklinik
St. Josef Hospital
Gudrunstrasse 56
D-44791 Bochum, Germany

Professor H. Tagami
Department of Dermatology
Tohoku School of Medicine
1-1 Seiryomachi, Aoba-ku
Sendai 980, Japan

Dr. H. Takiwaki
Department of Dermatology
School of Medicine
University of Tokushima
Tokushima 770, Japan

Mr. J.P. Taylor
Department of Dermatology
Leeds General Infirmary
Great George Street
Leeds LS1 3EX U.K.

Dr. R.A. Tupker
Department of Dermatology
State University Hospital
PO Box 30.001
NL-9700 RB Groningen, The Netherlands

Dr. M.A. Weinstock
Brown University Dermatoepidemiology Unit
Veterans Affairs Medical Center - 111D
830 Chalkstone Avenue
Providence, RI 02908 U.S.

Dr. R.C. Wester
Department of Dermatology
University of California School of Medicine
Box 0989, Surge Bldg. Room 110
San Francisco, CA 94143-0989 U.S.

Professor W. Westerhof
Department of Dermatology
Academic Medical Centre
University of Amsterdam
Meibergdreef 9
NL-1105 AZ Amsterdam, The Netherlands

R.R. Wickett, Ph.D
University of Cincinnati
Medical Center
College of Pharmacy
3223 Eden Avenue
Cincinnati, OH 45267-004 U.S.

K. Wårdell, Ph.D
Department of Biomedical Engineering
Linköping University
S-581 85 Linköping, Sweden

Dr. S.A. Yang
Department of Dermatology
Kaohsiung Medical College Hospital
100 Shih-Chuan 1st Road
Kaohsiung 807
Taiwan, R.O.C.

Professor H.S. Yu
Department of Dermatology
Kaohsiung Medical College Hospital
100 Shih-Chuan 1st Road
Kaohsiung 807
Taiwan, R.O.C.

Professor A. Zemtsov
Texas Technical University
Department of Dermatology
School of Medicine
3601 Fourth Street
Lubbock, TX 79430 U.S.

Dr. A. Zlotogorski
Department of Dermatology
Hadassah University Hospital
Jerusalem 91120, Israel

Professor B. Zweiman, MD
University of Pennsylvania
Division of Allergy and Immunology
School of Medicine
Philadelphia, PA 19104-6057 U.S.

Contents

Section A: Methodology and Techniques

1.0 Noninvasive Methodologies in General

1.1 Perspectives on Bioengineering of the Skin 3
   A.M. Kligman

1.2 Relevance, Comparison, and Validation of Techniques 9
   G.E. Piérard

2.0 Variables and Study Design

2.1 Prescription for a Bioengineering Study: Strategy, Standards, and Definitions 17
   J. Serup

2.2 The Skin Integument: Variation Relative to Sex, Age, Race, and Body Region 23
   N. Farinelli and E. Berardesca

2.3 Seasonal Variations and Environmental Influences on the Skin 27
   C.L. Goh

3.0 Technical Variation, Validation, and Statistics

3.1 Statistical Analysis of Sensitivity, Specificity, and Predictive Value of a Diagnostic Test 33
   N. Lange and M.A. Weinstock

3.2 Sample Size Calculation 43
   C. Bay and S. Møller

Section B: Epidermal Structure

4.0 Skin Surface Microscopy

4.1 Magnifying Lens — Noninvasive Oil Immersion Examination of the Skin 49
   H.I. Katz and J.S. Lindholm

4.2 Dermatoscopy 57
   W. Westerhof

4.3 Skin Replication for Light and Scanning Electron Microscopy 73
   B. Forslind

5.0 Skin Surface Contour Evaluation

5.1 Stylus Method for Skin Surface Contour Measurement 83
   J. Gassmüller, A. Kecskes, and P. Jahn

5.2 Skin Surface Replica Image Analysis of Furrows and Wrinkles 89
   P. Corcuff and L.L. Leveque

5.3 Laser Profilometry 97
   J. Efsen, H.N. Hansen, S. Christiansen, and J. Keiding

5.4 Three-Dimensional Evaluation of Skin Surface: Micro- and Macrorelief 107
   J. Mignot

5.5 Skin Surface Biopsy and the Follicular Cast 121
   R.P.R. Dawber

5.6 High-Resolution Ultrasound Examination of the Epidermis 125
   S. el-Gammal, T. Auer, K. Hoffmann, P. Altmeyer, C. Passmann, and H. Ermert

6.0 Nuclear Magnetic Resonance (NMR) Examination of the Epidermis

6.1 Nuclear Magnetic Resonance (NMR) Examination of the Epidermis *in vivo* 133
    B. Querleux

Section C: Epidermal Functions

7.0 Desquamation

7.1 Methods to Determine Desquamation Rate 143
    C. Edwards and R. Marks

7.2 Sticky Slides and Tape Techniques to Harvest Stratum Corneum Material 149
    D.L. Miller

7.3 Dry Skin and Scaling Evaluated by D-Squames and Image Analysis 153
    H. Schatz, P.J. Altmeyer, and A.M. Kligman

8.0 Epidermal Hydration

8.1 Measurement of Electrical Conductance and Impedance 159
    H. Tagami

8.2 Measurement of Epidermal Capacitance 165
    A.O. Barel and P. Clarys

9.0 Epidermal Barrier Functions

9.1 Measurement of the Transepidermal Water Loss 173
    J. Pinnagoda and R.A. Tupker

9.2 Comparison of Methods for Measurement of Transepidermal Water Loss 179
    A.O. Barel and P. Clarys

9.3 Measurement of Transcutaneous Oxygen Tension 185
    H. Takiwaki

9.4 Measurement of Transcutaneous P_{CO_2} 197
    C.N. Nickelsen

9.5 Noninvasive Techniques for Assessment of Skin Penetration and Bioavailability 201
    R.C. Wester and H.I. Maibach

10.0 The Skin Surface Microflora and pH

10.1 Sampling the Bacteria of the Skin 207
     E.A. Eady

10.2 Mapping the Fungi of the Skin 217
     J. Faergemann

10.3 Measurement of Skin Surface pH 223
     A. Zlotogorski and S. Dikstein

Section D: Structure of the Dermis

11.0 Dermatologic Digital Imaging

11.1 Overview of Dermatologic Digital Imaging 229
     D.A. Perednia

12.0 Ultrasound Examination of the Dermis

12.1 High-Frequency Ultrasound Examination of Skin: Introduction and Guide 239
     J. Serup, J. Keiding, A. Fullerton, M. Gniadecka, and R. Gniadecki

12.2 Ultrasound B-Mode Imaging and *in vivo* Structure and Analysis 257
     S. Seidenari

12.3 High-Frequency Sonography of Skin Diseases 269
   K. Hoffmann, A. Röchling, M. Stücker, K. Dirting, S. el-Gammal, A. Hoffmann, and P. Altmeyer

12.4 Ultrasound Examination of the Skin and Subcutaneous Tissues at 7.5 to 10 MHz 279
   B.D. Fornage

12.5 Ultrasound A–Mode Measurement of Skin Thickness 289
   T. Agner

12.6 Skin Thickness: Caliper Measurement and Typical Values 293
   A.D. Martin

13.0 Nuclear Magnetic Resonance (NMR) Examination of the Dermis

13.1 Nuclear Magnetic Resonance (NMR) Examination of the Skin 299
   A. Zemtsov

13.2 Nuclear Magnetic Resonance Examination of Skin Disorders 305
   S. el-Gammal, R. Hartwig, S. Aygen, T. Bauermann, K. Hoffmann, and P. Altmeyer

Section E: Function of the Dermis

14.0 Mechanical Properties of the Skin

14.1 Twistometry Measurement of Skin Elasticity 319
   P.G. Agache

14.2 Suction Chamber Method for Measurement of Skin Mechanical Properties: The Dermaflex® 329
   M. Gniadecka and J. Serup

14.3 Suction Method for Measurement of Skin Mechanical Properties: The Cutometer® 335
   A.O. Barel, W. Courage, and P. Clarys

14.4 Identification of Langer's Lines 341
   J.C. Barbenel

14.5 Levarometry 345
   V. Manny-Aframian and S. Dikstein

14.6 Indentometry 349
   V. Manny-Aframian and S. Dikstein

14.7 The Gas-Bearing Electrodynamometer 353
   C.W. Hargens

14.8 Ballistometry 359
   C.W. Hargens

Section F: The Cutaneous Vasculature

15.0 Visualization of Blood Vessels

15.1 Dynamic Capillaroscopy — A Sensitive Noninvasive Method for the Diagnosis of
   Conditions with Pathological Microcirculation 367
   H.S. Yu, C.H. Chang, G.S. Chen, and S.A. Yang

16.0 Measurement of Erythema and Skin Color

16.1 Spectrophotometric Characterization of Skin Pigments and Skin Color 373
   P. Bjerring

16.2 Measurement of Erythema and Melanin Indices 377
   H. Takiwaki and J. Serup

16.3 CIE Colorimetry 385
   W. Westerhof

17.0 Cutaneous Blood Flow, Vasomotion, and Vascular Functions

17.1 Laser Doppler Measurement of Skin Blood Flux: Variation and Validation 399
A.J. Bircher

17.2 Laser-Doppler Flowmetry: Principles of Technology and Clinical Applications 405
G. Belcaro and A.N. Nicolaides

17.3 Examination of Periodic Fluctuations in Cutaneous Blood Flow 411
R. Gniadecki, M. Gniadecka, and J. Serup

17.4 Laser Doppler Imaging of Skin 421
K. Wårdell and G. Nilsson

17.5 The ^{133}Xenon Wash-Out Technique for Quantitative Measurement of Cutaneous and Subcutaneous Blood Flow Rates 429
P. Sejrsen

17.6 Doppler and Duplex Scanning in Venous Disease 437
G. Belcaro and A.N. Nicolaides

17.7 Photoplethysmography and Light Reflection Rheography: Clinical Applications in Venous Insufficiency 443
G. Belcaro, D. Christopoulos, and A.N. Nicolaides

17.8 Strain Gauge Plethysmography: Diagnosis and Prognosis in Skin Lesions on the Feet and Toes and in Leg Ulcers 449
N.A. Lassen and P. Holstein

17.9 Air Plethysmography in the Evaluation of Skin Microangiopathy Caused by Chronic Venus Hypertension 453
D.C. Christopoulos

18.0 Skin Temperature and Thermoregulation

18.1 Thermal Imaging of Skin Temperature 457
E.F.J. Ring

19.0 Lymph Flow

19.1 Evaluation of Lymph Flow 473
P. Mortimer

Section G: Neural Supply

20.0 Sensory Function

20.1 Assessment of Skin Sensibility 483
P. Bjerring

20.2 Quantification of Cutaneous Pain 489
P. Bjerring

Section H: Sweat Glands

21.0 Sweat Gland Distribution

21.1 Techniques for the Localization of Sweat Glands 497
P. Dykes

22.0 Sweat Gland Activity

22.1 Methods for the Collection of Eccrine Sweat 503
J.H. Barth

22.2 Methods for the Collection of Apocrine Sweat 507
J.H. Barth

Section I: Sebaceous Glands

23.0 Distribution and Follicular Morphology

23.1 The Follicular Biopsy 511
O.H. Mills, Jr.

24.0 Sebum Excretion

24.1 Sebum-Absorbent Tape and Image Analysis 517
C. el-Gammal, S. el-Gammal, A. Pagnoni, and A.M. Kligman

24.2 Gravimetric Technique for Measuring Sebum Excretion Rate (SER) 523
W.J. Cunliffe and J.P. Taylor

Section J: Hair

25.0 Physical Properties of Hair

25.1 Hair Color 531
R.P.R. Dawber

25.2 Measurement of the Mechanical Strength of Hair 535
R.R. Wickett

26.0 Hair Growth

26.1 Measurement of Hair Growth 543
J.H. Barth and D.H. Rushton

26.2 Microscopy of the Hair — the Trichogram 549
U. Blume-Peytavi and C.E. Orfanos

Section K: Nails

27.0 Nail Thickness and Structure

27.1 Measurement of Nail Plate Thickness 557
G.B.E. Jemec

28.0 Nail Growth

28.1 Measurement of Longitudinal Nail Growth 561
J.S. Roth and R.K. Scher

Section L: Clincial Evaluation and Quantification

29.0 Clincial Scoring and Grading

29.1 Standard Schemes to Assess Skin Diseases 567
R.A. Logan

29.2 Clincial Grading of Experimental Skin Reactions 575
T. Agner

30.0 Evaluation of Lesional Extension

30.1 Measurement of Healing and Area of Skin Pathologies 581
S. Sarin

Section M: Experimental Test Procedures

31.0 Standard Testing of Skin Reactivity

31.1 Irritant Patch Test Techniques 587
P.J. Frosch and B. Pilz

31.2 Allergic Patch Test Techniques 593
T. Fischer

31.3 Type I Allergy Skin Tests 607
   L.J. Petersen

31.4 Ultraviolet Radiation Dosimetry 619
   B.L. Diffey

31.5 Phototesting: Phototoxicity and Photoallergy 627
   T. Horio

32.0 Special Experimental Techniques

32.1 Skin Chamber Techniques 633
   B. Zweiman

32.2 Skin Microdialysis 641
   L.J. Petersen

Section N: Legal and Ethical Aspects of Noninvasive Techniques

33.0 Regulatory Aspects

33.1 FDA and EEC Regulations Related to Skin: Documentation and Measuring Devices 653
   E.K. Seidenschnur

34.0 Ethical Aspects

34.1 Ethical Considerations 667
   P. Sohl and G.B.E. Jemec

35.0 Clincial Aspects

35.1 Good Clinical Practice 671
   E.F. Hvidberg

Index 681

Section A: Methodology and Techniques

1.0 Noninvasive Methodologies in General

1.1 Perspectives on Bioengineering of the Skin 3
A.M. Kligman

1.2 Relevance, Comparison, and Validation of Techniques 9
G.E. Piérard

Chapter 1.1
Perspectives on Bioengineering of the Skin

Albert M. Kligman
Department of Dermatology
University of Pennsylvania
Philadelphia, Pennsylvania

I. Overview

Bioengineering of the skin is coming of age as a scientific discipline. The International Society for Bioengineering of the Skin (ISBS) is barely 15 years old but already has an official journal and an international membership which numbers in the hundreds and meets regularly. It has a devoted leadership which spends considerable time and energy to advance the goals of the Society.

A committee on standardization of methodology chaired by Dr. Jorgen Serup is establishing with great diligence rigorous guidelines that will be universally applicable, eliminating controversies that emerge when investigators do not use exactly the same methods.[1,2] Standardization by consensus has already borne fruit in the measurement of transepidermal water loss by evaporimetry[1] and cutaneous blood flow by laser Doppler flowmetry.[2] There are other signs of a flourishing enterprise. The volume of published papers has increased exponentially in one decade. Bioengineering texts are appearing with increasing frequency with a growing number of contributing authors.[3-5] It is timely to take stock of the current posture, purposes, and hopes of the society.

Sometimes the leading practitioners of bioengineering seem to regard "non-invasiveness" as the commanding feature of the discipline. There are outstanding advantages to being able to study living skin in real time without violating its boundaries by surgery or other interventions. Nonetheless, this by itself should not become a religious mantra. It is not the be-all and end-all of bioengineering to replace and supplant time-proved intrusive procedures, such as biopsy. While that goal is approachable, it is probably not achievable and carries with it some mischief of its own. Some high-placed enthusiasts may think it within reach to match the reliability of histologic diagnosis by non-invasive technologies. They believe it possible to characterize the biochemical, cytologic, anatomic, and physiologic features of skin in health and disease. The grandiose ideas may engender dangerous practices. The diagnosis and treatment of malignant melanoma furnishes a telling example of the evils of overstatement. To be sure, the contours of the tumor can be visualized by ultrasound and thus might help in planning surgery. However, prognosis depends on many factors other than shape and volume of the tumor, viz., mitotic rate, degree of anaplasia, infiltration of individual tumor cells beyond the gross margins, antigenic markers, and other attributes which histologists take into consideration using multiple regression analyses.

Every new specialty requires a dedicated cadre of believers to create a curriculum that will attract the attention of the orthodox establishment. Enthusiasm is a necessary propelling force, as is enjoyment of new adventure. However, exaggerations and excessive zeal must be controlled if bioengineering is to gain respectability in the medical establishment, which is a destination to which we all fervently aspire.

On the other hand, ultrasound is an excellent way to follow the therapeutic responses of chronic dermatoses. Selecting the appropriate technique is at least as important as mastering the technology.

As a matter of fact, the Society faces a formidable task in educating dermatologists to understand and appreciate the benefits of bioengineering techniques in the clinical setting. Virtually every medical speciality except dermatology makes extensive use of sonography in the hospital practice of medicine. Sadly, I do not know a single department of dermatology in the U.S. where even a minimum of non-invasive instrumentation is available to the staff on a routine basis. This is deplorable and needs correction. A task force should be employed to inform academicians of the biophysical resources available not only for research but for patient care as well. The bioengineering establishment is too isolated. Two ways to correct this is to publish more articles in mainstream dermatologic journals and to become regular presenters, by lectures and posters, in national meetings. Dermatologists simply have no idea of the extent to which bioengineering methodologies could currently contribute to the study of skin. The potential applications are limitless in every area of the total skin enterprise.

An increasing number of devices are being developed. Still, we already have useful purchasable instruments, such as the evaporimeter to measure transepidermal water loss, Laser-Doppler for blood flow, the Skicon for electrical conductance, ultrasound for A- and B-scans, the gas-bearing dynanometer for viscoelasticity, and suction devices for measuring the mechanical properties of the dermis. Clinical trials of the comparative efficacy of new drugs will become more reliable using these devices.

I predicted 25 years ago that the time would come when a blind medical graduate would not be at a disadvantage if he or she decided to become a dermatologist.[6] "Sight" would come from instruments far more powerful than the eye. Today, there is reason to believe that this is not a preposterous prophecy of a senescent dreamer! Fantastic devices are waiting in the wings.

In my recent essay on "Invisible Dermatology", I tried to make the point that the finger and the eye are archaic instruments on which dermatologists have too long depended, under the sway of a seductive but specious imagery that the skin is a "window" through which one looks into the interior.[7] The fact is that the most important events that take place in the pathogenesis and expression of disease are subclinical and cannot be felt or seen. Healthy-looking skin may be quite abnormal. Appearance is deceiving. Bioengineering can penetrate this barrier and reveal occult events in the complete absence of signs and symptoms. A dozen examples come to mind. I will cite only three. (1) In psoriasis, no reliance can be placed on the clinical assessment of when the lesions have cleared. We know from histologic studies that lesional skin that has apparently healed remains abnormal for many weeks, even months. The disease has simply gone underground. This explains the frequent relapses at the very same sites. Accordingly, treatment should not be abruptly stopped, but gradually tapered, preferably under monitoring by bioengineering techniques. Spectrophotometry, ultrasound, Laser-Doppler velocimetry, and evaporimetry are far more reliable than the eye for establishing when to stop treatment. (2) The reading of patch tests for assessing irritants and allergens is primitive and unreliable. False-negative and false-positive reactions abound. The differentiation of irritant from allergic reactions is vexing, since clinicians cannot give sound advice regarding what chemicals to look for and avoid in the multitudinous world of cosmetics and drugs. It is a woeful fact that only about 50% of positive, putatively allergic reactions are relevant to the patient's problem. Seidenari and di Nardo of Modena, Italy, suggest that ultrasound technology, focusing on amplification of the area of interest, may rescue us from this diagnostic dilemma.[8] This would be a boon for students of contact dermatitis. A limitation in this and other efforts is oversimplification of the problem by relying on sodium lauryl sulfate as the prototype for irritants when in fact different irritants induce vastly different tissue reactions. So, conclusions must be guarded until comprehensive studies of diverse irritants have been completed. (3) There is currently a great interest in assessing the mildness of soaps and cleansers, since the daily use of surfactants over a lifetime may be damaging to skin. The eye cannot distinguish small gradations in erythema and is altogether useless when the reactions are subclinical. Histology may show appreciable pathology when the skin looks normal. Most soap bars contain anionic surfactants which are the principal culprits in soap-induced dry skin. It turns out that anionic injury begins with an assault on the integrity of the horny layer barrier resulting in an increase in transepidermal water loss (TEWL). Evaporimetry using the internationally acclaimed Servomed is the instrument of choice for detecting subtle changes in TEWL, enabling discrimination among soaps that differ only slightly in irritancy potential.[9] Then, too, there are agreeable advantages for the volunteers since sequential measurements after a single 24-hour exposure provide a kinetic plot of the induction and healing phases. The evaporimeter happily makes the Frosch-Kligman 5-day soap chamber test obsolete since the end-point of that procedure involved evocation of an unpleasant dermatitis, often accompanied by fissuring and exudation.[10] Evaporimetry also decides the tissue target of the injury. For example, cationic surfactants are severe irritants which can incite a brisk dermatitis within 24 hours. It is illuminating that this can happen without any increase in TEWL. Thus, bioengineering may be the royal road for understanding pathogenesis of disease states.

Physicians have always favored prophylaxis over treatment. The possibilities here for reducing the ravages of skin disease are nothing short of spectacular. Dermatologists need to be made aware that we already have techniques that enable identification of persons at high risk of persistent dermatoses years before these become clinically manifest. Chronic diseases do not spring up overnight. They often have occult signs which are below naked-eye detection. Internists know that monitoring serum lipids is valuable in preventing cardiovascular disease in predisposed persons. Externists (dermatologists) can be much more effective in the area of prophylaxis by recognizing early sub-clinical stages. A few unpublished examples will suffice. One is early identification of "flusher-blushers", some of whom will go on to develop full-blown inflammatory rosacea. Techniques such as spectrophotometry to monitor vascular responses to vasoactive agents can identify these particular patients, who can then be appropriately advised to avoid triggering factors such as sunlight, species, etc. A second example is identification of the high-risk acne-prone in pre-puberal children, especially girls 8 to 9 years of age. This is done by obtaining pore prints of sebum output by means of a sebum-trapping tape (Sebutape) applied for 1 hour.[11] A few oil-gushers is a harbinger of trouble in children whose parents had scarring acne in adolescence. Prophylactic treatment can then be commenced years before comedones and papulo-pustules develop. It is possible to identify individuals with sub-clinical atopic dermatitis who present only with dry skin. This mimics ordinary winter xerosis and is likely to be missed by clinicians. It is a lot more common than presently appreciated. Sticky slides that remove monolayers of horny cells will often reveal parakeratotic smaller corneocytes that may require changing the diagnosis to atopic dermatitis rather than simple xerosis.

II. The Proliferation of Non-Invasive Instruments: The Problem of Plethora

The devices described in the skin bioengineering literature have largely derived from the advanced fields of material science which have put man on the moon and have been indispensable for quality control in the production of metals, fabrics, etc. These technologies have endless possibilities for adaptation to the study of living tissues, especially the accessible skin. At the same time, there is a danger of excess.

Lévêque, in the introduction to his 1989 text, was impressed by the rapidly accumulating inventory of measuring devices, which at that time numbered about 20.[4] By the end of 1993, this list which I have culled from the literature, had swollen to more than 60 and that number is probably an underestimate! The remarks of R. Marks in his "Device and Rule" Dowling ovation are appropriate here.[12] "One could well argue that it is time for a pause. We are confronted by a plethora of instrumentation which is simply overwhelming."

Lévêque too warns that an "abundance of apparatuses with uncertain functionings and of measurements whose significance is often doubtful, debatable and contradictory will not serve the science of dermatology".[4]

An instinctual fascination with instruments for their own sake seems to have become an obsession for some bioengineering enthusiasts. Perhaps, it is ungenerous to restate that old down-putting maxim that "a fool with a tool is still a fool". Common-sense should not be abandoned in favor of clever mechanics. Not everything that can be measured is interesting. The converse is also true. Not everything that is interesting can be measured. In this connection, one can note among some bioengineers the dominance of a theme which maintains that only "objective" methods count in biological research. This school automatically discounts and denegrates every approach that is deemed to be "subjective". To these dogmatists, naked-eye assessments are neither credible nor useful, since they are idiosyncric subjective evaluations. This goes too far and does not take into account situations where subjective judgements may be advantageous and more revealing. An instructive example is the screening of corticosteroids for anti-inflammatory potency according to their ability to cause blanching (vasoconstriction) when applied to normal skin. This test, described 25 years ago, has certainly stood the test of time and does not seem to be in need of upgrading by "instrumentation".[13] It is fast, simple, inexpensive, convenient, and applicable by any person who can tell the difference between mild, moderate, and strong blanching. The subjective aspect is unacceptable to some who insist that Minolta colorimetry or Laser-Doppler velocimetry has greater accuracy. Nonetheless, the original test has not yet been surpassed for purposes of potency classification. Instrumentation in this case adds expense and technical complexities without increasing discriminating ability. In a witty letter to the editor, McKenzie, the originator of the test, opines that the use of objective methodologies are merely "cosmetic improvements".[14]

Again, measurements must not become a religion with beliefs that are heretical to challenge.

There is too little recognition that devices, no matter how ingenious and sophisticated, can give utterly false representations of reality. One telling example comes to mind. Some laboratories have made valiant attempts to obtain two-dimensional constructs of the microtopography of the skin's surface. This requires obtaining a replica with a silicone polymer which can then be scanned robotically by computerized image analysis to show surface patterns in exquisite detail.[15] There is only one difficulty. The final two-dimensional image does not look anything like skin, bearing a much greater resemblance to a mountain range with undulating hills separated by meandering valleys. In reality, the surface is a rather flat plateau, divided into neat geometric patterns by intersecting furrows, referred to as dermatoglyphics. The two-dimensional constructs are technical wonders which attest to impressive mathematical skills beyond the comprehension of the simple clinician. It should also be kept in mind that the eye can often detect and select out subtle patterns of a complex landscape which might be missed by recording devices. The eye, when properly connected to the brain, can sometimes see more surface detail than any instrument.

Along these lines, it is my sense that bioengineers tend to play down and ignore the value of photography, perhaps again because it is seen to be too subjective.

One always hears that photos can lie and therefore should be viewed with mistrust. The same charge can be leveled at objective devices which sometimes can be manipulated to produce a desired result. Photoskeptics would be amazed at the extraordinary textural details provided by high-resolution ultraviolet macrophotography. This technology requires little skill and uses an inexpensive camera in a standard lighting set up. Ultraviolet macrotopography is extremely valuable for visualizing fine wrinkles in the eye area and indeed is more revealing than the much-vaunted Silflo replica method.[16] Too many bioengineers put too much stock in replica measurements without examining confounding factors. Dorogi has emphasized that the naked figure itself doesn't always tell what is actually being measured, anatomically or physiologically.[17] This can be a serious snare and invites caution in interpretation. The question of the perturbing effect of the measuring probe is always with us, a reminder of the Heisenberg uncertainty principle. In our laboratory, we have shown that the Silflo replica distorts the surface and blurs the microtopographic patterns of the photoaged eye-area skin. As the polymer hardens, it contracts and effaces fine lines. It also generates some heat, followed occasionally by redness. One even finds scales clinging to the replica. These are all artifacts induced by the procedure. In the extensive literature which extols profilometry, they are only discussed sporadically.

III. Caveats and Cautions: The Commercial Connection

I am bold here to bring up some touchy subjects which impact on the respectability of the bioengineering fraternity. My comments fall into the category of conflicts of interest.

I submit that too much funding for bioengineering projects derives from industry. In particular, manufacturers of skin care products and cosmetics have been all too ready to exploit the resources of non-invasive technology for commercial advantages. In a competitive and largely unregulated marketplace where the shelves are crowded with alluring products that cater to the American fantasy of staying young forever, support of advertising claims becomes an important tactic for increasing sales. The temptation is great to satisfy the sponsor by selecting the right instruments to get the "right" answers. Instead of traditional advertising puffery to win market share for the latest breakthrough anti-aging product, one can now mobilize "scientific" proof to validate claims of efficacy.

Claims of objectively proved benefits are all the rage in fashion magazines. The guileless consumer learns, for instance, that the new Eternity formulation reduces wrinkles by 52% in two weeks, clarifies the complexion by 31% overnight, and enhances skin tone by 25% in a month! Impressive indeed! I do not doubt the accuracy of such measurements. However, their relevance and validity are often questionable and cannot be accepted at face value. We must forthrightly admit the limitations of non-invasive methodology. An instrumental measurement is unidimensional, yielding information on a single attribute. This underestimates the complexity of living tissue in which many different agencies are interacting simultaneously. Moreover, some measurements, like Laser-Doppler velocimetry, have no absolute dimension but are merely relative assessments, subject to great variations from day to day. Also, we cannot feel too comfortable with measurements that only indirectly reflect the physiologic attribute that is being putatively quantified. For example, non-invasively assessing the water content of the horny layer has received a great deal of attention since the softness and smoothness of the surface depends heavily on the level of hydration. There are at least three devices available that estimate water content by electrical measurements of capacitance, conductance, and impedance. Each oversimplifies the situation and provides only indirect measurements, which are highly susceptible to ambient conditions. First of all, there is a decreasing water gradient on the stratum corneum from the bottom to the top so that no single figure can be truly representative. Current methods reflect the state of hydration somewhere near the top. Then, there is the perturbing effect of the probe itself. Merely placing the sensing probe on the surface immediately blocks transepidermal water loss and leads to a build up of water. A true profile of the water gradient requires other technologies that are far more complex.

As mentioned above, it is important to understand just what is being measured. For example, firmness can be measured by ballistometry and some manufacturers use this approach to show that their latest anti-aging product props up sagging skin and imparts turgor. The fact is, however, that this change may not always be desirable. Turgor can be nonspecifically induced by any irritant which provokes a subtle edema. This can be accomplished in the absence of a visible inflammatory reaction. Fine wrinkles too can be effaced by a sub-clinical inflammatory reaction which in the long run can be harmful even though consumers quickly perceive a solid benefit. Profilometry can quantify wrinkle reduction but doesn't reveal mechanisms or tell anything about safety. This is a serious problem, which needs to be addressed seriously. Another example of the necessity for proper interpretation relates to measurement of epidermal cell turnover. "Anti-aging" advertisements often emphasize the speed-up of cell turnover obtainable by use of their product. It is certainly true that epidermal proliferation decreases in old age. What is left unsaid is that chemical irritation is the surest way to increase proliferative activity, moreover in direct proportion to the degree of injury. A cynical producer of cosmetics could easily reduce cell turnover by 50% in a few weeks by incorporating any of a wide assortment of irritants! Anionic surfactants now do this unwittingly. Thus, enhanced cell removal demonstrated by the non-invasive dansyl chloride fluorescence-extinction technique may mislead the consumer. Bioengineers must know their biology too or become gullible miscreants.

The unwillingness to concede limitations will undermine credibility if excessive claims are made regarding the sensitivity and discriminating powers of bioengineering technology. For example, it is not possible to visualize the epidermis with 20-MHz ultrasound instruments since the vertical resolving power is only about 80 μm. Nonetheless, otherwise beautiful images are marred by arrows pointing to the "epidermis" which turns out to be an entry echo. The "bio" must be put back into bioengineering. One needs to know a good deal about the anatomy and physiology of skin.

It is altogether inappropriate to use a technology to launch new products unless the measurements have been validated by at least one other coordinate. Validation must be made a high priority if misrepresentations are to be avoided. The following example is particularly invidious.

All corticosteroids have the potential to produce skin atrophy. Experience shows that atrophogenicity parallels potency. It would be highly desirable if the two could be dissociated, that is, a steroid that possessed both potency and minimal atrophy. So far this has not been achieved. The case in point involves a new midpotency steroid which was compared to triamcinolone acetonide, using ultrasound imaging. The authors concluded that the new steroid was a real advance since thinning of the dermal matrix was not observed by ultrasound. I take exception to this estimate for two reasons. The small decrease reported, about 10%, is at the level of the sensitivity of the instrument, and sophisticated image analysis of dermal texture was not performed. No effort was made to validate the finding by an alternative method, which is *de rigeur* for all non-invasive technology. I evaluated this same steroid histologically and found it to be as atrophogenic as triamcinolone!

IV. The Fabulous Future

It is already here! Instruments are coming on line that have unprecedented possibilities for non-invasive study of normal and diseased skin. These devices are wondrously ingenious though at present mostly not commercially available and frightfully expensive.

With these powerful tools, we shall witness stunning discoveries beyond our wildest dreams. I will mention a few that rival space-age technology in their reach and promise. Confocal microscopy is such a development, brilliantly described by Corcuff and Lévêque.[18] It is possible to visualize, sharply focused, 1-μm sections of living skin from the surface down to the dermis, in real time, with absolutely no disturbance to the tissue. Layer by layer we can cut through the stratum corneum optically, determining for the first time its thickness in the living state, descending successively to the hitherto elusive stratum lucidum, the stratum granulosum, on through to the living epidermis down to the germinative layer at its interface with the dermis. It is even likely that fine terminal neurons which are now known to penetrate the epidermis can be visualized! Red blood cells can be seen flowing through the capillary loops of the dermal papilla. The possibilities exist for studying the earliest events in the pathogenesis of disease and the response to therapeutic interventions. No front-line investigator who can afford the stiff price should be without a confocal microscope!

At last, magnetic resonance imaging (MRI) has caught up with the skin.[19] The resolving power of this fantastic instrument is very much greater than ultrasound B-scans. The dimensions and contours of the epidermis can now be precisely determined. Pilo-sebaceous units, formerly a blur, now come sharply into view, a boon to students of follicular disorders, such as acne and alopecias. Water distribution profiles can be delineated, compartment by compartment, viz., the stratum corneum, epidermis, the papillary and reticular dermis.[20,21] Lévêque's group at L'Oreal in France are the star performers in this area.

We are witnessing the upgrading of spectrophotometric techniques which will make it possible to determine quantitatively specific tissue components viz., melanin, hemoglobin, elastin, etc.

Stratum corneologists have long appreciated the influential role of water in keeping the horny layer soft and pliable, enabling it to conform to body movements. It has been known for a long time that water molecules have specific infrared absorption bands but the available techniques have not permitted accurate, reproducible measurements. This short-coming has now been overcome by using near-infrared wavelengths which reliably quantify water.[22] Finally, advances in imaging technology enable the construction of exacting three-dimensional models of lesions. El-Gammal and his group have brilliantly shown what a comedo really looks like.[23]

The excitement of the rapidly growing field of bioengineering is palpable at every meeting where the members of the Bioengineering Society sometimes appear to be hectic or slightly manic. Or, to use a different analogy, those on the cutting edge of this new discipline must feel like a rich child in a Vienna pastryshop!

References

1. Pinnagoda, J., Tupker, R.A., Agner, T., and Serup, J., A report from the Standardization Group of the European Society of Contact Dermatitis. Contact Dermatitis Guidelines for transepidermal water loss (TEWL) measurement. *Contact Derm.*, 22, 164, 1990.
2. Bircher, A., De Boer, E.M., Agner, T., Wahlberg, J.E., and Serup, J., Guidelines for measurement of cutaneous blood flow by laser Doppler flowmetry. A report from the Standardization Group of the European Society of Contact Dermatitis. *Contact Derm.*, 30, 65, 1994.
3. Frosch, P.J., Kligman, A.M., *Noninvasive Methods for the Quantification of Skin Functions.* Springer-Verlag, New York, 1993.
4. Lévêque, J.L., *Cutaneous Investigations in Health and Disease.* Marcel Dekker, New York, 1989.
5. Altmeyer, P., el-Gammal, S., and Hoffman, K., Eds., *Ultrasound in Dermatology,* Springer-Verlag, Berlin, 1992.
6. Kligman, A.M., Blind man dermatology, *J. Soc. Cosmet. Chem.,* 17, 505, 1966.
7. Kligman, A.M., The invisible dermatoses, *Arch. Dermatol.,* 127, 1375, 1991.
8. Seidenari, S. and DiNardo, A., Echographic evaluation with image analyses of allergic and irritant reactions, *Acta Derm. Venereol. Suppl.,* 175, 3, 1992.
9. Agner, T. and Serup, J., Sodium lauryl sulfate for patch testing: a dose-response study using bioengineering methods, *J. Invest. Dermatol.,* 95, 543, 1990.
10. Frosch, P.J. and Kligman, A.M., The soap chamber test. A new method for assessing the irritancy of soaps, *J. Am. Acad. Dermatol.,* 1, 35, 1979.
11. Kligman, A.M., Miller, D.L., and McGinley, K.J., Sebutape: a device for visualizing and measuring human sebaceous secretion, *J. Soc. Cosmet. Chem.,* 37, 369, 1986.
12. Marks, R., The Dowling Oration 1984. Device and Rule, *Clin. Exp. Dermatol.,* 10, 303, 1985.
13. Haigh, J.M. and Smith, E.W., Topical corticosteroid-induced skin blanching measurement: eye or instrument? *Arch. Dermatol.,* 127, 1065, 1991.
14. McKenzie, A.W., First blast of the trumpet against the monstrous segment of scientists, *Br. J. Dermatol.,* 115, 637, 1986.
15. Hoppe, U. and Sauermann, G., Quantitative analysis of the skin's surface by means of digital signal processing, *J. Soc. Cosmet. Chem.,* 36, 105, 1985.
16. Kligman, A.M. and Christensen, M., Wrinkles: pathogenesis and topical treatments, in *The Science of Skin Care,* de Lacharriere, O., Ed., 1994, in press.
17. Dorogi, P.L., Physical properties of the skin. *J. Dermal Clin. Eval. Soc.,* 1, 19, 1990.
18. Corcuff, P. and Lévêque, J.L., In vivo vision of the human skin with the Tandem Scanning Microscope, *Dermatology,* 186, 50, 1993.
19. Richard, S., Querleux, B., Bittoun, J., Jolivet, O., Idy-Peretti, I., de Lacharriere, O., and Lévêque, J.L., Characterization of the skin *in vivo* by high resolution magnetic resonance imaging: water behavior and age-related effects, *J. Invest. Dermatol.,* 100, 705, 1993.
20. Richard, S., Bernard, Q., Bittoun, J., Idy-Peretti, I., Jolivet, O., Cermakova, E., and Lévêque, J.L., *In vivo* proton relaxation times analysis of the skin layers by magnetic resonance imaging, *J. Invest. Dermatol.,* 97, 120, 1991.
21. Querleux, B., Richard, S., Bittoun, J., Jolivet, O., Idy-Perett, I., Bazin, R., and Lévêque, J.L., *In vivo* hydration profile in skin layers by high-resolution magnetic resonance imaging, *Skin Pharmacol.,* 7, 210, 1993.
22. de Rigal, J., Losch, M.J., Bazin, R., Camus, C., Sturelle, C., Descamps, V., and Lévêque, J.L., Near-infrared spectroscopy: a new approach to the characterization of dry skin, *J. Soc. Cosmet, Chem.,* 44, 197, 1993.
23. El Gammal, S., Aver, T., Hoffmann, K., and Matthes, U., High frequency ultrasound: a noninvasive method for use in dermatology, In *Noninvasive Methods for the Quantification of Skin Functions,* Frosch, P.J. and Kligman, A.M. Eds., Springer-Verlag, New York, 1993, 104.

Chapter 1.2
Relevance, Comparison, and Validation of Techniques

G.E. Piérard
Laboratory of Dermometrology
Department of Dermatopathology
CHU du Sart Tilman
University of Liège
Liège, Belgium

I. Introduction

Measuring in an objective way is always in need of additional breakthrough. Dermometrology and bioengineering have been and remain closely associated in the search for improvements of quantitative noninvasive assessments. The pre-bioengineering times and the descriptive phase of dermometrology are behind us. Ingenious researchers pioneered methods that may now look crude, time-consuming, and sometimes lacking in reproducibility.

Thanks to them, however, the noninvasive technology has made great advances in the study of skin during the last decade or so. This was paralleled by a greater knowledge of many aspects of skin biology. The result is now evident as we assist in the development and refinement of meaningful approaches used in cutaneous biology, cosmetology, and dermatology. Dermometrology has become a full-fledged science based on several disciplines. A healthy relationship exists among most professional societies, research institutes, and private commercial companies. A convergence of interests is clearly perceptible. Close working relationships among bioengineers, cosmetologists, clinical scientists, physiologists, and biologists are essential for the future of this discipline.

Reading these lines, the merest beginner could consider that skin physiopathology is now well studied by noninvasive dermometrology. He could also consider that the possibilities offered by the astonishing number of measuring devices are limitless. The situation is not so bright. At present, there are no guidelines with regard to the standardization of some techniques and no recognized quality control procedure is available for assessing uniformity of data collection and interpretation. This is the ransom of glory. The work of dermometrology covers a continually varying and ever-widening era of biological, medical, cosmetic, and skin care research. It is also used for commercial purposes by the drug and cosmetic industry. Some techniques are investigational, others are already present in many research units and even at the bedside for monitoring patients. Those that are now commercially available are in many hands everywhere. The danger resides in the apparent easiness of manipulation of these apparatuses. As already mentioned in the past, there may be a questionable use of bioengineering technologies. Sound methodology could be subverted for mercantile purposes and falsehoods promoted under the guise of scientific information. False claims can be so readily substantiated as worthy ones that the value of measurements lies in the strict application of standardized procedures and in the ability to interpret the data correctly. Inexpertise of the owner of the bioengineering device and credulity of the observer are the two most dangerous facets of dermometrology. Producing invalid "creative" advertising is another concern in the field of commercial marketing.

II. A Plea for Standardization and Quality Control

The scientific production is controlled in peer review journals. Should we discuss regulatory procedures before launching a dermometrology-supported claim about products? So far, manufacturers of cosmetics have perceived the great value of the continuously growing field of dermometrology. The drug industry and physicians have lagged behind for many years, but now have a similar focus of interest.

Optimization of noninvasive biophysical methods should benefit from strict standardization of measurements and frequent calibrations of devices. Constant developments in relevant links between dermometrology and biology should be pursued. Unfortunately, there have been only a handful of works devoted to standardization of measurements of skin properties. The problem is complicated by the fact that some laymen in the field of cutaneous biology who are owners of biophysical devices may prove to be brilliant in speculations and try to blur the borderline between claims and realities.

III. An Attempt to "Good Biometrological Practice"

Two distinct features should be considered, namely the characteristics of a device and the aspects of its application for measuring a biophysical property of skin.

Each main technological aspect of an apparatus has to be known by the scientist using it. The best conditions of reproducibility of measurements should be strictly evaluated. Frequent calibrations must be performed. Ideally, the procedure should be identical in all laboratories using the same device. Unfortunately the manufacturers of biophysical instruments do not always provide this information. Each researcher must therefore invent his own procedure, at least to keep a chance of reproducibility of data in his laboratory over the years!

The second facet of the problem concerns the application of technologies to skin-related biomedical problems. The basic knowledge of the biological aspect to be evaluated is of paramount importance. The choice of the technique of evaluation ensues. When possible, several methods of evaluation should be used in combination rather than one single type of measurement. This is crucial for the validity of interpretation of the data. In general, one device is designed to measure one single biophysical property, but the collected data may be influenced by parameters that are not evaluated by that apparatus. The association of diverse techniques may provide a better evaluation of complex interrelationships between cutaneous properties.

The experimental conditions should be correctly settled and controlled or monitored. In general, this concerns both the individuals tested and environmental conditions. For volunteers or patients, several characteristics should be choosen and/or recorded. This includes race, gender, age, region of the body, and any other parameter specific for the study. Seasonal, ovarian, and nyctohemeral cycles, diseases, previous manipulations such as preconditioning also clearly influence some measurements. The environmental conditions may significantly alter specific biophysical properties of skin. Every single biometrological evaluation benefits from a controlled environment where temperature and relative humidity are monitored. Exposure to irradiations, including ultraviolet light, total sunlight spectrum, infrared energy, strongly affect properties of skin, with sometimes a long remanence. It is also obvious that some drugs and cosmetics influence many measurements. Therefore, the choice of panelists or patients in a study proves to be very important. The same is true for any comparison group which should ideally concern both positive and negative controls.

The interpretation of data in biologic terms is often difficult for scientists, even if it is sometimes oversimplified in commercial strategies. It should always include adequate methods of statistics in combination with clear criteria of biologic relevance. Researchers should be aware of statistically validated data which prove to be not contributory on a biological basis. The reverse is also true, and sometimes it is simply by increasing the number of measurements that the validation is reached by statistical analysis.

IV. Selected Examples of Pitfalls

A. Evaluation of Dry Skin

Evaluations of impedance, conductance, and capacitance may be easily performed with some commercially available devices. When measurements are made on apparently "dry" skin, values are often decreased compared to normal skin. It is therefore tempting to conclude that "dry" skin lacks water. This is in fact a pitfall. So-called "dry skin" is a concept of laymen. This condition primarily corresponds to a rough skin surface (xerosis) which may or may not lack water. The decreased values of conductance or capacitance are in part due to a lack of close contact between the probes and the harsh surface of the skin. As quoted by Tagami,[1] "the recorded values tend to indicate a lower hydration state than the actual one when the measurements are performed on lesional skin covered by scales". No firm conclusion may therefore be drawn solely from such measurements with respect to the water amount in the stratum corneum of xerosis.

Other methods of evaluation should therefore be used in combination with electrical measurements in the evaluation of the hydration state of the skin surface. The relationship of conductance or capacitance with transepidermal water loss (TEWL) is complex. A significant positive correlation was found between these parameters when they were evaluated on normal-looking skin and when the stratum corneum barrier was just being altered by superficial strippings.[1] This relationship was no truer in some scaly disorders where TEWL increased in parallel with an actual or artefactual reduction in capacitance and conductance. From these foregoing data and according to our experience (Figure 1A) the relationship between electrical properties of the stratum corneum and TEWL is complex and does not prove to be always correlated with the severity of xerosis.[2] Another, and probably better, approach for evaluating xerosis is achieved by quantifying the amount of stratum corneum collected under standardized conditions by adhesive-coated disks

(D-Squames®).[3-5] There is no straightforward relationship between values of capacitance and of squamometry (Figure 1B).

From the aforementioned statements, we contend that capacitance, TEWL, and squamometry are three variables without any strong relationship between them. It appears that the specific pitfall in evaluating xerosis resides in the confusion between the aspect of "dry skin" and the water content of the stratum corneum. Data collected by a single biophysical method are at risk to be artefactual and wrongly interpreted. The combination of techniques is welcome.

B. Evaluation of Aging Through Changes in Mechanical Properties of Skin

The many different technologies used in the past to evaluate the mechanical properties of skin have yielded contradictory results. This was in part due to nonstandardized conditions of experimentations using self-made prototypes of measuring devices lacking precision. Even nowadays, the most recent sophisticated and electronically controlled apparatuses do not always provide similar information. Many features dramatically influence the measurements (Figure 2, Table 1). Hence, the interpretations of data that have been offered up to now are probably unsatisfactory on biological grounds.

When using the Twistometer, aging of the skin is best described as a sharp reduction in the elastic function (EF).[6] With the use of the Cutometer®, the largest changes with age concern the increased viscoelastic ratio (VER) and the decreased relative elastic recovery (RER).[7] The differences in data provided by the two distinct devices are probably related to the different operating modes of application of forces. The Twistometer works by torsion in the plane of the skin surface, whereas the Cutometer operates a suction perpendicular to the skin surface. If the general information provided by these two methods indicates a decreased overall elasticity with age, such information is no more precise than that gained by the historical tonometer.[8] The different but not contradictory subtle changes disclosed by the two modern electronic devices still defy accurate biological interpretation.

The lack of standardization is still an important problem even when laboratories are using the same device. For instance, by using the Cutometer, Cua et al.[7] measured the immediate distension (Ue = ED1) at 0.48 s of traction, whereas we found it more informative to measure at 0.15 s.[9] Hence, all biologically relevant ratios dealing with elasticity are not comparable between our laboratories. The same problem in the definition of terms arises with the concept of biologic elasticity (BE). We introduced it almost 20 years ago,[8] and we have not modified its calculation until now.[9] BE corresponds to $10^2(MD1 - RD1)MD1^{-1}$. Cua et al.[7] have altered its calculation to $10^2(MD1 - ER1)MD1^{-1}$; this ratio corresponding for us to the

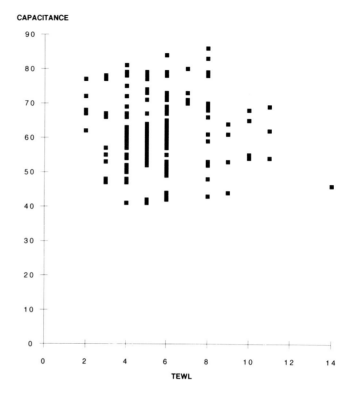

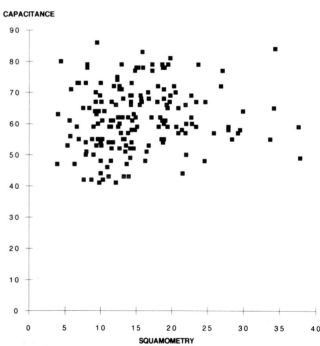

FIGURE 1 Biometrological evaluation of normal to dry skin of the forearms in a panel of 168 women. (A) The trend for an inverse relationship between skin capacitance and TEWL is not prominent and probably not biologically relevant. (B) Absence of correlation between capacitance and squamometry (see Reference 4).

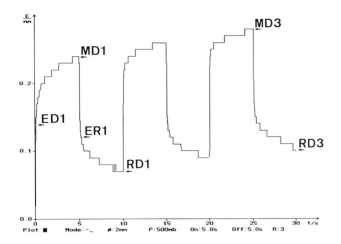

FIGURE 2 Recorded mechanical parameters. Variations in the elevation of skin (E/mm) in time (T/s). (A) Definition of parameters: ED1, elastic (immediate) deformation of the skin during the first traction (0.15 s); MD1, maximum deformation of the skin at the end of the first traction (5 s); ER1, elastic (immediate) retraction of the skin during the first cycle (5.1 s); RD1, residual deformation of the skin at the end of the first cycle (10 s); MD3, maximum deformation of the skin at the end of the third traction (25 s); RD3, residual deformation of the skin at the end of the third cycle (30 s).

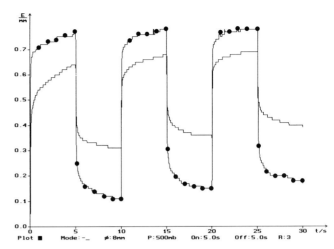

FIGURE 2 (C) Influence of preconditioning the test area. A 30-s preconditioning (dotted line) significantly alters the response of the skin to a similar mechanical stimulus.

TABLE 1 Calculated Mechanical Parameters

VER	Viscoelastic ratio = U_v/U_e = $10^2(MD1 - ED1)ED1^{-1}$
EF	Elastic function = U_r/U_e = $10^2(MD1 - ER1)ED1^{-1}$
BE	Biologic elasticity = $10^2(MD1 - RD1)MD1^{-1}$
RER	Relative elastic recovery = U_r/U_f = $10^2(MD1 - ER1)MD1^{-1}$
HY	Hysteresis (mm) = MD3 − MD1

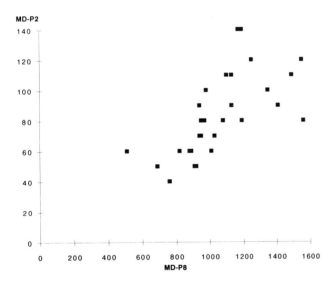

FIGURE 2 (B) Influence of the area of the tested site with a Cutometer. There is a positive linear correlation between the maximum deformations (MD, mm) of skin when the opening of the probe is 2 mm (MD-P2) and 8 mm (MD-P8) in diameter.

RER.[9] It is obvious that the same term encompasses two distinct evaluations and that two different names are given to the same biologically relevant measurement. This is representative and only an example of the multiple languages used in the Babel tower of dermometrology.

C. Sebum Output at the Skin Surface

The two most popular techniques for measuring sebum excretion are the Lipometer Sebumeter and the Sebutape® methods. They yield comparable information in many instances.[10] They are, however, not strictly equivalent.[11] The best sensitivity for detecting low sebum excretion is provided by the Sebumeter®, whereas the adhesive interposed between the lipid-absorbent tape and the skin limits the transfer of small amounts of sebum in the Sebutapes (Figure 3). A saturation effect may occur with both techniques, but manifests itself more readily with the Sebumeter method than with the Sebutape. The ideal sampling time is 1 h. The combined use of both evaluation methods of sebum excretion is therefore rewarding. Image analysis[12,13] and/or colorimetric assessments[13] of Sebutapes bring precision and provide information that is not accessible by other methods. For instance, the comparison of the mean and median of spot areas on Sebutape reveals two distinct groups of individuals (Figure 4). One is characterized by a Gaussian distribution (mean = median) of the amount of sebum delivered by individual follicles. The other one is defined by a distribution of the values following a Poisson law (any value of mean with a median close to 0).

The major pitfall with these techniques occurs when measuring the amount of sebum excreted in time. The

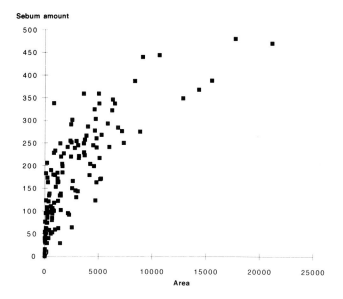

FIGURE 3 Comparison of the Sebumeter® and Sebutape®. The total amount of sebum collected by the Sebumeter ($\mu g/cm^2$) and the area of spots on Sebutapes (area, A.U.) are not linearly related in the extreme parts of the relationship.

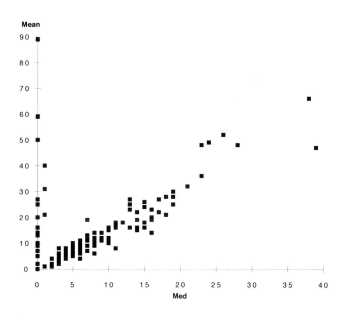

FIGURE 4 Relationship between means and medians of spot areas on Sebutapes applied in 100 unselected panelists. Two groups of individuals are distinguished according to the near-equality between mean and median of values or to the low value of the median irrespective of the mean.

calculated value of the rate of sebum excretion is in fact tremendously influenced by the care with which the sebum is cleaned before the measurements. There is obviously a need for standardization of that procedure. Simple wiping with a dry cotton wool gauze is not reliable (Figure 5A) and the situation is not ideal when excessive delipidization is done by contact with ether:acetone (1:1) in a cup for 1 min (Figure 5B).

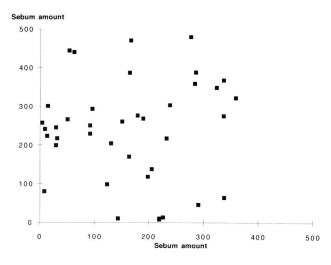

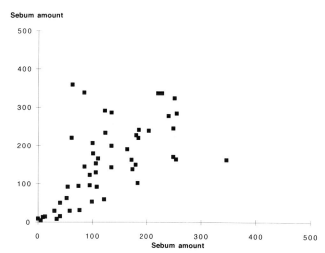

FIGURE 5 Inadequate evaluations of the sebum excretion rate ($\mu g/cm^2/3$ h) using a Corneometer. Comparison of two measurements on contiguous sites of the forehead in 38 panelists: (A) after wiping with a dry gauze; (B) after degreasing with ether and acetone (1:1 for 1 min).

Soaking a gauze in 80% ethanol and firmly wiping the skin three to five times is, in our hands, the best procedure because it removes the superficial lipids without altering the stratum corneum. Sebum present in the follicular reservoirs is preserved in large part.

V. Conclusion

Dermometrology is a fascinating discipline that needs the contribution of many researchers coming from diverse origins. Basic and applied research have still to explore many facets of biophysical properties of skin. Routine use of bioengineering devices may look simple but proves to be a field with multiple pitfalls for the inexperienced beginner. Even the skillful researcher is facing problems of relevance and interpretation of data. In every instance, emphasis should be placed on a strict respect of standardized and controlled conditions. We are still in a developmental phase of dermometrology where the brain of the researcher must control every single aspect of measurements. Devices only provide figures which may be relevant for a biological aspect of skin or may prove to be only distracting. If the skill of the experimenter or naive use by a layman influences the value of the data, the interpretation of the biophysical measurements also requires expertise. Should we need a license to manipulate biometrological devices as we need one to drive a car? Some regulatory procedures should perhaps be introduced to control claims and creative advertisements offered under the cover of dermometrology.

References

1. Tagami, H., Impedance measurement for evaluation of the hydration state of the skin surface, in *Cutaneous Investigation in Health and Disease.* Lévêque, J.L., Ed., Marcel Dekker, New York, 1989, 79.
2. Deleixhe-Mauhin, F., Piérard-Franchimont, C., Krezinski, J.M., Rorive, G., and Piérard, G.E., Biometrological evaluation of the stratum corneum texture in patients under maintenance hemodialysis, *Nephron,* 64, 110, 1993.
3. Serup, J., Winther, A., and Blichmann, C., A simple method for the study of scale pattern and effect of a moisturizer — qualitative and quantitative evaluation by D-Squame tape compared with parameters of epidermal hydration, *Clin. Exp. Dermatol.,* 14, 277, 1989.
4. Piérard, G.E., Piérard-Franchimont, C., Saint Léger, D., and Kligman, A.M., Squamometry: the assessment of xerosis by colorimetry of D-Square adhesive discs, *J. Soc. Cosmet. Chem.,* 47, 297, 1992.
5. Kligman, A.M., Schatz, H., Manning, S., and Stoudemayer, T., Quantitative assessment of scaling in winter xerosis using image analysis of adhesive-coated disks (D-Squames), in *Noninvasive Methods for the Quantification of Skin Functions,* Frosch, P.J. and Kligman, A.M., Eds., Springer-Verlag, Berlin, 1993, 309.
6. Escoffier, C., de Rigal, J., Rochefort, A., Vasselet, R., Lévêque, J.L., and Agache, P.G., Age-related mechanical properties of human skin: an in vivo study, *J. Invest. Dermatol.,* 93, 353, 1989.
7. Cua, A.B., Wilhelm, K.P., and Maibach, H.I., Elastic properties of human skin: relation to age, sex, and anatomical region, *Arch. Dermatol. Res.,* 282, 283, 1990.
8. Piérard, G.E. and Lapière, Ch. M., Physiopathological variations in the mechanical properties of skin, *Arch. Dermatol. Res.,* 260, 231, 1977.
9. Deleixhe-Mauhin, F., Piérard-Franchimont, C., Rorive, G., and Piérard, G.E., Influence of chronic haemodialysis on the mechanical properties of skin, *Clin. Exp. Dermatol.,* 19, 130, 1994.
10. Piérard, G.E., Follicle to follicle heterogeneity of sebum excretion, *Dermatologica,* 173, 61, 1986.
11. Serup, J., Formation of oiliness and sebum output — comparison of a lipid-absorbant and occlusive-tape method with photometry, *Clin. Exp. Dermatol.,* 16, 258, 1991.
12. Piérard, G.E., Piérard-Franchimont, C., Lê, T., and Lapière, Ch. M., Patterns of follicular sebum excretion rate during lifetime, *Arch. Dermatol. Res.,* 279, S104, 1987.
13. Piérard, G.E., Piérard-Franchimont, C., and Kligman, A., Kinetics of sebum excretion evaluated by the Sebutape-Chromameter technique, *Skin Pharmacol.,* 6, 38, 1993.
14. Piérard, G.E. and Piérard-Franchimont, C., Sebum analysis using a hydrophobic lipid-absorbent tape (Sebutape®), in *Non-Invasive Methods for the Quantification of Skin Functions,* Frosch, P.J. and Kligman, A.M., Eds., Springer-Verlag, Berlin, 1993, 83.

2.0 Variables and Study Design

2.1 Prescription for a Bioengineering Study: Strategy, Standards, and Definitions 17
J. Serup

2.2 The Skin Integument: Variation Relative to Sex, Age, Race, and Body Region 23
N. Farinelli and E. Berardesca

2.3 Seasonal Variations and Environmental Influences on the Skin 27
C.L. Goh

Chapter 2.1
Prescription for a Bioengineering Study: Strategy, Standards, and Definitions

Jørgen Serup
Department of Dermatological Research
Leo Pharmaceutical Products
Ballerup
Bioengineering and Skin Research Laboratory
Department of Dermatology
Bispebjerg Hospital
University of Copenhagen
Denmark

I. Introduction

A considerable amount of research is carried out employing noninvasive methods. In the daily work situation it is not possible to make a totally perfect study that is beyond any criticism. Hardly any method in medical research is totally validated and hardly any study is beyond improvement. To perform any study is to compromise at a certain time and place, with the state of the art at some level.

Researchers wish to do their very best and often because of this, even studies that might seem easy and simple often turn out to be more complicated than anticipated. However, there are a number of typical pitfalls, errors, or mistakes which can be avoided. Knowledge combined with sound clinical sense and the ability to analyze, clarify, and simplify are the essentials.

II. Purpose of the Study

A study may be planned to be descriptive and purely scientific or to be focused on a given piece of technology, a technique, the efficacy of a drug with registration purposes and a design to meet with bureaucratic demands, etc. Studies may be small with a few patients and one center or large and multicentric, planned to demonstrate or not to demonstrate a difference. There is obviously a wide range, and the appropriate design is entirely dependent on the situation. In any case it is important that the researcher uses sound judgment and remains flexible and cost efficient at all times. The crucial point is to master *the signal to noise ratio*.

Interindividual and intraindividual anatomical site variation are the most common sources of noise. Right/left comparison, regional control measurement, and use of the individual as his own control is therefore preferable whenever applicable.

III. The Problem of the Need for Good Clinical Definitions

The noninvasive method should be chosen so that it is relevant with regard to the disease or condition in question. However, clinical conditions are often poorly defined. Diagnosis in dermatology has strong elements of art and relies to a high degree on the talent and training of the dermatologist, although individual dermatologists have a clear concept of what they are dealing with. Common conditions such as "dry" skin, eczema, and irritancy still need a clear academic definition. Dry means literally with *no* water content, which is unlikely ever to happen under live conditions. The allergic patch test is a pragmatic red spot test and not a direct measure of type IV reactivity of the immune system; who dares to calculate the predictive value of a positive and a negative clinical reading?

Nevertheless, dermatologists do a superb job and help their patients. These points were only raised in order to illustrate the difficulty the bioengineer will meet in designing a study with a clinical reference point. It is a difficulty that is not related to the equipment but to the nature of the clinical reference point. It is, in a given study and under the given circumstances, of major importance to define a clear main purpose or a major clinical sign which the methods can then illustrate in an intelligent and useful way.

IV. Typical Pitfalls and Errors During a Study Employing Noninvasive Techniques

The possibility of the following typical pitfalls or errors should be considered when a study protocol is prepared:

1. study design (strategic error)
2. choice of variables (technological error)
3. the measuring device (technical error)
4. the use of the device (performance error)
5. measuring conditions (inadequate laboratory facilities)
6. selection and preconditioning of test subject (subject-related error)
7. data acquisition, storing, and handling (data error)
8. reporting and publication politics (explanatory mistake)

A. Pitfalls and Errors Related to the Phenomenon or Variable Being Measured

The following should be checked when a study protocol is prepared:

1. What information is expected?
2. What is the most relevant variable to be measured, and which variables serve for description, comparison, support, or exclusion?
3. What is the expected time course of variables, and when should the measurements be performed?
4. Are the variables expected to develop linearly or not?
5. What are the ranges of variables in relation to the expected phenomenon or structure being studied, including inter- and intraindividual variation and dependence of anatomical site, sex, and age?
6. What function or structure is actually being tested?
7. What is the measuring area and, if small, do more recordings need to be taken and averaged to overcome local site variation?
8. Are recordings with the equipment reproducible, and is the precision acceptable relative to variables being measured and their expected range?
9. Measuring standards and calibration procedures
10. Environmental influences (including season) and the need for special laboratory room facilities
11. Is it necessary to precondition the individuals before testing?
12. What keeps measurements from being performed?
13. Additionally, has the researcher or technician both the educational background and enough practical experience to conduct the study?

B. Definitions Directly Related to the Instrument and its Validation

To evaluate the technical variation related to the instrument the following definitions are often used:

Accuracy — Degree of similarity between the value that is accepted either as a conventional true value (in-house standard) or an accepted reference value (international standard) and the mean value found by performing the test procedure a certain number of times (provides an indication of systematic error)

Precision — Degree of similarity (degree of scatter) among a series of measurements obtained from multiple sampling of the same homogeneous sample under prescribed conditions, expressed as repeatability and reproducibility

Repeatability — Expresses the situation under the *same* conditions, i.e., same operator, same apparatus, short time interval, identical sample

Reproducibility — Expresses the situation under *different* conditions, i.e., different laboratories, samples, different operators, different days, different instruments from different manufacturers

Range — The interval between the upper and lower levels for which the procedure has been demonstrated as applicable with precision, accuracy, and linearity

Linearity — Ability of the procedure (within a given range) to obtain test results directly proportional to true values

Sensitivity — Capacity of the procedure to record small variations or differences within the defined range

Limit of detection — Lowest change above zero that is just detectable

Limit of quantification — Lowest change above zero that can be quantitatively determined (not only detected) with defined precision and accuracy under the stated experimental conditions

Ruggedness — Evaluates the effects of small changes in the test procedure on measuring performance

V. Monoinstrumental and Multi-Instrumental Designs

Bioengineering devices are typically based on one physical modality and therefore by nature only meant to assess one single fragment of a biological phenomenon. In contact dermatitis, for example, the evaporimeter measures the barrier disruption, the flowmeter the inflammation with vasodilation and increase of blood flow, the colorimeter the increase of blood volume under a given area of skin, and the ultrasound scanner the edema of the inflammatory process due to extravasation of water.[1] The evaporimeter is useful as an overall parameter in different types of dermatitis including atopy and irritant reactions to sodium lauryl sulfate, but other irritants such as nonanoic acid and calcipotriol create little barrier disruption and an increase in the blood flow, making measurement of transepidermal water loss (TEWL) less useful to the investigator. During the spontaneous healing phase of dermatitis the different variables have different time courses with redness and dryness of the skin surface being the last variables to normalize.[2]

The monoinstrumentalist therefore runs a great risk of missing the right variable unless he is well guided by previous studies describing the kinetics of the reactions studied.

It is generally much more conclusive to use a number of techniques in combination. Negative findings are, in the initial descriptive phase, equally important to positive findings. If the purpose of the study is quantification and dose ranging one selected method of known relevance may suffice and be more cost efficient.

VI. Standard Operating Procedure (SOP)

A standard operating procedure is a research laboratory's own written instructions for the proper use of its equipment. SOPs may concern validation, maintenance, performance, and the condition of the laboratory facilities employed. It may be both of practical use and a formal instrument in relation to regulatory approval and audits.

It may seem another frightening sign of approaching bureaucracy, but if practiced with a sense of proportion it is just a question of good order in the laboratory, and there are some straightforward formalities connected with this. Conformity with SOPs is mandatory if a study is to be given its full weight in a drug registration dossier. SOPs are essential in relation to Good Laboratory Practice (GLP), Good Clinical Practice (GCP), and Good Manufacturing Practice (GMP).[3,4] GLP is especially important in toxicity studies that include local tolerance, and GCP is of similar relevance when humans are studied. Although GCP by most national authorities is considered a guideline, it was nonetheless rapidly implemented by authorities and industries and determined the existing practice. Details about GCP and legal approval are found in other chapters of this handbook.

Although not applicable in every case the definition and validation requirements mentioned above may be useful when a SOP is established. In the field of bioengineering and the skin an in-house validation procedure can normally be defined and practiced, and normally the manufacturer's manual will contain some guidance. Recently, however, two authorized recommendations on validation of bioengineering techniques have been published.[5,6]

VII. Examples of Measurement Guidelines for Bioengineering Techniques

Measurement of skin surface hydration and water barrier function has become popular. The water barrier is believed to be located in the outer epidermis where a steep gradient of the water content is found (Figure 1). However, diffusional equilibrium between the skin surface and ambient air takes place within 5 min,[7-10] and we have immediately outside our skin a *water vapor mantel* approximately 10 mm in thickness. The sensors of the evaporimeter are mounted 3 and 6 mm above the skin surface. The water mantel, evaporimeter measurements, and the hydration of the stratum corneum are for obvious reasons quite sensitive to convection of air and other environmental factors. However, it is noteworthy that the water barrier is not simply an intra-epidermal structure but a far more complex function constantly depending on the close environment. Barrier function is even more complex in psoriasis and eczema where the water-holding capacity of the stratum corneum is reduced and the skin "dry" despite the exposure to an increased water vapor pressure. The clinical signs of

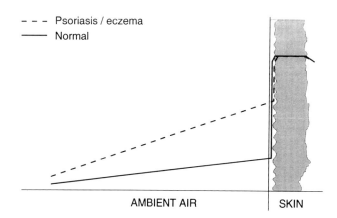

FIGURE 1 The water barrier function (schematic). The water barrier is not simply located within the epidermis. It also involves a mantel of water vapor outside the skin approximately 10 mm in thickness. The sensors of the Servo Med® evaporimeter are mounted in the probe chamber 3 and 6 mm over the skin.[5]

"dry" skin are minor scaling, greyish skin color, and microfissure with discomfort subjectively. The water molecule is not visible. Furthermore, thermoregulation and sweat gland function have a great influence on the water vapor mantel.

The different factors mentioned above were taken into account when the standardization group of the European Society of Contact Dermatitis recently developed the following guidelines for measurement of TEWL:

> Individuals should rest for 15–30 min before TEWL measurements, with the skin at the measuring site left uncovered. Measurement of skin temperature is recommended. Only TEWL values from the same anatomical area are expected to be comparable.
>
> Perform all TEWL measurements within a large "open-top" box, whenever possible. The ambient room relative humidity and temperature, during TEWL measurements, should be stated. If climate room facilities are available, the ambient room temperature should be regulated to 20–22°C, and the relative humidity to 40%. TEWL measurements of a single experiment should preferably be completed within one season. Do not measure TEWL under direct light sources.
>
> Do not hold the probe directly by hand. The probe should be handled with an insulating glove, or the calibration rubber stopper supplied with the equipment, or a burette clamp. The measuring surface should be placed in a horizontal plane, and the probe applied parallel to this surface. The contact pressure of the probe onto the skin should be kept low and constant. TEWL values should be registered 30–45 s after application of the probe to the skin, preferably using a pen-recorder. The filters should be used only after stabilization. Avoid using of the "offset" button for zeroing the instrument in between measurements, and allow it to zero on its own. The use of the protection covers should be clearly stated, since the cover with screen and grid influences the measurement.

For more detailed information, readers are referred to the original publication (Reference 5), or to the separate chapter in this book.

The standardization group of the European Society of Contact Dermatitis moreover developed guidelines for

TABLE 1 Factors and Variables with Effects on Cutaneous Blood Flow (CBF)

Age	Widely age independent
Sex	Minor or no difference
Menstrual cycle	Minor or no difference
Race	Minor or no difference
Anatomical site	Considerable variation
Position	Orthostatic dependence
Temporal, diurnal	Minor or no variation
Temporal, day to day	May be significant
Physical activity	Considerable effect
Mental activity	Considerable effect
Food and drugs	Considerable effect
Temperature	Very significant effect

laser Doppler flowmetry, which also illustrate the many factors that need be taken into account with a bioengineering technique (Table 1, Figure 2):

> Use the flowmeter in accordance with manufacturer's directions, and laser light safety directions, and with a defined operating procedure including validation (determination of resting cutaneous blood flow (CBF) and repeatability followed by 3 min arterial compression of the upper arm with a cuff and determination of biological zero and peak CBF on flexor side forearm skin of at least 3 healthy adults).
>
> Allow the flowmeter to warm up. Apply the probe to the measuring site with no special pressure. Make sure that the test subject has not taken any food or drugs which might influence CBF. Avoid any topical treatment of the test site prior to study unless it is a part of the experiment.
>
> Make sure that the test subject has not deliberately exercised, been exposed to unusual temperatures, or been under mental stress immediately before CBF measurements.
>
> Allow the test subject to rest for 15 min or more under quiet conditions, preferably in the laboratory room in the position in which recordings are going to be obtained, i.e., sitting or supine, and with the test site uncovered.
>
> Take three or more recordings and average them if the flowmeter operates within a small measuring field, i.e., 1–3 mm^2. Avoid movement of the optical cable.
>
> Use a pen-recorder or a computer with appropriate software for data collection. Store the basic data safely.

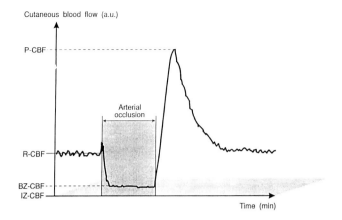

FIGURE 2 Cutaneous blood flow (CBF) before and during arterial occlusion and postocclusive reactive hyperemia (schematic). R-CBF, resting flow; BZ-CBF, biological zero; IZ-CBF, instrumental zero determined on white porcelain; P-CBF, Peak flow after occlusion and during reactive hyperemia. Latency period and recovery of hyperemia can also be defined. With the hyperemia experiment the range of operation of the instrument is described. In order to describe the repeatability of the instrument it is recommended that R-CBF is determined a number of times, and mean, SD, and coefficient of variation can be given. In the validation experiment a minimum of three healthy adults should be studied on flexor-side forearm skin. Results of validation experiments shall ideally appear in every publication including results obtained with any laser Doppler flowmeter. Details on standardization appear from Reference 6.

Control the laboratory room and the measuring conditions, in particular with respect to temperature, convection of air and noise. Avoid measurement under direct light, including sunshine, which might influence skin temperature. Perform measurements with the site studied at a standardized level relative to the level of the heart.

In reports and publications, clearly describe the instrument including details about the probe and the effective measuring area laterally and in depth, and also give information about instrument setting and validation, handling of the test subject and control of measuring conditions, in a way that provides precise and detailed information about the experiment, so that it can be reproduced.

Readers are referred to the original publication for more detailed information (Reference 6).

Official guidelines for ultrasound examination of the skin, measurement of skin elasticity, and measurement of skin color have not appeared, but readers can find practical guidance on these three methods in other chapters of this book.

References

1. Serup, J., Noninvasive techniques for quantification of contact dermatitis, in *Textbook of Contact Dermatitis,* Rycroft, R.J.G., Menné, T., Frosch, P.J., and Benezra, C., Eds., Springer-Verlag, Berlin, 1992, 323.
2. Wilhelm, K.P., Freitag, G., and Wolff, H.H., Surfactant-induced skin irritation and skin repair: evaluation of a cumulative human irritation model by non-invasive techniques, *J. Am. Acad. Dermatol.,* in press.
3. Hirsch, A.F., *Good Laboratory Practice Regulations,* 1st ed., Marcell Dekker, New York, 1989.
4. CPMP working party, Good clinical practice for trials on medical products in the European Community, *Pharmacol. Toxicol.,* 67, 361, 1990.
5. Pinnagoda, J., Tupker, R.A., Agner, T., and Serup, J., Guidelines for transepidermal water loss (TEWL) measurement, *Contact Dermatitis* 22, 164, 1990.
6. Bircher, A., de Boer, E.M., Agner, T., Wahlberg, J.E., and Serup, J., Guidelines for measurement of cutaneous blood flow by laser Doppler flowmetry: a report from the Standardization Group of the European Society of Contact Dermatitis, *Contact Dermatitis,* 30, 65, 1994.
7. Tagami, H., Impedance measurement for evaluation of the hydration state of the skin surface, in *Cutaneous Investigation in Health and Disease,* Lévêque, J.L., Ed., Marcel Dekker, New York, 1989, 79.
8. Blichmann, C.W., Serup, J., and Winther, A., Effects of single application of a moisturizer: evaporation of emulsion water, skin surface temperature, electrical conductance, electrical capacitance, and skin surface (emulsion) lipids, *Acta Derm. Venereol.,* 69, 327, 1989.
9. Serup, J., Urea revisited: including clinical uses and evaluation by bioengineering techniques, *Acta Derm. Venereol. Suppl.,* 177, 1, 1992.
10. Serup, J., A three-hour test for rapid comparison of effects of moisturizers and active constituents (urea). Measurement of hydration, scaling and skin surface lipidization by noninvasive techniques, *Acta Derm. Venereol. Suppl.,* 177, 29, 1992.

Chapter 2.2
The Skin Integument: Variation Relative to Sex, Age, Race, and Body Region

Nadia Farinelli and Enzo Berardesca
Department of Dermatology
University of Pavia
Pavia, Italy

I. Introduction

The skin is not a uniform sheath covering the body, but a specialized organ with several functions changing from site to site. These regional variations are of great importance because they can influence skin behavior and thus susceptibility to disease. The major anatomical differences related to site involve stratum corneum thickness, distribution of appendages and melanocytes, variation in the structure of the dermo-epidermal junction and of the dermis, and changes in blood supply.[1] Anatomical changes often induce functional changes that can be quantified with combined noninvasive techniques that allow the assessment of skin function relative to sex, age, and race.[2]

II. Sex

According to the HANES survey[3] skin pathology is consistently more prevalent among males than females. Most of this higher prevalence among males is accounted for by the higher prevalence of dermatophytoses and skin tumors. This difference, however, is felt to be related to differences in behavior (hygiene and occupation) and not in skin characteristics.

There is evidence of greater skin irritability in females than males, although this difference is not reported to be large.[4,5] The skin irritability to sodium lauryl sulfate in males and females has been studied by measurement of skin water vapor loss:[6] the study showed that mean values of unirritated skin in females were significantly lower than male volunteers. The differences seem due to a lower basal metabolic rate in females. There was no significant difference between the mean values of irritated skin of male and female volunteers and the irritation index was significantly lower in males than females. The study concluded that female skin is more prone to irritation. In spite of this, some other reports confirm that there is no difference in reactivity between the sexes.[5-7] However, variations in skin reactivity during the menstrual cycle seem to occur: changes in skin extensibility and increased proneness to develop strong irritant reactions are reported during the menstrual phase.[8,9]

III. Age and Body Region

Skin aging, more or less a physiological event, is characterized by several biological and histopatho-logical changes. Transepidermal water loss (TEWL) and skin hydration both decrease during the aging process, maintaining a directly proportional relationship.

The decrease of TEWL during life is conspicuous after the age of 60.[10] Several factors may be responsible. The increased size of corneocytes and the increased thickness of the stratum corneum due to the greater accumulation of corneocytes related to an impaired desquamation[11] are factors that should be considered. Similarly, stratium corneum hydration is decreased in elderly subjects.[12,13] Reduction in moisture content is more noticeable in exposed areas, where damage is the predominant factor accentuating aging.[13] The simultaneous decrease of TEWL and water content of the corneum is a distinct feature of elderly skin. It confirms the decreased stratium corneum hydration without impairment of the barrier function. Accordingly, "dry" skin in the elderly may be differentiated from pathologically dry skin since in the latter the barrier function is defective. One of the sites where xerotic changes occur more frequently in aged skin is the extensor aspects of the lower legs. Tagami and coworkers,[14] using an evaporimeter to measure TEWL and a skin surface hygrometer[15] to evaluate the hydration state of the skin, compared hydration of the lower legs in young and aged subjects. They found that skin of aged subjects is not particularly dry as compared with young adults.

A lowered TEWL in elderly subjects was confirmed by other groups.[10] A comparative study between aged individuals and children further substantiated that the skin surface in aged people is not necessarily dehydrated, indicating that aging of the skin itself does not induce any marked derangement in stratum corneum function.

On the other hand, stratum corneum moisturization detected by an impedance technique[13] revealed differences between chronically sun-exposed skin and unexposed skin in the same individuals. Electrical skin impedance varies greatly in the different sites investigated in relation to age. Levels in elderly subjects were higher than young controls both in exposed and unexposed sites, except for the palms. Statistically significant differences were recorded only in chronically exposed areas, such as forehead and neck.

The aging process varies greatly from site to site and from individual to individual, explaining the controversial data of the literature. Furthermore, senile xerosis represents a special pathologic condition affecting only certain subjects. Stratum corneum obtained from patients with senile dry skin showed a reduced capacity for secondary bound water, which plays an important role in maintaining corneum flexibility and suppleness.[14]

The skin of children is more easily irritated than that of adults.[4] The skin of elderly persons is also more reactive than that of younger adults. This is attributed to the dry skin factor which ensues with age.[5]

Many differences between a senile and young epidermis have been described, but a consistent interpretation of the aging process has proved difficult. In part, this is because the epidermis varies from site to site.[16]

Young skin from the back,[17] like that from the scalp and axilla,[18] has deep and complex rete ridges, whereas that of the face[18] has a fairly flat dermo-epidermal junction. It is widely agreed that in areas where the junction is corrugated in youth, it becomes flattened by age.[18-23] Similarly, there are differences in epidermal thickness even in young skin. On the face or on the dorsum of the hand, for example, it is considerably greater than on the arms, legs, or trunk.[18] In many areas the whole epidermis becomes thinner with age and the cells become less evenly aligned on the basement membrane and less regular in size, shape, and staining properties.[18,22-25] The thickness of the stratum corneum is obviously not uniform over the whole body surface. In particular, the differentiation of the stratum corneum of the palms and soles is unlike that of the rest of the skin.[26]

Hammarlund et al.[27] reported regional variations in newborns in the rate of TEWL. Comparing the abdomen, forearm, and buttocks they found twofold TEWL on the buttock. The findings were partially confirmed by Osmark et al.[28] using a Meeco analyzer to detect TEWL. The technique allowed more precise measurements, not biased by air turbulence and ambient relative humidity over the probe. They recorded similar TEWL on these three sites, but found a lower hydration level (obtained by inducing skin occlusion with plastic film for 1 h) on the abdomen, reporting a decreased water-holding capacity on this site.

In adults too, regional variations of moisturization reflecting differences of thickness and function of the corneum occur. Tagami et al.[29] reported drier skin on the extremities than on the trunk. This correlates with the fact that clinically dry skin tends to develop more frequently on the limbs during winter.

The reduction in TEWL with an increased thickness of the stratum corneum is not as much as would be expected.[5] In fact, in some regions, especially on the palms and soles, TEWL increases with the thickness of the stratum corneum. The increase in thickness of the stratum corneum seems to be compensated by a corresponding increase in its diffusivity,[5] resulting in skin with a relatively uniform steady-state TEWL over many parts of the body.[26] However, not all differences in thickness are compensated in this manner and some variations in the TEWL with the regional skin sites do exist. This regional variation in TEWL and permeability cannot be explained on the basis of differences in the chemical nature of the keratin molecule.[30] The regional variations in the total lipid concentrations of the stratum corneum, however, may be the most critical factor determining the regional variations in TEWL and permeability.[31]

Thus, barrier efficiency is not uniform over the whole body surface.[32] The scrotum has long been known to be particularly permeable.[33] The face, forehead, and dorsa of the hands may also be more permeable to water than the trunk, arms, and legs. The palms are practically impermeable, except to water and most water-soluble molecules. This may be the major reason why contact dermatitis is seen less often on the palms than on the dorsa of the hands.[32]

IV. Race

A. Physiological Differences

Although stratum corneum thickness is equal in blacks and whites,[34-35] more tape strippings are required to remove the stratum corneum in blacks. Weigand and associates[36] reported this is due to an increased number of corneocyte layers in black skin. Moreover, the same study revealed a great interindividual variability in the black race, whereas data from white subjects were more homogeneous. A correlation between the number of stratum corneum layers and the degree of pigmentation has not yet been demonstrated.

The increased number of cell layers in the stratum corneum and the increased resistance to stripping could be related to lipid molecules in the intercellular matrix that increase cell cohesion. Indeed, Rienertson and Wheatley,[37] investigating the stratum corneum lipid content, found higher values in blacks. Weigand and co-workers[36] confirmed this result. Other parameters investigated in different studies, such as skin electrical resistance,[38] were consistent with these findings.[39]

B. Percutaneous Absorption

In vitro penetration of fluocinolone acetonide through skin samples obtained from amputated black and white legs revealed an increased permeability in whites.[40] *In vitro* water permeation through human skin did not reveal the racial differences that had been reported by Bronaugh and coworkers.[41]

In vivo studies show different patterns of penetration depending on the molecules tested.[42] Tritiated diflorasone diacetate does not have different pharmacokinetics in blacks and whites.[43] Wedig and Maibach[44] applied C-labeled dipyrithione in different vehicles to stripped and unstripped skin of black and white volunteers and found 34% less absorption in blacks. A significantly lower penetration in blacks (47%) was also noted when a cosmetic vehicle (1:12:22:25:39 sodium lauryl-sulfate, propylene glycol, stearyl alcohol, white petrolatum, distilled water) was compared with methyl alcohol on the forehead and when the methyl alcohol vehicle was compared with the shampoo vehicle on the scalp. The penetration of intact vs. stripped skin by either the cosmetic cream or the shampoo vehicle was not different.

Racial differences in methylnicotinate-induced vasodilatation in human skin were studied by Guy and associates; they induced vasodilatation by applying the substance to the skin and monitored the response with laser Doppler velocimetry,[45] reporting statistically indistinguishable differences among the groups in the time-to-peak response, the area under the response-time curve, and the time from the response to 75% decay. Only the magnitude of the peak response revealed some significant differences, with increased levels in young white subjects. However, no important differences seem to exist between black and white skin when tested with this chemical model.[46]

C. Biomechanical Properties, TEWL, and Susceptibility to Irritants

These parameters have been measured in whites, Hispanics, and blacks to assess whether the melanin content could induce changes in skin biophysical properties.[47] Differences appear in skin conductance but are more marked in biomechanical parameters: skin extensibility, skin elastic modulus, and skin recovery. These relative variations of the parameters on dorsal and ventral sites are different according to the races and highlight the influence of solar irradiation on skin and the role of melanin in maintaining it unaltered.

Skin lipids may play a role in modulating the relationship between stratum corneum water content and TEWL, resulting in higher conductance values in blacks and Hispanics. Previously, equal baseline TEWL on the back was reported[48,49] among whites, blacks, and Hispanics. Moreover, TEWL revealed a different pattern of reaction in whites after chemical exposure to sodium lauryl sulfate with blacks and Hispanics developing stronger irritant reactions after exposure to 2% sodium lauryl sulfate. Skin color and race are an influential factor determining skin reactivity; black skin (Negroid) is the least susceptible to irritants.[50] Darkly pigmented individuals from the Mediterranean region are also less susceptible than light-complexioned individuals. It is thus reasonable to assume that fair-skinned persons of Celtic origin (Scottish-Irish-Welsh) have the most susceptible skin.[51]

All races show significant differences between the volar and dorsal forearms.[47] These results are in apparent contrast with TEWL recordings. Indeed, the higher the stratum corneum water content, the higher TEWL may be expected.[52] The data may be explained on the basis of different intercellular cohesion, lipid composition, or hair distribution. A greater cell cohesion with a normal TEWL could result in increased skin water content. The racial variability should be taken into account in terms of different skin responses to topical and environmental agents. Race provides a useful tool to investigate and compare the effects of lifetime sun exposure. It is clearly evident that melanin protection prevents sun damage; differences between sun-exposed and sun-protected areas are not detectable in races with dark skin.

References

1. Ebling, F.J.G., Eady, R.A.J., and Leigh, I.M., Anatomy and organization of human skin, in Rook/Wilkinson/Ebling, *Textbook of Dermatology*, Blackwell Scientific Publications, Oxford, 1992, 49.
2. Leveque, J.L., Ed., *Cutaneous Investigation in Health and Disease, Noninvasive Methods and Instrumentation*, Marcel Dekker, New York, 1989.
3. Stern, R.S., The epidemiology of cutaneous disease, in *Dermatology in General Medicine*, Vol. 1, 3rd ed., Fitzpatrick, T.B., Eisen, A.Z., Wolff, K., Freedberg, I.M., and Austen, K.F., Eds., McGraw-Hill, New York, 1987, 7.
4. Kligman, A.M. and Wooding, W.M., A method for the measurement and evaluation of irritants on human skin, *J. Invest. Dermatol.*, 49, 78, 1967.
5. Pinnagoda, J., Transepidermal Water Loss: Its Role in the Assessment of Susceptibility to the Development of Irritant Contact Dermatitis, Ph.D. thesis, London University, July 1990.
6. Goh, C.L. and Chia, S.E., Skin irritability to sodium lauryl sulphate — as measured by skin water vapour loss — by sex and race, *Clin. Exp. Dermatol.*, 13, 16, 1988.
7. Coenraads, P.J., Lee, J., and Pinnagoda, J., Changes in water vapour loss from the skin of metal industry workers monitored during exposure to oils, *Scand. J. Work Environ. Health*, 12, 494, 1986.
8. Berardesca, E., Gabba, P., Farinelli, N., Borroni, G., and Rabbiosi, G., Skin extensibility time in women. Changes in relation to sex hormones, *Acta Derm. Venereol.*, 69, 431, 1989.
9. Agner, T., Damm, P., and Skouby, S., Menstrual cycle and skin reactivity, *J. Am. Acad. Dermatol.*, 24, 566, 1991.

10. Leveque, J.L., Corcuff, P., de Rigal, J., and Agache, P., In vivo studies of the evolution of physical properties of the human skin with age, *Int. J. Dermatol.,* 23, 322, 1984.
11. Nicholls, S., King, C.S., and Marks, R., The Influence of Corneocytes Area on Stratum Corneum Function. Abstract, ESDR Annual Meeting, Amsterdam, 1980.
12. Berardesca, E. and Maibach, H. I., *Bioengineering and the patch test, Contact Derm.,* 18, 3, 1988.
13. Borroni, G., Berardesca, E., Bellosta, M., Bernardi, L., and Rabbiosi, G., Evidence for regional variations in water content of the stratum corneum in senile skin: an electrophysiologic assessment, *Ital. Gen. Rev. Derm.,* 19, 91, 1982.
14. Tagami, H., in *Cutaneous Aging,* Kligman, A.M. and Takase, Eds., University of Tokyo Press, Tokyo, 1988, 99.
15. Tagami, H., Kanamura, Y., Inoue, K., Suehisa, S., Inoue, F., Iwatsuki, K., Yoshikuni, K., and Yamada, M., Water sorption-desorption test of the skin in vivo for functional assessment of the stratum corneum, *J. Invest. Dermatol.,* 78, 425, 1982.
16. Graham-Brown, R.A.C. and Ebling, F.J.G., The ages of man and their dermatoses, in *Textbook of Dermatology,* Vol. 4, 5th ed., Rook, A.J., Wilkinson, D.S., and Ebling, F.J.G., Eds., Blackwell Scientific, Oxford, 1992, 2877.
17. Eller, J.J. and Eller, W.D., Oestrogenic ointments. Cutaneous effects of topical application of natural oestrogens with report of three hundred and twenty-one biopsies, *Arch. Dermatol. Syphilol.,* 59, 449, 1949.
18. Montagna, W., Morphology of the aging skin: the cutaneous appendages, in *Advances in Biology of Skin, Vol. 6, Aging,* Montagna, W., Ed., Pergamon Press, Oxford, 1965, 1.
19. Christophers, E. and Kligman, A.M., Percutaneous absorption in aged skin, in *Advances in Biology of Skin, Vol. 6, Aging,* Montagna, W., Ed., Pergamon Press, Oxford, 1965, 163.
20. Hill, W.R. and Montgomery, H., Regional changes and changes caused by age in the normal skin, *J. Invest. Dermatol.,* 3, 321, 1940.
21. Lavker, R.M., Zheng, P., and Dong, G., Morphology of aged skin, *Dermatol. Clin.,* 4, 379, 1986.
22. Montagna, W. and Carlisle, K., Structural changes in ageing skin, *J. Invest. Dermatol.,* 73, 47, 1979.
23. Montagna, W. and Carlisle, K., Structural changes in ageing skin, *Br. J. Dermatol.,* 122 (Suppl. 35), 61, 1990.
24. Gilchrest, B.A., *Skin and Ageing Processes,* CRC Press, Boca Raton, FL, 1984.
25. Lavker, R.M., Structural alterations in exposed and unexposed aged skin, *J. Invest. Dermatol.,* 73, 59, 1979.
26. Scheuplein, R.J. and Blank, I.H., Permeability of the skin, *Physiol. Rev.,* 51, 702, 1971.
27. Hammarlund, K., Nilsson, G., Oberg, A., and Sedin, G., Transepidermal water loss in newborn infants. Relation to ambient humidity and site of measurement and estimation of total transepidermal water loss, *Acta Paediatr. Scand.,* 68, 371, 1979.
28. Osmark, K., Wilson, D., and Maibach, H.I., In vivo transepidermal water loss and epidermal occlusive hydration in newborn infants: anatomical regional variations, *Acta Derm. Venereol.,* 60, 403, 1980.
29. Tagami, H., Masatoshi, O., Iwatsuki, K., Kanamaru, Y., Yamada, M., and Ichijo, B., Evaluation of skin surface hydration in vivo by electrical measurement, *J. Invest. Dermatol.,* 75, 500, 1980.
30. Tregear, R.T., The structures which limit the penetrability of the skin, *J. Soc. Cosmet. Chem.,* 13, 145, 1962.
31. Elias, P.M., Cooper, E.R., Core, A., and Brown, B.E., Percutaneous transport in relation to stratum corneum structure and lipid composition, *J. Invest. Dermatol.,* 76, 297, 1981.
32. Baker, H., The skin as a barrier, in *Textbook of Dermatology,* Rook, A., Ed., Blackwell Scientific, Oxford, 1986, 355.
33. Smith, J.G., Jr., Fisher, R.W., and Blank, I.H., The epidermal barrier. A comparison between scrotal and abdominal skin, *J. Invest. Dermatol.,* 36, 337, 1961.
34. Freeman, R.G., Cockerell, E.G., Armstrong, J., and Knox, J.M., Sunlight as a factor influencing the thickness of the epidermis, *J. Invest. Dermatol.,* 39, 295, 1962.
35. Thomson, M.L., Relative efficiency of pigment and horny layer thickness in protecting the skin of Europeans and Africans against solar ultraviolet radiation, *J. Physiol. (London),* 127, 236, 1955.
36. Weigand, D.A., Haygood, C., and Gaylor, J.R., Cell layers and density of Negro and Caucasian stratum corneum, *J. Invest. Dermatol.,* 62, 563, 1974.
37. Rienertson, R.P. and Wheatley, V.R., Studies on the chemical composition of human epidermal lipids, *J. Invest. Dermatol.,* 32, 49, 1959.
38. Johnson, L.C. and Corah, N.L., Racial differences in skin resistance, *Science,* 139, 766, 1963.
39. Berardesca, E. and Maibach, H.I., Skin color and proclivity to irritation, in *Exogenous Dermatoses,* Menne, T. and Maibach, H., Eds., CRC Press, Boca Raton, FL 1990, 65.
40. Stoughton, R.B., Bioassay methods for measuring percutaneous absorption, in *Pharmacology of the Skin,* Montagna, W., Stoughton, R.B., and Van Scott, E.J., Eds., Appleton-Century-Crofts, New York, 1969, 542.
41. Bronaugh, R.L., Stewart, F.R., and Simon, M., Methods for in vitro percutaneous absorption studies. VII. Use of excised human skin, *J. Pharm. Sci.,* 75, 1094, 1986.
42. Berardesca, E. and Maibach, H.I., Physical anthropology and skin: a model for exploring skin function, in *Models in Dermatology 4,* Maibach, H.I. and Lowe N., Eds., Karger, Basel, 1989, 202.
43. Wickema-Sinha, W.J., Shaw, S.R., and Weber, O.J., Percutaneous absorption and excretion of tritium-labelled diflorasone diacetate, a new topical corticosteroid in the rat, monkey and man, *J. Invest. Dermatol.,* 7, 373, 1978.
44. Wedig, J.H. and Maibach, H.I., Percutaneous penetration of dipyrithione in man: effect of skin color (race), *Am. Acad. Dermatol.,* 5, 433, 1981.
45. Guy, R.H., Tur, E., and Bierke, S., Are there age and racial differences to methylnicotinate-induced vasodilatation in human skin?, *J. Am. Acad. Dermatol.,* 12, 1001, 1985.
46. Berardesca, E. and Maibach, H.I., Contact dermatitis in blacks, *Dermatol. Clin.,* 6, 363, 1988.
47. Berardesca, E., de Rigal, J., Leveque, J.L., and Maibach, H.I., In vivo biophysical characterization of skin physiological differences in races, *Dermatologica* 182, 89, 1991.
48. Berardesca, E. and Maibach, H.I., Racial differences in sodium lauryl sulphate induced cutaneous irritation: black and white, *Contact Derm.,* 18, 65, 1988.
49. Berardesca, E. and Maibach, H.I., Sodium lauryl sulphate induced cutaneous irritation. Comparison of white and Hispanic subjects, *Contact Derm.,* 19, 136, 1988.
50. Kligman, A.M., Assessment of mild irritants, in *Principles of Cosmetics for the Dermatologist,* Frost, P. and Horwitz, S.N., Eds., C.V. Mosby, St. Louis 1982, 265.
51. Frosch, P.J. and Kligman, A.M., Recognition of a chemically vulnerable and delicate skin, in *Principles of Cosmetics for the Dermatologist,* Frost, P. and Horwitz, S.N., Eds., C.V. Mosby, St. Louis, 1982, 287.
52. Rietschel, R.L., A method to evaluate skin moisturizers in vivo, *J. Invest. Dermatol.,* 70, 152, 1978.

Chapter 2.3
Seasonal Variations and Environmental Influences on the Skin

Chee Leok Goh
National Skin Centre
Singapore

I. Introduction

The skin is subject to the influence of solar radiation, temperature, humidity, domestic contactants, occupational contactants, therapeutic agents, and a host of environmental agents. All of these environmental agents may have adverse effects on the skin.

The characteristic structure of the human skin places it as an important interface between human beings and their physical, chemical, and biological environment.[1] It is a primary organ of defence and adaptation. The skin is the largest organ of the human body. It is also the largest organ that is exposed to all elements of the external environment. It is a vulnerable target of environmental agents.[2] As a target organ, the skin is capable of responding in a variety of pathologic patterns.[1] It is also an important portal of entry for potentially hazardous agents.

The cell structure and appendages of the skin provide it with defences against environmental elements. It has protective elements against physical trauma, thermal stress, solar radiation, chemical agents, fluid loss, and antimicrobial agents. The stratum corneum provides a barrier against the various physical agents and biological agents, the sweat gland activities against thermal stress, and the pigmentary system, including the melanocytes and melanin, against ultraviolet radiation.

This chapter discusses the effect of seasonal variation and environmental influences on the skin.

II. Effect of Seasonal Variation on Skin Functions

The climatic and physical environmental conditions in different latitudes of the world differ. In temperate latitudes, seasonal variations occur during different periods of the year. Such climatic differences have an influence on the integrity and functional capacity of the skin. Certain skin disorders are more prevalent in different countries because of climatic differences and, similarly, certain skin disorders tend to occur during different seasons of the year.

The following environmental factors, which change with different seasons and which have an effect on the skin, will be discussed:

1. Solar radiation
2. Temperature
3. Humidity

A. Solar Radiation

Solar radiation, in particular ultraviolet light (UVL), has immediate and long-term effects on the skin. The flux of solar radiation varies during different seasons. Skin pigment provides protection against actinic or ultraviolet radiation which causes sunburn (ultraviolet erythema — immediate effects) and fragmentation and destruction of elastic tissue fibres, UV-induced aging of the skin, actinic keratoses, skin cancer, and alteration of the immune function of the skin (long-term effects).

1. Immediate Effects of Solar Radiation

Sunburn from UVL can be elicited in all human beings but photosensitivity is inversely related to the degree of melanin pigmentation. UVL exposure results in the immediate and delayed dilation of blood vessels in the dermis which is usually confined to the irradiated sites.[3]

UVB is the major cause of sunburn from sunlight. Cutaneous reactions from UVB is influenced by environmental conditions, season of the year, latitude and time of day, altitude, atmospheric pollution, and time of exposure and skin thickness and pigmentation, as well as other factors.[4] UVB erythema becomes visible within 2 to 6 h following irradiation, reaches a maximum at 24 to 36 h, fades in 72 to 120 h and is followed in most individuals by increased skin pigmentation (tanning).[4] Increased pigmentation provides protection against further damage from UVB, since the pigment is a remarkably effective absorber of UVB.

Although UVA is 1000-fold less potent than UVB in causing erythema, its predominance in the solar spectrum reaching the earth's surface (10- to 100-fold more than UVB) may account for its toxic effect on the skin. High-intensity UVA light sources, which may emit as much as five times more UVA than sunlight, widely used for cosmetic tanning, have effects on the skin and contribute to the long-term effects of UVL on the skin.

The effect of UVB and UVA on the skin has been studied in detail by Gilcrest et al.[5,6]

2. Long-Term Effects of Solar Radiation

UVA, in addition to UVB, is believed to contribute to the long-term effects of chronic sun exposure, including premature skin aging, actinic keratosis, and skin carcinogenesis.[7]

Solar radiation is the principal cause of skin cancer in humans. The most important wavelength responsible is UVB (290 to 320 nm). Recent studies have documented that the environmental flux of UVL radiation from the sun is increasing, especially over the North and South poles.[16-18] This is contributed by the liberation into the atmosphere of tons of chlorofluorocarbons by human activities which eventually removes the protective ozone shield in the stratosphere.[19,20]

UVB irradiation has been shown to induce immunosuppression.[21] Irradiation with high-dose UVB results in "systemic immunosuppression" while exposure to low-dose UVB produces "local immunosuppression".[22,23] There is substantial evidence to implicate the effect of UVL on the epidermal Langerhans cells as the cause of the change in immunosuppression.[21] The effect of UVL on Langerhans cells studies have also been demonstrated by studies which found lower Langerhans cells density in the non-sun exposed skin compared to density in chronically sun-exposed skin.[24]

The effect of seasonal variation in UVL flux, in particular UVB, may have an influence on the immunological response of the skin to contact allergens. The afferent and efferent limbs of allergic contact dermatitis in experimental animals may be suppressed by irradiation with UVB.[25] Bruze found fewer positive patch tests per tested patient in Sweden during the summer months of June, July, and August than the other months.[26] Similarly Veien et al. in Denmark also found significantly lower patch test reactivity during the same period when compared to other months.[27]

Epidemiological evidence has also shown the association of UVL exposure, which differs in different latitudes and different seasons and skin cancers. Studies have identified that in the white populations, there was an inverse relationship between latitude and the incidence of skin cancers.[28] Skin cancers showed a rising incidence with increasing dose of UVL exposure at different latitudes in North America.[29] A linear relationship was also observed on the incidence of non-melanoma skin cancers in countries of different proximity to the Equator. A linear relationship was observed between the incidence of non-melanoma skin cancers and the annual ultraviolet solar radiation.[30]

B. Effect of Temperature on the Skin

The skin is an important thermoregulation organ. The rate of blood flow and sweating controls body temperature. The body temperature is maintained at a very constant temperature with minimal variations. This is vital to the function of the various body organs.

One of the body responses to an increase in environmental temperature is increased rate of blood flow through dilatation of dermal capillaries and stimulation of the sweat glands to increase secretion of sweat. Sweating allows the evaporation process to occur, leading to loss of skin surface heat. Increased sweating is associated with increased hydration of the stratum corneum. An increased hydration of stratum corneum will enhance the penetration of chemical agents on the skin. Excessive sweating has clinical impact on contact dermatitis. Olumide et al. reported a high incidence in Nigerian workers of contact allergy to clothing dyes from work uniforms, caused by enhanced dissolution of dyes from clothing in a hot environment.[8]

Increased sweating secondary to high environmental temperature also tends to provoke workers to discard protective clothing and therefore expose workers' skin to irritants and allergens.[9] Excessive sweating in intertriginous areas leads to skin maceration and dermatitis. It predisposes the moist skin to colonization of fungus and superficial fungal infections.[9]

Excessive sweating due to high ambient temperature may lead to sweat duct swelling, resulting in obstruction, leading to miliaria. If severe, heat exhaustion and heat stroke may occur.

C. Effect of Humidity on the Skin

The effect of high ambient humidity on the skin is similar to that of high temperature. High ambient humidity prevents the skin surface sweat from evaporating and leads to an increased hydration of the stratum corneum.

Low humidity has been reported to cause skin symptoms. Rycroft et al.[10-12] described a phenomenon of "low-humidity occupational dermatoses" which affected office workers. Affected workers presented with itch and urticaria on covered parts of the body and scaly eczema on the face, scalp, and ears. The cause was believed to be due to low relative humidity in the work environment. The symptoms in these workers improved when the relative humidity of the work environment was raised above 45%.

Gaul and Underwood reported skin chapping on the hands, lips, and face and ichthyosis on lower legs and arms in subjects who were exposed to low environmental humidity.[13] Similar findings were also reported by Chernosky in patients living in air-conditioned houses when the environmental humidity was lowered drastically.[14]

However, the effect of low humidity on skin symptoms has been disputed. Andersen et al.[15] could not evoke similar symptoms in subjects exposed to 78 h in dry air. Subjects could not accurately judge whether they were exposed to low (20%) or high air humidity (70%) when the temperatures were held constant.[15]

III. Other Environmental Factors Influencing the Skin

Other factors influencing the skin include environmental contactants. The effects of skin contactants in different countries depend on several factors. They are influenced by the prevailing type of industry,[31] availability of topical medicaments and prescribing habits of physicians,[32] cultural and traditional habits of individuals in the country, and also the fauna of the country.[9]

The type of prevailing industrial activities in a country influences the prevalence of the type of contact dermatitis. For example, contact allergy to chromates from cement (representing about 6% of patients attending its patch test clinic) was prevalent in Singapore between 1980 and 1985 and declined (to less than 1% of patients attending its patch test clinic) after 1985 when the construction industry experienced a slump. Similarly, the prevalence of contact irritant dermatitis from solvents and cutting fluids which are widely used in the electronic and metal industries recorded an increase after 1985 as the two industrial activities took more prominence.[34]

Topical medicament allergy is common in developing countries where relatively cheap and more sensitizing medicaments are more widely used than in developed countries. In India contact allergy to nitrofurazone cream and in Singapore contact allergy to proflavine lotion were common causes of contact allergy to topical medicament, respectively.[33,35] These allergens are not known to cause problems in developed countries.

Preservatives allergy also varies in different regions of the world. Formaldehyde and formaldehyde releasers are common preservatives used in cosmetics and are the common cause of contact allergies from preservative in Europe. However, contact allergies to formaldehyde and formaldehyde releaser are relatively uncommon in Singapore and probably in South East Asia. One reason could be the widespread use of Japanese cosmetics in Singapore in which formaldehyde and formaldehyde releasers are not used as preservatives.[36]

Allergies from plants differ in different regions of the world, e.g., primin allergies from primulas which are common in Europe are uncommon in tropical countries. Rhus allergy is uncommon in South East Asian countries, whereas rengas allergy is very common in South East Asian countries.[37]

The characteristic warm and humid tropical climate is unique compared to the temperate climates. The tropical climate varies minimally throughout the year. The average ambient temperature of 30°C and humidity of more than 70% are about the same throughout the year. Other types of skin disorders associated with such climate and peculiar to the tropics include:

1. Acne estivalis, a condition described in patients who developed acneiform eruption after spending time in the tropical climate. The exact mechanism is unknown but heat is believed to play a role. Heat and high humidity have been known to affect the pilosebaceous unit.[37,38]
2. Miliaria, a common disorder in the tropics resulting from swelling of the keratin lining of the sweat ducts due to heat and high humidity.
3. Skin infections, high heat, and humidity in the tropics promote sweating and sweat retention, especially on skin folds, resulting in skin maceration. Secondary bacterial infection and fungal infections on macerated skin are common.

References

1. Suskind, R.R., The environment and the skin, *Environ. Health Perspect.*, 20, 27, 1977.
2. Suskind, R.R., Environment and the skin, *Med. Clin. North Am.*, 74, 307, 1990.
3. Farr, P.M. and Diffey, B.C., The vascular response of human skin to ultraviolet radiation, *Photochem. Photobiol.*, 44, 501, 1986.
4. In *Photosensitivity Diseases. Principles of Diagnosis and Treatment,* 2nd ed., Harber, L.C. and Bickers, D.R., Eds., B.C. Decker, Toronto, 1989, 112.
5. Gilchrest, B.A., Soter, N.A., Stoff, J.S. et al., The human sunburn reaction: histologic and biochemical studies, *J. Am. Acad. Dermatol.*, 4, 411, 1981.
6. Gilchrest, B.A., Soter, N.A., Hawk, J.L.M. et al., Histologic changes associated with ultraviolet A-induced erythema in normal human skin, *J. Am. Acad. Dermatol.*, 9, 213, 1983.
7. Staberg, B., Wulf, H.C., Klemp, P. et al., The carcinogenic effect of UVA irradiation, *J. Invest. Dermatol.*, 81, 517, 1983.
8. Olumide, Y., Oleru, G.U., and Enu, C.C., Cutaneous implications of excessive heat in the work-place, *Contact Derm.*, 9, 360, 1983.
9. Goh, C.L., Exogenous dermatoses in the tropics, in *Exogenous Eczema,* Menne, T. and Maibach, H.I., Eds., CRC Press, Boca Raton, FL, 1990, 351.
10. Rycroft, R.J.G., Occupational dermatoses among office personnel, *Occup. Med.*, 1, 323, 1986.
11. Rycroft, R.J.G., and Smith, W.D.L., Low humidity occupational dermatoses, *Contact Derm.*, 6, 488, 1980.
12. White, I.R., and Rycroft, R.J.G., Low humidity occupational dermatoses — an epidemic, *Contact Derm.*, 8, 287, 1982.

13. Gaul, L.E. and Underwood, G.B., Relation of dew point and barometric pressure to chapping of normal skin, *J. Invest. Dermatol.,* 19, 9, 1952.
14. Chernosky, M.E., Pruritic skin disease and summer air conditioning, *JAMA,* 179, 1005, 1962.
15. Andersen, I.B., Lundqvist, G.R., Jensen, P.L., and Proctor, D., Human response to 78 hour exposure to dry air, *Arch. Environ. Health,* 29, 319, 1974.
16. Callis, L.B. and Natarajan, M., Ozone and nitrogen dioxide changes in the stratosphere during 1979–1984, *Nature,* 323, 772, 1986.
17. Solomon, S., Garcia, R.R., Rowland, F.S., and Wuebbles, D.J., On the depletion of the Antartic ozone, *Nature,* 321, 755, 1986.
18. Stolarski, R.S., Kreuger, A.J., Shcoeberl, M.R., McPeters, R.D., Newman, P.A., and Alper, J.C., Nimbus and satellite measurements of the springtime Antarctic ozone decrease, *Nature,* 322, 808, 1986.
19. Farman, J.C., Gardiner, B.G., and Shanklin, J.D., Large losses of total ozone in Antarctica reveal seasonal CLOx/Nox interactions, *Nature,* 315, 207, 1985.
20. McElroy, M.B., Salawitch, R.J., Wofsy, S.C., and Longan, J.A., Reductions of Antarctic ozone due to synergistic interaction of chlorine and bromine, *Nature,* 321, 759, 1986.
21. Crus, P.D. and Bergstresser, P.R., The low-dose model of UVB-induced immunosuppression, *Photodermatology,* 5, 151, 1988.
22. Bergstresser, P.R., Ultraviolet B radiation induces "local immunosuppression", *Curr. Prob. Dermatol.,* 15, 205, 1986.
23. Kripke, M.L. and Morison, W.L., Modulation of immune function by UV radiation, *J. Invest. Dermatol.,* 85, 62s, 1985.
24. Czernielewski, J.M., Masouye, I., Pisani, A., Ferracin, J., Auvolat, D., and Ortonne, J.P., Effect of chronic sun exposure on human Langerhans cell densities, *Photodermatology,* 5, 116, 1988.
25. Sjovall, P. and Moller, H., The influence of locally administered ultraviolet light (UVB) on the allergic contact dermatitis in the mouse, *Acta Derm. Venereol.,* 65, 465, 1985.
26. Bruze, M., Seasonal influence on routine patch test results, *Contact Derm.,* 14, 184, 1986.
27. Veien, N.K., Hattel, T., and Laurberg, G., Is patch testing a less accurate tool during the summer months, *Am. J. Contact Derm.,* 3, 35, 1992.
28. Scotto, J. and Fraumeni, J., Skin cancer (other than melanoma), in *Cancer Epidemiology and Prevention,* Schotterfeld, D. and Fraumeni, J., Eds., W.B. Saunders, Philadelphia, 1982, 996.
29. Russell Jones, R., Consequences for human health of stratospheric ozone depletion, in *Ozone Depletion, Health and Environmental Consequences,* Russel Jones, R. and Wigley, T., Ed., John Wiley & Sons, New York, 1989, 207.
30. Gordon, D. and Silverstone, H., Worldwide epidemiology of premalignant and malignant cutaneous lesions, in *Cancer of the skin,* Andrade, R., Ed., W.B. Saunders, Philadelphia, 1976, 405.
31. Goh, C.L., Epidemiology of contact allergy in Singapore, *Int. J. Dermatol.,* 27, 308, 1988.
32. Goh, C.L., Contact sensitivity to topical antimicrobials. 1. Epidemiology in Singapore, *Contact Derm.,* 21, 46, 1989.
33. Goh, C.L., Contact sensitivity to topical medicaments, *Int. J. Dermatol.,* 28, 25, 1989.
34. Goh, C.L., Occupational dermatitis in Singapore. Changing Pattern: 1985–1989, *Hifu* (Skin Research), 33, 95, 1991.
35. Goh, C.L., Contact sensitivity to proflavine, *Int. J. Dermatol.,* 25, 449, 1986.
36. Goh, C.L., Allergic contact dermatitis from cosmetics, *J. Derm.,* 14, 248, 1987.
37. Goh, C.L., Occupational allergic contact dermatitis from Rengas wood, *Contact Derm.,* 18, 300, 1988.
38. Lobitz, W.C. and Dobson, R.L., Miliaria, *Arch. Environ. Health,* 11, 460, 1965.
39. Taylor, J.S., The pilosebaceous unit, in *Occupational and Industrial Dermatology,* Maibach, H.I. and Gellin, G.A., Eds., Year Book Medical, Chicago, 1982, 125.

3.0

Technical Variation, Validation, and Statistics

3.1 Statistical Analysis of Sensitivity, Specificity, and Predictive Value
 of a Diagnostic Test 33
 N. Lange and M.A. Weinstock

3.2 Sample Size Calculation 43
 C. Bay and S. Møller

Chapter 3.1
Statistical Analysis of Sensitivity, Specificity, and Predictive Value of a Diagnostic Test

Nicholas Lange
Analytical Biometrics Section
Biometric and Field Studies Branch
National Institute of Neurological Disorders and Stroke
National Institutes of Health

Martin A. Weinstock
Dermatoepidemiology Unit
VA Medical Center
Roger Williams Medical Center
Brown University

I. Introduction

In a perfect world, all of our diagnostic tests would be perfectly accurate — perfectly sensitive and specific — and 100% predictive of the disorder at issue. However, whereas perfect tests are all alike in that regard, every imperfect test is imperfect in its own way. Some will miss many cases, yet make few false diagnoses; others may miss few cases, yet falsely diagnose many. This chapter reviews appropriate methods for measuring the accuracy of tests.

The example we use to illustrate these principles is the use of the aspartate aminotransferase (AST) test to diagnose hepatic fibrosis in patients receiving methotrexate therapy. Methotrexate is a very effective treatment for severe recalcitrant psoriasis. However, its long-term use is limited by the occurrence of hepatic fibrosis, which can lead to cirrhosis of the liver and death. To avoid the adverse risk, a test is needed for the early stages of hepatic fibrosis so that the methotrexate can be stopped and clinical sequelae avoided. The present recommendation for monitoring includes periodic biopsies of the liver to determine the presence or absence of hepatic fibrosis. However, the biopsies themselves can have complications, so there is a need for a less invasive, safer procedure for determining whether hepatic fibrosis has developed. One such test is the AST, a determination of the levels of aspartate aminotransferase in the blood. High levels suggest injury to the liver.

The liver biopsy is viewed as the most accurate test for hepatic fibrosis. The AST test, either alone or in combination with other factors, is the alternative test for hepatic fibrosis used for the examples in this chapter. For simplicity, the numbers used in our examples are fictional. The reader is referred to an article by O'Conner and colleagues[1] for actual observational data pertaining to the issues presented here.

II. Basic Concepts, Definitions, and Methods

We begin by defining some basic terms and concepts. Many of these ideas are motivated by the *generic 2 × 2 table*. Table 1 gives an example of such a table, showing results for 50 fictitious patients cross-classified by their AST level and the presence of hepatic fibrosis. In order to understand more fully the properties of such a test, it is useful to abstract the clinical situation. Table 2 gives the generic form of the empirical cross-classification. In general, we use lowercase characters to denote observed quantities and events, and uppercase or Greek characters to denote theoretical quantities. For instance, $p(\cdot)$ is an observed probability or relative frequency, an empirical estimate of the theoretical probability $P(\cdot)$. In addition, the symbol d denotes a positive test result, $\bar{d}$ a negative test result, D a truly diseased state, and $\bar{D}$ a truly disease-free state.

The *observed prevalence* of the disease is defined as $p(D) = (a + c)/n$, 0.46 or 46% in our example. Thus, the observed prevalence is a *marginal probability*, for this measure sums over the rows of the table, ignoring the test result. If it can be assumed that the study population is *representative* of all such patient populations, then the observed quantities can serve as *valid, accurate*

TABLE 1 The Hepatic Fibrosis Example

Elevated AST Level?	Hepatic Fibrosis? Yes	No	Total
Yes	15	3	18
No	8	24	32
Total	23	27	50

TABLE 2 The Generic 2 × 2 Table

Diagnostic Test Result	True State Diseased (D)	Disease-Free ($\bar{D}$)	Total
Positive (d)	a	b	a + b
Negative ($\bar{d}$)	c	d	c + d
Total	a + c	b + d	n

estimates of the corresponding true, theoretical quantities. In representative samples, the observed prevalence can be interpreted as an estimate of the probability that a randomly selected individual will have the disease, i.e., $p(D) = P(D)$ on the average. If, on the other hand, the patient population under study cannot be assumed to be representative of all such patients, $p(D)$ is a *biased* prevalence estimate, i.e., $p(D) \neq P(D)$ on the average. *Selection bias* would be present, for instance, if patients were either included or excluded according to criteria not accounted for in the cross-classification. In the following, however, until Section III.A, we assume that the patients tested comprise a representative sample of all such patients. The *observed sensitivity* of the diagnostic test is defined as $p(d|D) = a/(a + c)$, 0.65 in our example, the observed proportion of *true positives* among the diseased.* Similarly, the *observed specificity* of the diagnostic test is defined as $p(\bar{d}|\bar{D}) = d/(b + d)$, 0.89 in our example, the observed proportion of *true negatives*.** Sensitivity and specificity are *conditional probabilities* as they include only those patients who are truly diseased or truly disease-free in their denominators, respectively. The overall *accuracy* of the diagnostic test is the sum of these two components weighted by the observed probabilities of the conditioning events, i.e., $p(d|D)p(D) + p(\bar{d}|\bar{D})p(\bar{D})$.

Inspection of the preceding conditional probabilities yields some useful relationships. A diagnostic test that is sensitive but not specific will correctly identify a large proportion of truly positive cases at the cost of labeling a large proportion of disease-free individuals as diseased. In such a case, the proportion $p(d|\bar{D}) = b/(b + d)$ of *false positives* will be large. Conversely, a diagnostic test that is specific but not sensitive will correctly identify a large proportion of truly negative cases at the cost of *not* labeling a large proportion of diseased individuals as diseased; the proportion $p(\bar{d}|D) = b/(b + d)$ of *false negatives* will be large. An extreme and unrealistic diagnostic "test" that declares every individual as diseased will not miss a single case and thus be completely sensitive, i.e., $p(d|D) = 1$, yet have a specificity of zero; similarly, a "test" that declares every individual as disease-free would be completely specific, i.e., $p(\bar{d}|\bar{D}) = 1$, yet have a sensitivity of zero! In the other extreme and more important case, a diagnostic test for which $P(d|\bar{D})$ and $P(\bar{d}|D)$ are *both zero*, or both *assumed* to be zero, is called a *gold standard* for the disease in question. When using a diagnostic test that does not qualify as a gold standard, trade-offs are required between increasing sensitivity at the cost of decreasing specificity, and vice versa, in order to develop a test with optimal properties; see the following Section IV on *receiver operating characteristic curves* for more detail on this point.

The *predictive value of a positive test result*, or simply the *positive predictive value*, is defined as $p(D|d) = a/(a + b)$. Similarly, the *predictive value of a negative test result* is defined as $p(\bar{D}|\bar{d}) = d/(c + d)$. In our example, the positive predictive value is 0.83 and the negative predictive value is 0.75. *Note the reversal of conditioning*: given that the diagnostic test is positive, the positive predictive value is the proportion of truly positive cases, and similarly for negative test results and true negatives. A simple form of *Baye's theorem* relates the two types of conditional probabilities:

$$p(D|d) = \frac{p(d|D) \cdot p(D)}{p(d|D) \cdot p(D) + p(d|\bar{D}) \cdot p(\bar{D})} \quad (1)$$

and similarly for $p(\bar{D}|\bar{d})$. This identity is verified trivially by inspection of Tables 1 and 2.*** In words, Bayes'

---

* Read $p(d|D)$ as "the observed proportion of patients testing positive given ('|') that they are truly diseased", and similarly for other conditional probability statements

** Note that the symbol d here denotes the number of disease-free patients that have a negative test result, not to be confused with the event d, a positive test result.

*** Equation 1 is only the simplest, discrete form of Bayes' theorem, however. When there are more than two possible true states (i.e., not only D and $\bar{D}$), the sum in Equation 1 gets more lengthy, until, in the limit, the state D is replaced by a *parameter* θ and the sum replaced by an integral over all possible parameter values, so that

$$p(\theta|d) = \frac{p(d|\theta) \cdot p(\theta)}{\int p(d|\theta) \cdot p(\theta)} \propto p(d|\theta)$$

Although we do not develop this idea further here, the preceding relationship suggests *Bayesian inference* for the types of problems discussed in this chapter; interested readers should see the accessible article by Breslow[2] on the subject.

theorem applied here shows that

$$\text{positive predictive value} \qquad (2)$$
$$= \frac{(\text{sensitivity}) \cdot (\text{prevalence})}{(\text{sensitivity}) \cdot (\text{prevalence}) + (1 - \text{specificity}) \cdot (1 - \text{prevalence})}$$

Equations 1 and 2 show exactly how the predictive value of a test depends on prevalence. Sensitivity and specificity, on the other hand, are conditional measures and thus do not depend on prevalence.

The positive predictive value, $p(D|d)$, is also called a *posterior probability*, as it is the probability estimate after the test result is known. It is thus an updated version of the *prior probability* $p(D)$, the prevalence estimate, which is the *a priori* probability of disease prior to any knowledge of the test result. Equation 1 shows that this updating of prior information is given explicitly as

$$\text{posterior} \propto (\text{likelihood}) \cdot (\text{prior})$$

with *likelihood* $p(d|D)$ and constant of proportionality the denominator of Equation 1. This relationship can also be expressed in terms of *odds ratios*. Indeed, an *odds ratio version of Bayes' theorem* is

$$\frac{p(D|d)}{p(\overline{D}|d)} = \frac{p(d|D)}{p(d|\overline{D})} \cdot \frac{p(D)}{p(\overline{D})}$$

with *prior odds* $p(D)/p(\overline{D})$, 0.85 in our example, and *likelihood ratio* $p(d|D)/p(d|\overline{D})$, 5.9 in our example, and hence a posterior odds of 5.0.

III. Limitations of Sensitivity, Specificity, and Predictive Value

In this section, we discuss the role of a definitive reference test, the clinical context, and the need to consider more than a simple normal/abnormal dichotomy.

A. The "Gold Standard"

Table 1 presupposes that we know who really has hepatic fibrosis, as Table 2 presupposes that, for each person, we know whether the disease is actually present. The standard for these determinations is commonly named the "gold standard" or "definitive reference test", i.e., the test that we assume is perfectly sensitive and specific. For hepatic fibrosis, the gold standard is generally the liver biopsy.

The requirement for a gold standard raises several problems. Typically, there is some difficulty with the gold standard that motivates the search for a sensitive and specific alternative. This difficulty may also impede study of the proposed diagnostic test, such as the AST, and the gold standard in the same large group of patients. With the liver biopsy, the difficulties include a small but nonzero risk of mortality or serious morbidity as well as discomfort and expense.

The gold standard itself may be an imperfect indicator of the disease being studied. This imperfection (inaccuracy) will decrease the observed sensitivity and specificity of the diagnostic test if the gold standard's inaccuracy is independent of the diagnostic test's result. However, inaccuracies in the gold standard may have opposite effects under other circumstances. Consider again our example of the AST test for hepatic fibrosis. If the pathologist who interpreted the liver biopsy was aware of the AST test when interpreting the histopathologic specimen, in equivocal cases he or she may have over-diagnosed hepatic fibrosis in the presence of an abnormal AST, and under-diagnosed the disease when the AST was normal, therefore increasing artificially the measured sensitivity and specificity of the AST test. A second type of bias would occur if, for instance, mild asymptomatic hepatitis caused both an increase in AST levels and a systematic over-diagnosis of hepatic fibrosis, and therefore an artifactual increase in measured sensitivity and specificity. Finally, it may be the case that the gold standard cannot be applied to all individuals subject to the diagnostic test due to logistical, cost, or ethical considerations, or due to comorbidity or risk of complications. If the result of the diagnostic test is used to determine which patients are subject to the gold standard evaluation, sensitivity and specificity estimates may be biased substantially; see Section IV.C.

A gold standard may be unavailable, in which case measurements of validity, including sensitivity and specificity, become problematic. Nevertheless, validity may be tentatively assessed by measuring a surrogate endpoint known to be associated with the diagnostic test. In addition, test-retest reliability and correlation with other diagnostic tests for the disorder support validity.

B. The Clinical Context

Sensitivity and specificity are used widely in part because these measures are independent of the prevalence of the disorder, as mentioned previously. However, this independence does not imply that sensitivity and specificity are constant. The diagnostic test may have different sensitivities in different stages of the disease, or in different forms of the disease, and these differences may vary geographically. Similarly, a variety of illnesses, treatments, or medications may affect the performance of the test. The test's validity may also depend on age, gender, socioeconomic status, ethnic background, and other demographic characteristics. Hence, when interpreting published measures of test validity, careful attention must be given to the clinical context in which such measures were determined. Test validity in a highly referred patient population may differ from that in a primary care setting; spatial and temporal factors may also affect validity. Replication of validity measures under diverse conditions is therefore quite helpful to establish the generalizability of results.

C. The Artificial Dichotomy

The terms "sensitivity" and "specificity" presume that a test result is simply "normal" (negative) or "abnormal" (positive). Typically, however, the test result is measured on a continuous scale and a *critical value* chosen to dichotomize the result. For instance, results of the AST test are reported initially in Système International units. Results above the critical value are labeled as abnormal, and results below, normal. This approach has the advantage of producing simple, easy to understand results, yet it is clearly not as informative as the actual result reported on a continuous scale.

IV. Receiver Operating Characteristic (ROC) Curves

Changing the critical value of a test almost invariably changes both its sensitivity and specificity. It is useful, therefore, to understand the consequences of choosing particular critical values in more detail and to use this understanding to help make optimal choices. To address the artificial dichotomy problem, a *receiver operating characteristic (ROC) curve* for a diagnostic test is often developed.*

A. Definition of the ROC Curve

The ROC curve displays the range of sensitivities and specificities that are possible for a corresponding range of choices for the critical value. Let us assume that an AST value greater than (>) a critical value z is labeled as a positive test result and that an AST value less than or equal to ($\leq$) z is labeled as a negative test result. Table 3 shows the range of AST values found in our example, providing classifications of finer resolution than that shown in Table 1. If the total number of possible test results is equal to t ($t = 13$ in our example), then each row j defines a different cross-classification, a different 2×2 table, for each $j = 1, ..., t$. The shaded cells in the fifth row of Table 3 ($j = 5$) define Table 1. Using the subscript j to indicate this 2×2 table, we thus find that $a_5 = 15$, $b_5 = 3$, $c_5 = 23 - 15 = 8$, and $d_5 = 27 - 3 = 24$, so that $p_5(d|D) = 15/(15 + 8) = 0.65$ as noted previously. The test becomes less conservative as the critical AST value decreases, since a lower AST value is then required for the result to be reported as positive. In other words, sensitivity increases and specificity decreases as one scans the third and fifth columns from top to bottom.

The ROC curve for these data is shown in Figure 1. This curve is a plot, for $j = 1, ..., t$, of the observed sensitivities $p_j(d|D) = a_j/(a_j + c_j)$ on the vertical axis against one minus the observed specificities $1 - p_j(\bar{d}|\bar{D}) = p_j(d|\bar{D}) = b_j/(b_j + d_j)$ on the horizontal axis. Scanning the rows in Table 3 from top to bottom corresponds to starting from the lower-left diagonal point (0,0) in Figure 1 and proceeding to the upper-right diagonal point (1,1) in the plot. As indicated, a critical value of AST equal to 34 U/l shows an increase in sensitivity at no cost to specificity when compared with those values attained using a critical value of 35 U/l for this test.

ROC curves help to remove the arbitrariness of clinical decision making by allowing one to investigate and control the critical value of a test to optimize decisions. ROC curves also facilitate the comparisons among different tests. Clearly, a diagnostic test with desirable properties will be one whose ROC curve lies in the upper left-hand region of the plot, where sensitivity and specificity are maximized jointly. A completely noninformative test, no better than a coin toss, will have a theoretical ROC curve that lies exactly on the diagonal between the lower left- and upper right-hand corners.

B. Area under the ROC Curve

Single-number summaries of ROC curves are useful for judging the validity of the test in question and also for making statistical comparisons of two or more ROC curves for competing diagnostic tests. Computing the area under an ROC curve is one way to reduce it to a single quantity. The area under an ROC curve (AUC), 0.8 for instance, can be interpreted as the probability (80%) that a randomly selected case from the diseased population will have a response to the diagnostic test worse (i.e., more "abnormal", more "positive") than that of a randomly selected individual from the disease-free population. For a completely uninformative test, $AUC = 0.5$; for a perfectly accurate test, $AUC = 1.0$.

Estimates of the AUC can be obtained by parametric and nonparametric methods. A parametric method to compute the AUC (Dorfman and Alf[7]) assumes that the test results for the disease-free population are distributed as standard Gaussian (Normal) with mean 0 and variance 1, and that the test results for the diseased population are distributed also as Gaussian with mean

---

* ROC curves were first developed in signal detection theory (Peterson et al.[3]). The "operating" end of this system (in this case, the true disease state) determines the signal sent to the "receiver" (the test result), together with noise; the task is then to determine what signal was actually sent. Each point on the curve describes the receiver's criteria for distinguishing between signal and noise, and is called an operating position on the curve. A classic text on ROC techniques is that by Green and Swets;[4] for medical applications, see, for instance, Metz.[5]

TABLE 3 Data Table Used to Construct the Empirical ROC Curve

Row (j)	Critical Value of the AST Test Result, z	Diseased Subjects With AST = z	Diseased Subjects With AST > z	Disease-Free Subjects With AST = z	Disease-Free Subjects With AST ≤ z
1	>38	9	0	0	27
2	38	2	9	0	27
3	37	0	11	2	27
4	36	4	11	1	25
5	35	3	15	0	24
6	34	2	18	1	24
7	33	1	20	3	23
8	32	0	21	2	20
9	31	1	21	1	18
10	30	1	22	4	17
11	29	0	23	9	13
12	28	0	23	2	4
13	<28	0	23	2	2
Total		23		27	

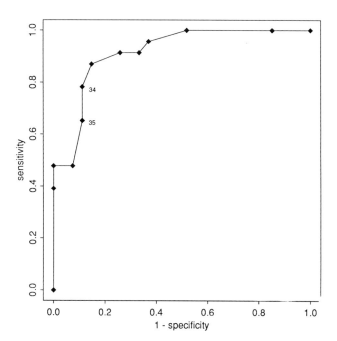

FIGURE 1 The empirical ROC curve for the hepatic fibrosis example; from Table 3.

μ and variance σ^2. Under these assumptions it has been shown that

$$AUC_1 = \Phi\left(\frac{a}{\sqrt{1+b^2}}\right),$$

with $a = \dfrac{\mu}{\sigma}$ and $b = \dfrac{1}{\sigma}$

where $\Phi(\cdot)$ is the standard Gaussian distribution function. If one makes the additional simplifying assumption that the points of the empirical ROC curve lie on the smooth curve defined by the Gaussian distribution functions, as they nearly do in our example, estimates of the parameters μ and σ^2 can be obtained by routine methods; otherwise, iterative maximization routines are required. The simplifying assumption in our case yields estimates of the mean and variance of the diseased group, which are simply the sample mean and sample variance of the distribution of test responses in the diseased group after these responses have been centered and scaled by the mean and the variance of the responses in the disease-free group. Under that simplifying assumption, our example yields estimates $\tilde{\mu} = 2.028$, $\tilde{\sigma} = 0.971$ and give an estimate $AUC_1 = \Phi(1.455) = 0.9272$. Maximum likelihood estimates, obtained from the program ROCFIT courtesy of C. Metz, are $\hat{\mu} = 2.283$, $\hat{\sigma} = 0.929$ and give an estimate $AUC_{ML} = \Phi(1.443) = 0.9255$.

One can use the trapezoidal rule to obtain a simple nonparametric estimate of the AUC. If the ROC curve has been evaluated at t points, then the AUC can be decomposed into $t-1$ trapezoidal "strips" ($t = 13$ in our example, 12 trapezoids). The area of each trapezoid is the product of the length of its base and its average height; the AUC is the sum of these areas. That is, for $j = 1, \ldots, t$, the horizontal coordinates are $x_j = 1 - p_j(d|\bar{D})$, i.e., the observed proportions of false positives at the jth level of the AST test, and the vertical coordinates are $y_j = p_j(d|D)$, i.e., the observed proportions of true positives at the jth level of the AST test. The area under the ROC curve by the trapezoidal approximation is thus

$$AUC_2 = \sum_{j=1}^{t-1}(x_{j+1} - x_j)\frac{y_{j+1} + y_j}{2}$$

The trapezoidal approximation to the area under the ROC curve shown in Figure 1 is $AUC_2 = 0.9147$. A trapezoidal approximation such as AUC_2 is smaller than the area under any smooth concave curve connecting the observed points, such as that assumed for AUC_1 and AUC_{ML}. It can be the case, however, that no smooth concave curve may exist, as for instance when diseased and disease-free populations are Gaussian with unequal variances not equal to 1. In such a situation, the ROC curve is sigmoidal in shape and $AUC_2 > AUC_{ML}$.

It should be noted that the trapezoidal approximation reflects the fact that the results have been "binned": there are ties among diseased and disease-free individuals at identical AST values. Without such binning, the ROC curve has a staircase appearance, taking discrete jumps connected by horizontal and vertical line segments throughout. The nonparametric estimate of the AUC is thus underestimated by the binned trapezoidal approximation (see, for instance, Zweig and Campbell[6,6a]). The magnitude of the underestimation depends of course on the number of ties and is an issue worthy of consideration when the results of a diagnostic test have been lumped into categories at a coarser resolution (e.g., "definitely abnormal", "probably abnormal", "equivocal", "probably normal", "definitely normal") than those reported originally.

Bamber[7a] and Hanley and McNeil[8] note the equivalence of the distribution of AUC_2 to that of the Wilcoxon-Mann-Whitney statistic:[9] both measure the probability of correctly ranking a pair of disease-free and diseased patients among all such pairs in the study population. This equivalence is important for it enables one to estimate the variance of AUC_2 explicitly. Specifically, suppose that there are n_D diseased and $n_{\bar{D}}$ disease-free patients. Let the symbol $A = AUC_2$ denote the area under the ROC curve as approximated by the trapezoidal rule. Denote by q_1 the estimated probability that two randomly selected and truly diseased patients are ranked below one randomly selected and truly disease-free patient. In addition, let q_2 denote the estimated probability that one randomly selected and truly diseased patient is ranked below two randomly selected and truly disease-free patients. Then, the estimated variance of AUC_2 is

$$\text{var}[AUC_2] = \text{var}[A] = \frac{1}{n_D n_{\bar{D}}}$$
$$\cdot \left[A(1-A) + (n_D - 1)(q_1 - A^2) + (n_{\bar{D}} - 1)(q_2 - A^2) \right]$$

Hanley and McNeil[8] approximate q_1 by $A/(2-A)$ and q_2 by $2A^2/(1+A)$. Weiand et al.[9a] estimate q_1 and q_2 directly. In our example, $A = 0.9147$, as stated previously. Application of the preceding approach to our example yields $q_1 = 0.8428$, $q_2 = 0.8740$. Hence, the estimate of the area under the ROC curve has an estimated standard error of $\text{SE}[A] = \sqrt{\text{var}[AUC_2]} = \sqrt{0.0019} = 0.0436$. A test of the null hypothesis that the true AUC is only 0.5 yields a z-score of $z = (0.9147 - 0.5)/0.0436 = 9.512$ using a standard Gaussian reference distribution and clearly rejects this null hypothesis.

C. Correction of the ROC Curve for Verification Bias

Note that ROC curve analyses may need to be corrected for *verification bias* in certain cases. Verification bias arises when the subjects used to assess the properties of the test have been selected in a nonrandom manner (Begg and Greenes,[10] Gray et al.,[11] Begg[12]). Proportions of subjects selected for test validation from a larger population of subjects may differ in some systematic, nonrandom manner across the levels of test results. In general, if the selection bias favors inclusion of more subjects with high "abnormal" test results, then the reported sensitivity of the test is inflated artificially. Conversely, if selection bias favors inclusion of more subjects with low "normal" test results then, in general, the reported specificity of the test would be inflated artificially. If selection bias favors higher proportions of subjects on both extremes of the test results, then both sensitivity and specificity are generally inflated.

Verification bias can be corrected if the sampling fractions across the levels of test results are available. In our example, suppose that one knows the total number of available subjects at each level of the AST test results, only a fraction of whom have been selected for test verification. Such a possibility is shown in Table 4, giving the cross-classifications of a larger fictional population of subjects from which Table 1 was constructed. Table 4 is Table 1 with the second-row entries increased by a factor of two. The subjects comprising Table 1 are a nonrandom sample from Table 4: those testing positive are sure to be included in the verification sample whereas subjects with negative test results have only a 50:50 inclusion probability. Sensitivity has been inflated artificially. (Note, however, that predictive value remains unchanged; both positive and negative predictive values are not altered when a single row is multiplied by a constant.) Correction of the ROC curve analysis for verification bias employs Bayes' theorem, as follows. Let the symbol s denote a selection indicator for positive and negative subjects, so that $p(s|d) = 18/18 = 1.00$ for those testing positive — sure inclusion — and only $p(s|\bar{d}) = 32/64 = 0.50$ for those testing negative. By definition, one does not know the true disease status for those subjects not selected for verification. One knows only that fraction testing positive out of the total number of subjects, i.e., $p(d) = 18/82$ or 22%, and, of the selected subjects testing positive, what fraction were found to be truly diseased, i.e., $p(D|d,s) = 15/18$ or 83% as given previously in Table 1. Similar

TABLE 4 The Cross-Classification of the Larger Population of Subjects from which the Verified Sample Shown in Table 1 was Drawn

	Hepatic Fibrosis?		
Elevated AST Level?	Yes	No	Total
Yes	15	3	18
No	16	48	64
Total	31	51	32

calculations apply for those testing negative and selected. One makes the additional assumption, plausible under the null hypothesis of no association, that true disease state and selection are *conditionally independent* given the test result. In other words, assume that $P(D,s|d) = P(D|d) \cdot P(s|d)$, implying $P(D|d) = P(D|d,s)$, and similarly for $P(D|\bar{d},s)$. Then, by Bayes' theorem, the observed sensitivity corrected for verification bias is

$$p*(d|D) = \frac{p(D|d,s) \cdot p(d)}{p(D|d,s) \cdot p(d) + p(D|\bar{d},s) \cdot p(\bar{d})}$$

$$= \frac{\left(\frac{15}{18}\right) \cdot \left(\frac{18}{82}\right)}{\left(\frac{15}{18}\right) \cdot \left(\frac{18}{82}\right) + \left(\frac{8}{32}\right) \cdot \left(\frac{64}{82}\right)}$$

$$= 0.484$$

matching the sensitivity indicated in the complete data Table 4. Similar calculations yield a corrected specificity of 0.941, also matching the specificity indicated in Table 4.

Table 5 is a separate example of verification bias at a higher resolution, at a finer level of detail. Table 3 has been extended in Table 5 to include two additional columns that give the total numbers of subjects, twice the number shown in Table 1, and fractions selected at each level of the test results. (Results given in Table 1 are again indicated by shading.) The fractions of selected subjects with high AST scores are consistently greater than the fractions of selected subjects with mid-level and low AST scores: the ROC curve analysis is thus again susceptible to verification bias. To see such biases in finer detail, we need a little more notation. Denote by $q_k(\cdot)$ that fraction of patients out of the total who have AST result at level k, and let $q_k(\cdot|\cdot,s)$ denote that fraction of patients giving positive or negative test results out of those included. For instance, at $k = 4$, $q_k(d) = 7/100$, $q_k(D|D,s) = 4/(4+1)$, $q_4(\bar{D}|d,s) = 1/(4+1)$, and, again, $q_4(d) = 7/100$. (This latter fraction is equal to $q_4(\bar{d})$: the fourth row in Table 5 represents the same fraction of available subjects regardless of their disease states. For each k, the fractions $q_k(d)$ and $q_k(\bar{d})$ are identical.) Observed sensitivities $p_j^*(d|D)$ corrected for verification bias are thus

$$p_j^*(d|D) = \frac{\sum_{k=1}^{j} q_k(D|d,s) \cdot q_k(d)}{\sum_{k=1}^{t} q_k(D|d,s) \cdot q_k(d)}, \quad j = 1,...,t$$

yielding corrected vertical coordinates for the ROC curve. Similar weighted partial sums over disease-free patients given observed negative test results and the inclusion of these patients in the verified sample yield corrected horizontal coordinates, $p_j^*(d|\bar{D})$, $j = 1, ..., t$.

TABLE 5 Augmented Data Table Used to Correct the ROC Curve for Verification Bias

		Verification Sample				Total Patient Population	
		Diseased		Disease-Free			
Row (j)	Critical Value of the AST Test Result, z	Subjects With AST = z	Subjects With AST > z	Subjects With AST = z	Subjects With AST ≤ z	Available	Fraction Selected
1	>38	9	0	0	27	9	1.00
2	38	2	9	0	27	2	1.00
3	37	0	11	2	27	2	1.00
4	36	4	11	1	25	6	0.83
5	35	3	15	0	24	4	0.75
6	34	2	18	1	24	4	0.75
7	33	1	20	3	23	6	0.67
8	32	0	21	2	20	4	0.50
9	31	1	21	1	18	5	0.40
10	30	1	22	4	17	14	0.36
11	29	0	23	9	13	28	0.32
12	28	0	23	2	4	7	0.29
13	<28	0	23	2	2	9	0.22
Total		23		27		100	0.50

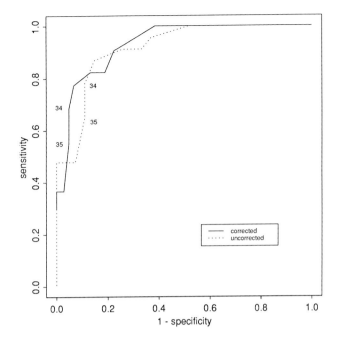

FIGURE 2 The corrected and uncorrected ROC curves for the hepatic fibrosis example; from Table 5.

Figure 2 shows the corrected and uncorrected ROC curves deriving from Tables 3 and 5. Although the corrected ROC curve appears better overall, the indicated AST values of 35 and 34 U/l demonstrate inflated sensitivities of the uncorrected results at the cost of deflated specificities.

D. Regression Methods

In addition to the ROC curve methods discussed here, there are several multivariable methods that should be mentioned, including ordinal regression methods (see McCullagh[13]) and logistic regression methods (see, for instance, Hosmer and Lemeshow[14] and Hunink and Begg[15]). The purpose of these methods is to calculate a single number that will predict the presence or absence of the disorder better than the diagnostic test alone. In order to do so, this single number is a function of the test result as well as other factors that are correlated with disease state, such as gender, age, and other intrinsic and extrinsic effects mentioned in Section III.B. The multivariable function that produces this single number is termed a *prediction rule*. The prediction rule can be subjected to ROC analysis and its AUC compared with that of the original diagnostic test alone. For instance, if gender, age, etc., are correlated with the outcome of the diagnostic test, then logistic regression methods allow for the inclusion of covariates x in a multivariable model that improves predictive performance. Estimated coefficients from the fit of such a model are combined with covariate values linearly to yield a prediction rule with superior performance, i.e., higher values of $p(D|d,x)$ than those attainable through ROC analyses that do not accommodate such factors.

V. Recommendations

We conclude the chapter with a brief discussion of additional elements of test evaluation and with our specific recommendations for future reports.

A. Considerations Other than Validity in Test Evaluation

Several considerations pertain to the assessment of tests beyond quantification of test validity. If the reasons for the observed inaccuracies are discovered, then it may be possible to enhance validity or to identify settings in which validity is greatest and poorest. It is important to assess both the reliability and validity of the test, since poor reliability may be an important source of inaccurate individual test results. It is also important to determine test performance outside of the research setting and under the conditions used by others.

Issues related to associated costs and health risks must also be considered. Results of cost-benefit and cost-effectiveness analyses may be crucial to the test's adoption in settings other than one's own. The test itself may have an impact on the clinical setting in which it is used; determination of appropriate circumstances for the test must be made. Devising a better test may not be worthwhile if test results will not affect significant therapeutic decisions. Finally, the consequences of inaccurate test results must also be considered. If a diagnostic test is less costly yet less accurate, then the consequences of errors that may follow (e.g., suffering, disability, and death) may outweigh the savings in monetary cost. In some circumstances, a formal decision analysis which takes into account all of the preceding issues may be worthwhile.

B. Specific Recommendations for Future Reports

The following 12 criteria may be useful in evaluation of future study reports that claim to validate diagnostic tests (from Weinstock[16]):

1. The test and other potential predictors of the disorder, the disorder to be diagnosed, and the gold standard for diagnosis of the disorder are defined with sufficient clarity and detail that an independent replication of the study may be conducted.
2. The population for which the test was validated is described. This includes the spectrum of disorder among those affected, the diagnoses among those not affected, demographic and medical data,

selection criteria for the test and evaluation with the gold standard, and any relevant referral patterns.
3. Statistical methods are described, applied appropriately, and cited in the literature clearly.
4. The test is interpreted blindly with respect to the gold standard diagnosis and also to the proposed test. In some circumstances, it may be reasonable to assess whether or not attempts to blind the observers were successful.
5. The reliabilities of the test and of the gold standard are estimated and reported.
6. The sensitivity, specificity, and predictive values of the test are calculated, and where appropriate, the dependence of these characteristics on other medical or demographic factors is estimated.
7. The ROC curve for the test is presented and the area under the ROC curve is also calculated and reported, if the test results are reported on an ordinal or interval scale.
8. The relation of the test to other predictors of the disorder, including results of other tests, is assessed. The incremental value of the test is determined and prediction rules are considered.
9. If a prediction rule is suggested, it is defined precisely, and its validity are evaluated and reported. If feasible, its validity is determined in a group not used to derive the prediction rule, or other statistical techniques are used to estimate its validity in such groups. The derivation of the prediction rule is described clearly, including variables considered and inclusion criteria.
10. Consideration is given to the generalizability and possible sources of bias.
11. Consideration is given to mechanisms that may account for the observed inaccuracies.
12. Consideration is given to the impact of the test in practice, i.e., effect on treatment decisions, consequences of inaccuracies, testing complications, costs, and benefits.

As diagnostic testing becomes more sophisticated, more costly, and scrutinized more intensively, proper attention to quantification of validity becomes crucial to test adoption, dissemination, and appropriate use.

References

1. O'Conner, G.T., Olmstead, E.M., Zug, K., Baughman, R.D., Beck, J.R., Dunn, J.L., and Lewandowski, J.F., Detection of hepatotoxicity associated with methotrexate therapy for psoriasis, *Arch. Dermatol.*, 125, 1209, 1989.
2. Breslow, N., Biostatistics and Bayes (with comments), *Stat. Sci.*, 86, 557, 1990.
3. Peterson, W.W., Birdsall, T.G., and Fox, W.C., The theory of signal detection, *Trans. IRE Professional Group on Information Theory*, PGIT-4, 171, 1954.
4. Green, D.M. and Swets, J.A., *Signal Detection Theory and Psychophysics*, revised edition, Krieger, Huntington, NY, 1974.
5. Metz, C.E., Basic principles of ROC analysis, *Semin. Nucl. Med.*, 8, 283, 1978.
6. Zweig, M.H. and Campbell, G., Receiver-operating characteristic (ROC) plots: a fundamental evaluation tool in clinical medicine, *Clin. Chem.*, 39/4, 561, 1993.
6a. Campbell, G.C., Advances in statistical methodology for the evaluation of diagnostic and laboratory tests, *Stat. Med.*, 13, 499, 1994.
7. Dorfman, D.D. and Alf, E., Maximum likelihood estimation of parameters of signal detection theory and determination of confidence intervals — rating method data, *J. Math. Psychol.*, 6, 487, 1969.
7a. Bamber, D., The area above the ordinal dominance graph and the area below the receiver operating characteristic curve, *J. Math. Psychol.*, 12, 387, 1975.
8. Hanley, J.A. and McNeil, B.J., The meaning and use of the area under a receiver operating characteristic (ROC) curve, *Radiology*, 143, 29, 1982.
9. Colton, T., *Statistics in Medicine*, Little, Brown and Company, Boston, MA, 1974.
9a. Wieand, S., Gail, M.H., James, B.R., and James, K.L., A family of nonparametric statistics for comparing diagnostic markers with paired or unpaired data, *Biometrika*, 76, 585, 1989.
10. Begg, C.B. and Greenes, R.A., Assessment of diagnostic tests when disease verification is subject to selection bias, *Biometrics*, 39, 206, 1983.
11. Gray, R., Begg, C.B., and Greenes, R.A., Construction of receiver operating characteristic curves when disease verification is subject to selection bias, *Med. Dec. Making*, 4, 151, 1984.
12. Begg, C.B., Biases in the assessment of diagnostic tests, *Stat. Med.*, 6, 411, 1987.
13. McCullagh, P., Regression models for ordinal data (with discussion), *J. R. Stat. Soc. Ser. B*, 42, 109, 1980.
14. Hosmer, D.W. and Lemeshow, S., *Applied Logistic Regression*, John Wiley & Sons, New York, 1989.
15. Hunink, M.G. and Begg, C.B., Diamond's correction method: a real gem or just cubic zirconium, *Med. Dec. Making*, 11, 201, 1991.
16. Weinstock, M.A., Validation of a diagnostic test, *Arch. Dermatol.*, 125, 1260, 1989.

Chapter 3.2
Sample Size Calculation

Claus Bay and Susanne Møller
Mathematical Statistical Department
Leo Pharmaceutical Products
Ballerup, Denmark

I. Introduction

Planning a scientific experiment consists of deciding on a number of equally important components: main purpose, design, variables to measure, methods of measurements, and hypotheses to test. The statistical planning is an integral part of this process. The statistical methods and tests of the specified hypotheses should be considered at an early stage in the process.

An important design feature is the number of observations on a particular response variable that the experiment should produce. The consequence of not considering this in relation to the questions the experimenter wants the study to answer may very well be that the size of the study is inadequate and therefore turns out to be inconclusive.

The sample size must be motivated by statistical considerations concerning *difference to detect* and the *power* of statistical test (e.g., as stated in the EEC guidelines for *Good Clinical Practice for Trials on Medical Products* [GCP][1] in Section IV.C).

The idea underlying the calculation of sample size of an experiment is to be able to detect, with a suitably high probability, an important difference between experimental units (e.g., treatment groups in a clinical trial) *if such a difference exists.*

Thus, the experimenter should decide on the size of the difference he would not like to overlook (often called the minimal clinical relevant difference) and how sure he would like to be on the decision made from the study, i.e., the probability of finding a difference if it really exists (= the power) and the probability of finding a difference when no true difference exists (= level of significance). Also, it should be decided how to measure the difference. When a number of variables are measured, it is preferable to name one variable or derived parameter to be of primary interest and consider the rest of secondary importance. The sample size is then determined based on this primary variable. If this procedure is considered infeasible, the sample size can be determined for all those variables considered of primary interest, and the study is dimensioned according to the largest of the calculated sample sizes, in this way assuring at least the specified power in respect of all variables.

The following sections will give a simple introduction to how sample size calculations can be done for most experiments and examples to illustrate the methods will be given. Only two-sided tests are considered, since one-sided tests are rarely used in practice. It is emphasized that the formulas are only approximate but for the vast majority of cases the gain in accuracy by applying the exact methods is negligible and need only to be considered for small sample sizes of less than about 20 for continuous data. It is beyond the scope of this chapter to go into a detailed discussion of the mathematical statistical theory from which the following formulas are derived. The interested reader is referred to Desu and Raghavarao[2] for a thorough introduction to *sample size methodology*. Altman[3] gives an excellent introduction to statistical methods in general for medical research and in Cox and Hinkley[4] the more theoretical aspects of test theory can be found. The special case of bioequivalence studies in general is dealt with in Chow and Liu.[5]

II. Sample Size and Power

To be able to understand the calculation of sample size it is important to understand the concept of *power* of a statistical test. Consider the situation where the means μ_1 and μ_2 of two normal distributed statistics are compared by a U-test. The test statistic $|U|$ is the numerical value of the difference between the means divided by its standard error, distributed as $N(0,1)$. The conventional null hypothesis is then: $H_0: \mu_1 - \mu_2 = 0$ against the alternative hypothesis $H_A: \mu_1 - \mu_2 \neq 0$.

The *P-value* of the test is *the probability of having observed the actual data (or more extreme data) when H_0 is true*, i.e., true means are equal, $P = P(|U| > u_{1-\alpha/2} | \mu_1 = \mu_2)$, where $u_{1-\alpha/2}$ is the $1-\alpha/2$ fractile of the standard normal distribution function (e.g., $\alpha = 0.05$: $u_{1-\alpha/2} = 1.96$ and $P(|U| > 1.96 | \mu_1 = \mu_2) = 0.05$). The cut-off point for statistical significance, *the significance level,* denoted by α, is equal to the probability of rejecting the null hypothesis when it is true, i.e., the probability of obtaining a "false-positive result". This value is also referred to as the type I error.

From the specification of the alternative hypothesis it appears that a whole range of values of μ_1 and μ_2 are contained in the set defined by H_A. For every pair of values the probability of accepting H_0 when it is false: $\beta = P(|U| < u_{1-\alpha/2}| \mu_1 - \mu_2 \neq 0)$, i.e., the probability of a "false-negative result" can be calculated. This value is called the type II error.

The power function $f(\alpha,\delta) = P(|U| > u_{1-\alpha/2}| \mu_1 - \mu_2 = \delta)$ is monotonously increasing with increasing value of δ. Note that the power function is equal to α for δ = 0. This follows from $f(\alpha,0) = P(|U| > u_{1-\alpha/2}| H_0) = \alpha$!

For a specific alternative H_1 defined by $\mu_1 - \mu_2 = \delta_0$ the type II error, β, is equal to $1 - f(\alpha,\delta_0)$. The quantity 1 – β is called the *power* of the test. When β is high, the power 1 – β is low, and it is unlikely that a true difference will be detected, i.e., yielding a statistically significant result of the statistical test, as the probability of failing to reject H_0 is high even when H_0 is false. This is obviously not desirable for the experimenter, so the aim is to have high power.

Let us now examine the general situation where the test statistic X is normal distributed as $N(0, \Sigma_0^2)$ under the null hypothesis, H_0, and normal distributed as $N(\delta, \Sigma_1^2)$ under the alternative hypothesis, H_1, where δ <0 or δ >0. Often Σ_0^2 and Σ_1^2 will depend on N: $\Sigma_0^2 = \sigma_0^2/N$ and $\Sigma_1^2 = \sigma_1^2/N$. With the significance level α and power 1 – β Lachin[6] shows that the sample size in this case is obtained by

$$N(\alpha,\beta,\delta) = \left[\frac{u_{1-\alpha/2}\sigma_0 + u_{1-\beta}\sigma_1}{\delta}\right]^2$$

In the following it will be shown how this simple expression can be used in the case of Student's *t*-test for equality of means of normal distributed variables with unknown variance and for chi-square tests for proportions. The corresponding u-values for commonly used values of α and β appear in Table 1 (e.g., α = 5%, $u_{1-\alpha/2}$ = 1.96 and β = 20%, $u_{1-\beta}$ = 0.84). With these and the above formula, the sample size can be determined for most experiments.

III. *t*-Test for Continuous Data

Let $(x_i)_{i=1,...,N}$ and $(y_i)_{i=1,...,N}$ be sets of mutually independent observations from normal distributions with means μ_1 and μ_2, respectively, and variance σ^2. The usual *t*-test statistic is $T = (\overline{x} - \overline{y})/s\sqrt{2/N}$ where $\overline{x}$ and $\overline{y}$ are the averages of the x_is and y_is, respectively, and s is the pooled estimate of the standard deviation σ

TABLE 1 Commonly Used u-Fractiles

Significance Level, α	$u_{1-\alpha/2}$	Type II Error, β	$u_{1-\beta}$
1%	2.58	20%	0.84
5%	1.96	10%	1.28
10%	1.65	5%	1.65

from the two samples. The difference $\overline{x} - \overline{y}$ is normal distributed as $N(0, 2\sigma^2/N)$ under H_0 and normal distributed as $N(\delta_0, 2\sigma^2/N)$ under H_1. In this case the sample size of each group necessary to have power 1 – β to detect a difference equal to or greater than δ_0 is

$$N = 2\left[\frac{(u_{1-\alpha/2} + u_{1-\beta})s}{\delta_0}\right]^2$$

The variances are equal under H_0 and H_1, and σ_0 and σ_1 are substituted by the estimate s. Because the true variance is unknown and replaced by an estimate of the variance the formula is only approximate since T is t distributed. The exact value of N is obtained by substituting the u-fractiles with the corresponding t-fractiles. N is then found by iteration.[2] For paired data, s is the estimate of the *intrasubject* standard deviation.

Example with independent data — Consider a clinical study in psoriatic patients. Two topical treatments are compared in a parallel group design. The response variable is the percentage reduction in Psoriatic Area and Severity Index (PASI, Frederikson and Petterson[7]) from start of treatment to end of treatment. The between-patient standard deviation of the percentage change in PASI is assumed to be 35 %-points. The clinician wants with 80% probability (power) to detect a true difference between treatments of 10%-points. The sample size calculation gives

$$N = 2\left[\frac{(1.96 + 0.84)35}{10}\right]^2 = 192 \text{ patients in each group}$$

Example with paired data — Consider an experiment in psoriatic patients where two topical treatments are compared in a right/left design. The response variable is the change in skin thickness measured by ultrasound scanning from start of treatment to end of treatment. The intrapatient standard deviation of the change in skin thickness is assumed to be 0.25 mm. The clinician wants with 80% probability to detect a true difference between treatments of 0.1 mm. The total sample size is then

$$N = 2\left[\frac{(1.96 + 0.84)0.25}{0.10}\right]^2 = 98 \text{ patients}$$

IV. Chi-Square Test for Dichotomous Data

Let $(x_i)_{i=1,...,N}$ and $(y_i)_{i=1,...,N}$ be sets of mutually independent observations from binomial distributions with means (probabilities) π_1 and π_2 respectively. These parameters are estimated by the observed frequencies p_1 and p_2. The usual chi-square test statistic is used to test the null hypothesis that $\pi_1 = \pi_2 = \pi_0$. The general formula for sample size can be applied by noting that the square root of a chi-square distributed variable with one degree of freedom is normally distributed. In this case the standard error of the difference $p_1 - p_2$ depends on the values of the proportions. The standard deviation of $p_1 - p_2$ under H_0 and under H_1 is estimated by

$$\sigma_0 = \sqrt{2\pi_0(1-\pi_0)} \quad \text{where } \pi_0 = (\pi_1 + \pi_2)/2$$

$$\sigma_1 = \sqrt{\pi_1(1-\pi_1) + \pi_2(1-\pi_2)}$$

Using these expressions, the sample size for each group necessary to have power $1 - \beta$ against the alternative given by (π_1, π_2) is

$$N = \left[\frac{u_{1-\alpha/2}\sqrt{2\pi_0(1-\pi_0)} + u_{1-\beta}\sqrt{\pi_1(1-\pi_1) + \pi_2(1-\pi_2)}}{\pi_1 - \pi_2}\right]^2$$

Extensive tabulations of the sample size using the exact distribution can be found in Fleiss.[8]

Example with independent data — Consider a clinical study in psoriatic patients. Two topical treatments are compared in a parallel group design. The response variable is the proportion of patients who achieve a marked improved or cleared status at the end of treatment according to the investigator's overall assessment of treatment response. It is assumed that the average proportion of response according to this criteria is about 60%. The clinician wants with 80% probability to detect a true difference between treatments of 20%-points corresponding to $\pi_1 = 0.5$ and $\pi_2 = 0.7$. The sample size calculation gives

$$N = \left[\frac{1.96\sqrt{2 \cdot 0.6(1-0.6)} + 0.84\sqrt{0.7(1-0.7) + 0.5(1-0.5)}}{0.7 - 0.5}\right]^2 = 93$$

patients in each group.

V. Discussion and Recommendations

We have seen from the previous examples that the calculation of sample size is technically very simple and can be done with the use of a pocket calculator. The difficulties lie in specifying the different elements of the calculations. Many statisticians have experienced that making an experimenter decide on the power and the difference to detect is very difficult. Thus, the emphasis in the previous section to explain the concept of power of a statistical test. The fixing of power at 80% or 90% is, of course, in many cases rather arbitrary, but it is recommended to carry out experiments with at least 80% power of the statistical test, the reason being that the power can be thought of as the receiver's (experimenter's) risk of getting a defective item (inconclusive result).

The decision about the *difference to detect* will depend on the purpose of the experiment. Suppose a novel drug to treat patients with a disease where no effective treatments are available, then even a small effect may be of clinical relevance. Conversely, in patients suffering from a disease with several efficacious treatments, a new drug may only be of interest if it shows a much better effect than placebo in a placebo-controlled trial. In any case the difference that one can detect with a realistic number of observations depends, as we have seen, on the standard deviation of the response variable.

How should the standard deviation be determined? The ideal situation is that it is well established from documented experiments with the same response variable, associated measuring technique, and comparable subject populations. However, this is often not the case. One solution is to perform a small pilot study to obtain an initial estimate of the standard deviation. If this is infeasible it is the author's experience that, as a rule of thumb the biological inter-individual coefficient of variation (CV) of many parameters is about 40%.

Another problem that often arises is that in published data only the interindividual standard deviation is reported. This is not of much help in an experiment where differences in responses are measured within individuals. Since the variance of the difference between related (paired) observations is equal to $2\sigma^2 - 2\tau$ where σ is the interindividual standard deviation and τ is the (unknown) covariance. Often the covariance will be at least half the size of the interindividual variance, so in many cases an upper limit for the relevant intrasubject variance σ_{intra}^2 is s^2, the between-individuals variance. Thus, the estimate of the interindividual variance can, in many cases, be used as an estimate of the unknown intraindividual variance.

References

1. CPMP Working Party on Efficacy of Medical Products, Note for guidance, Good clinical practice for trials on medical products in the European Community, Commission of the European Communities, 1990.
2. Desu, M.M. and Raghavarao, D., *Sample Size Methodology*, Academic Press, San Diego, 1990.
3. Altman, D.G., *Practical Statistics for Medical Research*, Chapman and Hall, London, 1991.
4. Cox, D.R. and Hinkley, D.V., *Theoretical Statistics*, Chapman and Hall, London, 1974.
5. Chow, S.-C. and Liu, J.-P., *Design and Analysis of Bioavailability and Bioequivalence Studies*, Marcel Dekker, New York, 1992.
6. Lachin, J.M., Introduction to sample size determination and power analysis for clinical trials, *Controlled Clin. Trials*, 2, 93, 1981.
7. Frederikson, T. and Petterson, U., Severe psoriasis — oral therapy with a new retinoid, *Dermatologica*, 157, 238, 1978.
8. Fleiss, J.L., *Statistical Methods for Rates and Proportions*, 2nd ed., John Wiley & Sons, New York, 1981.

Section B: Epidermal Structure

4.0 Skin Surface Microscopy

4.1 Magnifying Lens — Noninvasive Oil Immersion Examination
of the Skin 49
H. I. Katz and J.S. Lindholm

4.2 Dermatoscopy 57
W. Westerhof

4.3 Skin Replication for Light and Scanning Electron Microscopy 73
B. Forslind

Chapter 4.1
Magnifying Lens — Noninvasive Oil Immersion Examination of the Skin

H. Irving Katz and Jane S. Lindholm
Department of Dermatology
University of Minnesota
Minneapolis, Minnesota

I. Introduction

Visualization of the skin is perhaps the most important part of the dermatologic examination. The observation of a cutaneous sign reflects a pathologic change in the skin due to a derangement within the epidermis, dermis, and/or subcutaneous tissue. Recognition of cutaneous findings allows a definitive or differential diagnosis of a dermatological disorder.

Using naked-eye inspection the experienced observer detects many visible gross morphologic features, such as size, shape, color, or type of lesion. However, some subtle or ambiguous diagnostic features require further amplification of the finding. The use of skin surface magnification techniques assists an observer in visualizing such enigmatic findings.

Skin surface magnification or surface microscopy is not a new method to examine changes occurring on or within the skin. Hinselmann, Goldman, Gilje, and O'Leary, along with many others since, have reported the usefulness of skin surface microscopy methods for a variety of dermatologic conditions.[1-12] In addition in the past decade surface microscopy has been used to study and differentiate benign and malignant pigmented melanocytic neoplasms.[13-19]

Extremes of skin surface magnification that are available range from just slightly more than life size to hundreds of times the size of the actual image viewed, as summarized in Table 1. Although high-power resolution is obtainable, such applications are not within the reach of most practitioners. Therefore, only low-power methods of skin surface magnification, such as simple lens systems, are within the scope of practice of most observers.

Low-power skin surface magnification is considered less than 10× magnification for the purposes of this chapter. We have found that a good-quality single lens, hand-held 8× magnifier with adequate incident illumination is a practical technique, albeit not the most accomplished alternative, to assist the astute observer in the recognition of subtle morphologic features. We also have available a Wild M650 binocular surgical stereomicroscope with a 100- to 250-mm lens system and automatic 35-mm camera, which was used to obtain the photographs in this chapter (Wild Heerbrugg Ltd., Heerbrugg, Switzerland).[20] Further enhancement of the observer's ability to detect obscure morphologic findings occurs by pretreatment of the normally opaque stratum corneum with mineral oil.[2] Such enhanced low-power skin surface magnification allows noninvasive visualization of changes within the epidermis and superficial dermis. This chapter will describe the methodology and practical use of low-power skin surface magnification using a hand lens through an oil-glass interface for benign dermatoses. It is suitable for use by practicing clinicians for patients with a variety of dermatological disorders.

II. Objective

The major objective of noninvasive enhanced low-power skin surface magnification during the skin examination is to help delineate subtle morphological features occurring either on or within the superficial layers of the skin. Recognition of a subtle cutaneous morphologic feature can suggest a single condition or focused group of conditions, as summarized in Table 2. A nonpathognomonic finding integrated with the patient's history, gross examination, or other established dermatological examinations may possess diagnostic significance.

III. Methodological Principals

The simplest form of low-power skin surface magnification occurs when the observer moves closer to what he or she is trying to visualize. An individual's near-vision visual acuity diminishes after reaching the fourth or fifth decade of life (physiological presbyopia).

TABLE 1 Skin Surface Microscopy Techniques

Low-power surface magnification (generally <10×)
    Simple magnifying lens
    Magnifying loupes
    Otoscopes
    Ophthalmoscopes
    Commercial skin scopes
Medium-power magnification (generally <40×)
    Monocular surface microscope
    Colposcope
    Surgical microscope
High-power magnification (up to 1000×)
    Video microscopy systems

Therefore, further amplification of visual acuity is required as the mature observer reaches the age when bifocal lenses are necessary to see an object. Bifocal lenses increase the resolution of the eyes by slightly magnifying the size of the object within its field of view. Bifocals represent the simplest form of low-power skin surface magnification. We have found that an 8×lens magnifier (manufactured by Coil, 200 Bath Road, Berkshire, 5L14DW, England) is a practical option for low-power magnification in a clinical setting. It is possible using 8× magnification with an oil-glass interface and adequate lighting to discern occult detail of potential diagnostic significance.

TABLE 2 Selected Clinical Findings Amplified by Enhanced Skin Surface Magnification

Terminal Hair Shaft Alterations
1. Varying lengths of hairs (trichotillomania, dystrophic hair shafts, androgenetic alopecia)
2. Varying diameter of hairs (androgenetic alopecia)
3. Black dot hairs (tinea capitis)
4. Fractured hairs (dystrophic hair shafts, trichotillomania, tinea capitis)
5. Dystrophic hairs (dystrophic hair shafts)
6. Casts (inflammatory dermatoses)
7. Presence of foreign material (pediculosis, trichomycosis axillaris)
8. Exclamation point hairs (alopecia areata)

Upper Epidermis (Stratum Corneum) Alterations
1. Altered dermatoglyphics (neoplasia, atrophy, forms of sclerosis, scar formation)
2. Distorted or missing follicular orifices (neoplasia, scar, forms of sclerosis)
3. Follicular plugging (discoid lupus erythematosus, lichen sclerosus)
4. Surface irregularities (icthyoses, porokeratoses, actinic keratoses, basal cell nevus syndrome, pitted keratolysis)
5. Presence of foreign material (scabies, chromomycosis)
6. Scale formation (papulosquamous diseases)
7. Crust formation (inflammatory dermatoses, neoplasia, trauma)
8. Presence of blood (inflammatory dermatoses, neoplasia, trauma)
9. Black dots (nevocellular neoplasms, North American blastomycosis, tinea nigra palmaris)

Deeper Epidermis (Granular, Spinous, and Basal Cell Layers)
1. Reticulated white opacification (lichen planus, lichenoid dermatoses)
2. Diffuse white opacification (lichenoid dermatoses, lupus erythematosus)
3. Brown, tan, and dark discolorations (nevocellular neoplasms, seborrheic keratoses, photodamaged skin, lentigines, angiokeratomas)
4. Pigment network (nevocellular neoplasms)
5. Presence of foreign material (tattoos, perforating dermatoses)
6. Denundation (trauma, neoplasia, inflammatory dermatoses)
7. Appendageal pores (comedones, chromohydrosis, trichostasis, folliculoma)

Dermal Alterations
1. Distortion of capillary loop patterns (psoriasis vulgaris, inflammatory dermatoses, lupus erythematosus, scleroderma)
2. Visualization of the normally occult subpapillary vascular plexus (preatrophy)
3. Telangeictases (atrophy and atrophogenic skin conditions, neoplasms)
4. Brown, gray, blue, and other colorations (nevocellular neoplasms, foreign materials, hemangiomata)
5. Blanching erythema (inflammatory dermatoses)
6. Nonblanching erythema (petechial dermatoses, hemorrhage)
7. Presence of foreign material (foreign bodies, tattoos)
8. Ulcer formation (neoplasms, vascular abnormalities, trauma)
9. Altered dermal structure proliferations (neoplasms)

When used without an oil-glass interface, low-power skin surface magnification reveals only surface detail because the stratum corneum is opaque when viewed with incident light. The irregular surface of the stratum corneum reflects incident light in multiple directions rather than transmitting the light to the deeper layers of the epidermis and dermis.

Enhanced cutaneous visualization of the skin is achieved by placing a drop of mineral oil and a glass cover slip on the skin surface during low power skin surface magnification. Mineral oil acts as a conforming fluid that has a refractive index similar to that of the stratum corneum. It tends to smooth out the surface and prevent light from scattering. Thus, the incident light penetrates the epidermis, thereby allowing the skin to appear more translucent. The mineral oil renders the normally opaque epidermis translucent. The glass cover slip reduces reflective glare and stabilizes the oil in the visual field. The oil-glass interface provides the opportunity to view structures or findings located within the epidermis, at the dermal-epidermal junction, and in the papillary dermis. When used with an oil-glass interface, skin surface magnification with adequate illumination provides a suitable basis for an improved method to observe subtle changes. An external non-glare white light, preferably a portable cold light fiber optic source, held about 6 in. away from the skin, provides suitable illumination. A commercially available hand-held skin scope, somewhat similar in principle to an otoscope, with a built-in light source and a magnifying lens, is discussed in Chapter 4.2.

Normal skin demonstrates a varied but definitive dermatoglyphic surface pattern at 8× magnification. The dermatoglyphic pattern is made up of intersecting grooves that yield triangular, rhomboidal, and other geometric shapes that make up the architecture of the skin surface, as shown in Figure 1. In addition, depending upon the anatomic location, other normal skin surface characteristics, such follicular orifices, hairs, and sweat pores are visualized. In tanned skin delicate brown-colored lines form a grid-like uniform pigment network background. The fine reticulated brown-colored outline corresponds to the paucity of epidermal melanin at the tips of the dermal papillae and the apparent relative excess pigmentation due to a layering of melanin-containing basal cells of the rete ridges at the epidermal-dermal junction when viewed from the surface.[13-17,19] Alteration of the pigment network may occur if there is an inflammatory or neoplastic disorder involving the epidermal-dermal junction.

With practice enhanced low-power skin surface magnification examination will maximize the recognition of subtle morphologic detail. An example of such a subtle finding that was difficult to visualize with both naked-eye examination and with low-power skin

FIGURE 1 The dermatoglyphic pattern of normal skin surface with a faint pigment network at 8× through an oil-glass interface.

surface magnification without an oil-glass interface is demonstrated in Figure 2. In contrast, a white dermal foreign body opacification is easily seen after application of a drop of mineral oil and 8× magnification, as demonstrated in Figure 3. Both Figures 2 and 3 were taken at the same magnification! The only difference was that the mineral oil rendered the stratum corneum translucent in Figure 3, thereby allowing the incident light to penetrate to a deeper level and reflect the finding. With the exception of certain topographic and textural changes, such as surface dryness and scale formation, the oil-glass interface enhances visualization of alterations occurring either on or within the superficial layers of the skin.

The cutaneous morphologic signs observed using enhanced skin surface magnification are the same as those that are sought during the routine naked-eye dermatological examination. The finding may just be amplified and easier for the clinician to discern! Selected

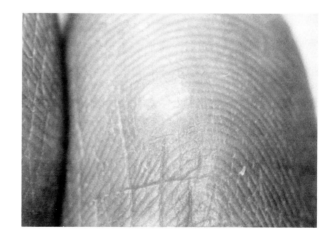

FIGURE 2 Relatively normal-appearing finger tip with an ill-defined white area at 8× without an oil-glass interface.

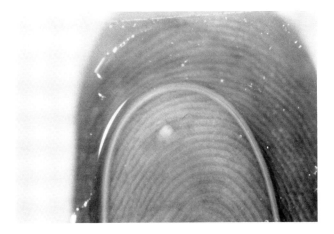

FIGURE 3 Small white opacification at site of a foreign body in the same finger depicted in Figure 2. However, it is viewed at 8× through an oil-glass interface.

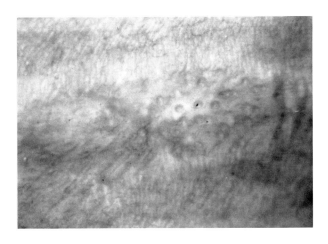

FIGURE 5 Follicular plugging from a plaque of lichen sclerosus at 8× through an oil-glass interface.

clinical findings observed with enhanced low-power skin surface magnification are summarized in Table 2. The findings are grouped according to their respective epidermal or dermal location. Examples of clinically significant changes occurring either on or within the stratum corneum include hair abnormalities, follicular plugging, and burrow formation.

Exclamation point hairs in a patient with alopecia areata above the skin surface are seen in Figure 4. In addition, hair shaft abnormalities such as the trichostasis spinulosa, tinea capitis, trichorrhexis nodosa, and other hair shaft abnormalities may be revealed using this technique. Follicular plugging in lichen sclerosus is demonstrated in Figure 5. A burrow occurring within the stratum corneum of a patient with scabies is demonstrated in Figure 6. Higher magnification is required to visualize the actual mite. Further examples of stratum corneum alterations that may be seen include surface depressions (basal cell nevus syndrome, porokeratosis of Mibelli, pitted keratolysis), discolorations (tinea nigra palmaris, chromomycosis), and contents of appendageal pores (comedones, chromhydrosis, trichofolliculoma).

Alterations within the epidermis are also demonstrated with enhanced low-power skin surface magnification. Reticulated delicate white opacification streaking (Wickham's striae) occurring within a lesion of lichen planus is shown in Figure 7. Other inflammatory dermatoses, such as various forms of dermatitis and psoriasis vulgaris, may demonstrate distinguishing features. Dermatitis may show superficial crust with dried blood and acanthosis with dilation of vertically and horizontally oriented blood vessels. Psoriasis vulgaris has scale, an alternating pattern of acanthosis, and thinning of the epidermis along dilated vertically oriented coiled capillary loops which penetrate the thinned portion of the checker board patterned acanthotic epidermis. In addition, delicate pin point bleeding from the thinned parts of the epidermis may be noted (Auspitz's sign), as is seen in Figure 8.

Enhanced low-power skin surface magnification is useful to detect occult alterations in the superficial

FIGURE 4 Exclamation point hairs in a patient with alopecia areata at 8× through an oil-glass interface.

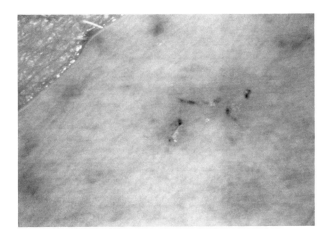

FIGURE 6 Burrows in a patient with scabies at 8× through an oil-glass interface.

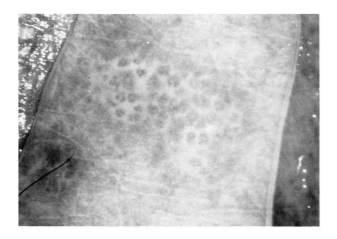

FIGURE 7 Reticulated white opaque streaking from a patient with lichen planus at 8× through an oil-glass interface.

FIGURE 9 Telangeictases in a patient with overt cutaneous atrophy at 8× through an oil-glass interface.

cutaneous vasculature as seen in atrophy of the skin. Telangeictases are a prominent finding in forms of cutaneous atrophy shown in Figure 9. An early sign of iatrogenic topical steroid atrophy is the observation of the normally obscure horizontally oriented subpapillary vascular plexus that has been termed preatrophy.[20] Preatrophy is due to steroid-induced shrinkage of the papillary dermis allowing the underlying structures, such as the subpapillary vascular plexus, to be nearer to the skin surface. Preatrophy change is shown with the oil-glass interface in Figure 10.

The presence of pigment within the epidermis or dermis is easily visualized with enhanced skin magnification. Alterations of the normal delicate pigment network can occur when there is a pathological process involving the dermal epidermal junction.[19] Mottled pigmentation in a person with extensive photodamaged skin is demonstrated in Figure 11. In addition, pigment changes are found in a variety of inflammatory

FIGURE 10 Horizontally oriented network of anastomosing delicate vascular channels in a patient with iatrogenic topical steroid cutaneous preatrophy at 8× through an oil-glass interface.

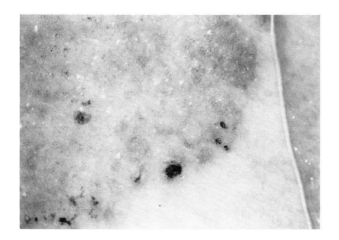

FIGURE 8 Well-demarcated erythema and patterned epidermal thinning containing coiled capillary loops and bleeding points in a plaque of psoriasis vulgaris at 8× through an oil-glass interface.

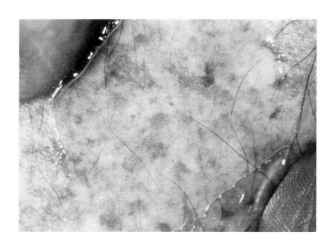

FIGURE 11 Mottled brown pigment change in a person with photodamaged forearm skin at 8× through an oil-glass interface.

dermatoses and neoplastic disorders. Skin surface microscopy changes found in pigmented neoplasms are discussed in the next chapter. Higher-power resolution may be necessary in order to distinguish benign from malignant nevocellular neoplasms. Extraneous discolorations depending on their age and/or depth, such as tattoo pigment, incontinence of melanin, hemoglobin, hemosiderin, along with other forms of foreign material, may be recognizable with experience.

IV. Sources of Error

Low-power skin surface magnification techniques can provide important diagnostic information. Some subtle findings may be diagnostic for cutaneous conditions, whereas others simply may be suggestive. Recognition and interpretation of such a finding(s) will depend on the competence of the observer. However, as with any type of clinical examination, a nonpathognomonic sign or a suspicious sign of malignancy should be further investigated with a biopsy. Subtle low-power skin surface magnification findings may be found in one or more dermatological conditions. A learning curve may be necessary for a morphologist to become familiar with the use of low-power skin surface magnification.

Visualization of certain surface changes, such as dryness, scale formation, and shadowing may be obscured when an oil-glass interface is used, as demonstrated in Figures 12 and 13. Such surface topographical detail is best appreciated without an oil-glass interface. Pigment network alterations, minute abnormal changes within capillary loops and other vascular channels may require medium or higher-power magnification to be fully appreciated. Unless a highly specific low-power skin surface magnification morphologic finding is found, further examination and clinical pathological correlation are necessary for it to be diagnostically meaningful.

V. Correlation with Other Methods

The findings using low-power skin surface magnification augment the routine dermatological examination. Low-power skin surface magnification does not supplant the gross examination of the skin.[19] Rather, low-power skin surface magnification is fine-tuning of observable morphologic skills that the dermatologist or other practitioner uses in everyday practice. Subtle cutaneous findings observed with low-power skin surface magnification of a benign dermatosis can be pathognomonic. However, further confirmation by clinical history, gross dermatological examination, a skin scraping, or a biopsy is necessary for a finding that is

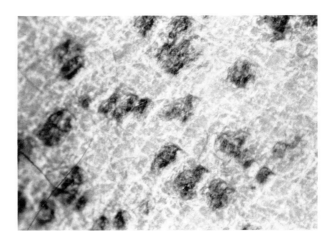

FIGURE 12 Surface dryness, adherent keratotic debris, and altered surface relief in a patient with X-linked ichthyosis at 8× *without* an oil-glass interface. See Figure 13.

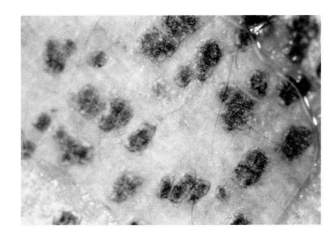

FIGURE 13 Same as Figure 12, except at 8× through an oil-glass interface. The dryness and surface relief are obscured by the oil-glass interface.

not pathognomonic. The combined use of traditional dermatological techniques along with low-power skin surface magnification fosters the recognition of clinical findings that strengthen the diagnostic acumen of a practitioner.

VI. Recommendations

Low-power skin surface magnification (8×) through an oil-glass interface is a useful technique that augments rather than supplants the routine naked-eye dermatological examination. It is noninvasive and easy to perform. The findings can assist the observer in making a diagnosis or developing differential diagnoses. Enhanced low-power skin surface magnification is a method of cutaneous examination that has a place in the dermatologist's diagnostic armamentarium.

References

1. Hinselmann, H., Die Bedeutung der Kolposkopie fur den Dermatologen, *Dermatol. Wochenschr.*, 96, 533, 1933.
2. Goldman, L., Some investigative studies of pigmented nevi with cutaneous microscopy, *J. Invest. Dermatol.*, 16, 407, 1951.
3. Gilje, O., O'Leary, P.A., and Baldes, E.J., Capillary microscopic examination of skin diseases, *Arch. Dermatol.*, 68, 136, 1958.
4. Goldman, L., Clinical studies in microscopy of the skin at moderate magnification, *Arch. Dermatol.*, 75, 345, 1957.
5. Goldman, L., A simple portable skin microscope for surface microscopy, *Arch. Dermatol.*, 78, 246, 1958.
6. Goldman, L., Direct microscopy of skin in vivo as a diagnostic aid and research tool, *J. Dermatol. Surg. Oncol.*, 6, 744, 1980.
7. Goldman, L. and Younker, W., Studies in microscopy of the surface of the skin, *J. Invest. Dermatol.*, 9, 11, 1947.
8. Gilje, O., Capillary microscopy in clinical dermatology, *Bibl. Anat.*, 1, 203, 1961.
9. Cunliffe, W.J., Forster, R.A., and Williams, M., A surface microscope for clinical and laboratory use, *Br. J. Dermatol.*, 90, 619, 1974.
10. Davis, M.J. and Lorincz, A.L., An improved technique for capillary microscopy of the skin, *J. Invest. Dermatol.*, 28, 283, 1957.
11. Epstein, E., Magnifiers in dermatology: a personal survey, *J. Am. Acad. Dermatol.*, 13, 687, 1985.
12. Tring, F.C. and Murgatroyd, L.B., Surface microtopography of normal human skin, *Arch. Dermatol.*, 109, 223, 1974.
13. Soyer, H.P., Smolle, J., Hodl, S., Pacernegg, H., and Kerl, H., Surface microscopy — a new approach for the diagnosis of cutaneous pigmented tumors, *Am. J. Dermatopathol.*, 11, 1, 1989.
14. Fritsch, P. and Pechlaner, R., Differentiation of benign from malignant melanocytic lesions using incident light microscopy, in *Pathology of Malignant Melanoma*, Ackerman, A.B., Ed., Masson, New York, 1981, 301.
15. Pehamberger, H., Steiner, A., and Wolff, K., In vivo epiluminescence microscopy of pigmented skin lesions. I. Pattern analysis of pigmented skin lesions, *J. Am. Acad. Dermatol.*, 17, 571, 1987.
16. Steiner, A., Pehamberger, H., and Wolff, K., In vivo epiluminescence microscopy of pigmented skin lesions. II. Diagnosis of small pigmented lesions and early detection of malignant melanoma, *J. Am. Acad. Dermatol.*, 17, 584, 1987.
17. Fritsch, P. and Pechlaner, R., The pigment network: a new tool for the diagnosis of pigmented lesions, *J. Invest. Dermatol.*, 74, 458, 1980.
18. Knoth, W., Boepple, D., and Lang, W.H., Differentialdiagnostische Untersuchungen Met dem Dermatoskop bei ausgewählten Erkrankungen, *Hautarzt*, 30, 7, 1979.
19. Bahmer, F.A., Fritsch, P., Kreusch, J., Pehamberger, H., Rohrer, C., Schindera, I., Smolle, J., Soyer, H.P., and Stolz, W., Terminology in surface microscopy, *J. Am. Acad. Dermatol.*, 23, 1159, 1990.
20. Katz, H.I., Prawer, S.E., Mooney, J.J., and Samson, C.R., Preatrophy: covert sign of thinned skin, *J. Am. Acad. Dermatol.*, 20, 731, 1989.
21. Lehmann, P., Zheng, P., Lavker, R.M., and Kligman, A.M., Corticosteroid atrophy in human skin. A study by light, scanning, and transmission electron microscopy, *J. Invest. Dermatol.*, 81, 169, 1983.

Chapter 4.2
Dermatoscopy

*Wiete Westerhof
Department of Dermatology
Academic Medical Center
University of Amsterdam
The Netherlands*

I. Introduction

Macroscopic morphological examination of skin disorders is, besides history taking, general physical examination, and laboratory investigation, one of the pilots of the diagnostic methods in dermatology. An extension of the morphological examination is the invasive technique of biopsy taking for histopathology, histochemistry, immune-histochemistry, and autoradiographic methods.

Dermatoscopy, which is a noninvasive micromorphological method, can be considered as a *trait d'union* between macroscopic morphological investigations and invasive histological techniques. From the historical point of view epiluminescence microscopy has been used for two different applications: capillary microscopy and dermatoscopy. The capillary microscopy of the nail fold and the nail bed will be dealt with in another chapter.

Dermatoscopy is a readily available diagnostic method, allowing us to appreciate the fine details not normally seen with the naked eye. It furnishes us with the overall detail, which is not the case in low-power scanning electron microscopy, the latter being, moreover, an elaborate and very expensive technique. The unexpected details seen with the dermatoscope often give us answers regarding the pathophysiology of various skin diseases, as well as the urge to pose questions requiring further investigation. As such, the dermatoscope can be applied as an objective tool in research, e.g., follow up of ultraviolet (UV)-erythema reaction, evaluation of sweat gland function, skin texture measurements, etc. With the existing techniques of UV, infrared (IR), and fluorescence photography, dermatoscopy is used to apply these methods to the fine detail of the skin. The possibilities of the technique are described in the following paragraphs.

Dermatoscopic photography provides the clinical instructor with a new valuable teaching aid that not only enhances what is seen clinically, but also creates a surprisingly pleasing visual image of color, shadows, texture, and form.

II. Object

The dermatoscope allows the examination of skin lesions at different magnifications. Based on these images differential diagnosis is possible. So far, relatively little has been published about the use of the dermatoscope in the diagnosis and follow up of general dermatological lesions. Several skin diseases constitute, even for experienced dermatologists, a diagnostic problem. This applies in individual cases to macroscopic detail as well as to the microscopic aspects. Often the diagnosis is based on the combination of the macroscopic clinical symptoms and the microscopic details. However, there are limits to this approach and other avenues have to be explored. In dermatology normally a magnifying lens is used to help in the diagnosis of macroscopic details. The magnification is maximally three times and, therefore, the link with the microscopic detail does not exist.

Basically the dermatoscope can be used for five groups of skin lesions:

1. Lesions having a vascular background
2. Lesions having a difference in color pattern
3. Lesions having a change in the skin texture
4. Lesions having abnormalities in the distribution of skin adnexae
5. Lesions with a malignant background

Before knowing what is abnormal it is necessary to study the normal features of the skin. In this context one has to remember that this depends on the age, sex, the site of the skin investigated, and the environmental conditions in which the investigation with the dermatoscope is carried out. Figure 1 gives an overview of the different locations of the body, where the skin detail is characteristically different from neighboring areas. It is clear from this table that "normal" skin does not exist. Furthermore, the skin of a baby cannot be compared with the skin of a 90-year-old person.

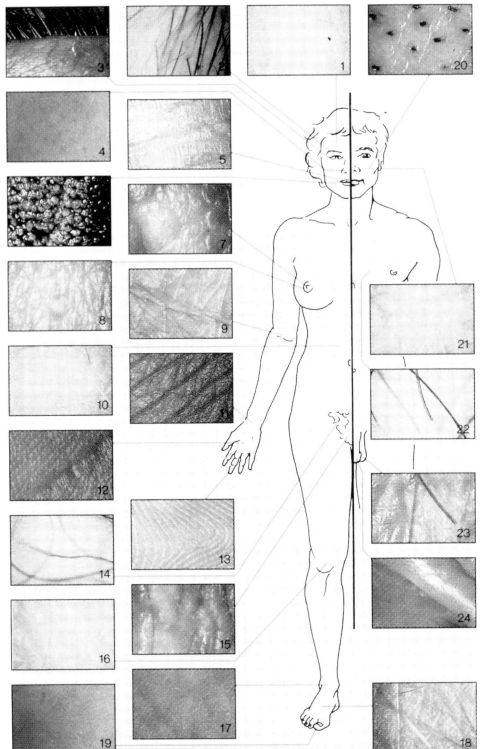

FIGURE 1 Map of the regional differences in skin texture.

III. Instrumentation and Technical Methods

A. Instruments

In this chapter different macroscopic techniques are described, which make use of similar optical instruments. These are dermatoscope, capillary microscope, epi-luminescence microscope, and surface microscope. Their optical properties are in essence equal to an operation or dissecting microscope. Sometimes the names of these instrument are exchanged in the text. Some of the most current instruments are described here. Many more are available on the market. However, it is not possible within the framework of this chapter to list them all.

- Zeiss dermatoscope (Zeiss, Germany):[1] The dermatoscope is supplied with a bendable arm fixed to the wall (Figure 2). For the examination of the patient it is not difficult to position the dermatoscope due to the flexibility of the arm. Some dermatoscopes are mounted on a movable tripod. For the handy manipulation of the tool the supporting arm is supplied with a counterweight, so that the dermatoscope remains in the working position. The Zeiss dermatoscope has a binocular and a monocular tube and one arm of the tube can also be fitted with a photocamera or film camera. The magnification system (motorized zoom) is stepless from 0.4 to 40 times. An objective is used with a focus of 175 mm. The ocular enlarges 10 times. The dermatoscope has a field cross section of 22.5 mm. The standard

In order to be able to make a diagnosis based on dermatoscopy, it is not possible to go by primary efflorescences as is used in dermatological practice. So new criteria describing skin changes at the macroscopic level have to be developed.

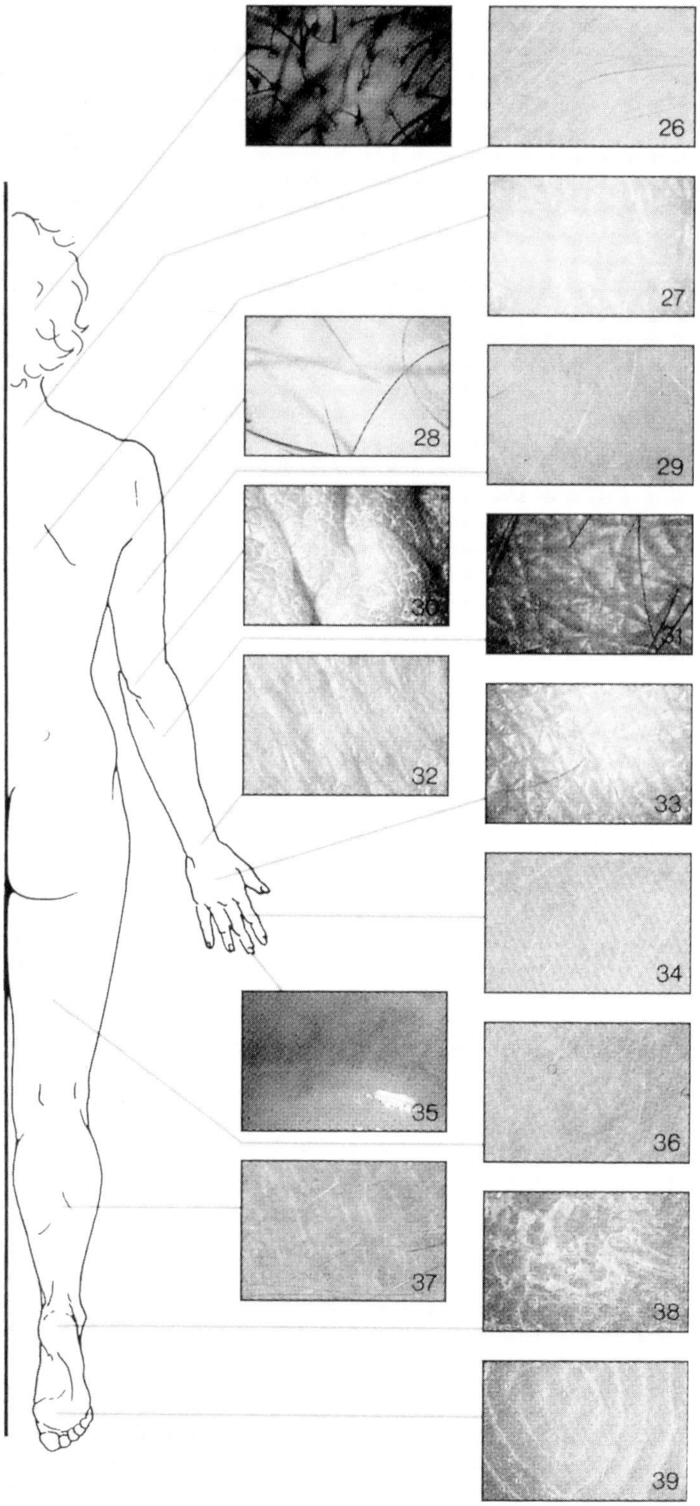

FIGURE 1 (continued).

equipment can be extended with different additional options. The binocular tube can be supplied with a microscope body with a magnification step times 3. Without problems also film cameras or a video camera can be coupled to the optical system. With a beemsplitter the visual image can be photographed when a camera is positioned on an additional tube. The stereoscopical observation by at least two doctors is an advantage.

- Wild M650 (Wild Heerbrugg AG, Switzerland): Epiluminescence microscopy (ELM) is a term often used in relation to the investigation of pigmented lesions but is in essence similar to dermatoscopy. ELM can be performed with a Wild M650 binocular surface microscope equipped with objectives of 91 mm working distance. Magnifications obtainable are ×6, ×10, ×16, ×25, and ×40. All pigmented skin lesions are first examined for surface structure. They are then covered with immersion oil and a glass slide that is applied with slight pressure; this renders the epidermis translucent and allows study of the dermoepidermal junction zone. Photographs of the pigmented skin lesions can be taken with an Olympus CM10 automatic camera, mounted on a side arm of the microscope.

- Delta 10 dermatoscope (Heine Optotechnik, Germany): With the Delta 10 dermatoscope after Braun-Falco, Billeck and Stolz, Heine Optotechnik developed an instrument[2] which, because of it small weight and easy handling, facilitates a quick analysis of pigmented skin lesions. The Delta 10 dermatoscope (Figure 3) is provided with an achromatic lens. It makes a 10-fold magnification possible of skin changes, when the skin is treated with emulsion oil. The illumination of the object is under an angle of 20° with a halogen lamp inside the dermatoscope and powered by a battery. With the Delta 10 only small areas can be studied, like the interdigital spaces, the nails, the eye corners, the genital, anal, and retro-auricular regions, as only a 8 mm diameter contact lens is illuminated by the halogen lamp via a special fiber optical system. Photographs cannot be made with this system and stereoscopic vision is also an omission of this instrument. Heine has developed a photographic system in which a reflex camera is supplied with a Dermaphot which is a special option for photographing contact dermatoscopical images (Figure 4). It has a built-in electron flash system. This Dermatophot might help to overcome the shortcomings of the Delta 10 dermatoscope. Although the contact photography is not as easy as the technique used with the Zeiss dermatoscope it is possible to make good pictures.

FIGURE 2 The Zeiss dermatoscope.

FIGURE 3 The Delta 10 dermatoscope (Heine).

- Microscan System (Fort, United Kingdom): The Fort UK Microscan MS 2500 (Figure 5) is a video microscope ideal for dermatological applications. The microscope consists of a video processing unit with an integral high-intensity cold light source. The miniature color camera has a built-in fiber optic ring light to which interchangeable objective lenses also having fiber optic ring lights are attached. A wide range of detachable lenses are available both in contact and noncontact formats with magnifications from 5 to 1000 times. The ring lights ensure even illumination of the

FIGURE 4 The Dermaphot (Heine).

subject under examination. The miniature camera with its lens has been designed for hand-held operation. The contact objective lenses have been specifically designed so that the contact face is smooth, rounded and comfortable for the patient under examination. When using the noncontact lenses the camera/lens assembly may be mounted into a special microscope stand which is also supplied by FORT UK. The cold light source is 150-W quartz halogen with electronic variable control of the lamp's intensity and features an exclusive, easy to operate filter wheel that has four neutral density filters for coarse setting of the illumination intensity and has red, orange, green, and blue colored filters. The use of colored light on the subject under examination can enhance features of interest. Also available are a microprocessor for image acquisition and analysis, a monitor, a mouse fitted to the PC for use as a pointing device, and image analysis software. The software is menu driven using a point-and-click-style user interface. The software provides facilities for storage and retrievel of records consisting of an image and associated notes and comments. The Fort Microscan outputs PAL composite video and can therefore be used with any picture monitor, monochrome or color. The Microscan is fully compatible with all currently available video printers for permanent hard copies and composite video tape recorders.

B. Infrared Photography and Dermatoscopy

The skin and superficial tissues reflect most of the infrared falling on and penetrating a short distance into the body, whereas the blood in the veins absorbs much of the infrared. This provides a tone separation. In an infrared color transparency the veins appear blue as well as dark, because they lay under a scattering layer.[3]

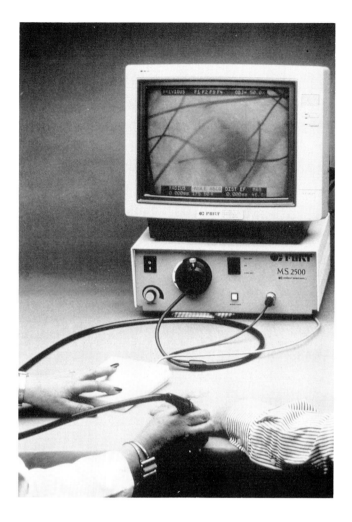

FIGURE 5 The Fort UK Microscan MS2500.

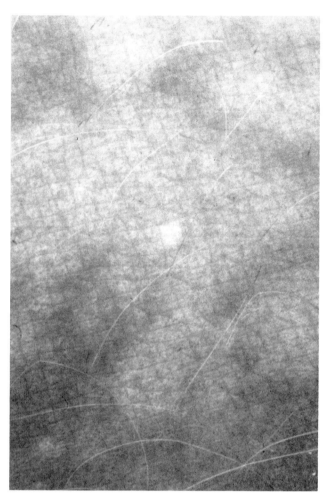

FIGURE 6 Pattern of dermal blood vessels of the arm using a Kodak high-speed infrared film 2481.

This technique introduces differences in color as well as tone (Figure 6). The pattern could be traced visually, but is much more apparent and detailed in the infrared photograph since it has become darkened. In general photographic practice it will be found that infrared radiation between 700 and 900 nm can penetrate the skin to a depth of about 3 mm. The translucency of tissues and particle size govern delineation and tone value in the reflected component. The venous circulation near the surface of the body is superficial and flows through larger vessels than its arterial counterpart. Venectasies, essential teleangiectasia, vascularization of tumors, and dermal pigmentation in the skin, are detectable earlier by infrared/black and white photography than by visual observation.[4] Incidentally, the more prominent superficial veins are recordable in black skin, which makes this type of photography of value in mapping venous patterns in dark-skinned people.

The absorption of melanin must be mainly in the visible region, because melanin granules on photomicrographic slides transmit actinic infrared. This is evidenced by the fact that melanotic tissue sections, and the black melanophores in frog skin, record red in infrared color transparencies made by transmission photomicrography, demonstrating translucency to infrared.[3,5] It has generally been assumed, rather without base, that black skin photographs are light-toned by infrared because melanin reflects infrared. However, the findings obtained by infrared color photography indicate that this is not so. The mechanism must be that of a high infrared transmittance to the underlying, or intermixed, tissue. These reflect the infrared back through the almost "transparent" melanin granules. Thus, as far as infrared photography of the skin is affected, the melanin particles might as well be a thin superficial layer of transmittant ground glass, with high absorption in the visible range of the spectrum, but little absorption in the infrared.

The false-color renditions of several skin lesions are described and discussed in Gibson.[3] In this paper, he demonstrated differentiation between a pigmented nevus of the fingernail and a splinter hemorrhage in the nail bed. The infrared translation color for melanin varies from a light reddish brown to a dark brown — quite similar to that observed visually except that it is

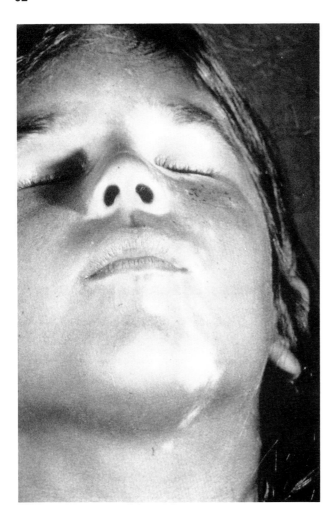

FIGURE 7 UV photograph (Kodak Panatomic-X film 135) enhances the contrast between pigmented and nonpigmented skin. Borders might reveal interesting dermatoscopic detail.

always more red. Skin not containing appreciable melanin appears as a pale, cold white.

With the basic translation colors for the normal pigments determined, several of the pigmentary disturbances encountered in human skin were explored. Argyria varied in skin coloration from deep brown to blue. In the patient with argyria, the greyness observed in the skin was intensified in the infrared record. But there was little added from the diagnostic point of view.

C. Ultraviolet Photography and Dermatoscopy

A source of ultraviolet light from which all visible rays have been excluded by a Wood's (nickel oxide) filter is an important investigative tool in the diagnosis and treatment of many dermatoses. Only the most important of the many uses to which ingenuity and imagination have contributed are mentioned here. Wood's Light[6] is a powerful UV-radiation source. The various dermatoses can also be photographed by dermatoscope.

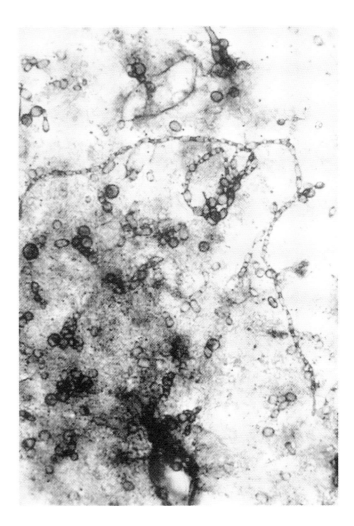

FIGURE 8 Yeast infection of the skin demonstrated by surface microscopy (courtesy of Dr. G.E. Pierard).

A flash light is used for the illumination, which can be filtered with colored filters, but can also be used to flash with parts of the ultraviolet spectrum. This enables us to visualize certain disease patterns, e.g., a disease in which fluorescence, evoked by ultraviolet illumination, is becoming visible. Special UV films are used to develop the image.

Examples of these are different fungus infections which give green fluorescence. Certain bacterial conditions like erythrasma give a coral-red fluorescence. A red fluorescence arises in the blister fluid of patients with porphyria cutania tarda.

Epidermal pigmentary disorders, such as vitiligo lesions (Figure 7) or *cafe au lait* spots, are better demonstrated with the help of ultraviolet illumination. Detection of fluorescing dies causing contact dermatitis can be demonstrated specifically. The remnants of mineral oils present in the pores of hair follicles persist even after washing. The presence of a black-light is therefore of importance to localize and to determine the size of a condition before photographing the condition. In the second place it is possible to combine the ultraviolet light with a fluorescent dye with fluorescein. This substance can be injected intravenously as a sodium salt. This is practiced, for example, in ophthalmology, which could also be applied in certain conditions of the skin. Under ultraviolet illumination the vasculature and the diseased skin can be studied with the dermatoscope when the passage of the fluorescing dye is maximal.[7] Characteristic changes can occur, for example, in chronic wounds, scabies, and in the case of malignant tumors. For the determination of the progress of relatively fast movements it would be handy to use a film camera instead of a photocamera. This is the so-called time-lapse cinematography.

D. Surface Microscopy[8-11]

Skin surface biopsy consists of sampling the superficial layers of the stratum corneum. There are two variants. One type is performed with a cyanoacrylate adhesive and a sheet of transparent plastic. The other one relies on the use of commercially available small adhesive discs. The method is noninvasive and painless. It may be sequentially repeated at the same site. The indications are multiple. They primarily concern the evaluation of xeroses and the diagnosis of squamous inflammatory dermatitis, superficial infections (Figure 8), and parasitoses, as well as pigmented neoplasms (Figure 9). Experimental studies concerning the biology of the stratum corneum and the pharmacokinetics of various drugs may also be undertaken.

IV. Examples of Dermatological Entities Studied with a Dermatoscope

A. Benign Conditions

Skin texture — The system of grading cutaneous microphotographs described by Beagley and Gibson[12] relates to changes in skin surface texture, which in normal skin is composed of a series of transverse and diagonal primary lines, which intersect to form quadrilaterals and triangles. Within these primary figures are sets of smaller secondary lines which often meet in the center of the figure, forming a star configuration (Figure 10). The six-step grading system devised by Beagley and Gibson, and based on alterations in these skin surface characteristics thought to reflect actinic damage, is summarized in Table 1.

Facial skin wrinkling or "crow's feet" are more common and more severe in cigarette smokers than in nonsmokers.[13] "Crow's feet" (wrinkles in the lateral periorbital area) have been thought also to be caused by sun exposure, the aging process, and other factors, such as massive weight reduction.[14] Daniell[13] inspected live subjects and reviewed photographs taken from the "crow's foot" area of their faces; the latter is called paraocular photography (POP).

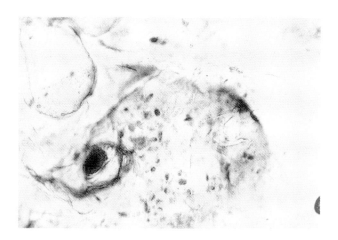

FIGURE 9 Melanoma cell in stratum corneum (surface biopsy) supporting the diagnosis by a noninvasive method (courtesy of Dr. G.E. Pierard).

Dermatoglyphics — Dermatoglyphics is a term applied to both the configurations of ridged skin, and the subject that deals with it. Characteristic ridge patterns of whorls and loops are found on the volar skin which are unique for any individual (Figure 11). The systematic classification of ridge patterns, as a means of personal identification or for use in studies of inheritance, requires image analysis as is possible with the Microscan System (Fort). Dermatoglyphic features may indicate an increased tendency to develop a particular condition and can be an aid to diagnosis in dermatology. For example, findings include, in alopecia areata, a decreased incidence of ulnar loops in the second left digit in both sexes, and in psoriasis an increased incidence of whorls in the fourth digit, more marked in the right hand.[15,16]

Ectodermal dysplasia — Hypoplasia of appendages occurs in the skin and mucus membranes. In the anidrotic form the skin is dry and there may be

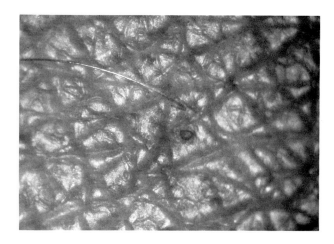

FIGURE 10 Primary, secondary, and tertiary skin lines.

FIGURE 11 Dermatoglyphics.

TABLE 1 The Beagley-Gibson System of Grading Cutaneous Microtopographs Taken From the Dorsum of the Hand

Grade	Features
1	Primary lines are all of the same depth. Secondary lines are all clearly visible, are nearly the same depth as the primaries, and often meet to form an apex of triangles (star formation).
2	Some flattening and loss of clarity of the secondary lines. Star formations are still present, but often one or more of the secondary lines making up the configuration are unclear.
3	Unevenness of the primary lines. Noticeable flattening of the secondaries with little or no star formation.
4	Macroscopic deterioration in texture. Coarse, deep primary lines. Distortion and loss of secondary lines.
5	Noticeable flat skin between the primary lines. Few or no secondary lines.
6	Large deep and widely spaced primary lines. No secondary lines.

palmoplantar keratoses. The most important feature is heat intolerance due to deficient sweat glands. This can be demonstrated in patients by rubbing the finger over a slightly rough surface, thereby provoking heat. In normal persons the sweat drops that appear due to profuse sweating at the orifices of the sweat ducts on top of the ridges can easily be seen with the dermatoscope. In patients with ectodermal dysplasia this feature is absent or highly decreased (Figure 12A and B). The decrease in number of sweat glands can also be demonstrated in female family members having the trait of this X-linked recessive inheritable disease (personal observation).

Parapsoriasis en plaque — The clinical evaluation of parapsoriasis en plaque with the dermatoscope gives an even so-called quiet image. The structure of the skin surface shows a regular pattern. The scales are discrete, not elevated or bizarre. The colors are evenly distributed in all the microscopic fields.

Seborrheic eczema — Seborrheic eczema has a variable pattern.[1] The image shows a certain succulence with yellow scales and an irregular surface. Also the colors are contrasting. In the first stages of mycosis fungoides the images arising from the dermatoscope resemble those of seborrheic dermatitis. Also here differentiating parapsoriasis en plaque from other eczemas is difficult.

Morphea — The morphea-like lichen sclerosis et atrophicus (LSEA) is in comparison to the small plaque type of circumscribed scleroderma especially remarkable with regard to the distinct follicular keratosis. This is not a new phenomenon; however, in the investigations of Knoth et al.[1] the regularity of the crater-like or plaque-like surface of the hyperkeratosis was very prominent in LSEA. In case of morphea the sur-

FIGURE 12 (A) Sweat droplets appear at the openings of the sweat ducts on the ridges, as seen in a normal person.

FIGURE 12 (B) Absence of dermatoglyphics in a patient with ectodermal dysplasia; no sweat secretion.

face pattern is more uniform and smooth. In lichen sclerosis et atrophicus of the glans of the penis or on the prepuce the follicular pattern is normally absent, which was unknown before the dermatoscopical investigation.

Lichen ruber planus — The various developmental stages of lichen ruber planus are especially remarkable in the early stage. The primary and secondary skin folds disappear. The papules are grouped, discrete, and slightly swollen. The development of hyperkeratosis and the Wickham striae comes later.

Corpora aliena — It is easily investigated with a dermatoscope, whether the particle is superficial or deep in case of corpora aliena in the face as a result of explosion of gun powder, which is an important implication with regard to the choice of treatment, for example excision or dermabrasion.

B. Alterations in Cutaneous Vessels in Various Diseases

Besides its usefulness in studying the anatomy and function of the superficial cutaneous vessels, capillary microscopy shows some promise in being used to differentiate skin diseases. At present, however, its use as a diagnostic aid is limited by the lack of complete data on the various changes in the superficial circulation of the skin in disease.[17,18]

Atopic skin — Capillary microscopy is particularly useful for studying vascular changes in atopic dermatitis in response to pharmacologic or physiologic stimuli. There is an enhanced vasoconstrictor tendency in atopic skin, which is demonstrable by the ease with which white dermographism is produced on stroking, by a more rapid rate of cooling in a cold environment and a slower rate of warming in warm environments, by exaggerated response to cold pressor tests, and finally by the delayed blanch phenomenon. In atopic skin there is a paradoxical reaction, which consists of the appearance of a white halo around the wheal formed by introducing dilute acetylcholine into the skin. This delayed blanch phenomenon begins between 3 and 5 min after the injection of acetylcholine and has been interpreted as vasoconstriction of the skin vessels in response to a vasodilator drug. This reaction was studied by dermatoscopy in a series of atopic patients[18] and direct observations were made of the capillary bed, after the introduction of acetylcholine. The capillaries at the periphery of the wheal remained dilated while the area appeared grossly blanched. Introduction of hyaluronidase into the test site before the injection of acetylcholine eliminated the blanching, and it appears that the whiteness seen grossly is due to the edema forming as a result of vasodilation of the capillaries in response to a vasodilator substance.

Atopic dermatitis — The capillaries in the involved skin of patients with atopic dermatitis are somewhat dilated. There is a regular distribution of the capillaries (which are all open) but acanthotic changes in the epidermis obscure the subpapillary plexus. The capillaries themselves are frequently obscured as a result of edema in the papillae and epidermis.[18] The arrangement of the vessels in atopic skin is somewhat reminiscent of the regular arrangement seen in psoriatic plaques; however, there is little tortuosity of the capillaries. The vascular bed in the uninvolved skin of atopic patients appears normal. The capillary changes seen in areas of localized neurodermatitis are not distinguishable from those of atopic dermatitis.

Psoriasis — Certain consistent alterations occur in the capillary bed in psoriatic lesion. Gilje[17] has described the regularly arranged distribution of capillaries in psoriatic plaques which, when viewed through the scales on the surface of the lesion, appear as puff balls. The end-capillaries in this disease are extremely tortuous and coiled. The subpapillary plexus is obscured, probably because of the acanthosis present. The capillaries are longer than normal but dilatation is not a prominent feature of the capillary changes in psoriasis. This is in contrast to the usual impression one gains from reviewing histological sections of psoriatic lesions. The apparent dilatation of the capillary vessels is probably an artifact which results from sectioning through a coiled vessel and unavoidably getting a tangential cut through at least some vessels. There is no increase in the number of capillaries. Flow through these capillaries appears normal and the capillaries can be seen to pulsate. The diameter of the venous and arterial limbs of these capillaries are approximately equal; this being another departure from normal. In early psoriatic lesions the capillaries are also abnormal in that the end-vessels are somewhat tortuous, but not nearly to the degree seen in old lesions. The "normal" skin of patients with psoriasis also shows some abnormally tortuous capillaries. This suggests that more attention should be directed to the role of the capillaries which may be involved in some way in producing the manifestations of psoriasis. The tortuosity and increased length of the capillaries in a psoriatic lesion represent a considerable increase in endothelial surface. This correlates well with the elevated oxygen consumption in psoriatic lesions. However, tortuosity of a capillary is not a finding limited to psoriasis; it has been found with some consistency in the nail folds of subjects who have neurasthenia and no evidence of skin disease.

Psoriasis pustulosa — In psoriasis pustulosa there is a similar picture, but it varies somewhat from the preceding picture of psoriasis. There is a ringlike arrangement of the capillaries around a prominent round or oval structure.

Eczema — The capillary microscopic picture in eczema varies a good deal from the clinical picture. There is a more irregular arrangement of the capillaries and these do not resemble rolled-up balls as in psoriasis. The tips of the capillary loops are always much smaller, and they may vary in size within the same field of vision. The group-like arrangement of the dilated capillaries corresponds to vesicles or papules. The dermal ridges are pronounced here. The subpapillary plexus is neither visible here nor in the pictures of psoriasis.

Lichen ruber planus — Lichen ruber planus shows an entirely different picture. The papule shows an irregular lighter central field, like a plateau, surrounded by a border of straight and slender, slightly twisted capillary loops, which slant inward and upward towards the central lighter region. In this condition there is a marked departure from the normal in the end-capillaries and in the subpapillary plexus. Some of the vessels are obliterated while others are deformed and dilated. There is a regular arrangement of the capillaries around the central area of severe changes, and these vessels are tortuous. At the periphery of the papule, there is noticeable palisading of the end-capillaries. This is seen not only in lichen planus but also in psoriasis and in other papular dermatoses. The violaceous hue, seen in lichen planus papules, is attributable to the engorgement of the capillaries and, to a greater extent, to the engorged venules in the subpapillary plexus seen through a hyperkeratotic stratum corneum. When the hyperkeratotic layer is removed, the lesions appear a dusky red.

Lupus erythematosus — In lupus erythematosus the most striking feature of the capillary microscopic picture is a round or oval formation which represents the follicular hyperkeratosis and the short, thick, usually horizontal loops or arcs of vessels. Some of these are arranged around the cornified plugs, others are distributed irregularly throughout the field of vision. There are almost no ordinary capillary loops here. There is almost a complete loss of the capillaries in lesions of chronic discoid lupus erythematosus. The vessels in the subpapillary plexus are deformed and dilated. A patient with lesions of several years' duration on the face, with no lesions in other areas, had atrophic changes grossly in the skin over the dorsum of the hands and forearms, sites which had never been involved with discoid lesions. This patient had no clinical or laboratory evidence of constitutional disease. The vascular beds on the dorsum of the hands and forearms were decidedly abnormal: there was an absence of end-capillaries and a dilated tortuous subpapillary plexus. The vascular bed on the volar surface of the forearm appeared normal. Gilje et al.[19] found abnormal capillaries in the nail folds of patients with discoid or disseminated lupus erythematosus.

Ichthyosis — The capillary microscopic picture in ichthyosis is entirely different from the foregoing diseases. The picture in ichthyosis usually shows capillaries which are slender, long and narrow with a narrow crest and with no particular differentiation of the arterial and venous branches. In some areas the capillaries branch out from the subpapillary plexus like plants in a bed or like the spray in a fountain. In some of the thin capillary loops it is possible to follow the blood stream, which has the appearance of beads on a string.

Hereditary hemorrhagic telangiectasia[20] — Examination of the grossly visible ectasias shows them to consist of markedly coiled, widely dilated superficial vascular channels. The flow of blood through these angiomatous vessels is slow. It is difficult to visualize the inflow and outflow vessels to the coiled mass; therefore, it is not known whether the mass consisted of one or more vessels. The normal skin of two patients with hereditary hemorrhagic telangiectasia also showed unusual vascular structures. These consisted of a circular vascular channel, about 0.3 mm in diameter, enveloping a second circular channel. The rate of flow of blood in these vessels was slow. Approximately two such structures per square millimeter could be seen. There were no end-capillaries in these areas and the circular channels may have been the precursors of the markedly coiled vessels in the involved skin of telangiectasia. These vessels superficially resembled the schematic representations of arteriovenous anastomoses. They were smaller than normal glomus bodies and were more superficially situated in the skin.

Senile skin and ectatic vessels — The vascular bed in senile skin shows certain distinguishing characteristics. The end-capillaries appear to be shorter than those in young subjects and the subpapillary plexuses appear more prominent. Since we do not yet have a method for accurately measuring the lengths of capillaries, because of their perpendicular orientation in the skin, the observation that the capillaries are shortened must remain only an impression. Shortening of the end-capillaries in aging skin correlates with the flattening of the dermal papillae. Thinning of the overlying structures accounts for the apparent prominence of the subpapillary vessels in senile skin. The ectatic and tortuous cutaneous vessels that characterize many diseases are dilated subpapillary venules, and possibly arterioles, but not capillaries (Figure 13). In areas of the skin in which ectatic vessels are found, a greatly reduced number of end-capillaries appear. The epidermis and dermal papillae above these ectatic vessels are usually atrophic. In skin diseases that are characterized by eventual atrophy of the skin, such as discoid lupus erythematosus, poikiloderma, or chronic radiodermatitis, there is usually complete destruction of the papillary capillaries and a dilatation of the

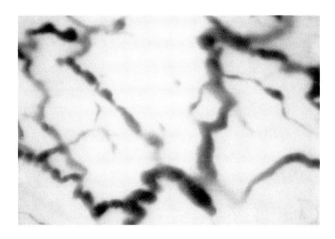

FIGURE 13 Ectatic tortuous blood vessels in senile skin.

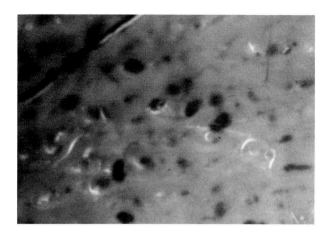

FIGURE 14 (B) Shakespeare type C, neavus flammeus on cheek, male 17 years old.

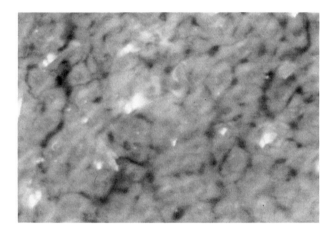

FIGURE 14 (A) Shakespeare type B, naevus flammeus on chin, female 2 years old.

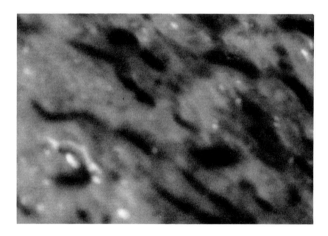

FIGURE 14 (C) Shakespeare type D, naevus flammeus on cheek, female 15 years old. (Courtesy of Prof. Dr. Ir, M.J.C. van Gemert, Dr. C.M.A.M. van der Horst, and Dr. S.D. Strackee.)

subpapillary vessels. The capillaries seem to be more susceptible to destruction in disease processes than are the subpapillary plexuses. Telangiectasias are better visualized when the horny layer is removed with formic acid.[21]

Nevus flammeus — In a study performed in the Academic Medical Center of the University of Amsterdam the dermatoscopic picture of nevus flammeus is investigated before argon laser treatment (Figure 14). The different appearances of capillaries in port wine stains of increasing severity are described in Table 2 (after Shakespeare and Carruth).[22]

Chronic ulcer — The capillary blood supply to chronic venous ulcers can be visualized by making use of dermatoscopy and even better in combination with intravenous perfusion of a fluorescent dye (Figure 15).[7,23]

C. Malignant Conditions

Before describing malignant conditions it is necessary to delineate a benign lesion, e.g., a pigmented melanocytic nevus. On dermatoscopic investigation a spider web-like superficial pigmentary pattern is observed (Figure 16). The color is brown to black and uniformly distributed. The greatest density of pigment is central and fading to the periphery. There are no dilated and abnormal capillaries seen. Consequently, there is no erythema.

Simple visual inspection is not always sufficient for the clinical diagnosis of melanocytic skin tumors.[24] This is particularly relevant for the differential diagnosis between dysplastic nevi and an early melanoma, as well as for the differentiation between melanomas and tumors of non-melanocytic origin. For this differentiation the epiluminescence microscope is being used with increasing frequency. Pehamberger et al.[25,26] and others[27-31] have previously reported that epiluminescent microscopy (ELM) definitely improves the diagnostic accuracy of pigmented skin lesions.

Goldman,[32] MacKie,[28] Fritsch and Pechlaner,[27] and many others[2,25,29] have published articles about the differential diagnostic possibilities of the epilumin-

TABLE 2 Transcutaneous Microscopic Classification of Port-Wine Stains by Blood Vessel Type

Type A	Capillaries of normal appearance, increased in number, are seen. The capillaries show the "pinpoint" ends visible in normal skin, and are analog ous to the "constricted type".
Type B	Capillaries have increased diameters but otherwise a normal appearance. They have a "pin-head" appearance, are very prominent, and are analogous to the "intermediate type".
Type C	Capillaries with thickened ends, resembling "ring doughnuts", are seen. The ends of the capillaries are flattened and turned to one side. These capillaries are very prominent, and are analogous to the "intermediate type".
Type D	Ectatic capillaries with ballooning, show "microcavernous hemangiomata". Thickening and ballooning of the capillary wall along the ascending and descending limbs is present. These capillaries are analogous to the "dilated type".
Type E	Very extended capillaries of reduced diameter and complex helical convul sions are seen. These are unusual, are generally seen on the limbs only, and have no analogue in Ohmori and Huang's description.
Type F	Formation of interconnections between capillaries and vessels in the subpa pillary dermal plexus with formation of a vascular network. Capillary "tufts" arising from small feeder vessels may be seen. These are analogous to the "deep-located type".

escence microscope. In these publications differential diagnostic microscopic criteria of a widespread nomenclature were described and evaluated. The work group Analytical Morphology of the Arbeitsgemeinschaft Dermatologische Forschung has organized a consensus meeting in 1989 in Hamburg about the nomenclature used in epiluminescence microscopy. With uniform and original description it will be easier to communicate between investigators and this will facilitate a better understanding of the method.

The technique is usually alluded to as epiluminescence microscopy or surface microscopy. In practice immersion oil is applied to the pigment lesion and covered with a cover slip. The oil makes the stratum corneum translucent, allowing a better observation of the underlying pigmented structures and patterns. Under epiluminescence illumination the lesion is then investigated with a stereomicroscope with magnifying range of 6 to 40 times.

The relations between the histological substrate and the diagnostic meaning of single epiluminescence microscopical criteria are summarized in Table 3. Other criteria of diagnostic importance which express the special arrangement of some of the properties inside a lesion are peripheral or central accentuation, regular or irregular distribution.

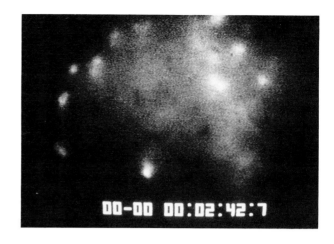

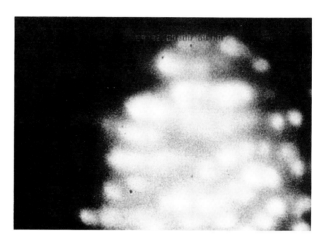

FIGURE 15 (A) Intravenous perfusion of fluorescent dye visualizing capillaries in border of granulating venous ulcer. (B) Same as (A) after transcutaneous electrical nerve stimulation. Observe that the newly formed capillaries are leaking. (Courtesy of Dr. M.J.H.M. Jacobs.)

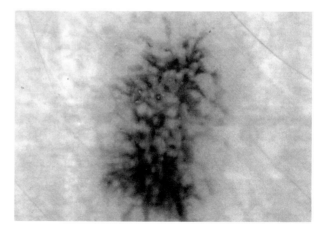

FIGURE 16 Dermatoscopic picture of junctional nevus.

TABLE 3 Diagnostic Criteria in Epiluminescence Microscopy (Proposal of the Consensus Meeting of the Work Group On Analytical Morphology, 1989, Hamburg)

ELM Criterion	Histological Substrate	Diagnostic Meaning
Normal pigment network	Pigmented retelists	Normal pigmented skin
Discrete pigment network	Slightly hypopigmented retelists	Likely to be a benign melanocytic lesion
Prominent pigment network	Hyperpigmented retelists	Likely to be a malignant melanocytic lesion
Regular pigment network	Regularly distributed retelists	Benign melanocytic lesion
Irregular pigment network	Irregularly distributed retelists	Dysplastic nevus malignant melanoma
Wide pigment network	Widely spaced retelists	Melanoma, e.g., melanoma *in situ*
Narrow pigment network	Tightly spaced retelists	Benign melanocytic lesion
Broad pigment network	Broad retelists	Melanoma, v.a. melanoma *in situ*
Delicate pigment network	Narrow retelists	Benign melanocytic lesion
Irregular extensions, pseudopods	Confluent pigmented juntional nests in the periphery	Malignant melanoma
Radial streaming	Radial distribution of pigmented junctional nests	Malignant melanoma pigment spindle-shaped cells
Brown globules	Superficial pigmented nests in the upper dermis	Regular dermal nevus: irregular malignant melanoma
"Black dots"	Aggregate pigmented melanocytes in the horny layer	Malignant melanoma
Whitish veil, "milky way"	Compact orthokeratosis and hypergranulosis	Malignant melanoma
White scar-like depigmented areas	Decrease of melanin and fibrosis	Regressive malignant melanoma
Greyish-blue areas	Superficial fibrosis melanophages	Regressive malignant melanoma
Hypopigmentation	Decrease of melanin	In many melanocytic lesions
Reticular depigmentation	Very wide retelists	Spitz-nevus
Milia-like cysts	Intraepidermal horn pearls in epidermis under the skin surface	Verruca seborrhoica papillomatous dermal nevus
Comedo-like openings	Intraepidermal horn pearls in epidermis	Verruca seborrhoica papillomatous dermal nevus
Telangiectasias	Dilated blood vessels in the upper dermis	Basal cell carcinoma
Reddish-blue areas	Telangiectasia in the upper dermis	Hemangioma
Maple leaf-like areas	Pigmented epithelial nests	Pigmented basal cell carcinoma

It should be emphasized that a diagnosis obtained with epiluminescence microscopy should never be based on a single criterion. In this respect it must be clear that under the term "diagnostic meaning" only frequently occurring features are mentioned. The demonstration of a single feature is not proof for the diagnosis and the absence of criteria is not exclusive of a diagnosis either.

An exact epiluminescence microscopic diagnosis can only be made in combination with careful simultaneous investigation of other criteria present. With the epiluminescence microscope a difficult problem is constituted by the over projection of skin levels so that it is difficult to recognize the structures and the colors, and it is also often impossible to associate a certain phenomenon with a corresponding specific histological feature. To support a diagnosis it is therefore important to include the histological as well as the epiluminescence microscopic diagnostic criteria in combination, rather than in isolation. The description of microscopic criteria by epiluminescence, for example, bizarre patterns and other multicomponent compositions, is often very complex and rich in variations, and not as exact as histopathology.

In the study of Steiner et al.[40] it was shown that better diagnostic accuracy could be achieved for pigmented Spitz nevi. Criteria have now been established but still have to be tested more extensively. This is important for small pigmented lesions that cannot be diagnosed by clinical criteria alone and in particular for the differentiation of pigmented Spitz nevi from malignant melanoma, dysplastic nevi, or common nevi. By applying ELM pattern analysis to Spitz nevi, we were able to increase the diagnostic accuracy from 56% to 93% (Table 4). This is a distinct improvement, particularly because some of these Spitz nevi had clinically been considered to be malignant melanoma. ELM correctly diagnosed three of five lesions as Spitz nevi. In two lesions the ELM pigment pattern was suggestive of malignant melanoma but histopathologically these lesions proved to be Spitz nevi. Schultz[41] developed a score system for dysplastic nevi, derived from a group of patients with junctional and dermal melanocytic nevi, which he compared with dysplastic nevi (Table 5).

The technology underlying epiluminescence microscopy or dermoscopy is based on the principle that oil reduces reflection from the skin surface, allowing light rays to penetrate more deeply. In general benign lesions exhibit consistent patterns of coloration and shape and have defined borders under epiluminescence microscopy, whereas malignant lesions tend to have irregular, less homogeneous patterns of development and color. With epiluminescence microscopy clear distinction can be made between melanocytic and pigmented non-

TABLE 4 Epiluminescence Microscopy Appearance of Pigmented Spitz Nevus and Melanoma

	Spitz nevi	Melanoma
General appearance	Monomorphous, starburst lesion, coffee bean-like appearance	Polymorphous, multiple pattern
Pigment network	Prominent, regular, stops abruptly at periphery or thins out, "negative" pigment network	Prominent, irregular, stops abruptly at periphery or thins out, no "negative" pigment network
Brown globules	Different size, regular throughout the lesion, rim of large brown globules at periphery of lesion	Different size, haphazardly spaced, no rim
Black dots	Center or throughout lesion, regular distribution	Often only at periphery, irregular distribution
Depigmentation	Bizarre, retiform in center	Irregular, often in periphery
Border	No pseudopods, no radial streaming	Often pseudopods, often radial streaming

melanocytic lesions as well as between malignant and benign melanocytic lesions.

Cutaneous pigmented lesions, when small and intensely pigmented, may provide difficulties in diagnosis. They may represent a lentigo, a vascular lesion such as angiokeratoma, a minute seborrheic keratosis, a pigmented basal cell carcinoma, or even foreign pigment such as a tattoo. With the aid of epiluminescence microscopy, these lesions can be differentiated. It was found that this technique was useful and even small darkly pigmented papules measuring 1 to 3 mm in diameter of intracutaneous metastatic melanoma could be recognized as such.

The cutaneous secondaries showed asymmetric brown-black globules and lacked the irregular pigment network, radial streaming, pigmented pseudopods, or black dots associated with primary melanoma. However, the epiluminescence microscopic appearance of these melanoma metastases is unlikely to be specific and their appearance would probably be shared by some forms of pigmented dermal nevi. Nonetheless, in the context of newly developing lesions, particularly when they are asymmetric, the finding can be helpful.[42]

Epiluminescense microscopy provides a practical and valuable diagnostic tool for the recognition of pigmented tumors of the skin.[38,39]

TABLE 5 Epiluminescence Microscopic Scoring Protocol for Dysplastic Nevi

	Criterium	Points
1.	Black dot on blue background	10
2.	Area with a regular distribution of capillaries	10
3.	Pseudospod-like border pattern	10
4.	Regressive remodeling of tissue with melanophages at the periphery	9
5.	Abrupt disappearance of pigment of trabeculae	7
6.	Dendrite-shaped blue grey trabeculae	6
7.	Bizarre network	5
8.	Multicomponent pattern	5
9.	Grey color	
10.	Grey blue globuli in center of papillae	2

Evaluation of score: 10–15 suspect for dysplastic nervus >15 highly suspect for dysplastic nevus

V. Validation of the Method

The dermatoscope is very useful in routine investigations of different diseases. The excellent stereoscopic image of the different observed skin areas are very realistically experienced by the investigator. The use of a simple instrument, for example as described by Braun Falco et al.,[2] is interesting but has draw backs because of the monocular vision and the poor illumination sources giving rise to very poor images.

Epiluminescence microscopy has proved 85% accuracy compared with 60% accuracy for clinical evaluation. In a study by Curley et al.[43] of 116 pigmented lesions, only 50% were diagnosed correctly by experienced dermatologists. To aid in diagnosis, Doppler sonography,[44] three-dimensional nevoscopy,[45] computerized digital imaging processing[46] for primary melanomas, and immunoscintigraphy with radioactive-labeled monoclonal antibodies[47] for metastatic disease have been developed. However, most of these methods are in the investigational stage, require experienced personnel, are not readily available, and are impractical for general use.

With defined morphologic criteria, skin surface microscopy was shown to improve the diagnostic accuracy of most pigmented lesions significantly.[25,26] In recent reviews,[25,26,48] the surface microscopic morphology of most pigmented lesions has been defined and illustrated, but the appearance of micropapular intracutaneous metastatic melanoma was not included.

VI. Archiving and Follow-Up

A new development is the use of a miniature video still camera with fiber optic illumination and optical lenses designed by various companies. They bring out the full structure of skin features, from intracutaneous examination to three-dimensional assessment of the surface of the skin. Full details are imaged quickly and easily to assure the highest level of diagnostic capability. The microscopic system comprises of a solid-state CCD-camera with illumination and camera controls housed in the compact supply box. The dermascope is compatible with a wide range of video recording and printing systems, overcoming problems associated with recording data over time. Hard copy prints of video-recorded information allow for progressive detailed study and analysis or presentation to teaching or research groups. The size and growth of skin melanomas, for example, can be progressed using the optional measurements scale located in the probe. Absolute measurements over time are easily recorded, unaffected by changing magnification and with no requirement for calibration.

Optional accessories include the high- and low-magnification adaptors which extend the capability of the microscopic system for more and fast investigation or where larger areas require referencing for achieving or recording purposes. A measurement graticule (1 cm in 1 mm divisions) easily locates on to the contact probe and is always in focus with the specimen.

References

1. Knoth, W., Boepple, D., and Lang, W.H., Differential diagnostische Untersuchungen mit dem Dermatoskop bei ausgewaehlten Erkrankungen, *Hautarzt*, 30, 7, 1979.
2. Braun-Falco, O., Stolz, W., Bilek, P., Merkle, T., and Lanthaler, M., Das Dermatoskop. Eine Vereinfachung der Auflichtmikroskopie von pigmentierten Haut-veränderungen, *Hautzartz*, 41, 131, 1990.
3. Gibson, H.L., Further data on the use of infrared color film, *J. Biol. Photogr. Assoc.*, 33, 155, 1965.
4. Aldis, A.S., and Marshall, R.J., Metastatic melanoma, detection by infrared recording, *Med. Biol. Illus.*, 13, 2, 1963.
5. Gibson, H.L., Medical infrared color photography, II, *Visual Med.*, 2, 43, 1967.
6. Rook, A., Wilkenson, D.S., and Champion, R.H., Principles of diagnosis, in *Textbook of Dermatology*, 4th ed., Rook et al., Eds., Blackwell Scientific, London, 1987, 61.
7. Jacobs, M.J.H.M., Breslau, P.J., Slaaf, D.W., and Reneman, R.S., Nomenclature of Raynaud's phenomenon: a capillary microscopic and hemorheologic study, *Surgery*, 101, 136, 1987.
8. Marks, R. and Dawber, R.P.R., Skin surface biopsy: an improved technique for the examination of the horny layer, *Br. J. Dermatol.*, 84, 117, 1971.
9. Cunliffe, W.J., Forster, R.A., and Williams, M., A surface microscope for clinical and laboratory use, *Br. J. Dermatol.*, 90, 619, 1974.
10. Cohen, P.R., Examination of the male genitalia with an ophthalmoscope: a rapid and simple approach to the detection of penile, venereal warts, *J. Am. Acad. Dermatol.*, 20, 521, 1989.
11. Pierard, G., Pierard-Franchimont, C., and Dowlati, A., The skin surface biopsy in clinical and experimental dermatology, *Rev. Eur. Dermatol.*, 4, 455, 1992.
12. Beagley, J. and Gibson, I.M., *Changes in Skin Condition in Relation to Degree of Exposure to Ultra Violet Light*, School of Biology, Western Australian Institute of Technology, Perth, 1980.
13. Daniell, H.W., Smoker's wrinkles. A study in the epidemiology of 'crow's feet', *Ann. Intern. Med.*, 75, 873, 1971.
14. Holman, C.D.J., Armstrong, B.K., Evens, P.R., Lumsden, G.J., Dallimore, K.J., Meehan, C.J., Beagley, J., and Gibson, I.M., Relationship of solar keratosis and history of skin cancer to objective measures of actinic skin damage, *Br. J. Dermatol.*, 110, 129, 1984.
15. Schamann, B. and Alter, M., *Dermatoglyphics in Medical Disorders*, Springer-Verlag, New York, 1976.
16. Verbov, J.L., Dermatoglyphic and other findings in alopecia areata and psoriasis, *Br. J. Clin. Pract.*, 22, 257, 1968.
17. Gilje, O., Capillary microscopy in the differential diagnosis of skin diseases, *Acta Derm. Venereol.*, 33, 303, 1953.
18. Davis, M.H. and Lawler, J.C., Capillary microscopy in normal and diseased human skin, *Biol. Skin*, 2, 1961.
19. Gilje, O., Kierland, R., and Baldes, E.J., Capillary microscopy in diagnosis of dermatologic diseases, *J. Invest. Dermatol.*, 22, 199, 1954.
20. Ryan, T.J., and Wells, R.S., Hereditary benign telangiectasia, *Trans. Rep. St John's Hosp. Derm. Soc. Lond.*, 57, 148, 1971.
21. Lehmann, P. and Kligman, A.M., In vivo removal of horny layer with formic acid, *Br. J. Dermatol.*, 109, 313, 1983.
22. Shakespeare, P.G. and Carruth, J.A.S., Investigating the structure of the port-wine stain by transcutaneous microscopy, *Lasers Med. Sci.*, 1, 107, 1985.
23. Jacobs, M.J.H.M., Jörning, P.J.G., Beckers, R.C.Y., Ubbing, D.T., Van Kleef, M., Slaaf, D.W., and Reneman, R.S., Foot salvage and improvement of microvascular blood flow as a result of epidural spinal cord electrical stimulation, *J. Vasc. Surg.*, 12, 354, 1990.
24. Bahmer, F.A., Fritsch, P., Kreusch, J., Pehamberger, H., Rohrer, C., Schindera, I., Smolle, J., Soyer, W.P., and Stolz, W., Diagnostische Kriterien in der Auflichtmickroskopie, *Hautarzt*, 41, 513, 1990.
25. Pehamberger, H., Steiner, A., and Wolff, K., In vivo epiluminessence microscopy of pigmented skin lesions. I. Pattern analysis of pigmented skin lesions, *J. Am. Acad. Dermatol.*, 17, 571, 1987.
26. Steiner, A., Pehamberger, H., and Wolff, K., In vivo epiluminescence microscopy of pigmented skin lesions. II. Diagnosis of small pigmented lesions and early detection of malignant melanoma, *J. Am. Acad. Dermatol.*, 17, 584, 1987.
27. Fritsch, P. and Pechlaner, R., Differentiation of benign from malignant melanocytic lesions using incident light microscopy, in *Pathology of Malignant Melanoma*, Ackerman, A.B., Ed., Masson, New York, 1981, 301.
28. MacKie, R., An aid to the preoperative assessment of pigmented lesions of the skin, *Br. J. Dermatol.*, 85, 232, 1971.
29. Bahmer, F.A. and Rohrer, C., Rapid and simple macrophotography of the skin, *Br. J. Dermatol.*, 114, 135, 1986.
30. Soyer, H.P., Smolle, J., Hoedl, S. et al., Surface microscopy: a new approach to the diagnosis of cutaneous pigmented tumours, *Am. J. Dermatopathol.*, 11, 1, 1989.
31. Stolz, W., Bilek, P., and Langthalen, M., Skin surface microscopy, *Lancet*, 2, 864, 1989.
32. Goldman, L., Some investigative studies of pigmented nevi with cutaneous microscopy, *J. Invest. Dermatol.*, 16, 407, 1951.
33. Haas, N., Ernst, T.M., and Stüttgen, G., Makrofotografie im transmittierenden Licht. Ein Beitrag zur horizontalen Strukturanalyse pigmentierter Hauttumoren, *Z. Hautkr.*, 59, 985, 1984.
34. Haas, N. and Ernst, T.M., Makrophotographische Korrelate zur Histologie bei Precursor-Nävi und SSm in Anlehnung an das Schema von McGovern, *Z. Hautkr.*, 61, 1535, 1986.
35. Hughes, B.R., Black, D., Srivastava, A., Dalziel, K., and Marks, R., Comparison of techniques for the non-invasive assessment of skin tumours, *Clin. Exp. Dermatol.*, 12, 108, 1986.
36. Kreusch, J. and Rassner, G., Struktur-analyse melanozytischer Pigmentmale durch Auflichtmikroskopie, *Hautarzt*, 41, 27, 1990.
37. Soltani, K., Use of the ophthalmoscope in the clinical diagnosis of cutaneous pigmented lesions, *J. Am. Acad. Dermatol.*, 17, 521, 1989.

38. Soyer, H.P., Smolle, J., Kerl, H., and Stettner, H., Early diagnosis of malignant melanoma using surface microscopy, *Lancet,* II, 8562, 1987.
39. Soyer, H.P., Smolle, J., Kresbach, H., Hoedl, S., Glavanovitz, P., Pachernegg, H., and Kerl, H., Zur Auflichtmikroskopie von Pigmenttumoren der Haut, *Hautarzt,* 39, 223, 1988.
40. Steiner, A., Pehamberger, H., Binder, M., and Wolff, K., Pigmented Spitz nevi: improvement of the diagnostic accuracy by epiluminescence microscopy, *J. Am. Acad. Dermatol.,* 27, 697, 1992.
41. Schultz, H., Auflichtmicroscopischer score zur differentialdiagnose dysplastischer Nâvi, *Hautarzt,* 43, 487, 1992.
42. Pang, B.K. and Kossard, S., Surface microscopy in the micropapular cutaneous metastatic melanoma, *J. Am. Acad. Dermatol.,* 5, 775, 1992.
43. Curley, R., Marsden, R., and Fallowfield, M., Diagnostic accuracy in the clinical evaluation of melanocytic lesions, *Br. J. Dermatol.,* 119(Suppl. 33), 345, 1988.
44. Srivastava, A., Hughes, L., Woodcock, J., et al., Vascularity in cutaneous melanoma detected by Doppler sonography and histology, *Br. J. Cancer,* 59, 89, 1989.
45. Dhawan, A., Early detection of cutaneous malignant melanoma by three-dimensional nevoscopy, *Comput. Methods Progr. Biomed.,* 21, 59, 1985.
46. Murray, A., Neill, S., Harland, C. et al., A new method for the quantification of physical change in unstable pigmented lesions using digital image processing techniques, *Br. J. Dermatol.,* 199(Suppl. 33), 45, 1988.
47. Mechl, Z. and Bauer, J., The detection of metastases, in *Cutaneous Melanoma Biology and Management,* Cascinelli, N., Santinami, M., and Veronesi, U., Eds., Masson, Milan, 1990, 218.
48. Bahmer, F., Fritsch, P., Kreusch, J. et al., Terminology in surface microscopy, *J. Am. Acad. Dermatol.,* 23, 1159, 1990.

Chapter 4.3
Skin Replication for Light and Scanning Electron Microscopy

Bo Forslind
Experimental Dermatology Research Group
Department of Medical Biophysics
Karolinska Institute
Stockholm, Sweden

Motto: Every honest researcher I know admits he is just a professional amateur. He is doing whatever he is doing for the first time. That makes him an amateur! He has sense enough to know that he is going to have a lot of trouble, so that makes him a professional!

Anonymous

I. Introduction

Replication is actually an old method within the realm of electron microscopy. Before the introduction of the scanning electron microscope (SEM), surface analysis was routinely performed by using multiple-stage replication techniques. A general feature of these techniques was the fact that they were all destructive, in the sense that the replicated material was lost in the preparation procedure. Hence, its applicability to investigation of the skin was limited to situations where skin (or in more general terms, integument) samples were taken by biopsy.

Recent reviews show that it is only to a minor extent that replication techniques have been used in clinical and experimental dermatology.[2,10] The advantages of this noninvasive technique, therefore, may still reveal unexplored areas of application. If analysis of the surface structure of skin and nails can be performed without the need for biopsy it will be possible to study processes of inflammation and/or infection sequentially, as well as the effect of topical drugs and direct physical wear and tear at the same site.

Sampling of topographical information for analysis of stratum corneum have been achieved through various techniques, including tape stripping,[16] plastic impression techniques,[1] skin surface biopsy,[4] besides the subject of this chapter. The plastic impression technique and, to a certain extent, the tape stripping technique (if it is to contain more than a few scattered stratum disjunctum cells), have the drawback that they require a thorough defattening of the skin surface. The surface biopsy strips off a thin surface layer of the stratum corneum but the surface that can be investigated in the SEM (or light microscope) represents the rift zone between the part of the stratum corneum *in situ* and the sample. This is then a surface that has not been directly exposed to the environment but represents the stratum corneum at a depth of one or more cell layers. Hence, the details of the surface structures such as villous projections[4] will be more pronounced than those of the actual surface of the stratum corneum.

In principle the replica of a surface can be a *negative* (or direct) one or a *positive*, essentially two-step, replica. Since the skin surface of a living subject contains moisture a number of replication media that require a dry surface are disqualified. Materials designed for replication in dental work have the required property of adhering tightly even to a wet surface and are therefore interesting candidates for skin replication molds. The silicon plastics have proven to allow detailed recording of surface characteristics of skin and nails, and have found extensive use in medical and biological scanning microscopy.[2,3,6,10-14]

One of the first papers to report the use of silicone rubber plastics to produce the negative mold for hard plastic replicas was given by Sampson[11] for use on plant material and insects. Although intended for light microscopic use the technique was soon adopted for SEM analysis of surfaces.

This chapter will, on a selective basis, review the surface replication technique as it has evolved in electron microscopy. We will not aim at providing the reader with a complete catalogue of materials but rather put forward some general and practical hints for the particular dermatological applications. The emphasis is put on simple and fast measures, suitable for use in the consultation office, which will produce replicas that can be analyzed at magnifications up to at least 2000×. Some practical aspects of producing and checking the result of a replication are forwarded. Readers with interests in high-resolution work are referred to current textbooks on electron microscopic preparation techniques and to the comprehensive reviews by Pfefferkorn and Boyd[9] and Pameijer,[7,8] which cover the literature up to 1978.

II. Carbon and Metal/Carbon Replication of Dry Solid Surfaces

A. Methodological Principle

This technique was developed for transmission electron microscopy (TEM) of surfaces before the SEM was available. It is mandatory that the object is completely free of water and any other substances that are volatile in the vacuum system used for evaporative coating of the specimen. For biological tissues this means that replication must be preceded by fixation, usually chemical fixation, but cryomethods are also possible. Subsequently a thorough dehydration of the specimen is required. This can be done according to the common protocol used for preparation of biological tissue for TEM. Alternatively critical point drying may be used.

The surface to be replicated should be cleaned by an appropriate method, e.g., ultrasonication, organic solvents, water, etc., and subsequently dried. The replication requires six steps (Figure 1):

1. The clean surface is covered with a thin carbon or metal (Cr, Pd/Pt, Pt, Au) film at an angle (often 6 to 10°) by evaporation in vacuum
2. The evaporated film is stabilized by application of a thin plastic film, e.g., formvar
3. The object is removed mechanically (which often disrupts the replica) or preferably by chemical dissolution
4. The (negative) replica is stabilized by a thick layer of carbon evaporated onto the replica in vacuum
5. The plastic film is removed by the appropriate solvent for the plastic in question
6. The positive carbon replica in now transferred to a conventional electron microscopic grid and viewed in the TEM

This technique has a resolution that is satisfactory at least down to 20 nm (200 Å) on metal surfaces. For biological specimens the resolution is somewhat less and very much dependent on the nature of the specimen surface and the preparation.

A similar process, omitting the plastic intermediate stage, is used in the production of freeze-fracture replication, which has been successfully used in membrane research.

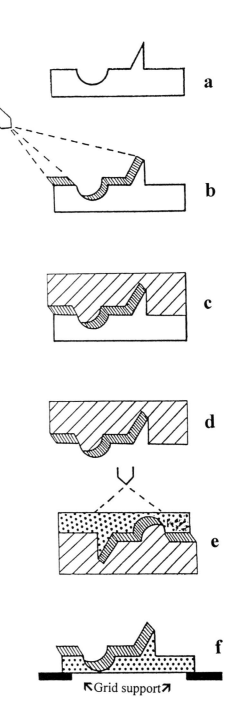

FIGURE 1 The steps in making a metal/carbon replica of surfaces. The clean surface (a) is covered with a thin carbon or metal (Cr, Pd/Pt, Pt, Au) film at an angle (often 6 to 10°) by evaporation in vacuum (b). The evaporated film is stabilized by application of a thin plastic film, e.g., formvar (c) (hatched area). The object is removed mechanically (which often disrupts the replica) or preferably by chemical dissolution (d). The (negative) replica is stabilized by a thick layer of carbon evaporated onto the replica (e) in vacuum. The plastic film is removed by the appropriate solvent for the plastic in question. The positive carbon replica is now transferred to a conventional electron microscopic grid (f) and viewed in the TEM.

B. Sources of Errors

It is to be noted that all fixation and drying procedures induce various degrees of surface structure artefacts due to linear or volume changes, shrinkage, etc.

C. Correlation with Other Methods

The main drawback of the technique is the loss of the original sample and the fragmentation of the replica on transferring to the electron microscopic grid, something that is virtually impossible to avoid. Thus, the technique is best when high resolution is required and therefore preferentially a TEM technique.

III. Plastic Impression Technique

A. Methodological Principle

Before introduction of the SEM, but even more recently, replicas for light microscopy have been made from plastic compositions that required a cleaned and dry, hairless area for the replication.[1,6] Consequently interest has been focused on low-resolution details of the skin, such as the cutaneous patterns of furrows that form patterns characteristic of a certain area of the integument.

The negative imprint is obtained simply by spreading the liquid plastic over the area to be sampled and allowing it to cure before mechanical removal. Household glues based on plastics in organic solvents that evaporate quickly can produce acceptable results at the low resolution required.

B. Sources of Errors

The preparation of the skin surface to be replicated, including shaving and degreasing with organic solvents, is likely to produce artefactual changes of fine surface structures. The gross features of the skin area replicated, i.e., the patterns of furrows and wrinkles, reproduce well even at low magnifications, e.g., >10×.

C. Correlation with Other Methods

There is some loss of fine details at the high magnifications attainable in the SEM.

IV. Silicone Elastomer Replication

A. Methodological Principle

The negative replica — The silicone-elastomer replication has its basis in the products developed for use in clinical dentistry.[3,14] There are five main requirements on the plastic used for producing the negative mold.

1. The silicone plastic should have a low viscosity to adhere closely even to the fine details of the surface
2. It should adhere well even to wet surfaces
3. After a fast, and complete, polymerization it should be released from the original specimen without leaving any material behind
4. It should possess an elastic memory to allow a complete return to the original status even when withdrawn from undercuts
5. The polymerization process should not produce heat, i.e., involve an exothermic reaction which may change surface properties of the object, and cause the discomfort of the subject.

The positive replica — The plastic used for producing the positive replica should cure at room temperature with as little release of heat as possible to prevent deformation of the negative mold.

In the SEM micrographs accompanying this chapter (Figure 2) molds were made from Provil-L® (Bayer Dental D-5090 Leverkusen, Germany), which is characterized as a low-viscosity, type I silicone meeting the requirements of ISO 4823, type (e) 3, category A (adhesion-induced polymerization). The silicone plastic is thoroughly mixed with an equal volume of catalyst and immediately applied to the surface to be replicated (Figure 3c). It is then allowed to set for ≥3 min before gentle removal from the (skin) surface (Figure 3d). The negative replica is subsequently covered with an Araldite® plastic (CIBA-Geigy) (Figure 3e) which cures within 3 to 5 h depending on the volume applied. Alternatively we have used a methacrylate designed for whole-mount embedding of insects, etc., which has longer curing time. The surface of the plastic, positive replica is subsequently made conductive by gold sputtering (Figure 3f and g).

B. Sources of Errors

The negative mold — A large negative imprint of the skin surface (i.e., >1 × 1 cm) tends to bend when loosened from the original surface and this large curvature remains when the positive replica is made (Figure 3f). Due to high total absorption of energy in the electron beam, a *larger-than-the-stub* specimen tends to be unstable in the beam, i.e., be subject to drift during viewing in the SEM.

The positive replica — When making the positive replica the amount of accelerator may be crucial to the final results. If the curing process occurs at too fast a rate, gas bubbles will accumulate at the replica surface.

C. Recommendations

The negative replica — The mixing of silicon base and curer is a critical stage in making a replication. The two components should be thoroughly mixed but agitation should not be so vigorous as to produce air bubbles. The drawbacks of manual mixing can be virtually eliminated when using a dual vessel ejector (Bayer Cartridge delivery dispensing gun) (Figure 3c). When making a negative replica for SEM care should be taken to cover a surface area no greater than the SEM specimen holder (the "stub") to avoid the unnecessary heating that follows from having a large specimen surface. When the negative replica has been removed after setting, its surface can be inspected

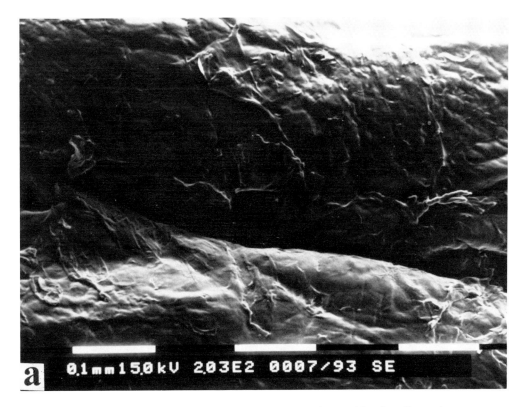

FIGURE 2 Replicas of a normal skin surface (volar aspect of lower arm) obtained by the silicon elastomer two-step method. (a) Skin surface after approximately 15 min of exposure to damp cloth saturated with distilled water, 203×.

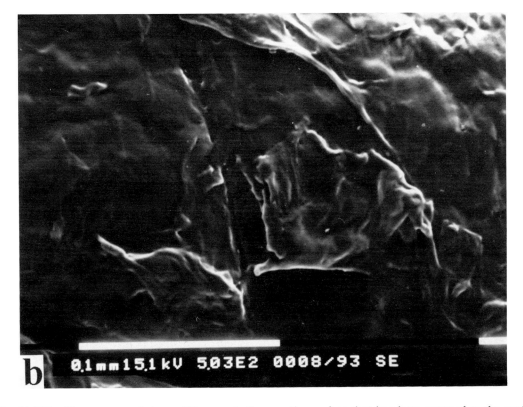

FIGURE 2 (b) 503×. There are no obvious *villiform projections* on the surface that has been exposed to the environment.

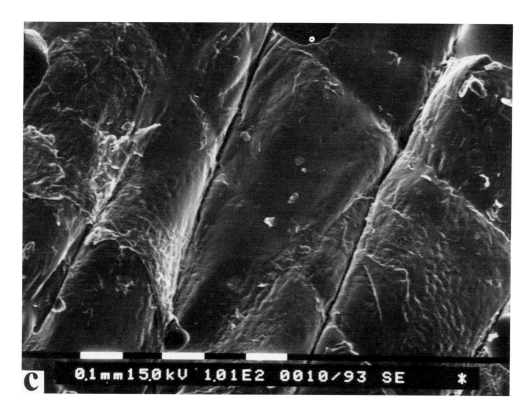

FIGURE 2 (c) corresponding area sampled dry on the following day, 101×.

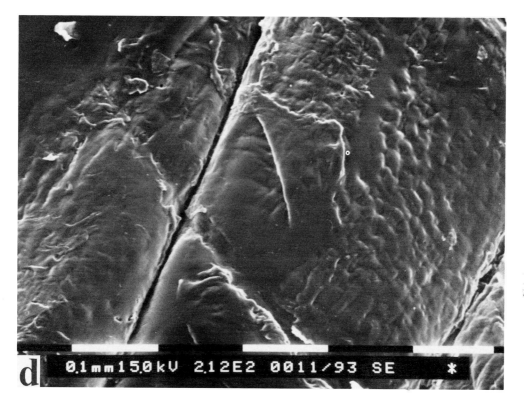

FIGURE 2 (d) 212×.

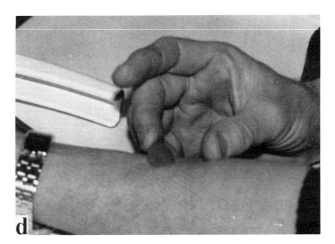
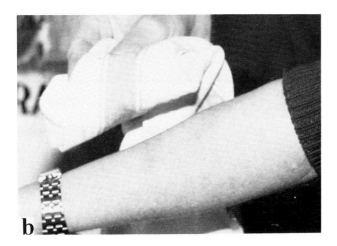

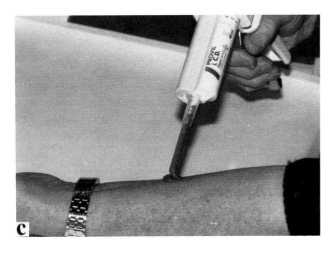

FIGURE 3 Silicon elstomer replication of an integument surface. (a) The skin surface is briefly cleaned by a quick rinse in cool tap water and (b) blotted dry. (c) The elastomer is applied and allowed to cure for about 3 min. (d) The negative mold produced is gently removed. (e) The mold is covered with a plastic to produce a positive replica. (f) After gold-sputtering the positive replica (left) and the negative mold can be inspected for surface defects by light microscopy.

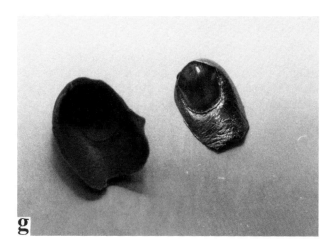

quick rinse under cool tap water (Figure 3a). The surface to be replicated is then blotted dry with soft, fluff-free tissue or dust-free cloth (Figure 3b) to avoid friction that could produce surface artefacts. Sometimes an improvement of resolution of details by the silicone is obtained through a quick cool water rinsing which undoubtedly removes water-soluble surface material. At low magnification (e.g., 40×) no swelling is apparent from this process. Rinsing may, however, introduce artefacts in lesional skin. It is then preferable to remove loose surface material by making two or three impressions from the same surface rather than making the lesion subject to tap water rinse. The sequentially obtained molds can be checked against each other for artifacts.

The negative imprint, *the mold,* is not suited for direct study in the SEM because it will melt and evaporate when hit by the electron beam. However, it can be used directly for light microscopic and photographic observations at low and moderate magnifications.[8] If this is the object, rather than a study in the SEM, large areas can be replicated, e.g., 2 × 2 cm (Figure 3g).

The positive replica — The making of a successful positive replica involves a good choice of plastic. We have used Araldite® (Ciba-Geigy) in a 1:1 mixture with accelerator. The manufacturer's advice on mixing proportions for plastic and accelerator should be tested for each batch of plastic as it may vary during aging of these materials. It is our experience that it is not unusual that the amount of accelerator should be somewhat reduced to get a curing rate that does not produce heat and solvent gas bubbles. However, it is easy to get an incomplete curing that results in a sticky surface which deforms on removal from the negative mold. An alternative way of reducing the risk of gas bubbles at the interface between the negative mold and the plastic is to moisten the surface of the silicon imprint with the solvent of the plastic (e.g., acetone for Araldite®) immediately before pouring the plastic onto the negative template. It is advised that positive casts are inspected for the presence of gas bubbles in the surface structures under a preparation microscope after gold sputtering. If bubbles are present they usually attain a size that allows them to be seen at a magnification of 40×. As an alternative to Araldite® we have used a methacrylate designed for embedding of large objects such as insects, e.g., a beetle. This methacrylate, which takes more than 24 h to cure even in thin sheets, tends to be very brittle. It reproduces the surface details well in our experience.

Pfister and Neukirchner[10] used a polystyrol granulate dissolved in toluol for the positive replica. The non-cured plastic has a syrup-like consistency. To

FIGURE 3 (g) Even large objects can be faithfully replicated. Left: fingertip of digit V of the author. (h) A stub machined to have a groove with undercut will prevent drift in large objects during SEM study.

under a light microscope at about 40× magnification to check that the surface is free from air bubbles. This simple measure saves much time and work.

If the skin surface is dry, *stratum disjunctum* cells may adhere to the negative imprint when it is released from the skin surface. Usually these scales can be removed by a jet-stream of dry, clean air. Alternatively the negative mold can be rinsed under running cool tap water and subsequently air dried. Since the stratum disjunctum cells may curl up or be deformed in other ways, a clean skin surface can be obtained by a avoid air bubbles in small crevices of the negative replica the authors "moisten" it with the solvent, toluol, before applying the plastic. The hardening time of this polystyrol plastic is comparatively long, approximately 24 h. The authors claim that magnifications up to 5000× are attainable with this technique.

Gold sputtering — The plastic material of the positive replica is an insolator. Gold sputtering of the surface provides a conductive film that distributes charges to ground potential but also contributes as a heat sink. The sputtering should be performed so as to get a continuous contact between the replica surface and the specimen stub. This is most easily achieved if the stub surface is cleared of the specimen at small point. When large objects are used, e.g., a replica of a fingertip with a nail, it is advantageous if the stub can be molded into the positive plastic replica. This can be achieved by making a groove with undercut (Figure 3h) with a milling cutter. Alternatively a cavity with undercut in the stub surface can be made using a dental drill. Through these means drift is virtually completely eliminated.

V. Present Status of Replication Techniques in Dermatology

Most areas of the human integument in health[1,2,6,12,15] have been described using replication techniques. In addition pathological conditions, including lesions of psoriasis,[13] superficial actinic porokerastosis, as well as more unusual conditions like Gorlin's syndrome,[6] have been documented. It is noteworthy that data presented in the literature on topographic data collected by replication (and corresponding) techniques on skin and its appendices in general merely have a descriptive character and provide little, if any, functional interpretation of the findings. Cosmetic industries have long utilized SEM studies of the effect of cosmetic formulations on the skin surface and the integument appendices, but details of this information have not been publicly available and cannot be scientifically evaluated. It is obvious that topographical methods of investigating the skin surface represent an interesting and potentially fruitful area of dermatological research. Combination with morphometric systems, image analysis systems, or other physical measurement systems[5] will allow quantitative analysis of changes in the surface structures as a result of the progress of a disease or a treatment of a disease. In addition to such applications a more extensive use of the excellent replication materials presently available will no doubt increase our knowledge of the dynamics of skin function in health and disease.

Acknowledgment

I am indebted to Ms. Eva Lundewall of the Swedish branch of Phillips Industrial Electronics AB for producing the SEM micrographs and to Ms. Margareta Andersson for photography and lay-out of illustrations.

The golden rule is that there are no golden rules.
George Bernhard Shaw

References

1. Chinn, H.D. and Dobson, R.L., The topographic anatomy of human skin, *Arch. Dermatol.*, 89, 155, 1964.
2. Forslind, B., Clinical applications of scanning electron microscopy and X-ray microanalysis in dermatology, *Scanning Electron Microsc.*, I, 183, 1984.
3. Jokstad, A. and Mjör, L.A., Assessment of marginal degradation of restorations on impressions, *Acta Odontol. Scand.*, 49, 15, 1991.
4. Marks, R. and Dawber, R.P.R., Skin surface biopsy: an improved technique for the examination of the horny layer, *Br. J. Derm.* 84, 117, 1971.
5. Marks, R. and Pearse, A.D., Surfometry. A method of evaluating the internal structure of the stratum corneum, *Br. J. Dermatol.*, 92, 651, 1975.
6. Nayler, J.R., Applications of the skin surface replica technique in dermatology, *J. Audiovis. Media Med.*, 21, 1984.
7. Pameijer, C.H., Replica techniques for scanning electron microscopy — a review, *Scanning Electron Microsc.*, II, 831, 1978.
8. Pameijer, C.H., Replication techniques with new dental impression materials in combination with different negative impression materials, *Scanning Electron Microsc.*, II, 571, 1979.
9. Pfefferkorn, G. and Boyde, A., Review of replica techniques for scanning electron microscopy, *Scanning Electron Microsc.*, I, 75, 322, 1974.
10. Pfister, T.C. and Neukirchner, A., Raster-elektronenmikroskopische Untersuchungen am kranken Nagel mittels Abdruck-Verfahren, *Fortschr. Med.*, 98, 1465, 1980.
11. Sampson, J., A method for replicating dry or moist surfaces for examination by light microscopy, *Nature*, 191, 932, 1961.
12. Tring, F.C. and Murgatroyd, L.B., Surface microtopography of normal human skin, *Arch. Dermatol.*, 109, 223, 1974.
13. Tring, F.C. and Murgatroyd, L.B., Psoriasis — changes in surface microtopography, *Arch. Dermatol.*, 111, 476, 1975.
14. Walsh, T.F., Waimsley, A.D., and Carrotte, P.V., Scanning electron microscopic investigation of changes in the dentogingival area during experimental gingivitis, *J. Clin. Periodontol.*, 18, 20, 1991.
15. Wagner, G. and Goltz, R.W., Human cutaneous topography. A new photographic technique: observation on normal skin, *Cutis*, 23, 830, 1979.
16. Wolf, J., Die innere Struktur der Zellen des Stratum desquamans der Menschlichen Epidermis, *Z. Mikr. Anat. Forsch.*, 46, 170, 1939.

5.0

Skin Surface Contour Evaluation

5.1 Stylus Method for Skin Surface Contour Measurement 83
    J. Gassmüller, A. Kecskes, and P. Jahn

5.2 Skin Surface Replica Image Analysis of Furrows and Wrinkles 89
    P. Corcuff and J.L. Leveque

5.3 Laser Profilometry 97
    J. Efsen, H.N. Hansen, S. Christiansen, and J. Keiding

5.4 Three-Dimensional Evaluation of Skin Surface: Micro- and
    Macro-Relief 107
    J. Mignot

5.5 Skin Surface Biopsy and the Follicular Cast 121
    R.P.R. Dawber

5.6 High-Resolution Ultrasound Examination of the Epidermis 125
    S. el-Gammal, T. Auer, K. Hoffmann, P. Altmeyer, C. Passmann,
    and H. Ermert

Chapter 5.1
Stylus Method for Skin Surface Contour Measurement

Johannes Gassmueller
BioSkin Institut für Dermatologische Forschung und Entwicklung GmbH
Hamburg, Germany

Andrei Kecskés
Schering AG, Dermatologie/Humanpharmakologie
Berlin, Germany

Peter Jahn
Schering AG, Diagnostika Koordination
Berlin, Germany

I. Introduction

Measuring surface texture requires a precise understanding of what is meant by surface. A surface is the boundary between two media. In medical terms, the surface of the skin is the boundary between the individual and the physical environment with all its diverse influences on the function and structure of the skin. In his handbook for surface texture analysis,[1] Mummery gives a very plastic description of what we have to deal with: "When specifying a surface profile, the examiner must be aware, that there are as many surface profiles as there are landscapes. The Himalayas and the Black Forest in Germany are both mountain ranges. This is where the similarity between them ends. The two mountain ranges differ not only in magnitude (height of the peaks) but also in their form (the shape of the peaks and valleys). Describing a surface is as complex as describing a mountain range."

The landscape of normal or impaired skin is determined by the cutaneous architecture. The arrangement and interlocking of adjacent keratinocytes, the epidermal rete ridges, the dermal papillary structure, and the cutaneous appendages all contribute to the surface form. Both internal and external processes such as aging, dehydration, hydration (cosmetic products), or atrophy (corticosteroids) continuously remodel the cutaneous landscape. Profilometry allows the objective measurement of the above-mentioned effects on skin surface and is of particular interest for the evaluation of the pharmaceutical or cosmetic action on the skin.

II. Skin Surface Measurement

The ideal approach for measuring the skin surface would be a touchless *in vivo* method. However, up to now no reliable and reproducible method has been reported which approaches the standard of "indirect" methods that rely on a negative replica of the skin. The main reason for the failure of direct measurements is the relative movement of the anisotropic skin surface while scanning the profile (pulsation of minor arteries, moving, trembling of subjects). Therefore, measuring the profile or surface texture of a replica remains the most common approach to describe surface characteristics related to the skin's surface geometry.

III. Methodological Principle

In our first attempts in 1980 to measure the influence of topical corticosteroids on skin surface texture, we adapted the stylus method. Established instruments for profilometry, like the Hommel Tester® (Hommelwerke GmbH, Schwenningen, Germany), were originally developed to measure tool traces that occur when working on a metallic surface. Later the method was modified for use in almost all areas of microgeometry. Accepted standardized parameters like roughness, waviness, total profile, spacing, and others serve to quantify surface texture. Therefore, surface texture analysis has grown into a field of its own with increasing importance not only for metalwork but also for many other research and production branches. This is especially true for experimental dermatology.

A. The Replica

Several materials have been used to obtain reliable impressions of the skin's surface. After having tested a number of materials Makki et al.[2] found silicone rubber, a dental impression material, to be most suitable for this purpose. After evaluating several materials ourselves, we found Provil® L C.D.* silicone rubber impression material to be the material of choice. Since the Hommel Tester automatically calculates an inverse profile of the measured negative impression, a second impression of the negative primary cast is not necessary. This eliminates possible errors introduced by secondary positive casting.

B. The Stylus Method

A diamond stylus is traversed across the replica surface. An electrical signal equivalent to the vertical displacement of the stylus is amplified and converted into a digital signal. The digital information is analyzed by computer according to selected parameters for roughness (Figure 1).

1. Technical Equipment

The pick-up — The pick-up has two functions: it supports the stylus and acts as a transducer converting the vertical movement of the stylus on the surface to an electrical signal. For dermatological purposes the pick-up consists of a diamond stylus that is cone-shaped with a 5 μm tip radius and a 60° tip angle. To protect the surface of the replica (to avoid "ploughing" through the ridges) a sliding Teflon** skid (Figure 2) with a round-shaped tip is mounted beside the stylus. This results in a more even and constant scanning of the surface. Furthermore, the use of a skid eliminates waviness due to mechanical filtering of the profile.

Transducer — The most common transducer used for dermatological purposes is a half-bridge inductive transducer. An inductive transducer is well suited for texture surface measurement because of its linearity and insensitivity to the surroundings (ambient temperature and humidity). Its compact size allows for packaging in a small housing. The resolution makes the measurement of displacements as small as 0.001 μm possible.

Traverse unit — The traverse unit furnishes the relative movement between the replica and the pick-up. The replica (Figure 1) is moved under the stationary pick-up mounted on a flexible pick-up holder.

---

* Registered Trademark of Bayer Dental Werke AG, Leverkusen, Germany

** Registered Trademark of E. I. du Pont de Nemours and Company, Inc., Wilmington, Delaware

FIGURE 1 Measuring set-up using a column-mounted linear traverse unit.

FIGURE 2 Special pick-up for use with a flexible replica.

Computer — Instrument control, data analysis of the digitized signal, and data output are all computer-controlled. Instruments are easily kept up to date by updating the software as standards change.

2. Profile Characterization

R_a, **average roughness** — R_a can be called the "grandfather" of all roughness parameters and is still young enough to do a good job. Numeric values in micrometers (μm) are obtained. It is commonly employed because of the ease of calculation when using simple analog devices. Although several other standard parameters have been applied to quantify the profile of the skin, we still prefer the mean roughness value R_a. The different parameters for roughness are discussed in detail by Cook.[3] From Figure 3 it can be seen that the quantity R_a is the average distance from the profile to the mean line over the traverse length of assessment. R_a is determined by the formula:

$$R_a = \frac{1}{l_m} \int_0^{l_m} |y| dx$$

l_m is the traverse (scan) length and $|y|$ is the absolute value of the location of the profile relative to the mean profile height (x-axis in Figure 3). R_a is a standardized roughness parameter according to relevant German and international industrial standards.

Filtering — A profile filter can be compared to a sieve. If a pile of rocks and stones is put through a sieve, it will be separated into two piles. One pile consist of rocks unable to pass through the sieve while the other is gravel able to pass through. The sieve hole size defines what is called rock and what is called gravel.[1] The filtering of surface profiles follows the same rules. A filter with a defined cut-off length divides roughness (gravel) from waviness (stones). The cut-off length is analogous to the hole size of the sieve. Filtering does not change the original profile, but the way of looking at it.

Statistical analysis — In addition to the average roughness R_a, the surface can be described by descriptive statistics such as variance, skewness, kurtosis, autocovariance, and autocorrelation functions, as well as Fourier analysis.[1] In order to go a step further and perform a three-dimensional analysis it is necessary to take a number of parallel measurements. This may be of interest for specific questions concerning nonhomogenic surfaces like the skin. In the future mathematical models describing regularities and irregularities will probably play an important role in the description of three-dimensional images.

3. A Recent Development — the Touchless Acoustic Stylus

A very recent development which promises to revolutionize the standard stylus principle is the new touchless acoustic pick-up with a resolution of 10 nm, reflecting a precise image of the measured surface without mechanical alteration (NANOSWING, Hommelwerke GmbH, Schwenningen, Germany). The new pick-up is fully compatible with the standard equipment.[4]

Principle — By means of an electronic sensor a standardized diamond tip is moved over the surface at a constant elevation. The vertical movement exactly corresponds to the surface profile.

Function — In order to maintain the diamond tip at a constant elevation, the distance to the surface has to be measured continuously. To achieve this oscillations with very small but constant amplitude are induced in the tip itself. The very low contact pressure of only about 0.2 µN does not result in a measurable impression on the elastic replica material. The power necessary to maintain the oscillation is measured by a precise electronic device completely integrated into the pick-up. The closer the oscillating diamond tip gets to the surface the more power is consumed as a result of the air friction between the tip and the surface. The power consumption of the oscillation serves as the input value for an electronic regulating circuit. The output is connected to an electric positioner for the diamond tip. Since the power consumption of the oscillation serves as a measure of the distance between the tip and the surface, maintenance of constant power results in a constant distance to the surface.

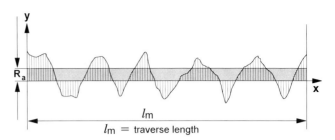

FIGURE 3 Mean roughness value R_a.

IV. Sources of Error

When employing profilometry in experimental dermatology there are a number of pitfalls which can lead to wrong conclusions: first, an adequate test site must be chosen; second, the replica must be made; and last, but not least, the equipment must be operated properly.

A. The Test Area

A suitable test site for measuring influences of topical treatments should have an even surface with a regular structure and few or no hair follicles. The volar side of the forearm meets most of the requirements. Subjects with too many hair follicles, scars, tatoos, visible dehydration (detergents), extensive sun exposure, and pathological skin conditions (scaling!) have to be excluded except under special circumstances.

B. Taking the Replica

For precise casting the replica material should be of low viscosity, fast hardening, and elastic without shrinking. The material should not produce heat while hardening. Large residues should not remain on the skin after removing the replica. The most commonly used material is silicone rubber impression material for dental purposes. With Provil® L C.D. all the above demands are covered. Other materials are listed by Cook.[3]

Provil L C.D. is a two-component material supplied in a ready-to-use cartridge, eliminating the necessity of further mixing. If a two-component material is

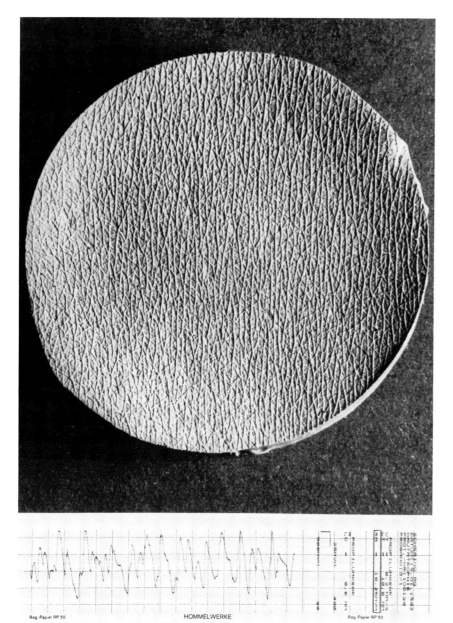

FIGURE 4 Skin replica before topical therapy with a corticosteroid ointment of medium strength.

before the start of every measurement series. Precise measurements require care and a stable environment. The regular use of a calibrated roughness standard ensures proper instrument function. Vibration is one of the main sources of measurement error and should be avoided as far as possible. In most buildings walking through the room while measuring causes distinct vibration! Protection from air currents is a matter of course (open windows!). With the standard stylus repeated measurements on the same replica should not be performed along the same line but have to be carried out at a sufficient distance from the previous measurement.

V. Correlation with Other Methods

A. Laser Profilometry

Laser profilometry[5] is a computer-assisted structural analysis of the skin surface which uses laser beams for touchless measurements with a very high resolution. A three-dimensional profile of the surface can be stored digitally. Different parameters of roughness can be determined. Additional mathematical and statistical procedures such as Fourier transformation and autocorrelation function complete the analysis. The touchless laser beam is said to allow more precise imaging of the peaks and valleys of a replica independent of its elastic properties (no bending of the peaks through contact) as compared to the conventional stylus method. However, one must be aware of the limitations of the method that are caused by the geometrical and optical properties of the object of interest. One aspect is the critical inclination (10 to 15°) of a profile where the laser system tends to overestimate the real depth of the profile. Furthermore, a high optical contrast (dark to bright areas) of the scanned surface may lead to a misinterpretation of the profile. Under unfavorable conditions a mistake of several hundred percent may result because the laser registers the optical contrast as well as the geometrical profile. The profile features produced by

used that has to be mixed by hand, it takes time and practice to mix the two components rapidly enough but at the same time without air bubbles. If the substance is already too hard, the replica will not be sufficient. For most purposes a mold with an inner marking that leaves an impression on the hardened replica is necessary for orientation.

C. Handling the Equipment

For comparable results the replica always has to be scanned in the same direction. The direction should be determined by the course of the tension lines. It is advisable to measure the replicas within a comparable time interval. The weight of the pick-up on the connecting surface always has to be the same (between 1 and 2 g) and should be controlled with a balance

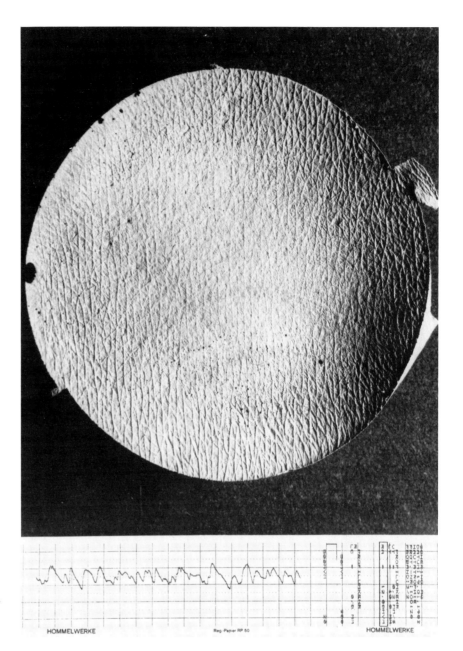

FIGURE 4 (continued) Skin replica 3 weeks after topical therapy with a corticosteroid ointment of medium strength.

the optical contrast are indiscernable from the geometrical profile. Further errors well known to the experienced investigator result from the entry of the beam into the replica, from porous surfaces, or structures with optic imaging properties.

These inevitable but not always predictable errors should be kept in mind when referring to the very high resolution of laser profilometry. In addition, to take advantage of the highest possible resolution over large areas of the replica surface, laser profilometry is very time consuming (requiring hours per replica). It should also be mentioned that the equipment for the laser method is considerably more expensive than that for the stylus method (both standard or acoustic pick-up). Despite the technical advantages of the laser method compared to the conventional stylus method, it remains to be seen whether this method is actually superior in practice and equal to the new touchless acoustic principle applied to the stylus method.

B. Image Analysis

The basic principle underlying image analysis is the measurement of shadows generated by incident lighting at the surface of a replica. Main target parameters are the number and mean depth of wrinkles. A major problem with this method is uneven lighting that may result from unlevel replicas that are true to the skin structure.

VI. Recommendations

A well-standardized procedure to assess the influence of different corticosteroids on the epidermal macropattern is the Duhring chamber test[6] in combination with profilometry. The loss of the detailed structure of the epidermis can be quantified (Figure 4). No residual cream should be present on the test site. In general it is advisable to stop treatment 24 h before any measurement. Skin replicas made of silicone rubber impression material have the "disadvantage" that their elastic properties may lead to incorrect measurement of the "mountain peaks". On the other hand, the material allows for repeated measurements by taking several replicas from the same area without substantial alteration of the skin surface. Cyanoacrylate replicas give a stable imprint of the skin surface but alter the test site by marked stripping when removing the replica.

One should keep an eye on the hydration of the skin because it influences the profile considerably. This is especially important when testing substances for their influence on aging skin. A reduced roughness may be due to temporarily enhanced hydration and not to reduced depth of the wrinkles because of changes in the elastic fibers.

Whichever roughness parameter is chosen by the experienced examiner, he/she must be aware of the method's limitations. It must always be kept in mind that while new technologies and parameters for skin surface texture analysis may reveal statistically significant differences between two skin conditions, the most important authority to validate significant data is clinical efficacy.

Acknowledgment

We thank B. Hughes for help with the English translation.

References

1. Mummery, L., *Surface Texture Analysis The Handbook,* Hommelwerke GmbH, VS-Muehlhausen, Germany, 1992.
2. Makki, S., Barbenel, J.C., and Agache, P., A quantitative method for the assessment of the microphotography of human skin, *Acta Derm. Venereol.,* 59, 285, 1979.
3. Cook, T.H., Profilometry of skin — a useful tool for the substantiation of cosmetic efficacy, *J. Soc. Cosmet. Chem.,* 31, 339, 1980.
4. Personal communication from Volk, R., Hommelwerke GmbH, Schwenningen, Germany.
5. Saur, R., Schramm, U., Steinhoff, R., and Wolff, H.H., Strukturanalyse der Hautoberfläche durch computergestützte Laser-Profilometrie, *Hautarzt,* 42, 499, 1991.
6. Frosch, P.J., Kligman, A.M., and Wendt, H., The Duhring chamber test for assaying corticosteroid atrophy in humans, in *Percutaneous Absorption of Steroids,* Mauvais-Jarvis, P., Vickers, C.F.H., and Wepierre, J., Eds., Academic Press, London, 1980, 185.

Chapter 5.2
Skin Surface Replica Image Analysis of Furrows and Wrinkles

Pierre Corcuff and Jean-Luc Leveque
Laboratoires de Recherche de L'OREAL
Aulnay-Sous-Bois, France

I. Introduction

Among the various noninvasive methods described in this handbook, the description and measurement of the geometric organization of the skin relief involve the same goal, i.e., the study of underlying biological and physiological phenomena based on their impact on the surface. The skin surface "messages" result mainly from the organization of the dermis and its collagen and elastin networks, but the state of the epidermis and stratum corneum can also play a role.

The problem is more complex than the simple measurement of an excretion (sebum, sweat, transepidermal water loss), as the structure is three-dimensional and needs a minimum of geometric parameters; it is therefore difficult to make a simple description and interpretation. A method developed in the 1970s to measure the roughness of metallic surfaces — mechanical profilometry — has since been adapted by several authors[1-3] to study cutaneous topography. Profilometry is still widely used today, as the replacement of the physical probe by a laser beam has overcome two major obstacles: contact between the instrument and the surface, and the time required for measurement. Given the early drawbacks of this method, we proposed in 1981 a technique based on image analysis of a skin replica, which was rapid and well adapted to the anisotropy of the skin relief, but which was prohibitively expensive.[4] With the fall in the cost of electronic components and computers, image analysis has now become more accessible, and an increasing number of teams have adopted this method which, as we shall see later in this chapter, requires impeccable technique and discipline.

II. Objective

At the beginning of the 1980s, the only method available was thus mechanical profilometry. It required a supple negative replica to be made, followed by a counter-replica made of hard resin (which resisted deformation by the probe). The main criticisms were the accumulation of artefacts by successive replication, the directional traces that were poorly adapted to the anisotropy of the skin surface, and the fact that the roughness parameters were difficult to interpret in terms of skin relief.

Image analysis was used by few teams at that time. We discovered that it was capable of scanning the first replica in several directions and that the geometric parameters provided by this technique (orientation, number, and depth) were easy to interpret. Finally, contact between the instrument and the study surface was avoided.

The objective was to describe the organization of the primary lines according to Hashimoto's classification[5] by using a minimum of parameters. When this objective had been reached, the method was extended to the study of wrinkles of the crow's feet area.

In order to obtain objective and reliable results, we opted for an entirely automated method, which disallowed all interactive intervention.

III. Basic Methodology

The principle of image analysis can be likened to an air passenger watching the shadows crossing the Alps as the sun traverses the winter sky. Each mountain has a dark side and a bright side; some valleys are illuminated in the morning, others in the afternoon. The brightness increases the contrast between the areas of shadow and light, just like topologic details.

A. Skin Surface Replica

Applied to the skin relief, this method carried the following constraints. The mold had to be white (as snow), matte, and opaque. Among the brands studied at the time, the silicon resin SILFLO from Flexico (England) fitted the bill.[6] It guaranteed reliable reproduction, the absence of further deformation, and documented artefacts.[7] The fact that the samples could be stored for about 2 years was an added advantage.

The negative replica is obtained in the following way. Two or three drops of Bayer catalyzer are added to 1 g of SILFLO resin and rapidly mixed in a cup with a spatula. The paste is then immediately applied to the study surface, which has first been delimited with an adhesive paper ring. The resin hardens within 2 or 3 min and the replica is removed gently by lifting from the tongue of the ring.

B. Shadowing Method

The cutaneous topography is such that the observer is obliged to "shadow" negative replicas rather than positive replicas, as is clearly shown in Figure 1. The sun in the above analogy is replaced by an optical fiber system providing illumination at a precisely defined angle relative to the plane of observation. For evident technical reasons, it is simpler to rotate the sample than the lighting system to simulate the movement of the sun. The negative replica with its ring is inserted into a metallic device covered with nonreflective black felt. This ensures that the sample is perfectly flat; it also means that the sample can be held perfectly horizontal and centered with regard to the light source and video camera.

C. Image Analysis
1. Image Analyzer

The basic principle of the image analyzer is the segmentation of an image according to shades of gray, and this is perfect for selecting areas of shadow created by the oblique light impinging on the sample (skin furrows).

In 1979, we selected the Quantimet 720 manufactured by Cambridge Instruments, as it had a number of technical features particularly adapted to this application and could be automated. We now use the Quantimet 970, which still has the essential elements of its predecessor; in this way, we can compare the results obtained in 1981 with those obtained in 1993. One important feature of this generation of system is the image format: 720 lines, 880 points per line: 605,000 pixels. To avoid introducing a bias into the measurements during sample rotation, the field of measurement must be circular, and this brings us down to 300,000 pixels (which is still a considerable amount). By way of comparison, a standard image formed of 512 × 512 pixels only provides a circular field of measurement of 180,000 pixels. Table 1 gives comparative data for the various measurement methods we have used. With the Quantimet, one can analyze 1 cm² of skin with a resolution of 10 µm in the horizontal plane and 8 µm in the vertical plane. One question that arises is whether more weight should be given to the field of measurement or to the resolution. Our experience shows that the area measured should be greater than 0.5 cm² and that a resolution of 10 µm is adequate to

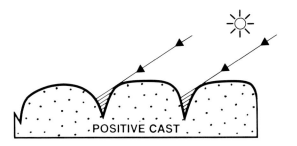

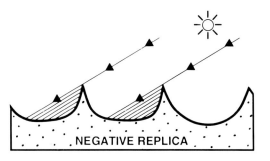

FIGURE 1 Shadows generated at the surface of negative and positive replicas by grazing lighting. On the negative imprint the width of the shadow is large enough to allow a correct estimation of the peak height.

study the primary lines. The most important factors are the sampling procedure and the reproducibility of the measurements. Obviously, the study of secondary furrows needs those higher resolutions provided by profilometry. But software should be capable of analyzing separately the various classes of skin furrows.

2. Analysis of the Image

The selection of the shadows is the most important phase of the operation. The 6-bit A/D converter provides a scale of 64 shades of gray (0 = black; 63 = white). Figures 2A and 2C show the results obtained with a threshold at level 45, i.e., selection of all the levels between 0 and 45.

The position of the light source relative to the video scan determines the significance of the measurement parameters: the light comes from the right of the screen. Subsequent operations consist of extracting only the shadows formed by peaks perpendicular to the light source, i.e., in a vertical direction. To do so, the binary image is opened by 15 pixels (erosion + dilation) with a linear, vertical structuring element (Figures 2B and 2D). This mathematical morphologic operation[8] has two consequences: to eliminate small events (noise, round objects, bubbles) and to emphasize the anisotropy of the skin furrows (suppression of oblique and horizontal shadows, joining of vertical shadow segments).

3. Measurement Parameters

The two parameters measured in binary images are known as field parameters. They correspond, in fact, to the very first stereological parameters available on

TABLE 1 Comparison of Methods Used for Three-Dimensional Analysis of Skin Surface Replica According to the Size of the Explored Surface, the Number of Points, Depth Resolution, and Recording Time

Instrument	Surface Area (mm²)	Picture Points (no.)	Depth Resolution (μm)	Time (min)
Quantimet 970	100	300,000	8	5
Standard I.A.	50	180,000	8	2
Mechanical Profilometer	25	62,500	1	90
Optical (laser) Profilometer	50	250,000	3	10
Confocal microscope (TSM)	1	250,000	1	10

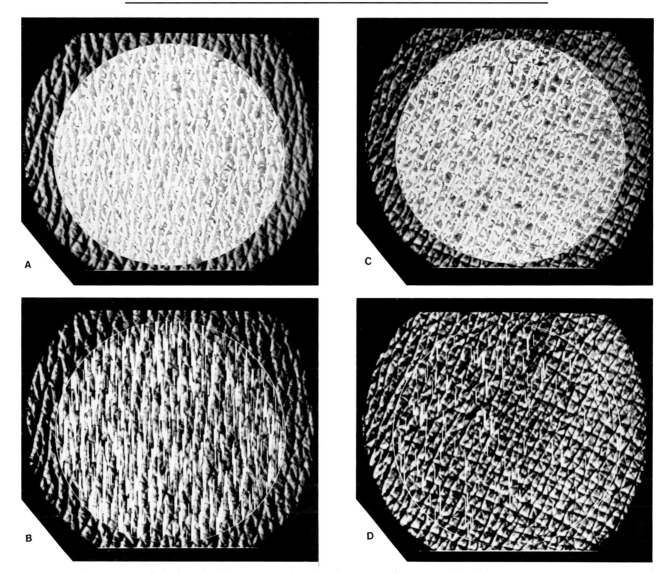

FIGURE 2 The selection of shadows by the image analyzer. In this series of pictures, white features correspond to binary images of shadows segmented at threshold level 45 (A) the principal orientation of furrows is perpendicular to the lighting; (B) binary image resulting from the vertical opening of A. (C) the principal orientation of furrows is not perpendicular to the lighting; (D) the vertical opening almost suppressed the nonperpendicular shadows.

image analyzers when the latter were incapable of identifying objects. The area fraction A_A is the percentage surface area occupied by shadows in the field of measurement. The intercept I is the horizontal projection of these shadows. Given the particular arrangement of the replica-lighting-camera ensemble, the I lines correspond exactly to the crests of the skin furrows, and thus represent the total length of furrows in the field of analysis (Figure 3).

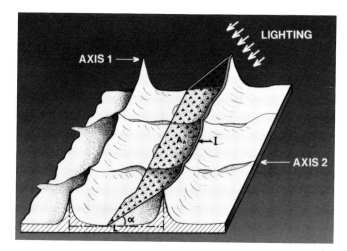

FIGURE 3 Parameters selected by the image analyzer. A_A is the fraction area of shadows, I is the intercept line drawing the crest line of negative furrows, α is the lighting angle.

On the basis of these two stereological parameters and the angle of incident light α, I can be used to estimate the number of lines per square centimeter (or per linear centimeter), while A_A, I, and α can be used to estimate their mean depth.

These parameters are measured at each step of the sample rotation (9° steps through 360°, giving 41 series of measurements). If the values of I are plotted in polar coordinates according to the angle of rotation, one obtains the rose of directions illustrated in Figure 4. This graph contains local maxima, with a 180° symmetry, which corresponds to the main orientations of the primary lines. The orientation of each network of parallel furrows can then be deduced relative to the reference axis (the tongue of the sample); the density of lines and their mean depth can also be deduced for each main axis of furrows. All these operations, i.e., the analysis of the orientations and parameters of each network, are carried out with a computer program written in Pascal. A supplementary parameter has been forwarded to characterize the relief by means of a single value. It is known as the coefficient of developed skin surface or CDSS, which represents the tissue reserve or deformation reservoir of the skin. On the basis of the mathematical cycloid arch model,[4] a *coefficient E* is calculated for each furrow axis. A simplified formula is provided below, which is valid when the field of analysis is close to 1 cm² (±10%).

Consider n the number of furrows per centimeter squared and p their mean depth in microns determined for a main axis:

$$k_1 = 2.10^{-4} \cdot \pi \cdot n \cdot p$$

$$k_2 = \sqrt{|1 - k_1^2|}$$

If $k_1 > 1$ then $c = \ln\left(\dfrac{k_1 + k_2}{k_1 - k_2}\right)$

If $k_1 < 1$ then $c = \pi - 2\arctan\left(\dfrac{k_1}{k_2}\right)$

Finally, $E = \dfrac{2 \cdot k_1}{\pi} + \dfrac{c}{\pi \cdot k_2}$

If E1 and E2 are the coefficients for each of the two axes mutually forming an angle β, CDSS = E1 [1 + (E2 − 1) sinβ]. The importance of this parameter has been shown in studies of chronological aging,[9,10] the effect of ultraviolet irradiation,[11] and the skin deformation process.[12]

4. Automation

Full automation of the image analysis procedure has been the subject of a previous publication.[13] A new "robot" has recently been designed to increase the analytical capacity without augmenting the time required for a full cycle. Eighty replicas instead of 40 can be analyzed in an 8-h period. Figure 5 shows details of the automated apparatus. A circular tray presents each sample which is translated by a magnetic arm for positioning under the camera. The replica is then raised by the autofocus motor of the Quantimet and rotated by a stepwise motor. At the end of the analysis, the replica is lowered, placed in its socket, and the next sample is presented for analysis. The time previously required to control the position of each sample is thus saved.

D. Choice of Lighting Angle

The lighting angle plays a crucial role in determining the nature of the shadows that will be analyzed: the

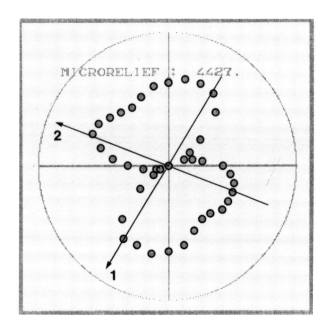

FIGURE 4 A rose of direction. Arrow 1 gives the orientation of the first axis of furrows and arrow 2 the second axis.

FIGURE 5 The "robot". **(a)** Objective lens, **(b)** optical fiber lighting, **(c)** the circular tray accepts 80 samples, **(d)** a replica in position for analysis, **(e)** magnetic arm, **(f)** autofocus motor.

lower the light source, the greater the detail. In this way, the lighting acts as a high-pass filter.

In the case of a fine and regular microrelief, such as that observed in a child, an angle of 17 or 20° will enable the observer to measure a large number of shallow furrows (Figure 6a), while an angle of 38 or 45° will select the few deeper furrows. In this type of topography, line density is the most sensitive parameter. In the case of aged skin (Figure 6B), from which the fine lines have disappeared, the lighting angle has little influence on the number of furrows, but more influence on the mean depth.

With a very low-angled light source, two risks arise. If the sample is not perfectly flat, large areas of shadow can be generated, but this artefact is generally easy to identify in the measurements. Very high peaks can mask lower peaks located in their shade, and this bias is, on the contrary, difficult to detect. A compromise solution has been based on the value of the coefficient E_1 as a function of the lighting angle (Figure 7). This coefficient passes through a maximum which depends on the surface topography. For the skin microrelief, the maximum is from 17 to 26°; we use the higher

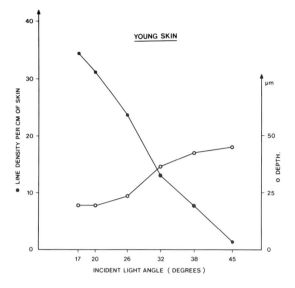

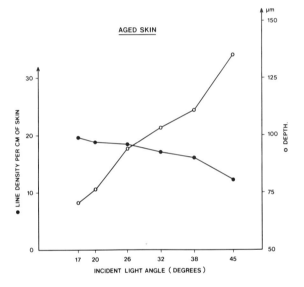

FIGURE 6 Density of lines and mean depth of furrows plotted vs. the lighting angle: (A) young skin replica of the volar forearm, (B) aged skin replica of the volar forearm.

value to avoid the above-mentioned risks. The optimum value when studying crow's feet is 38°.

IV. Sources of Error

A. Replica Artefacts

The artefacts created during the production of the skin surface replica arise from the preparation of the volunteer and the technician's experience. It is essential that the subject remain immobile during the polymerization phase. To this end, the room should be calm, with dim lighting and a temperature of 20°C; in addition, the subject should be comfortable, and given time to relax and adapt to the surroundings. These considerations are particularly important when taking replicas of the crow's foot area; in this case, the volunteer is placed in the lateral decubitus position, eyes closed and face relaxed.

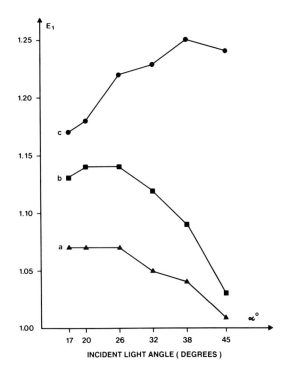

FIGURE 7 Evolution of the cycloid arch coefficient E_1 (see text) according to the lighting angle: **(a)** young skin of Figure 6A, **(b)** aged skin of Figure 6B, **(c)** crow's feet wrinkles.

The conditions necessary for obtaining negative replicas must be followed to the letter. The main sources of artefacts are as follows. Bubbles of sweat can form holes at the intersection of primary lines if the room is too warm or the subject stressed or emotional (Figure 8). Areas lacking any relief are due to inadequate mixing of the resin with the hardener (Figure 9). One of the most frequent causes of artefacts is polymerization of the resin before it is applied to the skin. To ensure that application has been carried out correctly, the underside of the replica should be inspected: if the surface is smooth and shiny, the application is correct; if the surface is irregular, matte and wrinkled, the resin had already started to harden before application (Figure 10).

B. Analysis of the Image

A frequent source of error is that the replica is not perfectly flat. Analysis through 360° permits this type of error to be identified simply, since the values for each 180° segment should be symmetrical. Normally, the rigid metal sample support enables most flatness errors to be corrected. All other errors are due to the image analyzer. The most frequent are time shifts and instability of the light source or electronic circuits of the camera. The uniformity of the incident light is also a critical factor: in general, there is a light gradient that has to be corrected by the image analyzer (background substraction). Interfering lights and changes in the ambient lighting influence the reproducibility of the measurements and it is best to work in total darkness. Incorrect focusing is also a source of error, and autofocusing is thus a major advantage. As a rule, all these types of error can be overcome by regular analysis of a standard replica during a series of measurements.

V. Correlation with Other Methods

There have been few publications comparing mechanical profilometry and image analysis by shadowing. In 1985, we reported a comparison of the effects of antiwrinkle product on the face, using mechanical profilometry for the wrinkles of the forehead and image analysis for the crow's foot wrinkles.[14] The percentage reductions were 27% for the profile area and 30% for the CDSS. The depth of the deeper lines (>50 µm) fell by 23% in the profilometric method and 16% in the image analysis technique. Schmidt et al.[15] compared the two methods on the same replicas taken from the crow's foot area. They found a good correlation between the depth given by image analysis and the height of the peaks given by profilometry, and also between the CDSS and peak surfaces. The best correlation was obtained with peak heights of between 50 and 100 µm, i.e., the optimal domain in image analysis with a lighting angle of 38°. Hayashi et al.[16] recently demonstrated that the fractions of shadow corrected for the lighting angle (RWA parameter) and the maximum depth of the wrinkles (V) were independent of the lighting angle. V was identical to the height given by a micrometer. Schrader and Bielfeldt[17] compared data from image analysis, mechanical profilometry, Corneometer,® and methylene blue staining of the skin, during cosmetic treatments. Linear regression study led to weak but significant correlations between the methods. Some authors have preferred a profilometric approach to a threshold method in image analysis: the optical profilometry traces curves corresponding to gray levels along a scanning line in the video image. The profiles are analyzed with the same parameters as those used to analyze a mechanical trace, and good correlations between the two types of profile have been reported.[18] Grove and Grove,[19] also using optical profilometry by image analysis, showed there was a good correlation between the roughness parameters (R_a, R_z) and Daniell's visual classification during treatment of the crow's foot area with retinoids.

VI. Recommendations

The study of the skin microtopography or facial lines by means of image analysis, whatever the technique used, is more a problem of sampling and reproducibility than one of sensitivity. It is thus best to use approaches that analyze as large an area as possible,

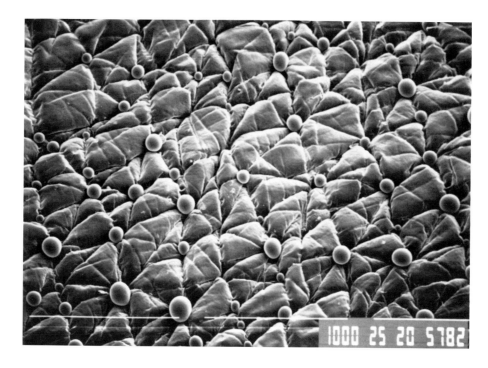

FIGURE 8 SEM picture of a skin surface replica showing droplets at the intersect of primary furrows.

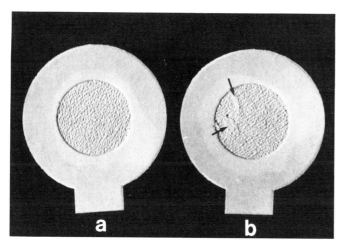

FIGURE 9 Replica artefacts: (a) an "academic" replica, (b) flat areas (arrows) obtained with a nonhomogeneous mixture.

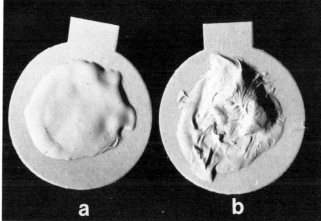

FIGURE 10 Replica artefacts: (a) the smooth verso of an "academic" replica, (b) aspect of a replica performed during the polymerization of the resin.

and a large number of samples. The time required for analysis is consequently important, and automation clearly has a role to play. In the special case of crow's foot wrinkles, the number of significant events (lines) is relatively small. Studies of changes in these lines during treatment must thus be based on strictly identical areas of analysis (before and after treatment). The image analyzer is a powerful tool in this setting, since it enables the study areas to be superimposed.

As we have seen, the production of the replica is a crucial step that requires the utmost care. This is an essential consideration because, despite the fact that the replica method permits samples to be studied some time after their collection, it is essential that the replicas truly reflect the state of the skin at the time they are made.

Replicas are durable and easy to store, and are also readily transportable. It is thus astonishing that so few multicenter studies comparing the different available methods have been published. A consensus on the effects of anti-aging treatments could be arrived at by this approach, which would finally convince the international scientific community of their efficacy.

References

1. Cook, T.H., Profilometry of skin. A useful tool for the substantiation of cosmetic efficacy, *J. Soc. Cosmet. Chem.,* 31, 339, 1980.
2. Makki, S., Mignot, J., and Zahouani, H., Statistical analysis and three dimensional representation of human skin surface, *J. Soc. Cosmet. Chem.,* 35, 311, 1984.
3. Hoppe, U. and Sauermann, G., Quantitative analysis of the skin surface by means of digital signal processing, *J. Soc. Cosmet. Chem.,* 36, 105, 1985.
4. Corcuff, P., de Rigal, J., and Lévêque, J.L., Image analysis of the cutaneous microrelief, *Bioeng. Skin Newslett.,* 4(1), 16, 1982.
5. Hashimoto, K., New methods for surface ultrastructure. Comparative studies of scanning electron microscopy, transmission electron microscopy and replica method, *Int. J. Dermatol.,* 13, 357, 1974.
6. Makki, S., Barbenel, J.C., and Agache, P., A quantitative method for the assessment of the microtopography of human skin, *Acta Derm. Venereol.,* 59, 285, 1979.
7. Gordon, K.D., Pitting and bubbling artefacts in surface replicas made with silicone elastomers, *J. Microsc.,* 134(2), 183, 1984.
8. Serra, J., in *Image Analysis and Mathematical Morphology,* Vol. 2, *Theoretical Advances,* Serra, J., Ed., Academic Press, London, 1988.
9. Corcuff, P., de Rigal, J., Makki, S., Lévêque, J.L., and Agache, P., Skin relief and aging, *J. Soc. Cosmet. Chem.,* 34, 177, 1983.
10. Corcuff, P., Lévêque, J.L., Grove, G.L., and Kligman, A.M., The impact of aging on the microrelief of peri-orbital and leg skin, *J. Soc. Cosmet. Chem.,* 82, 145, 1987.
11. Corcuff, P., François, A.M., Lévêque, J.L., and Porte, G., Microrelief changes in chronically sun-exposed human skin, *Photodermatology,* 5, 92, 1988.
12. Corcuff, P., de Lacharrière, O., and Lévêque, J.L., Extension induced changes in the microrelief of the human volar forearm: variation with age, *J. Gerontol. Med. Sci.,* 46(6), 223, 1991.
13. Corcuff, P., Chatenay, F., and Lévêque, J.L., A fully automated system to study skin surface patterns, *Int. J. Cosmet. Sci.,* 6, 167, 1984.
14. Corcuff, P., Chatenay, F., and Brun, A., Evaluation of anti-wrinkle effects on humans, *Int. J. Cosmet. Sci.,* 7, 117, 1985.
15. Schmidt, C., Camus, C., Candiu, H., Her, C., Soudant, E., and Bazin, R., Correlation d'une technique d'analyse d'images et d'une méthode profilométrique dans l'étude des rides de la patte d'oie, *Int. J. Int. Sci.,* 9, 21, 1987.
16. Hayashi, S., Matsuki, T., Matsue, K., Arai, S., Fukuda, Y., and Yoneya, T., Changes in facial wrinkles by aging and application of cosmetics, Proc. IFSCC Congress, Yokohama, 1992, 733.
17. Schrader, K. and Bielfeldt, S., Comparative studies of skin roughness measurements by image analysis and several in vivo skin testing methods, *J. Soc. Cosmet. Chem.,* 42, 385, 1991.
18. Marshall, R.J. and Marks, R., Assessment of skin surface by scanning densitometry of macrophotographs, *Clin. Exp. Dermatol.,* 8, 121, 1983.
19. Grove, G. and Grove, M.J., Effects of topical retinoids on photoaged skin as measured by optical profilometry, in *Methods in Enzymology,* Ed., Academic Press, New York, 1990, 360.

Chapter 5.3
Laser Profilometry

Jan Efsen, Hans Nørgaard Hansen, and Steen Christiansen
Technical University of Denmark
Institute of Manufacturing Engineering
Denmark

Jens Keiding
Department of Dermatological Research
Leo Pharmaceutical Products
Ballerup, Denmark

I. Introduction

The skin is a most versatile organ. It functions at the same time as a protective first-line defense for the body and an area of chemical communication between the body and the external world.

The *inside* part of the skin has been studied by a series of techniques. Histological, biophysical, and biochemical studies have led to a detailed understanding of the internal structure of the skin, i.e., organization of cell layers, fibers, and chemical constituents in different parts of the skin. More systematic studies of the *outside* part of the skin have been carried on since the 1930s. In the beginning the technique was used merely to describe features on the surface of the skin or a skin replica that could be observed in a microscope.

More quantitative studies of skin structure started about 30 years ago, and have included both two-dimensional and, more recently, three-dimensional studies. These methods can be subdivided into two categories:

1. Structure mapping based on analysis of an image of the surface. These systems include image analysis of videoscans, as in the Magiscan image analysis system,[1] and the fully automated system developed by Corcuff et al. in which the shadows cast by incident light on a replica may be used to determine primary and secondary direction and depth of furrows and also determine skin surface area.[2]
2. Topographical mapping of the structure based on scanning of surface height (two- or three-dimensional scanning). This category includes profilometers — mechanical or optical.

Normally skin replicas, i.e., impressions that are used as reproductions of the skin relief, are used as study objects in profilometric studies.

This chapter will give a short introduction to profilometry with an emphasis on the methodology of optical profilometry. The performance of a commercially available instrument is discussed, as is the skin structure data that can be obtained using this technique and how the dermatologist can make use of these data.

II. Object

The purpose of this chapter is to present the operation of an optical profilometer and how it may be calibrated. We also introduce the use of software for a three-dimensional surface description and give a practical example of the use of this system.

III. Methodological Principle

A. Preparation of Object — Making a Replica

The making of a replica is in principle very easy. In our routine we first attach an adhesive ring to the skin area to be studied. A silicone rubber (Silflo® Flexico) is mixed with a catalyst and the mixture is distributed as a thin layer covering the central opening of the ring and is allowed to harden for 3 to 5 min. The ring with the replica attached is gently removed and finally the replica is cut into an appropriate size with a cutting device. In principle this method leads to an accurate replica of the skin surface structure.[3] In practice, however, there are often problems. The skin under study may need pretreatment to remove scales, hair, etc. Due to the anatomy it may be difficult to obtain a replica of an appropriate structure and size. The process of mixing rubber and catalyst must be carefully controlled in order to avoid air bubbles getting stuck in the replica.

B. Control of the Optical Profilometer

The optical profilometer (Microfocus 1080^R) used in this investigation was provided by UBM Messtechnik, Ettlingen, Germany. The profilometer consists of four main parts: an optical sensor, an air bearing table, a control unit, and a computer. The optical sensor is fixed, and the object whose surface is to be measured is placed on the air bearing table. By means of the control unit the table can be translated in two perpendicular directions: an X-direction and a Y-direction, as illustrated in Figure 1. It is possible to move the table 150 mm in both directions.

Operation of this system is based upon the autofocus principle. An illustration of the system is shown in Figure 2 and described in Reference 6. Infrared light (wavelength 780 nm) emitted from a laser diode (**1**) is focused onto a small spot (diameter 1 μm) by a system of lenses (**8** and **9**). The light reflected from the object surface (**11**) is directed back into the sensor and is imaged as a pair of spots onto an arrangement of photodiodes (**5**). This is done in such a manner that both diodes are illuminated equally only when the objective lens (**9**) is precisely in its focal distance from the surface. If the distance to the object changes, the focus point is shifted too, and the illumination of the photodiodes becomes unequal. This unequal illumination of the photodiodes generates a focus error signal. A control circuit monitors the error signal and moves the objective lens accordingly. *This is the autofocus principle.*

The movement of the lens is accomplished by a coil (**6**) and magnet (**7**) arrangement. The vertical movement of the objective lens is registered by a light barrier measurement system (**10**) and corresponds to variations in the height of the object surface. The stand-off distance between object surface and optical sensor is approximately 2 mm. The optical sensor has two vertical measurement ranges: ±50 μm and ±500 μm. It is possible to obtain a vertical resolution of approximately 6 nm in the range ±50 μm and 60 nm in the range ±500 μm.

The surface of the object may be viewed during the measurements through a window (**4**) in the sensor by using a microscope (**13**) and a CCD-camera (**14**). All operations of the system are controlled by a computer.

C. Calibration Methods

Calibration and control of the optical profilometer is necessary to obtain reliable results. Our calibration procedure was divided into the following five parts:

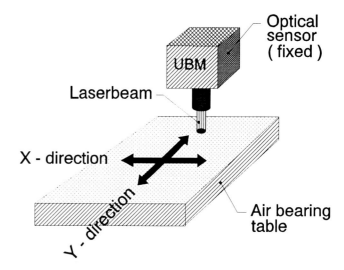

FIGURE 1 Optical profilometer system.

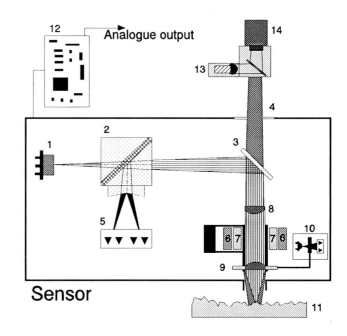

FIGURE 2 Autofocus principle.

- Control of optical sensor
- Control of air bearing table
- Control of software
- Calibration against standardized roughness specimens
- Determination of the frequency response of the optical profilometer

The tests were carried out at The Technical University of Denmark. The tests are described in detail in Reference 4.

1. Optical Sensor

The linearity of the sensor was tested with a sine bar. This is a metal bar mounted on two gaugeblocks to obtain a well-defined angle, see Figure 3. The sine bar was measured with the optical profilometer and the linearity determined from these results. The test was carried out both for the vertical measurement range ±50 μm and ±500 μm.

The ability of the optical sensor to measure inclined surfaces was tested with a roundness standard specimen. This is a glass ball with an almost perfect roundness. This specimen was measured with the optical profilometer as indicated in Figure 4, and the maximum allowable slope of the surface determined. A supplementary test of the sensor covered an investigation of which surfaces the sensor was able to detect. Furthermore, the control circuit of the sensor was tested by changing control parameters in the software.

2. Air Bearing Table

The flatness of the movement of the air bearing table was tested with an optical flat. The optical flat was measured in two areas: 1×1 mm^2 and 100×100 mm^2. The small area represents a typical three-dimensional area for investigation in mechanical engineering, and the large area covers almost the entire possible movement of the air bearing table. The positioning accuracy and repeatability in both the X- and the Y-axis direction was tested with laserinterferometry. Furthermore, the perpendicularity of the two axes was tested with an angle plate.

3. Software

The software was tested with a theoretic sine-profile. The results were compared to already existing and tested programs and calculations performed after the standard DS/ISO 4287/1.[7] The filter characteristics were determined by means of a signal analyzer (Brüel & Kjær, type 2032). A filter test as described in DIN 4777[8] was carried out.

4. Standard Roughness Specimens

The optical profilometer was calibrated with standard roughness specimens designed for mechanical stylus instruments. An optical flat was used to determine the background noise in the system during a measurement. A single grooved calibration standard specimen (DS/ISO 5436, type A[9]) was used to determine the static amplification of the profilometer (Figure 5). These measurements revealed overshoot of the sensor when measuring steep edges. A parameter specimen developed by Physikalisch Technische Bundesanstalt (PTB), Germany,[5] was used to test the ability to measure "real" surfaces (Figure 6). The PTB parameter specimens were not found suitable for calibration of the optical profilometer.

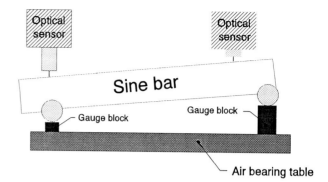

FIGURE 3 Sine bar.

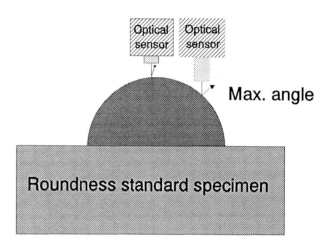

FIGURE 4 Roundness standard specimen.

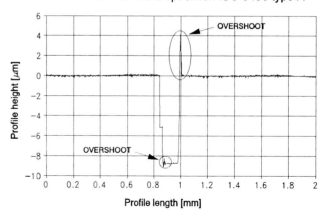

FIGURE 5 Measuring a calibration standard specimen (DS/ISO 5436 type A) with the optical profilometer.

5. Frequency-Response Analysis

By the term frequency response we mean the steady-state response of the optical sensor to a sinusoidal input. This analysis reveals the ability of the sensor to detect vertically oscillating surfaces. When measuring

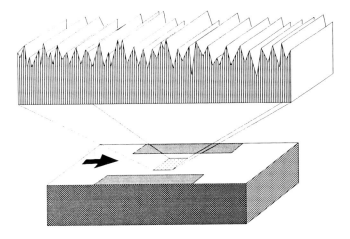

FIGURE 6 PTB parameter specimen.

surfaces with lateral waviness, the frequency response of the sensor determines which wavelength components it is possible to detect. The analysis showed that the vertical range of a surface influences the response of the optical sensor.

D. Summary of Calibration Methods

The tests found suitable for calibration of the optical profilometer are listed in Table 1.

E. Comparison of the Optical Profilometer with Mechanical Stylus Instruments

The performance of the optical profilometer was compared with that of mechanical stylus instruments (Rank Taylor Hobson Talysurf 5 and Rank Taylor Hobson Surtronic 3P). The comparison consisted in measuring a wide range of processed surfaces, i.e., turned, milled, and lapped surfaces. Also the well-defined surface of a PTB parameter specimen was measured. The standardized parameters calculated from profiles obtained

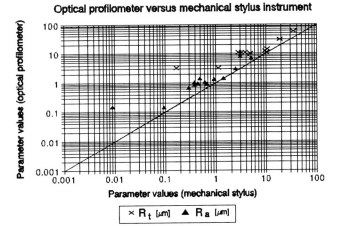

FIGURE 7 Measuring industrial surfaces with both an optical profilometer and a mechanical stylus instrument.

TABLE 1 Suitable Calibration Methods

Object	Property	Test Specimen
Optical sensor	Linearity	Sine bar
	Maximum allowable slope	Roundness standard specimen
Air bearing table	Flatness	Optical flat
	Positioning accuracy of axis	Laserinterferometry
	Repeatability	Laserinterferometry
	Perpendicularity of axis	Angle plate
Software	Parameter calculation	Theoretic sine profile
	Filter test (characteristics)	Signal analyzer, DIN 4777
Standard roughness specimens	Background noise	Optical flat
	Static amplification[a]	Calibration standard specimen (DS/ISO 5436, type A)
Frequency response analysis	Ability to measure oscillating surfaces	Determination of frequency response

[a] *Overshoot.*

by the optical profilometer were on all surfaces studied higher than the corresponding parameters from a mechanical stylus instrument. This is illustrated in Figure 7 where the R_a and R_t values are plotted. Values obtained from the mechanical stylus instrument are plotted along the X-axis and values obtained from the optical profilometer are plotted along the Y-axis. If both instruments would have given the same parameter values, all the points should have been placed on the marked line. It is obvious that parameter values obtained with the optical profilometer are higher than those obtained with the mechanical stylus instrument. This can be explained by the fact that the 1-µm laser spot detects more and deeper valleys in the surface than a mechanical stylus instrument (radius typically >2 µm), as shown in Figure 8. Furthermore, a mechanical stylus works like a mechanical filter (Figure 9). Finally, steep edges represent a problem, as illustrated by the overshoot phenomenon shown in Figure 5.

On a system calibrated as described it is possible to obtain a repeatability of 5 to 10% when measuring skin replicas.

1. Measuring and Data Collection with the Profilometer

The operation of the profilometer is controlled by an extensive software which is an integrated part of the equipment. It makes automatic scanning of a series of objects possible. Scanning takes place with the object positioned on the air bearing table. Several objects may be scanned in one measuring sequence. In a set-up table it is possible to predefine up to 99 positions on the table, and to each position a specific scan area

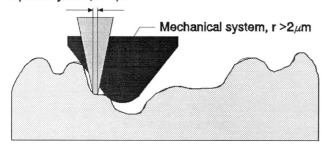

FIGURE 8 Comparison between optical stylus (laser beam) and a typical mechanical stylus.

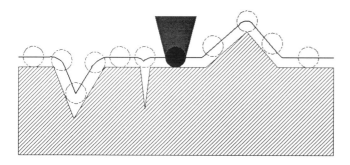

FIGURE 9 The stylus working as a mechanical filter.

may be defined. The information on position and area may be saved in the memory of the computer. When used together with information on speed and intensity of scanning it allows the user to set up automatic scan procedures for all these operations. It is possible also to automize the data analysis. An extended list of standard two-dimension roughness parameters and a few three-dimension parameters are included with the software. Power spectra and autocorrelation analysis in two and three dimensions and two-dimensional material ratio curves are some of the other parameters included. Software routines for three-dimensional material ratio parameters are, however, not included. In the following we will show how such parameters can be calculated and how they may be used for description of skin surface structure.

F. Characterization of Surfaces with Stratified Structure

This is a very common problem in engineering surfaces, since different layers in a surface have different functional properties. If the "top part" of a surface does not look like the "valley part" the surface cannot be characterized with commonly used parameters, such as R_a and R_z, as these parameters cannot distinguish "up" and "down" in the profile. This is also a problem in skin structure studies, as these parameters cannot distinguish a groove from a ridge. Therefore, other parameters that can distinguish stratified structures are needed.

1. Preprocessing the Parameter Calculations

A complete description of the skin surface is very complex since the replica contains many waves caused by the specific test conditions, such as the location where the replica has been taken and the replica preparation method. Grooves, ridges, and holes must be classified as "microstructure" and must be separated from the general form (the macroform), e.g., the form of the leg, the arm, or the face, before a reliable characterization of the microstructure alone can be made. Generally the true measured geometrical form — known as the raw set of data — may be classified into three categories:

- Microstructure, e.g., small holes, marks, and cracks
- Macrostructure, e.g., a cylinder, a plane, or a conus
- An arbitrary level, depending on the vertical range in which the measurement has taken place

It is difficult to distinguish between these categories since no exact limit between "micro" and "macro" has been defined yet. Different methods, e.g., fast Fourier transformation and digital cut-off filtering, have been tried,[10] and one of the most suitable ways to remove the macroform seems to be subtraction of the "Nth order least squares polynomial fit" from the raw set of data, as shown in this formula:

$$R(x,y) = Z(x,y) - F(x,y)$$

where $R(x,y)$ is the residual surface, $Z(x,y)$ is the original datapoint, and $F(x,y)$ is the least squares Nth order polynomial fit.

Different methods for definition of a reference plane have also been examined, e.g., the arithmetic mean plane and the least squares plane. The least squares plane method has been evaluated to be the best since this method is more robust than the arithmetic method. All parameters should be calculated with respect to this reference plane.

2. Algorithm

When the form element has been removed and the reference plane is defined the parameters can be computed after the following precept:[11]

1. The material ratio curve is computed as follows:
   (i) The surface is truncated into 4096 levels
   (ii) The relative volume of material in each level is calculated leading to the material distribution
   (iii) The material distribution is accumulated over the 4096 levels leading to the material ratio curve, which shows how much material is found above a specified surface level (this is illustrated in Figures 10 and 11).

2. The 40% secant is moved along the material-bearing curve until the 40% secant with least slope is found. These points are marked with A and B.
3. A line is projected through the A and B to the intersection with 0% and 100% material ratio. These points are labeled C and D as shown in Figure 12. The vertical distance between C and D is the core surface depth (S_k). S_{r1} and S_{r2} are the material ratios at the top and the bottom above and below the core surface.
4. The areas A_1 and A_2 are computed. S_{pk} is defined as the height in the triangle which has the area A_1 and the baseline S_{r1}. The S_{vk} is defined as the height in the triangle which has area A_2 and the baseline S_{r2}.

The S_k — the core surface depth — measures the height of the core material portion. It depicts the flattest part of the material ratio curve, i.e., the region with the greatest increase in material. A small S_k value generally indicates that the skin surface is very smooth, since the material volume is very large. This is seen, e.g., on the skin surface of a baby where the skin is virtually devoid of wrinkles.

The S_{pk} — the reduced surface peak height — denotes the height of the surface peak projecting beyond the core surface. A low S_{pk} value indicates that the surface does not include many ridges and may be free of extreme peaks.

The S_{vk} — the reduced surface valley depth — denotes the proportion of surface valleys extending into the material below the core surface. It provides useful information on grooves and holes in the surface. Generally it can be said that large S_{vk} values indicate that the surface contains a large amount of wrinkles.

Basically this set of parameters divides the surface into three categories: top, core, valleys. One parameter is related to each category and, therefore, the total set of parameters is required for a suitable characterization of surfaces with stratified structure.

3. Other Parameters

Other useful parameters for characterization of skin surfaces can be the arithmetic mean value S_a, the difference between maximum and minimum height S_t and skewness S_{sk}. Digitally the S_a is calculated as follows:

$$S_a = \frac{1}{MN} \sum_{i=1}^{MN} |Z_i|$$

where M is the number of points per profile, N is the number of profiles, Z_i is the numerical distance from least squares mean plane to the ith residual surface height. S_a is a mean value and gives an overall impression of the roughness properties of the object.

S_t is the maximum vertical distance between highest peak and deepest valley. This parameter gives information about extreme conditions, e.g., extremely deep holes.

S_{sk} is the skewness of the material distribution curve. This parameter can be used to describe the shape of the surface height distribution. It is given by the formula:

$$S_{sk} = \frac{1}{MNS_q^3} \sum_{i=1}^{MN} Z_i^3$$

where M is the number of points per profile, N is the number of profiles, Z_i is the distance from least squares mean plane to the ith residual surface height, and S_q is the root mean square deviation of the surface. For an asymmetric distribution of the surface heights, the skewness may be negative if the distribution has a longer tail at the lower side of the mean plane, which means that the surface consists primarily of valleys and holes. This is in contrast to positive skewness, which indicates a surface including many peaks and ridges. If skewness is zero no primary trend can be seen.

These three parameters have to be handled with care and conclusions based on these parameters alone must be avoided. The definition of S_a includes the numerical values, which means that this parameter

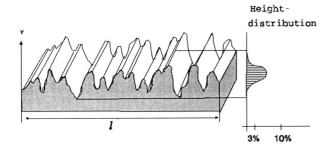

FIGURE 10 Illustration of height distribution curve.

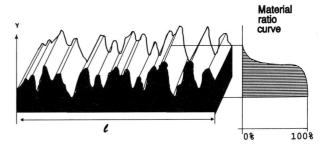

FIGURE 11 Calculation of the three-dimensional material ratio curve.

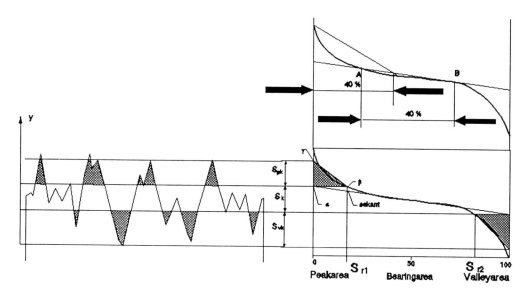

FIGURE 12 Calculation of the three-dimensional material ratio curve parameters.

cannot see the difference between "up" and "down". S_t is very sensitive for single-point values. One extreme value which may be caused by a vibration during the measurement can result in misleading conclusions. It can be seen from the formula for skewness S_{sk} that this parameter is dimensionless since it is normalized by S_q. Therefore, skewness can never give information on absolute properties of a surface, only relative properties can be characterized.

4. Using Three-Dimensional Parameters for Characterizing Skin Replicas

Sixteen skin replicas were investigated. An area of 5.6×5.6 mm² was measured with the optical profilometer and analyzed by means of the three-dimensional parameters described above. In this investigation only the parameters S_a, S_t, S_{vk}, and S_k were calculated.

The surface plots obtained from the optical profilometer were divided into two categories by means of visual inspection:

Category I: characteristic waviness with or without craters

Category II: no characteristic waviness but possibly craters

Category II could be subdivided further into these two subcategories:

Category IIa: no characteristic waviness but craters

Category IIb: no characteristic waviness and no craters.

Figure 13 shows an example of a profile plot from category I, and Figure 14 shows a replica from category IIb.

Table 2 shows the combination of parameters and parameter values found in this investigation. In this material it was possible based on a calculation of the parameters S_a, S_t, S_{vk}, and S_k to place each replica into one of these three categories. All four parameter

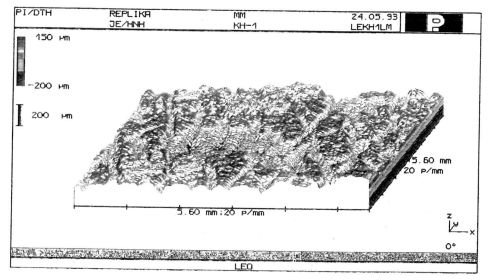

FIGURE 13 Example of a profile plot from category I.

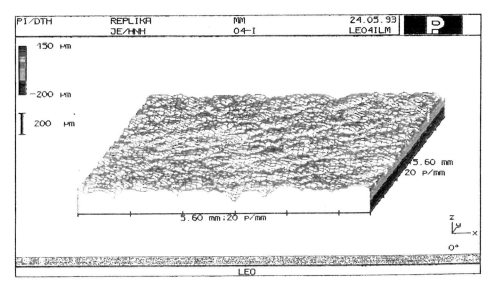

FIGURE 14 Example of a profile plot from category IIb.

values were needed to place the replica in the correct category.

5. What Has Quantitative Analysis of Skin Structure Been Used for?

Image analysis studies have been used by Corcuff and Lévêque[12] to describe development of skin structure as a function of age. They have shown that the regularity of the skin relief decreases with age, associated with the disappearance of secondary lines. They also showed that the aging process is a combination of physical age and external factors acting on the skin surface. Grove et al.,[13] among others, have used optical profilometry in clinical studies on the effect of retinoid treatment on wrinkles and have shown that roughness parameters R_a and R_z correlate with clinical gradings based on clinical ratings.[13] Zahouni et al.[14] have used profilometric studies to calculate volume and area of leg ulcers and have used the data for evaluation of wound-healing treatments.[14]

Image analysis of melanoma has recently been shown to be of help in the clinical diagnosis of pigmented lesions[15] and a profilometric study has shown that it is possible to distinguish between melanomas and nevocellular nevi from a profilometric analysis of replicas taken from these areas.

IV. Recommendations

In this chapter emphasis has been put on controlling and standardizing the operation of the optical scanner. However, it is equally important to have a standardized method of developing the material for study — the replica. The process of producing the replica and the quality of the replica should be controlled.

Sometimes the skin may need pretreatment prior to taking the replica, in order to remove scales or hair or clean the skin to remove residuals from skin surface treatments.

With respect to the use of the profilometer, it is necessary to specify the size of the area measured, the speed of the scanner, and the scanning intensity (i.e., points/mm). In addition to these specifications, which are included in the standard menu prior to starting the measurement, a series of hardware specifications are available. The actual set-values should also be specified in order to make reproducible measurements and for lab to lab comparisons.

In the test of the profilometer a test of the software was included. Only a small part of the extensive software was tested, however. An important part is the use of filtering methods in the determination of roughness parameters. If a digital filtering algorithm is used for removal of form error instead of a polynomial fit it is important to select the correct filter algorithm. A phase-correct filter (M-filter), not the common RC2 filter (which is not phase correct) should be used. A wrong choice of filter may introduce peaks where no such peaks can be identified in reality. Also the cut-off wavelength should be carefully chosen. In general the cut-off wavelength should be 0.2* the evaluation length.

TABLE 2 Categorization of Skin Replica Based on Three-Dimensional Surface Parameters

Category	Description	Suitable 3D Parameters (µm)	Value
I	Waviness, with or without craters	S_a	$S_a \approx 30$
		S_t	$210 < S_t < 300$
		S_k	$80 < S_k < 120$
		S_{vk}	$42 < S_{vk} < 49$
IIa	No waviness, with craters	S_a	$S_a \approx 20$
		S_t	$230 < S_t < 280$
		S_k	$60 < S_k < 100$
		S_{vk}	$30 < S_{vk} < 40$
IIb	No waviness, without craters	S_a	$S_a < 20$
		S_t	$170 < S_t < 195$
		S_k	$50 < S_k < 70$
		S_{vk}	$28 < S_{vk} < 33$

References

1. Grove, G.L. and Grove, M.J., Objective methods for assessing skin surface topography noninvasively, in *Cutaneous Investigation in Health and Disease*, Lévêque, J.-L., Ed., Marcel Dekker, New York, 1989, chap. 1.
2. Corcuff, P., Chatenay, F., and Lévêque, J.L., A fully automated system to study skin surface patterns, *Int. J. Cosmet. Sci.*, 6, 167, 1984.
3. Cook, T.H., Profilometry of skin — a useful tool for the substantiation of cosmetic efficacy, *J. Soc. Cosmet. Chem.*, 31, 339, 1980.
4. Efsen, J. and Hansen, H.N., *Optisk ruhedsmåling*, Institute of Manufacturing Engineering, Technical University of Denmark, Ed.MM.93.34, Copenhagen, Denmark, 1993.
5. Deutscher Kalibrierdienst, Physikalisch Technische Bundesanstalt, *Calibration of Stylus Instruments, Guideline DKD-R4-2*, Braunschweig, Germany, 1991.
6. UBM Messtechnik, *Microfocus Measuring System Manual*, Ettlingen, Germany, 1992.
7. *DS/ISO 4287/1*: Surface Roughness. Terminology. Part 1: Surface and its parametres, 1. ed., Dansk Standardiseringsråd, København, 1986.
8. *DIN 4777, Teil 1:* Oberflächenmesstechnik. Profilfilter zur Anwendung in elektrischen Tastschnittgeräten. Phasenkorrekte Filter, Berlin, Germany, 1988.
9. *DS/ISO 5436.* Calibration specimens — Stylus instruments — Types, calibration and use of specimens, 1.ed., Dansk Standardiseringsråd, København, 1987.
10. Stout, K.J., Sullivan, P.J., Dong, W.P., Manisah, E., Lou, N., Mathia, T., and Zahouani, H., The development methods for characterisation of roughness in 3 dimensions, Vol. 1 and 2; *BCR report, EC contract no. 3374/1/0/170/90/2;* Centre for metrology, University of Birmingham, Birmingham and L'Ecole centrale de Lyon, Lyon; 1993.
11. Christiansen, S., Function-Related 3-Dimensional Definition of Surface Microtopography, Ph.D thesis, Technical University of Denmark, Institute of Manufacturing Engineering.
12. Corcuff, P. and Lévêque, J.-L., Age-related changes in skin microrelief measured by image analysis, in *Aging Skin*, Lévêque, J.-L. and Agache, P.G., Eds., Marcel Dekker, New York, 1993, chap. 13.
13. Grove, G.L., Grove, M.J., Leyden, J.J., Lufrano, L., Schwab, B., Perry, B.H., and Thorne, G., Skin replica analysis of photodamaged skin after therapy with tretinoin emollient cream, *J. Am. Acad. Dermatol.*, 25 (2), 231, 1991.
14. Zahouni, H., Assoul, M., Janod, P., and Mignot, J., Theoretical and experimental study of wound healing: application to leg ulcers, *Med. Biol. Eng. Comput.*, 30, 234, 1992.
15. Busche, H., Connemann, B.J., Kreusch, J., and Wolff, H.H., Surface topography in the diagnosis of malignant melanoma, Poster session, 3rd Conference of the Int. Society for Ultrasound and the Skin, Elsinore, Denmark, 1993.

Chapter 5.4
Three-Dimensional Evaluation of Skin Surface: Micro- and Macrorelief

Jean Mignot
Laboratoire de Métrologie des Interfaces Techniques
Institut Universitaire de Technologie
Besançon, France

I. Introduction

Skin surface topography has been a matter of interest for dermatologists for 50 years. The first studies were carried out by Cummins and Midlo[1] and these were followed by many others by Wolf,[2] Tring and Murgatroyd,[3] and Marks and Saylan.[4] However, all these studies have been bi-dimensional, which means that the surface itself has not been studied, but one or several profiles of the surface in one or several directions have been. This simplification was necessary because of the capacities of the equipment available: no measurement device was able to analyze a sufficient number of data related to the surface analyzed within a short time.

Until recent years, these profilometric studies were made with equipment designed for measuring functional surfaces of parts of machines in the mechanical industry. The measuring part of these types of apparatus is a mechanical sensor or stylus in direct contact with the surface to be analyzed.[5,6] The detector of this sensor (Figure 1) is usually a thin diamond tip (radius of curvature from 5 to 10 µm) displaced at a constant speed on the surface (0.5 to 1 mm/s) according to a direction determined by the operator.

This type of device has several main drawbacks:

- It must not be in direct contact with the skin, because the pressure of the spheric head creates a major deformation of the skin surface.[7] This is why soft replicas of the surface are made,[8,9] from which solid replicas are obtained, using in most cases a polymerized material.[10]
- The scanning speed is limited in two-dimensional profilometry systems because of the mechanical contact with the analyzed surface, and this results in slow data acquisition.
- The results obtained with two-dimensional analysis (or roughness parameters) show the topography of the surface in one direction only. A top view examination of the skin surface (Figure 2), or the study of the changes of the various frequential components of the surface (two-dimensional Fourier transform) in all directions, are sufficient to show the anisotropy of this surface (Figure 3).

These main drawbacks have two consequences:

- Traditional surface measurement devices are not adequate.
- Because of its anisotropy, the analysis of the skin surface with profilometers is not reliable.

II. Aim of the Study

Because of the disadvantages of profilometry, it seems necessary to develop new devices that quickly acquire data and analyze skin topography, in three dimensions, and with a satisfactory speed. Nowadays, only non-contact sensors have the capacity to meet these requirements. Furthermore, the anisotropy of skin surface demands a three-dimensional treatment of data.

III. Improvements in the Apparatus

The study of the topography of a surface is usually carried out by scanning the surface with parallel profiles.[11]

In the case of the scanning of a surface of 512 ↔ 512 measurement points separated by a distance of 10 µm, data acquisition would take more than 2 h if a mechanical sensor were used under standard speed conditions (0.5 mm/s).[12] This length of time is unacceptable for assembly line monitoring.

The recent development of laser diodes and their miniaturization has given rise to the production of several devices based on two principles: focusing of the laser beam and optical triangulation.

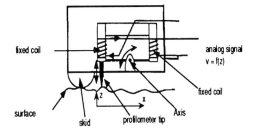

FIGURE 1 Principle of stylus profilometer.

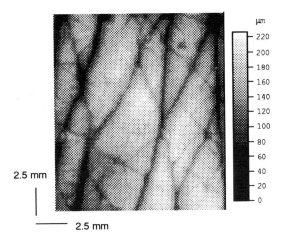

FIGURE 2 Cutaneous surface, top view.

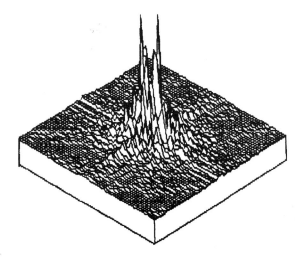

FIGURE 3 Two-dimensional Fourier transform of the cutaneous surface in a 25-year-old male subject.

A. Focusing System

The principle of this system, used in industry, is explained in Figure 4: the beam from the laser diode ($\lambda \approx 650$ to 800 nm) is focused on a point of the surface to be analyzed. This surface is displaced under the sensor at a constant speed, and any variation in height of each measured point enlarges the beam. This unfocusing induces a decrease of the energy received by the detector, which sends a signal to the linear motor that positions the optical system, thus regulating the convergence, and refocuses the beam. At each point of variation in the relief of the surface, a refocusing of the beam is necessary. The measurements, therefore, of the displacement of the optical system are equivalent to the variations of the relief.

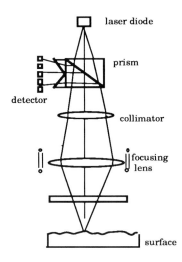

FIGURE 4 Optical focus profilometer.

This system is of interest because there is no contact with the analyzed surface; rapid measurements are theoretically possible. However, the mechanical displacement of the optical system, at each refocusing, requires time, and this waste of time diminishes the benefit that the principle of the system might bring. It does, however, have a very good vertical definition that reaches 0.1 μm.

For the measurement of skin relief made on silicon rubber negative replicas,[13] average measurement speeds of 0.5 mm/s in profilometry (two dimensions) and 0.3 mm/s in three dimensions have been obtained.

This system has another disadvantage: if it meets sudden changes in height on the relief, it has problems finding a new focusing point. These changes are often due to defects (bubbles) in the silicon rubber used to make the replicas.

B. Triangulation System

This kind of system, used in research laboratories only, is based on the classical principle of optical triangulation (Figure 5) utilized in non-coherent light. From a point of light projected on the surface being analyzed, a corresponding image is obtained on the surface of the detector: any variation of the image of the light on the detector corresponds to a change in the height. By using a laser diode ($\lambda = 850$ nm) and photodetectors with semiconductors (PSD or position sensing detector), the position of the image obtained on the surface of the PSD can be measured to the micrometer, thus giving a real sensitivity of about 3

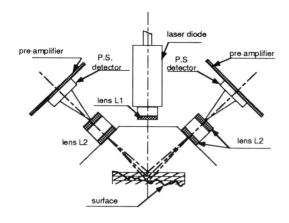

FIGURE 5 Optical profilometer: triangulation principle.

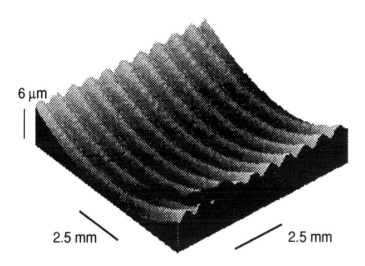

FIGURE 6 Roughness standard (6 μm amplitude).

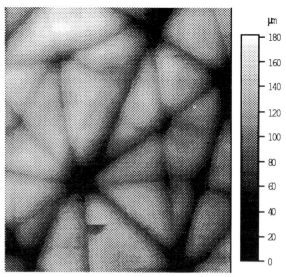

FIGURE 7 Mechanical detector.

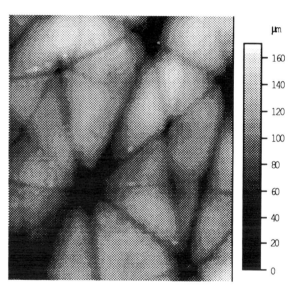

FIGURE 8 Optical detector.

μm. Figure 6 shows the image obtained with a roughness standard of a maximum amplitude of 6 μm, showing the present limitations of this type of system. Figures 7 and 8 compare the results obtained with two sensors of different types: one with mechanical contact, the other with a triangulation system. The results show the validity of the latter system, even for the measurement of very short displacements.

The main advantage of the triangulation sensor is its speed, which is limited only by the speed of precision translators. An average speed of 10 mm/s and accelerations of 8 mm/s^2 when starting each measurement line can be reached. With this sensor, the acquisition time of an image of 5.12 × 5.12 mm with one measurement point every 10 μm is only 5 min.

Another point of interest regarding this sensor is its wide measurement capability. It can measure a few tens of micrometers as well as variations in levels of up to 8 mm. Examples of the measurements of wounds showing wide surfaces and deep holes are given below.

IV. Quantification of Surfaces: Microrelief

The image of a surface analyzed by a sensor, whatever its principle, is represented by the law z = f(x,y), known for a discrete number of points (usually 512 × 512 or 256 × 256 points), with a vertical range between 12 and 14 bits. Since the height z = f(x,y) is known, the quantification of the surface is obtained using different methods.

A. Parameters Obtained from the Extension of Classical and Standard Roughness Parameters to Three Dimensions

Since standardization of three-dimensional measurement of surfaces has not yet been established,

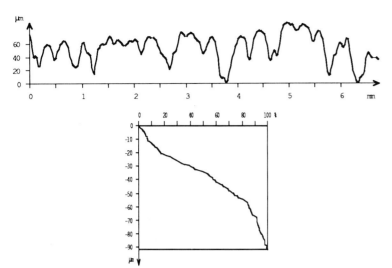

extending the known two-dimensional parameters is the obvious procedure to follow. A tri-dimensional parameter can correspond to a bi-dimensional parameter; for example, one of the most common two-dimensional parameters is R_a (arithmetic mean deviation of a profile) which gives the arithmetic mean of the absolute values of the profile departure, within a sampling length, i.e.:

$$R_a = \frac{1}{N}\sum_{i=1}^{N}|z(i)| \qquad (1)$$

where N is the number of experimental points. The same definition can be extended to surfaces:

$$SR_a = \frac{1}{N_1 N_2}\sum_{i=1}^{N_1}\sum_{j=1}^{N_2}|z(i,j)| \qquad (2)$$

A similar extension can be obtained for all the parameters quantifying heights.

The bearing curve of a profile is of special interest as it is a fundamental parameter in the study of friction and wear problems. The graph in Figure 9 illustrates the relationship between the values of the profile bearing length ratio to the profile section level. It shows the projected length at a determined height.

By applying this principle to studies on the surface of the skin, the surface of plateaus, and consequently their average size (Figure 10), are obtained.

B. Statistical Analysis

Another way to study the skin surface is by using statistical analysis that quantifies its anisotropy. It is possible to associate the autocovariance $C(\alpha, \beta)$ to the height $z(x,y)$ of each point of the surface:

$$C(\alpha,\beta) = \frac{1}{N_1 N_2}\sum_{i=0}^{N_1}\sum_{j=0}^{N_2}z(x,y)z(x+i\Delta x, y+j\Delta y)$$

with $\alpha = i\Delta x$

$$\beta = j\Delta y \qquad (3)$$

This value indicates the correlation between any point of the surface and another point distant from $i\Delta x$ and $j\Delta y$, with Ox and Oy being the experimental steps in the reference directions x and y, and i and j the integers of any value. The function $C(\alpha, \beta)$ shows possible periodicities; examples are provided showing the results obtained from a strictly regular surface (Figure 11) or from the skin surface (Figure 12).

Using $C(\alpha, \beta)$, the bi-dimensional Fourier transform

$$G(k_x, k_y) = \sum_{\alpha}\sum_{\beta}C(\alpha,\beta)\exp{-2i\pi(\alpha k_x + \beta k_y)} \qquad (4)$$

is the spectral density of the surface: it represents the stored "energy" of any part of the surface and shows the amplitude distribution of the different components of varied wavelengths that constitute the surface. Figures 13A and B illustrate the spectral density of the skin surfaces of two subjects of different ages. When $G(k_x, k_y)$ is known, it is possible to find the anisotropy of surfaces.[14]

The spectral moments described as:

$$m_{pq} = \sum_{k_x}\sum_{k_y}k_x^p k_y^p G(k_x, k_y) \qquad (5)$$

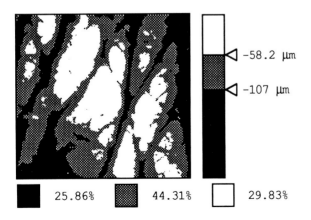

FIGURE 10 Surface of plateaus at different levels.

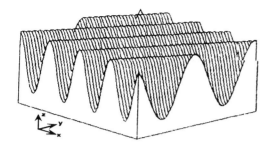

FIGURE 11 Autocovariance function of a sine surface.

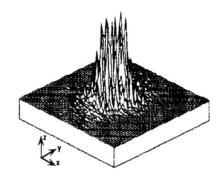

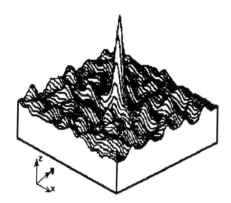

FIGURE 12 Autocovariance $C(\alpha, \beta)$ of a cutaneous surface in a 9-year-old child.

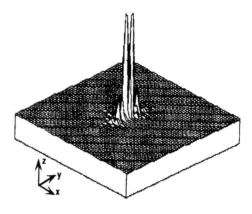

FIGURE 13 (A) Spectral density $G(k_x, k_y)$ function in a 9-year-old male subject; (B) spectral density $G(k_x, k_y)$ in an 86-year-old male subject.

and particularly the moments of second order m_{20}, m_{02}, and (zero order) m_{00}, give the anisotropy coefficient γ^2 (according to Longuet-Higgins[14]):

$$\gamma^2 = \frac{m_2 \min}{m_2 \max} \text{ with } \begin{cases} m_2\min=\frac{1}{2}\left[(m_{20}+m_{02})+\sqrt{(m_{20}+m_{02})^2+4m_{11}^2}\right] \\ m_2\max=\frac{1}{2}\left[(m_{20}+m_{02})+\sqrt{(m_{20}+m_{02})^2+4m_{11}^2}\right] \end{cases} \quad (6)$$

The main directions, i.e., the two directions on the surface in which the variations of m_2 are extreme, are obtained from the values of m_2 min and m_2 max. The examples in Figure 14 illustrate the correlation between the visual aspect of the surface and the value of its coefficient γ^2 of anisotropy.

In the case of a purely isotropic surface, $\gamma^2 \to 1$, and in that of a purely anisotropic surface, $\gamma^2 \to 0$. The values indicated in Figure 14 show that the influence of aging is well described by variations of this parameter.

C. Textural Analysis of the Skin Surface

The particular structure of the skin, made up of plateaus crossed by valleys, can be quantified by special operators applied to the image in its entirety. The techniques developed for image analysis make the quantification of furrows possible. Using a method proposed by Peuker and Douglas,[15] cutaneous furrows can be described by using techniques initially developed for earth relief study.

Any closed area, including a part of a furrow, shows particular variations of the local relief $z = f(x,y)$, characterized by two valleys (Figure 15). Thus, the height variations $(z_i - z_M)$ will show four changes of sign for any point M belonging to a cutaneous furrow, and the deepest furrows will be distinguished from the smallest by the value of the amplitude of the difference. Since any point selected belongs to a furrow (Figure 16), the density of furrows can be calculated, as well as their vertical distribution compared to the vertical distribution of the whole surface.

The ability to separate points on furrows from points belonging to the whole surface obviously induces greater sensitivity. For example, anti-wrinkle cosmetic products are developed to act mainly on deep furrows. An investigation of the whole surface would therefore drown the particular information needed in unnecessary experimental points. Furthermore, a selective study of the furrows of medium or large size (secondary or main furrows) can be obtained using discriminating thresholds.

Another method used in classifying cutaneous furrows into two families is the Fourier analysis or frequential analysis of the surface. The Fourier transform $F(u,v)$ of the surface $z(x,y)$ is given by discrete variables $x = k$ and $y' = k'$:

FIGURE 14 (A) The skin surface of a 9-year-old male subject; (B) the skin surface of an 86-year-old male subject.

$$F(n\Delta u, m\Delta v) = \sum_k \sum_{k'} z(k,k') \exp - 2i\pi(nk + mk')/N \quad (7)$$

where u = nDu and v = mDv in the frequency space.

One of the interesting features of this breakdown is that the entire surface can be reconstructed from only one part of the initial spectrum, i.e., after performing a low-pass filtering. The initial spectrum is composed of two parts: one with long wavelengths only (characteristic of plateaus separated by main furrows), the other with short wavelengths (characteristic of secondary furrows crossing the plateaus partially or completely), i.e.:

$$Z(n\Delta u, m\Delta v) = \sum_0^{k_1} \sum_0^{k_2} z(k,k') \exp - 2i\pi(nk + mk')/N +$$
$$\sum_{k_1}^{\infty} \sum_{k_2}^{\infty} z(k,k') \exp - 2i\pi(nk + mk')/N \quad (8)$$

A first approximation of the values of limits k_1, k_2 of wavelengths is given by calculating the average distance between plateaus when the cutaneous surface is cut by its mean plane. With this value, an approximate surface is reconstructed using the inverse transform on the part of the spectrum showing the low frequencies only:

$$Z_{LF}(k,k') = \sum_n \sum_m Z(n\Delta u, m\Delta v) \exp + 2i\pi(nk + mk') \quad (9)$$

In Figure 17, such a reconstruction applied to a skin profile can be seen: (a) the initial profile, (b) the corresponding total spectrum, (c) the profile reconstructed on the wavelength band $[k_1, \infty]$, and (d) the low-frequency part obtained from determining the mean distance Sm between plateaus (the limit k_1 = distance parameter Sm). The minima of profile (c) show the position of the main furrows.

The main furrows, detected by one or the other of the above-mentioned methods, are then subtracted from all the detected furrows, leaving only secondary furrows. Both families of furrows can then be quantified separately (Figure 18), resulting in a good description of the skin surface relief.

1. Directional Quantification of Furrows

Several methods of determining the linear forms of an image are available: one of them is the method proposed by Groch,[16] which first scans the image under study in two perpendicular directions so that the forms can be detected as precisely as possible, and then uses thinning and compression algorithms (Lu and Wang,[17] Holt et al.[18]) to reduce the calculation time. Such a method, used by Awajan et al.[19] in the detection of cutaneous furrows, provides good general results but poor directional sensitivity.

As proposed by Rosenfeld et al.[21] and Duda and Hart,[22] application of the Hough[20] transform to the extraction of linear forms is a more precise method. Hough uses a change of space: two points on a straight

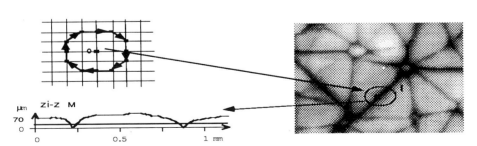

FIGURE 15 Detection of particular points (furrows): analysis of the height variations along a closed contour.

FIGURE 16 Recognition of the points in furrows.

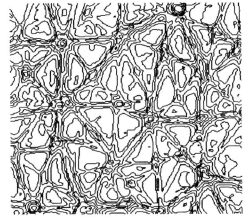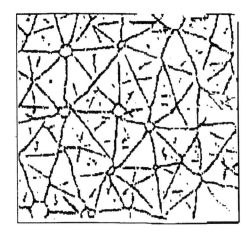

line of the image correspond to a point (r, υ) in the parameter space. In the space of parameters (r, υ), the resulting curve corresponds to each straight line intersecting a point $x_i\ y_i$ of the image space:

$$r = f(\theta, x_i, y_i) = x_i \cos\theta + y_i \sin\theta \qquad (10)$$

Therefore, a sine curve in the parameter space corresponds to each point $(x_i\ y_i)$ of the image space; the alignment of points on the image space will be expressed by a group of sine curves intersecting at the same point in the space of the parameters (Figure 19).

Several preliminary operations are necessary before applying this transform to the investigation of cutaneous furrows. The reference image must be simplified so that the points at the bottom of the furrows appear. As there are furrows of varied widths, it is necessary to develop a thinning technique for these linear forms so that an image can be obtained, which reproduces the original distribution of furrows and is composed of two different levels only: the first one classifying all the points belonging to furrows (level 1), the second one presenting all the external points (level 0).

The analysis of furrows is conducted as follows:

- Determination of cutaneous furrows
- Thinning of furrows
- Quantification

Furrows are localized by applying a local operator, which identifies a furrow on a point of the image. Several operators can be used, particularly those proposed by Awajan,[23] Cocquerez,[24] Groch,[16] Jimenez and Navalon,[26] and Montavert.[27]

Awajan uses a specific operator covering eight neighboring points, P_1 to P_8 (Figure 20):

$$\text{if } A(P) = \sum_i P_i \qquad (11)$$

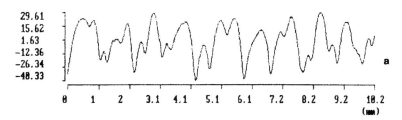
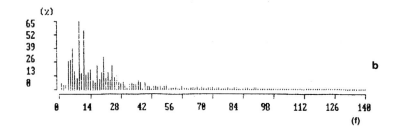

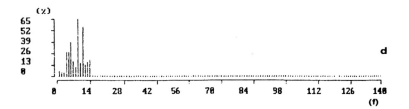

FIGURE 17 (a) Initial profile, (b) Fourier spectrum, (c) reconstructed profile from low frequencies, (d) Fourier spectrum of low frequencies.

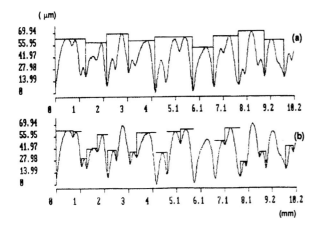

FIGURE 18 (a) Main furrows, (b) secondary furrows.

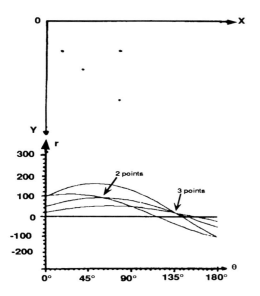

FIGURE 19 Line detection using the Hough transform.

then the points likely to be eliminated will be

$$2 \leq A(P) \leq 7 \tag{12}$$

The condition $A(P) > 2$ is necessary to retain the extremities of the skeleton, and condition $A(P) \leq 7$ states that the current point belongs to the edge.

All the points detected by the first scanning, thus belonging to the edges of the furrows, are analyzed by a second scanning and those satisfying the following relation are eliminated:

$$|\Sigma A_i - \Sigma B_i| = 3$$

with A_i and B_i defined in Figure 21. Examples of such a configuration are shown in Figure 22. The skeleton obtained after the application of this equation is not always of unit thickness, especially at the intersection of several furrows. To obtain a skeleton with a unit thickness, the point under discussion is eliminated as soon as one of the configurations shown in Figure 23 is found.

The result of the application of these different operators, shown in Figure 24, is compared to the original cutaneous surface. From this result, the Hough transform can be seen to express the directional distribution of furrows (density and directions).

V. Quantification of Surfaces: Macrorelief

Up to this point, we have studied only cutaneous microrelief. Two other sorts of skin surface characteristics, wrinkles and wounds, can be studied. They demand other methods of measurement.

A. Quantification of Wrinkles

The quantification of wrinkles is possible by using the silicon rubber replicas described earlier, and a sensor with a sufficient vertical range, under the condition that a suitable method be utilized. Usually the aim of this kind of study is to test the efficiency of an anti-aging product, and the investigation is therefore based on the comparison of two types of results: the evolution of a wrinkle without treatment (the reference), and the evolution of another wrinkle after treatment.

The results of the analysis depend on the different evolutions of both surfaces during the treatment (times t_0 and t_1).

To be reliable, this comparison study must be conducted on similar surfaces (same size and necessar-

a

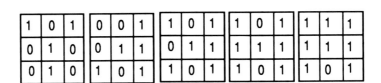
b

FIGURE 20 (a) Neighboring pixels, (b) masks used in the thinning algorithm (point elimination).

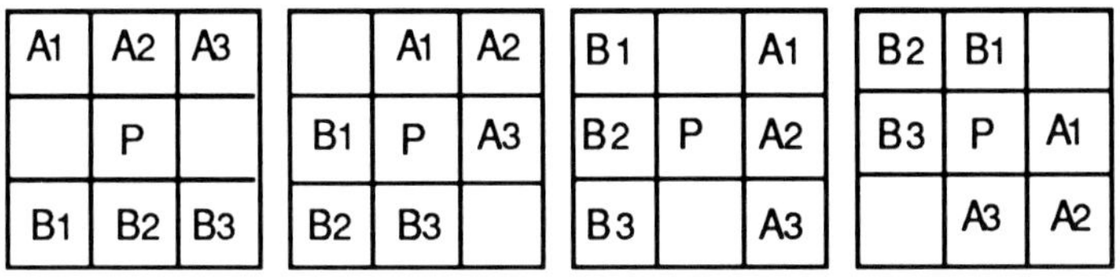

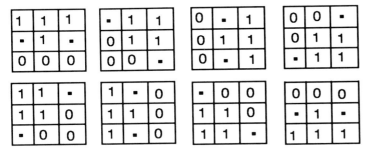

FIGURE 21 Configurations for pixel elimination.

FIGURE 22 Examples of application.

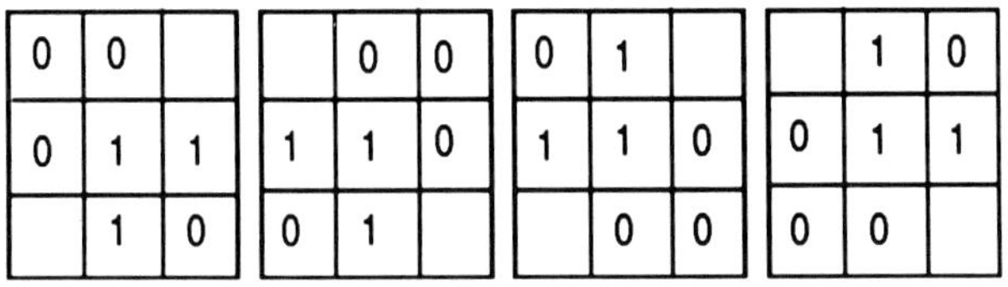

FIGURE 23 Masks used to obtain skeletons with a width of one unit.

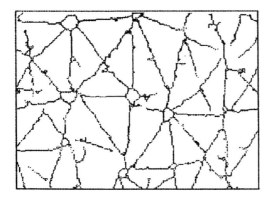

FIGURE 24 Skeleton of the area shown in Figure 16.

ily similar position); and be quantified by representative parameters.

To meet the first condition, replicas are made at t_0 and t_1 on similar parts of the body, the size of each replica being larger than the surface to be studied. Then the replica is scanned by a three dimension apparatus, either mechanical or optical, the scanned area being also larger than the wrinkle(s) to be examined. Both analyzed surfaces (Figure 25) are visualized: an area is selected on one of them and reproduced on the other one, either by a visual position control, or using the intercorrelation of both samples which consists in bringing to a maximum the product of correlation:

$$C(\alpha,n) = \sum_i \sum_\alpha z_1(x_i,y_i) z_2(x_i + \alpha, y_i + \alpha) \quad (13)$$

where z_1 is the height of a point on the first image with coordinates x_i, y_i and z_2 the height of a point on the second image. This product reaches a maximum when both images are correlated perfectly.

Once selected and carefully positioned, the areas are quantified as follows. Each profile is obtained by cutting the analyzed surface by a plane perpendicular to the mean plane of the surface. The real volume of the wrinkle is then the sum of all the elementary volumes.

An elementary volume is located between two parallel neighboring planes, the surface of the valley, and the two successive straight lines connecting the extreme points of each profile comprising the upper surface (Figure 26).

The maximum depth h, on average, of the whole surface is calculated:

$$h = \frac{1}{N} \sum_{i=1}^{N} h_i \quad (14)$$

B. Quantification of Wounds

The geometric characteristics of large wounds can be measured with one optical system because of its vertical range capacity and its wide precision translators. A typical example is a leg ulcer which can present differences in levels of several millimeters on a surface of a few square centimeters.

The healing progress of such wounds is monitored by making silicon rubber replicas of the surface at various times. It is useful to make such surface replicas for two main reasons:

- Keeping surface prints is important because they can be used later to check measurements and even to try new methods. These replicas can be stored over a long time.
- Only one measuring system is used, even if the prints come from other sources, thus allowing multicentric studies. Risk of error is reduced with just one system only.

Materials like SILFLO®* have been widely tested in dental surgery and in dermatology. This material is

---

* Registered Trademark of Flexico Developments Ltd.

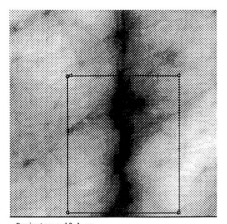

Perimeter : 12.1 mm
Surface : 8.07 mm^2
mean height : 123 µm
mean depth : 352 µm
Volume : 992 mm^2 x µm

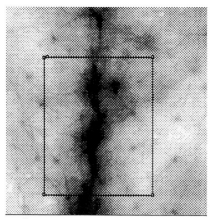

Perimeter : 12.1 mm
Surface : 8.13 mm^2
Mean height : 76.5 µm
Mean depth : 213 µm
Volume : 622 mm^2 x µm

FIGURE 25 Measurement of topographical parameters made on the same area, before and after treatment.

completely safe and causes no wound reaction; application to the wound is absolutely painless. It can reproduce the tiniest details: amplitudes of about a micrometer can be detected, which is sufficient for this kind of investigation. The white color of this material is accepted by the optical triangulation captor. In addition, it takes only 3 to 5 min to make a replica, even a large one.

1. Performance and Results

To analyze the surface of a wound (Figure 27), a scan of this surface is made line by line. The measuring step along a line and the distance between two neighboring lines are practically multiples of 10 µm. The result of this scanning is a table of N × L points, since the scanned area is generally rectangular.

The acquisition time depends on the size of the surface and the motor used (step by step motors); the step motor can be of 10, 50, 100 µm, producing measuring speeds of up to several tens of millimeters per second, the maximum frequency being 6000 steps per second.

The maximum scale of the vertical measuring range is 8 mm; it can be increased by using a large scale system.[28,29]

Once the surface z(x,y) is analyzed, the geometrical parameters are obtained in the following manner: flattening the surface by the least squared plane; and determining the edge of the wound, following it from a top view of the surface, with heights appearing as

FIGURE 26 Evaluation of the volume of a part of a furrow.

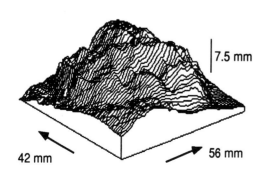

FIGURE 27 Ulcer replica scanning.

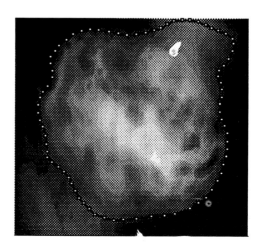

FIGURE 28 Determination of the boundary of the ulcer.

FIGURE 29 Elimination of the local form of the leg.

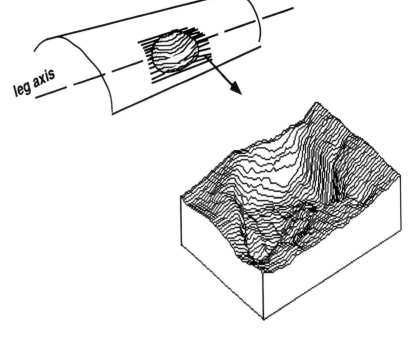

false colors (Figure 28). The perimeter of the boundary and the surface of the wound are the first parameters to be used, directly plotting the edges of the wound with transparent paper.[30] This method needs the operator's judgment and it can be replaced either by following up the contour on a screen with a mouse or by an automatic determination of neighboring points.[31]

Three-dimensional representation of wounds has been tried by photostereogrammetry,[32] but this method demands heavier apparatus and is not as accurate as the direct measurement of a relief by triangulation.

The exact plot of coordinates x, y, z of any point on the surface plays a part in the evaluation of the volume of the wound, which is a valuable parameter for the study of the healing process. However, the volume is significant only after elimination of the local form of the leg. As shown in Figure 29, this elimination is made by scanning the surface of the wound parallel to the longitudinal axis of the leg, and at a distance always greater than the real size of the wound. With this method it is possible to obtain, at the beginning of each profile, part of the healthy skin, which will be the reference.

For each profile, a segment of the straight line joins the reference parts and gives the general direction of the leg where there is no wound. The segments as a whole form a surface that reproduces the general shape of the leg without any wound; thus the real wound that will be analyzed is represented by the volume between the surface measured (included inside the boundary) and the "initial" surface of the leg.

Figure 30A shows the initial stage of an ulcer and Figure 30B the evolution of its boundary during the time of treatment, and the corresponding values of parameters: perimeter, surface, volume. Taking into account the volume of the wound avoids imprecise results in which the periphery and surface are likely to evolve in an unknown direction, whereas the volume decreases in accordance with the treatment duration.

2. Theoretical and Experimental Evolution of Healing

In view of this ability to trace the boundary of the wound, interesting future developments can be considered, such as theoretical research on its evolution and thus the prognosis of its evolution. Interesting theoretical research was carried out by Amiez[33] using geometric criteria combined with the results of a clinical investigation. She studied the displacement

FIGURE 30 Geometric parameters of a leg ulcer.

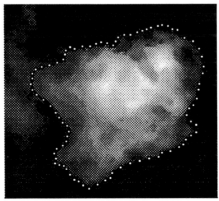

Perimeter : 639 mm
Surface : 297 cm^2
mean height : 817 μm
Volume : 24265 mm^3

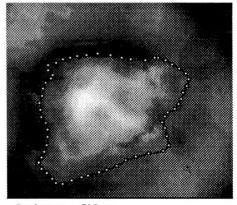

Perimeter : 516 mm
Surface : 185 cm^2
mean height : 658 μm
Volume : 14239 mm^3

of the contour of an ulcer over a specific period of time. If C1 and C2 are two contours at two very close moments t and t + dt (Figure 31), contour C2 is deduced from C1 by successive progressions defined by quantities r1, r2, r3 ... of points A1, A2, A3 ... of the contour at t. Any element such as Ai Ai + 1 scans the elementary surface defined by the quadrilateral Ai, Ai + 1, Bi + 1, Bi. With this method, the whole boundary can be split into several quadrilaterals. According to the clinical observations made by Agache, the change of the contour depends on its local curvature: an arc of the contour with a large curvature radius will change more quickly than an element with a smaller radius, so that the surface dSi scanned by the element di + 1 is given by dSi = Kdt di + 1 where K is a constant. Amiez links di + 1 to quantities ri, ri + 1, for the N points of the boundary, thus developing a system with N equations with N unknown.

Therefore, the resolution of the system is possible and, with the knowledge of the boundary at an instant t, knowledge of the contour at a further instant t + dt is produced. This new contour is then the basis for a new calculation, giving the solution at the next stage.

The successive healing stages of the contour of the wound can be foreseen with this iterative method. Figure 32 gives an example of the theoretical evolution of the real boundary of an ulcer applying the above method; Figure 33 shows the results of the theoretical calculation as well as the experimental results (broken lines) applied to a real boundary.

VI. Conclusion

The study of the microrelief of the surface of the skin is of interest to dermatologists, surgeons, and manufacturers of cosmetic products. The skin relief connected to the dermis and epidermis is dependent on numerous parameters that play a role in its evolution. Therefore, precise knowledge of skin relief can provide important information regarding the effects that certain illnesses, aging, and radiation have on

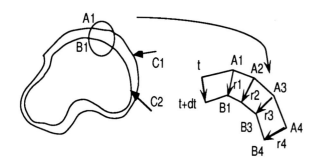

FIGURE 31 Contour of the ulcer at moments t and t + dt.

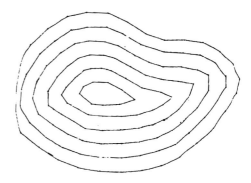

FIGURE 32 Theoretical ulcerous leg evolution as a function of time.

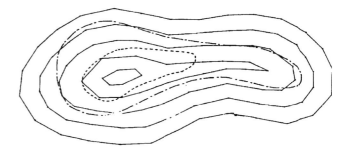

FIGURE 33 Ulcerous leg contour evolution. Theoretical contour, continuous lines; experimental contour, broken lines.

human beings. The capabilities of skin microrelief study have been greatly improved because of the existence of three dimensions in topographical studies.

Because of the use of both the real space $z = f(x,y)$ and the frequency (or wavelength) space, a large field of application has been opened up, where furrows can be separated into two families: one family that is connected to the dermis and one that is connected only to the stratum corneum.

Parameters specific to the cutaneous surface can take the place of traditional two-dimensional parameters, which were borrowed from the methods used in measuring industrial surfaces. The transfer to three dimensions opens up the field of study of texture, because the direction of the furrows can be detected and quantified, showing in particular the effects of the local mechanical deformation and aging. The macrorelief of the skin is composed of furrows of great amplitude, wrinkles, and wounds.

The study of wrinkles can be performed with much greater precision in three dimensions because the isolated detection of one and the same area at different stages of a treatment becomes possible. The quantification of geometric parameters such as surface, volume, and depth of the same part of a wrinkle offers a reliable way to test the efficiency of antiwrinkle products.

Wounds evolve, like wrinkles, according to different parameters. The healing process can be measured in an objective and quantitative manner by following the evolution of the geometric parameters: perimeter, surface, and volume of the wound. These are detected by making replicas which are then utilized for measuring the relief.

Analysed areas of large size demand the use of non-contact sensors, which are the only devices capable of working in three dimensions at satisfactory speeds.

The samples discussed throughout this chapter show the interest in and necessity for three-dimensional measurements of the geometric characteristics associated with the micro- and macrorelief of the skin, as well as all the potential possibilities of texture analysis.

References

1. Cummins, H. and Midlo, C., Palmar and plantar epidermal ridge configuration in European-Americans, *Am. J. Phys. Anthropol.*, 9, 471, 1926.
2. Wolf, J., Das oberflächenrelief der menschlichen Haut (Skin surface relief in man), *Z. Mikr. Anat. Forsch.*, 47, 351, 1940.
3. Tring, F.C. and Murgatroyd, L.B., Surface microtopography of normal skin, *Arch. Dermatol.*, 109, 223, 1974.
4. Marks, R. and Saylan, T., The surface structure of the stratum corneum, *Acta Derm. Venereol.*, 52, 119, 1972.
5. Thomas, T.R., Recent advances in the measurement and analysis of surface microgeometry, *Wear*, 33, 205, 1975.
6. Snaith, B., Edmonds, M.J., and Probert, S.D., Use of a profilometer for surface mapping, *Precis. Eng.*, 141, 87, 1981.
7. Xie, Y., in Quantification de la topographie des surfaces, Thesis, University of Besançon, France, no. 246, February 1992, chap 1.
8. Sampson, J., A method of replicating dry or moist surfaces for examination by light microscopy, *Nature*, 191, 932, 1961.
9. Sarkany, I., Method for studying the microtopography of the skin, *Br. J. Dermatol.*, 74, 254, 1962.
10. Cook, T.H., Craft, T.J., Brunelle, R.L., Norris, F., and Griffin, W.A., Quantification of the skin's topography by skin profilometry, *Int. J. Cosmet. Sci.*, 4, 195, 1982.
11. Williamson, J.P., The microtopography of surfaces, *Proc. Inst. Mech. Eng.*, 182, 21, 1968.
12. British Standard 1134, 1972, revised 1988.
13. Makki, S., Barbenel, J.C., and Agache, P., A quantitative method for the assessment of microtopography of human skin, *Acta Derm. Venereol.*, 59, 285, 1979.
14. Longuet-Higgins, M.S., The statistical analysis of a random moving surface, *Philos. Trans. R. Soc. London Ser. A*, 249, 966, 321, 1957.
15. Peuker, T.K. and Douglas, D.H., Detection of surface points by local parallel processing of discrete terrain evaluation data, *Comput, Graph. Imag. Process.*, 4, 373, 1975.
16. Groch, W.D., Extraction of line shaped objects from aerial images using a special operator to analyse the profiles of functions, *Comput. Graph. Imag. Process.*, 18, 347, 1982.
17. Lu, H.E. and Wang, P.S.P., An improved fast parallel thinning algorithm for digital pattern, IEEE Computer Soc. Conference on Computer Vision and Pattern Recognition, San Francisco, June 19 to 23, 1985, p. 364.
18. Holt, C.M., Stuart, A., Cunt, M., and Perrot, R.H., An improved parallel thinning algorithm, *Commun. ACM*, 30, 156, 1987.
19. Awajan, A., Rondot, D., and Mignot, J., Quick method of measuring the furrows distribution on skin surface replicas, *Med. Biol. Eng. Comput.*, 27, 379, 1989.
20. Hough, P.V.C., Method and means for recognizing complex patterns, US Patent N3, 069, 654, Dec. 18, 1962.
21. Rosenfeld, A., Thurston, M., and Lee, Y.H., Edge and curve detection for visual scene analysis, *IEEE Trans. Comput.*, C2D, 562, 1971.
22. Duda, R.O. and Hart, P.E., Use of the Hough transformation to detect lines and curves in pictures, *Commun. Assoc. Comput. Mech.*, 15 (N1), 11, 1972.
23. Awajan, A., Detection et analyse des structures linéaires d'une image. Applications biomédicales et industrielles. Thesis. University of Besançon, France, no. 78, 1988.
24. Cocquerez, J.P., Analyse d'images aériennes: extraction de primitives rectilignes et antiparallèles, Ph.D. thesis, Paris Sud (Orsay) University, France, 1984.
25. Cocquerez, J.P. and Devars, J., Détection de contours dans les images aériennes: nouveaux opérateurs. Traitement du signal, Vol 2, N1, 45, 1985.
26. Jimenez, J. and Navalon, J., A thinning algorithm based on contours, *Comput. Vision Graphics Imag. Process.*, 99, 186, 1987.
27. Montavert, A., Obtention d'une ligne médiane par connexion de l'axe médian. 5th Congress of AFCET-INRIA, Grenoble, France, November 27 to 29, 1985, p. 777.
28. Chuard, M., Mignot, J., Nardin, P., and Rondot, D., Range expansion and automation of a classical profilometer, *J. Manuf. Systems*, 6 (3), 223, 1987.
29. Zahidi, M., Assoul, M., Bellaton, B., and Mignot, J., A fast 2D/3D optical profilometer for wide range topographical measurement, *Wear*, in press.
30. Carrel, A. and Hartmann, A., Cicatrisation of wounds. The relation between the size of a wound and its rate of cicatrisation, *J. Exp. Med.*, 24, 429, 1916.
31. Zahouani, H., Assoul, M., Janod, P., and Mignot, J., Theoretical and experimental study of wound healing: application to leg ulcers, *Med. Biol. Eng. Comput.*, 30, 234, 1992.
32. Eriksson, G., Eklund, A.E., Torlegard, K., and Dauphin, E., Evaluation of leg wear treatment with stereophotogrammetry, *Br. J. Dermatol.*, 101, 2, 123, 1979.
33. Amiez, G., Cicatrisation des ulcères. National Meeting Numerical Analysis, Port Barcarès, France, 1988.

Chapter 5.5
Skin Surface Biopsy and the Follicular Cast

Rodney Dawber
Department of Dermatology
Churchill Hospital
Oxford, U.K.

I. Introduction

The technique of skin surface biopsy (SSB) was first described by Marks and Dawber.[1] It is a simple noninvasive method, removing only dead tissue, used to study the stratum corneum as a cohesive membrane (Figure 1), its constituent corneocytes and their relationship to each other, the many types of pathology within this compartment, and a vast array of microorganisms that may colonize or invade the layer.[1-3] It largely superseded the widely used sellotape stripping method.

II. Method

Essentially the method uses the known bonding properties of cyanoacrylate glue to proteins such as keratin — when the liquid adhesive is applied under pressure it solidifies by polymerization. Though the reaction is exothermic, it produces very little heat under these circumstances when applied to skin.

A drop of cyanoacrylate adhesive is applied to the area of skin to be studied. A clean transparent glass microscope slide is applied over the drop and pressure applied. The slide is removed after 20 to 30 s by slowly peeling it away from the skin; attached to this is a sheet of stratum corneum three to six cells thick. The bonded glue is transparent and of similar refractive index to glass. Thus, the attached sample can be viewed immediately by routine transmitted light microscopy,[1-5] preserved as present *in vivo*. The sampling method is less efficient and more inconsistent if the skin is either very dry or intrinsically thick, e.g., normal palms and soles.

Applications — The fact that the process involves no significant artefactual distortion means that conclusions drawn from studying the specimen can be justifiably related to the *in vivo* situation.

Light microscopy[1-5] — Studies carried out include (1) the relationship of corneocytes to each other in health and disease[2,4] and their surface morphology; and (2) commensal and pathological organisms, their qualitative and quantitative significance and destructive invasive characteristics.[8] The fact that the polymerization of the adhesive and its attachment to the stratum corneum generate heat (exothermic), does not compromise the histochemical reactivity with stains such as hematoxylin and eosin, periodic acid Schiff (PAS), Perls Prussian Blue reaction, silver impregnation for melanin granules (Fontana technique). Enzyme histochemical studies have also been undertaken, including nonspecific esterase, succinate dehydrogenase, lactate dehydrogenase, and glucose-6-phosphate dehydrogenase.

Scanning electron microscopy (SEM)[2,6,7] — Many SEM studies have been carried out in parallel with light microscopic ones described above. In studies on individual corneocytes in psoriasis (Figure 2) and other scaly and hyperproliferative skin diseases, surface morphological differences can be studied by magnifications up to 50,000. The method by which fungi proliferate and invade the corneocytes *in vivo* has been characterized in great detail.[3-5]

The follicular cast[9-11] — When the standard SSB is taken from sites with prominent hair growth or pilosebaceous openings, the act of taking the tissue typically removes a follicular cast and any associated hairs (Figure 3). Many detailed studies have been carried out, particularly in acne vulgaris, on the size of the follicular infundibulum (Figure 4) and the type and quantity of corneocytes using light and scanning electron microscopy and keratin biochemistry. Holmes and coworkers[9] found that even the normal skin of acne contained a greater degree of pilosebaceous keratin (Figure 5). Holland and Kearney[12] modified the SSB technique to quantitate the number of follicular microorganisms normally and in acne vulgaris. This method superseded many previous, more invasive, punch biopsy techniques.

FIGURE 1 Skin surface biopsy from flexor surface of forearm, showing the dermatoglyphic pattern and the continuity of the stratum corneum "sheet" as *in vivo* (scanning electron micrograph ×50).

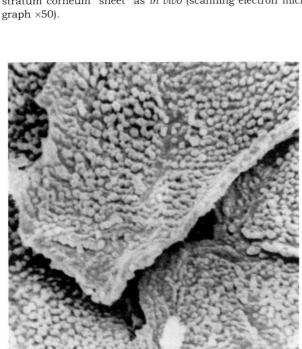

FIGURE 2 Psoriatic corneocytes — prominent villous folding of the plasma membranes. This reflects the rapid turnover and is not disease specific (scanning electron micrograph ×2000).

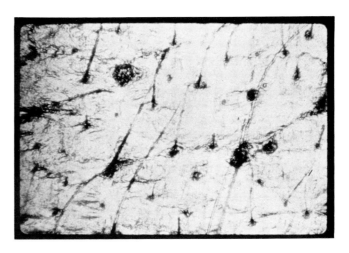

FIGURE 3 Skin surface biopsy from a hairy site. Note the attached "plucked" hairs and the surrounding, mainly infundibular follicular tissue (cast). (Light micrograph ×20.)

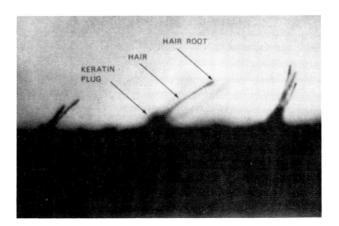

FIGURE 4 The keratin of the pilosebaceous duct sampled by skin surface biopsy (light micrograph). (Courtesy of Dr. W.J. Cunliffe, Leeds.)

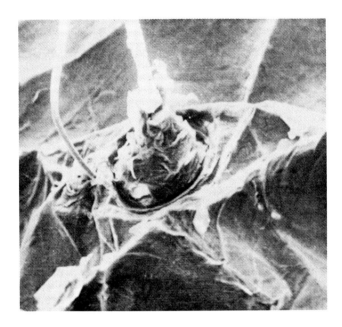

FIGURE 5 Scanning electron microscopic appearance of pilosebaceous keratin. The "whiteness" of parts of the image is due to charging of the specimen at these sites by the electron beam ×150).

References

1. Marks, R. and Dawber, R.P.R., Skin surface biopsy: an improved technique for the examination of the horny layer, *Br. J. Dermatol.*, 84, 117, 1971.
2. Marks, R., *Investigative Techniques in Dermatology*, 1st ed., Blackwell Scientific, Oxford, 1979, 243.
3. Whiting, D.A. and Bisset, E.A., The investigations of superficial fungus infections by skin surface biopsy, *Br. J. Dermatol.*, 91, 57, 1974.
4. Marks, R. and Dawber, R.P.R., In situ microbiology of the stratum corneum: an application of skin surface biopsy, *Arch. Dermatol.*, 105, 216, 1972.
5. May, R.J., Roberts, S.O.B., and MacKenzie, D.W.R., Laboratory diagnosis of the superficial mycoses, in *Textbook of Dermatology*, 5th ed., Chapman, R., Burton, J., and Ebling, F.J.G., Eds., Blackwell Scientific, Oxford, 1992, 1130.
6. Marks, R. and Griffiths, W.A.D., The significance of surface changes in parakeratotic horn, *J. Invest. Dermatol.*, 61, 251, 1973.
7. Dawber, R.P.R., Marks, R., and Swift, J.A., Scanning electron microscopy of the stratum corneum, *Br. J. Dermatol.*, 86, 272, 1972.
8. Marks, R., Histochemical applications of skin surface biopsy, *Br. J. Dermatol.*, 86, 20, 1972.
9. Holmes, R.L., Williams, M., and Cunliffe, W.J., Pilo-sebaceous duct obstruction in acne, *Br. J. Dermatol.*, 87, 327, 1972.
10. Cunliffe, W.J. and Cotterill, J.A., in *The Acnes*, 1st ed., W.B. Saunders, Philadelphia, 1975, 78.
11. Hughes, B.R. and Cunliffe, W.J., The effect of isotretinoin on the pilo-sebaceous duct in acne, in *Acne and Related Disorders*, Marks, R. and Plewig, G., Eds., Martin Dunitz, London, 1988, 227.
12. Holland, K.T. and Kearney, J.N., Microbiology of the skin, in *Methods in Skin Research*, 1st ed., Skerrow, D. and Skerrow, C.J., Eds., John Wiley & Sons, New York, 1985, 433.

Chapter 5.6
High-Resolution Ultrasound of the Human Epidermis*

S. el Gammal, T. Auer, K. Hoffmann, and P. Altmeyer
Dermatologic Clinic of the Ruhr-University
Bochum, Germany

C. Paßmann and H. Ermert
Institute for High Frequency Techniques of the Ruhr-University
Bochum, Germany

I. Summary

Using 50-MHz ultrasound, the healthy skin of the soles, dorsal forearm, and upper leg in 20 patients was examined and correlated to histology. To study hyperparakeratosis, 30 psoriasis plaques were examined in 10 patients. To evaluate the epidermis, low pre-amplification (digitization range 380 mV) was chosen in order to avoid overmodulation of the skin entry echo. Due to this low amplification of the echo signal, the dermis is not visible. The stratum corneum is displayed in orthohyperkeratosis as an echolucent band which is bordered by an echorich line on the top (skin entry echo), representing the water/stratum corneum interface and an echorich line downwards (interface to the stratum Malpighii). Sonometric measurements of the plantar stratum corneum thickness and the viable epidermis in all locations correlated well with histometry. The thin stratum corneum of the forearm and upper leg, however, could not be sufficiently resolved using 50 MHz and therefore sonometric and histometric thickness did not correlate. In hyperparakeratosis (psoriatic plaques), the stratum corneum is represented as an echorich, markedly widened, frequently interrupted band composed of spots varying in thickness, height, and echo density.

With a 100-MHz transducer, the skin of the palms and the dorsal forearm was studied in 15 patients (8 men, 7 women). Using a z-scan technique which composes the images of multiple stripes, a signal in-depth penetration of up to 2 mm is possible despite high resolution and a strongly focused transducer. With 100 MHz visualization of structural details within the epidermis is possible. At low pre-amplification the ridged shape of the palmar and plantar dermatoglyphics can be seen. In some of these crests there is a centrally located crater-shaped depression which corresponds to the orifice of the eccrine sweat gland duct. Further decrease of the pre-amplification also allows demonstration of the spiral-shaped sweat duct in the stratum corneum.

Our results indicate that ultrasound at 50 and 100 MHz is a promising method to study epidermal changes *in vivo*.

II. Introduction

In the last 10 years, high-resolution ultrasound instruments have been developed to study the skin architecture *in vivo*.[1-8] 25-MHz ultrasound technology is now widely available at reasonable costs and has proven to be particularly useful to define the dimensions of skin tumors[4,5,7,9-12] and to quantify the inflammatory activity in the course of diseases.[1,3,6,10,11,13,14] While pathological processes in the dermis can be easily visualized, many alterations of the epidermis and the stratum corneum cannot be demonstrated due to lack of resolution. Figure 1 shows that the axial resolution is mainly determined by the bandwidth. The lateral resolution is proportional to the center frequency and indirectly proportional to the focal length. To investigate the epidermis, both the axial and lateral resolution must be improved. By raising the center frequency and bandwidth of the ultrasound transducer, resolution increases but the signal penetration depth into the skin is reduced.[10,15] We modified the 100-MHz transducer technology in such a way that skin structures up to 2 mm depth can still be visualized. Using 50 and 100 MHz, we investigated the sonographic characteristics of normal epidermis and stratum corneum in different body regions.

---

* Supported by the Deutsche Forschungsgemeinschaft (ER 94/5) and the Deutsche Krebshilfe (70512/M3/92/AL1).

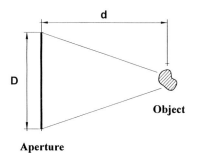

FIGURE 1 Physical parameters influencing the axial and lateral resolution in ultrasound.

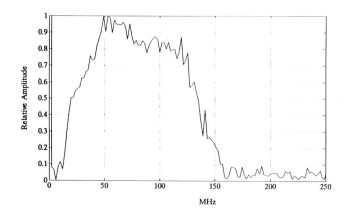

FIGURE 2 Frequency spectrum of the 100-MHz transducer in biological tissue.

III. Patients and Methods

An experimental ultrasound imaging unit was developed which can be operated with different transducers in a frequency range of 20 to 250 MHz. The ultrasound pulse shape is modulated by a programmable waveform generator and amplified. Using the reversed piezoelectric effect, this pulse is then transformed into an acoustic wave, which propagates through the coupling medium to reach the skin surface. Mainly longitudinal waves contribute to the acoustical echo signal.[15] The ultrasound wave is partly transmitted, partly reflected in different skin layers. The back-scattered echo signals are received by the same transducer, transformed into electric oscillations (receiver sensitivity 126 µV), amplified and digitized at 500 MHz with a transient recorder. Further signal processing is accomplished by an IBM-compatible 486 computer. It takes 1 to 16 s to record a single b-scan image, depending on the resolution chosen.

50-MHz sonography — The 50-MHz transducer has an axial resolution of 39 µm and lateral resolution of 120 µm within the focal area. Its piezoelectric layer is composed of polyvinylendifluoride (PVDF), a material that is characterized by a high flexibility making it particularly easy to focus this transducer.[16] The diameter of the transducer is 3 mm and its radius of curvature 8 mm. The in-depth signal penetration is 4 mm.[17] The healthy skin of the soles, dorsal forearm, and upper leg in 20 patients was examined with 50-MHz ultrasound. Seventeen of these patients had a tumor in one of the regions that were examined sonographically. These tumors (e.g., malignant melanoma) were excised with a tumor-free security margin. From this margin a histological section was taken after sonographic examination *in vivo* (dorsum of the arm 9×, proximal leg 5×, soles 3×). To study hyperparakeratosis, 30 psoriasis plaques were examined in 10 patients.

100-MHz sonography — The lateral resolution depends on the center frequency and focal length of the ultrasound field (see Figure 1). The 100-MHz transducer is built of ceramic material (PZT) and is strongly focused. Figure 2 shows the spectrum of this trans-

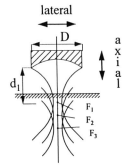

Center-Frequency:	f_0	= 80 MHz
Bandwidth:	Δf	= 120 MHz
Aperture:	D	= 3,2 mm
Focus Length:	d	= 4,3 mm

Resolution:	$\delta_{lateral}$ = 30 µm
	δ_{axial} = 11 µm
Penetration Depth: (using z-scan)	2 - 2,5 mm

FIGURE 3 (a) The "z"-scan technology makes it possible to use strongly focused transducers which have an excellent lateral resolution. (b) Physical properties of the 100-MHz transducer.

ducer. An acceptable in-depth signal penetration of 2 mm can be obtained by implementing a "z-scan" (Figure 3a), which eliminates the small focal depth of the transducer by decomposing the B-scan image into smaller image strips. The system characteristics have been summarized in Figure 3b. In the applicator, the transducer is moved by three independent motors in space (two lateral scan directions, "z"-scan). The skin of the palms and the dorsal forearm was studied in 15 patients (8 men, 7 women) using this 100-MHz transducer.

Quantification of sections — Distance measurements were performed on the ultrasound images and histological sections using image analysis (program AnalySIS, Soft Imaging Software GmbH, Münster, FRG). In each specimen at least five independent measurements were averaged. For statistical analysis the regression method was used.

IV. Results

A. 50-MHz Sonography

On the dorsal forearm and the upper leg, several zones can be differentiated from the skin surface down to the

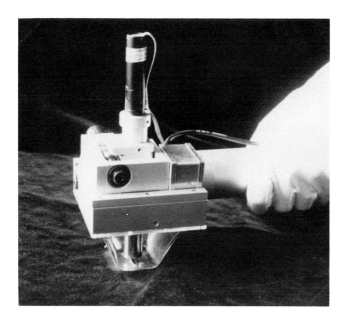

FIGURE 4 The applicator is equipped with two stepper motors and one linear motor to move the transducer. The cone-shaped plexiglass basin of the applicator is gently pressed to the skin surface and filled with water (coupling medium). Ultrasound scanning is done through a slit at the bottom of the plexiglass basin.

fatty tissue: above the skin surface, water, which is used as coupling medium, is represented as an echolucent area (Figure 4). Hairs lying above the horny layer are visualized as echorich ovaloid spots. The skin entry echo is an echorich line separating the water from a zone with multiple inhomogeneously distributed internal echoes. This zone corresponds to the dermis. The subcutaneous fat is seen as an echolucent area with obliquely oriented echorich lines (connective tissue bands). Many obliquely oriented echopoor ovaloid structures can be seen in the dermis (hair follicle complexes). In 50-MHz ultrasound images the thickness of the stratum corneum was 60 µm (range 40 to 88 µm; SD ± 19 µm), of the stratum Malpighii 78 µm (range 60 to 110 µm; SD ± 18 µm), and of the corium 1.50 mm (range 1.21 to 1.79 mm; SD ± 0.19 mm). In histological sections the stratum corneum measured 21 µm (range 10 to 40 µm; SD ± 18 µm), the stratum spinosum 81 µm (range 65 to 110 µm; SD ± 20 µm). The maximum corium thickness was 1.46 mm (range 1.2 to 1.7 mm; SD ± 0.19 mm) and minimum corium thickness was 0.79 mm (range 0.6 to 1.0 mm; SD ± 0.12 mm).

The palms and soles differ from this architecture. To evaluate the stratum corneum, the signal pre-amplification must be reduced. In 50-MHz ultrasound, on the soles the stratum corneum thickness was 568 µm (range 544 to 600; SD ± 23). In the histological sections the stratum corneum was 614 µm (range 650 to 680; SD ± 50). Figure 5 shows a longitudinal section of the palmar dermatoglyphics of a 26-year-old man. The distance between two small bars at the border of the image is 100 µm (referring to a mean sound speed of 1600 m/s in biological tissue). The upper echorich line corresponds to the skin entry echo, the lower to the stratum corneum/stratum Malpighii interface. The healthy orthohyperkeratotic stratum corneum and stratum Malpighii are echolucent in 50-MHz ultrasound images.

In hyperparakeratosis (psoriatic plaques, Figure 6), the stratum corneum is represented as an echorich, markedly widened, frequently interrupted band composed of spots varying in thickness, height, and echo density.

B. 100-MHz Sonography

At high echo signal preamplification, the skin of the dorsal forearm shows an echorich entry echo and an echodense corium (Figure 7). The image is composed of horizontal stripes due to the z-scan processing. The echopoor, sharply demarcated structure in the corium corresponds histologically to a hair follicle.

Low preamplification was used to study the epidermis. Figure 8 shows the ridged skin of the palms of a 26-year-old man. The ridges are cut perpendicular. Crests and valleys alternate. Some ridges show a crater-shaped central depression on the crest. This depression is sometimes connected to a more echorich

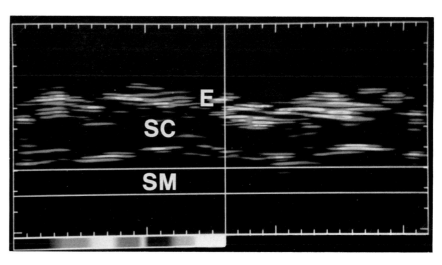

FIGURE 5 50 MHz B-scan of the palms at low pre-amplification. The distance between the smaller scaling marks at the border of the ultrasound image is 0.1 mm, between the larger ones 1 mm. The dermatoglyphics have been sectioned in longitudinal direction. Beneath the skin entry echo (upper echorich line) an approximately 200 µm thick echopoor zone is evident which corresponds to the orthohyperkeratosis. E, skin entry echo; SC, stratum corneum; SM, stratum Malpighii.

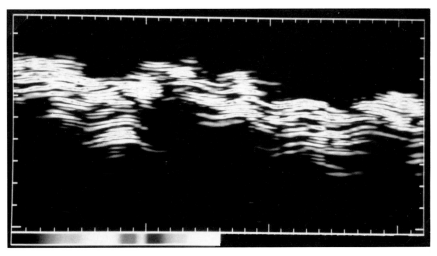

FIGURE 6 50-MHz B-scan of a psoriasis plaque at low pre-amplification. Beneath the skin entry echo multiple wide confluating echorich spots in an approximately 200 μm wide zone are observed. Histology reveals that these echorich spots correspond to focal hyperparakeratosis.

structure projecting into the stratum Malpighii (arrowhead). It corresponds histologically to the eccrine sweat gland duct. Visualization of this structure requires precise positioning of the applicator.

To study the fine architecture of the eccrine gland duct in the stratum corneum, the TGC-amplifier was disconnected. Due to the very low echo signal preamplification the slopes between the valleys and the crests of the dermatoglyphics disappear (Figure 9). The eccrine gland duct is now seen as a zig-zag-shaped double line.

V. Discussion

So far, high-resolution ultrasound has mainly been used to study the dermis. Only few attempts have been made to judge the epidermis in sonographic images. Using 10-MHz, Schwaighofer et al.[7] interpreted the skin entry echo as correlated to the stratum corneum. Their sonographic thickness measurements, however, did not correlate at all to values obtained from histological measurements, showing by far too high values. Others[18] equated the skin entry echo with the epidermis and felt that the echopoor line beneath the entry echo represented the papillary dermis and the subpapillary vascular plexus. The general consensus, however, was that the entry echo results from the impedance gap between water as coupling medium and stratum corneum and cannot be ascribed to a certain anatomical structure.[2,4,10] Failure to demonstrate a correlation between the epidermis in histology and the entry echo in sonography had lead to this commonly accepted hypothesis. Because of the low resolution at 20 MHz (100 μm axially, 200 μm laterally) epidermal substructures are only visible when they are greatly thickened (e.g., acanthosis and hyperkeratosis).

Using 50- and 100-MHz b-scan ultrasound we could very well correlate epidermal thickness histologically and sonographically. Moreover, we could demonstrate that both the stratum corneum and the viable epidermis are echopoor structures. These findings could only be obtained by optimizing two main factors (Figure 10): the echo signal pre-amplification and the resolution.

To visualize the epidermis, the pre-amplification of the echo signal has to be reduced so that the oscillation peaks of the echo signal are located within the digitization range of the transient recorder. During normal sonographic skin examination, the pre-amplification is usually chosen to particularly judge the dermis.[2,5,19] This leads to an overmodulation of the entry echo; the b-scan image shows a wide, echorich band[11] from which information about structural details cannot be obtained. Reduction of the overmodulated amplification goes parallel with a narrowing of the entry echo. When the entry echo no longer narrows and only fades, the optimal pre-amplification for its evaluation is reached. In images with such a low pre-amplification the dermis is no longer visualized.

The second crucial factor when studying the epidermis sonographically is the resolution of

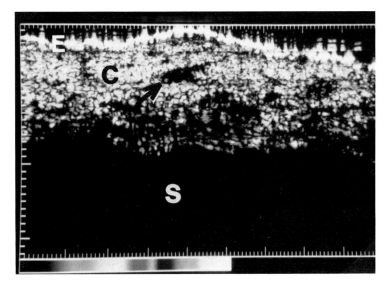

FIGURE 7 100-MHz B-scan of the dorsal forearm at high pre-amplification. Above the wide skin entry echo some echorich spots are visible which originate from hairs. The corium is echodense. A hair follicle is sharply demarcated within the corium (arrow). The subcutis is echopoor. E, skin entry echo; C, corium; S, subcutaneous fatty tissue.

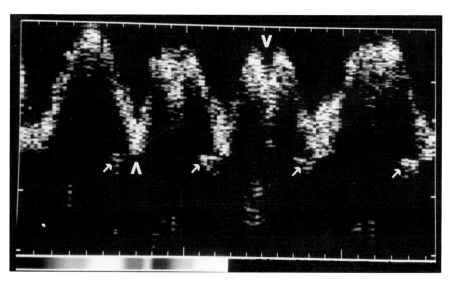

FIGURE 8 100-MHz B-scan of the palms at low pre-amplification. The dermatoglyphics are sectioned perpendicularly and exhibit crests and valleys. On the crests a central depression is often observed, from where an echorich structure sometimes leaves down to the stratum Malpighii. Down-pointing arrowhead, crest top (with central depression); up-pointing arrowhead, valley bottom; arrow, image stipe artifact.

the transducer. The axial resolution is inversely proportional to the bandwidth of the transducer, the lateral resolution to the center frequency and the focal length. Our experiences show that a structure has to be at least 1.5 times thicker than the axial resolution to be visualized proportional to its size.

At low pre-amplification, with 50 MHz we see a narrow skin entry echo, followed by a wider echolucent band which is bordered by an echorich line at its bottom. This line is less intense than the entry echo. There is strong evidence that the entry echo represents the water (coupling medium)/stratum corneum interface, the echopoor band in the middle the stratum corneum, and the deeper echorich line the stratum corneum/stratum Malpighii interface. This conforms to the general ultrasound principle that homogeneous structures are echolucent. "Homogeneous" in this case means that no scatterers are present in the tissue. Echorich regions emanate from interfaces between neighboring structures with markedly different acoustic impedance. For instance, the diaphragm is visualized in sonography of the abdomen as echorich. At higher resolution, however, it is represented as an echopoor structure.[20] The same effect leads to masking, in 100 MHz, of the echopoor stratum corneum by the wide echorich entry echo in 20 MHz sonography. We can conclude that the echo texture of a structure can only be judged when its thickness is significantly greater than the resolution of the transducer.

In contrast to orthokeratosis, hyperparakeratosis is echorich using 50 and 100 MHz. This is due to the cell nuclei, which strongly scatter the echoes.

With 100 MHz we were able to demonstrate the slopes between valleys and crests of the palmar dermatoglyphics and the central depression on the vertex, which corresponds to the meatus of the eccrine sweat gland duct. At reduced amplification, the spiral-shaped duct itself was also visible.

To study these fine structures of the epidermis, we used a new concept for 100-MHz ultrasound imaging. Presently narrow focused transducers are used in most dermatological sonographic equipment. It has been demonstrated that due to narrow focusing the reflected energy of an acoustic border is reduced to 6% at a 2-degree deviation from the main angle of reflection.[15] Narrow focused transducers therefore receive mainly signals from acoustic borders, which are oriented perpendicularly in relation to the ultrasound beam. We used a strongly focused 100-MHz

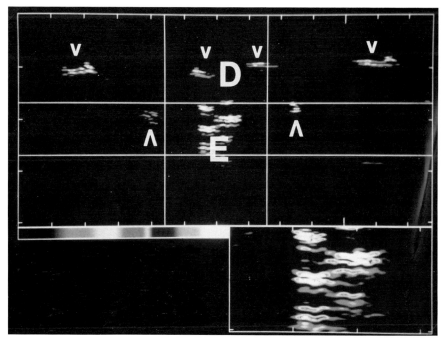

FIGURE 9 100-MHz B-scan of the soles at very low pre-amplification. The crest top and valley bottom are seen as echorich lines. The eccrine gland duct is observed as an echorich zig-zag structure in the stratum corneum. Down-pointing arrowhead, crest top; up pointing arrowhead, valley bottom; D, central depression; E, eccrine gland duct.

transducer to reduce the angular reflection artifacts discussed. To correct for the low echo signal depth penetration due to strong focusing, a "z"-scan was implemented. The final image therefore consists of stripes which are adapted by the computer. Slight deviations of these stripes sometimes lead to a blurred appearance of the image (Figure 8, arrow). In the future, correlation functions may further improve the adjustment of the stripes.

Our results from sonometric measurements of the stratum corneum and the viable epidermis illustrate the importance of optimal signal pre-amplification and resolution. Histologically, we found a mean stratum corneum thickness of 614 µm on the soles and of 21 µm on the dorsal forearm and upper leg. In the literature, data about the thickness of the stratum corneum (disregarding palmar and plantar skin) vary between 10[21] and 40 µm[22] Idson[23] measured a mean thickness of 15 µm for dry stratum corneum which increased to 48 µm after hydration. On the soles, sonometry of the stratum corneum at 50 MHz correlated well with histometry ($r = 0.99$, $p < .005$). On the dorsal forearm and upper leg, however, all sonometric values were above 40 µm; no correlation with the histology was possible ($r = 0.506$, $p = 0.06$). The reason is that the mean histological thickness of the stratum corneum (21 µm) is far below the axial resolution (39 µm at 50 MHz). Swelling of the horny layer, e.g., after application of emollients, can bring the stratum corneum thickness into the range of resolution and thus enable its visualization proportional to its size.[24]

With regard to the thicker stratum Malpighii there was good correlation between histometry and sonometry (mean thickness 81 resp. 78 µm, $r = 0.93$, $p < .005$) in all locations investigated.

An entirely exact correlation of histometry and sonometry cannot be expected as various artifacts influence the measurements in both methods. Histological processing leads to tissue shrinkage; the stratum corneum shows the characteristic basket-weave structure which does not correspond to the *in vivo* anatomy. Sonographic examination requires water as coupling medium which itself may lead to swelling of the horny layer. Sonometry is also influenced by the sound speed which is taken as the basis for distance calculations. In dermatological sonography, distance calculations from the echo signal time-lapse are usually based on the sound speed of the dermis (1580 m/s).[25,26] In the nail plate, however, Finlay et al.[27] found a sound speed of 2140 m/s comparing 20-MHz sonography and thickness measurements by a micrometer-screw. Jemec and Serup[28] divided the nail into two compartments with different speeds, an upper dry one (3103 m/s) and a lower humid inner one (2125 m/s). A similar sound speed could be postulated for the stratum corneum which consists of keratin and shows a similar structure to the nail.

We can conclude that high-resolution sonography is a promising technique which enables *in vivo* examination of fine structural details of the epidermis. In order to obtain images with significant information, understanding of the basic physics of ultrasound principles is indispensable. Only optimal pre-amplification, implementation of a z-scan to increase in-depth penetration at 100 MHz, and knowledge about the influence of resolution on the echo character enable correct interpretation of the phenomena.

References

1. Cole, C.W., Handler, S.J., and Burnett, K., The ultrasonic evaluation of skin thickness in sklerederma, *J. Clin. Ultrasound*, 9, 501, 1981.
2. Miyauchi, S., Tada, M., and Miki, Y., Echographic evaluation of nodular lesions of the skin, *J. Dermatol.*, 10, 221, 1983.
3. Serup, J., Decreased skin thickness of pigmented spots appearing in localized scleroderma (morphoea) — measurement of skin thickness by 15 MHz pulsed ultrasound, *Arch. Dermatol. Res.*, 276, 135, 1984.
4. Kraus, W., Nake-Elias, A., and Schramm, P., Diagnostische Fortschritte bei malignen Melanomen durch hochauflösende Real-Time-Sonographie, *Hautarzt*, 36, 386, 1985.
5. Breitbart, E.W., Hicks, R., and Rehpennig, W., Möglichkeiten der Ultraschalldiagnostik in der Dermatologie, *Z. Hautkr.*, 61, 522, 1986.
6. Myers, S.L., Cohen, J.S., Sheets, P.W., and Bies, J.R., B-mode ultrasound evaluation of skin thickness in progressive systemic sclerosis, *J. Rheumatol.*, 13, 577, 1986.
7. Schwaighofer, B., Pohl-Markl, H., Frühwald, F., Stiglbauer, R., and Kokoschka, E.M., Diagnostic value of sonography in malignant melanoma, *Fortschr. Röntgenstr.*, 146, 409, 1987.
8. de Rigal, J., Escoffier, C., Querleux, B., Faivre, B., Agache, P., and Léveque, J.L., Assessment of aging of the human skin by in vivo ultrasonic imaging, *J. Invest. Dermatol.*, 93, 621, 1989.
9. Mende, U., Petzoldt, D., Tilgen, W., and Schraube, P., Sonography: the ideal imaging method for staging and follow-up of malignant melanoma, in *Ultrasound in Dermatology*, Altmeyer, P., el-Gammal, S., and Hoffman, K., Eds., Springer-Verlag, Heidelberg, 1992, 119.
10. el-Gammal, S., Auer, T., Hoffmann, K., Matthes, U., Hammentgen, R., Altmeyer, P., and Ermert, H., High-frequency ultrasound: a non-invasive method for use in dermatology, in *Non-Invasive Methods in Dermatology*, Frosch, P., and Kligman, A.M., Eds., Springer-Verlag, Heidelberg, 1993, 104.
11. el-Gammal, S., Hoffmann, K., Auer, T., Korten, M., Altmeyer, P., Höss, A., and Ermert, H., A 50 MHz high-resolution imaging system for dermatology, in *Ultrasound in Dermatology*, Altmeyer, P., el-Gammal, S., and Hoffmann, K., Eds., Springer-Verlag, Heidelberg, 1992, 297.
12. Hoffmann, K., el-Gammal, S., Winkler, K., Jung, J., Pistorius, K., and Altmeyer, P., Skin tumours in high-frequency ultrasound, in *Ultrasound in Dermatology*, Altmeyer, P., el-Gammal, S., and Hoffmann, K., Eds., Springer-Verlag, Heidelberg, 1992, 181.

13. Akesson, A., Forsberg, L., Hederström, E., and Wollheim, E., Ultrasound examination of skin thickness in patients with progressive systemic sclerosis (scleroderma), *Acta Radiol. Diagn.,* 27, 91, 1986.
14. Hoffmann, K., el-Gammal, S., Gerbaulet, U., Schatz, H., and Altmeyer, P., Examination of circumscribed scleroderma using 20 MHz B-scan ultrasound, in *Ultrasound in Dermatology,* Altmeyer, P., el-Gammal, S., and Hoffmann, K., Eds., Springer-Verlag, Heidelberg, 1992, 231.
15. el-Gammal, S., Auer, T., Hoffmann, K., Altmeyer, P., Paßmann, C., and Ermert, H., Grundlagen, Anwendungsge-biete und Grenzen des hochfrequenten (20–50 MHz) Ultraschalls in der Dermatologie, *Zentralbl. Haut.,* 162, 817, 1993.
16. Höss, A.A., Ein hochfrequentes Ultraschall-Abbildungssystem hoher Bandbreite zur Tumordiagnostik in der Dermatologie, Dissertation, Bochum, Germany, 1991.
17. Höss, A.A., Ermert, H., el Gammal, S., and Altmeyer, P., High frequency ultrasonic imaging systems, in *Ultrasound in Dermatology,* Altmeyer, P., el-Gammal S., and Hoffmann, K., Eds., Springer-Verlag, Heidelberg, 1992, 22.
18. Querleux, B., Léveque, J.L., and de Rigal, J., In vivo cross-sectional ultrasonic imaging of the skin, *Dermatologica,* 177, 332, 1988.
19. Murakami, S. and Miki, K., Human skin histology using high-resolution echography, *J. Clin. Ultrasound,* 17, 77, 1989.
20. Lutz, H., Bauer, U., and Stolte, M., Ultraschalldiagnostik der Magenwand — experimentelle Untersuchungen, *Ultraschall,* 7, 255, 1986.
21. Holbrook, K.A., and Odland, G.F., Regional differences in the thickness (cell layers) of the human stratum corneum: an ultrastructural analysis, *J. Invest. Dermatol.,* 62, 415, 1974.
22. Blank, I.H., and Scheuplein, R.J., Transport into and within the skin, *Br. J. Dermatol.,* 81 (Suppl. 4), 4, 1969.
23. Idson, B., Hydration and percutaneous absorption, *Curr. Probl. Dermatol.,* 7, 132, 1978.
24. el-Gammal, S., Auer, T., Hoffmann, K., Matthes, U., and Altmeyer, P., Möglichkeiten und Grenzen der hochauflösenden (20 und 50 MHz) Sonographie in der Dermatologie, *Akt. Dermatol.,* 18, 197, 1992.
25. Alexander, H.D., and Miller, L., Determining skin thickness with pulsed ultrasound, *J. Invest. Dermatol.,* 72, 17, 1979.
26. Beck, J.S., Speace, V.A., Lowe, J.G., and Gibbs, J.H., Measurement of skin swelling in the tuberculin test by ultrasonography, *J. Immunol. Methods.,* 86, 125, 1986.
27. Finlay, A.Y., Moseley, H., and Duggan, T.C., Ultrasound transmission time: an in vivo guide to nail thickness, *Br. J. Dermatol.,* 117, 765, 1987.
28. Jemec, G.B.E. and Serup, J., Ultrasound structure of the human nail plate, *Arch. Dermatol.,* 125, 643, 1989.

6.0

Nuclear Magnetic Resonance (NMR) Examination of the Epidermis

6.1 Nuclear Magnetic Resonance (NMR) Examination
of the Epidermis *in Vivo* 133
B. Querleux

Chapter 6.1
Nuclear Magnetic Resonance (NMR) Examination of the Epidermis *in Vivo*

Bernard Querleux
Laboratoires de Recherche de L'Oréal
Aulnay-Sous-Bois, France

I. Introduction

The acquisition of sectional images of the epidermis *in vivo* is a difficult task, which, at the present time, may be achieved by three recent techniques: ultrasound, magnetic resonance (MR) imaging, and confocal microscopy. The oldest one, about 10 years old, is high-frequency ultrasound imaging. With an axial resolution in the order of 30 to 60 µm, it is well adapted for the visualization of the whole skin, but also differentiates epidermis from dermis in some cases.[1,2] The most recent technique is *in vivo* confocal microscopy, about 2 years old, which is characterized by an excellent spatial resolution of about 1 µm.[3,4] This technique is very efficient for visualizing the stratum corneum but optical improvements are still in progress in order to increase the signal-to-noise ratio on images of the inner layers of the epidermis. The first studies of skin by high-resolution MR imaging date from 1987–1988.[5,6] As with ultrasound, high-resolution MR imaging, obtained by modifying a standard whole-body MR scanner, is more adapted to the visualization of the whole skin, but with an in-depth resolution in the order of 35 to 70 µm, epidermis can be clearly delineated and thus analyzed.

II. Object

Over the last decade the nuclear magnetic resonance (NMR) technique has become a powerful method in medical diagnosis. This technique is of great interest because one can noninvasively obtain not only a spatial localization of the different tissues but also quantitative information on tissues by measuring their proton relaxation times T1 and T2, and proton density N(H). These are intrinsic parameters of each tissue, providing, for instance, useful information about correlation with water content or interactions of water protons with macromolecules and more generally in the understanding of the molecular organization.

The aim of this chapter is to present some applications of *in vivo* high-resolution MR imaging to epidermis examination.

III. Methodological Principle

A. *In Vivo* High-Resolution MR Imaging
1. Equipment

The main requirement for imaging the different skin layers is high spatial resolution, at least in the direction perpendicular to the skin surface, as the typical thickness of epidermis is less than 100 µm. In a conventional whole-body MR imaging system and in clinical use, the pixel size is limited to about 300 µm, which corresponds to a field of view of 8 cm. In the depth direction, it is insufficient to observe the epidermis. To increase the resolution in this direction, we designed a high-strength surface gradient coil of small dimensions allowing a decrease of the field of view in this direction to 18 mm corresponding to a pixel size of 70 microns.[7] A small surface radiofrequency coil was designed in order to improve the signal-to-noise ratio. This specific imaging module (Figure 1) is connected in place of the standard corresponding coils of the system allowing us to obtain high-resolution images of the skin on most parts of the body.

2. Application for Imaging of the Epidermis and Dermis

During the acquisition time, the skin area to be investigated lies on the imaging module (Figure 2) and motion is avoided by stabilizing the body with straps, and by surrounding skin area with a double-sided adhesive tape on the module. High-resolution MR images are obtained with an 18×50 mm² field of view, corresponding to a pixel size of 70×310 µm², and a slice thickness of 3 mm in two dimensional (2D) acquisition and 0.7 mm in 3D acquisition.[7] We are thus able to differentiate the skin layers: epidermis and its adnexae, dermis, hypodermis, and its fibrous septa, and even a thickened stratum corneum on the palm as well as on the heel (Figure 3a, b, and c). Such a high spatial resolution allows us to measure various thicknesses of the epidermis and to characterize shape, size, and density of the pilosebaceous units, according to the location.

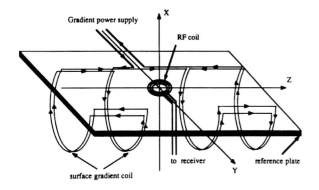

FIGURE 1 Schematic diagram of the specific imaging module comprising a high-strength surface gradient coil and a small surface radio frequency coil, 3 cm in diameter.

3. Application for Imaging of the Pathological Epidermis

If large cutaneous lesions can be studied with conventional MR scanners,[8-10] pathologic epidermis imaging must be performed *in vivo* with high-resolution imaging systems, and *in vitro* on biopsy samples with very high field microscopy NMR imaging systems.[11]

Our group has some preliminary results on pathological epidermis combined with inflammatory processes such as eczema or psoriasis. Figure 4a represents a psoriatic lesion before any treatment, located on the calf, and Figure 4b, the healthy contrelateral area. A thickened outer bright layer, about 1 mm in thickness, well differentiates the healthy area from the pathological one. However, studies are still in progress in order to establish whether this bright layer only represents a thickened epidermis, or whether it also represents the inflammatory process of the upper dermis.

The case of a squamous cell carcinoma on the heel gives an idea of the feasibility to delineate the exact border of epidermis tumors noninvasively. Morphological information about the shape, the depth, and the location of the tumor between high-resolution MR imaging (Figure 5a) and histology (Figure 5b) are in good agreement.

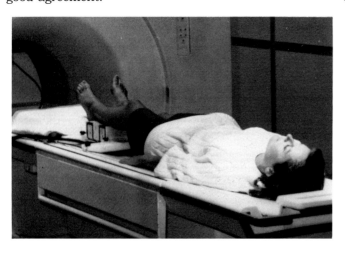

Consequently, with such a high resolution, MR imaging may be useful to discriminate tumors with regular or irregular boundaries, and diffused tumors.

B. In Vivo Measurement of the NMR Parameters in Skin Layers
1. Method

If high-resolution imaging is required to visualize in details both epidermis and dermis, the NMR imaging technique is also of great interest to obtain complementary information on the physicochemical properties of tissues by measuring their proton relaxation times T1 and T2,[12] and proton density N(H).

The received signal intensity can be approximately expressed as:[13]

$$S = k\, N(H)\, \exp^{(-TE/T2)}\, (1 - \exp^{(-TR/T1)}) \quad (1)$$

where k is a function of many parameters constant during the protocol, N(H) is proportional to the mobile proton density or free water content as determined by the MR imaging technique, T1 and T2 are MR tissue relaxation times, and TE and TR, parameters of the imaging sequence, are, respectively, echo time and repetition time.

T1 measurements are carried out by acquiring a set of 2D spin echo (SE) images only varying TR (TE = 16 msec, values of TR ranging from 100 to 4000 msec). T2 measurements were carried out, in a first step by acquiring a set of 3D gradient echo images, in order to achieve short echo time. At the present time, owing to many potential sources of artifacts related to the gradient echo sequence, T2 measurements are currently carried out by acquiring a set of 2D SE images (TR = 500 msec, values of TE ranging from 16 to 70 msec). T1 and T2 are calculated by fitting the signal intensities from a region of interest (ROI) to Equation 1.

2. Application for Characterization of the Epidermis and Dermis[14]

We measured T1 and T2 in the different skin layers on two different locations: calf and heel, which is characterized by a thickened stratum corneum of about 1 mm in thickness. Typical data on skin layers in the calf are presented in Figure 6 and on the heel in Figure 7.

Mean values on nine healthy volunteers (four women, five men, mean age ± standard deviation: 35 ± 6 years) are summarized in Table 1. Results have shown that epidermis and dermis are characterized by shorter T2 relaxation times than other biological

FIGURE 2 Whole-body MR imager with the specific imaging module for high-resolution imaging of the skin *in vivo*. For each acquisition, the subject lies on the examination cradle and the skin area of interest is centred on the surface of the module.

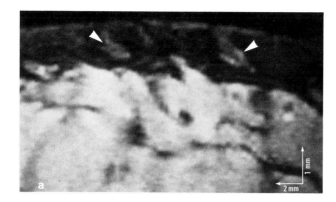

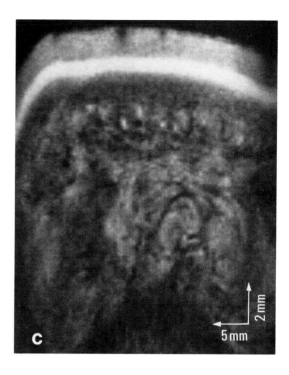

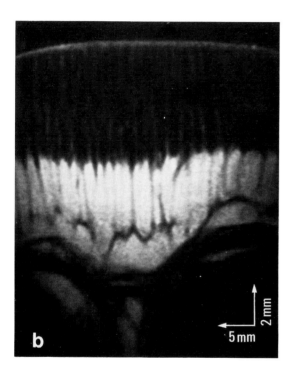

FIGURE 3 (a) MR image of skin on the calf. Layers of skin are clearly delineated: upper bright layer represents epidermis. The dermis appears as a thicker layer of hypointensity. Pilosebaceous units are seen as inclusions of epidermis inside dermis (arrowheads). The hypodermis and its fibrous septa are clearly visible. (b) MR image of skin on the back. The dermis is much thicker than on the calf and pilosebaceous units are more numerous and thinner. Hypodermis invaginations are also visible.

FIGURE 3 (c) MR image of skin on the heel. The upper layer is the stratum corneum, about 1 mm thick. The brighter layer between stratum corneum and dermis is epidermis, which is also particularly thick. The dermis appears more homogeneous and without visible epidermis annexe.

soft tissues, and that dermis and epidermis can be differentiated by their mean T2 values. These short T2 values may be essentially assigned to their fibrous protein content and particularly of the dermis. In contrast, the measured T1 values differentiate neither epidermis and dermis from other tissues, nor epidermis from dermis. Concerning the location dependence, heel epidermis T2 is lengthened and is closer to usual T2 values. This longer value, compared to the calf epidermis value, would be related to physiological differences induced by friction.[15] Finally, stratum corneum on heel presents short values not only of T2 but also of T1, characteristic of a low water content.

3. Application for Studies of Water Behavior in the Epidermis and Dermis[16]

In another study, and in addition to T1 and T2 measurements, we evaluated the relative proton density N(H). This parameter, which is proportional to the mobile proton density or free water content as determined by the MR imaging technique, allowed us to quantify the mobile water fractions in the epidermis and dermis. So a maximum signal intensity for each skin layer was computed according to Equation 1 using signal intensity for infinite TR value corrected by T2 decay for each layer. The quantity thus obtained was still dependent on the instrument's receiver gain, k, which was optimized for each subject. So the quantity N(H) had to be normalized in order to obtain comparable N(H) values between subjects. For this purpose an external reference constant was introduced for the entire experiment. This reference, mounted inside the radio frequency coil, was simultaneously scanned with each subject and every subject's relative proton density was normalized to that of the external reference.

After medical examination two groups of healthy volunteers were involved in the study: ten young women (25 to 40 years, mean ± standard deviation: 32 ± 4) and ten elderly women (70 to 81 years, mean ± standard deviation: 74 ± 4). Results have confirmed *in vivo* skin layer differentiation through relaxation times performed previously. Moreover, relative proton density quantification (Figure 8) has shown that epidermal

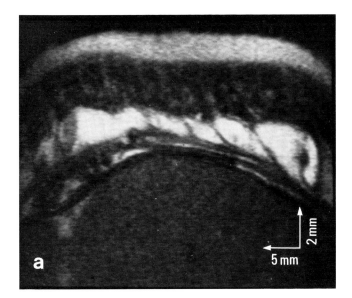

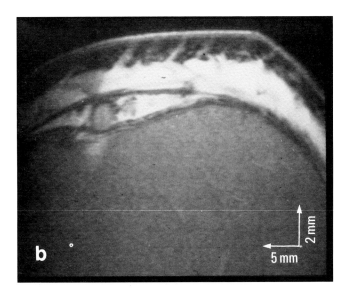

FIGURE 4 (a) MR image of a psoriatic plaque on the calf; (b) MR image of the healthy contralateral area. A thickened outer bright layer, about 1.1 mm thick, differentiates the pathologic epidermis from the healthy contralateral one, about 0.15 mm thick.

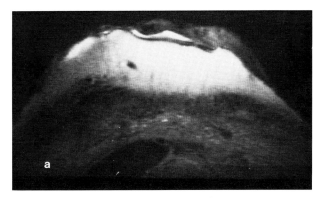

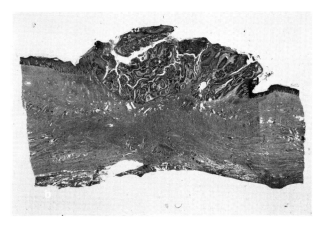

FIGURE 5 (a) MR image of a squamous cell carcinoma on the heel; (b) histology of the tumor. Dimensions, shape, and location are quite similar with the two methods.

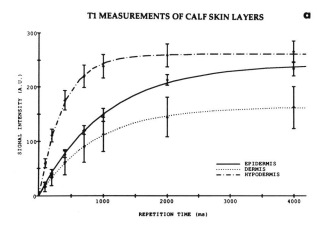

FIGURE 6 (a) Plots of signal intensities of the skin layers on the calf and the fitted curves obtained for T1 calculation.

mobile water is at least twice as abundant as dermal mobile water. So, as it is well established that epidermis and dermis have a total water content of the same order of magnitude, differences in N(H) values reflect differences in bound water content, much more important in dermis than in epidermis.

About differences between the two groups, if the main result concerns the upper part of the dermis (about 200 μm in thickness) which contains more mobile water protons in chronologically aged skin than in young adult skin, no clear difference could have been established in epidermis.

4. Application for Reconstruction of the Hydration Profile of the Stratum Corneum and Epidermis[17]

We applied the method of measuring mobile proton density by high-resolution MR imaging to analyze modifications of hydration of the stratum corneum *in vivo* under treatment. The images were acquired on

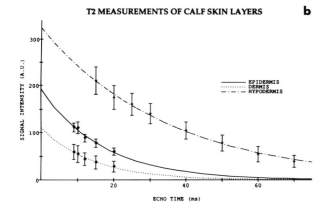

FIGURE 6 (b) plots of signal intensities and the fitted curves obtained for T2 calculation. Data presented are issued from one of the five ROIs for the dermis and hypodermis and from averaged 40 pixels in the epidermis. *Error bars,* standard deviation between pixels of the selected ROI.

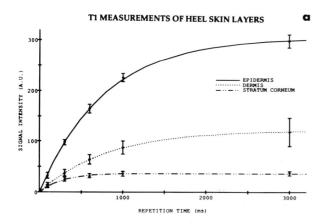

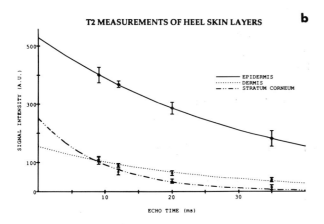

FIGURE 7 (a) Plots of signal intensities of skin layers on the heel and the fitted curves obtained for T1 calculation; (b) plots of signal intensities and the fitted curves obtained for T2 calculation. Data presented are issued from one of the five ROIs in each layer. *Error bars,* standard deviation between pixels of the selected ROI.

TABLE 1 Mean Values of the Relaxation Times T1 and T2 of the Different Skin Layers in Heel and Calf

	Heel		Calf	
	T1 (msec)	T2 (msec)	T1 (msec)	T2 (msec)
Stratum corneum	313 ± 47[a]	10.2 ± 2	—	—
Epidermis	720 ± 53	35.6 ± 5	887 ± 92	22.3 ± 7
Dermis	728 ± 104	17.9 ± 5	870 ± 143	13 ± 2.4
Hypodermis	—	—	393 ± 34	35.4 ± 3.6

[a] *Standard deviation calculated from nine subjects, five ROIs per subject.*

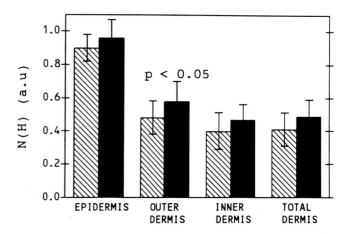

FIGURE 8 Relative proton density N(H) mean values of the skin layers on the thigh, normalized with the external reference and thus expressed in arbitrary units. The outer dermis corresponds to a subepidermal layer of 200 μm in thickness. The contribution of cutaneous appendages has been excluded from dermis measurements by semiautomatic image analysis. Relative proton density, i.e., mobile water fraction, is significantly higher in epidermis than in dermis ($p < .01$) and is significantly higher in the outer dermis than in the inner dermis ($p < .01$). N(H) is significantly higher in aged outer dermis than in young outer dermis ($p < .05$). *Hatched columns,* young subjects; *solid columns,* old subjects. *Error bars,* standard deviation.

the heel, where stratum corneum is particularly thick compared to our in-depth pixel size of 70 μm. The images were obtained by using a chemical shift imaging sequence,[18] which allows one to obtain two images: one related to the mobile water fraction within the tissues, the second one related to the lipid fraction. Signal intensity, extracted from the images related to the mobile water fraction, were measured for every depth increment equal to the pixel size. Then hydration profiles vs. depth were plotted after T1 and T2 fitting and normalization.

Hydration profiles of heel stratum corneum and epidermis, before and after immersion of the heel for 15 min in water at 30°C, are presented in Figure 9. An increase in hydration on the treated heel is clearly recorded in the outer layer of the stratum corneum. The main interest of this method lies in the fact that the physical signal is perfectly located, as spatial encoding is the basis of *in vivo* imaging. This method differs from other noninvasive methods which acquire an averaged signal from a nondelimited volume of interest.

This preliminary study tends to delineate two structures within the stratum corneum: an outer layer where hydration can be modified by external mechanisms and an inner layer where hydration is not altered. In the future this methodology will be available for *in vivo* measurement, in thick stratum corneum (palms and soles), of mobile water content vs. depth. Such a measure could be very useful for the follow-up of topical moisture product effects, and for the fundamental understanding of stratum corneum physiology in health or disease.

IV. Sources of Error

With *in vivo* imaging, whatever the technique used, ultrasound, X-rays, or magnetic resonance imaging, one major source of errors is spatial distortions induced by the encoding method. In MRI, these distortions are related to the use of gradients, nonlinear over the volume of interest. This problem is more important when using a high-intensity asymmetric surface gradient coil, because, with such a design, linearity cannot be as good as with a conventional symmetric gradient coil. Consequently, many experiments have to be carried out in order to specify the volume of interest in which the linearity of such a gradient is acceptable.

Another source of error is motion artifacts which are potentially critical in high-resolution imaging. This problem was well overcome by stabilizing the skin to investigate relative to the gradient coil by straps and double-sided adhesive tape, in such a way that no shift on images acquired at different times has been recorded.

Other sources of errors, specific to the MR technique,[19] have to be analyzed in order to assess their possible influence on high-resolution studies. These artifacts may be related to very short T2, susceptibility, and chemical shift. Very short T2 values induce a loss in spatial resolution, whereas susceptibility and chemical shift introduce spatial distortions. In fact, only the chemical shift artifact is visible on our MR images. It corresponds to a shift of 1.5 pixel, only localized at the boundary between water-rich and lipid-rich tissues, such as the dermo-hypodermis junction, and thus does not alter the visualization of a thin epidermis.

Finally, there is a source of error concerning epidermis thickness measurement. If the in-depth resolution corresponds to the pixel size, i.e., 70 μm, we have to keep in mind that the slice thickness is in the range of 0.7 to 3 mm. With such an isotropic voxel, measurement of epidermis thicknesses may be corrupted and overestimated due to partial volume effect, which could

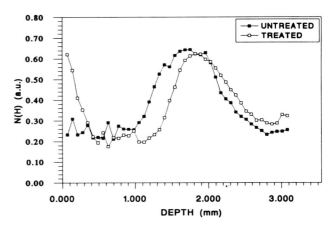

FIGURE 9 *In vivo* hydration profile of stratum corneum and epidermis by high-resolution MR imaging: effect of a bath. An increase in hydration is recorded in the outer layer of the thick stratum corneum on the heel.

happen if the direction of the slice thickness is not quite parallel to the skin surface.

In conclusion, even if partial volume effect seems to be the most important source of error, no clear degradation of the image quality arises from the potential sources of artifacts listed above.

V. Correlation with Other Methods

At the present time, no study has been published about a comparison of *in vivo* NMR examination of the epidermis and other imaging methods: ultrasound or confocal microscopy *in vivo,* and histology *in vitro*. In comparison to ultrasound, where the dermo-epidermis junction is more or less echogenic, in MRI the contrast is very important, so the interest in comparing it to other methods is less evident and should only concern epidermis thickness measurements in order to evaluate the overestimation probably induced by partial volume effect. Unfortunately, we have to keep in mind that the comparison with the "gold standard", histology, suffers from some difficulties: differences in slice thickness, effects of shrinkage, influence of dehydration, deparaffination. Thus, we cannot expect an exact correlation in epidermis thickness measurement or visualization of its annexe.

VI. Recommendations

If skin imaging may be performed in some cases with conventional wholebody imaging systems, in order to study large cutaneous tumors,[20] epidermis examination requires a modification of standard systems. We have proposed the use of not only a specific small surface radio frequency coil, but also of a high-strength small surface gradient coil which seems to be a very efficient method to obtain high-resolution images while maintaining a short echo time. We are thus able to obtain a pixel size in the range of 35 to 70 µm in the direction perpendicular to the skin surface, which is necessary for examining the epidermis *in vivo*. With such a high spatial resolution, morphological characteristics may give complementary information for the diagnosis of tumors. More particularly, differentiation between benign and malignant melanoma will be one of the most important challenges to assess the utility of this noninvasive technique.

Nevertheless we think that tissue characterization by quantitative measurements of the NMR parameters is the main interest of this technique. It already allows one to differentiate the skin layers, to quantify modifications of the physicochemical properties in normal and pathologic epidermis, and even to measure hydration processes in thick stratum corneum. Consequently, if microscopic examination of a biopsy specimen is, of course, a simple way to evaluate the epidermis, tissue characterization by high-resolution MR imaging is an *in vivo* noninvasive method which makes it possible to follow, on the same patient, physiologic processes or evolution of different pathologic conditions under treatment.

Acknowledgments

All the works presented are the results of collaborations between our laboratory with l'Institut d'Electronique Fondamentale-CNRS-Orsay, France and Centre Inter-Etablissement de Résonance Magnétique, CIERM, Hôpital de Bicêtre, Le Kremlin-Bicêtre, France.

References

1. Querleux, B., Lévêque, J.L., and de Rigal, J., In vivo cross-sectional ultrasonic imaging of human skin, *Dermatologica*, 177, 332, 1988.
2. el-Gammal, S., Hoffmann, K., Auer, T., Korten, M., Altmeyer, P., Höss, A., and Hermert, H., A 50-MHz high-resolution ultrasound imaging system for dermatology, in *Ultrasound in Dermatology*, Altmeyer, P., el-Gammal, S., and Hoffman, K., Eds., Springer-Verlag, Berlin, 1992, 297.
3. New, K.C., Petroll, W.M., Boyde, A., Martin, L., Corcuff, P., Lévêque, J.L., Lemp, M.A., Cavanagh, H.D., and Jester, J.V., In vivo imaging of human teeth and skin using real-time confocal microscopy, *Scanning*, 13, 369, 1991.
4. Corcuff, P. and Lévêque, J.L., In vivo vision of the human skin with the tandem scanning microscope, *Dermatology*, 186, 50, 1993.
5. Hyde, J.S., Jesmanowicz, A., and Kneeland, J.B., Surface coil for MR imaging of skin, *Magn. Reson. Med.*, 5, 456, 1987.
6. Querleux, B., Yassine, M.M., Darrasse, L., Saint-Jalmes, H., Sauzade, M., and Lévêque, J.L., Magnetic resonance imaging of the skin. A comparison with the ultrasonic technique, *Bioeng. Skin*, 4, 1, 1988.
7. Bittoun, J., Saint-Jalmes, H., Querleux, B., Darrasse, L., Jolivet, O., Idy-Peretti, I., Wartski, M., Richard, S., and Léêque, J.L., In vivo high-resolution imaging of the skin in a whole-body MR system at 1.5T, *Radiology*, 176, 457, 1990.
8. Zemtsov, A., Lorig, R., Bergfeld, W.F., Bailin, P.L., and Ng, T.C., Magnetic resonance imaging of cutaneous melanocytic lesions, *J. Dermatol. Surg. Oncol.*, 15, 854, 1989.
9. Schwaighofer, B.W., Fruehwald, F.X.J., Pohl-Markl, H., Neuhold, A., Wicke, L., and Landrum, W.L., MRI evaluation of pigmented skin tumors. Preliminary study, *Invest. Radiol.*, 24, 289, 1989.
10. Takahashi, M. and Kohda, H., Diagnostic utility of magnetic resonance imaging in malignant melanoma, *J. Am. Acad. Dermatol.*, 27, 51, 1992.
11. Aygen, S., el-Gammal, S., Bauermann, T., Hartwig, R., and Altmeyer, P., Tissue characterization and 3-D visualization of human skin tumors by high resolution proton NMR-microscopy at 9.4 Tesla, Eleventh Annual Meeting of the Society of Magnetic Resonance in Medicine, Berlin, 1992, 4605.
12. Bottomley, P.A., Hardy, C.J., Argersinger, R.E., and Allen-Moore, G., A review of ^{1}H nuclear magnetic resonance relaxation in pathology: are T1 and T2 diagnostic?, *Med. Phys.*, 14, 1, 1987.
13. Breger, R.K., Wehrli, F.W., Charles, H.C., MacFall, J.R., and Haughton, V.M., Reproducibility of relaxation and spin-density parameters in phantoms and the human brain measured by MR imaging at 1.5 T, *Magn. Reson. Med.*, 3, 649, 1986.
14. Richard, S., Querleux, B., Bittoun, J., Idy-Peretti, I., Jolivet, O., Cermakova, E., and Lévêque, J.L., In vivo proton relaxation times analysis of the skin layers by magnetic resonance imaging, *J. Invest. Dermatol.*, 97, 120, 1991.
15. MacKenzie, I.C., the effects of frictional simulation on mouse ear epidermis. I. Cell proliferation, *J. Invest. Dermatol.*, 87, 187, 1972.
16. Richard, S., Querleux, B., Bittoun, J., Jolivet, O., Idy-Peretti, I., de Lacharrière, O., and Lévêque, J.L., Characterization of the skin in vivo by high resolution magnetic resonance imaging: water behavior and age-related effects, *J. Invest. Dermatol.*, 100, 1993.
17. Querleux, B., Richard, S., Bittoun, J., Jolivet, O., Idy-Peretti, I., Bazin, R., and Lévêque, J.L., In vivo hydration profile in skin layers by high resolution magnetic resonance imaging, *Skin Pharmacol.*, 7, 210, 1994.
18. Dixon, W.T., Simple proton spectroscopic imaging, *Radiology*, 153, 189, 1984.
19. Bellon, E.M., Haacke, E.M., Coleman, P.E., Sacco, D.C., Steiger, D.A., and Gangarosa, R.E., MR artifacts: a review, *Am. J. Roentgenol.*, 147, 1271, 1986.
20. Zemtsov, A. and Dixon, L., Magnetic resonance in dermatology, *Arch. Dermatol.*, 129, 215, 1993.

Section C: Epidermal Functions

7.0 Desquamation

7.1 Methods to Determine Desquamation Rate 143
    C. Edwards and R. Marks

7.2 Sticky Slides and Tape Techniques to Harvest Stratum
    Corneum Material 149
    D.L. Miller

7.3 Dry Skin and Scaling Evaluated by D-Squames and
    Image Analysis 153
    H. Schatz, P.J. Altmeyer, and A.M. Kligman

Chapter 7.1
Methods to Determine Desquamation Rate

C. Edwards and R. Marks
Department of Dermatology
University of Wales College of Medicine
Cardiff, U.K.

I. Introduction

The integrity of the stratum corneum is vital to the constancy of the mammalian internal environment. Its barrier functions are quite remarkable when it is recognized that they are undisturbed by the continuous formation of the structure in its deeper parts as well as the loss of cells in its superficial zone.

The loss of stratum corneum at the skin surface (desquamation) is an integral and essential part of epidermal physiology which takes place in a controlled manner by the loss of single horn cells (corneocytes). It is clear that there is a relationship between epidermal cell production and desquamative loss, but the details of the link between these two facets of the population dynamics of the epidermis are obscure. Factors influencing cell production, including cytokine and mediator release, as well as the action of endocrine secretions, do so in a different time frame to factors influencing desquamation, such as the external mechanical stimulation of clothing, toilet, or social contacts. Not only do the two sets of factors differ as far as time frame is concerned but it appears likely that production and desquamation have separate "buffering" capacities. It seems highly likely that the epidermal cell population size stays constant over a period of several days but short-term inequalities may well occur in one or other of the compartments.

Regardless of the link between epidermal cell population and desquamation, measurement of that population at which the latter occurs is an important descriptor of this part of the skin. In particular determination of the rate of desquamation is helpful to the understanding of the vagaries of normal skin physiology and disease processes and in interpreting the actions of drugs on the skin.

One further comment before detailed descriptions of the methods are given is that it is important to distinguish desquamation from scaling. Scaling is due to the individual corneocytes failing to separate one from the other at the skin surface so that clumps of horn cells come off and is not directly related to the rate of desquamation. Measurement of the rate of desquamation can be direct, literally by counting the number of horn cells released at the surface, or indirect, in which the clearance rate of loss of a stain on the stratum corneum is estimated.

II. Measurement Techniques

A. Collection Techniques
1. The Chamber Technique

The passive or chamber technique of measurement of passive desquamation relies on the collection of all squames shed from the skin over a predetermined time period. Roberts and Marks[1] described the use of chambers consisting of perspex cylinders of 1 cm internal diameter, or 3.14 sq. cm area containing a bung with a glass fiber insert to allow free flow of water vapor, thus avoiding occluding the sample skin site. Nonirritant adhesive material was used to attach the chamber to the skin, and the whole assembly was covered with a foam pad. The enclosed volume was 2 ml. After 48 h the shed and loosely adherent squames were harvested and 2 ml of Triton X-100:0.06 M phosphate-buffered solution was carefully introduced into the chamber. The Triton X-100 is a nonionic surfactant which acts to prevent clumping of the squames and helps to keep them in suspension. The solution was then withdrawn and aliquots placed into a hemocytometer to count the corneocytes. Results were quoted as number of corneocytes collected per square centimeter per hour.

2. Forced Desquamation

This technique uses various gentle rubbing or scrubbing actions to collect free or loosely adherent squames from the skin surface. Early methods for quantitating desquamation were based on techniques first developed to collect cutaneous microflora.[2] These cup scrub techniques consisted of holding an open-ended cup onto the skin, placing a wash fluid in the cup, and agitating or scrubbing the skin surface to release loose material.

McGinley et al.[3] used a glass cylinder of 3.8 sq. cm held tightly against the skin. One milliliter of Triton X-100: phosphate-buffered solution was introduced and the skin surface rubbed vigorously with a smooth Teflon rod for 1 min. These authors stated that the scrubbing should be with firm pressure, and that experience is necessary before repeatability is obtained. The wash fluid was removed with a pipette, and the wash/scrub procedure repeated with clean wash fluid. The two wash samples were then pooled. The squames were then stained with a solution made from equal volumes of Hucker's crystal violet and basic fuschin.[4] This was done by dropping 0.05 ml of stain into 1 ml of wash sample, and the stained suspended squames were counted in a hemocytometer.

Nicholls and Marks[5] improved this method by building a motorized scrubber. A motor-driven smooth Perspex blade was held against the skin with a known and adjustable pressure in a modified cup apparatus. They used the same Triton X-100: phosphate-buffered solution described earlier, instilling this through a side arm, but did not stain the solution before counting suspended squames in a hemocytometer.

In a later report, Roberts and Marks[1] described a hand-held version of the motorized scrub apparatus, which they called the desquamator. Scrub application force was monitored as the torque required to rotate the blade against the skin and was kept constant by an electronic feedback circuit. Both the torque and the time of scrubbing were adjustable, and typical settings were 2 g-cm torque and a scrub time of 10 s. This latter study demonstrated a site variation for corneocyte shedding and a good relationship between the passive chamber technique and the forced desquamation technique (Table 1).

Corcuff et al.[6] used an air current generated by a turbine device to collect naturally shed corneocytes. In a later development, Corcuff and co-workers[7] described a similar device which incorporated a woolen pad as the friction element on a motor-driven stainless steel disc. The wool pad was chosen to resemble human hair in its biochemical nature and morphological organization, and was used to simulate the action of shampooing. The ratio of numbers of corneocytes collected by shampooing to the number of spontaneously shed corneocytes (24:1 in normal subjects, 32:1 in subjects suffering from dandruff) was similar to the ratio of wool scrubbed to spontaneous shed (22:1) or hair rubbed to spontaneous shed (24:1) The wool turbine device was also used to study the kinetics of desquamation after ultraviolet radiation-induced injury.[8]

B. Staining Techniques
1. Visual and Photometric Techniques

The cup or chamber scrub techniques, although very useful direct measures of numbers of corneocytes released, have the disadvantage of interfering in a physical or mechanical way with the "normal" environment and external forces experienced by the skin on different body sites. Staining methods aim to apply a substantive stain to the stratum corneum, the intensity of which can in some way be assessed or quantified. If the stain penetrates and is substantive the entire thickness of the stratum corneum is labeled and when the stained stratum corneum has been completely shed the stain will have disappeared, and the time taken is the stratum corneum turnover time. Applications of the method differ in the stains used and in the methods of measuring the amount or distribution of remaining stain.

Sutton[9] stained the horny layer with silver nitrate and observed the time required for disappearance of the black color. This method was modified by Roberts and Marks,[1] who used photographic developer to reduce the nitrate to metallic silver and then measured the intensity of staining with standardized photographic photometry. They found that the stain thus obtained was not wholly substantive, when sites washed in the normal way were compared to unwashed sites.

Baker and Kligman[10] noted that silver nitrate can cause irritant responses, and can also stimulate mitotic activity. They introduced the use of a fluorescent dye, tetrachlorsalicylanilide (TCSA) in a vehicle of ethylene glycol monomethyl ether (EGME). This was an antiseptic which bound strongly to the stratum corneum.

Their technique consisted of first establishing the time of penetration of the whole stratum corneum of each volunteer by applying the dye for 1, 1.5, and 2 h on different sites. These sites were then stripped down to the glistening layer (the stratum granulosum) to check penetration. The time of application to the first site to show dye penetration to, and including, the glistening layer was used in subsequent tests as the application time. After staining the sites to be investigated were examined daily using a Woods lamp (UV-A light of around 365 nm) to excite fluorescence. These authors checked that the dye was sharply localized to the stratum corneum by examination of transverse frozen sections of stained skin under the fluorescence microscope. The main problems occurred when it was found that about 50% of volunteers had an irritant reaction to the dye, which excluded them from any study. Contact sensitization occurred in about 15% of volunteers. However, despite these difficulties, stratum corneum turnover times could be estimated from the time taken for complete extinction of fluorescence. Reported times ranged from 6.3 (SD 1.4) d for forehead sites to 20.8 (SD 2.3) for the back-of-hand site.

Jansen et al.[11] introduced the use of another fluorescent dye, dansyl chloride (5-dimethyl-amino-1-naphthalene-sulfonyl chloride). They used a suspension of

TABLE 1 Comparison of Regional Variations in Rates of Corneocyte Loss Using Passive and Scrub Techniques with Dansyl Chloride (DC) Fluorescence Extinction Times

Site	Scrub Technique (Corneocytes/cm^2/10-scrub)	Chamber Technique (Corneocytes/cm^2/h)	DC Extinction Time (Days)
Forearm	104,243 (SD 15,570)	1,309 (SD 328)	18.5 (SD 4.6)
Upper arm	63,660 (SD 5,893)	795 (SD 141)	25.8 (SD 2.7)
Abdomen	55,623 (SD 8,215)	646 (SD 85)	25.9 (SD 5.4)
Thigh	67,638 (SD 4,736)	878 (SD 140)	26.5 (SD 5.4)
Back	81,962 (SD 12,929)	1,094 (SD 290)	23.6 (SD 5.2)

5% by weight of dansyl chloride in white petrolatum, which was applied on a cotton patch under occlusion for 24 h. They recommended lifting the patch after 6 to 8 h, rubbing the site and refastening the patch. The endpoint for measurement of turnover time was time elapsed until complete disappearance of fluorescence under a Woods lamp. Results confirmed earlier studies[12,13] that the forehead showed the shortest turnover time (about 8.5 d), with other body sites showing average turnover times of about 14 d. The authors also checked the binding of the dye to the stratum corneum, concluding that it must bind to the insoluble fibrous proteins which form the central keratin structure.

Finlay et al.[14] reported a fluorescence photographic photometric technique to measure objectively the intensity of fluorescence of dansyl chloride-stained skin. A fluorescent standard consisting of tiles of uniformly increasing fluorescence arranged in a circular fashion around a central aperture was placed over the area of unknown fluorescence. The site was photographed using a film holder with an integral calibrated step neutral density wedge.[15] The optical density of the standard tiles could be measured and plotted against their relative fluorescence on a log scale. A best fit slope was calculated and the unknown site fluorescence determined by reading off the log percentage fluorescence axis value at the optical density of the unknown area measured with reference to the internal step wedge standard.

This method was objective and accurate and the results were in good agreement with other reports. However, the fluorescent standards were not absolute or widely available (they were a selection of paper of different fluorescent qualities), and the method required considerable specialist photographic experience and equipment. It was not therefore easily accessible for more routine investigations.

Marks et al.[16] introduced a fluorescence comparator device for the assessment of skin fluorescence due to dansyl chloride staining. The device consisted of a viewing tube with an eyepiece at one end and an open aperture at the other. Half of the lumen of the tube was taken up by a linear varying optical density film strip (a gray wedge). The gray wedge was placed above a fluorescent standard positioned at the open end so as to be on the skin surface when the tube was placed on the skin. The standard was made from a skin surface biopsy (SSB)[17] from a site previously treated with a suspension of 5% dansyl chloride in white petrolatum. The gray wedge could be moved by a rack and pinion across the tube and its position accurately measured by reference to an accurate scale. In use the instrument is placed over the dansyl chloride-treated area of skin with the standard fluorescent SSB adjacent to it, while both are illuminated by a hand-held Woods lamp placed between 4 and 8 in. away from the site being examined. Through the simple eyepiece lens the test area and the fluorescent standard can be viewed simultaneously as equal halves of a circle divided by a septum in the tube. The gray wedge is then adjusted so as to attenuate the brightness of the standard until it appears to match the test site. The reading of the wedge position is then recorded as the arbitrary unit of fluorescence of the site. These authors checked the method against fluorescence photographic photometry.[14] They also measured inter- and intra-observer errors. Coefficients of variation for the intra-observer error and the means of the coefficients of variation for the interobserver error differences were between 4 and 6%. The extinction time of dansyl chloride fluorescence was determined by plotting the measured arbitrary fluorescence units for the dansyl chloride-treated site and an adjacent site (for a measure of "background" skin fluorescence) against time. Standard estimates of best fit allowed a mean slope to be fitted to the data and extrapolated to the time at which the slope of the treated site intercepted the background site slope. This was taken to be the extinction time.

Takahashi et al.,[18] using this fluorescence comparator device, demonstrated a circadian rhythm to the rate of decrease of dansyl chloride fluorescence and the effect of protection by a gauze pad and by a Finn chamber. A twofold difference was found in the rate of fluorescence loss between evening and morning readings (Tables 2 and 3).

Takahashi et al.[19] later developed an electronic fluorimeter containing sources of ultraviolet light, peaking at 338 nm, a photomultiplier detector and suitable filters to measure fluorescence of the skin at 446 nm (dansyl chloride fluorescence peak) without

TABLE 2 Stratum Corneum Renewal Time in Days Estimated Using Fluorescence Comparator Technique for Protected and Unprotected Flexor Forearm Sites

	Mean SC Turnover Time	Range
Unprotected site	14.5 (SD 3.8)	8.9–18.8
Gauze-protected site	23.7 (SD 6.7)	16.9–31.5
Ratio	1.75 (SD 0.19)	1.50–1.96
Unprotected site	13.2 (SD 1.7)	11.0–15.0
Chamber-protected site	29.1 (SD 12.2)	17.8–45.8
Ratio	2.21 (SD 0.77)	1.59–3.32

TABLE 3 Circadian Rhythm in Stratum Corneum Turnover Times as Estimated From Decreased Dansyl Chloride Fluorescence Measured in Arbitrary Units Using a Comparator Device

	Comparator Units/d	Comparator Units/h
Morning (9–10 a.m.) Period: 9–6 p.m.)	2.51 (SD 1.37)	0.17
Evening (5–6 p.m.) Period: 9–6 p.m.)	3.00 (SD 1.55)	0.33

effect from the UV light. Readings were presented on a digital display. This device was totally objective and easy to use.

2. Microscopic Techniques

Staining techniques tend to be easier to perform than direct estimates, but are probably less sensitive to small and transient changes.

Microscopic evaluation of the dansyl chloride staining of individual corneal layers has been reported.[20] This involved staining the skin with the standard 5% dansyl chloride preparation under occlusion for 24 to 48 h then taking biopsies at various intervals from 0 to 10 d after staining. Examination of transverse sections under the fluorescence microscope allowed identification of stained and new, unstained corneal layers. In this way the number of newly formed corneal layers was assessed and a rate of formation per 24 h calculated. It was claimed that increase in turnover rates greater than about 15 to 32% above normal were detectable using this method. An average rate of formation from a group of 10 normal volunteers was found to be 1.15 (SE 0.09) corneal layers per 24 h. Assuming the average stratum corneum to have 15 to 20 layers would translate to a turnover time of between 13 and 17 d, which is in accord with renewal times reported for the noninvasive dansyl chloride method.

More recently, Pierard[21] has reported a method for the assessment of dansyl chloride fluorescence of SSB samples from stained skin using a fluorescence microscope. In this study fluorescence on each test site used was clinically graded using a 0–10 analogue scale. Then an SSB was taken from that site. Photographs of the fluorescence microscope images were taken. The distribution of stain was reported to be at hair follicle openings, in primary and secondary lines, and on the plateaux. The extent of fluorescent and nonfluorescent areas was quantified using image analysis and the ratio of these areas used as a rate of fluorescence extinction.

Pierard reports dense and diffuse fluorescence immediately after staining, with labeling persisting in the adnexal openings, the primary and secondary lines, and on large patchy areas of the plateax, but with areas of focal extinction. Occlusion with some soaps and cosmetics revealed that some were capable of removing the dansyl chloride stain well before the stratum corneum could have been shed. The author concludes that dansyl chloride may not be suitable for testing the effects on stratum corneum turnover of all soaps, and that the reported increased desquamation measured using this test[22] could be due to removal of the dye by the products themselves.

3. Newer Agents

Pierard and Pierard-Franchimont[23] report on the use of fading of pigmentation induced by dihydroxyacetone (DHA) as a measure of stratum corneum renewal time. This product is widely used as a self-tanning agent. It reacts with stratum corneum proteins to form brown substances known as melanoids.[24]

For use as a turnover marker 10% DHA solution is applied for 2 h under occlusion. This is then repeated 6 h later. After this application the site is left uncovered. The evaluation of the intensity of browning of the skin is assessed using an industrial colorimeter based on the tristimulus method.[25] The CIE L* (lightness), a*, b* (chromaticity) color space parameters are used to measure differences between test sites and adjacent unstained skin.[26] The most sensitive and linear measures of fading of skin color are reported to be the change in b* chromaticity coordinate and the change in the overall color parameter (calculated from $[\Delta L^{*2} + \Delta a^{*2} + \Delta b^{*2}]^{1/2}$). These parameters are reported to be significant up to 18 d after application. In their attempt to measure the pigmentation effect throughout the stratum corneum the authors tape stripped the pigmented and adjacent skin using D-Squame tape and measured the color differences of subsequent pairs of D-Squame corneocyte collections. The progressive decrease in color found probably reflected a decreasing gradient of concentration of melanoids as well as a decreasing amount of collected corneocytes from lower and more cohesive stratum corneum layers. The DHA-induced pigmentation fading was compared to visual assessment of extinction of fluorescence from dansyl chloride-treated skin. Although the authors say that a good correlation exists between the two methods, they do not quantify this.

C. Cohesion

The binding forces between corneocytes at the skin surface are a function of the rate of desquamation. It must be remembered that it is the release of this intracorneal cohesion (ICC) that allows desquamation to occur. This binding force can be measured using a device known as a cohesograph. This employs a piston that is stuck to the skin surface with a cyanoacrylate adhesive and the force required to distract a segment of stratum corneum from the surface (about 2 cells thick) is measured electronically.

The first version of this device[27,28] used a manual piston retraction, but a motor-driven version followed. Changes in cohesion force were measured at increasing depths in the stratum corneum by measuring ICC after tape stripping. Mean force of cohesion of the forearm of a group of 10 normal subjects rose steeply from 100 g F at the surface to about 200 g F after 18 strips (about half the stratum corneum depth), when the rate of change slowed considerably.[29] There exists a (weak) relationship between the measured rate of forced desquamation and the ICC.[30] Also, an increase in ICC from skin from ichthyotic and other scaling disorders compared to normal can be measured with cohesography.[31]

D. Conclusion

The rate of loss of corneocytes from the skin surface is an important parameter of the population dynamics of the epidermis. Methods to measure desquamation have been devised and include direct counting techniques and the rate of clearance of substantive stratum corneum stains. The latter are the easier to perform but may be less sensitive than the more direct techniques.

References

1. Roberts, D. and Marks, R., The determination of regional and age variations in the rate of desquamation: a comparison of four techniques, *J. Invest. Dermatol.*, 74, 13, 1980.
2. Williamson, P. and Kligman, A.M., A new method for the quantitative investigation of cutaneous microflora, *J. Invest. Dermatol.*, 45(6), 498, 1965.
3. McGinley, K.J., Marples, R.R., and Plewig, G., A method for visualising and quantitating the desquamating portion of the human stratum corneum, *J. Invest. Dermatol.*, 53(2), 107, 1969.
4. Kolmer, J.A., Spaulding, E.H., and Robinson, H.W., *Approved Laboratory Technique*, 5th ed., Appleton-Century-Crofts, New York, 1951.
5. Nicholls, S. and Marks, R., Novel techniques for the estimation of intrecorneal cohesion in vivo, *Br. J. Dermatol.*, 96, 595, 1977.
6. Corcuff, P., Delasalle, G., and Schaefer, H., Quantitative aspects of corneocytes, *J. Soc. Cosmet. Chem.*, 33, 1, 1982.
7. Corcuff, P., Chatenay, F., and Saint-Leger, D., Hair skin relationships: a new approach to desquamation, *Bioeng. Skin*, 1, 133, 1985.
8. Corcuff, P., Chatenay, F., and Leveque, J.-L., Desquamation of the stratum corneum: kinetics following U.V. induced injury, *Acta Dermatol. Venereol. (Stockholm) Suppl.*, 134, 35, 1987.
9. Sutton, R.L., Early epidermal neoplasia, *Arch. Dermatol. Syphilol.*, 37, 738, 1938.
10. Baker, H. and Kligman, A.M., Technique for estimating turnover time of human stratum corneum, *Arch. Dermatol.*, 95, 408, 1967.
11. Jansen, I.H., Hojyo-Tomoko, M.T., and Kligman, A.M., Improved fluorescence staining technique for estimating turnover time of the human stratum corneum, *Br. J. Dermatol.*, 90, 9, 1974.
12. Weinstein, G.D. and Van Scott, E.J., Autoradiographic analysis of turnover time of normal and psoriatic epidermis, *J. Invest. Dermatol.*, 45, 257, 1965.
13. Rothberg, S., Crounse, R.G., and Lee, J.L., Glycine-14 incorporation into the proteins of normal stratum corneum and the abnormal stratum corneum of psoriasis, *J. Invest. Dermatol.*, 37, 497, 1961.
14. Finlay, A.Y., Marshall, R.J., and Marks, R., A fluorescence photographic photometric technique to assess stratum corneum turnover rate and barrier function in vivo, *Br. J. Dermatol.*, 107, 35, 1982.
15. Marshall, R.J., Infrared and ultraviolet photography in a study of the selective absorption of radiation by pigmented lesions of skin, *Med. Biol. Illus.*, 26, 71, 1976.
16. Marks, R., Black, D., Hamami, I., Caunt, A., and Marshall, R.J., A simplified method for measurement of desquamation using dansyl chloride fluorescence, *Br. J. Dermatol.*, 111, 265, 1984.
17. Marks, R. and Dawber, R.P.R., Skin surface biopsy: an improved technique for the examination of the horny layer, *Br. J. Dermatol.*, 84, 117, 1971.
18. Takahashi, M., Black, D., Hughes, B., and Marks, R., Exploration of a quantitative dansyl chloride technique for measurement of the rate of desquamation, *Clin. Exp. Dermatol.*, 12, 246, 1987.
19. Takahashi, M., Machida, Y., and Marks, R., A new apparatus to measure rate of desquamation using dansyl chloride fluorescence, *Arch. Dermatol. Res.*, 279, 281, 1987.
20. Johannesson, A. and Hammar, H., Measurement of the horny layer turnover after staining with dansyl chloride: description of a new method, *Acta Derm. Venereol.*, 58, 76, 1978.
21. Pierard, G.E., Microscopic evaluation of the dansyl chloride test, *Dermatology*, 185, 37, 1992.
22. Ridge, B.D., Batt, M.D., Palmer, H.E., and Jarett, A., The dansyl chloride technique for stratum corneal renewal as an indicator of changes in epidermal mitotic activity following topical treatment, *Br. J. Dermatol.*, 118, 167, 1988.
23. Pierard, G.E. and Pierard-Franchimont, C., Dihydroxyacetone test as a substitute for the dansyl chloride test, *Dermatology*, 186, 133, 1993.
24. Maibach, H.I. and Kligman, A.M., Dihydroxyacetone. A suntan-simulating agent, *Arch. Dermatol.*, 82, 505, 1960.
25. el-Gammal, S., Hoffmann, K., Steiert, P., Gassmuller, J., Dirschka, T., and Altmeyer, P., Objective assessment of intra- and inter-individual skin colour variability: an analysis of human skin reaction to sun and UVB, in *The Environmental Threat to the Skin*, Marks, R. and Plewig, G., Eds., Martin Dunitz, London, 1992, 99.
26. Robertson, A.R., The CIE color difference formulas, *Col. Res. Appl.*, 2, 7, 1977.

27. Nicholls, S. and Marks, R., Novel techniques for the estimation of intracorneal cohesion in vivo, *Br. J. Dermatol.,* 96, 595, 1977.
28. Marks, R., Nicholls, S., and Fitzgeorge, D., Measurement of intracorneal cohesion in man using in vivo techniques, *J. Invest. Dermatol.,* 69, 299, 1977.
29. Marks, R., Quantification of desquamation and stratum corneum cohesivity, in *Cutaneous Investigation in Health and Disease,* Leveque, J.-L., Ed., Marcel Dekker, New York, 1989, 33.
30. King, C.S., Nicholls, S., and Marks, R., Relationship of intracorneal cohesion to rates of desquamation in the scaling disorders, in *Bioengineering and the Skin,* Marks, R. and Payne, P.A., Eds., MTP Press, Lancaster, 1981, 237.
31. Nicholls, S., King, C.S., and Marks, R., Is there a relationship between corneocyte size and stratum corneum function in vivo?, in *Bioengineering and the skin,* Marks, R. and Payne, P.A., Eds., MTP Press, Lancaster, 1981, 227.

Chapter 7.2
Sticky Slides and Tape Techniques to Harvest Stratum Corneum Material

David L. Miller
Bionet Incorporated
Dallas, Texas

I. Introduction

The stratum corneum is both a barrier to the physical and chemical insults of the external environment and a record reflecting disease of the epidermis. In many cases examination of the skin surface and, particularly, the nature of the stratum corneum is more precisely accomplished by first separating a portion of the stratum corneum from the underlying tissue, then applying certain test modalities. Examples include techniques where the light absorbing/scattering properties of the surface are to be measured[1-3] (to estimate the scaliness), exfoliative cytology to elucidate certain features of the stratum corneum as a component of the diagnostic process,[4-7] and analytical methods for determining endogenous or exogenous chemical components of the barrier.[8,9] This chapter deals with the means of effecting convenient and reproducible samples of the desquamating stratum corneum.

II. Object

The object of these harvesting methods is to easily and reproducibly obtain a sample of the stratum corneum. To some extent, the ultimate fate of the sample places different requirements on the sampling technique.

III. Methodological Principle

Pressure-sensitive adhesives are high-molecular-weight organic substances that flow under applied pressure so as to form an intimate *mechanical* bond with a substrate, in this case the stratum corneum surface. As long as the adhesion of this interface, the internal cohesion of the adhesive material, and the adhesion to its support surface are all higher than the internal cohesiveness of the stratum corneum, some portion of the stratum corneum will be separated when the support surface is peeled away from the skin.

The use of pressure-sensitive adhesive tape (i.e., support surface = a flexible plastic film or fabric) to accomplish a sampling of the superficial stratum corneum for dermatological study probably originated with the publications by Wolf[10] beginning in 1939. Depending on the particular kind of adhesive and support employed, rather complex steps had to be taken to prepare samples for microscopy.[7] Many adhesives and their support materials are strongly stained by biological dyes. Goldschmidt and Kligman[11] coated an adhesive directly onto ordinary glass slides making a very rigid support. This eliminated the necessity of transferring the sample to a light-transmitting surface for observation, but required selection of an adhesive that did not interfere with the ultimate procedure. The very stiff nature of the support restricts sampling to flat areas. In any case, the preparation and storage of consistent coatings free of contamination from the air is an art in itself. Also, the nondrying adhesive film gradually loses tackiness due to oxidation.

Recently, pressure-sensitive adhesive discs with a moderately flexible plastic support specifically designed for dermatological use have become commercially available under the trade name D-Squame®* discs.[12] The discs are made from a very clear grade of polyester support film and an aggressive, super-clear adhesive. The combination of film and adhesive results in very high contrast between the optical properties of the adhering stratum corneum and the sampling medium. The clear carrier sheet on which the discs are supplied forms an effective barrier to dust and oxidation.

IV. Sources of Error

Contamination of the adhesive surface by dust and dirt prior to use can be a problem with sticky slides and adhesive tape. The design of the D-Squame® skin

---

* Registered trademark of CuDerm Corporation, Dallas, Texas.

sampling system tends to reduce this problem. Insufficient and/or inconsistent application pressure will result in samples that are difficult to compare. This error can be reduced by using a calibrated application pressure[2,3,13] and can be controlled by adopting a consistent application technique.[1,7]

Preparation of the skin surface is a critical step. While a clean, dry surface will provide maximal adhesion, samples taken for certain applications will be rendered useless by a cleaning step.

V. Correlation with Other Methods

There are only two other common methods of sampling the skin surface: scraping with a blade or other tool,[6] and the use of quick-setting cements, such as the cyanoacrylate type.[14]

Scraping the skin is not very useful in preparing a sample with any *quantitative value* as it is difficult both to control the extent of the area scraped and prevent loss of scraped material by air currents.

Cyanoacrylate cements have been used for preparing so-called skin surface biopsies.[14] Because the adhesion mechanism is based on a chemical reaction, the depth of skin removed will be determined by the depth of penetration of the adhesive before it hardens. Generally more stratum corneum is removed by this method than by the pressure-sensitive adhesives. This type of sampling technique is described in detail elsewhere in this handbook.

VI. Recommendations

The basic sampling technique is essentially the same regardless of the material used. The adhesive surface is pressed against the selected skin site for a few seconds, during which time sufficient pressure is applied to assure contact and flow of the adhesive substance. Then the adhesive support surface is peeled away from the skin and the sample set aside for further processing. Comments on the three approaches discussed follow.

Sticky slides — Preparation of the slides in the laboratory requires obtaining a source of adhesive solution then coating the solution onto glass slides and allowing the organic solvent to evaporate.[11] Producing an even, consistent coating will require skill and practice. Care must be taken in storing and handling the prepared slides to prevent contamination of the surface. The useful life of the slides after preparation is limited due to gradual air oxidation of the adhesive surface. The slides are available commercially.[15]

Adhesive tape — There are a number of commercially produced adhesive tapes which have been used for skin sampling,[6,7,13,16] but they are not necessarily well characterized with respect to component properties. In quantitative applications requiring a fixed sampling area,[8,9] tape should be precut under clean conditions.

D-Squame® discs — This skin sampling system eliminates many of the difficulties related to sticky slides and adhesive tape. It is a specially formulated adhesive system readily available worldwide directly from the manufacturer.[12] At the time of this writing, the discs are available in three sizes: the standard 22-mm diameter disc fits on an ordinary glass slide; the 13-mm diameter disc allows sampling from small areas such as the area covered by an experimental treatment patch, for example the large Finn chamber or Hill Top chamber, without including untreated skin; a 6-mm size allows a fixed area stratum corneum sample to be easily placed in the well of a microanalysis plate. All three sizes are produced on a clear carrier sheet, making storage and use of the discs easy. Figure 1 illustrates the appearance of skin samples made with the discs. The samples were obtained from the lower leg, then immediately placed on black background cards for viewing. The view is through the clear

FIGURE 1 Photographic comparison of D-Squame® skin surface samples representing three levels of scaliness: (a) fine scales.

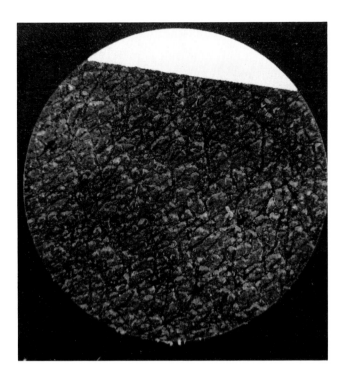

FIGURE 1 (b) medium scales.

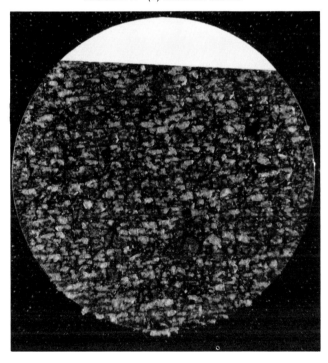

FIGURE 1 (c) coarse scales; 22-mm diameter discs against a black background viewed by scattered light.

adhesive disc, looking at the top surface of the skin. Note that differing scale thickness is represented in the photographs by differing degrees of scattered light intensity.

References

1. Prall, J.K., The scaliness of human skin, *Arch. Biochem. Cosmetol.*, 9, 87, 1966.
2. Piérard, G.E., Piérard-Franchimont, C., Saint Léger, D., and Kligman, A.M., Squamometry: the assessment of xerosis by cyanoacrylate surface biopsies and colorimetry of D-SQUAME adhesive discs, *J. Soc. Cosmet. Chem.*, in press.
3. Serup, J., Winther, A., and Blichman, C., A simple method for the study of scale pattern and effect of a moisturizer. Qualitative and quantitative evaluation by D-SQUAME tape in comparison with parameters of epidermal hydration, *Clin. Exp. Dermatol.*, 14, 277, 1989.
4. Piérard-Franchimont, C. and Piérard, G.E., Skin surface stripping in diagnosing and monitoring inflammatory, xerotic, and neoplastic diseases, *Pediatr. Dermatol.*, 2, 180, 1985.
5. Piérard-Franchimont, C. and Piérard, G.E., Assessment of aging and actinic changes by cyanoacrylate skin surface strippings, *Am. J. Dermatopathol.*, 9, 500, 1987.
6. Barr, R.J., Cutaneous cytology, *J. Am. Acad. Dermatol.*, 10, 163, 1984.
7. Jenkins, H.L. and Tresise, J.A., An adhesive-tape stripping technique for epidermal histology, *J. Soc. Cosmet. Chem.*, 1, 1969.
8. Pershing, L.K., New approaches to assess topical corticosteroid bioequivalence: pharmacokinetic evaluation, *Int. J. Dermatol.*, 38 (Suppl. 1), 14, 1992.
9. Rougier, A., Lotte, C., and Dupuis, D., An original predictive method for *in vivo* percutaneous absorption studies, *J. Soc. Cosmet. Chem.*, 38, 397, 1987.
10. Wolf, J., Das innere Struktur der Zellen des Stratum desquamans der menschlichen Epidermis, *Z. Mikrosk. Anat. Forsch.*, 46, 170, 1939.
11. Goldschmidt, H. and Kligman, A.M., Exfoliative cytology of human horny layer, *Arch. Dermatol.*, 96, 572, 1967.
12. CuDerm Corporation, Box 797686, Dallas, TX 75379.
13. Klaschka, F. and Norenberg, M., Individual transparency patterns of adhesive-tape strip series of the stratum corneum, *Int. J. Dermatol.*, 16, 836, 1977.
14. Marks, R. and Dawber, R.P.R., Skin surface biopsy, an improved technique for the examination of the horny layer, *Br. J. Dermatol.*, 84, 117, 1971.
15. Dermatologic Lab & Supply, 608 13th Avenue, Council Bluffs, IA 51501.
16. Prall, J.K., Theiler, R.F., Bowser, P.A., and Walsh, M., The effectiveness of cosmetic products in alleviating a range of dryness conditions as determined by clinical and instrumental techniques, *Int. J. Cosmet. Sci.*, 8, 159, 1986.

Chapter 7.3
Dry Skin and Scaling Evaluated by D-Squames and Image Analysis

Harald Schatz and Peter J. Altmeyer
Clinic of Dermatology
Ruhr-University of Bochum
Bochum, Germany

Albert M. Kligman
Department of Dermatology
University of Pennsylvania
Philadelphia, Pennsylvania

I. Introduction

A. Object

It is a well-known fact that the sampling and visual examination of tape strippings of the stratum corneum can reveal differences in the extent of dryness not noticeable by inspection of the skin surface. Following this simple method, a new image analysis technique was developed to objectively analyze the desquamating portion of the horny layer. Skin surface sampling discs (D-Squames) were employed to sample loose cells and scales from the superficial stratum corneum. Placing the discs against the black background of a storage card provides the maximum of contrast while evaluating the desquamation patterns. The discs are then illuminated in a light box and viewed by a CCD video camera attached to a stereo microscope. The video image of the sample is captured by a frame grabber in a personal computer and then processed with the aid of an image analysis program. Within seconds, the computer generates a few numbers, which represent the quantitative and qualitative properties of the sample. One of the values, introduced as "desquamation index", is a calculation of the disc area occupied with corneocytes and the individual thickness of the scales. The desquamation index provides exact, fast, and reproducible characterization of the D-Squames and is especially valuable for the assessment of dry skin.

B. Quantification of Dry Skin

The epidermis, the outer layer of the skin, is responsible for a wide range of properties which help to maintain the integrity of the body. It absorbs and excretes liquids, acts as a barrier against microorganisms, and provides cosmetic functions. Epidermal lipids and the sebum help to condition the stratum corneum and regulate the turnover of the horny layer. The shedding or accumulation of stratum corneum in visible flakes is called desquamation. Under normal circumstances, the epidermis is completely replaced every 25 to 30 d, depending on the skin site. At the end of the process of keratinization corneocytes get piled up to the stratum corneum. Abnormal shedding, produced by underlying conditions like parakeratosis (psoriasis) or by loss of lipids and hydration (xerosis) leads to squamous eruptions.

To quantify the rate of desquamation is one way to determine pathological changes in the keratinization of the epidermis. It has been used in the assessment of skin care products, to quantify xerosis, and to measure the degree of hydration of the stratum corneum.[1-4] Considering the high incidence of xerosis in the population and the growing interest in skin care, it was necessary to characterize and quantify dryness for clinical and commercial studies. Even though most efforts have been made in the field of cosmetic dermatology we realize more and more that the evaluation of the stratum corneum by means of noninvasive techniques becomes an important issue for clinical dermatology. In combination with other skin physiology techniques, like transepidermal water loss measurements, lipid assessments, pH and conductance measurements, the quantification of desquamation is important to characterize the condition of skin in health and disease.

There are three possible approaches to characterize the process of desquamation: the assessment of the stratum corneum turnover, the measurement of intracorneal cohesion, and the quantification of scaling.

Measuring the stratum corneum turnover can be achieved by application of dansyl chloride fluorescence dye on the skin.[5] The substance penetrates the entire thickness of the stratum corneum and can be detected easily under a Woods lamp. Consecutive observations under ultraviolet radiation up to the time of extinction of the fluorescence allow determination of the duration of one full turnover measured in days. A number of methods and modifications have been developed to measure the degree of fluorescence.[6] The measurement of stratum corneum cohesivity is much more complicated and requires a special apparatus to record the force that is required to rupture the stratum corneum after fixing a cyanoacrylate stub to the skin.[7] However, the method is very complicated and susceptible to environmental factors.

The most direct way to evaluate dryness is to look at the scale pattern itself. Since visual grading of scaliness is liable to environmental changes and subjectiveness, collecting corneocytes is a suitable approach towards an objective assessment of dry skin. Efforts were made to collect corneocytes by a washing technique. In a glass chamber placed on the skin, filled with water or a detergent (0.1% Triton X-100), loose scales and horny cells were sampled from the skin surface. The yield could be enhanced by using a spatula to scrub the skin manually.[8] The material was then ready for quantitative measurements (weighting, cell counter) or biochemical assessments (lipid extractions).

The tape stripping technique was first described over 50 years ago.[9] Readily available adhesive tape is used to strip the surface of the skin. After placement on a black background an experienced observer can grade the extent of scaliness on the tape. However, this elegant technique has never come into fashion. Later, adhesive-coated glass slides (sticky slides) were utilized to obtain a sample of the loose desquamating portion of the outer horny layer.[10,11] When appropriately stained, scales, which are really clumps of corneocytes, develop an intense color. Much depends on the nature of the adhesive and the thickness of the coating on the slide. The sticky slides themselves were highly variable, depending on the thickness of the adhesive, its source, and the uniformity of the adhesive coating.

A further step towards objectiveness was made by Serup et al. in 1989, who described a method to measure the amount of scales on adhesive tapes by the attenuation of transmitted light, using a projector and a light meter.[12] To utilize the full capacity of desquamation sampling we developed a rigidly controlled video and image analysis technique to evaluate the amount and structure of desquamating material on adhesive tapes.

II. Methodological Principle

A. D-Squame Image Analysis

Sampling desquamating horny cells on adhesive discs — The sampling device is a 22-mm crystal-clear, adhesive-coated disc (D-Squame®, CuDerm Corporation, Dallas, Texas) with a homogeneous adhesive layer. The medical-grade adhesive safely removes superficial corneocytes and provides optimum visibility of adhering skin cells. After peeling of the protective seal, the D-Squame is pressed to the skin surface. The degree of pressure can be precisely controlled, using the device described by Serup et al.[12] However, with experience, one can obtain a reliable sample manually with firm finger pressure. The disc is then peeled from the skin with tweezers and placed on a black storage card included in the D-Squame kit.

Obtaining the video image — An image analysis system consists of four components: the image source (live through video camera), a video screen that displays the image, a video digitizing board called a frame grabber, and a computer with monitor to run the software. In our setup, the video images of the samples were taken by a high-resolution black and white CCD video camera (Dage-MTI CCD72, Michigan City, IN) connected to a stereo microscope (ZEISS, West Germany). A separate video control panel with manual gain and black level controls, guarantees constant video processing under identical conditions. The image is then captured by an image analysis program. We use the Java (Jandel Scientific, CA) in combination with the frame grabber board (Truevision Targa-M8 Frame Grabber), both installed in a Unisys personal computer. However, any high-quality image analysis system available on the market can be used to grab the image. Through a sampling process the frame grabber translates the image into 512 × 480 picture elements (pixels). Each pixel will be given a numeric value according to its intensity on a gray level scale from 0 to 255.

Illumination — A crucial factor in video image analysis is a perfectly discriminating, homogeneous, and constant illumination. The best results are achieved using a reflecting white light box illuminated from two sides by means of two fiber optic light carriers. By scattering the light through white translucent glass an enhanced contrast of the corneocyte clusters on the discs can be accomplished. Due to their loose structure, the scales disperse the light and appear as shiny objects against the black background. The degree of brightness is proportional to the thickness of the scales. Before beginning the measurements, the illumination has to be calibrated using a reference gray card or by employing an automatic light meter which controls the power supply of the lamps by a feedback mechanism.

Image analysis procedures — The program applies a mask to the image, to define a measurement area of 200 mm². No contrast enhancement is necessary. The next step is to apply a lookup table to the image. This substitutes a new set of numeric values for the default gray scale so that ranges of gray levels are represented by single values. Each pixel is assigned to one of five arbitrary thickness levels of the corneocyte clusters. Transformations can be used to calculate the number of pixels in each thickness group as a percent value, as well as the total area occupied with cells. We also determine the percentage area occupied by corneocytes. These two functions are integrated to yield the desquamation index according to the following formula:

$$D.I. = \frac{2A + \sum_{n=1}^{5} T_n * (n-1)}{6}$$

D.I. is the desquamation index, A the percent area covered by corneocytes, T_n the percentage of corneocytes in relation to thickness, and n the thickness level (1–5). The macro facility of the program is used to record the sequences of all measurement functions and transformations and to perform them automatically with one keystroke.

Assessment of leg dryness by the D-Squame® method — Leg samples from non-dry, moderately dry, and severely dry skin were captured by the video camera (Figures 1a to c) and then processed by the image analysis (Figures 2a to c). Figures 1 and 2 illustrate the findings for these three levels of xerosis. The five thickness levels are reflected by five different colors. The computer outputs on these samples are presented in Figure 2. The differences are very striking. In non-dry skin the area occupied by scales is only 22%, in contrast to 97% for severely dry skin. Likewise, the thickness levels are very different, resulting in desquamation indices which for non-dry, moderately dry, and severely dry skin are 7.6, 28.3, and 70.0, respectively.

B. Correlation with Other Methods

The assessment of dry skin (xerosis), whether for purposes of classification, diagnosis, or therapeutic evaluations, has been notoriously handicapped by the traditional grading systems based on visual and tactile scoring. Such systems are highly subjective and suffer from unacceptable variability by inconsistencies from grader to grader and also from poor reproducibility. Environmental changes jeopardize such subjective grading systems since hydration swells the outer horny layer and leads to a camouflage of scaling and dryness.

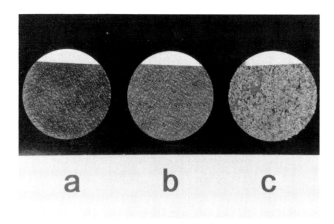

FIGURE 1 (a-c) Ordinary video camera illustrations of samples obtained from (a) nondry, (b) moderately dry, and (c) severely dry skin from the leg.

D-Squames® are certainly a technical advance in these respects, standardizing the collection of scales from the outermost portion of the horny layer, actually the "presumptive desquamating layer", where the loosened horny cells are ready to be shed.

To demonstrate the usefulness of the D-Squame® disc, we developed procedures based on computerized image analysis which enable quantification of the degree of scaling, i.e., the level of dryness. We measured both the percentage of area covered by corneocytes, and the distribution of scales according to five levels of thickness. From these one can calculate the desquamation index, which expresses the degree of xerosis in one integrated value.

C. Objectiveness and Reproducibility

With the computer analysis of adhesive tapes, we can add an objective method without the possible diversions in human judgment. According to their subjective nature, visual grading can only produce scores in the form of *ordinal values* with the restriction that the differences between the values are not consistent and therefore not meaningful. The data cannot be analyzed by parametric statistical tests and the use of mean scores is not permitted.

The image analysis provides *ratio measurements* with consistent differences between the values. The data can be analyzed with parametric statistical tests like t-tests or ANOVA and can easily be summarized in mean values. The desquamation index quantitatively and qualitatively characterizes the amount and pattern of desquamation with a high reproducibility.

III. Recommendations

We have developed a procedure that enhances the appearance of scaling by delipidization of the skin

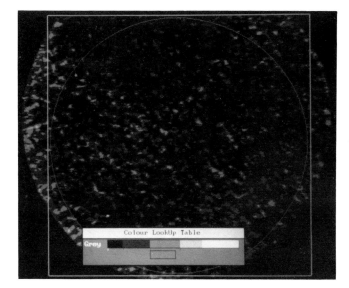
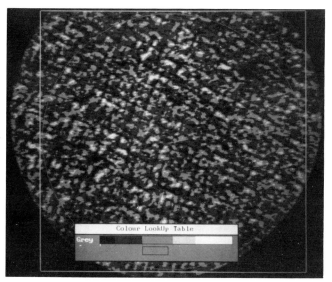
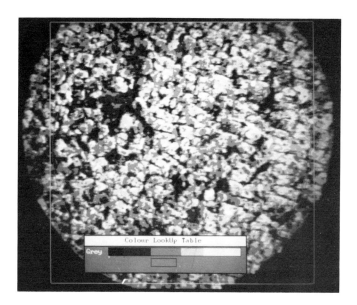

FIGURE 2 (a-c) Same samples as Figure 1, however, after image analysis with five scale thickness levels shown in five different colors (here presented in black and white). Note that the contrast and grading have been improved in comparison with Figure 1.

surface. Equal parts of ether:acetone are applied in a small glass cup, 2 cm in diameter, for 1 min. Even when scaling is not clinically apparent, the site becomes bright white, in proportion to the level of occult scaling. This enhancement is particularly notable on the face where sebum tends to mask scaling. We have demonstrated that the desquamation index increases when D-Squames® are applied to the delipidized surface. Stereophotographs of delipidized sites are very useful for clinical grading.

Finally, one can add another dimension to the D-Square® analysis by removing the scales from the tape, and after sonication to obtain a unicellular dispersion, determining the number of corneocytes per mm^2. We have found a good correlation between the desquamation index and the corneocyte count.

References

1. Boisits, E.K., Nole, G.E., and Cheney, M.C., The refined regression method, *J. Cut. Aging Cosmet. Dermatol.*, 1, 155, 1989.
2. Kligman, A.M., Regression method for assessing the efficacy of moisturizers, *Cosmet. Toilet.*, 93, 27, 1978.
3. Kligman, A.M., Lavker, R.M., Grove, G.L., and Stoudemeyer, T., Some aspects of dry skin and its treatment, in *Safety and Efficacy of Topical Drugs and Cosmetics,* Kligman, A.M., Leyden, J., Marks, R., Black, D., Hamami, I., Count, A., and Marshall, R.J., *Br. J. Dermatol.,* 111, 265, 1984.

4. Marks, R., Quantification of desquamation and stratum corneum cohesivity, in *Cutaneous Investigation in Health and Disease,* Leveque, J.-L., Ed., Marcel Dekker, New York, 1989, 33.
5. Jansen, L.H., Hoiyo-Tomoko, M.T., and Kligman, A.M., Improved fluorescent staining technique for estimating turnover of the human stratum corneum, *Br. J. Dermatol.,* 90, 9, 1974.
6. Finlay, A.Y., Marshall, R.J., and Marks, R., *Br. J. Dermatol.,* 107, 35, 1982.
7. Nicholls, S. and Marks, R., *Br. J. Dermatol.,* 96, 595, 1977.
8. McGinley, K.J., Marples, R.R., and Plewig, G., *J. Invest. Dermatol.,* 53, 107, 1969.
9. Wolf, J., Die innere Struktur der Zellen des Stratum Desquamans der menschlichen Epidermis, *Z. Mikrosk, Anat. Forsch.,* 46, 170, 1936.
10. Goldschmidt, H. and Kligman, A.M., Exfoliative cytology of human horny layer, *Arch. Dermatol.,* 96, 572, 1967.
11. Grove, G.L., Exfoliative cytological procedures as a nonintrusive method for dermatogerontological studies, *J. Invest. Dermatol.,* 73, 67, 1979.
12. Serup, J., Winther, A., and Blichmann, C., A simple method for the study of scale pattern and effect of a moisturizer — qualitative and quantitative evaluation by D-Squame tape compared with parameters of epidermal hydration, *Clin. Exp. Dermatol.,* 14, 277, 1989.

8.0

Epidermal Hydration

8.1 Measurement of Electrical Conductance and Impedance 159
    H. Tagami

8.2 Measurement of Epidermal Capacitance 165
    A.O. Barel and P. Clarys

Chapter 8.1
Measurement of Electrical Conductance and Impedance

Hachiro Tagami
Department of Dermatology
Tohoku University School of Medicine
Sendai, Japan

I. Introduction

The primary function of the epidermis is to produce the stratum corneum (SC) that protects our body from desiccation and invasion of various kinds of external attacks. It consists of about 15 to 20 tightly stacked layers of corneocytes, flattened dead bodies of epidermal cells. Although the SC is a thin membrane only about 20 µm thick, it is an efficient barrier to water and other substances. Hence, even in a dry environment, only a small amount of water is lost from the body. Surprisingly, just beneath it, hydrated living epidermal tissue functions to sustain our existence.

The SC plays another important role for human skin. It always keeps the skin surface soft and smooth. It allows free body movement without the skin surface becoming cracked or fissured. About 40 years ago, Blank[1] made interesting observations *in vitro*. He found that the isolated fragments of plantar SC became hard and brittle when dehydrated. Attempts to soften them with petrolatum or olive oil, which are clinically used for the treatment of rough scaly skin, completely failed. Only after absorption of water did they become soft and flexible and we can regard water as the ultimate moisturizer which improves subjective perception of the mechanical properties of human skin. By contrast, there is always a water supply from the underlying hydrated living tissue *in vivo* even in an atmosphere with extremely low relative humidity (RH) from which it is difficult for the SC to take up water. Despite the lack of water occlusive agents, such as petrolatum, exert a softening and smoothing clinical effect on the skin surface by preventing water loss. Thus, maintaining an appropriate water content in the SC is an important clinical and cosmetic concern.

II. Objective

It is easy to measure the water content of the SC *in vitro* by simple gravimetry.[1] In contrast, it is difficult to measure the absolute amount of water contained in the SC *in vivo* because of the presence of a concentration gradient of water within the SC.[2,3] Until lately we have lacked adequate methodologies to assess the state of skin surface hydration *in vivo*. Recent urgent demands for a practical technique to objectively evaluate the efficacy of different moisturizers have prompted development of various modalities of techniques that measure the skin properties that are influenced by the water content of the SC, e.g., electrical, mechanical, thermal, and spectroscopic properties. These techniques measure water content at poorly defined locations within skin and, with few exceptions, they have provided only qualitative information on changes in water content.

Most of all we should know the fact that there exists a concentration gradient of water within the SC *in vivo*, highest in the lowermost layer and lowest in the uppermost portion. The SC is the rate-limiting barrier between the water-saturated viable tissue and the dry outer environment and diffusion of water takes place as a purely passive process through the SC.[2,3] The superficial portion of the SC remains supple and flexible, as long as its water-holding capacity is intact. Small, water-soluble metabolites and proteinaceous structural components constitute the main components that bind water in the SC.[4] Ceramides, the main intercellular lipid component of SC, which is a key factor in the SC barrier function,[5] also play a crucial role in the water-holding capacity of the SC by preventing easy water passage through it.[6] By contrast, SC deficient in the water-holding capacity, such as that found in pathologic skin conditions, becomes brittle enough to break on flexing or stretching, which results in fissures and scaling, even under the normal ambient conditions of temperature and humidity.[7] The efficacy of various topical agents or cosmetics depends to some extent on their effectiveness in increasing the skin surface hydration state, which helps to recover softness and smoothness in dry skin. Thus, techniques are needed urgently to measure the hydration state of the exposed portion of the SC, the skin surface hydration state.

III. Skin Impedance

Among the diversity of techniques to evaluate the hydration state of the skin surface, the most widely used ones are those involving the measurement of skin impedance. Impedance *(Z)*, the total electrical opposition to the flow of an alternating current, depends on two components, resistance *(R)* and capacitance *(C)*, and their relationship may be formulated as: $Z = [R^2 + (1/2\pi fC)^2]^{1/2}$, where f stands for a frequency of an applied alternating current. In the past, for reasons of technical simplicity, many researchers studied the impedance of human skin. Tregear[8] speculated that when the skin surface is not deliberately hydrated, the reciprocal of specific impedance should be a measure of the hydration of its surface position. However, most measurements of the skin impedance have employed damp contact using electrode paste between the electrodes and the skin because of the high impedance of human skin, which is chiefly due to the properties of the SC. Certainly such an approach has a great influence on the water content of the SC.

We found that by employing dry electrodes, we can evaluate the hydration state of the skin surface quickly and quantitatively in a noninvasive way using high-frequency electric current.[9] Leveque and de Rigal[10] also reported good sensitivity to the water content of the skin surface of a similar instrument developed by them.

IV. Methodological Principle of the High-Frequency Method

The instrument being described operates at 3.5 MHz. It depends on the high-frequency apparatus developed by Masuda et al.,[11] which enables us to measure conductance $(=1/R)$ and capacitance separately. It consists of a main recording body and a long flexible cable, at the end of which a probe is attached. The skin probe consists of two concentric electrodes of 1 and 6 mm external diameter, respectively, separated by a dielectric. As soon as the probe is placed on the skin, both conductance and capacitance show a rapid initial increase for a few seconds followed by a gradual increase if the contact is maintained. The level of the initial increase represents the hydration state of the skin surface at the time of application of the probe and the later slow increase is due to accumulation of water beneath the probe resulting from insensible percutaneous water loss. Thus, we use the initial value for the evaluation of the hydration of the skin surface.

The Skin Surface Hygrometer (Skicon-200®; IBS Ltd., Hamamatsu, Japan) is a commercially available model of this apparatus. It automatically records measured values 3 s after application of a probe 6 mm in diameter (Figure 1). The reason for the relatively small size of the probe is that investigators can minimize the occlusive effect of the electrode placed on the skin. We can replace the conventional probe with a more sensitive one that contains a larger central electrode 2 mm in diameter; the sensitivity is three times greater than that of the conventional probe.

In general, skin conductance and capacitance show a very similar behavior (r = 0.95; $P < .01$). The only exception is the palmoplantar skin surface, where the values of capacitance are disproportionately low compared with conductance.[9] Thus, we can assess the hydration state of the skin surface with the measurement of conductance alone, in terms of micro-mho or micro-siemens.

V. Assessment of Accuracy

As mentioned above, there is a concentration gradient of water within the SC. The isolated piece of uniformly hydrated SC, which has been used in all experiments on the hydration state of the SC so far,[1] cannot be used as an *in vivo* model of SC. We have devised a simple simulation model of an *in vivo* SC, in which a concentration gradient of water exists between the surface and the lowermost portion.[11] It consists of an isolated sheet of SC that tightly occludes the underlying water-saturated filter paper placed as in a diffusion chamber. The filter paper is mounted on a glass slide and all free edges of the SC sheet are sealed to the glass with a removable frame of adhesive vinyl tape. The surface of the SC is exposed to the ambient atmosphere and passage of water is allowed only through this portion of the SC.

The underlying water-saturated filter paper, like fluid-saturated cutaneous tissues *in vivo*, is the water source of overlying SC and is also the conducting medium that allows the formation of an adequate electric field. Conductance values recorded with only one sheet of underlying filter paper were quite low. With an increase in the number of sheets of paper, the conductance value increased until five sheets of paper were in place. At this point the readings reached a plateau. The total thickness of five sheets of filter paper saturated with water was approximately 5 mm. Thus, to obtain an optimal reading, the high-frequency current of 3.5 MHz should extend at least 5 mm into the wet and electrically conductive substances. This condition is always attained *in vivo*.

We performed gravimetric determination of water content in the SC, together with the high-frequency conductometry. As a result, we confirmed that the recorded conductance values correlated well with the actual water content of the SC (r = 0.94). Moreover, by using a model consisting of five overlapped SC sheets

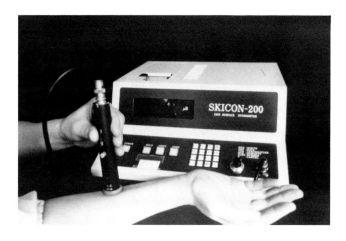

FIGURE 1 Skin surface hygrometer and a probe incorporated with a graduated spring mechanism.

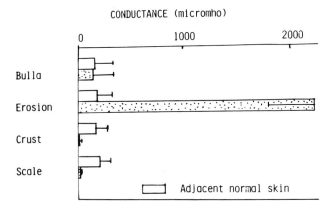

FIGURE 2 High-frequency conductance values measured at various skin lesions (dotted column) in comparison with those obtained at adjacent normal skin. Data are expressed as a mean and standard deviation (bar).

instead of one, we corroborated that there is a close correlation between the high-frequency conductance values and the water content in the uppermost SC sheet (r = 0.98).

VI. Sources of Error

In our instrument the high frequency of 3.5 MHz flows between the concentric electrodes via skin tissue. Thus, close contact between the probe and the skin surface is an important factor to obtain reproducible results. We have also found that even a small additional manual pressure on the probe greatly increases the observed values. To obtain reproducible results, the probe of the commercially available Skin Surface Hygrometer is now incorporated with a graduated spring mechanism to ensure the same pressure is applied to the skin each time (Figure 1). Even with it, however, sufficient fit with the irregularly contoured skin surface is difficult, particularly when the skin surface is dry and firm, and it should be kept in mind that the recorded values tend to indicate a lower hydration state than the actual one when the measurements are performed on lesional skin covered by scales (Figure 2). To obtain better contact even with a rough scaly skin surface, replacement of the probe with an interdigitated electrode has been suggested.[21]

Because the SC is exposed to the atmosphere, its surface hydration state is greatly influenced by the ambient relative humidity. Even simply breathing upon the skin induces an instantaneous increase in conductance. Thus, covered areas such as the trunk show much higher values than the exposed areas in the dry winter season due to the effect of thick, airtight clothes, when the measurement is performed just after removal of the clothes. At least 5 min should be allowed for the skin surface to adapt to the atmosphere.

Measured conductance values sometimes vary greatly even between sites only slightly apart from each other, particularly when the probe is applied to sites rich in sweat glands, such as the palmoplantar surface and the face. Generally the highest values are found on the forehead and palms and the lowest values on the abdominal wall and limbs on the body surface. We should avoid the facial and plamoplantar skin for comparative study, because these areas are always under the influence of mental sweating.

Our comparison made between summer and winter in the same subjects at 21 to 26°C (room temperature) showed that the high environmental relative humidity of the summer and possibly invisible sweating induced higher conductance values in the summer time.[13]

An exogenous supply of water on a skin surface results in a remarkable elevation in conductance value. However, this increase is not influenced whether we apply distilled water or a highly concentrated buffer solution; that is, it is not affected by the presence of other electrolytes in the applied water due to the fact that the skin surface is already rich in various electroconductive substances. (To totally remove such electroconductive substances from a fragment of SC, it was necessary to soak it in distilled water for at least 14 d.)

VII. Measurements in Normal Skin under Various Conditions

Removal of the SC from normal skin surface by stripping with adhesive cellophane tape demonstrates clearly that the principal hydration detected by this method is that in the outermost portion of the SC. The conductance value progressively increased as deeper layers of the SC were serially exposed by cellophane tape stripping, eventually reaching a certain high level, which presumably represents the water content of the fully hydrated viable epidermis.[9] It took about 14 d for the elevated conductance to return to the levels of the adjacent normal skin; it was the time point when the stripped skin finally resumed an almost nonscaly appearance.[7]

Removal of skin surface lipids with ethyl ether induces a marked decrease in conductance corresponding to the length of ether application. Extraction for 3 min was required to reach a nadir.[13] Recovery to the untreated levels was noted after 30 min on skin rich in sebaceous glands, such as the facial skin of adults, in contrast to other areas where it took much longer time. Thus, skin surface lipids plays a role in maintaining the skin surface hydration state. Children show low skin conductance levels among various age groups, probably reflecting the low sebum secretion rate.[14] Much lower levels are found in neonates.[15] In addition to the fact that they cannot sweat properly even in a warm environment, their SC, which has been under continuous exposure to the amniotic fluid, shows defective water-holding capacity.

Xerotic skin in elderly people shows decreased conductance in relation to clinical severity.[16] There was a significant correlation between the reduced amino acid content and skin conductance level. However, such age-associated alterations in the SC function seem to occur relatively uninfluenced by chronic sun exposure that causes skin photoaging on exposed skin areas.[17]

There is no marked sexual difference in conductance values when compared in the same age groups.

VIII. Measurements in Lesional Skin

Accumulation of tissue fluid beneath normal SC does not affect the readings made on the skin surface as long as the covering SC is intact. There is hardly any difference between bullous lesions covered by intact SC and adjacent normal skin (Figure 3).[9] In contrast the changes in the SC greatly affect the results. Scaly lesions noted in various dermatoses always show lower conductance values than those recorded in the adjacent normal skin (Figure 4). Even in the lesions of pityriasis alba, rough and hypopigmented macules on the face and trunk of children where it is difficult to visualize clearly the presence of very fine scales on the skin surface, we can find low conductance levels.[17] This is also the case with atopic xerosis, the dry, clinically noninflamed skin of patients with atopic dermatitis.[18] Such skin shows deficient water barrier function measured as transepidermal water loss (TEWL). By performing simultaneous measurements of TEWL and skin conductance in patients who had scaly lesions of various grades of severity, such as eczematous dermatitis or psoriasis, we obtained data indicating that there is an inverse relationship between these parameter, i.e., between the water barrier function of the SC and skin surface hydration state.[7]

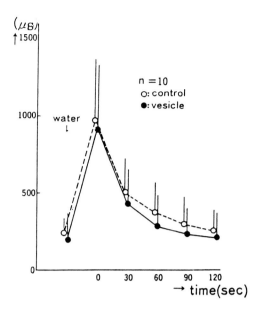

FIGURE 3 Water sorption-desorption test performed on bullous lesions covered by intact stratum corneum. No significant differences are noted before and after application of a water droplet for 10 s on the skin surface and at measurements conducted thereafter every 30 s up to 120 s. Data are expressed as a mean (circle) and standard deviation (bar).

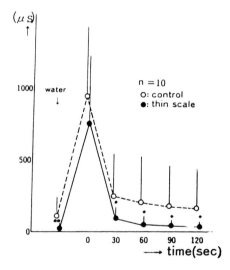

FIGURE 4 Water sorption-desorption test performed on skin covered by thin scale. Significantly lower conductance values are observed before and after application of a water droplet for 10 s and at measurements conducted every 30 s thereafter up to 120 s. Data are expressed as a mean (circle) and standard deviation (bar).*$P < .05$, **$P < .01$.

Similar to the removal of the SC by tape stripping to expose viable epidermis, even a small scratch wound in the test area results in a marked increase in conductance. Measurements in erosive lesions always yield a prominently high reading. Even in such lesions conductance becomes almost zero after the formation of dry crust.[9]

IX. Water Sorption-Desorption Test of the Skin Surface *in vivo*

The principal hydration detected by this method is in the outermost portion of the SC which is always under the influence of the relative humidity of the atmosphere. The superficial portion of the SC in normal skin takes up (= sorb) water quickly due to its hygroscopicity, but it releases water rather slowly, opposing a dehydration process (= desorption), due to its inherent water-holding capacity. In contrast, dry scaly skin surface can absorb only a small amount of water because of the deficient hygroscopicity of its SC and releases it quickly to the environment due to deficient water-binding capacity. Hence we have devised a rapid functional assay of the SC, the *in vivo* water sorption-desorption skin test, which furnishes information about the hygroscopicity and water-holding capacity of the surface SC in a short time.[19]

The test procedure is simple, consisting of serial electromeasurements before and after artificial hydration of the skin surface. Because the whole procedure takes only 2 min, we can easily repeat the test several times in nearby areas to confirm reproducibility as well as to obtain mean values. The hygroscopic property of the superficial portion of the SC is evaluated by the increase in conductance to attain a maximal value immediately after blotting a water droplet placed at the site for 10 s. The ability of the superficial SC to retain the absorbed water, i.e., its water-holding capacity, is measured from analysis of the subsequent desorption measurements of conductance at 30-s intervals for 2 min (Figures 3 and 4). The desorption curve is approximated to an exponential curve, from which the desorption rate constant for water can be calculated.[7] More accurately the entire area beneath the desorption curve is measured. Computer analysis is applicable for these procedures.

X. Assessment of the Efficacy of Skin Moisturizers

Immediately after application of moisturizing agents, the skin surface shows an increase in conductance value depending on the water content of the agents. This is followed by a rapid decrease due to evaporation of excess water from the skin surface. Thereafter, the conductance values are maintained at a certain increased level according to the efficacy of the agents, for several hours if undisturbed. In contrast, no initial increase is observed after application of emollients such as petrolatum that do not contain water. However, there is a gradual increase in conductance until a plateau is reached after 2 h due to the accumulation of water beneath it. To obtain reproducible results we apply 20 µl of the agent in a 4 × 4 cm skin area.

XI. Correlation with Other Methods

The Corneometer CM420; (Khzaka, Cologne, Germany), a commercially available electrical instrument that measures skin capacitance, is considered able to depict changes of hydration much deeper into the skin than the high-frequency method.[20] Thus, we have compared measurements with these instruments on various sites of involved and uninvolved skin of patients with psoriasis. Conductance revealed a wider range of distribution on uninvolved skin, whereas capacitance tended to show a wider range of distribution than high-frequency conductance when measured on dry skin, such as the involved psoriatic skin. Although there was a positive correlation between the values obtained (r = 0.60; $P < .001$), the X-Y intercept was clearly different from 0.[22] As mentioned above, in a simulation model of *in vivo* SC, the high-frequency conductance showed a close correlation with the hydration state of the surface SC (r = 0.99) compared with capacitance (r = 0.79). Both devices were insensitive to changes of hydration taking place in deeper viable skin tissue, e.g., the accumulated tissue fluids in suction blisters.

Recently we have had a chance to compare the results of the high-frequency conductometry with other parameters of the skin surface obtained with other techniques.[23] When the skin surface is observed under 200 times magnification with a special video camera we can observe the presence of small fragments of detaching corneocytes even on the normal skin surface. The number of such desquamating corneocytes is higher on dry skin surface. Image analysis for quantification of these detaching clumps of corneocytes showed an inverse correlation between them and high-frequency conduction levels. Firmness and elasticity of the skin surface is clinically evaluated by simply touching it with fingers. They can be measured with a new type of tactile sensor designed to detect a shift in resonant frequency generated in the piexoelectric element. We have also found that the data obtained with this instrument showed a close correlation with the high-frequency conductance.

XII. Recommendations

As mentioned above, care should be taken to apply the probe properly, perpendicularly to the skin surface. It is also recommendable to perform measurements at least three times in sites close together to get a mean value, rather than to record after only one, because even unnoticed poor contact between the probe and the skin surface causes a large decrease in obtained value. Furthermore, such a small series of readings can be completed in a short time.

The skin surface quickly absorbs water. Even invisible sweating or high relative humidity in the atmosphere can cause a great increase in conductance. Therefore, the skin of the forehead and palmoplantar surface is not suitable, because mental sweating occurs so easily on these sites, making it impossible to obtain reproducible results. A midportion of the flexor surfaces of the forearms is most suited for comparative studies.[9]

A marked increase in conductance also occurs when humidity rises above 60%. Rather constant readings are obtained between room temperature of 19 and 22°C but a steady increase in conductance is observed above 22°C.[12] Ambient conditions with temperature below 20°C and relative humidity below 40% are recommended for ordinary measurements.

References

1. Blank, I.H., Factors which influence the water content of the stratum corneum, *J. Invest. Dermatol.,* 18, 433, 1952.
2. Blank, I.H., Moleney, J., Emslie, A., Simon, I., and Apt, C., The diffusion of water across the stratum corneum as a function of its water content, *J. Invest. Dermatol.,* 82, 188, 1984.
3. Warner, R.R., Myers, M.C., and Taylor, D.A., Electron probe analysis of human skin: determination of the water concentration profile, *J. Invest. Dermatol.,* 90, 218, 1988.
4. Middleton, J.D., The mechanism of water binding in stratum corneum, *Br. J. Dermatol.,* 80, 437, 1968.
5. Elias, P.M., Lipids and the epidermal permeability barrier, *Arch. Dermatol. Res.,* 270, 95, 1981.
6. Imokawa, G., Akasaki, S., Hattori, M., and Yoshizuki, N., Selective recovery of deranged water-holding properties by stratum corneum lipids, *J. Invest. Dermatol.,* 87, 785, 1986.
7. Tagami, H. and Yoshikuni, K., Interrelationship between water barrier and reservoir functions of pathologic stratum corneum, *Arch. Dermatol.,* 181, 642, 1985.
8. Tregear, R.T., The interpretation of skin impedance measurements, *Nature,* 205, 600, 1965.
9. Tagami, H., Ohi, M., Iwatsuki, K., Kanamaru, Y., Yamada, M., and Ichijo, B., Evaluation of the skin surface hydration in vivo by electrical measurement, *J. Invest. Dermatol.,* 75, 500, 1980.
10. Leveque, J.L. and de Rigal, J., Impedance methods for studying skin moisturisation, *J. Soc. Cosmet. Chem.,* 34, 419, 1983.
11. Masuda, Y., Nishikawa, M., and Ichijo, B., New methods of measuring capacitance and resistance of very high loss materials at high frequencies, *IEEE Trans. Instrum. Meas.,* 29, 28, 1980.
12. Obata, M. and Tagami, H., Electrical determination of water content and concentration profile in a simulation model of in vivo stratum corneum, *J. Invest. Dermatol.,* 92, 854, 1989.
13. Tagami, H., Impedance measurement for evaluation of the hydration state of the skin surface, in *Cutaneous Investigation in Health and Disease. Noninvasive Methods and Instrumentation,* Leveque, J.-L., Ed., Marcel Dekker, New York, 1989, chap. 5.
14. Tagami, H., Aging and the hydration state of the skin, in *Cutaneous Aging,* Kligman, A.M. and Takase, Y., Eds., University of Tokyo Press, 1988, 99.
15. Saijo, S. and Tagami, H., Dry skin of newborn infants: functional analysis of the stratum corneum, *Pediatric. Dermatol.,* 8, 2, 1991.
16. Horii, I., Nakayama, Y., Obata, M., and Tagami, H., Stratum corneum hydration and amino acid content in xerotic skin, *Br. J. Dermatol.,* 121, 587, 1989.
17. Urano-Suehisa, S. and Tagami, H., Functional and morphological analysis of the horny layer of pityriasis alba, *Acta Dermatol. Venereol. (Stockholm),* 65, 164, 1985.
18. Watanabe, M., Tagami, H., Horii, I., Takahashi, M., and Kligman, A.M., Functional analyses of the superficial stratum corneum in atopic xerosis, *Arch. Dermatol.,* 127, 1689, 1991.
19. Tagami, H., Kanamaru, Y., Inoue, K., Suehisa, S., Inoue, F., Iwatsuki, K., Yoshikuni, K., and Yamada, M., Water sorption-desorption test of the skin in vivo for functional assessment of the stratum corneum, *J. Invest. Dermatol.,* 78, 77, 1982.
20. Blichman, C.W. and Serup, J., Assessment of skin moisture, *Acta Dermatol. Venereol. (Stockholm),* 68, 284, 1988.
21. Grove, G.L., personal communication.
22. Hashimoto-Kumasaka, K. et al., unpublished data.
23. Tanita, Y., unpublished data.

Chapter 8.2
Measurement of Epidermal Capacitance

A.O. Barel and P. Clarys
Laboratory of General and Biological Chemistry
Higher Institute of Physical Education and Physiotherapy
Vrije Universiteit Brussel
Brussels, Belgium

I. Introduction

The presence of an adequate amount of water in the stratum corneum is important for the following properties of the skin: general appearance of a soft, smooth, well-moisturized skin, in contrast to a rough and dry skin; of a flexible skin, in contrast to a brittle and scaly skin; and of an intact barrier function allowing a slow rate of transepidermal water loss (TEWL) under dry conditions.[1,2] As a consequence, the *in vivo* determination of the degree of hydration of the horny layer is an important factor in the characterization of normal and pathological situations of this layer, of an actinic aged skin, of irritated skin conditions, and, finally, in the assessment of the efficiency of various moisturizing topical products. As pointed out by Tagami,[2] the use of various dermatocosmetic products in order to restore softness, smoothness, and moisture in very dry skin is widely practiced in western countries.

It has been known for a long time that the electrical properties of the skin are related to the water content of the horny layer.[1,3] Therefore, the measurement of the impedance of the skin (the total electric resistance of the skin to an alternating current) has been studied extensively and is the most widely used technique to assess the hydration state of the skin surface.[1,2]

II. Object of this Study

Various experimental instruments have been developed in order to measure the water content of the horny layer (see the review of Lévêque and de Rigal[1]). Only a few commercially available instruments are used in dermatocosmetic research. There are two instruments based on the conductance method: Skicon® (version 100 and version 200) and Nova DPM 9003®, and one instrument based on the measurement of the capacitance: Corneometer CM 820 PC®*.

This chapter is concerned with the description of the measurements of the epidermal capacitance using the capacitance method (Corneometer®). Different aspects of the use of the capacitance apparatus, such as accuracy, reproducibility, range of measurements, and influence of external environmental factors on the measurements will be described. It will be shown that the capacitance apparatus is a reliable and very easy to use instrument which has been widely used for hydration measurements in healthy and pathological skin conditions and for quantitative assessment of the efficiency of various moisturizing products. As has been described earlier,[4] reliable and reproducible hydration measurements are obtained only if the experiments with the capacitance method are carried out under well-controlled standardized experimental conditions.

Previously extensive reviews on the electrical impedance methods for studying hydration of the skin have been published by Lévêque and de Rigal[1] and by Tagami.[2] Comparative studies about the measuring capabilities of the conductance method and the capacitance method have been published by Blichman and Serup[5] and Barel et al.[6]

III. Measuring System

A. Principle

The total impedance of the skin Z, when the skin is submitted to an applied alternating current of frequency F depends on the contribution of the resistance R and the capacitance C, according to the following relation:[2]

$$Z = (R^2 + 1/2\pi FC^2)^{1/2}$$

---

* Skicon® is a registered trademark of I.B.S. Co. Ltd., Shizuoka-ken, Japan. Nova® is a registered trademark of Nova Technology Corporation, Gloucester. Corneometer® is a registered trademark of Courage-Khazaka Electronic GmbH, Cologne, Germany.

In agreement with Mosely et al.,[7] by using an adequate design of the measuring electrodes and the oscillating electronic circuit, the apparatus measures primarily the capacitance contribution of the skin in contact with the measuring electrodes.[8,9]

B. Measuring Electrode

The measuring probe consists of an interdigital grid of gold-covered electrodes. The active part of the electrode covers a surface of 7 × 7 mm (Figure 1). The electrodes are 50 μm wide with an interdigital spacing of 75 μm. The active part of the electrode is covered by a low dielectric vitrified material of 20 μm thickness. The total probe (surface area 0.95 cm²) is applied with a constant pressure of 1,6 N/m² with a spring system. Since there is no direct galvanic contact between the gold-plated electrodes and the skin surface, no galvanic current occurs through the skin.[9] Only an electric field of variable frequency (40 to 75 kHz) is established in the upper parts of the skin (horny layer and epidermis).

C. Instrument

The form and the depth of the electric field present in the skin depends on the geometry of the electrodes and the dielectric material covering the electrodes (constant capacitance, Co) and of the capacitance of the biomaterial in contact with the electrode (variable capacitance, C).[1,6,9] As a consequence the whole system (electrode, horny layer, and upper parts of the epidermis) works as a variable capacitor. The total capacitance is only influenced by changes in the dielectric constant of the biomaterial in contact with the electrodes. A dry horny layer is a dielectrical medium. When the horny layer is hydrated a significant change in the dielectric properties of this medium is observed.[1] As a consequence the total capacity of this system is changed as a function of the degree of hydration of the skin (mostly in the horny layer[1]).

A resonating system in the instruments measures the shift in frequency (40 to 75 kHz) of the oscillating system which results from the changes in the total capacity.[9] The capacity of the skin surface is automatically measured after 1.5 s of application of the probe on the skin.

The variable total capacitance of the skin surface is converted in arbitrary units (a.u.) of skin hydration. The theoretical range of the apparatus varies from 0 to 150 a.u. As will be shown later in this chapter, calibration of the instrument with various synthetic materials of known thickness and water content is not possible. Practically the instrument gives values of hydration from 30 to 60 for very dry skin, 60 to 70 for dry skin, 70 to 90 for hydrated skin, and above 90 for very moist skin. The apparatus (Corneometer CM 820® PC version)

FIGURE 1 Schematic representation of the measuring probe of the capacitance apparatus (Corneometer CM 820® PC). Front and lateral views of the interdigital electrodes. The electrodes are covered with a vitrified material and are consequently not in direct contact with the skin surface.

can be connected through an RS-232 interface to an IBM-compatible personal computer using standard software developed by the company. Various information such as name of the test person, age, date, time, skin area to be tested, external temperature, and relative humidity can be stored in the computer. The instrument measures automatically the mean value of a maximum of seven measurements on the maximum ten skin areas selected. All the results are displayed on the screen or printed out.

D. Measurements
1. Accuracy and Sensitivity

Direct calibration of the capacitance instrument with various synthetic materials (cellulose, cellulose derivatives, plastics, etc.) of various thickness (from 10 to 200 μm) and of known water content is not possible. However, comparison of the hydration measurements obtained by the capacitance method with data obtained by the impedance method (Skicon-100)® which can be directly calibrated with a reference of known conductance, showed a very high degree of correlation (0.94) between both instruments over a broad range of hydration values.[6] The capacitance method shows a broad range of sensitivity from 20 to 110 a.u. In agreement with Blichman and Serup,[5] the capacitance method is very sensitive for measurements at low hydration and less sensitive in the range of very high hydration values (above 110 a.u.).[6]

2. Reproducibility

The reproducibility of the capacitance method was tested on different skin sites (forearm, forehead, crow's feet, abdomen, thigh, hand, leg, sole, etc.), on the same individual as well as on a large group of individuals of the same age group. In agreement with other studies[4-6] the reproducibility of the hydration measurements on the same individual varies from 4 to 5% for most skin sites under a broad range of skin conditions (very dry to moist skin). When measurements of reproducibility were carried out on a larger heterogeneous group of individuals of both sexes in the same age group, a coefficient of variability of 10% was observed for most of the skin sites. In accordance with other studies,[4,5] the results of the capacitance measurements are very reproducible.

3. Depth of the Detection of Hydration of the Skin

In order to investigate the depth to which the detection of hydration is possible by the capacitance method, experiments were performed on the forearm skin through increasing layers of tape (3M polyester tape, model 1516, 25 µm thickness). Polyester tape is a dielectric material which has almost zero value of hydration (arbitrary capacitance units). Measurements with the apparatus showed a sharp decrease in hydration a.u. after successive layers of tape were placed on the skin. Starting from a typical value of 90 a.u. of capacitance at the skin surface, at a depth of 125 µm (five layers of tape) the capacitance response of the instrument is lowered to 10 a.u.[6] In agreement with Blichman and Serup,[5] this result indicates that the capacitance method detects variations in the dielectric constant of the skin at a depth of about 100 µm.

4. Influence of External Environmental Factors

The effect of external factors such as temperature and relative humidity of the ambient air on the hydration values of the horny layer have been described before.[2,3] Figure 2 shows the results of the influence of external humidity on the hydration of the forearm skin. The hydration of the skin (volar part of the forearm) was examined under constant temperature (20 ± 2°C) as a function of an increase of relative humidity (from 37 to 87%). A linear relation (r = +0.98) was observed between the hydration as measured by the capacitance method and the external relative humidity (RH). This result is more or less in agreement with previous skin conductance measurements.[2-4,7,10] As a consequence, hydration measurements with the capacitance method must be performed in an experimental room where the relative humidity is kept more or less constant (50 ± 5% RH). When hydration measurements are carried out in variable conditions of relative humidity, the data obtained must be corrected with a correction factor taken from the slope of Figure 2. Similarly a steady increase of capacitance was observed by Rogiers et al.[4] as a function of temperature above 22°C, which corresponds to a higher hydration of the horny layer and the invisible beginning of sweating. The temperature of the experimental room must be kept constant and preferably below 22°C (ideally 20°C).

5. Seasonal Influences on Hydration

As a result of the higher external temperature and higher values of relative humidity occurring during

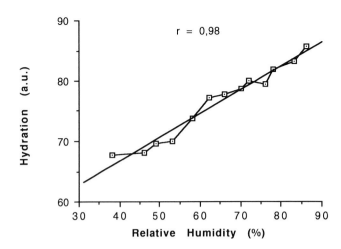

FIGURE 2 Effect of external relative humidity (%) on the capacitance hydration values of the forearm skin. Temperature = 22 ± 2°C (n = 15, age group 18 to 30 years).

summer, much higher hydration values are measured during this season on all skin sites.[2,11,12] In the winter, due to very low values of relative humidity, typical symptoms of very dry skin appear on all the exposed areas. With older individuals, capacitance hydration values as low as 30 to 40 are measured on the lower legs (winter xerosis).

Due to the interference of external high values of relative humidity and the possibility of small amounts of sweating on the skin surface observed in summer, it is very difficult to carry out reproducible and reliable hydration experiments in this period of the year. As a consequence quantitative efficiency experiments with moisturizing products are never carried out in the summer period in our laboratory.

In order to minimize the interferences of all these external factors, it is important that all the volunteers rest a minimum of 30 min in the experimental room prior to the experimental procedure.

6. Anatomical Skin Areas for Testing

The choice of the skin area selected for hydration measurements is important, since there are large variations in the hydration of the horny layer as a function of body region.[2-7] High hydration values are obtained at the forehead and the palm of the hand; lower values are observed at the abdomen, thigh, and lower leg. The hydration status of symmetrical sites of the body is generally identical, a situation that allows contralateral left-right comparative studies. In agreement with other researchers,[4] significant differences in hydration were observed when comparing skin sites located, respectively, at the distal part and the proximal part on the volar side of the forearm. It is important to point out that comparative contralateral hydration measurements must be performed exactly on identical anatomical skin sites.

7. Sex and Age

In agreement with other studies carried out with the conductance method[2,14] and the capacitance method,[4] both sexes, when compared within the same age group, show identical hydration values at all anatomical skin areas studied.

Different authors have shown that in adults (beginning at 18 to 25 years) there is a slow but steady decrease in hydration of the stratum corneum at all skin sites as a function of age.[14] One observes in elderly volunteers of both sexes a typical situation of xerosis on the lower legs (arbitrary units of skin capacitance around 30 to 40).

8. Influence of Stripping

Removal of the horny layer by stripping with an adhesive tape was examined by the skin capacitance method. The hydration values of the skin located at the forearm are progressively increased as deeper layers of the horny layer are serially removed by successive tape strippings.[2,6] Finally after 20 to 25 strippings a high value of 110 to 120 a.u. is obtained which corresponds to the fully hydrated horny layer located near the viable part of the epidermis.

E. Dermatocosmetic Applications

Besides the fundamental interest in studying *in vivo* the hydration properties of the superficial layers of the skin surface located at various anatomical sites in healthy and diseased skin situations, the determination of skin hydration is also important for assessing the therapeutic efficacy of topical skin products in various skin lesions and for the objective evaluation of the efficiency of various moisturizing preparations in cosmetics.

1. Skin Diseases and Lesions

Dry scaly lesions — Hydration measurements carried out on skin with dry scaly lesions (psoriasis, eczematous dermatitis) always reveal lower values of skin hydration (conductance measurements).[2,7] Similar results were obtained by Werner[15] and Berardesca et al.,[16] using the capacitance method on atopic dermatitis and psoriatic skin situations.

Irritation of the skin — When the skin is exposed to various chemical irritants, such as surfactants, one observes a complex phenomenon of skin irritation. In addition to typical irritation symptoms (swelling and redness), the barrier function of the stratum corneum is partially destroyed and the water content of this layer is significantly lower.[17] An objective assessment of the irritant character of some household detergent solutions can be carried out by following TEWL, skin color, and the hydration of the horny layer of the forearm and the dorsal part of the hands as a function of consecutive exposures to these products in the hand/forearm immersion test.[18] A significant decrease in hydration of the horny layer (forearm and hand) was observed after two and four consecutive exposures during 30 min at 40°C to two different household cleaning products (see Figure 3). When hydration measurements are carried out before and as a function of time after two consecutive exposures to the same cleaning products, the hydration of the skin returns progressively to normal values depending on the mildness of the surfactant (see Figure 4). When capacitance measurements are carried out on the skin in the immersion test under standardized experimental conditions, it is possible to discriminate between a mild surfactant (product 1) and a more irritant detergent (product 2), (see Figures 3 and 4 for comparison).[18-20]

2. Evaluation of the Efficiency of Cosmetic Moisturizing Products

Short-term effects — The short-term efficiency of moisturizing products can be readily analyzed from the increase of hydration after application of the product on the skin. Most experiments are carried out either on the volar side of the forearm or on the frontal side of the lower legs. Both anatomical sites allow contralateral, at-random comparison (untreated control skin areas vs. treated skin areas). The various hydrating products (O/W and W/O emulsions, hydro- and lipogels, occlusive creams, etc.) are gently applied by rubbing at a concentration of 1 to 2 mg of product per cm² skin area on the test area (4 × 4 cm). Recordings of the hydration of the control and the test areas are taken during a certain time (10 to 30 min) before application of the product on the test area in order to ascertain the hydration state of the skin (baseline value). Following application, recordings are performed on the treated and untreated skin areas every 10 to 15 min during a period of time lasting from 60 to 180 min, depending on the efficacy of the hydration product. Figure 5 shows the typical short-term effect of an O/W emulsion containing 2% urea as moisturizing factor on the hydration of the forearm skin. As has been previously described,[2,4] immediately after application of the O/W emulsion the hydration (capacitance arbitrary units) shows a significant increase which corresponds to the application of water present in the O/W emulsion on the skin surface. As a consequence of the evaporation of the excess of water present on the skin surface, a decrease in hydration of the horny layer is observed and after a certain time a constant increase in hydration is observed during a certain time. The level and the duration of the increased hydration of the horny layer is a measure of the efficacy of the moisturizing

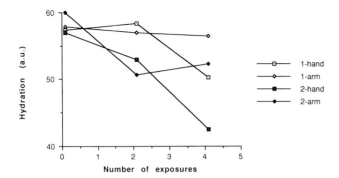

FIGURE 3 Effect of exposure of the skin (dorsal part of the hands and volar part of the forearm) to detergents (30 min, each exposure at 40°C). Hydration measurements (arbitrary capacitance units) were performed initially and 24 h after the second and the fourth exposure to detergent (n = 12, age group 18 to 25 years). Solution 1 (1% solution of a mild dishwashing liquid containing alkylethoxy sulfate and amphoteric surfactants) (open symbols) and solution 2 (1% solution of a more irritant dishwashing liquid containing linear alkylbenzenesulfonate and sodium lauryl sulfate surfactants (filled symbols).

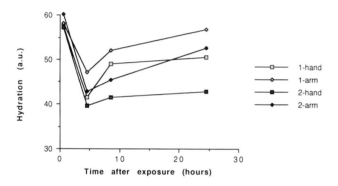

FIGURE 4 Recovery effect of the skin (dorsal part of the hands and volar part of the forearm) after two consecutive exposures to detergents (30 min, each exposure at 40°C). Hydration measurements (arbitrary capacitance units) were performed initially and as a function of time after two exposures. Same symbols as in Figure 3.

preparation. Figure 6 shows the effect of a single application of a W/O emulsion containing 4% urea as humectant, on the hydration of the forearm skin. One observes again a significant increase in hydration of the skin immediately after application of the product, followed by a decrease. A significant increase in hydration (plateau value) is maintained over a long time, reflecting the temporary occlusion effect on the skin surface of the W/O emulsion.

Long-term effects — Long-term hydration studies can also be quantitatively assessed by skin capacitance measurements. In a long-term study the moisturizing products are generally tested in a group (n = 12 to 30) of middle-aged to older women who present symptoms of dry to very dry skin (skin capacitance values of around 50 to 60 on the forearms and on the

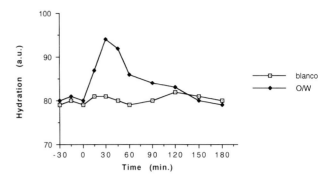

FIGURE 5 Changes in hydration (arbitrary capacitance units) of the forearm skin after a single application of an O/W emulsion containing 2% urea as humectant. Treated area (O/W, filled symbols) in comparison with untreated area (blanco, open symbols), (n = 15, age group 18 to 30 years).

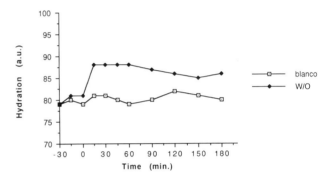

FIGURE 6 Changes in hydration (arbitrary capacitance units) of the forearm skin after a single application of a W/O emulsion containing 4% urea as humectant. Treated area (W/O, filled symbols) in comparison with untreated area (blanco, open symbols), (n = 15, age group 18 to 30 years).

lower legs). Testing on volunteers can be carried out in the laboratory or in an at-home study.[11] The skin areas (volar part of the forearms or frontal part of the lower legs) are treated two or three times daily with topical moisturizing products for 2 to 3 weeks. Contralateral skin sites are either untreated or treated with a placebo. Skin capacitance measurements are carried out before and, successively, after 1, 2, and 3 weeks of treatment with the products. Good correlations were obtained in our laboratory between clinical visual dryness scores, participant perception of skin dryness, and skin capacitance values during the progress of alleviating the dryness of the skin.[11,21]

IV. Conclusion

The capacitance apparatus is a simple-to-use, convenient, low-cost instrument which measures *in vivo*, noninvasively, the hydration of the skin down to a depth of about 100 µm (superficial layers of the epidermis).

The capacitance method, when used under well-controlled standard experimental conditions, is accurate and reproducible enough for skin hydration measurements in normal and pathological skin situations. Since the capacitance hydration values show a linear response as a function of external relative humidity, the capacitance data can be corrected as a function of the external relative humidity. The apparatus is suitable for quantitative evaluation of short- and long-term hydration effects on the skin after treatment with various topical dermatocosmetic moisturizing preparations. As a consequence, this instrument is widely used in dermatological, pharmacological, and skin care research.

Acknowledgments

The authors wish to thank Mr. W. Courage and Mr. G. Khazaka for their valuable technical support in this study.

The authors would also like to acknowledge the support of Dr. B. Gabard (Spirig, Switzerland) and wish to thank Prof. A. R. Barel (VUB, Brussels, Belgium) for valuable theoretical advice in this work.

References

1. Lévêque, J.L. and de Rigal, J., Impedance methods for studying skin moisturization, *J. Soc. Cosmet. Chem.*, 34, 419, 1983.
2. Tagami, H., Impedance measurements for evaluation of the hydration state of the skin surface, in *Cutaneous Investigation in Health and Disease, Noninvasive Methods and Instrumentation*, Lévêque, J.L., Ed., Marcel Dekker, New York, 1989, chap. 5.
3. Clar, E.J., Her, C.P., and Sturell, C.G., Skin impedance and moisturization, *J. Soc. Cosmet. Chem.*, 26, 337, 1975.
4. Rogiers, V., Derde, M.P., Verleye, G., and Roseeuw, D., Standardized conditions needed for skin surface hydration measurements, *Cosmet. Toilet.*, 105, 73, 1990.
5. Blichman, C.W. and Serup, J., Assessment of skin moisture. Measurement of electrical conductance, capacitance and trans epidermal water loss, *Acta Dermatol. Venereol. (Stockholm)*, 68, 284, 1988.
6. Barel, A.O., Clarys, P., Wessels, B., and de Romsée, A., Noninvasive electrical measurement for evaluating the water content of the horny layer: comparison between the capacitance and the conductance measurements, in *Prediction of Percutaneous Penetration — Methods, Measurements, Modelling*, Scott, R.C., Guy, R.H., Hadgraft, J., and Boddé, H.E., Eds., IBC Technical Services, London, 1991, 238.
7. Mosely, H., English, J.S., Coghill, G.M., and Mackie, R.M., Assessment and use of a new skin hygrometer, *Bioeng. Skin*, 1, 177, 1985.
8. Courage, W. and Khazaka, G., personal communication, 1992.
9. Barel, A.R., personal communication, 1992.
10. Batt, M.D., Davis, W.B., Fairhurst, E., Gerraid, W.A., and Ridgde, B.D., Changes in the physical properties of the stratum corneum following treatment with glycerol, *J. Soc. Cosmet. Chem.*, 39, 367, 1988.
11. Prall, J.K., Theiler, R.F., Bowser, P.A., and Walsh, M., The effectiveness of cosmetic products in alleviating a range of skin dryness conditions as determined by clinical and instrumental techniques, *Int. J. Cosmet. Sci.*, 8, 159, 1986.
12. Lévêque, J.L., Grove, G., de Rigal, J., Corcuff, P., Kligman, A.M., and Saint Leger, D., Biophysical characterization of dry facial skin, *J. Soc. Cosmet. Chem.*, 82, 171, 1987.
13. Saint Léger, D., François, A.M., Lévêque, J.L., Stoudemayer, T.J., Grove, G.L., and Kligman, A.M., Age-associated changes in stratum corneum lipids and their relation to dryness, *Dermatologica*, 177, 159, 1988.
14. Lévêque, J.L., Méthodes expérimentales d'étude du vieillissement cutané chez l'homme in vivo, *Acta Derm. Venereol.*, 144, 1279, 1987.
15. Werner, Y., The water content of the stratum corneum in patients with atopic dermatitis. Measurement with the Corneometer CM 420, *Acta Dermatol. Venereol. (Stockholm)*, 66, 281, 1986.
16. Berardesca, E., Fidelli, D., Borroni, G., Rabbrosi, G., and Maibach, H., In vivo hydration and water retention capacity of stratum corneum in clinically uninvolved skin in atopic and psoriatic patients, *Acta Dermatol. Venereol. (Stockholm)*, 70, 400, 1990.
17. Imokawa, G., Akasaki, S., Minematsu, Y., and Kawai, M., Importance of intercellular lipids in water-retention properties of the stratum corneum: induction and recovery study of surfactant dry skin, *Arch. Dermatol. Res.*, 281, 45, 1989.
18. Barel, A.O., Clarys, P., Wessels, B., and van de Straat, R., Quantitative biophysical measurements of the mildness properties of cleaning and detergent products in the hand immersion test, presented at the International Symposium on Irritant Contact Dermatitis, Groningen, October 3–5, 1991.
19. Clarys, P., van de Straat, R., Boon, A., and Barel, A.O., The use of the hand/forearm immersion test for evaluating skin irritation by various detergent solutions, presented at the First Congress of the European Society of Contact Dermatitis, Brussels, October 8–10, 1992.
20. Boon, A., Evaluatie van de irritatie van detergenten op de menselijke huid door middel van niet-invasieve biofysische metingen, B. Sc. Thesis, Vrije Universiteit Brussel, Belgium, 1992.
21. Barel, A.O., personal communication, 1992.

9.0

Epidermal Barrier Functions

9.1 Measurement of the Transepidermal Water Loss 173
J. Pinnagoda and R.A. Tupker

9.2 Comparison of Methods for Measurement of Transepidermal
Water Loss 179
A.O. Barel and P. Clarys

9.3 Measurement of Transcutaneous Oxygen Tension 185
H. Takiwaki

9.4 Measurement of Transcutaneous P_{CO_2} 197
C.N. Nickelsen

9.5 Noninvasive Techniques for Assessment of Skin Penetration
and Bioavailability 201
R.C. Wester and H.I. Maibach

Chapter 9.1
Measurement of the Transepidermal Water Loss

J. Pinnagoda
Singapore

R. A. Tupker
Department of Dermatology
State University Hospital
Groningen, The Netherlands

I. Introduction

Measurement of transepidermal water loss (TEWL) is used in many research centers for studying the water barrier function of the skin. Different methods for TEWL measurement from local skin sites have been described.[1] Unventilated chamber (closed chamber) methods are not capable of continuous measurement as they tend to occlude the skin. Ventilated chamber methods, using dry or moistened carrier gas, are capable of measuring TEWL continuously. However, both methods interfere with the microclimate overlying the surface of the skin, thereby influencing the water loss to varying extents. Thus, these methods have certain inherent drawbacks. The open chamber, gradient estimation method provides continuous measurement in ambient air, with little alteration of the microclimate overlying the skin surface.[1] This method is therefore preferable and has gained wide use in the evaluation of skin barrier function, and the quantification of patch test reactions. As a consequence of the different measuring principles, results obtained with different methods cannot be directly compared with any accuracy. In this chapter, attention is focused on the open chamber, gradient estimation method of the commercially available Evaporimeter EP1.

II. Object

The total amount of water vapor passing the skin can be divided into water vapor passing the stratum corneum by passive diffusion, and water vapor loss as a result of sweating.[2] Originally, the term "transepidermal water loss" was applied to indicate the amount of water vapor passing the stratum corneum by passive diffusion.[2] However, nowadays "TEWL" refers to the total amount of water loss through the skin. Therefore, it must be kept in mind that TEWL is a reflection of stratum corneum barrier function for water only when there is no sweat gland activity.

TEWL measurement is used to assess the barrier function of the stratum corneum in order to (1) perform predictive irritancy testing; (2) evaluate clinical conditions. Since TEWL measures total water vapor loss, another application of TEWL measurement may be to assess the degree of sweating in various diseases.

A. Predictive Irritancy Testing

The sensitivity of TEWL as a screening technique for early signs of irritancy was found to be superior as compared with visual scoring[3,4] as well as with laser Doppler flowmetry, colorometry, and skin thickness by ultrasound A-scan.[4] A wide variety of irritants, such as detergents, solvents, etc., exert their damaging influence on the skin by impairing the barrier function of the stratum corneum. In a multiple 3-week repeated exposure model we have demonstrated increasing TEWL values over time due to the cumulative irritating action of detergents and a solvent on the epidermal barrier.[3] This increase in TEWL was different for the various irritants. Thus, it was possible to rank the irritants according to their irritant potency.[3] However, for irritants that act largely on targets underneath the stratum corneum (for example, dimethyl sulfoxide and phenol), TEWL had a poor correlation with visual scoring.[5] Serup and Staberg studied the time course of TEWL after a single irritant exposure in an attempt to differentiate allergic from irritant skin reactions.[6]

B. Evaluation of Barrier Function in Clinical Conditions

A higher TEWL has been noticed on the involved skin in various types of dermatitis as compared with uninvolved sites.[7,8] Shahidullah et al. have observed an increased TEWL on the involved and uninvolved skin in dermatitis, being related to the severity of the disease.[7] Uninvolved skin of patients with manifest atopic dermatitis was demonstrated to have a higher TEWL than the same skin region of subjects without dermatitis.[9,10]

In psoriasis, TEWL is elevated on the plaques.[11] The clinical course showed a strict parallel with the TEWL values.[11] Also in patients with ichthyosis, the magnitude of increase in TEWL paralleled the severity of the scaling.[12]

TEWL measurement has been applied frequently for monitoring the epidermal repair process in (burn) wounds.[13,14] Soon after the burning, TEWL was extremely elevated, dependent on the degree of burn.[13] Evaporative water loss has been shown to be preeminently suited for the evaluation of epidermal regeneration of partial thickness wounds in guinea pigs which were covered with different artificial membranes or left uncovered.[15]

Hammarlund et al. have found a linear relationship between TEWL and ambient percent relative humidity (RH) in neonates.[16] By determining TEWL from different parts of the body and calculating the areas of the corresponding surfaces, the total cutaneous water loss could be obtained.[16]

C. Sweat Gland Activity in Clinical Conditions

Lower TEWL on the hands and forearms in patients with generalized scleroderma were observed.[17] Using TEWL measurement, a decreased sweat response after sympathomimetic stimulation was noted on the forehead in patients suffering from cluster headache.[18] In contrast, a higher TEWL was noticed in patients with Parkinson's disease than in healthy subjects, as a result of sweating.[19] In a patient with cholinergic urticaria, we have demonstrated an increase of TEWL on gustatory provocations of increasing strength.[20] These provocations were accompanied by erythematous macules and a sensation of warmth and itching of increasing severity, finally evolving into weals typical for cholinergic urticaria.[20]

III. Methodological Principle
A. The Theory

In the absence of forced convections, the human skin surface is surrounded by a water vapor boundary layer.[21] This layer, which forms a physical barrier against the environment, constitutes the transition zone for transportation of moisture and heat from the body to the ambient air. Considering the skin surface as a water-permeable surface, the process of water exchange through this zone can be expressed in terms of its vapor pressure gradient.[22] This gradient is approximately constant in the absence of forced convection and under steady-state conditions. It is, therefore, proportional to the amount of water vapor passing through this zone per unit time and area by the evaporation from the skin surface, i.e., TEWL (g/m^2h). The vapor pressure gradient is computed from the difference between the vapor pressures measured at two different fixed heights situated perpendicularly above the skin surface and within the zone of diffusion.[1] This estimation is valid only within this boundary layer,[1,23] its depth depending on the site, air speed, and convections, forced and/or free.[23] In the absence of convection currents or draughts, a mean depth of about 10 mm may be assumed for this boundary layer.[1,23]

The Evaporimeter EP1® (ServoMed, Stockholm, Sweden), consists of a detachable measuring probe connected by a cable to a portable main signal processing unit (see Figure 1). The sensor arrangement within the probe head is represented diagrammatically in Figure 2. The Teflon capsule of the probe head has a cylindrical measuring chamber, open at both ends, with a diameter of 12 mm and height of 15 mm. Within this chamber, relative humidity sensors (hygrosensors) are paired with temperature sensors (thermistors), at two fixed heights of 3 and 9 mm above the skin surface, i.e., within the boundary layer of diffusion. The distance between the centers of the two pairs of sensors is 6 mm.[1]

B. The Method

The measuring head of the probe is placed on the selected skin surface and a small area (1 cm^2) of skin is limited for measuring the TEWL. The actual vapor pressure at each fixed height of measurement above the skin is calculated from the formula $p = RH \cdot p_{sat}$, where p is the water vapor pressure (Pa), RH is the relative humidity (%), and p_{sat} is the saturation vapor pressure (Pa).

The RH is measured with a capacitive sensor based on an organic polymer with dielectric sensitivity to changes in RH, at each fixed height. The saturated vapor pressure, which is a function of the temperature alone, is calculated from the measured temperature value obtained with a fast thermistor, at each fixed height.[1] These sensors are mounted on the measuring probe head, as mentioned previously (see Figure 2). The calculated difference in the vapor pressure at the two fixed heights of measurement is the estimated vapor pressure gradient of the boundary layer of diffusion. From this gradient, the evaporative TEWL value, in g/m^2h, is calculated by the signal processing units in the probe handle and main unit, and digitally displayed. The electronic controls and digital display unit offer various measuring possibilities and ranges. Further details of the technical descriptions and accuracies of the instrument are given elsewhere.[1]

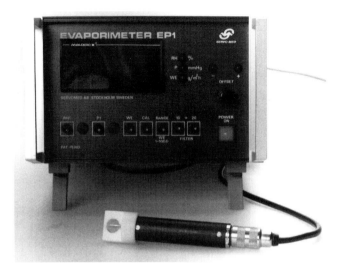

FIGURE 1 The Servo-Med EP1® Evaporimeter.

IV. Sources of Error

A. Validity of Method and Associated Variables

The measurement of TEWL with the Evaporimeter is valid only within the boundary layer of diffusion surrounding the human body.[1] The depth of this boundary layer is therefore crucial and depends on the environmental conditions (see Section III).

Thus, it is apparent that any environment- or instrument-related variables that influence the depth of this boundary layer would affect the gradient and, therefore, the measured TEWL value. Furthermore, due to the extreme sensitivity of the instrument, any variations in the microclimate, whether due to the instrument, environment, or individual, are immediately detected and instantly displayed as a fluctuation, in-

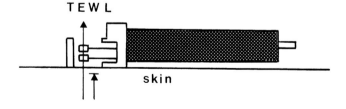

FIGURE 2 The probe of the Servo-Med EP1® Evaporimeter with an open measuring chamber and two sensors mounted 3 and 9 mm over the skin surface.

dicating an error in the measured TEWL level.[24] A detailed account of these influencing variables associated with the method of measuring TEWL using the Evaporimeter is given in the *Guidelines of the European Contact Dermatitis Society*.[25]

In this chapter, guidelines as to what should be considered "good laboratory practice" are given for the measurement of TEWL. To perform accurate and precise TEWL measurements, sources of variation due to the instrument, the environment, and the individual need to be known and taken into account.[25]

B. Instrument-Related Variables

The commonly occurring and important instrument-related variables are discussed in this section.

Start-up and use — The Evaporimeter should be turned "on" at least 15 min before measurements are performed, and, if the instrument is being used intermittently during the day, it should not be switched "off" between measurements.[26]

Zeroing — After the warm-up period, the instrument should be "zeroed" only if necessary.[25] Regularly calibrated and well-maintained instruments will not require this "zeroing" daily, if the "offset" knob is not used for zeroing in between measurements (see paragraph below on zero drift).

Measuring — The time the probe is applied to the skin should be as short as possible. Stabilization of the TEWL value is usually reached by 30 to 45 s after the start of measuring.[27,28] The skin area measured is approximately 1 cm^2, and the smallest amount measured corresponds to 0.00000006 g/cm^{2}s (2.16 g/m^{2}h).[29] Such precision also means that disturbances in the microclimate are immediately detected as a fluctuation in TEWL.[29] Therefore, further fluctuations after stabilization may occur, and as a rule it is easier to read the TEWL value on a pen recorder, because the average value is more readily comprehended from an analogue than from a digital reading. The built-in damping filters may also be used to smooth these fluctuations in TEWL.[26] Two filters giving time constants of 10 and 20 s are available, and thus the instrument can be operated with variable time constants, 0, 10, 20, and 30 s (10 s + 20 s). When using the filters, press filter button 10 after the 45-s stabilization period, wait about 5 s and then press button 20.[29] The TEWL value registered (recorded or displayed) during the 30-s period after stabilization is to be considered the measured value.[28] With the filters "on", the investigator obtains no information about fluctuations and artifacts. Therefore, direct recording of the TEWL values on a pen recorder is recommended.[25] The prescribed stabilization period of 30 to 45 s[27,28] is for baseline TEWL measurements only. If, however, measurements are made on excessively diseased or damaged skin sites, where high water evaporation rates are expected, or at high ambient relative humidities, a longer stabilization period may be necessary. This prescribed period is not a hard-and-fast rule, as it may also vary from instrument to instrument.[28]

Zero drift — Displacement of the water evaporation (WE) zero level, in between measurements, is attributed to the abrupt humidity changes as well as to the temperature changes of the probe, as a result of the measurement itself (see below).

Humidity changes — After a TEWL measurement has been made, condensation vapor will remain within the funnel of the probe for some time, causing a moisture gradient to persist.[26] Therefore, the instrument will for some time continue to indicate a "nonzero" WE value, which will of course disappear when the water vapor in the probe has evaporated.[26] To accelerate the evaporation of this moisture, the probe may be carefully waved, vertically up and down. A well-maintained instrument will return to "zero" within about 2 to 4 min post-measurement.[28] As this phenomenon is entirely a matter of a delayed response, and not a real "zero" displacement,[26] avoid using the "offset" button in between measurements for zeroing the instrument. Allow it to "zero" on its own, before the next measurement is made.

Temperature changes — The temperature-dependent variability of the sensors of the probe, and the amplifiers in the probe handle, is of importance when measuring TEWL.[26,29] During a measurement, the temperature of the probe increases as a result of heating from the measured skin surface, and also due to the operator's hand. The warmth from the hand influences the amplifiers in the handle of the probe, and thereby also the TEWL measurement.[26,29,30] A displacement in the WE "zero" level, of the order of ± 1 to 2 g/m^{2}h, may result due to a 5-min long measurement, in which the probe is hand-held.[26] Therefore, avoid holding the probe directly by hand, particularly in repeated measurements. The probe is best handled with an insulating glove, the holed rubber stopper of the Evaporimeter's calibration set, or with a lightweight (laboratory) burette clamp with rubber-covered ends.[25] Full control of the position of the probe and its contact with the skin surface should still be possible when using the above-mentioned accessories.

The surface plane — A standing person, warmer or cooler than the surrounding ambient air, will act as a chimney and cause an increased convection of air close to the surface.[29] To avoid this "chimney effect", place the measuring surface in a horizontal plane, and apply the probe parallel to this surface.[22,29] Moreover, with a horizontal plane, it is easier to control the probe position and pressure against the skin, thereby avoiding probe movement during the measurement.

Contact pressure — Variations in the contact pressure between the probe and the surface of the skin may cause alterations in the TEWL, due to changes in the distance between the skin and the sensors, and due to changes in the water permeability of the skin.[1] If the pressure is too light, gaps arise between the probe head and the skin surface. A constant light pressure should be applied when holding the probe against the skin.

Use of the probe protection covers — Although it is recommended that these protection covers, supplied with the Evaporimeter, be applied whenever possible during measurements,[26] it is important to recognize the variables that they introduce.[30] The use of the protection cover with the screen and the grid (no. 2107) elevates the probe, and therefore the sensors, above the water vapor boundary layer surrounding the skin (see Section III), due to the added height (6 to 7 mm) of the stainless steel screen. This will influence absolute TEWL measurement,[1,23,26,30] and has no importance only if the Evaporimeter is used for relative measurements.[26] However, measurements made with and without the screen (relative or absolute) cannot be directly compared.[26,30] With the screen, the TEWL values will be somewhat lower than without, the difference becoming greater as the TEWL rate increases.[30,31] Use of these protection covers should thus be stated clearly in reports and publications. For measuring absolute TEWL levels, the uncovered probe or the protection cover without the screen and grid (no. 2108) is thus recommended. If it is foreseeable that contaminations (ointments, oils, sweat, etc.) from the measuring skin surface may arise, or if a sterile probe surface is required, the protection cover without the stainless steel screen and grid (no. 2108), which can be sterilized, should be used.[30] The elevation of the probe above the measuring surface due to this attachment is only about 1 mm, and is negligible.[30] However, in hairy body regions, hair follicles may come into contact with the sensors. Dust and evaporated substances, including solvents, may also damage the sensors. In such conditions, the protection covers with the screen and grid (no. 2107) should be used.[25]

Intra- and interinstrumental variability — A high reproducibility of results was found for individual Evaporimeters, i.e., a low intra-instrumental variability.[28] However, large differences were found between the Evaporimeters, i.e., a high interinstrumental variability.[28] Interinstrumental variability is dependent on the age of the Evaporimeter.[28] The newer instruments respond faster, i.e., stabilization time is shorter. The older versions appear to measure much lower TEWL values.[28] This may be attributed to the sensors of the probe, which are reported to undergo a slight aging.[26] To keep this in check, the calibration of the instrument should be checked from time to time (see paragraph below on calibration).[26] To enable successful and reliable interlaboratory comparison of results, overcoming the effect of the interinstrumental variability, an additional calibration procedure incorporating a calibration for an actual *in vitro* measured water loss can be adopted.[25]

Calibration — As a general rule, calibration of the Evaporimeter, according to the manufacturer's specifications, should be performed at regular intervals.[26] For details of calibration and maintenance, refer to the operation handbook.[26] The Evaporimeter may also be checked against the standard constant water evaporation device periodically (see above),[25] to check the aging of the sensors.

Accuracy — It was reported that the Evaporimeter may underestimate the water evaporation rate.[23] At evaporation rates of about 20 g/m²h, the underestimation is only about 10%. When the evaporation rates exceed 80 g/m²h, however, the underestimation may be about 50%, even in still air.[23] It was concluded that the presence of the probe may restrict the flux of water vapor, particularly at the high evaporation rates.[23] This was also reported by Scott et al., who stated that this underestimation is at evaporation rates above 75 g/m²h.[32] Most of the changes in the barrier function that are caused by detergent damage[3] or diseased states, e.g., eczema,[7,8] psoriasis,[11] result in water evaporation rates within the range 20 to 60 g/m²h. Therefore, the consequences of a potential underestimation in water evaporation rates may be acceptable. However, more severe damage to the barrier, such as burns[13] or wounds,[15] gives rise to much higher evaporation rates (above 100 g/m²h), where the underestimation of TEWL is relevant, and should be considered in the interpretation of the results.

C. Environment-Related Variables

The effects of the most important environment-related variables are discussed in this section.

Air convections — This is the main source of disturbance resulting in rapid fluctuations of the measurements.[29] It is commonly produced by disturbances in the room, such as people moving about, opening and closing doors, breathing across the measurement zone, and air conditioners, etc. As these disturbances are difficult to avoid, some form of an enclosure (a measuring box) to serve as a draught shield is recommended.[25] A box or an incubator with an "open top" and sides preferably of Perspex (for visibility), with holes for the placement of the forearms (subject and investigator), will protect the measurement zone from rapid air movements.[25,28]

Ambient air temperature — The most important effect of the temperature of the ambient air is that it influences the skin temperature both directly (by convection) and indirectly (by central thermoregulatory effects).[2,33-35] With increasing ambient air temperature, the skin surface temperature increases, and the TEWL is almost double at the high ambient air temperature of 30°C, in comparison with that at the lower ambient air temperature of 22°C.[22] The ambient room air temperature is thus an important variable that must be controlled (see Section IV.D). A room temperature of 20 to 22°C is recommended.[25]

Ambient air humidity — Ambient relative humidity is a complex and important variable which influences TEWL measurements.[25] Ambient room relative humidity should be registered during TEWL measurements, reported in publications, and taken into consideration whenever TEWL results are compared. If climate room facilities are available, the relative humidity should be regulated to about 40%.[25]

Seasonal variation — This variable is mainly determined by variations in ambient air temperature and relative humidity. Provided that measurements are made at room temperatures between 20 and 22°C, a seasonal variation in TEWL is mainly attributable to the seasonal variation in ambient relative humidity and skin hydration state, the latter itself influenced by the former.[25] Publications should provide information about the season of the year in which the TEWL measurements were made.

Direct light — Direct light warms up the surface of the object, which in turn warms up the air close to the object, and an air convection is created.[29] TEWL measurements should not be made under direct light sources or close to windows with direct sunlight.[25]

D. Individual-Related Variables

The effects of the most important individual-related variables are discussed in this section. The effects of other common individual-related variables are given elsewhere.[25]

Sweating — Physical, thermal, or emotional sweating are important variables to control, in order to make accurate TEWL measurements. If the ambient air temperature is below 20°C, and the skin temperature is below 30°C, thermal sweat gland activity is unlikely, provided that the skin is not exposed to forced convection and no excessive body heat is produced (result of physical exercise).[7,36-38] Therefore, a premeasurement 15- to 30-min rest in a measurement room with an ambient air temperature regulated to about 20°C, possibly by an air conditioner, is best suited for accurate TEWL measurements.[38] Emotional sweating may be controlled by performing a couple of "dummy" TEWL measurements,[39] to put the test subject "at ease".

Skin surface temperature — Skin surface temperature is one of the essential factors dictating the rate of TEWL in normal skin,[22] and preconditioning of the test person is required (see above). At ambient room temperatures around 20 to 22°C, however, the normal range of skin surface temperature is between 28 and 32°C, which may be considered too little a variation to influence TEWL significantly.[22,25] Skin surface temperature should be measured and reported in publications, particularly if ambient room air temperature deviates from 20 to 22°C.[25]

References

1. Nilsson, G.E., Measurement of water exchange through skin, *Med. Biol. Eng. Comput.*, 15, 209, 1977.
2. Rothman, S., Insensible water loss, in *Physiology and Biochemistry of the Skin*, The University of Chicago Press, Chicago, 1954, 233.
3. Tupker, R.A., Pinnagoda, J., Coenraads, P.J., and Nater, J.P., The influence of repeated exposure to surfactants on the human skin as determined by transepidermal water loss and visual scoring, *Contact Derm.*, 20, 108, 1989.
4. Agner, T. and Serup, J., Sodium lauryl sulphate for irritant patch testing — a dose-response study using bioengineering methods for determination of skin irritation, *J. Invest. Dermatol.*, 95, 543, 1990.
5. Van der Valk, P.G.M., Kruis-de Vries, M.H., Nater, J.P., Bleumink, E., and De Jong, M.C.J.M., Eczematous (irritant and allergic) reactions of the skin and barrier function as determined by water vapour loss, *Clin. Exp. Dermatol.*, 10, 185, 1985.
6. Serup, J. and Staberg, B., Differentiation of allergic and irritant reactions by transepidermal water loss, *Contact Derm.*, 16, 129, 1987.
7. Shahidullah, M., Raffle, E.J., Rimmer, A.R., and Frain-Bell, W., Transepidermal water loss in patients with dermatitis, *Br. J. Dermatol.*, 81, 722, 1969.
8. Blichman, C. and Serup, J., Hydration studies on scaly hand eczemas, *Contact Derm.*, 16, 155, 1987.
9. Van der Valk, P.G.M., Nater, J.P., and Bleumink, E., Vulnerability of the skin to surfactants in different groups of eczema patients and controls as measured by water vapour loss, *Clin. Exp. Dermatol.*, 10, 98, 1985.
10. Tupker, R.A., Pinnagoda, J., Coenraads, P.J., and Nater, P.J., Susceptibility to irritants: role of barrier function, skin dryness and history of atopic dermatitis, *Br. J. Dermatol.*, 123, 199, 1990.
11. Rajka, G. and Thune, P., The relationship between the course of psoriasis and transepidermal water loss, photoelectric plethysmography and reflex photometry, *Br. J. Dermatol.*, 94, 253, 1976.
12. Frost, P., Weinstein, G.D., Bothwell, J.W., and Wildnauer, R., Ichthyosiform dermatosis. III. Studies on transepidermal water loss, *Arch. Dermatol.*, 98, 230, 1968.
13. Lamke, L.-O., Nilsson, G.E., and Reithner, H.L., The evaporative water loss from burns and the water-vapour permeability of grafts and artificial membranes used in the treatment of burns, *Burns*, 3, 159, 1977.
14. Frosch, P.J. and Czarnetzki, B.M., Effect of retinoids on wound healing in diabetic rats, *Arch. Dermatol. Res.*, 281, 424, 1989.
15. Jonkman, M.F., Molenaar, I., Nieuwenhuis, P., and Klasen, H.J., Evaporative water loss and epidermis regeneration in partial-thickness wounds dressed with by a fluid-retaining versus a clot-inducing wound covering in guinea pigs, *Scand. J. Plast. Reconstr. Surg.*, 23, 29, 1989.
16. Hammarlund, K., Nilsson, G.E., Öberg, P.Å., and Sedin, G., Transepidermal water loss in newborn infants. I. Relation to ambient humidity and site of measurement, and estimation of total transepidermal water loss, *Acta Paediatr. Scand.*, 66, 553, 1977.
17. Serup, J. and Rasmussen, I., Dry hands in scleroderma. Including studies of sweat gland function in healthy individuals, *Acta Dermatol. Venereol. (Stockholm)*, 65, 419, 1985.
18. Salvesen, R., de Souza-Carvalho, D., Sand, T., and Sjaastad, O., Cluster headache: forehead sweating pattern during heating and pilocarpine tests. Variation as a function of time, *Cephalalgia*, 8, 245, 1988.
19. Turkka, J.T. and Myllyla, V.V., Sweating dysfunction of Parkinson's disease, *Eur. Neurol.*, 26, 1, 1987.
20. Tupker, R.A. and Doeglas, H.M.G. Water vapour loss threshold and induction of cholinergic urticaria, *Dermatologica*, 181, 23, 1990.
21. Gates, D.M., in *Humidity and Moisture*, Vol. 2, Wexler, A. and Amdur, E.J., Eds., Reinhold, New York, 1965, 33.
22. Nilsson, G.E., On the Measurement of Evaporative Water Loss. Methods and Clinical Applications, Thesis, Linköping University Medical Dissertations, No. 48, Linköping, Sweden, 1977.
23. Wheldon, A.E. and Monteith, J.L., Performance of a skin Evaporimeter, *Med. Biol. Eng. Comput.*, 18, 201, 1980.
24. Pinnagoda, J., Occupational dermatitis risk estimated by transepidermal water loss measurements, Transepidermal Water Loss, Thesis, University of London, 1990, 141.
25. Pinnagoda, J., Tupker, R.A., Agner, T., and Serup, J., Guidelines for transepidermal water loss (TEWL) measurement, *Contact Derm.*, 22, 164, 1990.
26. ServoMed Evaporimeters, Operation Handbook, ServoMed, Vallingby, Stockholm, Sweden, 1981.
27. Blichman, C.W. and Serup, J., Reproducibility and variability of transepidermal water loss measurements, *Acta Dermatol. Venereol. (Stockholm)*, 67, 206, 1987.
28. Pinnagoda, J., Tupker, R.A., Coenraads, P.J., and Nater, J.P., Comparability and reproducibility of the results of water loss measurements: a study of 4 evaporimeters, *Contact Derm.*, 20, 241, 1989.
29. A guide to water evaporation rate measurement, ServoMed, Vallingby, Stockholm, Sweden.
30. Nilsson, G.E., personal communication, 1987.
31. Agner, T. and Serup, J., Transepidermal water loss and air convection, *Contact Derm.*, 22, 120, 1990.
32. Scott, R.C., Oliver, G.J.A., Dugard, P.H., and Singh, H.J., A comparison of techniques for the measurement of transepidermal water loss, *Arch. Dermatol. Res.*, 274, 57, 1982.
33. Grice, K.A., Transepidermal water loss, in *The Physiology and Pathophysiology of the Skin*, Vol. 6, Jarret, A., Ed., Academic Press, London, 1980, 2121.
34. Lamke, L.-O. and Wedin, B., Water evaporation from normal skin under different environmental conditions, *Acta Derm. Venereol.*, 51, 111, 1971.
35. Rothman, S., The role of the skin in thermoregulation: factors influencing skin surface temperature, in *Physiology and Biochemistry of the Skin*, The University of Chicago Press, Chicago, 1954, 258.
36. Baker, H. and Kligman, A.M., Measurement of transepidermal water loss by electrical hygrometry. Instrumentation and responses to physical and chemical insults, *Arch. Dermatol.*, 96, 441, 1967.
37. Blank, I.H., Factors which influence the water content of the stratum corneum, *J. Invest. Dermatol.*, 18, 433, 1952.
38. Pinnagoda, J., Tupker, R.A., Coenraads, P.J., and Nater, J.P., Transepidermal water loss: with and without sweat gland inactivation, *Contact Derm.*, 21, 16, 1989.
39. Pinnagoda, J., A pilot field study for the assessment of the practicability of the recommendations for the transepidermal water loss measurements and the skin patch testing technique, Transepidermal Water Loss, Thesis, University of London, 1990, 172.

Chapter 9.2

Comparison of Methods for Measurement of Transepidermal Water Loss

A. O. Barel and P. Clarys
Laboratory of General and Biological Chemistry
Higher Institute for Physical Education and Physiotherapy
Vrije Universiteit Brussel
Brussels, Belgium

I. Introduction

The measurement of transepidermal water loss (TEWL) is an important noninvasive method for assessing the efficiency of the skin as a protective barrier. The stratum corneum forms a barrier against diffusion of water through the epidermis and constitutes the main obstacle to the penetration of molecules coming in contact with the surface of the skin.[1-3]

As pointed out by Wilson and Maibach[2] and by Lévêque,[3] the efficiency of the horny layer depends on the integrity of the stacking of the corneocytes and the intercellular cement and on its water content. As a consequence, the measurement of TEWL provides information concerning the integrity of the epidermis in normal, irritated, and diseased skin situations, concerning the effects of chemicals on the surface of the skin and concerning the objective evaluation of occlusive pharmaceutical and cosmetic preparations.

II. Object of this Study

In the past different noninvasive methods and instruments have been developed to measure TEWL (see the review of Wilson and Maibach[2]). Until recently, the only commercially available TEWL instrument was the Evaporimeter®* (version EP1 or EP2) based on the open chamber evaporation gradient method. This widely used instrument measures the water evaporation gradient developed from the skin surface in an open chamber system. Hygrosensors coupled with thermistors measure the water evaporation at the skin surface, at two different distances from the skin surface.[2] Recently a new instrument based on the same principle of measurement of the water evaporation gradient in an open chamber, was developed and became commercially available: the Tewameter TM 210®*.

It is the purpose of this chapter to compare the two commercial instruments under identical experimental conditions. The following parameters will be comparatively analyzed and described: general technical description of the probes and the instruments; evaluation of the accuracy, reproducibility, and range of TEWL measurements; and a comparative study of some typical applications of TEWL measurements in dermatocosmetic research.

Recently an extensive study by Pinnagoda et al.[4] has been published giving valuable guidelines concerning the use of the Evaporimeter according "good laboratory practice" rules. Similarly the need for standardized TEWL measurements with the Tewameter was outlined by Rogiers.[5] A preliminary report of the comparative study of both TEWL instruments and some guidelines concerning the use of both instruments has been recently presented by Barel and Clarys.[6]

III. General Description of the TEWL Instruments

A. Description of the Measuring Probe
1. Evaporimeter®
The probe weighs 104 g and has a cylindrical open chamber measuring system, diameter 12 mm, height 15 mm, and skin area of 1.13 cm². Two sensor units are placed at 3 and 6 mm distance from the skin surface. The probe can be adapted with a chimney extension in order to reduce air turbulence and with a metallic shield. In this study the probe of the Evaporimeter was used without chimney extension and without shield.

2. Tewameter®
The probe weighs 25 g and has a cylindrical open chamber measuring system, diameter 10 mm, height

---

* Evaporimeter® is a registered trademark of ServoMed, Stockholm, Sweden. Tewameter® is a registered trademark of Courage-Khazaka Electronic GmbH, Cologne, Germany.

20 mm, and skin area of 0.79 cm². The two sensor units are placed at 3 and 8 mm distance from the skin surface. A special probe holder with clamp is furnished with the apparatus which allows a fixed positioning of the probe on the skin surface. Similarly, the probe of the Tewameter was used without chimney and without shield.

B. Description of the Instrument
1. Evaporimeter®

Two versions of this instrument are available: a single probe model and a dual probe model. The standard version of the instrument has a digital display of the following parameters: water evaporation (WE), (0 to 300 g/m² · h with two ranges of sensitivity), relative humidity (RH), (0 to 100%), and water vapor pressure (P), (0 to 50 mmHg). There is no display of mean value and standard deviation for the three parameters (WE, RH, and P). The temperature of the measuring probe as well as the temperature of the skin is not measured by this instrument. Calibration of the Evaporimeter with solutions of known vapor pressure is possible. The instrument must be manually adjusted to zero values of TEWL before each measurement. Built-in damping filters (time constants of 0, 10, 20, and 30 s) may be used during measurements in order to reduce fluctuations in TEWL data. Recently a software package, Evaporimeter Program®*, has been developed by the company to interface the instrument with an IBM-compatible XT/AT personal computer. This software allows automatic data collection, graphical display of desorption curves, calculation of evaporation rates, and display of the results on a spreadsheet program. Data about subjects, selected skin sites, date and time of measurements, and calibration can be collected and stored in the computer.

2. Tewameter®

The standard version of the instrument shows on a small LC screen the following experimental parameters: TEWL (0 to 90 g/m² · h with five ranges of sensitivity), relative humidity (0 to 100%), and temperature of the probe (0 to 50°C). The partial water pressure and the temperature of the skin are not measured by this instrument. Calibration of the Tewameter with a solution of known vapor pressure is possible. The automatic zero adjustment system of TEWL is carried out when the measuring probe is placed in an air turbulence-free cylindrical tube. There is an internal damping filter present in the instrument which cannot be manually adjusted. The instrument presents on the LC screen a graphical display of TEWL, relative humidity, and temperature of the probe as a function of time (kinetic curves). With TEWL vs. time curves it is possible to obtain during variable selected time intervals the mean value and the standard deviation of the data on the graphical display. A standard software STM201 which can be used on an IBM-compatible PC, was developed by the company for data collection.

C. Determination of the Accuracy, Reproducibility, and Sensitivity of Both Instruments
1. *In Vivo* Experiments

The comparative TEWL measurements were carried out on a variable number of volunteers ranging from 10 to 12 volunteers of both sexes, age group 18 to 30 years, with normal capacitance hydration values and normal TEWL measurements. Various anatomical skin sites were tested under normal conditions and after occlusion or stripping of the skin. Volunteers were requested not to use any moisturizers, body lotions, soaps, or occlusive cosmetic preparations on the tested skin areas 12 h before TEWL measurements were started. Unless otherwise described in the experimental conditions, all the TEWL experiments were performed in an experimental room where relative humidity and temperature were kept more or less constant (RH = 45 ± 5% and temperature = 20 ± 2°C). All the volunteers rested a minimum of 30 min prior to the experimental procedures.

2. Accuracy

The two instruments were calibrated with solutions of known vapor pressure as recommended by the respective companies. The determination of the real accuracy of both instruments is only possible when TEWL measurements are compared with the true gravimetric determinations of the amount of water that evaporates from the skin surface.[2,7,8] This experimental set-up was not available in our laboratory and consequently the determination of the "true value" of TEWL measurements for both instruments was not carried out in this study. It was reported previously that the Evaporimeter may underestimate the water evaporation rate, particularly at high TEWL values.[4,7] In agreement with previous studies[4,9] some instrumental variability (or inaccuracy) was observed for each instrument and when comparing the two instruments. As will be clearly demonstrated in the correlation study between both instruments (see correlation), the Tewameter used in this comparison measured systematically some higher TEWL values as compared with the Evaporimeter.

3. Reproducibility and Coefficient of Variability

The coefficient of variability of the two instruments as determined from repetitive measurements carried out with the same person on various anatomical skin sites (forehead, chest, abdomen, thigh, and forearm), are

---
* Evaporimeter Program® is a registered trademark of ServoMed, Stockholm, Sweden.

rather good for both apparatus. In agreement with previous data,[5] the range of coefficient of variation for both instruments varies from 3 to 8%. The interindividual variations in TEWL measurements were carried out with both instruments on various anatomical skin sites on a heterogeneous group of individuals of the same age group. In agreement with Pinnagoda et al.[4] and Rogiers,[5] interindividual variations in TEWL measurements for most skin sites were estimated to vary from 20 to 25%. But some skin sites such as the forehead and the palm of the hand show a higher coefficient of variation (from 30 to 50%) and these skin sites should be avoided when studying the barrier function of the skin with TEWL measurements.

4. Sensitivity Range and Correlation between Instruments

In order to assess the range of sensitivity of both instruments, TEWL measurements were carried out on various anatomical skin sites before and after 2 h of occlusion with surgical tape in order to obtain a broad range of TEWL values (see Table 1). A very good agreement was observed between TEWL data obtained from the Evaporimeter and the Tewameter (Figure 1). A high correlation (r = +0,97) was found between both instruments over a wide range of TEWL values.

IV. Influence of External and Environmental Factors

The following external and environmental factors, which are sources of variations, were considered: room temperature, external relative humidity, air turbulence, pressure of application of the probe on the skin, and temperature of the measuring probe.

A. Influence of External Temperature

The effect of external temperature on the TEWL of the forearm was examined with both instruments under constant relative humidity (RH = 45 ± 5%). A linear increase in TEWL as a function of temperature was observed with the two instruments (correlation factors of r = +0,88 for the Tewameter and r = +0,89 for the Evaporimeter®). In agreement with previous work,[4,10] the temperature of the ambient air influences the skin temperature, provoking an increase of TEWL. Thus, the ambient room temperature is an important factor for both instruments and must be strictly controlled (a constant room temperature of 20 to 22°C is strongly recommended).

B. Influence of the Temperature of the Probe

During the measurement, the temperature of the probe increases as a result of the heating effect from the measured skin surface.[4,5] The influence of the temperature of the probe on the TEWL was not investigated in

TABLE 1 TEWL Measurements on Different Anatomical Skin Sites

Anatomical Site	TEWL (g/m² · h)	
	Before Occlusion	After Occlusion
Tewameter		
Forehead	19.9 ± 2.6	30.7 ± 10.6
Chest	10.7 ± 1.9	19.1 ± 5.8
Abdomen	10.0 ± 1.4	22.0 ± 6.1
Thigh	9.5 ± 1.6	19.4 ± 4.5
Forearm	11.5 ± 2.7	22.7 ± 8.4
Evaporimeter		
Forehead	12.4 ± 2.9	26.4 ± 8.5
Chest	4.7 ± 3.4	14.2 ± 4.2
Abdomen	4.0 ± 2.5	12.3 ± 5.3
Thigh	4.5 ± 2.5	12.4 ± 4.7
Forearm	5.2 ± 1.9	16.9 ± 8.7

Note: Measurements made before and after 2 h of occlusion with surgical tape. Mean values and SD as measured with Evaporimeter and Tewameter (n = 12).

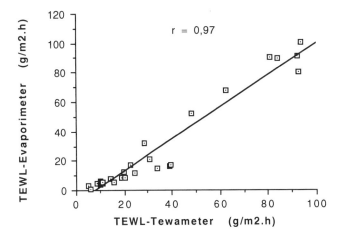

FIGURE 1 Correlation between TEWL measurements with Tewameter® and Evaporimeter® at different skin sites and under various experimental conditions (normal, stripping, and occlusion), n = 12.

the past, since the widely used TEWL instrument (Evaporimeter®) has no display of the temperature of the probe. The Tewameter® has the advantage that it shows the temperature of the measuring probe as a function of time. As has been shown by Rogiers[5] and confirmed in our laboratory,[6] the TEWL values measured with the Tewameter increase in a continuous way until the temperature of the probe has reached the skin temperature. Thermal equilibrium of the probe is generally reached after 10 to 15 min when the temperature of the probe is equal to skin temperature (around 30°C).

In agreement with the data of Rogiers,[5] Figure 2 shows the influence of the temperature of the probe on TEWL measurements on the forearm. Similar experiments were carried out with the Evaporimeter. Again

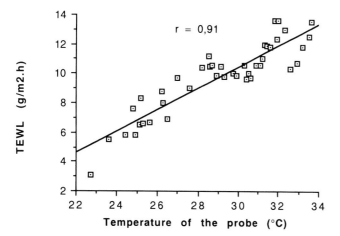

FIGURE 2 Influence of temperature of the probe on the TEWL of the forearm as measured by the Tewameter® (n = 10).

a continuous increase of TEWL was observed until after 10 to 15 min a constant level of TEWL was reached indicating thermal equilibrium with skin temperature. In order to have accurate TEWL data, the temperature of the probe must be equal to the skin temperature before carrying out TEWL measurements. This can be achieved along two different ways: (1) after application of the probe on the skin surface, follow the temperature of the skin on the LC display until thermal equilibrium is attained, or (2) place the probe on the skin surface on a place different from the test area and allow a warm-up period of 10 to 15 min before starting the real TEWL measurements.

C. Influence of External Relative Humidity

As pointed out by Pinnagoda et al.[4] and Lévêque,[3] changes in relative humidity modify the concentration of water in the horny layer and consequently the coefficient of diffusion of water through this layer, and also from Fick's law, modify the passive diffusion through a membrane. As a consequence there is more or less a general agreement that TEWL decreases as a function of an increase in relative humidity. We observed with both instruments a decrease in TEWL of the forearm when the relative external humidity was increased above 40%. These measurements with the Evaporimeter® and the Tewameter® are in agreement with previous work of Bettley and Grice,[11] Goodman and Wolf,[12] and Grice et al.,[13] but disagree with the findings of Petro and Komor[7] who found no dependence of TEWL on ambient relative humidity for the forearm. As a consequence ambient relative humidity is an important variable in TEWL measurements with both instruments. Either the ambient relative humidity must be reported in publications and taken into consideration when comparing TEWL results or better, use a constant relative humidity, preferably around 45 ± 5% by working in a climate room.

D. Influence of Pressure of Application of the Probe

Table 2 shows the influence of the pressure of the probe on the skin surface on TEWL of the forearm. TEWL measurements were carried out with both instruments with low pressure (100 g) and relative high pressure (300 g). In agreement with the previous work of Nilson,[8] Pinnagoda et al.,[4] and Rogiers,[5] there is a significant effect of the pressure of the probe on the skin. To obtain accurate TEWL data it is important to maintain with a constant light pressure (preferably around 100 g) the probe of both instruments on the skin surface during measurements.

E. Influence of External Air Turbulence

Air convection will influence the registered values by the sensors and consequently the TEWL measurements. It is known from previous studies with the Evaporimeter[4] that air turbulence produced in the room by people moving, by breathing near the opening of the probe, by opening and closing doors and by fans of air conditioners, is a main source of fluctuations. Consequently the use of a covering box was recommended in order to shield the probe from external fluctuations.[4] Similar experiments were carried out with both instruments under normal use and in a shielded box with open top, taking into account variable external air fluctuations such as breathing, opening and closing doors, a hair dryer, and a fan (Table 3). In agreement with previous work,[5] the Tewameter®, which has a smaller area of skin measurement site, is less sensitive to air turbulence. Breathing and moving doors have almost no influence on the TEWL measurements.

V. Applications of TEWL Measurements

As has been said in the introduction, TEWL is an excellent parameter for assessing the barrier function of the stratum corneum. Here follows a brief presentation of some typical areas of application of TEWL measurements which were carried out with both instruments.

TABLE 2 Effect of the Contact Pressure of the Probe on the Skin Surface on the TEWL of the Forearm as Measured with Both Instruments

	TEWL (g/m² · h)	
Pressure	Tewameter®	Evaporimeter®
Light (100 g weight)	11.5 ± 2.7	5.3 ± 1.9
High (300 g weight)	12.5 ± 2.2	5.6 ± 2.2

Note: Values are mean TEWL and SD (n = 10).

TABLE 3 Effect of External Air Turbulence on TEWL

	Shielded Box		Normal Use	
	Tewameter	Evaporimeter	Tewameter	Evaporimeter
Normal	11.4 ± 2.6	5.6 ± 1.5	11.5 ± 2.5	5.4 ± 1.9
Breathing	11.5 ± 2.5	5.7 ± 1.8	11.7 ± 2.6	5.6 ± 2.0
Moving doors	11.1 ± 0.9	5.4 ± 1.9	11.6 ± 2.7	5.2 ± 2.2
Hair dryer	10.6 ± 1.8	5.9 ± 1.4	8.8 ± 1.8	2.8 ± 2.0
Fan	9.7 ± 1.9	4.3 ± 2.1	5.1 ± 1.7	2.0 ± 2.1

Note: TEWL measured on the forearm in normal use and in a shielded box, measured with both instruments. Mean TEWL and SD (n = 10).

TABLE 4 TEWL After Exposure to a Detergent Solution

	TEWL (g/m² · h)	
Measurements	Tewameter®	Evaporimeter®
Hand		
Before exposure	18.6 ± 3.4	8.5 ± 1.4
After 2 exposures	26.3 ± 11.3	13.0 ± 5.1
After 4 exposures	34.1 ± 11.9	14.5 ± 6.3
Forearm		
Before exposure	12.8 ± 5.6	4.9 ± 1.8
After 2 exposures	21.7 ± 10.9	9.4 ± 5.2
After 4 exposures	28.8 ± 10.8	15.3 ± 8.2

Note: Measurements of TEWL (Evaporimeter® and Tewameter®) after successive exposures of the skin (volar part of the forearm and dorsal part of the hand) to a detergent solution (exposure time 30 min at 40°C). TEWL measurements were taken initially and 24 h after the second and the fourth exposure to a 1% solution of a conventional irritating dishwashing product containing sodium lauryl sulfate and alkyl benzene sulfonate surfactants. Mean TEWL and SD (n = 12).

A. Stripping

As previously reported by Lévêque,[3] the TEWL is inversely proportional to the thickness of the stratum corneum. To illustrate the barrier function of the stratum corneum, stripping experiments were carried out with adhesive tape (Scotch Magic Tape®). TEWL reaches very high values (above 100 g/m² · h) as measured with the Evaporimeter and Tewameter. As already found by many other researchers,[14-17] this significant increase in TEWL reflects the temporary destruction of the barrier function of the stratum corneum.

B. Occlusion

Occlusion of the skin at various anatomical sites (forehead, chest, abdomen, thigh, and forearm) with surgical tape for 2 h provokes a significant increase of TEWL (see Table 1). Similar values of increase in TEWL were observed with both instruments immediately after occlusion. After removal of the occlusion the TEWL returns rapidly (after 30 min) to normal values (see Figure 3).

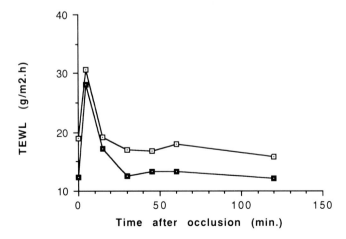

FIGURE 3 Changes in TEWL of the forehead (Tewameter® open symbols and Evaporimeter® filled symbols) as a function of time after 2 h of occlusion with surgical tape (n = 12).

C. Study of the Irritant Character of Detergents by TEWL in a Hand-Forearm Immersion Test

The forearm wash tests are more reliable and relevant than the chamber tests to differentiate the mildness-irritancy of cleaning products.[18] The immersion test is close to actual washing conditions and to prediction of consumers' skin conditions.[18] Of all the noninvasive biophysical techniques for characterization of irritation of detergent solutions, hydration and TEWL are objective and sensitive enough to discriminate between the irritation effects of mild and more irritant products.[19,20]

Table 4 shows the significant increase of TEWL as observed with the Evaporimeter® and the Tewameter® on the forearms and hands after two and four, respectively, consecutive short-term exposures to a typical irritant dishwashing solution.

VI. Conclusion

Under well-controlled standardized experimental conditions, the Evaporimeter® and the Tewameter® are adequate noninvasive commercial TEWL instruments for *in vivo* measurements of the barrier integrity function of the skin. Considering criteria such as accuracy, sensitivity, and reproducibility, both TEWL instruments are valid experimental tools in dermatocosmetic research.

In order to perform accurate and reproducible measurements, several external environmental factors must be controlled and taken into account with both instruments.

Considering the fact that the Tewameter® is a more recent apparatus which presents in the standard version possibilities of measuring the temperature of the probe and graphical display of TEWL in function of time, it appears that this apparatus is more complete and somewhat more convenient to use.

Acknowledgments

The authors wish to thank Mrs. I. Vanbeneden for her excellent technical assistance. The authors are grateful to Dr. B. Gabard (Spirig, Switzerland) and Dr. R. van de Straat (Procter & Gamble, The Netherlands) for technical support with the Evaporimeter®, and to Mr. W. Courage and Mr. G. Khazaka (Courage-Khazaka, Germany) for technical support with the Tewameter®.

References

1. Dupuis, D., Rougier, A., Lotte, C., Wilson, D.R., and Maibach, H., In vivo relationship between percutaneous absorption and transepidermal water loss according to anatomical site in man, *J. Soc. Cosmet. Chem.,* 37, 351, 1986.
2. Wilson, D.R. and Maibach, H., Transepidermal water loss: a review, in *Cutaneous Investigation in Health and Disease. Noninvasive Methods and Instrumentation,* Lévêque, J.L., Ed., Marcel Dekker, New York, 1989, chap. 6.
3. Lévêque, J.L., Measurement of transepidermal water loss, in *Cutaneous Investigation in Health and Disease. Noninvasive Methods and Instrumentation,* Lévêque, J.L., Ed., Marcel Dekker, New York, 1989, chap. 7.
4. Pinnagoda, J., Tupker, R.A., Agner, T., and Serup, J., Guidelines for TEWL measurement, *Contact Derm.,* 22, 164, 1990.
5. Rogiers, V., Capacitance and TEWL measurements: the need for standardisation, presented at Intensive Course in Dermato Cosmetics, Brussels, September 1992.
6. Barel, A.O. and Clarys, P., Study of the stratum corneum barrier function by trans epidermal water loss measurements. Comparison between two commercial instruments: Evaporimeter and Tewameter, presented at the 9th International Symposium on Bioengineering and the Skin, Sendai, October 19–20, 1992.
7. Petro, A.J. and Komor, J.A., Correction to absolute values of evaporation rates measured by the ServoMed® Evaporimeter, *Bioeng. Skin,* 3, 271, 1987.
8. Nilson, G.E., Measurement of water exchange through skin, *Med. Biol. Eng. Comput.,* 15, 209, 1977.
9. Pinnagoda, J., Tupker, R.A., Coenraads, P.J., and Nater, J.P., Comparability and reproducibility of the results of water loss measurements: a study of 4 evaporimeters, *Contact Derm.,* 20, 241, 1989.
10. Mathias, C.G., Wilson, D.M., and Maibach, H., Transepidermal water loss as a function of skin surface temperature, *J. Invest. Dermatol.,* 77, 219, 1981.
11. Bettley, F.R. and Grice, K.A., The influence of ambient humidity on transepidermal water loss, *Br. J. Dermatol.,* 78, 575, 1967.
12. Goodman, A.B. and Wolf, A.V., Insensible water loss from human skin as a function of ambient vapor concentration, *J. Appl. Physiol.,* 26, 203, 1969.
13. Grice, K., Salter, H., and Baker, H., The effect of ambient humidity on transepidermal water loss, *J. Invest. Dermatol.,* 58, 343, 1972.
14. Blank, I.H., Further observations on factors which influence the water content of the stratum corneum, *J. Invest. Dermatol.,* 21, 259, 1953.
15. Monash, S. and Blank, H., Location and reformation of the epithelial barrier to water vapor, *Arch. Dermatol.,* 78, 710, 1958.
16. Kermici, M., Dodereau, C., and Aubin, G., Measurement of biochemical parameters in the stratum corneum, *J. Soc. Cosmet. Chem.,* 28, 151, 1977.
17. Van Der Vak, P.G. and Maibach, H., A functional study of the skin barrier to evaporative water loss by means of repeated cellophane-tape stripping, *Clin. Exp. Dermatol.,* 15, 180, 1990.
18. Barel, A.O., Clarys, P., Wessels, B., and van de Straat, R., Quantitative biophysical measurements of the mildness properties of cleaning and detergent products in the hand immersion test, presented at the International Symposium on Irritant Contact Dermatitis, Groningen, October 3–5, 1991.
19. Clarys, P., van de Straat, R., Boon, A., and Barel, A.O., The use of the hand/forearm immersion test for evaluating skin irritation by various detergent solutions, presented at the First Congress of the European Society of Contact Dermatitis, Brussels, October 8–10, 1992.
20. Boon, A., Evaluatie van de irritatie van detergenten op de menselijke huid door middel van niet-invasieve biofysische metingen, B. Sc. thesis, Vrije Universiteit Brussel, 1992.

Chapter 9.3
Measurement of Transcutaneous Oxygen Tension

Hirotsugu Takiwaki
Department of Dermatology
University of Tokushima
Tokushima, Japan

I. Introduction

So-called cutaneous respiration, which means the absorption of oxygen and the elimination of carbon dioxide through the skin surface, was discovered over a century ago and was vigorously investigated in the early 1930s by Show et al.[1-3] They measured changes in oxygen content within a plethysmograph chamber into which a human subject put the arm,[3] and showed that oxygen diffuses from the ambient air into the skin even when the partial pressure of oxygen (P_{O_2}) of the air is as low as 3 to 4 mmHg. This fact implied that P_{O_2} in the viable part of the epidermis is lower than that value.

After 40 years, Evans and Naylor[4] and Huch et al.[5] measured the partial pressure of oxygen diffusing from the capillaries in the skin with a newly developed device consisting of a polarographic electrode set at the skin surface, and demonstrated that the oxygen tension, namely transcutaneous P_{O_2} (tcP_{O_2}), is close to zero. Since the direct oxygen supply from the atmosphere to the skin is excluded by the probe placed at the skin surface in this method, this finding indicates that almost all oxygen supplied by blood is consumed in the skin. It seems difficult, therefore, to detect oxygen diffusing from capillaries to the skin surface in the normal condition. However, vasodilatation induced by topically applied chemicals[4,5] or by heat[6] is found to increase the transcutaneous P_{O_2} value and eventually to correlate with the P_{O_2} in arterial blood (Pa_{O_2}).[6] By taking advantage of this, a compact polarographic electrode incorporated with a heating element was developed by Huch et al.[6,7] to monitor Pa_{O_2} transcutaneously.

Systems of this type were quickly commercialized and are now widely used in the management of premature or sick infants. From a dermatological viewpoint, however, they offer useful information about microcirculation and respiration of the skin as well, since tcP_{O_2} is determined not only by Pa_{O_2} but also by various cutaneous factors affecting oxygen flux to the skin surface. In this chapter, emphasis is not placed on the tcP_{O_2} measurement simply as a noninvasive monitoring of Pa_{O_2}, but rather as a method for the assessment of the respirocirculatory state of the skin.

II. Object

Transcutaneous P_{O_2} is the partial pressure of oxygen measured with an electrode placed on the skin surface, i.e., a reflection of P_{O_2} in the viable epidermis. As mentioned in the introduction, tcP_{O_2} measurement is usually performed at a heated site because the value is nearly zero in the normal condition. When the skin temperature rises up to about 43°C, tcP_{O_2} values become close to the Pa_{O_2}, especially in neonates. Therefore, tcP_{O_2} measurement is most advantageously utilized for a noninvasive monitoring to detect sudden changes in Pa_{O_2}. In practice, most of the commercially available instruments are designed for this utility with an alarm for a critical change in tcP_{O_2}. If the systemic Pa_{O_2} is normal, abnormal tcP_{O_2} values indicate that they are significantly influenced by peripheral or local factors, such as cutaneous blood flow, oxygen diffusibility, cellular respiration, and so on. In this sense, tcP_{O_2} is considered to be a parameter reflecting the extent of influence of various cutaneous factors, so that it can be utilized for the assessment of the skin on which the sensor is placed.

III. Methodological Principle

The tcP_{O_2} measurement is based on the polarographic technique using the Clark-type electrode, which consists of a platinum cathode, a silver anode, and electrolyte solution covered with a hydrophobic membrane (Teflon, polypropylene, etc.) permeable only to gases.[7] If a suitable voltage (polarizing voltage) is applied between the cathode and the anode, the following reactions take place at the electrodes:

$$O_2 + 2H_2O + 4e^- = 4OH^- \quad 4Ag + 4Cl^- = 4AgCl + 4e^-$$

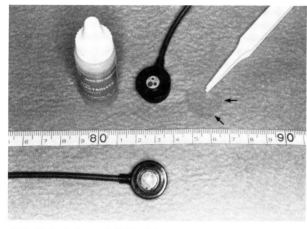

FIGURE 1 Instruments for transcutaneous P_{O_2} measurement. Upper: two types of sensors, electrolyte solution, and gas-permeable membrane covering electrodes (arrows). In the upper sensor, a glass electrode for tcP_{CO_2} measurement is incorporated. Lower: Control panel and recorder.

The resulting current is proportional to the concentration of oxygen. In order to transform the output voltage signal into the unit of pressure (mmHg or Torr or kPa, 1 mmHg = 1 Torr = 133 Pa), the sensor has to be calibrated with gases, usually air and pure nitrogen, the oxygen concentration of which is known. In order to supply heat to the skin and to regulate the temperature of the electrode, a heating element controlled by thermistors is set in the electrode. The sensor temperature can be optionally selected in the range of 37 to 45°C in most commercially available instruments, but the range depends on the design of each one. These elements are incorporated into a small probe head (diameter 15 to 20 mm) which can be applied to most body regions with double-sided adhesive ring tape (Figure 1). It is difficult, however, to attach it to fingers, toes, and the hairy scalp. Recent models contain both electrodes for oxygen and for carbon dioxide (pH-sensitive glass electrode) in one sensor, which enable us to obtain both types of information simultaneously. These instruments are commercially provided by several companies.*

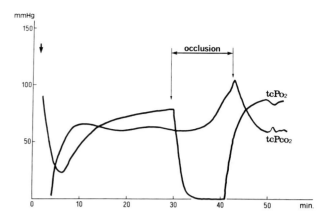

FIGURE 2 Simultaneous recording of tcP_{CO_2} and of tcP_{CO_2} monitored on the volar side of the forearm of an adult. Sensor temperature: 44°C. About 25 min were required to achieve the steady state of tcP_{O_2} after attachment of the sensor (first arrow). Suprasystolic pressure was maintained during cuff occlusion of the arm for 10 min.

By heating the skin, not only the cutaneous blood flow increases but the red blood cells also dissociate more oxygen owing to the shift of the oxygen dissociation curve to the right by the temperature effect,[8] both of which induce more flux of oxygen from capillaries to the skin surface. Transcutaneous P_{O_2}, therefore, depends on skin temperature. When the skin temperature rises up to about 43°C (sensor temperature 44 to 45°C), the tcP_{O_2} becomes close to the Pa_{O_2}[7,8] especially in newborn infants. Though the response time of the sensor is short enough to detect a rapid change in P_{O_2}, about 15 to 20 min are needed before we can determine the tcP_{O_2} value after placing the sensor on the skin, because some time is required to achieve maximal hyperemia. But once the steady state is achieved, tcP_{O_2} quickly responds to the change of P_{O_2} in the skin unless the sensor is placed on a hyperkeratotic region such as the palm. For instance, a rapid fall of tcP_{O_2} on the forearm is observed following a cuff occlusion of the proximal part of the arm with suprasystolic pressure, which indicates that an anoxic state takes place locally within a few minutes after the cuff inflation (Figure 2). The fall rate is regarded as the oxygen consumption rate of the skin.[9,10] Figure 3 shows the expected profile of P_{O_2} in the heated normal skin at 43°C during the steady state.[11] The P_{O_2} in capillaries is higher than Pa_{O_2} owing to the temperature effect; then, being consumed by epidermal cells, it decreases gradually through the epidermis. The diffusion resistance of the stratum corneum also contributes to the decrease in P_{O_2}, since the Clark-type electrode itself consumes oxygen and produces a diffusion field in front of it.[11,12] It finally becomes

---

* Kontron-Roche, Switzerland; Radiometer, Denmark; Novametrix, U.S.; Hewlett-Packard, U.S.; Hellige-Dräger, Germany; Kurare, Japan; etc.

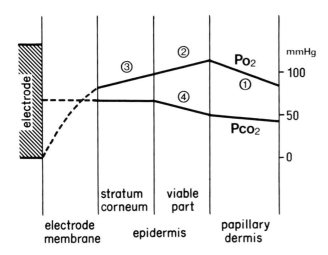

FIGURE 3 P_{O_2} and P_{CO_2} profiles of the skin during heat-induced hyperemia: **(1)** temperature effect; **(2)** oxygen consumption; **(3)** diffusion error by oxygen consumption of electrode; **(4)** CO_2 production. (Modified from Reference 11.)

close to the Pa_{O_2} value at the skin surface, and this value is the tcP_{O_2} discussed here. If measured in adults, however, tcP_{O_2} is usually lower than Pa_{O_2}, which is explained by insufficient hyperemia, higher diffusion resistance of the stratum corneum, and thicker epidermis of adult skin.[13]

IV. Sources of Error

The most serious "error" might be a wrong interpretation of a tcP_{O_2} value, especially when it is obtained in adults. As mentioned, tcP_{O_2} is a parameter which is determined by many factors related to the supply, diffusion, and consumption of oxygen. For example, significantly lower tcP_{O_2} values are found on the *facial and palmar skin* of adults than those on the other regions.[14,15] It is most likely explained by increasing oxygen consumption of the facial skin where pilosebaceous units are dense and highly developed, and by the low diffusibility of oxygen through the thick stratum corneum of the palmar skin.[15] If an examiner is not familiar with the histological structure of the skin he or she, or seeing low tcP_{O_2} monitored at these sites, might consider that Pa_{O_2} of the subject is low.

Many kinds of skin lesions, for example inflammatory skin diseases, also show low tcP_{O_2}, irrespective of their pathological differences.[16] Even if a therapeutic agent increases tcP_{O_2} on the lesion of a leg ulcer, no one can safely say that this drug improved cutaneous circulation because it might have improved only the inflammation that existed there. So the rigid conception that tcP_{O_2} is a parameter determined only by Pa_{O_2} and cutaneous perfusion, though it may be true when subjects are neonates, might consequently be misleading.

Since little skill is required for the tcP_{O_2} measurement, *technical errors* seldom occur, as long as the examiner handles the instrument properly according to the operation manual. Negligence of periodic exchange of electrolyte solution and a membrane of the sensor can be a source of errors. If one forgot calibration of the sensor prior to actual measurement, the data obtained are not reliable at all. Extraordinary tcP_{O_2} values higher than Pa_{O_2} suggest an influx of ambient air due to incomplete attachment of the sensor to the skin surface.

The *posture of subjects,* especially when the sensor is placed on the extremities, also affects tcP_{O_2},[17,18] so that a standardized position is desirable during the measurement. If the sensor is sited over bony prominences like ribs, sudden intermittent falls might take place during the monitoring probably due to the pressure on the skin-electrode.[19] When normal control subjects are needed, heavy smokers might be excluded because a significant difference in tcP_{O_2} is found between smokers and nonsmokers.[20] It is reported that the reproducibility of tcP_{O_2} was not necessarily good when values were compared on the right and the left lower extremities of normal adults[21] and when compared at adjacent sites on the dorsal foot of patients with arterial occlusive diseases, although the short-term (24 to 48 h) reproducibility of tcP_{O_2} was relatively good.[22]

Although the temperature of the test site is kept constant by the thermostatic probe, it is preferable to avoid the examination in a too hot or too cold room. Physical or mental stress on a subject before or during the examination also should be avoided, since it may affect the cutaneous blood flow or the respiration rate.

V. Correlation with Other Methods

Close correlations are reported between tcP_{O_2} at a sensor temperature of 44°C and Pa_{O_2} at 37°C in neonates,[23-25] except in the high Pa_{O_2} range.[23,25] A good correlation was also found in Japanese children of school age, with the exception of data for patients with atopic dermatitis.[26] In adults, however, a rather poor correlation is obtained, with far lower tcP_{O_2} values in the forearm relative to Pa_{O_2} values.[13] According to a multicenter study reported by Palmisano and Severinghaus,[25] which used a large number of patients with a wide age range, the tcP_{O_2} is useful only to indicate changes of Pa_{O_2} in an individual if he or she is older than neonates. The correlation becomes even poorer when P_{O_2} is greater than about 80 mmHg (Figure 4). The relationship between tcP_{O_2} and P_{O_2} of capillary blood from the ear lobe is relatively good in adults, but not so clear-cut.[27] The Pa_{O_2}-tcP_{O_2} gradient is unrelated to skinfold thickness, body surface area, and body mass indices determined by body weight and by height in adults.[28]

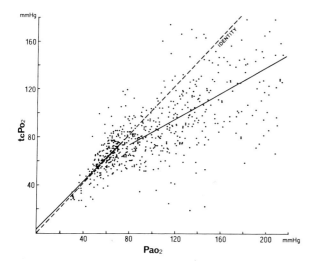

FIGURE 4 Relationship between tcP_{O_2} and Pa_{O_2} of 723 samples from 251 patients with wide age range. Sensor temperature is 44°C. Regression lines are shown for Pa_{O_2} below and above 80 mmHg. (Redrawn from Palmisano, B.W. and Severinghaus, J.W., *J. Clin. Monit.*, 6, 189, 1990. With permission.)

TABLE 1 tcP_{O_2} and tcP_{CO_2} (Mean ± SD) at Eight Body Sites of Ten Adult Males

Site	tcP_{O_2} (mmHg)	tcP_{CO_2} (mmHg)
Forehead	26.6 ± 21.0	67.2 ± 2.4
Cheek	29.6 ± 9.8	69.2 ± 2.8
Forearm, volar[a]	69.6 ± 5.3	62.6 ± 3.3
Abdomen	63.8 ± 6.4	63.7 ± 2.4
Back	60.6 ± 12.2	63.9 ± 3.1
Crus, anterior	66.6 ± 6.1	61.1 ± 4.1
Crus, posterior	67.3 ± 6.8	62.8 ± 2.9
Palm	26.4 ± 6.6	60.5 ± 3.1

Note: Probe temperature 44°C.
[a] *At 37°C, tcP_{O_2}: 8 ± 4, tcP_{CO_2}: 43 ± 6 (Haisjackl et al.[81]). The approximate values of P_{O_2} and P_{CO_2} (mmHg) in the airway and in the blood are as follows.[92]*

	P_{O_2}	P_{CO_2}
Tracheal air	149.2	0.3
Alveolar gas	104	40
Arterial blood	100	40
Venous blood	40	46

From Takiwaki, H. et al., Br. J. Dermatol., 125, 243, 1991. With permission.

Transcutaneous P_{O_2} is a parameter that is especially influenced by the cutaneous blood flow (CBF) at the test site. Since CBF depends on the heat supplied by the sensor,[29] tcP_{O_2} is positively correlated to both sensor temperature and CBF.[30-32] It is demonstrated that changes in tcP_{O_2} correspond to CBF variations more closely when the CBF is measured by laser Doppler flowmetry than by the analysis of heating power, which is a method for estimation of CBF, needed to maintain tcP_{O_2} sensor at constant temperature.[30,32] At a constant sensor temperature of 43°C, a positive correlation is reported between the change in tcP_{O_2} and that in CBF estimated from heating power during orthostatic changes of the leg.[18] In this case, however, the correlation becomes negative if the CBF is measured by the ^{133}Xe washout technique under unheated conditions, indicating that heating abolishes autoregulation of CBF following orthostatic changes.[33] Therefore, the relationship between tcP_{O_2} and CBF differs depending on whether the CBF is measured in the heated condition or in the unheated condition.

VI. Clinical and Experimental Applications

Besides practical use as a noninvasive Pa_{O_2} monitoring device for sick neonates, measurement of tcP_{O_2} has been utilized for various kinds of fundamental and clinical studies, as follows.

A. Fundamental Study

In healthy adults, mean tcP_{O_2} values are similar at various anatomical sites, with the exception of the face, the palm, and probably the sole,[14,15] although their standard deviations are not so small (Table 1). Women have significantly higher values than men.[14] The difference in the mean tcP_{O_2} between the cheek and forearm is very small in children (Figure 5), indicating that the low tcP_{O_2} value in the adult face is based on the anatomical characteristics of adult facial skin, i.e., probably related to the well-developed pilosebaceous units.[15] The oxygen consumption rate of the skin, which was found to be age dependent,[34] was estimated to be about 0.18 to 0.37 ml O_2/100 g/min at 43 to 45°C by some authors[10,34-37] from the reduction rate of tcP_{O_2} during arterial occlusion.

However, this rate measured on the forearm decreases after removal of the stratum corneum by tape-stripping,[34,36] while it markedly increases on the palm after removal of the stratum corneum with 50% salicylic acid plaster[15] (Figure 6). Although these results seem to be inconsistent, the former may be explained by the complex interaction of the significant resistance of the stratum corneum to oxygen diffusion with the oxygen consumption of the electrode.[34] A mathematical simulation using a simplified one-dimensional model, though oxygen consumption of the electrode is neglected in this model, can be applied for the explanation of the latter case where a far thicker stratum corneum exists[38] (Figure 7).

If a "real" oxygen consumption rate is required, therefore, the stratum corneum should be removed. In addition, tcP_{O_2} values increase after the removal of the stratum corneum by average values of 8 to 30 mmHg,[13,15,36,39] depending on the test region. Therefore, we should always be conscious of the interference of the stratum corneum with diffusion of oxygen when measuring tcP_{O_2} and examining dynamics of its changes.

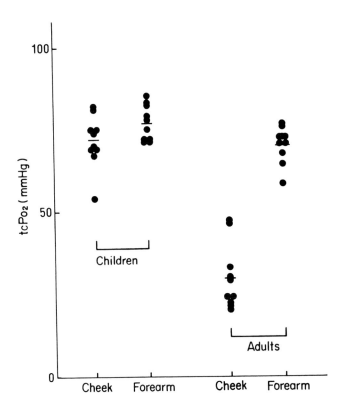

FIGURE 5 The tcP$_{O_2}$ values recorded on the cheek and forearm of healthy children (age range 3–9 years, mean age 5 years) and male adults (24–36 years, 29 years). (From Takiwaki, H. et al., *Br. J. Dermatol.*, 125, 243, 1991. With permission.)

In addition, the thickening of the viable layer of the epidermis also reduces tcP$_{O_2}$ by the increase in oxygen consumption, and the change of P$_{O_2}$ across the layer (from granular layer to basal layer) is estimated at 0.26 mmHg/μm by Falstie-Jensen and coworkers.[40] Thickening of the stratum corneum, namely hyperkeratosis, and that of the living part of the epidermis, namely acanthosis, frequently take place following chronic inflammation in the skin.[41] These changes are often noted at the edge of chronic ulcers and sometimes become prominent.[42] So we should remember the reduction of tcP$_{O_2}$ by epidermal changes when the measurement is performed on skin with an appearance different from normal.

B. Clinical Application

Transcutaneous P$_{O_2}$ measurement is reasonably applied for quantitative estimation of the efficient cutaneous circulation and for the evaluation of the severity of leg ulcers, since tcP$_{O_2}$ reflects the actual efficiency of oxygen supply by tissue perfusion and since oxygen availability is considered a limiting factor in wound healing.

In *arterial occlusive diseases*, several authors have reported that tcP$_{O_2}$ values of the lower leg or foot can be a good predictor of failure in wound healing after treatments or a method to determine amputation

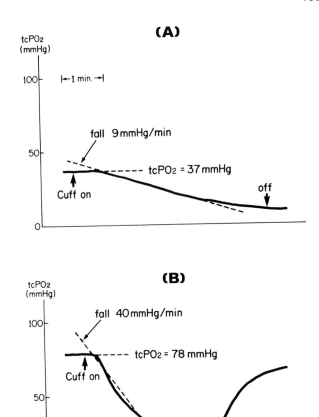

FIGURE 6 Actual records of the tcP$_{O_2}$ fall during arterial occlusion before (A) and after (B) removal of the stratum corneum of palmar skin. The occlusion was started at the point of "Cuff on" and released at "off".

level.[43-48] The validity, however, is still controversial.[49] It is also utilized for objective evaluation of the effect of nonoperative therapy.[50] On the other hand, tcP$_{O_2}$ levels at the edge of ulcers are not predictive of the response to therapy in *venous ulcers*[51] although it may be a valid indicator in the staging of chronic venous incompetence.[52] There are also discrepancies between the results of tcP$_{O_2}$ measurement in this disorder,[50,53-55] in which the extent of the decrease in tcP$_{O_2}$ differs considerably in each report. According to Franzeck et al.,[56] who performed simultaneous measurement of the tcP$_{O_2}$ and morphological examination of the nutritional capillaries at the monitoring site with a videomicroscope, tcP$_{O_2}$ varies markedly depending on capillary density in the test region. They reported that tcP$_{O_2}$ measured on "white atrophy" is almost zero while it is about 50 mmHg on average in the median ankle area without major trophic skin changes. Therefore, unlike arterial occlusive diseases, patients with venous insufficiency may show quite different tcP$_{O_2}$ values depending on the location selected for the measurement.

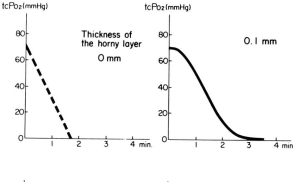
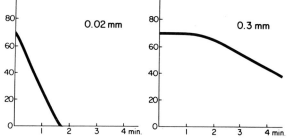

FIGURE 7 Calculated tcP_{O_2}-time curves during arterial occlusion using a one-dimensional simplified model, where the electrode consumes no oxygen and the O_2 consumption rate in the epidermis is constant. The diffusion equation $\partial P_{O_2}/\partial t = D\,(\partial^2 P_{O_2}/\partial x^2)$ is used for computation, where the diffusion coefficient of the stratum corneum (D) at 44°C is estimated to be 1.4×10^{-6} cm^2/s from the data of skin O_2 conductance.[90] (From Takiwaki, H. et al., *Nishinihon J. Dermatol.*, 53, 65, 1991. With permission.)

Some authors[53,54] applied the hypothesis of Browse and Burnand,[57] in which the cause of venous ulcer is a block of oxygen diffusion owing to pericapillary fibrin deposit caused by persistent venous hypertension, to the explanation for decreased tcP_{O_2} in venous insufficiency. On the contrary, Dodd et al.[58] reported that mean tcP_{O_2} was higher in patients than in controls in the recumbent position when the sensor temperature was set at 37°C, indicating that hypoxic state does not take place at this rather physiological temperature. These results imply that tcP_{O_2} at the lesions fails to increase by heating and that heating is regarded as a kind of load or stimulus to differentiate the lesions from normal skin, just like glucose in the glucose tolerance test for discrimination of diabetic patients. Another utilization of tcP_{O_2} is by plastic surgeons[59-62] for *intra- and postoperative management of flaps* and for determination of the optimal time for cutting pedicle flaps.

In the field concerning dermatology, sclerotic skin lesions, such as *systemic scleroderma* (PSS),[63-67] *morphea*,[66] and *hypertrophic scar*[64,68] have been especially examined, since there are few parameters available for the objective evaluation of sclerotic change. The results show significantly reduced tcP_{O_2} at sensor temperatures of 40.5 to 44°C in these lesions, except for the report by Kalis et al.[66] on PSS. Moreover, some authors[64,65,68] demonstrated a negative relationship between the severity of lesions and tcP_{O_2} values. At 37°C, however, no difference in tcP_{O_2} was found between PSS and the control.[67] The cause of reduced tcP_{O_2} is uncertain, but it is most likely explained by the reduced diffusibility of oxygen through the capillary wall and through the sclerotic dermis,[65,68] or by the poor response of the blood vessels to the heat stimulus, since laser Doppler blood flow measured simultaneously failed to be increased enough by heating.[64,67] However, low tcP_{O_2} under heating is never specific to sclerotic lesions, but is common to lesions of various skin diseases, such as *atopic dermatitis, psoriasis, discoid lupus erythematosus, cutaneous lymphomas, herpes, pemphigoid*, and so on,[16] all of which are familiar to dermatologists. It is probably because common pathological changes observed in various skin lesions, such as hyperkeratosis, acanthosis, cellular infiltrate or proliferation, fibrosis, and vascular changes, are all expected to reduce tcP_{O_2} from the theoretical point of view. We, therefore, should not use tcP_{O_2} value as a parameter representing a specified pathological or functional change in the skin unless the cause of the tcP_{O_2} change is clarified. It is a marker only indicating the extent of overall "interference" of the skin on oxygen flux from capillaries to the skin surface. Nevertheless, it might be a "nonspecific" marker indicating the severity of skin lesions if measured on the same lesion continually, since it becomes close to normal value as the lesion improves clinically (Figure 8).

Skin test reactions and their time course are also studied by analyzing tcP_{O_2} and/or transcutaneous values of P_{CO_2} (tcP_{CO_2}) at the test sites.[69-71] The tcP_{O_2} decreases in *positive tuberculin reactions* while tcP_{CO_2} increases, and the changes are greater in the strong reactors.[69] These authors concluded that the increased respiration of infiltrating lymphocytes and monocytes results in local hypoxia and hypercapnia, since the oxygen consumption rate increases significantly at positive test sites.[71] In *patch test reactions*, Prens et al.[70] found a significant correlation in negative fashion between the visual grading and the tcP_{O_2} values at 43°C, and mentioned that this method can be regarded as complementary to the laser Doppler technique. Unlike other noninvasive bioengineering techniques, however, it may be a weak point of this method that 10 to 20 min are required before obtaining a steady-state value. They speculated that the decrease in tcP_{O_2} results from the increase in the distance between the capillary lumen and the skin surface caused by the inflammation of test sites. This explanation, however, might be curious according to the solution of diffusion equation[72,73] for such a case. Indeed, the diffusion process is expected to be delayed as the distance increases but the distance itself has little influence on the value of P_{O_2} unless the total oxygen

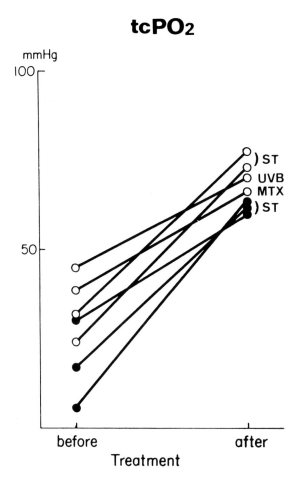

FIGURE 8 Recovery of tcP$_{o2}$ following treatment in plaques of psoriasis (○) and chronic dermatitis (•). Treatment: MTX, methotrexate; UVB, ultraviolet-B irradiation; ST, topical corticosteroid ointment. (From Takiwaki, H. et al., *J. Dermatol. (Tokyo)*, 18, 311, 1991. With permission.)

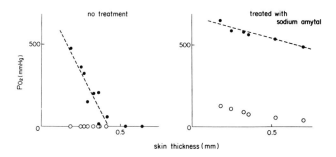

FIGURE 9 Cutaneous P$_{o2}$ against skin thickness during exposure to pure oxygen (•) and air (○) in an airtight chamber controlled at 37°C. The experiment was performed using rat skin cut in various thicknesses. P$_{o2}$ was measured at the dermal side of skin samples, which were directly attached to the sensor surface. To block cellular respiration, samples had been soaked in isotonic sodium amytal solution for 20 min. (From Takiwaki, H. and Nagai, R., *Jpn. J. Dermatol.*, 96, 497, 1986. With permission.)

consumption of cells varies. It is true that a medium with high diffusion resistance like the stratum corneum reduces tcP$_{o2}$ in cooperation with the oxygen consumption of the electrode.[11,34] However, tcP$_{o2}$ will not be dramatically reduced only due to an increase in extravasated plasma since the resistance of plasma to oxygen diffusion is far smaller.[74] In fact, the skin thickness does not have any crucial influence on the decrease in P$_{o2}$ through the skin when the respiration of cells is blocked *in vitro*[75] (Figure 9).

Moreover, by testing histamine and prostaglandin E$_2$-induced hyperemia and edema, Carnochan et al.[76] also demonstrated that tcP$_{o2}$ is little influenced by edema in the absence of cellular infiltrate although the oxygen consumption rate is apparently reduced by it, probably because the excess of interstitial fluid increases the distance between cells.

C. Analysis of Dynamics of Change in tcP$_{o2}$

Monitoring the dynamic response of tcP$_{o2}$ to some loads is more advantageous for the investigation of the functional state of the microcirculation and the respiration of the skin. Hyperkeratotic body sites should be avoided for this purpose. As already mentioned, the oxygen consumption rate of the skin can be estimated by the fall rate of tcP$_{o2}$ during transient arterial occlusion using a pressure cuff. It should be kept in mind that this fall rate is smaller during breathing in air than in 100% oxygen, since it is influenced by the nonlinear relationship between P$_{o2}$ and the oxygen content of blood. Although this fall rate (mmHg/min) is converted to the consumption rate (ml/100 g/min) by multiplying it by the oxygen solubility coefficient (ml O$_2$/ml water/atm at given temperature) in the high tcP$_{o2}$ range where hemoglobin is fully saturated, it becomes necessary to consider the oxygen fraction bound to hemoglobin when tcP$_{o2}$ is relatively low during air breathing.[36] Thus, the oxygen consumption rate is usually estimated from data obtained during pure oxygen inhalation. It is determined not only by the respiration of epidermal cells but also by that of inflammatory cells infiltrating the dermis.[71] A few minutes of arterial occlusion usually lead tcP$_{o2}$ to its minimum level close to zero, and the release of the cuff restores tcP$_{o2}$ exponentially to the previous level. The recovering time of tcP$_{o2}$ is regarded as a parameter reflecting the time required to refill tissue with blood flow, though influenced by the resistance of the skin to oxygen diffusion, and is found to be delayed in peripheral arterial occlusive disease.[21,43,77]

Regarding the *response of tcP$_{o2}$ to oxygen inhalation*, the pedal tcP$_{o2}$ hardly increases in severely ischemic limbs with arterial occlusive disease of the extremities.[47] Breuer et al.[78] reported that the maximum level of tcP$_{o2}$ during oxygen inhalation is significantly lower in diabetic patients than in controls, although similar time is required to achieve the level, and that it may be utilized for early detection of functional diabetic microangiopathy.

If the monitoring is carried out at a sensor temperature of 37°C, the physiological vasoreactions such as reflex vasoconstrictor response to posture change,[33,79] vasodilation response to exercise,[33,79] and reactive hyperemia after arterial occlusion[80-84] are reflected in the fast response of tcP_{O_2} following the stimuli or loads mentioned.

The sympathetic autoregulation of blood flow to thermal change is also detectable by examining the *response time of tcP_{O_2} to a sudden change in the sensor temperature from 45 to 37°C*.[85] These physiological vascular responses are never obtained under conditions of heat-induced maximal hyperemia at 44°C, where every kind of vasomotion is inhibited.[50] Since oxygen consumption and diffusibility of the skin do not vary abruptly in a test site, tcP_{O_2} changes in these examinations result from the change in blood flow or perfusion of the skin. Therefore, these examinations are valuable for the evaluation of weakened or absent vascular responses or reflexes, which are seen in diabetic patients,[82-84] in patients with severe venous incompetence,[33,79] and in critically ill patients.[81] By lowering the sensor temperature, however, the tcP_{O_2} value is so markedly reduced that any changes of it need to be interpreted with caution.[54] Moreover, the sensor responds more slowly to the change in P_{O_2} at 37°C than at 44°C. If "real-time" monitoring of the tcP_{O_2} change is required at 37°C, Tan et al.[86,87] recommend the use of an electrode without a gas-permeable membrane, though the output is expressed in arbitrary values because the system cannot be calibrated.

VII. Recommendation

After reviewing these reports, and from our experiences, it is especially recommended to measure not only resting tcP_{O_2} value but also its dynamic responses to some stimuli, and to measure *other parameters such as tcP_{CO_2} and CBF* simultaneously for more precise investigation of the respirocirculatory function of the skin, because the tcP_{O_2} value is a complex parameter influenced by many factors. The tcP_{CO_2}, which is measured on the heated skin surface just like tcP_{O_2}, is far less influenced by cutaneous factors than tcP_{O_2},[15] since CO_2 is readily eliminated owing to far more solubility in water than oxygen, and since the stratum corneum does not affect the tcP_{CO_2} value because the tcP_{CO_2} sensor (pH-sensitive glass electrode[88]) does not consume CO_2. Moreover, the P_{CO_2} can be kept stable by the buffering effect of the blood.

However, arterial occlusion induces gradual but continuous increase in tcP_{CO_2} (Figure 2). The extraordinarily high tcP_{CO_2} value at tcP_{O_2} close to zero, therefore, implies a severe disturbance of gas exchange in the skin. This finding is observed on blisters or severely edematous erythema, as seen in bullous pemphigoid and on prenecrotic skin of patients with arterial occlusive disease or necrotizing fasciitis.[16] Since these lesions frequently fall into necrosis of the epidermis and/or dermis (Figure 10), simultaneous measurement of tcP_{CO_2} provides more reliable information to predict necrotic change of the skin than tcP_{O_2} measurement alone.

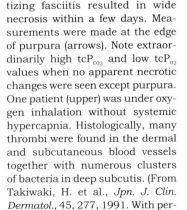
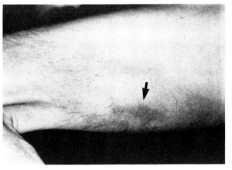
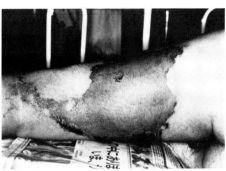
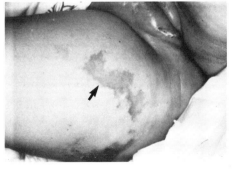
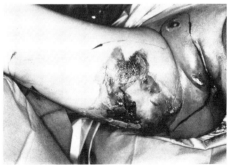

FIGURE 10 Two cases of necrotizing fasciitis resulted in wide necrosis within a few days. Measurements were made at the edge of purpura (arrows). Note extraordinarily high tcP_{CO_2} and low tcP_{O_2} values when no apparent necrotic changes were seen except purpura. One patient (upper) was under oxygen inhalation without systemic hypercapnia. Histologically, many thrombi were found in the dermal and subcutaneous blood vessels together with numerous clusters of bacteria in deep subcutis. (From Takiwaki, H. et al., *Jpn. J. Clin. Dermatol.*, 45, 277, 1991. With permission.)

Case 1 (under oxygen inhalation)
$tcPo_2/co_2$: 18 / 139 mmHg

after 72 hr.

Case 2 $tcPo_2/co_2$: 3 / 242 mmHg

after 24 hr.

The measurement of CBF is also preferable in order to get "pure" information on cutaneous microcirculation at the test site. Since some of the commercially available laser Doppler flowmeters have a thermostat probe holder by which the skin can be heated to a constant temperature, they are especially convenient for the assessment of blood flow under heated conditions corresponding to the circumstance of tcP_{O_2} measurement.

Laser Doppler flowmetry does not necessarily measure the perfusion involved in gas exchange since it appears to measure blood flow in arteriovenous anastomoses that does not participate in nutrition of the tissue,[89] but it is not influenced by cellular respiration and oxygen diffusibility. In contrast, tcP_{O_2} is more or less influenced by these factors, but skin perfusion affecting tcP_{O_2} is concerned directly with gas exchange. Consequently, simultaneous measurements of both most likely give us more precise information on the microcirculation of the skin.

References

1. Shaw, L.A., Messer, C., and Weiss, S., Cutaneous respiration in man. I. Factors affecting the rate of carbon dioxide elimination and oxygen absorption, *Am. J. Physiol.*, 90, 107, 1929.
2. Shaw, L.A. and Messer, C., Cutaneous respiration in man. II. The effect of temperature and relative humidity upon the rate of carbon dioxide elimination and oxygen absorption, *Am. J. Physiol.*, 95, 13, 1930.
3. Shaw, L.A. and Messer, A., Cutaneous respiration in man. III. The permeability of skin to carbon dioxide and oxygen as affected by altering their tension in the air, *Am. J. Physiol.*, 98, 93, 1931.
4. Evans, N.T.S. and Naylor, P.F.D., The systemic oxygen supply to the surface of human skin, *Respir. Physiol.*, 3, 21, 1967.
5. Huch, R., Lübbers, D.W., and Huch, A., Quantitative continuous measurement of partial oxygen pressure on the skin of adults and newborn babies, *Pflügers Arch.*, 337, 185, 1972.
6. Huch, A., Huch, R., Arner, B., and Rooth, G., Continuous transcutaneous oxygen measured with a heat electrode, *Scand. J. Clin. Lab. Invest.*, 31, 269, 1973.
7. Huch, R., Huch, A., and Lübbers, D.W., Transcutaneous measurement of blood P_{O_2} (tcP_{O_2}) — method and application in perinatal medicine, *J. Perinat. Med.*, 1, 183, 1973.
8. Lübbers, D.W. and Grossmann, U., Gas exchange through the human epidermis as a basis of tcP_{O_2} and tcP_{CO_2} measurement, in *Continuous Transcutaneous Blood Gas Monitoring*, Huch, R. and Huch, A., Eds., Marcel Dekker, New York, 1983, 1.
9. Naylor, P.F.D. and Evans, N.T.S., The action of locally applied barbiturates on skin oxygen tension and rate of oxygen utilization, *Br. J. Dermatol.*, 82, 600, 1970.
10. Severinghaus, J.W., Stafford, M., and Thunstrom, A.M., Estimation of skin metabolism and blood flow with tcP_{O_2} and tcP_{CO_2} electrodes by cuff occlusion of the circulation, *Acta Anaesthesiol. Scand. Suppl.*, 68, 9, 1978.
11. Lübbers, D.W., Cutaneous and transcutaneous P_{O_2} and P_{CO_2} and their measuring conditions, *Birth Defects: Orig. Artic. Ser.*, 15, 13, 1979.
12. Baumgärtl, H., Grunewald, W., and Lübbers, D.W., Polarographic determination of the oxygen partial pressure field by Pt microelectrodes using the O_2 field in front of a Pt electrode as a model, *Pflügers Arch.*, 347, 49, 1974.
13. Lemke, R., Klaus, D., and Lübbers, D.W., Experiences with the commercially available tcP_{O_2} electrode in adults, in *Continuous Transcutaneous Blood Gas Monitoring*, Huch, R. and Huch, A., Eds., Marcel Dekker, New York, 1983, 143.
14. Orenstein, A., Mazkereth, R., and Tsur, H., Mapping of the human body skin with the transcutaneous oxygen pressure method, *Ann. Plast. Surg.*, 20, 419, 1988.
15. Takiwaki, H., Nakanishi, H., Shono, Y., and Arase, S., The influence of cutaneous factors on the transcutaneous pO_2 and pCO_2 at various body sites, *Br. J. Dermatol.*, 125, 243, 1991.
16. Takiwaki, H., Arase, S., Nakanishi, H., and Takeda, K., Transcutaneous P_{O_2} and P_{CO_2} measurements in various skin lesions, *J. Dermatol. (Tokyo)*, 18, 311, 1991.
17. Svedman, P., Holmberg, J., Jacobson, S., Lindell, S.E., and Ponnert, L., On the relation between transcutaneous oxygen tension and skin blood flow, *Scand. J. Reconstr. Surg.*, 16, 133, 1982.
18. Eickhoff, J.H. and Jacobsen, E., Correlation of transcutaneous oxygen tension to blood flow in heated skin, *Scand. J. Clin. Lab. Invest.*, 40, 761, 1980.
19. Whitehead, M., Pollitzer, M., and Reynolds, E.O.R., Artefactual hypoxemia during estimation of Pa_{O_2} by skin electrode (letter), *Lancet*, ii, 157, 1979.
20. Workman, W.T. and Sceffield, P.J., Continuous transcutaneous oxygen monitoring in smokers under normobaric and hyperbaric oxygen conditions, in *Continuous Transcutaneous Blood Gas Monitoring*, Huch, R. and Huch, A., Eds., Marcel Dekker, New York, 1983, 649.
21. Lusiani, L., Visona, A., Nicolin, P., Papesso, B., and Pagnan, A., Transcutaneous oxygen tension (TcP_{O_2}) measurement as a diagnostic tool in patients with peripheral vascular disease, *Angiology*, 39, 873, 1988.
22. Rooke, T. and Osmundson, P.J., Variability and reproducibility of transcutaneous oxygen tension measurements in the assessment of peripheral vascular disease, *Angiology*, 40, 695, 1989.
23. Nilsson, E., Redham, I., Lagercrantz, H., and Olsson, P., The validity of transcutaneous P_{O_2} monitoring (tcP_{O_2}) as compared to intraarterial P_{O_2} monitoring in newborn infants, in *Continuous Transcutaneous Blood Gas Monitoring*, Huch, R. and Huch, A., Eds., Marcel Dekker, New York, 1983, 259.
24. Lofgrën, O. and Andersson, D., Transcutaneous carbon dioxide and transcutaneous oxygen monitoring in neonatal intensive care patients, in *Continuous Transcutaneous Blood Gas Monitoring*, Huch, R. and Huch, A., Eds., Marcel Dekker, New York, 1983, 413.
25. Palmisano, B.W. and Severinghaus, J.W., Transcutaneous P_{CO_2} and P_{O_2}: a multicenter study of accuracy, *J. Clin. Monit.*, 6, 189, 1990.
26. Oda, Y., Sano, Y., Sato, M., Ishizuka, T., Adachi, K., and Suzuki, H., Correlation between arterial P_{O_2} and transcutaneous P_{O_2} in children of school age, in *Continuous Transcutaneous Blood Gas Monitoring*, Huch, R. and Huch, A., Eds., Marcel Dekker, New York, 1983, 281.
27. Wimberley, P.D., Pedersen, K.G., Thode, J., Fogh-Anderson, N., Sørensen, A.M., and Siggaard-Andersen, O., Transcutaneous and capillary pco_2 and po_2 measurements in healthy adults, *Clin. Chem.*, 29, 1471, 1983.
28. Rafferty, T.D. and Morrero, O., Skin-fold thickness, body mass, and obesity indexes and the arterial to skin-surface P_{O_2} gradient, *Arch. Surg.*, 118, 1142, 1983.

29. Jaszczak, P. and Poulsen, J., tcP$_{O_2}$ dependence on a sufficient bloodflow, in *Continuous Transcutaneous Blood Gas Monitoring*, Huch, R. and Huch, A., Eds., Marcel Dekker, New York, 1983, 35.
30. Eberhard, P., Mindt, W., and Schäfer, R., Reliability of methods for skin blood flow measurement during transcutaneous P$_{O_2}$ monitoring, in *Continuous Transcutaneous Blood Gas Monitoring*, Huch, R. and Huch, A., Eds., Marcel Dekker, New York, 1983, 69.
31. Enkema, L., Holloway, G.A., Piraino, D.W., Harry, D., Zick, G.L., and Kenny, M.A., Laser Doppler velocimetry vs heater power as indicators of skin perfusion during transcutaneous O$_2$ monitoring, *Clin. Chem.*, 27, 391, 1981.
32. Beran, A.V., Tolle, C.D., and Huxtable, R.F., Cutaneous blood flow and its relationship to transcutaneous O$_2$/CO$_2$ measurements, *Crit. Care Med.*, 9, 736, 1981.
33. Dodd, H.J., Gaylarde, P.M., and Sharkany, I., Skin oxygen tension in venous insufficiency of the lower leg, *J. R. Soc. Med.*, 78, 373, 1985.
34. Jaszczak, P., Sejrsen, P., and Sørensen, P.R., The influence of epidermal membrane on percutaneous pO$_2$ and metabolic rate, *Scand. J. Clin. Invest.*, 48 (Suppl. 189), 17, 1988.
35. Evans, N.T.S. and Naylor, P.F.D., The oxygen tension gradient across human epidermis, *Respir. Physiol.*, 3, 38, 1967.
36. Jaszczak, P. and Sejrsen, P., Oxygen tension and consumption measured by a tcP$_{O_2}$ electrode on heated skin before and after epidermal stripping, *Acta Anaesthesiol. Scand.*, 31, 362, 1987.
37. Horio, H., Tamura, H., Hasegawa, T., and Ohminato, S., Oxygen transport model for transcutaneous P$_{O_2}$ measurement, *Med. Electr. Bioeng.*, 22, 31, 1984 (in Japanese).
38. Takiwaki, H., Oura, H., Utsunomiya, M., Kanno, Y., Arase, S., and Shigemi, F., The interference of the horny layer on the transcutaneous P$_{O_2}$ monitoring, *Nishinihon J. Dermatol.*, 53, 65, 1991 (in Japanese).
39. Fallenstein, F., Nef, W., Huch, A., and Huch, R., Effect of skin stripping on the level and variability of transcutaneous P$_{O_2}$, in *Continuous Transcutaneous Blood Gas Monitoring*, Huch, R. and Huch, A., Eds., Marcel Dekker, New York, 1983, 161.
40. Falstie-Jensen, N., Spaun, E., Brøchner-Mortensen, J., and Falstie-Jensen, S., The influence of epidermal thickness on transcutaneous oxygen pressure measurements in normal persons, *Scand. J. Clin. Lab. Invest.*, 48, 519, 1988.
41. Lever, W.F. and Schaumburg-Lever, G., Dermatitis (eczema), in *Histopathology of the Skin*, 6th ed., J.B. Lippincott, Philadelphia, 1983, 93.
42. Lever, W.F. and Schaumburg-Lever, G., Pseudocarcinomatous hyperplasia, in *Histopathology of the Skin*, 6th ed., J.B. Lippincott, Philadelphia, 1983, 505.
43. Franzeck, U.K., Talke, P., Bernstein, E.F., Golbranson, F.L., and Fronek, A., Transcutaneous P$_{O_2}$ measurements in health and peripheral arterial occlusive disease, *Surgery*, 91, 156, 1982.
44. Burgess, E.M., Matsen, F.A., Wyss, C.R., and Simmons, C., Segmental transcutaneous measurements of P$_{O_2}$ in patients requiring below-the-knee amputation for peripheral vascular insufficiency, *J. Bone J. Surg.*, 64-A, 378, 1982.
45. White, R.A., Nolan, L., Harley, D., Long, J., Klein, S., Tremper, K., Nelson, R., Tabrisky, J., and Shoemaker, W., Noninvasive evaluation of peripheral vascular disease using transcutaneous oxygen tension, *Am. J. Surg.*, 144, 68, 1982.
46. Christensen, K.S. and Klarke, M., Transcutaneous oxygen measurement in peripheral occlusive disease. An indicator of wound healing in leg amputation, *J. Bone J. Surg.*, 68-B, 423, 1986.
47. Bongard, O. and Krähenbühl, B., Predicting amputation in severe ischemia: the value of transcutaneous P$_{O_2}$ measurement, *J. Bone J. Surg.*, 70-B, 465, 1988.
48. Oishi, C.S., Fronek, A., and Golbranson, F.L., The role of noninvasive vascular studies in determining levels of amputation, *J. Bone J. Surg.*, 70-A, 1520, 1988.
49. Falstie-Jensen, N., Christensen, K.S., and Brøchner-Mortensen, J., Selection of lower limb amputation level not aided by transcutaneous pO$_2$ measurements, *Acta Orthop. Scand.*, 60, 483, 1989.
50. Stüttgen, G., Ott, A., and Flesch, U., Measurement of skin microcirculation, in *Cutaneous Investigation in Health and Disease*, Lévêque, J.L., Ed., Marcel Dekker, New York, 1989, 359.
51. Nemeth, A.J., Eaglstein, W.H., and Falanga, V., Clinical parameters and transcutaneous oxygen measurements for the prognosis of venous ulcers, *J. Am. Acad. Dermatol.*, 20, 186, 1989.
52. Mannarino, E., Pasqualini, L., Maragoni, G., Sanchini, R., Regni, O., and Innocente, S., Chronic venous incompetence and transcutaneous oxygen pressure: a controlled study, *VASA J. Vasc. Dis.*, 17, 159, 1988.
53. Clyne, C.A.C., Ramsden, W.H., Chant, A.D.B., and Webster, J.H.H., Oxygen tension on the skin of the gaiter area of limbs with venous disease, *Br. J. Surg.*, 72, 644, 1985.
54. Mani, R., Gorman, F.W., and White, J.E., Transcutaneous measurements of oxygen tension at edges of leg ulcers: preliminary communication, *J. R. Soc. Med.*, 79, 650, 1986.
55. Cheatle, T.R., Mcmullin, G.M., Farrah, J., Smith, P.D.C., and Scurr, J.H., Three tests of microcirculatory function in the evaluation of treatment for chronic venous insufficiency, *Phlebology*, 5, 165, 1990.
56. Franzeck, U.K., Bollinger, A., Huch, R., and Huch, A., Transcutaneous oxygen tension and capillary morphologic characteristics and density in patients with chronic venous incompetence, *Circulation*, 70, 806, 1984.
57. Browse, N.L. and Burnand, K.G., The cause of venous ulceration, *Lancet*, ii, 243, 1982.
58. Dodd, H.J., Gaylarde, P.M., and Sharkany, I., Skin oxygen tension in venous insufficiency of lower leg, *J. R. Soc. Med.*, 78, 373, 1985.
59. Achauer, B.M. and Black, K.S., Transcutaneous oxygen and flaps, *Plast. Reconstr. Surg.*, 74, 721, 1984.
60. Sonneveld, G.J., Kort, W.J., Meulen, J.C., and Smith, A.R., Transcutaneous P$_{O_2}$ monitoring in replanted extremities and free tissue transfer operations as an objective parameter for circulation control, in *Continuous Transcutaneous Blood Gas Monitoring*, Huch, R. and Huch, A., Eds., Marcel Dekker, New York, 1983, 715.
61. Keller, H.P. and Lanz, U., Transcutaneous P$_{O_2}$ measurement to evaluate blood flow in skin flaps, in *Continuous Transcutaneous Blood Gas Monitoring*, Huch, R. and Huch, A., Eds., Marcel Dekker, New York, 1983, 673.
62. Tuominen, H.P., Asko-Seljavaara, S., Svartling, N.E., and Härmä, A., Cutaneous blood flow in the TRAM flap, *Br. J. Plast. Surg.*, 45, 261, 1992.
63. Hiller, D., Kessler, M., and Hornstein, P., Vergleichende kutane Sauerstoffdruckmessung bei Gesunden und bei Patienten mit progressiver Sklerodermie, *Hautarzt*, 37, 83, 1986.
64. Takiwaki, H., Transcutaneous P$_{O_2}$ measurements on clinically sclerotic lesions in scleroderma and hypertrophic scar, *Nishinihon J. Dermatol.*, 49, 492, 1987 (in Japanese).
65. Silverstein, J.L., Steen, V.D., Medsger, T.A., and Falanga, V., Cutaneous hypoxia in patients with systemic sclerosis, *Arch. Dermatol.*, 124, 1379, 1988.
66. Kalis, B., De Rigal, J., Léonard, F., Lévêque, J.L., Riche, O., Le Corre, Y., and De Lacharriere, O., *In vivo* study of scleroderma by non-invasive techniques, *Br. J. Dermatol.*, 122, 785, 1990.

67. Valentini, G., Leonardo, G., Moles, D.A., Apasia, M.R., Maselli, R., Tirri, G., and Del Guercio, R., Transcutaneous oxygen pressure in systemic sclerosis: evaluation at different sensor temperatures and relationship to skin perfusion, *Arch. Dermatol. Res.*, 283, 285, 1991.
68. Berry, R.B., Tan, O.T., Cooke, E.D., Gaylarde, P.M., Bowcock, S.A., Lamberty, B.G.H., and Hackett, M.E.J., Transcutaneous oxygen tension as an index of maturity in hypertrophic scars treated by compression, *Br. J. Plast. Surg.*, 38, 163, 1985.
69. Spence, V.A. and Swanson, J.B., Transcutaneous measurement of P_{o_2} and P_{co_2} in the dermis at the site of the tuberculin reaction in healthy human subjects, *J. Pathol.*, 155, 289, 1988.
70. Prens, E.P., van Joost, V., and Steketee, J., Quantification of patch test reactions by transcutaneous P_{o_2} measurement, *Contact Derm.*, 16, 142, 1987.
71. Abbot, N.C., Spence, V.A., Swanson, B.J., Carnochan, F.M.T., Gibbs, J.H., and Lowe, J.G., Assessment of the respiratory metabolism in the skin from transcutaneous measurements of P_{o_2} and P_{co_2}: potential for non-invasive monitoring of response to tuberculin skin testing, *Tubercle*, 71, 15, 1990.
72. Hill, A.V., The diffusion of oxygen and lactic acid through tissues, *Proc. R. Soc. London Ser. B*, 104, 39, 1929.
73. Takahashi, G.H., Fatt, I., and Goldstick, T.K., Oxygen consumption rate of tissue measured by a micropolarographic method, *J. Gen. Physiol.*, 50, 317, 1966.
74. Thews, G., Die theoretische Grundlagen der Sauerstoffaufnahme in der Lunge, *Ergebnisse Physiol.*, 53, 42, 1963.
75. Takiwaki, H. and Nagai, R., Diffusion and consumption of oxygen in the rat skin *in vitro*, *Jpn. J. Dermatol.*, 96, 497, 1986 (in Japanese).
76. Carnochan, F.M.T., Abott, N.C., Beck, J.S., Spence, V.A., and James, P.B., The influence of histamine and PGE2-induced hyperaemia and oedema on respiratory metabolism in normal human forearm skin, *Agents Actions*, 29, 292, 1990.
77. Slagsvold, C.E., Stranden, E., Rosen, L., and Kroese, A.J., The role of blood perfusion and tissue oxygenation in the postischemic transcutaneous pO_2 response, *Angiology*, 43, 155, 1992.
78. Breuer, H.W.M., Breuer, J., and Berger, M., Transcutaneous oxygen pressure measurements in type I diabetic patients for early detection of functional diabetic microangiopathy, *Eur. J. Clin. Invest.*, 18, 454, 1988.
79. Sarkany, I., Dodd, H.J., and Gaylarde, P.M., Surgical correction of venous incompetence restores normal skin blood flow and abolishes hypoxia during exercise, *Arch. Dermatol.*, 125, 223, 1989.
80. Rooth, G., Ewald, U., and Caligara, F., Transcutaneous P_{o_2} and P_{co_2} monitoring at 37°C, *Adv. Exp. Med. Biol.*, 220, 23, 1987.
81. Haisjackl, M., Hasibeder, W., Klaunzer, S., Alterberger, H., and Koller, W., Diminished reactive hyperemia in the skin of critically ill patients, *Crit. Care Med.*, 18, 813, 1990.
82. Ewald, U., Tuvemo, T., and Rooth, G., Early reduction of vascular reactivity in diabetic children detected by transcutaneous oxygen electrode, *Lancet*, i, 1287, 1981.
83. Kobbah, M., Ewald, U., and Tuvemo, T., Vascular reactivity during the first year of diabetes in children, *Acta Paediatr. Scand. Suppl.*, 320, 56, 1985.
84. Haitas, B., Barnes, A., Shorgry, M.E.C., Weindling, M., Rolfe, P., and Turner, R.C., Delayed vascular reactivity to ischemia in diabetic microangiopathy, *Diabetes Care*, 7, 47, 1984.
85. Weindorf, N., Schultz-Ehrenburg, U., and Altmeyer, P., Plaqueförmige kutane Muzinose mit Teleangiektasien, *Hautarzt*, 39, 589, 1988.
86. Tan, O.T., Gaylarde, P.M., and Sarkany, I., Skin oxygen tension and blood flow changes in response to respiratory manoeuvres, *Clin. Exp. Dermatol.*, 7, 33, 1982.
87. Tan, O.T., Stafford, T.J., Sarkany, I., Gaylarde, P.M., Tilsey, C., and Payne, J.P., Suppression of alcohol-induced flushing by a combination of H_1 and H_2 histamine antagonists, *Br. J. Dermatol.*, 107, 647, 1982.
88. Eberhard, P. and Schäfer, R., A sensor for noninvasive monitoring of carbon dioxide, *J. Clin. Eng.*, 6, 35, 1981.
89. Engelhart, M. and Kristensen, J.K., Evaluation of cutaneous blood flow responses by ^{133}xenon washout and a laser-Doppler flowmeter, *J. Invest. Dermatol.*, 80, 12, 1983.
90. Eberhard, P. and Severinghaus, J.W., Measurement of heated skin O_2 diffusion conductance and P_{o_2} sensor induced O_2 gradient, *Acta Anaesthesiol. Scand. Suppl.*, 68, 1, 1978.
91. Takiwaki, H., Kanno, Y., Oura, H., Arase, S., and Shigemi, F., Clinical application of transcutaneous P_{o_2} and P_{co_2} measurements in dermatology, *Jpn. J. Clin. Dermatol.*, 45, 277, 1991 (in Japanese).
92. Comroe, J.H., *Physiology of Respiration,* Year Book Medical Publishers, Chicago, 1974, 9.

Chapter 9.4
Measurement of Transcutaneous P_{CO_2}

Carsten N. Nickelsen
Hvidovre Hospital
University of Copenhagen
Hvidovre, Denmark

I. Introduction

In 1793 John Albernethy[1] demonstrated for the first time carbon dioxide exchange through the intact skin by submerging his arm into mercury. He analyzed the gas bubbles accumulating above the mercury and found carbon dioxide. Sixty years later Gerlach[2] demonstrated changes in the gas mixture of a horse bladder glued to his chest. However, transcutaneous monitoring of the carbon dioxide tension was not possible until the construction of an electrochemical CO_2 electrode by Stow and Randall in 1954,[3] improved by Severinghaus and Bradley in 1956,[4] by Johns et al. in 1969[5] and modified by Huch et al. in 1973.[6]

II. Object

The acid-base balance is usually described by the pH, P_{CO_2} and the BE (base excess). The pH is defined as the negative logarithm of the hydrogen activity and is the resultant of two factors, a respiratory component and a metabolic component. The respiratory component is expressed as the P_{CO_2} in kPa (or in mmHg) being an index of dissolved CO_2. The metabolic component consists of all other acids or bases, which affects the pH, and can be expressed as base excess in meq/l or mmol/l.

According to these parameters, different clinical changes of human acid-base balance are defined. Whereas respiratory acidosis is characterized by accumulation of carbon dioxide in the organism, with elevated P_{CO_2}, decreased pH, and unchanged BE values, metabolic acidosis is characterized by accumulation of noncarbonic acid in the organism, with decreased pH and BE and primarily unchanged P_{CO_2} values. During respiratory alkalosis and metabolic alkalosis the changes are primarily in the opposite direction. These primary changes are followed by both chemical and physiological compensation in order to stabilize the pH. The compensation may be respiratory caused by hyperventilation or renal by means of increased excretion of bicarbonate. The chemical compensation is caused by the buffering system of the blood and the extracellular fluid. The major buffering system in the organism is the bicarbonate based in the changes in the equation:

$$CO_2 + H_2O \leftrightarrow H_2CO_3 \leftrightarrow H^+ + HCO_3^-$$

The same chemical reaction occurs when blood with low pH flows into the skin, leading to a P_{CO_2} increase in the skin; for instance, when metabolic acidosis is produced in another tissue during exercise, hypoperfusion, or in ketoacidosis, there may be an elevation of the carbon dioxide tension in the skin. Therefore, the P_{CO2}, being mainly an indicator for respiratory disturbances, when measured in the skin also can reflect metabolic disturbances.[7]

III. Methodological Principle

The electrode for transcutaneous carbon dioxide monitoring measures the partial pressure of CO_2 in a small lumen outside the skin. The lumen is usually filled with a contact liquid, an electrolyte solution, which very rapidly equilibrates with the carbon dioxide tension in the outer layer of the epidermis. The measured value is a reflection of the capillary blood P_{CO_2} value, but it is modified by the capillary blood flow, the temperature, and the production of CO_2 in the tissue between the capillaries and the electrode surface.

During normal conditions the P_{CO_2} of arterial blood is close to the P_{CO_2} of venous blood, the difference only being 1 kPa (7.5 mmHg).[8] At increasing capillary blood flow the capillary blood P_{CO_2} is decreasing towards arterial values, but at decreasing capillary blood flow the local capillary blood P_{CO_2} value may increase to much higher values, and even to values higher than central venous blood values. Therefore, a capillary blood flow above a certain limit is necessary in order to obtain transcutaneous values correlating to the arterial values. In order to obtain a satisfactory capillary blood flow, vasodilatation usually is induced using heated transcutaneous electrodes. Beside vasodilatation and a faster diffusion, heating of the electrode and the tissue below it will also induce a temperature-dependent elevation of the measured P_{CO_2} value, just as P_{CO_2} values increase with temperature in blood samples.[9] In blood samples this temperature-dependent elevation is described by the anaerobic temperature coefficient of P_{CO_2} in blood, α. When calculating the value at 37°C [P_{CO_2}(37)] from the measured value at t°C [P_{CO_2}(t)] the following formula is used:

$$P_{CO_2}(37) = P_{CO_2}(t)\frac{1}{10^{\alpha(t-37)}}$$

The anaerobic temperature coefficient of CO_2 in blood (α) is 0.021°C^{-1}.[9] According to that the correction factor is calculated for different temperatures in Table 1.

Besides correction for the elevated electrode temperature, the transcutaneous value must be corrected for the CO_2 production in the tissue between the capillaries and the electrode on the skin surface in order to correlate it to the capillary blood value of P_{CO_2}. This metabolic contribution to the transcutaneous value has been found to be approximately 0.5 kPa (3 to 4 mmHg).[10,11] The metabolic contribution may be changing when different electrode temperatures are used as the rate of metabolism increases with the temperature, but the increasing diffusion speed of CO_2 at increasing temperature will balance the change to some extent, making the metabolic contribution to the transcutaneous carbon dioxide value almost temperature independent.

Theoretically, the transcutaneous value of P_{CO_2} (tcP_{CO_2}) can be calculated as the product of the capillary P_{CO_2} (measured at 37°C) and the temperature factor (Table 1) added to the metabolic contribution (0.5 kPa), but a sufficient capillary blood flow is required for the use of this calculation.

Five different principles of transducers have been used for monitoring of tcP_{CO_2},[11] but the use of electrochemical electrodes has been predominant.

A. The Electrochemical Electrode

The electrochemical electrode used for tcP_{CO_2} monitoring is a thermostated Stow-Severinghaus electrode. This electrode consists of a glass pH electrode, a silver-silverchloride reference electrode placed around the pH electrode and a thermostated heating element. A CO_2-permeable membrane covers the electrodes, occluding an electrolyte chamber between the membrane and the electrodes. An O-ring secures the membrane to the electrode housing (Figure 1).

During monitoring the CO_2 released from the skin diffuses through a contact fluid through the gas-permeable membrane and into the electrolyte solution in the electrolyte chamber. The carbon dioxide reacts with water to form carbonic acid (H_2CO_3), which immediately

TABLE 1 Temperature Correction Factor for P_{CO_2}, Measured at Different Temperatures

t°C						
39	40	41	42	43	44	45
Correction from t° to 37°						
0.9078	0.8650	0.8241	0.7852	0.7482	0.7129	0.6792
Correction from 37° to t°						
1.1015	1.1561	1.2134	1.2735	1.3366	1.4028	1.4723

dissociates to hydrogen ions (H^+) and bicarbonate ions (HCO_3^-). Following the diffusion of CO_2 into the electrolyte the pH of the electrolyte will change according to the Henderson-Hasselbalch equation:

$$pH = pK + \log\frac{cHCO_3^-}{a\,P_{CO_2}}$$

where pK is the dissociation constant of carbonic acid, $cHCO_3^-$ is the concentration of HCO_3^-, a the solubility coefficient of dissolved CO_2, and P_{CO_2} the partial pressure of CO_2. The pH change (as a result of the P_{CO_2} change) in the electrolyte is measured as a change in the potential between the glass electrode and the reference electrode. The monitor therefore in principle is a pH-meter, but the display is usually changed and the value displayed is the P_{CO_2} value. In the monitor a microprocessor is usually integrated controlling the heating of the electrode, and often also performing the calibration of the electrode more or less automatically. In some cases the microprocessor modifies the measured value for the elevated electrode temperature and/or the metabolic contribution before displaying the result.

B. Electrode Calibration

The calibration procedure is usually recommended as a two-step calibration with two different gas mixtures (usually 5%/95% and 10%/90% CO_2/N_2 gas mixtures). Using this procedure the electrode and the monitor are adjusted (with the first gas mixture) and the actual sensitivity of the electrode is calculated (with the second gas mixture). The sensitivity of an electrode is, however, unchanged during most of its lifetime, and it is possible to perform a one-step calibration, in which a preset "standard electrode sensitivity" is used during monitoring. The possible measurement inaccuracy following one-step calibration is below 0.2 kPa when measuring in the range from 0 to 10 kPa (below 2%), and this was found acceptable.[13] Nevertheless, it is recommended to check the electrode sensitivity now and then by exposing the electrode to two different calibration gas mixtures, in order to avoid the use of old electrodes with low sensitivity. The one-point calibration is performed rapidly (approximately 5 to 10 min vs. 15 to 20 min by two-point calibration).

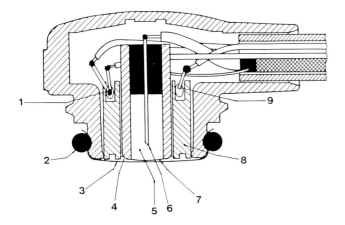

FIGURE 1 Electrochemical transcutaneous carbon dioxide electrode (E 5230, Radiometer Copenhagen). **(1)** Temperature sensor, **(2)** O-ring securing the membrane, **(3)** CO_2-permeable membrane, **(4)** electrolyte, **(5)** inner buffer solution, **(6)** inner reference Ag/AgCl electrode, **(7)** pH-sensitive glass membrane, **(8)** outer reference Ag/AgCl electrode, **(9)** heating element.

C. Electrode Application to the Skin

In the neonate and the adult the electrode application to the skin is performed using a fixation ring attached to the skin by adhesive tape. The fixation ring is attached to the skin, contact liquid is filled into the ring, and the electrode is locked into the fixation ring superseding some of the contact liquid. In this way the electrode membrane is positioned in direct contact to the small liquid-filled lumen on the skin. The same method is not possible for fetal monitoring, primarily because the adhesive tape does not attach to the wet fetal skin, but also because it is not possible to fill the contact liquid into the fixation ring when it is attached to the presenting part of the fetus (usually the fetal scalp) in the vagina. A special fixation ring for suction fixation is used for fetal monitoring.[13] The ring is made of soft plastic material and attached to the skin with a vacuum of 20 kPa. No contact liquid is used for fetal monitoring.

IV. Sources of Error

Insufficient fixation of the electrode will usually cause false low values. When applying the electrode to the skin, an area where the skin is thin and where movements causing dislocation of the electrode are few should be chosen. In the adult and the neonate the electrode is usually applied on the chest; in the fetus the electrode only can be applied on the presenting part, usually the fetal scalp. Insufficient blood flow in the skin below the electrode may cause false elevated values, only slowly reacting on changes in the P_{CO_2} values of blood in central blood values. The need for vasodilation is especially important in adult monitoring, but in healthy adults a short period of "preheating" to 45°C permits sufficient vasodilatation for monitoring at 37°C electrode temperature (in contrast to transcutaneous P_{O_2} monitoring).[14] However, in the critically ill patient a higher electrode temperature causing local vasodilatation is preferable. In the fetus, an electrode temperature of 44°C should be used.[15] The electrode must be placed in an area where mechanical pressure on the back of the electrode is avoided, as compression of the area under the electrode can cause false elevation of the P_{CO_2} value.

V. Correlation with Other Methods

In adults close correlations are found between transcutaneous carbon dioxide tension and capillary blood P_{CO_2}.[14] After correction for the elevated electrode temperature and the metabolic contribution from the skin, the values are almost identical. In the neonate a similar close correlation has been demonstrated even during respiratory insufficiency between transcutaneous P_{CO_2} and arterial P_{CO_2}.[16] During fetal monitoring transcutaneous P_{CO_2} has been correlated both to capillary blood P_{CO_2} values[17] and to P_{CO_2} values in umbilical artery blood[15] and close correlations were found in all cases.

VI. Recommendations

Although most studies have concluded that transcutaneous carbon dioxide tension measurement is both useful and sensitive for continuous respiratory monitoring, there are some limitations for the method. Especially in adults it seems that there are individual differences in either skin permeability or in central regulation of respiration,[18] which may limit the use of the method for the assessment of respiratory depression in extubated, spontaneously breathing patients recovering from general anesthesia.[19] The electrode temperature should be 42 to 43°C for adult monitoring.

Transcutaneous P_{CO_2} monitoring is especially useful as a continuous measurement for respiratory depression in the neonate, where the correlation to the arterial P_{CO_2} value is very close, and the need for repeated blood-gas measurements on arterial or capillary blood is thus diminished. The electrode temperature can be set as low as 38°C in the neonate, but as the electrode often is combined with a P_{O_2} electrode higher temperatures are used. In the fetus during labor transcutaneous P_{CO_2} monitoring gives valuable information on fetal distress, but the method cannot be used as the only method because of a low specificity. For fetal monitoring an electrode temperature of 44°C is necessary for accurate measurements.[15]

In newborns with arterial-venous shunts or compromised peripheral blood flow, it is possible by use of two transcutaneous electrodes placed in different areas noninvasively to diagnose the hemodynamic change. Further, the method is increasingly used in a variety of clinical situations, such as assessment of skin flap viability during evaluation of wound healing and selection of amputation level in peripheral vascular diseases.

References

1. Abernethy, J., An essay of the nature of the matter perspired and absorbed from the skin, in *Surgical and Physiological Essays*, Part 2, London, 1793, 107.
2. Gerlach, J.V., Über das Hautathmen, *Arch. Anat. Physiol. Leipzig*, 1851, 431.
3. Stow, R.W. and Randall, B.F., Electrical measurement of the pCO_2 of blood, *Am. J. Physiol.*, 179, 678, 1954.
4. Severinghaus, J.W. and Bradley, A.F., Electrodes for blood pO_2 and pCO_2 determination, *J. Appl. Physiol.*, 13, 515, 1958.
5. Johns, R.J., Lindsay, W.J., and Shephard, R.H., A system for monitoring pulmonary ventilation, *Biomed. Sci. Instrum.*, 5, 119, 1969.
6. Huch, A., Lübbers, A.W., and Huch, R., Patientenüberwachung durch transcutane pCO_2 Messung bei gleichzeitiger Kontrolle der relative lokalen Perfusion, *Anaesthesist*, 22, 379, 1973.
7. Rooth, G., Ewald, U., and Caligara, F., Transcutaneous pO_2 and pCO_2 monitoring at 37°C. Cutaneous pO_2 and pCO_2, in *Continuous Transcutaneous Monitoring, Advances in Experimental Medicine and Biology*, 220, Huch, A., Huch, R., and Rooth, G., Eds., Plenum Press, New York, 1986, 23.
8. Lübbers, D.W., Cutaneous and transcutaneous pO_2 and pCO_2 and their measuring conditions, *Birth Defects*, 15, 13, 1979.
9. Siggaard-Andersen, O., *The Acid-Base Status of the Blood*, 4th ed., Munksgaard, Copenhagen, 1974.
10. Hazinski, T.A. and Severinghaus, J.W., Transcutaneous analysis of arterial pCO_2, *Med. Instrum.*, 16, 150, 1982.
11. Severinghaus, J.W., Stafford, M., and Bradley, A.F., Tc-pCO_2 electrode design, calibration and temperature gradient problems, *Acta Anaesthesiol. Scand. Suppl.*, 68, 118, 1978.
12. Nickelsen, C., Thomsen, S.G., and Weber, T., Fetal carbon dioxide tension during human labour, *Eur. J. Obstet. Gynecol. Reprod. Biol.*, 22, 205, 1986.
13. Nickelsen, C. and Weber, T., Suction fixation of the tc-pCO_2 electrode for fetal monitoring, *J. Perionat. Med.*, 15, 383, 1987.
14. Wimberley, P.D., Pedersen, K.G., Olsson, J., and Siggaard-Andersen, O., Transcutaneous carbon dioxide and oxygen tension measured at different temperatures in healthy adults, *Clin. Chem.*, 31, 1611, 1985.
15. Nickelsen, C., Monitoring of fetal carbon dioxide tension during human labour, *Dan. Med. Bull.*, 36, 537, 1989.
16. Frederiksen, P.S., Wimberley, P.D., Melberg, S.G., Witt-Hansen, J., and Friis-Hansen, B., Transcutaneous pCO_2 at different temperatures in newborns with respiratory insufficiency: comparison with arterial pCO_2, in *Continuous Transcutaneous Blood Gas Monitoring*, Huch, R. and Huch, A., Eds., Marcel Dekker, New York, 1983, 227.
17. Schmidt, S., Langner, K., Gesche, J., Dudenhausen, J.W., and Saling, E., Correlation between transcutaneous pCO_2 and the corresponding values of fetal blood — a study at a measuring temperature of 39°C, *Eur. J. Obstet. Gynecol. Reprod. Biol.*, 17, 387, 1984.
18. Lehmann, K.A., Asoklis, S., Grond, S., and Huttarsch, H., Entwiklung eines kontinuierlichen Monitorings der Spontanatmung in der postoperativen Phase. I. Normalwertbereiche für kutane pO_2- und pCO_2-partialdrucke sowie pulsoxymetrisch bestimmte Sauerstoffsattigungen bei gesunden jungen Erwaachsenen, *Anaesthesist*, 41, 121, 1992.
19. Kavanagh, B.P., Sandler, A.N., Turner, K.E., Wick, V., and Lawson, S., Use of end-tital pCO_2 and transcutaneous pCO_2 as noninvasive measurement of arterial pCO_2 in extubated patients recovering from general anesthesia, *J. Clin. Monit.*, 8, 226, 1992.

Chapter 9.5
Noninvasive Techniques for Assessment of Skin Penetration and Bioavailability

Ronald C. Wester and Howard I. Maibach
Department of Dermatology
University of California
San Francisco, California

I. Introduction

Bioavailability is defined as the rate at and extent to which a compound is absorbed into the body to the target tissue. This is determined by assaying blood compound concentrations over a time course (invasive), or by monitoring compound excretion in urine and feces over a time course (invasive or noninvasive?). The skin is a visible target organ, so "direct" bioavailability can be done by assaying the compound in skin (invasive) or assaying some biological response from the surface of the skin (noninvasive). Assaying a biological response from an applied test article compared to a standard compound gives a "relative" bioavailability. This is a qualitative measurement, not quantitative. Measuring skin blood flow is a qualitative method.

Percutaneous absorption is the rate and extent of a compound passing through the skin and becoming systemically available. It can be determined from blood assay or urine assay. The first step in percutaneous absorption is the partitioning of chemical from vehicle to the stratum corneum, the outer layer of skin. Some speculate that this first step is the limiting step in percutaneous absorption.

This chapter will discuss and illustrate the use of a qualitative response, skin blood flow, and of a new technique, powdered human stratum corneum, to study the noninvasive interactions of compounds and skin.

II. Powdered Human Stratum Corneum (PHSC)

Powdered human stratum corneum is a product made from foot callus that a podiatrist will routinely remove and discard. The foot callus is cut into smaller pieces and then ground with dry ice to form a powder. Uniformity in particle size is achieved with sieving. In our laboratory, that portion of the powder that passed through a 40-mesh but not an 80-mesh sieve was used. The callus is human stratum corneum-derived and, thus, should retain some physical and chemical characteristics of human stratum corneum.[1]

Table 1 shows the partition of benzene, 54% PCBs, and nitroaniline between water and PHSC. The more lipophilic PCBs were the higher binding compound, followed by benzene and then the more hydrophilic nitroaniline. Thus, the binding/partitioning is able to distinguish between compounds. The 10-fold binding of nitroaniline concentrations also shows linearity. Table 2 shows that the binding of nitroaniline to PHSC correlated well with nitroaniline *in vitro* percutaneous absorption in human skin and *in vivo* percutaneous absorption in the Rhesus monkey.[1]

PHSC can be used to determine which chemicals/decontaminants might be able to remove (decontaminate) human skin. The procedure is to mix a chemical into the PHSC and then try to remove that chemical with a series of decontaminants. The liquid decontaminant is poured into the chemical/PHSC and mixed. After a set time period centrifugation will separate a solution from the PHSC. The content of the solution will show the decontamination potential of the solution. Table 3 shows the ability of water, soap and water, mineral oil, and ethanol to remove arachlor 1242 (42% PCBs) from PHSC. All but water can remove some PCBs from PHSC.[2] This is further illustrated in Table 4 and Figure 1, where alachlor readily binds to PHSC. Water alone will only remove a small portion of the alachlor. However, the introduction of 10% soap removes a larger portion of the alachlor and 50% soap removes most of the alachlor. Perhaps this is an elegant way to show that soap in water is necessary to wash one's hands. However, it does illustrate the adaptability of the PHSC to demonstrate skin decontamination.[3]

The PHSC will readily separate from any liquid with centrifugation. Thus, it is readily adaptable to a variety of liquid solutions. One of our laboratory's interests is the potential percutaneous absorption of contaminants from soil. Soil can be readily mixed with PHSC but

TABLE 1 Partition of Contaminants from Water Solution to Powdered Human Stratum Corneum Following 30-Min Exposure

Chemical Contaminant (µg/ml)		Percent Dose Partitioned to Skin
Benzene	21.7	16.6 ± 1.4
54% PCB	1.6	95.7 ± 0.6
Nitroaniline	4.9	2.5 ± 1.1
	1.6	3.7 ± 0.7
	0.49	3.9 ± 1.5

TABLE 2 In Vivo Percutaneous Absorption of p-Nitroaniline in the Rhesus Monkey Following 30-min Exposure to Surface Water: Comparisons to In Vitro Binding and Absorption

Phenomenon	Percent Dose Absorbed/Bound
In vivo percutaneous absorption, Rhesus monkey	4.1 ± 2.3
In vivo percutaneous absorption, human skin	5.2 ± 1.6
In vitro binding, powdered human stratum corneum	2.5 ± 1.1

TABLE 3 In Vitro Decontamination of Aroclor 1242 From Human Powdered Stratum Corneum

	Percent Dose Partitioned into Decontaminant			
Time	Soap and Water	Mineral Oil	Ethanol	Water
0 min	28.4 ± 5.9	63.3 ± 7.6	81.3 ± 6.1	
1 min	26.5 ± 3.5	48.8 ± 3.9	80.0 ± 6.1	<1%
10 min	31.5 ± 3.6	62.1 ± 3.5	80.2 ± 2.1	
1 h	34.8 ± 2.5	64.8 ± 2.3	85.7 ± 3.8	
4 h	38.3 ± 2.5	75.6 ± 1.8	92.2 ± 0.4	
8 h	40.2 ± 3.9	85.7 ± 5.2	92.1 ± 2.5	
Average	33.3	66.7	85.2	

TABLE 4 Partitioning: Alachlor in Lasso with Powdered Human Stratum Corneum

	[^{14}C]Alachlor (% Dose)
Stratum corneum	90.3 ± 1.2
Lasso supernatant	5.1 ± 1.2
Water only wash of stratum corneum	4.6 ± 1.3
10% soap and water wash	77.2 ± 5.7
50% soap and water wash	90.0 ± 0.5

Note: [^{14}C]Alachlor in Lasso EC formulation (1:20 dilution) mixed with powdered human stratum corneum, let set for 30 min, then centrifuged. Stratum corneum wash with (1) water only, (2) 10% soap and water, and (3) 50% soap and water.

TABLE 5A Partition Coefficient of Arsenic in Powdered Human Stratum Corneum/Water and Soil/Water

Test Material	Partition Coefficient[a]
Stratum corneum	1.1×10^4
Soil	2.4×10^4

[a] Partition coefficient =

$$\frac{\text{Concentration of arsenic-73 in 1000 mg PHSC (soil)}}{\text{Concentration of arsenic-73 in 1000 ml water}}$$

TABLE 5B Partition Coefficient of Cadmium Chloride in Powdered Human Stratum Corneum/Water and Soil/Water

Test Material	Partition Coefficient[a]
Stratum corneum	3.61×10^1
Soil	1.03×10^5

[a] Partition coefficient = Concentration of $CdCl_2$ in 1000 mg of powdered human stratum corneum (soil)/Concentration of $CdCl_2$ in 1000 mg of water.

Powdered human stratum corneum does have the potential for medical research application. For instance, a set of vehicles can be screened to determine which vehicle most readily releases a drug into the stratum corneum. This would help with drug delivery into the skin. Also, diseases involving the stratum corneum can be studied using PHSC. An example in Figure 2 is the partitioning of hydrocortisone from normal and psoriatic PHSC. In this case there was no difference in partitioning between normal and psoriatic PHSC.[6] It should be noted that there is no difference in the in vivo percutaneous absorption of hydrocortisone in normal volunteers and psoriatic patients.[7]

Powdered human stratum corneum is a new technique for which uses and applicability to science need to be explored. It also remains for the technique to be validated.

III. Skin Blood Flow

The laser Doppler velocimetry (LDV) method operates on the Doppler principle. Light from a 5-mW helium-neon laser is transmitted to the skin through a quartz optical fiber. The light is backscattered from stationary skin components and by red blood cells moving in the dermal capillaries, which are encountered as the radiation penetrates to a depth of 1 to 1.5 mm (approximately 0.05 in.) A second optical fiber collects the reflected light and the frequency-shifted component is converted to a single flow parameter.

Using the LDV as a noninvasive tool, a study was done to determine if skin blood flow is changed during shaving, and if this type of system can be used to compare the potential effect of razors on shaving. The study was done in human male volunteers to compare the potential effects or differences with two razors. Skin blood flow was measured on three face sites

centrifugation will not separate the two. Something that can be done is to determine separate partition coefficients relative to a common third liquid. This is shown for cadmium and arsenic in Table 5. These "relative" partitions can then be compared to those of other compounds and to skin absorption values.[4,5]

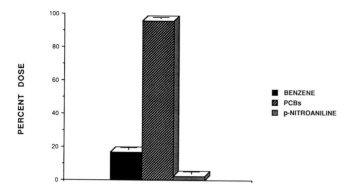

FIGURE 1 Powdered human stratum corneum is able to distinguish the partitioning of PCBs, benzene, and *p*-nitroaniline from water.

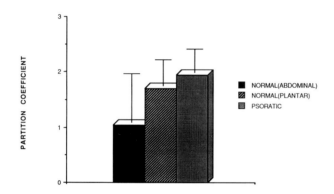

FIGURE 2 Hydrocortisone showed no difference in partitioning to powdered human stratum corneum from normal and psoriatic patients.

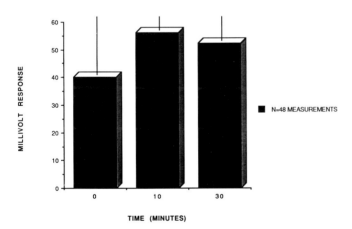

FIGURE 3 Skin blood flow significantly increased following razor shaving.

TABLE 6 Skin Blood Flow and Razor Shaving

	Millivolt Response (n = 8)		
Time	Site 1	Site 2	Site 3
	Razor 1		
Immediately	45.5 ± 34.3	31.2 ± 13.6	31.2 ± 13.6
10 min	68.5 ± 53.0	51.8 ± 24.5	51.4 ± 20.8
30 min	44.8 ± 25.2	42.0 ± 20.7	54.8 ± 28.2
	Razor 2		
Immediately	33.2 ± 26.3	50.2 ± 44.3	44.0 ± 28.3
10 min	57.6 ± 37.8	44.5 ± 20.1	62.6 ± 41.4
30 min	68.5 ± 48.5	48.0 ± 22.3	56.5 ± 26.0

TABLE 7 Human Skin Microcirculation Stimulation by Cosmetic Test Articles

Application	Skin Blood Flow[a] AUC (0–2 h)[b]
A. Methyl nicotinate (0.02%) (positive control)	117 ± 36
B. Base vehicle	73 ± 14
C. Test article I (2%) plus test article II (2%)	79 ± 27
D. Test article I (2%) plus test article II (20%)	85 ± 26
E. Untreated skin	72 ± 17

Statistical Summary AUCs (Student's *t*-test)
A > B ($p < .001$)*   D > B ($p = 0.005$)*
A > E ($p < .001$)*   D > E ($p = 0.005$)*
C ≠ B ($p = 0.22$)   B ≠ E ($p = 0.65$)
C ≠ E ($p = 0.09$)*

[a] Mean – SD for 20 subjects.
[b] AUC = area under time curve for millivolt response · h^{-1}.
* Statistically significant differences.

immediately after shaving and at 10 and 30 min after shaving. Table 6 gives the results. An analysis of variance (ANOVA) statistical analysis showed no difference in LDV response between razors or between skin face sites. However, there was significant difference ($p < .02$) in time. The "immediately after shaving" LDV response was statistically less than that at 10 min or at 30 min (Figure 3). Thus, from LDV response we have shown that shaving does increase face-skin blood flow. This change is not immediate but requires 10 to 30 min to be established.

Another example of using skin blood flow as a tool of investigation was in the search for a cosmetic chemical that could be used to stimulate human skin microcirculation. The first step was to screen a variety of chemicals looking for those which might alter LDV response. Two chemicals with potential LDV stimulation were designated test articles I and II. The study shown in Table 7 shows that the positive control chemical methyl nicotinate increased blood flow relative to base cream and untreated skin. Also the combination of test article I (2%) and test article II (20%) provided statistically significant stimulation to skin blood flow. The results created a new product that is now available in the marketplace.

Skin blood flow studies created a certain level of excitement when the possibility of hair growth in balding people was linked to minoxidil. The hypothesis of the time was that minoxidil stimulated the microcirculation around the hair shaft and that this vasodilation resulted in more nutrition and more

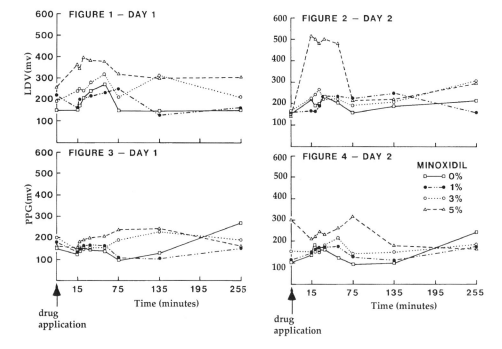

FIGURE 4 Skin blood flow changes *in vivo* by minoxidil application to balding male scalps.

hormonal support to the hair growing process. This was studied using both LDV and photopulse plethysmography (PPG) in balding male volunteers. PPG utilizes a diode-emitting infrared light and determines changes in blood volume passing through the microcirculation by the percentage of incident radiation absorbed.[8]

On day 1 before the drug was applied, blood pressure and pulse rates were obtained. Baseline measurement of blood flow in the scalp skin at the intended site of drug application was made with both techniques, LDV on one half of the scalp and PPG on the other half.

The topical formulations used were vehicle only, and minoxidil in 1, 3, and 5% concentrations in a vehicle of varying proportions of propylene glycol/ethanol/water. Volunteers were randomly assigned to one formulation. A volume of 0.25 ml of formulation was distributed uniformly over 100 cm² of the bald scalp area. A 15-min "drying period" after drug application was allowed before recording was started. The LDV and PPG probes were positioned on the scalp at the site of drug application. Cutaneous blood flow was recorded continuously for 40 to 60 min, and then intermittently for 4 h. Blood pressure and pulse rates were taken after the 4-h recording. Volunteers were asked not to wash the scalp before day 2.

On study day 2 (the following day), vital signs were again obtained and a baseline blood flow measurement obtained. This measurement was both the 24-h measurement for day 1 and the predrug measurement for day 2. The same randomly assigned formulation from day 1 was reapplied to the same scalp site and cutaneous blood flow recorded as on day 1. Vital signs were again obtained after the 4-h session.

The blood flow data were analyzed using ANOVA techniques and the paired *t*-test. Change from baseline (difference scores) were analyzed using the paired *t*-test within treatment groups. Differences between treatment groups were evaluated using ANOVA. A summary result was obtained by averaging the scores across the periods (excluding the predrug score). An ANOVA model with factors, time period at treatment (minoxidil solution), was used to obtain the significance levels for the summary results. Vital signs were analyzed using the paired *t*-test within treatment groups and ANOVA across treatment groups.

The results from the use of LDV were judged better than that with PPG; however, both techniques showed that minoxidil stimulated blood flow in male balding scalps (Figure 4). From a bioavailability standpoint, the LDV response was dose-related, with the 5% minoxidil having the greatest response. From a technical viewpoint the LDV response could be used to determine relative bioavailability of chemical doses in experimental vehicles, provided of course that the chemical stimulates skin blood flow. Other test articles have also been examined in male balding scalps and they show positive responses (Figure 5).

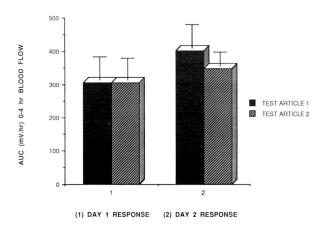

FIGURE 5 Various test articles can be screened as to their potential for increasing blood flow in balding human scalps.

IV. Discussion

Powdered human stratum corneum is a new technique to study the interactions of chemicals with the stratum corneum. Any interaction between a chemical and this outer layer of human skin can be studied. This includes chemical partitioning from vehicles to skin (the start of percutaneous absorption), skin decontamination, and quantitative structure-activity relationships (QSAR). This is a new technique which will need to be validated for any suggested correlation (such as to percutaneous absorption).

Skin blood flow is a biological response; thus, it is limited to chemicals or inflictions (razor shaving) that cause this biological response. Chemicals that stimulate skin blood flow show an increased response and thus can be measured. The vasoconstriction response decreases skin blood flow; thus, the response becomes self-limiting. In all cases the biological response must be considered qualitative and not quantitative.

References

1. Wester, R.C., Mobayen, M., and Maibach, H.I., *In vivo* and *in vitro* absorption and binding to powdered human stratum corneum as methods to evaluate skin absorption of environmental chemical contaminants from ground surface water, *J. Toxicol. Environ. Health,* 21, 367, 1987.
2. Wester, R.C., Maibach, H.I., Bucks, D.A.W., McMaster, J., Mobayen, M., Sarasan, R., and Moore, A., Percutaneous absorption and skin decontamination of PCBs: *in vitro* studies with human skin and *in vivo* studies in the Rhesus monkey, *J. Toxicol. Environ. Health,* 31, 235, 1990.
3. Bucks, D.A.W., Wester, R.C., Mobayen, M.M., Yang, D., Maibach, H.I., and Coleman, D.L., *In vitro* percutaneous absorption and stratum corneum binding of alachlor: effect of formulation dilution with water, *Toxicol. Appl. Pharmacol.,* 100, 417, 1989.
4. Wester, R.C., Maibach, H.I., Sedik, L., Melendres, J., Di Zio, S., and Wade, M., *In vitro* percutaneous absorption of cadmium from water and soil into human skin, *Fundam. Appl. Toxicol.,* 19, 1, 1992.
5. Wester, R.C., Maibach, H.I., Sedik, L., Melendres, J., and Wade, M., *In vivo* and *in vitro* percutaneous absorption and skin decontamination of arsenic from water and soil, *Fundam. Appl. Toxicol.,* 20, 1993 (in press).
6. Wester, R.C. and Maibach, H.I., Dermatopharmacokinetics in clinical dermatology, *Semin. Dermatol.,* 2, 81, 1983.
7. Wester, R.C., Bucks, D.A.W., and Maibach, H.I., *In vivo* percutaneous absorption of hydrocortisone in psoriatic patients and normal volunteers, *J. Am. Acad. Dermatol.,* 8, 645, 1983.
8. Wester, R.C., Maibach, H.I., Guy, R., and Novak, E., Minoxidil stimulates cutaneous blood flow in human balding scalps: pharmacodynamics measured by laser Doppler velocimetry and photopulse plethysmography, *J. Invest. Dermatol.,* 82, 515, 1984.

10.0

The Skin Surface Microflora and pH

10.1 Sampling the Bacteria of the Skin 207
   E. A. Eady

10.2 Mapping the Fungi of the Skin 217
   J. Faergemann

10.3 Measurement of Skin Surface pH 223
   A. Zlotogorski and S. Dikstein

Chapter 10.1
Sampling the Bacteria of the Skin

E. Anne Eady
Department of Microbiology
University of Leeds
Leeds, U.K.

I. Introduction

The bacterial flora of human skin resides in the upper layers of the stratum corneum and within the infundibula of pilosebaceous follicles.[1,2] These are lined with epithelium which is continuous with the epidermis so that the microorganisms that colonize them are sequestered external to the lining of the duct. The methods discussed in this chapter fall into two groups — those that are used to sample organisms from the skin surface and those that are used to extract organisms from within pilosebaceous ducts. It should be pointed out that there is no noninvasive method that will remove bacteria from healthy pilosebaceous units. The only method that will do this requires a biopsy as the starting material.[3,4] This article does not provide an exhaustive historical overview of bacterial sampling methods but instead highlights and describes in some detail those methods which should be of most use in a modern skin research laboratory and/or for routine patient diagnosis.

II. Factors Affecting the Choice and Efficacy of Sampling Methods

Human skin provides a variety of different habitats for the survival and multiplication of bacteria. For example, the environment in the axilla is moist and warm and contrasts sharply with that on the surface of the arms and legs, which is relatively cool and dry. These variations are reflected in differences in the density of the bacterial flora in the different locations. As might be expected, the axilla supports the growth of far higher numbers of bacteria than are found on the limbs (excluding the hands and feet). Exposed sites such as the hands and face are likely to carry higher numbers of transient organisms than nonexposed sites. The resident flora of any given individual varies little in both numbers and types of organism from puberty to old age, when the numbers slowly decline. Perturbation of the microflora results from many factors, such as disease, trauma, therapy (not necessarily antimicrobial), antisepsis, and increased hydration. Some organisms (e.g., group A beta-hemolytic streptococci) may assume resident status for short periods of time, especially in disease states, but will be eliminated once the skin returns to normal. For more information about the skin flora in health and disease, the reader is referred to the excellent book edited by W.C. Noble.[5]

Table 1 summarizes the main bacterial species resident on healthy human skin and the sites in which they are commonly found. Three bacterial groups — the propionibacteria, coagulase-negative staphylococci, and aerobic coryneforms (either *Corynebacterium* or *Brevibacterium* species) are widely distributed over the entire skin surface, although particular species may be associated with a specific habitat type. For example, *Propionibacterium acnes* is the commonest bacterium on sebaceous gland-rich areas of skin, but coagulase-negative staphylococci and aerobic coryneforms are dominant on moist skin. When the integrity of the skin barrier is interrupted by disease or trauma, a variety of nonresident bacteria can flourish. Table 2 provides a summary of the variables that affect the efficacy of skin sampling methods and gives suggestions how some, but not all, of the resulting problems can be overcome.

III. Methods Available

Faced with the degree of variation in numbers and types of bacteria and with the heterogeneity of the skin surface, it is obvious that no one sampling method can be universally applicable. Since the 1950s, when interest in skin microbiology began to increase, a variety of different sampling methods have been developed and subsequently modified so that those available today have resulted from many years of "evolution". It is convenient to divide the methods into four basic categories, which will each be discussed in turn.

A. Impression/Replica Methods
1. Contact Plates
The use of contact plates (see Figure 1) represents the easiest way of sampling skin bacteria. Contact plates

TABLE 1 Resident Bacteria Found at Various Anatomical Locations on Healthy Human Skin

Organism	Forehead	Axilla	Forearm	Hands	Perineum	Toe Webs
Staphylococcus aureus	−	(+)	−	(+)	(+)	−
Staphylococcus epidermidis and other coagulase-negative staphylococci[a]	+	+	+	+	+	+
Micrococcus spp.[b]	(+)	−	(+)	(+)	−	+
Propionibacterium acnes and other cutaneous propionibacteria[c]	+	+	(+)	(+)	+	+
Corynebacterium spp.	(+)	+	(+)	(+)	(+)	+
Brevibacterium spp.	(+)	+	(+)	(+)	(+)	+
Acinetobacter spp.	−	(+)	−	−	−	(+)
Anaerobic cocci[d]	+	?	(+)	?	?	?

Note: A plus sign indicates that the organism is carried by the majority of subjects and/or is usually present at that site in high numbers ($\geq 10^3$ cfu/cm^2). The brackets indicate that the organism may be present in a majority of subjects and/or in low numbers. A minus sign indicates that the organism is not usually found at that site.
[a] The various species of coagulase-negative staphylococci are not evenly distributed.
[b] M. sedentarius is commonest on the foot, whereas M. luteus is more common at other sites. The latter organism often occurs in low numbers on exposed skin and may not be a true resident.
[c] Propionibacterium avidum is most commonly found in intertriginous areas.
[d] Otherwise referred to as Staphylococcus saccharolyticus. No systematic study of the prevalence of these organisms on human skin has been carried out.

TABLE 2 Variables that Affect the Efficacy of Skin Sampling Methods

Variable	Associated or Resulting Problems
Uneven distribution of resident skin microflora	Difficult to define or obtain a representative sample
Heterogeneity of resident bacterial flora	Need to use a variety of selective and nonselective media to process sample
Variation in population density from site to site and person to person	Often need to serially dilute samples in order to achieve accurate quantification
Variation in surface contours of skin and presence of hairs at certain sites	Use of many sampling techniques impossible, especially quantitative methods
Damage to the skin by trauma or disease	Some sampling techniques too aggressive; bacterial flora modified, pathogens likely
Presence of antimicrobial substance (antiseptic, disinfectant, or antibiotic)	May need to add neutralizer to diluent and/or growth medium

are specialized Petri dishes which are filled with any desired culture medium until the agar surface is slightly concave. They are then simply pressed firmly onto the skin to remove surface bacteria. A grid is etched onto the base of the plates to facilitate colony counting. The method is not quantitative, although it does give an estimate of the number of microcolonies at the site sampled. Because the organisms are inoculated directly onto the agar surface, replicate samples cannot be transferred to different culture media or grown under different culture conditions. Therefore, the method is of most use when a specific organism is being sought, for example, the isolation of Staph. aureus from eczema lesions.[6] Further examples of the use of contact plates can be found in the work of Johnston et al.,[7] Brown et al.,[8] and Hendley and Ashe.[9]

2. Pads

The use of velvet pads to remove bacteria from the skin surface has the advantage over contact plates that sufficient organisms are removed to serially inoculate a number of different culture media.[10,11] The pile of the fabric picks up individual bacterial cells and then acts like a series of inoculating wires to transfer the organisms to the agar surface. The method is very inefficient and only a small proportion of organisms are successfully transferred from the velvet to the recovery medium.[12] Pads made from either 85% viscose and a 15% mixture of polyester and polyamide fibers or polyvinyl alcohol foam are now commercially available and represent a considerable improvement over the original material.[13] Both new types of pad can be attached to an applicator and are premoistened before use in a solution consisting of 2% polysorbate 80 (Tween 80) and 0.3% lecithin in phosphate-buffered saline (PBS). A template twice the size of the applicator is held on the skin surface and the pads are rubbed back and forth ten times. In this way organisms lying on and below the surface should be detached. Since Raahave demonstrated that bacterial recoveries could be considerably improved by mechanical rinsing,[12] the pads are no longer used directly to inoculate culture media.

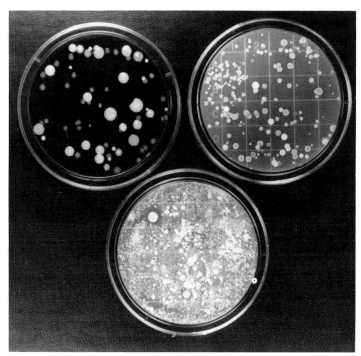

FIGURE 1 Contact plates from the forehead of one individual showing (a) colonies of aerobic coryneforms on fresh blood agar containing furazolidone (upper left), (b) colonies of coagulase-negative staphylococci on CLED (upper right), and (c) colonies of propionibacteria on Brain Heart Infusion containing furazolidone (lower middle). Note that the appropriate use of selective media has prevented the growth of unwanted bacteria and that the density of propionibacteria makes quantitation impossible.

Instead, organisms are released from the pads in a stomacher containing 50 ml of PBS. Aliquots of buffer are removed from the stomacher and plated onto one or more selective or nonselective media as required. The ability to detach the bacteria from the pads has made it possible to obtain a quantitative estimate of bacterial numbers.

3. Sellotape Stripping

Both velvet pads and contact plates only remove bacteria located on the skin surface. Sellotape stripping overcomes the problem of not reaching deeper organisms by facilitating serial sampling of successive layers of the stratum corneum, each just one cell thick.[14] In this way it was shown that the majority of aerobic bacteria reside in the upper part of the epidermis. In areas of skin containing numerous pilosebaceous follicles, some (but not all) organisms will be pulled out of the upper portion of each duct so that the numbers of bacteria may not decline in subsequent strips as they do in areas with few pilosebaceous units. The Sellotape strips are inverted onto the surface of the culture medium and may either be left in contact with the agar surface or removed before incubation. Both procedures are associated with problems. If the tape is left *in situ,* the oxygen tension beneath it is reduced so that growth of aerobic and facultatively anaerobic bacteria is inhibited. If the tape is removed, not all bacteria may have been successfully transferred to the culture medium. Care must be taken to ensure that the tape that is used is sterile and does not contain any bacteriostatic or bactericidal substances. Some authors have used a combination of Sellotape® stripping and contact plates for skin sampling, using the Sellotape® to remove successive sheets of epidermal cells and the contact plates to remove the exposed bacteria.[8,9]

B. Swabbing Methods

Methods for sampling skin bacteria based on the use of swabs generally have a poor reputation among skin microbiologists despite the fact that they have the best ratio of advantages to disadvantages of any currently available technique (see below and Table 3). It is the improper use of dry swabs and the associated poor recoveries of viable organisms that have relegated swabbing to a method of last resort among serious researchers. This is unfortunate, because, when properly used, moist swabs provide one of the most versatile of skin sampling techniques. It is important to stress that the use of swabs rarely provides accurate quantitative data but there are some situations in which the method can be used semiquantitatively, for example, when few organisms are present or when the area swabbed can be accurately defined and the bacteria of interest are known to reside superficially. There is disagreement in the literature as to whether virtually all, or only a small proportion, of the organisms collected on a swab are released during subsequent processing. This will depend on the type of swab used and what procedure is used to transfer organisms to the culture medium. Swabs are available in a variety of materials (cotton, rayon, calcium alginate) but all types should be moistened in either phosphate-buffered saline or Williamson and Kligmans' wash fluid (see below) before use. The area of the sample site can be standardized by holding a template onto the skin surface, but for many applications this is not necessary. The skin surface should be rubbed firmly and repeatedly for several seconds to ensure adequate removal of bacteria. If required, rubbing can be carried out for a fixed time. For semiquantitative work, the swab should be transferred to 1 ml of half-strength wash fluid and decimally diluted in the same. A fixed volume (usually 100 µl) of each dilution and the undiluted sample is then plated onto one or more suitable recovery media and spread with a sterile glass spreader. The appearance of the plates following incubation is shown in Figure 2. Alternatively, viable counts can be determined by the method of Miles and Misra.[15] For qualitative work, the swab can be used directly to inoculate one or more culture media (see Figure 3).

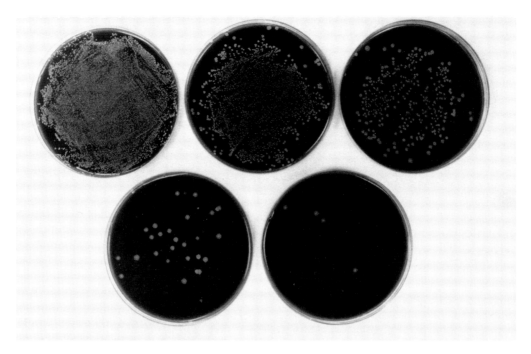

FIGURE 2 Enumeration of coagulase-negative staphylococci by plating fixed volumes from decimal dilutions of wash fluid onto heated blood agar. Note the decreasing number of colonies with increasing dilution (from left to right). Well-separated colonies have been obtained from the 10^{-3} dilution and this plate (bottom left) should be used to estimate the viable bacterial count.

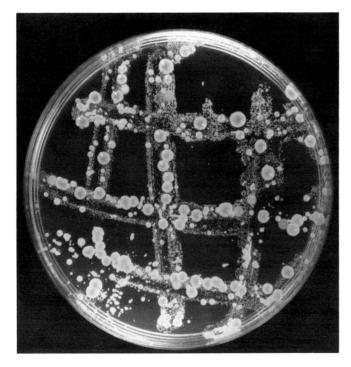

FIGURE 3 Appearance of a plate of Brain Heart Infusion inoculated with a moist swab taken from the forehead following anaerobic incubation for 7 d. The media contains 2 mg/l furazolidone to inhibit the growth of staphylococci and 5 mg/l tetracycline to select for tetracycline-resistant propionibacteria. The swab was taken and plated out by a nurse in a busy dermatology clinic.

There are many applications in which the correct use of swabs represents the best or only possible method of skin sampling. They can be used for the collection of samples for routine clinical bacteriology of patients with skin diseases and are equally useful on intact and broken skin, even in the presence of severe structural breakdown, as in burns or ulcers. Standardized swabbing techniques can be used for research purposes and provide the only means of sampling certain skin sites, such as the toe webs and ear canal. If the swab cannot be processed immediately it should be kept moist by immersion in wash fluid or transport medium. The viability of the collected bacteria will rapidly decline if the swab is allowed to dry out at any stage. Examples of the correct use of swabbing techniques can be found in the studies of Shaw et al.,[16] Keswick et al.,[17] Marshall et al.,[18] McGinley et al.,[19] Keyworth et al.,[20] and Harkaway et al.[21]

C. Washing Methods
1. The Detergent Scrub Technique

The bacterial skin sampling method that has achieved the greatest prominence and is the most widely used for research purposes is the detergent scrub technique of Williamson and Kligman.[22] The method is standardized, quantitative, reproducible, and efficient (i.e., removes over 95% of the aerobic bacteria present at the sample site). Various modifications and adaptations have appeared in the literature over the last 30 years but the technique in its simplest form remains the best and most universally applicable. A metal ring is held firmly against the skin surface and 1 ml of wash fluid (0.075 M sodium phosphate buffer, pH 7.9, containing 0.1% v/v Triton-X 100) is pipetted into it. The skin surface within the ring is rubbed firmly for 1 min with a Teflon policeman, which is lifted away from the skin every few seconds and then replaced. Care must be taken to ensure that the entire area within the ring is rubbed evenly. The wash fluid is collected into a suitable sterile container and the procedure is repeated. The two aliquots of wash fluid are pooled and decimally diluted in half-strength wash fluid. Aliquots of each dilution are then plated onto suitable recovery media, as described in Section III.B above.

There are several features of the scrub wash technique that are worthy of further comment. The wash fluid contains a mild detergent in order to facilitate dispersal of clumps of bacteria. Various authors have studied the survival of different skin bacteria in wash fluid with inconclusive results. It is best for individual investigators to estimate for themselves the survival time in wash fluid of those organisms of special interest to them. Alternatively, it is safe to assume no significant change in viable count if the samples are processed within 1 h of collection. In the original method, the sampling ring enclosed an area of 3.8 cm^2. This area appears to have been arbitrarily chosen. Rings of any size can be used depending upon the reason for sampling. Obviously, the smaller the diameter of the ring, the less representative the sample becomes. We use two ring sizes, 2.5 cm diameter enclosing 4.9 cm^2 for routine skin microbiology and for monitoring the efficacy of antibacterial therapies for acne and eczema, and a larger ring of 3.5 cm diameter enclosing 9.6 cm^2 for studies on antibiotic resistance in skin bacteria (see Figure 4). The cutaneous microflora is more heterogeneous in terms of resistance profiles than in terms of species so that it is necessary to remove bacteria from as large an area as possible to obtain a more representative sample of resistant strains. A smaller ring of 1.5 cm^2 enclosing an area of 1.8 cm^2 can be used to reveal the degree of variation in numbers of resident microorganisms at adjacent sites. The lower limit of detection of the method is form colony-forming units (cfu)/cm^2 using a 2.5 cm diameter ring and 2 ml of wash fluid.

A modification of this technique has been developed for assessing the efficacy of hand wash products (see also Section III.C.2 below).[23] Each fingernail region is immersed in 7 ml of collecting fluid and scrubbed for 1 min with an electric toothbrush. The sample site is chosen because the subungual space harbors large numbers of bacteria and is one of the most difficult sites to disinfect.

2. The Sterile Bag Technique

The bacterial flora of the hands consists of a much greater variety of species than is found elsewhere on the body, except the feet.[18,19] Many of these will not be true residents but transients which are picked up from the environment. The main interest in the microflora of the hand is how to get rid of it. The sterile bag technique is a simplification of the gloved hand method for assessing the efficacy of hand disinfection and provides a way of sampling the flora of the entire hand.[24,25] This is obviously an example of a specialized method developed to fulfil a particular need. Essentially, the hand is enclosed in a sterile plastic bag into which is poured sterile fluid. The bag is sealed at the wrist and the hand is rubbed vigourously for a standard length of time to release the bacteria. Aliquots of the fluid are used to estimate the viable count of bacteria using the most probable number method. Several different sampling solutions have been tested and wash fluid (see above) was found to yield the highest bacterial counts.[25] It may sometimes be necessary to add neutralizers to the sampling solution to inactivate any residual disinfectant.

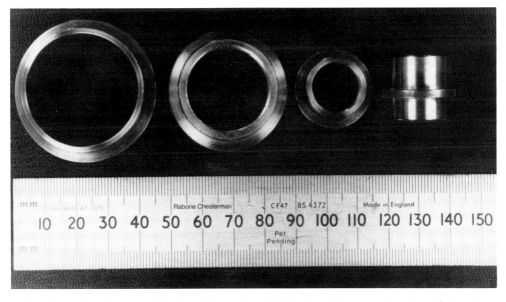

FIGURE 4 Examples of sampling rings used for the detergent scrub technique. For an explanation of when to use different-sized rings, see text.

D. Follicular Sampling Methods

Acne vulgaris is a disease of pilosebaceous follicles of the face, back, and chest in which the resident bacterium, *Propionibacterium acnes,* plays a role in the development of inflammatory lesions.[26] Researchers interested in the pathogenesis of acne need techniques that facilitate sampling of intrafollicular organisms. The methods that will be described below can be used to sample noninflamed lesions only and are not suitable for use with normal follicles or inflamed lesions. All the available data on the bacteriology of normal follicles have been obtained following their microdissection from skin biopsies.[3,4]

1. Comedone Extractor

Open comedones (blackheads) can be removed nontraumatically using a comedone extractor. These are available from chemists because of their cosmetic use. The skin surface is sterilized using an isopropanol swab and the comedone is removed by applying firm downward pressure with the extractor. The comedone is transferred to a sterile preweighed microcentrifuge tube with a sterile needle. After weighing, 200 μl of wash fluid is pipetted into the tube and the bacteria are dispersed from the comedone using a micro-tissue grinder. The fluid is then decimally diluted in half-strength wash fluid and plated onto one or more selective or nonselective media as required. Bacterial counts are expressed as cfu per milligram wet weight of comedonal material. The detection limit is 4 cfu/comedone. The method is unique in that it samples the entire bacterial flora of a single microhabitat. However, care must be taken to ensure that the whole of the comedone is removed from the skin because the distribution of bacteria in follicular ducts varies with depth.[27] The density and composition of the microflora also varies greatly between comedones from a complete absence of viable bacteria to >10^6 cfu/mg of propionibacteria alone, coagulase-negative staphylococci alone, or both.[28-30] Faced with such heterogeneity, it is essential to examine several comedones from each subject in order to obtain an accurate picture. This is especially important if the effects of antibacterial therapy are being studied.

2. Cyanoacrylate Glue

Rapidly polymerizing cyanoacrylate glue was first used to remove thin sheets of stratum corneum in a similar way to Sellotape® stripping.[31] It was quickly realized that follicular plugs were extracted from pilosebaceous follicles as the glue was pulled away from the skin surface. In the simplest form of the method, a drop of the glue is spread over an area of skin and left to polymerize for 1 min.[32] A second drop of glue is then applied on top of the first and spread uniformly by inverting a glass slide over it and pressing down firmly. After another minute, the slide is gently removed from the skin with its adherent sheet of adhesive and follicular casts. These casts represent the contents of microcomedones and consist of a mixture of corneocytes, sebum, and microbes. They can be microdissected from the glue and pooled or homogenized individually with micro-tissue grinders as described above. A more standardized version of this procedure uses a sterile glass sampler of known surface area instead of a slide and a sterile Teflon ring to delineate the sample area, which is extracted twice with glue.[33] The organisms are released by vortexing the end of the sampler in sterile medium in the presence of Ballotini beads. Exolift®* is a commercially available kit which includes a patented dermal tape and cyanoacrylate glue. It is easier to use than glass slides or samplers, but is much more expensive. Whichever procedure is followed, only follicular bacteria will be enumerated since surface organisms are sequestered between the glue and the thin sheet of stratum corneum. The main problems with the use of cyanoacrylate glue are the high frequency of incomplete "takes", when the glue fails to polymerize properly over part of the sample site, and the uncertainty of removing entire follicular casts. For obvious reasons, the method should not be used near the eyes.

IV. Correlation Between the Methods. Advantages and Disadvantages of Each

Of the methods described above, Sellotape stripping and the use of velvet pads are now outdated and have no place in modern skin bacteriology. They will not be discussed further in this article. Several research groups have studied the correlation between different sampling methods, with various outcomes.[6,13,16,20,34-36] It should be obvious that counts obtained with contact plates are consistently lower than those obtained by the other surface sampling methods because there is no dispersal step to break down aggregates of cells into smaller colony-forming units. Counts obtained by the detergent scrub technique are usually taken as the gold standard against which the efficacy of other methods is compared. No conclusive evidence has emerged from any of these studies to be able to state with certainty the relative efficacy of alternative surface sampling techniques. There is no doubt that none of the other methods is more efficient in terms of number of organisms recovered or in reproducibility than the detergent scrub technique. For many applications, the absolute number of organisms at the sample site is not of paramount importance but the investigator must be

---

* Registered trademark of Exovir Inc., Great Neck, New York 11021.

sure that the method chosen removes a constant proportion of the organisms present in order that comparisons between samples taken from different individuals, or sites, or at different times, can be validly made.

Correlations of the efficacy of follicular sampling methods would be meaningless because comedone extraction samples blackheads, whereas the use of cyanoacrylate glue samples predominantly microcomedones. In patients with numerous blackheads, these too will be removed by the glue method. Most investigators process blackheads individually whereas microcomedones are usually pooled because of their small size.

Table 3 summarizes the main advantages and disadvantages of the methods outlined above. Viscose and foam pads have been omitted because they possess no obvious advantages over moist swabbing of a defined area and are more cumbersome to use. Their large size (5 × 5 cm) means that they are unsuitable for use on many skin sites. The sterile bag technique is the accepted method of sampling hands and the only one which can be sensibly used for estimating the efficacy of hand-washing regimens but it is not applicable for bacterial sampling of other skin sites. When choosing a sampling method, the reader should consider which of the factors listed in Table 4 are the most important for a specific application. For example, it is no good selecting the detergent scrub technique if speed and simplicity are essential criteria.

V. Recommendations

For ease and versatility, there is no doubt that moist swabbing is the method of choice for surface sampling. It remains the most commonly used method for routine sampling of patients with diseases, infections, or wounds of the skin. The only practical alternative is the use of contact plates. However, these have severe limitations (see Table 3) and should only be used if a specific organism is being sought or if low numbers of bacteria (≤ 10 cm^2) are expected and they can all be cultured on the same medium. For research purposes, and when quantitative data are required, the detergent scrub technique should be chosen unless there are over-riding reasons why it cannot be used. The major limitation of this technique is that it is fairly aggressive and cannot be used on sensitive or damaged skin, although several groups have used it to sample bacteria from eczema lesions.[6,37,38] With all surface sampling methods, if temporal changes in the bacterial flora are being studied, adequate time must be allowed to elapse after sample collection to allow the bacterial flora to reestablish before a further sample can be taken from the same site. Some investigators overcome this problem by sampling from adjacent sites or from identical sites on the right and left side.

Comedone extraction is the best method of sampling intrafollicular bacteria and when the sample is correctly processed, generates data of very high quality. The limiting factor with this method is the supply of open comedones, which are only present in a minority of acne patients. Microcomedones are more common but the use of cyanoacrylate glue is far from easy. Open and microcomedones which can be obtained noninvasively must not be used as substitutes for normal follicles which can only be obtained following biopsy and which are less frequently colonized by bacteria.[39]

TABLE 3 Advantages and Disadvantages of Recommended Sampling Methods

Method	Advantages	Disadvantages
Contact plates	Easy and quick to use on intact and broken skin. Can be employed for routine patient sampling	Tedious to prepare. Limited to one recovery medium. Not quantitative. Cannot use if skin surface is uneven or hairy.
Moist swabbing	Easy and quick to use. Can be employed for routine patient sampling. Several recovery media can be inoculated immediately (in the clinic or laboratory). Can be used in intertriginous areas and on damaged skin.	Not quantitative.
Detergent scrub technique	Quantitative. Several recovery media can be employed.	Time consuming. Not suitable for routine patient sampling. Cannot be used on all skin sites or if skin is damaged. Specialized equipment needed which is not commercially available.
Comedone extraction	Samples intrafollicular bacteria. Simple and quick procedure. Can study microflora of a single pilosebaceous unit. Quantitative. Can inoculate several recovery media	Single sample not representative. Follicles are diseased. No similar method for normal follicles. Processing of sample tedious.
Cyanoacrylate glue	Samples intrafollicular bacteria. Can pool samples from multiple follicles or study singly. Quantitative. Can inoculate several recovery media.	Samples microcomedones only. Cannot use for normal follicles. Difficult and tedious procedure.

TABLE 4 Factors Important in Determining Choice of Sampling Method

Factor	Possible Alternatives/Important Considerations
Location of bacteria	Surface or follicle
Type of skin	Intact or damaged
	Hairy, smooth or uneven
Type of bacteria	Normal flora or pathogens
	Selection of appropriate culture media
Choice of sample site	One or several
	Pre-determined (e.g., lesion present) or selectable
Efficacy of technique	Qualitative or quantitative
	Number of microcolonies or individual bacteria
	Reproducibility
	Sensitivity
	Ease of use
Reason for sampling	Research or diagnosis
	Comparative or noncomparative

All methods that attempt to generate accurate viable count data include a dispersal step followed by serial dilution and inoculation of recovery media. Table 5 lists suitable media for the isolation of different types of skin bacteria, including major pathogens. It is a waste of time using a sophisticated sampling method if subsequent processing of the sample obtained is inadequate. All media should be incubated aerobically at 37°C for 24 to 48 h, except for the isolation of propionibacteria. These organisms are slow-growing anaerobes and media for their isolation must be incubated in an anaerobic jar or cabinet for 7 d.

It is unlikely that there will ever be a consensus among those who study skin bacteriology as to the most suitable methods for specific tasks. As a guide, some appropriate and inappropriate uses of the methods described in this chapter are shown in Table 6. There are many important unanswered questions in skin microbiology, such as the role of propionibacteria in acne. It is hoped that this article has demonstrated that the cutaneous microflora is easily accessible to study and that the effort may be very rewarding. The following chapter by Jan Faergeman deals with sampling techniques for fungi. All of the methods described above are suitable for the enumeration of *Malassezia furfur (Pityrosporum ovale)*, the yeast that is a member of the resident skin flora, if samples are plated out onto a selective growth medium such as that developed by Leeming and Notman.[40]

TABLE 5 Choice of Growth Media for Recovery of Resident Skin Bacteria and Primary Pathogens

Organism(s)	Recommended Medium	Other Resident Skin Bacteria Capable of Growth
Propionibacteria	Brain Heart Infusion or Reinforced Clostridial agar containing 6 mg/l furazolidone	None
Coagulase-negative staphylococci	Heated blood agar	Nonlipophilic coryneforms *Staph. aureus* *Micrococcus* spp. Gram-negative rods
Staph. aureus	Mannitol salt agar	Some coagulase-negative staphylococci
	Cysteine lactose electrolyte deficient (CLED) medium	Coagulase-negative staphylococci and non-lipophilic coryneforms
Aerobic coryneforms	Fresh blood agar containing 0.2% w/v glucose, 0.3% w/v yeast extract, 0.2% v/v Tween 80 and 6 mg/l furazolidone	*Micrococcus* spp.
Group A beta-hemolytic streptococci	Fresh blood agar containing 0.0002% crystal violet or 7.5 mg/l nalidixic acid and 17 units/ml polymyxin B	None

[a] Incubated anaerobically. All other media incubated aerobically.

TABLE 6 Examples of Appropriate and Inappropriate Uses of Sampling Methods

Reason for Sampling	Appropriate Method	Inappropriate Method
Isolation of *Staph. aureus* from suspected infected eczema.	Contact plate containing selective medium. Sample several lesions.	Dry swab used to inoculate non-selective medium only.
Detection of unknown pathogen in infected wound.	Moist swab transferred to laboratory in transport medium. Inoculate a range of selective and nonselective media immediately.	Dry swab used to inoculate non-selective medium only.
Estimation of the efficacy of a hand-washing regimen.	Sterile bag technique, followed by most probable number method of estimating viable bacteria using a nonselective medium.	Detergent scrub technique on one site.
Determination of the proportion of the resident staphylococcal flora resistant to an antibiotic.	Detergent scrub technique, followed by decimal dilution and plating onto antibiotic-containing and antibiotic-free medium. Sample site will affect result.	Contact plates using antibiotic-containing and antibiotic-free medium.
Identification of components of the aerobic skin flora of premature neonates.	Moist swabbing of many anatomical sites. Plating onto selective and non-selective media.	Contact plates.

References

1. Montes, L.F. and Wilborn, W.H., Location of bacterial skin flora, *Br. J. Dermatol.*, 81 (Suppl. 1), 23, 1969.
2. Wolff, H.H. and Plewig, G., Ultrastructur der mikroflora in follikeln und comedonen, *Hautarzt*, 27, 432, 1976.
3. Puhvel, S.M., Reisner, R.M., and Amirian, D.A., Quantification of bacteria in isolated pilosebaceous follicles in normal skin, *J. Invest. Dermatol.*, 65, 525, 1975.
4. Leeming, J.P., Holland, K.T., and Cunliffe, W.J., The microbial ecology of pilosebaceous units isolated from human skin, *J. Gen. Microbiol.*, 130, 803, 1984.
5. Noble, W.C., Ed., *The Skin Microflora and Microbial Skin Disease*, Cambridge University Press, London, 1992.
6. Williams, R.E.A., Gibson, A.G., Aitchison, T.C., Lever, R., and Mackie, R.M., Assessment of a contact-plate sampling technique and subsequent quantitative bacterial studies in atopic dermatitis, *Br. J. Dermatol.*, 123, 493, 1990.
7. Johnston, D.H., Fairclough, J.A., Brown, E.M., and Morris, R., Rate of bacterial recolonisation of the skin after preparation: four methods compared, *Br. J. Surg.*, 74, 64, 1987.
8. Brown, E., Wenzel, R.P., and Hendley, J.O., Exploration of the microbial anatomy of normal human skin by using plasmid profiles of coagulase-negative staphylococci: search for the reservoir of resident skin flora, *J. Infect. Dis.*, 160, 644, 1989.
9. Hendley, J.O. and Ashe, K.M., Effect of topical antimicrobial treatment on aerobic bacteria in the stratum corneum of human skin, *Antimicrob. Agents Chemother.*, 35, 627, 1991.
10. Holt, R.J., Pad culture studies on skin surfaces, *J. Appl. Bacteriol.*, 29, 625, 1966.
11. Gorril, R.H. and Penikett, E.J.K., New method of studying the bacterial flora of infected open wounds and burns, *Lancet*, 2, 370, 1957.
12. Raahave, D., Experimental evaluation of the velvet pad rinse technique as a microbial sampling method, *Acta Pathol. Microbiol. Scand. Sect. B*, 83, 416, 1975.
13. Hambraeus, A., Hoborn, J., and Whyte, W., Skin sampling — validation of a pad method and comparison with commonly used methods, *J. Hosp. Infect.*, 16, 19, 1990.
14. Updegraff, D.M., A cultural method of quantitatively studying the microorganisms in the skin, *J. Invest. Dermatol.*, 43, 129, 1964.
15. Miles, A.A. and Misra, S.S., The estimation of the bactericidal power of the blood, *J. Hygiene (Cambridge)*, 38, 732, 1938.
16. Shaw, C.M., Smith, J.A., McBride, M.E., and Duncan, W.C., An evaluation of techniques for sampling skin flora, *J. Invest. Dermatol.*, 54, 160, 1970.
17. Keswick, B.H., Seymour, J.L., and Milligan, M.C., Diaper area skin microflora of normal children and children with atopic dermatitis, *J. Clin. Microbiol.*, 25, 216, 1987.
18. Marshall, J., Leeming, J.P., and Holland, K.T., The cutaneous microbiology of normal human feet, *J. Appl. Bacteriol.*, 62, 139, 1987.
19. McGinley, K.J., Larson, E.L., and Leyden, J.J., Composition and density of microflora in the subungual space of the hand, *J. Clin. Microbiol.*, 26, 950, 1988.
20. Keyworth, N., Millar, M.R., and Holland, K.T., Swab-wash method for quantitation of cutaneous microflora, *J. Clin. Microbiol.*, 28, 941, 1990.
21. Harkaway, K.S., McGinley, K.J., Foglia, A.N., Lee, W.-L., Fried, F., Shalita, A.R., and Leyden, J.J., Antibiotic resistance patterns in coagulase-negative staphylococci after treatment with topical erythromycin, benzoyl peroxide, and combination therapy, *Br. J. Dermatol.*, 126, 586, 1992.
22. Williamson, P. and Kligman, A.M., A new method for the quantitative investigation of cutaneous bacteria, *J. Invest. Dermatol.*, 45, 498, 1965.
23. Mahl, M.C., New method for determination of efficacy of health care personnel hand wash products, *J. Clin. Microbiol.*, 27, 2295, 1989.
24. Michaud, R.N., McGrath, M.B., and Gross, W.A., Application of a gloved-hand model for multiparameter measurements of skin-degerming activity, *J. Clin. Microbiol.*, 3, 406, 1976.
25. Larson, E.L., Strom, M.S., and Evans, C.A., Analysis of three variables in sampling solutions used to assay bacteria of hands: type of solution, use of antiseptic neutralisers, and solution temperature, *J. Clin. Microbiol.*, 12, 355, 1980.
26. Webster, G.F., Inflammatory acne, *Int. J. Dermatol.*, 29, 313, 1990.
27. Kearney, J.N., Harnby, D., Gowland, G., and Holland, K.T., The follicular distribution and abundance of resident bacteria on human skin, *J. Gen. Microbiol.*, 130, 797, 1984.
28. Puhvel, S.M. and Amirian, D.A., Bacterial flora of comedones, *Br. J. Dermatol.*, 101, 543, 1979.

29. Leeming, J.P., Ingham, E., and Cunliffe, W.J., The microbial content and C3 cleaving capacity of comedones in acne vulgaris, *Acta Dermatol. Venereol. (Stockholm)*, 68, 468, 1988.
30. Ingham, E., Eady, E.A., Goodwin, C.E., Cove, J.H., and Cunliffe, W.J., Pro-inflammatory levels of interleukin-1 alpha-like bioactivity are present in the majority of open comedones in acne vulgaris, *J. Invest. Dermatol.*, 98, 895, 1992.
31. Marks, R. and Dawber, R.P.R., Skin surface biopsy: an improved technique for the examination of the horny layer, *Br. J. Dermatol.*, 84, 117, 1971.
32. Mills, O.H. and Kligman, A.M., The follicular biopsy, *Dermatologica*, 167, 57, 1983.
33. Holland, K.T., Roberts, C.D., Cunliffe, W.J., and Williams, M., A technique for sampling micro-organisms from the pilosebaceous ducts, *J. Appl. Bacteriol.*, 37, 289, 1974.
34. Noble, W.C. and Somerville, D.A., Methods for examining the skin flora, in *Major Problems in Dermatology*, Vol. 2, Rook, A., Ed., W. B. Saunders, Philadelphia, 1974, 316.
35. Ayliffe, G.A.J., Babb, J.R., Bridges, K., Lilly, H.A., Lowbury, E.J.L., Varney, J., and Wilkins, M.D., Comparison of two methods for assessing the removal of total organisms and pathogens from the skin, *J. Hygiene (Cambridge)*, 75, 259, 1975.
36. Whyte, W., Carson, W., and Hambraeus, A., Methods for calculating the efficiency of bacterial surface sampling techniques, *J. Hosp. Infect.*, 13, 33, 1989.
37. Aly, R., Maibach, H.H., and Shinefield, H.R., Microbial flora of atopic dermatitis, *Arch. Dermatol.*, 113, 780, 1977.
38. Nilsson, E., Henning, C., and Hjorleifsson, M.-L., Density of the microflora in hand eczema before and after topical treatment with a potent corticosteroid, *J. Am. Acad. Dermatol.*, 15, 192, 1986.
39. Leeming, J.P., Holland, K.T., and Cunliffe, W.J., The microbial colonisation of inflamed acne vulgaris lesions, *Br. J. Dermatol.*, 118, 203, 1988.
40. Leeming, J.P. and Notman, F.H., Improved method for isolation and enumeration of *Malassezia furfur* from human skin, *J. Clin. Microbiol.*, 25, 2017, 1987.

Chapter 10.2
Mapping the Fungi of the Skin

Jan Faergemann
Department of Dermatology
University of Gotenburg
Sahlgren's Hospital
Gothenburg, Sweden

I. Summary

Several different fungi may occasionally be isolated from the skin. However, it is only among the yeasts that we have true residents. The lipophilic yeast *Pityrosporum ovale* can constantly be cultured from the skin of adults. To culture the organism, it is important to know that it is lipophilic, grows best in the temperature range of 35 to 37°C and that it is very sensitive to drying. Two culture media, one with the addition of olive oil, glycerol monostearate, and Tween 80 and the other with addition of glycerol, glycerol monostearate, Tween 60, and cow's milk give the best results. Two techniques for quantitative culture are described. One is to use a modification of the Williamson-Kligman model for culturing skin bacteria. With this method a metal ring, a glass rod, and a phosphate-buffered solution containing 0.1% Triton X-100 is used to culture *P. ovale*. Samples are then transferred to a sterile tube, serially diluted, and plated onto the appropriate medium. Plates are read after 6 d of incubation in plastic bags at 37°C. With the contact plate a semiquantitative but easier technique is used. The plate is gently pressed against the skin for 15 s and then incubated in a plastic bag at 37°C for 6 d. A comparative study showed that the cow's milk medium was superior. Quantitative cultures could be used to follow the effect of antifungal treatment of various *Pityrosporum*-related diseases, to compare various diseases, or in other experimental situations.

II. Introduction

With the exception of yeasts, fungi colonize the skin only occasionally. Dermatophytes may be cultured in higher numbers in warm and humid environments and they may therefore be present on the skin due to a contamination instead of a colonization or real infection.[1] However, colonization with dermatophytes and signs of dermatophytosis are more commonly seen among soldiers than other groups due to the presence of predisposing factors.[2] Molds are no true residents of the skin but may now and then be found on the skin due to contamination.

Several yeasts can be cultured from the skin. However, even yeasts may be contaminating the skin. This is probably often so for *Candida albicans*. In healthy people it is not a member of the normal skin flora, but in immunocompromised patients it may be cultured from normal-looking skin.[3] *C. parapsilosis* and *Rhodotorula* sp. are often found on moist, warm, but otherwise normal-looking skin.[1,3] *Thrichosporum biegelii*, the etiological agent in white piedra, may often be cultured from normal-looking skin in areas such as the scrotum.

III. The Lipophilic Yeasts

The only fungus that constantly is cultured from several regions of the skin in almost 100% is the lipophilic yeast *Pityrosporum ovale*.[4-11] *P. ovale* was first cultured and described in skin scales from patients with dandruff and healthy subjects by Castellani and Chalmers in 1913.[12] A nonlipophilic member of the genus *Pityrosporum*, *P. pachydermatis*, was described in 1925 by Waldman.[13] It is a member of the normal skin flora in animals, especially dogs, cats, and rabbits. However, it has also occasionally been cultured from human skin, e.g., in patients with pustular psoriasis.[14]

A. Growth Characteristics of P. ovale
P. ovale is lipophilic and needs the addition of lipids to the culture medium for optimal growth.[4,6,7,12,15] Our previous standard medium for isolation and continuous growth of *P. ovale* is a glucose-neopeptone-yeast extract medium with the addition of olive oil, Tween 80, and glycerol monostearate.[7,16]

However, better results for isolation have been obtained with a culture medium primarily described by

Leeming et al.[11] This medium contained as lipid supplements glycerol, glycereol monostearate, Tween 60, and cow's milk.[11,15] P. ovale is able to grow in the temperature range of 28 to 38°C but in our studies the optimal temperature is 37°C. The yeast is very sensitive to drying and it is therefore of importance to increase the humidity in the incubator for optimal growth. In continuous cultures and when skin scales are scraped down onto the medium, growth is visible after 2 d of incubation at 37°C, and optimal growth is obtained after 3 to 4 d. Colonies are slightly raised, with irregular edges, white to creamy in color and 3 to 6 mm in diameter (Figure 1). The micromorphology of the round and oval forms of the organism is, according to the names, different but there are now several studies showing that the two forms may change into each other,[17-19] and today the majority of workers in the field believe that the different forms only are variations in the cell cycle of the same organism[16-19] (Figures 2 and 3).

B. Techniques for Culture of P. ovale

Qualitative cultures for P. ovale can be obtained by scraping the skin with a curette down onto one of the above-mentioned media.[15,16] The culture plate is then incubated at 37°C in a plastic bag and preferably in an incubator with increased humidity because P. ovale is very sensitive to drying. The plates are read after 4 d. Both of these media are very selective due to the addition of cyclohexamide, antibiotics, and lipids. However, C. albicans may sometimes be found.

Because P. ovale is a member of the normal human skin flora, *quantitative culture* is preferable. One method is a modification of the Williamson-Kligman scrub technique for culturing skin bacteria.[7,26] A stainless steel ring 2.6 cm in internal diameter and 2.0 cm deep covering 5.5 cm^2 area of the skin is used. The skin is gently rubbed with a blunt sterile glass rod and the fluid removed by a Pasteur pipette (Figure 4). The ring is held in place on the skin with moderate pressure from two fingers. One milliliter of sterile 0.075 M phosphate buffer, pH 7.9, containing 0.1% Triton X-100 was poured into the ring and the skin gently rubbed with the glass rod for 1 min. The fluid is removed by Pasteur pipette into a sterile glass tube. Serial dilutions are performed in phosphate-buffered saline (PBS), pH 7.4, containing 0.1% of Triton X-100. Samples (0.1 ml) from the dilutions are plated out. Plates are incubated in plastic bags at 37°C and read after 6 d.

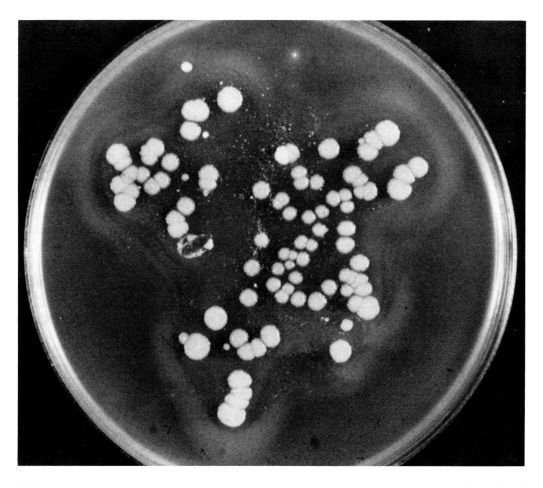

FIGURE 1 Macromorphology of *Pityrosporum ovale* (contact plate showing semiquantitative culture).

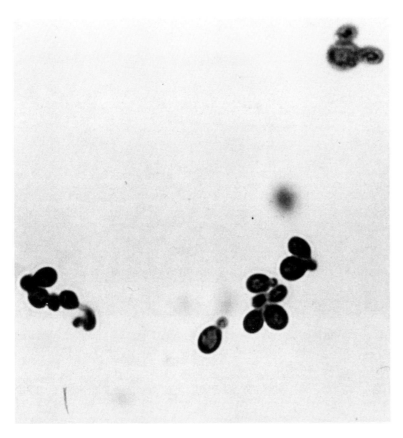

FIGURE 2 Microscopic picture of the round form of *Pityrosporum ovale* (formerly *P. orbiculare*) (× 400).

Other techniques for quantitative culture of yeasts, especially *P. ovale* are mentioned in the literature[27] but will not be mentioned here.

C. Skin Distribution of P. ovale

P. ovale is a member of the normal human cutaneous flora in adults.[16] However, there is great variation in its density and presence in various skin locations,[5,8,9,11] in children compared to adults,[9,20-23] and in normal skin compared to diseased.[7,16,21]

In a survey of children, from newborn to the age of 15 years, *P. ovale* was not cultured from normal-looking skin on the back before the age of 5 years, but was found in 93% of 15-year-old children.[9] In other studies *P. ovale* has been cultured from normal-looking skin in infants.[20,22] However, the incidence was much lower than in adults, although cultures were taken from the forehead[20] or from occluded skin in the diaper area.[22] The increase in colonization of normal skin with *P. ovale* is paralleled by the increase in sebum excretion in prepuberty and puberty. In a culture study in adults from the age of 30 to 80 years we found, using a modification of the Williamson-Kligman scrub technique for quantitative culture of *P. ovale*,[7] a parallel between a reduction in number of cultured *P. ovale* and an increase in age.[23] The reduction in number of cultured organisms with age may partially be explained by a reduction in lipid content of the skin in elderly individuals.[23]

Using direct microscopy or culture, Roberts found *P. ovale* on the normal scalp in 97% and on the chest in 92 to 100% of normal healthy adults.[5] McGinley and co-workers found, by direct count, the presence of *P. ovale*-like organisms on practically all of 112 normal scalps.[24] In a quantitative culture study, again using the modification of the Williamson-Kligman scrub technique, *P. ovale* was cultured from clinically normal skin on the chest, back, upper arm, lower leg, and dorsal aspect of the hand.[8] The highest count was found on the back (mean 333/cm^2) and the lowest count on the hand (mean 2/cm^2).

P. ovale is not only a member of the normal human cutaneous flora but also associated with several diseases, such as pityriasis

Another semiquantitative method is to use contact plates[10,11] (Figure 1). The plate is gently pressed against the skin for 15 s and then incubated into a plastic bag[11] or Bio-Bag CFj®[10] (Marion Laboratories, Kansas City) at 37°C and read after 6 d. A comparison between two contact plates showed that the plate using Leeming's medium[11] gave the best results. The results obtained with these contact plates are only semiquantitative but the plates are very easy to use and could be used in many comparative studies.

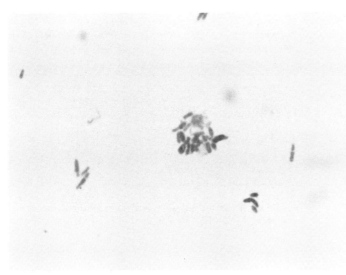

FIGURE 3 Microscopic picture of the oval form of *Pityrosporum ovale* (× 400).

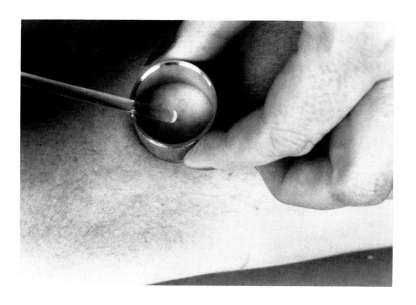

FIGURE 4 Quantitative culture of *Pityrosporum ovale* using a stainless steel ring and rubbing the skin gently with a glass rod.

versicolor,[16-18] *Pityrosporum* folliculitis,[16,24] seborrheic dermatitis,[25] and some forms of atopic dermatitis[21] and confluent and reticulate papillomatosis (Gougerot-Carteaud).[16] In pityriasis versicolor *P. ovale* change, under the influence of predisposing factors, from the normal round or oval blastospore form to the mycelial form.[16] In *Pityrosporum* folliculitis there is an increase in the number of *P. ovale* in the hair follicles, but only in a minority of patients does *P. ovale* change to the mycelial form.[24] In seborrheic dermatitis and probably also in atopic dermatitis the number of *P. ovale* on the skin is the same as in healthy individuals.[21,25] Several patients with seborrheic dermatitis have a slight defect in T-cell immunity and an increased amount of lipids on the skin and the disease is probably due to an abnormal reaction to *P. ovale* in predisposed individuals.[25] Several adult patients with a head and neck distribution of atopic dermatitis have a type I allergic reaction to *P. ovale*.[21]

There is a great variation in the number of *P. ovale* found in various individuals and although the number of *P. ovale* is significantly higher on the skin in patients with pityriasis versicolor compared to healthy subjects the intersubject variations are very high and even a quantitative culture is therefore of minor importance for the diagnosis of a *Pityrosporum*-related disease. In the other *Pityrosporum*-related diseases mentioned above, the number of *P. ovale* on the skin is the same as in normal individuals. However, quantitative cultures are still of importance in several investigations. They can be used to follow the effect of antifungal therapy, epidemiological studies, differences between various diseases, and in several other experimental situations.

References

1. Noble, W.C., *Microbiology of Human Skin*, 2nd ed., LLoyd-Luke, London, 1981.
2. Taplin, D., Fungous and bacterial diseases in the tropics: Final report to the U.S. army R and D Command. Contract DADA 17-71-C1084, 1978.
3. Odds, F.C., *Candida and Candidosis*, Leicester University Press, Leicester, 1979.
4. Gordon, M.A., The lipophilic mycoflora of the skin, *Mycologica*, 43, 524, 1951.
5. Roberts, S.O.B., *Pityrosporum orbiculare*: incidence and distribution on clinically normal skin, *Br. J. Dermatol.*, 81, 264, 1969.
6. Faergemann, J. and Bernander, S., Tinea versicolor and *Pityrosporum orbiculare*: a mycological investigation, *Sabouraudia*, 17, 171, 1979.
7. Faergemann, J., Quantitative culture of *Pityrosporum orbiculare*, *Int. J. Dermatol.*, 23, 110, 1984.
8. Faergemann, J., Aly., R., and Maibach, H.I., Quantitative variations in distribution of *Pityrosporum orbiculare* on clinically normal skin, *Acta Derm. Venereol.*, 63, 346, 1983.
9. Faergemann, J. and Fredriksson, T., Age incidence of *Pityrosporum orbiculare* on human skin, *Acta Derm. Venereol.*, 60, 531, 1980.
10. Faergemann, J., The use of contact plates for quantitative culture of *Pityrosporum orbiculare*, *Mykosen*, 30, 298, 1987.
11. Bergbrant, I.M., Igerud, A., and Nordin, P., An improved method for quantitative culture of *Malassezia furfur*, *Res. Microbiol.*, in press, 1993.
12. Castellani, A. and Chalmers, A.J., *Manual of Tropical Medicine*, Balliéré Cox, London, 1913.
13. Sloof, W.C., Genus *Pityrosporum*, in *The Yeasts*, 2nd ed., Lodder, J., Ed., North-Holland, Amsterdam, 1971, 1167.
14. Sommerville, D.A., Yeasts in a hospital for patients with skin diseases, *J. Hyg. (Cambridge)*, 70, 667, 1972.
15. Leeming, J.P. and Notman, F.H., Improved methods for isolation and enumeration of *Malassezia furfur* from human skin, *J. Clin. Microbiol.*, 25, 2017, 1987.
16. Faergemann, J., Lipophilic yeasts in skin disease, *Sem. Dermatol.*, 4, 173, 1985.
17. Faergemann, J. and Fredriksson, T., Experimental infections in rabbits and humans with *Pityrosporum orbiculare* and *P. ovale*, *J. Invest. Dermatol.*, 77, 314, 1981.
18. Faergemann, J., Aly, R., and Maibach, H.I., Growth and filament production of *Pityrosporum orbiculare* and *P. ovale* on human stratum corneum in vitro, *Acta Derm. Venereol.*, 63, 388, 1983.
19. Faergemann, J., A new model for growth and filament production of *Pityrosporum ovale* on human stratum corneum in vitro, *J. Invest. Dermatol.*, 92, 117, 1989.
20. Broberg, A. and Faergemann, J., Infantile seborrhoeic dermatitis and *Pityrosporum ovale*, *Br. J. Dermatol.*, 120, 359, 1989.

21. Broberg, A., Faergemann, J., Johansson, S., Johansson, S.G.O., and Strannegård, I.L., *Pityrosporum ovale* and atopic dermatitis in children and young adults, *Acta Derm. Venereol.*, 72, 187, 1992.
22. Ruiz-Maldonado, R., Lopez-Matinez, R., Chavarria, P. et al., *Pityrosporum ovale* in infantile seborrhoeic dermatitis, *Pediatr. Dermatol.*, 6, 16, 1989.
23. Bergbrant, I.-M. and Faergemann, J., Variations of *Pityrosporum orbiculare* in middle-aged and elderly individuals, *Acta Derm. Venereol.*, 68, 537, 1988.
24. Bäck, O., Faergemann, J., and Hörnqvist, R., *Pityrosporum* folliculitis: a common disease of the young and middle-aged, *J. Am. Acad. Dermatol.*, 12, 56, 1985.
25. Bergbrant, I.-M., Seborrhoeic dermatitis and *Pityrosporum ovale:* cultural, immunological and clinical studies, *Acta Derm. Venereol. Suppl.*, 167, 1991.
26. Williamson, P. and Kligman, A.M., A new method for the quantitative investigation of cutaneous bacteria, *J. Invest. Dermatol.*, 45, 498, 1965.

Chapter 10.3
Measurement of Skin Surface pH

Abraham Zlotogorski
Department of Dermatology
Hadassah University Hospital
Jerusalem, Israel

Shabtay Dikstein
Unit of Cell Pharmacology
School of Pharmacy
The Hebrew University of Jerusalem
Jerusalem, Israel

I. Introduction

The acidic nature of the skin surface was first determined by Hesus in 1892.[1] Since then numerous studies have confirmed this observation, which is found to be unique to human skin.[2] The normal distribution of skin surface pH on the forehead and cheek of adults was recently described by Zlotogorski[3] in a large study of 282 men and 292 women, aged 18 to 95 years (Figures 1 and 2). No difference was found between men and women regarding forehead and cheek pH distribution (Table 1). The "representative range" (5th to 95th percentile) for the population below the age of 80 is 4.0 to 5.5 on the forehead and 4.2 to 5.9 on the cheek. The values are more alkaline above age 80 (Figures 3a and 3b).

The skin surface pH on the cheek was higher than that on the forehead in 89% of the subjects measured. Forehead and cheek are correlated by the equation: pH cheek = 1.1 + 0.8 pH forehead.

Despite the 100 years that have passed since Hesus's original observation, several questions remain unanswered. What is the cause of this acidity? The lactic acid is the most reasonable source,[3] but this has yet to be confirmed. Why do we need this "acid mantle"? Does it have any protective mechanism? Until now no disease has been associated with increase or decrease of skin surface pH. However, several observations support the need to maintain the skin surface pH within the normal representative range. The skin is more alkaline on both the forehead and the cheek after age 80.[4] There is an increase in propionobacterial count on alkaline skin[5] and increased severity of irritant dermatitis.[6]

II. Object and Methodological Principle

Object — The aim is to measure the skin surface pH (representing the logarithmic reciprocal of the hydrogen ion concentration) in various areas of the body as a function of age and sex.

Instrument — Any commercial portable pH meter can be used. A planar electrode, which combines glass and reference electrodes into a single probe, is connected to the meter. The use of a planar electrode is mandatory in order to provide good contact with the skin.

Method — Place 1 or 2 drops of distilled water on the planar surface of the electrode and dry the surface with a filter paper. Attach the electrode to the area to be measured. The recommended areas are (1) the forehead in the midline, 3 cm above the nasion (Figure 4), and (2) the cheek below the zygomatic bone. However, if desired, every point over the skin can be measured. The electrode is attached to the skin for at least 10 s, until the reading has stabilized. Since the accuracy of the measurement is ±0.1 pH unit, only one decimal place is reported.

III. Sources of Error

Accurate measurement can be assured if the above guidelines are followed carefully. The electrode should always remain wet. The preferred storage solution is pH 4.0 buffer, with 1/100 part of saturated KCl added. Tap water will suffice for short-term storage. Do not soak in distilled water. The planar electrode has to be replaced once a year. The pH meter is calibrated daily

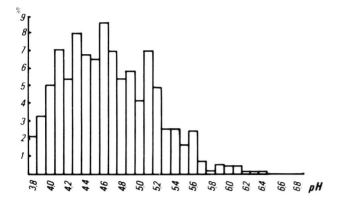

FIGURE 1 Histogram of forehead skin surface pH. The pH is recorded on the x-axis and the y-axis represents the percentage for each given pH grouping. Note that the histogram is skewed and, according to the Kolmogorov D statistics test, the distribution is not normal ($P < .01$; $n = 574$). (From Zlotogorski, A., *Arch. Dermatol. Res.*, 279, 398, 1987. With permission.)

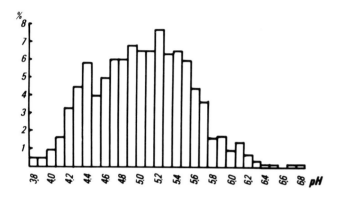

FIGURE 2 Histogram of cheek skin surface pH. See legend to Figure 1. (From Zlotogorski, A., *Arch. Dermatol. Res.*, 279, 398, 1987. With permission.)

TABLE 1 Mean and Percentiles of Forehead and Cheek Skin Surface pH by Sex

		Forehead		
Sex	n	Mean	Median	5th–95th Percentile
M	282	4.7	4.6	4.0–5.5
F	292	4.8	4.6	4.0–5.6
		Cheek		
Sex	n	Mean	Median	5th–95th Percentile
M	282	5.1	5.0	4.2–5.8
F	292	5.2	5.0	4.2–6.1

prior to measurement with standard solutions at pH 4.0 and 7.0. Sweat can influence the reading, so measurement is performed only below 23°C and relative humidity less than 65%.

Soaps, detergents, and creams applied to the skin can also change the results,[7] and readings should be taken at least 12 h after the last application of any soap or cosmetic product.

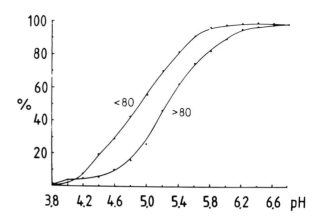

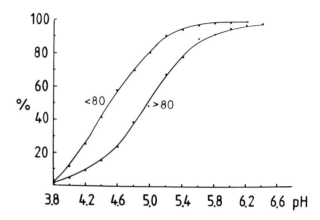

FIGURE 3 Cumulative distribution curves (Kolmogorov-Smirnov) for skin surface pH on (a) forehead and (b) cheek, for the population group under age 80 and for the group aged 80 years and over. Since no sex difference was evident, male (282) and female (292) populations have been combined. (From Dikstein, S. and Zlotogorski, A., in *Cutaneous Investigation in Health and Disease — Non Invasive Methods and Instrumentation*, Lévêque, J.L., Ed., Marcel Dekker, New York, 1989, 59. With permission.)

IV. Correlation with Other Methods

Before the availability of the electrometric procedures, other methods have been used to determine the pH of the skin surface. All methods confirmed the acidic nature of the skin. The colorimetric method is highly accurate when carried out *in vitro* using transparent aqueous solutions. Difficulties arise with colloidal solutions or suspensions, and when determinations are carried out on the skin of a living organism. Colorimetric pH measurements on the skin require the use of three different indicators, each covering a different pH range.[8] A large skin area is required for the application of the indicators. The accuracy is ±0.5 units (compared to ±0.2 units with the electrometric procedures).

Buffer capacity can also be measured by determining the time needed for neutralization of alkaline solution applied to the skin.[9]

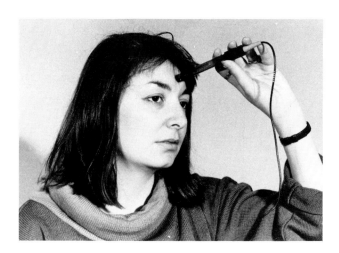

FIGURE 4 Measurement of forehead pH with a planar electrode.

The electrometric pH measurement using the glass electrode represents an improvement over the hydrogen, quinhydrone, or antimony electrodes.[10,11] Behrendt and Green[12] concluded that individual colorimetric and potentiometric measurements cannot be directly compared, and each requires its own reference range for valid interpretation of results. In contrast, trends established using both methods are similar and can be readily compared.[12] The planar electrode, in use since 1972, is superior in both sensitivity and reliability to other potentiometric procedures.[13] Among its advantages are better contact with the skin, inclusion of the active and reference electrodes in a single unit, and accuracy of the measurement within ±0.1 units.

The above-mentioned methods are reviewed in detail in our recent review.[4] A new method for pH measurement uses pH transistor technology, in which an ion-sensitive field effect transistor is used as the sensor.[14] This method is particularly suitable for measuring the pH of semisolid or very viscous materials. According to our preliminary observation this technique is suitable for skin pH measurement, providing the electrode is wet and good skin contact is maintained for 10 s.

V. Recommendations

Measurement of skin surface pH is a simple procedure and can be performed easily by an inexperienced examiner. It can even be done by the examiner himself on his own body (Figure 4). While checking, one should be aware of the sources of error mentioned previously. It is not clear whether or not improving sensitivity to match that of *in vitro* solutions measurements (±0.01) has any biological significance. We recommend that results be reported with only one decimal place precision. During measurement the second decimal place can be covered to avoid confusion due to measurement instability.

It is of great importance to find situations where the pH values differ significantly from the normal "representative range", and to try to correlate these abnormalities or disease states causing the change. This may help to define the factors regulating human skin surface acidity and also lead to an understanding of its biological significance. Attention should be given to the fact that localized alternations of skin surface pH may be due to external factors such as soaps or cosmetics.[7]

At present, however, the practical use of skin pH determinations is limited to evaluating the influence of various materials applied to the skin and possibly the effect of environmental factors on this parameter.

References

1. Hesus, E., Die Reaktion des Schweissen beim gesunden Menschen, *Monatsschr. Prakt. Dermatol.*, 14, 343, 1892.
2. Meyer, W. and Neurand, K., Comparison of skin pH in domesticated and laboratory mammals, *Arch. Dermatol. Res.*, 283, 16, 1991.
3. Zlotogorski, A., Distribution of skin surface pH on the forehead and cheek of adults, *Arch. Dermatol. Res.*, 279, 398, 1987.
4. Dikstein, S. and Zlotogorski, A., Skin surface hydrogen ion concentration (pH), in *Cutaneous Investigation in Health and Disease — Non Invasive Methods and Instrumentation*, Lévêque, J.L., Ed., Marcel Dekker, New York, 1989, 59.
5. Korting, H.Ch., Hubnen, K., Greiner, K., Hamm, G., and Braun Falco, O., Differences in the skin surface pH and bacterial microflora due to long term application of synthetic detergent preparations of pH 5.5 and pH 7.0, *Acta Derm. Venereol.*, 70, 429, 1990.
6. Wilhelm, K.P. and Maibach, H.I., Factors predisposing to cutaneous irritation, *Dermatol. Clin.*, 8, 17, 1990.
7. Bechor, R., Zlotogorski, A., and Dikstein, S., Effect of soaps and detergents on the pH and casual lipid levels, *J. Appl. Cosmetol.*, 6, 123, 1988.
8. Sharlit, H. and Scheer, M., The hydrogen ion concentration of the surface on the healthy intact skin, *Arch. Dermatol. Syphilol.*, 7, 592, 1923.
9. Olivetti, L., Meneghini, C.L., and Chislanzoni, Experimental observations on the skin compartments, its permeability and neutralization time in relation to age, *Gen. Gerontol.*, 14, 957, 1966.
10. Blank, I.H., Measurement of pH of the skin surface. I. Techniques, *J. Invest. Dermatol.*, 2, 67, 1939.
11. Dole, M., The theory of the glass electrode. I, *J. Am. Chem. Soc.*, 54, 3, 1932.
12. Behrendt, H. and Green, M., *Patterns of Skin pH from Birth Through Adolescence — With a Synopsis on Skin Growth*, Charles C. Thomas, Springfield, IL, 1971.
13. Peker, J. and Wahlbas, W., Zur Methodic der pH-Messung der Hautoberflache, *Dermatol. Wochenschr.*, 158, 572, 1972.
14. von Kaden, H., Oelssner, W., Kaden, A., and Schirmer, E., Die Bestimmung des pH-Wertes in vivo mit Ionensensitiven Feldeffecttransistoren, *Z. Med. Lab. Diagn.*, 32, 114, 1991.

Section D: Structure of the Dermis

11.0

Dermatologic Digital Imaging

11.1 Overview of Dermatologic Digital Imaging 229
D.A. Perednia

Chapter 11.1
Overview of Dermatologic Digital Imaging

Douglas A. Perednia
Department of Dermatology
Biomedical Information Communications Center
Oregon Health Sciences University
Portland, Oregon

I. Introduction

Human skin is a large, complex multilayered organ covering an area of approximately 1.75 m^2. Much of clinical and research dermatology depends upon our ability to draw relatively simple conclusions by direct examination of this large structure. Unfortunately, many important details are often obscured by the size and complexity of the skin itself. Are cancerous cells present or absent, and how deeply do they extend? Are the areas of affected skin larger or smaller than before? Which pigmented lesions were present or absent previously? The answers to these and other questions are often hidden within a surface which is irregular, opaque, continuous, and often covered with features which are difficult to measure and quantify. Still worse, the skin to be examined usually belongs to someone else and cannot be studied at the investigator's leisure! As a result, much of what is known in dermatology is based on measurements which are quick, but largely qualitative.

One powerful way to study complex objects is to create a replica or *image* which is simpler than the object, but still contains the information of interest. Images allow us to analyze complex objects such as the skin because they can be more easily examined and analyzed than the object itself. They are simplified versions of reality because they describe just one or a few attributes of the real object. As long as the attributes to be examined are well chosen and accurately portrayed, interpretation of images can provide a simple and direct means of finding answers to basic questions about the object.

The simplicity and versatility of images make them powerful tools for understanding the skin and changes occurring within it. High quality images are objective, easily measured, and provide a permanent record. The current revolution in digital computing has had the result of making the creation and interpretation of images faster, cheaper, and more reliable than ever before.

II. Images and Imaging

Like words, images provide a description of objects as they appear over time. Images differ from words, however, by being graphical representations derived from one or more measurable characteristics of the objects themselves. The process of forming an image involves mapping some property of an object into or onto an image space. This space is then used to visualize the object and its properties, and may be used to characterize its structure and/or function quantitatively. Imaging science is defined by efforts to develop better ways of creating, using, and understanding images.

Three important observations characterize imaging in general. First, any object, visible or invisible, large or small, static or changing, can be imaged as long as it possesses some property that can be measured and recorded. Thus, an internal organ, a mountain, and a star are all valid imaging objects; their common denominator is our ability to describe or map them by certain properties they possess, such as the reflectance of sounds waves, a change in elevation, or emission of light.

Second, virtually any property of an object can be imaged, but often only a few of the measurable properties will contain useful information. Moreover, each property can relay completely different information about the object. For example, an electrocardiogram (EKG), echocardiography, coronary angiography, and a photograph taken at the time of surgery each provide completely different information about the heart. The clinical value of an image is not necessarily related to complexity. Despite being more expensive and anatomically detailed, an echocardiogram is not nearly as useful as an EKG for making the diagnosis of early myocardial infarction. Thus, selection of the appropriate property to image therefore depends entirely on the information sought about the object in question.

Third, the image space itself can take any form. The only relevant consideration is whether it is able to represent the object's properties usefully. Thus, although we are accustomed to seeing objects represented by the effect of light on photographic paper or film, many other imaging spaces are possible. Sculptures, models, holograms, patterns of sound, and computer screens are all equally valid image spaces with their own unique advantages and disadvantages. Choice of a suitable image space therefore also depends upon what questions are asked about an object, and how the answers can best be visualized.

Images have long been used in dermatology. Sketches of skin lesions or surgical procedures are inexpensive, if subjective, representations of the skin onto a two-dimensional progress note. Photographs are more accurate objective representations of skin that currently serve as the accepted "standard" for transmitting objective information about the visual appearance of the skin. Three-dimensional models of skin, hair follicles, and other intricate help us understand the physical relationships between three-dimensional objects.

III. Digital Imaging

Just as letters are used as building blocks for descriptive words, numbers can be used as building blocks for descriptive images. Digital images are graphical displays of arrays of discrete numbers. Each number is a symbol that represents some property of a particular part of the image. In computerized optical imaging the property represented by each number is usually the amount of reflected light, but it could just as easily be transmitted light (e.g., for microscopic sections), reflected sound (ultrasound), or the magnetic spin properties of a material (magnetic resonance imaging, or MRI). Each number represents a part of the picture, or picture element (*pixel*). To form the image on a computer screen, a dot of appropriate brightness is substituted for each number in the picture array. The overall effect of viewing many pixels at once is to see the entire image.

The actual value each pixel takes is determined by convention. Black, for example, is typically represented by pixels with value 0, white by 255, and neutral gray by 128. Colors are represented by combinations of numbers representing red, green, and blue intensity values. The discrete, single-value nature of each pixel gives rise to at least two important types of resolution associated with each digital image, *spatial* and *depth* resolution. Spatial resolution refers to the number of pixels used to represent the image, and is expressed in two ways. The first is the number of pixels used to represent a given distance across the object, e.g. 10 pixels/mm of skin. This represents the magnification of the image, and can be varied to capture more or less detail about the object. If a resolution of 10 pixels/mm is chosen, each pixel represents a patch of skin 0.1 mm × 0.1 mm in size. This resolution is more than adequate for detecting features 0.5 mm in diameter, while a less detailed image with spatial resolution of 1 pixel/mm is likely to miss these structures completely.

The second type of spatial resolution is the display resolution or *dot pitch*. This is typically denominated as pixel width in millimeters (e.g., 0.25 mm/pixel) or, in the U.S., dots per inch (dpi). This measure is analogous to the graininess of film used in photography. Large numbers of dots per inch are desirable for viewing fine spatial details at any given level of magnification (see Figure 1). For example, an image taken of a 100 mm × 100 mm patch of skin at a resolution of 10 pixels/mm contains 1,000,000 (1000 × 1000) pieces of information about the brightness of the surface at different locations. A computer screen of fixed size with relatively low display resolution may only be able to show 300,000 pixels simultaneously. Given the relatively low display resolution, showing the entire

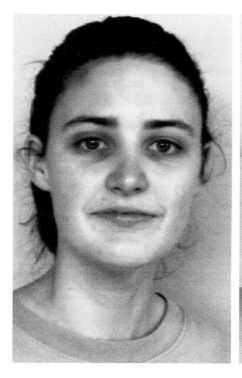

FIGURE 1 Effect of display resolution. Both images are identical with respect to magnification and spatial resolutions across the skin. Only the display resolutions differ. The image on the left has a linear display resolution 4 times higher than the right-hand image. Calculated by area, its resolution is 16 times higher.

image at once will result in either a marked loss of detail or the need to use a much larger monitor.

Depth resolution refers to the maximum size of the numbers that can be used to depict a shade of gray or color, and is expressed in "bits". Single-bit systems can depict only black or white (2^1 different shades), 8-bit systems can display up to 256 (2^8) colors or shades of gray, and 24-bit systems can be used to display nearly 17 million different colors (2^{24}). Depth resolution describes the accuracy with which real-world colors or gray scales can be duplicated. Because the human eye can distinguish about ten million colors, 24-bit pixel depth is sufficient to reproduce every visually perceived color.[1]

IV. Why Use Electronic Imaging?

Electronic digital imaging has four major advantages: (1) it is objective and quantifiable; (2) it is interactive and amenable to automated analysis; (3) it provides a permanent record with easy storage; and (4) it permits certain types of nonvisual imaging that would otherwise be impossible to use.

The first major advantage is that it is objective and quantifiable. Drawing, sketches, and clinic notes are inexpensive and convenient but relatively subjective. Photographs are objective if properly taken but difficult to quantify. Digital images are quantified by definition, and can be used to make exact measurements. If the spatial resolution of an image and the number of pixels used to represent a feature are known, that feature's area can be calculated directly. It is also possible to make quantitative determinations about shape, color, ultrasonic, and MRI thickness, and other observable features of the skin.

Second, electron digital imaging is interactive, and amenable to automated analysis. Unlike drawings, photos, or models, digital images can be queried, altered, and analyzed in real time. This allows the user to ask new questions about the image and sometimes have them answered without referring back to the original object. The computer's great advantage is its ability to process numbers (and thus symbols or portions of an image) by grouping, removing, enhancing, or rearranging them at a rapid rate. *Digital image processing* is the general term used to describe the process of manipulating images in an effort to obtain desired information or visual effects. Image processing algorithms try to automate the derivation of important information from images by highlighting important parts, eliminating noise, and classifying component parts. All digital imaging modalities, including ultrasound, MRI, and visible light imaging, rely heavily on image processing techniques.

In general, image processing algorithms can be divided into several different classes depending upon the functions they fulfill. Most algorithms require the use of computers because they require hundreds of thousands, or even millions, of calculations to implement. Many of these algorithms have been developed for use in military and space programs, where direct observation and measurements of objects is impossible.

Filters are mathematical sieves designed to remove image features or characteristics which may obscure important information about an object. Many common filters deal with the spatial characteristics of images, such as the presence of random "noise", the presence and clarity of feature boundaries, and problems caused by irregular backgrounds. Figure 2 shows an image of skin which is blurred and marred by random pixel values generated by the electronics of the imaging system, and

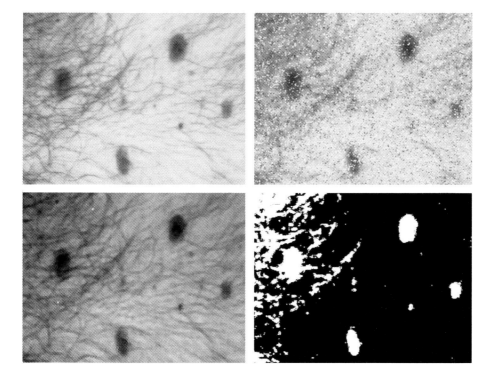

FIGURE 2 Effectiveness of basic image processing techniques. (Upper left) Original scene. (Upper right) Blurred image containing random pixel "noise". (Lower left) Results after applying noise reduction and sharpening filters to image at upper right. (Lower right) Segmentation of the image by taking a brightness threshold. Although the technique is simple and fast, it is sensitive to "noise" and overall illumination. Note that separate classification of image segments would be needed to distinguish hair from pigmented lesions.

the resulting image after application of noise reduction and sharpening filters. Filters are often specifically tailored for use in a given imaging application, and are frequently used in preprocessing images for subsequent examination by a human observer or other image processing algorithms.

Segmentation algorithms are processing steps which divide an image into distinct areas based upon some unifying characteristic(s) of the pixel values within the image. Segmentation is often a critical step in the measurement or interpretation of features within an image, especially when followed by a process of classification. Many different types of segmentation algorithms have been developed. One of the simplest and most frequently used is the process of *thresholding,* in which all pixels with numerical values at or above some given threshold value are assigned to one class, and all pixels with values below the threshold are assigned to a different class. The result is an image divided into distinct regions based on brightness, ability to reflect sound, or whatever other parameter was used to establish the threshold. Other segmentation routines make use of pixel connectivity, contrast, and a wide variety of other image characteristics, either individually or in combination.

Classification algorithms try to take the result of an image segmentation and assign meanings to the various components of an image. For example, after an image segmentation algorithm separates a skin image containing pigmented lesions into many light and dark areas, it still has not determined which of the dark areas represent pigmented lesions as opposed to hair, follicules, or shadows. Classification is often a very complex process, but is almost always needed before the results of area, color, and other measurements can be used effectively. The task of classification is called "interpretation" when done by a human observer, especially when classification is combined with knowledge about disease states and pathologic findings. Because of its complexity, most clinical classification is still done by human observers "reading" ultrasound, MRI, and visible-light images.

The third advantage is that electronic digital imaging provides a permanent record with easy storage. "Permanence" is relative in record-keeping. Photographs offer permanent storage if properly kept, but have some disadvantages in being relatively bulky, inconvenient to store or view, and relatively difficult to secure. New forms of digital media, such as optical disks, offer permanence, security, and easy and economical storage. Hybrid media, such as the Kodak Photo CD®, combine the use of film to acquire high-quality analog images, and then convert those images to a convenient digital format.

The fourth advantage is that it permits certain types of nonvisual imaging that would otherwise be impossible to use. Digital imaging makes a visual analysis of some nonvisible object properties possible, e.g., ultrasound and MRI. This can provide information that is otherwise hidden, such as skin thickness and depth of tumor invasion, although it is important to remember that this information may not be exactly equivalent to that derived by direct observation or other means. Nonvisual information derived by use of electronic imaging is always descriptive of reality as seen through the specific properties used for image acquisition. The images themselves are generated by complex algorithms which convert ultrasonic or MRI signals into a visual cross-section of the structures which must have been present to produce the signals obtained.

The most valuable characteristic of digital imaging is its extreme versatility. Its primary applications are outlined in the following section.

V. Nonvisible-Light Imaging

Only the most superficial structures of the skin are visible to the naked eye. Specific questions about deeper dermal or subdermal structures must be answered either invasively or by using imaging systems that do not rely on visible light. Until recently, few noninvasive means of examining deep cutaneous structures have been available. The skin's transparency to ionizing radiation limits the usefulness of most conventional radiology techniques. Xerography has been successfully adapted for determining skin thickness but is seldom used in practice because of expense, site limitations (generally only the extremities can be imaged), and the potentially harmful effects of X-rays.[2,3]

Recently, however, new nonvisual imaging modalities have become available with the development of cutaneous ultrasound and MRI. Both techniques rely heavily on digital imaging and their potential uses are the subject of active and ongoing investigation.

A. Cutaneous Ultrasound

Ultrasound's application to dermatology is a direct extension of its uses in diagnostic radiology. The principles involved are relatively straightforward.[4,5] Simply put, A- or B-scan ultrasound transducer directs high-frequency sound impulses into the skin. These signals are partially reflected whenever tissues with different acoustical properties are encountered. This generates a series of reflections that correspond to the different types of tissue in the skin. The pattern of these reflections is then plotted onto a screen to provide a map of the tissue interfaces the impulse encountered on its way through the epidermal, dermal, and subcutaneous layers (Figure 3). *Amplitude display,* or A-scan ultrasound maps the intensity of the reflections encountered along a single thin line through the skin surface. The result is displayed in the form of a graph that plots the size or amplitude of each reflection by the reflection's depth in millimeters. *Brightness display,* or B-scan ultrasound collects the data from many individual

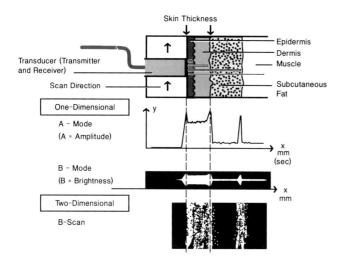

FIGURE 3 Steps used to produce a two-dimensional sonographic cross-section of the skin. A transducer is moved mechanically across the surface of the skin, transmitting ultrasonic pulses and receiving echoes from each transition zone where the acoustical properties tissues differ (Top). At each step, the returning echoes are timed. These return times are multipled by the speed of sound in tissue to determine the distance traveled before reflection. The strength of the echo at each point is then plotted against the time (or distance) traveled to create a one-dimensional A-scan plot (Middle). To create a one-dimensional B-scan image, the strength of the echo is converted to a brightness value between 0 and 255. Stronger echoes are given a higher brightness. When many of these individual readings are placed beside one another over the scan length of the transducer, a two-dimensional B scan is formed (Bottom). (From Perednia, D.A., *J. Amer. Acad. Dermatol.* 89–108, 1991. With permission from Mosby-Yearbook, Inc.)

A-scan soundings and presents it as a digitized two-dimensional picture; bright pixels represent strong reflections and dark pixels depict the absence of reflections. For most physicians B-scan displays are substantially easier to interpret than A scans because they represent a pictorial cross-section of the skin and subcutaneous tissues in a visual and intuitive way. Doppler ultrasound differs significantly from both A- and B-scan devices in using sound phase shift to track vascular blood flow. This system operates by monitoring the Doppler shift in the sound returning from red blood cells that flow toward or away from the transducer. In recent years, research and additional computing power have continued to change the way ultrasonic data is tabulated and displayed. Multiple adjacent B scans can be added together to describe a rectangular patch of skin, just as multiple one-dimensional B-mode scans can be added to form a two-dimensional B-scan image. C-scan ultrasound has been variously defined as different forms of B-scan addition. Just as A scan addition produces a two-dimensional B scan, B-scan addition can provide additional clues about the three-dimensional shape of an imaged object.[6,7]

In general, the sound reflectance of tissue (and thus its ultrasound picture) is a function of the tissue's physical composition and the frequency of the ultrasound signal. Collagen is an important source of echogenicity in any tissue, particularly in skin.[8,9] Cells and collagen fibers intertwined within the dermis produce a bright or "echogenic" layer within the B-scan map (see Section VI). Homogeneous regions, such as the subcutaneous fat, cystic structures, or masses of tumor cells contain little collagen and present relatively few areas in which tissue structure (and therefore sound impedance) changes. These structures appear dark or "echolucent" on the B-scan image. How finely tissue structures can be discriminated is determined by the resolving power of the system. *Axial resolution* refers to how precisely features can be distinguished in the vertical, or depth, dimension. *Lateral resolution* refers to the minimum width of the tissue swath sampled by the A- or B-scan instrument. Axial resolution is mainly a function of the transducer frequency used; higher frequencies allow the user to see at higher resolutions by sacrificing tissue penetration. Current dermatologic A- and B-scan ultrasound units operate at frequencies of 15 to 25 MHz or more, compared with the 2 to 10 MHz generally used in internal medicine applications. These high frequency machines are capable of distinguishing objects only 50 μm thick while surveying an area of skin 22.4-mm long and 0.3-mm wide down to a depth of 1 to 2 cm. Epidermis and adnexal structures can be seen with 20+ MHz devices, although cellular features such as the stratum corneum and basal-cell layers cannot be readily distinguished.[10] B-scan images have begun to resemble noninvasive histologic cross-sections of skin. However, the pattern of ultrasonic reflections do not necessarily bear any resemblance to the familiar cellular structures and staining artifacts seen with conventional histology. Although unfortunate for those desiring a noninvasive substitute for skin biopsy, it does provide a valuable new tool for investigators correlating newly discovered ultrasonic findings with changes in cutaneous biology.

Several important uses of cutaneous ultrasound have been investigated, including the measurement of skin thickness, assessment of mass lesions, and the general noninvasive evaluation of skin structure and function. The use of ultrasound to define skin tumors quantitatively and qualitatively is most valuable when the true extent or nature of the mass cannot be determined by direct examination or palpation. Ultrasound imaging has been successfully used to consistently and reproducibly measure cutaneous structures in a range of conditions.

After nearly 15 years of active research, ultrasound's potential as a useful clinical and research tool in dermatology is quite promising. One of the practical barriers to use in the past has been the cost of dermatologic ultrasound systems. Prices have fallen dramatically as the price of computing power continues to fall by approximately 50% every 18 months. In 1993, a portable clinical ultrasound system could be purchased for approximately $45,000. The less capable systems that it replaces carried prices of $70,000 to $90,000 just 2 years earlier. At the time of this writing, no guidelines had been established for reimbursing the cost of dermatologic ultrasound acquisition and interpretation in the U.S. This situation is likely to change if studies can demonstrate that ultrasound is a cost-effective way to improve the management of patient care.

Future improvements are already well underway in the form of faster scans, higher resolutions, more versatile displays of ultrasonic anatomy, and even attempts at definitive *in vivo* histologic diagnosis. High performance 50-MHz transducers with axial resolutions of 30 μm have been tested with some success.[11] Much higher resolution units, that operate at frequencies of ≥1 GHz (10^9 cycles/s) form the basis of acoustic microscopy.[12] These machines can produce ultrasound images with magnifications that approach that of conventional light microscopic images. Up to 2000× ultrasonic magnification is possible, with lateral resolutions of approximately 1.2 μm. One logical extension to previous work is three-dimensional sonographic modeling of superficial structures. el-Gammal and colleagues[13,14] have developed a three-dimensional modeling program called ANAT3D and reconstructed hair follicles and other structures from high-resolution ultrasound images.[13,14] The use of three-dimensional reconstruction has previously been limited by the computational intensity of the task.

B. Magnetic Resonance Imaging (MRI)

The range of clinical applications of MRI has expanded greatly over the past several years.[15,16] A detailed explanation of MRI is beyond the scope of this chapter, although many excellent works on the subject are available.[17-20] Several basic points deserve emphasis. Most MRI maps the density and behavior of tissue protons that have been stimulated by radio waves in the presence of a strong magnetic field. The protons seen are typically those of hydrogen atoms within water and other tissue molecules. Thus, tissues with markedly different hydrogen content will produce different MRI responses. This "proton density effect" is most prominent in air-filled tissues, such as lung, which appear black in MR images due to their low water content. It is much less useful in discriminating other structures, because differences in proton density between normal and abnormal tissues are usually quite small. The behavior of protons within various tissues often differs considerably, however, and provides the primary means by which structures can be distinguished. The two major proton responses that can be differentiated and quantified are commonly referred to as T_1 and T_2.

The application of MRI to dermatology has become practical with the advent of specialized surface coils that act as sensitive receivers for signals produced by superficial structures, and allow imaging of an extended planar region.[21] For some time, head and surface coils have been successfully used for detecting choroidal detachment, uveal melanomas, and other ophthalmologic tumors.[22,23] Their application to dermatologic imaging is a direct extension of this work.

The advantages of MRI include its ability to measure the depth or thickness of a given skin lesion, as well as its relation to bone, nerves, vessels, and other underlying structures. MRI also measures properties of tissue that cannot be discerned by any other means. The immediate disadvantages of MRI are both technical and monetary. Until recently the best available spatial resolution would detect structures measuring 0.5 mm × 0.5 mm × 2.5 mm. This resolution is low for many dermatologic applications. However, resolutions of 0.1 mm × 0.1 mm × 0.5 mm have been achieved in research devices.[24] A more serious concern is the high cost and complex nature of the necessary equipment. While considerable strides have been made in reducing the magnetic field strength and external shielding needed to obtain high-quality images, unit prices are still in the area of $1 million.

The clinical utility of MRI may be most limited by its high cost and cumbersome equipment requirements. Reimbursement schedules established for other uses of MRI have been used to pay for some dermatologic uses in the U.S., but use of MRI in dermatology has not yet become widespread. Continued attempts to lower the overall cost of health care services are likely to result in careful scrutiny of the costs of technologies such as MRI and competing modalities such as dermatologic ultrasound and direct biopsy.

VI. Visible-Light Imaging

Visible-light imaging records the appearance of the skin as we would ordinarily see it with our eyes. Applications of visible-light imaging range from the simple (recording the appearance of the skin for the medical record), to the complex (determining the diagnosis of a particular skin lesion by use of artificial intelligence). These methods have great potential for *in vivo* examination of the skin as computing power, storage media, and image acquisition methods have become faster and cheaper. Virtually all the benefits of digital imaging and analysis can be applied equally well to images derived from microscopic as well as macroscopic observation.

The technology of visible-light image acquisition is evolving rapidly. Digital images of skin can be acquired either directly or indirectly. Indirect acquisition uses conventional film to first capture an analog photograph of the skin. An example of this technology is

Eastman Kodak's Photo CD process. Using this method, standard color film is developed and rapidly digitized with a high level of both spatial and color resolution — as many as 3072 × 2048 24-bit pixels per frame. The digitized images are then placed on CD-ROM disks which can be used by any personal or workstation computer equipped with a standard CD-ROM drive. The cost of this conversion is quite low; in 1993, 24 photographs could be converted to digital form for less than $20. Indirect imaging permits immediate electronic use of both archival and newly acquired photographic images on a massive scale. Moreover, the technique takes advantage of the large installed base of photographic equipment present in both research and clinical dermatology.

Most direct acquisition on digital images is done by use of solid-state charged-coupled device (CCD) video cameras. CCD cameras contain an array of electronic photosensors which become electrically charged when exposed to light. The magnitude of the charge is then measured and converted to a digital value by a device known as an analog-to-digital converter, or *framegrabber* within a computer, thus forming a digital image. CCD devices are small, lightweight, and increasingly less expensive. A high-quality camera with separate red, green, and blue CCD chips can now be purchased for less than $5,000, making it practical for both research and some clinical use.

A. Clinical Documentation

Of all applications for visible-light imaging, clinical documentation and communication are the simplest and most readily understood. Photographs have long been the "gold standard" of clinical recording. Digital imaging provides certain advantages, however, including: (1) improved storage and access in the form of electronic databases capable of mixing images with text and other information; (2) simplified duplication and transmission of information; and (3) the ability to rapidly measure, analyze, and even interpret the information in images by computer.

The computer's ability to file, relocate, and cross reference data with remarkable speed makes it an obvious choice for storing unwieldy amounts of information. Fully integrating text and photographs in computer-readable form has major advantages: access to images such as photographs could be improved considerably and the physical storage space needed could be reduced to a small fraction of that currently required. With data compression, a single optical disk can be used to store over 2000 full-color images.[25]

Image quality is currently the single most important and difficult issue in this field. There are several factors that determine the quality and "useability" of an image, including fidelity, informativeness, and aesthetics.[26] Conventional photographs have an effective resolution which is high in almost every respect. Photographic spatial resolution is approximately 4000 lines, or dots, per inch. Color resolution is equally high because virtually any color can be represented accurately with careful processing. Despite these high standards, it is not necessarily true that digital images must duplicate them to relay the same amount of clinically useful information.[27] Studies of the image quality necessary for dermatologic applications are still underway. Recent experiments suggest that 24-bit color RGB images with resolutions within the range of current CCD cameras are as informative as photographs for diagnosing a wide range of clinical conditions.[28] These studies will eventually be extended to gauge the effects of image compression and other space-conserving techniques.

One important future use of high-quality digital images is their projection over distances, or *teledermatology,* for diagnostic purposes. The primary application of teledermatology is in improving patient care in medically underserved areas, and obtaining expert opinions in medically difficult cases. Note that this technique can be used to an advantage with any digital imaging modality, including direct imaging, ultrasound, and MRI.

B. Quantitative Measurement

The detection and characterization of change forms the basis of most research in clinical dermatology and is central to the assessment of therapy. Even experienced clinical observers vary when assessing the clinical severity of skin conditions.[29] Fortunately, computers are ideally suited to the task of counting, measuring, and comparing. Once images are archived in digital form, image-processing algorithms can extract objective and quantitative information of clinical interest with great efficiency. Computer imaging has been applied to the quantitative assessment of scalp hair, wrinkles, scale, scars, and pigmented nevi.[30–34] Most macroscopic efforts have concentrated on finding and tracking multiple moles or dysplastic nevi, although similar technology should be applicable to recording and measuring psoriatic plaques, vitiligo, or almost any other visible condition.[35–37] Many microscopic applications have concentrated on measuring diffuse or nonuniformly distributed features, such as elastic tissue, cell cultures, or autoradiography, in slide preparations.[38–41] The speed, accuracy, objectivity and ability to automate digital techniques make them especially useful in obtaining tedious, repetitive, or frequent measurements. Once features are found, a number of automated and semi-automated image processing techniques can be used to measure area, color, border regularity, and shape.[42–46] One of the most important qualities a computer system can offer is *consistency* of measurement. Once a system is consistent, conversion to real-world measurements (such as area in millimeters) can usually be made rather easily. For many clinical applications in screening or therapeutics, it is far more important to know that a change has definitely occurred than to know some "absolute" magnitude of that change.

Once a lesion is found and characterized, change evaluation/comparison techniques can determine whether a predefined "allowable" amount of change in area, color, or shape has occurred. The system can then bring these abnormalities to the attention of the human observer.

C. Digital Image Interpretation

Computer-based image interpretation and diagnosis is the logical conclusion to an automated process of image archiving, feature measurement, and change detection. Such systems have traditionally asked the physician to make observations and enter them into a computer program.[47,48] The program is then asked to analyze these observations, compare them with the features of known diseases, and propose the most likely diagnoses. Over time, the subjectivity of human input and tediousness of the process has spurred efforts to remove humans from the early stages of diagnosis.

The descriptive inputs needed for clinical diagnosis are considerably more complex than those required for most feature characterization and change detection applications. Nevertheless, several groups have performed promising studies into the automated microscopic diagnosis of skin diseases, including cutaneous lymphomas and melanomas.[49-54] Progress has also been made in the area of macroscopic diagnosis. At least two groups have reported overall accuracies of 92 to 98% in the automated diagnosis of malignant melanoma by use of visible-light images, although predictive values are still relatively low.[55,56] The ultimate limitation of these and other computer-driven diagnostic algorithms appears to be the limited amount of information contained within the visible-light, ultrasonic or other appearance of pigmented and other skin lesions.[57]

VII. Summary

Digital imaging is a general tool which is valuable for its great versatility and speed. Techniques used for dermatologic image acquisition and processing are essentially the same as those used in satellite remote sensing, industrial MRI and ultrasound, and many other applications. Customization of these techniques for dermatologic applications has produced promising results in many areas, including easy acquisition and storage of high-quality photographic images, quantitative assessment of skin disease, noninvasive examination of deep skin structures, and early attempts at automated or semiautomated diagnosis. Several of these applications are discussed in detail in the following chapters.

References

1. Hunt RWG. *Measuring Color.* Chichester, England: Ellis Harwood, 1987, 17.
2. Black MM. A modified radiographic method for measuring skin thickness. *Br J Dermatol* 81: 661–666, 1969.
3. Marks R, Dykes PJ, Roberts E. The measurement of corticosteroid induced atrophy by a radiological method. *Arch Dermatol Res* 253: 93–95, 1975.
4. Payne PA. Applications of ultrasound in dermatology. *Bioeng Skin* 1: 293–320, 1985.
5. Payne PA. Ultrasonic methods for skin characterization. *Bioeng Skin* 3: 347–357, 1987.
6. Hoffmann K, el-Gammal S, Altmeyer P. [B-scan sonography in dermatology]. *Der Hautarzt* 41: 11–20, 1990.
7. Perednia DA. What dermatologists should know about digital imaging. *J Amer Acad Dermatol* 25(1), 89–108, 1991.
8. Price RR, Jones TB, Goddard J, James AE. Basic concepts of ultrasonic tissue characterization. *Radiol Clin N Amer* 18: 21–30, 1980.
9. Rosenfeld AT, Taylor KJW, Jaffe CC. Clinical applications of ultrasound tissue characterization. *Radiol Clin N Amer* 18: 21–30, 1980.
10. Querleux B, Lévêque JL, de Rigal J. In vivo cross-sectional ultrasonic imaging of human skin. *Dermatologica* 177: 332–337, 1988.
11. Höß A, Emert H, el-Gammal S, Altmeyer P. A high-frequency ultrasound system for the examination of skin disorders and tumor diagnosis in dermatology. *Biomedizinische Technik* 34: 142–143, 1989.
12. Buhles N, Altmeyer P. Ultrasonic microscopy of skin sections. *Z Hautkr* 63(11): 926–934, 1988.
13. el-Gammal S. Experimental approaches and new developments with high-frequency ultrasound in dermatology. Congress Report, International Congress on Ultrasound in Dermatology, March 15–17, 1990: 7–8.
14. Altmeyer P, el-Gammal S, Hoffman K. Looking within the skin without incision and biopsy. *Münch med Wschr* 132(18): 14–22, 1990.
15. Kaplan PA, Helms CA. Current status of temporomandibular joint imaging for the diagnosis of internal derangements. *AJR* 152: 697–705, 1989.
16. Hasso AN, Christiansen EL, Alder ME. The temporomandibular joint. *Radiol Clin North Am* 27(2): 301–314, 1989.
17. Longmore DB. The principles of magnetic resonance. *Brit Med Bull* 45(4): 848–880, 1989.
18. Kean DM, Smith MA. *Magnetic Resonance Imaging, Principles and Applications.* Baltimore: Williams & Wilkins, 1986.
19. Sigal R, Doyon D, Halimi Ph, Atlan H. *Magnetic Resonance Imaging, Basis for Interpretation.* Berlin-Heidelberg: Springer-Verlag, 1988.
20. Council of Scientific Affairs. Fundamentals of magnetic resonance imaging. *JAMA* 258: 3417–3423, 1987.
21. Hyde JS, Tesmanowicz H, Kneeland BJ. Surface coil for MR imaging of the skin. *Magn Reson Med* 5: 456–461, 1987.
22. Peyman GA, Mafee MF. Uveal melanoma and similar lesions: the role of magnetic resonance imaging and computed tomography. *Radiol Clin North Am* 25: 471–486, 1987.

23. Mafee MF, Peyman GA, Grisolano JE et al. Malignant uveal melanoma and stimulating lesions: MR imaging evaluation. *Radiology* 160: 773–780, 1986.
24. Margulis AR, Crooks LE. Present and future status of MR imaging. *AJR* 150: 487–492, 1988.
25. Cookson J, Sneiderman C, Colianni J, Hood A. Image compression for dermatology. Proc SPIE Medical Imaging IV 1990; 1232: in press.
26. Smith WJ. Viewing computer color images for medical applications. *M.D. Computing* 5(4): 58–70, 1988.
27. Kundel HL. Visual perception and image display terminals. *Rad Clin North Am* 24(1): 69–78, 1986.
28. Perednia DA, Gaines JA, Butruille TW. Using multiple-choice receiver operating characteristic (ROC) analysis to compare the informativeness of clinical imaging media. In review, 1993.
29. Marks R, Barton SP, Shuttleworth D, Finlay AY. Assessment of disease progress in psoriasis. *Arch Dermatol* 125: 235–240, 1989.
30. Gibbons RD, Fielder-Weiss VC. Computer-aided quantification of scalp hair. *Dermat Clin* 4(4): 627–640, 1986.
31. Gibbons RD, Fiedler-Weiss VC, West DP et al. Quantification of scalp hair — a computer-aided methodology. *J Invest Dermatol* 86(1): 78–82, 1986.
32. Gormley DE. Computer models and images of the cutaneous surface. *Dermatol Clin* 4(4): 641–649, 1986.
33. Barton SP, Marks R. Image analysis as a tool for measuring biological phenomena of the skin. *Int J Cosmet Sci* 10: 137–144, 1988.
34. Murray AE. A routine method for the quantification of physical change in melanocytic naevi using digital image processing. *J Audiov Media Med,* 11: 52–57, 1988.
35. Devereux DF. Melanoma-pigmented lesion center. *N J Med* 86(5): 401–403, 1989.
36. Devereux DF. Diagnosis and management of dysplastic nevus syndrome and early melanoma. *Primary Care Cancer* 10(5): 19–31, 1990.
37. White RG, Perednia DA, Schowengerdt RA. Automated feature detection in digital images of skin. *Comput Methods Prog Biomed* 1991, in press.
38. Flotte TJ, Seddon JM, Zhang Y, Glynn RJ, Egan KM, Gragoudas ES. A computerized image analysis method for measuring elastic tissue. *J Invest Dermatol* 93(3): 358–362, 1989.
39. Uitto J, Brockley K, Pearce RH, Clark JG. Elastic fibers in human skin: quantitation of elastic fibers by computerized digital image analyses and determination of elastin by radioimmunoassay of desmosine. *Lab Invest* 49: 499–505, 1983.
40. Stolz W, Scharfetter K, Abmayr W, Köditz W, Krieg T. An automatic analysis method for in situ hybridization using high-resolution image analysis. *Arch Dermatol Res* 281: 336–341, 1989.
41. Smolle J, Helige C, Soyer H-P, Hoedl S, Popper H, Stettner H, Kerl H, Tritthart HA, Kresbach H. Quantitative evaluation of melanoma cell invasion in three-dimensional confrontation cultures in vitro using automated image analysis. *J Invest Dermatol* 94(1): 114–119, 1990.
42. Umbaugh S. Computer vision in medicine: color metrics and image segmentation methods for skin cancer diagnosis. Ph.D. Thesis, University of Missouri-Rolla, 1990.
43. Umbaugh SE, Moss RH, Stoecker WV. Automatic color segmentation of images with application to detection of variegated coloring in skin tumors. *IEEE Med Biol* 8(4): 43–52, 1989.
44. Golston JE, Moss RH, Stoecker WV. Boundary detection in skin tumor images: an overall approach and a radial search algorithm. M.S. Thesis, University of Missouri-Rolla, 1989.
45. Li WW. Computer vision techniques for symmetry analysis in skin cancer diagnosis. M.S. Thesis, University of Missouri-Rolla, 1989.
46. Golston JE, Moss RH, Stoecker WV. Boundary detection in skin tumor images: an overall approach and a radial search algorithm. *Pattern Recognition* 23(11): 1235–1247, 1990.
47. Stoecker WV. Computer-aided diagnosis of dermatologic disorders. *Dermat Clin* 4(4): 607–625, 1986.
48. Stoecker WV. Computer diagnosis in dermatology. Presentation to the 47th annual meeting of the American Academy of Dermatology, December 3, 1988.
49. Stolz W, Schmoeckel C, Burg G, Braun-Falco O. Circulating Sézary cells in the diagnosis of Sézary syndrome (quantitative and morphometric analysis). *J Invest Dermatol* 81: 314–319, 1983.
50. Stolz W, Vogt T, Braun-Falco O, Abmayr W, Eckert F, Kaudewitz P, Vieluf D, Bieber K, Burg G. Differentiation between lymphomas and pseudolymphomas of the skin by computerized DNA-image cytometry. *J Invest Dermatol* 94(2): 254–260, 1990.
51. Stolz W, Abmayr W, Schmoeckel C, Braun-Falco O. High-resolution image analysis: a new tool for the recognition of malignant melanocytic nuclei in light microscopy. *J Invest Dermatol* 89(4): 448 (Abst.), 1987.
52. Abmayr W, Stolz W, Korherr S, Wild W, Schmoeckel C. Chromatin texture of melanocytic nuclei: correlation between light and electron microscopy. *Appl. Optics* 26(1): 3343–3348, 1987.
53. LeBoit PE, Fletcher HV. A comparative study of spitz nevus and nodular malignant melanoma using image analysis cytometry. *J Invest Dermatol* 88(6): 753–757, 1987.
54. Stolz W, Groß J, Schmoeckel C, Abmayr W, Braun-Falco O. Which is the best ultrastructural morphometric parameter differentiating between intraepidermal melanocytic cells of benign nevi and malignant melanomas? In: Burger G, Ploem JS, Goerttler K, Eds., *Clinical Cytometry and Histometry.* London: Academic Press, 1987: 523–525.
55. Cascinelli N, Ferrario M, Bufalino R, Zurrida S, Galimberti V, Mascheroni L, Bartoli C, Clemente C. Results obtained by using a computerized image analysis system designed as an aid to diagnosis of malignant melanoma. *Melanoma Res* 2: 163–170, 1992.
56. Schindewolf T, Stolz W, Albert R, Abmayr W, Harms H. Classification of melanocytic lesions with color and texture analysis using digital image processing. *Anal Quant Cytol Histol* 15(1): 1–11, 1993.
57. Perednia DA, Gaines JA, Rossum AC. Variability in physician assessment of lesions in cutaneous images and its implications for skin screening and computer-assisted diagnosis. *Arch Dermatol* 128: 357–364, 1992.

12.0

Ultrasound Examination of the Dermis

12.1 High-Frequency Ultrasound Examination of Skin: Introduction and Guide 239
J. Serup, J. Keiding, A. Fullerton, M. Gniadecka and R. Gniadecki

12.2 Ultrasound B-Mode Imaging and *In Vivo* Structure and Analysis 257
S. Seidenari

12.3 High-Frequency Sonography of Skin Diseases 269
K. Hoffmann, A. Röchling, M. Stücker, K. Dirting, S. el Gammal, A. Hoffman, and P. Altmeyer

12.4 Ultrasound Examination of the Skin and Subcutaneous Tissues at 7.5 to 10 MHz 279
B. D. Fornage

12.5 Ultrasound A-mode Measurement of Skin Thickness 289
T. Agner

12.6 Skin Thickness: Caliper Measurement and Typical Values 293
A.D. Martin

Chapter 12.1
High-Frequency Ultrasound Examination of Skin: Introduction and Guide

Jørgen Serup, Jens Keiding, Ann Fullerton,
Monika Gniadecka, and Robert Gniadecki
Department of Dermatological Research
Leo Pharmaceutical Products
Ballerup
Bioengineering and Skin Research Laboratory
Department of Dermatology
University of Copenhagen
Bispebjerg Hospital
Copenhagen, Denmark

I. Introduction

Ultrasound examination of the skin integument is relatively new. Since the introduction in the late 1970s this field has, however, been in a phase of exponential growth. Computer technologies and digital imaging techniques became widely used during the same period.[1]

In this chapter we wish, mainly based on our own experience, to present an introduction and practical guidance to clinicians and researchers, who are starting up with the ultrasound technique. Our group has participated in the development from the very beginning with uncertainty and prototypes until the more advanced level of technical and clinical experience of today.[2]

The fundamental physical principle of a dermatological ultrasound scanner is the emission of high-frequency ultrasound (>10 MHz) from a transducer. The sound emission is not continuous but pulsed, i.e., the equipment automatically and very rapidly switches between emission of sound and registration of the sound coming back to the same transducer (the echo) from objects being studied. The time lag between emitted and reflected sound waves depends on the physical distance between the surface of the object and the different layers of the object, which might reflect the sound. On the screen a line with peaks representing echoes from different layers, i.e., an *A-mode scan* is seen (Figure 1). The distance between peaks within the object is easily calculated when the intraobject velocity of sound is known. In *B-mode scanning* the transducer is automatically moved tangentially over the object and a number of A scans are depicted and processed electronically, resulting in a cross-sectional image of the object in two dimensions shown on the screen (Figure 2). In *C-mode scanning* a horizontal picture is depicted. The transducer is automatically moved in the horizontal level over an area of skin both along an X-axis and a Y-axis. In three-dimensional *(3D)-scanning* a cube of data and a true 3D image is obtained (Figure 3). In *real time scanning* the scan speed allows visualization of motile structures such as arterial walls. *M-mode scanning* is a special procedure in which such structures and their motility pattern may be characterized.

Using different modes of scanning tissue parameters such as *in vivo* distance, *in vivo* cross-sectional area, and *in vivo* volume can be calculated on a purely noninvasive basis. These facilities, including documentation on a quantitative and widely observer-independent basis, makes ultrasound examination an attractive and powerful tool for both research purposes and for diagnostic purposes in dermatology.

B-mode scanning and cross-sectional imaging is the mode of major and more widespread interest. The more advanced ultrasound techniques are mainly used in special centers of excellence.

The present review will not include specialized techniques such as ultrasound microscopy and Doppler ultrasonography but will focus on the techniques directly relevant for dermatological purposes.

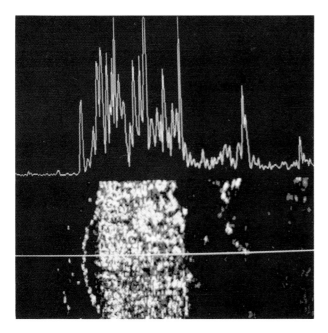

FIGURE 1 A-mode scan (upper half) and B-mode scan (lower half) of skin. An entrance echo is seen (left) corresponding to the epidermal surface. There is an echolucent area just underneath the epidermis. The dermis is rich in ultrasound reflections, i.e., echodense in structure. The interphase to the subcutaneous fat, which is echopoor, is well defined. The scan was obtained from a positive allergic reaction, and the band underneath the epidermis corresponded to inflammatory edema.

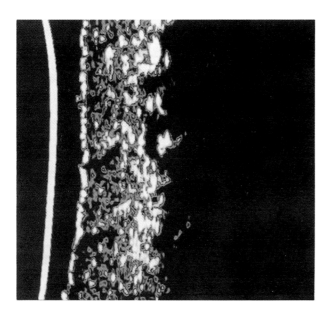

FIGURE 2 The normal skin echogram as shown by B-mode scanning. The membrane between water (coupling between membrane and transducer) and gel (coupling between membrane and skin surface) is seen in the left side. The epidermis is slightly irregular in particularly corresponding to the hair follicles, which are seen as echolucent structures within the dermis and under an oblique angle. The reticular dermis (right side) creates stronger reflectance than the papillar dermis. The interphase to the echo-poor subcutaneous space (right side) is irregular.

FIGURE 3 Three-dimensional scan showing the bifurcation of the cubital vein, the cubital artery and the level of the dermis.

II. Physical Principles and Techniques

The velocity of longitudinal sound waves in a tissue is determined by its elasticity and density. The acoustic characteristic impedance of a tissue is defined as the product of its density and the velocity of sound in the tissue. It is the difference in acoustic impedance between two adjacent media which determines interface echogenicity. Ultrasound reflection and refraction follows optical laws. Thus, the character of a tissue interface and the incident angle of the ultrasound beam are also important for an echo to be registered. Since the same transducer emits and records sound the interface preferably should be perpendicular to the beam unless the interface is somewhat uneven and creates scattering of sound.

It is well known from practical use of ultrasound equipment that the axial resolution is related to the center frequency of the transducer. In theory, however, resolution is not directly determined by the center frequency but by the bandwidth (half-value range around center frequency) of the system. A thorough theoretical discussion is outside the scope of this chapter. An important consequence of these considerations is that the center frequency alone gives only limited information about the resolution of the system. With respect to resolution a high center frequency and a bandwidth of 10 to 15 MHz is optimal for skin examination. General experience shows that in the field of dermatological ultrasonography a center frequency of 20 MHz provides a good compromise between resolution and viewing depth, and 50 MHz or higher frequencies may only be suitable for scanning of the epidermis, where a great number of other methods are easily applicable. It is often forgotten that with high frequency, the viewing field in depth becomes too small.

Resolution and usefulness of a system can only be partly deduced from a list of technical specifications, and skilled use of equipment to solve real problems is the final test. It should be kept in mind that images are qualitative and open to subjective or biased evaluation unless special evaluation techniques such as digital image analysis is applied. Thus, modern principles of objective and blinded comparison should be employed whenever possible. Technical specifications to consider are

Bandwidth and center frequency
Resolution (axial and lateral)
Scan speed (images per s or s per image or real time)
Swept gain (fixed or adjustable)
Scanning field (B-mode or C-mode)
Viewing field in depth (fixed or adjustable)
Scan modes (A, B, C, or M)
Measuring facilities and image analysis
Image storage facilities
Hard copy facilities
Selection of probes and transducers
Selection of display modes (color scales or split screen or zoom, etc.)

III. Ultrasound Velocity in Skin

Estimates of ultrasound velocity are stratum corneum, 1550 m/s; epidermis, 1540 m/s; dermis, 1580 m/s; and subcutaneous fat, 1440 m/s.[3] The average for normal full-thickness skin is 1577 m/s. Ultrasound velocity of 1580 m/s is commonly used for the calculation of total skin thickness. A study showed that ultrasound velocity of skin depended on body region (average: 1605 m/s).[4] Previous studies based on oral mucosa suggested a velocity of 1518 m/s, while studies based on human abdominal skin and porcine skin suggested a velocity of 1710 m/s.[5,6] From a practical point of view, a minor deviation of ultrasound velocity from the true value of a particular location will not influence significantly the result of the thickness measurement, expressed in millimeters to one decimal point. The ultrasound velocity of the entire nailplate is 2459 m/s, and of the dorsal plate and nail matrix 3101 m/s and 2125 m/s, respectively.[7-9]

IV Correlation Between Ultrasonography and Histology

Histology and electron microscopy are, with some limitations, important comparative techniques. One important difference in ultrasonography and histology is that microscopy cannot determine tissue elasticity, and *in vivo* elasticity is an important factor in the acoustic behavior of tissues. Histological staining of

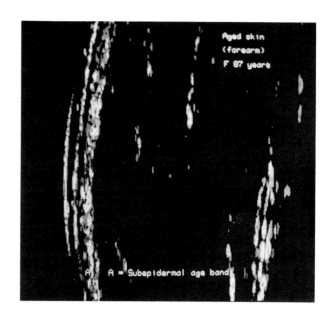

FIGURE 4 Low echogenic subepidermal band of aged skin of the forearm.

tissue specimens is a kind of desirable artifact which need not visualize significant alterations of structure demonstrable by other techniques. Thus, ultrasonography is a separate modality not directly comparable to microscopy. Some structural features are better visualized by ultrasound than histology, and vice versa. An example is the age band of the papillary dermis seen in Figure 4.[10] In scleroderma the collagen may stain normally in histology but it may be severely degraded in electron microscopy, and ultrasound may show an echolucent band in accordance with electron microscopy.[11] Punch biopsies are of fairly limited value for comparative studies since they may undergo retraction and gross change of dimension on cutting depending on the different circumstances.[12] Nevertheless, with respect to the correlation between histological thickness, typically based on surgical biopsy, and ultrasonographic measurement of tumor thickness in malignant melanoma (Breslow thickness), a number of studies showed a remarkably high correlation between the two techniques (see Section V). *In vivo* 20-MHz ultrasound examination does not have the resolution of histology. The advantage of ultrasound is its noninvasiveness and immediate result. A large *in situ* tissue block can be examined with a good presentation of the tissue microanatomy, and with free choice of body region. Thus, ultrasound is essentially a method somewhere in between subjective clinical examination and microscopy based on biopsy, depending on the problem to be evaluated. This intermediate position is also the actual situation in major and very well-established fields such as abdominal and ophthalmological scanning.

TABLE 1 Skin Thickness (Full-Thickness) and Acoustic Density of the Dermis Relative to Anatomical Site

	Females (n = 10)		Males (n = 8)	
	Skin Thickness (mm)	Acoustic Density (a.u.)	Skin Thickness (mm)	Acoustic Density (a.u.)
Forehead	1.79 ± 0.28	16.5 ± 8.6	2.19 ± 0.23	8.9 ± 3.3
Cheek	1.49 ± 0.17	23.7 ± 10.7	1.83 ± 0.15	13.4 ± 7.10
Neck, anterior	1.34 ± 0.17	31.4 ± 9,3	1.61 ± 0.38	28.0 ± 10.1
Neck, posterior	1.92 ± 0.40	17.3 ± 8.9	2.09 ± 0.21	13.4 ± 4.6
Pectral area	1.77 ± 0.21	24.2 ± 6,8	1.92 ± 0.14	22.5 ± 7.8
Abdomen	1.62 ± 0.20	23.4 ± 6,7	1.88 ± 0.12	18.3 ± 5.9
Back, upper	2.33 ± 0.32	14.8 ± 5.9	2.62 ± 0.44	10.1 ± 2.6
Back, lower	2.09 ± 0.34	16.0 ± 7.8	2.21 ± 0.35	12.1 ± 4.8
Upper arm, anterior	1.45 ± 0.34	28.0 ± 10.8	1.53 ± 0.18	22.5 ± 8.1
Upper arm, posterior	1.05 ± 0.09	30.6 ± 6.2	1.21 ± 0.12	26.1 ± 8.0
Forearm, extensor	1.36 ± 0.25	27.4 ± 8.5	1.42 ± 0.14	26.6 ± 7.4
Forearm, flexor	1.12 ± 0.19	29.3 ± 7.1	1.31 ± 0.13	27.0 ± 5.3
Hand, dorsal	1.26 ± 0.18	25.2 ± 8.4	1.50 ± 0.14	23.5 ± 5.9
Palm	1.50 ± 0.52	14.3 ± 7.8	1.48 ± 0.45	9.7 ± 2.7
Thigh, anterior	1.42 ± 0.12	30.3 ± 7.4	1.59 ± 0.21	25.6 ± 9.3
Thigh, posterior	1.46 ± 0.16	25.9 ± 5.0	1.51 ± 0.21	25.8 ± 4.3
Crus, anterior	1.34 ± 0.20	29.0 ± 7.5	1.42 ± 0.19	27.1 ± 8.5
Crus, posterior	1.30 ± 0.12	29.9 ± 7.3	1.34 ± 0.14	24.4 ± 4.5
Foot, dorsal	1.49 ± 0.31	27.2 ± 4.9	1.74 ± 0.26	24.5 ± 9.8
Ankle	2.08 ± 0.38	13.9 ± 9.4	2.01 ± 0.44	6.7 ± 2.3
Heel	—	—	—	—
Sole	1.53 ± 0.29	10.0 ± 5.9	1.60 ± 0.28	9.0 ± 3.3

Note: Mean ± SD; a.u. = arbitrary units. Results were obtained in 18 healthy individuals (ages 24 to 41) with the Dermascan C® and the inbuild image analysis program of this equipment.[26] Imaging of the dermis was difficult in the palm and the sole, and not possible on the heel. In some individuals the face might also be difficult to scan.

V. Correlation Between Ultrasonography, Skinfold Caliper, and Radiography

With skinfold caliper measurement some subcutaneous tissue is inevitably included depending on the anatomical site, and with xeroradiography the dermis-fat interface is somewhat vague.[13,14] Previous studies on the correlation of the different methods for evaluation of skin structures were based on A-mode ultrasonography only. There was, generally, a fair correlation between the methods, but also discrepancies in, for example, inflamed skin.[15–23] There is a general trend that ultrasound shows lower thickness values than the other methods, as might be expected.

There are no studies directly comparing thickness measurement of normal skin based on B-mode 20-MHz ultrasound with thickness measurement using the previous methods because researchers of today find the ultrasound method based on B-mode scanning clearly much more accurate than the previous methods. In our laboratory the reproducibility is 0.05 mm. Three A-scan lines from the top, middle, and bottom of the B-scan image are selected and the average calculated. A theoretically more precise but more cumbersome way of thickness measurement is to outline the area of the block of skin on the image with the region of interest (ROI) function and divide with the width of the block.

VI. Ultrasound Structure of Normal Skin

The epidermis and the dermis reflect ultrasound variably, but with well-defined interface echoes toward ambient air or coupling medium and toward the underlying subcutaneous fat. The normal skin echogram (Figure 2 and Table 1) was described in previous publications and in the different contributions in a recent book.[24–27]

Epidermal echoes may be disturbed by air contained within scales (particularly psoriatic scales), and by the keratotic material of seborrheic keratoses, which causes heavy reflections and shadows that are characteristic or even pathognomonic. The epidermis-dermis interface is obviously very uneven but observations in psoriasis and acanthosis indicate that the ultrasound interface between epidermis and dermis is mainly determined by the top of the rete papillae.[28] This is probably also the case in normal skin.

Epidermis itself is low-reflectant in its internal structure. By 20-MHz A mode scanning one internal epidermal echo is seen close to the entry echo. This profile is obvious in palms and soles, where the epidermis is thick. The internal epidermal echo probably represents the water barrier zone of the skin, comparable to the internal echo found in the nailplate.[7] Reflections from peaks and valleys of dermatoglyphic ridges may be seen with 50-MHz transducers.[29,30] The epidermal thickness is easily measured on palms and soles, where the interface to the very low-reflectant dermis of these regions can be reliably defined.

Dermal echoes are, in most body regions, many and variable (Figure 2 and Table 1). They originate from the well-organized fiber network of the dermis, which is also responsible for the tensile properties of skin and the Langer lines. Affections which erode or disturb this network cause low reflectancy. Subepidermal increase of interstitial water in edema is a common cause of low reflectancy (Figure 1). Dermal echoes may be influenced by the distensibility state of the skin and thus by the position of joints. Hair follicles and sebaceous glands are sometimes seen, depending on the body region (Figure 2). Thus, the normal and undisturbed regular fiber network of the dermis is a kind of natural contrast medium in which different pathologies can be outlined if they cause low reflectancy, or disturbance of interfaces and dimensions. Palms, soles, and to some extent the face and the scalp are exceptions. In these regions the fiber orientation is variable, and ultrasound reflectancy is consequently less. In *neonatal skin,* particularly in premature infants, the whole dermis is low reflectant or echolucent.[31] In early infancy the dermal echo pattern changes toward a normal or adult pattern. In *aged skin* a well-defined subepidermal band of low reflectancy appears[10,32] on sun-exposed sites, such as the forearm (Figure 4). The skin becomes thin in old age, particularly on distal extremities and the dorsum of the hand unless sun damage and repair with actinic elastosis results in thickening. Bleeding and bruises in senile skin progress in this subepidermal zone of low mechanical resistance. In advanced corticosteroid atrophy a similar alteration is seen.

The *subcutaneous space* is normally low reflectant. Low reflectancy depends, however, on the equipment and the gain. With high gain, subcutaneous veins are seen as dark structures. Other anatomical structures such as hypoechogenic tendons can also be visualized by proper adjustment of the gain. On the neck, chest, and back the subcutaneous fascia is often visible.

The *muscle fascia* is easy to define with smooth surfaces, especially toward the muscle, while it may have attachments of retinacula toward the fat. Muscles have few internal echoes. Bone causes heavy reflection.

Ultrasound measurement of *in vivo* distances can be no more reproducible than the actual biology. Obviously, rete papillae do not constitute a line or a plane, and the interface between dermis and subcutaneous fat is far from smooth due to attachments of subcutaneous retinacula (Figure 5). The anatomical thickness of the subcutaneous fatty layer is even more variable. Thus, the biology itself is so "noisy" that an extremely precise distance measurement can never be attained

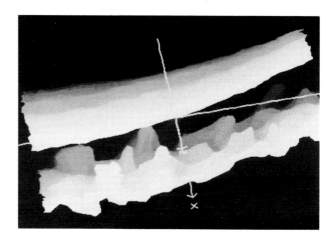

FIGURE 5 3D-ultrasound reconstruction of the skin surface (upper part) and the interphase between the reticular dermis and the subcutaneous tissue (lower part) demonstrating the irregularity of the latter.

even if the ideal scanner existed. There is a popular but obviously incorrect view that, if the frequency of the ultrasound equipment is high enough, all problems of variation and precision are overcome.

VII. Ultrasound Image Analysis

S. Seidenari and A. DiNardo of Modena, Italy developed the field of ultrasound image analysis of skin. This group has in a series of publications demonstrated the utility of the technique for the characterization of a number of clinical conditions including allergic and irritant contact dermatitis, corticosteroid effects, and psoriasis.[33–40] The Modena group works with the Dermascan C® scanner and the Dermavision 2D® dedicated software (Cortex Technology, Hadsund, Denmark) enabling the selection of amplitude range of interest and the transformation into a binary color system (Figure 6). By attributing one color to a selected amplitude range of particular importance, part of an image can be highlighted and characterized by a value corresponding to the number of amplitudes within that range. In the image analysis system amplitudes are represented by pixels (picture elements), which may vary from 0 to 255 in intensity. By selecting a bandwidth from 0 to 30 the hypoechogenic part of the image typical for edema and inflammation may be highlighted. In psoriasis a band between 0 and 10 appears more discriminative. By means of a band of 201 to 255 the echodense areas in the tissue may be selectively quantified. Texture in tissue echogram, speckle formation, and clinical relevance were reviewed in a recent paper.[41] (Seidenari will give a detailed description in Chapter 12.2.)

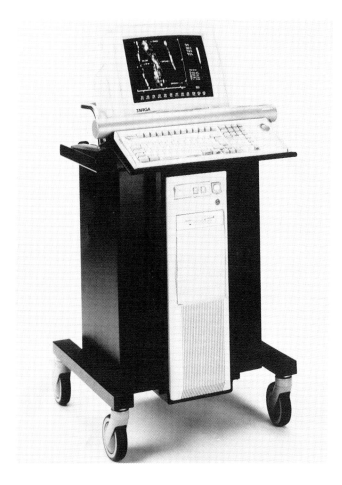

FIGURE 6 The Dermascan C® originally developed in our laboratory as a prototype.[2]

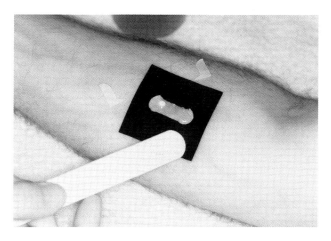

FIGURE 7 Spacing device used in our laboratory to ensure scan location, direction and standard thickness (1 mm) of the gel layer.

VIII. Ultrasound Examination, Variables, and Practical Guidance

A. Equipment, Laboratory Facility, and Examiner

To perform state of the art echography of skin a specialized scanner is needed (Figure 6). Our experience is primarily based on the Dermascan C® (Cortex Technology, Hadsund, Denmark), which was originally developed as a prototype in our laboratory and described in a separate paper.[2] A center frequency of 20 MHz generally represents the best compromise between resolution and the need for a certain viewing field in depth. The resolution of the system is as previously mentioned mainly determined by the bandwidth rather than the center frequency. Generally, 20-MHz scanners have an axial resolution in skin of 0.05 mm and a lateral resolution of 0.15 to 0.35 mm. Thus, the axial resolution is better than the lateral. The viewing field in depth is typically 15 to 25 mm.

It is important that the *laboratory facility* is adequately equipped as a diagnostic facility for ultrasound evaluation of skin. It is typically a problem that dermatological clinics have no tradition for this type of diagnostic procedure. Obviously, the needs for space, assistance, and time must be the same as it is at other hospital departments with routine functions in diagnostic radiology or echography.

The *examiner* must be familiar with the equipment, the way it is operated, the background literature, and must have adequate knowledge about skin structure, dermatological application, and interpretation. There is no formal training or education in dermatological echography, and the examiner has to get the insight and the routine in operation on his/her own. The field is covered by a number of recent reviews.[1,2,27,42–51]

It is a fundamental rule that the body site, which is examined, must be kept in a *fixed position* during scanning. Examinations on the trunk are best performed with the patient resting in the supine position. Next to the fundamental rule about the fixed positioning of the body site being examined the significance of the scan *angle* and scan *direction* and awareness of the importance of *gain setting* (amplification) of the equipment, *gel layer thickness,* and the *axial positioning* is important. In image analysis of acoustic density of the skin these prerequisites must be carefully standardized. We routinely mark the examination site and apply a spacing device to ensure scan location and direction as well as a 1-mm gel layer thickness defined by the thickness of the spacing device (Figure 7). We also work with two ink marks on the monitor screen (one for the probe membrane, another for the epidermal surface) to ensure axial positioning relative to the transducer. The more the scan angle deviates from the right angle the less sound will be reflected back to the scanner, and the dermis will appear of lower acoustic density (Figure 8). The dermis may be measured more thick unless the change of density and delineation of dermis-fat interface happens to compensate for the

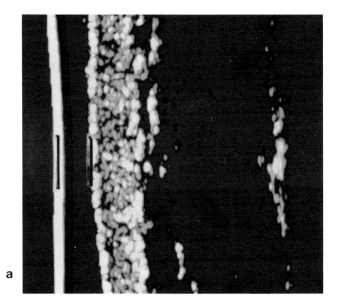

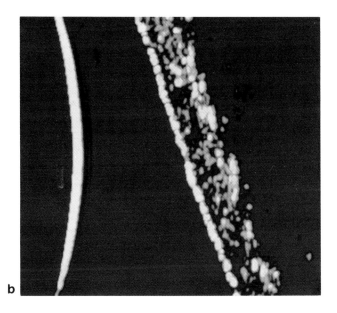

FIGURE 8 (a) B-mode scan of forearm skin. To ensure constant axial positioning the image is oriented vertically and placed so that the probe membrane and the epidermal surface covers their respective ink marks on the monitor screen. (b) B-mode scan of same examination site but under an oblique angle. Note the decrease of acoustic density of the dermis.

deviation from the right angle. The epidermal surface must be oriented strictly vertically on the monitor screen.

The scan direction in the examination field should always be the same since the image may be influenced by differences in the fiber orientation within the dermis as represented by the Langer lines. There are, in most body regions, systematic differences even within relatively small local distances. If a skin condition is monitored over time the examination field must be carefully marked to permit follow-up examination at exactly the same position.

There is at the moment no ultrasound phantom for 20-MHz scanners available except a rubber plate with a surface texture close to the epidermal microrelief (available through Cortex Technology, Hadsund, Denmark). With the rubber phantom the gain may be adjusted to a level where the entrance echo reaches the upper margin of the monitor screen. For follow-up check of speckles and internal structure the examiner may use his own flexor side forearm.

There is a certain interinstrument variation. It may be of limited value to operate with the same gain in every situation. For the study of forearm skin we operate our two Dermascan C® scanners with linear fixed gain ranging from 16 to 22 dB. In micro pigs we have used 19 dB, in the rat 25 dB, in mice 20 dB, and in hairless guinea pigs 23 dB (but for image analysis a gain of 18 or 19 dB). It is generally the most convenient and reproducible way to operate with a linear gain setting.

When groups are studied or monitored over time the equipment, as mentioned, must be ensured to remain constant (the fixed gain method). However, the chosen gain setting is not optimal for any skin lesion or any body site. If scanning is performed as a diagnostic procedure in dermatological clinics to obtain a maximum of information in individual cases whatever they might suffer from corticosteroid atrophy, scleroderma, or nodular malignant melanoma it is far more fruitful not to operate with a fixed gain setting but to perform the examination in real time with adjustment of the gain on the live image (the live image gain adjusted method). A pigmented seborrheic wart will, at a gain setting normally used, create a heavy shadow but after gaining up a normal dermis without tumor erosion is visualized, differentiating this lesion from a nodular melanoma (personal observation).

IX. Biological Variables

Biological variables of ultrasonographic significance include variations relative to sex, age, body site, (Table 1 and Figures 9a, b, and 10) and diurnal variation related to orthostatic position as well as endogenous biorhythms. Moreover, bodyweight and physical activities may be influential.

- Females have more thin skin than males[25,26,52]
- Extremity skin is more thin that truncal skin[25,26,52]
- Facial skin is low echogenic[26]
- Skin of the palm and sole is echopoor or echolucent, however, with a thick epidermis[26,29,30,52]
- The dermis-subcutaneous tissue interphase is better defined in females than it is in males since females have a more echodense dermis[26]
- Thin skin is generally more echogenic (female > male, extremity skin > truncal skin)[26]
- Reticular dermis is more echogenic than papillar dermis[10,27,32]
- Children have more thin skin[31]

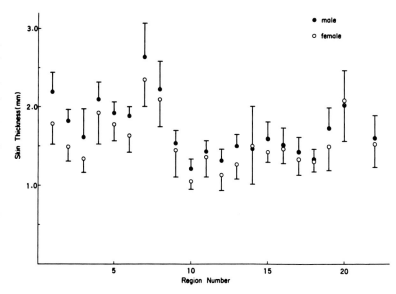

FIGURE 9 (a) Skin thickness in males (closed circles) and females (open circles) vs. anatomical site (region by number, see Table 1). Results were obtained with the Dermascan C®.[26] Females have more thin skin. (b) Ultrasound density of full thickness dermis in males (closed circles) and females (open circles) vs. anatomical site (region by number, see Table 1). Results were obtained with the Dermascan C® and the inbuild software of this equipment.[26] Females have more dense skin.

- Senile skin is more thin, particularly on extremities; a subepidermal low-echogenic band appears[10,32,47]
- Thin (and more transparent) extremity skin of old people may correlate with osteoporosis[53,54]
- Persons with overweight may have more thick skin[55]
- Sportsmen and persons with heavy physical activity and sun exposure may have more thick skin[56]
- Hot weather may be associated with cutaneous hyperemia, extravasation, and generalized cutaneous edema which may influence ultrasound scanning significantly[140]

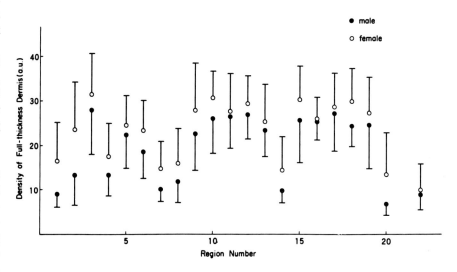

The echogenicity of the skin shows *diurnal variation* (Figures 11 and 12) originating from two different mechanisms.[57–59] The skin becomes generally slightly more echodense during the daytime irrespective of orthostatic position, body site, and age. The skin on the legs of aged individuals on the other hand become less echogenic during daytime. This orthostatic decline in density probably related to water accumulation may also be observed in a minor part of the population of young individuals, however, the echogenicity of their skin remains stable if they stay supine during the whole day (Figure 13). The adaptation to orthostatic position mainly takes place the first 2 h after standing up in the morning.

The above-described variables need to be taken into account when studies including ultrasound examination are performed. Controlling these variables is to control noise, and to go for precision on a minimum study sample.

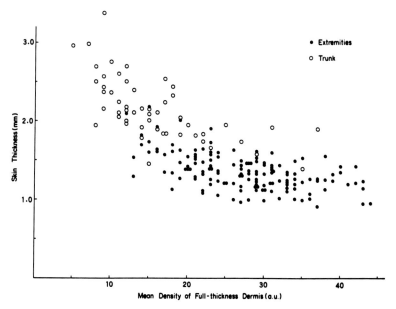

FIGURE 10 Skin thickness vs. mean density of full thickness dermis on extremities (closed circles) and truncal skin (open circles) demonstrating the echodense structure of the thinner extremity skin. Results obtained with the Dermascan C® and the inbuild software of this equipment.[26] 21 different body sites were examined.

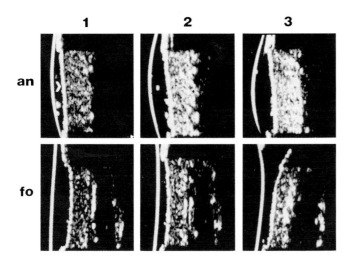

FIGURE 11 Diurnal variation of ultrasound density in a young individual in the morning (1), 2 h after standing (2), and after 12 h (3) in the ankle region (an) and on flexor side forearm skin (fo).[57]

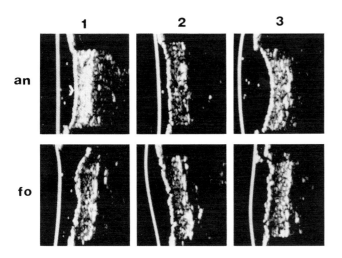

FIGURE 12 Diurnal variation of ultrasound density in an aged individual in the morning (1), 2 h after standing (2) and after 12 h (3) in the ankle region (an) and on flexor side forearm skin (fo).[57]

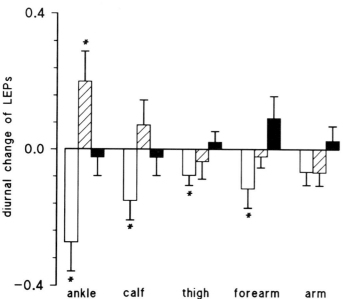

X. Applications in Clinical and Experimental Dermatology

A. Inflammatory Skin Diseases

Ultrasound imaging shows that the process of inflammation does not spread evenly throughout the dermis but concentrates in the outermost part corresponding to the papillary dermis where swelling and a *subepidermal band* is formed[60] see Figure 1. This is a common trait in any kind of inflammation. The structural background is disturbance of the fibrillar network due to edema formation. The papillar dermis is more easily distended under the pressure of edema as compared to the reticular dermis. The skin becomes thicker at the same time, and the skin surface may become folded as it is seen in lichenification in atopic dermatitis, where the reticular dermis at the same time remains straight and undisturbed, according to ultrasonography.[61,62]

Skin thickening was previously described in *positive allergic patch test reactions* to the classical contact allergens (Figure 1), tuberculin and "recall" allergens[63-68] as well as *irritant reactions* elicited by sodium lauryl sulfate (SLS) and as a number of other primary irritants including dimethylsulfoxide (DMSO).[64,68-74] The Modena group showed, however, that ultrasound image analysis is a more powerful tool to quantify those reactions than simple thickness measurement although thickness measurement has the advantage of being easier to do.[33-40] The degree of edema formation after SLS depends on variables such as body site, complexion of the skin, menstrual phase, etc. *Wheals of urticaria* (Figure 14) tend to involve the dermis more deeply, however, in pseudopodia, a protrusion of the edema at the subepidermal level, where band formation normally takes place can be seen. Thickening of wheals of urticaria does not develop linearly.[75] When a wheal has reached a diameter of approximately 5 mm the progression in thickness stops, and further development takes place laterally. At the same time the wheal changes in shape from globoid to flat. This dynamic behavior is a manifestation of the mechanical behavior of the papillar dermis and its ability to resist the pressure of edema as previously outlined. Effects of anti-

FIGURE 13 Average change of low echogenic pixels during the day (mean ± 2 SE). Open bars represent a group of young individuals (n = 22, age 17–27), hatched bars a group of aged individuals (n = 22, age 75–100), and black bars a group of subjects (n = 10, age 17–83) who remained in the supine position in bed during the whole day. The LEP (low echogenic pixels, Dermavision software, Cortex Technology, amplitude range 0–30) represent differences between values in the morning and 12 h later. Asterisk indicates significant changes (p <0.05). Young individuals (open bars) showed an increase in acoustic density (decrease in LEP) during the day while aged individuals (hatched bars) showed a decrease in echo density particularly in the ankle region. Individuals who remained supine (black bars) were constant in echo density during the daytime.[58]

FIGURE 14 Wheal of urticaria with elevation and flattening of the skin surface and a decrease of echo density of the dermis.

histamines can be documented.[76] The Modena group also demonstrated the utility of image analysis for characterization of wheals.[77]

Inflammatory lesions of *acne* are echopoor with deformation of the skin surface contour around the hair follicle. Interestingly, underneath atrophic scars a hyperechogenic massive probably representing a subcutaneous scar is quite often seen[141] (Figure 15). This is an example how ultrasonography can provide unexpected information not hitherto available by ordinary techniques such as histology, which cannot be practiced routinely in the face.

In *psoriasis,* thickening and a subepidermal band is also seen (Figure 16) representing acanthosis and inflammation at the same time.[28,39,78-82] Scales may create shadows. Pretreatment with petrolatum may be necessary. In pustular psoriasis the same type of abnormality is seen, however, focal in distribution. Skin thickness measurement by ultrasound is superior to color measurements as regards following the healing of psoriasis as exemplified by the psoriasis plaque test.[80] At end of treatment the skin thickness normalizes while redness often persists, and this is reflected in the parameters measured.

Special diseases such as nodular prurigo exhibits echographic alterations in accordance with inflammatory skin diseases in general with focal band formation, and shadow due to scratch and hyperkeratosis.

B. Connective-Tissue Diseases of the Skin

Scleroderma was one of the first applications of dermatological ultrasonography.[83-86] The skin surface to bone distance over the digits, and the skin thickness on the dorsum of the hand and the forearm was measured (Figures 17 and 18). Both the acrosclerotic and the localized (morphea) type of scleroderma as well as scleroderma have been extensively studied.[87-93] Acroscleroderma first clears on proximal parts. It is important that monitoring includes both forearm skin thickness measurement as well as measurement of skin-phalanx distances. In localized scleroderma, spontaneous regression of thickening takes place and the final outcome may be thinning and pigmentation.[85,86] Thickness measurement has generally turned out to be more useful than measurement of skin elasticity for the purpose of following therapy of scleroderma because the stiffening of the skin and the contractures tend to be permanent once established.[85] In cases of Pasini-Pierini atrophy and in selected cases of localized scleroderma the skin is thin from the very beginning, and there may be loss of subcutaneous fat and even bony depression. Subepidermal band formation can occasionally be seen in morphea, which can also be bullous in rare cases.[11] If subepidermal band formation is observed in a patient with acroscleroderma during treatment with penicillamin, particularly over body sites exposed to pressure and accompanied with bruises, penicillamin dermatopathy or elastosis perforans serpiginosa should be suspected and the treatment taped out. It might be unethical to institute risky and expensive treatments such as photopheresis without proper monitoring and quantification to demonstrate efficacy and justify long-term treatment. With ultrasound the thickening of the muscle fascia in fascitis and the

FIGURE 15 (a) Early inflammatory lesion of acne with decrease of acoustic density of the papillar dermis and the dermis corresponding to the hair follicle.

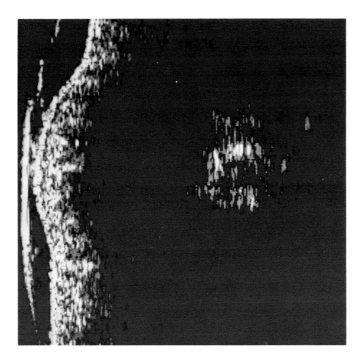

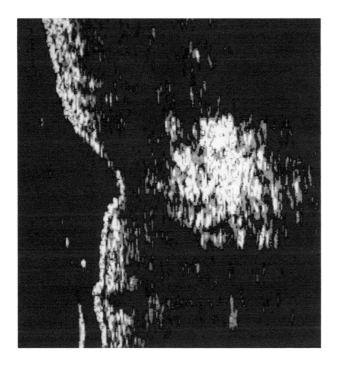

FIGURE 15 (b) Advanced acne lesion with deformation of the skin contour and a pathological echodense area in the subcutaneous space (right side of image). (c) Atrophic acne scar with no sign of active inflammation. A hyperechogenic mass is seen in the subcutaneous space possibly representing the scar formation after a previous inflammatory process in the subcutaneous tissue.

dermal thickening and serrate fat interface corresponding to dimpling clinically in chronic graft vs. host reaction may be demonstrated[141]

Dermal thinning during development of *corticosteroid atrophy,* typically 0.2 to 0.3 mm after 4 weeks irrespective body site, is easy to measure following potent corticosteroid treatment but the decline in thickness may not allow a detailed grouping of any corticosteroid.[19,42,95–100] Open application of the corticosteroid is preferable since occlusive application may result in edema formation, which will interfere with the corticosteroid induced suppression of ground substance and skin thinning. Posttreatment normalization of thickness takes place within a few weeks.[42] It was recently observed that ultrasound image analysis offers an extra opportunity for detailed study of the effects of corticosteroid on dermal connective tissue, i.e., with a drop in low amplitude echoes.[142] Ultrasound is the first line quantitative method for documentation of corticosteroid effect using the psoriasis plaque test.[80]

Traumatic scars including hypertrophic burn scars normally are irregular and relatively echopoor due to the poor organization of the connective tissue fibrils of such tissues.[101] Underneath atrophic acne scars a hyperechogenic massive may be found as previously mentioned.

Thermal injury and ultrasound determination of the burn-nonburn interface in the dermis was evaluated in animal as well as human experiments.[101–103] The skin swells the first hour postburn but this does not directly allow the determination of the vitality of the dermis at a certain level. *Burns* were not studied by state of the art 20-MHz ultrasound skin scanners and image analysis. Logically the critical question should be whether the echogenicity of the reticular dermis decreases significantly during the first 48 to 72 h postburn as a sign of vital reaction with inflammation, or not. A necrotic reticular dermis is not expected to undergo this change of echostructure.

C. Cutaneous Neoplasms

Cutaneous neoplasms invade and erode the connective tissue of the dermis and they are consequently seen as echopoor or echolucent structures on ultrasound (Figure 19). The literature on cutaneous and subcutaneous neoplasms is extensive and also conclusive with respect to the precision and usefulness of ultrasonography in preoperative evaluation and monitoring.[104–131] There is a wide overlap between the internal echostructure of *basal-cell carcinoma, squamous-cell carcinoma* and *melanoma,* and definite differentiation between these tumors on the basis of internal echostructure is not possible, although findings may be characteristic or typical. Moreover, *benign naevi* are typically echopoor as well, and distinction from *dysplastic naevi* by ultrasonography is uncertain or impossible. However, this distinction may also be difficult in histology and cause controversy.

Ultrasonography is, nevertheless, helpful in the tumor clinic. Ultrasonographic tumor thickness and depth evaluation is rational prior to cryosurgery. It is impor-

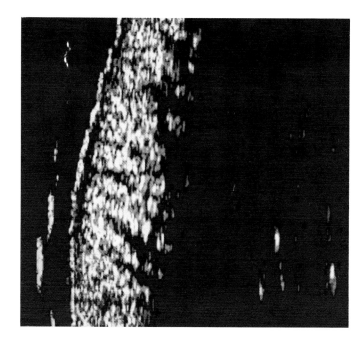

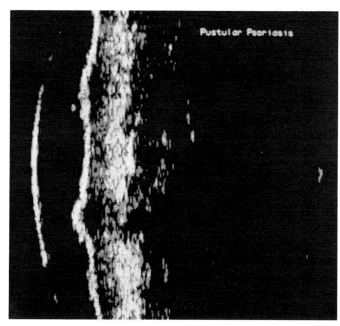

FIGURE 16 (a) Psoriatic plaque with a subepidermal low echogenic band. (b) Pustular psoriasis with a subepidermal low echogenic band and areas of more advanced alteration of the echo structure.

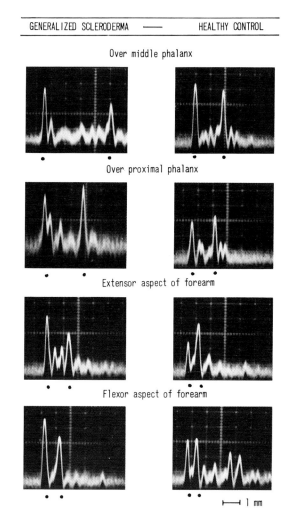

FIGURE 17 Ultrasound A-mode measurement of skin surface-phalanx distance and skin thickness (0–0) on the third finger and the forearm. A patient with generalized scleroderma (acrosclerosis) is compared with a healthy control. Note that the thickening becomes more advanced on the acral examination sites.

tant to know if a basal-cell carcinoma has broken through the dermis into the subcutaneous space, and if Moh's surgery is indicated. The studies on *malignant melanoma* performed with 20-MHz state-of-the-art scanners have shown a high (>90%) correlation between ultrasound measurement of melanoma thickness and histological thickness despite the known changes of dimension of small biopsies during operation and histological precessing.[12] It is useful to know whether the patient should be prepared for a small biopsy or a wide excision. The 20-MHz ultrasonography will never have the resolution to directly replace histology for the final diagnosis but ultrasonography can be helpful in the planning of the strategy of treatment, including more precise information of the patient early in the treatment phase. Distinction from pigmented seborrhoeic keratosis is easy since these create heavy shadows in ultrasonography.

In clinical dermatology, however, there is no real tradition for more complicated instrumentation directly used by the clinicians to support diagnosis and planning. This is very much in contrast to other specialties such as ophthalmology.

Kaposi sarcoma has a special ultrasound structure (Figure 20) with a lump of separate islands of tumor-tissue echopoor. The depth of the Kaposi lesions may be of interest when laser treatment is planned and the modality of laser light selected.

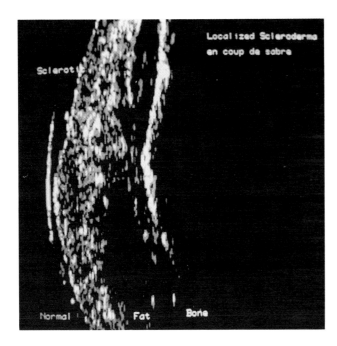

FIGURE 18 Horizontal scanning in the front of a patient with localized scleroderma en coup de sabre. The sclerotic skin (upper part) is thickened, the subcutaneous fat is reduced, and there is a depression of the bony contour of the frontal bone. The skin of the perilesional area (lower part of the picture) appears normal.

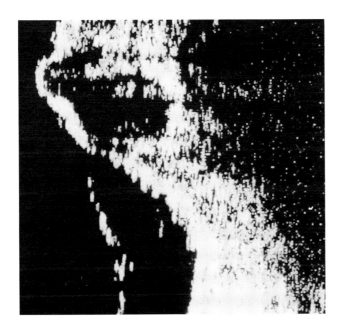

FIGURE 19 (a) Basal-cell carcinoma in the face. The tumor has *not* eroded through the reticular dermis and reached the subcutaneous tissue.

Infiltrates of *cutaneous lymphoma* spread primarily in the subepidermal space as any infiltration. The subepidermal space may be enormously distended, however, with a continuous but deformed reticular dermis underneath.[141] It is useful to know the thickness of the infiltrate prior to radiotherapy.

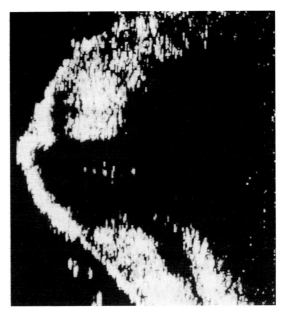

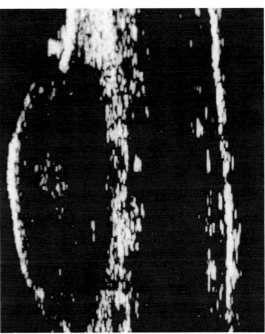

FIGURE 19 (b) Basal-cell carcinoma in the face. The tumor has reached the subcutaneous space and it also has spread laterally into the dermis at a superficial and more profound level. (c) Basal-cell carcinoma on the leg. Despite the size of the tumor, there is no erosion through the reticular dermis into the subcutaneous tissue.

D. Leg Ulcers and the Vascular System

In *arterial ischemia* the skin of the distal leg becomes transparent and thin. In *venous insufficiency* and *venous leg ulcer* the skin becomes thicker, and a prominent low echogenic subepidermal band (Figure 21) appears, especially in skin surrounding the ulcer.[47,57,58,59,133] This indicates a severe disorganization of structures at the level of the superficial vascular plexus of the skin with possible consequences for the nutritional supply of the epidermis. Dilated intradermal vessels may also be

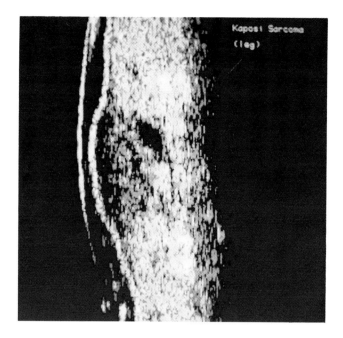

FIGURE 20 Kaposi sarcoma with irregular internal echostructure.

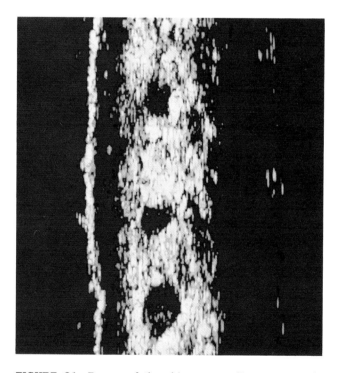

FIGURE 21 B-scan of the skin surrounding a venous leg ulcer. A pronounced subepidermal low echogenic band is seen, and dilated vessels within the dermis may be observed.

seen as a pathological observation. Ultrasonography is important for characterization and delineation of lipodermatosclerosis. In *lymphedema* the total skin becomes thicker, however, in this condition no special band formation is observed, and there is no major tendency to leg ulceration. Patients with venous leg ulcer seems to suffer from a reduced mechanical resistance of the skin of the lower leg and distension and accumulation of edema especially at the level of the papular dermis with the consequences mentioned.[47,57,133] Compressive bandages and stockings were shown to diminish the ultrasonographic findings described.[56,57] The intradermal edema is a key factor in venous leg ulcer, and ultrasound is, obviously, an advanced and powerful tool for this application.

Small-size arteries such as the temporal artery and the radial artery can be demonstrated by ultrasound, and the diameter measured. With M-mode scanning the pulse waves can be followed. The effect of antimigraine therapy was evaluated by ultrasonographic studies of the temporal artery, and also the degree of dilation relative to clinical symptoms was described.[134-137]

E. The Nail

The *nail thickness* is easily measured by ultrasound, particularly with the small and more convenient A-mode scanners.[7,8] The correct ultrasound velocity in nail tissue needs to be used.[7] There is a quite strong internal echo in the nail between the dorsal plate and the matrix, and there is a weak echo in the nailbed between the underlying epidermis and the connective tissue of the distal phalanx, which is echopoor. This echopoor zone between the nail and the phalanx bone extends proximally under the nailbed with a relatively sharp delineation toward extensor side skin and the distal interphalangeal joint.

XI. Ultrasound Examination of Experimental Animals

Ultrasound examination may be possible in animals covered with hair, however, it is easier and more reliable in the hairless species. The researcher has to be familiar with the histology and microanatomy of the animal as a prerequisite to interpret ultrasound images. According to our experiences, animal skin is generally less echogenic than human skin. The interphase between dermis and subcutaneous fat is acoustically well defined both in hairless mice and in guinea pigs and rats. It was recently shown that ultrasound image analysis is useful for quantitation of irritant reactions to SLS in hairless mice.[138] Minipigs and domestic pigs have a low echogenic or echopoor dermis, and it is often impossible to measure the thickness of the skin. In the Yucatan micropig the skin thickness is easily measured, and ultrasound is a useful tool to follow corticosteroid atrophy in experimental studies with this animal. Ultrasound imaging of the hamster flank organ was shown to correlate with histology and to be convenient for monitoring of hormone effects on this specialized sebaceous gland.[139]

References

1. Perednia DA. What dermatologists should know about digital imaging. *J Am Acad Dermatol* 25:89–108, 1991.
2. Serup J. Ten years' experience with high-frequency ultrasound examination of the skin: development and refinement of technique and equipment. In: *Ultrasound in Dermatology* (P. Altmeyer, S. el-Gammal, K. Hoffmann, Eds.), Springer-Verlag, Berlin-Heidelberg, 1992, 41–54.
3. Edwards C, Payne PA. Ultrasound velocities in skin components. International Society for Bioengineering and the Skin: ultrasound in dermatology. Symposium, Liege. November 9th, 1984.
4. Escoffier C, Querleux B, de Rigal J, Lévêque J-L. In vitro study of the velocity of ultrasound in the skin. *Bioeng Skin* 2:87–94, 1986.
5. Daly CH, Wheeler JB. The use of ultra-sonic thickness measurement in the clinical evaluation of the oral soft tissues. *Int Dental J* 21:418–29, 1971.
6. Dines KA, Sheets PW, Brink JA, Hanke CW, Condra KA, Clendenon JL, Gross SA, Smith DJ, Franklin TD. High frequency ultrasonic imaging of skin: experimental results. *Ultrasonic Imaging* 6:408–34, 1984.
7. Jemec GBE, Serup J. Ultrasound structure of the human nailplate. *Arch Dermatol* 125:643–46, 1989.
8. Finlay AJ, Western B, Edwards C. Ultrasound velocity in human fingernail and effects of hydration: validation of in vivo nail thickness measurement techniques. *Br J Dermatol* 123:365–73, 1990.
9. Jemec GBE, Agner T, Serup J. Transonychial water loss: relation to sex, age and nail-plate thickness. *Br J Dermatol* 121:443–46, 1989.
10. de Rigal J, Escoffier C, Querleux B, Faivre B, Agache P, Lévêque J-L. Assessment of aging of the human skin in vivo. Ultrasound imaging. *J Invest Dermatol* 93:621–25, 1989.
11. Kobayasi T, Willeberg A, Serup J, Ullman S. Morphea with blisters. *Acta Derm Venereol (Stockh)* 70:454–56, 1990.
12. Serup J. Punch biopsy of the skin by electric power drill: effect of speed. *Dan Med Bull* 32:189–91, 1985.
13. Tanner JM, Whitehouse RH. The Harpenden skinfold caliper. *Am J Physiol Antropol* 13:743–46, 1955.
14. Black MM. A modified method for measuring skin thickness. *Br J Dermatol* 81:661–66, 1969.
15. Dykes PJ, Francis AJ, Marks R. Measurement of dermal thickness with the Harpenden skinfold caliper. *Arch Derm Res* 256:261–63, 1976.
16. Dykes PJ, Marks R. Measurement of skin thickness: a comparison of two in vivo techniques with a conventional histiometric technique. *J Invest Dermatol* 69:275–78, 1977.
17. Alexander H, Miller DL. Determining skin thickness with pulsed ultrasound. *J Invest Dermatol* 72:17–19, 1979.
18. Tan CJ, Roberts E, Stathan B, Marks R. Reproducibility, validation and variability of dermal thickness measurement by pulsed ultrasound. *Br J Dermatol* 105:25–26, 1981.
19. Tan CJ, Marks R, Payne PA. Comparison of xeroradiographic and ultrasound detection of corticosteroid induced dermal thinning. *J Invest Dermatol* 76:126–28, 1981.
20. Newton JA, Whitaker J, Sohail S, Young MMR, Harding SM, Black MM. A comparison of pulsed ultrasound, radiography, and micrometer screw gauge in the measurement of skin thickness. *Cur Med Res Opin* 9:113–18, 1984.
21. Lawrence CM, Shuster S. Comparison of ultrasound and caliper measurements of normal and inflamed skin thickness. *Br J Dermatol* 122:195–200, 1985.
22. Stautner-Brückmann C, Reitmeier K, Gressner U, Zöllner N. Vergleichende Messing der Hautdicke des Handrückens mit Caliper und Ultraschall — erste Ergebnisse einer prospektiven Untersuchung. *Blindgebund* 57:67–69, 1990.
23. Martin AD, Drinkwater DT, Clarys JP, Daniel M, Ross WD. Effects of skin thickness and skinfold compressibility on skinfold thickness measurement. *Am Hum Biol* 4:453–60, 1992.
24. Miyauchi S, Miki J. Normal human skin echogram. *Arch Dermatol Res* 275:345–48, 1983.
25. Fornage BD, Deshayes J-L. Ultrasound of normal skin. *J Clin Ultrasound* 14:619–22, 1986.
26. Olsen LO, Takiwaki H, Serup J. High-frequency characterization of normal skin. Skin thickness and echographic density of 22 anatomical sites. *Skin Res Technol*, in press.
27. Altmeyer P, el-Gammal S, Hoffmann K. *Ultrasound in Dermatology*. Springer-Verlag, Berlin-Heidelberg, 1992.
28. Olsen LO, Serup J. High-frequency scanning for non-invasive cross-sectional imaging of psoriasis. *Acta Derm Venereol (Stockh)* 73:185–87, 1993.
29. Murakami S, Miki J. Human skin histology using high-resolution echography. *J Clin Ultrasound* 17:77–82, 1989.
30. el-Gammal S, Hoffmann K, Auer T, Korten M, Altmeyer P, Höss A, Ermert H. A 50 MHz high-resolution ultrasound imaging system for dermatology. In: *Ultrasound in Dermatology* (P. Altmeyer, S. el-Gammal, K. Hoffmann, Eds). Springer-Verlag, Berlin-Heidelberg, 1992.
31. Petersen JR, Petersen S, Serup J. High-frequency ultrasound measurement of subcutaneous fat thickness in newborn infants. *Skin Res Technol* in press.
32. Richard S, Querleux B, Bittoun J, Folivet O, Idy-Perelli I, Lachaniere O, Lévêque J-L. Characterization of the skin in vivo by high resolution magnetic resonance imaging: water behaviour and age-related effects. *J Invest Dermatol* 100:705–709, 1993.
33. Seidenari S, di Nardo A. A new image analysis system for the assessment of allergic patch test reactions recorded by B-scanning. *Contact Dermatitis* 25:329, 1991.
34. Seidenari S, di Nardo A, Pepe P, Gianetti A. Ultrasound B-scanning with image analysis for assessment of allergic patch test reactions. *Contact Dermatitis* 24:216–222, 1991.
35. Seidenari S, di Nardo A. Cutaneous reactivity to allergens at 24-h increases from the antecubital fossa to the wrist: an echographic evaluation by means of a new image analysis system. *Contact Dermatitis* 26:171–76, 1992.
36. Seidenari S, di Nardo A. Echographic evaluation of corticosteroid inhibition of allergic patch test reactions. *Contact Dermatitis* 26:212–13, 1992.
37. Seidenari S, di Nardo A. B-scanning evaluation of allergic reactions with binary transformation and image analysis. *Acta Derm Venereol (Stockh)* (Suppl.) 175:3–7, 1992.
38. Seidenari S, di Nardo A. B-scanning evaluation of irritant reactions with binary transformation and image analysis. *Acta Derm Venereol (Stockh)* (Suppl.) 175:9–13, 1992.
39. di Nardo A, Seidenari S, Gianetti A. B-scanning evaluation with image analysis of psoriatic skin. *Exp Dermatol* in press.
40. Seidenari S, di Nardo A, Gianetti A. Assessment of topical corticosteroid activity on experimentally induced contact dermatitis: echographic evaluation with binary transformation and image analysis. *Skin Pharmacol* 6:85–91, 1993.

41. Thijssen JM, Oosterveld BJ. Texture in tissue echograms. Speckle or informations. *J Ultrasound M* 9:215–229, 1990.
42. Søndergaard J, Serup J, Tikjøb G. Ultrasonic A- and B-scanning in clinical and experimental dermatology. *Acta Derm Venereol (Stockh)* (Suppl.) 120:76–82, 1985.
43. Querleux B, Lévêque J-L, de Rigal J. In vivo cross-sectional ultrasonic imaging of human skin. *Dermatologica* 177:332–37, 1988.
44. Pugliese P. Use of ultrasound in evaluation of skin care products. *Cosmetics Toiletries* 104:61–76, 1989.
45. Hoffmann K, el-Gammal S, Altmeyer P. B-scan Sonographie in der Dermatologie. *Hautarzt* 41:7–16, 1990.
46. Payne PA. Skin thickness measurement and cross-sectional imaging. In: *Cutaneous Investigation in Health and Disease.* Noninvasive methods and instrumentation (J-L Laaevêque, Ed.), Marcel Dekker, New York, 1989, 183–213.
47. Serup J. High-frequency ultrasound examination of aged skin: intrinsic, actinic and gravitational aging, including new concepts of stasis dermatitis and by ulcer. In: *Aging Skin, Properties and Functional Changes* (J-L Lévêque Ed.) Marcel Dekker, New York, 1993, 69–85.
48. el-Gammal S, Auer T, Hoffmann K, Matthes U, Hammentgen R, Altmeyer P, Ermert H. High-frequency ultrasound: a noninvasive method for use in dermatology. In: *Noninvasive Methods for the Quantification of Skin Functions* (PJ Frosch and AM Kligman, Eds.), Springer-Verlag, Berlin, 1993, 104–129.
49. Agner T, Serup J. Ultrasound — an update on methodology and application with special references to inflammatory reactions. In: *Noninvasive Methods for the Quantification of Skin Functions* (PJ Frosch and AM Kligman, Eds.), Springer-Verlag, Berlin, 1993.
50. Gropper CA. Diagnostic high-resolution ultrasound in dermatology. *Int J Dermatol* 32:243–50, 1993.
51. Fornage B, Duvic M. High-frequency sonography of the skin. *U Eur Acad Derm Venereol* 3:47–55, 1994.
52. Southwood WFN. The thickness of the skin. *Plast Reconstr Surg* 15:423–29, 1955.
53. McConkey B, Fraser GM, Bligh AS, Whiteley H. Transparent skin and osteoporosis. *Lancet* i:693–95, 1963.
54. Schatz H, Staudemayer T, Kligman AM. Applications in the study of human skin disorders and the response to treatment. In: *Ultrasound in Dermatology* (P Altmeyer, S el-Gammal, K Hoffmann, Eds.), Springer-Verlag, Berlin, 1992, 256–263.
55. Petersen H, Agner T, Storm T. Skin thickness in patients with osteoporosis and controls quantified by ultrasound A-scan. *Skin Pharmacol* submitted.
56. Lévêque J-L, Porte G, de Rigal J, Corcuff P, Francois AM, Saint Leger D. Influence of chronic skin exposure on some biophysical parameters of the human skin: an in vivo study. *J Cut Aging Cosmet Dermatol* 1:123–27, 1988.
57. Gniadecka M, Serup J, Søndergaard J. Influence of gravitational stress on skin echogenicity in young vs. aged individuals. High-frequency ultrasonography and digital image analysis: novel techniques for evaluation of dermal water distribution. *Br J Dermatol* submitted.
58. Gniadecki M, Gniadecka R, Serup J, Søndergaard J. Ultrasound structure and digital image analysis of the sub-epidermal low echagenic band in aged human skin: diurnal changes and interindividual variability. *J Invest Dermatol* 102:362–65, 1994
59. Gniadecka M, Gniadecki R, Serup J, Søndergaard J. Skin mechanical properties present adaptation to man's upright position. In vivo studies in young and aged individuals. *Acta Derm Venereol (Stockh)* 74:188–90, 1994.
60. Serup J. Noninvasive techniques for quantification of contact dermatitis. In: *Textbook of Contact Dermatitis* (RJG Rycroft, T Menné, PH Frosch, C Benezra, Eds.), Springer-Verlag, Berlin, 1992, 323–338.
61. Serup J. Characterization of contact dermatitis and atopy using bioengineering techniques. A survey. *Acta Derm Venereol (Stockh)* (Suppl.) 177:14–25, 1992.
62. Hoffmann K, Schwarzt M, Dirschka T, el-Gammal S, Hoffmann A, Altmeyer P. Non-invasive evaluation of inflammation in atopic dermatitis. *J Eur Acad Derm Venereol,* in press.
63. Serup J, Staber B, Klemp P. Quantification of cutaneous oedema in patch test reactions by measurement of skin with high-frequency pulsed ultrasound. *Contact Dermatitis* 10:88–93, 1984.
64. Serup J, Staber B. Ultrasound for assessment of allergic and irritant patch test reactions. *Contact Dermatitis* 17:80–84, 1987.
65. Brazier S, Shaw S. High-frequency ultrasound measurement of patch test reactions. *Contact Dermatitis* 15:199–201, 1986.
66. Beck JS, Spence VA, Lowe JG, Gibbs JH. Measurement of skin swelling in the tuberculin test by ultrasonography. *J Immunol Methods* 86:125–30, 1986.
67. Eun HC, Marks R. Dose-response relationships for topically applied antigens. *Br J Dermatol* 122:491–499, 1990.
68. Berardesca E, Maibach HI. Bioengineering and the patch test. *Contact Dermatitis* 18:3–9, 1988.
69. Agner T, Serup J. Quantification of the DMSO-response — at test for assessment of sensitive skin. *Clin Exp Dermatol* 14:214–17, 1989.
70. Agner T, Serup J, Handlos V, Bastberg W. Different skin irritation abilities of different qualities of sodium lauryl sulphate. *Contact Dermatitis* 21:184–88, 1989.
71. Agner T, Serup J. Skin reactions to irritants assessed by non-invasive bioengineering methods. *Contact Dermatitis* 20:352–59, 1989.
72. Agner T, Serup J. Individual and instrumental variations in irritant patch-test reactions — clinical evaluation and quantification by bioengineering methods. *Clin Exp Dermatol* 15:29–33, 1990.
73. Agner T, Serup J. A dose-response study of SLS irritation evaluated by different bioengineering methods. *J Invest Dermatol* 95:543–47, 1990.
74. Agner T, Serup J. Seasonal variation of skin resistance to irritants. *Br J Dermatol* 121:323–28, 1989.
75. Serup J. Diameter, thickness, area, and volume of skinprick histamine weals. *Allergy* 39:359–64, 1984.
76. Shall L, Newcombe RG, Lush M, Marks R. Dose-response relationship between objective measures of histamine-induced weals and dose of terfenadine. *Acta Derm Venereol (Stockh)* 71:199–204, 1991.
77. di Nardo A, Seidenari S. Echographic evaluation with image analysis of histamine induced weals. *Skin Pharmacol* in press.
78. Serup J. Non-invasive quantification of psoriasis plaques-measurement of skin thickness with 15 MHz pulsed ultrasound. *Clin. Exp. Dermatol* 9:502–508, 1984.
79. Hermann RC, Ellis CN, Fithing DW, Ho VC, Voorhees JJ. Measurement of epidermal thickness in normal skin and psoriasis with high-frequency ultrasound. *Skin Pharmacol* 1:128–36, 1988.
80. Broby-Johansen U, Karlsmark T, Petersen LJ, Serup J. Ranking of the anti-psoriatic effect of various topical corticosteroids applied under a hydrocolloid dressing, skin thickness, blood-flow and colour measurements compared to clinical assessments. *Clin Exper Dermatol* 15:343–48, 1990.

81. Rogers S. Skin thickness in psoriasis. *Clin Exper Dermatol* 17:324–27, 1992.
82. Krieg PHG, Bacharach-Buhlers M, el-Gammal S, Altmeyer P. The pustule in palmoplantar psoriasis: transformed preside or mature microabscess? A three-dimensional study. *Dermatology* 185:104–12, 1992.
83. Serup J. Quantification of acrosclerosis: measurement of skin thickness and skin-phalanx distance in females with 15 MHz pulsed ultrasound. *Acta Derm Venereol (Stockh)* 64:35–40, 1984.
84. Serup J. Localized scleroderma (morphea): thickness of sclerotic plaques as measured by 15 MHz pulsed ultrasound. *Acta Derm Venereol (Stockh)* 64:214–19, 1984.
85. Serup J. Localized scleroderma (morphea). Clinical, physiological, biochemical and ultrastructural studies with particular reference to quantification of scleroderma. *Acta Derm Venereol (Stockh)* (Suppl.) 122:1–61, 1986.
86. Serup J. Decreased skin thickness of pigmented spots appearing in localized scleroderma (morphea). Measurement of skin thickness by 15 MHz pulsed ultrasound. *Arch Dermatol Res* 276:135–37, 1984.
87. Rodman GP, Lipinski E, Luksick J. Skin thickness and collagen contact in progressive systemic sclerosis and localized scleroderma. *Arthritis Rheum* 22:130–40, 1979.
88. Cole GW, Handler SJ, Burnett K. The ultrasonic evaluation of skin thickness in scleroderma. *J Clin Ultrasound* 9:501–3, 1981.
89. Åkesson A, Forsberg L, Hedeström E, Wollheim F. Ultrasound examination of skin thickness in patients with progressive systemic sclerosis (scleroderma). *Acta Radiol Diagn* 27:91–94, 1986.
90. Myers SL, Cohen JS, Sheets PW, Bies JR. B-mode ultrasound evaluation of skin thickness in progressive systemic sclerosis. *J Rheumatol* 13:577–80, 1986.
91. Kalis B, de Rigal J, Leonard F, Lévêque J-L, Riche O, le Corre Y, de Lacharriere O. In vivo study of scleroderma by non-invasive techniques. *Br J Dermatol* 122:785–91, 1990.
92. Hoffmann K, Gerbaubt U, el-Gammal S, Altmeyer P. 20-MHz B-mode ultrasound in monitoring the course of localized scleroderma (morphea). *Acta Derm Venereol (Stockh)* (Suppl.) 164:3–16, 1991.
93. Levy JJ, Gassmüller J, Andring H, Brenke A, Albrecht-Nebe H. Darstellund der subkutanen Atrophie der Zirkumskripten Sklerodermie im 20 MHz-B-scan Ultraschall. *Hautarzt* 44:446–51, 1993.
94. Gomez EC, Berman B, Miller DL. Ultrasonic assessment of cutaneous atrophy caused by intradermal corticosteroids. *J Dermatol Oncol Surg* 8:1071–72, 1982.
95. Serup J, Holm P, Stender IM, Pichard J. Skin atrophy and telangiectasia after topical corticosteroids as measured non-invasively with high frequency ultrasound, evaporimetry and laser Doppler flowmetry. Methodological aspects including evaluations of regional differences. *Bioeng Skin* 3:43–58, 1987.
96. Lubach D, Grütter H, Behl M, Nagel C. Investigations on the development and regression of corticosteroid-induced thinning of the skin in various parts of the human body during and after topical application of amcinomide. *Dermatologica* 178:93–97, 1989.
97. Korting HC, Vielluf D, Kerscher M. 0.25% Prednicarbate cream and the corresponding vehicle induce less skin atrophy than 0.1% betamethasone-17-valerate cream and 0.05% clobetasol-17-propionate cream. *Eur J Clin Pharmacol* 42:159–61, 1992.
98. Kerscher MJ, Korting HC. Topical glucocorticoids of the non-fluorinated double-ester type. Lack of atrophygenicity in normal skin as assessed by high-frequency ultrasound. *Acta Derm Venereol (Stockh)* 72:214–16, 1992.
99. Lubach D, Rath J, Kietzmann M. Steroid-induced dermal thinning: discontinuous application of clobetasol-17-propionate ointment. *Dermatologica* 185:44–48, 1992.
100. Katz SM, Frank DH, Leopold GR, Wachtel TL. Objective measurement of hypertrophic human scar: a preliminary study of tonometry and ultrasonography. *Ann Plast Surg* 14:121–27, 1985.
101. Cantrell JH, Goans RE, Roswell RL. Acoustic impedance variations in burn — nonburn interfaces in porcine skin. *J Acoust Soc Am* 64:731–36, 1978.
102. Goans RE, Cantrell JH, Meyers FB. Ultrasonic pulse-echo determination of thermal injury in deep dermal burns. *Med Physics* 4:259–63, 1977.
103. Bauer JA, Sauer T. Cutaneous 10 MHz ultrasound B-scan allows the quantitative assessment of burn depth. *Burns* 15:49–51, 1989.
104. Goldberg B. Ultrasonic evaluation of superficial masses. *J Clin Ultrasound* 3:90–94, 1975.
105. Rukavina B, Mohar N. An approach of ultrasound diagnostic techniques of the skin and subcutaneous tissue. *Dermatologica* 158:81–92, 1979.
106. Miyauchi S, Tada M, Miki J. Echographic evaluation of modular burns of the skin. *J Dermatol* 10:221–27, 1983.
107. Breitbart EW, Rekpenning W. Möglichkeiten und Grenzen der ultraschalldiagnostik zur in vivo Bestimmung der Invasionstiefe des maglignen Melanoms. *Z Hautkr* 58:975–87, 1983.
108. Kraus W, Schramm P, Hoede N. First experiences with a high-resolution ultrasonic scanner in the diagnosis of malignant melanomas. *Arch Dermatol Res* 275:235–38, 1983.
109. Brenner S, Ophir J, Weinraub Z. Thickness of basal cell epithelioma measured preoperatively by ultrasonography. *Cutis* 34:509, 1984.
110. Shafir R, Itzchak Y, Heyman Z, Azizi E, Tsur H, Hiss J. Preoperative ultrasonic measurements of the thickness of cutaneous malignant melanoma. *J Ultrasound Med* 3:205–208, 1984.
111. Kraus W, Nake-Elias A, Schramm P. Diagnostsische Fortschritte bei malignen Melanomen durch die hochauflösende Real-time-sonographie. *Hautarzt* 36:386–392, 1985.
112. Rodriques JCF, Carlos MJ, Féria R. Squamous-cell carcinoma. Echotomographic study and treatment by chemosurgery without systematised microscopic control. *Skin Cancer* 2:135–40, 1987.
113. de Ascensao AC. Echotomographic study of skin tumours. *Skin Cancer* 2:41–45, 1987.
114. de Ascensao AC. Ultrasonography of skin tumours — basic principles and main procedures. *Skin Cancer* 2:5–12, 1987.
115. Schwaijhofer B, Pohl-Markl, H, Frühwald R, Stiglbauer R, Kokoschka EM. Der Diagnostische Stellenwert des Ultraschalls beim malignen Melanom. *Forschr Röntgenstr* 146:409–11, 1987.
116. Hughes BR, Black D, Srivastava A, Dalzid K, Marks R. Comparison of techniques for the noninvasive assessment of skin tumours. *Clin Exper Dermatol* 12:108–11, 1987.
117. de Ascensao AC. The importance of echography in the treatment of basal-cell carcinoma. *Skin Cancer* 3:237–53, 1988.

118. Breitbart EW, Müller CE, Hicks R, Vieluf D. Neue Entwichlungen der Ultraschalldiagnostik in der Dermatologie. *Aktuelle Dermatologie* 15:57–61, 1989.
119. Edwards C, Al-Aboozi MM, Marks R. The use of A-scan ultrasound in the assessment of small skin tumours. *Br J Dermatol* 121:297–304, 1989.
120. Reali UM, Santucci M, Paoli G, Chiarugi C. The use of high resolution ultrasound in preoperative evaluation of cutaneous malignant melanoma thickness. *Tumori* 75:452–55, 1989.
121. Hoffmann K, el-Gammal S, Matthes U, Altmeyer P. Digitale 20 MHz-Sonographie der Haut in der präoperativen Diagnostik. *Z Hautkrankh* 64:851–58, 1989.
122. Nessi R, Betti R, Bencini PL, Crosti C; Blanc M, Uslenghi C. Ultrasonography of nodular and infiltrative lesions of the skin and subcutaneous tissues. *J Clin Ultrasound* 18:103–9, 1990.
123. Gassenmaier G, Kieseweller F, Schell H, Zinner M. Wertigkeit der Hochauflösenden Sonographie für die Bestimmung des vertikalen Tumordurch-messers beim malignen Melanom der Haut. *Hautarzt* 41:360–64, 1990.
124. Hoffmann K, Stücker M, el-Gammal S, Altmeyer P. Digitale 20-MHz-Sonographie des Basalioms im b-scan. *Hautarzt* 41:333–39, 1990.
125. Betti R, Neosi R, Blanc M, Bencini PL, Galimberti M, Crosti C, Uslenghi C. Ultrasonography of proliferative vascular lesions of the skin. *J Dermatol* 17:247–51, 1990.
126. Neosi R, Blanc M, Bosco M, Dameno S, Venegoni A, Betti R, Bencini PL, Crosti C, Uslenghi C. Skin ultrasound in dermatologic surgical planning. *J Dermatol Surg Oncol* 17:38–43, 1991.
127. Beeckman P, de Clerck S, Jong B, de Maeseneer M. The ultrasound aspect of the skin and subcutaneous fat layer in various benign and malignant breast conditions. *J Belge Radiol* 74:283–88, 1991.
128. Hoffman K, Jung J, el-Gammal S, Altmeyer P. Malignant melanoma in 20 MHz B-scan sonography. *Case Rep Dermatol* 185:49–55, 1992.
129. Nitsche N, Iro H, Hoffmann K. Ultraschalldiagnostik der ausseren Nase. *Hals Nase Ohrin HNO* 40:181–85, 1992.
130. Harland CC, Bamber JC, Gusterson BA, Mortimer PS. High frequency, high resolution B-scan ultrasound in the assessment of skin tumours. *Br J Dermatol* 128:x–x, 1993.
131. Hoffmann K, Winkler K, el-Gammal S, Altmeyer P. A wound healing model with sonographic monitoring. *Clin Exper Dermatol* 18:217–25, 1993.
132. Maurad MM, Marks R. Assessment of disease severity and outcome in the gravitational syndrome using pulsed A-scan ultrasound to measure skin thickness. *Clin Exper Dermatol* 15:200–05, 1990.
133. Maurad MM, Edwards C, Marks R. Skin extensibility in the gravitational syndrome. *Bioeng Skin* 4:199–215, 1988.
134. Nielsen T, Iversen H, Tfelt-Hansen P. Dermascan A. A high-frequency acoustic scanner for non-invasive studies of the luminal diameter of superficially situated medium-sized arteries. *Cephalalgia* (Suppl.) 10:72–73, 1989.
135. Iversen H, Nielsen T, Tfelt-Hansen P, Olesen J. Headache and changes in the diameter of the radial artery during 7 hours intravenous nitroglycerin infusion. *Cephalalgia* (Suppl.) 10:82–83, 1989.
136. Iversen HK. N-Acetylcysteine enhances nitroglycerine-induced headache and cranial arterial responses. *Clin Pharmacol Ther* 52:125–33, 1992.
137. Iversen KH, Nielsen TH, Olesen J, Tfelt-Hansen P. Arterial responses during migraine headache. *Lancet* 336:837–39, 1990.
138. Seidenari S, Zanella C, Pepe P. Echographic evaluation of sodium lauryl sulfate (SLS)-induced irritation in mice. *Contact Dermatitis* 30:41–42, 1994.
139. Combettes C, Durand-Seme V, Querleux B, Saint-Leger D, Lévêque J-L. Imaging the hamster flack organ by an ultrasonic technique: a new approach to animal tests. *Br J Dermatol* 121:689–99, 1989.
140. Fullerton A, personal observation.
141. Serup J, personal observation.
142. Serup J, Keiding J, Fullerton A, Gniadecka M, and Gniadecki R, personal observation.

Chapter 12.2
Ultrasound B-Mode Imaging and *In Vivo* Structure Analysis

Stefania Seidenari
Department of Dermatology
University of Modena
Modena, Italy

I. Introduction

The recent introduction of high-frequency ultrasound in dermatology has opened new perspectives concerning skin physiology and diagnosis of dermatological diseases. The two-dimensional sonographic methods are called B-scan procedures. A two-dimensional ultrasound image consists of single lines (A-scan lines) whose amplitudes are represented by means of a grey or color scale, which are added to form a sectional image.

As other diagnostic procedures used in dermatology, based on bioengineering methods, ultrasound B scanning is an effective means for the objective assessment of healthy and diseased skin. The difference between ultrasound B-mode imaging and other noninvasive techniques, providing digital displays or numbers, lies in the possibility of obtaining a pictorial representation corresponding to a cross-sectional image of the skin. The problem of looking under the surface of the skin has so far been dealt with by microscopy, which is the standard reference for morphologists. However, in respect to invasive procedures like biopsying for histological examination, an *in vivo* noninvasive assessment enables the evaluation of a skin site without altering the connections with the surrounding tissue, and without changing tissue orientation, tension, and thickness.[1]

A considerable amount of data can be immediately obtained by looking at an echographic image. This information is perceived by intuition and evaluated by the observer according to his/her experience. A second phase, consisting of the computerized elaboration and quantification of data, turns the echographic procedure into an objective diagnostic means. In fact, if an immediate evaluation of an image or a comparative judgment over a short time span is conceivable without the need for instrumental elaboration, then interpretation and comparison of pictorial data showing slight variations in space and time are only possible by means of an appropriate software for image analysis.[2]

This chapter will provide a simplified nonmathematical description of image-processing methods, which can be employed for the study of echographic pictures taken from the skin. At present, little experience on analytical methods suitable for 20-MHz ultrasound images is available. However, studies employing different image analysis procedures and different software packages are underway in many centers, in order to identify the most suitable processing equipment and techniques for analyzing high-resolution ultrasound images of the skin. For a technical description of the different image analytical techniques, please refer to References 3 through 5.

Image-processing methods were first employed to improve various types of picture distortion and to restore images for human interpretation. One of the first applications was in correcting digitized newspaper pictures sent by submarine cable between London and New York in the early 1920s. In addition to applications in the space program, digital image-processing techniques are now used for solving a variety of problems relating to methods to enhance pictorial information for human interpretation and also for machine perception and analysis. The latter case deals with procedures for extracting information, which is suitable for computer processing, to end up with an easier interpretation or a specific application.[3] A major problem regarding image-processing methods applied to medicine is that all of these procedures provide information, which, although quantitative in a relative sense, is highly instrument dependent, making data collected from groups working with different equipment incompatible and incomparable.

II. Ultrasound Equipment

A. The Instrument

All the procedures which will be exemplified have been achieved on images recorded by the same instrument (Dermascan C®, Cortex Technology, Hadsund,

Denmark). Because of the specificity of the information, as previously mentioned, this instrument will be briefly described.

The instrument consists of a probe with a 20-MHz transducer fixed onto an articulated self-balanced arm, a central unit for the elaboration of the data, a visualization system, and a memorizing and data storing system. The scanner displays four images per second, sufficient for easy orientation in tissue and for swept gain regulation on the live image. The ultrasound transducers used in all the scanning heads are standard Cortex 20 MHz with a bandwidth of 15 MHz, a focal distance of 30 mm, and a 6-dB focal length of approximately 13 mm. Using this transducer, the axial system resolution is 50 µm, the lateral system resolution 350 µm, and the typical usable tissue penetration 10 mm. The field of view in depth can be set in four steps going from a minimum of 1.7 mm full screen to a maximum of 13.4 mm full screen. All four fields of view can be panned live to a maximum scanning depth of 30 mm. Our evaluations have been performed by employing the zoom function in the axial direction at the first magnification (at factor 2) which enables exploration of the tissue within a depth interval of 6.71 mm.

The images are presented either in a 256-shade pseudocolor scale or a gray scale. The scales are coded to the A-scan amplitude. The amplitudes displayed are proportional to the recorded amplitudes up to the value 200 within the 0 to 256 scale. Above this value the amplitudes are displayed compressed. The images are computed from 224 A-scan lines recorded along the transducer movement, each A scan having 256 sampling points. The live images can be instantly frozen and can either be saved in the system memory or permanently in a personal computer (PC). All parameter settings are saved simultaneously. Any A-scan line can be displayed, and its position marked on top of a frozen image, for closer analysis and measurement of distances. Further calculations like area and mean echo amplitude within a region of interest as defined by the user can be made directly on the scanner.

The instrument has a fully documented data interface available for connection to PCs for image analysis packages. All the pictures undergoing image analysis have been recorded by the counter-balanced probe capable of scanning within a 22.4 mm × 22.4 mm area, with a built-in closed water path, using a water-comparable ultrasound gel as contact medium.

System calibration can be carried out using a standard phantom (Cortex Technology). This consists of a selected natural rubber compound which produces an entrance echo very similar to human skin and a 1-mm thick spacing plate which has an open slot for the ultrasound transmission contact medium (Cogel®, Comedical, Trento, Italy). The plate thickness is used as the distance between the probe membrane and the phantom when calibrating the gain, which is set to give A-scan peaks up to the maximum limit of the screen. In our instrument this adjustment is equal to 22 dB, used in all our recordings. We have tested the amplitude variations relating to the probe-to-skin distance in a water tank using a plane PVC plate as reflector. By measuring the entrance echo as a function of the depth, we found that the variations in amplitude can be neglected in the interval of 1.5 to 3.5 mm in front of the nominal zero of the instrument (F = 30 mm). Accordingly, during recordings, the distance between the zero point and the skin is kept at 1.7 ± 0.2 mm. This can be achieved by making the epidermis echo coincide with a point marked on the screen frame, corresponding to a 1.7-mm distance from the zero.

B. Technical Aspects of Echographic Recording for Image Analysis

Some technical aspects should be considered when performing recordings of 20-MHz B-scan pictures for image analysis. In fact, in order to obtain reproducible data, images should be recorded under the same experimental conditions.

1. Probe to Skin Distance

When recording echographic images for computerized analysis, one should consider, that the received ultrasound signal is influenced by attenuation due to absorption and/or scattering and reflection. While reflection takes place mainly at the boundaries of tissue compartments with different impedance, absorption is high when the relative amount of proteins and collagen in the tissue increases, and low if the liquid content is elevated. One result of attenuation is that the signal coming from a given tissue structure not only varies according to acoustic tissue parameters, but also to the distance between the structure and the transducer.[6] For this reason the probe-to-skin distance should be kept constant. This can be achieved as previously described or by using a punched spacing plate of constant thickness to be put between the skin and the probe. The quantity of gel between the probe membrane and the skin will be standardized as well, corresponding to the amount filling the plate's slot.

2. The Swept-Gain Curve

The time-gain compensation is used to correct for the echo attenuation of the signals coming from greater depth and can be adjusted as required. Interactive regulation of the swept-gain curve can be useful for a direct observation of the skin site under examination in order to improve the contrast and provide further information, but it eliminates the possibility of performing image analysis. In fact, if the swept-gain curve is not constant the echoes' amplitudes are variably modified, not depending on the characteristics of the tissue and depth, alone.

3. Other Parameters

Other parameters should be considered and kept constant during recording. For example, acoustic properties of the skin rely also on elasticity, tension, pressure, and temperature.[6] The orientation of the collagen fibers, which strongly influences the acoustic properties of the dermis changes under different tension forces, as shown by scanning electron microscopy.[7] This observation can explain the variations in the echogenicity of the skin under different conditions of tension during ultrasound examination, which can be determined by stretching or by increasing the pressure of the probe onto the skin.[8] Moreover, an increase in the water temperature in the applicator seems to be a method of improving the amplitude of the received echo signal.[5]

III. Ultrasound Tissue Characterization

Ultrasound tissue characterization can be defined as the process of describing tissue structure in terms of information, deriving from computer processing of echographic images.[2] Its major goal consists in extracting image features, which may not be available to the human observer, and transforming subjective feature analysis, into a quantitative process, so as to obtain objective and comparable results, which are not dependent on the diagnostician. Many acoustic parameters, such as elasticity, density, attenuation coefficient, and speed of sound, represent suitable features for tissue characterization.[2] However, only image-processing techniques for the evaluation of the distribution of amplitude variations of echographic pictures of the skin are available so far. A simple approach to tissue differentiation is based on the evaluation of mean grey-level values (with standard deviation) within a region of interest. A more specific procedure consists of the use of image segmentation. *Segmentation* is the process which subdivides an image into areas of uniform appearance. Regions or objects of interest are subsequently extracted for further processing, such as description or recognition.[9] When texture characteristics corresponding to different body regions or diseases have to be defined for the first time, the segmentation attributes should be found during processing, and regions of interest must be established interactively by the user, specifying the number and location of the thresholds by trial and error.

A. The Software for Image Analysis

Software for image analysis on Dermascan C® images will be briefly described. The program ascribes an arbitrary scale, ranging from 0 to 255, to the amplitude values of the echoes. It enables the interactive selection of amplitude bands, whose width and positioning on a numerical scale can be chosen. The choice of the band of interest is a function of the problem being considered, which consists in highlighting the picture areas formed by *pixels* (picture elements) characterizing structures or functional aspects of skin reactions, which should be quantitatively evidenced and assessed. In order to identify a meaningful band of interest, the recorded image can be scanned systematically, using different amplitude intervals. This can be achieved manually, by selecting and highlighting a band, corresponding to an amplitude interval of interest, on the scale appearing in the upper part of the screen. At the same time only those parts of the image, whose pixels are reflecting within the selected amplitude interval will be evidenced. After defining a region of interest, it is possible to assess quantitatively the areas, in which the echoes' amplitude is included within the selected values, by calculating its extension in square millimeters or in number of pixels.[10]

In Figures 1 through 4 different parts of the same echographic image are enhanced by using different amplitude bands. The epidermis and the lower part of the dermis are marked with amplitude bands covering the upper part of the scale; intermediate and low amplitude intervals characterize different parts of the dermis.

B. Present Possibilities and Perspectives

Indications for the use of image analysis on 20-MHz B-scan recordings comprise: studies on normal skin and skin aging; determination of the intensity of allergic and irritant patch-test reactions and differentiation of

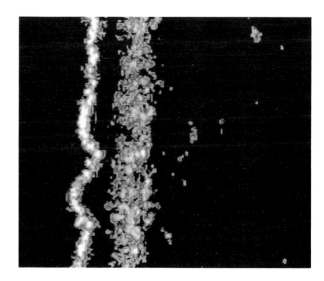

FIGURE 1 Echographic aspect of dorsal forearm skin in a woman aged 86. The image is represented by the pseudocolor scale.

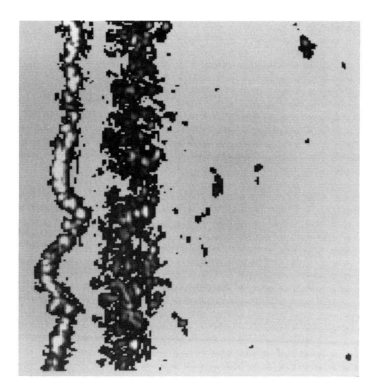

irritant substances; studies of the effect of steroids on the skin; and, finally, evaluation of the course of psoriasis and of the therapeutic effects of drugs used for the treatment of this disease.

Studies on skin tumors and other inflammatory skin diseases are underway. Most skin tumors appear as hyporeflecting areas without any particular acoustic characteristics. Attempts have been made to differentiate skin tumors by quantifying their reflectivity. The main difficulty consists in obtaining quantitative evaluations, regardless of actual attenuation, both in front and inside the tumor itself.

FIGURE 2 Segmentation of the image in Figure 1, after selection of a 0 to 30-amplitude band. Hyporeflecting parts of the dermis are highlighted.

Other skin diseases appearing with a characteristic echographic pattern of the dermis, such as scleroderma, cannot be differentiated from normal skin on the basis of image segmentation. More elaborated image descriptors will probably be needed to characterize this and other types of tissue alterations.

C. Evaluation of Normal Skin

Variations of normal skin according to site, age, and sex have been studied by means of numerous noninvasive methods, including A scanning, which enables the determination of skin thickness at different sites.[11-13] Elaboration of B-scan images allows the quantification of aspects, which can also be appreciated by the simple observation of the images regarding the different distribution of the echogenicity according to site, age, and sex (Figures 5 through 8).[14]

Tables 1 and 2 illustrate the results of a study performed on 48 men and 48 women at six different skin sites.[15] Four age groups (27–31, 60–70, 70–80, and 80–90) were studied. For the evaluation of the images, two bands with low (0–30) and intermediate (50–150) amplitude values were used (Figures 2 and 3). Since skin-

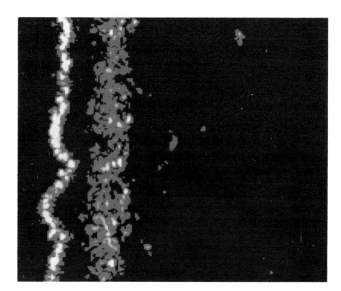

FIGURE 3 Segmentation of the image in Figure 1, after selection of a 50- to 150-amplitude band.

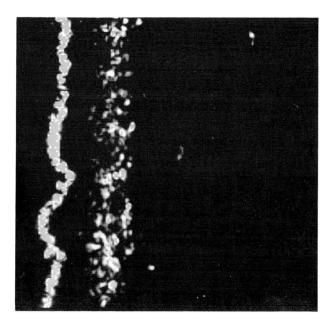

FIGURE 4 Segmentation of the image in Figure 1, after selection of a 201- to 255-band. The entrance echo and hyperreflecting parts of the lower dermis are visible.

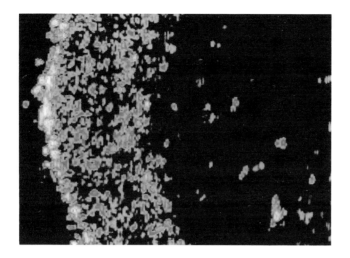

FIGURE 5 Cheek skin in a woman aged 25.

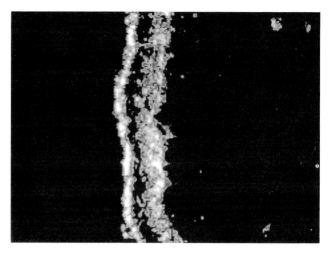

FIGURE 8 Normal volar forearm skin of an elderly subject. A hypoechogenic subepidermal band is clearly visible.

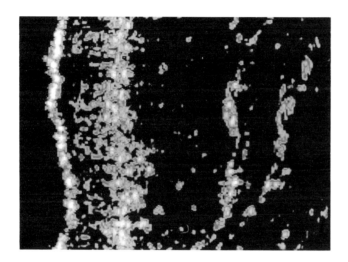

FIGURE 6 Cheek skin in a woman aged 81.

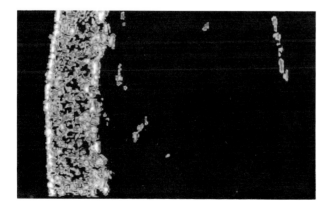

FIGURE 7 Normal volar forearm skin in a young subject.

thickness variations are high according to skin site, age, and sex, echogenicity values are not expressed in absolute numbers, but as the ratio between pixel values (extension of areas) and thickness values.

Low numbers in Table 1 correspond to a limited extension of hyporeflecting (0–30 band) areas and thus, to higher reflectivity in respect to other skin sites: volar and dorsal forearm skin are most echogenic, and forehead skin shows the lowest reflectivity. Accordingly, intermediate amplitude values (50–150) are not so marked on the forehead (Table 2). Regarding variations related to sex, we can observe that at forehead and cheek, echogenicity of the dermis is higher in women. No significant differences can be seen on abdomen and back skin, whereas for volar and dorsal forearm skin the extension of the hyporeflecting area corresponding to the hypoechogenic band in the upper part of the dermis is higher in men. Finally, concerning modifications of the dermis with the aging process, we can see that the reflectivity of the dermis increases at forehead, cheek, abdomen, and back in elderly subjects. Volar and dorsal forearm skin show a decrease of the dermal echogenicity due to the appearance of the hyposonic subepidermal band, which in some subjects occupies most of the dermis (Figure 8). Figure 9 represents the values of the extension of the hyporeflecting (0–30) area at volar forearm skin in three different age groups in elderly subjects. Whereas in men the thickness of the hyposonic band seems constant, in elderly women it increases progressively with aging. Values referring to dorsal forearm skin show the same trend.

To summarize, we can say that there is a great regional variation in the behavior of ultrasound reflection of the skin according to skin site, sex, and age. Regarding skin aging, it is not possible to identify a general tendency for the variations of the acoustic

TABLE 1 Echogenicity of the Dermis According to Skin Site, Sex, and Age

	Forehead	Cheek	Volar Forearm	Dorsal Forearm	Upper Abdomen	Lower Back
YM	67	58	5	9	46	56
YW	48	41	5	10	49	63
EM	58	53	18	25	32	41
EW	41	33	14	19	27	34

Note: A 0- to 30-band evaluation. YM = men aged 27–31, YW = women aged 27–31; EM = men aged 60–90; EW = women aged 60–90. Numbers are obtained by dividing the mean amplitude value by the skin thickness value and by 100.

TABLE 2 Echogenicity of the Dermis According to Skin Site, Sex, and Age

	Forehead	Cheek	Upper Abdomen	Lower Back
YM	15	19	29	16
YW	24	30	27	14
EM	16	18	33	27
EW	27	35	37	32

Note: A 50- to 150-band evaluation. YM = men aged 27–31; YW = women aged 27–31; EM = men aged 60–90; EW = women aged 60–90. Numbers are obtained by dividing the mean amplitude value by the skin thickness value and by 100.

properties of elderly skin: at four skin sites examined, a shift from ultrasound echoes of low intensity, characteristic of the dermis of young subjects, to intermediate reflection amplitudes, are visible in the elderly. On the contrary, forearm skin generates echoes of lower intensity.[16] The variations of dermal reflectivity may be correlated to structural and biochemical diversities of collagen and elastic bundles and also to water content and to composition of the ground substance, which may alter density, homogeneousness, and spatial organization of structural elements which are responsible for the acoustic behavior and the reflectivity to ultrasound. These changes can only be appreciated by echography and quantitatively assessed by performing segmentation of the image, which represents an innovative means in the study of skin aging.

1. Echographic Characteristics of Normal Volar Forearm Skin

Forearm skin is the most suitable site for experimental allergic and irritant patch testing. Yet, great regional variations in the response to irritation, medication with corticosteroids, vasodilation, and reactivity to allergens have been described.[10,17-19] In fact, by performing image analysis on 20 MHz echographic pictures taken at various sites of forearm skin, great differences can be appreciated in the distribution of the reflectivity of the dermis. To carry out patch tests in a reproducible way, and to overcome intra- and interindividual variations, the same skin sites should be assessed using accurate spacing references. To this aim we use flexible rubber foils with precise measures, having holes 2.5 cm in diameter corresponding to the testing areas (Figure 10). Before each evaluation, the upper edge of the rubber foil is applied to the elbow crease and test sites are drawn on the skin. Thus, each successive measurement can refer to its baseline value. Table 3 illustrates values of pixels reflecting within a 0–30 amplitude interval at the five different sites indicated in Figure 10. Significant differences can be established

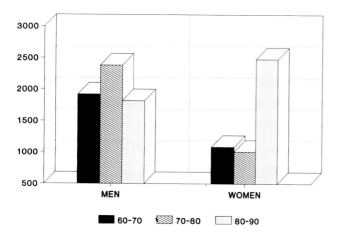

FIGURE 9 Values of the extension of the hyporeflecting area at volar forearm skin in three different age groups, are expressed in number of pixels (picture elements).

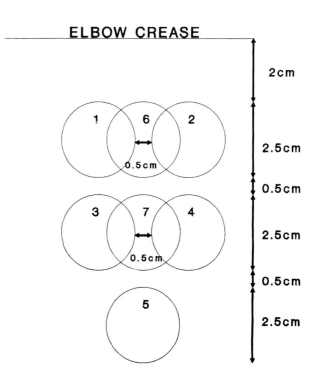

FIGURE 10 Sites and measures of testing areas.

TABLE 3 Regional Differences of Volar Forearm Skin

Area	1	2	3	4	5	6	7
Mean pixel values	1368.27	1258.71	1466.43	1440.8	1208.69	989.29	1015.72
SD	851.8	554.82	780.6	792.09	581.86	551.28	497.45
No. of cases	56	56	53	55	58	80	79

Note: A 0- to 30-band evaluation.

between areas 6 and 7 and the other sites, and between area 3 vs. 5 to 8, pointing to the necessity of referring observations on allergic and irritant reactions to precise baseline data. Moreover, experimental irritant patch testing on forearm skin should be performed in subjects under 45 years of age, without a subepidermal band of lower echogenicity. In fact, edema, deriving from an inflammatory reaction and appearing as a hypoechogenic band or area, can be difficult to recognize on the forearm of elderly subjects (Figures 7 and 8).

D. Evaluation of Allergic Patch-Test Reactions

In clinical practice allergic reactions can be evaluated by visual and palpatory examination. However, in some circumstances, such as the monitoring of drug-induced or spontaneous variations of reactivity to contact allergens, a quantitative assessment is required, enabling an objective collection of data, and their statistical evaluation. The echographic assessment of patch-test reactions was first performed by Serup and co-workers,[20] who evaluated doubtful and positive patch-test readings by A-scan measurements of skin thickness, and found that this parameter, which can be considered an expression of allergic edema, increases according to the intensity of the reaction.

In a B-scan image, a positive patch-test reaction appears with an increased thickness of the skin, and with a greater homogeneity and a decreased echogenicity of the dermis (Figures 11 and 12). At the same time, we can observe a rise in the extension of hyposonic areas, a decrease in the quantity of objects reflecting within intermediate amplitude values, and the disappearance of the hyperreflecting parts of the dermis at volar forearm skin. No difference can be noticed among reactions induced by different contact allergens. The quantification of the inflammatory response can be achieved by processing the image by segmentation using a band covering the low echoamplitude values of the scale. Intervals like 0–20, 0–30, 0–40, 5–20, and 3–40 provide results that can be usefully employed in the quantification of the intensity of the reactions, correlating well with the clinical observation. In order to quantify allergic reactions of different intensities, we performed patch tests on 12 nickel-sensitized women with nickel sulfate at different concentrations (0.5, 1, 2, 3, and 5%), with a 72-h

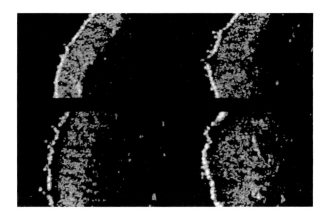

FIGURE 11 Patch-test reactions of different degrees of positivity.

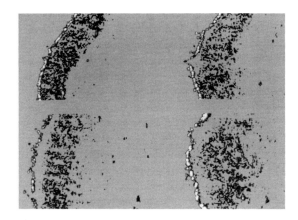

FIGURE 12 Same images as in Figure 11, after selection of a 0–30 band. The hypoechogenic part of the dermis corresponding to edema and inflammatory infiltration is highlighted.

observation period, employing a 0–30 and a 201–255 band to assess the reflectivity of the dermis. A progressive increase in the number of low reflecting pixels, and a decrease in the values assessing hyperreflectivity, according to the nickel patch-test concentration and the elapsing of time was demonstrated (Figure 13).[21]

E. Evaluation of Subclinical Allergic Patch-Test Reactions

Image segmentation enables the quantification of allergic patch test reactions, which are regarded clinically as negative or doubtful. This has been demonstrated by patch testing 70 nickel-sensitized patients with 0.05%

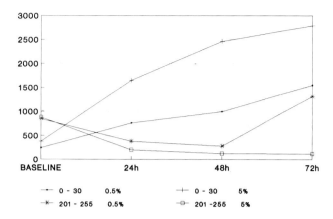

FIGURE 13 Image processing of 20-MHz B-scan recordings of 0.5 and 5% nickel sulfate pet. patch-test reactions on volar forearm skin. Segmentation after selection of a 0–30 and a 201–255 interval. Numbers express mean values of pixels reflected within the chosen amplitude bands.

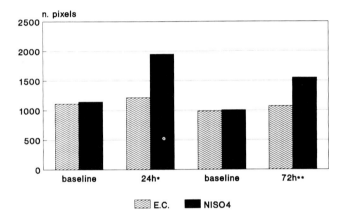

FIGURE 14 Echographic evaluation of doubtful and negative allergic patch-test reactions. Image analysis by the 0–30 band.

nickel sulfate, and by recording the echographic images at 24 and 72 h.[22] There were 22 doubtful 72-h reactions and 12 negative 24-h responses at test sites which showed positive reactions at 72 h; they were processed by the image analysis software. A significant increase in the extension of the hyporeflecting areas in the dermis at clinically negative and doubtful readings was clearly shown (Figure 14). Thus, this method can be particularly useful when evaluating substances with low sensitizing potential, or allergens contained at low concentrations in complex preparations, or subjects with a high sensitization threshold.

When performing echographic quantification of allergic reactions, one should consider the differences in the reactivity to allergens on forearm skin. By patch testing 17 nickel-sensitive women at four different sites on the volar forearm, a greater response to patch tests in skin areas near the wrist crease was demonstrated (area 4

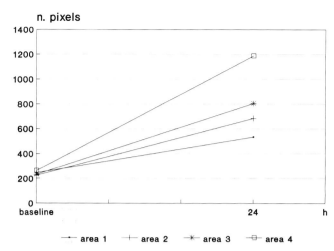

FIGURE 15 Extension of the 0–30 area at four sites on volar forearm skin, before and after a 24-h application of a 5% nickel sulfate patch test. Mean values at test sites near the wrist crease are higher.

in Figure 15).[10] This is a clear indication for the use of rotating patterns or symmetrical sites in quantitative and comparative investigations concerning the evaluation of contact allergy on volar forearm skin.

F. Evaluation of the Activity of Topical Corticosteroids

The echographic quantification of allergic patch-test reactions can be employed as an evaluation method for experimental procedures specifically concerning the anti-inflammatory action of corticosteroids. In fact, epicutaneous testing, which reproduces eczematous reactions in a controlled manner in sensitive subjects, seems to be an ideal procedure for studying the inhibitory activity of these drugs.[23] The procedure consists in patch testing sensitized subjects with nickel sulfate 5% at different skin sites on the volar forearm, and in applying medications with corticosteroids of diverse potency 16 and 40 h after the induction of the allergic reactions. The inhibition of a patch-test reaction after pharmaceutical treatment is proportional to the potency of the steroid used for topical application, as evaluated by segmentation of the echographic recording and determination of the extension of the hypoechogenic area of the dermis (Figure 16).

G. Evaluation of Irritant Reactions
1. Assessment of Sodium Lauryl Sulfate Induced Irritation

The inflammatory component of irritant patch-test reactions can be assessed by image analysis on 20-MHz B-scan recordings using the same method as for evaluating allergic responses (Figure 17).[24] At volar forearm patch-test sites, sodium lauryl sulfate (SLS)-induced inflammation appears echographically with an edema, which is more superficial than the one caused by nickel sulfate in sensitive subjects. Fre-

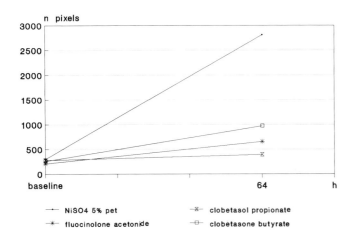

FIGURE 16 Evaluation of the anti-inflammatory effects of corticosteroids. Extension of the 0–30 areas of nickel sulfate test sites treated with different preparations.

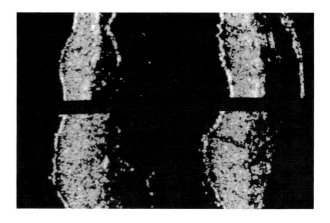

FIGURE 18 Echographic aspect of SLS-induced irritant reactions of different intensity.

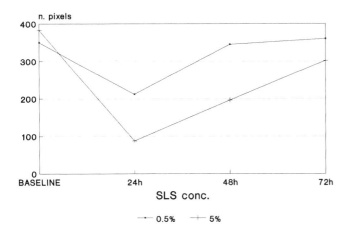

FIGURE 17 Echographic evaluation of SLS-induced irritation. Assessment was by a 0–30 and 201–255 band.

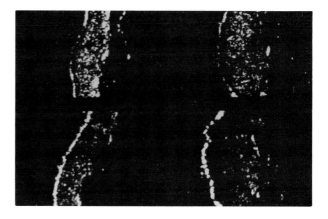

FIGURE 19 SLS-induced reactions are represented in the upper part of the picture; nickel sulfate responses are represented in the lower part. Segmentation after selection of a band highlighting the entrance echo.

quently, a subepidermal hypoechogenic band represents the first sign of irritation at the dermal level. Subsequently, hypoechogenicity of the remaining part of the dermis appears, but, even if the lower dermis is involved, hyperreflecting areas at this location do not disappear completely as they do at allergic patch-test sites (Figure 18). While at positive nickel patch-test sites the entry echo shows only a slight attenuation even for very intense reactions, at SLS patch-test sites, a clear decrease in the echogenicity of the epidermis which is hyporeflecting or interrupted or has totally disappeared, can be appreciated at 24-h evaluations (Figure 19). Going on the above, it is possible to distinguish between an allergic and an SLS-induced reaction. Whereas data calculated by the 0–30 evaluation show a fair correlation to clinical scoring, reflectivity of the epidermis, as assessed by the 201–255 band, is inversely proportional to transepidermal water loss (TEWL) values.[24] Thus, at SLS patch-test sites, loss of superficial hypoechogenicity can be considered as a visual equivalent of impaired barrier function of the skin.

2. Evaluation of Subclinical Irritation

Image analysis on 20-MHz B-scan recordings also represents a sensitive method for evaluating subclinical irritation. In fact, SLS-treated skin areas, clinically evaluated as negative tests, show echographic modifications which are significant in comparison to normal skin values.[25]

3. Evaluation of Sodium-Lauryl-Sulfate Induced Irritation in Hairless Mice

SLS-induced irritation can be echographically assessed also in hairless mice. Segmentation can be performed by using the same amplitude intervals as in humans, whereas other amplitude intervals can be employed for the determination of the dermis-hypodermis boundary.[26]

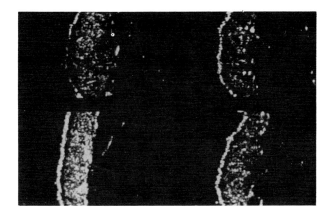 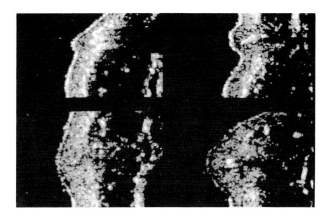

FIGURE 20 Echographic images of skin reactions induced by different irritant substances: SLS in the upper left corner, sodium hydroxide in the upper right, nonanoic acid in the lower left, and hydrochloric acid in the lower right. A band highlighting the entrance echo has been chosen.

FIGURE 21 Echographic aspect of histamine induced wheals of different size.

4. Evaluation of Other Irritant Substances

While the inflammatory component of irritant reactions has a fairly uniform expression, superficial reflectivity corresponding to epidermis enables the characterization of different irritant substances from an echographic point of view. In contrast to hypoechogenicity induced by SLS, at nonanoic acid and hydrochloric acid test sites a progressive increase in the epidermal reflectivity is visible, whereas propylene glycol and sodium hydroxide do not bring about any statistical variations in comparison to baseline skin (Figure 20).[27]

5. Indications for the Use of Units for Patch Tests to be Evaluated by Echography

When performing patch tests with irritant substances, we must bear in mind that pressure caused by the adhesion of rigid chambers to the skin can induce an edema which is not always assessable by clinical evaluation but is echographically evident. Since the presence of edema represents a positive element in the evaluation of irritancy, it is advisable to perform patch tests with irritant substances by using units which do not exert too much pressure on the skin.[28]

H. Evaluation of Wheals

Besides measurement of superficial wheal extension by planimetry, and assessment of skin thickness by A scanning, determination of the dermal hypoechogenic area by image analysis can be employed in the evaluation of skin responses to wheal-inducing substances (Figure 21). By measuring histamine-induced wheals, we were able to demonstrate that edema of the dermis, corresponding to the extension of the hyposonic area, increases in time according to the histamine concentration. However, for the same concentration, a more

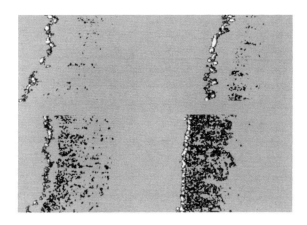

FIGURE 22 Echographic aspect of a psoriatic plaque: before treatment, in the upper left corner, after 7 d of treatment with dithranol in the upper right corner, and after 14 d of treatment in the lower left corner. Normal contralateral skin in the lower right corner. Images are represented after selection of a 0–30 band.

intense response with a greater wheal extension and vaster dermal hypoechogenic area, was observed at proximal sites in respect to distal ones.[29]

I. Evaluation of Psoriatic Skin

The efficacy of antipsoriatic drugs is generally evaluated clinically. However, even if well-standardized criteria are used, reproducible data are difficult to obtain, and a comparison between studies performed in different centers is seldom possible. Therefore, different noninvasive methods have been proposed, with a view to standardizing parameters of potential relevance in following the evolution of the disease.[30] Echographic evaluation enables the identification of three aspects which characterize psoriatic lesions: an increase in skin thickness, the presence of a thick hyposonic subepidermal band, and an enhanced entrance echo

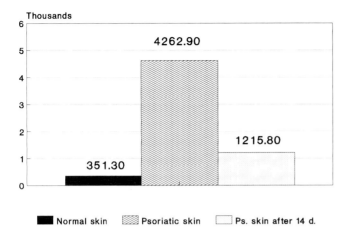

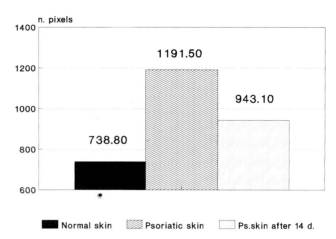

FIGURE 23 Echographic assessment of psoriatic skin. Evaluation by a 0–30 band. Histograms express values of hyporeflecting areas at normal contralateral skin, psoriatic untreated skin, and psoriatic skin after 14 d of treatment.

FIGURE 24 Echographic assessment of psoriatic skin. Evaluation by a 201–255 band. Histograms express values of hyperreflecting areas at normal contralateral skin, psoriatic untreated skin, and psoriatic skin after 14 d of treatment.

(Figure 22). Since echoes coming from tissues are determined by differences in acoustic impedence, we can presume that the hyposonic band is caused by the presence of a homogeneous component, which in psoriatic lesions could be represented by inflammatory infiltrate and vasodilatation. The enhancement of the entrance echo, on the other hand, corresponds to a thickening and a modification of psoriatic epidermis. Figures 23 and 24 show the results of processing 20-MHz B-scan images performed on 10 psoriatic plaques, before, and after 14 d of treatment with anthralin. Both the extension of the hyposonic subepidermal band and that of the hyporeflecting dermal echo gradually and constantly decrease according to clinical observations.[31] Thus, processing of 20-MHz B-scan images allows the characterization and quantification of echographic parameters, which vary during the course of the disease according to treatment.

IV. Conclusion

Superficiality of the skin, enabling high resolution ultrasound, represents a unique opportunity for analyzing echographic images. In fact, changes in skin thickness are very slight and recording conditions can be standardized geometrically by fixing probe-to-skin distance, whereby the intensity variations caused by the focusing of the transducer will remain very small.

New developments on high-resolution ultrasound image analysis methods are underway. Segmentation can represent the first step for further computer processing and description of images. In fact, echographic images of some skin conditions do not present any quantitative variation in the echoes' amplitudes, the total area covered by pixels reflecting within a certain amplitude range are the same. However, the spatial distribution of the amplitudes differs consistently, characterizing what we call a different texture of the dermis. The aim of future computer processing is to find a way of using descriptors that highlight essential differences between objects or classes of objects, regardless of changes in factors such as location, size, and orientation: boundary descriptors such as contour length, diameter, and curvature, or regional descriptors such as area, perimeter or compactness could be used in the future for describing ultrasound images of the skin.

References

1. Serup J., Ten years' experience with high frequency ultrasound examination of the skin: development and refinement of technique and equipment, in *Ultrasound in Dermatology*, Altmeyer P., el Gammal S. and Hoffmann K., Eds., Springer Verlag, Berlin Heidelberg, 1992, 41.
2. Bamber J.C. and Tristam M., Diagnostic ultrasound, in *The Physics of Medical Imaging*, Webb S., Ed., Adam Hilger, Bristol, 1990, chap. 7.
3. Gonzalez R.C. and Wintz P., *Digital Image Processing*, Addison-Welsey, New York, 1987.
4. Webb S., *The Physics of Medical Imaging*, Adam Hilger, Bristol, 1990.
5. el Gammal S., Hoffman K., Hoss A., Hammentgen R., Altmeyer P., and Ermert H., New concepts and developments in high resolution ultrasound, in *Ultrasound in Dermatology*, Altmeyer P., el Gammal S., and Hoffmann K., Eds., Springer Verlag, Berlin Heidelberg, 1992, 397.
6. Mc Dicken W.N., Ultrasound in Tissue, in *Diagnostic Ultrasonics. Principles and Use of Instruments*, Mc Dicken W. N., Ed., Churchill Livingstone, Hong Kong, 1991, chap. 4.
7. Brown I.A., A-scanning electron microscopic study of the effects of uniaxial tension on human skin, *Br J Dermatol*, 89, 383, 1973.
8. el Gammal S., Experimental approaches and new developments in high frequency ultrasound in dermatology, *Zentralbl Haut Geschlechtskr*, 157, 327, 1990.
9. Gonzalez R.C. and Wintz P., Image segmentation, in *Digital Image Processing*, Gonzalez R.C. and Wintz P., Eds., Addison-Welsey, New York, 1987.

10. Seidenari S. and Di Nardo A., Cutaneous reactivity to allergens increases from the antecubital fossa to the wrist: an echographic evaluation by means of a new image analysis system, *Contact Dermatitis*, 26, 171, 1992.
11. Berardesca E., Farinelli N., Rabbiosi G., and Maibach H.I., Skin bioengineering in the non-invasive assessment of cutaneous aging, *Dermatologica*, 182, 1, 1991.
12. Alexander H. and Miller D.I., Determining skin thickness with pulsed ultrasound, *J Invest Dermatol*, 72, 17, 1979.
13. Sondergaard J., Serup J., and Tikijob G., Ultrasonic A and B scanning in clinical and experimental dermatology, *Acta Derm Venereol (Stockh)*, (Suppl.) 120, 76, 1985.
14. De Rigal J., Escoffier C., Querleux B., Faivre B., Agache P., and Lévêque J.-L., Assessment of skin aging of human skin by in vivo ultrasonic imaging, *J Invest Dermatol*, 93, 622, 1989.
15. Seidenari S., Pagnoni A., Lasagni C., and Giannetti A., Age-dependent variations of normal skin: image processing of 20 MHz B scan recordings. Abstracts of the 9th Int Symp. of Bioengineering and the Skin, Sendai, Japan, October 19–20, 1992, *Skin Pharmacol*, 5, 230, 1992.
16. Seidenari S., Pagnoni A., Di Nardo A. and Giannetti A., Echographic evaluation with image analysis of normal skin: variations according to age and sex. *Skin Pharmacol*, in press.
17. Van der Valk P.G.M. and Maibach H.I., Potential for irritation increases from the wrist to the cubital fossa, *Br J Dermatol*, 121, 709, 1989.
18. Kirsch J., Gipson J.R., Darley C.R., Barth J., and Burke C.A., Forearm skin variation with the corticosteroid vasoconstrictor assay, *Br J Dermatol*, 106, 495, 1982.
19. Tur E., Maibach H.I., and Guy R.H., Spatial variability of vasodilatation in human forearm skin. *Br J Dermatol*, 113, 197, 1985.
20. Serup J., Staberg B., and Klemp P., Quantification of cutaneous oedema in patch test reactions by measurements of skin thickness with high frequency pulsed ultrasound, *Contact Dermatitis*, 10, 88, 1984.
21. Seidenari S. and Di Nardo A., B scanning evaluation of allergic reactions with binary transformation and image analysis, *Acta Derm Venereol (Stockh)*, (Suppl.) 175, 3, 1992.
22. Seidenari S., Echographic evaluation of subclinical allergic patch test reactions, *Contact Dermatitis*, in press.
23. Seidenari S., Di Nardo A., and Giannetti G., Assessment of topical corticosteroid activity on experimentally induced contact dermatitis: echographic evaluation with binary transformation and image analysis, *Skin Pharmacol*, in press.
24. Seidenari S. and Di Nardo A., B scanning evaluation of irritant reactions with binary transformation and image analysis, *Acta Derm Venereol (Stockh)*, (Suppl.) 175, 9, 1992.
25. Seidenari S. and Pepe P., Valutazione ecografica dell'irritazione subclinica indotta da laurilsolfato di sodio, *G Ital Dermatol Venereol*, in press.
26. Seidenari S., Zanella C., and Pepe P., Echographic evaluation of sodium lauryl sulfate induced reactions in mice, *Contact Dermatitis*, submitted for publication.
27. Seidenari S., Di Nardo A., Schiavi E., and Pepe P., Echographic evaluation of irritant reactions induced by nonanoic acid, hydrochloric acid and sodium lauryl sulfate: a comparison with TEWL assessment. Abstracts of the 9th Int Symp. of Bioengineering and the Skin, Sendai, Japan, October 19–20, 1992, *Skin Pharmacol*, 5, 240, 1992.
28. Seidenari S., Turnaturi C., Motolese A., and Pepe P., Echographic evaluation of edema induced by patch test chambers, *Contact Dermatitis*, 27, 331, 1992.
29. Di Nardo A. and Seidenari S., Echographic evaluation with image analysis of histamine induced wheals, *Skin Pharmacol*, submitted for publication.
30. Berardesca E. and Maibach H.I., Non-invasive bioengineering assessment of psoriasis, *Int J Dermatol*, 28, 157, 1983.
31. Di Nardo A., Seidenari S., and Giannetti A., B-scanning evaluation with image analysis of psoriatic skin, *Exp Dermatol*, 1, 121, 1992.

Chapter 12.3
High-Frequency Sonography of Skin Diseases

Klaus Hoffmann, Anja Röchling, Markus Stücker, Kay Dirting,
Stefan el Gammal, Andrea Hoffmann, and Peter Altmeyer
Dermatology Department
Ruhr University Bochum
St. Josef's Hospital
Bochum, Germany

I. Introduction

In dermatology imaging techniques are of increasing interest. For many reasons we observe more patients with skin diseases, e.g., malignant skin tumors, than in the past.[27,32] Especially for this reason diagnostic imaging methods have found wide application. When making a diagnosis the dermatologist uses a number of fixed rules. His primary diagnosis is based on a subjective assessment of the skin disease. In spite of this unsatisfactory situation no procedure has yet been developed which can assist the dermatologist in making a differential diagnosis with a similar degree of reliability, for example, of the electrocardiography used by the internist in the diagnosis of cardiac diseases.

In other fields of medicine the search for noninvasive diagnostic methods fulfilling the above postulates led to the development of ultrasound several decades ago. In dermatology ultrasound initially meant a step forward in the search for metastases, particularly of malignant melanomas. The descriptions of ultrasound images of melanoma metastases vary considerably. A fundamental change was brought about by the "high-resolution" (high frequency) scanners which are now available in over 150 dermatologic centers all over the world. These scanners are able, for the first time, to produce a differentiated two-dimensional image even of individual layers of the skin.

Ultrasound examination of skin diseases was made possible by the fundamental work of Alexander.[1] Initially the skin was examined mainly with A-mode (A = amplitude) scanners.[1,5,6,9,11,30,40–42] This mode involves direct visualization of the voltage oscillation curves on an oscilloscope. For distance measurement the elapse time between the acoustic pulse and the returning echo is transformed into spatial distances, assuming a constant speed of sound in different skin tissues (≈1480 m/s). We know and agree that the A scan is useful to examine skin thickness.[49,50] However, examination of the skin with the A scan is not unproblematic as the investigator obtains only a one-dimensional amplitude information. As far as differentiation of histological structures is concerned it is very difficult to determine where the individual reflections have arisen. In particular the echopoor hair follicles and finger-like projections of islets of adipose tissue into the corium, which are frequently found on the posterior aspect of the thighs and the buttocks, cannot be differentiated with certainty in the A scan.[21] As almost all skin tumors appear more or less echolucent in the ultrasound scan and the method was thus not of genuine differential diagnostic assistance, some working groups began to experiment with scanners which we today would describe as low to middle frequency but which permitted so-called B-scan imaging (B = brightness).[9,45] By moving the transducer laterally in one direction (e.g., X axis) roughly parallel to the skin surface, multiple A scans can be obtained. Every A scan is "demodulated" by detecting the envelope of the voltage oscillation curve. The amplitude of the envelope curve is then transformed into a brightness modulation.[4] At the beginning only specialized engineering departments and laboratories were able to design, construct, and use such high resolution ultrasound imaging systems. The latest result in the development is the C scan (Computerized). If two motors are used to move the transducer over a defined area of skin, a tomographic horizontal section from every part of the skin can be computed by using a "time window" (only echoes from a certain skin depth [defined as running time of the signal] are selected for this tomographic section).

Today two commercially available systems are on the market which provide excellent B-scan images (Dermascan C®, Cortex, Denmark, and DUB 20 S®, Taberna pro medicum, Germany). Both scanners work with 20 MHz, the axial resolution is ≈80 µm (defined by the bandwidth of the ultrasound signal) and the lateral resolution (defined by the shape of the ultrasound beam) is ≈200 µm. Looking at the advertisements of the companies there is some confusing information regarding resolution. Resolution is defined as the minimum distance between two point sources, which can be discerned as being distinct. Fundamental physical laws apply, there are no mathematical tricks (algorithms) that can increase the resolution. The resolution is directly dependent on the quality of the transducer. Both commercially available scanners use nearly the same one. We should also take into account that resolution can be improved by increasing the center frequency of the transducer but in-depth signal penetration diminishes, due to frequency-dependent signal attenuation (in biological tissue about 1dB/MHz/cm). For routine work the 20-MHz system with an in-depth penetration of $\approx$13 mm is very suitable. We developed a 100-MHz scanner. It might be, that a system with transducers of 20 and 100 MHz will be the standard, because the 100-MHz system provides additional information about the epidermis.[14] The main differences between the two commercially available scanners can be found in the signal processing and the software. A very important message is that the prices for the scanners were cut by 50 to 70% within the last 2 years.

Exact correlation of the sonograms with histological sections or typification of specific reflex patterns are possible with high-frequency scanners.[16] Our working group believes that only investigators with experience in the direct correlation of about 500 direct comparisons between ultrasound and histological sections should interpret B-scan images. Training in a specialized center should be completed and a detailed knowledge about the physical basis of ultrasound must exist. Another very important basic requirement is the comparability of ultrasound scans both the images within and between scanners and different groups. As comparable sonograms can only be obtained at constant scanner settings all examinations for tumor diagnosis should be made with the same settings (amplification of echo amplitudes, color scales). Only then one can generate comparable ultrasound scans suitable not only for measurement of tumor thickness but also for differential diagnosis. In nearly all standards the light colors (white, blue) represent strong reflection and the dark colors (black, dark green) represent weak reflection, however if we look on the images of different scanners they look very different. We should not confuse the differences of those images with a higher or lower resolution of a scanner, as is said by the scanner distributors.[18] The international society for skin imaging will define standards for ultrasound scanners in the near future to ensure comparable results. At the moment training in a center that provides different scanners seems to be the only solution of the problem for beginners in the technique. Fortunately, there are more than 100 papers and a book about high frequency ultrasound to be read in the meantime.

II. Examination Procedure

In order to permit later follow-up studies a well-ordered standardized filing system for collection of information should be set up early. Our working group uses a data base developed for this purpose with the software dBASE IV-2.0®. We recommend the following examination procedure:[4,18,24]

- The patient should be well informed about the nature, aim, and procedure of the examination as his cooperation will be required. This is especially true for the DUB 20® System. The water path for the ultrasound signal is not closed, so that water is frequently lost during the examination. It can be avoided for management of this problem by applying pressure to the skin to close the system.
- Standardized examination conditions must be observed and documented at all times. As far as possible no changes should be made in the settings for signal intensity, reflex amplitude amplification, etc. If such changes are necessary a sonogram in the "normal setting" should always be made for reference.
- The examiner should accustom himself as early as possible to a standardized posture for examining his patients. It is advisable for the patient to be lying down in a prone position during the examination. The skin to be examined should be under moderate tension. For better visualization of structures it may be necessary to examine the skin under various degrees of tension. This can be achieved with the help of an assistant.
- For correlation between the sonogram and the histological section it is imperative to mark the planes to be examined. The histological examination should be performed in exactly the marked areas.
- The examiner must on all account avoid exerting pressure on the skin with the scanner during the examinations. He should always hold the scanner head vertical to the skin surface (90°).
- As much as possible, a scan of the contralateral healthy skin should also be made, in order to have even more information for interpreting the various echo phenomena.

There should exist a *description of routine diagnostic findings*. Apart from the customary patient data and the date of examination the report of the ultrasound findings should cover the following fundamental points:[3,4,18,24]

- General
    - Clinical diagnosis of the structure to be examined by ultrasound
    - Condition of the skin (elastosis, surrounding inflammation, tension of the skin)

- Exact anatomical description of the area to be examined
- How many planes of section were examined in which axes?
- Description of findings — We propose the use of a standardized nomenclature for describing ultrasound images so as to avoid misunderstandings. A standard nomenclature is of high importance (e.g., hypoechoic is a term that can be used for echopoor and echolucent areas). A proposal for this standardization can be found in Tables 1 and 2.
  - Reflex (echo) characteristics and thicknesses of entry echo (epidermis), corium, subcutis, and muscle fascia must be described in detail.
  - Any structures of interest (STOI) or regions of interest (ROI) should be described (whether pathological or physiological!). It is better to describe the structure in its fundamental characteristics than to name it anatomically or pathologically.
  - Location and size of ROI and STOI should be described.
  - Evaluation of a tumor area (ROI) should include a description of the density, structure, and distribution of the internal echoes. A description should also be given of the demarcation (sharp vs. indistinct), border zone (present vs. not present), any echopoor or echorich processes of the ROI and attenuation or intensification of echoes below the ROI (Tables 1 and 2).
- Evaluation — The final evaluation of the findings should also take into consideration the clinical and gross morphological information. We must strictly warn against overrating sonograms, on which doctors inexperienced in the method too quickly base their diagnosis.

III. Normal Skin

When we look at the ultrasound image of normal skin we find a typical zoning of reflex pattern. Four main structures can be described: (1) entry echo, (2) corium, (3) subcutaneous fatty tissue, and (4) bands of connective tissue in the subcutis.

A. Entry Echo

On every high-frequency ultrasound scan of the skin a strong band-like echo is seen at the boundary between the water path and the skin, the so-called *entry echo*. In the past the entry echo was equated with the epidermis.[40,41] In our opinion this is incorrect. The highly reflective entry echo originates from the uppermost portion of the epidermis, probably as a result of the change in impedance from the coupling medium (water) to the stratum corneum.[8,14,51] Exact information can be found in Chapter 5.6.

TABLE 1 Nomenclature for General Ultrasound (Echo/Reflex) Phenomena in B-mode Evaluation

Area/region	Echorich; same as surrounding tissue (scarcely distinguishable); echopoor
Margins of area/region	Sharp (abrupt), indistinct (partly abrupt, partly indistinct)
Echoes below an area/region of interest	Attenuation/intensification of dorsal echoes dorsal acoustic shadow

TABLE 2 Nomenclature for General Description of Ultrasound (Echo/Reflex) Phenomena

Intensity of echoes	Echorich (strong), Echopoor, and without any echoes = echolucent
Structure of echoes	Regular, irregular, ill-defined (weak)
Distribution of echoes	Homogeneous, inhomogeneous

B. Corium

Apart from the entry echo it is the corium (and within it the collagenous fibers) which is the main echogenic structure of the skin.[37,43,44] Fields and Dunn could demonstrate that the supporting tissue has a thousand times higher bulk modulus or "stiffness" than the surrounding parenchyma. Depending on the plane of section, the collagenous bundles of collagenous fibers appear as plaque-like or band-like, moderately or highly reflective structures (B scan x or y direction, C scan). In skin cancer the tumor tissue can replace normal collagen tissue thereby forming regions of reduced echogenicity.[7,9] In a fresh skin scar, which is accompanied by an increase of collagen fibers, one could expect more echo reflexes. However, the disarranged and very small fibers cause an echopoor region.[20] In connection with this finding we should not forget why a signal reflection occurs. A signal reflection is caused at the border of two tissues with different impedance. The impedance is defined by the stiffness and the density of the tissue. We talk about an *impedance jump* when the signal hits a border between tissues with different impedance characteristics. Due to this impedance jump one part of the signal is reflected and one part continues its way to the depth. If we keep this in mind we can understand why a scar (increased amount of ground substance[29]) or inflammation (edema and lymphocytic infiltration) are echopoor to echolucent. The differences in impedances are faded.[10,20,23] The same is true for densely packed tumor cells. Here again we do not find a difference in stiffness or density, the area does not cause reflexes and therefore we find an echopoor area.[4,5,6,14,16,18,24,26,36,48] In contrast to this observation we find more reflexes in skin with high tension as compared to skin with lower tension. This is of immense practical importance when investigating an echopoor tumor in an echopoor corium (e.g., a basal-cell carcinoma in an actinic elastosis). If an assistant tightens the elastotic skin the echogenicity of the skin affected solely by actinic elastosis can be increased, while the area of the basal-cell carcinoma remains echopoor. Thereby the border of the tumor can be detected in many cases.[28,48]

The forearm is a very suitable region for experimental studies in dermatology. In this area we can describe another tension phenomenon. We believe that in this region the skin has two *tension arches*. The first one at the upper border of the cutis is represented by the entry echo (upper tension arch) and the lower one can be found at the lower border of the corium (margin corium and subcutaneous fatty tissue). This "lower tension arch" or "echorich corium" presents itself as an echo rich band comparable to the entry echo.

C. Subcutaneous Fatty Tissue and Connective Tissue Bundles

Unlike its appearance in middle frequency scanners (5 to 10 MHz) the fatty tissue does not produce reflexes.[35] It is depicted completely in black as an echolucent area. In this echolucent area we find highly reflective spots caused by bands of connective tissue.[45] These bands connect the muscle fascia and the lower margin of the corium. They are of special interest in morphea.[25] It may well be that these bundles on becoming sclerotic after an inflammation are at least partly involved in the reduction of the subcutaneous fatty tissue.

IV. Skin Tumors

Of all skin diseases skin tumors have received the greatest attention from the different research groups using high-resolution ultrasound. An experienced dermatologist can diagnose and categorize a skin tumor with a specificity of 90%.[17] Two problems remain: how we can increase the score of hits in inexperienced doctors and describe the uncertain 10% of diagnoses.

Different techniques were introduced: epiluminescence microscopy, nuclear magnetic resonance (NMR)-microscopy, ultrasound, image analysis, skin surface quantification, and others. All of these techniques are imaging of techniques and all of them improve the score of hits but they are never ever able to meet all of our requirements for making a differential diagnosis without clinical information about the tumor. At the moment the information we can gather from the images is not sufficient. This will be different in the future. With image analysis and "intelligent" computer software (neuronal network) the situation will change completely.

Nearly all skin tumors appear echopoor. The most important information we get from ultrasound is the *in vivo* tumor thickness. The tumor thickness can tip the balance between different therapies.[22] We treat a basal-cell carcinoma with an in-depth penetration up to the middle of the corium with cryosurgery while basal-cell carcinomas with ill-defined margins or with an in-depth penetration down to the lower parts of the corium are treated with other techniques like mohs-surgery.[19] The importance of the tumor thickness as regards to planning the surgical procedure need not be explained. In the treatment of Kaposi's sarcoma we use the tumor thickness to decide about laser therapy (thin tumor) and/or combined therapy excision and interferon. Kaposi's sarcoma and metastases are classical examples for the possible use of ultrasound as a noninvasive method for follow-up investigations. However all high-frequency scanners (20 to 50 MHz) are unable to precisely distinguish between tumor parenchyma and peritumoral infiltrate and depict both as a single echopoor area. This phenomenon is well described and is also found in other tumors which are accompanied by peritumoral infiltration.[20,22] It might be that the recently available 100-MHz sonography will solve this problem.

Comparison between histometry and computer-aided determination of tumor thickness in the sonogram, which we call "sonometry", shows the fundamental problem of distinguishing between the tumor itself and similarly echopoor structures situated very close to the tumor, which also explains the differences in the measurements obtained from the same image by different investigators. In our experience this problem arises mainly from hair follicles and bands of adipose tissue as finger-like projections into the corium.[26] In the A scan a distinction is possible in exceptional cases only. High-frequency B-mode scanners of the first generation were likewise not always capable of producing a topographically correct image of these structures. With the Dermascan C® or the DUB 20®, however, the experienced examiner can obtain good differentiation of follicles and bands of fatty tissue. An argument which can be held against this procedure, which we have termed *weighted sonometric measurement* of the ultrasound scan, is that precisely in the case of malignant melanomas (e.g., Lentigo maligna melanoma) agglomerates of tumor cells are sometimes found in the hair follicles. Nevertheless, on account of its better correlation and the resulting greater clinical relevance, weighted measurement is currently the sonometric method of choice.[26]

It should be noted that the skin thickness measured in the histological sections is often greater than that measured *in vivo*. This is attributed to the fact that the biopsy is under less tension after excision. Due to the work-up of the histological material involving dehydration, deparaffinization and the associated shrinkage processes we would, however, expect a reduction in skin thickness. The increased thickness of the biopsy due to the diminished tension evidently overcompensates for the shrinkage artefacts resulting from the histological work-up.[9] It should also be kept in mind that the histological section is 7 µm thick. The scanner on the other hand, in its lateral resolution, subsumes structures along a width of 200 µm to form a sectional image. We thus cannot expect an absolute correlation between the ultrasound scan and the histological section with respect to either measurement of thickness or the visualized structures.

Typical echo phenomena can be seen in various skin diseases. However we should not confuse "typical" with "pathognomonic". For example typical acoustic characteristics of seborrhoeic keratosis in high-frequency ultrasonography have been described.[5,6,37] They do not apply to all forms of seborrhoeic keratoses, however we were only able to confirm a typical pattern for the hyperkeratotic (thick)-type. As a rule this type shows massive attenuation of dorsal echoes sometimes even a complete acoustic shadow. Such is typical in more than 3-mm-thick tumors pathognomonic acoustic pattern is not always found in purely acanthotic or adenoid types. The reflex characteristics are not influenced by the pigmentation of the seborrhoeic keratosis but are influenced by the frequently found horny cysts. Sometimes patients present with a seborrhoeic keratosis that is irritated, rubbed smooth, or even bleeding. Skin cancer must be ruled out. These lesions usually display the reflex characteristics described in the above case and can thus be clearly distinguished from other skin tumors. In this subgroup of seborrhoic keratosis high-frequency ultrasound can provide a reliable basis for diagnosis. The contrast of attenuation of dorsal echo pattern is the intensification (enhancement) of dorsal echoes. We described this phenomenon typical for basal-cell carcinomas. In up to 75% of nodular basal-cell carcinomas and 60% of all basal-cell carcinomas we found a subtumoral echo enhancement.[28,48] The tumor areas were echopoor with inhomogenously distributed internal echoes. We have to point out that only in exceptional cases internal echoes in the tumor areas can be correlated with particular structures in histology.[9,28,48]

The most likely explanation for the internal echoes often found in the nevi is that they are produced by the highly echogenic normal connective tissue lying between the rete ridges and the nevus cell nests.[18,24] Ramified rete ridges lying at 90° to the signal can also produce echoes. A very important question is what information ultrasound may provide for making a differential diagnosis of pigmented lesions. The first answer has been given: it is very helpful if we have to differ pigmented seborrhoeic keratosis and malignant melanoma. Another question refers to the differential diagnosis between pigmented nevi and malignant melanoma. We know from different publications that malignant melanoma should appear echolucent and in the echopoor area of nevi we should find a weak reflex pattern.[5–7] Unfortunately this pattern of echogenicity is absolutely not pathognomonic or typical. In 75% of the malignant melanomas and 55% of nevi we found irregular borders to the sides and to the depth. In both tumor types, the margin to the entry echo is sharply delimited. In the tumor areas we find an echopoor to echolucent area in 74% of malignant melanoma and the same pattern in 65% of nevi. Underneath nevi and malignant melanomas both echo enhancement and attenuation can be seen.[18,24,46,47] Sonometrically determined tumor thickness can be correlated significantly with the histometric results in all types of tumors ≈r = 0.98.[6,16,26] However sonometry produces higher values than histometry on average ≈0.27 mm. There are various reasons for the marked tendency towards such higher values. One of the main problems is comparison of an *in vivo* (sonometry) measurement with an *in vitro* (histometry) technique (shrinkage and tension problems). Histomorphological problems increase the mistake. For example hair follicles and/or glands lying next to the tumor can lead to a considerable exaggeration of the tumor thickness in the ultrasound scan. Other problems are muscle-fiber bundles and projections of subcutaneous fatty tissue into the dermis. More frequently overestimates of thickness are caused by the presence of subtumoral inflammatory infiltration, as this often lies immediately below the tumor parenchyma and cannot be distinguished from it sonographically due to the limited resolution of standard high-frequency scanners. In basal-cell carcinomas the echopoor stroma is another source of higher sonometric values.[48] At present such overestimates of tumor depth are unavoidable. However we should underline that the tumors are overestimated, in our therapeutical conclusions this *adds* to security.

The information contained in the ultrasound scan is available in the form of digital information. It has been shown that, using three-dimensional reconstructions, it is possible to calculate the surface area and volume of the tumor.[12,13,15,38,39] These two parameters can then be used to calculate the *invasive tumor mass* as a new prognostic criterion. It is not yet clear what significance this criterion will have in the last instance. But even today there is no doubt that ultrasound decisively improves the planning of surgery for malignant melanoma.[16,22,37,43]

The following is a case study of a 25-year-old woman with malignant melanoma on her left thigh. The tumor was discovered by the patient 3 months prior to our treatment. Histology showed a superficial spreading melanoma, Clark Level III, and a Breslow's tumor thickness of 0.6 mm.

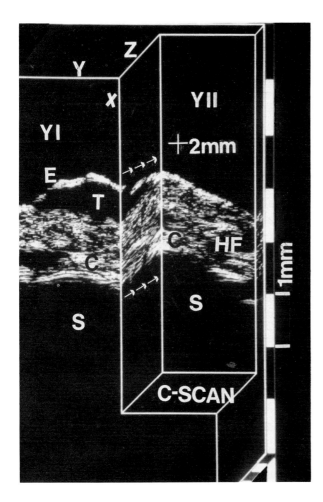
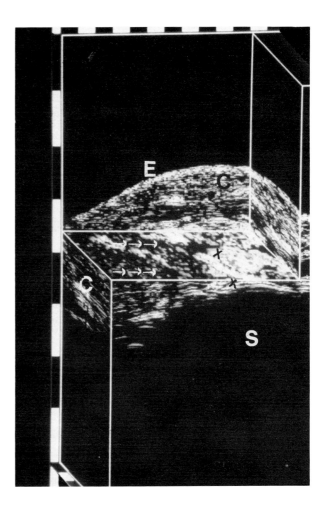

FIGURE 1 Scan of malignant melanoma. X,Y,Z direction in the ultrasound image. Note that in all images normally the skin surface is on the right side of the image. All published images are leaned at an angle of 90°. E = Entry echo; YI = view of the foreground; YII = section 2mm behind Y1; T = tumor, C = corium, S = subcutis, HF = hair follicle (made with DUB 20 S®).

FIGURE 2 Ultrasound scan of the same tumor as shown in Figure 1. C = corium, E = entry echo, S = subcutis (made with DUB 20 S®). In this image the area directly under the tumor is depicted as a C scan. We find a dorsal echo enhancement in the tissue underneath the tumor.

In malignant melanoma we routinely use the 3D options of our scanner. This option differs in the DUB 20 S® and the Dermascan C®. The DUB 20 provides a pseudo 3D image. The images consist of serial sections that are displayed together without gaps. With this trick we create the impression of a 3D voxel reconstruction. To improve this impression the image can be rotated and any wanted tomographic section can be selected. Figure 1 shows the tumor (T) as an echolucent, sharply delimited area directly underneath the entry echo (E). The corium (C) is "normal" echorich and the subcutis (S) again echolucent. In the right lower corium we find a cut through an echo poor hair follicle (HF). The pseudo 3D image provides the possibility to select every section in every axis we wish to see. In this image there is a difference of 2 mm from the left part of the image (YI) to the right part of the image (YII). Both tomographic sections are connected with the according part from the Z axis. With this technique we can minimize measurement problems caused by benign echopoor structures. Another advantage is the improved possibility to find the section with the highest tumor thickness. Additional information can be obtained by the use of the C-scan (plane corresponding to a horizontal cut as marked).

In Figure 2 the area directly underneath the tumor is depicted as C scan. The tumor itself is not shown. We found a marked echo enhancement under the echopoor area. With this technique the study of echo phenomena will be improved markedly. It might be, that due to the section planes not being strictly defined (vertical angle of examination) we fail to discover all occurring echo phenomena. It may also be that our statistics concerning echo phenomena must be reevaluated. Unfortunately this technique has only been available very recently (August 1993) so that more studies have to prove the sense and possibilities of these new software options. True 3D reconstruction which can be obtained by the Cortex scanner Dermascan C® are well known.

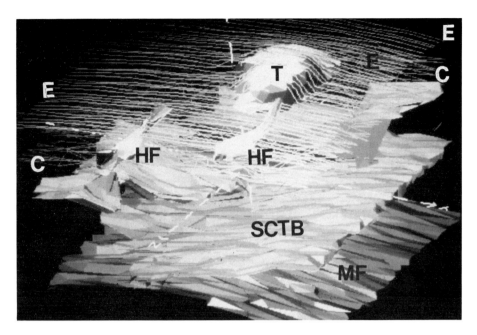

FIGURE 3 True 3D reconstruction obtained from the Dermascan C®. T = tumor, E = entry echo, C = corium, HF = hair follicle, SCTB = subcutaneous trabeculae, MF = muscle faszia. In contrast to Figure 1 and 2 or a voxel reconstruction the line and surface reconstruction offered by the dermascan software is better suited to describe structures and better controllable concerning the calculated surface and volume data.

The reconstruction with this technique are true 3D reconstructions. The images can be depicted as line or surface model. In Figure 3 we again show the malignant melanoma. This reconstruction consists of 40 serial equidistant sections. The software for this reconstruction has mostly been developed in our study group. The technique is based on the idea of reducing the data prior to reconstruction. We therefore talk about a structure boundary reconstruction. We prefer this technique due to the fact that in dermatology only the interaction of a few biological structures is of interest. In ultrasound, structure-boundary 3D models are superior to voxel-reconstruction models (which are very widespread) because ultrasonic artefacts due to scattering, absorption, and attenuation can be limited. In our example (Figure 3) we combine false color codation, with line and surface reconstruction to improve the information. The blue colored entry echo (E) is reconstructed as well as the green lower border of the corium (C) in a line modus. Directly underneath the entry echo the yellow tumor (T) is reconstructed in a surface mode. As previously mentioned in the ultrasound scan it is nearly impossible to differentiate small echopoor regions in the lower corium. If we reconstruct these areas we can easily classify them. In Figure 3 we find two hair follicles (HF), the left one breaking through the lower part of the corium in the subcutis. The right one is completely embedded in the corium. In the surface reconstruction mode we found two subcutaneous connective tissue trabeculae (SCTB, blue colored) connecting the border of the lower corium with the muscle fascia (MF). Conventional prognostic staging of malignant melanoma is based on the depth of invasion of the tumor. The Clark level and the tumor thickness according to Breslow are generally used as prognostic parameters for malignant melanoma. Neither accounts for the fact that the computed surface area and volume of a tumor are also of prognostic significance. We therefore propose that for malignant melanoma these two computed values should be combined to form a new prognostic parameter which we term *invasive tumor mass*. So far we have investigated about 100 cases, the average follow-up time being 1.1 years. Our data are therefore too small to prove the actual prognostic value of our proposed parameter. It may well be that the tumor volume is of greater relevance than the surface area as it is widely independent of resolution. In other words, the surface value of a tumor is directly dependent on the resolution used for the reconstruction. Therefore the data for the surface area resulting from ultrasound reconstruction (20 MHz) and histology are not comparable, whereas the data we obtain for the volume are comparable.

V. Inflammatory Diseases

Due to the limited possibilities being offered to quantify inflammatory skin diseases objectively, ultrasound is of fast growing importance. We have to differentiate between two applications: routine diagnostics and scientific questions. The two most important inflammatory diseases with regard to their investigation by ultrasound as a true routine application are localized (morphea) and progressive systemic scleroderma. In patients with morphea the skin thickness was increased up to 300%.[25] We calculate this change by comparing affected skin with healthy skin from the corresponding contralateral side of the body. We found that the increase in corium thickness depends on the site affected and varies from region to region. The degree of increase in corium thickness also depends on the original thickness of healthy skin. In the groin, where the skin is thinnest as compared to other regions, the relative increase in corium thickness is particularly marked. With increasing thickness of healthy skin the average scleroderma-induced percentage increase in corium thickness decreases in the order: groin, lower leg, thigh, abdomen, chest, back. A further feature of morphea is the increased occurrence of echorich bands which originate at the corium subcutis border and transverse the subcutaneous fatty tissue. In earlier studies we were able to show that follow-up investigations are possible by using the criteria skin thickness, echogenicity (densitometry), and subcutaneous trabeculae. The same criteria were suitable to investigate patients with progressive systemic sclerosis. However measurement of the skin-bone distance of the fingers of these patients is often very difficult. In patients with edema of the distal part of the finger the in-depth penetration of a 20-MHz signal is often not high enough. It might be that a 15-MHz A scanner is better suited to solve this problem. We believe that the ultrasound technique especially in the case of the follow-up evaluation of morphea is superior to all other bioengineering techniques.

Another interesting field are scientific questions that can be answered with ultrasound techniques. In inflammatory dermatoses we typically find an echopoor area (ELA) underneath the entry echo, the echodensity (echogenicity) of which increases under therapy and which completely disappears with the resolution of a lesion.[23] However, differential diagnosis of various inflammatory skin diseases based on 20-MHz sonography is not possible.[20] In atopic dermatitis the ELA corresponds to inflammatory skin changes in the form of infiltration and edema and in psoriasis it corresponds to infiltration, edema, and acanthotic epidermis. The ELA must not be confused with the so-called echolucent band (ELB) which is present in actinically damaged skin (actinic elastosis).[10] The ELB shows a serrated border towards the corium and inhomogeneously distributed weak internal echoes which account for higher densitometry values than those seen in inflammation (ELA). The ELA is mostly well demarcated to its upper and lower border. While the depth and density of the ELB remain constant dynamic changes in thickness and echodensity are characteristic for the ELA.

Considering the methodical problems (e.g., resolution and staining effects) there is a good correlation between the ELA and the histological infiltrate/edema (acanthosis). The skin thickness which is increased in the inflammatory stage of inflammatory diseases by about 40% constantly diminishes under therapy parallel to the increasing densitometry (echogenicity) values. Reduction of the acanthosis and subsidence of the inflammatory infiltrate are responsible for these sonographic phenomena. Normally we do not need objective information with an exactness of 80% for our routine treatment of psoriasis and atopic dermatitis. However, in multicenter studies this technique is used as objective criterion for evaluating therapeutic effects. Based on the knowledge of the phenomena in inflammatory skin diseases ultrasound can be used in the evaluation of artificially caused type-IV reactions. This is explained in Chapter 12.2.

VI. Outlook

High-frequency ultrasound is still a very young method which will no doubt undergo further, very rapid development. Its principal advantage is that it is a noninvasive procedure which can provide the examiner with valuable information about skin diseases. As we still have too little experience with the currently available scanners, which in addition require modification in some points, some important problems such as reliable differential diagnosis have not yet been solved. Scanners with higher frequencies (50 to 150 MHz) or acoustic microscopes (1 to 2 GHz)[33,34] operating *in vitro* can be expected to provide additional information. Recently, scanners with very high frequencies >100-MHz center frequency were developed. For the *in vivo* use this generation is well suited to answer more questions. Magnetic resonance tomographs specially designed for the skin also appear promising.[44] It can be expected that self-learning, "intelligent" computer programs will soon be available for image evaluation.

There is, however, no doubt that high-frequency sonography is a meaningful procedure for the measurement of tumor thickness/wideness and for monitoring the progress of treatment (measurement of volume of metastases). The same is true for the evaluation of progression and regression of scleroderma. In answering these questions the method is at present unmatched so that, at least in hospitals working in the oncological/scleroderma field, its use must be strongly recommended.

Acknowledgment

The work presented in this chapter was supported by a grant from the *Deutsche Krebshilfe,* project number 70512/M3/92/AL1.

References

1. Alexander H, Miller DL (1979), Determining skin thickness with pulsed ultrasound, *J Invest Derm* 72:17–19.
2. Altmeyer P, Hoffmann K, el Gammal S (1989), Dermatologische Ultraschalldiagnostik — gegenwärtiger Stand und Perspektiven (Editorial), *Z Hautkr* 64:727–728.
3. Altmeyer P, Hoffmann K, el Gammal S (1990), Sonographie der Haut, *MMW* 132:14–22.
4. Altmeyer P, Hoffmann K, Stücker M, Goertz S, el Gammal S (1992), General phenomena of ultrasound in dermatology, in: Altmeyer P, el Gammal S, Hoffmann K (Eds), *Ultrasound in Dermatology,* Springer, Berlin, 55–79.
5. Breitbart EW, Hicks R, Rehpennig W (1985), Möglichkeiten der Ultraschalldiagnostik in der Dermatologie, *Z Hautkr* 61:522–526.

6. Breitbart EW, Rehpennig W (1983), Möglichkeiten und Grenzen der Ultraschalldiagnostik zur in vivo Bestimmung der Invasionstiefe des malignen Melanoms, *Z Hautkr* 58:975–987.
7. Breitbart EW, Müller CH, Hicks R, Vieluf D (1989), Neue Entwicklungen der Ultraschalldiagnostik in der Dermatologie, *Akt Dermatol* 15:57–61.
8. Buhles N, Altmeyer P (1988), Ultraschallmikroskopie an Hautschnitten, *Z Hautkr* 64:926–934.
9. Dines KA, Sheets PW, Brink JA, Hanke CW, Condra KA, Clendenon JL, Goss SA, Smith JS, Franklin TD (1984), High frequency ultrasonic of skin: experimental results, *Ultras Imag* 6:408–434.
10. Dirschka T, Hoffmann K, Stücker M, Altmeyer P (1993), Bewertung der aktinischen Elastose mittels 20 MHz Sonographie *Akt Dermatol* 19:224–228.
11. Fornage DW, Deshaynes JL (1986), Ultrasound of normal skin, *J Clin Ultras* 14:619–622.
12. el Gammal S (1987), ANAT3D: A computer program for stereo pictures of three-dimensional reconstructions from histological serial sections. In: Elsner N, Creutzfeld O (eds). *New Frontiers in Brain Research.* Thieme; Stuttagard, New York; p 46.
13. el-Gammal S, Altmeyer P, Hinrichsen K (1989), ANAT 3D: Shaded three-dimensional surface reconstructions from serial sections. Applications in morphology and histopathology. *Acta Stereol Suppl* 8:543–550.
14. el-Gammal S, Hoffmann K, Auer T, Korten M, Altmeyer P, Höss A, Ermert H (1992), A 50 Mhz high-resolution ultrasound imaging system for dermatology. In: Altmeyer P, el Gammal S, Hoffmann K (eds). *Ultrasound in Dermatology.* Springer, Berlin, pp 297–325.
15. el-Gammal S, Hoffmann K, Kenkmann J, Altmeyer P, Höss A, Ermert H (1992), Principles of three-dimensional reconstructions from high-resolution ultrasound in dermatology. In: Altmeyer P, el Gammal S, Hoffmann K (eds). *Ultrasound in Dermatology.* Springer, Berlin, pp 355–385.
16. Gassenmeier G, Kiesewetter F, Schell H, Zinner M (1990), Wertigkeit der hochauflösenden Sonographie für die Bestimmnug des vertikalen Tumordurchmessers beim malignen Melanom der Haut, *Hautarzt* 41:360–364.
17. Grin CM, Kopf AW, Welkovich B, Bart RS, Levenstein MJ (19xx), Accuracy in the clinical diagnosis of malignant melanoma, *Arch Dermatol* 126:763–766.
18. Hoffmann K, Altmeyer P, el Gammal S, Winkler K, Stücker M, Dirschka T, Matthes U (1991), Ultraschallphänomene in der Dermatologie, *Therapiewoche* 41:1088–1102.
19. Hoffmann K, Dirschka T, Stücker M, Rippert G, Hoffmann A, el Gammal S, Altmeyer P (1993), Ultrasound and cryosurgery, *Dermatologische Monatsschrift* 179:270–277.
20. Hoffmann K, el-Gammal S, Altmeyer P (1989), 20 MHz B-scan Sonographie an Händen und Füßen in Handsymposium: Dermatologische Erkrankungen der Hände und Füße, Altmeyer P et al., *Edition Roche:* 285–300.
21. Hoffmann K, el Gammal S., Altmeyer P (1990), Ultraschall in der Dermatologie, *Hautarzt* 41:W7–W15.
22. Hoffmann K, el Gammal S, Matthes U, Altmeyer P (1989), 20 MHz Sonographie der Haut in der präoperativen Diagnostik, *Z Hautkr* 64:851–858.
23. Hoffmann K, el-Gammal S, Schwarze H, Dirschka T (1992), Examination of Psoriasis vulgaris using 20-Mhz B-scan ultrasound In: Altmeyer P, el Gammal S, Hoffmann K (eds). *Ultrasound in Dermatology.* Springer, Berlin, pp 207–220.
24. Hoffmann K, el Gammal S, Winkler K, Stücker M, Hammentgen R, Schatz H, Altmeyer P (1991), Hauttumoren im Ultraschall, *Hautnah* 7:4–24.
25. Hoffmann K, Gerbaulet U, el Gammal S, Peter Altmeyer (1991), 20 MHz B-mode ultrasound in monitoring the course of localised scleroderma (Morphea) *Acta Derm Venerol (Stockh) Suppl* 164:3–16.
26. Hoffmann K, Jung J, el Gammal S, Altmeyer P (1992), Malignant melanoma in 20 MHz b-scan sonography, *Dermatology* 185:49–55.
27. Hoffmann K, Matthes U, Stücker M, Segerling M, Altmeyer P (1990), "Prevention week" in Bochum 1989 for Malignant Melanoma Information, *Öff Gesundh Wes* 52:9–13.
28. Hoffmann K, Stücker M, el Gammal S, Altmeyer P (1990), Digitale 20- MHz-Sonographie des Basalioms im b-scan *Hautarzt* 41:333–339.
29. Hoffmann K, Winkler K, el Gammal S, Altmeyer P (1993), A wound healing model with sonographic monitoring, *Clin Exp Dermatol* 18:217–225.
30. Kirsch JM, Hanson ME, Gibson JR (1984), The determination of skin-thickness using conventional ultrasound equipment, *Clin Exp Derm* 9:280–285.
31. Kraus W, Nake-Elias A, Schramm P (1985), Diagnostische Fortschritte bei malignen Melanomen durch die hochauflösende Real-Time-Sonographie, *Hautarzt* 36:386–392.
32. Koh HK, Lew RA, Prout MN (1989), Sreening for melanoma/skin cancer: theoretical and practical considerations, *J Am Acad Dermatol* 20:159–172.
33. Kolosov OV, Levin VM, Myev RG, Senjushkina TA (1987), The use of acoustic microscopy for biological tissue charakterization, *Ultras Med Biol* 13:477–483.
34. Matthes U, Höxtermann S, Hoffmann K, el Gammal S, Bruschke E, Altmeyer P (1992), Acoustic microscopy in dermatology: normal skin structures and tumours. In: Altmeyer P, el Gammal S, Hoffmann K (eds). *Ultrasound in Dermatology.* Springer, Berlin, pp 207–220.
35. Miyauchi S, Murikami S, Miki Y (1988), Echographic studies of superficial lymphadenopathies, *J Derm* 15:263–267.
36. Miyauchi S, Tada M, Miki Y (1983), Echographic evaluation of nodular lesions of the skin, *J Derm* 10:221–227.
37. Murikami S, Miki Y (1989), Human skin histology using high-resolution echography, *J Clin Ultras* 17:77–82.
38. Nitsche N, Strasser GM, Hoffmann K, Hilbert M, Arnold W (1993), Neue Aspekte der Hochfrequenzsonographie, *Otorhinolaryngol Nova* 3:155–160.
39. Pawlak F, Hoffmann K, el Gammal S, Altmeyer P (1990), Three-dimensional reconstruction of ultrasonic images of the skin, *Zbl Haut* 157:330.
40. Payne PA (1983), Non-invasive skin measurement by ultrasound, *RNM Images* 13:24–26.
41. Payne PA (1985), Medical and industrial applications of high resolution ultrasound, *J Phys E:Sci Instrum* 18:465–473.
42. Price R, Jones TB, Goddard Jr. J, James AE (1980), Basic concepts of ultrasonic tissue charaterization, *Radiol Clin N Amer* 18:21–30.
43. Querleux B, Lévêque J-L, de Rigal J (1988), In vivo cross-sectional ultrasonic imaging of human skin, *Dermatologica* 177:332–337.
44. Querleux B, Yassine MM, Darasse L, Saint Jalmes, Sauzade M, Lévêque J-L (1988), Magnetic resonance imaging of the skin: a comparison with the ultrasonic technique, *Bioeng Skin* 4:1–14.
45. Rukavina B, Mohar N (1979), An approach of ultrasound diagnostic techniques of the skin and subcutaneus tissue, *Dermatologica* 158:81–92.
46. Schwaighofer B, Pohl-Markl H, Frühwald F, Stiglbauer R, Kokoschka EM (1987), Der diagnostische Stellenwert des Ultraschalls beim malignen Melanom, *Fortschr Röntgenstr* 146:409–411.

47. Strasser W, Vanscheidt W, Hagedorn M, Wokalek H (1986), B-scan Ultraschall in der Dermatologie, *Fortschr Med* 25:495–498.
48. Stücker M, Hoffmann K, el Gammal S, Altmeyer P (1992), The acoustic characteristics of the basal cell carcinoma in 20 MHz ultrasonography In: Altmeyer P, el Gammal S, Hoffmann K (eds), *Ultrasound in Dermatology*. Springer, Berlin, pp 207–220.
49. Tan CY, Marks R, Payne P (1981), Comparison of xeroradiographic and ultrasound detection of corticosteroid induced dermal thinning, *J Invest Derm* 76:126–128.
50. Tan CY, Statham B, Marks R, Payne PA (1982), Skin thickness measurement by pulsed ultrasound: its reproducibility, validation and variability, *Brit J Derm* 106:657–667.
51. Tanaka M (1985), The development and medical application of an ultrasonic microscope, *Nippon Rinshi* 43:2713.

Chapter 12.4
Ultrasound Examination of the Skin and Subcutaneous Tissues at 7.5 to 10 MHz

Bruno D. Fornage
Department of Radiology
The University of Texas
M. D. Anderson Cancer Center
Houston, Texas

I. Introduction

Despite the fact that the skin is virtually always traversed during body imaging, very few radiologists have been interested in imaging the skin. Although promising advances in magnetic resonance imaging (MRI) of the skin have been reported with the use of dedicated surface coils,[1] this application remains investigational. Currently, the most practical modality for imaging the skin is ultrasound (US). US examination of the skin itself is best performed using dedicated equipment with 20-MHz transducers.[2] However, well-known limitations of such equipment include the narrow field of view (less than 2 cm in width) and the inability to visualize the subcutaneous tissues beyond a depth of approximately 1.5 cm. These limitations emphasize the need for lower frequency (7.5 or 10 MHz) transducers — which are part of the armamentarium of the radiologist — to image those structures that lie beyond the scope of the dedicated machines. This chapter describes the capabilities and limitations of 7.5- and 10-MHz US in demonstrating skin lesions and emphasizes its role in imaging the subcutaneous tissues.

II. Technical Considerations

Because of their better near-field resolution and wider field of view, linear-array transducers are preferred to mechanical sector probes of similar frequency. The US beam of the former is perpendicular to the skin surface and to most underlying interfaces in the subcutis, thus minimizing scattering artifacts. The field of view of most commercially available 7.5- and 10-MHz linear-array probes is about 3 to 4 cm wide and 4 to 6 cm deep. Examination of the skin and superficial subcutaneous tissues with this type of equipment is improved by use of a thin standoff pad, which results in the skin being closer to the focal zone.[3]

Doppler US has already been used to improve the diagnosis of skin lesions.[4,5] Continuous-wave high-frequency Doppler instruments are the most sensitive but cannot determine the depth of the vessels. In contrast, pulsed color Doppler imaging allows for convenient cross-sectional mapping of blood flow.

A major strength of US is its real-time imaging capability. The use of a standoff pad allows the operator to palpate the skin under continuous real-time US monitoring by sliding one or two fingers between the skin and the pad. With the use of the 7.5- or 10-MHz transducers, percutaneous needle biopsy of subcutaneous lesions is readily performed under continuous real-time monitoring.[6]

III. Normal Ultrasound Anatomy

At 7.5 or 10 MHz, the skin appears as a thin, regular layer of tissue that is more echogenic than the underlying fat (Figure 1).[7,8] The epidermis, even where it is thickest, i.e., at the sole or at the hypothenar eminence, cannot be resolved, and skin appendages cannot be visualized. The interface between the echogenic skin and the hypoechoic subcutis is clearly depicted in most areas of the body, allowing measurement of the skin thickness — although with less accuracy than at higher frequencies. The thickness of the skin as measured at 10 MHz has been reported to range from 1.4 to 4.8 mm, depending on the examination site;[7] slightly different values have been reported recently with the use of 20-MHz US.[2]

The subcutaneous tissues are composed mostly of hypoechoic fat with a few scattered echogenic strands of connective tissue.[9] US has been used to measure the thickness of subcutaneous fat at various sites in normal subjects.[10] Occasionally, anechoic veins are seen. They are readily confirmed by their collapse under compression. State-of-the-art, color Doppler US

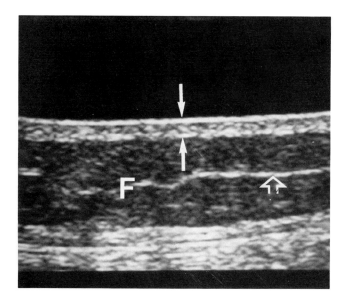

FIGURE 1 Normal skin. Scan of the anterior thigh shows the echogenic skin (arrows), which is sharply demarcated from the underlying hypoechoic subcutaneous fat (F). Note an echogenic strand of connective tissue (open arrow).

equipment can demonstrate blood flow in the subcutaneous fat (Figure 2). Because of the easy identification of subcutaneous veins, US has been proposed to guide venipuncture for venography in edematous legs.[11] Superficial benign fat-infiltrated lymph nodes are often demonstrated. They appear typically as elongated, oval-shaped structures with an echogenic central fatty component and a peripheral hypoechoic rim of residual lymphoid tissue.[12]

IV. Abnormalities of the Skin

Imaging skin lesions is feasible with 7.5- or 10-MHz transducers. However, the resolution cannot match that of 20-MHz transducers, and thin superficial lesions, such as junctional nevi, cannot be demonstrated. On the other hand, these transducers demonstrate larger lesions and ascertain their location within the skin. The vast majority of skin tumors, benign and malignant, appear as hypoechoic masses that contrast with the surrounding echogenic dermis.

A. Benign Lesions

Sebaceous cysts are round to oval with an anechoic or markedly hypoechoic central cavity. Occasionally, a sebaceous cyst in the skin of the breast may be at the origin of an indeterminate density on mammo-grams. In such a case, US examination of the skin can confirm that the mass indeed arises from the skin (Figure 3).

Only thick nevi can be visualized at 7.5 to 10 MHz. Scars appear as focal hypoechoic areas involving the dermis (Figure 4), which may continue in the subcutaneous tissues as an area of distortion. US can demonstrate the deep infiltration of large dermatofibromas.

US is rarely performed to evaluate inflammatory or infectious skin lesions. On 7.5- or 10-MHz sonograms, edematous skin is markedly thickened and its echogenicity usually decreased (Figure 5).[12] Furthermore, the interface between the dermis and subcutaneous fat is often blurred. Color Doppler can demonstrate increased vascularity in the edematous skin at the level of the deep plexuses (Figure 6). US at 10 MHz has been used successfully to evaluate the depth of burns,[13] although sonograms at higher frequencies yield better results.[14]

B. Malignant Lesions

Malignant tumors of the skin are hypoechoic, with melanomas being nearly anechoic. The thickness of melanomas can be measured at 10 MHz,[15] although the best results are obtained at 20 MHz (Figure 7).[2,16] Color Doppler studies can demonstrate the increased vascularity associated with malignant tumors, in particular with malignant melanomas (Figure 8). Cutaneous metastases from various primary tumors can be visualized if they are large enough. US can confirm that such a lesion is located within and confined to the dermis (Figure 9).[12]

V. Abnormalities of the Subcutaneous Tissues

US has been used with success to evaluate changes in the thickness of subcutaneous fat associated with various clinical conditions in humans[17-19] as well as to evaluate the thickness of the subcutaneous fat in cattle in the meat industry.[20] US is highly sensitive in demonstrating abnormal masses in the subcutaneous tissues.[21]

Trauma to subcutaneous tissues occurs frequently but is rarely brought to the attention of the sonographer. Subcutaneous hema-tomas have a wide spectrum of US appearances:[22] a recent hemorrhage generally appears as a focal hyperechoic area; as the hematoma liquefies and becomes organized, a complex fluid collection can develop, leading on rare occasions to a purely cystic collection (serohematoma).

Subcutaneous tissues are a common site for traumatic foreign bodies. Real-time US has proved to be highly sensitive in the detection of foreign bodies, including those that are not radiopaque.[23-25] The vast majority of foreign bodies appear as brightly echogenic reflectors (Figure 10) and are associated with shadowing, a trail of reverberation echoes ("comet-tail" artifact), both, or none, depending on the object's physical nature; the comet-tail artifact is virtually pathognomonic of metallic objects.[26] When present, inflammatory reaction to the foreign body appears as a hypoechoic area surrounding the echogenic foreign body. US provides

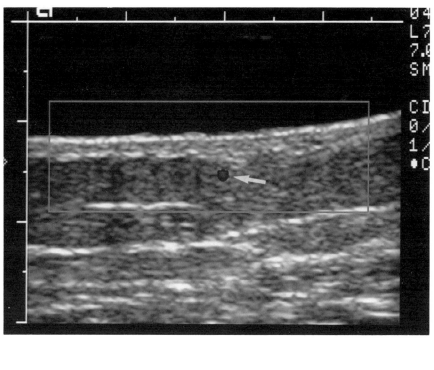

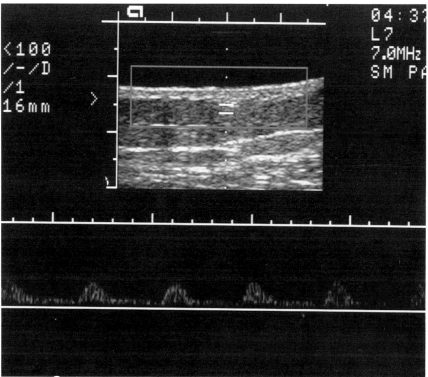

FIGURE 2 Color Doppler examination of normal skin and subcutaneous tissues. (A) Scan shows a small color dot (arrow) in the subcutaneous fat. (B) Spectral analysis confirms that the flow recorded at this site is arterial.

accurate preoperative three-dimensional localization of foreign bodies, which is greatly appreciated by the surgeon.[27,28]

A common clinical problem in evaluating inflamed subcutaneous tissues is how to differentiate between cellulitis and abscess. In cellulitis, US scans show a diffuse increase in the echogenicity of the subcutaneous fat, which often cannot be delineated from the skin (Figure 11A). Hypoechoic streaks can be seen, representing prominent blood and lymphatic vessels. Color Doppler examination identifies flow in blood vessels, whereas dilated lymphatics remain void of flow (Figure 11B). Abscesses appear as fluid-filled collections, often with internal echoes or a complex appearance, irregu-

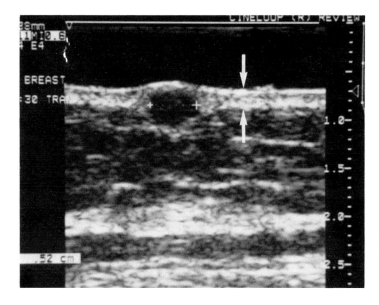

FIGURE 3 Sebaceous cyst of the breast appearing as an indeterminate density on mammograms. Sonogram shows a markedly hypoechoic 0.5-cm mass developed from the dermis (calipers). Arrows indicate the normal skin.

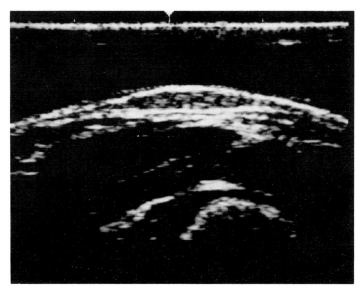

FIGURE 4 Cutaneous scar. Scan shows the marked hypoechoic thickening of the skin (arrows).

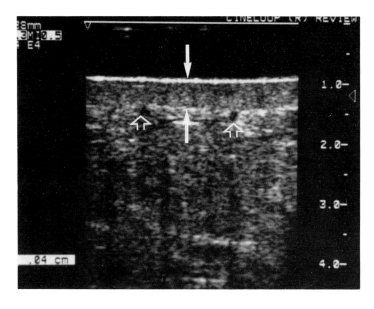

FIGURE 5 Post-irradiation changes of the skin of the breast. Sonogram shows marked thickening and decreased echogenicity of the skin (arrows). Note the blurred interface between the skin and subcutis and the prominent veins (open arrows).

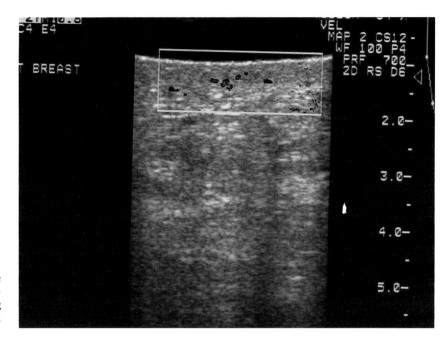

FIGURE 6 Post-radiation changes of the skin of the breast. Color Doppler examination shows several color dots representing increased vascularity in the thickened and hypoechoic dermis.

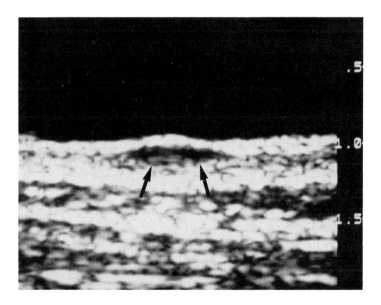

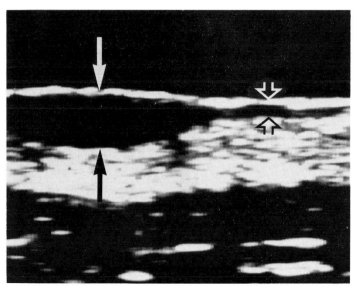

FIGURE 7 Malignant melanoma. (A) Scan obtained at 10 MHz shows a hypoechoic lesion involving the upper half of the skin (arrows). (B) Scan of the same lesion obtained at 20 MHz shows the margins of the lesion better (arrows) and demonstrates superficial spreading of tumor (open arrows) not seen at 10 MHz.

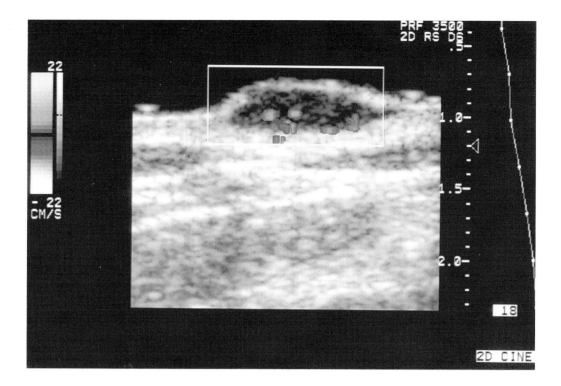

FIGURE 8 3-mm-thick malignant melanoma. Color Doppler examination shows marked vascularity at the deep margin of the tumor.

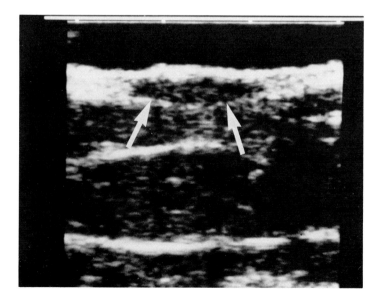

FIGURE 9 Cutaneous metastasis from a malignant fibrous histiocytoma. Scans obtained at 7.5 MHz shows that the lesion (arrows) is confined to the skin.

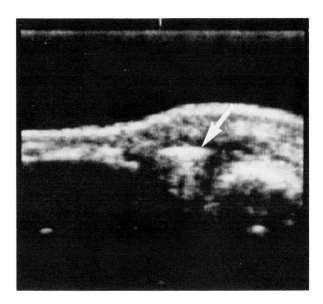

FIGURE 10 Metallic foreign body in the hand. Sonogram shows a bright linear echo (arrow) associated with a comet-tail artifact.

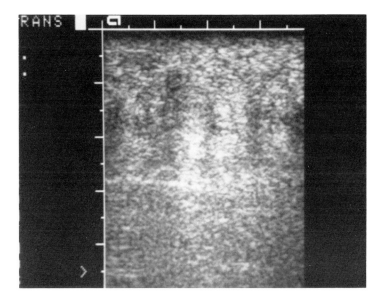

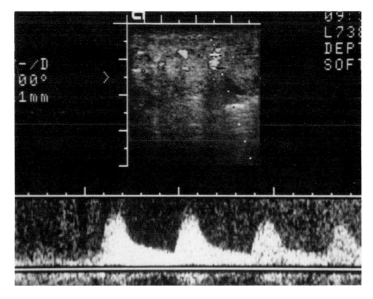

FIGURE 11 Cellulitis of the anterior abdominal wall. (A) Sonogram shows a marked diffuse increase in echogenicity of the subcutaneous fat and loss of demarcation between the skin and subcutis. (B) Color Doppler scan shows increased vascularity and spectral analysis at one site indicates arterial flow.

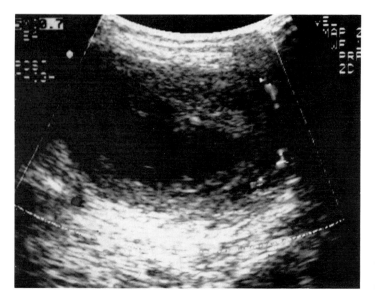

FIGURE 12 Abscess in subcutaneous tissues. Sonogram shows fluid-filled collections with internal echoes and ill-defined margins.

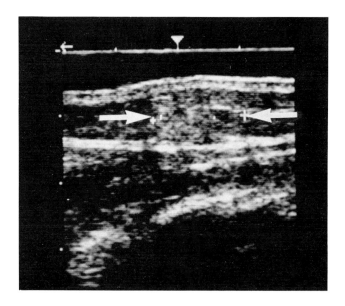

FIGURE 13 Fat necrosis. Sonogram shows an ill-defined area of increased echogenicity in the subcutaneous fat (arrows). Note the overlying normal skin.

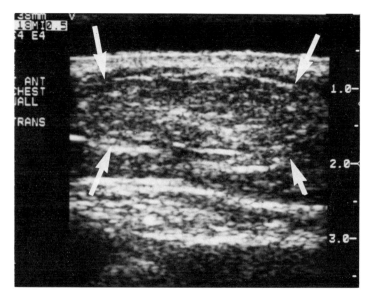

FIGURE 14 Subcutaneous lipoma. Sonogram shows an ill-defined area of mildly increased echogenicity (arrows) in the subcutaneous fat.

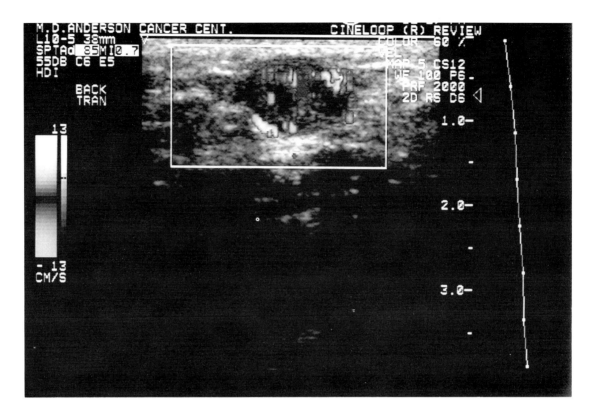

FIGURE 15 Subcutaneous hemangioma. Color Doppler scan shows marked vascularity inside the hypoechoic mass.

lar walls, and increased vascularity at the periphery of the lesions on color Doppler scans (Figure 12). If the abscess is small or not easily defined by palpation, US can be used to guide diagnostic fine-needle aspiration and subsequent percutaneous drainage. On sonograms, focal fat necrosis and panniculitis appear as areas of increased echogenicity in the subcutaneous fat (Figure 13).

The most common benign subcutaneous tumors include lipomas and hemangiomas. Both types of lesions display a wide spectrum of echogenicity. Lipomas may be hypoechoic, hyperechoic, or isoechoic relative to the surrounding fat (Figure 14).[29] Because isoechoic lipomas may be difficult to identify, when a lipoma is suspected, it is good practice to use palpation under real-time sonoscopy, which allows delineation of the mass on the scan with certainty. Hemangiomas are usually — but not always — hypoechoic. Bright echogenic foci indicate the presence of phleboliths, an important clue to the diagnosis of hemangiomas. Color Doppler may demonstrate significant blood flow within the lesion (Figure 15), unless it is thrombosed.

The development of a hypoechoic mass in subcutaneous tissues in patients with malignant melanoma should be considered as suspicious for local recurrence or in transit metastasis until proven otherwise. US can demonstrate lesions that are only a few millimeters in diameter (Figure 16). Metastases from melanoma are highly vascularized and usually exhibit significant flow on color Doppler studies. US has also been shown to be as accurate as MRI in the detection of early recurrence in patients treated for soft-tissue

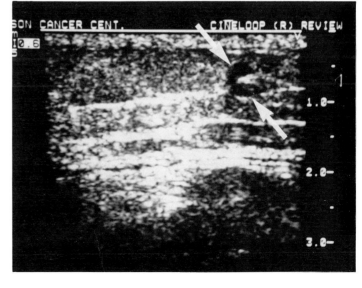

FIGURE 16 Metastasis from melanoma in subcutaneous fat. Sonogram obtained during US-guided, fine-needle aspiration shows a nearly anechoic 0.5-cm mass (arrows) and the bright tip of the biopsy needle in the center of the lesion.

sarcoma.[30] US-guided, fine-needle aspiration can be performed within minutes to confirm and document the malignant nature of a suspicious lesion (Figure 16).[6]

VI. Summary

While US examination of the skin is feasible with the use of transducers of 7.5- and 10-MHz, it is best done with transducers of 20 MHz or more. However, the limited penetration of these very-high-frequency transducers does not allow adequate examination of the subcutaneous tissues. That is best achieved using 7.5- or 10-MHz probes. US of subcutaneous tissues provides unsurpassed spatial resolution, and it is more versatile and quicker to perform than any other cross-sectional imaging modality currently available. In addition, the unique real-time capability of the US allows the pinpoint placement of a biopsy needle in lesions as small as a few millimeters in diameter. Because of its low cost and wide availability, 7.5- or 10-MHz US is the ideal modality with which to image the subcutaneous tissues, and its wider use should be encouraged.

References

1. Hyde JS, Jesmanowicz A, Kneeland JB. Surface coil for MR imaging of the skin. *Magn Reson Med* 5:456–461, 1987.
2. Fornage BD, McGavran MH, Duvic M, Waldron CA. Imaging of the skin with 20-MHz US. *Radiology* 189:69–76, 1993.
3. Fornage BD, Touche DH, Rifkin MD. Small parts real-time sonography: a new "water-path". *J Ultrasound Med* 3:355–357, 1984.
4. Srivastava A, Hughes BR, Hughes LE, Woodcock JP. Doppler ultrasound as an adjunct to the differential diagnosis of pigmented skin lesions. *Br J Surg* 73:790–792, 1986.
5. Srivastava A, Laidler P, Hughes LE, Woodcock J, Shedden EJ. Neovascularization in human cutaneous melanoma: a quantitative morphological and Doppler ultrasound study. *Eur J Cancer Clin Oncol* 22:1205–1209, 1986.
6. Fornage BD, Lorigan J. Sonographic detection and fine-needle aspiration biopsy of nonpalpable recurrent or metastatic melanoma in subcutaneous tissues. *J Ultrasound Med* 8:421–424, 1989.
7. Fornage BD, Deshayes JL. Ultrasound of the normal skin. *J Clin Ultrasound* 14:619–622, 1986.
8. Miyauchi S, Miki Y. Normal human skin echogram. *Arch Dermatol Res* 275:345–349, 1983.
9. Fornage BD. Ultrasonography of muscles and tendons. Examination technique and atlas of normal anatomy of the extremities. New York: Springer-Verlag, 1988.
10. Maruyama Y, Iizuka S, Yoshida K. [Ultrasonic observation on distribution of subcutaneous fat in Japanese young adults with reference to sexual difference.] *Ann Physiol Anthropol* 10:61–70, 1991.
11. Johns CM, Sumkin JH. US-guided venipuncture for venography in the edematous leg. *Radiology* 180:573, 1991.
12. Fornage BD. Echographie des membres [Sonography of the extremities]. Paris, Vigot, 1991.
13. Bauer JA, Sauer T. Cutaneous 10 MHz ultrasound B scan allows the quantitative assessment of burn depth. *Burns Incl Therm Inj* 15:49–51, 1989.
14. Brink JA, Sheets PW, Dines KA, Etchison MR, Kanke CW, Sadove AM. Quantitative assessment of burn injury in porcine skin with high-frequency ultrasonic imaging. *Invest Radiol* 21:645–651, 1986.
15. Shafir R, Itzchak Y, Heyman Z, Azizi E, Tsur H, Hiss J. Preoperative ultrasonic measurements of the thickness of cutaneous malignant melanoma. *J Ultrasound Med* 3:205–208, 1984.
16. Jung J, Hoffman K, El-Gammal S, Altmeyer P. Sonographic characteristics of malignant melanoma in 20-MHz B-scan sonography (abstract). *Zentralblatt Haut Geschlechtskrankheiten* 157:325, 1990.
17. Heckmatt JZ, Pier N, Dubowitz V. Measurement of quadriceps muscle thickness and subcutaneous tissue thickness in normal children by real-time ultrasound imaging. *J Clin Ultrasound* 16:171–176, 1988.
18. Koskelo E-K, Kivisaari LM, Saarinen UM, Siimes MA. Quantitation of muscles and fat by ultrasonography: a useful method in the assessment of malnutrition in children. *Acta Paediatr Scand* 80:682, 1991.
19. Gooding GAW, Stess RM, Graf PM, Moss KM, Louie KS, Grunfeld C. Sonography of the sole of the foot. Evidence for loss of foot pad thickness in diabetes and its relationship to ulceration of the foot. *Invest Radiol* 21:45–48, 1986.
20. Butler LG, Head GM. The medium frequency (7.5 MHz) ultrasound image characteristics of cattle skin. *Aust Vet J* 70:344–347, 1993.
21. Nessi R, Betti R, Bencini PL, Crosti C, Blanc M, Uslenghi C. Ultrasonography of nodular and infiltrative lesions of the skin and subcutaneous tissues. *J Clin Ultrasound* 18:103–109, 1990.
22. Wilson DJ. Ultrasonic imaging of soft tissues. *Clin Radiol* 40:341, 1989.
23. Fornage BD, Schernberg FL. Sonographic diagnosis of foreign bodies of the distal extremities. *AJR* 147:567–569, 1986.
24. Gooding GAW, Hardiman T, Sumers M, Stess R, Graf P, Grunfeld C. Sonography of the hand and foot in foreign body detection. *J Ultrasound Med* 6:441–447, 1987.
25. Banerjee B, Das RK. Sonographic detection of foreign bodies of the extremities. *Br J Radiol* 64:107, 1991.
26. Fornage BD, Rifkin MD. Ultrasound examination of the hand and foot. *Radiol Clin North Am* 26:109–129, 1988.
27. Fornage BD, Schernberg FL. Sonographic preoperative localization of a foreign body in the hand. *J Ultrasound Med* 6:217–219, 1987.
28. Fornage BD. Preoperative sonographic localization of a migrated transosseous stabilizing wire in the hand. *J Ultrasound Med* 6:471–473, 1987.
29. Fornage BD, Tassin GB. Sonographic appearances of superficial soft-tissue lipomas. *J Clin Ultrasound* 19:215–220, 1991.
30. Choi H, Varma DGK, Fornage BD, Kim EE, Johnston DA. Soft-tissue sarcoma: MR imaging vs sonography for detection of local recurrence after surgery. *AJR* 157:353–358, 1991.

Chapter 12.5
Ultrasound A-mode Measurement of Skin Thickness

Tove Agner
Department of Dermatology
Gentofte Hospital
Hellerup, Denmark

I. Introduction

Recent technical developments have made ultrasonographic studies of superficial structures possible. High-frequency ultrasound for skin examination was introduced in 1979 by Alexander and Miller.[1] Today, A-scan for one-dimensional, B-scan for two-dimensional, and C-scan for three-dimensional study of the skin are commercially available.[2] In some fields the areas that B and C scan with its increased possibilities of imaging the skin structures has surpassed A scan, but in other fields the A-mode scanning is still an exact and useful tool. The object of this paper is to give the basic principles of A-mode ultrasound scanning, and to review the potentiality of the instrument.

II. Methodological Principle

Ultrasound is sound waves in frequencies above what is audible to the human ear, i.e. >15–20 kHz. A mode (A for amplitude) indicates that the amplitudes reflected from the surfaces are measured as series of peaks. The ultrasound source is a probe including electrical equipment and a transducer. The latter emits the ultrasound beam as well as receives the reflected ultrasound echoes. When received by the probe the echoes produce an electrical signal in the transducer, which is presented as amplitudes on the oscilloscope.

The utilized sound frequency is decided as a compromise between image definition and penetration depth. Tissue penetration is better with lower frequencies, while higher frequencies issue better-defined images. Generally, ultrasound frequencies <10 MHz are used to study profound localized structures like abdominal organs, while for the study of superficial structures such as the skin, ultrasound frequencies >10 MHz provide better results. At 10 MHz the subcutaneous area has a high resolution, while at 50 MHz the epidermis can be studied more in detail. With respect to the compromise between resolution and depth of the viewing field, 15 to 20 MHz for ultrasonographic studies of the skin has been established as a good solution.[2]

The velocity of ultrasound waves depends on the medium through which they are propagated. As it passes through varying tissues in the body, the velocity of the ultrasound wave varies slightly, being estimated for practical purposes as 1540 m/s.[3] The velocity depends on the intensity as well as the elasticity of the medium through which it is propagated, and is much slower in air (350 m/s) than in bone (4000 m/s).[3] In human skin the velocity has been estimated as 1580 to 1605 m/s.[4,5]

When an ultrasound beam meets an interface between two media of different acoustic impedance it is reflected. The degree of reflection will be greater if the acoustic impedance, i.e., the velocity with which the sound is propagated, between the media is accentuated. Thus, high amplitudes indicates a considerable difference in acoustic impedance from one area in the tissue to another. In the skin, sound velocity in the subcutaneous fat is considerably lower than in the epidermis and dermis, which means that the interface between the dermis and the subcutaneous fat normally is sharply outlined and clearly demarcated. However, the amplitude of the signal is also influenced by the distance from the medium to the ultrasound source. The intensity of the echoes decreases proportionate to the distance it has passed through the tissues. The angle of reflection of the ultrasound beam will be identical to the angle of the emission of the beam. It is consequently necessary to keep the probe exactly perpendicular to the skin surface, not to miss capturing of the reflected echoes. For A-mode ultrasound it may for the study of superficial structures like the skin be advantageous to use a probe immersed in water, which ensures contact and may provide a better visualization.

By presenting the echo signals directly on the oscilloscope as received by the probe, A-mode scanning gives an unidimensional picture of the structures passed by the ultrasound beam on its way through the tissues. The distance between the echoes indicates the thickness of each medium passed. This distance can be calculated in millimeters, from knowledge on the sound velocity in the tissue.

In practice the measurement is easily performed; the probe head is held gently toward the skin, and the echoes are displayed on the oscilloscope. The interval between the echo from the stratum corneum and the echo from the interface between the dermis and subcutaneous fat is a measure of the thickness of the skin (i.e. epidermis and dermis). In most parts of the body the differentiation between epidermis and dermis by ultrasound is difficult. An exception is the palm, where the interface between the epidermis and the dermis can actually be detected. In most body regions, the dermis is highly echogenic. It is the regularly ordered fiber network which reflects ultrasound. Disturbances of the regularly arranged fiber network is normally followed by a decrease in reflectance. Edema in the dermis leads to a zone with very few and low echoes. The velocity of the ultrasound wave through the subcutaneous fat is markedly lower than in the dermis. This gives rise to high amplitude echoes from the dermis/subcutaneous fat interface, which facilitates interpretation of full skin thickness.

III. Examples of Applications in Clinical Dermatology

A. Contact Dermatitis — Reading of Patch Tests

Allergens, irritants, and chemical or physical trauma may elicit inflammation. One of the main features in the inflammatory response in the skin, or elsewhere, is edema formation. Parametrically quantification of edema formation, reflecting the strength of the inflammatory response, can be obtained by ultrasound A-scan.[6] Quantification of the response is simple, since edema formation leads to a low-echogenic zone within the dermis, while the interface between dermis and subcutis still gives rise to a distinct echo. Ultrasound A-scan was found useful for objective measurement of edema formation in experimentally induced inflammation in several studies. Measurement of skin swelling in the tuberculin test by ultrasonography was proposed in a study using A-mode scanning.[7] Skin-prick histamine weals were also successfully studied by A-mode ultrasound scanning, and significant differences in weal formation between three different concentrations of histamine could be detected.[8] Quantification of response to the DMSO-test (dimethylsulfoxide test), used for determination of the sensitivity of the skin,[9] is also possible by ultrasound,[10] thus improving the objectivity of the test. Among other noninvasive bioengineering methods A-mode ultrasound scanning was used to characterize the skin responses to a number of different primary irritants.[11] Quantification of irritant patch-test reactions has been used in experimental studies for determination of skin susceptibility under varying physiological and pathophysiological conditions. Seasonal variation in response to irritants were studied by A-mode ultrasound, as well as by other noninvasive methods,[12] and the inflammatory response to sodium lauryl sulfate (SLS), as quantified by ultrasound A-scan, was found to be modified by hormonal variations during the menstrual cycle.[13] Quantification of SLS-induced edema in patients with acute, chronic, and atopic eczema was studied and compared.[14-16] A-mode ultrasound scanning has not proved useful for differentiation between allergic and irritants reactions, although in one study, including allergic as well as irritant patch-test reactions, clinically similar reactions were claimed to have an increased edema formation when elicited by an allergen than by an irritant.[17]

Although most papers report the use of A-mode ultrasound scanning for quantification of edema as valuable, one study reports less optimistically. Brazier and Shaw[18] studied positive allergic patch-test reactions and found that they could not by A-mode scanning differentiate weak positive reactions from normal skin.[18] However, the equipment used in that study was less precise than the scanners which are available today.

B. Scleroderma

Measurement of skin thickness in patients with systemic sclerosis may be helpful in the clinical evaluation of patients, as well as for quantification of sclerodermatous skin changes during medical treatment. Measurements are easily performed, and ultrasonography provides a method by which skin thickness can be monitored over time. In 22 patients with acrosclerosis skin-phalanx distance as well as skin thickness on the forearms were measured by 15-MHz A-mode ultrasound, and compared to skin thickness of age-matched healthy controls.[19] Skin-phalanx distance of the digits and skin thickness of the forearm was increased and patients with acrosclerosis, and was concluded to be appropriate for the diagnosis of this disease.

In another study skin thickness of the phalanges of both second fingers was measured in 40 patients with systemic sclerosis with 10-MHz ultrasound. The skin thickness was significantly increased in patients with scleroderma compared to controls.[20] A-mode scanning was also found helpful for evaluation of localized scleroderma, demonstrating the skin thickness in the plaques to be increased more than 300%.[21]

C. Tumors

Ultrasonic evaluation of cutaneous tumors including malignant melanoma has been reported in a number of studies.[22,23] Although promising results have been presented, ultrasonography cannot at the present time replace the traditional histological examination. In general, A-scanning of tumors will only provide a one-dimensional picture, while B and C scanning present cross-sectional imaging which is more valuable for this purpose. Thus, A-mode scanning is not very helpful in the diagnosis of skin tumors, and should not be used for that purpose.

D. Atrophy

The atrophic potential of local steroids has been evaluated by ultrasonographic studies. Amount of dermal thinning was studied using different corticosteroid creams, and A-mode ultrasound scanning was concluded an accurate and safe method for determining dermal thickness.[24] A-mode scanning is useful for quantification of skin thickness and is, therefore, helpful in comparative studies of skin atrophy.

E. Other Areas Where A-mode Scanning May Be Useful

In some experimental set-ups insertion of a needle into either the dermis or the subcutaneous fat is accomplished. Microdialysis of the skin is an example of this.[26] To obtain valuable information it is important that the needle is localized in the right layer of the skin. A-mode ultrasound scanning may be helpful to situate the needle correctly.

Until recently many physiological properties of the skin had not been thoroughly studied. Examples of areas which have only briefly been looked into are relation between skin thickness and age, or[11,27] skin thickness in relation to body mass index and in relation to osteoporosis and physical capacity.[28] Changes in skin thickness in relation to different endocrine diseases is also an area which lacks investigation. A-mode ultrasound examination of animals included in experimental studies has been used very little up to now. It is the author's personal experience that measurement of skin thickness by A-mode scanning in mice is very difficult. C-scanning may be more helpful for this purpose.

Recently, A-mode ultrasound scanning was reported valuable for examination of the nail plate.[29,30] The human nail plate was studied using a 20-MHz ultrasound scanner. A superficial dry compartment with the ultrasound velocity 3103 m/s, and deep humid compartment with the ultrasound velocity 2125 m/s was identified.[30] No significant correlation was found between nail-plate thickness and measurement of transonychial water loss.[31] Since changes in the nail may reflect dermatological or sometimes internal diseases, introduction of ultrasound for nail examinations may reveal a new area for investigation.

IV. Sources of Error

The reproducibility of skin thickness measurement by ultrasound A scan is high. The coefficient of variation is reported as 2.2% on normal skin and 2.4% on inflamed skin.[32] However, the reproducibility of the obtained results depends on the examiner. The main drawback of the method is that a certain experience in interpretation and identification of the interface is needed until precise measurements can be obtained. Normally the measuring technique is mastered within a few months.

The variation of the biology itself limits the precision of thickness measurements. The interface between the dermis and the tela subcutanea is far from flat. However, if three or more sites are examined and averaged, the above-mentioned reproducibility of the measurements can be obtained.[32]

The ultrasound technique has the advantage compared to most other bioengineering methods used for quantification of inflammation, that modest changes in ambient temperature, relative humidity, air convection, and emotional status will not influence the result of the measurement. Skin thickness, edema, etc. are relatively stable parameters, and no preconditioning of the subjects is necessary before measurement.

V. Correlation with Other Methods

The sensitivity of A-mode ultrasound scanning for quantification of experimentally SLS-induced skin damage was compared to other bioengineering methods (measurement of transepidermal water loss [TEWL], laser Doppler flowmetry, and colorimetry). In this study the sensitivity of ultrasound was found consistently better than colorimetry, and almost as high as measurement of TEWL.[33]

A-mode ultrasound scanning is far superior to measurement of skin fold thickness[34] because the latter method inevitably includes the subcutaneous fat. Ultrasound B- and C-scan both include the facility of A scan.[2] This equipment is therefore more advanced and includes more far more potentialities than does the A scan, although the risk of observer bias is also increased for these more advanced procedures. In choice of equipment, however, economical cost is also a factor. It is the author's experience that averaged triple-site A-mode ultrasound measurement of skin thickness is highly accurate.

VI. Recommendations

A-mode ultrasound scanning is a simple, highly reproducible technique for measurement of skin thickness. As opposed to B- and C-mode scanning the A-mode technique is not image producing. The result is presented in a diagram as amplitudes, and some skill is needed to acquire correct interpretation and identification of the interfaces presented. A-mode ultrasound scanning will serve very well when a simple measurement of skin thickness is required. A-mode scanning should not be employed in more advanced procedures such as tumor diagnosis.

References

1. Alexander H, Miller DL. Determining skin thickness with pulsed ultrasound. *J Invest Dermatol* 72: 17–19, 1979.
2. Serup J. Ten years of experience with high frequency ultrasound examination of the skin: development and refinement of technique and equipments. *Proc Int Symposium on Ultrasound and the Skin,* Bochum, FRG, March 15–17, 1990 (Ed. P. Altmeyer) Springer-Verlag, New York.
3. Cabral de Ascensao A. Ultrasonography of skin tumours — basic principles and main procedures. *Skin Cancer* 2: 5–12, 1987.
4. Edwards C, Payne PA. Ultrasonic velocities in skin components. Abstract presented in International Society of Bioengineering of the Skin, 9 November 1984; Liege, Belgium.
5. Escoffier C, Querleux B, De Rigal J, Lévêque J-L. In vitro study of the velocity of ultrasound in the skin. *Bioeng Skin* 2: 87–94, 1986.
6. Serup J, Staberg B, Klemp P. Quantification of cutaneous oedema in patch test reactions by measurement of skin thickness with high-frequency pulsed ultrasound. *Contact Dermatitis* 10: 88–93, 1984.
7. Swanson Beck J, Spence VA, Lowe JG, Gibbs JH. Measurement of skin swelling in the tuberculin test by ultrasonography. *J Immunol Meth* 86: 125–130, 1986.
8. Serup J. Diameter, thickness, area and volume of skin-prick histamine weals. *Allergy* 39: 359–364, 1984.
9. Frosch PJ, Duncan S, Kligman AM. Cutaneous biometrics. I. The response of human skin to dimethyl sulphoxide. *Br J Dermatol* 102: 263–274, 1980.
10. Agner T, Serup J. Quantitation of the DMSO-response: a test for sensitive skin. *Clin Exp Dermatol* 14: 214–17, 1989.
11. Agner T, Serup J. Skin reactions to irritants assessed by non-invasive bioengineering methods. *Contact Dermatitis* 20: 352–359, 1989.
12. Agner T, Serup J. Seasonal variation of skin resistance to irritants. *Br J Dermatol* 121: 323–328, 1989.
13. Agner T, Damm P, Skouby SO. Menstrual cycle and skin reactivity. *J Am Acad Dermatol* 24: 566–570, 1991.
14. Agner T. Basal transepidermal water loss, skin thickness, skin blood flow and skin colour in relation to sodium-lauryl- sulphate-induced irritation in normal skin. *Contact Dermatitis* 25: 108–114, 1991.
15. Agner T. Skin susceptibility in uninvolved skin of hand eczema patients and healthy controls. *Br J Dermatol* 125: 140–146, 1991.
16. Agner T. Susceptibility of atopic dermatitis patients to irritant dermatitis caused by sodium lauryl sulphate. *Acta Derm Venereol (Stockh)* 71: 296–300, 1991.
17. Serup J, Staberg B. Ultrasound for assessment of allergic patch test reactions. *Contact Dermatitis* 17: 80–84, 1987.
18. Brazier S, Shaw S. High-frequency ultrasound measurement of patch test reactions. *Contact Dermatitis* 15: 199–201, 1986.
19. Serup J. Quantification of acrosclerosis: measurement of skin thickness and skin-phalanx distance in females with 15 MHz pulsed ultrasound. *Acta Derm Venereol (Stockh)* 64: 35–40, 1984.
20. Åkeson A, Forsberg L, Hederström E, Wollheim F. Ultrasound examination of skin thickness in patients with progressive systemic sclerosis (scleroderma). *Acta Radiol Diagn* 27: 91–94, 1986.
21. Serup J. Localized scleroderma (morphoea): thickness of sclerotic plaques as measured by 15 MHz pulsed ultrasound. *Acta Derm Venereol (Stockh)* 64: 214–219, 1984.
22. Shafir R, Itzchak Y, Heyman Z, Azizi E, Tsur H, Hiss J. Preoperative ultrasonic measurement of thickness of cutaneous malignant melanoma. *J Ultrasound Med* 3: 205–208, 1984.
23. Kraus W, Schramm P, Hoede N. First experiences with a high-resolution ultrasonic scanner in the diagnosis of malignant melanoma. *Arch Dermatol Res* 275: 235–238, 1983.
24. Tan CY, Marks R, Payne P. Comparison of xeroradiographic and ultrasound detection of corticoid induced dermal thinning. *J Invest Derm* 76: 126–128, 1981.
25. Serup J, Holm IP, Stender IM, Pichard J. Skin atrophy and telangiectasia after topical corticosteroids as measured noninvasively with high frequency ultrasound, evaporimetry and laser Doppler flowmetry. Methodological aspects including evaluations of regional differences. *Bioeng Skin* 3: 43–58, 1987.
26. Andersson C. Microdialysis. *Acta Derm Venereol (Stockh)* 71: 389–393, 1991.
27. de Rigal J, Escoffier C, Querleux B, Faivre B, Agache P, Lévêque J-L. Assessment of ageing of the human skin in vivo. *J Invest Dermatol* 93: 621–625, 1989.
28. Pedersen H, Agner T, Storm T. Skin thickness and osteoporosis. Submitted.
29. Finlay AY, Moseley H, Duggan T. Ultrasound transmission time: an in vivo guide to nail thickness. *Br J Dermatol* 117: 765–770, 1987.
30. Jemec GBE, Serup J. Ultrasound structure of the human nail plate. *Arch Dermatol* 125: 643–646, 1989.
31. Jemec GBE, Agner T, Serup J. Transonychial water loss: relation to sex, age and nail plate thickness. *Br J Dermatol* 121: 443–446, 1989.
32. Agner T, Serup J. Individual and instrumental variations in irritant patch-test reactions — clinical evaluation and quantification by bioengineering methods. *Clin Exp Dermatol* 15: 29–33, 1990.
33. Agner T, Serup J. Sodium lauryl sulphate for irritant patch testing — a dose-response study using bioengineering methods for determination of skin irritation. *J Invest Dermatol* 95: 543–547, 1990.
34. Andersen KE, Staberg B. Quantitation of contact allergy in guinea pigs by measuring changes in skin blood flow and skin fold thickness. *Acta Derm Venereol (Stockh)* 65: 37–42, 1985.

Chapter 12.6
Skin Thickness: Caliper Measurement and Typical Values

Alan D. Martin
School of Human Kinetics
University of British Columbia
Vancouver, B.C., Canada

I. Introduction

Few direct measurements of the physical dimensions and weight of human skin have been reported in the literature. The surface area of the skin is a measure with a wide variety of applications ranging from normalizing metabolic test functions and the determination of appropriate drug dosage,[1] to hydrodynamics and exercise physiology.[2] Despite this, few direct measurements of surface area have been reported, resulting in the development of a variety of surface-area estimation equations based on height and weight.[3] Other measures of skin size have received little attention. Even data on normal skin weight are sparse. A survey of the limited available cadaver data indicated adult skin weights in the range 3.2 to 5.9 kg, and our recent dissections of 25 older-adult cadavers revealed a range of 2.5 to 5.5 kg.[4] Knowledge of skin thickness is useful in a variety of medical conditions as well as in areas such as thermoregulation[5,6] and body fatness.[7] However, prior to the use of ultrasound, there have been few reports on the measurement of skin thickness. Several studies have estimated skin thickness from radiographs,[8,9] but only at a single site on the forearm. A further limitation of these two studies was the poor precision of the method; radiographs cannot provide the resolution required to accurately assess skin thickness. The first report of directly measured skin thickness investigated forearm skin thickness in an autopsy study of 35 Chinese adults.[10] A second autopsy study by the same group was more comprehensive in that it reported thickness at nine sites on the body.[11] However, both autopsy studies suffered the severe limitation of the measurement procedure, which consisted of a plastic ruler inserted into an incision in the skin, giving a precision of only 0.5 mm. There is thus a surprising lack of comprehensive direct data on skin thickness. As a part of our cadaver analysis study investigating tissue masses of the human body, skin thickness was measured at 25 sites in 13 cadavers. The technique and results are the subject of this chapter.

II. Materials and Methods

As part of a comprehensive cadaver study of body composition, 13 unembalmed cadavers ages 55 to 94 years (6 male and 7 female) were subjected to extensive anthropometry prior to complete separation by dissection of skin, adipose tissue, muscle, bone, and organs. No cadavers showing emaciation or signs of compositional changes due to chronic illness were selected. Two thirds of those cadavers available were rejected on this basis. Since one of the aims of the study was to investigate the contribution of skin thickness to skinfold thickness as measured by skinfold caliper, the sites selected were those typically used for skinfold assessment of body fat.[12]

Measurements were made at 12 sites bilaterally and at one central site (abdomen). The selected sites were triceps (at the mid-acromiale-olecranon line on the posterior surface of the arm), biceps (at the level of the triceps but on the anterior surface), forearm (on the anterior surface of the forearm, at the level of maximum girth), subscapular (1 cm below the inferior angle of the scapula), chest (on the anterior axillary line at the level of the xyphoid process), waist (on the anterior axillary line at the level of the waist narrowing), supraspinale (5 to 7 cm above the anterior iliac spine on a line to the axilla), abdominal (2 cm below and lateral to the umbilicus), front thigh (on the anterior surface of the leg midway between the inguinal fold and the midpatella), medial thigh (on the medial surface, midway between the inferior ramus of the pubic bone and the inferior border of the medial femoral condyle), rear thigh (on the posterior surface, midway between the gluteal fold and the kneefold), patella (2 cm proximal to the the proximal tip of the patella)

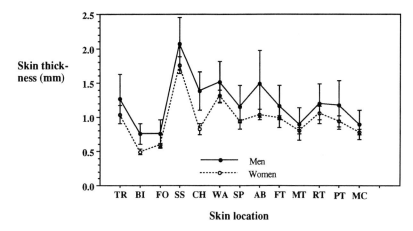

FIGURE 1 Skin thickness (mm) at 13 sites in six male and seven female cadavers, showing standard deviations. Values are the means of right and left side measurements. (Skinfold sites as defined in text and abbreviations as defined in Table 1.)

III. Results

Descriptive data for the cadavers are shown in Table 1. The subjects ranged in age from 59 to 86 years, and in body weight from 48.2 to 88.9 kg. Tissue masses have been reported elsewhere,[4] but individual skin weights ranged from 2.5 to 5.5 kg, with a mean of 3.9 kg in the men and 3.2 kg in the women. A separate study demonstrated that the anthropometric measures on the cadavers did not differ significantly from those taken on a group of living Belgians of similar mean age.[14] These cadavers were thus representative of their living peers in terms of physique. Skin thicknesses for both right and left sides at all sites are shown in Table 2 for the male cadavers and Table 3 for the females. The tables also show the means for each subject, the means for each site, and the overall means for men and women. Skin thickness varied by subject and site with males having thicker skin than females at all individual sites and overall (1.19 mm compared to 0.96 mm). The mean of 25 sites ranged from 0.81 to 1.43 mm in the six males and from 0.73 to 1.10 mm in the seven females. The site with the thinnest skin was the biceps: 0.76 mm in the men and 0.49 mm in the women. The thickest skin was at the subscapular site: 2.07 mm in men and 1.76 mm in women. There was considerable overlap between men and women: one of the men had a mean

and the medial calf (on the medial surface at the level of the maximum girth). After skinfold caliper readings of skin plus subcutaneous adipose-tissue thickness had been taken, incisions were made at all sites using a scalpel to measure adipose-tissue thickness directly; this enabled rapid site location for skin-thickness measurements which were made after dissection[13] (see Figure 1).

The cadavers were then dissected and all skin was completely removed in a number of pieces. Any adipose tissue adhering to the skin was removed by gentle scraping. Skin thickness at each site was measured by folding the skin at the incision and applying Harpenden skinfold calipers (British Indicators, Ltd., St. Albans, U.K.) to the double layer. These values were then halved and all values reported here are for a single layer. Skin thickness at each site was taken as the mean of five measurements around the incision, measured to a precision of 0.05 mm.

TABLE 1 Description of Subjects

Cadaver Identification No.	Age (yr)	Cause of Death	Weight (kg)	Stature (cm)	Skin Weight (kg)
Males					
17	65	heart disease	54.8	166.1	3.2
20	59	heart disease	76.8	173.2	4.3
23	81	unknown	61.0	176.9	2.7
25	73	heart disease	85.1	172.0	4.6
26	73	heart disease	57.7	163.6	2.9
27	55	suicide	88.9	186.5	5.5
Mean (SD)	68(10)		70.7(14.8)	173.0(8.2)	3.9(1.1)
Females					
15	79	heart disease	58.9	159.7	3.1
18	83	heart disease	74.2	172.3	3.3
19	82	renal insuff.	48.2	161.7	2.5
21	77	heart disease	71.6	151.3	3.5
22	68	unknown	69.0	153.6	3.4
24	86	leukemia	61.2	156.5	3.6
28	82	heart disease	68.8	163.5	3.1
Mean (SD)	80(6)		64.6(9.0)	159.8(7.0)	3.2(0.3)

TABLE 2 Right and Left Side Skin Thicknesses By Site in Male Cadavers

ID No.	SIDE	TR	BI	FO	SS	CH	WA	SP	AB[a]	FT	MT	RT	PT	MC	All Sites Mean	
17	R	1.10	0.80	0.65	1.85	1.40	1.45	1.15	1.20	1.10	0.95	1.40	1.35	0.95	1.18	0.32
	L	0.80	0.70	0.65	1.85	1.10	1.25	0.95	1.20	0.95	0.95	1.35	1.20	1.00	1.07	0.31
20	R	1.40	0.90	1.05	2.55	1.60	1.80	1.75	1.55	1.40	1.25	1.55	1.25	1.00	1.47	0.43
	L	1.55	0.80	1.10	2.15	1.75	1.95	1.20	1.55	1.20	1.30	1.35	1.00	1.05	1.38	0.39
23	R	0.65	0.50	0.45	1.55	1.15	1.30	0.70	—	0.70	0.70	0.70	0.75	0.50	0.80	0.34
	L	0.90	0.75	0.60	1.45	1.00	1.05	0.60	—	0.75	0.60	0.65	0.85	0.60	0.82	0.26
25	R	1.95	0.85	1.00	—	1.65	1.85	1.45	1.80	1.65	0.75	1.20	1.70	0.80	1.39	0.44
	L	1.40	0.80	0.65	—	1.50	1.60	1.35	1.80	1.55	0.75	1.40	1.50	0.75	1.25	0.40
26	R	1.30	0.55	0.70	2.15	1.05	1.25	1.15	1.15	1.05	0.75	1.05	0.80	0.85	1.06	0.40
	L	1.50	0.60	0.55	2.50	1.10	1.25	1.00	1.15	0.85	0.65	1.00	0.70	0.80	1.05	0.52
27	R	1.25	1.00	0.75	2.25	1.50	1.85	1.40	1.75	1.35	1.05	1.35	1.45	1.25	1.40	0.39
	L	1.40	0.85	0.95	2.40	1.75	1.55	1.10	1.75	1.40	1.15	1.35	1.60	1.10	1.41	0.41
Mean		1.27	0.76	0.76	2.07	1.38	1.51	1.15	1.49	1.16	0.90	1.20	1.18	0.89	1.19	
SD		0.36	0.15	0.21	0.38	0.28	0.30	0.32	0.29	0.31	0.24	0.29	0.35	0.21		0.35

Note: Measurements are reported in millimeters. Data are the means and standard deviations (SD) for each side, each male subject, each site, and all male subjects. Abbreviations are (TR) triceps, (BI) biceps, (FO) forearm, (SS) subscapular, (CH) chest, (WA) waist, (SP) supraspinale, (AB) abdominal, (FT) front thigh, (MT) medial thigh, (RT) rear thigh, (PT) patellar, (MC) medial calf, — no data.

[a] *Not taken bilaterally; the single value is included in both right and left sides to make the means comparable.*

TABLE 3 Right and Left Side Skin Thicknesses By Site in Female Cadavers

ID No.	SIDE	TR	BI	FO	SS	CH	WA	SP	AB[a]	FT	MT	RT	PT	MC	All Sites Mean	
15	R	1.45	0.65	0.75	1.70	1.20	1.65	1.00	1.25	1.10	0.95	1.05	1.15	0.90	1.14	0.32
	L	1.05	0.50	0.60	2.10	1.05	1.55	1.10	1.25	1.05	0.90	0.95	0.85	0.90	1.07	0.41
18	R	0.45	0.35	0.50	1.25	—	1.05	1.00	0.75	1.00	0.55	0.60	0.90	0.75	0.76	0.28
	L	0.50	0.40	0.45	1.25	—	0.85	0.55	0.75	0.75	0.60	1.05	0.70	0.60	0.70	0.265
19	R	1.20	0.35	0.55	1.70	0.60	1.00	1.05	—	0.85	0.85	0.90	0.85	0.70	0.88	0.35
	L	1.05	0.40	0.50	1.85	0.65	1.20	1.00	—	1.10	1.05	1.15	1.10	0.90	1.00	0.38
21	R	1.40	0.60	0.65	2.20	0.75	1.30	0.90	1.30	1.00	0.75	1.00	1.10	0.70	1.05	0.43
	L	1.10	0.65	0.60	2.00	0.80	1.00	0.90	1.30	0.80	0.70	1.00	1.25	0.80	0.99	0.37
22	R	0.95	0.50	0.65	1.70	0.75	1.30	0.95	1.05	1.10	0.70	1.30	0.65	0.85	0.96	0.33
	L	0.95	0.40	0.60	1.55	0.90	1.55	1.10	1.05	0.95	0.75	1.35	0.70	0.65	0.96	0.36
24	R	1.20	0.45	0.55	2.10	0.75	1.50	0.90	0.95	1.10	0.95	1.45	1.25	0.95	1.08	0.43
	L	1.05	0.45	0.75	2.10	0.80	1.50	0.90	0.95	1.10	0.95	1.30	0.80	0.80	1.03	0.41
28	R	1.05	0.55	0.55	1.50	0.85	1.45	0.90	0.95	1.00	0.75	0.95	1.05	0.70	0.94	0.29
	L	1.10	0.65	0.55	1.60	0.85	1.40	0.85	0.95	0.90	0.75	0.80	0.80	0.75	0.92	0.29
Mean		1.04	0.49	0.59	1.76	0.83	1.31	0.94	1.04	0.99	0.80	1.06	0.94	0.78	0.96	
SD		0.28	0.11	0.09	0.31	0.16	0.25	0.14	0.20	0.12	0.15	0.23	0.21	0.11		0.30

Note: Measurements are reported in millimeters. Data are the means and standard deviations (SD) for each side, each female subject, each site, and all female subjects. (TR) triceps, (BI) biceps, (FO) forearm, (SS) subscapular, (CH) chest, (WA) waist, (SP) supraspinale, (AB) abdominal, (FT) front thigh, (MT) medial thigh, (RT) rear thigh, (PT) patellar, (MC) medial calf, — no data.

skin thickness (0.81 mm) that was smaller than all but one of the women. The overall range in skin thickness was wide: two women had skin of thickness 0.35 mm at the biceps site, while one male had a thickness of 2.55 mm at the subscapular site.

IV. Discussion

Data on skin thickness are sparse. In an autopsy study on 35 Chinese subjects,[10] forearm skin thickness ranged from 0.82 to 1.82 mm. At the same site, Shephard and Meema,[9] using radiography, reported a mean thickness of 1.43 mm in male and 1.34 mm in female Caucasians. A third study,[8] using a similar procedure, found that (1) for a given age and weight, males had a thicker skin than females and (2) for each sex, skin thickness decreased with age, but these findings were based only on radiographs of the forearm site. In a more comprehensive study,[11] skin thickness was measured directly at nine typical skinfold sites in 35 adult Chinese cadavers. The mean values at each site ranged from 0.96 (biceps) to 3.41 mm (subscapular) with somewhat smaller values in females. The measurement technique, however, which utilized a plastic ruler inserted in an incision in the skin, only gave readings to the nearest 0.5 mm, a value larger than some of the thicknesses being measured.

The data reported demonstrate considerable variability in skin thickness both from site to site within a single subject as well as between subjects. For example, in the six males, skin thickness at the triceps ranged from 0.65 to 1.95 mm (Table 1). The corresponding range in the seven females was 0.45 to 1.45 mm. In interpreting these data the nature of our sample should be kept in mind. These were all older subjects of one nationality. The normal range of skin thickness may be considerably greater than this if there are significant ethnic differences. The effects of aging on skin thickness are not well known, but it seems likely that skin thins with age. Thus the highest values reported here may be lower than those found in a healthy young population.

References

1. Siber, G.R., Smith, A.L., and Levin, M.J., Predictability of peak serum Gentamicin concentration with dosage based on body surface area, *J. Pediatr.*, 94, 135, 1979.
2. Nadel, E.R., Holmer, I., Bergh, U., Astrand, P.O., and Stolwijk, J.A., Energy exchanges of swimming men, *J. Appl. Physiol.*, 36, 464, 1974.
3. Martin, A.D., Drinkwater, D.T., and Clarys, J.P., Human body surface area: validation of formulae based on a cadaver study, *Hum. Biol.*, 56, 475, 1984.
4. Clarys, J.P., Martin, A.D., and Drinkwater, D.T., Gross tissue weights in the human body by cadaver dissection, *Hum. Biol.*, 56, 459, 1984.
5. Vendrik, A. and Vos, J., A method for the measurement of the thermal conductivity of human skin, *J. Appl. Physiol.*, 11, 211, 1957.
6. Ducharme, M. and Tikuisis, P., In vivo conductivity of the human forearm tissues, *J. Appl. Physiol.*, 70, 2682, 1991.
7. Martin, A.D., Ross, W.D., Drinkwater, D.T., and Clarys, J.P., Prediction of body fat by skinfold caliper: assumptions and cadaver evidence, *Int. J. Obesity*, 9, 31, 1985.
8. Bliznak, J. and Staple, T.W., Roentgenographic measurement of skin thickness in normal individuals, *Radiology*, 118, 55, 1975.
9. Shephard, R.H. and Meema, H.E., Skin thickness in endocrine disease, a roentgenographic study, *Ann. Int. Med.*, 66, 531, 1967.
10. Lee, M., Physical and structural age changes in human skin, *Anat. Rec.*, 129, 473, 1957.
11. Lee, M. and Ng, C., Post-mortem studies of skinfold caliper measurements and actual thickness of skin and subcutaneous tissue, *Hum. Biol.*, 37, 91, 1965.
12. Ross, W.D. and Marfell-Jones, M.J., Kinanthropometry, in *Physiological Testing of the Elite Athlete.*, McDougall, J.D., Wenger, H.A., and Green, H.J., Eds., Mutual, Ottawa, 1982, 75.
13. Martin, A.D., Drinkwater, D.T., Clarys, J.P., Daniel, M., and Ross, W.D., Effects of skin thickness and skinfold compressibility on skinfold thickness measurement, *Am. J. Human Biol.*, 4, 453, 1992.
14. Clarys, J.P., Martin, A.D., Drinkwater, D.T., and Marfell-Jones, M.J., The skinfold: myth and reality, *J. Sports Sci.*, 5, 3, 1987.

13.0

Nuclear Magnetic Resonance (NMR) Examination of the Dermis

13.1 Nuclear Magnetic Resonance (NMR) Examination of the Skin 299
A. Zemtsov

13.2 Nuclear Magnetic Resonance Examination of Skin Disorders 305
*S. el-Gammal, R. Hartwig, S. Aygen, T. Bauermann,
K. Hoffmann, and P. Altmeyer*

Chapter 13.1
Nuclear Magnetic Resonance (NMR) Examination of the Skin

Alexander Zemtsov
Department of Dermatology and Biochemistry
Texas Tech University School of Medicine
Lubbock, Texas

I. Introduction

There are two magnetic resonance (MR) techniques used today in clinical medicine and biomedical sciences. Magnetic resonance imaging (MRI) also known as proton resonance imaging and magnetic resonance spectroscopy (MRS). MRI scans are two- and three-dimensional radiological pictures that are similar to radiological images produced by computed tomography (CT). MRS is a nondestructive technique to noninvasively study a metabolism of intact biological system such as a human body. MRI in the U.S. has become a standard radiological technique that already had a truly revolutionary impact on many areas of clinical medicine.[1] MRS is still considered by the FDA (Food and Drug Administration) an investigational device and the Human Subject Institutional Review Boards' approval is required for its use in a clinical setting.

In 1938 Professor Rabi described nuclear magnetic resonance phenomenon.[2] Since then, MR has been extensively used in physics, chemistry, and in the past decade also in biomedical sciences. For their contribution to this field Rabi (in 1944) and, later, Bloch and Purcell (in 1952) were awarded a Nobel Prize in Physics.[3] In 1971, Damadian show that tumors and normal fissure have different relaxation times.[4] This discovery led to the eventual development of MRI as a radiological imaging technique that differentiates between tissue of different pathological types. In the past few years, software was developed that performs human in *in vivo* MRS on a standard hospital-based MRI scanner.[5]

Zemtsov et al.,[6] Schwaighofer et al.[7] and later Takahashi and Kohda[8] described the MRI scan of skin lesions. Hyde's[9] original design of MRI skin surface coil was perfected by a French group headed by Bittoun.[10] Plastic surgeons were first to perform *in vitro* ^{31}P-MRS experiments.[11,12] Zemtsov was first to collect human *in vivo* ^{31}P-MRS data and confirmed it by high-pressure liquid chromatography (HPLC) analysis.[13–14] Recently Zemtsov wrote an editorial in *Archives of Dermatology* on the use of MR techniques in dermatology.[15]

II. Object and Methods

The use of MR techniques in dermatology can be subdivided into three areas:[15]

1. Radiological evaluation of skin neoplasms by standard commercially available surface coils
2. Use of specialized skin surface coils to study biophysics of skin disorders and to detect malignant changes in melanocytic lesions
3. Ability of MRS to noninvasively study metabolism in cutaneous tissue (see Table 1).

Development of MR surface coils allowed radiological visualization and evaluation of small superficial structures such as skin.[16] The purpose of MRI in dermatology of using commercial or experimental skin-surface coil is similar to high-frequency ultrasound techniques. It also allows radiological examination of skin and subcutis. Zemtsov showed that tumor depth measured by MRI correlates well with postoperative histological measurements.[17] MRI can define and detect skin cancer recurrences under skin flaps and grafts.[18] MRI is also a very reliable technique to determine preoperatively the extent of underlying tissue involvement.[17] For example, Figure 1 shows large basal-cell carcinoma in the orbital area. One can clearly see the tumor does not invade orbital structures. Various researchers proposed radiological criteria to differentiate between benign and malignant lesions.[6–8,17] Finally, MRI is of help[32] in monitoring the response of skin tumors, facial hemangiomas, and other skin lesions to radiation, chemotherapy, and interlesional steroid injections. In summary, MRI utilization of standard commercially available surface coils is becoming a radiological technique of choice for evaluation of large and/or aggressive skin tumors in important anatomical locations such as orbital area and genitalia.

TABLE 1 MR in Dermatology

	Standard coils	Skin surface coils	Spectroscopy
Purpose	Radiological evaluation of skin tumors	Detection of malignant changes in melanocytic lesions Biophysical evaluation of skin layers	Monitor skin metabolism
Application	Measure tumor depth Extent of involvement Malignant vs. benign Recurrences of tumors under skin grafts and flaps Monitor response to therapy	Study radiological changes that correlate with melanocytic dysplasia Effect of disease and aging on skin layers T1 and T2 relaxation times.	^{31}P MRS measures intracellular pH, rate of keratinocyte turnover, and skin bioenergetics

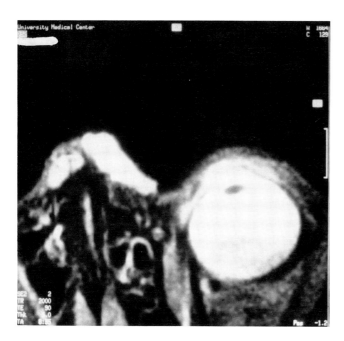

FIGURE 1 MRI scan of basal cell carcinoma in the periorbital area. The tumor is superficial and does not invade orbital structures.

A French group described their experience with skin surface coils in measuring skin layers T_1 and T_2 relaxation times in normal, diseased, and aged skin. They correlated these findings with underlying biophysical processes.[20] Zemtsov believes that MR skin surface coils will play a major role in defining dysplastic nevi syndrome by studying radiological changes that correlate with melanocytic dysplasia.[15]

Except for one paper dealing with *in vitro* proton spectroscopy,[21] all other investigators working in the area of cutaneous MRS concentrated their efforts on the ^{31}P atom. Days or even weeks before any clinical changes occur, ^{31}P MRS can noninvasively detect biochemical changes. ^{31}P MRS monitors and measures cutaneous tissue bioenergetics, intercellular pH, and keratinocytes' turnover rate. ^{31}P MRS is expected to be widely used in noninvasively monitoring survival of skin flaps and grafts,[11,12] and in monitoring therapeutic response of psoriasis, mycosis fungoides, skin tumors, and leg ulcers to various forms of therapy.[15-22]

Finally, ^{31}P MRS can better help to understand the pathophysiology of skin disorders.

A. Methodological Principle

There are dozens of excellent review articles and books written on the subject of MRI and MRS biophysical and engineering principles.[23,24] In this section, only a brief discussion is included. For more detailed explanations the reader is referred to References 23 and 24.

Atoms with an odd number of protons and neutrons possess a nuclear spin. When these atom are placed in a strong magnetic field (Bo), they develop different energy states. When an energy of specific frequency (specific for each particular atom) in radiowaves' range is applied (called Larmour frequency) the atoms will populate a higher energy state. Once this radiofrequency is turned off and as atoms return to the ground state, they release energy that is also in radiowave range. This energy is called free induction decay (FID). This signal is picked up by a surface coil and is used either for an MRI image or MRS spectrum reconstruction.

In order to obtain an image, one needs to have a high concentration of atoms. Hydrogen atoms (protons) present in all biological tissue, but carbon-12 has no nuclear spin (even number of protons and neutrons). Therefore only protons can be used to get radiological images. However, MRS spectra can be obtained from ^{13}C or ^{31}P atoms. Until recently, most work in biomedical sciences was done with ^{31}P MRS because spectra are simple and easily interpretable. ^{31}P MRS provides information about intercellular pH, tissue turnover rate, and tissue bioenergetics. Figure 2 is a ^{31}P-MRS spectrum of a normal skin. One sees beta, alpha, and gamma peaks of ATP molecule, phosphocreatine peak, and inorganic phosphate peak. During skin ischemia, ATP and phosphocreatine peaks disappear and inorganic phosphate peak increases.[11-13] Phosphomonester concentration reflects phospholipid biosynthesis. On the other hand, phosphodiesters are the products of membrane catabolism. Therefore, phosphomonester to phosphodiester ratio is related and is increased in a disease characterized by increased production of new cells.[22-25] As expected, preliminary results indicate that phosphomonester to phosphodiester ratio is

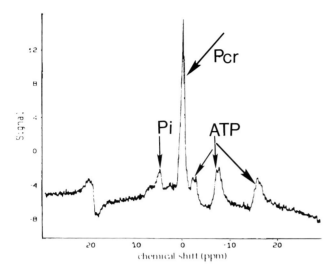

FIGURE 2 ^{31}P MRS spectrum of normal nonischemic skin. Arrows point to phosphocreatine (Pcr), ATP, and inorganic phosphate (Pi) peaks. In hyperprolific states such as psoriasis one also observes PME and PDE peaks (data not shown).

increased in patients with psoriasis. Intercellular pH is calculated from the chemical shift of inorganic phosphate.[26] Zemtsov's preliminary data indicates that pH of many chronic leg ulcers is acidic and as the ulcers are starting to heal the pH returns to the normal tissue range.

As we have discussed, in MRS, chemical shift and metabolites' ratio are the most important parameters. In MRI, the parameters that determine the quality of pictures, are protein density (concentration of proton atoms) and proton relaxation times. These relaxation times are called T_1 (spin-lattice) and T_2 (spin-spin). T_1 and T_2 determine the time frame for the excited protons to return to the ground state. T_1 and T_2 differ widely within tissues and between normal and pathological tissues. As was mentioned, the discovery by Damadian explains a unique ability of MRI to differentiate between tissue of different pathological types.[4] The train of radiofrequencies sent to the tissue at Larmour frequency is called a pulse sequence. Pulse sequences determine the speed of image acquisition and most importantly whether the image will be mostly formed as the result of T_1 or T_2 relaxation times. (In MRI language these images are called T_1- or T_2-weighted images). There are over 200 pulse sequences that were described in MRI literature. Recently a book describing these various pulse sequences was published.[27] However, in a standard pulse sequence, called spin echo, T_1-weighted images as a rule emphasized anatomy of a lesion while T_2-weighted images enhanced a contrast between normal and pathological tissue. This contrast between tissues can be further enhanced in skin tumors by an MRI contrast agent called gadolinium.[18]

Zemtsov published his recommendations of what coil to use for a skin lesion depending on its size and anatomical location.[17] In general, commercially available surface coils are adequate for almost all skin lesions except for the very small and totally flat ones. For small and flat skin lesions, one needs to use experimental skin surface coils. Skin surface coils will be discussed in greater detail in Recommendation, Section III.

Computer function involved in MRI image and MRS spectrum reconstruction is far more complex than in any other method discussed in this book. For example in high-frequency digital ultrasound techniques for image reconstruction, the computer merely performs digital to analog conversion, but in MR techniques a computer also controls timing and the duration of radiofrequency pulses and the timing and intensity of magnetic-field alterations.[28] Good resolution skin images can be obtained at any 1.5-T supraconducting magnets.

B. Sources of Error

The best skin MRI images are obtained on the extremities and in the head and neck region. Skin MRI images on the chest, and most importantly on the abdomen may appear blurry because of the breathing motion artifact. To eliminate this motion breathing artifact, one can perform a so-called gated study in which the computer averages the patient's number of breaths per minute so they can be eliminated from MRI data. Alternatively, one can simply place a tight bandage or an ace wrap around the patient's abdomen, and thus decrease the breathing motion.

Another factor that needs to be taken into consideration is the relationship between field of view and the resolution. In comparison to commercially available surface coils, specialized skin-surface coils are supposed to provide, as a rule, better resolution and visualization of skin lesions. However, they have a very limited field of view. Skin surface coils will be useless in determining depth and the extent of underlying tissue involvement of large tumors, such as basal-cell carcinoma shown in Figure 1. This issue has been discussed in detail by Zemtsov.[17]

In MRS of skin as in other types of spectroscopy, the tissue localization is the key factor. One has to be sure that the signal is coming primarily from the skin and not from the underlying tissue. In skin ^{31}P MRS, a skeletal muscle that has a high concentration of ATP and phosphocreatine, is the most common source of data contamination. To prevent data contamination phantom experiments are performed to insure proper localization of ^{31}P-MRS signal. In these phantom experiments, the ^{31}P-MRS coil sensitivity is tested versus depth of penetration.[13] Ideally, in ^{31}P-MRS skin experiments the coil should not pick up a signal from 1 cm below the skin surface. The slotted cross-over coil for ^{31}P-MRS examination of skin was developed by Nagel.[29-30] This coil was designed to have a very limited depth of penetration. It was very effective when acquiring skin ^{31}P-MRS data in rodents' experiments. However, phosphometabolite concentration in rodents' skin is approximately 10 times higher than in a human skin.[31] (In rodents' skin there is a layer of skeletal muscle called panniculus carnosus that cannot be separated from other skin layers.) It remains to be seen if this coil is sensitive enough for human *in vivo* ^{31}P-MRS skin experiments. The experience of our research group with this coil is disappointing so far.

C. Correlations with Other Methods

MRS noninvasively provides a unique biochemical information that cannot be obtained by any other method. However, much of the information obtained by MRI can also be obtained by high-frequency ultrasound techniques. Advantages of ultrasound techniques are obviously lower cost, portability, and speed of obtaining images. Ultrasound equipment can be placed in any dermatology office, does not require a special building and can be used in patients with pace makers and other ferromagnetic objects in their bodies. However, MRI provides a much better resolution, helps to differentiate between benign and malignant skin lesions, and has contrast agents available. Furthermore, ultrasound has a limited tissue penetration. On the other hand, MRI skin-surface coils are not commercially available and commercially available MR skin-surface coils do not provide good resolution of epidermis and dermis layers. Therefore because of speed and portability, high frequency ultrasound techniques are probably a technique of choice for radiological evaluation of processes that primarily affect epidermis and dermis (monitoring skin thickness with patients with progressive systemic sclerosis, differentiating between second and third degree burns, etc.).

Zemtsov believes MRI and ultrasound will become complimentary radiological techniques in dermatology the same way ultrasound, CT scans, and MRI are in general diagnostic radiology. Because of portability and speed, ultrasound may become useful for initial evaluation of skin lesions while MRI will be primarily reserved for larger, complex skin tumors, and special applications. Figure 3 shows histology, MRI, and ultrasound scans of unusual hypertrophic lichen planus tumor on the shin. MRI obviously provides a better resolution, but ultrasound is adequate to determine that the growth is superficial and does not invade soft tissue and muscle.

FIGURE 3 (A) Histology of verrucous hypertrophic lichen planus growth on a shin.

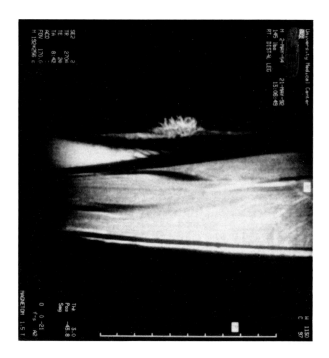

FIGURE 3 (B) MRI scan of this lesion.

III. Recommendations

The next most important task for MR application techniques in dermatology is to make MR skin-surface coils for MRI and MRS widely available. Probably the best MRI skin-surface coil called (Skin Imaging Modele) is available only from one research group in France.[10] Hopefully General Electric, Siemans, or other MR equipment manufacturers will consider producing it commercially. Dr. Salter (Sambrook, U.K.) at the American Academy of Dermatology meeting in San Francisco presented a poster of his skin MR surface coil. This coil had a better resolution than the

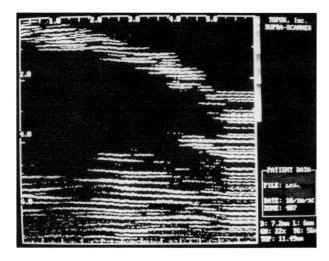

FIGURE 3 (C) 20-MHz ultrasound B-scan of the lesion.

coil produced by the French group. However, the MRI scan of skin obtained by this coil had two obvious disadvantages: nonstandard (2T) magnetic field (1.5-T used commercially), and a small diameter of a magnetic bore limiting skin examination only to the hands and feet. Therefore, it remains to be seen whether this coil can provide a better resolution of skin structure if it is connected to a standard 1.5-T supraconducting magnet system.

Most MRI scans are obtained with 256 × 256 digital matrix. However, software for a 512 × 512 matrix is becoming commercially available. The 512 × 512 matrix should probably provide a much better resolution of skin structure. The obvious tradeoff will be that to acquire images, more time will be needed. Finally, almost all MRI scans of skin lesions were obtained by spin-echo pulse sequences. It remains to be seen if fat-suppression pulse sequences will allow a better contrast and resolution.

The MRS of skin is still in its early pioneering stages. There are only six or seven research groups in the world working in this area. In my view, there is room for development of the ^{31}P MRS coil that can both adequately detect skin signal and localize a signal primarily to the depth of 1 cm from the skin surface. The potential application of ^{31}P MRS in dermatology is truly limitless. It can be applied to noninvasively study metabolism of any skin lesion from leg ulcers, to psoriasis, mycosis fungoides, skin tumors, and skin grafts and flaps.

References

1. Council on Scientific Affairs. Magnetic resonance imaging. Prologue. *JAMA*, 258, 3283, 1987.
2. Rabi II, Zacharias JR, Millman S, Kursch P. A new method of measuring nuclear magnetic moment. *Phys Rev.* 53, 318, 1938.
3. Hurwitz R. Milestones in magnetic resonance: a new method of measuring nuclear magnetic moment. *JMRI*, 2, 131, 1992.
4. Damadian R. Tumour detection by nuclear magnetic resonance. *Science,* 171, 1151, 1971.
5. Ng TC, Majors AW, Meany TF. In vivo MR spectroscopy of human subjects with 1.4T whole-body MR imager. *Radiology,* 158, 517, 1986.
6. Zemtsov A, Lorig R, Bergfeld WF, Bailin PL, Ng TC. Magnetic resonance imaging of cutaneous melanocytic lesions. *J Dermatol Surg Oncol,* 15, 854, 1989.
7. Schwaighofer BW, Fruehwald FXJ, Pohl-Markl H, Neuhold A, Wicke L, Landrum WL. MRI evaluation of pigmented skin tumors. Preliminary study. *Invest Radiol,* 24, 289, 1989.
8. Takahashi M, Kohda H. Diagnostic utility of magnetic resonance imaging in malignant melanoma. *J Am Acad Dermatol,* 27, 51, 1992.
9. Hyde TS, Tesmanovwicz H, Kneeland BJ. Surface coil for MR imaging of the skin. *Magn Reson Med,* 5, 456, 1987.
10. Bittoun J, Saint-James H, Querleux BG. In vivo high-resolution MR imaging of the skin in a whole-body system at 1.5T. *Radiology,* 176, 457, 1990.
11. Cuono CB, Armitage IM, Marquetand R, Chapo GA. Nuclear magnetic resonance spectroscopy of skin: predictive correlates for clinical application. *Plast Reconstr Surg,* 81, 1, 1988.
12. Klein HW, Gourly IM. Use of magnetic resonance spectroscopy in the evaluation of skin flap circulation. *Ann Plast Surg,* 20, 547, 1988.
13. Zemtsov A, Ng TC, Xue M. Human in vivo ^{31}P spectroscopy of skin: potentially a powerful tool for noninvasive study of metabolism in a cutaneous tissue. *J Dermatol Surg Oncol,* 15, 1207, 1989.
14. Zemtsov A, Cameron G, Stadig B, Martin J. Measurement of phosphocreatine concentration in skin by high pressure liquid chromatography. *Am J Med Sci* in press.
15. Zemtsov A, Dixon L. Magnetic resonance in dermatology. *Arch Dermatol* in press.
16. Fisher MR, Barker B, Ampero EG, Brandt G, et al. NMR imaging using specialized coils. *Radiology,* 157, 443, 1985.
17. Zemtsov A, Lorig R, Ng TC, et al. Magnetic resonance imaging of cutaneous neoplasms: clinicopathologic correlation. *J Dermatol Surg Oncol,* 17, 416, 1991.
18. Zemtsov A, Reed J, Dixon L. Magnetic resonance imaging evaluation helps to delineate a recurrent skin cancer present under the skin flap. *J Dermatol Surg Oncol,* 18, 508, 1992.
19. Richard S, Querleux B, Bittoun J, et al. In vivo proton relaxation times analysis of the skin layers by magnetic resonance imaging. *J Invest Dermatol,* 97, 120, 1991.
20. Richard S, Querleux B, Bittoun J, et al. Characterization of the skin in vivo by high resolution MRI: water behavior and age-related effects. *J Invest Dermatol* in press.
21. Kim YH, Oreberg ER, Faull KF, Wade-Jardetzky NG, Jardetzky O. ^{1}H NMR spectroscopy: an approach to evaluation of diseased skin in vivo. *J Invest Dermatol,* 92, 210, 1989.
22. Redmond OM, Bell E, Stack JP. Tissue characterization and assessment of preoperative chemotherapeutical response in musculoskeletal tumors by ^{31}P MRS. *Magn Res Med,* 27, 226, 1992.
23. Andrew ER, Bydder G, Griffiths, Iles R, Styles P. *Clinical Magnetic Resonance: Imaging and Spectroscopy.* John Wiley & Sons, Chichester, 1990.
24. Kramer DM. Basic principles of MRI. *Radiol Clin North Am,* 22, 765, 1984.

25. Breit A. "In vivo magnetic resonance spectroscopy", in *Magnetic Resonance in Oncology.* Springer-Verlag, Berlin, 1990.
26. Dhasmana JP, Digerness SB, Geckle JM. Effect of adenosine deaminase inhibitors on the heart. *J Cardiovasc Pharmacol,* 5, 1040, 1983.
27. Parikh AM. Magnetic resonance imaging techniques. Elsevier, New York, 1992.
28. Kuni CC. *Introduction to Computers and Digital Processing in Medical Imaging.* Year Book Medical Publishers, Chicago, 1988, chap. 7.
29. Nagel T, Alderman DW, Schoenborn RS, et al. The slotted crossover surface coil: a detector for in vivo NMR of skin. *Magn Res Med,* 16, 252, 1990.
30. Chen Y, Richards TL, Izenberg S, et al. In vivo phosphorus NMR spectroscopy of skin using a crossover surface coil. *Magn Res Med,* 23, 46, 1992.
31. Zemtsov A. Presence of phosphocreatine in rat testis. *Magn Reson Med,* 27, 198, 1992.
32. Zemtsov A. Unpublished data.

Chapter 13.2
Nuclear Magnetic Resonance Examination of Skin Disorders*

S. el-Gammal,[1] R. Hartwig,[1] S. Aygen,[2] T. Bauermann,[2] K. Hoffmann,[1] and P. Altmeyer[1]

[1]*Dermatologische Klinik der Ruhr-Universität Bochum*
[2]*Institut für Zentrale Analytik und Strukturanalyse der Universität Witten/Herdecke*
Germany

I. Introduction

A characteristic feature of the dermatological specialty is that the whole spectrum of diseases, from slight irritations to malignant transformations, lies directly before ones eyes. Clinical examination by inspection and palpation thus plays a crucial role in dermatological diagnosis. Moreover in any doubtful case the easy accessibility of the skin allows biopsies to be taken without much discomfort for the patient. Many diagnoses can be confirmed by their typical histological picture. The diagnosis of a skin disease is therefore usually based on clinical and histological criteria.

In the diagnosis of malignant melanoma, where the prognostic classification of the tumor depends on the level of invasion into the dermis[8] and the vertical tumor thickness,[6,7] preoperative evaluation of these factors is extremely helpful to determine the therapeutic excision margin and judge the risk of metastasis. Pretherapeutic information would also be valuable in differential diagnosis improving diagnostic accuracy.

During the last years the development of noninvasive techniques has opened new possibilities of preoperative diagnosis. Lateral and axial tumor expansion can be determined by means of high resolution 20-MHz[2,22] or 50-MHz sonography.[16] However this method does not always allow definition of the tumor parenchyma exactly.[2,15,16,22] Our investigations show that it is especially difficult to differentiate the tumor from the concomitant inflammatory infiltrate.[15-17] Attempts to evaluate sonographic pictures with image-analytical methods could not solve this problem.[15,17]

Other noninvasive imaging techniques like computer tomography and xeroradiography require X-ray radiation and are therefore less suitable for dermatological diagnosis considering the benefit-risk ratio for the patient. A noninvasive method which became increasingly important in the whole medical field throughout the last years is the magnetic-resonance (MR) technique. The investigation and description of the magnetic properties of the atomic nuclei laid the base for this method. In the years 1924 to 1927 the physicist Pauli analyzed the detailed structure of the atomic nuclei (he described the anomal Zeeman effect and the exclusion principle).[34] Then 20 years later Bloch[5] and Purcell,[31] independently from each other, discovered the nuclear-spin resonance signal. The nuclear-spin resonance signal provides information about the chemical composition of the investigated object. The idea to use nuclear magnetic resonance (NMR) for investigations of human tissue has its origin in studies by Jackson in 1967.[39] A milestone in the history of NMR was the publication of a section of water-filled capillaries by Lauterbur in the year 1973.[26] A new imaging technique (magnetic resonance imaging, MRI) had thus been created which he himself named zeugmatography and from which today's MR tomography was developed. In MRI different sequence protocols are used to get images of two-dimensional (2D) sections. For spin-echo images a nonselective 90° rectangular pulse of 3 to 20 µs (hard- or high frequency (HF)-pulse) and a magnetic gradient field are combined with a slice-selective pulse of 4000 µs (so-called soft pulse) in such a way that the tissue characteristics can be assessed in very small volumes nondestructively. With the current MR tomographs for clinical applications, which

---

* Essential parts of this publication are part of the dissertation of R. Hartwig.

use a homogeneous magnetic field between 0.23 and 2.4 T and work with magnetic gradient fields of up to 1 G/cm to assess internal organs or bones, voxel (volume element)-resolutions of 5 × 5 × 5 mm³ at best can be obtained.[25]

Recently time imaging MR systems have become available which work with very high magnetic ground and gradient fields and thus allow examination of the fine structure of tissues at a microscopic level. Of particular diagnostic value is the differentiation of various tissues and pathological alterations using localized T1- and T2-measurements. Since the construction of this strong homogeneous magnetic ground field and high gradient fields is technically difficult for greater probe volumes, it is possible to study with our equipment small living animals *in vivo*, whereas human skin must be investigated at present *ex vivo*.

II. Materials and Methods

Various skin tumors, among them 15 nevocellular nevi, 20 basal cell carcinomas, and 6 malignant melanomas were investigated. Immediately after excision the tissue was either rinsed in 0.9% physiologic NaCl solution or briefly fixed in 5% formalin. The tissue specimen was placed in the center of a 5 or 10 mm proton-resonant frequency-coil and examined with a supraconducting nitrogen-cooled MR-spectroscopy/microscopy unit (AM 400 WB NMR, Bruker GmbH, Rheinstetten, FRG) equipped with an advanced microimaging accessory. A vertical 50-mm diameter gradient coil was used to generate gradients between 5.7 and 11.7 G/cm. The thickness of the MR slices in the specimen was calculated according to the following formula: slice thickness = selective pulse width/gyromagnetic ratio × gradient strength.

Our specimens had a spacial resolution in the range of 20 to 39 µm and a slice thickness between 300 and 500 µm. We used an image processing workstation (X32, Bruker GmbH, Rheinstetten, FRG) for off-line image retrieval and to plot relaxation curves from consecutive images. Finally tissue-specific T1- and T2-relaxation times were calculated.

After MR imaging the tissue was processed for histology (fixation in formalin 5%, dehydration in increasing ethanol concentrations, embedding in paraffin). In order to obtain histological sections which correlate to the MR-multislice images the tissue has to be precisely oriented in the paraffin block. The whole paraffin block was cut in 7-µm-thick sections collecting every 18th section and abandoning the rest until the next MR-slice has been reached (calculations see above). The specimens were stained with H & E and examined and photographed using an axiophot-microscope (Zeiss, Oberkochen, FRG).

A. MR Principle

When entering a specimen in a homogeneous magnetic ground field the spins of the atomic nuclei within the specimen statistically rearrange in the magnetic field in such a way that the sum vector of all spins points to the direction of the magnetization. The length of the sum vector is proportional to the strength of the magnetic field. When a second magnetic field alternating at a certain frequency is applied at a 90° angle to the ground field, the sum vector (overall magnetization vector) changes are spatial orientation. This temporal change of the magnetic sum vector perpendicular to the ground field induces in the coil used for signal detection a HF alternating current. The resulting curve is characteristic for the composition of the specimen. The energy previously absorbed by the specimen is now emitted causing an exponential decrease of the received nuclear resonance signal, while the spins of the atomic nuclei relax.

This relaxation process can be divided into two components, the longitudinal and the transversal relaxation. The longitudinal relaxation describes the process in which the momentum of all nuclei due to the interaction with their surrounding atoms and molecules (called grid) return into their original position parallel to the field. This phenomenon is also called spin-grid relaxation. The time it takes for 63% of the momentum of all nuclei to reach their original state in relation to the magnetic ground field is referred to as longitudinal relaxation time T1.[28]

When the high frequency impulse is switched off all nuclei first point to the same direction and precess uniformly. The precession movements are, however, increasingly influenced by the inhomogeneity of the applied magnetic ground field on the one hand and interactions between neighboring nuclear moments on the other hand. This process is named spin-spin interaction or cross relaxation, the according time constant being the transversal relaxation time T2.

With MR spectroscopy the entire material is measured as integral signal assuming an equal distribution of all substances within the specimen. The use of additional magnetic gradient fields makes it possible to measure selectively within small volumes of the specimen as well. With MR imaging the specimen is analyzed in slices and the qualities of the received signal within the voxel are presented as grey-level modulation of one pixel (picture element) in the 2D image (Figure 1).

B. MR Microscopy

The NMR-microscopy/spectroscopy-unit Bruker AM 400 WB NMR works with a helium-cooled supraconducting magnetic coil of 9.4 T and 3 orthogonal gradient coils which produce linear gradient fields of up to 75 G/cm.

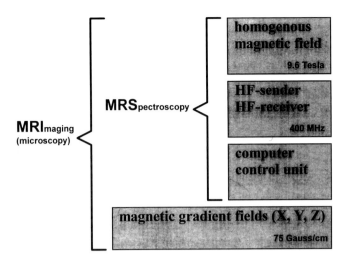

FIGURE 1 MR spectroscopy and MR microscopy. For MR microscopy additional gradient fields are applied to obtain a spin-echo signal from small voxels (volume elements).

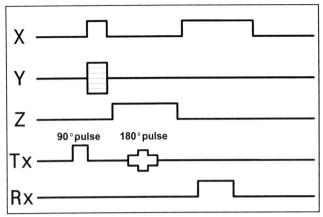

FIGURE 2 Sequence protocol of the single-slice spin-echo procedure. The nonselective 90° pulse is a rectangular "hard" pulse, the slice selective 180° pulse is displayed as a symbol standing for different "soft"-pulse shapes (e.g., Hermite or Gauss). X, Y, and Z refer to the gradient fields. A high-frequency signal is emitted (T_x) and received (R_x) by the same coil. The strength of the Z-gradient determines the slice thickness. The axial resolution is proportional to the strength of the X gradient.

In a tissue block of approximately 1 cm³ a resolution of 20 to 40 μm can be achieved in the x/y-plane when selecting a slice thickness of 300 to 500 μm. In smaller tissue specimens theoretically a pixel resolution of 10 μm² should be achievable. We used the following imaging techniques for our investigations:

- Spin-echo sequence for single slices
- Spin-echo sequence for multiple slices
- Inversion-recovery sequence to determine the longitudinal relaxation time, T1
- Saturation-recovery sequence to determine the longitudinal relaxation time, T1
- Multiecho sequence to determine the transversal relaxation time, T2

C. Spin-echo Sequence

Figure 2 shows the sequence protocol to record spin-echo images. The gradient fields are designated by X, Y, and Z. The emitted high frequency pulse is referred to as Tx, while Rx represents the position of the so-called interval of signal reception in the time sequence. In the beginning the whole specimen is stimulated by a nonselective, rectangular (90°) hard pulse. Switching on the Y gradient with a specific incrementation phase codes the MR signal. Switching on the Z gradient during the selective 180° soft pulse (e.g., Hermite- or Gauss-shaped pulse) causes the refocusing of the spins in only one particular slice. The X gradient is the readout gradient and determines the axial resolution. The received nuclear resonance signal provides information about the spatial location in x and y direction through the frequency and the phase of the signal. This spin-echo sequence can be used for either a single slice (single-slice imaging) or for several slices (multislice imaging with 4, 8, 16, and 32 slices). For the latter technique the stimulation sequence with slice-selective pulses has been combined in such a way that overlap effects are kept to a minimum. While one already stimulated slice relaxes, another is stimulated. This allows a more efficient use of the extinction time periods which occur because the nuclei have to relax completely in longitudinal magnetization prior to stimulation. For multislice imaging with eight slices for example the sequence 1, 3, 5, 7, 2, 4, 6, 8 is chosen[25] in order to reduce extinctions due to overlapping of the slice profiles.

D. Determination of the Longitudinal Relaxation Time, T1

The relaxation time, T1, characterizes the speed with which the longitudinal magnetization returns to its resting value after stimulation with a 90° pulse. This time is determined by the following measurement techniques.

1. Inversion-Recovery Sequence

At the beginning a 180° pulse is applied which causes an inversion of the magnetization sum vector (Figure 3). This 180° pulse is therefore also called inversion pulse. The sum vector pointing in a negative z direction returns to its original orientation within the magnetic ground field during relaxation. To measure the amplitude of the magnetization vector after a particular relaxation time period, the z magnetization is turned in the x-y plane by a 90° pulse. The interval between the two applied pulses is called inversion time. The received signal is used for grey-level modulation according to the amplitude of the z-magnetization sum vector.

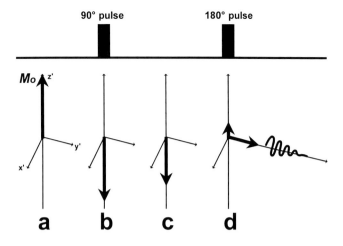

FIGURE 3 Inversion-recovery sequence to determine T1-tissue constants: (a) balance, (b) inversion, (c) recovery, (d) measurement. The time elapse between the 180° pulse and the 90° pulse is called inversion time. The 90° pulse turns the magnetic sum vector in transversal direction to measure its amplitude (d).

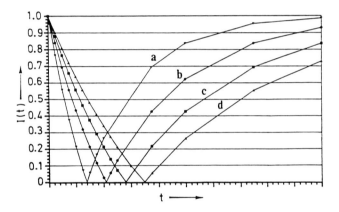

FIGURE 4 Substance-specific temporal (x axis) changes of the signal amplitude of different human tissues (y axis), assessed by inversion-recovery sequence: (a) adipose tissue, (b) white matter of the brain, (c) grey matter of the brain, (d) liquor cerebri. (Modified from Schmiedl, U., Kölbel, G., and Griebel, J., Röntgen praxis, 38, 352–356, 1985. With permission.)

By evaluating the mean grey value within the same area (corresponding to the amplitude of the magnetization sum vector in a particular direction) of an image series registered at different inversion times, a tissue-specific T1 relaxation curve can be plotted. In Figure 4 the x axis represents the different inversion times and the y axis represents the grey value within the same area of the tissue. As only the absolute value of the z magnetization, but not its sign (e.g., spacial orientation) is measured, the tissue first appears darker with increasing inversion time and then becomes whiter while rising asymptotically to a fixed value (Figure 4). If the z magnetization was displayed with regard to its sign, the first part of the curve in Figure 4 would have to be mirrored along the abscisse (e.g., Figure 11).

2. Saturation-Recovery Sequence

The second method which we used to measure the longitudinal T1 relaxation time is the aperiodic saturation-recovery technique. Fast application of several 90° hard pulses and a slice-selective 180° hermitian-shaped pulse (echo time 20 ms, repetition time 100 to 5000 ms) leads to a saturation of the magnetization sum vector. To measure amplitude the magnetization sum vector in a particular direction (i.e., the inversion-recovery technique) a rectangular pulse is emitted to turn the magnetization sum vector into the xy plane. The received signal is processed in a similar way as in the inversion method. The main advantage of the saturation-recovery in comparison to the inversion-recovery procedure is that due to the saturation of the magnetization no waiting time is needed between the successive pulses and thus the measurement time can be markedly reduced.

E. Determination of the Transversal Relaxation Time, T2

Like in the spin-echo imaging technique a 180° pulse is superimposed on a previously applied 90° pulse. A spin-echo signal is emitted during refocusing. In order to obtain an increasing T2 weighting in multi-echo images, the 180° pulse is repeatedly emitted several times after the echo time. In the practice the single echoes are not registered separately, but integrated while the sequence 180° pulse — echo time — 180° pulse — echo time is applied in succession.

F. Voxel Reconstruction

We used a 3D spin-echo sequence to acquire an isotropic 128^3 data cube. The pixel resolution was between $(40\ \mu m)^3$ and $(20\ \mu m)^3$. The echo time was 5.5 ms, the repetition time 60 ms. The 3D spin echo method makes it possible to shorten the echo time by the use of hard pulses. This minimizes the signal loss due to rather short $T2_{eff}$ relaxation times often observed in small probes.

The 3D visualization software enables reconstruction of sectional 2D images in oblique planes. 3D-image presentation provides the possibility of spacial reconstructions by extracting different structures from the data cube using an image-processing workstation (X32, Bruker GmbH, Rheinstetten, FRG).

III. Results

As mentioned before, MR images are nondestructive sections through the whole tissue block. Using a defined slice thickness, the MR slices could be located within the tissue block. After having precisely oriented the histological section plane in comparison to the MR images, the histological sections were used to compare the morphological aspect and the visual appearance of different skin structures.

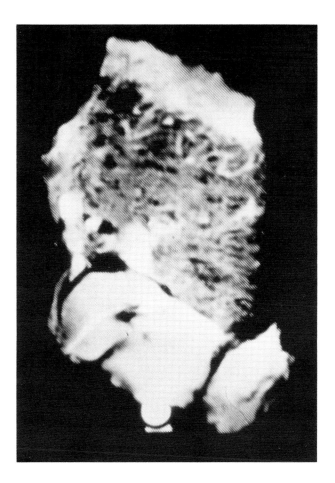

FIGURE 5 Superficial basal-cell carcinoma from the back of an 81-year-old man. (a) One out of 16 images from a spin-echo-sequence multi-slice imaging. Due to the relatively thick slice there is superposition of basal cell carcinoma nests.

A. Basal-cell Carcinoma

Figure 5a shows the MR image of a sklerodermiform basal cell carcinoma from the back of an 81-year-old patient. Figure 5b exhibits the corresponding histology. In the spin-echo image (Figure 5a) three skin regions of different texture are evident. In the upper part of the picture there is a light zone which is distinctly demarcated from a darker, more inhomogeneous, broad area in the middle part. From the upper light zone several round globules connected with it project into the darker area underneath. The lower part of the image consists of an irregular, very light region which is partly separated by dark band-like structures. The correlating histological section demonstrates that the the regions shown in the spin-echo image correspond to the epidermis, dermis, and subcutaneous tissue. Especially along the right upper edge of the histological picture epidermal buds of basaloid cells are apparent which project into the papillary dermis. They correspond to the globules in the spin-echo image which project from the light upper into the darker middle zone. In the lower part of the histological picture the subcutaneous fat is visible with its typical lobular structure and connective tissue septae inbetween.

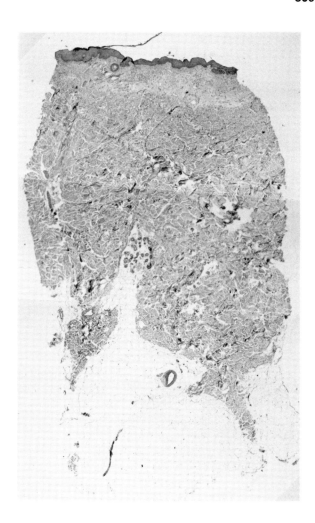

FIGURE 5 (b) Correlating histology exhibits an only moderately thickened epidermis with basal cell carcinoma nests discharging into the corium.

The T1 relaxation times were determined in this basal-cell carcinoma by an inversion-recovery sequence. Figure 6 shows the tumor at eight different inversion times. Remarkably, epidermis, dermis, and subcutaneous tissue go at different times through the "dark point", i.e., the intersection point of their tissue-specific T1 relaxation curve with the abscissae. In the third picture of the eight-image sequence the subcutis is running through the dark point while the dermis and epidermis have the same lightness as in the spin-echo image. In the seventh picture epidermis and dermis are not visible due to an amplitude of zero for the z magnetization while the subcutis which has already fully relaxed appears as a light region. The last picture of the image sequence shows all tissue components in complete relaxation. By determining the intensity of the signal in the same picture location at different inversion times a tissue-specific relaxation curve can be plotted and the absolute T1 relaxation time calculated. In this basal-cell carcinoma we found a longitudinal relaxation time of 1783 ms for the tumor tissue, of 1030 ms for the corium, and of 437 ms for the subcutaneous fat.

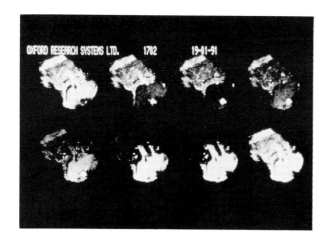

FIGURE 6 Inversion-recovery-sequence with 8 different inversion times to determine the longitudinal relaxation time, T1, of a superficial basal-cell carcinoma.

B. Nevocellular Nevus

Figure 7a shows the spin-echo image of a nevocellular nevus from the back of a 25-year-old patient. The corresponding histological section is shown in Figure 7b. The MR image exhibits an upper, wide, light area with several dimple-like indentations of its upper edge (Figure 7a, arrowheads) sharply demarcated from a lower, dark, inhomogeneous-appearing region. A cone-shaped structure with a dark loop inside projects from the upper into the lower part of the picture. The corresponding histology shows a dense aggregation of nevus cells within the thickened papillary dermis, correlating to the upper white part of the MR image. The surface of the epidermis shows several indentations. These are filled with retained horny material which is not visible in the spin-echo image due to its protone deficiency. The nevus extends into the dermis along a dilated hair follicle. The loop-shaped structure in the MR image represents a longitudinally cut hair and retained cornified material.

Figure 8 displays the time course of the longitudinal relaxation at 16 different measurement times. The T1 times were determined using a saturation-recovery sequence. As the specimen was oriented upside down during the measurement, the image was turned, so that the relaxation-curve values must be determined in opposite direction in the image sequence (from the right bottom to the left top). Like in the inversion technique the different lightness levels allow calculation of the absolute T1 relaxation times of different tissue areas. In this nevocellular nevus the values were 1365 ms for the tumor tissue, 1057 ms for the corium, and 441 ms for the subcutaneous fatty tissue lobules.

C. Malignant Melanoma

Figure 9 shows the spin-echo image and the correlating histological picture of a superficial spreading

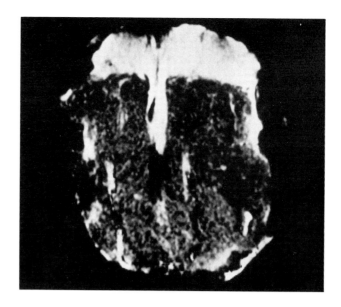

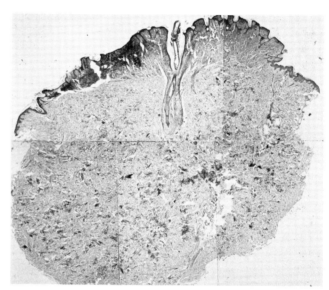

FIGURE 7 Nevocellular nevus from the back of a 25-year-old man. (a) Spin-echo-image from a multislice series. The nevus cell nests extended into the deeper dermis along a hair follicle as displayed also in the corresponding histological section (b).

amelanotic malignant melanoma from the scalp of a 74-year-old patient. The MR image exhibits three distinct regions. The left upper part contains a hemispheric structure which is lighter than its surroundings and reaches almost the middle of the picture. An incomplete, sickle-shaped, dark border separates this structure from a relatively dark region extending to the right side of the picture. This area contains several irregular, light, small nests in its right part. The whole lower part of the image shows a light zone in which irregular dark thread-like bands are embedded.

The corresponding histology (Figure 9b) shows underneath the ulcerated epidermis on the left a hemispherical tumor which extends in irregular formations into the subcutis. The tumor is surrounded by an

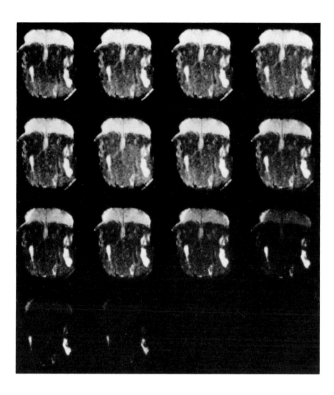

FIGURE 8 Saturation-recovery-sequence (16 different saturation times) in order to determine the longitudinal relaxation time T1 of the nevocellular nevus. Since the specimen was upside-down during data acquisition, the picture has been turned. Relaxation-curve values must therefore be determined from the right bottom image to the left top image.

FIGURE 9 (b) The histology exhibits that the hemispheric structures correspond to tumor formations.

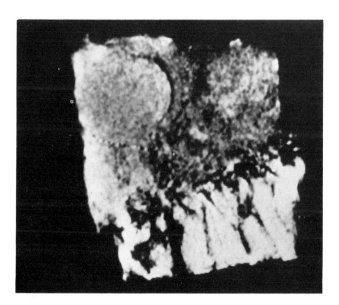

FIGURE 9 Amelanotic malignant melanoma from the scalp of a 74-year-old man. (a) Spin-echo image from a 16-slice sequence.

inflammatory infiltrate around dilated blood- and lymph-capillaries and by edematously swollen connective tissue. The dermis in the right side of the section is interspersed with tumor proliferates extending discontinually from the epidermis to the deep dermis. The subcutis underneath is separated by irregular fibrous septae. Approximately in the middle of the subcutis a cross-section through a blood vessel is apparent.

The longitudinal relaxation times for this malignant melanoma were calculated with the saturation-recovery sequence. Figure 10 shows the tumor at 16 different saturation times. The subcutaneous fat is the first structure to reach its final lightness in the image sequence which means it relaxes first. The partly preserved epidermis relaxes next and then follow tumor conglomerates and the rest of the corium. The following T1 relaxation times were evaluated: 1672 ms for the tumor tissue, 1229 ms for the epidermis, 1122 ms for the corium, and 427 ms for the subcutaneous fatty tissue. The measured relaxation curves for the epidermis, the dermis, the subcutaneous fat and the malignant melanoma are displayed in Figure 11.

Figure 12 exhibits the same malignant melanoma at different transversal relaxation time periods, which were registered using a multiecho sequence. While epidermis, tumor, and dermis are not visible after 66.7 ms (third picture of the image sequence), the subcutaneous fatty tissue is apparent even in the last image. Evaluation of the T2 relaxation times of the different tissues results in 29.75 ms for the epidermis, 23.5 ms for the dermis, 57.5 ms for the subcutaneous fat, and 29.3 ms for the malignant melanoma.

T1 and T2 relaxation times were also measured in normal melanotic malignant melanoma and have been integrated in Table 1.

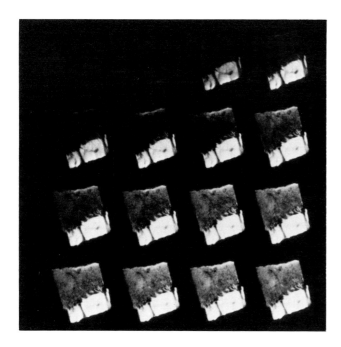

FIGURE 10 Image sequence at 16 different saturation times to determine the longitudinal relaxation time T1 of a malignant melanoma.

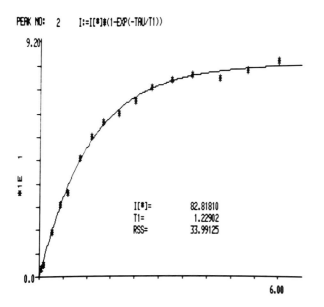

FIGURE 11 T1-relaxation curves of different skin layers. (a) epidermis.

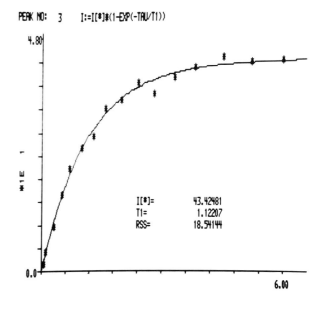

FIGURE 11 (b) Corium.

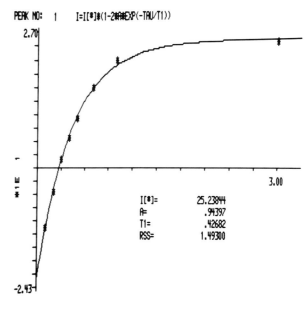

FIGURE 11 (c) Subcutaneous fatty tissue.

IV. Discussion

The figures demonstrate the very good correlation between MR image and histological picture. This comparison enables interpretation of MR images to obtain basic tissue data of different skin structures. It has to be considered however that the thickness of the MR sections is 300 μm as opposed to the 7-μm thick histological sections. In MR microscopy, therefore, skin structures appear more blurred, for example, basal-cell carcinoma nests are represented as globules. Moreover artifacts cannot be fully avoided both with MR due to technical difficulties and with histology due to varying shrinkage of the tissue during the paraffin embedding and tissue processing. Our results indicate that especially the subcutaneous fatty tissue lobules can be well differentiated from adjacent structures and therefore can serve as an orientation mark in the MR image. Earlier investigations by other research groups revealed that in computer tomograms there are no sex differences concerning the adipose tissue[11] and that this tissue type does not change with age.[19] Dooms et al. concluded that the adipose tissue would be suitable as a reference tissue in MR imaging and proved this assumption on a group of 78 patients.

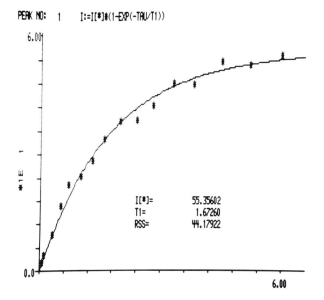

FIGURE 11 (d) Tumor tissue of the amelanotic malignant melanoma. The curves display different ascents, different intersection points with the X-axis and different end values at complete relaxation.

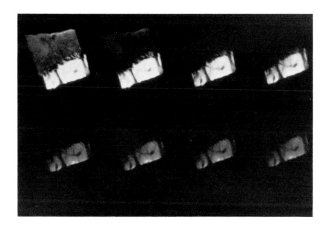

FIGURE 12 Multiecho sequence with eight images to calculate the transversal relaxation time T2 of the amelanotic malignant melanoma.

It has been relatively difficult to evaluate the epidermis in normal skin,[13] which is only partly due to thickness variations of the different skin layers depending on the body region. The physiological skin surface water evaporation particularly reduces the spin-echo signal in addition to the rather low proton content of the normal stratum corneum. This phenomenon was already observed in the verruca seborrhoica using MR microscopy.[13] Salter et al.[36] also had difficulties in visualization of the epidermis *in vivo*. They solved the problem by occluding the upper skin surface with petroleum gel to reduce skin surface evaporation significantly prior to MR-imaging. Using this procedure, they were able to visualize the upper layers of the stratum corneum. They also observed significant changes in the appearance of the stratum corneum during the evaporation process *in vivo*. Salter et al.[36] concluded, that the stratum corneum exhibits two different layers, which appear to be different, when the skin is hydrated. We were not able to visualize these two different layers. Since we did our examinations *ex vivo*, it is possible that this barrier had already broken down.

Some skin tumors increase the thickness or change the properties of the epidermis to such an extent that the epidermis can be easily visualized *ex vivo*. The basal-cell carcinoma, a tumor which develops from the epidermis is represented as a broad light zone in the upper corium. The single basal cell tumor nests appear as globule-shaped structures connected to the epidermis and can be quite well delimited from the adjacent corium. These aggregates of tumor cells specific for a basal-cell carcinoma could be exactly localized in the MR image and correlated with histological findings in recent investigations.[13,14] Likewise, the nevocellular nevus can be precisely localized in the spin-echo image. Correlating with the histological picture it is clearly demarcated from the corium below. The expansion of the tumor along the hair follicle is represented in the MR image as a cone-shaped structure and the hair follicle with the sebaceous gland can be separated from the surrounding corium. Protone-poor structures like hairs, retained horny material[13] are not visible. That the hyperkeratotic stratum corneum is not displayed has also been observed in investigations of verrucae seborrhoicae.[13]

In MR-microscopical images the normal corium has a quite irregular texture due to the skin appendages.[13] T1 and T2 values of the normal healthy corium vary depending on the region evaluated (e.g., due to density variability of skin appendages in different body regions). This can make it difficult to differentiate small tumorous processes in the corium. Homogenous tumor aggregates, such as the amelanotic malignant melanoma, are on the other hand easily detected as hemispheric regions with a quite homogenous texture in spin-echo images (Figure 9). In this situation, texture analysis could be of help to determine the morphological borders of the tumor.[17] Despite the differences in lightness and texture it was difficult in some melanocytic tumors to exactly determine the border between tumor masses and the surrounding corium in the MR image. In these cases it was sometimes helpful, that the peritumoral inflammatory infiltrate with edematous swelling of the connective tissue was visualized as a darker sickle around the tumor in the MR image.

A. MR Imaging and High-Resolution MR Microscopy

In 1973 Lauterbur published a procedure based on nuclear spin resonance, which enabled him to display two water-filled capillaries.[26] One year later he succeeded in visualizing a cross-section image of the

TABLE 1 Comparison of the Longitudinal Relaxation Time T1 (In Milliseconds) As Obtained by Different Research Groups

	Damadian et al.[10]	Yassine et al.[40]	Richard et al.[33]	Our results
Epidermis		200 ± 20	887 ± 92	1180 ± 100
Corium		200 ± 15	870 ± 143	1100 ± 80
Subcutis		±	393 ± 34	400 ± 40
Total skin	616 ± 19			±
Basal cell carcinoma	1236			1800 ± 100
Nevocellular nevus				1300 ± 70
Malignant melanoma	738			±
Amelanotic				1650 ± 80
Melanotic				1350 ± 70

Note: Important is the relation between the measured values within every column of the table, since the absolute values are specific to the instrument setup (e.g., calibration) and depend on the intensity of magnetic ground field.

thorax of an anesthetized mouse, this being the first MR image of a living organism.[27]

The first microscopic images using the MR technique were presented by Aguayo et al. in 1986.[1] Application of MRI *in vivo* in dermatology has only been possible since special surface coils have become available.[23] These coils have already been known in ophthalmology, where they were used to detect for example uvea melanomas. Querleux et al.[32] tried to analyze the different skin layers with MRI using a magnetic ground field of 0.1 T. They stated that this low magnetic ground field is not very well suitable to obtain images with high resolution. Bittoun et al.[4] used particularly strong surface coils for skin analysis. Consistent with our results they found that the epidermis is visualized lighter than the corium in spin-echo images. The subcutaneous fat can be distinguished from other tissues by its lightness due to its short-T1 and long-T2 relaxation time. According to our observations we cannot agree with the assumption that the multiple, almost parallel located, light inclusions in the dermis represent hair follicles,[4] neither could we differentiate the stratum corneum within the epidermis. There has been no confirmation of these hypotheses neither by histology nor any other established imaging technique.

Recently MRI has also been used to detect metallic foreign bodies in skin wounds, taking advantage of their ferromagnetic qualities.[3] The first application of MRI to assess pathologic skin changes was described by Jorizzo et al.[24] who investigated the blue-rubber-bleb-nevus syndrome. In general the examination of pathological skin alterations has focused mainly on pigmented skin tumors, especially malignant melanomas and their metastases. Mafee et al.[29] found that uvea melanoma metastases in the brain and the eye behave heterogeneously, being light in T1-weighed and dark in T2-weighed images compared to the adipose tissue. This is consistent with Zemtsov's observation concerning pigment tumors of the skin that melanin-containing lesions appear very light in T1-weighed images.[41] Schwaighofer et al.[37] in contrast described melanoma metastases in T1-weighed images as dark and in T2-weighed images as light compared to the reference tissue.

We can conclude that currently there are different points of view about the appearance of malignant melanomas and their metastases in MR images, possibly due to different melanin content of the tumors on the one hand and the different technical setup on the other hand.

B. Tissue Differentiation by Determination of Relaxation Times

Table 1 shows a comparison of our results with the values published in the literature. Damadian observed that the T1 relaxation times in tumor tissue (Novikoff-hepatoma and Walker-sarkoma of the laboratory rat) are longer than in tumor-free tissue,[9] which motivated him to investigate the T1 relaxation time of 106 different tumors using MR spectroscopy.[10] Yassine et al.[40] measured the relaxation times for the different skin layers on the forearm and finger using NMR imaging with a magnetic ground field of 0.1 T. Richard et al.[33] obtained measurements of the lower leg skin using 1.5 T.

One should compare only the relation of the T1 measurements between the different columns as the absolute values are specific for the instrument and depend on the intensity of the magnetic ground field (Figure 1). Despite the high resolution and the rather small examined volume our results are consistent with the measurements of other investigators. T1 relaxation time differences between different malignant melanoma types are rather striking. Amelanotic malignant melanomas revealed with 1650 ms rather long longitudinal relaxation time, while melanotic malignant melanomas exhibited T1 values of 1350 ms which lie only slightly above the values measured in nevocellular nevi. These findings suggest, that the relaxation time is influenced by the pigment concentration of the tumor.

Our *ex vivo* measurements were accompanied with difficulties due to the fact that the determination of the T1 relaxation time, especially the inversion-recovery

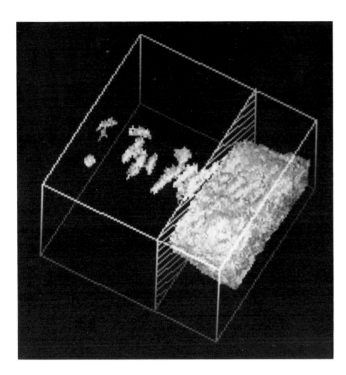

FIGURE 13 3D spin echo of normal skin (128^3 isotropic voxel data set). Surface reconstruction of the skin sample (right) and of different skin appendages (left).

experiment, needed several hours. To delay autolytic processes which cause a prolongation of the T1 and T2 relaxation time,[20] we briefly fixed the specimens prior to the measurement. However it has been described that fixation influences the relaxation times as well, causing a decrease in T1 time.[20]

Despite these difficulties we cannot confirm the statement of Zimmer et al.[42] that a differentiation between malignant and benign tumors according to their relaxation times is impossible. We therefore prefer to modify this statement in that sense, that MR imaging can deliver some new pieces (T1 and T2 values, shape, and texture) for the puzzle making up the diagnosis.

3D reconstruction procedures finally reveal additional information about skin structures and their orientation in space.[18] Fast 3D spin-echo methods can be used to register 3D data cubes for voxel reconstruction. This additional information can be of particular interest to identify and separate normal skin structures, such as skin appendages, from pathological processes in the corium and subcutaneous fatty tissue using MRI. Figure 13 demonstrates the extraction of the outer surface of a biopsy of normal skin (right) and of skin appendages (mainly hair follicles) within the biopsy (left) from such a 3D data cube using MR microscopy *ex vivo*.

V. Outlook

According to Zemtsov et al.[41] an appropriate imaging technique for the diagnosis of skin tumors does not exist currently. However he judges MR imaging to be a good candidate. Perednia[30] is more reserved about the rentability of MR imaging in its present stage for dermatology. Apart from the high technical requirements and the time necessary for picture acquisition as compared to the much faster histological diagnosis, the enormous costs have to be considered as well.[30] On the other hand, this imaging technique has, like sonography, the advantage that no ionizing radiation is applied and thus all health risks connected with this are avoided.[21] Roth judges the NMR technique as completely harmless for the human health according to our current knowledge.[35]

We believe that MR microscopy using local surface coils and gradient fields placed directly on the skin could potentially become an important noninvasive diagnostic tool in order to confirm the diagnosis in doubtful cases and to facilitate in difficult cases the exact preoperative staging of skin tumors.

References

1. Aguayo JB, Blackband SJ, Schoeniger J, Mattingly MA, Hintermann M (1986) Nuclear magnetic resonance imaging of a single cell. *Nature* 322: 190–191.
2. Altmeyer P, Hoffmann K, el-Gammal S (1990) Allgemeine dermatologische Ultraschallphänomene. *Hautarzt* 41 (Suppl.) 10: 124–129.
3. Bajanowski T, Hüttenbrink KB, Brinkmann B (1991) Detection of foreign particles in traumatized skin. *Int J Leg Med* 104: 161–166.
4. Bittoun J, Saint-Jalmes H, Querleux BG, Darrasse L, Jolivet O, Idy-Peretti I, Wartski M, Richard SB, Lévêque J-L (1990) In vivo high-resolution MR imaging of the skin in a whole-body system at 1,5 T. *Radiology* 176: 457–460.
5. Bloch F (1946) Nuclear induction. *Phys Rev* 70: 460–474.
6. Breslow A (1970) Thickness, cross-sectional areas and depth of invasion in the prognosis of cutaneous melanoma. *Ann Surg* 172: 902–908.
7. Breslow A (1975) Tumor thickness, level of invasion and node dissection in stage I cutaneous melanoma. *Ann Surg* 182: 572–575.
8. Clark Jr. WH, From L, Bernardino EA, Mihm MC (1969) The histogenesis and biologic behavior of primary malignant melanomas of the skin. *Cancer Res* 29: 705–726.
9. Damadian R (1971) Tumor detection by nuclear magnetic resonance. *Science* 171: 1151–1153.
10. Damadian R, Zaner K, Hor D (1973) Human tumors by NMR. *Physiol Chem Physics* 5: 381–402.
11. Dixon AK (1983) Abdominal fat assessed by computer tomography: sex difference in distribution. *Clin Radiol* 34: 189–191.
12. Dooms GC, Hricak H, Margulis AR, de Geer G (1986) MR imaging of fat. *Radiology* 158: 51–54.
13. el-Gammal S, Aygen S, Hartwig R, Bauermann T, Hoffmann K, Altmeyer P (1992) NMR-Mikroskopie der menschlichen Haut. *H G Z Hautkr* 67: 114–121.
14. el-Gammal S, Hartwig R, Aygen S, Bauermann T, Hoffmann K, Altmeyer P (1993) NMR-Mikroskopie und Gewebsdifferenzierung von Hauttumoren am Beispiel des Basalioms. in: Petres D and Lohrisch (eds.) *Das Basaliom und verwandte Tumoren*. Springer-Verlag, Berlin, pp. 99–114.

15. el-Gammal S, Hoffmann K, Auer T, Höβ A, Altmeyer P, Ermert H (1990) Computergestützte sonographische (50 MHz) Gewebsdifferenzierung von Hauttumoren. *Zbl Haut Geschlkr* 158: 105.
16. el-Gammal S, Hoffmann K, Auer T, Korten M, Altmeyer P, Höβ A, Ermert H (1992) A 50-MHz high-resolution ultrasound imaging system for Dermatology. In: Altmeyer P, el-Gammal S, Hoffmann K (eds.) *Ultrasound in Dermatology.* Springer-Verlag, Berlin, pp. 297–322.
17. el-Gammal S, Hoffmann K, Höβ A, Hammentgen R, Altmeyer P, Ermert H (1992) New concepts and developments in high-resolution ultrasound. In: Altmeyer P, el-Gammal S, Hoffmann K (eds.) *Ultrasound in Dermatology.* Springer-Verlag Berlin, pp. 399–442.
18. el-Gammal S, Hoffmann K, Kenkmann J, Altmeyer P, Höss A, Ermert H (1992) Principles of three-dimensional reconstructions from high-resolution ultrasound in dermatology. In: Altmeyer P, el-Gammal S, Hoffmann K (eds.) *Ultrasound in Dermatology.* Springer-Verlag, Berlin, pp. 355–387.
19. Giloteaux S, Linz MH (1983) Histology of aging: adipose tissues. *Gerontol Geriatr Educ* 4: 107–111.
20. Grodd W, Schmitt WGH (1983) Protonenrelaxationsverhalten menschlicher und tierischer Gewebe in vitro, Änderungen bei Autolyse und Fixierung. *Fortschr Röntgenstr* 139: 233–240.
21. Hausser KH, Kalbitzer HR (1989) *NMR für Mediziner und Biologen Strukturbestimmung, Bildgebung, In-vivo-Spektroskopie.* Springer-Verlag Berlin, p. 175.
22. Hoffmann K, el-Gammal S, Altmeyer P (1990) B-scan Sonographie in der Dermatologie. *Hautarzt* 41: W7-W16.
23. Hyde JS, Jesmanowicz A, Kneeland JB (1987) Surface coil for MR Imaging of the skin. *Magn Res Med* 5: 456–461.
24. Jorizzo JR, Amparo EG (1986) MR imaging of blue rubber bleb nevus syndrome. *J Comput Ass Tomogr* 10: 686–688.
25. Kuhn W (1990) NMR-Mikroskopie — Grundlagen, Grenzen und Anwendungsmöglichkeiten. *Angew Chem* 102: 1–20.
26. Lauterbur PC (1973) Image formation by induced local interactions — examples employing nuclear magnetic resonance. *Nature* 242: 190–191.
27. Lauterbur PC (1974) Magnetic resonance zeugmatography. *Pure Appl Chem* 40: 149–157.
28. Longmore DB (1989) The principles of magnetic resonance. *Brit Med Bull* 45: 848–880.
29. Mafee MF, Pegman GA, Grisolano JE, Fletcher ME, Spigos DG, Wehrli FW, Rasouli F, Capek V (1986) Malignant uveal melanoma and simulating lesions: MR imaging evaluation. *Radiology* 160: 773–780.
30. Perednia DA (1991) What dermatologists should know about digital imaging. *J Am Acad Dermatol* 25: 89–108.
31. Purcell EM, Torrey HC, Pound RV (1946) Resonance absorption by nuclear magnetic moments in a solid. *Phys Rev* 69: 37–38.
32. Querleux B, Yassine MM, Darrasse L, Saint-James H, Sauzade M, Lévêque J-L (1988) Magnetic resonance imaging of the skin. A comparison with the ultrasonic technique. *Bioeng Skin* 4: 1–14.
33. Richard S, Querleux B, Bittoun J, Idy-Peretti I, Jolivet O, Cermakowa E, Lévêque J-L (1991) In vivo proton relaxation times analysis of skin layers by magnetic resonance imaging. *J Invest Dermatol* 97: 120–125.
34. Richter S (1976) Wolfgang Pauli und die Entstehung des Spin-Konzepts. *Gesnerus* 33: 253–270.
35. Roth K (1984) *NMR-Tomographie und -Spektroskopie in der Medizin — Eine Einführung.* Springer-Verlag, Berlin, p. 108.
36. Salter DC, Hodgson RJ, Hall LD, Carpenter TA, Ablett S Moisturization processes in living human skin studied by magnetic resonance imaging microscopy. IFSCC (Yokohama, Japan), pp. 587–595.
37. Schwaighofer BW, Fruehwald FXJ, Pohl-Markl H, Neuhold A, Wicke L, Landrum WL (1989) MRI evaluation of pigmented skin tumors. *Invest Radiol* 24: 289–293.
38. Schmiedl U, Kölbel G, Griebel J (1985) Begriffe der medizinischen Kernspintomographie, Teil 3: Die Kontrastmechanismen und der Einfluß der biologischen Parameter auf das MR-Bild. *Röntgenpraxis* 38: 352–356.
39. Wehrli FW (1988) Principles of magnetic resonance. in: Stark DD, Bradley WG (eds.) *Magnetic Resonance Imaging.* The CV Mosby Company, St. Louis, pp. 3–23.
40. Yassine MM, Darrasse L, Saint-James H, Sauzade M, Querleux B, Lévêque J-L (1987) In vivo skin imaging at O,1 Tesla. Book of abstracts, *Soc Magn Res Med* 1: 466.
41. Zemtsov A, Lorig R, Bergfield WF Bailin P L, Ng TC (1989) Magnetic resonance imaging of cutaneous melanocytic lesions. *J Dermatol Surg Oncol* 15: 854–858.
42. Zimmer WD, Berquist TH, McLeod RA, Sim FH, Pritchard DJ, Shives TC, Wold LE, May GR (1985) Bone tumors: magnetic resonance imaging versus computed tomography. *Radiology* 155: 709–718.

Section E: Function of the Dermis

14.0

Mechanical Properties of the Skin

14.1 Twistometry Measurement of Skin Elasticity 319
     P.G. Agache

14.2 Suction Chamber Method for Measurement of Skin Mechanical
     Properties: The Dermaflex® 329
     M. Gniadecka and J. Serup

14.3 Suction Method for Measurement of Skin Mechanical Properties:
     The Cutometer® 335
     A.O. Barel, W. Courage, and P. Clarys

14.4 Identification of Langer's Lines 341
     J.C. Barbenel

14.5 Levarometry 345
     V. Manny-Aframian and S. Dikstein

14.6 Indentometry 349
     V. Manny-Aframian and S. Dikstein

14.7 The Gas-Bearing Electrodynamometer 353
     C.W. Hargens

14.8 Ballistometry 359
     C.W. Hargens

Chapter 14.1
Twistometry Measurement of Skin Elasticity

Pierre G. Agache
Department of Functional Dermatology
University Hospital
Besançon, France

I. Introduction

For the two last decades, numerous attempts were made to noninvasively assess the human-skin mechanical behavior *in vivo*. This was felt to be a useful way to allow a more precise follow-up of the numerous diseases or skin states characterized by an abnormal skin induration of softening and to estimate the efficacy of treatments or cosmetics. To attain this goal devices were constructed which operate either by inducing a skin deformation and recording the resisting force or by putting a load and assessing the resulting deformation. Four directions of loads are conceivable: a vertical pressure, a vertical suction, a linear horizontal traction, and a torsion in the horizontal plane.

Mechanical stimuli perpendicular to skin surface have the disadvantage of involving the subcutis at least in part. This layer has wide differences in thickness or in fat content and consequently would have widespread mechanical properties. Stimuli in the horizontal plane, on the other hand, use probes stuck on the skin surface and, if the displacement is small, may be expected to only involve epidermis and dermis. Unidirectional stresses should consider the skin mechanical anisotropy. But Langers' lines are not easy to find and most of them are oblique to the body or limb axis and thereby difficult with which to comply.[1] Also the sampled area is poorly delimited with such devices. All these reasons prompted some authors to use torsional devices made of a central rotating disc and a peripheral fixed ring, thus allowing a skin narrow annulus to be twisted.

Vlasblom[2] from Utrecht was the first to use such a device and made a theoretical study of forces and deformations implicated. Finlay[3,4] from the Strathclyde group (Glasgow) extended the investigation and applied the technique to the assessment of changes in mechanical behavior with aging. Since then the l'Oreal group (Paris) conducted by Jean-Luc Lévêque, has been using torsional devices extensively and contributed a great deal to the development of the technique.

II. Technical and Theoretical Considerations

Before using or interpreting the results of a torsional test, some elements of the relationships between stress and deformation should be recalled and also the requirements necessary for an experiment to be mechanically valid.

A. Equipment

Torsional equipment acts through a disk glued to the skin, which is rotated by a motor powered by a controlled voltage, thereby loading the peripheral skin with a torque, the value of which can be adjusted. Under this torque the skin glued under the disc moves with it, supposedly without any brake from the subcutaneous tissue. The skin around the disc is elongated in a twisting way. This is the mechanical behavior of this part of the skin which is assessed.

There are two types of twists to be considered whether there is or is not a peripheral guard ring concentrical to the disc, also glued to the skin and immobile during the disc rotation, thereby delimiting an annular area of skin submitted to elongation (Figure 1). When there is no guard ring,[5] the skin peripheral to the disc is implicated up to an unknown distance. As it is much less extensible than the subcutis, the latter will provide for most of the twisting strain. By contrast when there is a guard ring the shape and limits of the twisted area are known, the sliding of the skin over the subcutis is limited, and the essential part of the twisting strain takes place within the skin itself.

Finlay's[4] and Jaskowski and Maceluch's[6] devices were constructed to give repeated twists at adjustable frequencies and adjustable increasing or decreasing rates. The first one was also equipped by a strain gauge which recorded the torque generated by the skin under a constant twist and assessed force relaxation. Lévêque et al.'s[7] device (Figure 2) has a guard ring integrated to the body of the apparatus. It delivers an analogue signal directly proportional to the disc rotational angle. In a

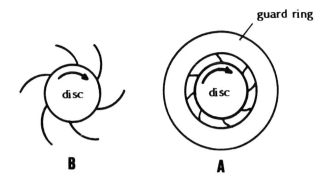

FIGURE 1 Skin deformation produced at the periphery of a rotating disc with (A) and without (B) a guard ring.

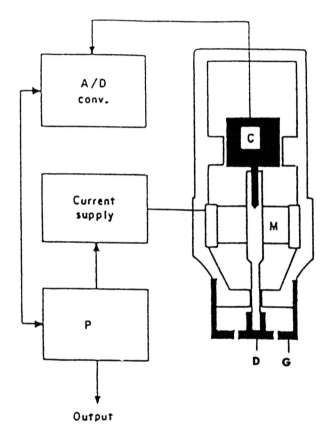

FIGURE 2 Lévêque's twistometer diagram. (C) Rotational sensor, (M) motor, (D) disc, (G) guard ring, (P) microprocessor. Both guard ring and disc are removable.

newer version the signal is digitalized and a microprocessor both computes the main parameters and controls the measurement phases. The applied torque can be chosen between 4 and 57 mN · m, the width of the crown of skin submitted to torque is either 1, 3, or 5 mm, and the disc radius is 18 or 25 mm. The equipment is now commercially available (Dermal Torque Meter®, Diastron, Ltd., Andover, GB).

B. Mechanical Testing

Finlay's protocol[4] included a progressive increase in disc rotation, then a standstill to assess the skin torque

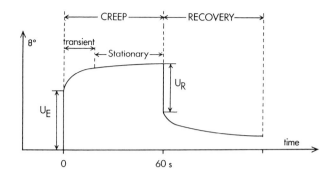

FIGURE 3 Skin angular deformation (Θ) versus time upon application of a constant torque. Ue : immediate deformation, Ur : immediate recovery upon torque switching off.

relaxation, then a progressive return of disc position to zero. He recorded both the strain and the resisting force induced in the skin during the first run and the following ones. Jaskowski and Maceluch[6] used repeated torque application up to a resonance frequency and recorded the torque oscillating amplitude, the optimum frequency, and the attenuation of torque by the skin.

The simple way to use torsional tests is to apply a constant torque for a couple of seconds and record the skin angular deformation (Figure 3). Upon torque application there is an immediate elastic deformation (Ue) followed by a creeping viscoelastic deformation (Uv). Torque suppression is associated with an immediate recovery (Ur) which is always incomplete. This is the way Lévêque's device works.[8] One or several runs can be programmed and preconditioned.

C. Mechanical Parameters

In vivo mechanical tests on the skin have two purposes. The first and main one is to quantitatively assess changes which are usually detectable by palpation but not measurable otherwise. The second aim is to get an access to the skin intrinsic structure as far as mechanical components are concerned and to the structure-function relationships of these components. The absolute parameters concerning the strength of skin elasticity and viscosity need to be derived from the assumptions that it is an homogeneous layer and is uniform in thickness, which is obviously wrong but is a much rewarding approximation.

1. Elasticity Parameters

In simple elongation tests the well-known equation of Young's modulus (E) is a commonly used way to express the stiffness at elastic phase.

$$E = \sigma (1 - \nu)/\varepsilon \qquad (1)$$

where σ is the stress (ratio of force to the section area submitted to force), ε is the strain (elongation-to-initial sample length ratio) and ν is the ratio of relative narrowing to strain (Poisson's ratio).

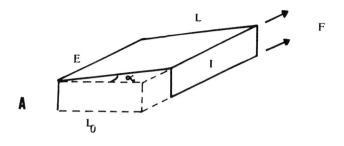

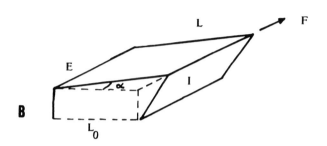

FIGURE 4 A diagram of skin annulus deformation in torsional experiments. (A) Deformation homogeneous through thickness. (B) Deformation predominant on the skin surface. Deformation gradients are supposed linear: (L_0) initial length (width of skin annulus); (L) length after elongation; (I) internal aspect of skin annulus, facing disc; (E) external aspect of skin annulus, facing guard ring, (F) force; (α) deformation angle.

In a torsional experiment the deformation is more complex because elongation is replaced by shear and is rotational. The skin can be deformed through its full thickness to the same extent (Figure 4). This occurs when the force is high enough to act in depth while only applied on the surface, and when the skin annulus is wide enough to allow the force gradient within the skin to fully reach the deeper layers. In that case, the skin is supposedly isotropic in torsion tests and the mechanical parameter is the shear modulus μ (Lamé's coefficient) given by the formula

$$\mu = \sigma (1 - \nu)/\alpha \qquad (2)$$

where α is the deformation angle, ν the Poisson's ratio, and σ the stress as calculated by the ratio of force (torque/radius of disc) to area submitted to torsion (skin thickness × disc perimeter). Agache et al.[9] have proposed the following formula

$$\mu = \frac{M}{2\pi e\, r_1 r_2 \Theta} \qquad (3)$$

where M is the torque momentum, r_1 the disc radius, r_2 the guard ring inner radius, e the skin thickness, θ the rotational angle of the disc at equilibrium. The Lamé's coefficient is usually close to 0.4 E therefore it is proposed to compute E by simply dividing the above formula by 0.4. Hence

$$E = \frac{1}{0.4} \times \frac{M}{2\pi e\, r_1 r_2 \Theta} \qquad (4)$$

If only E variation is considered, the same formula gives the following equation by differentiation:[10]

$$\Delta E/E = -(\Delta Ue/Ue + \Delta e/e) \qquad (5)$$

When the torque is low and mostly when the skin annulus is narrow, the superficial layers of the skin are more implicated than the deeper ones because the tangential force is applied on the skin surface (Figure 4). In that case the deformation has two components, a visible and measurable one in the superficial plane, and an invisible and immeasurable one in the vertical plane (the shear strain gradient in depth). It could be possible, from the deformation seen at the skin surface, to calculate both of these strain components and consequently the relevant more complex shear coefficient and corresponding Young's modulus equivalent. But in the present state of experimental conditions which are submitted to large variability in the results, obtaining such precision seems unrealistic or at least unnecessary.

2. Viscosity Parameters

The experimental protocol designed by Finlay[4] includes a typical stress-relaxation step. The observed half-time of skin torque relaxation during this step is inversely related to skin viscosity for a given initial torque. Accordingly, this time would change for another torque value at the start of the relaxation step.

As in an elongation test, in torsional experiments where a constant torque is applied, the deformation versus time, as shown in Figure 3, can be described by the following equation (Vlasblom).

$$U(t) = Ue + Uv(1 - e^{-t/\tau}) + At \qquad (6)$$

where Ue is the immediate deformation, Uv the viscous deformation (transient creep), and At a linear deformation following a longer time of torque application (stationary creep). Vlasblom showed that in the stationary creep the deformation is not proportional to time and suggested a constant term should be added to At. As this term is small compared to elastic deformation he thought possible to neglect it. Pichon, de Rigal and Lévêque,[11] using the finite differences method, undertook to determine the creep law without any influence of Ue and proposed the following equation for the creep deformation ε

$$\varepsilon = \varepsilon_0\, At^m \qquad (7)$$

In an experiment on six subjects with a 1 mm wide skin annulus they found m = 0.33 ± 0.03 and 0.335 ± 0.06 for 9 × 10⁻³ and 12 × 10⁻³ Nm torques respectively. Increasing the width of the annulus did not change the m value although increasing the interindividual variation. Accordingly they proposed the following equation for the creep.

$$\varepsilon = Uv(1 - e^{-t/\tau}) + At^{1/3} \qquad (8)$$

This law was also validated for the recovery part of the curve, and allowed a correct determination of the immediate recovery.

3. Skin Rheological Model

Both for better intuitive understanding of what occurs within the skin during traction and easier computing of mechanical parameters, rheological models of skin have been proposed. Wijn et al.[12] used Bürger's model for their uniaxial elongation experiments (Figure 5) Pichon et al. used the same model for skin torsional experiments under a constant force and using the above-quoted creep law (Equation 7). Accordingly the equation for the model is

$$\varepsilon(t) = \frac{\sigma}{k_0} + \frac{\sigma}{k_1}\left(1 - e^{-tk_1/\eta_1}\right) + \frac{\sigma t^{1/3}}{\eta_0} \qquad (9)$$

The three phases of experimental strain (Ue, transient creep, and stationary creep) correspond to each of the three terms of the equation. The characteristic parameters of the springs (k_0, k_1) and dashpots (η_0, η_1) should correspond to those of special structures or arrangements within skin, e.g., the elastic resistance of elastic fibers to elongate or collagen network to deform, and the viscous resistance to displacement.

All parameters of Bürger's model can be obtained from the *in vivo* experiment, as follows:

- Stress: σ = force/area submitted to stress (i.e., disc perimeter × skin thickness)
- Young's modulus E (i.e., first spring strength at the casual level of first spring tension): E = k_0 = σ/Ue.
- Second spring strength: k_1 = σ/Uv curvilinear (i.e., transient creep)
- Viscosity associated with the first spring: η_0 = σ/U_v $t^{1/3}$
- Viscosity associated with the second spring: η_1 = τk_1

This model looks satisfactory but has been used in only one published study.[11] Accordingly in the next sections the calculations relative to creep used only the second member of Equation 6.

D. Requirements for Test Validity and Correct Interpretation

Barbenel and Payne[13] in a report to the International Society for Bioengineering and the Skin have presented the requirements for a torsional testing to give interpretable and reliable results. Most of them should be recalled along with some additional warning.

1. Geometry of the Applied System

The area submitted to torque should be delimited by a guard ring in order (1) to allow the skin itself to be deformed and reduce to a minimum the skin sliding over the subcutis, and (2) to calculate the strain (i.e., deformation/initial length). The width of the twisted skin annulus should be narrow enough to prevent torsion of subcutis. In any case the disc diameter and inner guard-ring diameter should be quoted, and also the torque value.

The rotational angle θ should be small (inferior to 10°) and the radius r of the rotating disc should be large enough to allow the displacement rθ which is circular to be considered as linear.

2. Absence of Pressure Perpendicular to Skin Surface

As shown in Figure 6, such a pressure would change the mechanical behavior of the skin. In practice the applied pressure, if any, should always be mentioned.

3. Attachment of Disc to the Skin

Slipping or deformation of the attachment system should be avoided. Accordingly the mechanical behavior of the attachment system used should be tested beforehand at the same torque as in the experiment. This was the case for Lévêque's device, as tested on a steel plate,[7,8] and in Finlay's experiments.[4] Anyway the type of adhesive used should be indicated.

4. Rate of Torque Application

As skin is viscoelastic any application of the torque at low rate would allow both an elastic and viscous deformation to take place at the same time. Very low rate (quasistatic experiment) would reduce the viscous resistance to almost zero and consequently assess only

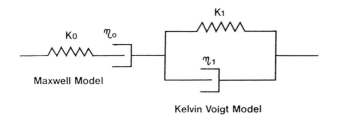

FIGURE 5 Bürger's model of skin mechanical behavior, and corresponding extension (k_0, k_1) and viscosity (η_0, η_1) parameters.

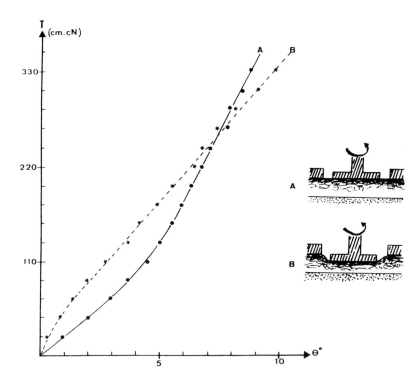

FIGURE 6 Experimental plot of Ue (Θ°) versus applied torque (T) under 11.25 kPa pressure (A) and under 24.52 kPa pressure (B) onto skin surface. Higher pressure induced an artifactual curve where the curvilinear relationship between stress and strain was no longer visible. (From Lévêque, J-L, et al., Arch. Dermatol. Res., 269, 221–232, 1980. With permission.)

the elastic resistance. On the other hand a very high deformation rate would overcome viscous resistances lower than the applied stress, and assess the elastic resistance through the measurement of immediate deformation. In any case the rate of applied torque should be mentioned.

III. Skin Mechanical Aging

A. Intrinsic Aging

The skin mechanical behavior over the life span, as assessed by torsional experiments *in vivo*, was first studied by Finlay[4] in the lateral aspect of forearm on a 4-mm-wide skin annulus using a 15-mm-diameter rotating disc inducing on the skin an 11-kPa pressure (Figure 6). The run consisted of progressively twisting the skin annulus (rotation of the disc 2° per second) up to about 20 mN · m torque, then maintaining the same twist level for 1 min before relaxing the twist down to zero at the same rate. During such a run the torque first rises up to a peak then subsides partially while the deformation is maintained (force relaxation) then falls to a negative value, i.e., the skin would have remained deformed if it had not been forced to come back to baseline.

The angular deformation associated with a 2 mN · m torque was felt by the author a good parameter of "what can be sensed by palpation". Other chosen parameters were the ratio peak-to-final torque (i.e., the end of relaxation) and the time for half relaxation indicating the relaxation intensity within 1 min, and rate, respectively. Only the angular deformation at low torque was found to significantly decrease with aging, indicating an increase in skin stiffness.

Three other runs were done after the first, each one at 1-min interval, and the same parameters pooled. Again only the angular deformation at low torque was significantly depressed with aging. Finally a fifth run in the reverse direction was done 1 min after the fourth run. Only the ratio peak-to-final torque was found significantly lower with aging (p = 0.05), indicating a decrease in force relaxation within the extended skin during this time interval.

As the skin components can be grossly compared to springs and dashpots connected both in parallel and in series (Figure 5), the stiffer skin and less relaxing forces inside skin may denote either a weakening of springs or strengthening of dashpots (viscosity) or both. Unfortunately the panel of subjects comprised a bulk of middle-aged adults and very few children or aged people, which may explain the small number of significant correlations found in this study.

Sanders,[5] using Vlasblom's apparatus and formulas, computed the torsional modulus of the outer forearm skin in 19 healthy subjects aged 6 to 61 years. The torque was 0.83 mN · m through a 8.7-mm-diameter disc without any guard ring, so that the twisted area kept undefined. The pressure on the skin induced by the disc was not specified. The immediate elastic deformation (Ue) was found to increase almost linearly with aging, and the delayed viscoelastic deformation (Uv) seemed to increase only beyond 40 years of age. From these data a torsional elastic modulus of 20 to 100 kPa was computed and found to decrease with aging. As the torque was very weak only elastic fibers were supposed to be implicated. Accordingly the decreased skin stiffness with aging was tentatively ascribed to the well-known decay of elastic fibers over the life span. However as said previously, the absence of guard ring let the subcutis bear the major (if not the entire) part of the deformation. As this layer is loose, this explains the very low Young's modulus observed and the increase with aging. These observed parameters were probably those of the subcutis, not of the skin.

Jaskowski and Maceluch[6] investigated the dynamic torsional behavior of forearm, forehead, and abdomen skin in 380 normal subjects aged 5 to 80 years. The crown of skin between disc and guarding was made vibrate by the rotating oscillation of the disc, up to a maximum amplitude (resonance). The observed parameters were the resistance, the frequency, and the attenuation. The skin stiffness (Nm/rad) rose steadily from 20 to 60 years then sharply beyond 60 years and attenuation (Nms/rad) followed the same increase. The resonance frequency and resistance slightly rose over the life span.

An investigation using a torsional device similar to Finlay's, i.e., with a guard ring, was undertaken by Lévêque's[8] and Agache's[9] groups in a series of experiments in 1980. In the two first studies on 141 subjects in the age range 3 to 89 years, they used a device made of a central disc 25 mm in diameter and a skin annulus 5 mm wide, and exerted a 12.6 kPa pressure on the skin. The protocol consisted of abruptly putting (within 15 ms) a set torque for 2 min on the volar forearm and recording the angular rotation (24 mV per degree). Two stresses were applied at some distance on the same forearm: 9 and 28.6 mN · m. Under the lower stress the immediate deformation Ue decreased by about 30% during the first two decades, kept stable until 60 years, then rose by about 10%. But the product of immediate deformation by half skinfold at the same site (as assessed by a caliper) showed a steady and significant decrease with aging. Surprisingly at almost any age this product was lower in females than in males indicating a higher Young's modulus under this tensile stress and accordingly a stiffer skin. Under the higher stress the immediate deformation Ue strongly decreased until the age of 30 then rose moderately, while the Young's modulus, as calculated by the above-quoted formula slightly decreased to a minimum in the twenties then rose progressively with aging. Again with this high torque the Young's modulus, i.e., the skin stiffness, was greater in females in all age classes beyond 20 years. The absolute values of this parameter lies in the same range for both stresses, between 0.6 and 2.9 MPa, but were statistically higher with the lower torque. The direction of the difference would favor a technical origin rather than a physiological one, as the higher torque was too slight to induce any damage to the skin and anyway gave no unpleasant sensation upon application.

This work was undertaken again 9 years later[14] using the same torsional equipment, but owing to the newly available technique of assessing skin thickness by ultrasound high-resolution imaging (25 MHz). The measurements were done on volar forearm set in the same position as in the previous paper and on a panel of 123 subjects. Ultrasound studies showed a thinner skin before 10 years and after 70 years of age, with nonsignificant variation in between and a thickness significantly greater (+16%) in males than in females in the whole range of ages. The skin annulus submitted to twist was reduced to 3-mm wide in order to minimize the possible implication of subcutis. Also the

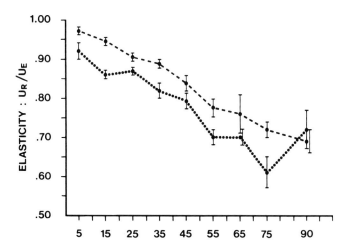

FIGURE 7 Skin elasticity (Ur/Ue) over the life span with 2.6 mN/m torque (dotted line) and 10.4 mN/m torque (dashed line). Bars indicate SEM. (From Escoffier C, et al., *J. Invest. Dermatol.*, 93, 353–357, 1989. With permission.)

applied torque values were reduced to 2.3 and 10.4 mN · m as compared to 9 and 28.6 mN · m in the previous paper, using a central disc of 18 mm diameter instead of 25 mm. No significant change over the life span nor differences between sexes were found for either immediate extension or creep. Only did the product Ue × thickness significantly ($p < 0.01$) decrease by about 85% beyond 70 years of age. By contrast the immediate recovery steadily decreased with aging and also the ratio Ur/Ue (Figure 7). This trend was highly significant ($p < 10^{-4}$). It confirms a current observation that skin folding keeps longer as age advances. The relaxation time of the creep decreased with aging; this was only significant with the higher torque. That means a less viscous dermis with advancing age, a data in accordance with the well-known decline in mucopolysaccharide content.[15]

The results of all above-quoted studies, as presented in Table 1, show an agreement for a decreased skin immediate extensibility with aging mostly beyond 60 years, and an increase in the viscous delayed extension. As the skin thickness decreases in the sixties, the result is an increase of the elastic modulus with aging beyond this age. Only Sander's data are discrepant with this trend. As stressed previously the main reason seems to lie in the absence of guard ring and this experiment was probably an assessment of the subcutis distensibility.

Other discrepancies include to the force relaxation time, i.e., the ability of structural elements inside the skin to move relative to each other. As this time is reduced, as shown in Agache's,[9] Lévêque's,[16] and Escoffier's[14] experiments, their mobility would increase with aging. The reverse was observed by Finlay.[4] An explanation could lie in the difference in the type of experiments. While Finlay investigated the mobility of elements during a maintained stretch (for relaxation

TABLE 1 Literature Data On Intrinsic Aging As Assessed By Torsional Tests

Author	Disc Diam (mm)	Torque (mN/m)	Width of Skin Annulus (mm)	Ue Range (mm)	Deformation with Aging		Elastic Modulus Range (MPa)
					Elastic	Viscous	
Finlay	15	2	4		↓		
Finlay	15	20	4		→		
Sanders	8.7	0.83	no	0.4–1.7	↑	↑	0.02–0.1
Lévêque	25	9	5	0.1–0.4	↓		0.85–2.94*
Agache	25	28.6	5	0.6–1.5	↓	→	0.64–2.04*
Escoffier	18	2.3	3	0.4–0.6	↓	↑	0.16–0.36*
Escoffier	18	10.4	3	1.1–1.2	↓	↑	0.34–0.62*

Note: Ue is immediate elongation. Asterisks indicate figures not produced by the authors but computed from Equation 4.

time), the two other papers dealt with the mobility during a maintained stress (deformation time). The former would implicate the strength of an interior spring attempting to recoil against viscosity, and the latter the resistance to a deformation through associated interior spring and viscosity. Accordingly the strength of the interior spring would diminish with aging while spring resistance to extension plus viscosity also diminish. So the discrepancy would be only apparent, not real.

B. Actinic Aging

A more specific study of the influence of sun exposure of skin mechanical aging, was conducted by Lévêque's group in 1988.[17] This witty investigation was made on a panel of 35 professional cyclists at the end of the period of intensive training in spring and early summer in often sunny countries (Spain, Italy, France) before entering the "Tour de France" competition. Almost all of them wore a short sleeved shirt dividing the outer aspect of their arm into a sun-covered and a sun-exposed area. This was evident by the suntan limit. As UVA rays do not penetrate the skin beyond the papillary dermis, the authors used a narrower ring (1 mm wide) with a 18-mm-diameter disc, in order to restrict the shear stress to the superficial layers of the skin. A 9-mN · m torque was applied for 10 s both on exposed and covered areas on the same arm each at 1 cm distance from the suntan limit. In exposed areas the rotation angle was reduced by 33% ($p < 0.0001$), and the skin thickness increased by 22.5% ($p < 0.0001$). As calculated by Equation 5 the mean Young's modulus moved up from 541 to 698 kPa, thus demonstrating a skin stiffening following repeated sun exposure. Also the relative immediate recovery was reduced by 22% ($p < 0.0001$) denoting a substantial loss of elasticity.

This work was extended by a study in 3 racial groups (15 blacks, 12 white, 12 hispanics) made on both ventral and dorsal forearm with a 15 mN · m torque, an 18-mm-radius disc and a free skin crown 3 mm wide.[18] The torque was set abruptly, maintained for 60 s, then abruptly suppressed. While no difference between races were observed in skin thickness (as assessed by ultrasound A-scan, 15 MHz), in all subjects the dorsal forearm skin was thicker. The skin elastic modulus E on the volar forearm was the same in all groups whereas in dorsal forearm it was statistically lower in blacks. Also in blacks there was no difference in E values between the two sides while the difference was significant in other groups. Viscoelasticity was lower in dorsal forearm but significantly only in whites and hispanics. Finally the clinical elasticity (recovery/extensibility) was lower on dorsal forearm except in blacks, while irrespective of groups and sites it steadily decreased with aging. In summary the volar forearm skin had the same mechanical behavior in all races. Differences appeared on dorsal forearm where blacks differ by the absence of alterations related to chronic sun exposure such as increase in Young's modulus (i.e., stiffness) and decrease in skin extensibility. Also a black-hispanic-white trend was noted on dorsal side for such alterations.

The results of the these two studies show that, mechanically speaking, long lasting and repeated sun damage resemble intrinsic aging in reducing skin elasticity and increasing Young's modulus. But the former is associated with a thicker skin while it is the contrary for the latter. Accordingly chronic actinic damage cannot be simply assimilated to a premature aging.

IV. Stratum Corneum and Skin Biomechanics

From a mechanical standpoint the skin is a stratified composite material whose superficial layer is the stiffest but represents only 1/100 of the total thickness. For that reason its role in whole skin mechanical behavior is often overlooked. Torsional experiments offered Lévêque and de Rigal[7] the possibility to investigate *in vivo* the mechanical behavior of stratum corneum (SC). Using a 9 mN · m torque they first studied the direct effect of pouring water on a skin crown either 1, 3, or 5 mm wide in volar forearm of 10 volunteers. The result was a highly significant increase in Ue parameter, 80, 40, and 15% for crown widths of 1, 3, and 5 mm, respectively. Indirect SC hydration by 15, 30, and 120 min occlusion with a plastic sheet in six subjects caused Ue to rise progressively and significantly. The rise was also significantly steeper with a 1 mm than with a 5 mm wide crown.

These Ue variations are inversely related to Young's modulus ones. On the other hand only SC mechanical properties can be modified by pouring water on the skin surface. Therefore this experiment demonstrated that: (1) SC takes an important part in the mechanical behavior of whole skin, (2) using a narrow crown of twisted skin the contribution of SC to the assessed mechanical parameters is strongly increased, and (3) the SC moisturization can be accurately assessed *in vivo* by torsional experiments.

In the same paper, the efficacy of cosmetic formulations on SC hydration was assessed using the same technique in 13 volunteers. The significantly increased Ue found 1 h following application of either O/W or W/O creams devoid of moisturizers, or petrolatum, partially or totally subsided 1 h later, but at a lower rate with petrolatum and with W/O emulsion. Addition of moisterizers such as 10% lactic acid and/or PCNa or urea had the same effect, not greater, by the first hour and unexpectedly no sustained effect was found by the second hour following application. Only 10% glycerol had a protracted hydration effect although less marked than with lactic acid. In a long-term study carried out on the legs of 3 groups of 14 volunteers, selected for showing dry skin patches, two O/W preparations containing 10% glycerol or 10% urea were applied twice daily for 3 weeks. By the first, second, and third weeks both preparations brought a significant rise in Ue while the vehicle used as a blank did not differ from the control site. Only the effect of glycerol was still measurable ($p \leq 0.05$) 1 week after cessation of treatment.[7]

The efficacy of other cosmetic products on SC compliance was also investigated by the l'Oreal's group with the same device and experimental conditions in 10 subjects.[19] The compliance parameter Ue was raised by $34 \pm 0.9\%$ 2 min following total forearm immersion in hot water (30°C) for 3 min. Also Ur and Ur/Ue significantly rose, by 40 ± 6 and 5%, respectively. The increased skin elasticity could be related to the rise in skin temperature together with SC hydration. In the same paper the variations of the same parameters over 24 d of treatment are presented, each measurement was done at time intervals 24 h after the last application. All three parameters showed a steady and highly significant increase. These experiments also demonstrated the daily individual variations probably related to climatic changes (relative humidity and temperature).

The dryness of facial skin was assessed on forehead and cheeks in 55 selected subjects using both an ordinal scale (0 to 3) and torsional experiments.[20] On cheeks there was a striking inverse relationship between Ue and the clinical dryness score ($r = -0.66$, $p < 0.001$). This was also significant although less marked on forehead ($r = -0.41$, $p < 0.002$).

This body of data on the clear-cut influence of SC hydration effect on whole skin mechanical parameters stress the role of SC in overall skin mechanical behavior, which often had been underestimated. The conclusive evidence, i.e., modification of skin extensibility and elasticity upon SC removal, was brought again by l'Oreal's group in the following experiment.[19] After 10, 15, and 20 SC strippings with an adhesive tape stuck for 15 s under 1 kg cm^{-2} on forearm skin in 8 volunteers, Ue and Ur and Ur/Ue steadily and significantly rose relative to baseline. This simple experiment confirmed unequivocally that SC takes a substantial part in the skin stiffness and recovery following extension and should be accounted for as a functional as well as a structural component in any interpretation of data concerning the mechanical properties or behavior of the skin. This is the confirmation that from a mechanical standpoint skin is a composite material made up of three layers of different thickness, stiffness, and elasticity, working in parallel.

V. Medical Applications

A. Effect of Topical Retinoic Acid

Because they are supposed to take place within SC and papillary and subpapillary dermis, the changes induced by topically applied retinoic acid (RA) might modify the mechanical behavior of the skin. Consequently torsional devices which are stuck on the skin and put force onto the superficial layers are particularly well suited for assessing this effect. Pichon et al.[11] in a group of 17 subjects evaluated the extensibility and viscosity parameters as described in the preceding chapter, using both 3 and 1 mm wide skin crowns. With the wider annulus they found a significant decrease in both types of parameters while only extensibility parameters (k_0, k_1) were significantly decreased with the narrow (1 cm wide) annulus. The difference could be ascribed to the remodeling effect of RA on both epidermis and subpapillary dermis.

The same drug was tentatively used to reduce the skin side-effects of long-term systemic corticosteroid treatment in 27 kidney-graft recipients.[21] Each patient's volar forearms were treated either by RA 0.5% (or 0.025%) in O/W emulsion, or vehicle alone, at random, for 180 d. Skin thickness, as determined by an ultrasound A-scan device working at 25 MHz, was found intensively decreased at start and significantly increased by the 60th day of treatment onward, mostly in women. Torsional tests used a 3 mm free-skin crown, and a 18 mm disk radius. The skin elasticity (Ur/Ue) was increased in the treated versus control areas by a factor of 3 to 20% by the 60th day and further on, but this was only significant in women. As no experiment was done with the narrower skin crown (1 mm) the results are difficult to interpret in term of location of structural change.

B. Scleroderma

Sclerodermas are diseases characterized by induration of the lower dermis and uppermost subcutis, either in localized and well-defined plaques or bands as in morphea or diffuse as in progressive scleroderma. Increase in thickness and lower extensibility have been noninvasively documented using ultrasound echography and a suction extensometer. Kalis et al.[10] confirmed these findings by using l'Oreal's torsional device with a 3 mN · m torque applied through an 18-mm-diameter disc and a 3-mm-wide free-skin annulus. The skin thickness was measured by A-scan echography.

In active plaques of morphea (n = 5) the results were highly significant both for increased skin thickness (60 ± 5%) and decreased extensibility (–67 ± 11%) relative to symmetrical healthy control area. In regressing lesions (n = 12) the skin thickness did not differ from the control area but the extensibility was still lower although to a less extent. Using Equation 5, the authors concluded to a negligible and not significant difference in the elastic modulus E in case of active plaques ($\Delta E = -6\%$) whereas in the case of regressing lesions E was significantly higher ($\Delta E = +49\%$). From these data one can infer that the excess of collagen synthesized during the sclerotic phase retains normal mechanical properties whereas at the regressing phase the progressive dermal atrophy is associated with a stiffer collagen.

Measurements in progressive systemic scleroderma were made on volar forearm. The 11 patients were compared to 10 age-matched controls. The skin thickness was about twice that of controls (+106 ± 18%) and extensibility (Ue) was reduced by 68 ± 8%. Both these differences were highly significant. As a surprising result E was decreased by 38%. But the skin elasticity was also reduced.

Humbert et al. used the same device and a 15-MHz ultrasound A scan for skin thickness measurement in a patient with morphea treated with 1,25 dihydroxyvitamin D_3. After 6-months therapy Ue rose by 211% while skin thickness kept unchanged (1.41 and 1.35 mm, respectively). Accordingly the Young's modulus E decreased by 204%, a finding in accordance with clinical scoring. This result was confirmed in a series of five patients with morphea whose duration of disease had varied from 2 to 10 years.[23] They received oral 1,25 (OH) 2 vit D_3 (mean dose: 1.75 µg/d). In line with clinical improvement a significant decrease in the Young's modulus (–62%, $p < 0.01$) appeared after a 2- to 24-months therapy. In an open uncontrolled study Humbert et al.[24] included in the same protocol 11 patients suffering from systemic scleroderma. A significant decrease in the Young's modulus (–36%, $p < 0.01$) was also noted at the end of the study. Furthermore these investigations demonstrated the possibility to use this device for assessing the effect of a treatment over time in a single patient.

C. Inherited Connective-Tissue Diseases

Torsional measurements on volar forearm were performed on five patients with type 2 Ehlers-Danlos syndrome, using the l'Oreal's device, and the results compared to those of age and sex-matched healthy subjects (five for each patient).[25] The torque was acted by a 18-mm-radius disc with a free skin annulus 3 mm wide. The skin thickness was considered half of skinfold in the same area, as measured by a caliper. While a significantly lower skinfold thickness was found in patients than in controls (1.1 ± 3 mm vs. 1.5 ± 0.2 mm, respectively), the skin extension (Ue) was significantly increased (40 ± 11 mmV vs. 25 ± 5 mmV, respectively). By combining these data no change in the elastic modulus was found in these patients. Also the ratio Ur/Ue (elasticity) was found identical to that of controls. The conclusion was that only skin thinning could account for the clinically observed extensibility in these patients.

By contrast in three cases of Marfan's syndrome the skinfold thickness and distensibility Ue of each patient were higher than that of his (her) five age-and-sex-matched controls. The product Ue × thickness was significantly higher by 46 and 53% on left and right forearms, respectively, and the Young's modulus as calculated by Equation 5 was decreased by 60%. On the other hand skin elasticity was found slightly and nonsignificantly increased. From these data the authors suggested that Marfan's syndrome was associated with "a true decrease in tissue stiffness". Some years later this was confirmed by an abnormality in the fibrillin network in that disease.[26]

VI. Conclusion

Torsional experiments as used mostly with l'Oreal's device proved a reliable and easy-to-handle tool for assessing skin mechanical behavior in current practice. As in any use of instrument however, great care should be taken by the clinician to comply with the above-cited requirements for the validity of the experiment. Over the last 12 years this allowed significant advances to be made concerning skin physiology and *in vivo* mechanical properties. By contrast only a few investigations have been made in disease, at least for two reasons. The first one is the dimension of the guard ring which initially was 54 mm and now is reduced to 40 mm in outer diameter, thus requiring a rather large flat area of skin, which is uncommon in pathology. The second reason is that the device, named "twistometer" was a l'Oreal's prototype, not commercially available, and consequently usable by only a few groups. This is no more the case today when the Dermal Torque Meter has come to the market. Dermatologists whether they are interested in skin physiology or pathology, have now an easy-to-use technique to investigate one of the main skin functions and possibly have an insight into the functioning of skin major components. They feel indebted to the cosmetic industry to have created the device, designed the protocol, and made the experimental studies needed to validate clinically and structurally relevant skin mechanical parameters.

Acknowledgment

The author would like to acknowledge Jean-Luc Lévêque for his outstanding contribution to our knowledge in skin biomechanics physiology.

References

1. Meirson D, Goldberg LH. The influence of age and patient positioning on skin tension lines. *J Dermatol Surg Oncol* 19: 39–43, 1993.
2. Vlasblom DC. Skin elasticity. Ph.D. Thesis, University of Utrecht 1967.
3. Finlay B. Dynamic mechanical testing of human skin "in vivo". *J Biomechanics* 3: 557–568, 1970.
4. Finlay B. The torsional characteristics of human skin in vivo. *J Biomed Eng* 6: 567–573, 1971.
5. Sanders R. Torsional elasticity of human skin in vivo. *Pflügers Arch* 342: 255–260, 1973.
6. Jaskowski J, Maceluch J. Nowe mozliwosci badan wlasciwosci mechanicznych skory czlowoieka. *Wiad Lek* 35: 1149–1155, 1982.
7. Lévêque J-L, De Rigal J. In vivo measurement of the stratum corneum elasticity. *Bioeng Skin* 1: 13–23, 1985.
8. Lévêque J-L, De Rigal J, Agache P, Monneur C. Influence of ageing on the in vivo extensibility of human skin at a low stress. *Arch Dermatol Res* 269: 127–135, 1980.
9. Agache P, Monneur C, Lévêque J-L, De Rigal J. Mechanical properties and Young's modulus of human skin in vivo. *Arch Dermatol Res* 269: 221–232, 1980.
10. Kalis B, De Rigal J, Leonard F, Lévêque J-L, Riche O, Le Corre Y, De Lacharriere O. In vivo study of scleroderma by non invasive techniques. *Br J Dermatol* 122: 785–791, 1990.
11. Pichon E, De Rigal J, Lévêque J-L. In vivo rheological study of the torsional characteristics of the skin. 8th Int Symp Bioenginering and the Skin, Stresa (Italy) 13–16 June 1990.
12. Wijn PFF. The alinear viscoelastic properties of human skin in vivo for small deformations. Ph.D. thesis, University of Nijmegen, Holland, 1980.
13. Barbenel JC, Payne PA. In vivo mechanical testing of dermal properties. *Bioeng Skin* 3:8–38, 1981.
14. Escoffier C, De Rigal J, Rochefort A, Vasselet R, Lévêque J-L, Agache P. Age-related mechanical properties of human skin: an in vivo study. *J Invest Dermatol* 93: 353–357, 1989.
15. Fleischmajer R, Perlish JS. The vascular inflammatory and fibrotic components in scleroderma skin. *Monogr Pathol* 24:40–54, 1983.
16. Lévêque J-L, Corcuff P, De Rigal J, Agache P. In vivo studies of the evolution of physical properties of the human skin with age. *Int J Dermatol* 23:322–329, 1984.
17. Lévêque J-L, Porte G, De Rigal J, Corcuff P, Francois AM, Saint Leger D. Influence of chronic sun exposure on some biophysical parameters of the human skin: an in vivo study. *J Cutaneous Aging Cosmet Dermatol* 1: 123–127, 1988/89.
18. Berardesca E, De Rigal J, Lévêque J-L, Maibach HI. In vivo biophysical characterization of skin physiological differences in races. *Dermatologica* 182: 89–93, 1991.
19. Aubert L, Anthoine P, De Rigal J, Lévêque J-L. An in vivo assessment of the biomechanical properties of human skin modifications under the influence of cosmetic products. *Int J Cosmet Sci* 7: 51–59, 1985.
20. Lévêque J-L, Grove G, De Rigal J, Corcuff P, Kligman Am, Saint Leger D. Biophysical characterization of dry facial skin. *J Soc Cosmet Chem* 82: 171–177, 1987.
21. De Lacharriere O, Escoffier C, Gracia AM, Teillac D, Saint Leger D, Berrebi C, Debure A, Lévêque J-L, Kreis H, De Prost Y. Reversal effects of topical retinoic acid on the skin of kidney transplant recipients under systemic corticotherapy. *J Invest Dermatol* 95: 516–522, 1990.
22. Humbert P, Dupond JL, Rochefort A, Vasselet R, Lucas A, Laurent R, Agache P. Localized scleroderma — response to 1,25-dihydroxyvitamin D3. *Clin Exp Dermatol* 15: 396–398, 1990.
23. Humbert P. Unpublished data.
24. Humbert P, Dupond JL, Agache P. Treatment of scleroderma with oral 1,25-dihydroxy vitamin D3. An open study. *Acta Dermato Venereol* 1993, in press.
25. Bramont C, Vasselet R, Rochefort A, Agache P. Mechanical properties of the skin in Marfan's syndrome and Ehlers-Danlos syndrome. *Bioeng Skin* 4: 217–227, 1988.
26. Mollister J, Godfrey M, Sakai LY, Pyeritz RE. Immunohistologic abnormalities of the microfibrillar fiber system in the Marfan syndrome. *N Engl J Med* 323: 152–159, 1990.

Chapter 14.2
Suction Chamber Method for Measurement of Skin Mechanical Properties: The Dermaflex®

Monika Gniadecka* and Jørgen Serup*,**
*Bioengineering and Skin Research Laboratory
Department of Dermatology
University of Copenhagen
Bispebjerg Hospital
Copenhagen, Denmark
**Department of Dermatological Research
Leo Pharmaceutical Products
Ballerup, Copenhagen, Denmark

I. Introduction

Mechanical properties of the skin have been studied noninvasively by a number of methods, using uniaxial stretching, ballistometric techniques, torsion, indentation, and, finally, suction.[1–7] The principle of the suction method is the measurement of skin elevation caused by the suction force exerted over a defined area of the skin.[8–10] Measuring skin elevation in the function of the suction time or the suction force calculates various parameters of skin mechanical properties. From the physiological and pathophysiological point of view it is crucial to determine at least two properties: *stiffness* (resistance to change of the shape) and *elasticity* (the ability to recover shape after stretch). Additional parameters may be calculated with the use of different types of equipment.

From the mechanical point of view skin may be considered as a *five-layered structure* consisting of:

- The stratum corneum and outer portion of the epidermis with a fluctuating water content and continuous adaptation to environmental humidity conditions
- The internal portion of the epidermis with the basement membrane zone, desmosomes, and elastic fiber anchoring to deeper structures
- The papillary dermis with a relatively loose connective tissue defining the microrelief of the skin
- The reticular dermis built-up of tight connective tissue spatially organized as expressed by the Langer lines. Coarse wrinkles involve this layer
- The subcutaneous space with attachments to deeper structures, such as fascia

Selection of the method for measuring of skin elasticity depends on the expected abnormality to be characterized, region of the body under consideration, and technical specifications of the piece of equipment, in particular the measuring probe, the area of the skin which is being measured, and the type of stress (vertical, horizontal, or torsional). Methods can be divided into those exerting a *proportional full-thickness strain,* useful mainly for dermatological and medical applications, and those exerting a *disproportional superficial strain* mainly useful for cosmetological purposes. The Dermaflex® with its chamber diameter of 10 mm is an example of a proportional method, whereas the Cutometer® (chamber diameter of 1 mm) and twistometers are examples of disproportional methods, which highlight the influence of the outer mechanical compartment of the skin.

The *relevance* of mechanical parameters is a complex issue. In scleroderma the correlation between stiffness and collagen excess is logic; the increased stiffness of chronically inflamed skin can also be easily understood. However, using methods based on disproportional deformation to verify effects of cosmetic creams may be more difficult to understand, since this type of strain is quite far from common practices among users, who assure skin elasticity by hand movements parallel to the skin surface.

II. Equipment and Determination of Skin Mechanical Properties

The prototype of the suction machine for the investigation of skin mechanical properties was described by Grahame.[10] He adopted the principle of the diaphragm method of Dick[11] and Tregear[12] which was successfully used for the determination of the mechanical properties of skin explants *in vitro*. Later, more sophisticated electronic instruments became commercially available; the most well known are the Dermaflex A® (Cortex Technology, Hadsund, Denmark) developed in our laboratory, and the Cutometer SEM 474® (Courage and Khazaka, Cologne, Germany) described extensively in the laboratory of Maibach.[9]

The Dermaflex A®[7,25] consists of three main parts: (1) the generator of the vacuum which is transduced to the suction probe placed directly on the skin; (2) sensor of skin elevation inside the probe; and (3) data elaboration and visualization system. The suction probe is equipped with a steel ring (diameter of 10 mm) that sharply demarcates an area of the skin where suction is applied (Figure 1). The double adhesive ring around the steel ring additionally prevents skin creeping during suction. The elevation of the skin is determined electronically by measuring electric capacitance between skin surface and the electrode placed in the top of the suction chamber. The accuracy of this measurement has been shown to be 0.01 mm.[8] The instrument allows adjustment of the vacuum strength, the length of the suction period, and the number of suction cycles. The most often used set of parameters was: suction 300 mbar or 450 mbar, suction period 4 s or 20 s, number of cycles 1 or 5-6.

The parameter of skin stiffness (distensibility) is a value of skin elevation or strain (in millimeters) at the end of the first suction; residual skin elevation after the release of the first suction is named resilient distension (Figure 2). The relative elastic retraction (RER), reflecting biological elasticity, can be calculated from the formula:

$$\text{RER} = \frac{\text{distensibility - resilient distention}}{\text{distensibility}} \times 100\%$$

A 100% value represents the perfect recovery of skin shape after stretch. An additional parameter, hysteresis, represents the creeping phenomenon and is defined as the distance that the skin is stretched beyond the distensibility when the suction is repeated over the same area (Figure 2).

Another suction instrument for assessment of skin mechanical properties is the Cutometer SEM 474®. The main difference from the Dermaflex A® is the considerably smaller size of the suction chamber in the standard version, 2 vs. 10 mm in Dermaflex A.® It is conceivable that with the Cutometer® one measures the mechanical properties of the epidermis, papillary dermis, and to a lesser degree deeper layers of the dermis and subcutis.[9,13] The results of the measurements with the Dermaflex®, which is equipped with a larger suction probe, are mainly influenced by the mechanical properties of the dermis. Thus equipment with a small-size suction probes would be more suit-

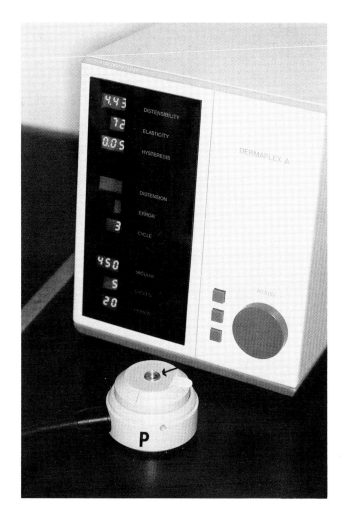

FIGURE 1 Dermaflex A® for the measurement of skin mechanical properties. (P) Suction probe of the instrument, arrow shows a guard steel ring demarcating the area of measurement.

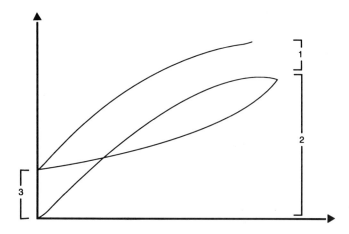

FIGURE 2 Example of skin elevation plot obtained from Dermaflex A® for determination of skin mechanical properties (a series of nine cycles): (2) distensibility, (3) resilient distension. Hysteresis (1) is observed on repeated suctions.

able for detection of alterations of mechanical properties due to the changes in epidermis and papillary dermis, whereas with the larger suction probes the mechanical functional status of the whole integument would be measured.

The correlation of the separate parameters of skin mechanical properties with the structural elements in the skin has not been fully elucidated. According to the mechanical model, skin is a sponge-like viscoelastic structure, composed of fibers and the colloidal ground substance, covered with a relatively rigid epidermal sheet, and filled with a free-movable (Newtonian, colloid-poor) fluid.[14] Evidence has been presented that elastic fibers in the dermis are principally responsible for elastic properties of the skin under minor loads. When a *minor stretch* of the skin takes place elastic fibers are primarily responsible for the return of the skin to the previous shape.[15,17] The loss of elastin, e.g., in cutis laxa, is connected with diminished skin elasticity.[18] The relative architecture of these fibers is important. It has been shown, that in normal conditions elastic fibers in the immediate subepidermal area form a delicate framework perpendicular to the epidermal surface.[19] These fibers are attached to the epidermal-dermal junction from one side, and to the layers of elastic fibers that run parallel to the surface of the skin from the other. Therefore, the structure and shape of the epidermal-dermal junction may influence skin elasticity (for example, in psoriasis, where the rete ridges of the epidermis are elongated and skin elasticity is compromised).[8] The importance of the structural relationship of the elastic fibers for skin elasticity can be exemplified in elastoderma, where excessive and disorganized elastin proliferation in the skin takes place with complete loss of its elastic recoil.[20]

In the situation of mild suction, collagen bundles unfold and do not resist skin stretching. However, when *higher traction forces* are applied, stiff and inelastic collagen bundles become totally stretched and counteract mechanical deformation of the skin. The quantitative deficiency of collagen and the disorganized collagen architecture results in excessive skin extensibility, laxity, and fragility to mechanical stress.[21,22] Besides elastin and collagen other fibrillar components in the skin, such as fibrilin, may be of importance in determination of skin elasticity.[23] Suction chamber pressure above 50 to 100 mbar will normally represent a higher traction force, and the measured distensibility represent resiliency of the collagen fiber network.

An important, but overlooked component determining skin elasticity is epidermal and dermal water. Jemec et al.[24] showed conclusively with the Dermaflex A® that skin distensibility and hysteresis increase slightly after epidermal moisturizing with tap water. Elasticity of the skin decreases slightly upon epidermal hydration. Similarly, in psoriasis and histamine wheals, where inflammatory edema in the dermis is present, hysteresis of the skin increases.[8,25,26] This suggests that water, which represents approximately 70% of skin mass, acts as an 'oil' both in the epidermis and in the interstitium, increasing the distensibility and creeping, while dampening elastic retraction. Moreover, decreased water content in the skin is associated with decreased skin elasticity, exemplified by the phenomenon of inelastic skinfold which is a reliable clinical sign of dehydration.

III. Variables, Prerequisites, and Practical Guidance to Measurements

Variables may be divided into technical variables related to the equipment, biological variables, and variables related to the laboratory, the investigator, and its environment. The manufacturer's manual together with the literature on the equipment should provide sufficient *technical information* including suggestions of presettings. *Biological* and *environmental variables* which need be considered are

- Sex
- Age
- Race
- Anatomical site, the vertical vector of skin stiffness
- Endocrine factors (menstrual cycle, hormonal treatment)
- Water balance including treatment with diuretics
- General and localized diseases, including gravitational syndrome
- Diurnal variation of water accumulation in the dependent parts of the body
- Consequences of sun exposure (photodamage, *PUVA* photosclerosis)
- Seasonal variations in humidity of the stratum corneum
- Previous physical stress to the skin (stress 'memory' lasts for at least 1 h)
- Topical treatment

The following points should be considered to standardize measurements:

- Standardized positioning of the body region, in particular with respect to positioning of adjacent large joints including supination/pronation
- Careful selection and marking of the site to be measured; ideally adjacent sites should be measured and the mean value should be calculated
- Standardized placement of the probe with respect to load and angle
- Be especially aware of the importance of the *zero situation* immediately at initiation of the measurement; deviation from the natural position will strongly influence the result. In contrast, once the suction has been applied, the probe can be moved vertically in the millimeter range with no influence

on the final result, thus demonstrating a marginal importance of the subcutaneous septa bindings
- Standardize with respect to axial orientation and Langer's line position if a noncircular suction chamber is used or a method sensitive to axial orientation is applied
- Observe that the surrounding skin is fixed and remains fixed during measurement with no creeping into the suction probe (this phenomenon may manifest as a sudden increase in distention during repeated measurement)
- Avoid repeated measurements in the same site, at least for 1 h; the initial recording will temporarily modify the viscoelastic behavior within that site.

Skin mechanics is *not* a static parameter but depends heavily on a complex interaction between tissue fluid and solid components. Skin is clearly an anisotropic and viscoelastic system. Skin mechanical properties may depend on skin thickness. However, it does not seem to be justified to adjust mechanical parameters of the skin-to-skin thickness, as is typically done in traditional engineering. Precise characteristics of the skin specifically in the measured site is more useful.

IV. Mechanical Properties of Normal Skin

Mechanical properties of normal human skin have been investigated in detail by Grahame,[10] Cua et al.,[27] Malm et al.,[28] and Gniadecka et al.[29] The authors found that these properties depend on the *anatomical site*. A vertical vector of distensibility has been described,[28,29] where the skin in the acral sites (malleoli, forearms) is less distensible (more stiff) than in the more proximal sites (thigh, arm, trunk) (Figure 3). In the study, where higher suction pressure has been applied,[29] the similar vertical vector of elasticity (RER) has been detected, skin being less elastic in the acral sites. In addition Malm et al. detected the vertical vector of hysteresis. Acral female skin is more distensible than male skin, whereas the distensibility of the trunkal skin does not differ significantly between men and women.[10] It is not known whether genetic or *endocrine factors* are responsible for modulation of the elasticity of the skin. It is likely, however, that female hormones play some role in modulation of skin mechanical properties, since skin elasticity in women differ throughout the menstrual cycle.[30]

Age is an important factor influencing mechanical properties of the skin. It is generally accepted that in humans skin elasticity (RER) diminishes with age, whereas the distensibility may remain relatively unaffected.[28,39] This finding was confirmed with other methods for assessment of skin mechanical properties.[31-36] In aged

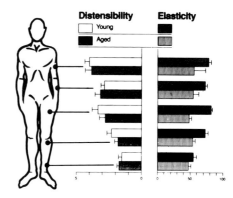

FIGURE 3 Vertical vector of skin distensibility and elasticity in young and aged individuals in different sites of the upper and lower extremity (means with 2 SE). Distensibility is measured in millimeters (elongation after suction), biological elasticity is in percent of skin retraction after stretch. (Reproduced from Gniadecka, M., Gniadecka, R., Serup, J., and Søndergaard, J., *Acta Derm. Venereol.,* 74, 188, 1994. With permission.)

people it has been demonstrated with the use of the Cutometer® that the indices of elasticity diminish with age both in sun-exposed and sun-protected areas.[9,13,36,37]

It is crucial to distinguish between the intrinsic aging within sun-protected areas, characterized by skin thinning and epidermal and dermal atrophy, and actinic aging of the sun-exposed areas, where epidermis is hyperplastic, and dermis is thickened. Lévêque et al.[38] studied the effects of chronic sun exposure on skin extensibility and elastic recovery. These authors found that in the areas exposed to long-term *sun irradiation* extensibility and elasticity are decreased. In another study Berardesca et al.[39] found decreased skin elasticity after ultraviolet irradiation. The impairment of skin elastic properties could be prevented with topical sunscreens and reversed with topical retinoic acid.[40] It has been proposed that diminished elasticity in aged skin is due to the damage, disintegration, or changes of the structure of elastic fibers[41-43] that appear both in intrinsic and solar aging. Glycosaminoglycans that accumulate in the papillary dermis of the actinically damaged skin may interfere with the elastic-fiber system located in this region.

The effect of aging on skin stiffness has not been fully elucidated. Some authors claim that photoaging is associated with diminished extensibility,[38] whereas others found no differences between aged and young individuals[28,39,40] but in photodamaged and photoprotected sites. The explanation of the contradictory findings may be the different degree of photodamage of the subjects enrolled in the studies and the differential effect of UVA and UVB irradiation. Borroni et al.[44,45] showed that experimental irradiation of photoprotected sites with UVA causes diminishing of both elasticity and distensibility. Our own studies showed that skin distensibility in different sites of the body of the aged individuals was not different from that in the corresponding regions in youths. However, the vertical gradient of distensibility in the extremities was more weak in the aged group.[29]

Decreased vertical vector of stiffness in the aged extremity skin may have serious pathophysiological consequences. Studies in tall animals, such as the giraffe, revealed that skin, fascia, and veins in the dependent areas are more stiff to directly compensate for the increased hydrostatic pressure of blood in the standing position.[46] This congenital 'antigravity suit' can prevent venous distention and leg edema formation in the upright position, in a manner comparable to external compression. The findings of the vertical vector of skin stiffness provide evidence of a similar *antigravity suit* in humans.[29] Impairment of this antigravity suit in aged individuals may enhance formation of the varicose veins and leg edema in the standing position in old individuals.

Studies of Gniadecka et al.[29] showed that *distensibility* and *elasticity* (RER) are not constant but they *fluctuate during the day*. In young persons RER and distensibility increases in the afternoon (Figure 4). Although the mechanism of the diurnal variation of skin elasticity has not been elucidated, this phenomenon may present an adaptation to the standing position. It has been postulated that stretching of the skin enhances the time-dependent dermal accumulation of fluid while restoration of previous shape is associated with fluid removal.[14] Skin elastic forces contribute to the recovery of skin shape after stretch and thus enhance clearance of the intercellular fluid. Thus, evening increase of elasticity of the acral skin is advantageous because strong recoil forces assist in the removal of the edematous fluid that accumulates in the extremities during the day.

V. Skin Mechanical Properties in Pathological Conditions

A full review of dermatological applications is outside the scope of this chapter but a few illustrative examples of the use of Dermaflex® are provided below (Table 1). In the *acute edema* in histamine wheals the distensibility and hyseresis is increased.[8] The excess of water acts as oil and the fiber network of the dermis is further extended. Repeated suction at the same measuring site results in a relatively minor strain at a higher stress. The initial maneuvre brought the fibers into a relatively extended and locked position for a period at least 1 h. *Psoriasis* representing inflammation and protracted edema with a zero situation of a certain strain of fibers manifests with decreased distensibility, but increased hyseresis due to inflammatory edema.[8] In *scleroderma* the distensibility is decreased

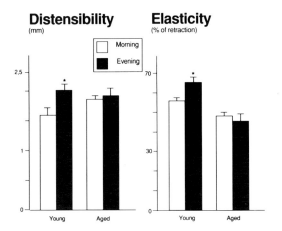

FIGURE 4 Diurnal changes of ankle skin distensibility and elasticity (RER) in young vs. aged individuals. *significant $p = 0.002$, paired t-test. Reproduced from Gniadecka, M., Gniadecka, R., Serup, J., and Søndergaard, J., *Acta Derm. Venereol.*, 74, 188, 1994. With permission.

TABLE 1 Summary of Principal Changes in Parameters of Skin Elasticity in Selected Pathological Conditions[8,25,26]

Condition	Distensibility	Hysteresis
Cutaneous edema	Increased	Increased
Previous mechanical stress	Increased	Decreased
Psoriasis	Decreased	Increased
Scleroderma		
Sclerotic plaque	Decreased	Decreased
Early remission	Slightly decreased	Increased
Late remission	Decreased	Slightly decreased

in every phase,[25,26] and increased skin stiffness is a constant feature, in contrast to thickening. In Ehlers-Danlos syndrome, in contrast, the distensibility increases particularly on the cheek and the arms.[48] The subcutaneous fat and connective tissue bindings have no major influence on skin mechanics.

References

1. Marks R, Payne PA. *Bioengineering and the Skin*. MTP Press Ltd., Lancaster, 1981.
2. Marks R. Mechanical properties of the Skin. In: *Biochemistry and Physiology of the Skin*. ed. Goldsmith LA. Oxford University Press, Oxford 1983, 1237–1254.
3. Christensen MS, Hargens CW, Nacht S, Gans EH. Viscoelastic properties of intact human skin, instrumentation, hydration, effect and the contribution of the stratum corneum. *J Invest Dermatol* 69:282, 1977.
4. Kijd WL, Daly CH, Nansen PD. Variation in the response to mechanical stress of human soft tissues in the elderly and the young. *J Prosthet Dent* 32:493–500, 1974.
5. Wan Abas WAB, Barbenel JC. Uniaxial tension test of human skin *in vivo*. *J Biomed Eng* 4:65, 1982.
6. Adhoute H, Berbis P, Privat Y. Ballistometric properties of aged skin. in: *Aging Skin. Properties and Functional Changes*. eds. Lévêque J-L, Agache PG. Marcel Dekker, New York, 1993:39–48.

7. Pierard GE. Mechanical properties of aged skin: indentation and elevation experiments. in: *Aging Skin. Properties and Functional Changes.* eds. Lévêque J-L, Agache PG. Marcel Dekker, New York, 1993: 49–56.
8. Serup J, Northeved A. Skin elasticity in psoriasis. *In vivo* measurement of tensile distensibility, hysteresis and resilient distension with a new method. Comparison with skin thickness as measured with high-frequency ultrasound. *J Dermatol* 12:318–324, 1985.
9. Elsner P, Wilhelm D, Maibach HI. Mechanical properties of human forearm and vulvar skin. *Br J Dermatol* 122:607–614, 1990.
10. Grahame R. A method for measuring human skin elasticity *in vivo* with observations on the effects of age, sex, and pregnancy. *Clin Sci* 39:223–238, 1970.
11. Dick JC. The tension and resistance of stretching of human skin and other membranes with results from a series of ormal and oedematous cases. *J Physiol* 112:102–113, 1951.
12. Tregear RT. *Physical Function of the Skin.* Academic Press, London, 1966.
13. Agache P, Monneur C, Lévêque J-L, de Rigal J. Mechanical properties and Young's modulus of human skin *in vivo*. *Arch Dermatol Res* 269:221–232, 1980.
14. Oomens CWJ, van Campen DH, Grotenboer HJ. A mixture approach to the mechanics of skin. *J Biomechanics* 20:877–885, 1987.
15. Oxlund H, Manschot J, Viidik A. The role of elastin in the mechanical properties of skin. *J Biomech* 21:213–218, 1988.
16. Daly CH. The role of elastin in the mechanical behaviour of human skin. Digest 8th ICMBE, Chicago, IL, 1969:7.
17. Oxlund H. Changes in connective tissues during corticotrophin and corticosteroid treatment. *Dan Med Bull* 31:187–206, 1984.
18. Fitzsimmons JS, Gilbert G. Variable clinical presentations of cutis laa. *Clin Genet* 28:284–295, 1985.
19. Bravermann IM, Fonferko E. Studies in cutaneous aging. The elastic fiber network. *J Invest Dermatol* 78:434–443, 1982.
20. Kornberg RL, Hendler SS, Oikarinen A, et al. Elastoderma-disease of elastin accumulation within the skin. *N Engl J Med* 312:771–772, 1985.
21. Burton L. Disorders of connective tissue. In: *Textbook of Dermatology.* eds. Champion RH, Burton JL, Ebling FJG. Blackwell Scientific Publications, Oxford. 1992:1791–1797.
22. Silence DO, Senn A, Danks DM, et al. Genetic heterogeneity in osteogenesis imperfecta. *J Med Genet* 16:101–116, 1979.
23. Hollister DW, Godfrey MP, Kene DR, et al. Immunohistologic abnormalities of the microfibrillar-fiber system in the Marfan syndrome. *N Engl J Med* 323:152–159, 1990.
24. Jemec GBE, Jemec B, Jemec BIE, Serup J. The effect of superficial hydration on the mechanical properties of human skin in vivo: implications for plastic surgery. *Plast Reconstruct Surg* 85:100–103, 1990.
25. Serup J, Norheved A. Skin elasticity in localized scleroderma (morphoea). Introduction of a biaxial *in vivo* method, and the measurement of tensile distensibility, hysteresis and resilient distension of diseased and normal skin. *J Dermatol* 12:52–62, 1985.
26. Serup J. Localised scleroderma (morphoea): clinical, physiological, biochemical and ultrastructural studies with particular reference to quantitation of scleroderma. *Acta Dermatol Venereol* 66 (suppl. 122): XX, 1986.
27. Cua AB, Wilhelm KP, Maibach HI. Elastic properties of human skin: relation to age, sex, and anatomical region. *Arch Dermatol Res* 282:283–288, 1990.
28. Malm M, Bartling MS, Serup J. *In vivo* skin elasticity of twenty-two anatomical sites. The vertical gradient of skin extensibility and implications in gravitational aging. Submitted.
29. Gniadecka M, Gniadecki R, Serup J, Søndergaard J. Skin mechanical properties present adaptation to man's upright position. *In vivo* studies in young and aged individuals. *Acta Dermatol Venereol (Stockh),* 74, 188, 1994.
30. Berardesca E, Gabba P, Farinelli N, Borroni G, Rabiosi G. Skin extensibility time in women. Changes in relation to sex hormones. *Acta Dermatol Venereol (Stockh)* 69:431–433, 1989.
31. Berardesca E, Farinelli N, Rabbiosi G, Maibach HI. Skin bioengineering in the noninvasive assessment of cutaneous aging. *Dermatologica* 182:1–6, 1991.
32. Pierard GE, Lapiere CM. Physiopathological variations in the mechanical properties of skin. *Arch Dermatol Res* 260:231–239, 1977.
33. Pierard GE. A critical approach to *in vivo* mechanical testing of the skin. In: *Cutaneous Investigation in Health and Disease,* ed. Lévêque J-L. Marcel Dekker, New York, 1989:215–240.
34. Lévêque J-L, de Rigal J, Agache PG, Monneur C. Influence of ageing on the *in vivo* extensibility of human skin at a low stress. *Arch Dermatol Res* 269:127–135, 1980.
35. Dikstein S. *In vivo* mechanical properties of the skin measured by indentometry and levarometry. *Bioeng Skin Newsl* 2:23, 1979.
36. Pierard GE. Evaluation des proprietes mechanique de la peau par les methodes d'indentation et de compression. *Dermatologica* 168:61, 1984.
37. Pierard GE, Lapiere CM. Physiopathological variations in the mechanical properties of skin. *Arch Dermatol Res* 260:231–239, 1977.
38. Lévêque J-L, Porte G, de Rigal J, Corcuff P, Francois AM, Saint Leger D. Influence of chronic sun exposure on some biophysical parameters of human skin: an *in vivo* study. *J Cutan Aging Cosmet Dermatol* 1:123–127, 1988/89.
39. Berardesca E, Vignoli GP, Borrni G, Rigano L, Gaspari F. Acute effects of UV rays on mechanical properties of the skin *in vivo*. In: Proceedings of the 16th IFSCC. New York, October 1990.
40. Berardesca E, Gabba P, Farinelli N, Borroni G, Rabiosi G. *In vivo* tretinoin-induced changes in skin mechanical peoperties. *Br J Dermatol* 122:525–529, 1990.
41. Fazio MJ, Olsen DR, Uitto JJ. Skin aging: lessons from from cutis laxa and elastoderma. *Cutis* 43:437–444, 1989.
42. Bouissou H, Pieraggi M, Julian M, Savit T. The elastic tissue of the skin: a comparison of spontaneous and actinic (solar) aging. *Int J Dermatol* 27:327–335, 1988.
43. Balin AK, Pratt LA. Physiological consequences of human skin aging. *Cutis* 43:431–436, 1989.
44. Borroni G, Zaccone C, Vignati G, et al. Assessment of biomechanical changes induced by long-term PUVA treatment (>1000 J/sqcm) in psoriatic patients (abstract). 8th International Symposium on Bioengineering and the Skin. Stresa, Italy. June, 1990:59.
45. Borroni G, Vignai G, Vignoli GP, et al. PUVA-induced viscoelastic changes in the skin of psoriatic patients. *Med Biol Environ* 17:663–671, 1989.
46. Hargens AR, Millard RW, Pettersson K, Johansen K. Gravitational haemodynamics and oedema prevention in the giraffe. *Nature* 329:59–60, 1987.
47. Wickman M, Olenius M, Malm M, Jurell G, Serup M. Alterations in skin properties during rapid and slow tissue expansion for breast reconstruction. *Plast Reconstr Surg* 90:945–951, 1992.
48. Serup J, unpublished observations.

Chapter 14.3
Suction Method for Measurement of Skin Mechanical Properties: The Cutometer®

A. O. Barel*, W. Courage**, and P. Clarys*
*Laboratory of General and Biological Chemistry
Higher Institute for Physical Education and Physiotherapy
Vrije Universiteit
Brussels, Belgium

**Courage & Khazaka Electronic GmbH
Cologne, Germany

I. Introduction

The mechanical properties of the human skin have been extensively studied in the past most *in vitro* and less *in vivo*.[1] Skin is a complex organ which as many other biologicals, presents in a combined way the typical properties of elastic solids and viscous liquids.[2] As a consequence the mechanical properties of the skin are called viscoelastic. Typical properties of viscoelastic materials are nonlinear stress-strain properties with hysteresis (the stress-strain curves obtained on loading will not be superposed on the curves obtained by unloading).[2-4] Furthermore the deformation of the skin is time-dependent with a typical phenomenon of creep. The creep is characterized as an increasing deformation of the skin in function of time when a constant stress is applied on this material. The viscoelastic properties of the skin are due to the components of the skin: collagen fibers and elastin fibers impregnated in a ground substance of proteoglycans.[2-4]

II. Object of this Study

The general purpose consists of noninvasively measuring the biomechanical properties of the human skin *in vivo* by means of a simple, reliable, and reproducible biophysical technique. There have been in the past many investigations of the mechanical properties of human skin using various equipment which measure the deformation of the skin after application of uniaxial or biaxial oriented forces. In these studies normally healthy and diseased skin situations must be considered. The influence of physiological factors such as age, sex, normal and actinic ageing, diseases, and changes in biomechanical properties induced by various topical treatments can also be examined using mechanical measurements.

Many experimental instruments and devices have been developed in research laboratories but only very few instruments are commercially available: a torsion method (Torque Meter®*) and two instruments using the suction method (Dermaflex®* and Cutometer®*). This report describes the use of the Cutometer SEM 474® based on the suction method. The instrument measures the vertical deformation of the skin surface when the skin is pulled in the circular aperture of the measuring probe after application of a vacuum.

III. Methodological Principle

A. Description of the Measuring Probe

Figure 1 shows a schematic representation of the measuring hand-held probe which is attached to the main apparatus with an electric cable and an air tube. A variable vacuum (ranging from 50 to 500 mbar) is applied on the skin surface through the opening of the probe. The skin surface is pulled by vacuum in the aperture of the probe. The depth of the skin penetration is measured by an optical system which measures in function of skin penetration the diminution of light intensity of an infrared light beam. Calibration of this optical system is carried out in the factory with a micrometer (from 0 to 3.0 mm). The skin adjacent to the opening of the probe is maintained in position by an external guard ring attached to the probe shield (external diameter of 25 mm).

---

* Torque Meter® is a registered trademark of Dia-Stron Ltd., Andover, United Kingdom. Dermaflex® is a registered trademark of Cortex Technology, Hadsund, Denmark and Cutometer® is a registered trademark of Courage-Khazaka, Cologne, Germany.

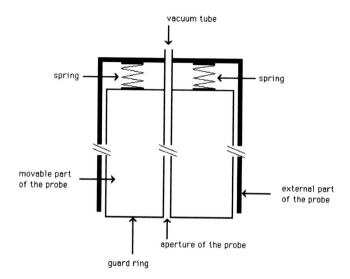

FIGURE 1 Schematic view of the measuring probe of the Cutometer®. Probe with a small aperture (2mm diameter).

The measuring probe, weight ±90 g, is applied vertically on the surface of the skin with a constant pressure by means of a spring (50 g/cm² or 2×10^3 N/m²). The skin deformation can be measured by this optical system up to an accuracy of 0.010 mm. The standard measuring probe has an aperture diameter of 2 mm (test area of about 3 mm²). Optional probes with an aperture of 4, 6, and 8 mm are available for studying the mechanical properties of larger skin areas. With the larger measuring probes, deeper layers of the skin (dermis and perhaps hypodermis) are deformed by suction.

The instrument is linked through an RS232 serial interface to an IBM compatible personal computer, version AT. A standard control software directing the Cutometer® is provided by the company. In this study the most recent available software program (Cuto,® version 5.3) was used.

The standard software allows storage of various data concerning the volunteer, date and time of experiment, skin area, external temperature and relative humidity, type of probe used, and mode of measuring technique. In addition graphic display of the obtained experimental curves allows calculations of individual values on the curve by means of a variable cursor.

B. Description of the Measuring Modes

Essentially two different measuring techniques are available: stress-vs.-strain mode and strain-vs.-time mode. In the stress–strain mode the deformation of the skin (strain) is displayed as a function of the stress (load-vacuum). In the strain–time mode the deformation of the skin is shown as a function of time (see Figure 2). In both experimental modes the choice of vacuum (from 50 to 500 mbar), the duration of the measurements (from 0.1 to 320.0 s), and the number of measurement cycles (from 1 to 163) can be preset.

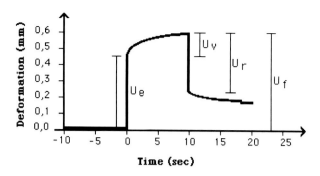

FIGURE 2 Graphical representation of a strain–time curve obtained for forearm skin. Aperture 2 mm, time of application 10 s, relaxation time 10 s, no pretension applied, load 500 mbar. Skin deformation in mm.

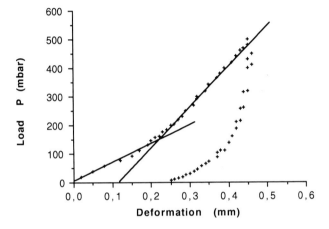

FIGURE 3 Graphical representation of a stress–strain curve obtained for forearm skin. Aperture 2 mm, linear increase and decrease in vacuum (50 mbar/s), total application time 20 s.

The strain–time mode which is mostly used in the mechanical studies on human skin, offers a choice of four measuring techniques. The four measuring techniques are the result of different combinations of choices of application rates and release rates of vacuum on the skin: immediate or slow linear increase and decrease of vacuum during a chosen period of time varying from 0.1 to 60 s.

In measuring system two where the deformation of the skin is measured during a linear increase in vacuum from 0 to maximal 500 mbar and subsequently during the linear decrease of vacuum, the resulting graphical display can be automatically replotted as a stress–strain curve (see Figure 3).

C. Pretension of the Skin

The skin immediately adjacent to the opening of the measuring probe is normally held in position by the guard ring of the probe on application of the probe on the skin surface. It appears, however, that during suction some lateral displacement of the skin adjacent to the opening may occur during the suction. It is the purpose of this instrument to measure only the vertical displacement of the skin during suction without major contribution of lateral displacement of the skin. In order to minimize this lateral displacement, it is possible to pretension the skin before the measurements are carried out. Pretension of the skin is possible by applying a preliminary suction on the surface of the skin during a short time (mostly short times of 0.1 to 0.2 s) before the real vertical deformation measurements are executed. The calibration of the optical system which measures the skin deformation is automatically preset by the instrument after pretension before the real measurements start.

D. Analysis of the Measuring Systems
1. Strain-Vs.-Time Curves

In the strain–time mode which is mostly used in viscoelastic studies on human skin, the vacuum is applied for a period of time varying from 2 to 10 s followed by a relaxation for 3 to 10 s. Figure 2 shows the results of a typical experiment carried out on the volar part of the forearm. The deformation parameters used in most studies in order to describe the different parts of the curve are those proposed by Aubert et al.[5] and Escoffier et al.[6]

Ue is the immediate deformation-skin extensibility. Uv is the deviation which reflects the viscoelastic contribution of the skin. Uf is the total deviation of the skin. Due to the steady increase of deformation of the skin in function of time after application of a given load (creep), the total deformation Uf does not reach a constant plateau value. As a consequence the maximal deformation Uf is usually determined after a constant time period of suction (usually after 10 s of application of suction). Ur is the immediate recovery of the skin after removal of vacuum. Finally due to typical viscoelastic properties of the skin, the deformation returns very slowly to the original state. After application of suction and removal of vacuum, values of 0.11 s were systematically observed for Ue and Ur. The deformation-versus-time curves obtained for the second, third, and subsequent deformation–relaxation cycles are similar to the first curve. But the curves progressively shifted vertically upwards as a consequence of the slow return of the skin to the original state.

All the deformation parameters Ue, Uf, Uv, and Ur are dependent of skin thickness. Since there are significant differences in skin thickness between women and men and that skin thickness varies with age, the use of these extrinsic deformation parameters for comparative studies is not adequate.[6] Consequently either intrinsic skin deformation parameters are used (deformation × skin thickness) or ratios of the extrinsic parameters are considered. The ratio Ur/Ue, ratio between immediate recovery and immediate deformation, is independent of skin thickness. This ratio is considered as a biological important factor for the characterization of elasticity of the skin.[6,7] The ratio Ur/Uf which closely resembles Ur/Ue is also used as a measure of elastic recovery.[8]

2. Stress-Vs.-Strain Curves

In the stress–strain mode, Uf is measured as a function of vacuum during a linear increase of vacuum followed by a linear decrease of suction (Figure 3). Generally nonlinear curves are obtained with hysteresis. The stress–strain curve with loading will not be superposed with the unloading curve. Furthermore, the values of strain in the relaxation procedure do not return to origin and will return to zero values only after a long period of relaxation.

Strictly speaking because of the nonlinearity of the stress–strain curves for human skin, the modulus of Young is not applicable. One can always define a coefficient of elasticity which corresponds to the derivative of the stress to the strain $d\sigma/d\varepsilon$ at a given stress value.[9] However in the experimental ascending curves (see Figure 3), a linear part in the stress–strain mode is clearly present between 150 and 500 mbar. From the linear part of the curves a modulus of Young E, in principle, can be calculated.[1,10]

For the practical calculation of the modulus of Young a theoretical model for the deformation of the skin in the suction aperture as proposed by Agache et al.[11] can be used to calculate the strain and stress. In this theoretical model one assumes that the initial flat surface of the circular test area is transformed by suction in a curved surface of a segment of a sphere.

IV. Measurements

A. General Results

The stress–strain and strain–time curves measured at different skin sites (forearm, face, thigh, etc.) by different researchers[12–17] using the suction method, are in good agreement with the results obtained by other methods (tensile, torsional, elevation, and indentation).[1] It is obvious that the deformation parameters (Ue, Uf, and Ur) vary in function of the load (vacuum) and the aperture of the suction probe. With the 2-mm suction probe, typical skin deformation data (Uf) ranging from 0.1 to 0.6 mm are recorded as a function of vacuum (from 100 to 500 mbar). These values of vertical skin elevation correspond typically to deformations of the dermis and with perhaps some contribution of the hypodermis when large deformation values are measured.

The values obtained for the elasticity ratio Ur/Ue are a function of anatomical sites, age, and other physiological factors. However, typical values of elasticity ratio ranging from 0.4 to 0.9 are recorded for the different anatomical skin sites. These elasticity recovery values are in good agreement with the results obtained by the torsional method.[6]

From the linear part of the stress–strain curves typical values of the modulus of Young ranging from 13×10^4 N/m^2 to 26×10^4 N/m^2 were computed for the different anatomical skin sites. These *in vivo* values obtained by the suction method are in agreement with previous data obtained by the torsional system (Modulus of Young = 42×10^4 N/m^2).[1,18] As pointed out previously by Piérard,[1] the determination of the extent of the linear part of the stress–strain curve is not always obvious and is generally rather subjectively determined by the researchers. This explains the much larger variations generally observed in the reported Young's modulus. As a consequence depending on the values of the applied stress and the mechanical system used in the study, the modulus of Young varies in a large range from 10^4 to 10^6 N/m^2.[1]

B. Accuracy and Reproducibility of the Measurements

Based on the calibration of the optical system with a micrometer, the deformation of the skin is measured with an absolute accuracy of 0.010 mm. The relative accuracy of the vacuum varies from 10% for small vacuum values (from 20 to 100 mbar) to 5% for higher vacuum values (from 100 to 500 mbar).

C. Reproducibility of Skin Deformation Parameters and of the Modulus of Young

Repetitive Uf deformation measurements without skin pretension carried out on the same person show a coefficient of variability ranging from 4 to 6% depending on the skin sites examined. With skin pretension the coefficient of variability is generally lower (around 4%). Similar results were obtained by other researchers using the suction method on different skin sites.[7,8,19] Similar repetitive measurements carried out on the same individual in the stress–strain mode show a coefficient of variability for the modulus of Young around 10%.

D. Factors Influencing the Measurements

1. Influence of Load

As would be logically expected and dependent on the opening of the suction probe, all the skin deformation parameters are nonlinearly increased as a function of load (vacuum). Due to limitations of the instrument, skin deformation measurements of 400 to 500 mbar are not possible with the 8-mm suction probe. In agreement with Elsner et al.[8] the elasticity ratio and other deformation ratios are independent of load for most anatomical sites (with the exception of vulvar skin).

Due to the curvature of most of the stress–strain curves and depending on the choice of the linear part in the stress–strain curves used for the calculation, the values of the modulus of Young are different and consequently load dependent. The values of modulus of Young obtained in this study are of the same order of magnitude as those recently reported by Agache et al.[11] In agreement with Agache,[11] the modulus of Young is more or less independent of the vacuum between 150 to 500 mbar.

2. Influence of the Diameter of the Suction Device

The maximal deformation Uf of the skin increases linearly in function of the diameter of aperture of the suction device. Maximal vertical deformation of 1.2 mm can be observed with a 6-mm suction probe. As previously mentioned, skin deformation measurements are not possible at vacuum loads of 400 to 500 mbar with the 8-mm suction probe.

3. Influence of the Orientation of the Probe

The vertical deformation of the skin is recorded by the optical measuring system along a well-defined direction which is indicated on the surface of the measuring probe. As a consequence the eventual anisotropy in the mechanical properties of the skin can be readily evaluated by measuring the vertical deformation of the skin under different orientations of the probe.

The mechanical properties of the forearm skin were evaluated in two directions: parallel and perpendicular to the primary lines of the skin microtopography of the skin surface on the forearm. No significant differences were observed in our laboratory for the elasticity ratio Ur/Ue or for the modulus of Young when measurements were parallel and perpendicular to the primary lines (preferential directions of the skin surface pattern) in young and middle-aged individuals.[20,21] These results seem to indicate that the suction method measures in an isotropic way the vertical deformation of the skin located at the forearm. In the future, studies concerning the influence of the orientation of the suction probe on the viscoelastic properties of the skin located at other anatomical sights (crow's feet for example) and with older individuals should be performed using the suction method. In all experiments it is recommended to maintain the probe exactly vertical to the skin surface.

4. Influence of Pressure of Application on the Skin

With the spring system the hand-held probe is always applied with constant pressure on the skin surface. However in order to reduce as much as possible the small variations in pressure of application, we have found it more convenient to perform the experiments under the following experimental set-up: the skin surface to be measured is always placed in a horizontal position and the suction probe is applied with the help of a stable probe holder.

5. Preconditioning of the Skin Surface

The skin adjacent to the aperture of the probe is immobilized by a guard ring in order to reduce as much as possible lateral displacement of the skin towards the opening of the suction device. In the strain–time mode it is possible to pretension the skin by applying a suction for a short time before the real deformation measurements are carried out (pretime setting on the instrument). It has been shown in our laboratory and also by others[19] that under pretension the values of the skin deformation parameters are more reproducible and more accurate. Measurements carried out at different skin sites (forearm, forehead, and crow's feet) show that with pretension the elastic recovery parameter Ur/Ue is systematically higher (typically for forearm skin in young individuals, Ur/Ue values change from 0.82 without pretension to 0.89 with pretension). This result indicates that with preconditioning the skin regains, to a greater extent, the initial position after deformation.

V. Applications

Due to the versatility of the measurements, the suction method is well suited to study *in vivo* the fundamental viscoelastic properties of the dermis in normal and diseased skin. In addition to this fundamental approach of the properties of the dermis, the influence of various factors such as sex, normal and actinic ageing, anatomical skin sites can be readily evaluated by this technique. Furthermore the efficiency of various dermatocosmetic treatments can be evaluated quantitatively by the suction method.

A. Influence of Ageing

Biological and photodamage ageing of the dermis have been extensively studied by several research groups using different experimental techniques.[3–7,8,16,17,20–24] Using the suction method we were able to show that the skin elasticity Ur/Ue was continuously and significantly decreased with age at various anatomical sites for both sexes.[20–22] Typically the elastic recovery ratio changes from 0.92 (age group 2 to 15 years), 0.71 (age group 18 to 25 years), 0.64 (age group 40 to 55 years) to 0.50 (age group 70 to 95 years). Significant increases of the modulus of Young were observed as a function of age at most skin areas (forearm, forehead, and thigh).[20,21] These results on the influence of ageing on the modulus of Young and on the Ur/UE parameter, are in good agreement with other studies obtained by the same suction method and by the torsional technique.[6–8]

B. Anatomical Skin Sites

The mechanical properties of the dermis located at different skin sites (forearm, forehead, temporal side, thigh, and vulvar skin) have been examined using the suction method.[7,8,13,15,16] Significant differences in viscoelastic properties have been observed reflecting differences in the structure of the papillary and reticular dermis. Ur/Ue varies in a range from 0.48 (vulvar skin), 0.50 (crow's feet), 0.60 (forehead), to 0.80 (forearm). Despite differences in the viscoelastic parameters, changes related with age seemed to be very similar for all the selected skin sites.[7,8,20–22]

C. Influence of Sex

In agreement with many other authors,[6–8] no significant differences were detected in the intrinsic skin deformation parameters (skin elasticity ratio Ur/Ue and the modulus of Young) between men and women using the suction method.

D. Objective Assessment of Topical Dermatocosmetic Treatments and Various Diseases

Skin diseases (scleroderma and Ehlers-Danlos Syndrome) and the therapeutic effect of retinoids in the treatment of photodamaged skin were examined on the level of the viscoelastic properties of the dermis.[24,25]

VI. Conclusions

The skin elasticity meter, Cutometer SEM 474® based on the suction method, allows us to measure in a simple way *in vivo* the viscoelastic properties of the skin. Under well-controlled experimental conditions where various parameters such as load (vacuum), aperture of the suction device, position and pressure of application of the probe, time of application and relaxation, and pretension of the skin are kept constant, reproducible and accurate stress–strain and strain–time curves are obtained which give quantitative information concerning the purely elastic and viscoelastic properties of the dermis. These parameters are used to study the properties of various anatomical skin areas in normal and diseased skin situations. Finally the influence of physiological parameters such as ageing, anatomical skin sites, and the efficacy of topical dermatocosmetic treatments can be quantitatively examined by this suction method.

References

1. Piérard, G., A critical approach to in vivo mechanical testing of the skin, in *Cutaneous Investigation in Health and Disease, Noninvasive Methods and Instrumentation,* Lévêque, J.-L., Ed., Marcel Dekker, New York, 1989, chap. 10.
2. Larrabee, W., A finite element model of skin deformation, *Laryngoscope,* 96, 399, 1986.
3. Daly, C.H. and Odland, G.F., Age-related changes in the mechanical properties of human skin, *J. Invest. Dermatol.,* 73, 84, 1979.
4. Vogel, H.G., Age dependence of mechanical and biochemical properties of human skin, *Bioeng. Skin,* 3, 141, 1987.
5. Aubert, L., Anthoine, P., de Rigal, J., and Lévêque, J.-L., An in vivo assessment of the biomechanical properties of human skin modifications under the influence of cosmetic products, *Int. J. Cosmet. Sci.,* 7, 51, 1985.

6. Escoffier, C., de Rigal, J., Rochefort, A., Vasselet, R., Lévêque, J.-L., and Agache, P., Age-related mechanical properties of human skin: an in vivo study, *The Journal of Investigative Dermatology*, 93, 353, 1989.
7. Cua, A.B., Wilhelm, K.P., and Maibach, H.I., Elastic properties of human skin: relation to age, sex, and anatomical region, *Arch. Dermatol. Res.*, 282, 283, 1990.
8. Elsner, P., Wilhelm, D., and Maibach, H.I., Mechanical properties of human forearm and vulvar skin, *Br. J. Dermatol.*, 122, 607, 1990.
9. Manschot, J.F. and Brakkee, A.J., The measurement and modelling of the mechanical properties of human skin in vivo. I. The measurement; II. The Model, *J. Biomech.*, 19, 511, 1986.
10. Piérard, G.E. and Lapière, C.M., Structures et fonctions du derme et de l'hypoderme, in *Précis de Cosmétologie Dermatologique*, Pruniéras, M., Ed., Masson, Paris, 1981, chap. 2.
11. Agache, P., Varchon, D., Humbert, P., and Rochefort, A., Non-invasive assessment of Biaxial Young's modulus of human skin in vivo, presented at The 9th International Symposium on Bioengineering and the Skin, Sendai, October 19–20, 1992.
12. Barel, A.O. and Clarys, P., Noninvasive measurements of the viscoelastic properties of the human skin with the suction method, presented at The 8th International Symposium on Bioengineering and the Skin, Stresa, June 13–16, 1990.
13. Malm, M. and Serup, J., In vivo skin elasticity of different body regions: the vertical vector, presented at The 8th International Symposium on Bioengineering and the Skin, Stresa, June 13–16, 1990.
14. Anfossi, T., Bosio, D., and Emanuelle, G., Influence of environment factors on skin elastometric patterns, presented at The 8th International Symposium on Bioengineering and the Skin, Stresa, June 13–16, 1990.
15. Barbanel, J.C., A suction method for determining the direction of Langer's lines, presented at The 8th International Symposium on Bioengineering and the Skin, Stresa, June 13–16, 1990.
16. Nishimura, M. and Tsuji, T., Measurements of skin elasticity with a new suction device: relation to age, sex and anatomical regions in normal skin and its comparison with some diseased skin, presented at The 9th International Symposium on Bioengineering and the Skin, Sendai, October 19–20, 1992.
17. Barel, A.O. and Clarys, P., In vivo evaluation of skin ageing. Relations between visco-elastic properties and skin surface parameters, presented at The 9th International Symposium on Bioengineering and the Skin, Sendai, October 19–20, 1992.
18. Agache, P., Monsieur, C., Lévêque, J.-L., and de Rigal, J., Mechanical properties of Young's modulus of human skin in vivo, *Arch. Dermatol. Res.*, 269, 221, 1980.
19. Vandenbussche, K., Meting van elasticiteit en viscoelasticteit van de huid, Pharmacy Thesis, Vrije Universiteit Brussel, Belgium, 1991.
20. Van Den Eynde, A., Invloed van de leeftijd op de viscoelastische eigenschappen van de huid, B. Sc. Thesis, Vrije Universiteit Brussel, Belgium, 1990.
21. VanWonterghem, M., Mechanische eigenschappen van de menselijke huid: invloed van verschillende factoren, B. Sc. Thesis, Vrije Universiteit Brussel, Belgium, 1991.
22. De Schrijver, K., Evaluatie van de visco-elastische parameters van de huid, B. Sc. Thesis, Vrije Universiteit Brussel, Belgium, 1992.
23. Berardesca, E., Borroni, G., Borlone, R., and Rabbiosi, G., Evidence for elastic changes in aged skin revealed in an in vivo extensometric study at low loads, *Bioeng. Skin*, 2, 261, 1986.
24. Sparavigna, A. and Galbiati, G., Strain-time curve in the assessment of topical retinoin as an anti ageing agent, presented at The 8th International Symposium on Bioengineering and the Skin, Stresa, June 13–16, 1990.
25. Dempo, K., Sasaki, T., and Nakajima, H., Skin elasticity of systemic scleroderma, presented at The 9th International Symposium on Bioengineering and the Skin, Sendai, October 19–20, 1992.

Chapter 14.4
Identification of Langer's Lines

J. C. Barbenel
Bioengineering Unit
University of Strathclyde
Glasgow, U.K.

I. Introduction

Langer's lines reflect systematic directional variation in the mechanical behavior of the skin. Dupuytren[1] reported that a stilleto, which has a blade of circular cross-section, may produce an oval wound. The phenomenon was systematically investigated by Langer[2] who punctured the skin of a series of cadavers with an awl which had a blade which was circular in cross-section. The resulting wounds were generally *clefts,* i.e., oval holes. Langer placed the stab wounds as closely together as possible and showed that the long axes of the wounds formed a regular pattern, now called Langer's lines (Figure 1). The work has been repeated by Cox[3] who confirmed the systematic nature of the direction of the long axes of the wounds, although the direction of the lines obtained by Cox did not wholly agree with those of Langer.

There are two basic causes of the phenomenon which characterize Langer's lines. Even under passive, resting, conditions the skin has a series of built-in internal tensions which are, in part, due to the fact that the skin acts as container for its contents. If a circular incision is made in the skin, then the tensions produce an oval defect with major and minor axes which are larger than the diameter of the original circular incision. The circular piece of skin within the incision will shrink into an oval with axes smaller than the diameter of the incision. The long axis of the oval of the defect will generally be at right angles to the long axis of the skin circumscribed by the incision. These effects are all due to the release of the resting tension due to cutting through the skin. The skin also appears to be mechanically anisotropic even after the removal of resting tensions by excising strips of skin and testing them *in vitro,* as was done by Langer.[1]

The relationship between the anisotropic biomechanical behavior of skin and Langer's lines make the latter of considerable practical importance. Their direction influences the placement and orientation of surgical incisions because incisions in the direction of Langer's lines generally gape less and produce better quality and less conspicuous scars than incisions across Langer's lines. The results of those methods of noninvasive testing of the mechanical properties of skin which extend the tissue may also be strongly influenced by the relationship between the test direction and the local Langer's lines, particularly if the test loading is uniaxial. The orientation of the lines must be known if useful and meaningful comparisons are to be made between the results of the tests at similar sites on different subjects, or of the secular changes at the same site on a single subject.

Unfortunately the directions of Langer's lines are not constant but show significant variations between people and may not remain constant at a single site for a specific subject. It is necessary, therefore, to develop test methods for the detection of the direction of Langer's lines.

II. Object

The technique used by Langer is the only one that clearly and directly identifies the local directions of Langer's lines, but it is certainly not a widely applicable method. In order to develop a noninvasive method it is important to understand those underlying biomechanical properties of the skin which are related to Langer's lines and to identify those features which can be measured.

A. Skin Biomechanics

The skin shows nonlinear load extension of stress strain when subjected to simple tension[4] (Figure 2). Langer[1] himself carried out such tests on a large number of excised strips of skin, and identified the nonlinearity of the load-extension response, commenting that "skin does not stretch proportionally with applied load, indeed the amount of extension grows steadily smaller so that the course of progressive extension cannot be represented by a straight line but by a curve." The nonlinearity has been repeatedly demonstrated by both *in vitro*[5] and *in vivo*[6] studies. The load-extension response

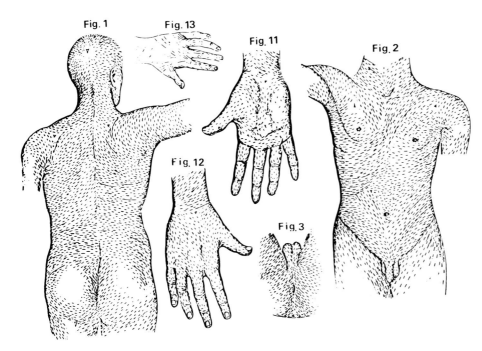

FIGURE 1 Langer's lines. (From Langer, K., *Br. J. Plast. Surg.*, 31, 3, 1978. With permission.)

can generally be divided into three phases. The first phase of initial extension is characterized by the high compliance of the tissues when small loads produce large extensions. The second phase represents a progressive, but nonlinear, stiffening of the tissue with increasing extension. The third, terminal, phase is one in which the stiffness of the tissues is greater and the load-extension response becomes linear.

Repeated load or extension cycling produces a progressive change in the load-extension response of the skin, with the three phases of behavior becoming more marked in the second and subsequent cycles compared to the initial cycle. With an increasing number of cycles the extensions obtained by the application of a specific load increases, but the difference between cycles decreases until a stable, reproducible, response is obtained. This behavior is usually called preconditioning and occurs both *in vivo* and *in vitro*. It is particularly marked if the skin has been stretched in a direction other than the test direction.

The skin also shows time-dependent behavior. The load required to maintain the tissue at a constant extension decreases with time, a phenomenon known as stress relaxation. Conversely the application of a constant load produces a time-dependent increase in extension, which is known as creep. These effects imply that constant rate testing is desirable, but rate dependence is not marked in the range of extension rates possible for *in vivo* tests.

B. Qualitative Correlates of Langer's Lines

The load-extension curves described previously are directionally dependent (Figure 2) and this is usually discussed in connection with the directional dependence of Langer's lines. It must be realized, however, that such *in vivo* comparisons are not precise because it is impossible to measure the actual direction of Langer's lines using a penetration technique and comparisons can only be drawn between specific noninvasive test results and what are believed to be the general form and direction of Langer's lines at the test site.

The most striking features of tests carried out in the direction of, or normal to, Langer's lines is that within the second and third phase of the skin's stress-strain behavior, the extension obtained at a specific load is always least in tests made in the direction of the local Langer's lines (Figure 2).

The slope of the third, linear, phase is rather more variable, but generally is greatest when the skin is stretched in the direction of Langer's lines. The slope of the initial, compliant, phase is more difficult to investigate *in vivo*. *In vitro* tests made on strips of skin aligned either along or across Langer's lines suggested that the differences in extensibility at a specific load shown in phases 2 and 3 also occurred in the initial phase; there was, however, no significant or consistent relationship between the stiffness and the direction of Langer's lines.[7]

III. Methodological Principle

The relationship between the anisotropy of skin extensibility and the local orientation of Langer's lines

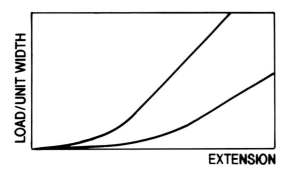

FIGURE 2 Results of uniaxial tension tests *in vivo*. The anisotropy of the skin appears as a difference in the terminal, linear, slope, and the extensions produced by the same load/unit width.

provides a method of detecting the latter. There are two techniques which can be applied for the purpose of detecting differences of the skin extensibility.

A. Multiple Simple Tension Tests

Several devices have been described to carry out uniaxial tension tests on skin *in vivo*. These can be employed to make uniaxial tests in multiple directions, and the correlation between the direction of Langer's lines and the biomechanical property of the skin previously described may be used to detect the local direction of the lines.

The only major systematic study[8] to validate the correlation between simple tension tests and Langer's lines utilized a constant speed extension device which was coupled to the skin by double-sided adhesive tape. The load-extension response was obtained for three extension tests performed in a single direction at a given site. The tests were then repeated in three other directions such that the four test axes were at 45° to their nearest neighbor. The tangent of the linear portion of the third phase of the load-extension curve in each direction was extrapolated and the extension at which this extrapolated tangent cut the zero load axis was defined as the limit strain. The values of the limit strain in the four test directions were then used to identify the direction of minimum limit strain which was strongly correlated with the direction of Langer's lines at the sites investigated.

The technique has several drawbacks. Multiple tests to obtain the complete load-extension curve, with three replicates in each direction, is extremely time consuming. Limitations of the number of tests to reduce the duration of the program reduces the accuracy with which the direction of the minimum limit strain can be defined. In part, the need to investigate the terminal, linear, phase of extension which is required to define the limit strain increases the amount of preconditioning in other directions and the number of repetitions required.

A simpler method of using uniaxial tension tests would be to measure the directional variation of extension produced by a load large enough to be well within the second phase or the start of the third phase of extension. The definition of the necessary load is not simple because uniaxial tension tests carried out *in vivo* produce inhomogeneous deformations[9] which depend on such variables as the widths of the loading tabs and their initial separation.

Stark[10] investigated the angular variation of extension produced by a defined load in order to detect the orientation of Langer's lines. He used an initial length, equivalent to the separation between the loading tabs, of 30 mm and a load of 14.2 g/mm. He found that the extension reached a relatively stable value 1.5 s after the application of load and that repeated testing was not required because of the lack of preconditioning. The rapidity with which tests could be completed allowed the investigations of eight directions at each site. The published results show an impressive correlation between the direction of minimum extension and the classical direction of Langer's lines. Stark produced his constant load using a simple plastic instrument similar to a pair of dividers, but containing a constant tension coiled flat strip spring.[11]

B. Suction Chamber Methods

The suction/inflation of a circular disc of skin with a fixed periphery has formed the basis of devices to assess the mechanical properties of the skin (see Chapters 14.2 and 14.3). Imaging the shape of an inflated disc with a fixed periphery showed that anisotropic tissue produced an inflated dome which was not axisymmetric. The directions of the axes of symmetry of the dome defined the direction of maximum and minimum extensibility of the tissue.[12] The method has not been applied to skin *in vivo* but has the potential to determine the direction of minimum extensibility, and hence the direction of the local Langer's lines from the results of a single test.

The suction test devices currently used to assess skin mechanics do not image the shape of the inflated skin disc within the test apparatus but only indicate the maximum height of the inflated dome. Hence they cannot be used to investigate anisotropy. In principle, a parallel-sided slit may be used instead of a circular orifice and variation of dome height with orientation could be used to determine the direction of least extensibility, in a way analogous to multiple simple tension tests. The technique appears, however, to offer few advantages over repeated uniaxial tension tests.

It has been suggested that the suction method can be used to detect Langer's lines if the skin is allowed to slip under the periphery of the suction device, rather than being constrained at the boundary.[13] The suction device consisted of a transparent plastic cylinder which was smooth and rounded where it was applied to the skin surface in order to reduce the interface friction. The cylinder housed a dome-shaped piston which defined the shape and amount of skin sucked into the cylinder. The extent of skin slip and translation was determined by drawing a line on the skin around the circular periphery of the suction cylinder. When the suction was discontinued and the skin allowed to return to its resting configuration the original circle became an oval. Tests were made in the abdominal midline, where both the direction of Langer's lines and anisotropy are well established. Load-extension tests made at the test site confirmed that the long axis to the oval was in the direction of minimum skin extensibility and hence of the local Langer's line. Qualitative investigation of other thoracic and abdominal sites showed that the load axes of the ovals produced by the suction device coincided with the directions of Langer's lines.

IV. Sources of Error

Body posture may change the extension and tension in the skin and therefore modify the directions of Langer's lines. There may be a direct effect due to the skin being stretched, particularly close to, or overlying, mobile joints. Changes in muscle size due to shortening or contraction will also stretch the skin and alter Langer's lines at areas distant from the joints. It is thus necessary to relate the direction of Langer's lines to posture. A standard posture should be maintained throughout the test program if the determination of the direction of the local Langer's lines is to precede other tests of the mechanical properties of the skin.

Short-term changes in body shape due to such events as pregnancy, rapid slimming or muscle building, may produce significant alterations in Langer's lines by stretching the overlying skin.

V. Correlation with Other Methods

The usefulness and relevance of measurements of skin extensibility previously outlined depend on the correlation between the anisotropy of these mechanical properties and the direction of Langer's lines. There also appears to be a correlation between the limit strain and the orientation of skin surface roughness caused by the intersecting grooves and ridges which produce the characteristics of skin surface patterns.[16] The measurement of the skin surface roughness is even more complicated than the measurement of skin extensibility, and it cannot be considered as a means of determining Langer's lines.

VI. Recommendations

There is no commercially available equipment designed specific to detect the local orientation of Langer's lines. There are, however, devices for carrying out uniaxial tension tests *in vivo* (see Chapters 14.2 and 14.3) which can be used to determine the load-extension behavior of the skin in multiple directions. Devices which produce a force indication can be used in the constant force mode described by Stark,[10] which is simpler than the determination of limit strain.

The use of a suction device which allows skin slip appears to have considerable promise but has not yet achieved general use and must, at present, be considered as having undergone only limited validation.

References

1. Dupuytren, G., *Theoretisch-praktische Vorlesungen uber Verletzungen durch Kriegswaffen,* Vert, Berlin, 1836.
2. Langer, K., On the anatomy and physiology of the skin, (from Sitzungsberichte der mathematisch-naturwissenschaftlichen Klasse der Kaiserlichen Akademie der Wissenschaften, 44, 19, 1861), translated by Gibson, T., *Br. J. Plast. Surg.,* 31, 3, 1978.
3. Cox, H.T., The cleavage lines of the skin, *Br. J. Surg.,* 29, 234, 1941.
4. Kenedi, R.M., Gibson, T., Evans, J.H., and Barbenel, J.C., Tissue mechanics, *Phys. Med. Biol.,* 20, 699, 1975.
5. Fung, Y.C., *Biomechanics in Mechanical Properties of Living Tissues,* Springer-Verlag, New York, 1981, chap. 7.
6. Barbenel, J.C., Skin biomechanics, in *Concise Encyclopaedia of Biological and Biomedical Measurement Systems,* Payne, P.A., Ed., Pergamon Press, Oxford, 1991, 347.
7. Wan Abas, W.A.B. and Barbenel, J.C., The response of human skin to small tensile loads *in vitro, Eng. Med.,* 11, 43, 1982.
8. Gibson, T., Stark, H.L., and Evans, J.H., Directional variation in extensibility of human skin *in vivo, J. Biomech.,* 2, 201, 1969.
9. Wan Abas, W.A.B. and Barbenel, J.C., Uniaxial tension test of human skin *in vivo, J. Biomed. Eng.,* 4, 65, 1982.
10. Stark, H.L., Directional variations in the extensibility of human skin, *Br. J. Plast. Surg.,* 30, 105, 1977.
11. Stark, H.L., The surgical limits of extension and compression of human skin, Ph.D. thesis, University of Strathclyde, Glasgow, U.K., 1977.
12. Zioupos, P., Barbenel, J.C., and Fisher, J., Mechanical and optical anisotropy of bovine pericardium, *Med. Biol. Eng. Comput.,* 30, 76, 1992.
13. Barbenel, J.C., A suction method for obtaining the direction of Langer's lines, in *The 8th International Symposium on Bioengineering and the Skin,* 1990, 68.
14. Ferguson J.M. and Barbenel, J.C., Skin surface patterns and the directional mechanical properties of the dermis, in *Bioengineering and the Skin,* Marks, R.M. and Payne, P., Eds., MTP Press, Lancaster, 1983, 833.

Chapter 14.5
Levarometry

Vered Manny-Aframian and Shabtay Dikstein
Unit of Cell Pharmacology
School of Pharmacy
The Hebrew University of Jerusalem
Jerusalem, Israel

I. Introduction

The mechanical properties of the human skin are known to change with age: the skin becomes lax and wrinkled with the loss of elasticity and turgor. A scientific approach to the prevention and treatment of these changes requires a noninvasive method of assessing structural changes in the skin with age. Such a method would enable us to quantify the changes in the mechanical properties of the skin with age, as well as changes due to environmental causes such as solar damage. It would also help us to quantify the effects of different treatments. Levarometry is a method of evaluating skin slackness.

II. The Measuring System

One of the main characteristics of wrinkled skin is that it stretches more readily than smooth skin when force is applied, i.e., wrinkled skin is more "slack" than smooth skin. To measure this "skin slackness" one has to apply a perpendicular pull to the skin without a guard ring, thus enabling the skin to respond freely to the applied force. Levarometry works on this principle.

III. Methodological Principle

To measure skin "slackness", we developed a device called the levarometer[3] which is a modified Schade instrument[1,2] with increased sensitivity. The increased sensitivity was achieved by using modern electronic techniques, allowing the skin to be analyzed *in vivo* in the low range of the stress-strain curve shown by Daly[4] to be the most sensitive parameter in differentiating between young and old skin elasticity measurements.

The levarometer (Figure 1) operates by applying a perpendicular pull to the skin without a guard ring. A circular piece of Perspex® with a diameter of 0.5 cm is attached to the skin by double-sided adhesive tape. This is then attached to a counterbalanced measuring rod so that the net pressure of the system is less than 1 g/cm^2. Different weights can be applied to the rod, thereby providing any desired elevating force. The elevating forces used are usually low; in the range of 5 to 40 g/cm^2. The measuring rod is connected to a linear variable differential transformer (LVDT), the output of which is recorded graphically. The precision of the measuring system is 2 μm. The main problem is in ensuring that there is no friction between the measuring rod and the LVDT.

The most sensitive measurement is the immediate levarometric response without measuring the creep. As a standard, we use a 4-g weight and since the area of the probe is $0.25^2\pi \cdot$ cm^2 = 0.2 cm^2, the pull is 20 g/cm^2 (see validation). A theoretical treatise of levarometry has been published by Lanir et al.[5]

IV. The Sensitivity and Reproducibility of the Levarometry Measurements

1. When measuring the same area at different times the variance of the measurements is less then 5.5%.[6]
2. When measuring the dispersal of measurements taken immediately one after the other over an area of 4.5 × 4.5 cm^2, the deviation from the mean is 6%. This deviation is very small considering that the skin is not a homogeneous tissue.[6]
3. Topical application to the skin of a humectant (5% glycerine) did not affect the measurements. This implies that the measurement is not affected by environmental conditions or treatments that affect the stratum corneum.[6]

V. Validation

Does levarometry differentiate between young and old skin? In order to answer that question the skin of two

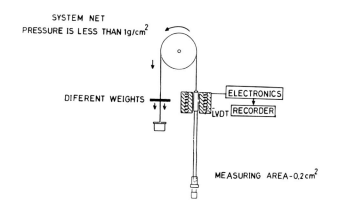

FIGURE 1 Schematic view of a levarometer.

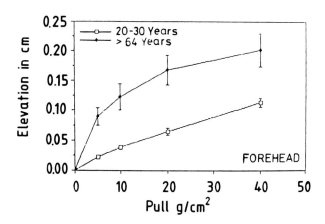

FIGURE 2 The dependence of levarometric measurements on age. N = 15 in each age cohort.

groups was compared: young women 20 to 30 years of age vs. older women 65 to 80 years old. There were 15 healthy volunteers in each group. The measurements were done on the forehead and four elevating pulls were applied: 5, 10, 20, and 40 g/cm².[6] The results (figure 2) show that levarometry distinguishes between the two age groups. The difference was significant for all the elevation forces ($p <0.02$ for 5 g/cm² and $p <0.01$ for 10, 20, and 40 g/cm²).

On the basis of the forehead measurements we decided that the standard pull for all other measurements should be 20 g/cm². This pull was chosen because it is the highest pull within the linear range. With a lower pull there is greater risk of human error, since small pulls are more difficult to handle.

Can levarometry differentiate between male and female skin? The inner forearm skin in two groups: young subjects 20 to 30 years of age and older subjects 61 years old was measured at the standard pull of 20 g/cm².[6] There were 22 women and 22 men in each age group. The inner forearm was selected as the measuring site because it is usually less exposed to sun.

The results of the forearm measurements show (Figure 3) that levarometry differentiates between the two age groups and also between men and women. Most of the men tested in both age groups had less skin elevation than women. This difference was more marked in the older age group; in the young age group it was 40%, while in the older group it was 75% ($p <0.01$).

What is the difference between young and old skin by levarometry at 20 g/cm² pull? The measurements of the female forehead skin show that the change between the young and old cohort is 160% ($p <0.01$), (Figure 2).[6] The measurement on the volar forearm skin shows a change of 60% for the young versus old male cohort ($p <0.01$), whereas in the young versus old female cohort the change is 100% ($p <0.01$), (Figure 3).[6]

VI. Correlation with Other Methods

There are two research groups using similar methods[7,8] with which the results of the levarometry were compared.

A. Pull, Guard Ring, and Weight of Probe

The elevating forces used by Pierard[7] are 16 to 128 g/cm² with a measuring disc diameter of 14 mm, and 64 to 512 g/cm² for a diameter of 7 mm. Gartstein[8] uses 10 g elevating force with a measuring disc diameter of 3 mm.

Pierard[7] uses guard rings of different diameters. We found that if the ratio between the diameter of the measuring disc and the diameter of the guard ring is at least 6, then the ring has minimal influence on the levarometry. Otherwise, one is measuring "skin elasticity" rather then skin slackness. As pointed out, we do not use a guard ring.

In Pierard's method,[7] the disc is glued to the skin by cyanoacrylate; whereas Gartstein et al.[8] use a vacuum. We use double-sided adhesive tape instead of cyanoacrylate to avoid the difficulty of applying the cyanoacrylate only on the precise area of the measurement. It is important that the starting net pressure of the measuring probe on the skin should be minimal, so that it does not indent the skin before elevating it. There is no published information on the net pressure of the systems described above. Our system's net pressure is about 1 g/cm².

B. Elevations

Pierard[7] measures loading deformation, which is a measurement of the change 20 s after loading the force, i.e., measuring immediate levarometry with the immediate creep. Gartstein et al.[8] uses 0.25 s for loading time thereby measuring immediate levarometry, allowing the same time for recovery with a total cycle time of 0.50 s, recorded by a computer. Gartstein et

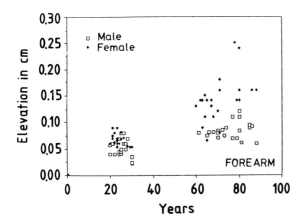

FIGURE 3 Change of standard levarometric measurement according to age for women and men. N = 22 in each age and sex cohort.

al.[8] also measure skin elasticity as the percentage of recovery after deformation. We measure the immediate elevation without the creep. The reproducibility of Gartstein's system is 4%,[8] Pierard's is lower than 8%,[10] and ours is lower than 5.5%.[6]

C. Conclusions

Gartstein et al.[8] found that young, undamaged skin stretches less and recovers more completely than aged or solar-damaged skin. We agree that young skin is less slack than aged skin (see Section V, Validation). A comparison of levarometry with other mechanical methods such as torsion, elevation with guard ring, indentometry and ballistometry,[6,11–16] suggests that levarometry without a guard ring is highly discriminating between old and young skin, and also between old male and old female skin. Levarometry is, therefore, eminently suited to the study of aging skin.

VII. Future Studies

A portable, easy to handle, commercially available levarometer with a fixed elevating force of 20 g/cm^2, should be developed. Such a device could rapidly generate data from many researchers and hopefully broaden knowledge about hormonal and other factors influencing skin slackness.

References

1. Schade, H., Die Elasticitatsfunction des Bindegewebes und Die Intravitale Messung Ihrer Strorungen, *Z. Exp. Pathol. Ther.*, 11, 369, 1912.
2. Kirk, E. and Koorning, S.A., Quantitative measurements of the elastic properties of the skin and subcutaneous tissue in young and old individuals, *J. Gerontol.*, 4, 273, 1969.
3. Dikstein, S. and Hartzshtark, A., In-vivo measurement of some elastic properties of human skin, in *Bioengineering and the Skin*, Marks, R. and Payne, P.A., Eds., MTA Press, England 1979, p. 43.
4. Daly, H.C. and Odland, G.F., Age related changes in the mechanical properties of human skin, *J. Invest. Derm.*, 73, 84, 1979.
5. Lanir, Y., Manny, V., Zlotogorski, A., Shafran, A., and Dikstein, S., The influence of ageing on the in-vivo mechanics of the skin, *Skin Pharmacol.*, 6:223–30, 1993.
6. Manny, V., In-vivo deformation by small forces as a criterion for assessing skin ageing, M. Pharm. thesis, Hebrew University Jerusalem, 1989.
7. Pierard, G.E., Investigating rheological properties of skin by applying a vertical pull, *Bioeng. Skin Newsl.*, 2, 31, 1980.
8. Warren, R., Gartstein, V., Kligman, A.M., Montagna, W., Allendorf, R.A., and Ridder, G.M., Age, sunlight and facial skin: a histologic and quantitative study, *J. Am. Acad. Dermatol.*, 25, 751, 1991; Gartstein, V. and Elsau, W.H., Measurement of skin elasitity: tonometry method, *8th International Symposium on Bioengineering and the Skin*, Stresa, 1990, p. 70.
9. Pierard, G.E. and Lapier, Ch. M., Physiopathological variations in the mechanical properties of skin, *Arch. Derm. Res.*, 260, 231, 1977.
10. Pierard, G.E., Structure et properties mechaniques descompartiment adventitiel et reticulaire du derme, thesis for Agrege de l'Enseignement Superieur, University of Liege, 1984.
11. Brozek, J. and Warren-Kinzey, W., Age changes in skinfold compressibility, *J. Gerontol.*, 15, 45, 1960.
12. Graham, R. and Holt, P.J.L., The influence of ageing on the in vivo elasticity of human skin, *Gerontologia*, 15, 121, 1969.
13. Tosti, A., Compagno, G., Fazzini, M.L., and Villardita, S., A ballistometer for the study of the plastoelastic properties of skin, *J. Invest. Dermatol.*, 69, 315, 1977.
14. Cook, T., Alexander, H., and Cohen, M.C., An experimental method for determining the two-dimensional mechanical properties of living human skin, *Ned. Biol. Eng. Comput.*, 15, 381, 1977.
15. Lévêque, J.-L. and Corcuff, P., In vivo studies of the evaluation of physical properties of human skin with age, *Int. J. Dermatol.*, 23, 322, 1984.

Chapter 14.6
Indentometry

Vered Manny-Aframian and Shabtav Dikstein
Unit of Cell Pharmacology
School of Pharmacy
The Hebrew University of Jerusalem
Jerusalem, Israel

I. Introduction

When asked which of various body parts are soft or hard, people generally unwittingly demonstrate the principle of indentometry. They first prod their faces, palms, etc., with the tips of their fingers or a pencil, etc. When asked what they were attempting to determine, a typical remark would be "to see how deep the pencil would go." Industrial measurements of hardness of materials are based on a similar principle. By the same token, in order to measure "skin softness" one has to apply a perpendicular force to the skin, which will then indent in response to the applied force to a degree related to its softness. The various indentometry methods are based on this principle.

II. Various Measuring Systems

The first indentometer was built by Schade[1] in 1912. All other indentometry measuring systems are modifications of his system. Schade's measuring system,[1] the elastometer, consists of a registering lever and a rotating drum. One end of the lever is connected to a vertical metal rod terminating in a hemispherical brass knob with an area of 50 mm² that is applied to the skin. The rod also carries a small platform upon which a weight (ranging from 5 to 75 g) may be placed. A pen point and ink is attached to the other end of the lever, which plots the curve on paper on the rotating drum. The measuring pressure is 10 to 150 g/cm².

The measuring system of Kirk et al.,[2,3] first published in 1949, uses a similar apparatus, the only difference being that movements of the lever are traced by a celluloid point on smoked paper instead of by a pen point and ink. The standard measuring weight was 50 g, and the measuring pressure was 100 g/cm².

The measuring system of Tregear et al.,[4] first published in 1965, consists of a weighted metal rod with a measuring area of 0.1 to 5 cm²; the rod is free to slide in a vertical glass tube, and can be loaded with weights ranging from 100 to 2000 g. The rod is placed on a skinfold area, the undersurface of which rests on a thin metal disc of the same diameter as the rod. The vertical position of the rod is registered by a spring loaded dial micrometer, accurate to 2 µm. The measuring pressure is 20 to 400 g/cm².

In 1969, Robertson et al.[5] first described a measuring system consisting of a measuring rod and linkage system with a low moment of inertia. The probe is a hollow aluminum rod mounted vertically on low-friction bearings. The lower, measuring end, which is applied to the skin, has a diameter of 0.5 cm. An aluminum plate at the upper end bears the weights. This probe has a simple linkage to a rotating transducer, the outer case of which rotates in an arc around a stationary inner core in response to vertical movement of the rod. The inner core contains a photosensitive element. The output of the photoresistive cell is related to the vertical displacement of the probe, and is measured on a recording voltmeter. The instrument has a response time of 50 msec and can record movements of up to 1 cm with a resolution of 0.01 cm. The internal friction is approximately 1 dyn. A 50-g weight was used in all his experiments. The measuring pressure is 250 g/cm².

The measuring system of Daly et al.,[6,7] first published in 1974, consists of an ultrasonic transducer connected to a lead rod 4.5 mm in diameter and 20 mm long. Loads are applied to the tissue by a servocontrolled loading system. A semiconductor strain-gauge bridge measures the load directly at the transducer. The load used is 5 g, and the measuring pressure is 71 g/cm².

The instrument of Pierard et al.,[8] first described in 1984, uses an aluminum rod, the force is created by a manometer, and the measurement is carried out using a comparator. The measuring pressure used ranges from 1 to 4 N per 320 mm². The precision of the measurement is 10 µm. The measuring pressure used was 28 to 111 g/cm².

The measuring system of Dikstein et al.,[9] first published in 1981, has a measuring area of 0.2 cm² (Perspex®), connected to a light metal measuring rod that is counterbalanced so that the net pressure of the system is less than 1 g/cm². The measuring rod can be loaded with different weights. We decided on a standard load of 2 g, since the resultant pressure is still in the linear part of the stress-strain curve. The measuring rod is connected to a linear variable differential transformer (LVDT), and the output is recorded graphically by an electronic recorder. The measuring rod is adjusted so as to just touch the skin. Once the baseline is stabilized, the weight is suddenly applied. Measuring the same area at different times using this method the variance of the measurements is less than 6%.[10] Measuring the dispersal of measurements taken immediately one after the other over an area of 4.5 × 4.5 cm², the deviation from the mean is 10%.[10] The standard measuring pressure used was 10 g/cm².

III. Indentometry Measurements Using Different Methods

Indentometry measurements were used to assess skin softness in a variety of physiological conditions: old and young skin, male and female skin, and in various edematous conditions.

The measurements of Schade et al.[1] were carried out on the leg below the knee and on the forearm below the elbow in various edematous conditions. The effect of age on the initial indentation and on elastic recovery was also studied. It was found that in edema, both the indentation and the elastic recovery decrease. With age, it is mainly the elastic recovery that changes (decreases).

The measurements of Kirk et al.[2,3] were carried out on the medial surface of the tibia, using a weight of 50 g. The measurements were performed in women aged 20 to 101 years and in men aged 18 to 104 years. The observations showed a definite decrease in indentation with age and an even more marked decline in the degree of immediate resiliency (rebound — called by the author elasticity) of the skin following removal of the weight. Women's skin in general was found to possess higher elastic properties than that of men of similar ages; the elasticity values recorded for women were usually of the same order of magnitude as those exhibited by men 10 years younger.[3]

Indentometry measurements were also carried out on the skin covering the lower and upper tibia. The weight used was again 50 g. The measurements were done on old (60 to 86 years) and young (18 to 22 years) skin. The results again showed a marked difference between the two groups: the depth of indentation in the young skin was much greater than in the older age group; the immediate rebound of the tissue after removal of the weight was also much greater in the young than in the old age group.[2]

A force of 100 to 2000 g acting over 0.1 to 5 cm² for up to 20 min. was used by Tregear et al.,[4] to compress human skin *in vivo*. The subjects were young white male adults. The area tested was the skin of one leg over the tibia, and the dorsal aspect of one forearm. The maximum compression produced was 0.6 mm. The time course of the compression was not exponential; there was a long-continued "tail" of deformation, with the compression remaining after the force had been removed. The speed of compression was increased by increasing the force. Tregear did not carry out comparative physiological studies.

The measurements of Robertson et al.,[5] were made on the dorsum of the hand, forearm, biceps, triceps, knee, calf, and foot. Measurements were carried out on three groups of women: nonpregnant, pregnant with no clinical edema in the third trimester, and pregnant women with clinical edema. In addition, measurements were performed on one pregnant woman who developed widespread edema in late pregnancy. The indentation measurements produced a similar pattern to the compressibility measurements by a modified caliper: increasing depth of indentation was accompanied by a lowered index of compressibility.

Daly et al.,[6] measured the anteromedial surface of the tibia. There were 20 volunteers, male and female, ranging in age from 8 to 86 years. Age did not have a dramatic effect on the response of compressive loading. In subjects aged 8 to 10 years, there was some indication of a tendency toward greater compression. Age was, however, shown to have its most significant effect on the recovery phase. Children aged 8 to 10 years showed almost immediate recovery to about 97%; adults 15 to 23 years old showed 90% recovery in the same period; those aged 72 to 86 showed recovery to about 67% in 10 min., and took 4.5 h for complete recovery.

The areas used for testing by Pierard et al.,[8] were the volar surface of the forearm and the tibia. The effect of age and sex on the measurements is not stated.

Dikstein et al.,[11] measured the forehead in females aged from 2 to 70 using a pressure of 10 g/cm². In 20-year-old patients the indentation was found to be 0.043 cm, and at the age of 70 it was 0.054 cm — a difference of 26%. The elastic recovery at the age of 20 was 80.5%, compared to 65.5% at the age of 70. From those measurements it can be seen that the skin of older females has less resistance to the applied force and less elastic recovery after removing the force than does that of younger subjects.

Manny,[10] using the apparatus of Dikstein, measured the indentation on the forehead and the indentation of the back of the hand. The indentation of the back of the hand was measured both in the case of open palm

edematous conditions, and altered water handling of the dermis.[1,5]

References

1. Schade, H., Untersuchungen Zur Organfunktion des Bindegewebes, *Z. Exp. Pathol. Therap.*, 11, 369, 1912.
2. Kirk, E. and Kvorning, S.A., Quantitative measurements of the elastic properties of the skin and subcutaneous tissue in young and old individuals, *J. Gerontol.*, 4, 273, 1949.
3. Kirk, J.E. and Chieffi, M., Variation with age in elasticity of skin and subcutaneous tissue in human individuals, *J. Gerontol.*, 17, 373, 1962.
4. Tregear, R.T. and Dirnhuber, P., Viscous flow in compressed human and rat skin, *J. Invest. Dermatol.*, 45, 119, 1965.
5. Robertson, E.G., Lewis, W.Z., Billewicz, W.Z., and Foggett, I.N., Tow devices for quantifying the rate of deformation of skin and subcutaneous tissue, *J. Lab. Clin. Med.*, 73, 594, 1969.
6. Kydd, W.L., Daly C.H., and Nansen, D., Variation in the response to mechanical stress of human soft tissues as related to age, *J. Prosthet. Dent.*, 32, 493, 1974.
7. Daly, C.H. and Odland, G.F., Age related changes in the mechanical properties of human skin, *J. Invest. Dermatol.*, 73, 84, 1979.
8. Pierard, G.E., Evaluation de proprietes mecaniques dela peau pa les methodes dindentation et de compression, *Dermatologica*, 168, 61, 1984.
9. Dikstein, S. and Hartzshtark, A., In-vivo measurement of some elastic properties of human skin, in *Bioengineering and the Skin*, Marks, R. and Payne, P.A., Eds., MTA Press, England 1981, p. 43.
10. Manny, V., In-vivo deformation by small forces as a criterion for assessing skin ageing, M. Pharm. thesis, Hebrew University Jerusalem, 1989.
11. Dikstein, S., Hartzshtark, A., and Bercovici, P., The dependence of low-pressure indentation, slackness and surface pH on age in forehead skin of woman, *J. Soc. Cosmet. Chem.*, 35, 221, 1984.
12. Robert, C., Blac, M., Lesty, C., Dikstein, S., and Robert, L., Study of skin ageing as a function of social and professional conditions: modification of the rheological parameters measured with a noninvasive method — indentometry, *Gerontology*, 34, 284, 1988.
13. Dikstein, S. and Hartzshtark, A., What does low-pressure indentometry measure?, *Arztliche Kosmetologie*, 13, 327, 1983.
14. Ogston, A.G. and Stanier, J.E., The physiological function of hyaluronic acid in synovial fluid, *J. Physiol.*, 119, 244, 1953.
15. McLean, D. and Hale, C.W., Studies on diffusing factors, *Biochem. J.*, 35, 159, 1941.

and closed fist. The volunteers were women aged 20 to 30 years and over 65 years. There was no difference in the indentation between the two age groups in the above experiments. There was, however, difference between the open-palm and closed-fist measurements — the indentation in the case of open palm was higher.

Using the same method Robert et al.[12] assessed the influence of social and professional factors on the rheological properties of aging skin. The measurements were carried out on the forehead, using a pressure of 10 g/cm^2. Three different populations were studied: nuns (females), white collar workers (males and females), and blue collar workers (males and females). For the purposes of the study subjects were defined as "young" if they were less than 55 years old, and "old" if they were over 55. They found that the elastic rebound (after removing the weight) decreased steadily with age; this effect was always more marked in females. Working women lost their elasticity more rapidly than did the nuns, and the male blue collar workers lost their elasticity more rapidly than did the male white collar workers.

IV. What Does Indentometry Measure?

A basic question is: Which skin layer is the most important for indentation measurements? In the stratum corneum, 4% lactic acid (pH 4.2) applied to the skin of four volunteers with dry skin of high pH was found to have no effect on indentation. Also, 5% glycerin, which is a strong humectant, was tried as a moisturizer on six volunteers with dry skin, with no effect on indentation.[13] In an another experiment the stratum corneum was removed by stripping techniques, monitoring the surface pH to ensure that the stratum corneum was indeed removed. It was found that stripping did not affect indentation measurements.[13] From those experiments it is clear that the stratum corneum does not affect indentation.

In the dermis the forehead skin of three male volunteers was injected intradermally with 0.2 cm^3 saline, and on a subsequent occasion (at least 10 d later) the same volume of saline containing 300 IU hyaluronidase, 4 U elastase, or 4 U collagenase was injected. The results showed that saline decreased indentation, whereas hyaluronidase, and to a lesser degree elastase, increased indentation. Collagenase had no significant effect. The decrease in indentation caused by saline suggests, therefore, that the water moisture content in the dermis decreases indentation. Elastase, and even more so hyaluronidase, increases indentation in spite of the presence of saline. The interpretation of those experiments is that in the dermis, the state of the ground substance-elastin network is responsible for the observed changes in indentometry.[13]

More evidence of the effect of the ground substance on indentation was found in rat skin.[4] A cross-section cut through compressed rat skin showed that most of the compression occurred in the dermis. Since liquids are virtually incompressible, for the dermis to be compressed, the water must have flowed out sideways from the compressed area. The dermal fluid, the ground substance, is known to contain enough mucopolysaccharide to make it highly viscous.[14] Diffusion through dermis is greatly increased when the polysaccharide is depolymerized by hyaluronidase.[15]

V. General Recommendations for Using the Indentometry Method

1. In order to compare different physiologic conditions — young vs. old skin, male vs. female, and various edematous conditions — one must work within the linear area of the stress-strain curve of the skin area being measured. The standard indentation force used by Dikstein et al. is 10 g/cm^2; this force is in the low linear area of the stress-strain curve of the forehead, the area on which the measurements are made.[9]
2. The pressure of the measuring apparatus itself should not exceed 20% of the total measuring pressure. The weight of Dikstein et al.'s apparatus creates an initial pressure of 1 g/cm^2 which is 10% of the final measuring pressure. Unfortunately, no other article specified the exact initial pressure of the apparatus on the skin.
3. During the measurement there should not be any slip between the skin and the apparatus.
4. The reproducibility of the measurements should be greater than 90%.[10]

VI. General Conclusions

Indentometry is a technique for measuring skin softness. In effect, it actually measures the water status of the dermis, which is largely determined by the amount of mucopolysaccharides and their ability to hold water in the measured area. The immediate indentation measurement using indentometry was found in most studies not to discriminate sufficiently between young and old skin.[3,6,11-12] The difference between the two age groups can be better differentiated by measuring the elastic rebound, which is the immediate rebound of the skin after unloading the force. It was found that in young skin, after unloading the force, the immediate rebound comprised a greater percentage of the initial indentation.[1-3,6,10-12] Although some studies showed age dependence and differences between the skins of males and females with the indentometry method, it seems to us that indentometry is of most use to evaluate

Chapter 14.7
The Gas-Bearing Electrodynamometer

C.W. Hargens
Philadelphia, Pennsylvania

I. Introduction

The present chapter deals with the fundamental mechanical properties of the skin and their quantitative measurement. This is basic to an evaluation of existing dermal conditions and/or the efficacy of any treatment.

The method is direct and not "implied" by inference from some other test or property such as electrical conductivity or sonic propagation. It is the method a mechanical or materials engineer would use, the determination of stress (applied force) versus strain (resulting deformation), to describe the characteristics influencing the performance of any structural substance.

The skin is one of the most important "structural" substances holding the body together. We can further demonstrate that the skin's principal strength resides in a few outer cell layers of the stratum corneum. This can be easily shown with the gas-bearing electrodynamometer (GBE) instrumentation if one examines an area where these thin cell layers are stripped away by just a few repeated applications of Scotch® tape (adhesive). A glistening layer is quickly reached which the instrument will show has no strength of containment. In fact such a test reveals that the mechanical modulus of the stripped area diminishes from a high value to almost nothing. This is the same phenomenon occurring in burns of the skin which have a similar destructive effect.

The term dynamometer may be somewhat foreign to the biomedical field, but technically it implies a device for measuring power, force, electrical current, or voltage. In this case we are interested in a convenient and accurate way to measure force. A dynamometer involves the interaction of electrical and mechanical quantities, in essence a kind of transducer. Electric current in a conductor can be measured, and, through its interaction with a magnetic field, mechanical force thereby will be determined; or the reverse, measure force and derive the value of current in an electrical conductor. Usually force is determined through an arrangement of coils, magnetic fields, and a current measurement. The adjustable nature of these quantities and devices allows one great latitude in designing measuring systems to suit various purposes. Also, the precision and accuracy can be of a high order, because the calibrating forces, currents, and voltages involved can be fundamentally verified through standards.

In the measurement of the skin's mechanical properties stress-strain (force-displacement) modulus the principles just mentioned have been utilized. To further understand what is involved it must be recognized that the skin, like other body tissues, is a viscoelastic substance, i.e., it is partly spring like and partly viscous. Thus, the mechanical moduli one seeks to determine are significantly rate related due to the large viscous component of the modulus. In other words, the observed deformation will depend upon the speed with which a force is applied. This is why one cannot be satisfied with static measurements that were done in earlier times, for they will not explain dynamic behavior which influences our subjective perception of skin quality.

The fundamental reason for there being a complex, rate-related modulus to describe skin behavior is the same as in the case of all large molecule polymers. In all of these materials there is a finite time required for a stress-imposed molecular rearrangement to take place. To this phenomenon we assign the simplifying term viscosity, and it is present in soft solid substances as well as in fluids which have been studied extensively in this regard.[1]

Skin is composed of a well-known and complicated architecture consisting of such viscoelastic materials. As just mentioned, the main mechanical influences of this kind, as far as skin measurements are concerned, reside in the cells of the stratum corneum and the natural cements that hold the aggregate together. The GBE is sensitive enough to show that the mechanical integrity of these thin membranes is strongly influenced by small external effects such as moisture and other topically applied reagents.

Returning to the matter of feasible instrumentation for skin tests, one needs equipment capable of dynamic measurement. The interesting phenomena relating to skin involve time constants on the order of 1 s. If one has equipment sufficiently lightweight, with agile components that can be quickly accelerated, these features of elasticity and viscosity can be recorded. One can apply small forces, actually a periodic forcing function such as a sine, at the 1-s rate and observe by suitable recording means the resulting deformations as a phase-displaced signal. Details of the application of these principles are explained in Section II.

Another point of initial consideration concerns the best direction in which to apply the test stresses in the skin to clearly enhance the desired responses. It turns out that the most sensitive indications of modulus changes in a membrane such as the stratum corneum will be obtained with the shear mode, i.e., the force vector applied directly in the plane of the surface. Techniques which indent the surface or attempt to raise a bell-shaped deformation normal to the surface with suction have been used but are far less sensitive. This is because the measurement relies upon only the smaller vector components of the principal stresses.

Finally, one finds that a stress-strain characteristic taken in the shear mode gives a very adequate measurement. Plotted in real time (approximately a 1-s cycle) the diagram will be a smooth ellipse in four quadrants. It is best to keep within the linear parameters of the skin if a simple numerical modulus is to be obtained. Beyond these limits the tissue's response becomes very nonlinear, and analysis becomes much more complicated, involving a Fourier analysis of the individual stress and strain waveforms. The consequent interpretation of results is not worth the trouble, since the linear response involving a simple phase shift between sinusoidal stress and strain waveforms is quite adequate for the comparative evaluations one is seeking. The ellipse is of course just a display suitable for calculating the phase angle between X and Y waveforms and the modulus or dynamic spring rate of the test surface. This will be explained further in Section IV.

II. Instrumental Application

Having introduced the basic objective of the noninvasive protocol, we now present the method that has been developed over several decades and used successfully for as long.[2-9] The protocol is to adhere a probe to a chosen test site on the skin by means of an accepted medical adhesive similar to products used in surgical procedures. The surface area contacted by the probe need be only a few square millimeters.

The probe attaches in turn to the GBE as shown in Figure 1. Here a facial site is being examined. The probe moves with the reciprocating action of the GBE. The amplitude of motion is usually about 1 mm, so the active area of the skin under test might be said to extend up to 10 mm beyond the attachment point depending upon the type of skin involved and its pretension state.

A more detailed view of what takes place around the probe in the plane of the skin during the stressing of the surface would seem to be the following. Stratum corneum is in itself so stiff that bodily actions are only possible because there is a certain excess of skin. That is, there are microscopic as well as macroscopic folds, creases, and ridges in its texture. As the skin moves, these function in accordion fashion to relieve the stress that would otherwise occur. Therefore, what the probe measures through the resulting strain determination, the "stretch", is actually the bending of the stratum corneum within these folds, i.e., the unfolding process brought about by the stretching forces.

This reality in no way diminishes the utility of the measurement, because the apparent stretchability and dynamic behavior of the skin are what one seeks to know. Regardless of the precise mechanism of dynamic response, the influence of the various skin treatments upon the measurable and subjectively sensed stiffness or dryness will become apparent from the data obtained.

As the probe of the GBE applies specific forces in the plane of motion, its resulting displacement, also monitored, is controlled by the elastic and viscous moduli of the outer skin layer. It is a simple matter to display these two quantities, force and displacement, as a typical stress-strain diagram for the material.

A plotting device with sufficiently rapid movement, such as a storage oscilloscope, analog or digital, can display the complete elliptical diagram every second as the probe drives the skin back and forth in repeated cycles. If rapid changes in skin condition occur, as with the introduction of moisture or drying, they will appear as corresponding alterations of the diagram. For example, dryness and stiffening of the surface will appear as an elevated slope of the major axis of the ellipse; softening will cause a decrease in slope.

When only a few tests are to be performed, the dynamic stress-strain plots can be measured graphically and the moduli calculated, as will be explained, from their geometry. On the other hand, in laboratories where very large testing programs continue day after day, it is best to computerize the process, taking data directly through analog-to-digital conversion of the stress and strain signals.

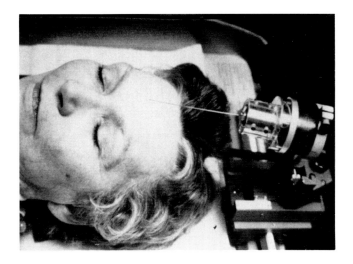

FIGURE 1 Dynamometer attached to a facial site.

III. Instrumentation

So far the GBE has been described only as to its capabilities, whereas little has been said about the mechanism itself. The probe just referred to in Section II is attached to the skin in a novel way so that the test site can be precisely preserved while the GBE is disconnected for use elsewhere. Thus, the subject can be released for other activities during the course of the test period. To facilitate this a small plastic button serves to contact the skin, adhered by the adhesive film previously mentioned. However, the actual probe wire fits tightly into a tiny hole in the top of the button and remains thusly attached during the test. Disconnection is achieved by simply popping the probe wire out of the hole in the button. A supply of buttons and film adhesive is furnished with the instrument, along with a detailed application procedure.

The GBE is visible at the top of Figure 1 and is mounted on a small optical bench immediately adjacent to the subject whose head rests on the same base so as to minimize relative motion. The optical bench provides through rack and pinion manual adjustments, precise but rapid positioning along the three coordinate axes. The attachment button secured to the probe wire is moved about over the subject by means of these controls until the desired test location is attained. The GBE is lowered until the tip of the probe (button) touches the skin and adheres to it.

It is, of course, desirable that the subject not move for the several seconds of the test in order to avoid disturbing the recording process. This is usually not a problem even without any particular stabilizing means for the head or other body part under study. A relaxed subject can hold quite still even to the extent required by the rather large displacement magnification of the electronic and display systems involved. That is, 1 mm of skin displacement at the test site can correspond to ≥25 mm spot movement on the oscilloscope screen or plotter (a magnification of at least 25).

The displacement of the probe just discussed is induced by the reciprocating armature of the GBE to which it is attached. The probe is held by a setscrew in a small plastic chuck fitted into a protruding sleeve of the armature. This chuck with probe attached can be removed as an assembly owing to its tapered fit. This is an added convenience as will be discovered when moving the GBE.

The armature carries also a coil of fine copper wire which moves in a strong, radial magnetic field. The field is created by a permanent magnet whose flux can be depended upon to remain constant. As the GBE's name implies, the armature "floats" on a coaxial air bearing. Therefore, there is no metal-to-metal contact in the bearing, and frictional forces are reduced to those caused by the viscosity of air. This means that they are so small as to be orders of magnitude less than the skin moduli and hence negligible. To a slightly lesser extent this is true of the hair-like leads carrying current to the coil.

The force produced by the coil and magnetic field combination is in accordance with the standard electric motor equation:

$$\text{Force} = B \cdot l \cdot i \qquad (1)$$

where force is in dynes, B (magnetic flux) is in gausses, l (length of conductor) is in centimeters, and i (current) is in abamperes (1 abampere = 10 amperes).

Thus, if B and l are constant, as in the dynamometer, the force is directly proportional to the current, and this fact allows one to use current to determine the force exerted by the GBE probe.

Displacement (strain) is determined through another electrical device built into the GBE. This tracks the armature's movements without exerting any force upon it. It is referred to as a linear variable differential transformer (LVDT). It consists of a primary and two secondary coils, all magnetically coupled together by a movable core. As the core moves axially with the armature through the three coils' windings, it couples flux differentially from the primary into the two secondaries. If an electronic system is associated with the LVDT to supply current to the primary coil and rectify the combined secondaries' output voltages (demodulate), one obtains a DC voltage whose polarity and magnitude give an accurate record of the armature's instantaneous position. This electronic system associated with the LVDT is referred to as a signal conditioner.

The stress-strain diagram is thus created by applying a voltage corresponding to the force coil's current to the oscilloscope's vertical deflection system and at the same time connecting the LVDT (signal conditioner output) to the horizontal deflection system. For a viscoelastic material, whose deformation lags in time the impressed sinusoidal forcing function, an ellipse is formed as in Figure 2.

The GBE has several important advantages for measuring the mechanical properties of soft, viscoelastic substances such as living tissue. One advantage is its ability to exert small dynamic forces and record sizeable displacements that ensue in these soft materials. Other transducers by contrast are very stiff, such as strain gauges and load cells, or they impose frictional elements in parallel with the test specimen. They also are usually rather bulky and inconvenient for clinical testing, or they cannot respond at high frequencies. The construction of a GBE on the other hand is similar to the transducer in an audio loudspeaker, but without the stiffness.

For completeness it should be mentioned that a basic air-bearing dynamometer in an evolutionary form has been used at audio frequencies. However, at audio rates it operates in a slightly different fashion, because the mass of the armature, although small, and in fact negligible at the 1-Hz rate discussed earlier, would now become significant. However, it can be shown that by combining the electrical motor equation (Equation 1) with the corresponding generator equation involving velocity (E = Blv) one does not need the LVDT or any other measure of displacement. Instead, it is possible to employ an electric bridge circuit to obtain precise electrical impedance analogs of the specimen's mechanical moduli. From these electrical quantities one can calculate the actual mechanical characteristics of the material.[10,11] These would combine to yield the so-called mechanical impedance.

Returning to the low-frequency operation of the GBE described here, one does measure both force and displacement to produce the elliptical stress-strain diagram directly. From the diagram's dimensions which are calibrated in appropriate units, such as grams and millimeters, it is a simple matter to compute the moduli.

The most useful modulus is the dynamic spring rate (DSR), which is the slope of the ellipse's major axis, B/A, as seen in Figure 2. As mentioned earlier, it correlates well with subjective descriptors such as dryness, softness, etc. used by trained dermatological graders.[12-14]

Additional information about the combined elastic and viscous parameters can be obtained by noting the openness of the elliptic loop. For example, a strictly elastic element like a steel spring would show no openness. The correct interpretation of this would be to say that there is no energy loss if the loop degener-

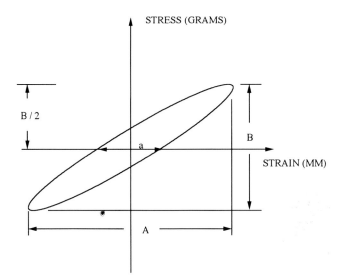

FIGURE 2 Dynamic stress-strain diagram plotted by the GBE apparatus showing the principal measurable quantities.

ates to a straight line, i.e., the integrated area under the curve in Figure 2 (the energy or work done) is the same during stretch as during return. Energy is conserved in the elastic case but not in the case involving viscous elements.

There should be no mystery about the elliptical display. It is simply the Lissajous figure used to measure the phase angle between two sinusoidal waves, in this case, stress and strain. If they were shown as sine functions on two parallel time axes, one would observe the time lag or phase angle lag of the strain behind the applied stress. In our case it is called the loss angle.

In Figure 2 the numerical value of the loss angle is computed from the ratio of the loop opening to the total displacement. The sine of the loss angle equals this ratio.

$$\sin \theta = A/a \qquad (2)$$

The fundamental reasons for this time lag in a viscoelastic material have been previously discussed.[1]

What has the loss angle to do with tissue properties in dermatological terms? It has been observed that young, healthy tissue has the least loss. This was found to be true in the case of ocular tissue when ophthalmic experiments were conducted, and it is true of skin. While speaking of the loss factor we should interject that there is of course more to tactile sensation, self-comfort, and appearance than the dynamic shear moduli of the stratum corneum as measured by the GBE. The underlying dermal tissues are important as well, and their contribution can be conveniently measured using ballistometry.

To illustrate dermal loss factor further in a very homely analogy we are speaking about the "upholstery" of the body. That is to say, subjectively tactile sensation responds to all of the above, plus frictional effects, and there is a noticeable difference between a strictly "springy" surface and one with a gradual yield, even though their static stress-strain deformation coefficients may be identical. Again, note the differences between a down-filled cushion, one of foam rubber, and one packed with cotton batting. The dynamic response counts strongly.

IV. Data Reduction

Graphical data reduction methods were employed when the GBE was first introduced. A Polaroid camera photographed each loop which was later scaled and the calibrated measurements tabulated. The force calibration of the GBE is straightforward and positive. First, the instrument is simply tipped up and fixed in a vertical position. The wire probe is removed from the chuck, and the dynamometer becomes a platform balance on which gram weights can be placed, a few at a time. The armature, now vertical, is balanced against gravity by means of the DC offset control of the function generator which drives the GBE. The various balance currents introduced into the force coil of the instrument will cause a series of corresponding vertical displacements of the trace along the force axis of the oscilloscope or other plotter. The vertical amplification setting of the latter can be chosen to give a convenient scale for the force range being used in any particular experiments. Once set, of course it should not be changed until the system is recalibrated.

The displacement calibration is equally simple and positive in its overall inclusion of the system component variables. It is done with the GBE in a horizontal position and with a micrometer or other accurate scale indicating armature movement. The corresponding scale can thus be generated on the horizontal (displacement) axis of the display, again choosing an appropriate adjustment of the amplifier.

When a large number of experiments must be carried out on a daily basis, a computer can be connected to the system. This will greatly facilitate recording of data, automatic calculation of DSR, and other quantities as well as trend or statistical data summaries. The connection is made directly from the oscilloscope or recorder terminals using the signal voltages produced by the GBE system components.

Each experimenter will establish his or her own protocols for equilibrating subjects' skin when they arrive for tests. For extremely precise work an environmental chamber has been used, although most modern laboratories have sufficiently standardized atmospheres in their general working spaces.

References

1. Ferry, J.D., *Viscoelastic Properties of Polymers,* John Wiley & Sons, New York 1961.
2. Hargens, C.W., Glaucoma and vibration tonometry, *J. Franklin Instit,* 207, 143, 1960.
3. Hargens, C.W., Viscoelastometry of biomaterials, in *Proc. 26th. Annual Conf. Engineering in Medicine and Biology* (IEEE), Minneapolis, 1973, 203.
4. Hargens, C.W., Instrumentation to measure the viscoelastic properties of intact human skin, in *Proc. 28th. Annual Conf. on Engineering in Medicine and Biology* (IEEE), New Orleans, 1975, 17, 179.
5. Christensen, M.S., Hargens, C.W., Nacht, S., and Gans, E.H., Viscoelastic properties of intact human skin: instrumentation, hydration effects, and the contribution of the stratum corneum, *J. Invest. Dermatol.,* 69, 282, 1977.
6. Hargens, C.W., Measurement of dynamic moduli and loss factor in viscoelastic materials using the gas bearing electrodynamometer (GBE), *J. Acoust. Soc. Am.,* (Suppl. 1) 67, S25, 1980.
7. Hargens, C.W., The gas bearing electrodynamometer (GBE) applied to measuring mechanical changes in skin and other tissues, in *Bioengineering and the Skin,* Marks, R. and Payne, P.A., Eds., MTP Press, Lancaster, 1981, chap. 14.
8. Hargens, C.W., Instrumented testing of human skin in vivo, in *Proc. 35th Ann. Conf. Engineering in Medicine and Biology,* (IEEE), Philadelphia, 1982, 12.
9. Missel, P.J., Bowman, M.J., Benzinger, M.J., and Albright, G.B., An in vitro method for skin preservation to study the influences of relative humidity and treatment on stratum corneum elasticity, *Bioeng. Skin,* 2, 203, 1986.
10. Hargens, C.W. and Keiper, D.A., Tonometry — challenge for electronics, in Digest 1961 Int. Conf. Med. Electronics (IEEE), 77.
11. Keiper, D.A., Dynamic mechanical properties tester for low audio and subaudio frequencies, *Rev. Sci. Inst.,* 33, 1181, 1962.
12. Christensen, M.S., Nacht, S., and Packman, E.W., Facial oiliness and dryness: correlation between instrumental measurements and self-assessment, *J. Soc. Cosmet. Chem.,* 34, 241, 1983.
13. Maes, D., Short, J., Turek, B.A., and Reinstein, J.A., In vivo measuring of softness using the gas bearing electrodynamometer, *Int. J. Cosmet. Sci.,* 5, 189, 1983.
14. Cooper, E.R., Missel, P.J., Hannon, D.P., and Albright, G.B., Mechanical properties of dry, normal and glycerol-treated skin as measured by the gas bearing electrodynamometer, *J. Soc. Cosmet. Chem.,* 36, 335, 1985.

Chapter 14.8
Ballistometry

C.W. Hargens
Philadelphia, Pennsylvania

I. Introduction

The ballistometer is a neat and easy-to-use device for measuring *certain properties* of human skin *in vivo*. This statement implies the ballistometer's usefulness, but at the same time, the expression "certain properties" is a caution. One should be very specific in all claims regarding the determination of skin properties because there are so many. One might do well to first decide which ones tell something about the subjectively sensed skin condition to be evaluated, then see which "property" is the best indicator and measure it with the correct instrument.

An example of two measurements which determine entirely different skin properties concerns the role of the ballistometer in contrast to the GBE (gas-bearing electrodynamometer). The first measures certain elastic parameters below the surface, while the GBE is best for quantifying stiffness in the surface plane of the skin, the stratum corneum.

Here we will speak about the ballistometer for skin studies, and it should be appreciated that this is a narrow application of the concept. Our discussion will differentiate between a strictly practical device that has served well as an experimental tool and consideration in some detail of its extended capabilities.

Ballistometry, the classical use of impacting masses to measure their material properties through their interaction, is not new. Sir Isaac Newton (1642 to 1727) is credited with certain relevant philosophical propositions and experiments involving impacting bodies.[1] His observations led to the conclusion that the relative velocity of two bodies after they have impacted each other is in a constant ratio to their relative velocity before impact and in the opposite direction. This constant ratio has been called the "coefficient of restitution", usually designated by e.[2]

The ballistic method of investigating materials' mechanical properties in modern times was initially applied to homogeneous, usually hard, substances such as metals.[3] On the other hand, the use of the concept to examine relatively soft, viscoelastic matter is still more recent, i.e., in the middle of this century. Hollinger and Thelen[4] extensively studied asphalts in the 1950s using an impacting pendulum to find the storage (elastic) and loss (viscous) moduli.

The ball rebound tests of natural and synthetic rubber stocks given in handbooks show the considerable influence of temperature on polymers.[3] For example, in these ballistic measurements the percent rebound of a 1.9-cm steel ball dropped 100 cm onto a natural rubber (Hevea) sample 1.9 cm thick is 45% at 20°C and 71% at 100°C. For polychloroprene (neoprene) the change was from 35 to 67% for temperature change of 20 to 100°C, respectively.

Arbitrary scales have been of necessity the practice in all of these tests, and the specification of equipment details, such as the geometry of the impacting components, is essential if uniformity of results is expected. For metals the diameter of the indentation made by a small hardened steel sphere (Brinnel hardness) or the height of rebound of a small hammer (Shore sclerosope) serve as arbitrary measures of hardness.

In the case of skin ballistometry the depth of penetration to particular layers of the dermis will depend to some extent upon the geometric sharpness of the impacting mass. The sharpness of course determines the instantaneous pressures exerted upon the tissue. Some standardization eventually will be helpful.

The ophthalmic applications go farther back to a ballistic tonometer devised in 1930 by Vogelsang[5] to assess ocular tension. The method involved photographing the rebound oscillations of a small hammer striking the cornea of the eye. Wigersma[6] in 1955 apparently discovered the benefit of a lighter hammer in a method referred to as elastometry. Eventually Mamelok and Posner[7] published their conviction that these measures had more to do with the mechanical properties of the cornea itself than with intraocular pressure, i.e., more in accordance with Newton's original contention.

In more recent times others have used the method to study skin properties. Here the deeper dermal structures would be influential. Tosti[8] made a ballistometer for this purpose. However, one should note that the published analysis must be viewed with caution, because e is defined as an energy ratio instead of the accepted momentum ratio which Newton originally postulated. Thus, without noting this difference, one could be confused by the numerical results of the experiments. The differences result from energy being proportional to the square of the velocity, whereas momentum is mass times the first power of velocity.

More specifically, the example given in the reference is of a free-falling body striking a horizontal surface. Actually a pendulum was used, and we will have more to say about pendulums. Basically it is true that kinetic energy of whatever mass system is involved should equal the starting potential energy however it is created. Then the rebound kinetic energy should equal the next potential energy peak. This reference defines e as H'/H, the ratio of rebound height, H', to the previous starting height, H, as previously discussed.

The coefficient of restitution is an important element in any analysis of the motion ensuing after the collision of two bodies. In the most elementary case it is assumed that as soon as contact begins there will be a certain period of time referred to as the "period of compression". After maximum compression and deformation, recovery to some extent will occur during a time called the "period of restitution"; hence the name of the coefficient. These times are of course fundamentally dependent upon the period required by the materials' molecular structures to rearrange themselves. Thus one observes the differences in behavior between those substances which are highly elastic and those that are more viscous.

Experimental determination of e is traditionally done by measuring the rebound height of one object falling upon another. The expression for e is derived simply as follows:

$$e = \frac{v_2 - v_0}{v_0 - v_1} \quad (1)$$

If it can be assumed that the second body does not move, v_0, their common velocity at maximum compression, equals zero, and hence we will have

$$e = -\frac{v_2}{v_1} \quad (2)$$

In these equations v_1 is the velocity of the falling mass before and v_2 its velocity after impact and separation. Since these velocities are defined by the starting and final rebound heights, i.e., potential energies, one may apply the following relationships to calculate e.

$$v_1 = \sqrt{2gH} \text{ and } v_2 = \sqrt{2gh} \quad (3)$$

Then, substituting in Equation 1,

$$e = \sqrt{\frac{h}{H}} \text{ or } e^2 = \frac{h}{H} \quad (4)$$

Thus, instrumentally one measures the height, H, from which the mass falls and the rebound height, h, to which it rises. The task of the experimenter is to find a practical way of doing this.

II. Mechanics of Ballistometry

This elementary determination of the coefficient of restitution assumes that the falling mass and the surface struck are of the same material, that the surface is rigidly supported, its velocity is at all times zero, and hence the velocity of both masses at the instant of greatest compression will be equal to zero. These assumptions are not completely satisfied when ballistometry is applied to the skin. Now we must consider the details of the process further to see how valid our interpretation of the results will be.

In general, when the impacting mass strikes the skin surface, several things happen. First, the elastic component of the skin begins to store some of the kinetic energy of the falling object. The subsequent release of this stored energy provides the rebound. The processes of both compression and restitution are slowed however by the viscous component whose reaction force, like a shock absorber, is proportional to velocity. It will be at a maximum at first and then decrease as indentation proceeds, energy is dissipated as viscous internal friction, and the area of contact simultaneously enlarges. The elastic component of force is not dependent upon velocity.

A second deviation from simple assumptions is that all of the involved skin does not have zero velocity throughout the impact process. Instead, some parts will have been accelerated, and an exchange of momentum will occur as well as potential energy storage. This momentum initiates an acoustic wave in the skin, however rapidly attenuated, propagating in indeterminate directions. The acoustic energy will be mostly lost, although a small amount may be returned to the rebounding mass and contribute to or detract from the reaction depending upon its time phase. This is similar to a diver bouncing up and down on the end of a diving board; if he gets out of synchronism with the board, unpleasant results ensue.

With skin much of the indentation energy will be lost in shear viscosity within the tissues, conversion to heat. The greatest rebound will of course occur where the ratio of elastic to viscous effect is largest. Hence the expression that the skin is "more elastic" in such cases is a fair statement and can be used to describe the observed accentuated characteristic in younger skin and for certain beneficial dermal treatments. The coefficient of restitution theoretically ranges from 0 for inelastic impact, i.e., zero after-impact velocity, to 1 for completely elastic impact (which of course never occurs). On this basis ballistometry has come to be useful for evaluating youthful and more mature skin and its response to skin treatment products.

For a viscoelastic, nonlinear substance like skin, especially in compression during impact, one observes diminished times between successive rebounds. This is to be expected because the time for a mass to fall is a function of its starting height. Thus, impacts will be closer and closer in time. (See Figure 1 for typical rebound recordings.)

One additional effect of the decreasing fall height is that the diminished impact velocity produces shallower indentation, involving essentially a different substance composed of the more superficial skin layers. This explains in part why a constant rebound ratio is not maintained. Also, it should be appreciated that, with indentation of soft substances, the area of contact increases rapidly, so that, depending upon the shape of the impacting mass, an altered stress distribution in the region occurs with depth.

These last comments direct our attention to the influence of the shape of the impacting mass. This subject interested mathematicians as far back as the 19th century. Heinrich Hertz, the discoverer of wireless waves, solved the problem of impact between solid spheres and presented an extended theory of impact between a solid sphere and an elastic plate, the latter being more like our situation. These relations in mathematical form showed the dependence of the coefficient of restitution upon the material constants, Young's modulus, density, and Poisson's ratio.[9] More recent works on this topic are the classical analyses of Timoshenko in his Theory of Elasticity in which precise relationships for stress distributions in various geometries are presented.[10,11]

A pendulous mass that impacts on the test surface is a better way to get control of the several ingredient parameters than by dropping weights and attempting to observe their rebounds. A compound pendulum provides a means of distributing several mass elements to increase its moment of inertia and so control the impact velocity. We can show how this is better than a simple, pivoted hammer which, because of its uncontrolled impact velocity, produces only minimal results, i.e., unsatisfactory rebounds for soft, viscoelastic substances. Large rebounds are necessary to provide a high degree of discrimination among the varying viscoelastic parameters.

Let us illustrate graphically the reason for controlling impact velocity when applying ballistometry to the study of viscoelastic materials. This can be shown with a graph of the force, and hence the work relationships accompanying the periods of compression and restitution.

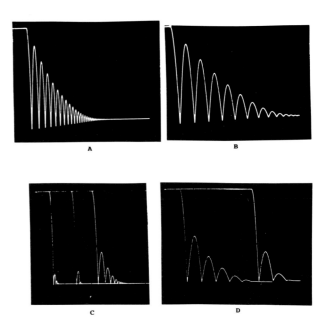

FIGURE 1 Sample recordings: (A and B) Reduced period with lower impact on highly elastic surface; (C) improved rebound with lower velocity impact on more viscous surface; (D) comparison of more elastic with less elastic material.

Figure 2 shows two plots of force acting on the indenting mass as it compresses the skin downward along the velocity axis at several speeds, one high and the other more slow. In both cases we will assume identical kinetic energy inputs to the same test site. Fundamentally it is the maximum indentation that stores the most elastic, potential, energy that will be returned as a large rebound.

Identical energy input means that the total work done by the viscous and by the elastic force, i.e., the area under these curves, must be equal in both cases. It is unimportant in making our point that the exact functional relationship between these forces and the indentation of the skin cannot be known. However, the graphs show the greater apportionment of viscous energy (loss) in the highspeed case. This is of course because the viscous force is proportional to the speed, whereas the elastic force is not. Note also that the slope of the elastic force-displacement characteristic is everywhere the same as determined by the skin involved. The conclusion is that the slow-speed case stores more energy in the elastic element and to do so must mean greater indentation, as shown. Hence, the greater stored energy will occur, and the rebounds will be higher. The approximate curvature of these force characteristics is introduced to account for the varying area of contact, hence pressure, during compression and restitution. If the contact is assumed to have a conical geometry, the force will vary somewhere between a square and cubic function of the indentation.

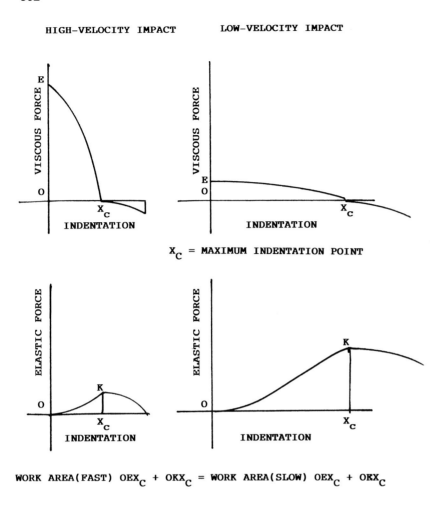

FIGURE 2 Force vs. indentation (work) diagrams of high and low velocity impact on Viscous and elastic substances showing greater rebound potential for the low-velocity impact.

III. Practical Ballistometry for Skin Studies

The qualitative measures presented in the previous section are sufficient to provide ballistometer design criteria. A practical embodiment of the ballistometer is a form of pendulum in which the pivotal angle is measured instead of mass heights. This may be done conveniently by connecting its shaft to some low-friction angle transducer such as a rotary variable differential transformer which imposes no force on the pendulum. In Reference 8 a stationary coil with its inductance varied by a moving iron core attached to the pendulum was associated with an electronic circuit to indicate angle. To directly obtain numerical values for computer entry high-resolution, digital, optical angle encoders are available. In most cases the ballistometer will operate with large angles of swing, so that some trigonometric processing will be required as will be discussed in more detail. Still another approach is to employ an angular accelerometer whose output is integrated to give angular velocity. Programmed with a computer to deliver information precisely at the moments of impact and rebound, this arrangement can directly calculate the coefficient of restitution. Other useful functions of the computer are the establishment of a time base for release of the pendulum from its raised position and graphical display of the rebound pattern.

Release of a pendulum type of moving mass will result in a certain velocity of normal impact, R dΘ/dt, upon the test surface, where R is the effective length of the pendulum (pivot-to-impacting point). The analysis of the pendulum's motion must now recognize certain practical design features based upon the theory presented in the previous section. The desirability of a low-velocity, compound (physical) pendulum with a considerable moment of inertia has been discussed in terms of improved rebound. Figure 1 shows this improvement in measurable rebound amplitude, achieved simply by adding an adjustable counterweight to a pivoted beam assembly. Such a ballistometer is pictured in Figure 3 showing a bar with impacting mass and counterweight pivoted on an angle transducer.

The angular acceleration of a compound pendulum determines its ultimate velocity. For any rotating body the angular acceleration is equal to the torque applied to it, in this case gravitational, depending upon its instantaneous angular position, divided by its moment of inertia about the pivot point. Thus the angular acceleration and hence the impact velocity can be easily adjusted by changing the positions of the impacting mass and the counterweight to alter the net gravitational torque and the moment of inertia of the system. The latter is the moment of inertia of the pivoted beam plus the contributions of the two masses. The moment of inertia of the system can be measured by timing its natural period of oscillation about the pivot and using the following relationship:

$$I = \left[\frac{T}{2\pi}\right]^2 mgL \quad (5)$$

I is the moment of inertia, T is the period, m is the total mass, g is gravitational acceleration, and L is the distance from the pivot to the center of mass. L can be found by balancing the beam and masses on a knife edge. The differential equation of motion for the system is

FIGURE 3 Ballistometer with compound pendulum and angle transducer. Impacting mass is on the right, counterweight on the left.

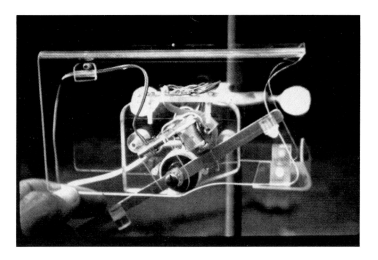

$$d^2\Theta/dt^2 - (mgL/I) \cos \Theta = 0 \quad (6)$$

where Θ is the pivotal angle referenced to the horizontal plane of impact.

Solution of the kinetic equation of motion is not absolutely necessary to the understanding, design, and use of a practical ballistometer. We can analyze the compound pendulum to see how the coefficient of restitution can be derived from energy and moment considerations.

In Figure 4 the center of mass of the pendulum is shown at a distance L from the pivot. If the pendulum is to rotate clockwise when released, and the beam is pivoted at its midpoint, the center of mass must be to the right of the pivot. Taking moments about the center of mass to find its location one obtains

$$L = \frac{M_1 r_1 - M_2 r_2}{M_1 + M_2} \quad (7)$$

If we can make the two masses equal, L becomes further simplified. When $M_1 = M_2 = mg$, then

$$L = \frac{1}{2}(r_1 - r_2) \quad (8)$$

We will use L to calculate the effective height to which the mass of the system is raised for a given starting angle of the pendulum, Θ_1, relative to the horizontal. Thus the potential energy will be obtained as

$$PE = 2mgL\sin\Theta \quad (9)$$

This potential energy can be equated to the kinetic energy to find the velocity of impact as well as the rebound velocity, both of which are needed to find the coefficient of restitution. The kinetic energy of such a rotating system is $1/2 \, I \, (d\Theta/dt)^2$. Solving for velocity gives

$$v = r_1(d\Theta/dt) = r_1 \sqrt{2mg(\sin\Theta)/I} \quad (10)$$

Considering this relationship for v_1 where Θ_1 is the starting angle of the pendulum and v_2 where Θ_2 is the rebound angle, the value of e may be calculated exactly as

$$e = \frac{v_2}{v_1} = \sqrt{\frac{\sin\Theta_2}{\sin\Theta_1}} \quad (11)$$

This is the basic equation that should be used to derive the coefficient of restitution in a system which measures pendulum angles. This is in conformity with the definition of e in Equation 2. Equation 11 can be verified by showing it to be identical with Equation 4. This is done by substituting under the radical in Equation 11 the geometric expressions for $\sin\Theta_2$, which is h/L, and the equivalent of $\sin\Theta_1$, which is H/L. The L cancels, and one is left with the square root of h/H or Equation 4.

We have shown how the potential energy transfer to rotational energy and back to potential is accomplished and does not alter the impact conditions. The point is made also that the angular measure is a simple way to register these processes, and the necessary trigonometric computation is not complicated. In doing this one gains control over the impact velocity which is critically important in applying ballistometry to viscoelastic materials such as the skin.

IV. Data Obtained with Ballistic Measurements

The fundamental quantity sought in ballistometrics is the coefficient of restitution. Tests on the skin as a specific extension of the measurement from its use on other materials have been reported to clearly distinguish elastic modulus differences between young and old, various body sites, as well as similar changes after pharmaceutical treatments.[8,12]

Ballistometry is attractive for several reasons. It is noninvasive and easy to use. No probes have to be attached to the skin. The instrument is not as expensive as for example the dynamometer. Although one cannot obtain data on the status of the stratum corneum as one does with shear measurements, it provides a practical indication of underlying tissue changes, i.e., for example, expansion from topical retinoid or other similar treatment.

As has been pointed out, the depth of tissue responding to ballistic impact will depend upon the

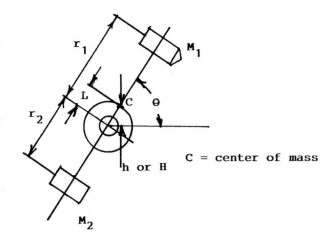

FIGURE 4 Compound pendulum mechanics.

height to which the falling mass is raised. Thus specific skin layers will be responsive in each case. Skin structure variations between body sites and between individual subjects will of course have an effect. However, in any series of tests planned to study response to a greater or lesser depth it is advisable to select an impact velocity and not change. The moment of inertia of the pendulum, although variable, is usually not altered after the performance of the apparatus has been optimized.

An impressive amount of data *(in vivo)* can be obtained with a ballistometer in a short time, particularly if the information is processed on-line via computer.[12] The computer may also control a solenoid release of the ballistometer's pendulum and start the rebound angle recording process. This may be done also with a manual electronic system. Images such as those shown in Figure 1 may be recorded on the screen of a storage oscilloscope whose sweep is synchronized to begin with the pendulum release. Alternatively, an X-Y plotter with good response and writing speed might be used to record raw bounce angle data. These would have to be corrected trigonometrically in accord with Equation 11 to obtain accurate coefficients of restitution.

Experiments should be done under the same atmospherically controlled conditions as would be applied to skin studies. Subjects should be allowed to equilibrate prior to testing.

References

1. Timoshenko, S. and Young, D.H., *Engineering Mechanics*, McGraw Hill, N.Y., 1940.
2. Loney, S.L., *A Treatise on Elementary Dynamics*, Cambridge Univ. Press, N.Y., 1900.
3. *Handbook of Chemistry and Physics*, Chemical Rubber Co., Cleveland, Ohio, 1946, pp. 30, 1302, 2328.
4. Hollinger, R. and Thelen, E., Design and development of an impact tester for use with asphalt, Report R-11, *National Asphalt Research Center,* Franklin Institute, Philadelphia, 1956.
5. Vogelsang, K., Ueber mechanische Gewebsprüfung am Auge, *Arch. f. Augenh.*, 108, 714, 1934.
6. Wiegersma, G., Elastometry of the eye, *Am. J. Ophth.*, 39, 811, 1955.
7. Mamelok, A.E. and Posner, A., Measurements of corneal elasticity, *Am. J. Ophth.*, 39, 817, 1955.
8. Tosti, A., Giovanni, C., Fazzini, M. L., and Villardita, S., A ballistometer for the study of the plasto-elastic properties of skin, *J. Invest. Dermatol.*, 69, 315, 1977.
9. Hertz, H., *J. Math. (Crelle)*, 92, 1881.
10. Timoshenko, S., *Theory of Elasticity*, McGraw Hill N.Y., 1934, p. 390.
11. Lamb, H., *Proc. London Math. Soc.*, 35, 141, 1902.
12. Fthenakis, C.G., Maes, D.H., and Smith, W.P., In vivo assessment of skin elasticity using ballistometry, *J. Soc. Cosmet. Chem.*, 42, 211, 1991.

Section F: The Cutaneous Vasculature

15.0

Visualization of Blood Vessels

15.1 Dynamic Capillaroscopy — A Sensitive Noninvasive Method for the Diagnosis of Conditions with Pathological Microcirculation 367
H.S. Yu, C.H. Chang, G.S. Chen, and S.A. Yang

Chapter 15.1
Dynamic Capillaroscopy — A Sensitive Noninvasive Method for the Diagnosis of Conditions with Pathological Microcirculation

Hsin Su Yu, Chung Hsing Chang, Gwo Shing Chen, and Sen Ann Yang
Department of Dermatology
Kaohsiung Medical College Hospital
Taiwan

I. Introduction

Cutaneous microcirculation can be divided into thermoregulatory shunt vessels and nutritive skin capillaries. Laser doppler flowmetry, transcutaneous oxygen-pressure measurement, and capillary microscopy are known noninvasive methods for assessing cutaneous microcirculation. Flex in nonnutritional shunt vessels dominates the signal recording of the laser doppler flowmetry. Transcutaneous oxygen pressure most probably primarily reflects the function of the nutritive skin capillaries. Presently capillary microscopy is the best choice for studying the nutritional status of a certain skin area.[1-2] The skin capillaries can be studied directly with an ordinary light microscope. In 1919 Basler described a sophisticated mechanical set-up for the actual measurement of the velocity of the blood cells in human nailfold capillaries.[3] In 1964 Zimmer and Demis demonstrated a microscope-television system for studying dynamic blood flow in human skin capillaries.[4] Bollinger and co-workers further refined this method and adapted a frame-to-frame analysis to measure both blood-flow velocity and vessel diameters in nailfold capillaries.[5] The application of the cross-correlation technique[6] for measuring the velocity of blood cells in the capillaries greatly improved this measurement. Recently capillary microscopy has been coupled with a videophotometric system and used with software to analyze the capillary blood-cell velocity (CBV).[7] This technique makes it possible to noninvasively study human skin capillaries under physiological and pathophysiological conditions. Some applications have been quickly developed to evaluate the dynamic microcirculatory status in peripheral vascular disorders including arterial occlusive diseases, collagen vascular diseases, and diabetes. It is the purpose of this chapter to introduce recent advances in dynamic capillaroscopy in the study of diseases (namely Tetralogy of Fallot and Raynaud's syndrome) and diabetes mellitus; a focus on clinical applications will be presented.

II. Methodological Principle

A. Computerized Videophotometric Capillaroscopy

During the past few years, a fully computerized system for CBV measurement has been developed. This system includes a microscope, television camera, monitor, and software (CapiFlow®, Stockholm) (Figure 1). For CBV, capillaries in the nailfold area are suitable for this purpose because they are parallel to the skin surface and can be visualized rather nicely in their full length (Figure 2). The details of dynamic capillaroscopy have been reported elsewhere. In brief, the finger or toe is placed on the investigation plate, and a small bracket is allowed to lightly touch the distal end of the nail. A drop of immersion oil is applied to make the nailfold transparent. The image can be seen on a TV monitor. All recordings are stored on videotape and analyzed by the CapiFlow® computerized analysis system. With this system the whole process of CBV calculations is performed automatically either by temporal correlation or spatial correlation velocimetry. Observations consist of: recording the number, caliber, length, relative area, visibility of the capillary loops, intercapillary distances, and CBV measurements, measured and analyzed as described in detail.

B. Dynamic Capillaroscopy with Load — Pressure and Temperature

Computerized videophotometric capillaroscopy is the most sensitive noninvasive method for evaluating both morphological change and rest CBV (rCBV) in capillaries. However, in some pathological conditions, physical load is necessary to display the abnormality in capillaries. The 1-min arterial occlusion and the cold provocation are useful methods for this purpose. An

FIGURE 1 Equipment for dynamic capillaroscopy: (1) Leitz capillary microscope; (2) Ikegami video camera; (3) Toshiba video tape recorder; (4) JVC video monitor; (5) CapiFlow® computer system.

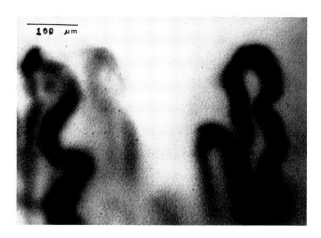

FIGURE 3 Extremely dilated, torturous capillary loops observed in patient of Tetralogy of Fallot with hemoglobin 22 g/dl.

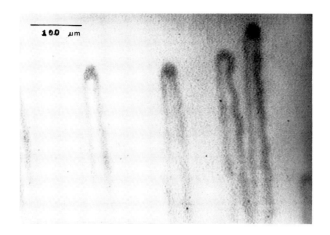

FIGURE 2 Normal hairpin-like capillary loops in regular arrangement.

arterial occlusion of digital arteries results in a reactive hyperemia, which indicates the myogenic response of the precapillary sphincter. The important parameters for postocclusive reactive hyperemia response include peak CBV (pCBV), time to pCBV (tpCBV), and the percent increase of CBV above rCBV during postocclusive reactive hyperemia (%PRH). The cold provocation test (5 to 15°C) has been used for studying the disturbances of skin microvascular reactivity in different types of Raynaud's syndrome.

III. Application

A. Tetralogy of Fallot

Tetralogy of Fallot (TF) is recognized to be the most common of the congenital cyanotic cardiac malformations, consisting of a ventricular septal defect, pulmonary stenosis, aortic override, and right ventricular hypertrophy. Patients with TF usually suffer from generalized cyanosis, clubbing digits, and a high risk of thrombosis due to secondary polycythemia. Cutaneous microcirculation was studied and we found that the nailfold capillaries in TF patients became dilated, torturous, and branching with an increase of total length and vascular area (Figure 3). The degree of dilation and vascularity was closely related to the hemoglobin (Hb) concentration. CBV declined with the increase of hematocrit (Hct), significant decreases were noted during Hb > 19 g/dl or Hct > 60%, i.e., 0.08 ± 0.05 mm/s vs. normal control 0.60 ± 0.32 mm/s ($p < 0.01$). TF patients are good models for studying the effect of long-term hypoxia. Capillary dilation and vascularity increase are compensatory reactions and are reversible. The capillary patterns became normal after successful operation, which were observed in one of our patients. We can use capillary microscopy to evaluate the compensatory status of TF patients before operation and the dynamic changes after operation.[8]

B. Raynaud's Syndrome

Raynaud's syndrome is the paroxysmal constriction of small arteries of the extremities, usually precipitated by cold. When exposed to low temperatures the digits become white (ischemic), then blue (cyanotic), and finally red (hyperemia).[9] Over 90% of the cases are primary Raynaud's syndrome. The principal clinical challenge is to distinguish idiopathic cases of primary Raynaud's syndrome from secondary to underlying disease of connective tissues, obstructive arterial disease, blood dyscrasias, drug toxicity, or artery injury.

Initially, abnormal capillary morphology was described in connective tissue disease, especially in scleroderma by Maricq et al.[10,11] It is just a qualitative method to differentiate primary from secondary Raynaud's syndrome. Then, a quantitative morpho-

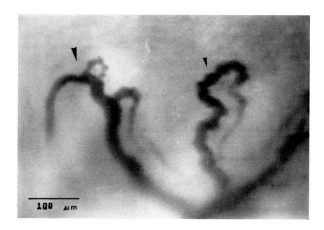

FIGURE 4 Loss of normal capillary loop and abnormal branching from superficial plexus (▶), blood flow in granular pattern (▸) observed in PSS patient.

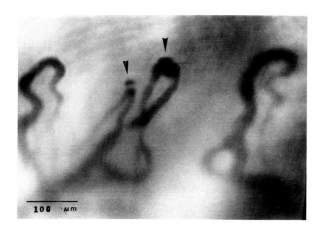

FIGURE 5 Aneurysm formation (▶) in capillary loop observed in PSS patient.

logical analysis of nailfold capillary microscopy was developed by Lefford et al.[12] They found that enlarged and deformed capillary loops surrounded by relative avascular areas in progressive systemic sclerosis (PSS) and mixed connective tissue disease (MCTD) patients. Finally, dynamic blood-flow analysis was performed after local cold exposure.[13] It clarifies the role of cold temperatures in Raynaud's syndrome. Prior to our study a comprehensive study utilizing all of dynamic capillaroscopy's potential had yet to be undertaken.

Morphologic changes of primary Raynaud's disease are approximately normal but the CBV decreased markedly after cold exposure.[14] Abnormal morphologic changes were noted in secondary Raynaud's syndrome, including an increase in intercapillary distance and ratio of torturous capillary loops, and a decrease in capillary loops, number, and total length.[15,16] Besides, it was also found that significant dilation of arterial and venous limbs and apex.[17] The degree of abnormal morphologic changes of PSS patients is more severe than SLE patients (Figures 4 and 5). In both of the groups, the CBV is slow even at room temperature. The results of a 2-year study utilizing dynamic capillaroscopy revealed its accuracy in differentiating primary from secondary Raynaud's syndrome and the clinical progressions of underlying diseases.

C. Diabetic Mellitus

Microcirculation is known to be disturbed in many organs of diabetic patients. A previous study revealed that prevalence of coiled and slightly enlarged microvessels is significantly increased in long-term diabetes at the nailfold of fingers[18] and toes.[19] It was also found that rCBV did not differ significantly in patients with diabetes as compared to controls.[20] However, the tpCBV after 60 s of arterial occlusion was significantly ($p < 0.01$) increased in the patients with both short and long durations.[21] Recently we studied the correlation between cutaneous microcirculation and retinopathy in diabetes mellitus (DM) patients. We found that DM patients with retinopathy had a significant increase of capillary tortuosity and tpCBV, and significant decrease of rCBV and pCBV in comparison with controls and DM without retinopathy.[22] Therefore skin dynamic capillaroscopy in concert with ophthalmoscopy and fundus photography could give a matchable, comprehensive, and complete assessment of DM vascular system. The morphologic changes of cutaneous microcirculation are not sufficiently pronounced to be used diagnostically for diabetic microangiopathy in single cases. However, they may be useful for characterizing patient samples. Correlation between morphologic and functional changes will be important in the future.

References

1. Fagrell B. Vital capillary microscopy — a clinical method for studying changes of the nutritional skin capillaries in legs with arteriosclerosis obliterans. *Scand J Clin Lab Invest* Suppl 133, 1973.
2. Fagrell B, Intaglietta M, Tsai AG, Ostergren J. Combination of laser doppler flowmetry and capillary microscopy for evaluating the dynamics of skin microcirculation. *Prog Appl Microcirc* 11:125–138, 1986.
3. Basler A. Uber die Bestimmung der Stromungsgeschwindigkeit in den BlutKapilallen der menschlichen Haut. *Muench Med Wochenschr* 13:347–348, 1919.
4. Zimmer JG, Demis DJ. The study of the physiology and pharmacology of the human cutaneous microcirculation by capillary microscopy and television cinematography. *Angiology* 15:232–235, 1964.
5. Bollinger A, Butti P, Barras JP, Trachsler H, Siegenthaler W. Red blood cell velocity in nailfold capillaries of man measured by a television microscopy technique. *Microvasc Res* 7:61–72, 1974.
6. Intaglietta M, Silverman NR, Tompkins WR. Capillary flow velocity measurements in vivo and in situ by television methods. *Microvasc Res* 10:165–179, 1975.

7. Fagrell B, Eriksson SE, Malmstrom S, Sjolund A. Computerized data analysis of capillary blood cell velocity. *Int J Microcirc Clin Exp* 7:276, 1988.
8. Chang CH, Yu HS. Study of the cutaneous microcirculation in Tetralogy of Fallot. *J Dermatol Sci* 4(2):110, 1992.
9. Coffman JD. Raynaud's phenomenon. *Hypertension* 17:593–602, 1991.
10. Maricq HR, LeRoy EC. Patterns of finger capillary abnormalities in connective tissue disease by wide-field microscopy. *Arthritis Rheum* 16:619–628, 1973.
11. Maricq HR, LeRoy EC, Dangelo WA, Medsger TA, Rodnan GP, Sharp GC, Wolfe JF. Diagnostic potential of in vivo capillary microscopy in scleroderma and related disorders. *Arthritis Rheum* 23:183–189, 1980.
12. Lefford F, Edward JCW. Nailfold capillary microscopy in connective tissue disease: a quantitative morphological analysis. *Ann Rheum Dis* 45:741–749, 1986.
13. Mahler F, Saner H, Bose Ch, Annaheim M. Local cold exposure test for capillaroscopic examination of patients with Raynaud's syndrome. *Microvasc Res* 33:422–427, 1987.
14. Jacohs MJHM, Breslau PJ, Slaaf DW, Reneman RS, Lemmens JAJ. Nomenclature of Raynaud's phenomenon — A capillary microscopic and hemorheologic study. *Surgery* 101:136–145, 1987.
15. Ohtsuka T, Ishikawa H. Graphic analysis of nailfold capillary in patients with collagen disease, especially in those with systemic sclerosis. *Jpn J Clin Dermatol* 45:637–644, 1991.
16. Caspary L, Schmees C, Schoetensack I, Hartung K, Stannat S, Deicher H, Creutig A, Alexander K. Alternations of the nailfold capillary morphology associated with Raynaud's phenomenon in patients with systemic lupus erythematosus. *Rheumatol* 18:559–566, 1991.
17. Liu CG, Su W, Luo Y. Changes in cutaneous microcirculation, hemorheology and platelet aggregation function in dermatomyositis. *J Dermatol Sci* 2:346–352, 1991.
18. Rouen LR, Terry EN, Doft BH, Clauss RH, Redisch W. Classification and measurement of surface microvessels in man. *Microvasc Res* 4:285–292, 1972.
19. Chazan BI, Balodimos MC, Larine RL, Koncz L. Capillaries of the nailfold in diabetes mellitus. *Microvasc Res* 2:504–507, 1970.
20. Tooke JE, Lins PE, Ostergren J, Fagrell B. Skin microvascular autoregulatory responses in type I diabetes: the influence of duration and control. *Int J Microcirc Clin Exp* 4:249–256, 1985.
21. Tooke JE, Ostergren J, Lins, PE, Fagrell B. Skin microvascular blood flow control in long duration diabetes with and without complications. *Diabetes Res* 5:187–192, 1987.
22. Chang CH, Yu HS. Using dynamic capillaroscopy with NaF fluorescence to assess cutaneous microcirculation of diabetic mellitus patients. Abstract of 18th Pacific Skin Research Club meeting, Nara, Japan, Oct 22–24, 1993.

16.0

Measurement of Erythema and Skin Color

16.1 Spectrophotometric Characterization of Skin Pigments and Skin Color 373
P. Bjerring

16.2 Measurement of Erythema and Melanin Indices 377
H. Takiwaki and J. Serup

16.3 CIE Colorimetry 385
W. Westerhof

Chapter 16.1
Spectrophotometric Characterization of Skin Pigments and Skin Color

Peter Bjerring
Department of Dermatology
Marselisborg Hospital
University of Aarhus
Aarhus, Denmark

I. Introduction: Spectrophotometric Characterization of Skin Pigments and Color

Skin color is of vast importance for the diagnosis of skin diseases and for the evaluation of progress of therapy of most dermatological diseases. Objective measurements of skin color using registration of the proportion of reflected light from the skin surface (reflectance spectroscopy) offers an accurate method to describe skin color. The method has been refined for more than a century, starting with the studies of Bunsen in 1861.[1,2] The earlier investigations lacked both sensitivity and reproducibility, but the development of electronics during the 1930s allowed construction of the first reliable reflectance spectrophotometers.[3] Reflectance spectrophotometry has since that time been widely used for the study of skin color during physiological changes and in disease.[4-16]

The use of reflectance spectrophotometry has often been compared with visual readings of skin color for evaluation of skin color changes, and it has been claimed that the human eye is able to differentiate as accurately as any apparatus when presented with two test samples. This may be true for direct comparisons or scoring of skin color, e.g., in tests of skin blanching induced by topical steroids, but only provided that the areas that are compared can be observed simultaneously.[9,15,17] For quantitative or repetitive investigations or investigations performed in different laboratories, only objective measurements using objective reflectance spectroscopic methods can provide reproductive results.[7,18-21]

II. Instrumentation

Instruments based on measurements of a few bands of reflected visible light from the skin are described in Chapters 16.1 and 16.2.[22-24] In this section, instruments based on scanning reflectance spectrophotometry of broader bands of the spectrum will be described.

A. Spectrum of Operation

Reflectance spectrophotometers have been designed for operation in either the ultraviolet (UV), visible, or infrared part of the optical spectrum.[25-30] However, the vast majority of clinical experimental data have been obtained from scanning in the visible spectrum where characteristic absorption peaks can be found for a series of important molecules, e.g., melanins, hemoglobins, bilirubin, and carotene.[17,27,31-35]

B. Construction Principles

Basically, scanning reflectance spectrophotometers may be constructed according to one of the following principles:

1. Broad band irradiation of the skin with filtering of the reflected light before detection
2. Irradiation of the skin with filtered light and detection of light with a broad band detector

Instruments with either single or double (reference) beams have been constructed. Double beam instruments continuously perform measurements not only on the skin, but also on a reference surface, and are therefore able to correct for alterations in lamp spectrum and other optical changes during the measurement procedure. This technology will in theory maximize both sensitivity and reproducibility of a reflectance spectrophotometric system compared to a single beam apparatus.

Also instruments have been specially constructed in order to reduce the effect of ambient light on the measurements. These instruments irradiate the skin surface with chopped light. The light detector is coupled (gated) to the chopper and only registers reflected light during periods of irradiation. The ambient light is measured during the dark periods and these measurements can be used for back light compensation.

The energy source for broad band irradiation of the skin was originally a tungsten lamp, but lately broadband Xenon arc lamps or Xenon flash lamps have been employed as sources of light.[30] The part of the instrument which transmits incident light and picks up reflected light from the skin surface may be subject to many different designs. Both the incident beam and the detector for reflected light can be mounted at specific angles to the skin surface. Some devices illuminate the skin surface at a right angle (90°) and detect at angles between 45° and 90°.[9,17,30] The light collecting device may also be constructed as an integrating sphere, which accumulates reflected light from all directions, or by a black chamber allowing only light emitted in the direction of the detector to be measured.[36]

The type of detector for measurement of visible light depends on the intensity of the reflected light and on the basic construction of the apparatus. Newer constructions incorporate flashlamps for broad-banded high-intensity irradiation of the skin combined with a detection system consisting of linear array of 512 to 2048 photodiodes covered with a wedge filter. These systems can perform measurements in only fractions of a second whereas the monochromator-based detector systems may use several seconds to minutes in order to perform a mechanical scan of the visible spectrum based on movement of a holographic filter plate.[30]

Flashlamp-based systems may incorporate both a flashlamp and a detector system in an integrated sphere. This eliminates any light loss in fiber-optic light guides. However, combination of light source, filter, and detector in one device results in a relatively heavy apparatus. When such a device is placed on the skin surface, it may compress skin capillaries during the measuring procedure if not suspended by a spring arrangement or mounted on a pivot system.[30,37,38] In fact, lack of control of skin surface pressure exerted by the measuring device may be one of the most significant sources of methodological errors in scanning reflectance measurements.

Calibration of a reflectance spectrophotometer is performed on a standard white surface consisting of either barium sulfate, magnesium carbonate, or magnesium oxide. These chemicals reflect visible light nearly 99%.

Some spectrophotometers have some internal leakage of light from one wavelength band to the next. This stray light or "cross talk" will also occur if the wavelength scans are extended from one wavelength to more than the double wavelength of the initial due to lack of absorption of "harmonic" wavelengths. Additional filters are necessary for blocking this effect. Calibration on a black hollow "infinite" tube can compensate for this phenomenon.

C. Incorporation of Computers

Microcomputers or built-in controllers have provided versatility during both measuring procedures and subsequent data processing and read-out. Computers enable easy standardization, calibration, and on-line monitoring of all critical components of a reflectance spectrophotometer, i.e., the lamp, fiber optics, filter (monochromator), and detector.[37,38]

D. Number of Measurements and Skin Area

Reflectance spectra from human skin are often presented as a mean of a number of consecutive measurements or scans of the same skin area. Averaging of measurements are often performed in order to reduce effects of small unavoidable movement of the measuring device relative to the skin and to reduce the effect of pulsating blood flow through dermal vessels during measurements. Also, if the measuring area is small, a series of measurements should be performed within a larger test area in order to reduce variation due to the uneven distribution of skin vessels.

E. Computerized Calculation of Chromophore Content

The content of specific chromophores present in the skin can be calculated by a computer using standard spectra of each chromophore as a standard. Either *in vivo* or *in vitro* spectra of each chromophore have been used. A specific mathematical method for convenient computerized calculation, the Choleski algorithm, can be used for rapid determination of skin chromophore concentrations.[17] The calculations are based on *a priori* knowledge of a standard spectrum of all significant chromophores present in the skin.

Both natural and artificial chromophores which may be present in the skin should be known *a priori*. The specific reflectance and transmission spectra of each chromophore form templates on which the computer superimposes a reflectance scan from the skin. Each specific chromophore spectrum is then applied onto the actual skin spectrum in order to make the best fit of all chromophores to the actual reflectance spectrum. As not all reflected light from the skin originates from "real" chromophores, also the reflectance spectra of the structural elements of the skin should be known to the computer before reliable calculations can be performed.[17] The spectrum of vitiligo skin can thus be used as a "negative" standard for melanins, and suction-blister-separated epidermis serves as a standard for epidermal attenuation.[31]

The computerized calculations of chromophore content can be calculated in absorbance units. Assuming a uniform distribution within the skin this can be converted to concentration of chromophores in the skin. The same principle has been used on other organs than the skin: a special reflectance spectrophotometer was developed for determination of concentrations of mitochondrial enzymes in living liver cells as a measure of intracellular oxidative status, and also the intracellular oxidative metabolism of the myocardium can be monitored by this principle.[39]

III. Application of Reflectance Spectrophotometry in Dermatology

This noninvasive method for quantification of cutaneous pigments has many potential applications. Among these are spectroscopically determined cutaneous oxyhemoglobin and deoxyhemoglobin, which can be used in the study of homeostatic control, to evaluate the potency of topically or systemically administered vasoactive compounds, to quantify treatment effects of, e.g., laser treatment of port wine stains, to compare the effects of different UV sources, and for evaluation of sunscreens after standardized UV-irradiation.[17,40,41] Also quantification of skin irritancy and allergy after topical application of different products can be performed by reflectance spectroscopy.[42,43]

Evaluation of melanin pigments has been used for quantification of immediate pigmentation, and formation of new pigments, and for monitoring of treatment effects of pigmentary disorders as melasma and vitiligo.[17,30–32] Determination of other pigments as betacarotene or bilirubin can be used for treatment monitoring in uremic patients, in light sensitive patients, and during phototherapy for neonatal hyperbilirubinemia.[33–35] Reflectance spectroscopy in the UV or the infrared spectra may be used for monitoring of water content in the skin or for determination of absorption, binding, and elimination kinetics of topically applied drugs with specific absorption peaks in these regions.[25–29,44–48]

References

1. Abel JJ, Davis WS. On the pigment of the negro's skin and hair. *J Exp Med* 1: 360–401, 1896.
2. Breul L. Üer die Verteilung des Hautpigments bei verschiedenen Menschenrassen. *Morphol Arb* 6: 691–719, 1896.
3. Hardy AC. History of the design of the recording spectrophotometer. *J Opt Soc Am* 24: 360–364, 1938.
4. Edwards EA, Duntley SQ. The pigments and color of human living skin. *Am J Anat* 65: 1–33, 1939.
5. Banerjee S. Fluctuation pigmentary and hormonal changes. *J Genet Hum* 32: 345–349, 1984.
6. Anderson RR, Parrish JA. Optical properties of human skin. In: The science of photomedicine. Reagen JD, Parrish JA (eds). Plenum Press, New York, pp: 147–193, 1982.
7. Ryatt KS, Feather JW, Dawson JB, Cotterill JA. The usefulness of reflectance spectrophotometric measurements during psoralens and ultraviolet A therapy for psoriasis. *J Am Acad Dermatol* 85: 558–562, 1985.
8. Wan S, Jaenicke KF, Parrish JA. Comparison of the erythemogenic effectiveness of ultraviolet-B (290–320 nm) and ultraviolet-A (320–400 nm) radiation by skin reflectance. *Photochem Photobiol* 37: 547–552, 1983.
9. Feather JW, Ryatt KS, Dawson JB. Reflectance spectrophotometric quantification of skin colour changes induced by topical corticosteroid preparations. *Br J Dermatol* 106: 436–440, 1982.
10. Breit R, Kligman AM. Measurement of erythemal and pigmentary responses to ultraviolet radiation of different spectral qualities. In: *The Biologic Effects of Ultraviolet Radiation.* Urbach F (ed). Pergamon Press, Oxford 1969; 267–275.
11. Wan S, Parrish JA, Jaenicke KF. Quantitative evaluation of ultraviolet induced erythema. *Photochem Photobiol* 37: 643–648, 1983.
12. Kollias N, Baqer AH. Quantitative assessment of UV-induced pigmentation and erythema. *Photodermatology* 5: 53–60, 1988.
13. Tang S, Gilchrest B. Spectrophotometric analysis of normal, lesional and treated skin of patients with port wine stains (PWS). *J Invest Dermatol* 78: 340–345, 1982.
14. Farr PM, Diffey BL. A quantitative study of the effect of topical indomethacin on cutaneous erythema induced by UVB and UVC radiation. *Br J Dermatol* 115: 453–454, 1986.
15. Queille-Roussel C, Poncet M, Schaefer H. Quantification of skin colour changes induced by topical corticosteroid preparations using Minolta Chroma Meter. *Br J Dermatol* 124: 264–270, 1991.
16. Argenbreit LW, Forbes PD. Erythema and skin blood content. *Br J Dermatol* 107: 569–574, 1982.
17. Andersen PH, Bjerring P. Noninvasive computerized analysis of skin chromophores in vivo by reflectance spectroscopy. *Photodermatol Photoimmunol Photomed* 7: 249–257, 1990.
18. Bjerring P, Andersen PH. Skin reflectance spectrophotometry. *Photodermatology* 4: 167–171, 1987.
19. Brunsting LA, Sheard C. The color of the skin as analysed by spectrophotometric methods. *J Clin Invest* 7: 575–593, 1929.
20. Buckley WR, Grum F. Reflection spectrophotometry. *Arch Dermatol* 83: 249–261, 1961.
21. Puttnam NA, Baxter BH. Spectroscopic studies of skin in situ by attenuated total reflectance. *J Soc Cosmet Chem* 18: 469–472, 1967.
22. Robertson AR. The CIE 1976 color difference formulas. *Color Res Applic* 2: 7–11, 1977.
23. Feather JW, Hajizadeh-Saffar M, Leslie G, Dawson JB. A portable scanning reflectance spectrophotometer using visible wavelengths for the rapid measurements of skin pigments. *Phys Med Biol* 34: 807–820, 1989.
24. Diffey BL, Oliver RJ, Farr PM. A portable instrument for quantifying erythema induced by ultraviolet radiation. *Br J Dermatol* 111: 663–672, 1984.
25. Gloor M, Hirsh G, Willebrandt U. On the use of infrared spectroscopy for the in vivo measurement of the water content in the horny layer after application of dermatological ointment. *Arch Dermatol Res* 27: 307–314, 1981.
26. Gloor M, Willebrandt U, Thomer G, Kuperschmidt W. Water content of the horny layer and skin surface lipids. *Arch Dermatol Res* 268: 221–233, 1980.
27. Koelmel K, Mercer P. Determination of the moisture of the horny layer by means of infrared reflection at three different wavelengths. *Arch Dermatol Res* 267: 206, 1980.

28. Gloor M, Heymann B, Stuhler T. Infrared spectroscopy of the water content of the horny layer in healthy subjects and in patents suffering from atopic dermatitis. *Arch Dermatol Res* 271: 429–436, 1981.
29. Comaish S. Infrared studies of human skin in vivo by multiple internal reflection. *Br J Dermatol* 80: 522–528, 1968.
30. Bjerring P, Andersen PH. Skin reflectance spectrophotometry. *Photodermatology* 4: 167–171, 1987.
31. Kollias N, Baqer A. Spectroscopic characteristics of human melanin in vivo. *J Invest Dermatol* 85: 593–601, 1985.
32. Kollias N, Baquer A. A method for the non-invasive determination of melanin in human skin in vivo. In: *The Biological Effect of UV-A Radiation,* F. Urback and R.W. Gange, Eds., Praeger, New York, 1986: 226–230.
33. Hannemann RE, Dewitt DP, Hanley EJ, Schreiner RL, Bonderman P. Determination of serum bilirubin by skin reflectance: effect of pigmentation. *Pediatr Res* 13: 1326–1329, 1979.
34. Hanley EJ, Dewitt DP. A physical model for the detection of neonatal jaundice by multispectral reflectance analysis. In: *Proceedings of the Sixth New England Bioengineering Conference,* D.O.V. Jaron, Ed. Pergamon Press, New York, 1978: 346–349.
35. Barbour RL, Benes L, Brown CD, Lee J, Lundin AP. Characterization of transcutaneous reflectance spectra in normal and uraemic patients. *Clin Chem* 31: 907 (abstr.), 1985.
36. Tang S, Wan S. The spectrophotometer and measurement of skin color. In: *Cutaneous Laser Therapy: Principles and Methods.* Arndt KA, Noe JM, Rosen S, Eds., John Wiley & Sons, New York, 1983: 27–39.
37. Minolta Camera Co. Osaka, Japan. *Precise Color Communication — Color Control from Feeling to Instrumentation.* Publication no: PCC S109 (E)-F1, 1990.
38. Carl Zeiss, Oberkochen, Germany. MCS Diode Array Spectrometer (publication no: CM-TS VIII/91 UToo). 1991.
39. Frank KH, Kessler M, Appelbaum, Dummler W. The Erlangen micro-lightguide spectrophotometer EMPHO 1. *Phys Med Biol* 34: 1883–1900, 1989.
40. Wendt H, Frosch PJ. Clinicopharmacological models for the assay of topical corticoids. S. Karger AG, Basel, 1982.
41. Andersen PH, Bjerring P. A comparative evaluation of sun protective properties of a shea butter by reflectance spectroscopy, laser Doppler flowmetry and visual scoring. *Skin Pharmacol Toxicol* 181: 267–276, 1990.
42. Agner T, Serup J. Sodium lauryl sulphate for irritant patch testing — a dose-response study using bioengineering methods for determination of skin irritation. *J Invest Dermatol* 95: 543–547, 1990.
43. Andersen PH, Nangia A, Bjerring P, Maibach HI. Chemical and pharmacologic skin irritation in man. A reflectance spectroscopic study. *Contact Dermatitis* 25: 283–289, 1991.
44. Bendit EG. Infrared absorption spectra of keratin. I. *Biopolymer* 4: 539–559, 1965.
45. Bendit EG. Infrared absorption spectrum of keratin. II. *Biopolymer* 4: 561–577, 1966.
46. Klimish HM, Chandrag G. Use of Fourier transform infrared spectroscopy with attenuated total reflectance for in vivo quantitation of polydimethylsiloxanes on human skin. *J Soc Cosmet Chem* 37: 73–87, 1986.
47. Kuppenheim HF, Heer RR. Spectral reflectance of white and negro skin between 440 and 1000 mu. *J Appl Physiol* 4: 800–806, 1959.
48. Kuppenheim HF. Spectral reflectance of the human skin in the ultraviolet, visible, and infrared range and its interpretation. In: Proceedings of the 1. International Photobiological Congress. Veemon and Zonen, Amsterdam. pp: 228–232, 1956.

Chapter 16.2
Measurement of Erythema and Melanin Indices

Hirotsugu Takiwaki* and Jørgen Serup**
*Department of Dermatology
University of Tokushima
Tokushima, Japan
**Bioengineering and Skin Research Laboratory
Department of Dermatology
University of Copenhagen
Bispebjerg Hospital
Copenhagen, Denmark

I. Introduction

In the field of dermatology or dermatopharmacology, inspection of skin-color change is often used to determine the degree of a skin-test reaction or to evaluate the efficacy of a drug such as topically applied corticosteroid which induces vasoconstriction. Although human eyes are quite an excellent sensor to distinguish the subtle difference between two colors, we cannot tell the "absolute" difference in a quantitative manner. It is true that visual scoring or grading of color change is a method to quantify the difference between skin colors, but it may differ from examiner to examiner. In addition, the score obtained is not a true physical quantity but a roughly graded number, so that it is unsuitable for the detailed analytic study by mathematical means. Moreover, since we cannot memorize a color precisely we may fail to tell the difference between two similar colors if each of them is shown to us at separate times. It is required, therefore, to quantify skin color objectively with instruments. Up to now, reflectance spectrophotometers and colorimeters have been mainly utilized for such a requirement. Although the former provides fine and precise information of the intensity of reflected light, namely reflectance, throughout the wavelengths of visible light, it may be troublesome to analyze data obtained with it unless the user is a specialist in color research. Additionally, this equipment is expensive and cumbersome for clinical use. The latter is easier to use due to its portability. Most of the portable instruments adopt the system in which a color is numerically expressed as coordinates in the standard color space defined by CIE (Commission Internationale de l'Eclairage).[1] Although this system must be suitable for the quantitative expression of numerous kinds of color in the field of industry, it seems to be difficult for dermatologists to understand its conception of "color space". Since hemoglobin and melanin are two well-known major pigments in the skin, our need might be said to be the quantification of selected colors which are mainly determined by the content of the two chromophores. Simple photoelectric reflectance meters designed to quantify erythema have been available for some years,[2,3] but they were prototypes and not appropriate instruments for routine clinical use due to limitations in technology.[4] More recently, portable optoelectronic instruments are developed for the quick and quantitative measurement of the degree of erythema and/or pigmentation. The principle and clinical application of these instruments will be discussed in this chapter.

II. Object and Methodological Principle

The quantity of hemoglobin, which corresponds directly to the extent of erythema, is expressed as the erythema or hemoglobin index, and that of melanin as the melanin index. The erythema index is suitable for the quantitative evaluation of anemic or hyperemic color changes of the skin, and the melanin index for hypo- or hyperpigmented skin conditions.

A. Theoretical Aspect
Both erythema and melanin indices are derived from the reflectance of the object in a selected band of the spectrum. In order to understand the meaning of the erythema and melanin indices, it is helpful to explain it using a multilayered skin model. We assume a simplified skin model composed of three layers of which the uppermost layer contains only melanin, the middle layer only hemoglobin, and the bottom layer

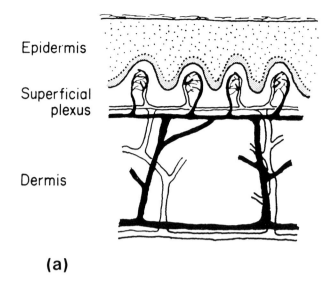

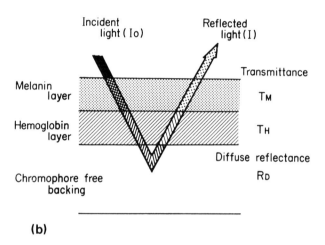

FIGURE 1 (a) Schematic structure of the skin. (b) Optical skin model of three-layered structure with an outer melanin layer, an inner hemoglobin layer, and a backing representing chromophore-free dermis.

neither chromophore (Figure 1). They represent the epidermis, the plexus of blood vessels in the upper dermis, and the dermis below them, respectively. In this model, we assume that there is no regular reflectance from the surface and that the diffuse reflectance of each of the upper two layers is nearly zero while that of the bottom layer is high, when each layer is placed separately on an ideal black background. According to Dawson et al.[5] and Diffey et al.,[6] the total reflectance, R, of this model at a given wavelength is roughly expressed as

$$R = I/I_0 \approx T_M^2 T_H^2 R_D \quad (1)$$

where I_0 is the intensity of incident light, I is that of reflected light, T_M is the transmittance of the uppermost layer, T_H is that of the middle layer, and R_D is the diffuse reflectance of the bottom layer. By taking the logarithm to the base 10 of the inverse reflectance (LIR) from Equation 1 we get

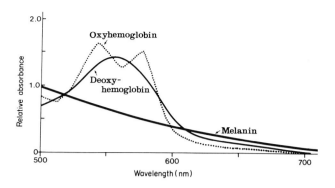

FIGURE 2 Absorbance spectra of oxyhemoglobin, deoxyhemoglobin, and melanin. (Redrawn from Feather, J.W., Ellis, D.J., and Leslie, G., *Phys. Med. Biol.*, 33, 711–722, 1988. With permission.)

$$\log(1/R) \approx 2\log(1/T_M) + 2\log(1/T_H) + \log(1/R_D) \quad (2)$$

Since the scattering within the upper two layers is extremely small from the above-mentioned assumption, the Beer-Lambert law of radiation absorption can be applied when the concentration of melanin and hemoglobin is low. Then Equation 2 is converted to

$$\log(1/R) \approx d_M E_M C_M + d_H E_H C_H + \log(1/R_D) \quad (3)$$

where d_M, E_M, and C_M are the thickness of the melanin layer, the coefficient determined by the absorbance of melanin at a given wavelength, and the concentration of melanin, respectively, and d_H, E_H, and C_H are their counterparts in the hemoglobin layer. Writing A = log (1/R), $m = d_M E_M$, $h = d_H E_H$, and $d = \log(1/R_D)$, Equation 3 is written as

$$A_\lambda \approx m_\lambda C_M + h_\lambda C_H + d_\lambda \quad (4)$$

where λ equals the given wavelength.

Figure 2 shows the absorbance spectra of reduced or oxygenated hemoglobin and that of melanin. Oxyhemoglobin has a peak of light absorption at around 560 nm (green light), and it absorbs little light in the wavelengths band of 650 to 700 nm (red light) indicating $h = d_H E_H = 0$ in this band. Therefore if selecting a few narrow bands of spectrum corresponding to red light (r, or r_1 and r_2) we get

$$A_r \approx m_r C_M + d_r \quad (5)$$

or

$$A_{r_1} - A_{r_2} \approx \left(m_{r_1} - m_{r_2}\right) C_M \quad (5')$$

since human dermis has similar reflectance values at wavelengths in these bands.[7] Moreover, when selecting a green band (g) and a red band (r) we get

$$A_g - A_r \approx (m_g - m_r)C_M + h_g C_H + (d_g - d_r) \quad (6)$$

M_g does not equal m_r because green light is more absorbed than red light by melanin. But if the concentration of melanin is very low and the thickness of the layer is thin enough

$$A_g - A_r \approx h_g C_H + (d_g - d_r) \quad (7)$$

Equations 5 and 5' indicate that LIR in the red band is a linear function of the melanin content (concentration × thickness) in this three-layered model irrespective of its hemoglobin content. This value, therefore, is suitable for the melanin index. Equation 7 also indicates that (LIR in the green band - LIR in the red band) is a linear function of the hemoglobin content if the melanin content is low, indicating that this value is suitable for the erythema (hemoglobin) index.

Since the skin is an optically turbid material with nonnegligible scattering, the assumption used in this model cannot be adapted in practice.[8] Nevertheless, it is found that the LIR of normal skin is dominated by the summed absorbances (log (1/T)) of hemoglobin and melanin[5] and that the erythema index has almost a linear relationship with the concentration of diluted hemoglobin solution *in vitro*.[9] We also confirmed that the melanin index correlated linearly with the concentration of diluted melanin solution.[10] The results derived from this model, therefore, are expected to show roughly estimated reflectance of the skin unless it contains large amount of hemoglobin or melanin.

B. Instruments

We would like to introduce only portable instruments which adopt the principle previously mentioned, although "full range" reflectance spectrophotometers are also utilized to obtain different types of the erythema and melanin indices.[5,11,12] Several kinds of equipment have been developed and some of them are commercially available.[4] One type has a halogen lamp as the light source that emits white light to the object, and the reflected light is selected within the two narrow bands centered at 546 and 672 nm using two interference filters.[6] The other type (Figure 3) have two[13,14] or three[9] light-emitting diodes (LEDs) that emit green light centered at 565[13] or 568 nm[14] and red light at 635[13] or 655[14] in the former, and 566, 640, and 670 nm in the latter. Since these LEDs emit narrow-band radiation no wavelength-selective filters are needed in the optical system. These instruments, irrespective of their types, thereby measure the reflectance of the skin within the narrow bands of the spectra corresponding to green and red light. After detection and photoelectric conversion of reflected light by photodiodes, the reflectance signals are processed by analogue electronics or built-in microcomputers, then the

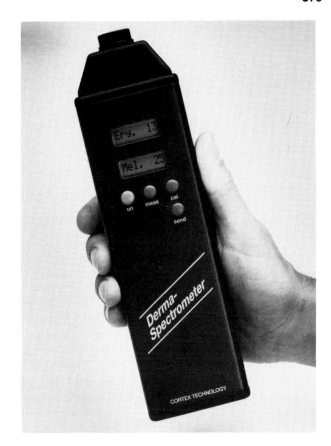

FIGURE 3 Example of a handheld instrument for measuring the erythema and melanin indices (Dermaspectrometer®, Cortex Technology, Hadsund, Denmark).

erythema[6,13,14] (hemoglobin[9]) index and/or the melanin[9,14] index are displayed on the digital panel meters. The size of the opening in the measuring head depends on instruments (approximately 20 to 150 mm²). For the zero calibration of the indices (i.e., 100% reflectance), measurements on a white standard surface (magnesium oxide, etc.) have to be made prior to each series of experimental procedures.

III. Sources of Error

Although these instruments offer convenient indices, there are some critical points that examiners should keep in mind. First, the erythema index is influenced by melanin index.[9] Equation 6 indicates that $A_g - A_r$, namely erythema index, is not exclusively a linear function of the hemoglobin content but it also depends on the melanin content. It is true that the term $(m_g - m_r) C_M$ in Equation 6 can be neglected if the subject has fairly light skin but it cannot be any more if the skin is pigmented. Figure 4 shows the relationship between the erythema index and the melanin index measured at the sites with various extent of pigmentation induced by ultraviolet-B (UVB) irradiation 14 d before the measurement (n = five Japanese subjects). This result surely indicates that the erythema index apparently increases

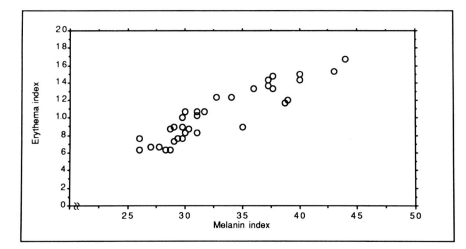

FIGURE 4 Dependence of erythema index on melanin index. Both indices were measured with the Dermaspectrometer® at test sites showing various degrees of pigmentation induced by UVB irradiation of different doses 14 d before the measurement.

in a linear fashion as the melanin index increases. We therefore should avoid the comparison of erythema indices between two sites showing quite different degrees of pigmentation. Second, the melanin index is influenced by the oxygen saturation level of hemoglobin. Since reduced hemoglobin absorbs more red light than oxyhemoglobin (Figure 2), the term $h\lambda\, C_H$ in Equation 4 cannot be neglected if the concentration of reduced hemoglobin becomes considerably higher than that of oxyhemoglobin. This implies that melanin index apparently increases when congested or cyanotic conditions take place. In fact, reduced hemoglobin easily increases during arm lowering,[15] which results in the increase in the melanin index.[9,16] Figure 5 shows the difference of the erythema and melanin indices at three different arm positions measured at the center of inner forearms of 15 Caucasians. Both indices significantly increased even if the arm was lowered for a short period of a few minutes, indicating that the posture of a subject during the measurement affects not only the erythema index but also the melanin index.

These errors result inevitably from the definition of the erythema and melanin indices. In addition, errors due to incorrect techniques may occur when the measuring head is placed with too much pressure, or when the measurement area is too hairy. Measurement in direct sunlight should be avoided as well. Sources of variation and pitfalls in the measurement of the erythema and melanin indices are summarized in Table 1.

IV. Correlation with Other Methods

There are few reports regarding the comparison between this type of instruments and other methods. According to our assessments[17,18] of the color on normal skin and lesions of psoriasis using the Dermaspectrometer®* and the Minolta Chromameter CR-200®,** a strong linear correlation was shown between the erythema index and the a* representing the red-green axis in the CIE-L*a*b* space (Figure 6). The melanin index also correlated negatively with L* values representing brightness (Figure 7). This correlation, however, does not indicate that the melanin index is equivalent to L*, because L* values decreased significantly on erythematous lesion while the melanin index did not.[18] It is because L* is mainly determined by the reflectance of green light[1] while the melanin index by that of red light. Therefore, the melanin index is considered more specific to the intensity of pigmentation.

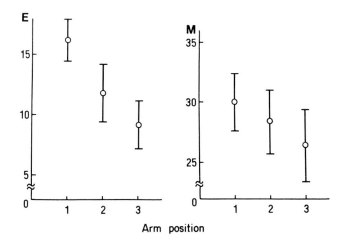

FIGURE 5 Variations of erythema (E) and melanin (M) indices (mean ± SD) of forearm skin at three different arm positions in 15 Caucasians. Positions: (1) lowering; (2) at heart level; (3) elevation. Measurements were carried out with the Dermaspectrometer® after keeping each position for 1 to 2 min. The differences between mean index values at two different positions are statistically significant in all cases.

TABLE 1 Summary of Sources of Variation and Pitfalls

Erythema index and heavy pigmentation (tan, race)
Erythema/melanin indices and blood deoxygenation
Orthostatic position
Regional variation
Diurnal variation
Moderate/heavy scaling
Direct light including sunshine
Physical contact between probe and skin

---

* Registered Trademark of Cortex Technology, Hadsund, Denmark.

** Registered Trademark of Minolta Camera Co., Tokyo, Japan.

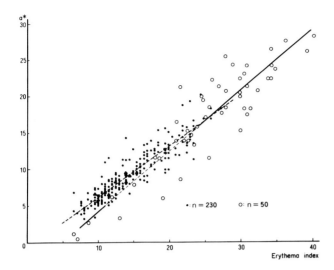

FIGURE 6 Correlation between erythema index and a* measured at 23 anatomical sites of 10 healthy subjects (•) and that in 50 plaques of psoriasis of 10 patients (○). Correlation coefficient r = 0.92 and r = 0.91, respectively. Regression lines are shown for each group. Subjects are all Caucasians. Cortex Dermaspectrometer® (erythema index) and Minolta Chromameter CR-200® (a*) were used.

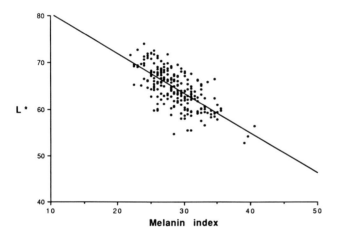

FIGURE 7 Correlation between melanin index and L* measured at 23 body sites of 10 Caucasians: r = −0.56, $p < 0.001$.

Concerning the correlation between erythema index and laser-Doppler flow (LDF), it seems to depend upon what is the cause of erythema. Gawkrodger et al.[19] performed both measurements in irritant and allergic patch-test reactions, and reported that a disproportionately greater increase in erythema index than in LDF was seen in mild irritant reactions compared to allergic reactions. The erythema index also correlates positively with LDF in UVB-induced erythema, whereas LDF decreases despite the increase in erythema index during cuff occlusion of the arm at diastolic pressure.[20] Figure 8 shows the negative relationship between the change in erythema index and that in LDF measured on the forearm skin following position changes of the arm. Considering that a congested change takes place during the arm lowering, this is most likely explained by the reason that erythema index reflects only the blood volume while LDF reflects the quantity multiplied by the blood flow rate.[21]

V. Clinical and Experimental Applications

In comparison with the conventional method such as reflectance spectrophotometry or tristimulus colorimetry, this type of instrument has advantageous points as follows:

1. Intensity of erythema or that of pigmentation can be separately quantified in the convenient description, index.
2. The erythema index is expected to have a roughly linear relationship with the content of red blood cells in the upper dermis, and the melanin index with that of melanin in the epidermis, unless the extent of erythema or pigmentation is intense.
3. Instruments are suitable for clinical use because of their portability and easy operation.

Since the measured reflectance is limited within two or three narrow bands concerning the absorption spectra of hemoglobin and melanin, these instruments are not suitable for the color change due to other chromophores such as jaundice, for example. Moreover, these indices may be inadequate for the quantification of the hemoglobin and melanin seated deeply in the dermis in such lesions as cavernous hemangioma, subcutaneous hemorrhage, Ota's nevus, and blue nevus because spectral reflectance of them is influenced by the optic characteristics of the dermis, and differs from that of the usual erythema and pigmentation.[22,23] The most suitable application, therefore, is the quantification of color change due to the increase or decrease in the content of red blood cells in the upper dermis and that of melanin in the epidermis. Measurement of the erythema index on black skin is found to be unsuccessful.[24]

Using the erythema index in practice, Farr and Diffey[25] found that the sensitivity to UV radiation shows considerable variation over the back and the measured increase in erythema index is linearly related to the logarithm of the radiation dose from the minimal erythema dose,[25] with different slopes depending on the wavelengths.[26] They also reported similar time courses of UVB and UVC erythemas[27] and similar effects of topical indomethacin on them,[28] though different in sensitivity. They concluded that the difference in erythemal response to UVB and UVC radiation is not due to the formation of different mediators at these wavelengths. In addition, they found that some

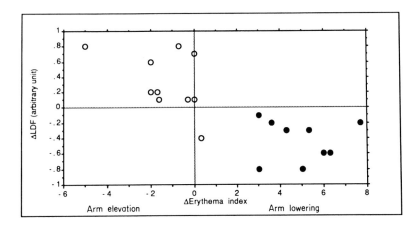

FIGURE 8 Relationship between the change in erythema index and that in laser-Doppler blood flow following arm elevation (○) and lowering (●), where the change is expressed by the difference (Δ) between the value at heart level and that in each position.

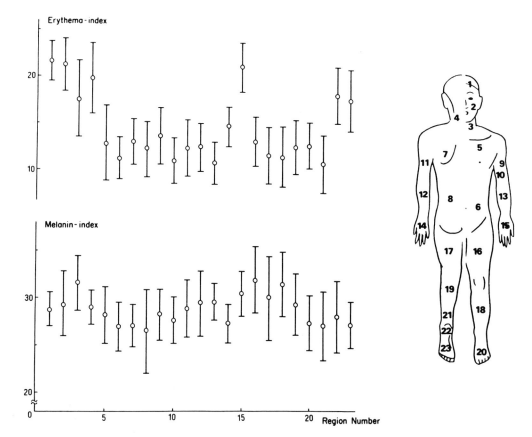

FIGURE 9 Erythema index (upper) and melanin index (lower) against 23 anatomical sites of 10 Caucasians measured with the Dermaspectrometer®. Region numbers correspond to those in body figures.

patients with polymorphic light eruption showed abnormal augmentation of UVB- and UVA-induced erythema by indomethacin.[29] Moreover, in the assessment of the onset of UVB and UVC erythema, Diffey and Oakley[30] reported that vasodilation occurs in the "latent period" before the erythema becomes visible. A series of these works might be impossible to have been completed without reliable means to quantify erythema. The erythema index was also used successfully for the objective evaluation of the efficacy of some drugs on UVB-induced erythema[31] and on sundamaged skin,[32] but was found to be unsuitable for the assessment of the effect of antihistamines against histamin flare reaction.[33] Response of port-wine stains, or nevus flammeus, to the laser treatment and its time course were readily quantified as well.[9] We measured the variation of both indices at different body sites in winter time,[17] and confirmed that subtle differences between the locations are readily detected (Figure 9). The erythema index also can be a marker which reflects the extent of erythema and the amount of scales on psoriatic plaques[18] (Figure 10). Unfortunately, it cannot be utilized as a marker of severity of the plaque because it becomes

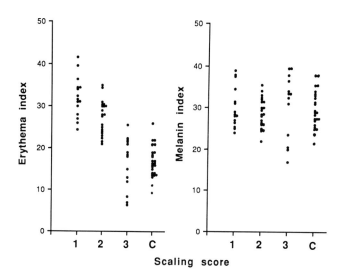

FIGURE 10 Erythema and melanin indices of psoriatic plaques with various amounts of scales. Scaling score definition: (1) no or slight powder-like scaling; (2) moderate scaling (small and some large flakes); (3) heavy scaling (large flakes); (C) regional controls adjacent to lesions.

rather closer to that of normal skin as erythema is covered by thicker scale. However, it will provide helpful information to quantify the differences between lesions by using it with clinical records.

VI. Recommendation

Considering how the erythema and melanin indices are defined, the following is recommended in order to make the best use of these instruments:

1. Subjects should take a standardized position for each series of the measurement, especially if it is performed in the extremities. Otherwise, not only erythema index but also melanin index is influenced by their positions.
2. Measuring head should be placed very softly and perpendicularly onto the skin. Holding it too long at a test site should be avoided. Much or incorrect pressure makes the site anemic or congested. Measurements should not be performed under direct light including sunshine.
3. Avoid comparing erythema indices between two sites where the levels of pigmentation (melanin index) are considerably different. If obtainable, empirical corrections for the indices[9] are desirable.
4. Skin color shows spontaneous diurnal variations[34] and regional differences. When the intensity of skin-test reaction is successively examined in a wide area, the data for the control should be obtained in the normal skin adjacent to the test site at each measurement.

References

1. Robertson, A., The CIE 1976 color-difference formulae, *Color Res. Appl.*, 2, 7–11, 1977.
2. Daniels, F. and Imbrie, J.D., Comparison between visual grading and reflectance measurements of erythema produced by sunlight, *J. Invest. Dermatol.*, 30, 295–303, 1958.
3. Frank, L., Rapp, Y., and Bergman, L.V., An instrument for the objective measurement of erythema, *J. Invest. Dermatol.* 38, 21–24, 1962.
4. Diffey, B.L. and Farr, P.M., Quantitative aspects of ultraviolet erythema, *Clin. Phys. Physiol. Meas.*, 12, 311–325, 1991.
5. Dawson, J.B., Barker, D.J., Ellis, D.J., Grassam, E., Cotterill, J.A., Fisher, G.W., and Feather, J.W., A theoretical and experimental study of light absorption and scattering by *in vivo* skin, *Phys. Med. Biol.*, 25, 695–709, 1980.
6. Diffey, B.L., Oliver, R.J., and Farr, P.M., A portable instrument for quantifying erythema induced by ultraviolet radiation, *Br. J. Dermatol.*, 111, 663–672, 1984.
7. Anderson, R.R. and Parrish, J.A., The optics of human skin, *J. Invest. Dermatol.*, 77, 13–19, 1981.
8. Wan, S., Anderson, R.R., and Parrish, J.A., Analytical modeling for the optical properties of the skin with *in vitro* and *in vivo* applications, *Photochem. Photobiol.*, 34, 493–499, 1981.
9. Feather, J.W., Ellis, D.J., and Leslie, G., A portable reflectometer for the rapid quantification of cutaneous haemoglobin and melanin, *Phys. Med. Biol.*, 33, 711–722, 1988.
10. Takiwaki, H., Watanabe, Y., Utsunomiya, M., Shirai, K., Nakagawa, K., and Kanno, Y., Analysis of skin color using a videomicroscope with computer, *Skin Pharmacol.*, 5, 242 (abstr.), 1992.
11. Feather, J.W., Ryatt, K.S., Dawson, J.B., Cotterill, J.A., Barker, D.J. and Ellis, D.J., Reflectance spectrophotometric quantification of skin colour changes induced by topical corticosteroid preparations, *Br. J. Dermatol.*, 106, 437–444, 1982.
12. Andersen, P.H. and Bjerring, P., Spectral reflectance of human skin in vivo, *Photodermatol. Photoimmunol. Photomed.*, 7, 5–12, 1990.
13. Pearse, A.D., Edwards, C., Hill, S., and Marks, R., Portable erythema meter and its application to use in human skin, *Int. J. Cosm. Sci.*, 12, 63–70, 1990.
14. DermaSpectrometer operation manual, Cortex Technology, Hadsund, 1991.
15. Hajizadeh-Saffar, M., Feather, J.W., and Dawson, J.B., An investigation of factors affecting the accuracy of *in vivo* measurements of skin pigments by reflectance spectrophotometry, *Phys. Med. Biol.*, 35, 1301–1315, 1990.
16. Takiwaki, H. and Serup, J., Variation in color and blood flow of the forearm skin during orthostatic maneuver, *Skin Pharmacol.*, 7, 226–230, 1994.
17. Takiwaki, H., Overgaard, L., and Serup, J., Comparison of narrow-band reflectance spectrophotometric and tristimulus colorimetric measurement of skin color — 23 anatomical sites evaluated by the DermaSpectrometer® and the Chroma Meter CR-200®, *Skin Pharmacol.*, 7, 217–225, 1994.
18. Takiwaki, H. and Serup, J., Measurement of color parameters of psoriatic plaques by narrow-band reflectance spectrophotometry and tristimulus colorimetry, *Skin Pharmacol.*, 7, 145–150, 1994.
19. Gawkrodger, D.J., McDonagh, A.J.G., and Wright, A.L., Quantification of allergic and irritant patch test reactions using laser-Doppler flowmetry and erythema index, *Contact Dermatitis*, 24, 172–177, 1991.

20. Shirai, S., Utsunomiya, M., and Takiwaki, H., Measurement of skin color and cutaneous blood flow in UVB-induced erythema and in the skin under congestive state, presented at 10th Asian-Australasian Conference of Dermatology, Manila, November 15 to 19, 1992.
21. Stüttgen, G., Ott, A., and Flesch, U., Measurement of skin microcirculation, in *Cutaneous Investigation in Health and Disease,* Lévêque, J.-L., Ed., Marcel-Dekker, New York, 1989, pp. 363–366.
22. Findlay, G.H., Blue skin, *Br. J. Dermatol.,* 83, 127–134, 1970.
23. Parrish, J.A., Responses of skin to visible and ultraviolet radiation, in *Biochemistry and Physiology of the Skin,* Goldsmith, L.A., Ed., Oxford University Press, New York, 1983, pp. 717–722.
24. Takiwaki, H., personal communication, 1991.
25. Farr, P.M. and Diffey, B.L., Quantitative studies on cutaneous erythema induced by ultraviolet radiation, *Br. J. Dermatol.,* 111, 673–682, 1984.
26. Farr, P.M. and Diffey, B.L., The erythemal response of human skin to ultraviolet radiation, *Br. J. Dermatol.,* 113, 65–76, 1985.
27. Farr, P.M., Besag, J.E., and Diffey, B.L., The time course of UVB and UVC erythema, *J. Invest. Dermatol.,* 91, 454–457, 1988.
28. Farr, P.M. and Diffey, B.L., A quantitative study of the effect of topical indomethacin on cutaneous erythema induced by UVB and UVC radiation, *Br. J. Dermatol.,* 115, 453–466, 1986.
29. Farr, P.M. and Diffey, B.L., Effect of indomethacin on UVB- and UVA-induced erythema in polymorphic light eruption, *J. Am. Acad. Dermatol.,* 21, 230–236, 1989.
30. Diffey, B.L. and Oakley, A.M., The onset of ultraviolet erythema, *Br. J. Dermatol.,* 116, 183–187, 1987.
31. Trevithick, J.R., Xiong, H., Lee, S., Shum, D.T., Sanford, S.E., Karlik, S.J., Norley, C., and Dilworth, G.R., Topical tocopherol acetate reduces post-UVB, sunburn-associated erythema, edema, and skin sensitivity in hairless mice, *Arch. Biochem. Biophys.,* 296, 575–582, 1992.
32. Eskelinin, A. and Santalahti, J., Special natural cartlage polysaccharides for the treatment of sun-damaged skin in females, *J. Int. Med. Res.,* 20, 99–105, 1992.
33. Lever, L.R., Hill, S., Marks, R., Rosenberg, R., and Thompson, D., Effects of cetirizine, loratadine, and terfenadine on histamine weal and flare reaction, *Skin Pharmacol.,* 5, 29–33, 1992.
34. Queille-Roussel, C., Poncet, M., and Schaffer, H., Quantification of skin-colour changes induced by topical corticosteroid preparations using Minolta Chroma Meter, *Br. J. Dermatol.,* 124, 264–270, 1991.

Chapter 16.3
CIE Colorimetry

*Wiete Westerhof
Department of Dermatology
Academic Medical Center
University of Amsterdam
Amsterdam, The Netherlands*

I. Introduction

Color information may be acquired and communicated in various ways. However in a scientific context there are requirements for consistency independent of time, distance, and language.[1] Although the average person may be able to distinguish several thousands of colors, the description of visual observations by the use of general color terms is the least satisfactory in terms of precision. The range of names upon which people reliably agree is very limited and there is no simple way of using visual observations to describe the differences between colors.[2]

Color assessment based on comparison with sets of colored samples, known as color-order systems, can increase the range and reliability of color designations significantly.[2] The Munsell color-order system[3] is the oldest and perhaps the most widely recognized. Color comparisons may involve metamerism, that is: the actual physical composition of the colored lights might be different but these are perceived as the *same* color. Therefore color-order systems are most reliable when used under fixed conditions of illumination to match colors of relatively flat surfaces that are uniformly pigmented. They are less appropriate for samples that are heterogeneous in surface structure. A particular limitation of color-order systems is that they only identify single colors and do not provide a way of specifying the nature or magnitude of color differences.

The definitive method of measuring skin color uses the recording spectrophotometer adapted for reflectance readings. However these readings are not communicable, as they do not conform to an international standard. The limitations of visual observations may, in principle, be overcome by the instrumental evaluation of colors and color differences according to the system of color measurement established by the CIE (Commission Internationale de l'Eclairage).[2,4-6] This system has become widely used with the availability of reliable reflectance spectrophotometric instrumentation that conforms to the recommendation of the CIE for the measurement of the color of surfaces.[7-9] This method does not give information about the substances generating the color but it is highly appropriate for color matching (e.g., grading of erythema and grading of melanin pigmentation, etc.). In fact, the use of the CIE tristimulus values is a concise and reliable way of approximating an actual normal color observer. In other words, by using these units one can imitate what is normally done by visual inspection but with greater reliability and reproducibility than a given human observer could achieve.

II. Aim

The objective of this chapter is to present qualitative and quantitative measurement methods of color of the skin, mainly erythema and melanin pigmentation. This is done with a tristimulus computer-controlled color analyzer which measures reflected object color in an accurate and reproducible way utilizing CIELAB (CIE 1976 L*a*b*) color space values.

These CIE color space parameters are proposed for the unambiguous communication of skin-color information that relates directly to visual observation of clinical importance or scientific interest.

III. Methods

A. CIE Color System

The perceived color of objects depends on: (1) the nature of the illuminating light, (2) its modification by interaction with the object, and (3) the characteristics of the observer response. The CIE system defines these conditions as follows: (1) The relative spectral energy distributions of various illuminants, known as CIE standard illuminants, are specified and available as published tables,[2,4-9] (2) the modification of an illuminant by interaction with the object is measured with a reflectance spectrophotometer having an optical configuration that conforms CIE recommendations,[4] and provides a visible spectrum expressed as the fractions of incident light intensity reflected in the wavelength range 400–700 nm; (3) the nature of human color vision has been quantified for the purpose of color measurement in terms of three color-matching functions x, y, and z (Figure 1). Three are required because color vision has been found to be trichromatic: a single perceived color may be regarded as resulting from the effect of three separate stimuli on the visual cortex.[6] Their numerical values are available as published tables and are known collectively as a CIE standard observer.[2,4,5,9] They may be regarded simply as a numerical description of average human color vision. From a practical viewpoint their use has been made more convenient by incorporation of the tabulated values within the software provided with color-measuring reflectance spectrophotometers. Similarly, the tabulated values of the relative spectral-energy distributions of various CIE illuminants are provided within the software because they do not exist as actual physical sources of light within instrumentation.

Colors are measured in terms of their tristimulus values X, Y, and Z by combining a selected table of illuminant data, the measured values of reflectance, and a selected table of color-matching functions with three summations, each having the form

$$\Sigma \text{ ER } x\text{-} = X$$

At selected intervals in the wavelength range 400 to 700 nm the relative energy (E) of the chosen illuminant is multiplied by the fraction reflected (R) and the numerical value of the standard observer (x or y or z). A wavelength interval of 10 nm, which requires 31 terms in each summation, gives adequate precision for most purposes. Modern color-measurement instrumentation normally incorporates microcomputer hardware and software so that the spectral measurements and subsequent calculations are integrated so as to produce a copy of the results within a few seconds. However the tristimulus values of colors are difficult to relate to the experience of seeing them. In addition, in any study involving comparisons, contrasts, or changes, tristimulus values do not directly enable measurement of the difference between two colors. This concern has now been overcome by using the tristimulus values to calculate the CIE 1976 L*a*b* (CIELAB) color space values.[2,4,7,9] The mathematical manipulations that convert tristimulus values to CIELAB color space values enable colors to be regarded as existing in an approximately uniform three-dimensional space in which each particular color has

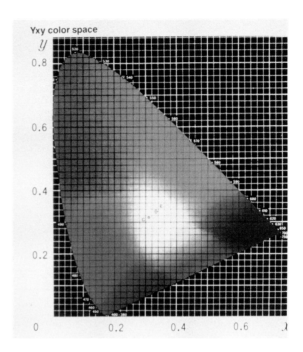

FIGURE 1 X, Y, and Z color coordinates.

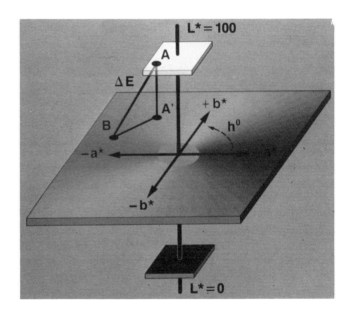

FIGURE 2 Color space; CIELAB.

a unique location defined in terms of its cartesian coordinates with respect to the axes L*, a*, and b* as shown in Figure 2. Modern computer-controlled reflectance spectrophotometers provide for automatic calculation of CIELAB values from the spectral data they produce.

The measured value of a color has been recommended by the CIE as the psychometric correlate of the visually perceived color attribute of "lightness",[4,7] to which the descriptive terms assigned might include the words "light", "dark", etc. In other words, L* would measure the change along a grey scale from black to white that visually varied in a perceptually uniform manner. The L* scale, which ranges from 0 for a theoretical black to 100 for white, corresponds to the notion of the value attribute in the Munsell color system.

The a* and b* coordinates may be conceptually related to Hering's opponent color theory,[8] which was based on the proposition that the retina of the eye contains opponent color channels that distinguish colors according to their red-versus-green and yellow-versus-blue attributes. In CIELAB space they are more useful when converted into polar coordinates. This facilitates the definition of a hue angle, h* = arctan (b*/a*) (Figure 2), which is recommended by the CIE as the psychometric correlate of the visually perceived attribute of hue (e.g., red, orange, yellow, etc.).[4,8] Measured hue angles make the use of visually assigned hue terms unnecessary, although it is simple and often convenient to relate them in a general way. The general angular position of some of the main generic hues are shown in Figure 1. CIE hue angle corresponds conceptually to the attribute of hue in the Munsell color-order system but no simple relationship has been found between measured hue-angle values and Munsell hue designations.[10] Colors for which both a* and b* are zero, and therefore lie on the L* axis, are termed achromatic and would be perceived as grey, white, or black. The visually perceived color attribute of "saturation", which might be described by the use of the terms "weak", "strong", etc., may be measured in terms of its distance away from the L* axis in the a*b* plane. This is the length of the line C in the diagram. It is termed the CIE(1976)a*b* chroma and is calculated using coordinate geometry as

$$C = [(a^*)^2 + (b^*)^2]^{1/2}$$

It corresponds conceptually to the attribute of chroma in the Munsell system but the measured values do not relate in any simple way to Munsell designations.[10] Thus the use of CIELAB coordinates enables measurement of the three attributes of a color by which it is visually distinguished.

CIELAB space is not only more convenient than tristimulus values with respect to its conceptual relationship to the actual experience of seeing colors but it has the important advantage of providing a means of measuring the differences between any two colors.[2,8] Their color difference (E) is calculated, using coordinate geometry, as the length of the line joining their coordinate positions:

$$E = [(L^*)^2 + (a^*)^2 + (b^*)^2]^{1/2}$$

In some circumstances the differences between two colors may also be considered in terms of differences in hue angle and/or chroma.

Modern computer-controlled reflectance spectrophotometers used for color measurement often include the software necessary for automatic calculation of color differences from two sets of spectral data.

For more details the reader is referred to the appendix of the excellent paper by Weatherall and Coombs.[1]

B. Technical Details of the Colorimeters

Several colorimeters have been manufactured for medical and scientific use. Here we discuss three of the most commonly used instruments.

1. Labscan 6000® (Hunter Associates Inc., USA) scanning reflectance visible spectrophotometer has 0° illumination and 45° viewing geometry with the specular component excluded. The instrument is calibrated with a supplied white standard traceable to the National Bureau of Standard's perfect white diffuser. The spectrophotometer is controlled by an IBM-XT microcomputer, which performs all color calculations from the digitized spectral data by means of a menu-driven set of programs supplied with the instrument. The skin is measured over a 30-mm diameter open circular port with a 26-mm illuminated area in the horizontal upper surface of the sensor module. Reflectance spectra over the wavelength range of 400 to 700 nm, requiring about 3 s for measurement, can be obtained. Tabulated data for CIE illuminant D65 and the CIE 1964 10′ standard observer are selected under software control and combined with the spectral data at 10-nm intervals to compute the CIELAB L*a*b* values. The latter two are then further converted to CIELAB color space and chroma. The instrumental set up is not suitable for field conditions and routine clinical applications because of its volume and weight.
2. The Minolta Chromameter CR 200® (Osaka, Japan) is a light-weight and compact tristimulus color analyzer for measuring reflected object color (Figure 3). Utilizing high-sensitivity silicon photocells, filtered to match CIE Standard Observer Response, the Chromameter CR 200® assures good accuracy and reproducibility (measuring error ≤1%). Readings, taken through the measuring head, are processed by a built-in microcomputer and then presented digitally on a liquid-crystal display. The measuring head also contains its own standard light source: a high-power xenon arc lamp which provides diffuse illumination from

FIGURE 3 Chromameter.

a controlled angle for vertical viewing and constant, even lighting on the object. The illumination chamber of the Minolta equipment is cylindrical, with a conical opening pointing toward the skin surface. These two parts are separated by a diffusing plate with an inner circular aperture 4 mm in diameter. The outer aperture is 11 mm, and the diameter of the estimated measuring area is 8 mm. The photo receiver is centered in the top of the chamber, and the tristimulus optical analysis unit with six silicon photocells is located directly in the probe unit (in the CR series), and in previous equipment in the main body. The meter's precise double-beam feedback system, using six photocells, detects any slight deviations in the xenon light's spectral distribution, and the microcomputer compensates for them, thus ensuring the utmost accuracy in measurements. The measuring area is circular and 8 mm in diameter. The Chromameter offers five measuring modes: two chromaticity-measuring modes — Y^{xy} and $L^*a^*b^*$ — and three color-deviation measuring modes — ± (Y^{xy1} ± ($L^*a^*b^*$) and Eab). The weight of the Minolta CR-200 is 1951 g.

3. The Dr. Lange Micro Color® (Dr. Bruno Lange GmbH, Dusseldorf, FRG) is computerized with a number of facilities. L, a*, and b* values are presented on a digital display. It is based on illumination with a xenon flash light. In the Lange equipment the chamber is spherical with a circular aperture, 5 mm in diameter, directed toward the skin surface. Measuring and reference photo receivers are mounted on the top of this sphere, but eccentrically aligned by 6° and 8°. Optical fibers lead the signal to the main body containing six silicon photocells for tristimulus analysis (red, green, blue) of both signal and reference light, according to DIN 5033. The Lange equipment has a short cable (600 mm) between probe and main body. The weight of the Lange equipment is 6 kg. It works at a slightly slower rate than the Minolta equipment. The technical reproducibility of the two colorimeters as given by manufacturers is the same, i.e., 0.15 E* to white. Both are calibrated against their respective white calibration tiles before use.

For the practical handling of the colorimeters the reader is referred to the instruction manuals of the respective apparatus.

IV. Sources of Error

A. Instruments

It is not possible to go into much technical detail within the framework of this chapter. A source of error could be the illumination. A light source comparable to the sun does not exist. Most of the artificial light sources emit incontinuous bands out of the whole light spectrum. This may give rise to metamerism. The filters used in this tristimulus chromameter are not exactly monochromatic. Especially at the periphery of the wave band of absorbtion, deviations in the filtering capacity exist which can lead to incorrect color measurement. Also the liquid or glass fiber light guide can contribute to inhomogeneous loss in light spectrum constituents.

The angle of illumination of the object being measured, in our case the skin, determines the degree of specular reflection which negatively influences the color analysis of the reflected light. The lack of standardization in the production of different colorimeters will lead to incomparable results of measurement. One given instrument is usually very reproducible in measurements of the same object.[5]

B. Measurement Method

The skin is not a homogeneous surface structure. It differs in primary, secondary, and tertiary skinfolds depending on the site of the body. Also the greasiness and humidity of the skin differ from site to site. This will affect the glossiness of the skin. Similarly the hairs, blood vessels, blemishes spots, and scars may determine the color aspects of the skin. The reflection from the skin is influenced by all the above factors leading to incorrect color measurement and significant site-related variations in reflectance. It is therefore imperative to carefully select the skin sites to be measured and to take the measurements in duplicate or triplicate.

If erythema is to be measured it has to be borne in mind that this is a function of blood vessel size and flow. These parameters can be influenced among other things, by compression. Therefore the measuring head with the aperture must not be pressed too hard against the skin. The measuring head must also be kept in a perpendicular fashion to the skin surface. Otherwise differences in reflectance of the light source will occur.

C. Reproducibility

Another source of error lies in the method of provoking the color change. In UV-induced erythema a 3-plus visual grading with edema may be less red due to capillary compression and extracellular fluid accumulation.

The time lapses of UVA and UVB induction is different. Therefore the moment of measurement is critical. If follow-up measurements are required from one and the same measuring site it is necessary that care is taken to meticulously place the aperture over exactly the same place as chosen for the previous measurements. As was already explained a few milimeters away from the measuring site the characteristics of the skin may be completely different.

We have devised measuring bracelets which allowed us to exactly measure the UV-irradiated sites of the skin of the lower arm for the development of erythema and pigmentation (Figure 4). For other sites of the body similar appliances could be devised. These devices help to increase the reproducibility of the measuring technique. In Section VI, *Fields of Application* the methodology of inducing color changes or measuring color changes over time need to be standardized in order to be able to make chromametric measurements in a reproducible way.

V. Correlation with Other Methods

Reflectance chromameters have not been compared with spectrophotometers. However Weatherall et al.[1] used a spectrophotometer, which converted the reflected wavelengths taken at 10-nm intervals into CIELAB values.

Serup et al.[11] evaluated two commercially available colorimeters (Minolta chromameter and Dr. Lange microcolor) and compared them with laser-Doppler flowmetry, which is widely used for quantification of erythema and cutaneous inflammation. Irritant reactions after application of sodium lauryl sulfate (SLS) were studied.

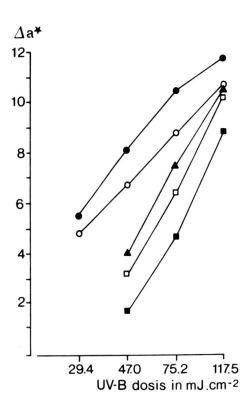

FIGURE 4 UV sensitivity of the skin related to body site. Diagram of mean Δ a* vs. irradiation doses given to 5 body regions of 8 volunteers: •-• right scapular back, o-o right lumbar back, ▲-▲ distal arm, ▫-▫ medial arm, □-□ proximal arm.

Evaluations of technical reproducibility using white tiles and a red standard simulating erythema are presented. Repeated measurements (n = 10) were performed with the two colorimeters on their respective white calibration tiles and on a red color standard simulating moderate erythema[12] The difference between the two standard tiles was measured by the Lange equipment. The linearity of the colorimeters was examined using different color scales from the color-standard book.[12] Red (erythematous scale) was registered as mainly parallel a*-axis curves both in the lower end, relevant for erythema measurement, as well as in the upper end. The Lange equipment showed a steeper increase in values in comparison with the Minolta equipment. The Lange equipment appeared more sensitive for determination of yellow, with a steeper increase in values on the b* axis. The equipment determined blue with parallel curves. However, the Lange equipment measured at a substantially lower level, i.e., about 40 U lower on the b* axis. Green was also determined at a lower level, on the a* axis, by the Lange equipment, especially at high intensities. In the three axes only a small part of the large dynamic range of the colorimeters was relevant for skin color and erythema measurement.

Color measurement by colorimeters using the CIE system was useful for the characterization of erythema, confirming previous studies.[3,4] In this paper dealing with light-induced erythema a more detailed evaluation of erythema was performed. Serup et al.[11] concentrated on the technical aspects of color measurement.

Laser-Doppler flowmetry proved useful for the characterization of erythema as already shown, with a positive correlation to clinical scoring and colorimetry (a*-axis movement toward red).[15,16] The colorimetry and the flowmeter probably measure somewhat different features of cutaneous circulation, i.e., the colorimeters mainly measures the capillary accumulation of blood and the flowmeter measures the total blood flow, mainly determined by the arteriolar tone.[17] Thus, the methods are, to some extent, complementary in the evaluation of erythema.

The technical reproducibility of the two colorimeters was very good with low coefficients of variation, although higher for a* than for the other axes. The variation in color in the group of subjects studied was remarkably low with respect to the L value, and limited with respect to the a* value. The high coefficient of variation of b*-axis values recorded with the Dr. Lange equipment was partly attributable to the low mean values. Standard deviations might indicate a similar reproducibility of colorimeters in control skin and erythematous skin in contrast to laser-Doppler flowmetry, which showed a five times higher standard deviation in erythematous skin.

This is probably due to the small area of skin illuminated by the laser light in contrast to the colorimeters, which operate with a far larger area and thus provide a better overall assessment of the circulation. Thus, with the laser-Doppler it is of special importance that results are obtained through averaging a number of recordings of the irritant reaction studied. In contrast, a single colorimeter measurement may suffice, depending on the situation or the purpose. It is well known that laser-Doppler flowmetry, recording a dynamic situation, is sensitive to factors such as noise and talking. Colorimetry is not affected by such environmental sources of variation, making the method more suitable for routine work.

The two colorimeters were essentially equally suited as tools for quantification of erythema. However they do not give identical values, and the technical part of the study concluded that they record some differences on the different color axes. Thus, a simple transformation factor cannot be applied.

The CIE color system has the advantage of being internationally accepted, and equipment using this system is available. Modifications were developed by Munsell and more recently by a Swedish group, the latter system being named NCS (natural color system).[18] This system was suitable for assessment of skin color during Argon-laser treatment of port wine stains. However, for the wider exchange of information it is advantageous if a uniform system is used, although experts may not find it ideal in every situation.

It should not be forgotten that inflammatory responses start with vasodilatation, while edema formation takes over in advanced reactions, compressing the vasculature. Laser-Doppler flowmetry and high-frequency ultrasound measurement of skin thickness of histamine wheals, irritant reactions after application of SLS, and allergic patch-test reactions show that grading of reactions relative to vasodilatation is only possible in light and moderately severe reactions.[16,19,22] However, ultrasound examination also shows the development of edema within irritant reactions is less in comparison with allergic reactions. Thus, in studying irritancy, colorimetry is likely to be suitable for the grading of a relevant spectrum of reactions, which previous studies also demonstrated.[13,14] In conclusion, Serup et al,[11] found that colorimetry using the CIE system is reproducible and useful for quantification of erythema, even under busy laboratory conditions. In group studies (winter) using untanned skin, the influence of melanin pigmentation appears negligible. However seasonal and regional differences in skin color need to be studied in more detail. The relevance and potential areas of application in experimental and clinical dermatology are many since this color system takes into account the actual color perception of the human eye, which spectrophotometric recording of light absorbency and reflectance does not do.

VI. Fields of Application

The application of reflectance chromameters in clinical practice and dermatological research is wide and still expanding. The most important applications reported in the literature are measurement of:

1. UV radiation-induced erythema
2. erythema due to irritants and contact allergens
3. the blanching effect of corticosteroids
4. skin color
5. UV-induced pigmentation
6. dose-response curves of UV-induced erythema and pigmentation
7. the bleaching effect of depigmenting agents

A. Ultraviolet Radiation-Induced Erythema Measurement

The assessment of sensitivity of human skin to UV radiation is important to photochemotherapy, diagnosis of photodermatosis, photoaging, photocarcinogenesis, and photoprotection. As an example we use the method described by Westerhof et al.[5]

Eight healthy volunteers (four males, four females, 22 to 26 years old) with skin type II and III were irradiated on seven different regions of the body with a solar simulator. The regions were chosen following the reports of Farr and Diffey[23] and Olson et al.[24] On the ventral side of the left forearm, the three sites were: proximal near the elbow, distal near the wrist, and in between. On the subject's back the four sites chosen were related to the vertebral level: upper dorsal (T3) — a spot to the left and a spot to the right of the spine, and for lumbar (L2) — a spot to the left and a spot to the right of the spine at 1 cm from the midline. Care was taken to select the spots in such a way they were without terminal hairs and without nevi or scars, etc.

The test areas were then exposed to a series of four simulated solar radiation dosed increasing in delivered energy dose by a factor $2^{2/3}$ of the foregoing dose. The radiation dose was chosen in such a way that one or two test spots (out of four) received a dose equal to one minimal erythema dose (MED).

The solar simulator consisted of a 1000-W xenon arc in a lamphouse fitted with a quartz collimating lens (f/0.7) and a water filter with quartz windows. A dichroic ('cold type') mirror which reflects most of the radiation below 500 mm was placed into the beam to minimize the infrared radiation further. A liquid light guide conducted the radiation from the exit port of the lamp optics to the volunteer's skin. An applicator which housed a quartz lens was attached to the proximal end of the light guide and produced a uniform beam of radiation, 10 mm in diameter, on the skin. The stimulated solar emission curve was determined with the aid of a Zeiss M4 QIII® monochromator and an EG&G 550® radiometer at a distance of 25 cm from the xenon-arc source. The UVB output was measured with an EG&G 550® radiometer fitted with a calibrated UV-enhanced silicon detector probe, on top of which a Scott UVB filter was mounted. This filter matches the erythematogenic action spectrum of normal skin at the long wave side as presented by Berger.[25] The monitored UVB output of the solar simulator per second was 2.35 mW/cm².

After 24 h all irradiated sites were visually scored by two independent investigators according to the following erythema grading scale:

 0 = no perceptible erythema
 $1/2$ = slight or partial erythema
 1 = minimal erythema with defined borders (MED)
 $1^1/2$ = three arbitrarily increased steps of erythema
 2 = formation without edema or vesiculation
 $2^1/2$ = (2+ erythema in the commonly used scale)
 3 = erythema with induration and blisters

Thereafter the erythema was measured with the Chromameter II Reflectance.® The same site was measured twice on the L*a*b* mode. Of the three chromaticity values L*a*b*, the a* value is the best index for the erythema measurement. This is to be expected since a* stands for chromaticity coordinates ranging from perceived red to green in the three-dimensional color space diagram (CIE 1976). After measuring the erythematous spot, a piece of nonirradiated adjacent skin was measured for comparison. The difference between the two gives the erythematous color aspects of the skin (a*).

The Chromameter was always held perpendicular to the skin surface, hardly touching it. The aperture of the measuring unit is fitted with an applicator so that the cutaneous capillaries are not compressed.

Angle errors in the vertical measuring position of the instrument's head did not show any influence on the a* and b* chromaticity coordinates and gave a slight deviation of L* (coefficient of variation 6, 8%; n = 20). The differences between duplicate readings showed a SD of 0.30 in case of L* in the range 57.0 to 67.9; a SD of 0.14 in case of a* in the range 6.5 to 20.7 and a SD of 0.21 in case of b* in the range 11.8 to 22.2.

The L* value decreased when erythema developed, indicating some skin darkening but to a relatively smaller extent than the increase of a*. The b* values did not change significantly in our erythema measurements. Westerhof et al.[5] tried to relate the visual ratings (0, $1/2$, 1, $1^1/2$, 2, $2^1/2$, 3) to the Chromameter measurements.

In Figure 2 the mean visual grading is plotted as a function of a*. It is easy to see that in the range of 1 to 11 there is a linear relation between visual grading and the Chromameter reading. Above a Chromameter reading of 11 a saturation of the visual grading can be seen. Westerhof et al.[5] were able to confirm the findings of Olson et al.[24] and Farr and Diffey.[23] The sensitivity to erythema formation increases when one compares the distal part of the flexor side of the forearm with the proximal part near the elbow while keeping the energy dose of UVB radiation constant (Figure 5). The same is true on the subject's back when comparing the thoracal region (T3) with the lumbar region (L2). It is also remarkable that the back is more sensitive to UV radiation than the flexor side of the forearm.

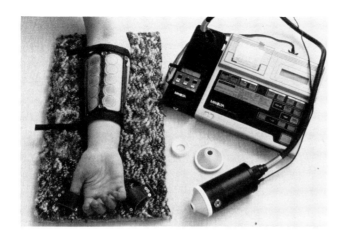

FIGURE 5 Bracelet to make follow-up measurement with chromameter possible. The measuring head exactly fits into the holes of the bracelet, which can be applied to the lower arm in only one way as a notch should exactly overlay an ink marking made on the skin.

B. Measurement of Erythema Due to Irritants and Contact Allergens

In dermatology and the cosmetic sciences tristimulus colorimetric measurements of erythema offer an advantage in quantifying skin reactions to allergic reactions evoked by patch tests and irritation responses. We studied the role of melanin in inflammatory reactions by provoking erythema with the potent irritant anthralin in vitiliginous skin versus normal pigmented skin.[26] The locations chosen for anthralin application were the trunk, arms, and legs under the condition that normal control skin was to be as close as possible to lesional skin. These precautions were taken because of the known existence of a site-dependent inflammatory reaction to anthralin.[27] Anthralin was used in concentrations of 0.1, 0.5, 1.0, and 5.0% dispensed in a cream containing 1% salicylic acid as a stabilizer; the vehiculum served as a control. The serial applications of the four concentrations of anthralin were done with patch-test chambers. The patch-test chambers were fixed with hypoallergic cloth tape. The patch-test chambers were removed after 24 h and the irritation response was read after the next 24 h.

For evaluation of the irritation response the following semiquantitative scale was used: 0, no erythema; 1, erythema (+, slightly visible erythema, ++, moderate erythema, +++, intense erythema, ++++, extreme erythema); 2, erythema with induration; and 3, erythema with induration and blisters. To avoid an optical illusion, a white paper was used to cover the test area so that the surrounding reference area was always the same.

For an accurate quantitative evaluation of the anthralin-induced erythema we used the Chromameter Reflectance II® (Minolta). The control substance did not induce any inflammatory reaction. The erythematous reactions with 0.1% anthralin were weak, those with 0.5 and 1.0% tended to be stronger, and those with 5.0% showed a very intense erythema. The irritation response never reached beyond the margins of the test chamber and at the time of reading, neither induration nor blisters were observed. With the visual grading the irritation response was considered to be more intense in normal skin than in vitiliginous skin. In contrast to the visual estimations the chromameter readings indicated that the changes in redness due to anthralin were more pronounced in vitiliginous skin than in normal pigmented skin. The staining produced by anthralin appeared to be regular and of the same intensity when vitiliginous and neighboring pigmented test sites were compared. This fact was confirmed by colorimetric measurements.

The relation between the degree of pigmentation and the irritation response to anthralin could only be evaluated making use of a sensitive colorimetric method. Visual estimation of the erythema can be misleading as the redness observed in the pigmented skin is known to be comprised of genuine erythema and a red component of the complex brown color. The human eye cannot discriminate these two different sources of redness. Colorimetric measurements can help to overcome this difficulty.

From this study it was concluded that melanin pigment had an inhibitory effect on the anthralin-induced irritation response probably by scavenging free radicals and reactive oxygen species.

C. Measurement of the Blanching Effect of Corticosteroids[28]

The vasoconstrictor assay[29] (human skin-blanching assay) is a well-established method for ranking topical steroids. Cornell and Stoughton[30,31] gave a rank for U.S. formulations with seven orders of potency. In Europe, the current classification for topical corticosteroids has four ranks,[32,33] with group I being the most potent. In this assay the visual assessment of blanching generally gives good results, but under certain circumstances is not a very satisfactory method, e.g., nonoccluded tests for weak or moderately potent formulations. Many studies using quantitative instrumental methods have attempted to improve the objectivity of this test. Some have completely failed[34,36] and those that succeeded were too cumbersome and time consuming for general use.[37,39] The only study using a colorimetric system[40] was not convincing and this method was abandoned. Queillo-Roussel et al.[28] used the Minolta tristimulus colorimeter to quantify the blanching effect of topical corticosteroids in a nonoccluded vasoconstriction test. To investigate the influence of time on variations in colorimetric parameters, an initial series of measurements was performed on day one on six predetermined sites on the ventral surface of the forearm of six healthy volunteers every 2 h over a 12-h period. The colorimetric values were shown to be site related but hourly variations occurred with similar profiles for all sites. On day two, four topical corticosteroid creams, representative of their potency group, as well as a cream base were applied in a randomized double-blind manner on five predetermined sites. Visual gradings and colorimetric measurements were carried out every 2 h over the following 12-h diurnal period and were continued on day three. The colorimetric parameters L* (luminance) and a* (color hue ranging from green (–) to red (+)) gave a rank order correlated to corticosteroid potency that showed superior discrimination compared to simple visual grading. In this study L* was a more discriminative parameter than a*.

D. Measurement of Skin Color[1]

Reflectance colorimetric methods have been widely applied in studies of the chromatic characteristics of skin in the context of anthropologic studies of human population genetics.[41-43] CIELAB color space parameters also have potential cosmetic applications, such as the design of prosthetic devices and the use in color matching of skin grafts in plastic surgery (e.g., after skin tumor excision).

For complete color measurement it has been shown that filter colorimetry can give results that differ systematically from those based on scanning spectrophotometry.[2,44] Evaluation of the ventral upper arm skin color in volunteers of various ethnic origin were measured by applying reflectance spectroscopy with CIELAB color space parameters. The volunteers classified themselves as being of European, Chinese, Indian, or Polynesian ethnicity, or of mixed race. Measurements of the visible reflectance spectrum of skin and the calculation of CIELAB values provided a practical numerical basis for quantifying the perceived color of human skin. The excess of data in a full spectrum can be reduced to a set of color space parameters that relate directly to the appearance of the color as it would be clinically observed and without making any assumptions about the nature of the pigments involved. However the same spectral data could be used if there was a requirement beyond the simple numerical specification of appearance as in the determination of the concentration of chromophores in physiologic studies.

E. Measurement of Ultraviolet-Induced Pigmentation

Tanning is a cardinal response of human skin to UV radiation. A spectrophotometric technique for measuring changes in the rate of transition between the first erythematous response and delayed pigment formation has important applications for evaluating the efficacy of sunscreens and potential pigment enhancers (e.g., psoralens with UVA and tyrosine).[44,46,47] Unfortunately, the clinical usefulness of early spectrophotometers for measuring this color transition of skin has been limited by technical problems including the stability of the instrument light source, the need for external references, and the use of multiple filters to obtain the exact skin color.[48,49] Furthermore, the wide spectrum of human skin colors has made comparative treatment studies with these devices virtually impossible since there is no means of recording baseline data on an individual subject for later paired comparison.[50] With a handheld tristimulus colorimeter these problems have been overcome. Natural skin tones can be stored in the colorimeter memory from a subject for direct comparison to the solar exposed skin color. The capability of the tristimulus colorimeter to simultaneously evaluate the hue and saturation of skin color affords an improved opportunity to quantitate the transition from cutaneous erythema to tanning.

Seitz et al.[51] tested areas on the back, which were UV irradiated at the exposed sites previously estimated to produce 1, 1.5, and 2 MED for each individual. 24 h later erythema and tanning were scored by a dermatologist. The subjects received repeated UV exposures once every 48 h for 14 d. The Minolta chromameter showed subtle, continuous transition between the primary erythamous response and delayed tanning of the skin. The initial changes were below the visual threshold for detection as seen in the dermatologist's scores and in the instrument validation responses. Linear regressions comparing the dermatologist's scores with the meter values indicated that erythema to UVA and UVB exposure was perceived as a pure red shift in skin color, darkening in intensity with continued exposure until tanning started to appear. In contrast, tanning was perceived as an intensifying yellow hue superimposed on the existing erythema. However, since the human eye integrates all visual stimuli, any admixtures of color or other variables, e.g., skin blemishes, background skin tone, quality of lighting, etc., would partially explain the variability in the dermatologist's erythema scores. These results also emphasize the difficulty in visually distinguishing between erythema and tanning.

F. Measurement of Dose-Response Curves of Ultraviolet-Induced Erythema and Pigmentation[52]

The assessment of sensitivity of human skin to UV radiation is important to photochemotherapy, photodermatoses, photoaging, photocarcinogenesis, and photoprotection. Today time, the most frequently used means of attempting to predict UV sensitivity, without phototesting, is classification of the persons's skin type. This Working Classification of Sun Reactive Skin Types, as introduced by Fitzpatrick, is based on the history of an individual's tendency to sunburn and to tan, along with some racial parameters. When more objective determination of sensitivity of the skin to UV radiation is desired, is determined experimentally.

A number of investigators have sought to compare and correlate the various means of predicting and measuring UV sensitivity. Olson et al.[35] reported that the minimal erythema dose (MED) correlated well with melanosome size, quantity, density, and distribution in various skin colors. Shono et al.[36] also found a good correlation between MED and skin color, but a less clear one between the minimal melanogenic dose (MMD) and either skin color or MED. Sayre et al.[37] studied the erythema action spectrum of 30 Caucasians and concluded that several factors, among which was constitutional skin color, had no significant effect on MED. Haake et al.,[38] studying a population consisting of Fitzpatrick's types I to IV, found no correlation between the UV sensitivity and color of the skin. It is inconclusive from these studies whether or not the MED correlates with skin types or skin color.

Most researchers dealing with photo testing recognize that, within a so-called skin type, there is a larger interindividual variation in UV sensitivity, as assessed by MED measurements. Since a value such as the MED is merely a point on a dose-response curve, it can be expected that the full curve will yield much more information than does any arbitrarily chosen single point such as the MED.

In studies of the responses of human skin to UV radiation, it is convenient to have a means of estimating the reactivity of an individual's skin without performing phototesting. The most frequently used means at present is classification of skin into one of the six skin reactive types, based primarily on the history of previous responses of the skin to sunlight. This has proved to be a convenient method because it requires no measurements and can be done rapidly. When measurement of skin responses to UV radiation is performed, the usual method is the measurement of a MED. The MED is only an estimate of the amount of UV radiation required to produce detectable erythema and does not reveal the incremental increases in erythema with increasing UV doses. Dose-response data for erythema would more accurately measure UV responses of human skin,[23] but obtaining such data has been difficult. The availability of a sophisticated chromameter interfaced with a computer has now made possible the easy, objective measurement of erythema and pigmentation responses to UV radiation and the obtaining of dose-response data. Westerhof et al.,[52] performed comprehensive photo testing, including the determination of MED and MMD values and the measurement of dose responses for erythema and pigmentation, on a large number of volunteers and attempted to assess the value of the skin type and objectively measure skin color in predicting the skin sensitivity to UV radiation.

Westerhof et al.,[52] confirmed the results of Stern and Momtaz[53] and those of Azizi et al.,[54] who found a greater than threefold difference between the highest and the lowest MED values in each skin type, reflecting the large variation in individual MED values. The conclusion from this study[52] is that the skin type does not correlate well with the MED. The reason for this may be that the skin type does not accurately estimate UV sensitivity or that the MED is not a sensitive measurement of UV responses of human skin. The lack of a close relationship of skin type with the dose-response curves for erythema and pigmentation would point to the skin type as an inadequate predictor of UV responses.

With the reflectance measurement (Y) obtained with the chromameter prior to phototesting, we were able to objectively estimate skin color. The constitutional skin color did not correlate well with the skin type and neither with measured MED or MMD values. However, we found the objectively measured skin color to be the best predictor of the dose-response measurements for both erythema and for pigmentation. In lightly pigmented skin, the dose-response curves were steep, whereas in darkly pigmented skin the curves were much flatter. From these studies[52] it appears that dose-response data best measures the sensitivity of human skin to UV radiation. Because the technology to obtain dose-response information is not widely available, MED measurements will continue to be the most convenient means of measuring the response to UV radiation. However, for sophisticated photo testing in the future, such as in measuring the protection provided by sunscreens, the use of dose-response determinations might better reflect changes in UV sensitivity. The dose-response curves for erythema, which can be easily measured 24 h after irradiation of the skin, should become the standard for modern photo testing.

In summary, skin color is a more valid predictor than skin type for the measurement of UV sensitivity, which is best expressed by dose-response curves. Although skin typing will continue to be used because of its convenience, one must be aware that it has severe limitations as a predictor of UV sensitivity. Perhaps if a means of easily estimating constitutional skin color at the site of anticipated UV exposure becomes readily available, this will prove a more valid means of predicting UV sensitivity.

G. Measurement of Bleaching Effect by Depigmenting Agents

There is a need to show efficacy in the treatment of hyperpigmentary disorders such as melasma, lentigo solaris, cafe au lait spots, postinflammatory hyperpigmentation, etc. Many products have been advocated; few of them work after months of application, some have serious side effects such as depigmentation or even more severe hyperpigmentation.

Duteil and Ortonne[55] aimed to assess the activity of 20% azelaic acid cream in light-induced skin pigmentation in subjects. There were five test zones, all located on the middle of the back: two were treated with azelaic acid cream, two others with the vehicle, and one was left untreated. Each product was applied twice daily, 5 d a week, for 4 weeks on one zone, and for 5 weeks on the other. In the middle of the fourth week, the test zones were exposed to UVB + UVA + visible light, with a total of three times the minimal erythema dose distributed progressively over 3 consecutive days. For 7 and 10 d after the last irradiation, the induced photopigmentation was assessed by colorimetric and visual means. Compared with its vehicle, the azelaic acid cream had neither a depigmenting effect nor a preventive effect on the light-acquired skin pigmentation. Moreover, interrupting or continuing azelaic acid treatment after skin irradiation had no influence on the resulting pigmentation.

VII. Conclusions

As already stated, the complexity of the erythematogenic response in the optically complex multilayered system of the skin is difficult to approximate by a formula. Erythema indices based on such equations are bound to be oversimplifications, so that the computed mathematical interpretations might lead to inappropriate correlation with the degree of erythema. The advantage of measuring the erythema of the skin by using the CIE system of tristimulus values is that one does not need to make any assumptions about the processes underlying erythema formation.

In clinical situations, e.g., when only the degree of sensitivity of the skin to UV radiation is required, one can simply estimate the degree of erythema while comparing it to normal neighboring skin. Therefore, especially in the early stages of erythema formation, it is not necessary to account for the different translucent layers of the skin, in particular the pigments in each of these layers and the degree of light scattering due to differences in turbidity. These other qualities are leveled out when comparing erythematous skin with normal skin. For scientific investigations, however, colorimetric measurements are indispensible.

Colorimeters using the CIE system allow for quantitative comparison of erythema and pigmentation in individuals and between individuals in a way that is consistent with visual judgements, but with greater reliability and reproducibility than would be possible by human observers. In this way, reliable dose response curves can be constructed. Colorimeters able to quantify reflected colors from the skin using the CIE system appear to be precise, quick, and handy in their operation.

Acknowledgments

This work was supported by the Dutch Pigment Cell Foundation. The constructive advise of Professor Oscar Estevez Uscanga and Dr. Henk E. Menke greatly increased the readability of this chapter.

References

1. Weatherall IL, Coombs BD. Skin color measurements in terms of CIELAB color space values. *J Invest Dermatol* 99:468–473, 1992.
2. Billmeyer FW, Saltzman M. *Principles of Color Technology*, 2nd ed., Wiley-Interscience, New York, 1981.
3. Feather JW, Ryatt KS, Dawson JB, Cotteril JA, Barker DJ, Ellis DJ. Reflectance spectrophotometric quantification of skin color changes induced by topical corticosteroid preparations. *Br J Dermatol* 106:437–444, 1982.
4. Babulak SW, Rhein LD, Scala DD, Simion FA, Grove GL. Quantitation of erythema in a soap chamber test using the Minolta Chroma (Reflectance) Meter: comparison of instrumental results with visual assessments. *J Soc Cosmet Chem* 37:475–479, 1986.
5. Westerhof W, van Hasselt BAAM, Kammeyer A. Quantification of UV-induced erythema with a portable computer controlled chromameter, *Photodermatology* 3:310–314, 1986.
6. Hacham H, Freeman SE, Gange RW, Maytum DJ, Sutherland JC, Sutherland BM. Do pyridine dimer yields correlate with erythema inducation in human skin irradiated in situ with ultraviolet light (275–365nm)? *Photobiology* 53(4):559–563, 1991.
7. Hunter RS, Harold RW. *The Measurement of Appearance*, 2nd ed., Wiley-Interscience, New York, 1987.
8. Marchesini R, Brambilla M, Clemente C, Maniezzo M, Sichirollo AE, Testori A, Venturoli DR, Cascinelli N. In vivo spectrophotometric evaluation of non-neoplastic skin pigmented lesions. I. Reflectance measurements. *Photochem Photobiol* 53(1):77–84, 1991.
9. Feather JW, Hajizadeh-Saffiar M, Leslie G, Dawson JB. A portable scanning spectrophotometer using visible wavelengths for the rapid measurements of skin pigments. *Phys Med Biol* 34:807–820, 1989.
10. Kollias N, Bager A. Quantitative assessment of UV-induced pigmentation and erythema. *Photodermatology* 5:53–60, 1988.

11. Serup J, Agner T. Colorimetric quantification of erythema — a comparison of two colorimeters (Lange Micro Color and Minolta Chroma Meter CR-200) with a clinical scoring scheme and Laser-Doppler flowmetry. *Clin Exp Dermatol* 15:267–272, 1990.
12. Kollias N, Bager A. Spectroscopic characteristics of human melanin in vivo, *J Invest Dermatol* 85:38–42, 1985.
13. Babulak SW, Rhein LD, Scala DD, Simion A, Grove GL. Quantitation of erythema in a soap chamber test using the Minolta Chroma (reflectance) Meter: comparison of instrumental results with visual assessments. *J Soc Cosmet Chem* 37:475–479, 1986.
14. Westerhof W, van Hasselt BAAM, Kammeyer A. Quantification of UV-induced erythema with a portable computer controlled chromameter. *Photodermatology* 3:310–314, 1986.
15. Nilsson GE, Wahlberg JE. Assessment of skin irritancy in man by laser Doppler flowmetry. *Contact Dermatitis* 8:401–406, 1982.
16. Staberg B, Serup J. Allergic and irritant reactions evaluated by laser Doppler flowmetry. *Contact Dermatitis* 18:40–45, 1988.
17. Engelhart M, Kristensen JK. Evaluation of cutaneous blood flow responses by 133 Xenon washout and a laser Doppler flowmeter. *J Invest Dermatol* 80:12–15, 1983.
18. Malm M, Jurell G, Tonnquist G. Natural color system — a new method for evaluating skin colors during Argon laser treatment of port-wine stain. *Ann Plast Surg* 20:317–321, 1988.
19. Serup J. Quantification of wheal reactions with laser Doppler flowmetry. *Allergy* 40:223–237, 1985.
20. Staberg B, Klemp P, Serup J. Patch test responses evaluated by cutaneous blood flow measurements. *Arch Dermatol* 120:741–743, 1984.
21. Serup J, Staberg B, Klemp P. Quantification of cutaneous oedema in patch test reactions by measurement of skin thickness with high-frequency pulsed ultrasound. *Contact Dermatitis* 10:88–93, 1984.
22. Serup J, Staberg B. Ultrasound dor assessment of allergic and irritant patch test reactions. *Contact Dermatitis* 17:80–84, 1987.
23. Farr PM, Diffey BL. Quantitative studies on cutaneous erythema induced by ultra violet radiation. *Br J Dermatol* 111:673–682, 1984.
24. Olson RL, Sayre RM, Ecerett MA. Effect of anatomic location and time on ultraviolet erythema. *Arch Derm* 93:211–215, 1966.
25. Berger DS. The sunburn ultravioletmeter: design and performance. *Photochem Photobiol* 24:587–593, 1976.
26. Westerhof W, Buehre Y, Pavel S, Bos JD, Das PK, Krieg S, Siddiqui AH. Increases anthralin irritation response in vitiliginous skin. *Arch Dermatol Res* 281:52–56, 1989.
27. Dawson JB, Barker DJ, Ellis DJ, et al. A theoretical and experimental study of light absorption and scattering by in vivo skin. *Phys Med Biol* 25:695–209, 1980.
28. Queillo-Roussel C, Poncet M, Schaefer H. Quantification of skin color changes induced by topical corticosteroid preparations using Minolta Chroma Meter. *Br J Dermatol* 124:264–270, 1991.
29. Wan S, Parrish JA, Jeaniche KF. Quantitative evaluation of ultraviolet induces erythema. *Photochem Photobiol* 37:643–648, 1983.
30. Cornell RC, Stoughton RB. Correlation of the vasoconstriction assay and clinical activity in psoriasis. *Arch Dermatol* 121:63–7, 1985.
31. Robertson AR. The CIE 1976 color difference formulas. Color Research and Application 2:7–11, 1977.
32. Polano MK, August PL. In: Polano MK (ed.). *Topical Skin Therapeutics.* Churchill Livingstone, Edinburgh, 1984, p. 101.
34. Fitzpatrick TB, Pathak MA, Parrish JA. Protection of human skin against the effects of the sunburn ultraviolet (290–320nm). In: Fitzpatrick TB et al. (eds). *Sunlight and Man-Normal and Abnormal Photobiological Responses.* University of Tokyo Press, Tokyo, 1974, p. 751.
35. Olson RL, Gaylor J, Everett MA. Skin Color, malonin and erythema. *Arch Dermatol* 108:541–544, 1973.
36. Shono S, Imura M, Ota M, Ono S, Toda K. Relationship of skin color, UVB-induced erythema and melanogenesis. *J Invest Dermatol* 84:265–267, 1985.
37. Sayre RM, Desrocher DL, Wilson CJ, Marlowe EL. Skin type, minimal erythema dose (MED), and sunlight acclimatization. *J Am Acad Dermatol* 5:429–443, 1981.
38. Haake N, Buhles N, Altmeyer P. Sensitivity of human skin to UV-light-practicability and limits of clinical diagnostics. *Z Hautkr* 62:1505–1509, 1987.
39. Berger DS. Design of a solar simulator. *J Invest Dermatol* 53:192–199, 1969.
40. Pathak MA, Fitzpatrick TB, Greiter F, Kraus EW. Preventive treatment of sunburn, dermatoheliosis, and skin cancer with sunprotective agents. In: Fitzpatrick TB et al. (eds). *Dermatology in General Medicine.* McGraw-Hill, New York, 1987.
41. Little MA, Wolf ME. Skin and hair reflectance of women with red hair. *Ann Human Biol* 8(3):231–241, 1981.
42. Clarke P, Stark AE, Walsh RJ. A twin study of skin reflectance. *Ann Human Biol* 8:529–541, 1981.
43. Towne B, Hulse FS. Generational change in skin color variation among Habbani Yemini Jews. *Human Biol* 62(1):85–100, 1990.
44. Stevenson JM, Weatherall IL, Littlejohn RP, Seman DL. A comparison of two different instruments for measuring venison CIELAB values and color assessment by trained panel. *N Zeal J Agr Res* 34:207–211, 1991.
45. Jarnecke-Munster H. Uber die Zusammenhange der am Hauteigenstoffwechsel beteiligten Aminosauren, ins besondere Histiden und Tyrosin. *Arch Dermatol Syph* 180:290–293, 1940.
46. Cripps DJ. Natural and artificial photoprotection. *J Invest Dermatol* 76:154–157, 1981.
47. Shigeaki S, Imura M, Ota M. The relationship of skin color, spectrophotometric technique. *J Invest Dermatol* 84:265–267, 1985.
48. Buckley WR, Grum F. Measurement of skin color, spectrophotometric technique. *J Soc Cosmet Chem* 15:79–85, 1964.
49. Farrington D, Imbrie JD. Comparison between visual grading and reflectance measurements of erythema produced by sunlight. *Br J Dermatol* 111:295–304, 1984.
50. Wasserman HP. The colour of human skin. *Dermatologica* 143:166–173, 1971.
51. Seitz JC, Whitmore CG. Measurement of erythema and tanning responses in human skin using a tri-stimulus colorimeter. *Dermatologica* 177:70–75, 1988.
52. Westerhof W, Estevez-Uscanga O, Meens J, Kammeyer A, Durocq M, Cairo I. The relation between constitutional skin color and photosensitivity estimated from UV-induced erythema and pigmentation dose response curves. *J Invest Dermatol* 94:812–816, 1990.
53. Stern RS, Momatz K. Skin typing for assessment of skin cancer risk and acute response to UVB and oral methoxalen photochemotherapy. *Arch Dermatol* 120:869–872, 1984.

54. Azizi E, Lusky A, Kushelevsky AP, Shewach-Millet M. Skin type, hair color, and freckles are predictors of decreased minimal erythema ultraviolet radiations dose. *J Am Acad Dermatol* 19:32–38, 1988.

55. Duteil L, Ortonne JP. Colorimetric assessments of the effects of azelaic on light-induced skin pigmentation. *Photodermatol Photoimmunol Photomed* 9:67–71, 1992.

17.0

Cutaneous Blood Flow, Vasomotion, and Vascular Functions

17.1	Laser Doppler Measurement of Skin Blood Flux: Variation and Validation .... 399 *A.J. Bircher*	
17.2	Laser-Doppler Flowmetry: Principles of Technology and Clinical Applications .... 405 *G. Belcaro and A.N. Nicolaides*	
17.3	Examination of Periodic Fluctuations in Cutaneous Blood Flow .... 411 *R. Gniadecki, M. Gniadecka, and J. Serup*	
17.4	Laser Doppler Imaging of Skin .... 421 *K. Wårdell and G. Nilsson*	
17.5	The ^{133}Xenon Wash-Out Technique for Quantitative Measurement of Cutaneous and Subcutaneous Blood Flow Rates .... 429 *P. Sejrsen*	
17.6	Doppler and Duplex Scanning in Venous Disease .... 437 *G. Belcaro and A.N. Nicolaides*	
17.7	Photoplethysmography and Light Reflection Rheography: Clinical Applications in Venous Insufficiency .... 443 *G. Belcaro, D. Christopoulos, and A.N. Nicolaides*	
17.8	Strain Gauge Plethysmography: Diagnosis and Prognosis in Skin Lesions on the Feet and Toes and in Leg Ulcers .... 449 *N.A. Lassen and P. Holstein*	
17.9	Air Plethysmography in the Evaluation of Skin Microangiopathy Caused by Chronic Venus Hypertension .... 453 *D.C. Christopoulos*	

Chapter 17.1
Laser Doppler Measurement of Skin Blood Flux: Variation and Validation

Andreas J. Bircher
Department of Dermatology
University Hospital
Basle, Switzerland

I. Introduction

The cutaneous microcirculation plays an outstanding role in physiologic processes and is also involved in many pathologic conditions. Therefore, there is considerable interest to objectivate changes in the cutaneous blood flow (CBF) in physiologic conditions and pathologic disorders as well as to pharmacological stimuli. Blood flow has been measured on virtually all animal and human organ surfaces such as the nasal and gastrointestinal mucosa, in the eye, and on bone and muscular tissue. The skin, however, remains the most readily accessible organ and therefore a vast literature on numerous aspects of CBF have been published in the last years.

Several methods to identify and quantitate CBF fluctuations have been developed. Among these are more direct methods such as photoplethysmography, venous occlusion plethysmography, and ^{133}Xenon (^{133}Xe) clearance, as well as indirect techniques such as the determination of skin temperature and transcutaneous pO^2. A newer noninvasive technology is laser Doppler velocimetry or flowmetry (LDF), which is based on optical principles and which allows a more direct measurement of the cutaneous microcirculation. LDF has some advantages over the other methods in that it allows noninvasive, continuous recording of CBF in relatively superficial skin layers on virtually any reachable skin or mucous membrane surface.

The application of the Doppler phenomenon and the use of the unique properties of laser light to detect the motion of macromolecules was suggested by Cummins et al. and first used in a biological system by Riva et al. Later the approach was improved and applied to human retinal vessels by Tanaka et al. Stern[9] was the first to determine blood flow in the intact cutaneous microvasculature of humans.[1] Several instruments which were usable also in clinical settings were then developed. The first, the laser Doppler velocimeter, was designed and later improved by Holloway and Watkins.[2] Subsequently a modified version, the LDF, was developed by Nilsson and colleagues.[3] Both instruments use basically the same principles based on light bearing spectroscopy with some technical variations to improve the yield of the measurements. Detailed descriptions of the theoretical aspects of LDF and the available instruments which have been later designed have been published.[1]

II. Anatomical and Physiological Factors

A. Skin Microvasculature

The human epidermis contains no vessels and is therefore nourished by diffusion from the dermis. Its thickness, which is dependent on the anatomical site, varies from 50 to 200 µm. In the dermis, supplied by vessels from the subcutaneous tissue, there is a complex vascular network to perform the particular tasks such as thermoregulation, nutrition, and metabolism. In the upper dermis a superficial plexus is present with a special capillary network extending into the dermal papillae, the so-called capillary loops. They have a mean length of approximately 0.2 to 0.4 mm and each supplies an average of 0.04 to 0.27 mm^2 to the skin surface. A deeper plexus is situated at the dermal-subcutaneous border. The microcirculatory blood flow is regulated by smooth muscle cells which are mainly located in the walls of the small arteries and arterioles and to a lesser extent in venules and small veins. The true capillaries are free of smooth muscle cells, however, the pericyte cells located in the capillary vessel walls may induce to some extent capillary contractions. Arteriovenous anastomoses are present in some locations, particularly in the face and in the acral areas. They play an outstanding role in the thermoregulation of the organism.[4]

B. Cutaneous Blood flow

Blood flow or flux implies the movement of blood, which is a heterogeneic mixture of liquid, soluble, and cellular elements, through the preformed channels of the arterial and venous system which is linked by the above-mentioned capillary network or by arteriovenous anastomoses. Blood flow is far from being constant, on the contrary, it is dynamically regulated to fulfill the requirements of the organism. Dependent on the measuring technique, its resolution, and the investigated tissue volume, any method of measurement expresses blood flow only in relation to the type and the localization of the examined vessels. In optical methods further variables that influence the measurements are the structure of the skin surface, the skin thickness, and the pigmentation of the epidermis. Due to the small measuring area and the anatomical architecture of the dermal microvasculature considerable variations in blood flow are present.[5] Therefore, the regional variations of CBF have a considerable magnitude and have to be taken into account when using such measuring methods.

Most of the current laser Doppler devices use helium-neon lasers with a wavelength, $\lambda = 632.8$ nm and small probes. It is estimated that the tissue volume in which CBF is measured with such devices, has an approximate surface area of 1 mm^2, extending to an estimated depth of 1 mm resulting in a theoretical total measured tissue volume of 1 mm^3. The penetration depth of laser radiation, however, is rather variable and dependent on the above-mentioned factors. The incident laser light is absorbed, scattered, and only to a small extent reflected by the skin tissue structures. Stationary tissue, such as fibers, macromolecules, and vessel walls, scatter and reflect the incident radiation at the same frequency. Typically the largest fraction of the diffusely reflected laser light waves comes from stationary tissues. Tissue components such as red blood cells moving with a mean estimated speed of 1 mm/s, reflect the light with a shifted frequency (optical Doppler effect). The blood flow signal measured by laser Doppler instruments is an indicator of the cutaneous perfusion, however, due to the complex structure of the skin texture and somewhat random orientation of the cutaneous vessels, only semiquantitative, relative measurements of CBF can be made[6] with these techniques. Commonly the CBF results are given in millivolts or in arbitrary units.

Laser Doppler measurements of CBF from normal relaxed subjects have three major characteristics (Figure 1): (A) pulsatile flow synchronized with the cardiac cycle, (B) vasomotor waves of lower frequency (approximately 4 to 6 per minute) and (C) skin blood flow, i.e., the deviation from the instrument baseline.[7]

III. Methodological Principles: Technical Aspects

In laser Doppler measurements, a laser light source and the Doppler effect are used to generate an output proportional to the red blood cell movement through the skin vessels under investigation. A prerequisite for this technique are the characteristics of laser radiation such as monochromaticity, spatial, and temporal coherence. In typical laser Doppler instruments, a 2 or 5 mW powered helium-neon laser ($\lambda = 632.8$ nm) is usually the light source. Other instruments make use of an infrared laser ($\lambda = 780$ nm). The laser radiation is guided through an optical fiber to the probe head. The probes are usually held in contact with the skin by a double-sided adhesive ring-shaped tape. The emitted radiation enters the skin and is reflected by stationary and moving tissue components (Figure 2). Stationary tissue scatters and reflects the incident radiation at the same frequency. Red blood cells moving with a certain speed, reflect the light with a shifted frequency, the Doppler effect. Thus the light waves returning to the instrument are composed of two components: the frequency-modulated, spectrally broadened light, which is directly related to the number of erythrocytes times their velocity, and the nonshifted fraction, which has been reflected from nonmoving tissue. The reflected light is transmitted through one or more receiving optical fibers and the nonshifted reference and the Doppler shifted beam are then mixed on a photodetector and processed by optical heterodyning. The generated "beat frequency" is then converted into an electrical output. The signal is usually measured as a fluctuating voltage, expressed in millivolts. This flow signal measured by laser Doppler instruments is an indicator of cutaneous perfusion, however, due to the complex structure of the skin texture and the somewhat random orientation of the cutaneous microcirculation, only semiquantitative, relative measurements of CBF are obtained.[1]

IV. Factors Influencing Cutaneous Blood Flow

A. Subject-Related Variables

Recently a position paper on *guidelines for LDF* has been published by the *standardization group of the European Contact Dermatitis Society*.[8] In this paper variables influencing CBF have been extensively reviewed.

A broad spectrum of aspects may influence the measurements of CBF. Such variables include individual, interindividual, environmental, and technical factors. The latter variables are discussed in Reference 8. Individual-related variables include age, gender, and ethnic background. Selectable individual-related variables are the anatomical location of the measurements, the skin temperature, the subject's position, physical and mental activities, previous consumption of food, beverages, drugs and nicotine, the menstrual cycle, possibly some laboratory values, and to some

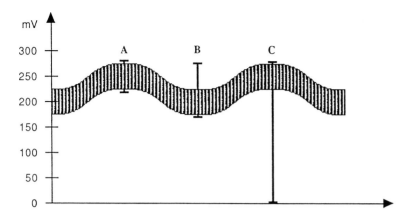

FIGURE 1 Schematic response pattern in LDF measured blood flow: (A) pulsatile flow synchronized with the cardiac cycle, (B) vasomotor waves of lower frequency, and (C) relative blood flow, i.e., the deviation from the instrument baseline. (Redrawn from Karanfilian, R. et al., *Am. Surg.*, 50, 641, 1984. With permission.)

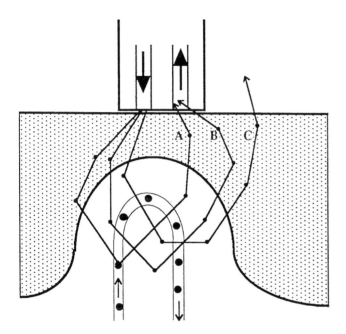

FIGURE 2 Schematic presentation of laser-light interaction with tissue components. Transmitted light is reflected and shifted in frequency by (A) moving erythrocytes or (B) reflected without frequency shift from stationary tissue or (C) does not reach the receiving probe fiber.

extent the considerable temporal variation of repeated measurements. Environmental variables include air convection, ambient temperature, and probably air humidity (Table 1).

B. Instrument-Related Variables

In the above-mentioned position paper also guidelines for LDF have been published.[8] Therefore these guidelines will only be briefly summarized. Due to the wide range of instruments available which work with several techniques and principles, a universal standard procedure cannot be developed. However, some general recommendations for validation can be given.

The instrument should be used in accordance with the recommendations of the manufacturer, particularly with regard to instrument set-up and safety precautions. The operating procedure of every laboratory should be clearly stated and a standardized procedure which includes validation under the laboratory's own specific conditions is proposed.

Some recommendations concerning the adjustments of the zero levels are briefly outlined. The instrument zero is the value obtained when the probe is held against a white surface, e.g., white porcelain. It may slightly deviate from the electrical zero of the instrument. The biological zero, that is higher than the instrumental zero, is obtained in an anatomical site, e.g., the forearm, under arterial occlusion. In the same procedure the dynamic range of the blood flow can be measured. This basic standard reactive hyperemia experiment should be performed in at least three healthy adults. These recommendations should facilitate the comparison of measurements obtained at different research facilities with different laser Doppler instruments.

Naturally, the individual and environmental variables described, also should be controlled and mentioned. Subjects should be managed with regard to smoking, food, and drug intake, as well as physical and mental stress. When small probes are used several blood flow determinations which may be averaged are recommended. The study environment should be controlled with respect to temperature, air movements, and other factors. Instrumental factors to be controlled include warming up of the instrument and particular precautions with the optical fiber and the application pressure of the probe.

V. Correlation With Other Methods

A. ^{133}Xenon Clearance

^{133}Xe clearance was compared to LDF in ultraviolet-stimulated CBF in forearm skin.[2,9] In both studies ^{133}Xe was delivered by injection. To minimize the spatial and temporal variation in LDF measurements, the average of five and ten measuring sites, respectively, was determined. In these experiments a linear relationship (r = 0.9), between the two methods was found. In a study using an atraumatic delivery technique for ^{133}Xe in four subjects fingertip and finger web blood flow was examined.[10] With both methods, parallel changes of CBF in skin without arteriovenous anastomoses were observed, however, in areas with shunt

TABLE 1 Individual and Environmental Variables Affecting Cutaneous Blood Flow

Variable	Influence on Blood Fow	Remarks
Age	Mostly minor	Dependent on location
Gender	Minor or none	—
Menstrual cycle	Minor or none	—
Ethnic background	None or minor	Reflection of pigmented skin
Subject position	Major	Orthostatic dependence
Anatomical location	Considerable variation	Also age related
Skin temperature	Major	Also environment related
Laboratory hematologic values	None or minor	Only when major pathology is present
Temporal (same day)	Minor	—
Temporal (day to day)	Considerable	—
Physical activity	Considerable	Short influence
Mental activity	Considerable	Short influence
Food and beverages	Considerable	e.g., Caffeine, alcohol
Systemic drugs	Major	Vasoactive compounds
Topical drugs	Major	Vasoactive compounds
Nicotine	Major	Vasoconstriction
Ambient temperature	Major	—

Modified from Bircher, A.J. et al., Contact Dermatitis, *30, 65, 1994. With permission.*

vessels, no relation was present. Also in healthy skin and in uninvolved skin of psoriatic patients[11] a poor correlation between LDF and xenon washout was observed. A better correlation was obtained in psoriatic lesions leading to the conclusion that, although LDF allows a rough estimate of CBF at high perfusion rates, it is not reliable in areas with low blood flow. A comparison of LDF and ^{133}Xe washout in skin lesions of localized and generalized morphea showed a good correlation between the methods. The sclerotic plaques and the perilesional inflammatory lilac rings, had a significantly greater CBF than normal skin. Older plaques had higher CBF than early lesions. However, in the most advanced 'burnt out' plaques, CBF was not different from that in normal skin.[12] Very similar results were obtained in another study of scleroderma.[13] Again in ultraviolet B-induced erythema of the forearm a significant correlation between ^{133}Xe washout and LDF measurements of CBF was found. Surprisingly LDF measurements showed a fourfold greater increase over ^{133}Xe determinations. The importance of the use of the biological zero instead of the instrumental electrical zero was also emphasized.

B. Venous Occlusion Plethysmography

Venous occlusion or strain gauge plethysmography is an established method to determine total blood flow particularly in extremities. In thermally-stimulated CBF by direct heating of the whole body surface[14] a good correlation between CBF measured by LDF and total forearm blood flow determined by occlusion plethysmography (r = 0.94 to 0.98) was found. However the correlation varied considerably between and within subjects. A nonlinear relationship between occlusion plethysmography and LDF-CBF measurements, stimulated by exercise and heat stress, was reported suggesting that the two methods measure different blood flow parameters.[15] In a comparison of LDF, occlusion plethysmography, and thermal clearance, considerable variability in the results was found, which was explained on the basis of the different skin volumes and types of blood flow measured by the methods.[16]

C. Photoplethysmography

The resting blood flow levels as a function of the anatomic position were determined in 52 body sites in ten subjects[17] by photoplethysmography and LDF. Photoplethysmographic and LDF measurements agreed well in the areas with low CBF but differed considerably in highly perfused areas. This was interpreted to be the consequence of the different parameters measured by the two methods (blood volume in photoplethysmography, volume times velocity in LDF).

D. Thermographic Methods

Several thermal methods which have been compared to LDF were found to give acceptable correlations. The thermal measurements, however, were usually slower in the speed of the response and more dependent on ambient temperature.[18] Also upon an intravenous injection of naftidrofuryl, similar blood flow values as measured by thermal conductivity and LDF were observed. The skin temperature also decreased over time, however, with a temporally delayed response.[19]

E. Other Lasers

In a study investigating histamine-induced CBF a helium-neon (λ = 633 nm) and an infrared laser (λ = 780 nm) were compared.[20] Very similar results of CBF changes were found with the two lasers, although the infrared laser radiation has a greater penetration. This implies that probably only superficial vessels were mainly affected by the histamine effect.

VI. Conclusions

Laser Doppler flowmetry requires a sophisticated technology, however, the instruments handling is simple. It allows noninvasive measurements of skin blood flow on virtually any area of the body and provides a continuous recording of the actual mostly superficial blood flow. It is particularly suited to measure and follow stimulated blood flow. Major fields of interest and application have been the determination of the qualitative and quantitative effects of systemically and topically applied vasoactive drugs, the study of inflammatory mediators, the investigation of allergic and irritant skin reactions, particularly in combination with other noninvasive bioengineering methods, and the evaluation of vascular phenomena in skin and other diseases, and also effects of therapeutic interventions on skin disorders.

Because the instrument is easy to use, some precautions should be taken when performing measurements. One major limitation is the rather small measured tissue volume which may be overcome by future technical improvements, e.g., of scanning probes, as described in the next chapter. A variety of factors including the instrument, the subject, and the environment influence the measurements. Because of the high variability of skin blood flow such as inter- and intraindividual subject variability and environmental parameters, these factors have to be taken into consideration in the planning and realization of experiments.

References

1. Shepherd, A. and Öberg, P., *Laser-Doppler Blood Flowmetry*, Kluwer Academic Publishers, Boston, 1990.
2. Holloway, G. and Watkins, D., Laser Doppler measurement of cutaneous blood flow, *J. Invest. Dermatol.* 69, 306, 1977.
3. Nilsson, G., Tenland, T. and Öberg, P., A new instrument for continuous measurement of tissue blood flow by light beating spectroscopy, *IEEE Trans. Biomed. Eng.*, 27, 12, 1980.
4. Tenland, T., On Laser Doppler Flowmetry. Thesis, Linköping University, Linköping, 1982.
5. Braverman, I., Keh, A. and Goldminz, D. Correlation of laser Doppler wave patterns with underlying microvascular anatomy, *J. Invest. Dermatol.*, 95, 283, 1990.
6. Holloway, G. Laser Doppler measurement of cutaneous blood flow, in *Non-invasive Physiological Measurements*, Rolfe, P., Ed., Academic Press, New York, 1983, 219.
7. Karanfilian, R., Lynch, T., Lee, B., Long, J. and Hobson, I. R., The assessment of skin blood flow in peripheral vascular disease by laser Doppler velocimetry, *Am. Surg.*, 50, 641, 1984.
8. Bircher, A. J., de Boer, E., Agner, T., Wahlberg, J. and Serup, J., Guidelines for the measurement of cutaneous blood flow by laser Doppler flowmetry, *Contact Dermatitis*, 30, 65, 1994.
9. Stern, M., Lappe, D., Bowen, P., Chimosky, J., Holloway, G., Keiser, H. and Bowman, R., Continuous measurement of tissue blood flow by laser Doppler spectroscopy, *Am. J. Physiol.*, 232, H441, 1977.
10. Engelhart, M. and Kristensen, J., Evaluation of cutaneous blood flow responses by 133 Xenon washout and laser Doppler flowmeter, *J. Invest. Dermatol.*, 80, 12, 1983.
11. Klemp, P. and Staberg, B. The effect of antipsoriatic treatment on cutaneous blood flow in psoriasis measured by 133Xenon washout method and laser Doppler velocimetry, *J. Invest. Dermatol.*, 85, 259, 1985.
12. Serup, J. and Kristensen, J. Blood flow of morphoea plaques as measured by laser Doppler flowmetry, *Arch. Dermatol. Res.*, 276, 322, 1984.
13. De Lacharrière, O. and Kalis, B. Measurement of cutaneous microcirculation in dermatology and dermatopharmacology, in *Cutaneous Investigation in Health and Disease*, Lévêque, J.-L., Ed., Marcel-Dekker, New York, 1989, pp. 385–420.
14. Johnson, J., Taylor, W., Shepherd, A. and Park, M. Laser Doppler measurement of skin blood flow, comparison with plethysmography, *J. Appl. Physiol.*, 56, 798, 1984.
15. Smolander, J. and Kolari, P. Laser Doppler and plethysmographic skin blood flow during exercise and acute heat stress in the sauna, *Eur. J. Appl. Physiol.*, 54, 371, 1985.
16. Saumet, J., Dittmar, A. and Leftheriotis, G. Non-invasive measurement of skin blood flow: comparison between plethysmography, laser-doppler flowmeter and heat thermal clearance method, *Int. J. Microcirc. Clin. Exp.*, 5, 73, 1986.
17. Tur, E., Tur, M., Maibach, H.I. and Guy, R.H., Basal perfusion of the cutaneous microcirculation: measurements as a function of anatomic position, *J. Invest. Dermatol.*, 81, 442, 1983.
18. Johnson, J., The cutaneous circulation, in *Laser-Doppler Blood Flowmetry*, Shepherd, A. and Öberg, P., Eds., Kluwer Academic Publishers, Boston, 1990, pp. 121–139.
19. Dittmar, A., Skin thermal conductivity, in *Cutaneous Investigation in Health and Disease*, Lévêque, J.-L., Ed., Marcel-Dekker, New York, 1989, pp. 323–358.
20. Coulsen, M., Hayes, N. and Foreman, J., Comparison of infrared and Helium-Neon lasers in the measurement of blood flow in human skin by the laser Doppler technique, *Skin Pharmacol.*, 5, 81, 1992.

Chapter 17.2
Laser-Doppler Flowmetry: Principles of Technology and Clinical Applications

Gianni Belcaro and Andrew N. Nicolaides
Irvine Laboratory for Cardiovascular Investigation and Research
St. Mary's Hospital Medical School
London, U.K.

I. Introduction

Noninvasive optical methods to evaluate skin flow have been used for many years. The most used method is photoplethysmography (PPG) which records variations in skin flow by the evaluation of the absorption characteristics of the skin that are related to its blood-flow content. By PPG qualitative data may be obtained but quantitative data are difficult to obtain and standardize. Fluctuations in venous volume, i.e., due to postural changes or motion artifacts may alter the signal and make the interpretation of skin flow variations difficult. Therefore, clinical applications of PPG in the assessment of arterial or venous disease have never been defined with standard and conclusive diagnostic methods of evaluation. The use of PPG has been mainly limited to complementary methods able to evaluate some aspects of vascular disease, i.e., the pulsatility of skin flow in Raynaud's disease or the refilling time after venous emptying following an exercise test. Furthermore PPG evaluates skin flow for a variable depth (2 to 6 mm) including in this layer different skin circulatory elements without separating them from the more superficial, nutritional skin flow. Also the PPG probes are not easily usable to assess flow in cavities (such as intestine) or parenchymatous organs (liver, spleen, muscle, etc.)

Laser Doppler flowmetry (LDF) a theoretically comparable method has been developed in the last 15 years from a pure scientific instrument to a research tool and in the last few years to a clinical diagnostic technique to assess tissue viability and perfusion.[1–4] The fundamental principle of LDF as applied to the measurements of blood flow have been described in detail in several publications.[4–8,15] Measurement of blood velocity in single, large vessels and the measurement of skin perfusion can be both obtained using LDF. However, the technology required for these two applications are completely different and in this chapter we refer only to the measurement of microvascular perfusion.

II. Theory of Laser Doppler Flowmetry

A detailed technical description of the method is beyond the aim of this chapter. However some simple concepts are needed to understand the method, its applications, and possible limitations. Most commercially available laser-Doppler instruments utilize helium-neon gas or gallium-aluminum arsenide elements to produce a weak laser beam with low tissue penetration. This type of laser light does not alter the tissues under evaluation or produce an increase in tissue temperature. Most tissues (i.e., skin) are relatively opaque as they contain matters which refract (scatter) light in various and random directions. Excluding blood within large vessels, blood itself constitutes only a small fraction of tissue volume and most light scattering is due to small particles and to stationary tissue elements. Only moving parts in the sample volume (blood cells) will cause a Doppler shift in the light frequency. The scattering angles and red cell velocities are variable and they can be determined only in a statistical sense.[22] Therefore the LDF signal is a stochastic representation of the number of cells in the sample volume multiplied by their velocities. The technology using coherent laser light overcomes some problems observed with other optical and nonoptical methods used to record skin blood perfusion. The helium-neon laser which emits a red light and is used in many LDF instruments, detects very small changes in the wavelength of the laser light as a result of red blood cell movements, well below the resolution of the optical spectroscope.[1,2] The frequency distribution of the signal is defined by computerized spectral analysis of the output that produces a power spectrum with a separation between the noise and the signal due to blood cell motion. Some components of the LDF signal are due to external biological or instrumental elements (e.g., vibration) and some to internal (mainly electronic noise) factors. The LDF photodetector signal contains all the Doppler frequencies arising from the laser interaction with moving particles in the tissue and static components. An elaborate electronic processing and filtering is needed to transform the LDF signal into a physiologically reproducible, meaningful, and useful parameter so that the LDF output varies linearly with the blood flow within the sample volume. Considering the LDF output, as seen above, the total power in the signal depends on the number of the moving particles producing a laser Doppler shift of frequency. Therefore an LD flowmeter must be capable of measuring the mean frequency shift in the signal. Some LD flowmeters digitize the signal and analyze it with a fast Fourier transform from which the mean frequency shift can be measured. Other systems employ analogue signal processing circuitry. Electronic components are needed to normalize the signal and to compensate for noise. In most equipment the noise level remains relatively stable and theoretically it can be extracted from the LDF output by subtracting a constant offset. Different technical solutions have been applied in low noise lasers or laser diodes.[4,5,13,14]

This simple preliminary outline of LDF technology may be useful to comprehend the basic concepts. More specific technical information and details on the structure of LD flowmeters can be found in other reports.[1,2,5,6]

III. Terminology

The output from the LDF is referred to as flux instead of flow. This may cause some confusion. The difference in the terminology is explained by an example reported by Almond.[2] Replacing, by hypothesis, blood with saline which does not contain particles generating Doppler shift, a method which measures volume flux (flow) of the fluid such as venous occlusion plethysmography would give approximately the same value of flow. In these conditions an LD flowmeter would give a zero output as there are no scattering particles to produce a Doppler shift. However it is reasonable to suppose that in most physiological and clinical situations even with a variable relationship — due to the fact that blood cells are not homogeneously distributed — LDF volume flux and flow have a good correlation.

IV. Calibration

Many studies demonstrate a good correlation between LDF measurements and other blood flow measurements[2,13,23] while some other studies have shown an irregular correlation between LDF and isotopic methods — i.e., xenon clearance[7] where arteriovenous shunts are present[8] or in the bone.[10]

The major problem at this stage is that there is no gold standard against which LDF can be compared particularly for skin flow. Also for the medical operator and vascular technologist it is still confusing to consider that any single LDF instrument gives its own values and it is difficult to compare results obtained with an instrument to results obtained with a different flowmeter. Therefore relative changes (i.e., before and after treatment, after thermal or postural changes or after arterial occlusion to cause reactive hyperemia) are clinically better accepted than absolute flux measurements.

The validity of an universal calibration does not appear possible at the moment as in different tissues, several factors (i.e., light penetration, diffusion, and reflection) make the calibration relevant only to that single tissue.

Even with these limitations LDF has many positive aspects which have progressively enlarged its fields of application from experimental and clinical physiology to clinical practice.

A. The Zero LDF Calibration and the Biological Zero

Flux zero calibration is obtained by positioning the probe against a white surface. An 'instrumental zero' is different from the 'biological zero', which can be obtained from the skin[9–20] by complete occlusion of the arterial supply (Figure 1). When blood flow in the skin is completely abolished, the LDF signal decreases to 20 to 50% of the tissue flux measurement. The relative ratio between the normal resting flux state and the occlusion state values are variable from region to region and also vary in different organs. It has been observed that in a fingertip the ratio is 5 to 7% and that in closely related areas the biological zeroes are very similar.[20] The origin and physiology of the biological zero is not clearly understood. Studies on excised tissue have shown an elevated baseline from the instrumental zero even several hours after excision. The elevated biological zero disappeared after a few days. Freezing tissues immediately abolish the biological zero baseline lowering the values to a level very similar to the instrumental baseline.[20] In situations of inflammation of human skin it has been shown that the biological zero is increased up to approximately 50 to 70%.[9] The clinical implications of these findings are still not clear and further studies in this field are essential. However in the practical clinical evaluation of limbs, particularly in low perfusion states, it is recommended that each measurement should be associated with an estimation of the zero baseline and of the biological zero.

V. Problems of LDF in Clinical Practice

LDF is noninvasive and does not interfere with the microcirculation when measuring local blood flow.[22] LDF is also particularly useful as it produces a continuous output which can be used for prolonged monitoring of tissue viability — i.e., after plastic surgery or to record skin flow perfusion during sleep or in newborns, etc. — a technique which is impossible to

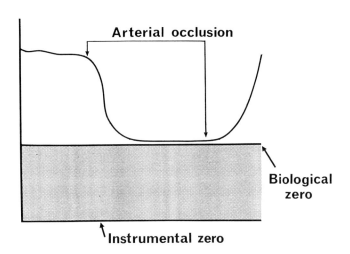

FIGURE 1 The concept of the instrumental zero (obtained placing the probe against a white, neutral surface) and the biological zero obtained with arterial occlusion.

obtain with any other noninvasive or invasive techniques. LDF monitoring is also relatively stable, reproducible, and easy to learn to apply. The technology of clinically usable LDF instruments is continuously improving. Multiple channel systems capable of measuring distant tissue areas at the same time are now produced by different manufacturers. LDF imagers[20] and multiwavelength systems[2] are currently being developed and may be used for more accurate and interesting, physiological clinical applications.

LDF has been extensively used in clinical physiological evaluations and in clinical practice in humans. The technique is easy, noninvasive, and has become very popular but at times the interpretation of clinical results has been uncritical.[9] LDF, as seen above, measures minute particle motion. In living tissue such particles include mainly blood cells which will constitute the major component of the LDF output. However it is obvious that in situations of altered microcirculation or alteration of the blood hematocrit (i.e., in inflammation or leukemia), the number of particles can change, determining a different output. The LDF signal can be considered a stochastic representation of the motion of all particles in the sample volume. This measurement cannot be a physiologically correctly defined flow although it has been shown that LDF values in certain tissues and conditions are proportional and closely related to flow[9] measured with isotopic methods. Also, in the skin the precise sampling volume is not always easy to determine and volume flow cannot be measured.

Therefore the LDF output signal recorded has been defined flux but possibly the most correct expression in clinical application is *relative perfusion units*.[9]

Other terms used to express output signal such as volt, millivolt, arbitrary units, and other terms need to be unified. This appears to be difficult at the moment as there is no dialogue concerning a common standard

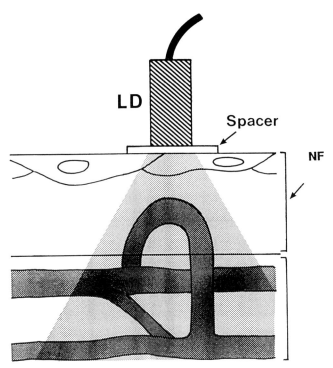

FIGURE 2 The sampling volume of a standard laser-Doppler (LD) flowmeter includes the more superficial nutritional flow (NF) and the deep thermoregulatory capillary layer.

among manufacturers. However, when referring to flow in certain contexts and considering the above limitation, the term flow may be used to express a concept which is more familiar to most physicians.

A. Depth of Measurements and Volume in the Skin

The sensitivity of the reflected LDF measurements decreases exponentially with the distance from the probe.[22] It has been estimated that the theoretical average depth of sampling in the skin is 0.14 mm but in artificial models it has been estimated that it is 1.5 mm. However these values have little importance in clinical applications. There is also evidence that in intestinal models the measuring depth can be greater than 6 mm and that by placing a mirror on the opposite intestinal wall the signal output can be increased from 85 to 100%. The above observations indicate that the measuring depth is not a fixed value for the skin but most probably a continuous variable as in other tissues.

As reported by Fagrell[9] the microcirculation — particularly skin microcirculation — can be considered divided in a small, more superficial, thin layer characterized by nutritional capillaries and in a deeper layer with thermoregulatory vessels. Figure 2 shows that the nutritional capillaries are the most superficial ones (0.1 to 0.05 mm from the skin surface) and in normal conditions supply only a very small amount of skin blood (5 to 10%). In the subpapillary mostly thermoregulatory bed (0.05 to 2.0 mm in depth) the most common (95%)

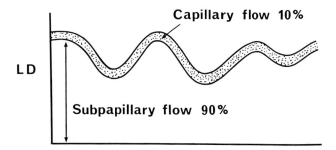

FIGURE 3 The percentage of nutritional flow in the LDF tracing is possibly less than 10% as most of the signal component is due to the non-nutritional subpapillary flow.

vessels are venules with only a small portion of arterioles. In this bed at least 95% of skin blood flow can be responsible for the LDF output while in the nutritional bed only a small percentage of flow (5 to 10%) contributes to the global output (Figure 3). Therefore LDF measurements in the skin measuring 1 to 2 mm of skin depth record a signal in which the thermoregulatory flow component is predominant (usually >89%) and the nutritional flow is a small component. New microprobes and new LDF flowmeters may reduce the measuring depth and selectively evaluate the most relevant superficial nutritional capillaries, but so far clinically relevant data are not available.

Using 1-mm thick plastic spacers with the same reflectance properties and light transmission characteristics of the skin, it is theoretically possible to evaluate the most superficial skin flux. However the more distal LDF signal seems to contain a high proportion of noise components and it appears to be weaker. Early results from our study in which skin spacers are used[24] indicate that there is a significant important difference in skin flux values and in flux responses, i.e., venoarteriolar response, flow increase with local temperature increase, etc. This difference is noted both with and without the spacers but with a different, not correlatable extent in *low perfusion microangiopathy* — peripheral vascular disease, hypertension associated with vasospasm, and Raynaud's disease and phenomenon. However, in venous disease and in diabetic microangiopathy, (these conditions can be defined as *high perfusion microangiopathy*) the reflex responses and skin flux measurements are reduced with the spacers but comparable and parallel to measurements obtained without them.

This possibly indicates that the differentiation between the nutritional and the thermoregulatory component is more important in states of low perfusion microangiopathy. In low perfusion microangiopathy the increase in skin flux as measured by LDF, i.e., following administration of a vasoactive drug or superficial revascularization, may only reflect an increase in the thermoregulatory, non-nutritional flow. This often could be irrelevant in the evaluation of skin flow changes due to the treatment and produce misleading conclusions in relation to the healing or development of skin necrosis.

VI. Vasomotion

As LDF measures flux continuously flux motion can also be easily evaluated. During LDF monitoring in normal conditions the skin vessels in the microcirculation fill with blood rhythmically as a consequence of pressure and flow changes due to cardiac action respiration and vasomotion. The LDF output therefore shows a continuous variation and the sample volume varies continuously. Flux motion patterns are markedly changed in patients with peripheral vascular disease who frequently present a high frequency flux motion component. The prevalence of high frequency motion waves is significantly increased in low perfusion states and it is proportionally more evident with increased levels of ischemia. The relationship between the presence of high-frequency flux motion waves at the forefoot has been evaluated by Hoffman[11] before and after percutaneous transluminal angioplasty. In successfully treated patients a significant decrease in high-frequency flux waves was observed after angioplasty. However the prevalence of such waves after angioplasty indicated that high-frequency waves are related to severe chronic ischemia. Different patterns of vasomotor waves were detected by Hoffman[11] in different conditions of perfusion. Flux motion in normals was characterized by low frequency and pulsatile flux waves. Occasionally additional high-frequency wave components appeared in the recording. Flux motion patterns, in severe ischemia, showed almost no pulsatile flux

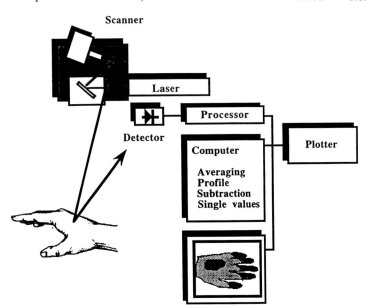

FIGURE 4 A diagram of the LDF scanning system.

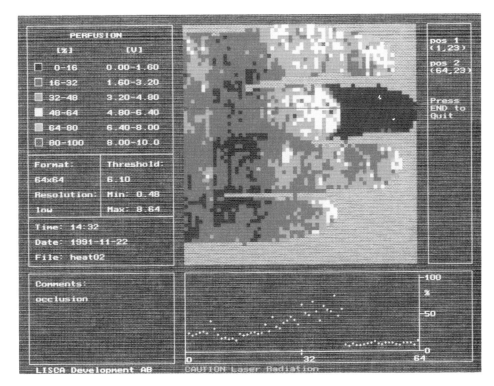

FIGURE 5 An example of LDF scanning. The low flow in the second finger from below is due to occlusion with a rubber band and is indicated in the color-coded perfusion scan by the dark-blue color.

waves whereas high-frequency waves were frequently observed in more severe ischemia. In severe cases of ischemia no flux motion was observed. By contrast patients with intermittent claudication showed variable patterns (excluding the last pattern with no vasomotion).

Therefore using frequency analysis it appears possible to qualitatively differentiate degrees of ischemia. However while there is little doubt that the alteration of vasomotion in low or high perfusion states is clinically relevant to indicate microcirculatory disturbances, no definite clinically usable answer has been provided about the analysis of vasomotion.

Sophisticated frequency analysis, not commercially available with the instrumentation but usable in postprocessing with the signal output, is needed to extract clinically meaningful data from the analysis and thus far limits the application of the method. There is, however, little doubt that qualitative analysis of vasomotion may offer new and interesting concepts to evaluate microcirculatory disturbances.

VII. New Technology

LD imaging is a new method of evaluating skin perfusion (other tissue can also be studied). A diagram of the system is shown in Figure 4. The multiple skin sampling produces a color perfusion image of the tissue under evaluation and a perfusion profile (Figure 5). The color-coded perfusion scale clearly indicates absence of perfusion in the second finger from the bottom in the image due to occlusion with a rubber band (dark blue).

The system has been developed very recently and some clinical application of the system appears very promising.

VIII. Conclusions

The use of LDF is progressively increasing in physiological, pharmacological, and practical clinical evaluation of vascular disease. The monitoring of the effects of treatment on the microcirculation appears to be one of the most promising fields of application of LDF. The technical development of LDF combined with extensive clinical research application and coupled with frequency analysis systems will make this method one of the future's most interesting non-invasive fields of investigation in vascular disease.

At this stage, there are some problems concerning standardization of the different systems and calibration problems. However, the very recently published guidelines of the European Contact Dermatitis Society may overcome this problem.[25]

References

1. Almond NE, Jones DP, Bowcock SA, and Cooke ED. A laser Doppler blood flowmeter used to detect thermal entrainment in normal persons and patients with Raynaud's phenomenon. In: Spence VA, Sheldon CD (Eds), *Practical Aspects of Skin Blood Flow Measurement*, Biological Engineering Society, London: 31, 1985.
2. Almond NE. Laser Doppler flowmetry: Instrumentation theory and practice. In: Belcaro G, Hoffman U, Nicolaides AN, Bollinger A (Eds), Med-Orion Publishing Co. (In publication).
3. Almond NE, Jones DP, and Cooke ED. Noninvasive measurement of the human peripheral circulation: relationship between laser Doppler flowmeter and photoplysmograph signals from the finger. *Angiology* 39: 819, 1988.
4. Boggett D, Blond J, and Rolfe P. Laser Doppler measurement of blood flow in skin tissue. *J Biomed Eng* 7: 225, 1985.
5. Bonner RF and Nossal R. Model for laser Doppler measurements of blood flow in tissue. *Appl Optics* 20: 2097, 1981.
6. Bonner RF, Clem TR, Bowen PD, and Bowman RL. Laser Doppler continuous realtime monitor of pulsatile and mean blood flow in tissue microcirculation. In: Chen S, Chu B, Nossal R (Eds), *Nato Advanced Study Institute, Series B, Physics, Scattering Techniques Applied to Supramolecular and Non-equilibrium Systems*. Plenum Press, New York: 685, 1981.

7. Borgos JA. TSI's LDV blood flowmeter. In: Shepherd A, Oberg P (Eds). *Laser Doppler Blood Flowmetry,* Kluver Academic Publ., Boston: 73, 1990.
8. Engelhart M and Kristensen JK. Evaluation of cutaneous blood flow responses by ^{133}Xenon washout and a laser Doppler flowmeter. *J Invest Dermatol* 80: 12, 1983.
9. Fagrell B. Problems using laser Doppler on the skin in clinical practice. In: Belcaro G, Hoffman U, Nicolaides AN, Bollinger A (Eds). Med-Orion Publishing Co. (In publication.)
10. Hellem S, Jacobsson LS, Nilsson GE, and Lewis DH. Measurement of microvascular blood flow in cancellous bone using laser Doppler flowmetry and ^{133}Xe clearance. *Int J Oral Surg* 12: 165, 1983.
11. Hoffman U, Seifert H, Baider E, and Bollinger A: Skin blood flux in peripheral arterial occlusive disease. In: Belcaro G, Bollinger A, Franzeck U, Hoffman U, Nicolaides AN (Eds). *Laser-Doppler Flowmetry Experimental and Clinical Applications.* Med-Orion Publishing Co (in press).
12. Holloway GA and Watkins DW. Laser Doppler measurement of cutaneous blood flow. *J Invest Dermatol* 69: 306, 1977.
13. Johnson JM, Taylor WF, Shepherd AP, and Park MK. Laser Doppler measurement of skin blood flow: comparison with plethysmography. *J Appl Physiol Respirat Environ Exercise Physiol* 56: 798, 1984.
14. Nilsson GE, Tenlan T, and Oberg PA. Evaluation of a laser Doppler flowmeter for measurement of tissue blood flow. *IEEE Trans Biomed Eng* 27: 597, 1980.
15. Obeid AN, Boggett DM, Barnett NJ, Dougherty G, and Rolfe P. Depth discrimination in laser Doppler skin blood flow measurement using different lasers. *Med Biol Eng Comput* 26: 415, 1988.
16. Obeid AN, Barnett NJ, Dougherty G, and Ward G. A critical review of laser Doppler flowmetry. *J Med Eng Technol* 14: 178, 1990.
17. Riva C, Ross B, Benedek GB. Laser Doppler measurement of blood flow in capillary tubes and retinal arteries. *Invest Ophthalmol* 11: 936, 1972.
18. Stern MD. In vivo evaluation of microcirculation by coherent light scattering. *Nature* 254: 56, 1975.
19. Stern MD, Lappe DL, Bowen BD, Chimosky JE, Holloway GA, Keiser HR, and Bowman RL. Continuous measurement of tissue blood flow by laser Doppler spectroscopy. *Am J Physiol* 232: H441, 1977.
20. Tonneson KH and Pederson LJ. Laser Doppler flowmetry: problems with calibration. In: Belcaro G, Hoffman U, Nicolaides AN, Bollinger A (Eds). Med-Orion Publishing Co. (In Press.)
21. Wardell K, Jakobbson A, and Nilsson GE. A laser Doppler imager for microcirculatory studies. 1st European Laser Doppler Users Meeting, Oxford, March 1991.
22. Weis GH, Nossal R, and Bonne RF. Statistics of penetration depth of photons re-emitted from irradiation tissue. *J Mod Optics* 36: 349, 1989.
23. Winsor T, Haumschild DJ, Winsor DW, Wang Y, and Luong TN. Clinical application of laser Doppler flowmetry for measurement of cutaneous circulation in health and disease. *Angiology* 10: 727, 1987.
24. Belcaro G and Nicolaides AN. Article in preparation.
25. Bircher AJ, de Boer E, Agner T, Wahlberg J, and Serup J, Guidelines for the measurement of cutaneous blood flow by laser Doppler flowmetry, *Contact Derm* 30: 65, 1994.

Chapter 17.3
Examination of Periodic Fluctuations in Cutaneous Blood Flow

Robert Gniadecki,[1] Monika Gniadecka,[2] and Jørgen Serup[1,2]
[1]Department of Dermatological Research
Leo Pharmaceutical Products
Ballerup, Denmark
[2]Bioengineering and Skin Research Laboratory
Department of Dermatology
University of Copenhagen
Bispebjerg Hospital
Copenhagen, Denmark

I. Physiology of Vasomotion

Small arteries and arterioles in the skin and the subcutaneous tissue exhibit *vasomotion*, i.e., rhythmic changes of vessel diameter due to a series of contractions and relaxations. The principal features of vasomotion have been described by Nicoll and Webb[1,2] who observed rhythmic changes of diameter of small vessels in bat wing *in vivo*. These authors reported that vasomotion induced variations of blood blow in the vessels (flow-motion). Since that time vasomotion has been directly observed in various animal and human tissues: in the skin (hamster skinfold window preparation[3] and cheek pouch[4]), nailfold capillaries in man,[5] in skeletal muscle,[6] testicle,[7,8] conjunctiva,[9] retina,[10] brain,[11] and heart.[12]

The origin of vasomotion has been studied mainly in the rabbit tenuissimus-muscle model.[6] It has been found that vasomotion is elicited by the rhythmic activity of smooth muscle pacemaker cells in which the spontaneous depolarization occurs.[13-15] These cells are located in cushion-like thickening of vessel wall near the branching points and are supposed to provide the original trigger for vasomotion, that is eventually propagated downwards to the larger arterioles (Figure 1).[16-18] Frequency of vasomotion changes abruptly at bifurcation points and gradually increases in the downstream direction, from 0.5 to 18 cycles per minute (cpm: 60 cpm = 1 Hz) in the larger arteries to 9 to 21 cpm in terminal arterioles. This downstream propagation of contractions and dilations causes superposition of waves and the final vasomotion pattern in the distal elements of the vascular tree is the composite effect of signals that originate at various branching points in the microvasculature.[18]

The major role of arteriolar vasomotion and cyclic oscillations of blood flow in the microcirculation is probably the enhancement of blood passage in the capillaries. Poiseuille's law, that describes flow of the fluid in the cylinder-like vessel, states that conductivity of the vessel is proportional to the fourth power of its radius. Therefore, vessel of the oscillating diameter will be much more conductive than the vessel of the same mean but constant diameter.[19,20] Additionally, vasomotion plays an important role in the control of vascular resistance. Secomb et al.[19] and Secomb and Gross[21] predicted that the increase of vascular resistance to four times of the initial value would be impossible without the participation of vasomotion, because in this instance the vascular diameter must have been controlled within very tight and unrealistic limits, and the diameter of the vessel would be smaller than the critical minimum diameter for passage of red blood cells. Therefore, vasomotion enables red blood cell transfer in the conditions of increased vascular resistance. Other functions of vasomotion are listed in Table 1.

II. Methods of Detection and Analysis of Vasomotion

The ideal method for the examination of cutaneous vasomotion would be the direct observation of the changes of the diameter of the arterioles in the skin. Since this cannot be easily accomplished in humans, methods have been developed that detect temporal variations of blood flow (flow-motion) or red blood cell flux* (flux-motion) that appear secondarily to arteri-

---

* Flux is the product of the red blood cell concentration and speed.

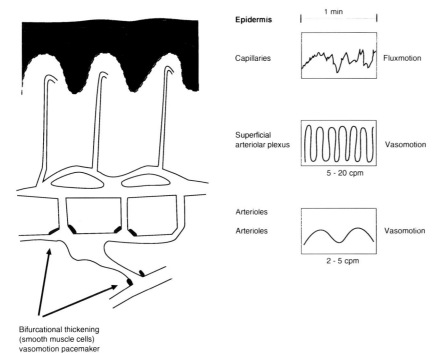

FIGURE 1 Hypothetical origin of vasomotion in human cutaneous microcirculation. (Based on References 13–18, 75.) Vasomotion is triggered by pacemaker smooth muscle cells located at arterial bifurcations. The dominant vasomotor frequency (5–10 cpm) is generated at the origin of ascending arterioles, whereas vasomotion in the larger arteries has lower frequency. The complex pulsating flow in the superficial arteriolar plexus and the capillaries is a result of superposing of rhythms from several pacemakers.

TABLE 1 Role of Vasomotion in Microcirculation

Function	Ref.
Enhancement of vessel conductivity	19
Control of vascular resistance	21
Improvement of oxygen delivery to the tissues	19
Maintainance of blood fluidity	91
Removal of tissue edema	92,100
Stimulation of lymph flow	93,101
Detachment of cells adherent to endothelium	90

olar vasomotion. Flow-motion in skin capillaries can be analyzed with dynamic capillaroscopy, whereas laser Doppler technique is used for the analysis of fluxmotion in the microvasculature.

A. Intravital Microscopy

Intravital dynamic capillaroscopy is based on obtaining a magnified image of skin capillaries and dynamic recording of red blood-cell velocity in these vessels. The most often used system has been developed by Bollinger et al.[5] and Butti et al.[22] and further improved by Fagrell et al.[23,24] In this system vessels are visualized with an *in vivo* microscope and the image is recorded with the aid of the closed-circuit television.[25] Nailfold capillaries (in the finger or toe) have been commonly investigated, but some studies were also done in a titanium chamber system.[26] The alterations of the velocity of red cells are analyzed with the video densitometric techniques with correlation of the photometric signals.[27] Wayland and Johnson[28] developed a dual slit method which is based upon the measurement of the delay between the two signals elicited by the same configuration of red blood cells as they pass two photosensors separated by a fixed distance. More recently Intaglietta et al.[29] described a video dual-window technique for measuring of blood velocity. The video signal passes through two independent square areas (windows) fitted in size to the capillary width. The velocity of blood is determined by measuring the intrawindow transit time. Simultaneously with blood velocity other parameters may be recorded, such as arterial pulsations in the finger, respirations, ECG, etc.[23] However, the real-time on-line determination of blood cell velocity with these systems is complicated, especially when high velocities are to be measured. Slaaf et al.[30] proposed an easy to operate system based on a three-stage prism grating technique. A microscopic image of a microvessel is projected on a grating of alternating transparent and opaque lines and the light that passes through the grating is focused on a photosensor via the transfer lens. Moving erythrocytes modulate the intensity of light that is recorded by the photosensor. This allows the on-line determination of the velocity and the direction of flow in the capillary.

Direct observation of blood flow velocity changes in human nailfold capillaries revealed spontaneous fluctuations, in most cases at the frequency of 6 to 10 cpm that were not related to normal respiration.[5,23,31] These fluctuations of flow velocity are considered to be a result of arteriolar vasomotion[31-33] and are correlated with changes of pressure of blood in a capillary, increased pressure being associated with the most rapid speed of flow.[34]

The principal drawback of examining vasomotion by direct observation of blood vessel is that capillary microscopy is restricted to nailfold capillaries. Examination of vasomotion in other regions requires implantation of titanium chamber and cannot be considered fully noninvasive.

B. Laser Doppler Technique

The laser Doppler method has recently become an attractive alternative to intravital dynamic capillaroscopy for studying cutaneous vasomotion. The laser Doppler technique enables simple, real-time monitoring of relative changes of red blood cell flux in the cutaneous microvascular bed in any region of the body. Pioneering laser Doppler studies of Holloway and Watkins,[35] Tenland et al.,[36] and Salerud et al.[37] demonstrated spontaneous oscillations of blood flux (flux-motion) in normal human skin in resting conditions. The average frequency of the oscillations recorded by these authors was 8.6 cpm and this value was comparable with rhythmic variations of capillary blood pressure,[34,38] capillary blood velocity,[5,39,40] and vasomotion.[3] Oscillations of flux could not be suppressed by a proximal nervous blockade, implying a local, myogenic mechanism. It therefore has been concluded that the flux-motion recorded with the laser Doppler technique reflects changes of microcirculatory blood flow due to arteriolar vasomotion.[37]

It is not fully understood what actually generates the laser Doppler signal in the skin[32,41–44] and there is much debate to what extent flux-motion recorded with laser Doppler technique reflects flow-motion in the microvasculature due to arteriolar vasomotion. Tooke et al.[43] compared the laser Doppler method with video dynamic capillaroscopy and found that with both techniques it was possible to record oscillations of 4 to 6 cpm. When Pearson product momentum correlation coefficient, which predicts the relationship between two sets of paired data, was calculated, a linear relationship between laser Doppler signal and red blood cell velocity was seen in 11 out of 14 sets of records. In 8 of the 14 recordings the best correlation of flux-motion and flow-motion was found at no time delay. Spectral analysis showed that in 14 out of 16 recordings the flow-motion pattern had similar frequency spectra to flux-motion. The amplitudes of flow oscillations were significantly higher than the amplitudes of flux-motion. Therefore, although laser Doppler measures blood flux not only in the nutritive capillaries but also in deeper elements of the skin vascular tree,[42-45] this is a useful tool for evaluation of cutaneous vasomotion. The issue of the correlation of the origin of laser Doppler signal with the microvascular segments of human skin has been further investigated by Braverman et al.[46] They found that with a commonly used laser Doppler probe the maximum amplitude of rhythmic oscillations could be obtained when the probe was placed directly over ascending elastic arteriole and its immediate branches. On the other hand, when the probe was moved to the site between ascending arterioles the nonpulsatile laser Doppler signal of the flow flux value was recorded.

The dependency of the laser Doppler flux pattern on the position of the probe on the skin may explain low reproducibility in the detection of flux-motion among different authors. In some studies cyclic changes of cutaneous flux could not be detected in normal conditions in humans.[47,48] Because of these difficulties methods for amplification of vasomotion have been devised. Wilkin[47] observed that cutaneous flux-motion is provoked in the early phase of *postocclusive reactive hyperemia*. The hyperemia response has been obtained after occlusion of the brachial artery with a sphygmomanometer cuff for 6 min. After release of the occlusion a prompt increase of flux is seen and blood flux oscillations can be recorded from the forearm skin during the return of blood flux to normal values (Figure 2). The average period of these oscillations is 9.6 ± 0.3 s (SE) is in accordance with the reported rhythmic variations of blood flow observed microscopically in human nailfold capillaries[23] and flux-motion in the forehead[37] and leg[49] obtained in the normal resting conditions. It is conceivable that amplified laser Doppler flux oscillations are due to the local synchronicity of oscillatory blood flow in a group of cutaneous capillaries in the period of hyperemia. The synchronicity is limited to a small area of the skin. Loss of synchronicity has been reported in the sites only 0.5 to 2 cm apart implying a local origin of the oscillations.[37,43,47]

Besides postocclusive hyperemia, cutaneous flux oscillations can be induced by other stimuli that provoke a local increase in skin blood flow (Table 2). Wilkin[50] applied topically a 5 M aqueous solution of propionaldehyde and reproduced a tenfold increase of cutaneous blood flux. During the recovery to the resting flux level marked oscillations of flux could be detected. A similar hyperemic response and augmentation of flux-motion in the recovery phase could be induced with a topical 2.2×10^{-2} M nitroglycerine.[51] Kastrup et al.[52] reported that during local skin heating to 42°C flux-motion is induced in the posthyperemic phase. The oscillations, with a mean frequency of 6.9 cpm (range 5.2 to 10.4), were not suppressed during local or central nervous blockade, with lidocaine or trimethaphan

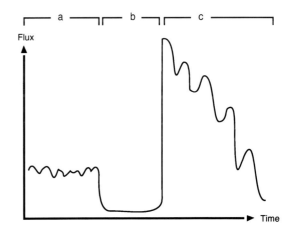

FIGURE 2 Enhancement of cutaneous blood flow oscillations in the early phase of postocclusive hyperemia. (a) Baseline flux, (b) arterial occlusion, and (c) reactive hyperemia. Note prominent oscillations of blood flux on the descending arm of hyperemic response (c).

TABLE 2 Modulation of Oscillatory Blood Flow in Cutaneous Circulation

Condition	Oscillatory Blood Flow Amplitude	Oscillatory Blood Flow Frequency	Ref.
Dependent position	↓	0	49
(increased venous pressure)	↓	↓	57
Post-thrombotic	↓	↓	70[a],66,67
syndrome	↓	ND	68
	↑	↑	69,71
Arterial insufficiency	0	↑	56,73
Sickle cell disease	↑	0	48,77
Diabetes mellitus	↓	↓	94[b],95[c],96
General anesthesia	↓	0	51,97
Postocclusive reactive hyperemia	↑	ND	47,48,51
Topical application of:			
Propionaldehyde	↑	ND	50
Nitroglycerine	↑	ND	51
Carbon dioxide	↑	ND	53
Thermal challenge	↑	0	52
Hyperventilation	↑	0	54
Smoking	0	↓	26
Hypertension	↑	0	98[d]
Aging	↓	0	99

Note: ↓-decreased, ↑-increased; 0-no change; ND-not done.
[a] In legs with edema, vasomotion restored after edema removal.
[b] Based on the thermal clearance method (detection of arteriovenous anastomotic blood flow).
[c] Based on venous occlusion pletysmography.
[d] Studies in experimental animals.

campsylate, respectively. Some authors were also able to induce vasomotion by a local application of carbon dioxide[53] and during hyperventilation.[54]

C. Analysis of Waveform Patterns

Cyclic oscillations in the blood flow or flux tracings may be analyzed by manual determination of wave frequency, amplitude, and prevalence. It has been realized that the frequency and amplitude of fluxmotion waves is heterogeneous and may be further subdivided into discrete components. Kastrup et al.[52] divided flux-motion into two constituents: high-frequency regular oscillations (mean 6.8 cpm) of the non-neurogenic (probably myogenic) origin, and irregular low-frequency oscillations (mean 1.5 cpm) that were caused by periodic changes in the autonomic tone. Similarly, Scheffler and Rieger,[55] Seifert et al.,[56] and Bongard and Fagrell[57] found two categories of oscillations: large waves of large amplitude and low frequency (<10 cpm), and small waves of low amplitude and the frequency higher than 15 cpm. The prevalence of the small waves in the laser Doppler signal has been reported to be between 5 and 50%.

For the objective analysis of vasomotion the spectral methods have been recently introduced. The most widely used techniques, fast Fourier transform, autoregressive algorithms, and Prony spectral line estimation (PSLE), will be briefly reviewed in Sections II.C.1–3.

1. Fast Fourier Transform

Fourier transform isolates the repetitive harmonic (sine and cosine) parts of the time domain data and calculates the frequency and power of each repetitive component.[58] Rather than the classical Fourier transform, which is extremely computational intensive, its 'fast' version is usually used. The requirement for the fast Fourier transform (FFT) is that the number of sample points in a time domain signal must be a power of two. A FFT of an oscillatory tracing gives two sets of frequency domain data: real and imaginary. The first set shows the values of component sine waves at different frequencies, while the latter shows cosine harmonic components. For example, if the original data is a simple sine wave with a frequency of 1 Hz and an average amplitude of 10 U the FFT will give value of 10 at frequencies +1 Hz (real) and −1 Hz (imaginary). The power of the component frequencies is usually shown in a power spectrum. Each power spectrum value is the sum of the squares of the values of the real and imaginary FFT values (in our example $10^2 + 10^2 = 200$ U at 1 Hz). Another presentation of FFT is the power spectral density graph. Each value for power spectral density is calculated as power spectrum value/frequency bandwidth per point and thus the area under the power spectral density curve will be in units of power. Power spectral density allows for easier comparison of the data independently on the sample length and the rate of recording.

The most prominent limitation of FFT is the resolution of the spectral components of similar frequency. A second limitation is due to the windowing of the data that occurs when processing with FFT. In practice one deals always with finite sets of time-domain data and this introduces error to Fourier analysis, since FFT tries to find repetitive signals from the edges of the data block (a 'leakage' phenomenon).[58] Due to the 'leakage' phenomenon, the power of the main lobe 'leaks' to the side lobes distorting other spectral responses that are present so that weak components can be masked by higher sidelobes from stronger spectral responses. Leakage may be substantially reduced by mathematical modification of the input data (*windowing*[59,60]). Such transformed data give more 'clean' power spectra, however at the expense of reduced frequency resolution. Data windowing should be principally used when the analysis of a weak repetitive component, that may be easily obscured by leakage, is required. The typical laser Doppler recording with the visible cyclic oscillations and the result of its FFT analysis is shown in Figure 3. Usually, besides the heart beat component at about 1 Hz (60 cpm) two or three low frequency spectral components are obtained at approx. 2.5 cpm, 5 cpm, and 12 cpm, the first two being the most prominent. It is conceivable that the very low frequency components (2.5 cpm) arise in the

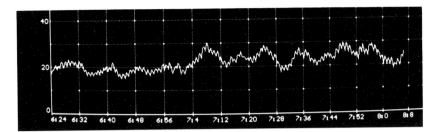
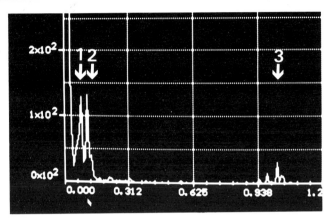

FIGURE 3 Fast Fourier transform (FFT) analysis of cutaneous flux-motion in humans. (a) Laser Doppler recording taken from the lower extremity (horizontal position), x axis = time (min:sec), y axis = units of flux. (b) FFT analysis — PSD graph. x axis = frequency (Hz), y axis = power of harmonic components. Note two distinct vasomotion-specific peaks at 2.34 (1), and 4.68 (2) cpm, in addition to a smaller peak at 61 cpm (3), characteristic for pulse-related flux changes in the skin.

larger arterioles whereas the higher frequency components are produced in the ascending arterioles and subcapillary arteriolar plexus in the skin. The origin of high frequency components (>10 cpm) that are sometimes seen in PSD may reflect either vasomotor activity or, alternatively, the ventilatory effect on venous return, respiratory sinus arrhythmia, or periodic blood pressure waves — all were shown to reside in the 9.6 to 15 cpm frequency band.[61]

2. Prony Spectral Line Estimation (PSLE)

PSLE technique has been proposed for analysis of vasomotion by Colantuoni et al.[18] and Meyer and Intaglietta.[62] The basic algorithm has been devised by de Prony[63] and later by Burkhardt.[64] A finite block of time-domain data are modeled as the sum of finite numbers of nonharmonically related sinusoids in white (uncorrelated) noise and the original waveform pattern is reconstructed by the iterative procedure. A first order of size approximations contains one sinusoid and the correlation coefficient is computed at each increment of order size. If the correlation coefficient is larger than the previous order, then this solution is kept and the previous is discarded. Thus a maximal correlation will be seen at a specific order.[65] If the correlation is greater than the predetermined level (usually the correlation coefficient >0.8 to 0.95), a solution has been found. In certain circumstances PSLE is more powerful than FFT, because the resolution is not dependent upon the data length.[58] The drawback of the PSLE method is computational complexity and the necessity to determine the order.[58] Moreover, the PSLE graph may be difficult to interpret for an uninitiated individual due to a relatively large number of spectral lines.

3. Autoregressive Modeling

Autoregressive PSD estimation employ the autocorrelation method to find the repetitive components in the input data.[58] Autoregression gives good results for strong sinusoidal components, however the power of amplitude of the components cannot be calculated.[58] Other limitations of autoregression methods involve the degrading effect of observation noise, presence of multiple spurious peaks and shifting of main frequencies towards higher values at low signal-to-noise ratio.[58] The resolution of autoregressive modeling is often better than that of FFT. Moreover, this procedure is not dependent on the number of samples, and the results are not affected by windowing.

III. Vasomotion in Pathologic Conditions

Vasomotion is easily modulated in a variety of physiologic and pathologic conditions (Table 2). Changes of vasomotion in arterial insufficiency, venous hypertension, and sickle cell disease were investigated in some detail.

A. Increased Venous Pressure, Chronic Venous Insufficiency, and the Postthrombotic Syndrome

The influence of the increased venous pressure in the lower extremity, caused either experimentally by leg lowering or by the pathologic process of venous thrombosis and valve injury (a postthrombotic syndrome), has been recently investigated by several groups with the laser Doppler flowmetry.[49,57,66-69] Increased leg venous pressure led to postural vasoconstriction and decrease of amplitude of vasomotion[49,57] (Figure 4). It is not clear whether increased venous pressure causes changes of the frequency of vasomotion. Bongard and Fagrell[57] found a decrease of vasomotion frequency from 3.5 to 2.1 cpm, whereas the Fourier spectral analysis performed by Gniadecki et al.[49] did not reveal significant changes of the frequency of the dominant harmonic component of the laser Doppler flux signal. The postural attenuation of vasomotion could be suppressed by the application of leg compression,[49] a finding providing further proof that increased venous

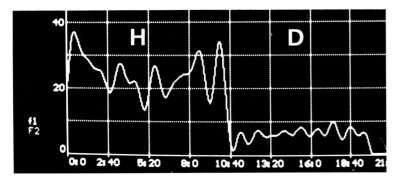
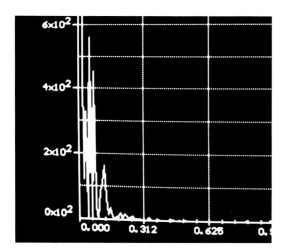
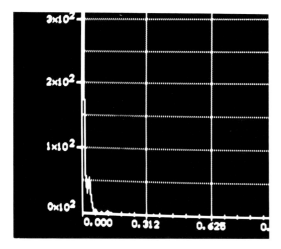

FIGURE 4 Effects of the increased venous pressure, caused by leg lowering, on cutaneous blood flux oscillations: (a) original laser Doppler recordings (H = horizontal position, D = dependent position), (b) power spectral densities (FFT) of blood flux in the horizontal position, and (c) in the dependent position. Note the marked decrease of power amplitudes in (c).

chronic venous insufficiency, with the long-term (4 weeks) compressive therapy, restored normal cutaneous vasomotion. Therefore, it is likely that limb edema itself inhibits vasomotion in lipodermatosclerosis. Similar findings were reported by Pekanmäki et al.[66] These authors were able to restore vasomotion (5 cpm, range 2 to 8 cpm) in the patients with lipodermatosclerosis by a single treatment with intermittent pneumatic compression (Ventipress®, Lemi, Finland). However, in this study, the position of the patients during laser Doppler measurement has not been reported and leg edema has not been assessed. Therefore, it is difficult to conclude whether the restoration of vasomotion due to intermittent pneumatic compression of the leg could be attributed to the removal of leg edema or to other mechanisms. Decrease of the amplitude of blood flux oscillations in the post-thrombotic syndrome has also been observed by Belcaro et al.[68] In a marked contrast, Chittenden et al.[69] and Cheatle et al.[71] reported an enhancement of vasomotion in patients with chronic venous insufficiency in the areas of lipodermatosclerosis. These authors recorded statistically significant higher baseline laser Doppler flux, higher vasomotion frequency (3.2 vs. 1.92 cpm in the control), and higher vasomotion amplitude (8.5 vs. 1.2 mV, control) in the lipodermatosclerosis. Increased frequency and amplitude of vasomotion was unlikely to be attributed to high flux because increasing of flux with pilocarpine did not induce changes of vasomotion. These results are difficult to compare with the studies showing decreased vasomotion in chronic venous insufficiency, because the authors did not assess leg edema in their patients. In the absence of edema, vasomotion in lipodermatosclerosis and chronic venous insufficiency could be enhanced secondarily, due to capillary plugging by white blood cells,[72] a similar mechanism to that seen in sickle cell disease where cutaneous vasomotion is induced by stiff sickle erythrocytes blocking blood flow in the microvessels. (See Sections III.B and III.C.)

B. Arterial Insufficiency

The effects of leg ischemia due to peripheral atherosclerosis on vasomotion were studied by Yanar et al.[73] and Seifert et al.[56] Cutaneous vasomotion in the second and third toe were recorded with laser Doppler in the four groups of patients with different stages of the disease (Table 3). The authors found increased prevalence of the high-frequency vasomotion laser Doppler waves proportional to the severity of the peripheral ischemia (Table 3). Increased vasomotion frequency was directly linked to peripheral hypoperfusion since the prevalence of high-frequency waves decreased after successful restoration of peripheral circulation by means of angioplasty or thrombolysis.[74] The pathophysiologic basis of the changes of vasomotion in

pressure and venous distention are responsible for the suppression of vasomotion.

The vasomotion pattern in patients with chronic venous insufficiency are not consistent. Gniadecka et al.[70] reported that in these patients vasomotion could not be detected in the region of lipoderma-tosclerosis neither in the horizontal nor in the dependent position of the lower extremity. Removal of the leg edema, that coexists with

TABLE 3 Fluxmotion in Patients with Peripheral Ischemia

Degree of Limb Ischemia	Ankle/arm Pressure Ratio[a]	Walking Distance	High Frequency Waves[a]		
			Frequency (Hz)	Amplitude[b]	Prevalence (%)
No (control)	1.22 (0.2)	—	26	0.24	8
Light	0.81 (0.1)	>200 m	23.0 (5.3)	0.27 (0.1)	33
Moderate	0.62 (0.2)	<200 m	19.9 (3.9)	0.22 (0.1)	75
Severe	0.38 (0.1)	Rest pain	22.5 (4.0)	0.19 (0.1)	92

[a] Mean with (SE).
[b] Laser Doppler blood flux units.
Modified from Bollinger, A., Hoffmann, U., Seifert, H., in Progress in Applied Microcirculation, Vol. 15, Intaglietta, M., Ed., Basel, Karger, 1989, 87–92. With permission.

peripheral ischemia was studied by Bertuglia et al.[75] in a hamster skinfold window model. These authors divided skin microvasculature according to Strahler classification,[75,76] so that Order 0 was capillaries and Order 4 the largest skin arterioles. In the baseline situation the dominant vasomotion frequency was found in Order-1 arterioles (4 to 18 cpm, mean 9.1 ± 3.9 cpm) and decreased with the increasing arteriole order, a finding in accordance with earlier observations of Colantuoni et al.[18] However, during experimentally evoked tissue hypoxia caused by the inspiration of 8 and 11% O_2 mixture, the frequency of vasomotion increased, in all arteriolar branches and the dominant frequency was generated in Order-3 arterioles (25.5 ± 4.5 cpm at 8% O_2 vs. 3.4 ± 1.8 cpm in the control). Increased frequency of vasomotion was accompanied by a decreased mean and effective vessel diameters and reduced capillary blood flow. It is probable that the laser Doppler findings of increased flux oscillation frequency during limb ischemia in humans could be explained by the increased activity of vasomotion pacemaker in larger cutaneous arteries. Such a phenomenon is likely to reduce resistance in the skin microcirculatory network and ensure adequate blood supply to the tissue.

C. Sickle Cell Disease

Cutaneous vasomotion in the patients with sickle cell disease has been investigated with the laser Doppler fluxmetry by Rodgers et al.[48] and Gniadecka et al.[77] These authors found prominent oscillations of blood flux of the period 7 to 10 s and peak-to-trough magnitudes about half the mean flow. The oscillations were apparently associated with the presence of the pathological hemoglobin S, since in two patients studied by Rodgers et al.[48] blood transfusion resulted in the diminution or disappearance of the rhythmic variations in blood flux.

The origin of enhanced vasomotion is not clear. One possibility is that sickle cells, which are more stiff and resistant to deformation than normal erythrocytes,[78,79] become trapped at the entrance to capillaries, as shown in animal models and in human retina.[80-84] This phenomenon may lead to the local increase of intravascular pressure[85] that stimulates contraction of smooth muscle in the precapillary sphincters and 'myogenic' oscillation pattern in the arterioles.[86-88] Alternatively, cyclic skin hypoperfusion and hypoxia caused by the plugging of capillaries with sickle cells may stimulate vasomotion as shown in animal models by Zweifach and Lipowsky[89] and Bertuglia et al.[75] Mechanical stimulation of endothelium by sickle cells and release of vasoactive mediators cannot be excluded.[90]

As discussed above, vasomotion significantly improves tissue perfusion. Therefore, it is likely that enhancement of vasomotion in sickle cell disease is the compensatory mechanism to upregulate nutritive blood flow in the skin. The finding that, unlikely in the normal situation, vasomotion does not disappear after postural vasoconstriction caused by the lowering of the leg[77] (Figure 4) further supports the hypothesis for the compensatory role of vasomotion in sickle cell disease.

References

1. Nicoll PA, Webb RL. Blood circulation in the subcutaneous tissue of the living bat's wing. Ann NY Acad Sci 46: 697–709, 1946.
2. Nicoll PA, Webb RL. Vascular patterns and active vasomotion as determiners of flow through minute vessels. Angiology 6: 291–310, 1955.
3. Colantuoni A, Bertuglia S, Intaglietta M. Quantification of rhythmic diameter changes in arterial microcirculation. Am J Physiol 246: H508–H517, 1984.
4. Duling BR, Gore RW, Dacey RC, Damon DN. Methods for isolation, cannulation and in vitro study of single microvessels. J Appl Physiol 241: H108–H116, 1981.
5. Bollinger A, Butti P, Barras JP, Traschler H, Siegenthaler W. Red blood cell velocity in nail fold capillaries of man measured by a television microscopy technique. Microvasc Res 7: 61–72, 1974.
6. Lindbom L, Tuma RF, Arfors KE. Blood flow in the rabbit tenuissimus muscle. Influence of preparative procedures for intravital microscopic observations. Acta Physiol Scand 114: 121–127, 1982.
7. Collin O, Bergh A, Damber JE, Widmark A. Control of testicular vasomotion by testosterone and tubular factors in rats. J Reprod Fertil 97: 115–121, 1993.
8. Damber JE, Maddocks S, Widmark A, Bergh A. Testicular blood flow and vasomotion can be maintained by testosterone in Leydig cell-depleted rats. Int J Androl 15: 385–393, 1992.

9. Bachir D, Maurel A, Portos JL, Galacteros F. Comparative evaluation of laser Doppler flux metering, bulbar conjunctival angioscopy, and nail fold capillaroscopy in sickle cell disease. *Microvasc Res* 45: 20–32, 1993.
10. Braun RD, Linsenmeier RA, Yancey CM. Spontaneous fluctuations in oxygen tension in the cat retina. *Microvasc Res* 44: 73–84, 1992.
11. Morita Tsuzuki Y, Bouksela E, Hardebo JE. Vasomotion in the rat cerebral microcirculation recorded by laser Doppler flowmetry. *Acta Physiol Scand* 146: 431–439, 1992.
12. Iversen PO. Evidence for long-term fluctuations in regional blood flow within the rabbit left ventricle. *Acta Physiol Scand* 146: 329–339, 1992.
13. Casteels R, Droogmans G, Himpens B. Excitation-contraction coupling in vascular smooth muscle cells and perivascular nerve stimulation. *J Cardiovasc Pharmacol* 6 (Suppl.): S9–S12, 1985.
14. Mulvany MJ. Functional characteristics of vascular smooth muscle. *Progress in Applied Microcirculation,* Vol. 3, Karger, Basel, 1983, 4–18.
15. Colantuoni A, Bertuglia S, Intaglietta M. The effects of alpha- or beta-adrenergic receptor agonists and antagonists and calcium entry blockers on the spontaneous vasomotion. *Microvasc Res* 28: 143–158, 1984.
16. Meyer JU, Lindblom L, Intaglietta M. Coordinated diameter oscillations at arteriolar bifurcations in skeletal muscle. *Am J Physiol* 253: H568–H573, 1987.
17. Meyer JU, Borgström P, Intaglietta M. Is vasomotion due to microvascular pacemaker cells? In *Vasomotion and Flow Modulation in the Microcirculation* (ed. Intaglietta M). *Progress in Applied Microcirculation* Vol. 5, Basel, Karger, 1989, 41–48.
18. Colantuoni A, Bertuglia S, Intaglietta M. Variation of rhythmic diameter changes at the arterial microvascular bifurcations. *Pflügers Arch* 403: 289–295, 1985.
19. Secomb TW, Intaglietta M, Gross JF. Effects of vasomotion on microcirculatory mass transport. In Vasomotion and Flow Modulation in the Microcirculation (ed. Intaglietta M). *Progress in Applied Microcirculation,* Vol. 15, Basel, Karger, 1989, 49–61.
20. Wilkin JK. Poiseuille, periodicity, and perfusion: rhythmic oscillatory vasomotion in the skin. *J Invest Dermatol* 93: 113S–118S, 1989.
21. Secomb TW, Gross JF. Flow of red blood cells in narrow capillaries. Role of membrane tension. *Int J Microcirc Clin Exp* 2: 229–240, 1983.
22. Butti P, Intaglietta M, Reimann H, Hollinger CH, Bollinger A, Anliker M. Capillary red blood cell velocity measurements in human nail fold by videodensitometric method. *Microvasc Res* 10: 1–8, 1975.
23. Fagrell B, Frontek A, Intaglietta M. A microscope-television system for studying flow velocity in human skin capillaries. *Am J Physiol* 233: H318–H321, 1977.
24. Fagrell B, Frontek A, Intaglietta M. A microscope-television system for studying flow velocity in human skin capillaries. *Am J Physiol* 246: H508–H517, 1984.
25. Bloch EH. A quantitative study of the haemodynamics in the living microvascular system. *Am J Anat* 110: 125–153, 1962.
26. Asano M, Bränamark P-I. Cardiovascular and microvascular responses to smoking in man. *Adv Microcirc* 3: 125–158, 1970.
27. Wood E, Strum RE, Sanders JJ. Data processing in cardiovascular physiology with particular reference to roentgen videodensitometry. *Mayo Clin Proc* 39: 849–865, 1964.
28. Wayland H, Johnson PC. Erythrocyte velocity measurement in microvessels by two-slit photometric method. *J Appl Physiol* 22: 333–337, 1967.
29. Intaglietta M, Silverman NR, Tompkins WR. Capillary flow velocity measurements in vivo and in situ by television methods. *Microvasc Res* 10: 165–179, 1975.
30. Slaaf DW, Rood JPSM, Tangelder GJ, Jeurens TJM, Alewijnse R, Reneman RS, Arts T. A bidirectional optical (BDO) three-stage prism grating system for on-line measurement of red blood cell velocity in microvessels. *Microvascular Res* 22: 110–122, 1981.
31. Fagrell B. Capillary dynamics in man. In *Vasomotion and Quantitative Capillaroscopy.* (eds. Messmer, Hammersen). *Progress Applied Microcirculation.* Vol. 3, Karger, Basel, 1983, 119–130.
32. Fagrell B. Microcirculation in the skin. In *Physiology and Pharmacology of the Microcirculation* (ed. Mortillaro NA). Vol. 2, Academic Press, New York, 1984, 133–180.
33. Funk W, Endrich B, Messmer K, Intaglietta M. Spontaneous arteriolar vasomotion as a determinant of peripheral vascular resistance. *Int J Microcirc Clin Exp* 2: 11–25, 1983.
34. Mahler F, Muheim MH, Intaglietta M, Bollinger A, Anliker M. Blood pressure fluctuations in human nailfold capillaries. *Am J Physiol* 236: H888–H893, 1979.
35. Holloway GA, Watkins DW. Laser Doppler measurement of cutaneous blood flow. *J Invest Dermatol* 69: 306–309, 1977.
36. Tenland T, Salerud EG, Nilsson GE, Öberg PÄ. Spatial and temporal variations in human skin blood flow. *Int J Microcirc Clin Exp* 2: 81–90, 1983.
37. Salerud EG, Tenland T, Nilsson GE, Oberg PA. Rhythmical variations in human skin blood flow. *Int J Microcirc Clin Exp* 2: 91–102, 1983.
38. Wiederhielm CA, Weston BV. Microvascular, lymphatic and tissue pressures in the unanesthesized mammal. *Am J Physiol* 225: 992–996, 1973.
39. Fagrell B, Intgalietta M, Östergren J. Relative hematocrit in human skin capillaries and its relationship to capillary flow velocity. *Microvasc Res* 20: 327–335, 1980.
40. Fagrell B, Intgalietta M, Tsai AM, Östergren J. Combination of laser Doppler flowmetry and capillary microscopy for evaluating the dynamics of skin microcirculation. *Prog Appl Microcirc* 11: 125–138, 1986.
41. Caspary L, Creutzig A, Alexander K. Biological zero in laser Doppler fluxmetry. *Int J Microcirc Clin Exp* 7: 367–371, 1988.
42. Nilsson GE, Tenland T, Öberg PÄ. Evaluation of a laser Doppler flowmeter for measurement of tissue blood flow. *IEEE Trans Biomed Eng* 27: 597–604, 1980.
43. Tooke JE, Östergren J, Fagrell B. Synchronous assessment of human skin microcirculation by laser Doppler flowmetry and dynamic capillaroscopy. *Int J Microcirc Clin Exp* 2: 277–284, 1983.
44. Svensson H, Jönsson BA. Laser Doppler flowmetry during hyperaemic reaction in the skin. *Int J Microcirc Clin Exp* 7: 87–96, 1987.
45. Hales JRS, Westerman RA, Roberts RGD, Fawcett AA, Stephens FRN. Evidence for laser Doppler discrimination between skin AVA & capillary perfusion. European Laser Doppler Users Group (ELDUG), plenary session. London, 1992.
46. Braverman IM, Keh A, Goldminz D. Correlation of laser Doppler wave patterns with underlying microvascular anatomy. *J Invest Dermatol* 95: 283–286, 1990.

47. Wilkin JK. Periodic cutaneous blood flow during postocclusive reactive hyperaemia. *Am J Physiol* 250: H767–H768, 1986.
48. Rodgers GP, Schechter AN, Noguchi CT, Klein HG, Nienhuis AW, Bonner RT. Periodic microcirculatory flow in patients with sickle-cell disease. *N Engl J Med* 311: 1534–1538, 1984.
49. Gniadecki R, Gniadecka M, Kotowski T, Serup J. Alterations of skin microcirculatory rhythmic ascillations in different positions of the lower extremity. *Acta Derm Venereol (Stockholm)* 72: 259–260, 1992.
50. Wilkin JK. Periodic cutaneous blood flow during aldehyde-provoked hyperemia. *Microvasc Res* 35: 287–294, 1988.
51. Wilkin JK. Vasomotion in the cutaneous circulation. In *Vasomotion and Flow Modulation in the Microcirculation* (ed. Intaglietta M). *Progress in Applied Microcirculation* Vol. 15, Basel, Karger, 1989, 62–74.
52. Kastrup J, Bülow J, Lassen NA. Vasomotion in human skin before and after local heating recorded with laser Doppler flowmetry. A method for induction of vasomotion. *Int J Microcirc Clin Exp* 8: 205–215, 1989.
53. Erdl R, Schnizer WE, Schöps K. Untersuchungen zur Wirkungsweise von CO_2-bädern. Messungen an der Mikrocirkulation der Haut mittels eines Laser-Doppler-Flowmeters. *Herz/Kreisl* 18: 387–391, 1986.
54. Smits TM, Aarnoudse JJ, Geerdink JJ, Zijlstra WG. Hyperventilation-induced changes in periodic oscillations in forehead skin blood flow measured by laser doppler flowmetry. *Int J Microcirc Clin Exp* 6: 149–159, 1987.
55. Scheffler A, Rieger H. Signalverlaufmuster des Laser-Doppler-Fluxes bei peripherer arterieller Verschlusskrankheit und ihre Beziehung zum systolischen Knöcheldruckindex. In: Proceedings of the 12th Annual Meeting of the Schweizerischen Gesellschaft für Microzirkulation, Bern, 1988, pp. 73–77.
56. Seifert H, Jäger K, Bollinger A. Analysis of flowmotion by laser Doppler technique in patients with peripheral arterial occlusive disease. *Int J Microcirc Clin Exp* 7: 223–236, 1988.
57. Bongard O, Fagrell B. Variations in laser Doppler flux and flow motion patterns in the dorsal skin of the human foot. *Microvasc Res* 39: 212–219, 1990.
58. Kay MS, Marple SL. Spectrum analysis. A modern perspective. *Proc IEEE* 69: 1380–1419, 1981.
59. Harris FJ. On the use of windows for harmonic analysis with the discrete Fourier transform. *Proc IEEE,* 66: 51–83, 1978.
60. Nuttall AH. Some windows with very good sidelobe behavior. *IEEE Trans Acoust Speech Signal Process* 29: 84–89, 1981.
61. Schmid-Schönbein H, Ziege S. Attractors and quasi-attractors and the assessment of fluctuations in laser-Doppler signals by spectral analysis. European Laser Doppler Users Group (ELDUG), plenary session. London, 1992.
62. Meyer JU, Intaglietta M. Measurement of the dynamics of arteriolar diameter. *Ann Biomed Eng* 14: 109–117, 1986.
63. Prony GRB. Essai experimental et analytique, etc. *Paris J de l'Ecole Polytechnique,* 1: 24–76, 1795.
64. Burkhardt P. Modification of the Prony spectral line estimator with applications to vasomotion. Ph.D. Dissertation, University of California, San Diego, 1983.
65. Hildebrand FF. Introduction to numerical analysis. McGraw-Hill, New York, 1956.
66. Pekanmäki K, Kolari PJ, Kiistala U. Laser Doppler vasomotion among patients with post-thrombotic venous insufficiency: effect of intermittent pneumatic compression. *Vasa* 20: 394–397, 1991.
67. Rowell LB. Human circulation: regulation during physical stress. Oxford University Press, Oxford, 1986, p. 416.
68. Belcaro C, Rulo A, Vaskedis S, Williams MA, Nicolaides AN. Combined evaluation of postphlebitic limbs by laser Doppler flowmetry and transcutaneous PO_2/PCO_2 measurements. *Vasa* 17: 259–261, 1988.
69. Chittenden SJ, Shami SK, Cheatle TR, Scurr JH, Coleridge-Smith PD. Vasomotion in the leg skin of patients with chronic venous insufficiency. *Vasa* 21: 138–142, 1992.
70. Gniadecka M, Gniadecki R, Serup J. Vasomotion, posture, and leg ulceration (Abst.) 2nd Annual Meeting of the European Tissue Repair Society, Malmö, Sweden. August 26–28, 1992.
71. Cheatle TR, Shami SK, Stibe E, Coleridge Smith PD, Scurr JH. Vasomotion in venous disease. *J Royal Soc Med* 84: 261–269, 1991.
72. Coleridge-Smith PD, Thomas P, Scurr JH, Dormandy JA. Causes of venous ulceration: a new hypothesis. *Br Med J* 296: 1726–1727, 1988.
73. Yanar A, Hoffmann U, Geiger M, Franzeck UK, Bollinger A. Laser-Doppler-Fluxmotion bei peripherer arterieller Verschlußkrankheit (PAVK). *Vasa* suppl. 8: 48–50, 1987.
74. Bollinger A, Hoffmann U, Seifert H. Flux motion in peripheral ischemia. In *Vasomotion and Flow Modulation in the Microcirculation.* (Intaglietta M. ed.). *Progress in Applied Microcirculation* Vol. 15, Basel, Karger, 1989, 87–92.
75. Bertuglia S, Colantuoni A, Coppini G, Intaglietta M. Hypoxia- or hyperoxia- induced changes in arteriolar vasomotion in skeletal muscle microcirculation. *Am J Physiol* 260: H362–H372, 1991.
76. Ellsworth ML, Liu A, Dawant B, Popel AS, Pittman RN. Analysis of vascular pattern and dimension in arteriolar networks of the retractor muscle in young hamsters. *Microvasc Res* 34: 168–183, 1987.
77. Gniadecka M, Gniadecki R, Serup J. Söndergaard J. Microvascular reactions to postural changes in patients with sickle cell disease. *Acta Derm Venereol* 74:191–93, 1994.
78. Kaul DK, Fabry ME, Windisch P, Baez S, Nagel RL. Erythrocytes in sickle cell anemia are heterogeneous in their rheological and hemodynamic characteristics. *J Clin Invest* 72: 22–31, 1983.
79. Chien S. Rheology of sickle cells and erythrocyte content. *Blood Cells* 3: 283–303, 1977.
80. Nagpal KC, Goldberg MF, Rabb MF. Ocular manifestations of sickle hemoglobinopathies. *Surv Ophthalmol* 21: 391–411, 1977.
81. Lipowsky HH, Usami S, Chien S. Human SS red cell rheological behavior in the microcirculation of cremaster muscle. *Blood Cells* 8: 113–126, 1982.
82. Goldberg MF. Natural history of untreated proliferative sickle retinopathy. *Arch Ophthalmol* 85: 428–437, 1971.
83. Serjeant GR. The clinical features of sickle cell disease. Elsevier, New York, 1974, pp. 208–219.
84. Klug PP, Lessin LS. Microvascular blood flow of sickled erythrocytes: a dynamic morphologic study. *Blood Cells* 3: 263–272, 1977.
85. Bonner RF, Rodgers GP, Schechter AN. Laser-Doppler measurements (LDV) of skin blood flow and number density of flowing RBCs in sickle cell patients. *Int J Microcirc Clin Exp* 3: 432, 1984.
86. Bayliss WM. On the local reactions of the arterial wall to changes of internal pressure. *J Physiol (London)* 28: 220–231, 1902.
87. Folkow B. Intravascular pressure as a factor regulating the tone of the small vessels. *Acta Physiol Scand* 17: 289–310, 1949.

88. Patterson GC. The role of intravascular pressure in the causation of reactive hyperaemia in the human forearm. *Clin Sci* 15: 17–25, 1956.
89. Zweifach BW, Lipowsky HH. Pressure flow relations in blood and lymph microcirculation. In *Handbook of Physiology* Sec. 2. (eds. Ranking EM, Michel CC). *The Cardiovascular System.* Vol. 4, Part 1. American Physiological Society, Bethesda, MD. 1984, 251–307.
90. Chien S. Rheology of sickle cells and the microcirculation. *N Engl J Med* 311: 1567–1569, 1984.
91. Schmid-Schönbein H, Klitzman B, Johnson PC. Vasomotion and blood rheology: maintenance of blood fluidity in the microvessels by rhythmic vasomotion. *Bibl Anat* 20: 138–143, 1981.
92. Papenfuss HD, Gross JF. Vasomotion and transvascular exchange of fluid and plasma proteins. *Microcirc Endothel Lymphatics* 2: 577–596, 1985.
93. Intaglietta M, Gross JF. Vasomotion, tissue fluid flow and the formation of lymph. *Int J Microcirc Clin Exp* 1: 55–56, 1982.
94. Corcoran JS, Yudkin JS. Loss of spontaneous variability of fingertip anastomotic blood flow in diabetic autonomic neuropathy. *Clin Sci* 72: 557–562, 1987.
95. Christensen NJ. Spontaneous variations in resting blood flow, postischemic peak flow and vibratory perception in the feet of diabetics. *Diabetologia* 5: 171–178, 1969.
96. Le Dévéhat C, Khodabandehlou T, Testu N. Effects of Buflomedil* on cutaneous blood flow, vasomotion, and transcutaneous oxygen pressure in diabetic patients in an open study. European Laser Doppler Users Group (ELDUG), plenary session. London, 1992.
97. Colantuoni A, Bertuglia S, Intaglietta M. Effects of anaesthesia on the spontaneous activity of the microvasculature. *Int J Microcirc Clin Exp* 3: 13–28, 1984.
98. Boegehold MA. Enhanced arteriolar vasomotion in rats with chronic salt-induced hypertension. *Microcirc Res* 45: 83–94, 1983.
99. Weiss M, Milman B, Rosén B, Eisenstein Z, Zim Lichman R. Analysis of the diminished skin erfusion in elderly people by laser Doppler flowmetry. *Age Ageing* 21: 237–241, 1992.
100. Intaglietta M. Vasomotor activity, time-dependent fluid exchange and tissue pressure. *Microvasc Res* 21: 153–164, 1981.
101. Skalak TC, Schmid-Schönbein GW, Zweifach BW. New morphological evidence for a mechanism of lymph formation in skeletal muscle. *Microvasc Res* 28: 95–112, 1984.

Chapter 17.4
Laser Doppler Imaging of Skin

Karin Wårdell and Gert Nilsson
Department of Biomedical Engineering
Linköping University
Linköping, Sweden

I. Introduction

Skin blood flow is an important parameter to record and assess in a large number of clinical and experimental settings. Peripheral vascular disease often manifests itself as a disturbance in cutaneous microcirculation, while an early result of agents irritating the skin is an elevation in its perfusion. Measurement of skin blood flow, therefore, constitutes an important diagnostic procedure in the vascular laboratory as well as in the evaluation of consumer products and potential skin irritants.

Methods for measuring skin blood flow should preferably be noninvasive, analytical, versatile, easy to use and cost effective. The noninvasiveness should, if possible, also be extended to imply that the measuring device does not have to be in physical contact with the tissue, because even the weakest external stimuli may disturb the flow conditions of the microvascular network under study. The methods should be analytical in the sense that the recorded blood flow signals may be stored on a hard disk for later analysis by means of dedicated software packages. Versatility implies that the same device should be applicable to studies of microvascular perfusion under clinical as well as laboratory conditions. Ease of use and cost-effectiveness are important features especially in clinical situations where the measurements need to be done on a routine basis. Very few of the methods described in the literature over the years fulfill all these requirements. Laser Doppler flowmetry (LDF) and laser Doppler perfusion imaging (LDI), influence the microvascular network only to a minimal extent during measurements and are possible to apply to studies of many tissues of the body. Since these technologies also are easy to use and are supported by analytical software packages, they have become increasingly important for studies of microcirculation both in the laboratory and in clinical settings.

II. Object

Skin blood flow generally possesses both substantial temporal and spatial variations. The temporal variations can be rhythmic in nature or show a more fluctuating and stoichastic pattern.[1] Depending on the architecture of the underlying microvascular network, skin blood flow shows a characteristic granular speckle pattern and large variations in perfusion can be demonstrated even at adjacent skin sites.[2] In addition, the blood flow in the skin is generally known to be compartmentalized. The superficial capillaries are perfused by slow-speed red cells that supply the tissue with oxygen and nutritive substances and remove waste metabolites. Deeper lying arteriovenous anastomosis take an active part in body temperature regulation, while small arteries and veins constitute routes for the supply and drainage of blood.

Taking all this into account, the ideal method for assessing skin perfusion should be able to capture both the temporal and spatial variations in skin blood flow as well as have a potential to separate the signals generated by the different compartments of the microvascular network. Conventional LDF can readily track fast changes in perfusion, but the small measuring volume prevents assessment of the spatial variability in skin blood flow. In order to overcome this limitation, LDI was developed in the late 1980s. With this method it is possible to create two-dimensional flow maps of a specific tissue and to visualize the spatial variation of its perfusion.

III. Methodological Principle

A. Operating Principle

The laser Doppler perfusion imager[3] is a data acquisition and analysis system that generates color coded images of the tissue perfusion (PIM 1.0 Laser Doppler perfusion imager®, Lisca Development AB, Linköping,

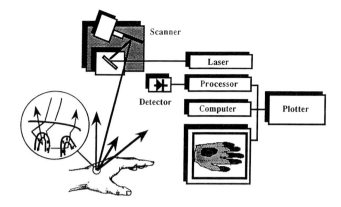

FIGURE 1 Block diagram of the LDI system. (Reproduced from Wårdell, K., Jakobsson, A., and Nilsson, G. E., *IEEE Trans. Biomed. Eng.*, 40, 309, 1993. With permission.)

FIGURE 2 The laser Doppler perfusion imaging system.

Sweden). An optical scanner comprising two mirrors, the positions of which are controlled by two stepping motors, guides a low-power He-Ne laser beam to the tissue surface (Figure 1). The mirrors move the laser beam sequentially over the tissue step by step throughout a maximum area of 12 cm × 12 cm. A scanning procedure with a maximum image format of 4096 measurement sites is completed in about 4.5 min. At each measurement position, the beam illuminates the tissue to a depth of a few hundred micrometers.[4] At each point of interaction with the moving blood cells, the light becomes spectrally broadened due to the Doppler effect. A fraction of this Doppler-broadened light is backscattered and detected by a photodetector positioned in the scanner head. The instantaneous light intensity is converted into an electrical signal and processed to form an output value proportional to the perfusion, defined as the product of the blood-cell speed and concentration. For each measurement position, the perfusion value is stored in the computer memory for further signal processing, image generation, and data analysis.

B. Instrumentation

The LDI system utilizes a 386 or 486 personal computer equipped with an analog to digital converter and a signal processing unit. The scanner head, which is mechanically fixed to the laser tube, is mounted on a flexible stand and can thereby readily be positioned over the tissue for investigation. A color plotter (HP Paint Jet, Hewlett Packard, USA) is connected to the system for hard copy documentation of perfusion images. All operations are controlled by the user from the keyboard through menu-driven software. The system is shown in Figure 2.

C. File Handling and Image Generation

During a measurement procedure, the total light intensity (TLI) and the Doppler components are sampled and stored in two separate matrices. The TLI value is used by the system to automatically determine whether the light level is adequate enough for recording of a Doppler value. If so, the Doppler signal is sampled and stored in the Doppler matrix at the position corresponding to the actual measurement site. If not, a dummy value (–1) is stored. In practice, this arrangement makes discrimination between the object and the background possible (assuming that a light absorbing material is used as the background). When a measurement is terminated and the user wants to store the captured data on the hard drive or diskette, the matrix-values are stored as files with the extensions *.dc* (TLI values) and *.dop* (Doppler values). A perfusion image is then generated by masking the Doppler values with the TLI values. The result is stored as an *.img* file which is used for image presentation and data analysis. In addition, text files containing data about the measurement are stored in connection with the measurement (*.txt*) and with the image data (*.doc*). Altogether one complete measurement generates five files which for a full format recording require about 25 kByte of memory.

D. Color Coding

During a measurement procedure each captured perfusion value is immediately color coded and an image is continuously generated on the monitor in *absolute* mode. When the measurement is completed the image is presented in *relative* mode next to the absolute mode image (Figure 3). A fixed range is used for color coding in absolute mode. While recording, the image is simultaneously processed with this fixed range set to between 0 and 10 V (the entire range of the system). In relative mode the highest of the captured perfusion values is set to 100%, and all other values are scaled relative to this maximum value. The color scale generated in both absolute and relative modes is divided into six intervals, each given a specific color in the image. All values ranging from 80 to 100% are coded

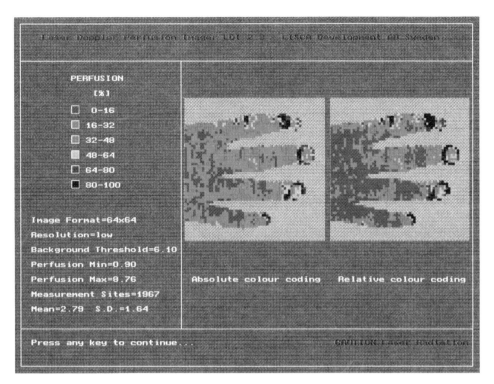

FIGURE 3 Perfusion image recorded from the dorsum of the hand in absolute and relative color coding as presented on the monitor after terminated measurement.

as red. The remaining intervals are divided into five intervals of equal width. The background is coded light gray. Information about the recording, such as the mean, standard deviation, number of captured values, time, date, and filename are presented in an information box to the left of the images. For later image processing and for the comparison of individual images in a set of recorded images, the operator may set a fixed upper and lower limit for color coding. This gives the images the same color scale and they can consequently be compared. This does not, however, change the original data.

E. Data Analysis

To convert the set of perfusion values into more quantitative parameters, data analysis functions such as perfusion value profile generation, statistics and image subtraction have been incorporated as an integral part of the system. In the perfusion value profile mode, the perfusion values along a horizontal or vertical line (the endpoints of which are set by using the cursor) can be displayed in a separate window. By using the cursor, two rectangular regions of interest can be selected from the image in the statistical mode. The mean, standard deviation, and number of samples are automatically calculated and a color frequency histogram is presented on the monitor. In Figure 4 this is exemplified by a recording from the ventral forearm. Approximately 10 min before the recording, the skin was stroked twice with a pointed instrument causing a local vasodilatation. The image subtraction mode allows for subtraction of one image from another, thereby facilitating separate visualization of the vascular response to a stimuli-response experiment.

F. Exporting Images to Other Software Packages

Images recorded by the LDI system may also be converted to TIFF or ASCII formats for further exportation to more powerful image and data analysis software packages. The TIFF format allows either gray scale or the use of the same colors as in the original perfusion image and is combined with the option of selecting whether to interpolate between measurement sites or not. Interpolation increases the resolution by a factor of three on a linear scale between two consecutive measurement sites. The TIFF formatted data set can be read by several commercially available image analysis software packages such as Global Lab® Image for Windows (Data Translation, Inc., USA) or NIH Image (Version 1.47 or higher) for the Macintosh (NIH (shareware) USA). In addition, various word processing packages such as WORD® (Microsoft Inc., USA) can read TIFF formatted files. Figure 5 demonstrates a recording performed on the dorsal side of the hand 15 min after a vasodilatating cream was applied. The same image is presented with three different fixed color scales, namely the actual minimum and maximum values (0.89 to 7.10 V) of this image (A), the entire range of the system, 0 to 10 V (B) as well as the range 0 to 5 V (C). In D through F the corresponding interpolated images are displayed. All images shown were TIFF formatted, exported to WORD® for windows and plotted on an HP Paint Jet plotter.

After converting an image to ASCII format, it can be exported to, e.g., EXCEL® (Microsoft Inc., USA), or other spread-sheet software packages. Values converted to ASCII code appear in the original format ranging from 0 to 4095. Conversion to the format used in the PIM 1.0 laser Doppler perfusion imager® (0.00 to 10.00 V) is easily performed by multiplying all values by a factor of 10/4095. When the images have been converted to ASCII format and exported to a spread-sheet software package, there are virtually an unlimited number of ways in which to analyze the images. For instance, the program can automatically analyze user-specified subsets of the image corresponding to low-perfusion and high-perfusion areas.

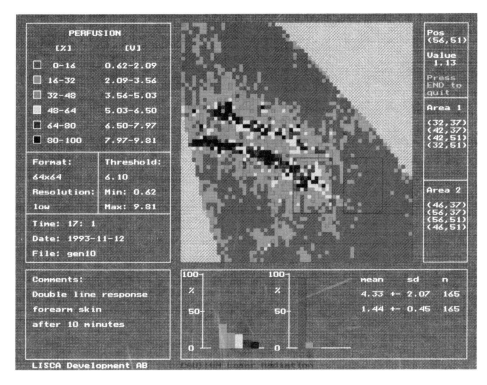

FIGURE 4 Perfusion image recording demonstrating statistical mode. The recording was performed on the ventral forearm 10 min after the skin was stroked with a pointed instrument.

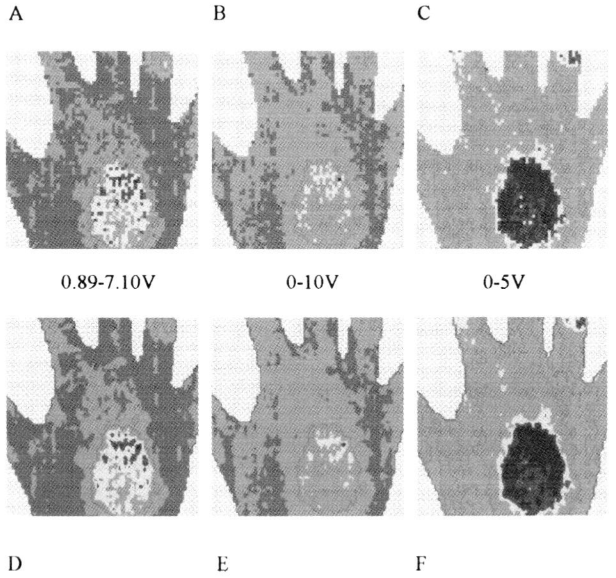

distributed over a circular skin area with a diameter of approximately 1 mm² resulting in an intensity of less than 1 mW/mm². This can be compared with the upper limit of visible light to which skin may be exposed, accumulated over a time period of 8 h during each 24-h interval. This limit has been set to 2000 W/m² (Swedish radiation protection and occupational health ordinance). Direct exposure of the retina to the laser light may, however, be harmful. Therefore the operator and the test subject should use protective glasses when there is a risk for direct eye exposure.

IV. Sources of Error

A. Design of a Measurement Procedure

Before a measurement is initiated it is advisable to establish a well-organized protocol. Such a protocol simplifies the analysis and evaluation which often take place at several points in time following image capturing. Parameters to take into consideration are: the filename, the image format, the resolution, the subject data, and the distance between the tissue and the lower part of the scanner head. The latter two may be entered as comments via the keyboard and will then appear on the screen together with the other parameters which are automatically stored after a measurement is completed. It is, however, recommended that notes be taken and kept in a separate protocol.

The test subject should be sitting or laying down in a comfortable position during the entire measurement procedure. Since the ambient temperature is known to have a substantial influence on skin blood flow, it is recommended that this be recorded. When imaging of the skin perfusion of an extremity is to be performed, this extremity should preferably be attached to, e.g., a sandbag, or an arm- or leg-rest with tape or Velcro® straps, in order to keep the tissue in the same position throughout the scanning procedure. Treat the skin area under investigation carefully and avoid scratching or applying unintentional pressure or mechanical stimuli, since even the slightest stimulation may substantially influence the perfusion.

FIGURE 5 The same image presented in three individual color ranges (0.89–7.10 V, 0–10 V and 0–5 V) (A, B, and C, respectively). The corresponding interpolated images are presented below the corresponding color ranges (D–F).

G. Safety

The LDI system design (PIM 1.0 Lisca Development AB, Sweden) complies with IEC 601-1 safety standards for medical equipment and FDA standards (21 CFR 1040–10 and 1040–11), which for laser equipment allows a total maximum output power of 1 mW. Skin exposure at this low laser intensity is considered harmless. The power emitted from the scanner head is

In order to be able to differentiate the backscattered light as either tissue or background, the object should be placed on a light-absorbing material. To simplify the identification and orientation of the perfusion image, it is useful to mark at least two to three reference points on the skin. If the measurement situation allows, use either black or green ink pen, dark paper, or tape which can be removed after the scanning procedure is completed. The advantage of using ink markers is that they can be used as a reference in long-term experiments and can easily be filled out to ensure identification of the same region of interest during the entire experimental period. When the laser beam hits the dark background or markers, the light is absorbed instead of being backscattered to the detector. The markers will appear as gray pixels (the background color) in the image. The size of the markers must be at least twice the current spatial resolution (distance between two adjacent measurement sites).

The spatial resolution depends on the distance between the scanner head and the tissue surface as well as on the system parameter "resolution" which can be set by the operator. Having the "resolution" set to "low" implies that the beam is moved two steps instead of one, prior to signal sampling. By varying these parameters, the spatial resolution may be changed within the range of approximately 0.5 mm (the distance 8 cm, resolution set to "high") to 2 mm (the distance 20 cm, resolution set to "low"). Combining these parameters with the image format, which can be varied from 2×2 up to 64×64 measurement sites, the total scanning area and thereby the total scanning time can be set to a value ranging from approximately 1 mm^2 to 150 cm^2 and 0.5 s up to 4.5 min, respectively (Table 1).

An image capturing procedure starts by positioning the lower side of the scanner head in parallel with the tissue surface of interest. This can best be achieved by placing the test subject in a comfortable position and carefully positioning the scanner head over the tissue surface while adjusting the stand. The position of the light spot on the tissue surface always indicates the center of the image. By keeping the scanner head parallel to the tissue surface, geometrical distortion of the image is reduced. Some distortion, however, is inevitable if imaging is performed on a surface with sharp curvatures.

When all parameters are set, the extension of the tissue area to be imaged may be marked by moving the beam along its boundaries. This procedure indicates the area to be imaged and allows the operator to adjust the scanner head or any of the parameters before a measurement is started. It is also recommended that the distance between the tissue and scanner head be measured and recorded. Even though the system is designed to give perfusion values independent of the distance between the scanner head and tissue within the range 12 to 22 cm, a minor dependence on this distance is inevitable in practice. If minor changes in tissue perfusion are to be studied it is therefore important that the distance be kept the same for successively recorded images.

TABLE 1 Typical Values of Scanning Parameters for Two Different Settings

	Range	
Image format	2×2	64×64
Scanning area	1 mm^2	150 cm^2
Distance	8 cm	20 cm
Spatial resolution	0.5 mm	2 mm
Scan time	0.5 s	4.5 min.

To avoid optical interference with the laser light, the ambient light must be switched off during a measurement procedure. Alternatively, the scanner head and tissue area of interest can be covered with a dark cloth or a nontransparent box. Excessive ambient light levels may render the light-absorbing reference markers placed on the skin invisible in the image. Furthermore, since ambient light levels add to the light level caused by the backscattered laser light, underestimation of the perfusion value due to the normalization procedure in the signal processor cannot be neglected if the ambient-light level is not low in comparison to the laser-light level on the detector surface.

Recordings of dark lesions or pigmented skin perfusion may require a reduced distance between the scanner head and tissue in order to increase the backscattered light level to above the threshold level for tissue-background discrimination. Alternatively this threshold level can be reduced to values below the default "background threshold" (TLI level). Reduction of the threshold level is, however, always made at the expense of a lower signal-to-noise ratio of signals from the darkest tissue spots. The lowest "background threshold" allowed in the system is set to a value above what is considered to be the acceptable noise limit. If the lesion or skin remains too dark despite optimization of the distance and "background threshold", the laser light will be absorbed in the tissue as if it were the background, and the corresponding pixels will show up as gray areas in the image.

B. Temporal Changes in Perfusion and Movement Artifacts

In LDI it is assumed that the perfusion of the tissue to be imaged is stationary and does not show any temporal variations during the time required to capture an image. If this is not the case, temporal variations will manifest themselves as a false spatial heterogeneity in tissue perfusion. Rhythmical variations in tissue perfusion such as vasomotion or a temporary reduction in flow, due to, e.g., taking a deep breath, generally show up as isolated stripes of perfusion values different from the perfusion in adjacent spots.

Isolated gross movements of the tissue will likewise appear as stripes of falsely elevated and sometimes saturated values. To avoid such artifacts it is important that the test subject sit as still as possible during the scanning procedure or in the case of animal experiments, to apply anesthesia to prevent muscle contraction and shivering.

Rhythmical movements of the whole tissue caused by breathing may give rise to a periodic pattern in the image which is partly caused by tissue movement artifacts and partly by respiratory-related changes in tissue perfusion. The respiratory-tissue movement artifacts may be particularly substantial in small animals, and a careful selection of skin areas that are only minimally affected by these movements is recommended. A practical way of evaluating the influence of tissue movements on the perfusion values which have been recorded is to place a piece of excised and nonperfused tissue on top of the normal skin and capture an image. This image will then show only the false perfusion values caused by tissue movement artifacts and a comparison with an image recorded from living tissue with the same amount of movement will reveal how severe the movement artifacts actually are.

C. Distance, Angular, and Reflection Errors

By utilizing a somewhat divergent laser beam, the system has been designed to be virtually independent of the distance between the skin surface and the scanner head. In practice the residual distance dependence within an interval of 12 to 22 cm is limited to less than 1% change in perfusion value for each centimeter change in distance between the skin surface and scanner head. At tissue-scanner head distances of less than 12 cm, the perfusion value is underestimated and at distances greater than 22 cm, measurements are generally not possible because of a too deficient light level on the detector surface. The focal point of the divergent laser beam is positioned at a distance of about 4 cm below the lower edge of the scanner head. At this point the beam diameter in air is approximately 0.3 mm. Consequently, images with the highest resolution (using the high resolution mode) and best signal-to-noise ratio can be recorded at a distance of 4 cm, albeit at the expense of an impaired compensation for distance dependence in the perfusion value.

The dependence of the angle between the light beam incident on the detector surface and a line perpendicular to the detector surface for the calculated perfusion value is automatically compensated for by the system for each measurement site. When measuring curved surfaces, the geometrical shape of the object as such does not influence the perfusion value. Surfaces with a pronounced curvature, however, have a tendency to scatter away the light due to pure surface reflection from the object, at least at the image boundaries.

A small fraction of the light beam is scattered directly on the surface of the skin due to a mismatch in the refractive indexes of air ($n = 1$) and tissue ($n = 1.5$). This surface reflection is, however, generally limited to about 5% in normal skin. If the skin is covered by an optical semitransparent material, images of the underlying tissue can still be captured, but the sensitivity is generally reduced due to scattering and absorption in the material or reflections from the surface. Imaging of skin surfaces immersed in water, where the water has been thermostated to a well-defined temperature, can be performed. A geometrical distortion due to the differences in the refractive indexes of water and air is, however, inevitable. Care must also be taken, to ensure that the water surface is kept still. If it is not, the light reflected from the water surface may include Doppler components that are difficult to separate from perfusion-generated Doppler shifts in the final perfusion image.

D. Noise Level, Electrical and Biological Zero

The noise limited resolution of the LDI system has been determined to be within 0.5% of maximum value (full scale). This resolution limit corresponds to a perfusion value of 0.05 V. Perfusion values of individual pixels below 0.05 V can therefore not be separated from noise with any statistical significance. By utilizing the system's ability to record the perfusion in a large number of adjacent points in the same image together with the averaging option in the integrated statistics mode, far smaller changes in tissue perfusion can be proved with statistical significance.

Due to the statistical nature of the measurement procedure, the "zero" level of the system is not identical for all pixels. If a static light scattering surface is imaged, the standard deviation is in the order of 0.02 to 0.04 V for a full format image. In order to render the majority of the "zero" values positive, an offset of about 0.04 V is therefore automatically added to the image.

A "biological-zero" level caused by residual movements of the arrested blood cells in the occluded tissue is superimposed on the electrical "zero" level. This is equivalent to the "biological-zero" level to which the conventional laser Doppler flowmeter output signal is reduced after occlusion.

E. Summary

The most important steps in preparing a successful image of skin tissue are

1. Use a well-designed protocol
2. Position the test subject such that gross movement artifacts can be avoided
3. Use light absorbing material as the background
4. Mark reference points on the skin surface
5. Position the scanner head in parallel with the skin surface

6. Mark a measurement area and adjust the scanner head if necessary
7. Protect the scanner head from ambient light

V. Correlation with Other Methods

LDF and LDI use a similar signal processing algorithm for the calculation of the perfusion value. LDF,[5] however, is intended for continuous recordings at a single site, whereas LDI records the perfusion within a specific tissue area. Combining the two methods, therefore, facilitates studies of both the temporal and spatial variations in skin perfusion. Thus, it is not surprising that Sefalian et al.[6] as well as Harrison and colleagues[7] have been able to demonstrate a good agreement between the two methods when studying skin perfusion. This agreement was demonstrated despite the spatial variations that exist in normal skin perfusion between adjacent skin areas. However, if the LDF-probe is moved to an adjacent site, the slope of the LDI/LDF curve can be expected to change due to the heterogeneity in skin perfusion. This has been demonstrated by several studies.[1,8] By the use of biopsy, and a special probe holder for the generation of LDF topographic maps, Braverman et al.[2] correlated different LDF perfusion patterns with vessel type. Furthermore, this LDF-mapping technique has been compared to LDI with the modification using a longer sampling time at each measurement site than in the ordinary set up.[9] Perfusion values recorded from ventral forearm skin areas with consistent high and low perfusion, coincided when measurements were made by both systems.

Harrison et al.[7] also compared LDI with thermographic mapping of the skin and found that the regional temperature profile resembled the perfusion values but the extreme heterogeneity picked up as large pixel-to-pixel variations by LDI could not be demonstrated by thermography. Thermography, however, records a parameter (skin temperature) that is only indirectly related to skin perfusion, while LDI directly senses the speed and concentration of the blood cells in the microvascular network. Diverging results may therefore be expected especially when measurements are made at the fingertips or over ulcers with a high evaporative water loss that significantly reduces the tissue temperature.

VI. Recommendations

From the experience with LDI that has been gained so far, major potential fields of application include dermatology (skin irritant and consumer product testing, evaluation of portwine stain therapy), plastic surgery (flap two-dimensional mapping), general surgery (peroperative organ imaging), vascular surgery (peripheral skin perfusion imaging), neuropathy assessment (axon reflex response), and vasoactive drug evaluation.

One of the most important features of the system is its ability to generate many different images from the same data. When the image is captured, the corresponding Doppler signals and perfusion values are calculated and stored in the computer memory. Depending on how the configuration file is set up the corresponding image can be generated in *absolute* mode (with user-selected upper and lower limits) or in *relative* mode (default). The relative mode image is a good starting point for inspecting the heterogeneity in tissue perfusion. If several images are to be compared, however, they should all have the same color scale. This is achieved by generating new images based on the same data, but with identical color scales. A good rule of thumb is to select zero as the lower limit and the highest perfusion value observed in any of the images to be compared as the upper limit.

By successively reducing the upper limit in a series of images generated from the same data, more detailed information of the low perfusion areas is revealed. This option is useful, for example, when the extent of the hyperperfused area around a point stimuli is to be investigated. At the point of stimulation the perfusion values may be so high that other areas will all be coded in more or less the same color. By reducing the upper limit, the perfusion values at the stimulation point will all be saturated, and the entire color scale will be utilized to visualize the heterogeneity of the blood flow in the surrounding skin area. Whatever the manipulation of the color scale may be, the perfusion value behind each pixel is left unchanged. Statistical procedures such as the average perfusion calculation are therefore not influenced by changing the color scale.

In many applications the vascular response to a stimuli is investigated separately. By subtracting the image captured prior to stimulation from the image captured after stimulation, changes in the two-dimensional perfusion map may be isolated and analyzed. In order to study the intensity, extension, and time course of the axon reflex induced by electrical skin stimulation, subtraction of images recorded before and after stimulation has been performed.[10] To minimize "noise" due to temporal variations in skin blood flow, all values less then 10% of the maximum value were set to the background value in the resulting image. The average perfusion in different selected areas in the resulting image was calculated and, by counting the number of nonbackground pixels, the extension of the axon reflex was computed. The method was then adapted to compare the axon reflex response on the dorsum of the feet of a group of healthy volunteers to that of a group of insulin dependent diabetics. A significant difference between the healthy and diabetic group ($p < 0.001$) was observed.[11]

In order to study dynamic changes in skin perfusion a *sequential* mode has been introduced as an integral part of the system. In this mode the image format can be selected to be 4 × 4, 8 × 8, 16 × 16, or 32 × 32 measurement sites. During the scanning procedure sequential images from the same area are successively captured, shown one after the other on the monitor, and eventually stored in the same file. The trade-off between the image format and the number of images captured per unit time allows, for instance, 48 images/min with the format set to 4 × 4 and about 1 image/min with the format extended to 32 × 32 measurement points.

In addition to being an imaging system, LDI can be regarded as a laser Doppler multipoint measuring system that allows the user to specify the number of measuring points and how frequently consecutive readings are to be made. This option implies that the primary limiting factor of conventional LDF, the small measuring volume in combination with the localized flow variations in tissue, can be effectively circumvented by the LDI system. Instead of performing a large number of time-consuming individual recordings with conventional LDF, the laser-Doppler perfusion imager simply captures a single image including both normally perfused and stimulated tissue. This image comprises a sufficient number of measurement points with which to prove differences in perfusion via statistical significance.

Since no physical contact is necessary between the scanner head and the skin, both the potential risk of influencing the skin perfusion mechanically as well as the problem with the additional risk of infection and discomfort to the patient have been effectively eliminated. Furthermore, as no dyes or tracers need to be injected in order to capture the perfusion images and no probes need to be positioned on the skin, the system is user-independent and there is virtually no way of influencing, intentionally or unintentionally, the two-dimensional perfusion value data recorded with LDI.

References

1. Tenland, T., Salerud, G., and Nilsson, G.E., Spatial and temporal variations in human skin blood flow. *Int J Microcirc Clin Exp,* 2, 81, 1983.
2. Braverman, I.M. and Schechner, J.S., Contourmapping of the cutaneous microvasculature by computerized laser Doppler velocimetry. *J Invest Dermatol,* 97, 1013, 1991.
3. Wardell, K., Jakobsson, A, and Nilsson, G.E., Laser Doppler perfusion imaging by dynamic light scattering. *IEEE Trans Biomed Eng,* 40, 309, 1993.
4. Jakobsson, A. and Nilsson, G.E., Prediction of sampling depth and photon pathlength in laser Doppler flowmetry. *Med Biol Eng Comput,* 31, 301, 1993.
5. Nilsson, G.E., Tenland, T., and Öberg, P. Å., Evaluation of a laser Doppler flowmeter for measurement for tissue blood flow. *IEEE Trans Biomed Eng,* 27, 597, 1980.
6. Seifalian, A. M., Stansby, G., Jackson, A., Howell, K., and Hamilton, G. Comparison of laser Doppler perfusion imaging, laser Doppler flowmetry, and the thermographic imaging for the assessment of blood flow in human skin. *Eur J Vasc Surg,* in press, 1993.
7. Harrison, D.K., Abbot, N.C., Swanson Beck, J., and McCollum, P.T., A preliminary assessment of laser Doppler perfusion imaging in human skin using the tuberculin reaction as a model. *Physiol Meas,* 14, 241, 1993.
8. Braverman, M.I., Keh, A., and Goldmintz, D., Correlation of laser Doppler wave patterns with underlying microvascular anatomy. *J Invest Dermatol,* 3, 283, 1990.
9. Wårdell, K., Braverman, I.M., Silverman, D.G., and Nilsson, G.E., Spatial and temporal skin perfusion studied with laser Doppler perfusion imaging. CNVD 5th international symposium, Aachen, Germany, Dec. 3–5. 1993.
10. Wårdell, K. Naver, H.K., Nilsson, G.E., and Wallin, B.G., The cutaneous vascular axon reflex in humans characterized by laser Doppler perfusion imaging. *J Physiol,* 460, 185, 1993.
11. Naver, H.K., Wårdell, K., Nilsson, G.E., and Wallin, B. G., Vascular axon reflex responses in diabetes mellitus. 17th European Conference on microcirculation, London, July, 5–10, 1992.

Chapter 17.5
The ^{133}Xenon Wash-Out Technique for Quantitative Measurement of Cutaneous and Subcutaneous Blood Flow Rates

Per Sejrsen
Department of Medical Physiology
The Panum Institute
University of Copenhagen
Copenhagen, Denmark

I. Introduction

Measurement of blood-flow rates in cutaneous and subcutaneous tissues is of interest in human physiology, pathophysiology, and in control of therapeutic effect. It is especially of interest in the understanding of the distribution of cardiac output to the skin during test, orthostatic maneuvers, and dynamic exercise and in thermoregulation. Most of the methods developed for this purpose have been qualitative of nature. The introduction of the ^{133}Xe washout method after epicutaneous labeling has made it possible to measure the cutaneous and subcutaneous blood-flow rates quantitatively during atraumatic conditions.[1]

II. Object

The purpose of the present chapter is to describe the measurement of cutaneous and subcutaneous blood-flow rates by the washout of ^{133}Xe after atraumatic local, epicutaneous labeling using external residue detection.

III. Methodological Principles

A. Physical Principles

^{133}Xe is a radioactive inert gas isotope with a physical half-life of 5.3 d. The radiations emitted by disintegration of ^{133}Xe are X-ray, and β- and γ emission. By a NaI (T1) scintillation detector coupled to a γ spectrometer, it is possible to register the ^{133}Xe γ emission of 81 keV, with an incidence of 35.5% and the X-ray of about 40 keV, with an incidence of 64.5% by setting the window to include these two energy peaks. The distance between the ^{133}Xe deposit and the detector shall be kept constant throughout the total period of registration to measure the relative washout rate. The collimation shall be so wide, that registration is obtained from the total ^{133}Xe depot area, also when this expands by diffusion to the surrounding tissue area. A suitable distance between deposit and detector is 15 to 20 cm. This distance will minimize the effect on the counting efficiency of the expansion of the ^{133}Xe deposit by diffusion, which will increase the distance between the detector and the labeled area.

Another detector suitable for registration of ^{133}Xe activity is a Cadmium Telluride (Chloride) detector (Cd Te (Cl)).[2] As this detector type can be fixed to the region by adhesive plaster keeping the counting geometry almost constant, it is a portable solution of the registration. It is important to note that the labeled area, by the short distance used with this detector type, either shall be so small that ^{133}Xe cannot leave the counting area by diffusion or so large, that a constant concentration is present in an area somewhat larger than the area of registration. The short counting distance used by this detector type has been from 1 to 20 mm, which makes it very important to correct for the elimination rate due to expansion by diffusion if present. This can be done by subtracting the elimination rate measured during blood flow cessation, from that obtained with intact blood flow.

As ^{133}Xe is a gas it is freely diffusible in the tissues, and equilibrium between tissue and blood is obtained for ^{133}Xe during the passage of blood through the tissue. This has been shown in experiments on semiisolated, autoperfused gastrocnemius muscles in cats, where the ^{133}Xe washout method was compared to the directly measured out-flow rate of blood.[3] On this basis ^{133}Xe is a suitable indicator for measurement of blood flow rates, as ^{133}Xe washout rate is proportional to blood-flow rate.

B. Atraumatic Local Labeling

Atraumatic local labeling of the tissue with ^{133}Xe can be done by application of ^{133}Xe gas or a ^{133}Xe in saline solution on the skin surface for a few minutes, e.g., 3 min.[4] In practice this is performed by the following technique. A deposit of ^{133}Xe is placed on the skin surface in a chamber formed by the skin surface and a circular gastight Mylar® membrane 3 to 8 cm in diameter and 20 μm thick. The membrane is attached to the skin surface by a ring-shaped, 0.7 to 1.5 cm wide, adhesive membrane with adhesive material on both sides, see Figure 1.[1] The dimension of the central chamber will then be from 1.6 to 5 cm in diameter. A thin injection needle is placed between the Mylar and the adhesive membrane leading from the outside into the chamber. From a syringe it is then possible to introduce a deposit of ^{133}Xe gas or ^{133}Xe in isotonic saline solution into the chamber. After labeling by diffusion from the deposit on the skin surface into the skin for about 3 min, the deposit is redrawn to the syringe. The membrane with adhesive ring, needle, and syringe is removed, the region is dabbed with a piece of soft tissue paper, and the surplus of ^{133}Xe blown away.

C. Registration and Data Management

The registration of the ^{133}Xe activity is then performed as a residue detection by external counting in time intervals of e.g., 20 s, (from 1 s to 1 min) dependent on the purpose of the measurement and the level of activity. The data obtained is then plotted in a semilogarithmic diagram after subtraction of the background activity. The x axis is time in a linear scale, and the y axis is the activity in a logarithmic scale.

The washout of ^{133}Xe after atraumatic local labeling of a skin area shows a biexponential course, see Figure 2. This is due to diffusion of ^{133}Xe from the cutaneous venous blood out through the walls of the venous vessels during the passage of this blood stream through veins located in the subcutaneous tissue. ^{133}Xe has about 10 times higher solubility in subcutaneous adipose tissue than in blood. This has the effect that the subcutaneous tissue acts as a zinc for ^{133}Xe, resulting in an accumulation of ^{133}Xe in the subcutaneous tissue with time. This is illustrated in Figure 3a and b, showing the distribution of ^{133}Xe in the tissue after atraumatic local labeling by autoradiographic technique. After 2 min of *in vivo* washout the ^{133}Xe is located almost exclusively in the cutaneous tissue and after 70 min almost exclusively in the subcutaneous tissue. The result of the very limited transport by diffusion alone without blood flow after 70 min is shown in figure 3c.[5] A diffusion of ^{133}Xe directly from cutaneous to subcutaneous tissue over the contact area between these two tissues thus seems of lesser importance than the above-mentioned transport by convection with venous blood flow combined with diffusion out through the venous vessel walls and into the surrounding subcutaneous tissue. A transport in the opposite direction from subcutaneous tissue to cutaneous tissue later in the washout process seems to be negligible due to the following reasons. The very high solubility of ^{133}Xe in subcutaneous tissue compared to that in blood and cutaneous tissue will counteract an exchange by diffusion from the subcutaneous tissue. The higher linear velocity of blood in the arterial vessels, and the lower contact area between blood and tissue in these vessels compared to that of the venous vessels, will also minimize the exchange between the subcutaneous and the cutaneous tissues.

The *biexponential washout of ^{133}Xe* is, on the abovementioned conditions, a combined washout curve including an initial, fast washout component from the cutaneous tissue, and an accumulation in the subcu-

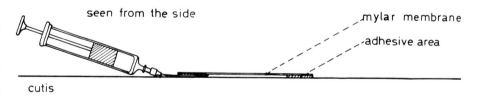

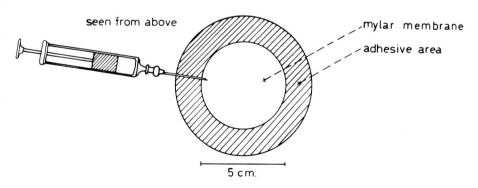

FIGURE 1 Technique for epicutaneous application of ^{133}Xe gas or ^{133}Xe dissolved in isotonic saline. (From Sejrsen, P., *J. Appl. Physiol.*, 24, 570, 1968. With permission.)

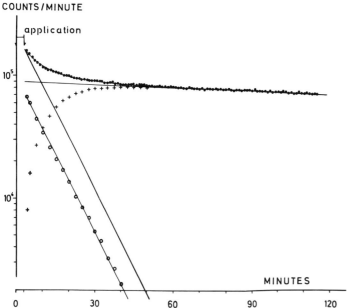

FIGURE 2 Washout curve after epicutaneous labeling with ^{133}Xe for 3 min on the lateral side of crus. **Solid circles** show the registered activity with time. The **open circles** show the result of graphic curve resolution. The mathematical expression of the two exponentials obtained by the graphic curve resolution are presented in Equation 8. The washout curve separate for the cutaneous tissue is constructed by drawing a line parallel to the straight line through the open circles from the top of the registered curve. By subtracting the values given by this line from those of the registered curve the separate curve for the subcutaneous tissue is constructed, here denoted by crosses. The mathematical expressions of these separate washout curves for the two tissues are given in Equation 7. (From Sejrsen, P., *Circ. Res.*, 25, 215, 1969. With permission.)

taneous tissue, followed by a washout from this tissue. The accumulation in the subcutaneous tissue is determined by the convective transport with the cutaneous venous blood. Thus the registered curve contains only two washout rate constants: the cutaneous and the subcutaneous, and by a graphic curve resolution these two rates can be obtained, see Figure 2.[5]

In a special region of the skinfold between the extended thumb and the forefinger it has been possible to measure the washout of ^{133}Xe separately from cutaneous tissue. This was done after atraumatic local labeling with ^{133}Xe gas of the region and a shielding of the rest of the hand by a 3 mm thick lead shield. By registration of the activity from the distal, unshielded 3 to 4 mm of the skinfold, being solely cutaneous tissue, a monoexponential washout curve was obtained over 3.5 decades, see Figure 4. A similar result has been obtained for a skinfold raised on the back of the hand.

A monoexponential washout of ^{133}Xe from subcutaneous tissue has also been observed. After an atraumatic local labeling of subcutaneous fatty tissue in an autoperfused inguinal fat pad preparation in cats a monoexponential washout was followed over 2.5 h, see Figure 5.[5]

D. The Washout Model

On the basis of these observations of monoexponential washout of ^{133}Xe from cutaneous and subcutaneous tissues the following combined in-series and in-parallel washout model is described.[5] The model assumes that under steady state conditions, a constant fraction, **E,** of the ^{133}Xe in the cutaneous venous blood is extracted, as it passes through the subcutaneous tissue, due to the tenfold higher solubility of ^{133}Xe in this tissue than in blood.

The model consists of two homogeneous compartments symbolized by C and S for the cutaneous and

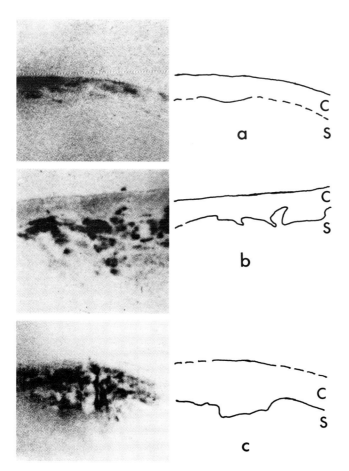

FIGURE 3 Radioautograms of cutaneous (C) and subcutaneous (S) tissues. The tissue boundaries are illustrated in the schematic drawings to the right, (a and b) were taken after epicutaneous labeling with ^{133}Xe gas for 3 min. (a) was taken after 2 min of *in vivo* washout, (b) after 70 min. (c) is the distribution of ^{133}Xe in the tissue after intracutaneous injection of 0.1 ml of ^{133}Xe in isotonic saline taken 70 min later demonstrating the minimum of exchange by diffusion between the cutaneous and the subcutaneous tissue without blood flow. The exposure time of the film emulsion was 20 h in a, and 120 h in b and c. (From Sejrsen, P., *Circ. Res.*, 25, 215, 1969. With permission.)

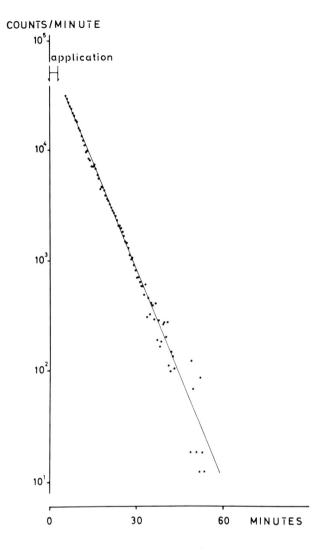

FIGURE 4 Washout curve separate from cutaneous tissue after epicutaneous labeling with ^{133}Xe gas in 3 min. (From Sejrsen, P., *Circ. Res.*, 25, 215, 1969. With permission.)

the subcutaneous compartments, respectively. The input to the system is in the form of an impulse into C. With the initial amount of 1 U the output from C is divided into two fractions: (1) the extracted fraction, E, and (2) the transmitted fraction, $1 - E$. The extracted fraction, E, of the output from C is reaching the subcutaneous compartment, and thus C is in-series with S for this fraction. The complementary fraction, $1 - E$, is transmitted solely through the vascular volume by the flowing blood. The transport of this fraction is thus an output from C, which is in-parallel with the output from S, see Figure 6.

The total amount 1 is initially located in compartment C and the cumulative output from C at time t is called H_1. The amount retained in compartment C at time t, R_c, can then be written as $R_c = 1 - H_1$. It is observed, that compartment C is a well-mixed compartment with an exponential washout. The expression

$$R_c = 1 - H_1 = e^{-k_c \cdot t} \quad (1)$$

can then be written for compartment C, where k_c is the elimination rate constant for C.

At time zero there is no indicator in compartment S. The rate of input of tracer to S is a constant fraction, E, of the output rate from C. This can be obtained by differentiation of H_1 with change of the sign. The input rate to S, I_s, is then

$$I_s = E \cdot k_c \cdot e^{-k_c \cdot t} \quad (2)$$

It is assumed, that the output from S follows a monoexponential function, which can be described by $k_s \cdot e^{-k_s \cdot t}$ for 1 U of indicator. Corresponding to the input rate to S, given by Equation 2, the output, O_s, from S is the input rate, I_s, convoluted by the impulse response for S:

$$O_s = \int_0^t E \cdot k_c \cdot e^{-k_c \cdot \tau} \cdot k_s \cdot e^{-k_s \cdot (t-\tau)} d\tau \quad (3)$$

$$O_s = \frac{E \cdot k_c \cdot k_s}{k_c - k_s} \cdot \left(e^{-k_s \cdot t} - e^{-k_c \cdot t} \right) \quad (4)$$

A combined expression for the amount of indicator contained in C plus S, R_c plus R_s is called $Q(t)$:

$$Q(t) = R_c + R_s \quad (5)$$

By inserting Equation 1, 2, and 4 in Equation 5 the following expression is obtained:

$$Q(t) = R_c + \int_0^t I_s dt - \int_0^t O_s dt \quad (6)$$

$$Q(t) = e^{-k_c \cdot t} + \left[\frac{E \cdot k_c}{k_c - k_s} \cdot \left(e^{-k_s \cdot t} - e^{-k_c \cdot t} \right) \right] \quad (7)$$

$$Q(t) = \left(1 - \frac{E \cdot k_c}{k_c - k_s} \right) \cdot e^{-k_c \cdot t} + \frac{E \cdot k_c}{k_c - k_s} \cdot e^{-k_s \cdot t} \quad (8)$$

Thus it is possible by graphic curve resolution to determine the rate constants for the cutaneous and the subcutaneous components. Furthermore it is possible to calculate the E fraction from the intercepts of the two curves and their rate constants. The rate constants are the blood-flow rate to partition coefficient ratios for the two tissues. On average E was observed to be 0.50 in ten washout experiments on the lateral side of the lower leg.

E. Calculation of Blood-Flow Rates

Blood flow rates can be calculated from the rate constants using the equation introduced by Kety[6,7]

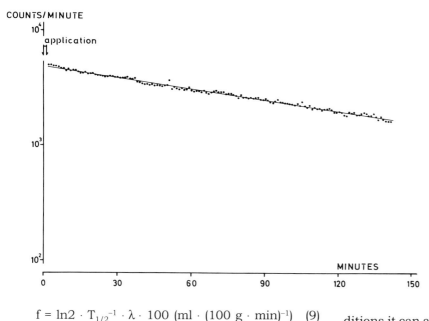

FIGURE 5 Washout curve separate from subcutaneous tissue after local labeling with ^{133}Xe gas in 1 min. (From Sejrsen, P., *Circ. Res.*, 25, 215, 1969. With permission.)

$$f = \ln 2 \cdot T_{1/2}^{-1} \cdot \lambda \cdot 100 \ (\text{ml} \cdot (100 \ \text{g} \cdot \text{min})^{-1}) \quad (9)$$

where ln2 is the natural logarithm to 2, λ is the tissue to blood partition coefficient for ^{133}Xe in milliliters per gram being 0.7 for cutaneous and 10 for subcutaneous tissue, respectively.[1] $t_{1/2}$ is the half-time of the two monoexponential components as obtained by graphic curve resolution for the two tissues. The factor 100 is introduced to give the results per 100 g of tissue which is the conventional term.

F. Loss of ^{133}Xe from the Skin Surface

The question concerning loss of ^{133}Xe by diffusion out through the intact skin surface during the washout period has been elucidated by experiments with blood-flow cessation. This was done by a cuff placed on the upper arm and inflated to a pressure of 230 mmHg, a pressure chosen well above the systolic pressure. Under this condition the very slow elimination rate is solely due to a loss of ^{133}Xe out through the epidermal membrane, see Figure 7.[4] This has been demonstrated by placing a gastight Mylar membrane, 20-μm thick, over the deposit area with a drop of water interposed. After this gastight sealing of the surface the curve has an almost horizontal course with a decline equal to the physical decay. Without this gastight sealing of the surface the observed very slow elimination rate is about 1 to 2% of that observed during intact blood flow at normal, thermoneutral conditions (Figure 7). However under sweating conditions it can account for as much as 20 to 25% of the measured elimination rate of ^{133}Xe from a cutaneous deposit, as measured just after the end of the labeling period. In such situations a correction can be performed to give the cutaneous blood flow rate. This can be done by subtraction of the elimination rate measured during blood flow cessation from that obtained for cutaneous tissue during intact blood flow after curve resolution.

Thus the elimination rates are in the order of 10^{-4} min^{-1} due to the physical decay of ^{133}Xe, 10^{-3} min^{-1} due to diffusional loss of ^{133}Xe out through the intact epidermal membrane, 10^{-2} min^{-1} due to profuse sweating, and from 0.07 to 0.7 min^{-1} due to cutaneous blood flow, corresponding to blood-flow rates of about 6 to 50 ml · (100 g · min)$^{-1}$ at normal conditions and at local heating to 45°C, respectively.[5,8] On this basis the errors due to physical decay and loss of ^{133}Xe out through the intact epidermal membrane are considered negligible when calculating blood-flow rate from ^{133}Xe washout curves. Blood-flow rates in subcutaneous tissue are at normal conditions and during heating of the skin surface to 45°C measured to about 3 and 50 ml · (100 g · min)$^{-1}$, respectively.[1,8]

IV. Sources of Error

The loss of ^{133}Xe out through the skin surface is, as mentioned, negligible with an intact epidermal membrane, but correction can be necessary during sweat secretion. The gastight epidermal membrane can be removed by 40 to 50 times of stripping with adhesive plaster. This procedure removes the dry part, of the epidermis, stratum corneum, which normally has a water content of about 4%. By this procedure the basal, living cell layers in stratum germinativum is left *in situ*. After 40 or 50 times of stripping with adhesive

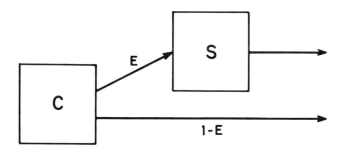

FIGURE 6 Schematic diagram of the washout model. (C) is the cutaneous and (S) the subcutaneous component. (E) is the extracted fraction, which is washed out via the subcutaneous tissue. (1 − E) is the complementary fraction, which is washed out from the cutaneous tissue and transmitted by convection through the vascular volume. (From Sejrsen, P., *Circ. Res.*, 25, 215, 1969. With permission.)

FIGURE 7 Washout of ^{133}Xe after epicutaneous labeling with and without blood-flow cessation. (From Sejrsen, P., *Circ. Res.*, 25, 215, 1969. With permission.)

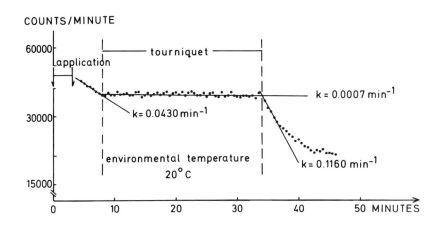

plaster the loss of ^{133}Xe from the skin surface will increase about 30 to 40 times.[9]

Epidermal desquamation or denudation due to pathological processes therefore can give rise to a severe loss of ^{133}Xe out through the skin surface. However, by placing a 20-μm thick Mylar membrane on the skin with a drop of water interposed, it is possible to reestablish a gastight sealing of the surface. Steady state of blood flow is an assumption for the method. This is the reason for a demand of constant thermal conditions during a measurement. Also traumatic influence has to be avoided. The reason for not using a labeling technique with intracutaneous injection of ^{133}Xe dissolved in isotonic saline is just the trauma effected by the injection leading to hyperemia in the following 10 to 30 min.

An uptake of ^{133}Xe in rubber and plastic materials has been observed. When a cadmium telluride detector mounted with a rubber cap is placed in contact with the labeled skin area, then the ^{133}Xe uptake in the rubber invalidates the method. The use of standard values for the tissue to blood partition coefficient, λ, for cutaneous and subcutaneous tissues are presumably acceptable in many regions. However in regions with a thin subcutaneous layer the lipid contents can be reduced, and a lower value has to be used for this tissue.

In such a region it is possible to make an estimate of the relative contents of lipid, water, and protein in a tissue biopsy, and from these values, and the corresponding values for the blood, to calculate a λ value for the tissue in question.

V. Correlation with Other Methods

Other methods have been employed in attempts to measure cutaneous blood-flow rates. ^{85}Krypton washout has been used with registration of the β activity by a Geiger-Müller tube after an intraarterial injection of ^{85}Kr dissolved in isotonic saline.[10] This method is invalidated by diffusion processes in the tissue due to the short half-value thickness in the tissue of the emitted β radiation, only 0.25 mm.[11]

Other radioactive isotopes have been used such as ^{24}Na,[6,7,12] ^{131}Iodine,[13] ^{125}I-antipyrine, and ^{131}I-antipyrine[14] given as local injections with the isotopes dissolved in isotonic saline. The problems have been the trauma of injection, and that equilibrium between tissue and the flowing blood cannot be obtained for most of these indicators, as they are not freely diffusible in the tissues.

Helium uptake through the skin has been used on the extremities. This method underestimates cutaneous blood-flow rate due to the existence of the epidermal diffusion barrier to gases.[15,16] In accordance with this valuation, values obtained by this method have been low, about 3 to 4 ml · (100 g · min)$^{-1}$.

Heat conduction has been employed as a qualitative measure of cutaneous blood-flow rate.[17,18] The loss of heat to the surroundings invalidates this method as a quantitative method. From the amount of heat dissipated from the skin a blood-flow rate in the order of magnitude 2 to 10 ml · (100 g · min)$^{-1}$ seems likely.[19,20]

Venous occlusion plethysmography can only give a rough evaluation of the cutaneous blood-flow rate, as it is based on measurements of blood-flow rates before and after iontophoresis of epinephrine into the skin combined with complicated subtraction procedures. These procedures are necessary to correct for blood flow in subcutaneous and muscle tissue.[21,22] By using values of the ratio between the weight of the cutaneous and subcutaneous tissues and the blood-flow rate in subcutaneous tissue a rough estimate gives a cutaneous blood-flow rate in the order of 6 to 9 ml · (100 g · min)$^{-1}$.

Measurements with laser Doppler technique are determined both by the velocity and the contents of the red blood cells in the tissue causing this method to be of a semiquantitative nature. The change in blood cell contents in the vessels during orthostatic maneuvers and heat stress limits use of this method.[23]

VI. Recommendations

It is recommended to make control experiments with blood-flow cessation to exclude loss of ^{133}Xe from the skin surface or change in counting geometry during the registration. By this type of measurement it is possible to estimate a rate constant for loss of ^{133}Xe to the surrounding air and thereby to make a correction for this non-blood-flow dependent elimination. As the ^{133}Xe washout method with graphic curve resolution is based on the assumption of a steady-state blood flow rate it is important to maintain temperature in the body and in the surroundings constant during the registration. Also the body and the region under study shall be kept in a constant position during the measurement to obtain steady state conditions.

Changes in blood-flow rate in cutaneous tissue during the measurement can be registered quantitatively by the ^{133}Xe washout method in the skinfold on the hand. A similar possibility is present for the subcutaneous tissue in the later part of the washout curve, which is separate from this tissue. It is important to use a sufficiently high ^{133}Xe activity and to follow the washout for a sufficiently long time (1.5 to 2 h) to get an acceptably low standard deviation of the subcutaneous washout rate. This is a necessary basis for a reasonable graphic curve resolution and determination of the cutaneous washout rate.

References

1. Sejrsen, P., Measurement of cutaneous blood flow by freely diffusible radioactive indicators, *Dan Med Bull*, (Suppl.) 18, 1, 1971.
2. Bojsen, J., Staberg, B., and Kølendorf, K., Subcutaneous measurements of ^{133}Xe disappearance with portable CdTe(Cl) detectors: elimination of interference from combined convection and diffusion, *Clin Physiol*, 4, 309, 1984.
3. Sejrsen, P., and Tønnesen, K.H., Inert gas diffusion method for measurement of blood flow using saturation techniques: comparison with directly measured blood flow in isolated gastrocnemius muscle of the cat, *Circ Res*, 22, 679, 1968.
4. Sejrsen, P., Atraumatic local labelling of skin by inert gas: epicutaneous application of xenon-133, *J Appl Physiol*, 24, 570, 1968.
5. Sejrsen, P., Blood flow in cutaneous tissue in man studied by washout of xenon-133, *Circ Res*, 25, 215, 1969.
6. Kety, S.S., Quantitative measurement of regional circulation by the clearance of radioactive sodium, *Am J Med Sci*, 215, 352, 1948.
7. Kety, S.S., Measurement of regional circulation by the clearance of radioactive sodium, *Am Heart J*, 38, 321, 1949.
8. Jaszczak, P. and Sejrsen, P., Determination of skin blood flow by ^{133}Xe washout and by heat flux from a heated tc-P$_{o2}$ electrode, *Acta Anaest Scand*, 28, 482, 1984.
9. Sejrsen, P., Epidermal diffusion barrier to Xenon-133 in man and studies of clearance of Xenon-133 by sweat, *J Appl Physiol*, 24, 211, 1968.
10. Jacobsson, S., Studies of the blood circulation in lymphoedematous limbs, *Scand J Plast Reconstruct Surg*, (Suppl.) 3, 1, 1967.
11. Sejrsen, P., Diffusion processes invalidating the intra-arterial krypton-85 beta particle clearance method for measurement of skin blood flow in man, *Circ Res*, 21, 281, 1967.
12. Braithwaite, F., Farmer, F.T., and Herbert, F.I., Observations on the vascular channels of tubed pedicles using radioactive sodium, III, *Brit J Plast Surg*, 4, 38, 1951.
13. Alpert, J.S. and Coffman, J.D., Effect of intravenous epinephrine on skeletal muscle, skin, and subcutaneous blood flow, *Am J Physiol*, 216, 156, 1969.
14. Kövamees, A., Skin blood flow in obliterative arterial disease of the leg, *Acta Chir Scand*, (Suppl.) 397, 1, 1968.
15. Behnke, A.R. and Willmon, T.L., Cutaneous diffusion of helium in relation to peripheral blood flow and the absorption of atmospheric nitrogen through the skin, *Am J Physiol*, 131, 627, 1940–41.
16. Klocke, R.A., Gurtner, G.H., and Farhi, L.E., Gas transfer across the skin in man, *J Appl Physiol*, 18, 311, 1963.
17. Hensel, H., Messkopf zur Durchblutungsregistrierung an Oberflächen, *Arch ges Physiol*, 268, 604, 1959.
18. Golenhofen, K., Die Hautdurchblutung des Menschen — Möglichkeiten zur Objektivierung von Hautreaktionen, *Fette. Seifen Anstrichmittel*, 3, 177, 1968.
19. Hardy, J.D. and Soderstrom, G.F., Heat loss from the nude body and peripheral blood flow at temperatures of 22 to 35°C, *J Nutr*, 16, 493, 1938.
20. Stewart, H.J. and Evans, W.F., The peripheral blood flow under basal conditions in normal male subjects in the third decade, *Am Heart J*, 26, 67, 1943.
21. Cooper, K.E., Edholm, O.G., and Mottram, R.E., Blood flow in the skin and muscle of the human forearm, *J Physiol* (London), 128, 258, 1955.
22. Kontos, H.A., Richardson, D.W., and Patterson, J.L., Blood flow and metabolism of forearm muscle in man at rest and during sustained contraction, *Am J Physiol*, 211, 869, 1966.
23. Klemp, P. and Staberg, B., The effect of antipsoriatic treatment on cutaneous blood flow in psoriasis measured by ^{133}Xe washout method and laser Doppler velocimetry, *J Invest Dermatol*, 85, 259, 1986.

Chapter 17.6
Doppler and Duplex Scanning in Venous Disease

Gianni Belcaro and Andrew Nicolaides
Irvine Laboratory, Vascular Surgery
St. Mary's Hospital Medical School
London, U.K.

I. Introduction

Chronic venous insufficiency (CVI) may be the result of outflow obstruction, reflux, or a combination of both. The first aim of the clinical examination is to detect whether obstruction or reflux is present. The second aim is to define the anatomic localization of the abnormality and finally the problem of quantification of the reflux or obstruction must be addressed.

In evaluating venous stasis, noninvasive tests combine physiologic and imaging techniques. These tests are widely available, simple, quick, and cost effective. Therefore, they are the methods of choice for initial objective evaluation.

It should be noted that different tests provide answers to different questions. Many of these have been established during the past decade. Most recently, however, technologic advances and current thinking indicate that the optimum useful information can be obtained using only three instruments:

1. Pocket Doppler
2. Duplex scanning device, preferably with color-flow imaging
3. Air plethysmography

II. Tests for Venous Reflux

Venous reflux is the result of gravity drawing the venous bloodstream distally. Therefore, reflux testing should be performed with the patient standing. Recent studies have determined that venous reflux detected in the supine position is frequently abolished when the patient is standing. This is because valve closure occurs only after reflux exceeds a critical flow velocity. This is more easily achieved with the patient standing rather than supine. When the patient is standing, it is important to avoid muscular contractions. Therefore, the patient should be examined holding onto a frame or table. The leg to be examined should be relaxed with the knee slightly flexed, with the weight on the opposite leg. Studies have shown that during full knee extension, an occlusion of the popliteal vein may occur in some 20% of healthy people.

After the initial physical examination, a pocket Doppler instrument is used. Pocket instruments are satisfactory to complement the physical examination as a screening test for outpatients.[1-3] The continuous-wave instrument provides information about reflux at the saphenofemoral and saphenopopliteal junctions.[2]

The knee of the leg to be examined should be slightly flexed to relax the muscles and the skin over the area under examination, i.e., the popliteal fossa. Manual calf compression produces cephalad flow and reflux generally occurs when the compression is released (Figure 1). Abolition of the reflux by compression of the superficial veins — with a finger or with a tourniquet — just below the probe suggests that reflux is confined to the superficial system. Failure to abolish reflux by such a maneuver indicates that the reflux is in the deep system.[4]

In experienced hands even a pocket Doppler provides clear answers regarding the presence or absence of reflux at the saphenofemoral or/and saphenopopliteal junctions in more than 90% of patients.[5] Abnormal anatomy in the popliteal fossa (Figure 2) is responsible for most of the errors (8%).

In some cases reflux in the gastrocnemius veins may be interpreted as reflux in the popliteal vein or in the short saphenous vein.

Furthermore the continuous-wave Doppler venous examination is not accurate in localizing incompetent perforating veins.

Duplex scanning supplements the physical examination and the preliminary evaluation with the pocket continuous-wave Doppler instrument. The duplex scan provides information about reflux in the specific veins the examiner wants to evaluate. For example, the common and superficial femoral, popliteal, deep calf veins, and perforating veins can be individually sampled. The use of color has made duplex scanning much faster and more accurate especially in areas such as the popliteal fossa where many different venous trunks may be responsible for reflux. Sampling, with

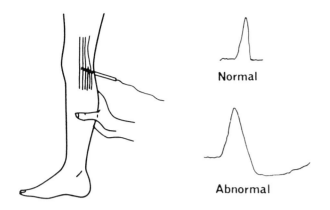

FIGURE 1 Pocket Doppler evaluation of the popliteal fossa. The patient is standing. A compression release maneuver reveals monodirectional venous flow (above tracing) or a bidirectional venous flow (below tracing) indicating reflux on the release of compression. If compressing the superficial system, the reverse flow is abolished and the reflux is localized in the superficial veins. Failure to abolish reflux by superficial compression indicates the presence of deep reflux.

duplex, each vein for reflux may be very time consuming and even small movements of the leg may displace the focus of the ultrasound sampling beam. With color duplex a simple manual calf compression-release maneuver may show, in real time, all venous structures and reveal the sites of significant incompetence.

As in the examination of a patient with the continuous-wave instrument, the patient is examined standing. The non-weight-bearing lower extremity is evaluated and the sites to be interrogated are imaged with a 5 or 7.5 MHz probe. The saphenofemoral junction, the popliteal venous anatomy, and the perforating veins are visualized. Manual calf compression or, ideally, compression by a rapidly deflatable cuff is used. The cuff inflation produces cephalad flow. Rapid release of the compression is essential to test for reflux and valve closure and can be documented.

Figure 3 shows the sapheno-popliteal junction. The blue (cephalad flow) is seen on calf compression while the reverse red flow indicating reflux and incompetence is seen — i.e., at the level of the short saphena or, if incompetent, within the popliteal vein.

Figure 4 shows an example of testing for perforating veins. Localization of calf perforating veins and reflux in them is time consuming with conventional duplex, but color-flow imaging has made this examination practical. The varicose vein is followed proximally or distally with the color probe to the point where the communicating vein crosses the fascia connecting the superficial and deep systems. At this point gentle distal compression will show the bidirectional flow in the communicating vein.

Duplex and color scanning for localization of sites of reflux are particularly useful in patients with recurrent varicose veins after previous surgery. Such examination also confirms the normal function of deep veins and the extent and site of venous reflux when it is present. Both localized and generalized reflux (e.g., whether it is present throughout the deep venous system) can be identified.

Although quantification of reflux with duplex in individual veins is possible, this is very time consuming.[6] Accurate and reproducible results are obtained more easily for the whole leg using air plethysmography which has become the test of choice for quantitating reflux. As an alternative ambulatory venous-pressure measurements can be used to quantify he degree and level of venous reflux.

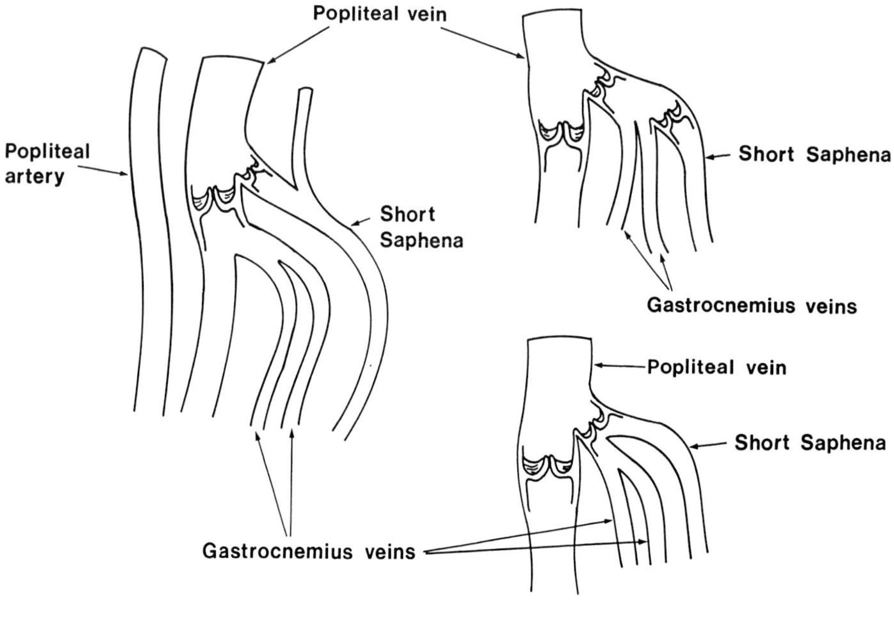

III. Ultrasound Evaluation of Venous Obstruction

Ascending venography remains the standard method of delineating persistent venous obstruction. There are several noninvasive tests that determine the presence and quantity the degree of outflow obstruction (Figures 5 and 6).

However, the simple evaluation with continuous-wave Doppler

FIGURE 2 Three different but common anatomic variations of the veins in the popliteal fossa. Duplex scanning reveals the anatomy and enables testing of each individual vein for reflux. Color duplex makes this examination faster.

FIGURE 3 Example of saphenopopliteal reflux. Blue color on calf compression indicates cephalad flow. Reflux is indicated by the red color on release of compression.

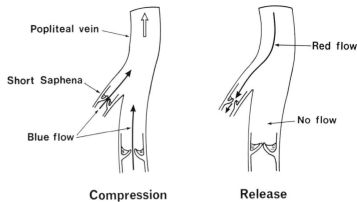

can be used as a screening device in outpatients. A history of deep-venous thrombosis or persistent leg and ankle swelling indicates the need for such an examination.[14]

Although the best test to evaluate chronic venous obstruction is the arm-foot pressure differential developed by Raju,[15] other non-invasive tests based on the measurements of venous outflow by various techniques using different instrumentation (i.e., strain-gauge and air plethysmography) can be used. Air plethysmography is now considered the noninvasive method of choice.

A. Ultrasound Techniques: Continuous Wave Doppler and Duplex Scanning

The patient is first examined with the legs horizontal and the knee slightly flexed. The trunk should be at 45° and the ultrasound probe is held over the femoral vein.[16] Flow velocity is normally phasic with respiration. If this is found, the finding indicates a normally patent iliocaval segment. Absence of phasic flow or the finding of flow that is continuous and not affected by respirations suggests obstruction or venus compression. If flow is diminished or abolished by compression of the contralateral groin or suprapubic area, the presence of obstruction and collateral circulation is suspected. Augmentation of the velocity in the common femoral vein by calf compression indicates absence of popliteal and femoral venous obstruction. This maneuver can be repeated with occlusion of the long saphenous vein at the knee by external pressure. This double checks the patency of the popliteal vein. Augmentation of the velocity in the popliteal vein produced by digital compression of each venous compartment in the leg suggests patent axial deep calf vein flow.[17]

Duplex scanning and color-flow imaging detect with great accuracy, the particular veins containing organized thrombus that are not compressible by probe pressure. The transversal compression of the vein (particularly the common and superficial femoral, popliteal, and soleal sinuses) indicates absence of thrombi within the lumen. More experience and scanning skill is required to evaluate the iliac veins and the cava. This can be easily achieved with a 2.5 to 3.5 abdominal scanning probe. Such duplex visualization of the deep veins may also reveal irregular vein walls with abnormal echo and partially recanalized lumens.[18,19]

A B-mode scanner is very often good enough to evaluate the compressibility of the major deep venous trunks. The use of a color duplex makes the test faster. It also shows very clearly nonocclusive parietal thrombi and more clearly the distal venous bed (i.e., tibial veins) and the abdominal venous system.

In our experience we perform the test with a pocket Doppler first and then with a color duplex scanner. The test is performed both with the patient in the supine position and in a standing position as described.

IV. Conclusion

It is now possible to detect the presence or absence of reflux and/or obstruction in the superficial or deep venous circulation noninvasively using Doppler and duplex ultrasound scanning.

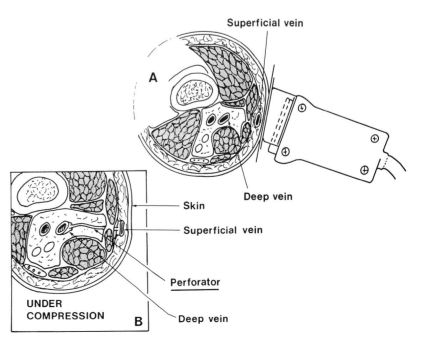

FIGURE 4 Transverse scan of superficial and deep veins (A) by moving the probe up or down the limb with continuous visualization of the two veins, the presence and level of communicating vein is defined (B). The direction of flow with calf compression and release can then be tested.

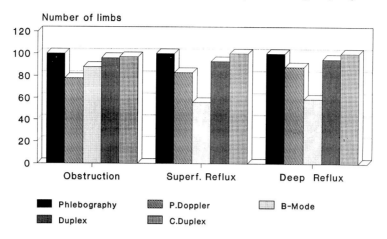

FIGURE 5 Ultrasound test and phlebography detected problems: obstruction, superficial (superf.), or deep reflux). B-mode ultrasound is mainly used for the diagnosis of deep venous thrombosis. Color (c.) duplex, in expert hands, may detect the same number of problems detected with phlebography.

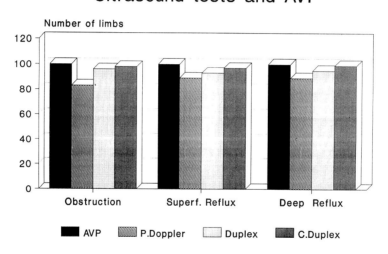

FIGURE 6 Comparison between ultrasound tests and signs of obstruction, superficial (superf.), or venous incompetence detected with ambulatory venous pressure (AVP) measurements. Color (c.) duplex is almost as effective as phlebography in detecting both reflux and obstruction.

By these methods it is also possible to define the anatomic site of flow abnormality and obtain some quantitative measurements of the severity of reflux and/or obstruction.

Despite the plethora of investigations that have emerged during the past years it is possible to obtain optimum information using simple tools such as the pocket Doppler, the duplex scanner (with or without color-flow imaging) and the air plethysmograph to quantitate the degree of obstruction.

References

1. Folse R, Alexander RH. Directional flow detection for localising venous valvular incompetence. *Surgery* 67: 114–121, 1970.
2. Hoare MC, Royle JP. Doppler ultrasound detection of saphenofemoral and saphenopopliteal incompetence and operative venography to ensure precise saphenopopliteal ligations, *Aus NZ J Surg* 55: 411–6, 1985.
3. Nicolaides AN, Hoare M, Miles CR et al. Value of ambulatory venous pressure in the assessment of venous insufficiency. *Vasc Diagn Ther* 3: 41, 1981.

4. Sigel B, Felix RW Jr, Popky GL. Diagnosis of lower limb venous thrombosis by Doppler ultrasound technique. *Arch Surg* 104: 174–9, 1972.
5. Lewis JD, Parsons DCS, Needham T et al. The use of venous pressure measurements and directional Doppler recordings in distinguishing between superficial and deep valvular incompetence in patients with deep venous insufficiency. *Br J Surg* 60, 312, 1973.
6. Vasdekis SN, Clarke H, Nicolaides AN. Quantification of venous reflux by means of duplex scanning. *J Vasc Surg* 10: 670, 1989.
7. Nicolaides AN, Zukowski AJ. The value of dynamic venous pressure measurements. *World J Surg* 10: 919, 1986.
8. O'Donnell J, Burnand KG, Clemenson G et al. Doppler examination vs clinical and phlebographic detection of the location of incompetent perforating veins. *Arch Surg* 112: 31, 1977.
9. Bydgeman S, Aschberg T. Venous plethysmography to the diagnosis of chronic venous insufficiency. *Acta Chir Scand* 137: 423, 1971.
10. Nicolaides AN, Miles C. Photoplethysmography in the assessment of venous insufficiency. *J Vasc Surg* 5: 405.11, 1986.
11. Christopoulos D, Nicolaides AN. Noninvaisive diagnosis and quantitation of popliteal reflux in the swollen and ulcerated leg. *J Cardiovasc Surg* 29: 535, 1988.
12. Christopoulos D, Nicolaides AN, Szendro G. Venous reflux: quantitation and correlation with the clinical severity of chronic venous disease. *Br J Surg* 75: 352, 1988.
13. Christopoulos D, Nicolaides AN, Cook A et al. Pathogenesis of venous ulceration in relation to the calf muscle pump function. *Surgery* 106: 829, 1989.
14. Holmes ML. Deep venous thrombosis of the lower limbs diagnosed by ultrasound. *Med J Aust* 1: 427, 1973.
15. Raju S. New approaches to the diagnosis and treatment of venous obstruction. *J Vasc Surg* 4(1): 42, 1986.
16. Bendick PJ et al. Pitfalls of the Doppler examination for venous thrombosis. *Am Surg* 49: 320, 1983.
17. Nicolaides AN, Christopoulos D, Vasdekis S. Progress in the investigation of chronic venous insufficiency. *Ann Vasc Surg* 3: 278, 1989.
18. Rollins DS, Ryan TJ, Semrow C et al. Characterisation of lower extremity chronic disease using realtime ultrasound imaging. In: Negus D, Jantet G (eds). *Phlebology 85*. J Libbey and Co., London, 1985, 576–579.
19. Stallworth JM, Talbot SR. Use of realtime imaging in identifying deep venous obstruction, a preliminary report. *Bruit* 6: 41, 1982.
20. Belcaro G, Christopoulos D, Nicolaides AN. Venous insufficiency: noninvasive testing. In: Bergan J, Kistner R (eds). *Atlas of Venous Surgery,* Saunders, Philadelphia, 1992, 9–24.

Chapter 17.7
Photoplethysmography and Light Reflection Rheography: Clinical Applications in Venous Insufficiency

G. Belcaro, D. Christopoulos, and A.N. Nicolaides
Irvine Laboratory
Vascular Unit
Academic Surgical Unit
St. Mary's Hospital Medical School
London, U.K.
Microcirculation Laboratory
Cardiovascular Institute
Chieti University
Italy

I. Introduction

Photoplethysmography (PPG) and light-reflection rheography (LRR) are simple noninvasive techniques that detect changes in the blood content of tissues. Both methods have been widely used to screen and evaluate venous insufficiency. This chapter will provide some basic information concerning the practical clinical applications of PPG and LRR in the assessment of venous insufficiency.

In PPG and LRR instruments light from the emitting probe is directed into the skin and the detector records changes in the intensity of the light emerging from the surrounding skin. Skin tissues produce considerable attenuation of the light and additional attenuation is produced by the blood present in the dermal circulation. The attenuation in the signal is proportional to the blood contact in the skin and therefore a high quantity of blood causes a great attenuation of the PPG and LRR signals. Other factors such as type of skin tissue, wavelength of the light used and the packing, orientation, and oxygen saturation of the red cells present in the sample volume may also affect the light attenuation.

A linear relationship between the PPG pulse volume (amplitude) and skin blood flow when both are measured simultaneously using venous occlusion plethysmography or the method of thermal conductance has been shown in the past.[1] However attempts to calibrate PPG and LRR and to express results in units proportional to absolute units of flow (milliliters per unit of volume per time) have been unsuccessful. This is due to the fact that any linear relationship found between pulse amplitude and blood flow applies only to the particular bed studied at the time. Reapplication of the probe to another area of tissue or even to the same area at different times will produce a regression line with a different slope. Therefore it is considered that PPG and LRR are reliable expressions of the changes in skin blood flow but that measurements are not quantifiable in absolute standard units. Also PPG and LRR cannot be used to compare flows at different sites.

In recent years with better definition of the basic requirements for photoplethysmography and the development of better PPG and LRR probes the signal has been found to be of clinical value in the study of venous problems. The venous system can be evaluated by recording changes in dermal blood volume induced by a standard exercise of the calf muscle pump simulating deambulation. When the probe is applied on the dorsum of the foot following an exercise test a decrease in dermal blood volume occurs (Figure 1) as a result of a decrease in venous pressure. At the end of the exercise text the time taken for the dermal plexus to refill is then calculated. It is interesting to observe that a comparable curve has been obtained using a laser-Doppler flowmeter probe applied on the same skin region (Belcaro G, data in preparation).

The exercise test should be ideally performed to reach a plateau of dermal/venous emptying avoiding a hyperemic response which follows strenuous or

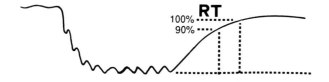

FIGURE 1 The photoplethysmograph curve with the deflection due to venous emptying and the refilling time (RT).

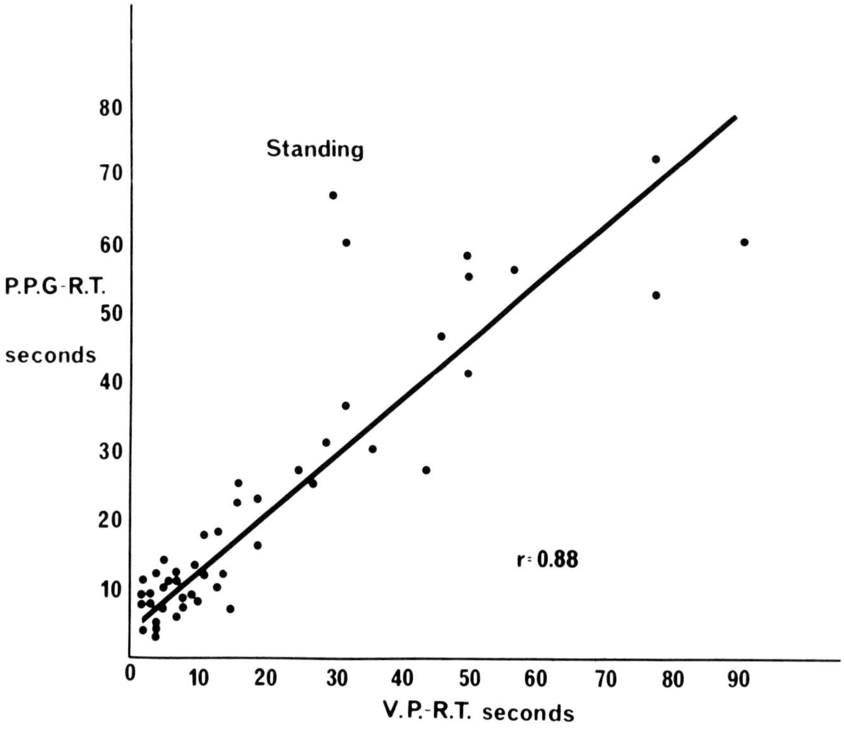

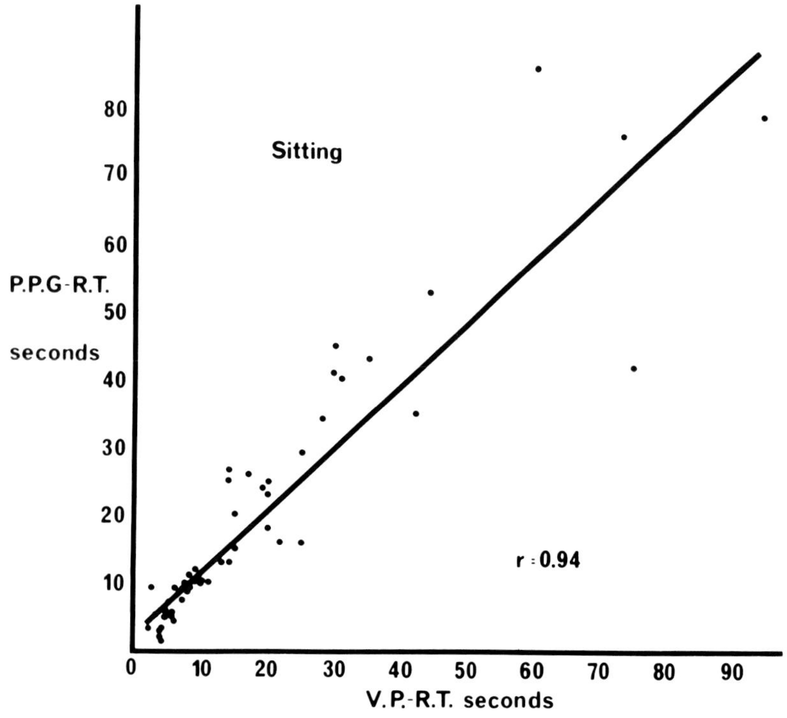

prolonged exercise. The PPG and LRR probes should be applied flat on the skin so that direct ambient light will not reach the detector. Ideally the probe should also have an overall surface area greater than that of the detector in order to prevent ambient light from being scattered through the skin onto the detector. No pressure should be applied to the skin — to avoid alterations in capillary flow — by the probe and therefore double adhesive tape with a central hole for light transmission is generally used.

II. Photoplethysmography and Light-Reflection Rheography in Venous Insufficiency

The value of PPG and LRR in the functional assessment of venous insufficiency must be evaluated comparing PPG and LRR measurements with ambulatory venous pressure (AVP) measurements which are considered the gold standard in the dynamic assessment of venous insufficiency.[1-3] In a previous study,[1] PPG (Medasonics, USA) was compared to AVP in 62 limbs (42 patients). Venous pressure measurements were recorded by inserting a 23-gauge butterfly needle into a vein of the dorsum of the foot. A 2.5-cm wide pneumatic cuff was placed around the ankle so that when it was inflated, reflux in the superficial veins was abolished without occluding the deep venous system. All subjects in this group were evaluated both in the standing and sitting position. In the standing position the exercise consisted of alternative plantar flexion and dorsiflexion of the foot; in the sitting position it consisted of repeated tiptoeing. Each exercise was performed at the rate of one per second for 30 s unless a plateau of dermal emptying and venous pressure was reached.

In this series each exercise was repeated three times without and with inflation of the ankle cuff and the measurements presented are the mean of three tests. In the standing position the patient was holding onto an orthopedic frame to prevent mo-

FIGURE 2 The correlation between photoplethysmography refilling time (PPG-RT) and ambulatory venous pressure refilling time (AVP-RT) in the standing (A) and in the sitting (B) position.

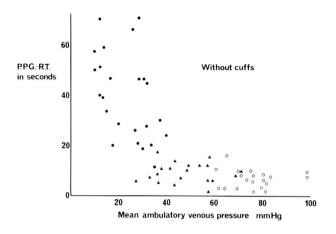
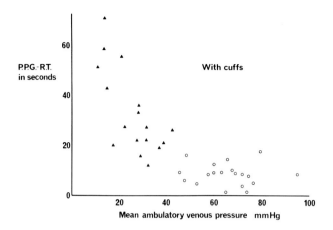

FIGURE 3 The relationship between ambulatory venous pressure and photoplethysmography refilling time (PPG-RT) without a cuff (A) and with an ankle cuff excluding the superficial system (B).

tion artifacts in the AVP and PPG recordings. The limbs were classified as: (a) normal limbs (24); (b) limbs with superficial venous incompetence (primary varicose veins) (20 limbs); and (c) limbs with deep venous incompetence (18). This classification was based on AVP confirmed by the evaluation of the presence or absence of reflux in the superficial or deep veins assessed with Doppler-duplex scanning.

The PPG refilling time (PPG-RT) and AVP refilling time (AVP-RT) were defined as the time taken from the end of the exercise test to the point where recording reached 90% of the distance between the plateau during exercise and the postexercise plateau measured on the vertical axis. The plot of PPG-RT against AVP-RT in both the standing (Figure 2A) and sitting position (Figure 2B) are almost equivalent. A linear relationship was observed between the PPG-RT and AVP-RT both in the standing and sitting position with correlation coefficients (r) of 0.88 and 0.9 respectively.

The severity of venous hypertension in venous insufficiency may be expressed by the mean AVP (the lowest venous pressure observed during AVP measurements) in the standing position. Therefore AVP was plotted against the PPG-RT (Figure 3A and B). The relationships between PPG-RT and mean AVP when the patients were standing without and with the ankle cuffs inflated to occlude the superficial system are shown in Figure 3 (A without cuffs and B with cuffs). Normal limbs had an AVP lower than 40 mmHg and with the exception of one limb the PPG-RT was longer than 18 s. When the exercise was repeated with the ankle cuff inflated AVP, AVP-RT, and PPG-RT did not change. In all limbs with venous valvular incompetence AVP was greater than 40 mmHg and PPG-RT shorter than 18 s with the exception of six limbs with mild superficial venous incompetence. There was a moderate overlap between PPG-RT in limbs with superficial incompetence and in those with deep venous incompetence when the ankle cuff was not inflated. After the cuff was inflated — abolishing the superficial reflux — a complete separation of the limbs with superficial reflux from those with deep reflux was obtained. The PPG-RT increased to more than 18 s in all but two limbs (10%) while it remained shorter than 18 s in all limbs with deep venous incompetence. A marked similarity between the AVP-RT and PPG-RT was observed in the standing position but not in the sitting position. In the sitting position a large overlap between limbs deep and superficial venous incompetence was observed. Therefore the diagnostic accuracy of PPG-RT is higher when the test is performed with the patient standing.[1-3]

Table 1 shows the data concerning 100 normal limbs, 100 patients with superficial venous incompetence and 100 with deep venous incompetence tested in our laboratories. The test was repeated according to the experimental designs and with the same equipment previously described.[1] The only differences were that: (1) the test consisted in a simplified ten-tiptoe exercise test (one per second) (the 90% RT was considered); and (2) patients were selected on the basis of color duplex scanning as having normal veins or superficial or deep venous incompetence.

Results show comparable values (Table 1). The standard deviation in all groups of patients was significantly higher in the sitting position than in the standing position indicating a better reproducibility of the test when the patient was standing.

In conclusion exercise in a normal limb results in a sudden reduction in venous pressure and dermal blood content as demonstrated by AVP, PPG and LRR. The time taken for postexercise refilling depends on: (a) the efficiency of the calf muscle pump; (b) the competence of the valves in the deep and in the superficial venous systems; (c) on the arterial inflow; and (d) in the case of PPG and LRR, the dermal vascular tone during the test.

In normal limbs the postexercise refilling time depends mainly on the arterial inflow and when standing it ranges

TABLE 1 PPG-RT and AVP-RT in 100 Normal Limbs, 100 Limbs with Superficial Venous Incompetence, and 100 Limbs with Deep Venous Incompetence

		AVP-RT		PPG-RT	
		No Cuffs	Cuffs	No Cuffs	Cuffs
Normal limbs	Sit	56 ± 13	55 ± 15	64 ± 16	63 ± 14
	Sta	48 ± 10	46 ± 11	53 ± 8	51 ± 10
Superficial incompetence	Sit	10 ± 4	24 ± 6	11 ± 5	26 ± 6
	Sta	9 ± 2	25 ± 3	9 ± 2	24 ± 4
Deep incompetence	Sit	8 ± 3	12 ± 5	8 ± 3	12 ± 4
	Sta	7 ± 2	11 ± 2	7 ± 1	13 ± 2

Note: Ambulatory-venous-pressure refilling time = AVP-RT and photoplethysmography refilling = PPG-RT. Results (in seconds; mean ± SD) shown are relative to the test performed both in the sitting (sit) and standing (sta) position and without and with an ankle cuff to exclude the superficial venous system.

from 19 to 80 s. In the case of venous incompetence the volume of blood emptied from the calf is deficient and emptying reaches a plateau within three to four calf contractions as shown by AVP and PPG. An early plateau is also reached in limbs with large varicose veins with marked reflux approximating the calf pump-stroke volume. At the end of the exercise test refilling is faster (shorter than 18 s) following reflux in the incompetent veins.

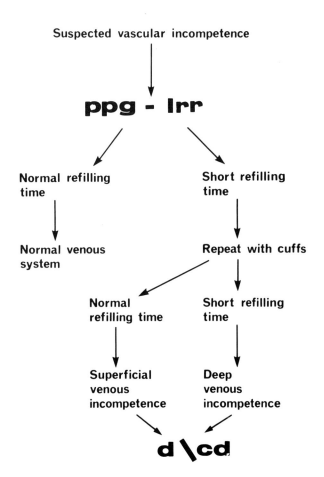

FIGURE 4 Flow chart summarizing the applications of photoplethysmography and light-reflection rheography in the evaluation of venous insufficiency.

The PPG-RT in the standing position immediately separates limbs without valvular incompetence from those with valvular incompetence. This separation occurs even in subjects with moderate to mild incompetence. In limbs with PPG-RT <18 s the test is repeated with the ankle cuff inflated. When RT is >19 s (with an ankle cuff inflated) deep venus incompetence is present (in 95% of limbs). This separation is not as good when the test is made in the sitting position.

An RT shorter that 18 s may be considered normal and indicates that AVP is lower than 40 mmHg. A PPG-RT shorter than 18 s is indicative of abnormal AVP (any value between 40 and 100 mmHg) and therefore of venous incompetence. It is generally considered that PPG or LRR cannot grade exactly venous incompetence and the methods are mainly used as a simple, inexpensive and not time-consuming screening method. When quantitative results need to be considered AVP or air plethysmography (APG) should always be used.[3-5] Therefore PPG (and LRR) classifies patients into one of the following three clinically important groups: (a) normal valvular function; (b) superficial venous incompetence; and (c) deep venous incompetence with or without superficial incompetence. However the severity of venous incompetence cannot be clearly graded. The diagnostic flow-chart used in many vascular laboratories for the screening of patients is shown in Figure 4. Subjects with a PPG-RT >18 s are considered normal. In subjects with PPG-RT <18 s the exercise is repeated with a cuff excluding the superficial system. Prolongation of PPG-RT to >18 s indicates superficial venous incompetence only (primary varicose veins). The test can be repeated with the cuff below and above the knee to localize proximal or distal venous incompetence. When the cuff is placed distally (ankle) the most distal site of deep to superficial venous reflux — particularly incompetent calf perforating veins — can be localized. Doppler ultrasound will confirm the presence of reflux at more proximal sites and at the major junctions (saphenofemoral, short saphenapopliteal vein).

A PPG-RT shorter than 18 s with an ankle cuff excluding the superficial system suggests incompetence of the deep venous system. The introduction of duplex scanning and color duplex Doppler has definitely modified the scheme first proposed by Miles and Nicolaides in 1981.[1,5] However in many venous clinics good results may be obtained in the evaluation of most patients with venous insufficiency with two simple, noninvasive instruments such as a pocket Doppler and a PPG.

III. Light-Reflection Rheography

In the last few years an improvement in the application of PPG for the evaluation of venous insufficiency has been made with the LRR. A recent review by Fronek[6] states that there is a very good correlation of LRR-RT with AVP-RT. A less close relationship was found for the LRR signal change (LRR deflection during the exercise test) when compared to absolute AVP changes. The Hemodynamics AV 1000 (Laumann Medizintechnik GMBH, Germany) has an improved sensor head weighing only 10 g and 2 cm in diameter that contains three light-emitting diodes (LEDs) around a central photodetector. The light emission is close to the infrared range and is focused into the skin at a depth range of 0.3 to 2.3 mm. The sensor head contains also a thermistor sensitive to changes in skin temperature. The test is performed when skin temperature is between 28 and 32°C. Also a metronome is inserted in the instrument and a beeping sound and a red light indicate to the patient under examination the time to tiptoe and — with a more prolonged sound — when to stop and rest. This ensures a standard tiptoe exercise test.

In our series we have evaluated, both with AVP and LRR measurements, normal subjects and patients with superficial and deep venous incompetence. The probe was placed in the perimalleolar region (5 cm above the medial malleolus). As with LRR an ankle cuff is more difficult to apply above the probe (due to the dimension of the sensor head) in comparison with PPG a specially designed cuff (12 cm wide) should be applied to the ankle (Figure 5). This cuff including and holding the probe without compressing it onto the skin is useful to exclude even distal perforators in the perimalleolar region. It also excludes reflux from perforators just below the area of examination. The pressure of occlusion of the cuff was maintained at 70 mmHg just for the time of the test. The exercise test was performed only on the standing position as a previously published study had shown that tiptoe exercise with the patient standing as defined by Seycek[7] is more reproducible and reliable in the standing position. In this position the maximum hydrostatic load on the insufficient veins is present. As already observed for PPG the test on standing reveals venous insufficiency better than the exercise program in the sitting position. The data relative to the three groups of limbs are shown in Table 2. The good, global (considering all three groups) correlation between LRR-RT and AVP-RT (r = 0.81) and the friendly use of the instrument make this method a fast and simple screening method to be used even by nonvascular staff. As an example in the diabetic clinic and in the microcirculation laboratory in our institution LRR is used as a screening method in any case of leg swelling associated with some degree of pain or when some veins are present. If the test is normal (RT >28 s) a hemodynamically relevant venous disease is ruled out. If the test is abnormal the patients are sent to the venous clinic for a more complete (duplex/color duplex) assessment.

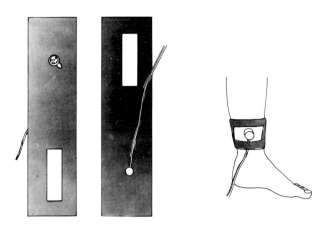

FIGURE 5 The special light-reflection rheography ankle cuff developed to exclude venous reflux form distal perforators.

The screening of diabetic patients with leg swelling and/or pain (a series of 300 patients) has screened out 53% of the patients (defined as normals) cutting the number of referrals and allowing more time for patients with actual venous problems. A retrospective evaluation of limbs evaluated as normals by LRR was shown by color duplex that all limbs (123 in 112 subjects) were effectively free from significant venous disease. A less precise but still clinically good separation may also be obtained with the LRR exercise performed in the sitting position.

Recent developments[5] indicate that it is possible (at least theoretically) to attempt calibration. This may be obtained by evaluating the changes in optical density.

TABLE 2 Basic Data of the LRR Study

		AVP-RT		LRR-RT	
		No Cuffs	Cuffs	No Cuffs	Cuffs
Normal limbs	Mean	31.5 ± 9	30 ± 8	35 ± 12	34 ± 13
	Range	15–29	16–27	22–75	24–68
Superficial Incompetence	Mean	9.3 ± 3	28 ± 7	12 ± 5	32 ± 12
	Range	5–15	19–50	7–23	21–65
Deep incompetence	Mean	9 ± 3	10 ± 4	14 ± 6	12 ± 5
	Range	3–13	5–14	6–25	7–26

Note: Ambulatory-venous-pressure refilling time = AVP-RT and light-reflection-rheography refilling time = LRR-RT. LRR-RT and AVP-RT in 52 normal limbs, in 30 limbs with superficial venous incompetence and in 14 limbs with deep venous incompetence. In normal limbs the average AVP was 21 ± 3.7 without the ankle cuff and 20 ± 4 with the cuff. The AVP range was 15 to 29 mmHg without the cuff and 16 to 27 with the ankle cuff. In limbs with superficial venous incompetence the mean AVP was 61.2 ± 9.79 mmHg (with the ankle cuff it was 22 ± 6). The range was 40 to 71 mmHg without the cuff and 10 to 28 with the cuff. In deep venous incompetence the mean AVP was 74 ± 15 without the ankle cuff and it was 70 ± 16 with the cuff. The range was between 58 and 88 mmHg without cuff and 55 and 79 mmHg with the cuff.

'Calibrated PPG' (Fronek[6]) is based on a standardized change in light source intensity (e.g., 5%). All other changes either exercise- or tilt-induced can be expressed proportionally. A second, more complex system automatically controls the light source intensity in order to deliver a constant input voltage to the amplifier. The first clinical experience has been recently presented[6] and instrumentation is not commercially available at present.

No serious attempt has been made to correlate the PPG and LRR deflection during exercise with alterations in venous outflow. Therefore PPG and LRR are **not recommended** in the evaluation of deep venous thrombosis. A recent European Consensus Meeting on deep venous thrombosis held in Windsor (UK, Nov. 1991) has found no documented application of PPG or LRR in the diagnostic assessment of DVT (European Consensus Meeting on DVT, book in publication, Med Orion Press, 1992).

Limitations to the use of PPG and LRR are muscular and bone/joint problems as any other problem affecting the ability of patients to perform the test. As for any other microcirculatory test we perform the PPG and LRR evaluation in a room at constant temperature (22 ± 2°C) after 20 min acclimatization and supine or sitting rest.

Some major skin changes, liposclerosis and ulcerations also affect the possibility of testing limbs with PPG and LRR (particularly LRR). In our experience less than 10% of the patients referred to our venous clinic were unable to correctly perform the exercise test.

In conclusion PPR and LRR are effective, fast, easy to learn and simple noninvasive tests to screen patients with venous insufficiency. The low cost of the test and the possibility of being performed by staff without vascular experience make these methods valid to select patients with venous insufficiency and to classify them as normal or abnormal — separating limbs with superficial and/or deep venous insufficiency.

References

1. Miles C, Nicolaides AN. Photoplethysmography: principles and development. In: Nicolaides AN, Yao JST (Eds.) *Investigation of Vascular Disorders,* Churchill Livingstone, New York, 1981, pp 501–515.
2. Flinn WR, Queral LA, Abramowitz HR, Yao ST. Photoplethysmography in the assessment of chronic venous insufficiency. In: Nicolaides AN Yao JST (Eds). *Investigation of Vascular Disorders,* Churchill Livingstone, New York, 1981, pp. 516–531.
3. Nicolaides AN, Zukowski AJ. The value of dynamic venous pressure measurements. *World J Surg* 10: 919–924, 1986.
4. Belcaro G, Christopoulos D, Nicolaides AN. Lower extremity venous hemodynamics. *Ann Vasc Surg* 5: 305–310, 1991.
5. Belcaro G, Christopoulos D, Nicolaides AN. Venous insufficiency: noninvasive testing. In: Bergan JJ and Kirstner RL (Eds.) *Atlas of Venous Surgery,* WB Saunders Company, Philadelphia, 1992, pp 9–24.
6. Fronek A. Recent development in venous photoplethysmography. 6th San Diego Symposium on vascular diagnosis 15–21 Feb 1992, pg 327.
7. Seycek J. Licht-Reflections-Reographie und blutige Venendruckmessung mbeim Arbeitstest im Stehen. *Vasa* 18: 18–23, 1989.
8. Nuzzaci G, Mangoni N, Tonarelli PA, Lucente E, Righi D, Borgioli F, Lucarelli F. Our experience on light reflection rheography (LRR): a new non-invasive method for lower limbs venous examination. *Phlebology,* I: 231–242, 1986.

Chapter 17.8
Strain Gauge Plethysmography: Diagnosis and Prognosis in Skin Lesions on the Feet and Toes and in Leg Ulcers

N.A. Lassen and P. Holstein
Department of Clinical Physiology and Nuclear Medicine
Department of Thoracic and Vascular Surgery
Vascular Division
Bispebjerg Hospital
University of Copenhagen
Copenhagen, Denmark

I. Introduction

A stretch sensitive resistor is a type of strain gauge widely used in clinical physiology. The mercury-in-rubber or mercury-in-silastic strain gauge is simple and convenient. When stretched its resistance increases and this can be accurately recorded by a Wheatstone bridge. When used for plethysmography (volume recording) the strain gauge is recording volume changes of a limb segment distal to a conventional cuff encircling the limb.

II. Blood Flow

Sudden inflation of the proximal cuff to pressure levels somewhat below the diastolic blood pressure (usually 50 mmHg) will cause the limb segment with the strain gauge to swell at the rate of the inflow. This indicates the unobstructed (prevenous-stasis) flow provided the local venous capacitance vessels are relatively empty at the moment of cuff inflation so that for a short time venous pressure does not build up.

The measured flow is total flow, i.e., flow in muscle, tendon, fascia, bone, subcutis, and skin. Therefore strain gauge plethysmography, and also other types of plethysmography, is not well suited for measuring skin blood flow. Forearm skin blood flow can be estimated by measuring total flow before and after adrenaline iontophoresis to arrest skinflow.[1] This gives a value of 9 ml/100 g/min, given certain assumptions about the mass of skin present and the depth of the penetration of the adrenalin. It agrees fairly well with other estimates of this flow. But the method is clearly not very accurate and at the same time quite cumbersome. If one applies the same principle to a finger or a toe, then the arteriovenous anastomoses will interfere so markedly at thermoneutral temperatures as to completely invalidate the approach for measuring skin flow exclusively.

III. Systolic Blood Pressure

The strain gauge affords a simple and very accurate way of recording the distal blood pressure in a limb. The pressure in the occlusion cuff is first rapidly inflated to suprasystolic levels and then slowly deflated just like when measuring blood pressure in the arm in the conventional manner. Because the strain gauge is sensitive to the nonpulsatile signal of limb swelling the approach is *much more accurate for measuring very low pressures,* levels of 0, 10, 20, or 30 mmHg, than auscultation or Doppler probes. This crucial aspect, and the fact that it demands little technical skill or medical knowledge to perform routine measurements, accounts for the widespread use of strain gauge for measuring distal systolic blood pressure in occlusive arterial disease.

The pressure recorded is the highest systolic pressure at the level where the cuff is placed. The sensor, the strain gauge, may be placed at any convenient level distal of the cuff. Most often used is *ankle blood pressure* and *toe blood pressure,* in both cases with the strain gauge encircling the tip of the big toe. Due to the many distal anastomoses and to the low-pulse pressure amplitude in chronic arterial occlusive arterial disease, there are good reasons to believe that no gross differences in blood pressure exists between the various arteries at a given level, say the ankle. This concept agrees with the results obtained by Doppler recording of pressure on anterior and posterior tibial arteries in such patients.

One drawback in systolic blood-pressure measurements distally on the legs is of particular importance:

the *transmission* of the cuff pressure through the arterial wall may be *compromised by medial sclerosis* causing the measured pressures to be falsely high. This invalidates the reliability of ankle pressures in diabetic patients. Medial sclerosis also exists in the pedal and digital arteries but in toe pressure measurements it is, strangely enough, not a problem.

Thus the toe systolic blood pressure is the pressure in the arteries supplying the skin of this most vulnerable site for ischemic ulcerations. In order to emphasize this point a brief account of technique and results of this measurement will be given.

IV. Crucial Problems in Diagnosis and Treatment

In occlusive arterial disease of the legs ischemic skin lesions are characteristic manifestations of the pathophysiology. However, such lesions on the feet and toes may be confused with nonischemic lesions as for instance diabetic neuropathic lesions. And in leg ulcers it may be very difficult to judge by clinical examination alone whether ischemia from arterial occlusion plays a significant role in the pathogenesis. The problem is whether a vascular reconstructive procedure is required or not. This problem can be noninvasively settled by measurement of the toe blood pressure.

V. Technique

A. The Toe Pressure

The systolic blood pressure in small arteries in fingers and toes is in normal subjects close to the systolic arm blood pressure.[2,3] This gives a proper wide scale for quantitative evaluation. The blood pressure can be measured by a miniature blood pressure cuff, 2.4 cm wide and long enough to encompass the digit. Various physical systems can be used for picking up the endpoint where blood penetrates under the cuff during deflation. The mercury in silastic strain gauge as introduced in pressure measurement by Strandness[4] is a most suitable device for recording the volume expansion.

Some important points in the technique should be emphasized. Measurements on the big toes are made with the subject in the supine position after about 20 min of rest (Figure 1). The temperature in the laboratory must be comfortable (22 to 25°C) and the temperature of the skin on the toes must be at least 25°C.[5] Failure to achieve these requirements will result in low recordings which are not relevant in prognostic studies. Cold feet should be gently heated in a tempered bath 38 to 40°C for about 10 min. If skin lesions are present a plastic bag should be drawn over the foot during heating (Figure 2). When measuring the toe blood pressure the toe should be carefully emptied

FIGURE 1 Measurements on the big toes are made with the subject in the supine position after about 20 min of rest.

from blood by squeezing it with the fingers *before* cuff inflation (Figure 3). This detail greatly facilitates the reading of the curves.

It is a drawback that patients with toe amputations or extensive gangrene cannot be measured but a minor skin lesion on the big toe does not necessarily impede the procedure. Also measurements on the second toe with a more narrow cuff can alternatively be measured.

B. The Ankle Pressure

In the daily routine we also measure the ankle blood pressure with the strain gauge technique. The ankle pressure — often measured with Doppler-ultrasound — is today the most widely used subset in examining patients with peripheral arterial disease. It is useful in the quantitative diagnosis of occlusive arterial disease and in following the response to treatment. It has some prognostic value as to the risk of progression[6] and to the risk of amputation in case of gangrene.[7] Besides the problems with medial sclerosis, however, ankle blood pressures are measured too far away from the lesions on the toes and forefoot to be valid in the prognosis.

In many leg-ulcer patients ankle blood pressures cannot be measured due to bandages and pains.

VI. Skin Lesions on the Toes and on the Feet

A. Nondiabetics

In a series of investigations the toe blood pressure has been shown to correlate excellently with the fate of lower limbs with ulcers or gangrene on the feet and toes, i.e., the potential for spontaneous healing and

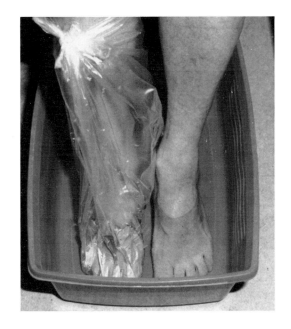

FIGURE 2 If skin lesions are present, a plastic back should be drawn over the foot during heating.

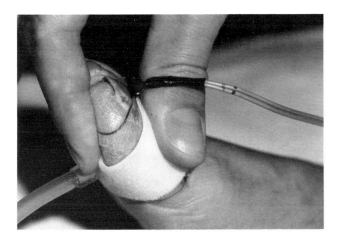

FIGURE 3 When measuring the toe blood pressure, the toe should be carefully emptied of blood by squeezing it with the finger *before* cuff inflation.

the risk of major amputation.[7-10] A toe blood pressure of less than 30 mmHg indicates a risk of amputation of about 60 to 70% and even more when it is below 20 mmHg. On the other hand a toe blood pressure of more than 30 mmHg indicates that ischemia is not a problem in wound healing. Thus patients requiring vascular arterial intervention are identified.

B. Diabetic Foot Lesions

The 30 mmHg mark is also valid in diabetics.[7-10] This is particularly useful due to the special problems in these patients. Foot lesions in diabetic patients which are neuropathic or neuroischemic are possibly complicated with more or less invasive infection. The neuropathic lesion is typically a chronic ulceration in a foot with normal pulsations localized to a skin area covering a protruding substructure and exposed to repetitive stress during walking. The neuroischemic lesion is typically a chronic ulceration or a necrosis in a foot with nonpalpable pulsations and possibly, a history with intermittent claudication. Peripheral neuropathy is nearly always present when diabetics present with ischemia; the term neuroischemic lesion is adequate at this point. However, the clinical examination is in many cases inadequate for classification of the lesion. This is, in particular, the case when pulsations are lacking and infection is superimposed. The toe pressure is again the subset which solves the classification problems. Low pressures, i.e., below 30 mmHg, mean that the patient requires the attention of a vascular surgeon. Toe pressures above 30 mmHg mean that conservative measures can be relied upon, i.e., infection has to be controlled and a proper organization for the treatment of neuropathic ulcers must be at hand. It should be remembered in wound healing that failure at adequate circulation in most cases indicates that the conservative treatment is inadequate.

VII. Leg Ulcers

Patients with stasis ulcers of long duration may end up with occlusive arterial disease not only further compromising the healing, but perhaps causing progressive necrosis, pain, and threatening amputation. Moreover, the use of compression bandages and stockings, which are otherwise indispensible in the conservative treatment of stasis ulcers, may become dangerous due to the risk of pressure necrosis.

Surprisingly the toe pressures have proven valuable in this situation. In two recent investigations[11,12] a toe blood pressure of below 30 or 35 mmHg indicated a significant risk of amputation, i.e., of about 40 to 70%. Since then the toe pressures are measured in all patients with leg ulcers in this hospital and values below the critical level are the main indication for arterial reconstruction surgery or arterioplasty in these cases.

VIII. Comments on the Method

In the above considerations toe pressures of 30 mmHg have been indicated as the crucial point below which vascular intervention should be considered. In pathology there are of course no such clearcut borderlines. About 5% of our patients with gangrene undergoing vascular reconstruction actually have pressures a little bit higher, i.e., on the order of 30 to 35 mmHg. One must always take into consideration the accuracy of the method; at low pressures the standard deviation is on the order of 5 mmHg.[8] On the other hand spontaneous healing at pressures below 30 mmHg occurs, in particular at pressures between 20 and 30 mmHg. Thus a certain toe pressure value can only indicate the *chance* of healing or *risk* of amputation (if not treated).

In both nondiabetics and nondiabetics with an ankle pressure below 50 mmHg reliably indicates severe ischemia with threatening major amputation. But in most cases with ischemic lesions the ankle blood pressure is more than 50 mmHg and unfortunately there is no reasonably clearcut ankle pressure level above which healing most certainly will take place. Also for this reason ankle pressures should be supplemented with toe pressures.

Toe pressures (and ankle pressures) measured by strain gauge technique have been the basic noninvasive procedure in evaluating patients with peripheral arterial disease for more than 20 years in vascular laboratories in Denmark. A series of other methods have been introduced over the years and the majority of these methods have been investigated in our laboratory. The skin perfusion pressure measured on the dorsum of the foot by isotope washout[7,10] may be valuable when the toes are lacking but so far an adequate reason to choose other methods in the routine evaluation has not been found. Recently, moreover, the toe blood pressure has been introduced in the European Concensus Document on Chronic Critical Leg Ischaemia[13] and in the United States the toe blood pressure has just been incorporated in the recommendations for noninvasive evaluation of diabetic patients with vascular leg problems by a workshop organized by the American Heart Association and the American Diabetes Association.[14]

References

1. Rowell LB. The cutaneous circulation, In *Physiology and Biophysics*. Vol. 2, Ruch TC and Patton HD (Eds.). WB Saunders, Philadelphia.
2. Gundersen J. Segmental measurements of systolic blood pressure in the extremities including the thumb and the great toe. *Acta Chir Scand* (suppl.) 426:40, 1972.
3. Nielsen PE, Bell G, Lassen NA. The measurement of the digital systolic blood pressure by strain gauge technique. *Scand J Clin Lab Invest* 29:371–379, 1972.
4. Strandness DE, Bell JW. Peripheral vascular disease: diagnosis and objective evaluation using a mercury in silastic strain gauge. *Ann Surg* 161:3–35, 1965.
5. Carter SA. The effect of cooling on toe systolic pressure in subjects with and without Raynaud's syndrome. *Clin Physiol* 11:253–261, 1991.
6. Jelnes R, Gaardsting O, Hougaard Jensen K, Bækgaard N, Tønnesen KH, Schroeder T. Fate in intermittent claudication:outcome and risk factors. *Br Med J* 293:1137–40, 1986.
7. Holstein P, Lassen NA. Healing of ulcers on the feet correlated with distal blood pressure measurements in occlusive arterial disease. *Acta Orthop Scand* 51:995–1006, 1980.
8. Paaske WP, Tønnesen KH. Prognostic significance of distal blood pressure measurements in patients with severe ischaemia. *Scand J Thor Cardiovasc Surg* 14:105–108, 1980.
9. Ramsey DE, Manke DA, Sumner DS. Toe blood pressure. A valuable adjunct to ankle pressure measurement for assessing peripheral arterial disease. *J Cardiovasc Surg* 24:43–49, 1983.
10. Holstein P. The distal blood pressure predicts healing of amputations on the feet. *Acta Orthop Scand* 55:227–233, 1984.
11. Sindrup JH, Danielsen L, Karlsmark T, Gikjøb G, Jensen BL, Kristensen JK, Holstein P. Prognostic significance of digital blood pressure in leg ulcer patients. *Acta dermatovenerologica (Stockh)* 70:259–261, 1990.
12. Nielsen G, Mølby L, Nielsen JA. Relationship between distal systolic blood pressure and wound healing. A retrospective investigation of patients with crural ulcers. *Ugeskr Laeger* 154:3658–3661, 1992.
13. Second European Consensus Document on Chronic Critical Leg Ischaemia. *Eur J Vasc Surg* 6(Suppl. A):1–32, 1992.
14. Anon. The assessment of peripheral vascular disease in diabetes. Recommendations of an international workshop sponsored by the American Heart Association and the American Diabetes Association. Submitted for publication in Circulation and Diabetes Care.

Chapter 17.9
Air Plethysmography in the Evaluation of Skin Microangiopathy Caused by Chronic Venous Hypertension

D.C. Christopoulos
B2 Surgical Unit
Hippokrateion Hospital
University of Thessaloniki
Greece

I. Introduction

Chronic venous hypertension is related to microangiopathy of the skin. The latter is mainly responsible for the sequelae of venous disease such as edema, skin changes (lipodermatosclerosis, pigmentation, eczema), and ulceration.[1] Although the latter is the result of poor skin oxygenation and nutrition, arterial inflow to the lower limb of these patients seems to be increased and several authors have been postulating the presence of arteriovenous shunts or reduced capillary resistance.[2-4] Controversies in the measurement of arterial inflow in these limbs with segmental plethysmographic devices, i.e., strain-gauge, indicated the necessity of an accurate and reproducible method to evaluate this condition. Air plethysmography, with an air chamber surrounding the whole leg may prove to be the method of choice for the evaluation of arterial inflow to the lower limb.[5]

II. Object

Air plethysmography has been used to measure the resting arterial inflow in normal limbs and limbs with various grades of venous malfunction. This is in order to correlate these measurements to the clinical severity of venous disease and also to determine the effect of elastic compression on the abnormally increased blood flow.

III. Methodological Principle: Air Plethysmography

The air plethysmograph (APG-1000, ACI/Medical) consisted of a 35 cm long polyurethane tubular air chamber which surrounded the whole leg. This was inflated with air at 6 mmHg and connected to a pressure transducer, amplifier, and recorder (Figure 1). The air chamber was fitted with the patient in the supine position, the leg on external rotation with the knee slightly flexed and the heel resting on a support, 15 cm in height.

Calibration was performed by depressing the plunger of the syringe (Figure 1), reducing the volume of the air chamber and tubing by 100 ml, and observing the corresponding pressure change. After calibration the plunger was pulled back to its original position until the pressure in the air chamber returned to 6 mmHg.

Arterial inflow was measured in the whole leg (foot excluded) with venous occlusion air plethysmography using an 11 cm wide inflatable tourniquet, which was placed just above the knee and connected to a manometer (Figure 1). Ten minutes after the air-chamber was inflated, when a stable leg/air chamber/room temperature gradient was achieved and resting arterial inflow to the leg was ensured, the tourniquet was inflated to 50 mmHg. An increase in volume was recorded for 20 seconds and then the tourniquet was deflated. This increase represented the arterial inflow to the limb in milliliters per minute (Figure 2).

The above method was applied to 20 normal volunteers (N) (20 limbs), 32 patients (40 limbs) with primary varicose veins (PVV) without skin changes or edema, and 62 patients (72 limbs) with chronic edema and/or skin changes (lipodermatosclerosis, pigmentation, eczema) and/or ulceration. Of the 62 patients with edema and skin changes, 26 (32 limbs) had PVV with normal deep venous system (PVV/S) and 34 (36 limbs) had deep venous disease (DVD). The patients were classified as having PVV. PVV/S and DVD according to the results of clinical examination. Doppler ultrasound, Duplex scanning, and ascending venography.[5]

The air-plethysmographic measurements of arterial inflow (AI) in milliliters per minute are shown in Figure 3a. The AI was not dissimilar in normal limbs (N)

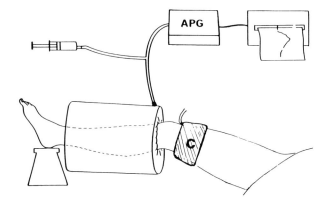

FIGURE 1 The air plethysmograph consists of a tubular air chamber surrounding the whole leg connected with a pressure transducer amplifier (APG) and recorder. Calibration is performed with the 100-ml syringe connected with the tubing. The above knee cuff (C) is inflated to 50 mmHg. (From Christopoulos, D. et al., *J Dermatol Surg Oncol* 17: 809–813, 1991. With permission.)

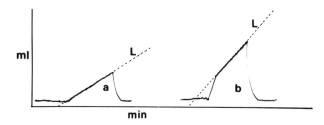

FIGURE 2 A typical recording from a normal subject (a) and from a patient with gross saphenofemoral incompetence (b). Arterial inflow is evaluated in milliliter per minute according to the slope L. The rapid increase in (b) is an artifact due to reflux of venous blood because of the compression of the tourniquet. (5) (From Christopoulos, D. et al., *J Dermatol Surg Oncol* 17: 809–813, 1991. With permission.)

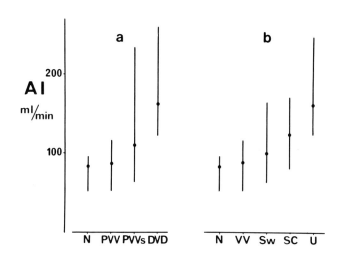

FIGURE 3 (a) The measurements of arterial inflow (AI) expressed as 95% range and median in normal limbs (N), limbs with primary varicose veins (PVV), limbs with primary varicose veins with chronic edema and/or skin changes/ulceration (PVV/S) and limbs with deep venous disease (DVD). (b) The same measurements according to the clinical appearance of the limbs. (N) Normal, (VV) varicose veins only, (Sw) chronic swelling, (Sk) Skin changes (lipodermatosclerosis, eczema, pigmentation), (U) ulceration. (From Christopoulos, D. et al., *J Dermatol Surg Oncol* 17: 809–813, 1991. With permission.)

IV. Sources of Error

It should be pointed out that the measurements should be performed 5 to 10 minutes after the application of the air chamber, in order to have resting arterial inflow and no temperature influences.

During the inflation of the tourniquet, there is a rapid increase in volume in limbs with venous incompetence (Figure 2). This is the result of venous reflux. Arterial inflow should be measured after this artifact.

V. Correlation with Other Methods

The air plethysmograph was designed to investigate venous macrocirculation[6] by measuring venous reflux, obstruction, and ejecting capacity of the calf muscle pump. Recently, it has been demonstrated that it is able to provide reproducible measurements of arterial inflow in the whole leg,[5] in contrast to other segmental plethysmographic devices which provided measurements in a part of it not necessarily representative and leading to controversial results.[7] Another advantage is that the measurements in milliliters per minute are not influenced by decrease or increase of edema. However, the air-plethysmographic results expressed in milliliter per minute instead of milliliters per 100 ml of tissue per minute may produce difficulties when comparing legs of different sizes. Nevertheless because of the high reproducibility of the method and the great difference between normal and abnormal, significant differences have been shown.

compared with limbs with PVV, but it was markedly increased (up to 4 times) in the limbs PVV/S and DVD.

Figure 3b shows the arterial inflow of the same limbs according to their clinical appearance. An increase in arterial inflow has been observed with increasing clinical severity. Limbs with ulceration had a median arterial inflow which was about double the arterial inflow of normal limbs, irrespective of whether the underlying venous insufficiency was in the superficial or the deep venous system. Also ulceration occurred only when the arterial inflow was greater than 120 ml/min.

Arterial inflow was measured five times in five limbs in order to determine the reproducibility of air plethysmography. The coefficient of variation was in the range of 3.2 to 7.6%.

In 20 limbs of patients with PVV/S the measurements of arterial inflow were repeated after a knee-length graduated compression stocking was applied (Futuro Surgical, with 25 mmHg compression). The mean arterial inflow decreased from 98 ml/min to 82 ml/min ($p < 0.01$, sum rank test for paired samples).

Laser Doppler measurements of skin flux have the advantage of being direct.[5] However, the information involves only a small skin area, the method of calibration is not clear and reproducibility is not very good.

Arteriography has demonstrated a faster passage of contrast media from the arteries to the veins in limbs with severe venous disease.[3] Histology of liposclerotic tissue revealed the existence of an abnormally great number of dilated capillaries surrounded by excessive amounts of fibrin.[4] Despite increased blood flow in these capillaries $TcPO_2$ of liposclerotic tissue is low.[1] In contrast PO_2 in the veins of limbs with chronic venous disease is higher than normal.[8–10] These observations indicate that increased blood flow through abnormal capillaries is associated with decreased oxygen release to the tissue which is responsible for skin alterations and ulceration.[11]

Air-plethysmographic blood-flow measurements are in agreement with these observations and provides an objective measurement of this condition. As shown in Figure 3 the greater the increase in blood flow the higher is the increase in clinical severity of skin alterations, ulceration occurring when blood flow is superior to 120 ml/min.

VI. Recommendations

Air plethysmography is a noninvasive technique, easy to perform, requiring minimal training by the user and with excellent tolerance by the patient.

Although it measures the blood flow of the whole leg, the good correlation with the clinical picture of the skin indicates its usefulness as a screening method in this field. Also, the good reproducibility could make it a technique of choice to evaluate therapy.

References

1. Partch H. Hyperhaemic hypoxia in venous ulceration. *Br J Derm* 110:249–51, 1984.
2. Shalin L. Arteriovenous communications in varicose veins localised by thermography and identified by operative microscopy. *Acta Chir Scand* 147: 409–21, 1981.
3. Haimovici H. Role of precapillary arteriovenous shunting in the pathogenesis of varicose veins and its therapeutic implications. *Surgery* 101: 515–22, 1987.
4. Browse NL, Burnand KG. The cause of venous ulceration. *Lancet* ii: 243–245, 1982.
5. Christopoulos D, Nicolaides AN, Belcaro G, Kalodiki E. Venous hypertension microangiopathy in relation to clinical severity and effect of elastic compression. *J Dermatol Surg Oncol* 17: 809–813, 1991.
6. Nicolaides AN. Christopoulos DC. Methods of quantitation of chronic venous insufficiency. From Bergan JJ and Yao JST (Eds.). *Venous Disorders* WB Saunders, New York, 1991: 77–90.
7. Strandness and Sumner. Venous valvular incompetence. In: *Hemodynamics for Surgeons* 1977: 409–410.
8. Blalock A. Oxygen content of blood in patients with varicose veins. *Arch Surg* 19: 898–9, 1929.
9. Holling HE, Beecher HK, Linton RR. Oxygen tension of blood contained in varicose veins. *J Clin Invest* 17: 555–561, 1938.
10. Guis JA. Arteriovenous anastomosis in varicose veins. *Arch Surg* 81: 299–310, 1960.
11. Partch H. Investigations on the pathogenesis of venous ulcers. *Acta Chir Scand* (Suppl.) 544: 25, 1988.

18.0

Skin Temperature and Thermoregulation

18.1 Thermal Imaging of Skin Temperature 457
    E.F.J. Ring

Chapter 18.1
Thermal Imaging of Skin Temperature

E. F. J. Ring
Department of Clinical Measurement
Royal National Hospital for Rheumatic Diseases
Bath, U.K.

I. Historical Background

The association between temperature and disease can be traced to the earliest records in history. Fever was observed as a natural phenomenon, which could be detected by touching the skin. It was also claimed that, if wet mud was applied to the skin, rapid drying out in a localized area may indicate the presence of an underlying disease. The oldest known record may be that of a papyrus dating back to the 17th century B.C. This alludes to practices which could be as old as 29 B.C., when suppurating wounds were described as hot, and constantly issuing heat detectable by the human hand. It is now known that human fingers can only discriminate approximately 1°C under certain favorable conditions.

Throughout the 17th century A.D., the thermometer evolved as an indicator of temperature level. While the application to medicine had previously received little attention, a few pioneers had experimented with the Galileo thermoscope. The Florentine glass blowers were ingenious in their variety of designs, and had produced a density thermometer which could be tied to the body surface. Beads, suspended in a liquid within the sealed bulb of the instrument, were observed to rise or fall at given levels of heat! Considerable progress was made by Carl Wunderlich who introduced a clinical thermometer for axillary temperature in 1871. Working systematically on his sick patients in Leipzig, he recorded the course of temperature changes on paper with over 10,000 observations.[1] The hitherto suspicion of temperature observation which was independent of touch and clinical judgment was finally laid. Over 100 years later, medical practitioners throughout the world regard temperature measurement to be a first line clinical observation.

Attempts to measure skin temperature per se had not however been so successful. Progress ultimately came from the work of Seebeck, who in 1821 established the principles of thermoelectricity. Four years later, he constructed the thermocouple by the fusion of two dissimilar metals, which produced a current relative to the temperature of the coupled elements. It was many years later, closer to the turn of the 19th century, that medical and physiological researchers began to study temperature of the skin of man and animals. Multiple thermocouples and (later) chart recorders were used, but their data were difficult to interpret. Thermopiles with increased sensitivity were found to be of value for non-contact skin thermometry as late as the mid-20th century. Important progress was made in 1934 by J. D. Hardy, an American physiologist, who determined that human skin behaves as a near black body radiator, with an emissivity close to unity.[2]

In the late 1950s, thermal imaging devices which had been developed during World War II were declassified in Europe and the U.S. These systems produced an image of skin temperature, which opened up a new era in the functional study of the skin and human thermoregulation. The original pioneering work in thermal radiation can be attributed to Sir William Herschel in 1800.[3] He detected the presence of heating rays beyond the visible red of the spectrum in a classic experiment with a prism and a set of thermometers (Figure 1). His son, John Herschel, in 1840, substantiating his father's observations, created the first thermal image, using a suspension of carbon in alcohol and a lens placed in a beam of natural sunlight.[4] He introduced the word "thermogram" (Figure 2). It is interesting to note that John Herschel's closest and lifelong friend was Charles Babbage, a founding father of the computer. Modern thermograms are produced with the aid of computer false color and resulting temperature measurements — the ultimate product of three famous men who worked together.

The development of thermal imaging from the original military systems has been dramatic. The advent of solid-state circuits, high-definition oscilloscope and liquid crystal displays, and finally small image processing computers have resulted in reliable compact cameras for passive infrared measurement. It is therefore possible to study skin temperature distribution from many aspects of health and disease using accurate non-contact methods.

FIGURE 1 Sir William Herschel.

FIGURE 2 Sir John Herschel.

Contact thermal imaging is also possible, now superseding cumbersome arrays of thermocouples. In 1877 Lehmann discovered "liquid crystals", a liquid crystalline state of certain cholesteric crystals. In the early 1960s a few experimenters had applied suspensions of these materials to the skin, which was previously coated with black paint. Changes in color were visible with a well-focused light beam set at 45° to the line of vision. Developments in this technology have resulted in encapsulated materials which are laid on flexible latex sheets. The process was patented in 1968, and has found applications throughout the 1980s as an inexpensive qualitative method for observing limited areas of skin temperature.

Infrared thermography has been received with mixed enthusiasm by medical authorities. The original hope was that this technique might be used for breast cancer screening. This proved not to be so, with many academic centers rejecting the technique, despite its unchallenged applications in other areas of medicine. The concept that skin surface temperature could be of clinical value was also challenged. In due course, attention turned to thermal detection at microwave frequencies. Experimental studies indicated that microwave measurements at certain wavelengths were able to detect heat from deeper tissues, up to several centimeters, according to the dielectric properties of the organic media. Interesting results have been obtained from this research, although microwave detector technology is far less advanced than that of infrared. Furthermore, the skin emits microwave energy at considerably lower levels than at the infrared wavelengths. There are, however, good prospects for the future in this area of research as technology improves.

II. Human Body Temperature and the Skin

The skin plays a vital role in the thermal ecology of man. As the human body is homeothermic, it constantly self-regulates to maintain a steady "core" temperature. The skin functions as a thermal interface between the body and its environment. Skin temperature is therefore influenced by both internal and external conditions. Only a very small quantity of heat is produced by the skin itself, and this is evenly spread over the total surface. In most situations, a temperature gradient can be measured over the skin surface; the temperature is usually maximal at the head and minimal at the lower extremities. This difference in temperature, which becomes increasingly marked as ambient temperature falls, is the result of a complex mechanism based on the balance required by the body for thermal regulation. This may be overridden by specific local demands, particularly in the extremities. The hyperemic flush in the fingers after handling a cold substance is an example of this. Blood convection is the main warming agent of the skin, transferring heat from the core. Localized changes in blood flow can therefore induce changes in skin temperature. Some heat is transferred to the skin by conduction from the underlying organs and tissues. The skin temperature at a given site is therefore dependent on (1) the vascular supply to the area (at that time) and (2) the thermal conductivity of the subcutaneous structures. In certain pathological conditions, increased metabolism (for example, of a tumor) may generate extra heat to a localized area.

External influences, i.e., ambient temperature or air or water velocity, must also be considered in the study of skin temperature. In most clinical studies, skin temperature is measured in air, not in water. However, applications of pharmaceutical or cosmetic products can provide a barrier layer which may disturb the thermal state to be measured. Ideally, skin temperature should be measured under constant conditions. These relate to both clothing and to the temperature, moisture, and air flow surrounding the subject. Clothing serves to insulate the body, creating a microclimate layer or layers to slow down direct heat exchange. Most body heat leaves the body through the skin into the environment. Of all heat loss, 25% is by conduction, and slightly less by evaporation in neutral temperature (still air at 50% humidity). In such conditions, 45% of heat is exchanged by radiation.

The degree to which each area of the skin surface will exchange heat with the environment partly depends on skin anatomy. The forehead is different from the intermedial aspect of the thighs, where the surfaces are radiating to one another. Body position will also modify the thermal distribution of the skin, by increasing exposure to some areas, and decreasing others (Figure 3). Heat loss can be increased when the subject is lying on a bed or other flat surface. If that surface becomes an open mesh, increased heat exchange with the ambient will occur. Conversely, a subject in the crouching position will lose less heat than in the normal sitting position on a chair. The postural position should therefore be standardized for all skin temperature measurements on a given subject, even when measurement is confined to one specific area.[5]

Since heat is also generated by exercise, skin temperature is usually measured after a period of rest, to minimize the effects of physical activity. Food and drink are usually quoted as having an effect on skin temperature. In practice, a moderate meal taken without alcohol or a hot drink may have only transient effects. Alcohol flushing, particularly of the face and extremities, can be marked, for 1 hour or more. In certain physiological states, a hot drink, if taken quickly and in sufficient quantity, can also have a measurable effect on these areas.

Thermal neutrality of the human body occurs when the subject is at rest and thermoregulatory mechanisms are inactive. Basal heat production and heat loss are then equal. This state can usually be achieved at an ambient temperature of 30°C at low humidity, <30%, and air movement of less than 0.5 ms^{-1}. At all other temperatures, the skin will be at a different thermal state for a given site in a given ambient temperature. In "temperate countries" skin temperature is often measured in a cold environment. If the temperature is too low, <17°C, vasoconstriction occurs and shivering may be induced. At higher temperatures, e.g., >25°C, sweating commences, thus creating surface heat loss by evaporation. The optimal tempera-

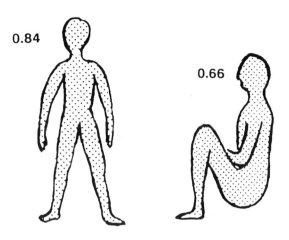

FIGURE 3 The ratio of effective radiating surface area of skin to total area is affected by postural position.

ture must therefore lie between these observable points. They differ with climatic and racial variation and may change in a given subject due to acclimatization. In Northern Europe most subjects are "comfortable" for skin temperature measurement in an environment of 20 to 22°C when partially clothed. Higher temperatures are adopted for thermal comfort levels in warmer climates.

A. Mean Skin Temperature

Physiologists have for some time debated the question of how to obtain a meaningful single value representing the state of thermal balance in a subject. This is a theoretical concept that has many inaccuracies. Mean skin temperature is usually based on a series of skin temperature measurements. There is little agreement over the optimal sites selected. To relate skin temperature variation to the underlying structures, a coefficient is used for each measurement site. A large simple anatomic surface, such as the abdomen, requires a single measurement with a high coefficient. The extremities, which are more complex, require a greater number of measurements with a lower coefficient. Some of the different methods were reviewed by Mitchell and Wyndham.[6] They concluded that each of the published methods, though different, could be used for most applications. Later studies, using techniques such as infrared thermography, have shown that relatively few sites can be used: upper abdomen, medial forearm, sacroiliac area, and posterior knee, for a reasonable estimation of mean skin temperature.[7]

III. Skin Temperature Measurement

Skin temperature is often required to assess the change induced by an external or internal stimulus within a given area. It may also be required as a comparison with a similar contralateral or reference point.

There are three fundamental methods of heat transfer, all of which are used in the study of body temperature.

A. Conduction

The earliest sensors applied to the skin were able to demonstrate changes in fluid density by heat conduction. Fluid thermometers, and later more popular thermistors and thermocouples, placed on the skin, record by conducted heat. The latter are well suited to long-term temperature recordings from one site over several minutes to hours. Temperature distribution of a small area may be demonstrated with liquid crystal sheets, which change color according to temperature.

B. Convection

Schlieren photography is a technique used to study the heat flow from the body surface. It is used by physiologists to study heat loss in high convection currents, e.g., industrial machines, helicopters, and Arctic survival studies.

C. Radiation

The skin is a highly efficient emitter of infrared and microwave radiation. Radiation thermometers and infrared imaging systems provide high-speed two-dimensional temperature recordings. Microwave radiometers may be used for temperature measurement from "deeper" tissues. Most thermal radiation systems are non-contact and therefore do not interfere with the skin itself.

D. General Principles

Any form of temperature measurement can affect the heat exchange between the skin and its surroundings. In contact thermometry, the intrinsic heat of the sensor can influence the skin. In radiation detection, the sensor must be distanced from the skin to avoid a direct thermal influence. In practice, all temperature measurement techniques should be so designed that the influences introduced by the method are smaller than the acceptable levels of error. There are particular problems with surface contact sensors, for differences in pressure on the skin not only affect the thickness and position of the tissue layers, but also affect the pattern of local blood flow. Invasive techniques, e.g., needle thermocouples, are even more disturbing, with changes created by the release of histamines, etc., by hematoma or by healing tissues with modified thermal properties.

Reproducible technique is essential for all measuring procedures, particularly for temperature. It is important that all parameters which can be altered should be noted and repeated for each measurement. The effect of changing parameters on the technique should be known by the investigator. The errors introduced by the detector or instrument should also be defined before conclusions are drawn from the measurement data. These technical conditions may appear to be obvious, but are frequently overlooked. As the magnitude of change decreases, so the requirements for the scientific methodology become more critical. In many cases, skin temperature changes may be small, so that good technique is of great importance.[8]

E. Skin Temperature Measured by Conduction

Contact thermometry requires the detector and the skin surface to equilibrate, usually from the warm skin to the cooler detector. Passive systems using thermistor and thermocouple probes work in this way. The measurement probe thus cools the skin on contact until an intermediate temperature is reached between that of the detector and the skin surface. In most cases, the mass of detector is very much less than that of the skin, so the resulting temperature measured is close to the initial skin temperature prior to contact. The temperature/time function of the equilibrating process is exponential and not linear. The final stable temperature is reached very slowly; therefore, the exact "end point" cannot be precisely defined. Most instrument response times are given as a fraction or percentage, i.e., 90% of the total response. Some thermocouples designed for skin temperature measurement are actually calibrated in water. The response time in water will be less than that of surface contact with the skin. It is generally considered that thermistors are more suited to long-term temperature measurements at a given site. Thermocouples with fast response times are preferred for short-term local use. A thermally conducting paste may be required as a coupling medium, to improve stability of long-term recording (e.g., intensive care monitoring).

1. Thermocouples

Certain metals produce a thermoelectric voltage difference when brought together in a simple circuit. Two junctions can be used, one making contact with a reference temperature, the other with the unknown (e.g., skin). When the voltage is amplified the temperature difference can be accurately shown (Figure 4). However, the relationship between thermocouple voltage and temperature is not linear. For this reason, the selection of suitable metals for a specific temperature range is limited. The most commonly used are iron, copper, nickel, and a copper-nickel alloy, constantan. Copper-constantan produces a voltage charge of 42 $\mu V/°C^{-1}$ within the body range of 30 to 40°C. Solid-state amplifiers and digital displays have greatly reduced the size and improved the convenience of these devices (Table 1).

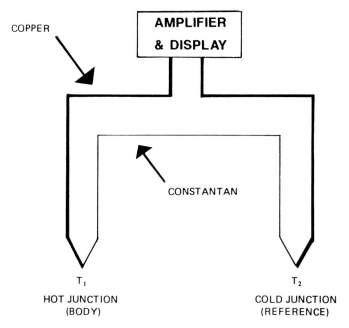

FIGURE 4 Thermocouple circuit where one junction is placed on the skin T_1 and compared to a reference junction T_2.

TABLE 1 Voltages Produced by a Copper Constantan Junction at Temperatures from 10 to 45°C

Temperature (°C)	Millivolts — reference at 0°C			
	10	20	30	40
0	0.389	0.787	1.194	1.610
1	0.429	0.827	1.235	1.652
2	0.468	0.868	1.277	1.694
3	0.508	0.908	1.318	1.737
4	0.547	0.949	1.360[a]	1.779
5	0.587	0.990	1.401	1.821
6	0.627	1.030	1.443	
7	0.667	1.071	1.485	
8	0.707	1.112	1.526	
9	0.747	1.153	1.568	

[a] *Example: at 34°C the voltage is 1.360 mV.*

2. Thermistors

These detectors depend on the electrical resistance or conductance changes of certain semiconductors with temperature. It is now possible to obtain very small sensors which can be made exchangeable. Thermistors may be formed in a variety of shapes and sizes, and are often manufactured for specific probes to be used in body cavities as well as on the skin surface. The effect of aging on these detectors can produce sizeable errors. It is, therefore, important that stability tests are made against a known temperature standard at regular intervals. Resistance thermometers require an external power supply so that the current and voltage changes can be measured. There are two categories of resistance thermometry sensors, metal resistors and semiconductor resistors. Metal resistors, normally platinum or nickel wire, have a constant high resistance and prove stable over long periods of time. The electrical resistance of metals rises in proportion to the square of the temperature.

Semiconductors (thermistors) are often inhomogeneous in chemical composition. They are therefore characterized by an exponential dependence of resistance on temperature. Most of the sensors used in the medical range show reduced resistance with increasing temperature. Very small bead resistors of 0.2 mm upward can be constructed. They are often fused to a platinum-iridium lead and protected by a glass cover. There is necessarily a flow of current, which may have a thermal effect on the probe. For this reason they are not recommended for long-term monitoring of temperature.

The response time of a thermometer may be important for particular applications. The adjustment of heat between the skin and the sensor is reflected in an exponential change of temperature with time. The time constant involved depends on the conditions of the measurement, which may not be defined by a manufacturer. It is obvious that the smaller the probe, the more rapidly it will equilibrate with the skin.

Calibration of contact thermometers used on the skin should be made by surface coupling rather than immersion. The most suitable method will be against radiation measurement of a surface at known emissivity.[9] It should also be noted that measurements made on the living human skin surface will be subject to the microclimate around the area to be measured. This in turn will be affected by the ambient temperature and the air movement within the room.

The manufacture of semiconductors has reached an advanced stage. It is now possible to have inexpensive "disposable" sensors for clinical use. A successful Swedish instrument for oral temperature uses disposable thermistors in a paper strip which can be clipped into a power and display handle (Figure 5). This device, produced to overcome the safety and hazard problems of cross-infection and mercury spillage, is capable of reliable skin temperature measurement. It has, however, a limited range suitable for oral temperatures, but indicates that advances in the semiconductor field will lead to further developments in this area.

3. Active Contact Thermometry

The general principles described above refer to the passive measurement of skin temperature. It is possible to learn more of the skin's thermal behavior with a heated probe. These probes, for which there are several designs, can give an approximation of skin microcapillary blood flow, based on thermal conductivity.

The probe usually consists of an annular array of thermocouples around a circular heater which raises the skin temperature by several degrees. The system is

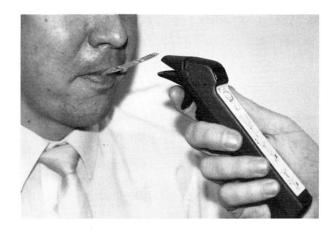

FIGURE 5 An oral thermistor thermometer with disposable sensor.

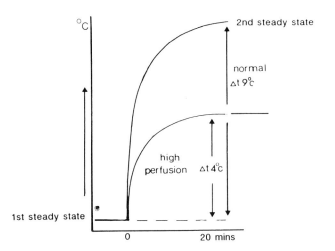

FIGURE 6 Principle of thermal clearance, where temperature increase beneath the probe is measured between first steady state (unheated) and second (heated stage).

first allowed to stabilize with the skin, then the heater current is applied. A new steady state is reached after a few minutes and the difference in temperature between the first and second steady states are recorded. In principle, the more blood perfusion present beneath the probe, the more heat is conducted away, resulting in a small temperature difference between steady states. The converse is true with higher temperature differences showing lower thermal conductivity (Figure 6). There are a number of problems with the technique which should be carefully defined. The heater and sensors must be as far apart as possible (normally 1 cm) to avoid lateral diffusion of heat. The site chosen must not be varied for comparative studies. Britton et al.[10,11] have shown through a mathematical model that both fat and blood flow are major factors which alter the thermal characteristics of the skin for thermal clearance techniques. The diameter of the probe appears to be one of the major determinants in the estimated depth of this particular measurement. A probe design based on a ceramic potentiometer base has been published by Ring et al.[12] (Figures 7A and B). Recent interest in tcPO^2 (transcutaneous oxygen tension) of the skin has resulted in a wider use of this technique on different anatomical sites. The probe includes a heater to dilate the superficial capillaries and a thermocouple to act as a regulating sensor. It is designed to measure the heat required to raise the skin to a fixed temperature of 40°C in order to assess oxygen tension. In principle, this information can also be used to indicate the thermal clearance of the sample area of skin.

4. Liquid Crystal Contact Temperature Measurement

Cholesteric liquid crystals are substances which exhibit some liquid properties, e.g., low viscosity, combined with other properties of crystals. Their optical properties are temperature dependent, hence their application for thermography. Other forms of liquid crystals exist, such as nematic and sinetic forms which have different physical properties. The characteristic optical properties of the cholesterics are explained by their helical structure, circular dichroism, and rotating power. When an incident beam of light is directed at a thin film of liquid crystals double polarization occurs. Linearly polarized light with a specific wavelength is reflected, while another component is transmitted to the next layer. Each molecular layer reflects a part of the incident white light with the same specific wavelength. When that beam is within the spectrum of visible light, the liquid crystals appear to be colored. The actual wavelength is related to the pitch of the helix representing the structure of the molecule. As temperature increases, the pitch decreases because the molecular layers move toward each other. On cooling, the layers move apart, thus changing the wavelength again. Within the visible spectrum, the cooler colors appear at the longest wavelengths, i.e., red. At higher temperatures shorter wavelengths, green and blue, are visible. Liquid crystal formulations for medical use are prepared by mixing several esters of cholesterol. The range of temperature of the given mixture is dependent on the proportions of the component esters. To prepare a mixture of a specified temperature range is difficult and therefore expensive.[13]

Little use is now made of the original paint and spray mixtures, which are largely unsuitable for clinical application. Microencapsulated liquid crystals have advantages in that they are protected from exposure to the environment which causes rapid decay of their properties. When prepared in polymer microcapsules of around 20 µm, they become stable for long periods (approximately 4 to 5 years), and can be packaged within elastic membranes for reusable contact application. The simplest form of plastic foils used in medi-

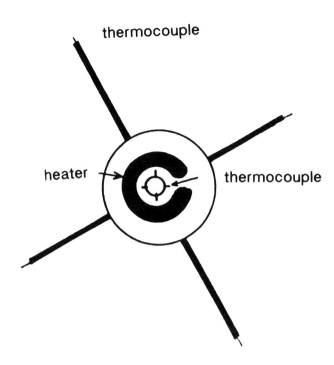

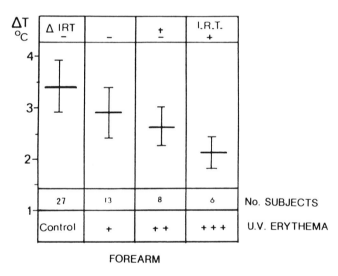

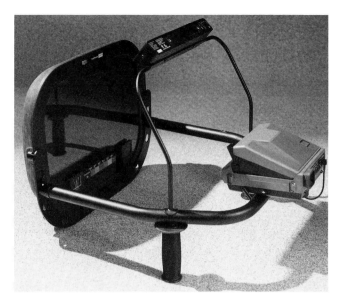

FIGURE 8 The Novatherm® liquid crystal detector system, with a detector unit in place for recording.

FIGURE 7 (A) Thermal clearance probe with annular heater and four thermocouple bridge circuits. (B) Results obtained with probe (A) on 27 subjects given experimental UV erythema at constant dosage. Slight erythema in 13 subjects did not show thermographic skin temperature change. Marked erythema in six subjects produced thermographic skin temperature changes in excess of 2°C. Bars indicate mean thermal clearance temperatures ± S.E.M.

cine is held in an open rigid frame. A number of these sheets are required, since each may have a specified range of some 3°C. The shape and size of the detector sheets varies with the manufacturer. All are supplied with a fixed focus camera and oblique electronic flash system. Ideally, a video CCD camera should be used, so that the warming on contact and cooling of the detector sheet are recorded. Lighting for a video system is, however, much more difficult. Most high-intensity lamps have a heating effect, and room lighting is rarely sufficient to generate sufficiently high lux values from the surface of the liquid crystal sheets.

Improved contact with the skin surface is achieved by an air-filled cushion. The detector surface is laid on an elastic material which is thermally conductive. The active crystal layer is inside the air-filled cushion and viewed from outside through a rigid clear plastic window. The cushion can be supported by a hand-held stand to facilitate photographic recording[14] (Figure 8).

It is essential that a strict protocol is adopted when recording temperature patterns by this technique. It is also important to note that each cushion has a narrow specified range of temperature. All areas of the skin that fall outside this range may register black or show as bright blue. Only when several intermediate colors, typically brown, orange, and green, are observed has the correct temperature range been selected. The examination must also be repeated with the next lower and next higher detectors to observe any detail beyond the range of the "optimal" level. Care should also be taken when using older detectors, since the calibration and stated range can change with time. For this reason precise temperature measurement cannot be guaranteed by this method. The spatial resolution of a given detector may be observed by lightly placing a point thermal source and imaging the chromatic halo around the hot spot. Since the membranes on which these liquid crystals are mounted can be damaged by this procedure (also by fingernails), great care is required. A good detector will show spatial resolution of around 1 mm. The thermal inertia and color response time of a detector will also vary with manufacturer and age. Good detectors will exhibit a return to base color (black) in less than 10 seconds.[15]

Liquid crystal or contact thermography is a relatively low-cost method for imaging small areas of skin temperature, but has limitations. Small localized and thermally contrasting features of skin temperature distribution may be adequately recorded by good technique; for example, vascular patterns, varicosed veins, etc. can show very clearly. These detectors are also suitable for outdoor use to identify localized sports injuries. However, there are now small portable electronic imagers available, which are less cumbersome and more reliable.

IV. Principles of Thermal Radiation

Infrared radiation is a particular form of electromagnetic energy. It occurs beyond the visible spectrum at wavelengths from 1 μm (near infrared) to 15 μm in the far infrared. From 15 μm to 1 mm an extreme infrared band occurs, before the radio wavelengths at extra high frequency lead into the UHF and VHF band, commonly used in sound and television broadcasting.

Infrared or thermal radiation is generated by the motion of charged atomic particles in any material whose temperature is above absolute zero, i.e., 0 K or −273°C. In addition to the natural properties of waves, electromagnetic radiation also behaves in the manner of a stream of particles or photons.

The amount of energy an object radiates is dependent on two factors, temperature T and emissivity ε. Emissivity is an expression of the ratio of the rate of energy emitted to that emitted from an ideal "black body" or perfect radiator at the same temperature.

An important principle was established by Stefan in 1879, who found that the total power emitted by a black body of emissivity 1.0 is proportional to the fourth power of temperature. In practice, surfaces usually emit and absorb less than an ideal "black" surface.

Emissive power or emittance of a surface at temperature T is defined as

$$M_e = \varepsilon \sigma_e T^4 \ (W/cm^2)$$

The symbol σ represents Stefan-Boltzmann's constant and has the value of $5.67 \times 10^{-8} \ Wm^{-2}K^{-4}$.

If the emissivity of a surface is known, the radiant power is a measure of its temperature.

When the spectra emitted by a black body at different temperatures are recorded, the spectral radiant emittance at each wavelength interval can be seen to pass through a maximum, which moves to shorter wavelengths as the temperature of the body is increased. When the temperature of the body reaches 1000 K, the radiant emittance in the visible part of the spectrum becomes luminous, with color changes according to the mixtures of wavelengths present. For example, a white hot or yellow iron bar taken from a furnace will be at a temperature higher than if it is red in color.

In a theoretical perfect radiator the assumption is made that the surface does not impede the emission of energy from any part of the electromagnetic spectrum. However, a perfect radiator not only emits its total energy without impedance, but also absorbs energy with the same freedom.

This theoretical situation does not exist for real bodies or surfaces. The characteristics of the surface do actually exert some degree of influence over the absorption and emission of radiant energy. The emissive properties of a substance can be strongly dependent on wavelength, so that a high emissivity substance at one wavelength may be lower at another.

The human body is a living radiator, with the skin as a dynamic interface between the internal body organs and the environment surrounding the subject. One of the highly organized functions of living skin is to have efficient radiative properties in the infrared region, with a high emissivity. This property is unaffected by color or pigmentation of the skin in the visible range of the spectrum (Figure 9).

A. Skin Emissivity

An American physiologist, J. D. Hardy, showed by experiment that living skin on a normal subject will exhibit an average emissivity over the whole spectrum of 0.97 to 0.98. A number of workers have reexamined this phenomenon, and have given general agreement. It is important that the emissivity of skin should be known, for without such data it is not possible to determine skin temperature by the radiation emitted from its surface. There are many advantages of radiation thermometry as a means of measuring skin temperature. One major advantage is that the sensor does not make contact with the skin, which as a living organ is very reactive to physical contact.

Investigations into skin emissivity have been conducted by a number of scientists. These include Buttner,[16] Mitchel et al.,[17] Watmough and Oliver,[18] and Stekettee.[19,20]

The results of these studies are in approximate agreement, but show many inconsistencies which result in a range of emissivity values of some 8%. In all these studies, infrared radiation was measured in a thermal steady state. The conditions used provided for a heat flow across the surface of the skin, due to the temperature difference between the skin and the ambient space. When both temperatures are equal, the reflection and emission at the surface have the same energy spectrum. As a result it is not possible to discriminate between these functions, and therefore to accurately measure emission per se.

When a steady-state flux does exist across the skin, a temperature gradient is built up in the tissues. In this case, it is difficult to measure the exact temperature of the skin with a contact thermometer. In

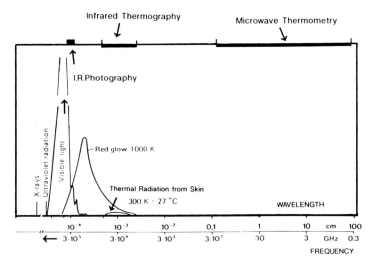

FIGURE 9 A comparative representation of thermal radiation from the skin, which is a fraction of the intensity of heat emitted by a red glowing object at 1000 K. The relative spectra used for IR thermography and microwave thermometry are shown.

reflectance measurement, in a steady state, incident radiation causes a change in heat flow across the skin. This has the resulting effect that the actual temperature of the skin will be influenced. Even when these effects are minimal, they can cause a significant effect on the calculated skin emissivity. When the emissivity value is close to unity, as in the case of human skin, the small difference between radiant energy from the skin and from a black body at the same temperature must be determined.

Togawa in 1985 proposed a measurement method for determining skin emissivity in a transient state.[21] The method used is a zero-heat-flow thermometer described by Fox and Solman in 1971 to achieve isothermal conditions in the skin.[22] Radiation from the skin was measured immediately after the probe was removed from the surface. This technique presented a problem in that an accurate calibration was needed between the heat-flow thermometer and the radiation detector. It also suffered from the disadvantage that some 15 min was required to achieve isothermal conditions beneath the heat-flow probe. The improved technique by Togawa in 1989 addressed these problems very effectively.[23] In this later method, they measured the infrared radiation emitted by the skin at a stepwise increment in ambient radiation temperature. As a result they were able to remove the necessity for absolute calibration between the contact and non-contact devices. At the same time the measurements could be made in 1 min only. Togawa was thus able to show that skin emissivity could be calculated from the radiometric measurements before and after switching the ambient radiation, as long as the temperature of the latter was known.

The ingenious device used two conical shades to create a specific ambient over the test skin surface. The radiometer sensed the infrared energy emitted from the skin through an aperture in the peak of the cone. A mechanical device allowed the two cones, one of which was heated to 40°C, to be alternately moved into position between the radiometer and the skin surface within 0.5 second. The radiometer, a HgCdTe (77 K) detector with a uniform sensitivity within the 8- 14-μm range, was sampled every 100 ms. By applying this device to different areas of the human body surface in male and female subjects, an average emissivity value was obtained. Neither the different sites nor the male and female differences were found to be statistically different. Their observed emissivities remained in a range between 0.961 to 0.981, with an overall average and standard deviation of 0.971 ± 0.005.

Togawa has discussed in detail why these results differ from some of the earlier published data. The advantages of the Togawa method are improved accuracy of measurement and rapid acquisition of data, making it highly adaptive for clinical studies in pathological conditions. Little is known of the possible changes, if any, in emissivity of human skin in certain disease states. It is not unreasonable to question the effects of some disruptive dermatological conditions on the physical properties of the skin surface. Ulcerated skin lesions are clearly abnormal, once the intact surface of the epidermis is destroyed.

B. Radiative Heat Transfer from the Skin

The human body is a thermal organism and is described as homeothermic or self temperature regulating. Changes in the skin occur to regulate the superficial blood flow necessary to balance the heat flow to and from the ambient. This is necessary to maintain a specific "core" temperature around the vital organs of the head and thorax. The skin thus forms a dynamic interface between the body and its (normally) cooler environment. As such it is the principal source of distributing excess heat into the immediate surroundings of the body. It does this by radiating heat to form a microclimate or thin layer of warmth over the body surface. Ideal conditions for measuring skin temperature occur when the skin is in a thermal steady state. In clinical practice some compromise is usually adopted.

Light in the near infrared up to 6 μm is partially reflected by skin tissues. The slight differences in reflectance of skin of different pigmentation and color in the near infrared become negligible in the far infrared. However human skin is moderately transparent to the infrared wavelengths between 0.7 and 1.5 μm down to a depth of 2 mm.[16] Veins near the skin surface absorb infrared radiation and may appear dark in reflected light. This can be demonstrated by viewing superficial veins beneath the skin surface through red glass or taking infrared photographs by reflected red light.

Some thermal radiation passes through the superficial horny layer of the skin. Infrared temperature measurements of the skin are therefore not entirely dependent on the actual surface temperature but also from the immediate subsurface of the skin.

C. Thermal Imaging Systems

Electrooptical systems for passive non-contact imaging of temperature are now highly developed. The current technology used is based on the detection of infrared radiation emitted by the skin. The detectors commonly used cadmium mercury telluride, or indium antimonide, which generates an electrical signal which is amplified and displayed. A scanning optical arrangement allows the rapid construction of a two-dimensional image.[24] In its simplest form this may be displayed as white = hot, black = cool, or vice versa. Increased sensitivity for the narrow temperature range of the human skin is obtained by cooling the detector, either by liquid gas (N_2) or by some form of cooling engine. The speed of image capture is dependent on the type of detector and scanning mechanism. The simplest are single element detector systems, while those operating at higher speed use multielement dectors.[25] One high-resolution high-speed detector (Sprite-Mullard, U.K.) uses a multiple array of strips of mercury cadmium telluride.[26,27] Infrared transmitting lenses of silicon or germanium are now used, with many operational advantages similar to optical lenses used in visible light optics[28] (Figure 10).

Thermal imaging systems for medical use are normally equipped with basic controls over a temperature range of 2 to 40°C and may be calibrated by an internal or external thermal reference. Temperature measurement facilities vary, but may include selected spot temperatures, temperature difference between two or more spots or regions of the image. Isotherms, which link points at the same temperature in the electronic image, and selectable rectangular or irregular regions of interest are also standard. Calculations from this region may be for maximum, minimum, or mean temperature or a defined index calculated by a chosen formula. Computer image processing is now almost universal, with provision for color coding of temperature (color isotherms). A conventional color thermogram will use a spectral color range, where blue is cold and red-white is hot. Intermediary temperatures are shown in shades of green, yellow, orange, etc.

D. The Examination Room

Modern thermal imagers are small and portable and therefore usable in a clinic or ward. However, unless good environmental conditions can be provided, the

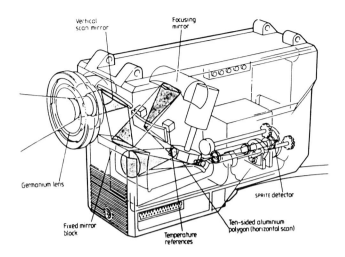

FIGURE 10 Schematic diagram of an infrared thermal imaging system (Agema Infrared Systems) using a cooled detector and IR transmitting lens.

image obtained will be of limited value. Ideally, the ambient temperature must be controlled, and not subject to variation, or achieved by high air speeds. Direct sunlight on the subject must be avoided. The patient should, where possible, be positioned in a standard and reproducible way, at least 15 cm from the nearest wall, to avoid reflection of heat. Humidity should be optimal at around 45% RH, since high humidity has a marked effect on increasing sweat function, which will affect the resultant image. Recommendations for the ideal conditions were described in 1979[29] and 1984.[30] Good quantitative and reproducible technique is only achieved when the instrumentation is of proven stability and when the physiological conditions are also correct for the patient. Examination of inflammatory areas of the skin requires a cool environment — 20°C in Europe — whereas neurological and sympathetically mediated reactions should be tested in a warmer ambient temperature to avoid vasoconstriction; 22 to 24°C is used in Europe, while higher temperatures have been used in Japan and "warmer" countries.

Preparation of the patient is important. The areas to be examined should be unclothed. A period of 15 min is often used to achieve some form of equilibrium with the environment. The thermal contrast on the skin, temperatures measured over the body surface from head to foot, will be dependent on ambient temperature and time of exposure.[31] These conditions should be determined and standardized for any given investigation using thermal imaging techniques.

V. Applications of Thermal Imaging

A. Thermal Patterns of the Skin

The human body is subject to a wide range of "biological" variants, many of which influence skin temperature. However, practical experience with infrared

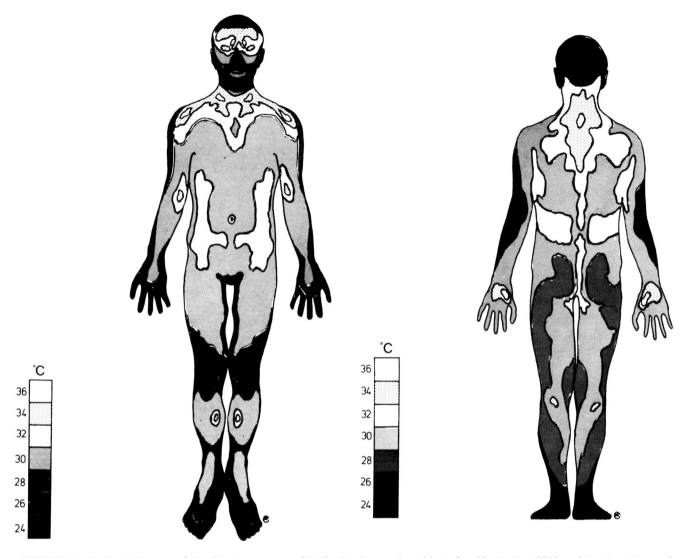

FIGURE 11 Stylized diagram of the skin temperature distribution in a male subject after 20 min in a 20°C ambient (anterior and posterior).

thermography shows that under given conditions human skin temperature will exhibit a predictable pattern from which pathological effects can be measured (Figure 11). There is a definite relationship between skin perfusion and temperature.[32] In general well-perfused tissues are warm and will be less subject to temperature reduction in a cooler ambient. Conversely, tissues with low blood perfusion will lose heat more dramatically in a colder ambient. In the extremities, sympathetic control of the vascular system may be triggered by external stimuli, which may be physical (hot or cold) or chemical (vasoactive drugs). Inflammation with increased local perfusion may be present on or below the skin. The congested vascular network may show little or no direct response to humoral or neurological influence.

B. Increased Skin Temperature

There are many practical applications of this natural phenomenon in clinical medicine which can be monitored by infrared imaging. Inflammation has been recognized for centuries by heat, redness, pain, swelling, and loss of function. Thermal imaging is a reliable technique for monitoring local heat and therefore to identify the extent, degree, and response to treatment.[33–36] Under strictly controlled conditions, inflammatory lesions such as arthritis may be measured and the efficacy of antiinflammatory drug treatment or surgery (e.g., synovectomy) can be quantified.[37,38] Unlike the many subjective parameters used to assess arthritis severity, thermal imaging will reveal the temperature independent of local pain (Figure 12). Therefore, analgesic doses of some agents may improve function, but only true antiinflammatory doses of a drug will produce a reduction in temperature over the affected area.[39] This observation made in 1974[40] has been used extensively in clinical trials of new nonsteroidal, steroid, and second line treatments (penicillamine, methotrexate, colloidal gold therapy, etc.). A thermal index which is based on the distribution of isotherms and mean temperature of a defined region has been used to compare known analgesic and antiinflammatory

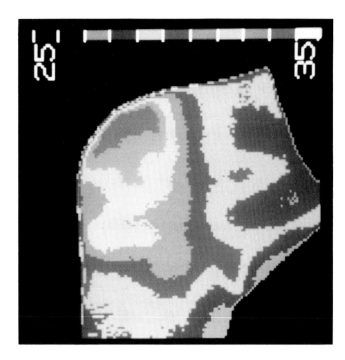

FIGURE 12 Thermal image of lateral knee joint, showing inflammation over the joint and high temperature over the patella area.

agents. Different analogues of prednisolone have been tested on groups of inflamed joints to compare plasma drug levels with local and systemic effects.[41] The thermal index has also been used to demonstrate dose response curves of antiinflammatory drugs in groups of patients and in experimental arthritis in animals.[38] It is one of the few objective tests of inflammation which can be equally applied to animals and man.

Increased skin temperature has been quantitatively monitored in joint infection[42] and in Paget's disease of bone.[43,44] Certain anatomical sites where bone is close to the skin surface, e.g., forehead and tibiae, may show large temperature increase in active disease (Figure 13). Figure 14 illustrates the thermal indices measured over active painful Paget's disease of the tibia compared with non-painful disease and controls. Increasing temperature often precedes increased bone pain and vice versa. Temperature may be shown to fall during calcitonin and bisphosphonate therapy, rebounding when the treatment is discontinued and the disease is still active.

Studies on erythema and urticarial eruptions have shown that quite large skin temperature increases can be measured. However, the area of visible skin color change may be smaller than the surrounding skin temperature increase.[45] A comprehensive collection of thermograms relating to dermatological problems is given by Stuttgen and Flesch.[46]

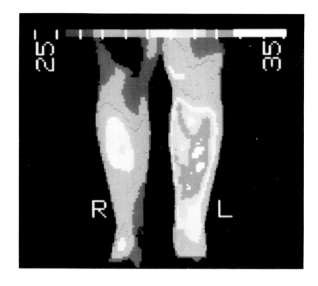

FIGURE 13 Paget's disease of the tibia, showing increased heat over left anterior leg (also synovitis of left knee).

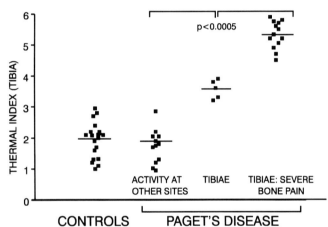

FIGURE 14 Thermal index recorded over the tibiae of controls and patients with Paget's disease. Increased temperature is associated with bone pain.

C. Decreased Skin Temperature

A reduction in skin temperature may occur in a number of conditions where there is decrease in blood perfusion. On the extremities, this is readily detectable in a unilateral condition, where comparison with the unaffected arm or leg can readily be made. In certain situations, e.g., venous ulceration, the hypothermic area may be localized, revealing the extent of decreased skin perfusion surrounding the lesion. It may however be necessary to apply a thermal or chemical stress to provoke a temperature difference. This is commonly applied to the measurement of severity of Raynaud's phenomenon, where recovery from a brief immersion of the hands in water is significantly delayed (Figure 15).

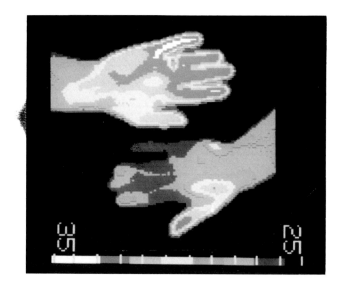

FIGURE 15 Dorsal hands of a patient with Raynaud's phenomenon after cold stress thermography. Fingers on one hand are cold 10 min after 20°C water immersion.

1. Raynaud's Phenomenon

Thermal imaging is a very efficient technique for non-contact assessment of vasospasticity.[47] The generally accepted procedure is for the patient to rest in a chair at 22 to 23°C ambient temperature for at least 10 min. A thermogram of the dorsal surface of the hands is then recorded. This is followed by a 1-min 20°C waterbath immersion of both hands up to wrists. The hands are kept dry by large disposable plastic gloves (or bags). At the end of the stress period, the hands are allowed to rest for 10 min and are then rethermogrammed. Measurement of the mean temperature of two separate regions, fingers to MCP joints and MCPs to wrist, can be used to indicate the mean temperature gradient (strictly mean temperature difference) from the wrist to fingertips. This gradient records a positive value when reactive hyperemia occurs. In Raynaud's phenomenon, a negative gradient before cold stress becomes even more negative after stress. The degree of negativity is a measure of severity.[48] Some centers have modified this method, either by varying the temperature of the waterbath, e.g., 10 to 15°C, or in continuously monitoring fingertip temperatures during recovery. The conclusions are largely similar, but results are only comparable when the same strict protocol has been followed. While absolute temperatures over the hand may be influenced by external climatic conditions, the stress test overcomes external influences and is a consistent and objective indicator of Raynaud's phenomenon (Figure 16). The stress test has also been used to test for neurological effects from cervical sympathectomy, brachial plexus injury, stroke, hemiplegia, and reflex sympathetic dystrophy.

2. Neurological Dysfunction

In the absence of disease or injury and under controlled conditions, the human body shows a high level of thermal symmetry. Independent studies on control subjects have shown that in many cases this may be of the order of 0.3°C, from left to right over much of the trunk and limbs.[49,50] Subjects who are regularly involved in sports may show increased temperature difference due to higher perfusion of the dominant arm or leg. However, in certain cases where peripheral nerve injury or congestion has occurred, a decrease in temperature at the extremity may be evident. This hypothermic region does not necessarily fall into a dermatome distribution, but will sometimes be quite characteristic.[51] Critics of this application of thermography rightly point out that this is not a perfect sign for nerve root compression or injury. However, it has been used effectively in conjunction with other investigations such as CT, MRI, EMG, etc., and can be repeated more frequently to monitor progress. In reflex sympathetic dystrophy decreased temperatures on the affected limb can be marked — up to 5 or 6°C in ambient temperatures of 22°C.[52] Here also, a very localized hypothermia can be demonstrated, which cannot be attributed to simply "disuse" of the limb. Thermal imaging does offer objective evidence of such peripheral skin temperature changes which cannot be influenced by the patient (Figure 17). For this reason, it is a good procedure to use when investigating possible psychosomatic complaints, where normal temperature distribution may preclude a genuine pathology.

In post-trauma follow-up, an abnormal response in the suspect lesion may be revealed by thermal imaging. Where relevant, (e.g., injury to one or more digits) cold stress tests can reveal an abnormal reaction which was not indicated by the resting thermogram.

Thermal imaging has also been used successfully to confirm the level of limb amputation prior to surgery.[53] A poorly vascularized region, showing cold on a thermogram, is likely to result in delayed or impaired healing of scar tissues. A warm well-perfused area is more likely to heal successfully after surgery. Limited use of the technique has also been made in assessing burn injuries and subsequent skin grafting. A thermogram may give early indications of graft rejection by persistent low skin temperature.

D. Conclusion

Thermal imaging, particularly infrared thermography, is a highly developed technology. Its use in medicine has proved controversial in some areas. Nevertheless, it is indisputably the most efficient means of studying skin temperature since the birth of medicine. Used correctly, under controlled conditions, it can far outweigh the information which is so universally accepted from the

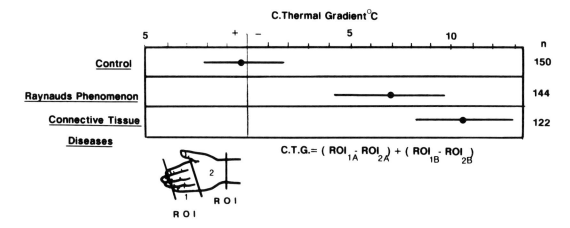

FIGURE 16 Scale of values obtained from the dorsal hand temperatures measured before and after cold stress at 20°C. Increasing negativity on this scale indicates severity of the vasospastic reaction.

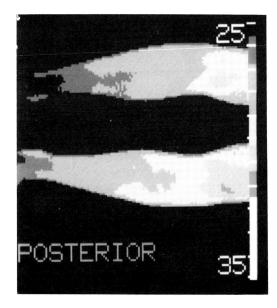

FIGURE 17 Decreased temperature over the lower limbs in a patient with Sudeck's algodystrophy.

clinical or oral thermometer. Since these modern technologies have become available, the clinical knowledge of skin temperature patterns has been extended and clarified. The ability to image discrete cooling effects of sweat pore secretion in real time is one of many new areas for study by thermography. Unfortunately, the highest resolution thermal imaging is expensive at the present time. Nevertheless, the new technical possibilities with high-resolution thermal imaging and prospects for low-cost portable thermal imaging cameras are proving of value in medical research.[54] With the improved knowledge of their potential and better understanding of thermal physiology, they are more likely to find a valuable role in medicine than when the first systems were used in the early 1960s. Unlike the early days of thermal imaging, it is possible to extend the observation of thermal patterns on the skin to fast and efficient two-dimensional measurement of skin temperature.

When you can measure what you are speaking about, and express it in numbers, you know something about it. But when you cannot measure it, when you cannot express it in numbers, your knowledge is of a meagre kind. It may be the beginning of knowledge, but you have scarcely in your thoughts advanced to the stage of science.

Lord Kelvin (1824–1907)

References

1. Wunderlich, C., *On the Temperature in Disease: A Manual of Medical Thermometry,* translated into English, Bathhurst Woodman W, New Sydenham Society, London, 1972.
2. Hardy, J.D., The radiation of heat from the human body. III. The human skin as a black body radiator, *J. Clin. Invest.,* 13, 615, 1934.
3. Herschel, J.F.W., Investigation of the power of the prismatic colours to heat and illuminate objects, *Philos. Trans.,* 90, 255, 1800.
4. Herschel, J.F.W., On the action of the rays of the solar spectrum on preparation of silver and other substances etc. Note 1 on the distribution of the calorific rays in the solar system, *Philos. Trans.,* 51, 1840.
5. Houdas, Y. and Ring, E.F.J., Man and his environment, in *Human Body Temperature: Its Measurement and Regulation,* Plenum Press, New York, 1982, chap. 4.
6. Mitchell, D. and Wyndham, C.H., Comparison of weighting formulas for calculating mean skin temperature, *J. Appl. Physiol.,* 26, 616, 1969.
7. Houdas, Y., Sauvage, A., Bonaventure, M., and Guieu, J.D., Control of heat exchange: an alternative concept for temperature regulation, in *Regulation and Control in Physiological Systems,* Iberall, A.S. and Guyton, A.C., Eds., International Federation for Automatic Control, Pittsburgh, 1973, 217.
8. Schonecht, G., Electrical contact thermometry, in *Thermological Methods,* Engel, J.M., Flesch, U., and Stuttgen, G., Eds., Verlag Chemie, Weinheim, 1985, 35.
9. Magdesburg, H., Regulations for the certification of contact thermometers, in *Thermological Methods,* Engel, J.M., Flesch, U., and Stuttgen, G., Eds., Verlag Chemie, Weinheim, 1985, 203.

10. Britton, N.F., Barker, J.R., and Ring, E.F.J., An assessment of the thermal clearance method for measuring perfusion, in *Recent Advances in Thermology,* Ring, E.F.J. and Phillips, B., Eds., Plenum Press, New York, 1984, 327.
11. Britton, N.F., Barker, J.R., and Ring, E.F.J., A mathematical model for a thermal clearance probe, *IMA J. Math. Appl. Med. Biol.,* 1, 95, 1984.
12. Ring, E.F.J., Watson, C., and Barker, J., Infrared thermography and thermal clearance of the skin, in *Thermological Methods,* Engel, J.M., Flesch, U., and Stuttgen, G., Eds., Verlag Chemie, Weinheim, 1985, 133.
13. Flesch, U., Techniques for liquid crystal thermography in medicine, in *Thermological Methods,* Engel, J.M., Flesch, U., and Stuttgen, G., Eds., Verlag Chemie, Weinheim, 1985, 45.
14. Ring, E.F.J., Skin temperature measurement, *Bioeng. Skin,* 2, 15, 1986.
15. Gautherie, M., Techniques of liquid crystal thermography, in *Atlas of Breast Thermography,* Padusa s.r.i., Milan, 1989, 10.
16. Buttner, K., *Physikalische Bioklimatologie,* Akademische Verlagsgesellschaft, Leipzig, 1938.
17. Mitchel, D., Wyndham, C.H., and Hodgson, T., Emissivity and transmittance of excised human skin in its human emission, *J. Appl. Physiol.,* 23, 390, 1967.
18. Watmough, D.J. and Oliver, R., Emissivity of human skin in-vivo between 2.0 µ and 5.4 µ measured at normal incidence, *Nature,* 218, 885, 1968.
19. Stekettee, J., Spectral emissivity of skin and pericardium, *Phys. Med. Biol.,* 18, 686, 1973.
20. Stekettee, J., The influence of cosmetics and ointments on the spectral emissivity of skin, *Phys. Med. Biol.,* 21, 920, 1976.
21. Togawa, T., Skin emissivity measurement using unsteady state immediately after removal of a zero-heat-flow thermometer probe, Proc. 14th Int. Congress of Medical and Biological Engineering, ESPOO, 1985, 1016.
22. Fox, R.H. and Solman, S.J., A new technique for monitoring the deep body temperature in man from the intact skin surface, *J. Physiol.,* 212, 8, 1971.
23. Togawa, T., Non-contact skin emissivity: measurement from reflectance using a step change in ambient radiation temperature, *Clin. Phys. Physiol. Meas.,* 10, 39, 1989.
24. Putley, E.H., The development of thermal imaging systems, in *Recent Advances in Medical Thermology,* Ring, E.F.J. and Phillips, B., Eds., Plenum Press, New York 1984, 151.
25. Lloyd, J.M., Thermal imaging system types, in *Thermal Imaging Systems,* Plenum Press, New York, 1975, chap. 8.
26. Elliott, C.T., New detector for thermal imaging systems, *Electron. Lett.,* 17, 312, 1981.
27. Alderson, J.K.A. and Ring, E.F.J., Sprite high resolution thermal imaging systems, *Thermology,* 1, 110, 1985.
28. Ring, E.F.J., Technical advances in thermal imaging, in *Thermal Assessment of Breast Health,* Gautherie, M., Albert, E., and Keith, L., Eds., MTP Press, Boston, 1983, 3.
29. Anglo-Dutch Thermographic Society, Thermography in locomotor diseases: recommended procedure, *Eur. J. Rheumatol. Inflammation,* 2, 299, 1979.
30. Ring, E.F.J., Engel, J.M., and Page Thomas, D.P., Thermological methods in clinical pharmacology, *Int. J. Clin. Pharmacol. Ther. Toxicol.,* 22, 20, 1984.
31. Ring, E.F.J., Standardisation of thermal imaging in medicine: physical and environmental factors, in *Thermal Assessment of Breast Health,* Gautherie, M., Albert, E., and Keith, L., Eds., MTP Press, Boston, 1983, 29.
32. Houdas, Y. and Ring, E.F.J., Temperature distribution, in *Human Body Temperature, Its Measurement and Regulation,* Plenum Press, New York, 1982, 96.
33. Collins, A.J. and Ring, E.F.J., Measurement of inflammation in man and animals by radiometry, *Br. J. Pharmacol.,* 44(1), 145, 1972.
34. Bacon, P.A., Collins, A.J., Ring, F.J., and Cosh, J.A., Thermography in the assessment of inflammatory arthritis, *Clin. Rheum. Dis.,* 2, 1, 51, 1976.
35. Ring, E.F.J., Thermographic and scintigraphic examination of the early phases of inflammatory disease, *Scand. J. Rheumatol.,* Suppl. 65, 77, 1987.
36. Engel, J.M., Thermographische Diagnostik, *Rheumatol., Aktuel Rheumatol.,* 4, 25, 1979.
37. Bacon, P.A., Ring, E.F.J., and Collins, A.J., Thermography in the assessment of anti rheumatic agents, in *Rheumatoid Arthritis,* Gordon, J.I. and Hazleman, B.I., Eds., Elsevier, Amsterdam, 1977, 105.
38. Ring, E.F.J., Thermal imaging and therapeutic drugs, in *Biomedical Thermology,* Gautherie, M., Ed., Alan R. Liss, New York, 1982, 463.
39. Bacon, P.A. and Ring, E.F.J., Thermal imaging in assessment of drugs in rheumatology, in *Recent Advances in Medical Thermology,* Ring, E.F.J. and Phillips, B., Eds., Plenum Press, New York, 1984.
40. Ring, E.F.J., Collins, A.J., Bacon, P.A., and Cosh, J.A., Quantitation of thermography in arthritis using multi-isothermal analysis. II. Effect of nonsteroidal anti-inflammatory therapy on the thermographic index, *Ann. Rheum. Dis.,* 33(4), 353, 1974.
41. Bird, H.A., Ring, E.F.J., and Bacon, P.A., A thermographic and clinical comparison of three intra articular steroid preparations in rheumatoid arthritis, *Ann. Rheum. Dis.,* 38, 36, 1979.
42. Bird, H.A. and Ring, E.F.J., Thermography and radiology in the localisation of injection, *Rheumatol. Rehab.,* 17, 103, 1978.
43. Ring, E.F.J. and Davies, J., Thermal monitoring of Paget's Disease of bone, *Thermology,* 3, 167, 1990.
44. Ring, E.F.J., Quantitative thermal imaging, *Clin. Phys. Physiol. Meas.,* 11 (Suppl. A), 87, 1990.
45. Stuttgen, G., Dermatology and thermography, in *Thermological Methods,* Engel, J.M., Flesch, U., and Stuttgen, G., Eds., Verlag Chemie, Weinheim, 1984, 257.
46. Stuttgen, G. and Flesch, U., *Dermatological Thermography,* Verlag Chemie, Weinheim, 1985.
47. European Association of Thermology Report, Raynaud's phenomenon assessment by thermography, *Thermology,* 3, 69, 1988.
48. Ring, E.F.J. and Elvins, D.M., Quantification of thermal images, *J. Photogr. Sci.,* 37, 164, 1989.
49. Uematsu, S., Symmetry of skin temperature comparing one side of the body to the other, *Thermology,* 1, 4, 1985.
50. Uematsu, S., Edwin, D.H., and Jankel, W.R., Quantification of thermal asymmetry. I. Normal values and reproducibility, *J. Neurosurg.,* 69, 552, 1988.
51. Green, J., Leon Barth, C.A., Hickey, S.T., and Dieter, J., Efficacy of neurodiagnostic studies in patients with lumbosacral and single-leg pain of sciatic distribution of 90 days or more, *Pain Digest,* 2, 213, 1992.
52. Will, R. K., Ring, E.F.J., Clarke, A.K., and Maddison, P.J., Infrared thermography: what is its place in rheumatology in the 1990's?, *Br. J. Rheumatol.,* 31, 337, 1992.
53. Spence, V.A., Walker, W.F., Troup, I. M., and Murdoch, G., Amputation of the ischaemic limb: selection of the optimum site by thermography, *Angiology,* 32, 155, 1981.
54. Ring, E.F.J., Video thermal imaging, in *Thermological Methods,* Engel, J.M., Flesch, U., and Stuttgen, G., Eds., Verlag Chemie, Weinheim, 1985, 101.

19.0

Lymph Flow

19.1 Evaluation of Lymph Flow 473
 P. Mortimer

Chapter 19.1
Evaluation of Lymph Flow

P.S. Mortimer
Consultant Skin Physician
St. George's and Royal Marsden Hospitals
London, U.K.

I. Introduction

Few techniques exist for the functional assessment of skin lymphatics; yet, it is in this area of microcirculation research that most questions relating to the role of lymphatics in disease remain unanswered.

The lymphatic vessels provide an important "limb" to the microcirculation of the skin and, together with the blood vessels, cater for the constant recirculation of protein- and lymph-borne cells, e.g., Langerhans cells[1] and T lymphocytes. It is the essential function of the lymphatic system to return to the vascular compartment extravascular protein molecules, colloids, and particulate matter too large to reenter the blood capillaries directly.[2] The rate at which labeled protein molecules or colloids are removed from the interstitial tissues has been regarded as an index of lymphatic function.[3-5]

Measurement of skin and subcutaneous lymph flow has employed the same principle of isotope clearance as measurement of skin and subcutaneous blood flow.[6,7] However, the interpretation of the clearance of tracers from the skin in disease states appears difficult and unreliable.[8,9]

II. Background

A. Lymphangiography

In vivo visualization of lymphatic vessels **(lymphangiography) using X-ray contrast medium**[10] remains the gold standard for lymphatic vessel abnormalities. The technique, however, is invasive, difficult to perform, and provides only anatomical detail with no functional information. Only subcutaneous lymphatics as large, or larger than, collectors can be opacified except in pathological circumstances when dermal backflow occurs and small skin lymphatics become visible.

Intravital dyes, e.g., patent blue, used to delineate subcutaneous lymphatics prior to direct cannulation for X-ray lymphography, can be used to visualize dermal lymphatics but not capillaries.[11] The results are however transitory.

Two new methods of lymphangiography have been developed in recent years.

1. Fluorescence Microlymphangiography[12]
This technique enables the superficial lymph capillary network of the skin to be seen under the vital microscope by means of fluorescing macromolecules (FITC-DEXTRAN, SIGMA) injected subepidermally and cleared exclusively by lymphatics. Information regarding the morphology of lymphatic capillaries and precollectors (initial lymphatics) and the extent of tracer propagation within the dermal lymphatic network can be recorded on video for analysis.

2. Indirect Lymphography[13]
Indirect lymphography employs water-soluble non-ionic X-ray contrast media that can be administered via an interstitial injection without recourse to direct access to lymphatics. Iotralan® or Iotasol® (Schering AG, Berlin) is infused by a motor pump into the skin; 2 to 3 ml injected intradermally leads to considerable local skin distention and is not without discomfort. Dermal and subcutaneous collec-ting lymphatics can be visualized by X-ray using the mammography film method. In the presence of incompetent valves and dermal backflow, initial lymphatics can also be seen.

All lymphangiographic methods are limited in their ability to evaluate lymph flow as the techniques are essentially for demonstrating the anatomy of lymph vessels.

B. Lymphoscintigraphy
The development of lymphoscintigraphy was aimed mainly at imaging the lymphatic system and in particular the lymph nodes.[14,15] Lymphoscintigraphy has proved more useful in the determination of lymph flow. Because of the close inter-relationship between lymph formation and flow, lymphoscintigraphy theoretically provides a much more comprehensive and functional examination of lymph drainage than does X-ray lymphography.

1. Historical Perspective

The earliest studies involved measurement of lymph flow by external counting following the **subcutaneous** injection of ^{131}I-labeled plasma protein into the hind limb of healthy dogs.[16] The author, using an external scintillation counting technique, concluded incorrectly that the major route of the removal from the tissue spaces of crystalloid and protein molecules is via the blood capillary bed. Taylor et al.,[3] using ^{131}I-human serum albumin (HSA), concluded that the behavior of radioactive proteins injected subcutaneously was consistent with removal by the lymphatic route and that the rate of removal was slower in patients with lymphedema than in normal subjects. Hollander et al.[4] studied patients with and without edema by ^{131}I-HSA clearance from the subcutaneous tissue and concluded that the rate of removal was significantly reduced in edema caused by lymphatic obstruction but significantly increased in edema caused by venous obstruction, congestive cardiac failure, or hypoproteinemia. Similar results were obtained by Sage et al.[17] using radioactive gold (^{198}Au) but the absorbed dose of radioactivity was unacceptably high. Emmett et al.[18] stated that ^{131}I-protein clearance studies are of little value for the initial assessment of individual patients with swollen limbs but may well prove to be the most sensitive method for evaluating the response to treatment. Although further studies on lymphatic tracer clearance were performed, their usefulness for clinical measurement was of some doubt.

2. Lymph Transport Kinetics

Lymphoscintigraphy has proved to be a sensitive and specific method for the study of lymph transport kinetics[19,20] and useful for repeat examinations.[21] Computerization has allowed data on depot clearance, colloid transit, and nodal uptake to be correlated. Investigation times have been reduced owing to extrapolation of data collected over 1 to 2 hours.

a. Lymph Node Uptake

Lymph node uptake has proved to be a more reliable measurement than tracer clearance.[22] Most studies have employed a subcutaneous injection. **Intradermal** administration of tracer seems to encourage increased migration of tracer possibly due to higher interstitial pressures in the dermis than the subcutis.

b. Transit Times

The speed of passage of tracer along main lymphatic vessels immediately following the administration of tracer can be measured. Such **transit times** indicate patentcy of lymph drainage routes as well as being an indirect measurement of lymph flow.

c. Fractional Removal Rate (Depot Clearance)

The amount of an injected tracer deposited in a tissue principally decreases along an exponential curve, the slope of which is expressed as a clearance constant.[23] The tracer must be freely diffusible through the tissue and to equate with lymphatic clearance the tracer must be removed solely through the lymphatic route.

C. Skin Lymph Flow

Measurement of lymphatic function specifically in the skin by the disappearance rate of ^{131}I albumin from the dermis was first reported in 1970;[8] 50 patients were investigated and radioactivity declined exponentially giving a linear plot over 50 hours. Radioactivity fell more rapidly during the first 4 hours giving an initial curve on a semilogarithmic plot. Results were reproducible and clearance rates varied according to the injection site.

The only further study of skin lymph flow examined albumin clearance from psoriatic skin.[9] Again small quantities (0.1 ml) of ^{131}I-HSA containing 10 µCi of radioactivity were administered intradermally and the radioactivity of each depot measured by sodium iodide scintillation detectors. Half clearance times were calculated by regression analysis by the least squares method. Clearance was shown to be monoexponential and increased in involved psoriatic skin indicating increased lymph drainage.

III. Object

If interstitial protein clearance is the essential function of the lymphatic system, can skin lymph flow be measured reliably using the principle of isotope clearance and give meaningful results?

A measure of lymphatic function is the efficiency of protein removal from the tissues. Lymph flow is considered here as equivalent to the movement of protein and accompanying fluid for the purpose of lymph drainage. (Flow refers to bulk transport per unit volume of tissue per unit time. Strictly, it is not possible to measure absolute lymph flow *in vivo*.)

Skin lymph flow relies on several interdependent steps which include lymph formation, its entry into lymph capillaries followed by transit through noncontractile lymphatics, and then propulsion by subcutaneous contractile lymphatics. Movement of solid matter need not necessarily relate to that of fluid and lymph flow is driven by many extrinsic forces. Colloidal (or protein) clearance from the dermis does involve every step of skin lymph flow and is theoretically the ideal test. Perhaps a more appropriate term would be skin lymph drainage instead of lymph flow. For this reason "lymph flow" was expressed as a half clearance time ($t_{1/2}$) and calculated from the slope of the exponential clearance curve.

A. Radiolabeled Tracers

Macromolecules and colloids of a certain size are transported exclusively in the lymph. The ideal tracer is one which migrates freely from the injection site through the tissues and away in the lymphatics. It was discovered that the optimal colloid was one with a particle size of a few nanometers with a small dispersion around that value.[24] Too large a particle resulted in poor absorption from the injection depot and too small a particle risked blood clearance.

^{99m}Tc-labeled agents offer significant advantages in terms of radiation dose and energy characteristics. Much lower doses of radioactivity are possible with external scintillation counting but image formation using scintiscanner or gamma camera demands higher radioactivity.

In a study[25] comparing ^{131}I-HSA, ^{198}Au, and ^{99m}Tc-colloid (TCK 17, Cis), $t_{1/2}$ values for ^{198}Au were extremely long and variable. The clearance of ^{99m}Tc-colloid was slightly faster, but not significantly so, than ^{131}I-HSA and more consistent (Figure 1). To reduce leakage the needle was inserted obliquely through the skin. Tracer was injected slowly with minimal force. The needle entry site was wiped once with a cotton wool swab after withdrawal of the needle.

B. Procedure

Injections of tracer were made into the dermis at a superficial level (subepidermal). An injection volume of 0.03 ml was used using a 30-gauge needle. Albumin may be more physiological but the main function of the lymphatics is the removal of macromolecules, including exogenous colloids, from the interstitium. A well-collimated sodium iodide detector was positioned over the injection site with the collimator surface 10 mm from the skin. Additional detectors can be used to monitor other injection sites when paired studies are to be performed or to monitor uptake in the regional lymph nodes. Each detector was connected to a dual-channel interface analyzer operating in the multiscaler mode to give a digital output of radioactive counts with time. Counts were integrated over periods of 10 to 50 seconds and data recorded on a computer for a total time of 30 min. Observations up to periods of 90 min were not found to give any greater consistency.

C. Analysis of Data

Changes in radioactive counts were plotted as the percentage of the maximum count rate against time. The resulting clearance curves were analyzed using a single exponential equation.[25] The data points were fitted to this equation using a nonlinear least square method. The rate of removal was corrected for the decay of the isotope (^{99m}Tc = 6 hours) and lymph flow expressed as a half clearance time ($t_{1/2}$) in hours.

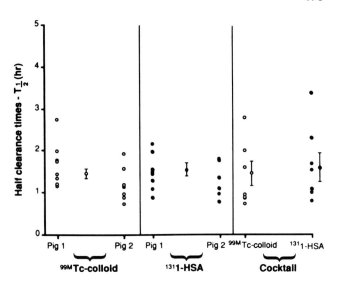

FIGURE 1 Variations in $t_{1/2}$ values for the clearance of ^{99m}Tc-colloid with ^{131}I-HSA (•) when injected as single tracers or together as a cocktail into the flank skin of the pig. The individual results of separated measurements on two pigs are given along with the mean ± SE for each set of results.

D. Reliability and Reproducibility

1. Animal Studies

The ICT proved to be a reliable and reproducible method when performed as a group test under controlled laboratory conditions in anesthetized large white pigs.[25] Data were found to fit best to a monoexponential equation with good correlation coefficients (>0.86), indicating that the clearance was a mono rather than biexponential function. This is in keeping with the previously published human work.[8,9] Long investigation times are not only impractical but fluctuations in lymph flow may be expected, particularly from movement in an unanesthetized animal.

Differences were observed in lymph flow between the two pigs studied (Figure 1) and between skin sites within each animal. Differences in lymph vessel density and distribution and in tissue compliance were the likely explanations. Age was not a major factor in these studies but has been recently shown to result in a decline in limb lymph drainage in humans.[26]

2. Human Studies

As in the controlled studies performed in pig skin,[25] similar studies performed in human skin revealed essentially monoexponential clearance.[27] There was, however, very little consistency of $t_{1/2}$ in repeat basal studies (basal = without interference, e.g., lymph flow enhancement) and serious doubts must be raised regarding the value of single lymph flow determinations at rest (basal). Fractional removal rates were not significantly different in pre-tibial skin compared with the skin of the thigh or foot. The wide range of $t_{1/2}$ values witnessed in normal human skin differed from the reproducible results seen under controlled conditions in pig skin. This was considered to indicate real differences in lymph flow rather than technical error, particularly as results varied in repeat studies in the same subject on consecutive days whereas right and left legs showed similarity when examined together.

Lymph flow at rest is slow and subject to instant fluctuations and its measurement is clearly prone to error unless carried out under strictly controlled conditions. All components of lymph movement depend upon changes in local tissue and hydrostatic pressure. These changes are produced by external compression,[28] muscular activity,[29,30] skin surface massage,[31,32] passive movements,[33] and local arterial pulsation.[34] Movement of macromolecules from interstitium toward, into, and through peripheral lymphatics would appear to be a predominantly passive process dependent on many extrinsic forces rather than an active process generated by the lymphatic itself.

A study of lymph flow enhancement comparing vibration with local massage demonstrated greater colloid clearance from massage. The response to vibration was disappointing. A possible explanation for this failure would be that the rapid movement of vibration was too fast to allow adequate lymph vessel filling.

Local massage significantly enhanced colloid clearance both in normal pig[25] (Figure 2) and human skin (Figure 3). Massage performed some distance away in the leg from the injection depot according to the principle of manual lymphatic drainage[35] invoked a significant increase in clearance superior to that generated by pneumatic compression therapy (Talley Medical Equipment Ltd.) (Figure 4). The proposed theory is that such massage has a "milking" or "siphoning" effect on distal lymph. Stimulation of the intrinsic contractility of the main lymphatic collecting vessels in the limb would pump lymph proximally so generating a pressure gradient which draws lymph from peripheral lymphatics including in the skin. The lack of efficacy from pneumatic compression therapy was a surprise as the ten-chamber inflatable garment produced a pressure wave moving repeatedly up the limb that was far stronger than the massage. These machines are widely used for the treatment of lymphedema but may do little to mobilize protein via lymphatics.[36]

Comparison of colloid clearance from normal skin on the dorsum of the foot with the same site in a lymphedematous leg showed no difference under basal conditions with the subject supine. Clearance of ^{99m}Tc-colloid from the dermis, as a measure of skin lymph flow, could only differentiate normal subjects from patients with lymphedema by the response to massage. Single lymph flow determinations using ^{99m}Tc-colloid clearance from the dermis over a short investigation time are therefore only meaningful when attempts are made to enhance lymph flow and so test lymph trans-

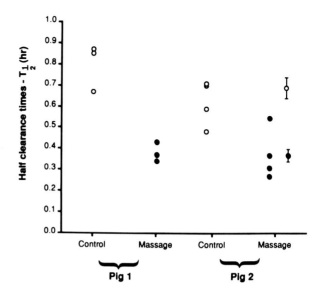

FIGURE 2 The effect of local massage on the half clearance time for ^{99m}Tc-colloid from skin on the flank of pigs. The individual results are given for separate measurements on two pigs (•, unmassaged sites; ○, massaged sites) along with the mean values ± SE.

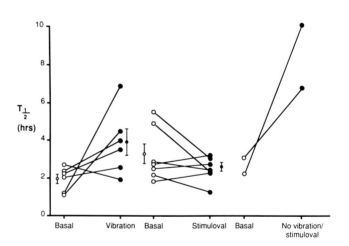

FIGURE 3 The clearance of ^{99m}Tc-colloid from normal skin of the lower leg expressed as $t_{1/2}$, in normal subjects at rest (basal), and following a period of lymph flow enhancement (vibration or surface massage).

port capacity. Massage by stimulating lymph flow exposes the deficiency in lymph transport which examination at rest may miss. Only then can lymphatic insufficiency be distinguished from normal function.

3. Studies in Pathological Skin

As edema and connective tissue changes are well known sequelae of lymphatic damage and as such changes occur following radiation to the skin, lymph flow studies using the I.C.T. (^{99m}Tc-colloid) were undertaken in pig skin following single doses of X-rays.[37] Paired estimates of lymphatic clearance were performed in irradiated and unirradiated sites on the

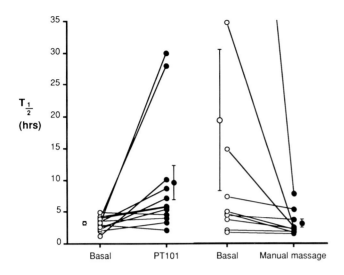

FIGURE 4 The clearance of ^{99m}Tc-colloid, expressed as $t_{1/2}$, from the skin of the dorsum of the foot during rest, basal, (0 to 30 min post-injection) and following a period (30 to 60 min) of lymph flow enhancement with either pneumatic compression (PT101) or manual massage.

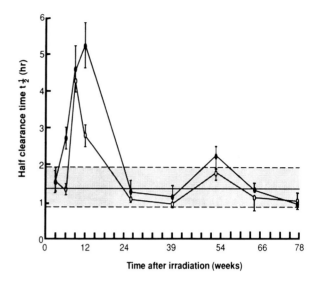

FIGURE 5 Time-related changes in the clearance of ^{99m}Tc-colloid from the dermis of the flank of the pig following irradiation with a dose of 18Gy of X-rays. Results are expressed as the mean half clearance times (± SE) for irradiated (•) and nonirradiated (○) fields. The shaded area represents the mean $t_{1/2}$ value (± SE) in normal skin. Error bars indicated.

flank skin of anesthetized large white pigs at 3, 6, 9, 12, 26, 39, 52, 64, and 78 weeks after a single dose of 18 Gy of X-rays. The results demonstrated good consistency of results relative to site and time examined. The results demonstrated two waves of impaired lymphatic clearance with time which correlates well with the gross morphological changes observed (Figure 5). It was concluded that impaired lymphatic drainage probably contributes to the gross and histological changes observed in the skin following X irradiation. The study demonstrated the value of the ICT for lymph flow measurements even in pathological skin providing experimental conditions are well controlled.

IV. Sources of Error

The advantage of the ICT is that it utilizes the principle function of the lymphatic, namely the removal of protein or colloid from the interstitium, and examines it dynamically. As such it explores the capacity of the lymphatic system to absorb material from the interstitial space and transport it to the regional node. This corresponds to the route taken by protein leaked from blood capillaries and subsequently drained from the interstitium by the lymphatics. Of the whole circulating plasma protein pool, 50 to 100% leaves the circulation daily. It is the lymphatic system which maintains this "extravascular circulation of plasma proteins".[38]

A. Tracer Migration

Flow can be calculated from clearance provided that (1) the tracer leaves only by one route and (2) reaches an instant diffusion equilibrium between lymph and tissue. The rate-limiting step for measuring lymph flow is almost certainly the poor migration of tracers from the injected site. Injected proteins behave differently from native plasma proteins. When an isotope in its colloidal form is injected into the tissues, approximately 90% is precipitated on the local tissue proteins. Only a small proportion is attached to the mobile proteins and to phagocytes and so taken up by the lymphatics.[39] This obviously limits the "measurable" quantity of ^{99m}Tc-colloid available for clearance. In the studies described an average of 15% of colloid injected was absorbed by lymphatics in the first 30 min.

B. Blood Clearance

Isotope clearance, as a measure of lymph flow, has been criticized because of the risk of significant amounts of radioactivity escaping by the blood stream thus invalidating clearance values interpreted as lymph flow. This problem was investigated by comparing blood clearance of ^{99m}Tc-colloid with its total (lymphatic) clearance.[25] Results revealed that the percentage of tracer cleared by the bloodstream was never more than 1.5% of total clearance. Lymphoscintigraphic studies confirmed that blood clearance was negligible.

C. Injection Trauma

Injection into the skin is obviously traumatic and nonphysiological. The insertion of a needle will undoubtedly on occasions disrupt a lymph capillary mesh 600 to 1000 μm wide. Disruption of blood capillaries is more likely but bleeding at the entry point can be minimized if care is taken. Obvious bleeding should lead to termination of the measurements. Laceration of the lymph capillary network effectively results in direct injection into lymphatics with increased uptake and filling of the vessels.[11] Nevertheless, this clearly does not prevent continuation of satisfactory clearance and with good injection techniques the disturbance is both minimal and consistent.[40]

D. Injection Depth

Studies of injection depth in pig skin demonstrated the importance of an accurate injection for dependable results.[25] Subepidermal localization of the tracer produced faster and more consistent clearance (Figure 6). Similar results were observed in blood flow studies using the same technique[7] where clearance rates correlated with local vascular density. The much denser network of lymphatic capillaries existing in the subpapillary region of pig and human skin[41] provides a greater surface area for absorption and would satisfactorily explain the faster subepidermal clearance of colloid compared with deep dermal and subcutaneous sites.

E. Volume of Distribution

Consideration must also be given to injection pressure and volume when interpreting results. A change in injection volume did not significantly influence $t_{1/2}$.[25] Clearance rates were slower with a larger volume. However, the clearance rate (K_T) is only proportional to lymph flow (F_L) when the volume of distribution (V_i) remains the same:

$$K_T = \frac{F_L}{V_i}$$

Therefore, no change in $t_{1/2}$ despite an increase in volume suggests an increased lymph flow. This is possibly the most serious source of error when interpreting clearance rates as lymph flow. Clearly, therefore, it is not possible to compare directly clearance from normal skin with lymphedematous skin because V_i remains unknown.

F. Extrinsic Forces

The differences in the consistency of lymph flow measurements performed in pig skin compared with the wide variation seen in human skin was considered a reflection of the influences of extrinsic forces. By controlling through the laboratory conditions for ambient temperature, active and passive movements, and pulse rate, these problems were largely overcome. This is not easily possible in human studies unless the extrinsic forces are specifically used to enhance lymph flow. Only by increasing lymph flow in response to standardized stimuli, e.g., massage, could impaired lymph drainage from the skin be detected.[25]

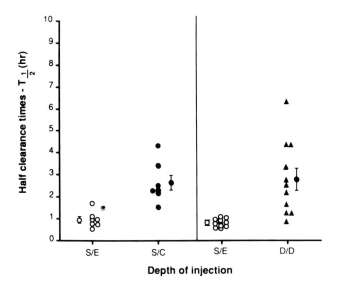

FIGURE 6 Variations in the $t_{1/2}$ values for the clearance of ^{99m}Tc-colloid from the normal flank of the pig following injections at different depths. Injections were subepidermal (S/E, ○) subcutaneous (S/C, •), and deep-dermal (D/D, ▲). The individual results of separate measurements are given along with the mean ± SE for each set of results.

V. Correlation with Other Methods

The scintillation detector system with simultaneous measurement of two comparative sites, or depot clearance and nodal uptake, provides a portable and low radiation method for the functional assessment of peripheral lymphatics in humans. The relative and simple and inexpensive equipment permits the technique to be used at the bedside. This has benefits in centers without gamma camera or whole body scintillation scanners.

An increased number of external detectors connected in series at intervals along the lymph drainage route from the injection site could be used to improve the sensitivity of the technique particularly in relation to speed of lymph movement. Small detectors strapped to the limb in a similar method to pressure transducers would be a possibility.

Isotope clearance is the only method currently available which provides objective and dynamic information on skin lymph flow. Most clinical methods which examine the lymphatic system — histology and electron microscopy, lymphangiography, and lymphoscintigraphy — either focus on large lymphatic vessels outside the skin or provide static, anatomical, or structural detail.

Lymphoscintigraphic studies supported the results of the external detector studies and demonstrated that clearance of approximately 99% of the tracer was lymphatic. Blood clearance and tracer diffusion were negligible in normal skin. Results using external scintillation detectors were in broad agreement with the published findings using a gamma camera, although most examinations employed a subcutaneous injection of tracer. Gamma camera studies demand a 20-fold greater dose of radioactivity but nevertheless still fall within category I for radiation risk. External detector studies permit multiple repeat lymph flow measurements with safety.

Gamma camera measurements of isotope clearance from the injection site have been considered unreliable but Pecking[21,42] in extensive studies has shown significant differences between lymphedema and normal limbs, according to clearance, with good reproducibility.

VI. Recommendations

Physiological and clinical measurement of the microcirculation is important for understanding the dynamic changes that occur with pathology. So often functional questions are answered incorrectly by extrapolation of data from static studies. The ICT has the major advantage that it utilizes the principle function of the lymphatic, namely the removal of protein or colloid from the interstitium, and examines it dynamically. Skin lymph flow can only be reliably measured when conditions are controlled. For lymph flow at rest this means controlling for all extrinsic influences as well as inter-site and inter-subject variations. Extrinsic factors such as massage strongly influence lymph flow. Greater sensitivity in detecting lymphatic insufficiency may be achieved if a standardized stimulus to lymph flow is administered. The response in clearance to the lymph flow enhancement will be the best indicator of lymph drainage abnormalities.

References

1. Silberberg-Sinakin, I., Thorbecke, G.J., Baer, R.L., Rosenthal, S.A., and Berezowsky, V., Antigen bearing Langerhans cells in skin, dermal lymphatics and in lymph nodes, *Cell. Immunol.*, 25, 137, 1976.
2. Drinker, C.K. and Field, M.E., The protein content of mammalian lymph and the relation of lymph to tissue fluid, *Am. J. Physiol.*, 97, 32, 1931.
3. Taylor, G.W., Kinmonth, J.B., Rollinson, E., and Rotblat, J., Lymphatic circulation studied with radioactive plasma protein, *Br. Med. J.*, 1, 133, 1957.
4. Hollander, W., Reilly, P., and Burrows, B.A., Lymphatic flow in human subjects as indicated by the disappearance of ^{131}I-labelled albumin from the subcutaneous tissue, *J. Clin. Invest.*, 40, 222, 1961.
5. Fernandez, M.J., Davies, W.T., Owen, G.M., and Tyler, A., Lymphatic flow in humans as indicated by the clearance of ^{125}I-labelled albumin from the subcutaneous tissue of the leg, *J. Surg. Res.*, 35, 101, 1983.
6. Engelhart, M. and Kristensen, J.K., Evaluation of cutaneous blood flow responses by 133-Xenon washout and a laser Doppler flowmeter, *J. Invest. Dermatol.*, 80, 12, 1983.
7. Young, C.M. and Hopewell, J.W., The evaluation of an isotope clearance technique in the dermis of pig skin: a correlation of functional and morphology parameters, *Microvasc. Res.*, 20, 182, 1980.
8. Ellis, J.P., Marks, R., and Perry, B.J., Lymphatic function: the disappearance rate of ^{131}I albumin from the dermis, *Br. J. Dermatol.*, 82, 593, 1970.
9. Staberg, B., Klemp, P., Aasted, M., Worm, A.M., and Lund, P., Lymphatic albumin clearance from psoriatic skin, *J. Am. Acad. Dermatol.*, 9, 857, 1983.
10. Kinmonth, J.B., Lymphangiography in man, *Clin. Sci.*, 11, 13, 1952.
11. Hudack, S.S. and McMaster, P.D., Lymphatic participation in human cutaneous phenomena, *J. Exp. Med.*, 57, 751, 1933.
12. Bollinger, A., Jager, K., Sgier, F., and Seglias, J., Fluorescence microlymphography, *Circulation*, 64, 1195, 1981.
13. Partsch, H., Wenzel-Hora, B., and Urbank, A., Differential diagnosis of lymphoedema after indirect lymphography with Iotasul, *Lymphology*, 16, 12, 1983.
14. Battezzati, M. and Donini, I., The use of radioisotopes in the study of the physiopathology of the lymphatic system, *J. Cardiovasc. Surg.*, 5, 691, 1964.
15. Anghileri, L.J., Lymph nodes distribution of several radio colloids: migration ability through the tissues, *J. Nucl. Biol. Med.*, 11, 180, 1967.
16. Jepson, R.P., Simeone, F.A., and Dobyns, B.M., Removal from skin of plasma protein labelled with radioactive iodine, *Am. J. Physiol.*, 175, 443, 1953.
17. Sage, H.H., Sinha, B.K., Kizilay, D., and Toulon, R., Radioactive colloidal gold measurements of lymph flow and functional patterns of lymphatics and lymph nodes in the extremities, *J. Nucl. Med.*, 5, 626, 1964.
18. Emmett, A.J., Barron, J.N., and Veall, N., The use of ^{131}I-albumin tissue clearance measurements and other physiological tests for the clinical assessment of patients with lymphoedema, *Br. J. Plast. Surg.*, 20, 1, 1967.
19. Kleinhaus, E., Baumeister, R., Hahn, D., Siuda, S., Bull, U., and Moser, R., Evaluation of transport kinetics in lymphoscintigraphy. Follow-up study in patients with transplanted lymphatic vessels, *Eur. J. Nucl. Med.*, 10, 349, 1985.
20. Stewart, G., Gaunt, J., Croft, D.N., and Browse, N.L., Isotope lymphography: a new method of investigating the role of lymphatics, *Br. J. Surg.*, 72, 906, 1985.
21. Pecking, A., Cluzan, R., Desprez-Curely, J.P., and Guerin, P., Functional study of the limb lymphatic system, *Phlebology*, 1, 129, 1986.
22. Mostbeck, A., Kahn, P., and Partsch, H., Quantitative lymphography in lymphoedema, in *The Initial Lymphatics*, Bollinger, A., Partsch, H., and Wolf, J.H., Eds., Thieme Verlag, Stuttgart, 1985, 123.
23. Kety, S.S., Measurement of regional circulation by the clearance of radioactive sodium, *Am. Heart J.*, 38, 321, 1949.
24. Strand, S.E. and Persson, B.R., Quantitative lymphoscintigraphy. I. Basic concepts for optimal uptake of radiocolloids in the parastemal lymph nodes of rabbits, *J. Nucl. Med.*, 20, 1038, 1979.
25. Mortimer, P.S., Simmonds, R., Rezvani, M., Robbins, M., Hopewell, J.W., and Ryan, T.J., The measurement of skin lymph flow by isotope clearance — reliability, reproducibility, injection dynamics, and the effect of massage, *J. Invest. Dermatol.*, 95, 677, 1990.

26. Bull, R.H., Gane, J., Evans, J., Joseph, A., and Mortimer, P.S., Abnormal lymph drainage in patients with chronic venous leg ulceration, *J. Am. Acad. Dermatol.*, in press.
27. Mortimer, P.S., Measurement of Skin Lymph Flow by an Isotope Clearance Technique, M.D. thesis, University of London, London, 1990.
28. Miller, G.E. and Seale, J.L., The mechanics of terminal lymph flow, *J. Biomech. Eng.*, 107, 376, 1985.
29. Yoffey, J.M. and Courtice, F.G., *Lymphatics, Lymph and Lymphomyeloid Complex*, Academic Press, New York, 1970.
30. Barnes, J.M. and Trueta, J., Absorption of bacteria, toxins and snake venoms from the tissues, *Lancet*, I, 623, 1941.
31. Calnan, J.S., Pflug, J.J., Reis, N.D., and Taylor, L.M., Lymphatic pressures and the flow of lymph, *Br. J. Plast. Surg.*, 23, 305, 1970.
32. Olszewski, W.L., *Peripheral Lymph: Formation and Immune Function*, CRC Press, Boca Raton, FL, 1985.
33. Jacobsson, S., Lymph flow from the lower leg in man, *Acta Chir. Scand.*, 133, 79, 1967.
34. Parsons, R.J. and McMaster, P.D., The effect of the pulse upon the formation and flow of lymph, *J. Exp. Med.*, 68, 353, 1938.
35. Stijns, H.J. and Leduc, A., The contribution of physical therapy in the treatment of lymphoedema, in *Lymphoedema*, Clodius, L., Ed., Thieme, Stuttgart, 1977, 27.
36. Partsch, H., Mostbeck, A., and Leitner, G., Experimentelle untersuchungen zur wirkung einer druckwellen-massage (lymphapress) bein lymphodem, *Lymphologie*, Band V, Heift 1, 1981, 35.
37. Mortimer, P.S., Simmonds, R.H., Rezvani, M., Robbins, M.E., Ryan, T.J., and Hopewell, J.W., Time related changes in lymphatic clearance in pig skin after a single dose of 18Gy of X rays, *Br. J. Radiol.*, 64, 1140, 1991.
38. Mayerson, H.S., The physiologic importance of lymph, in *Handbook of Physiology*, Vol. 2, Hamilton, W.F. and Dows, P.G., Eds., Waverly, Baltimore, 1963 1035.
39. Haagensen, C.D., Methods of study of the lymphatic system, in *The Lymphatics in Cancer*, Haagensen, C.D., Ed., W.B. Saunders, Philadelphia, 1972, 14.
40. Courtice, F.C., Lymph and plasma proteins: barriers to their movement throughout the extracellular fluid, *Lymphology*, 4, 9, 1971.
41. Mortimer, P.S., Jones, R.L., and Ryan, T.J., Human skin lymphatics: regional variation and relationship to elastin, in *Immunology and Haematology Research: Progress in Lymphology*, Vol. 2, Heim, L., Ed., Immunology Research Foundation, Inc., Nearburgh, 1984, 59.
42. Pecking, A., Cluzan, R., Desprez-Curely, J.P., and Guerin, P., Indirect lymphoscintigraphy in patients with limb oedema, *Phlebology*, 1, 211, 1986.

Section G: Neural Supply

20.0

Sensory Function

20.1 Assessment of Skin Sensibility 483
    P. Bjerring

20.2 Quantification of Cutaneous Pain 489
    P. Bjerring

Chapter 20.1
Assessment of Skin Sensibility

Peter Bjerring
Department of Dermatology
Marselisborg Hospital
University Hospital of Aarhus
Aarhus, Denmark

I. Introduction

Clinical assessment of skin sensibility is normally performed in order to detect abnormalities in peripheral sensory end organs, afferent neurons, central pathways, or cerebral areas, and the cutaneous distribution of altered skin sensibility may thus provide information on the anatomical localization of neurologic lesions. The clinical evaluation of sensory functions normally involves very simple instruments and provides only information of gross sensory loss with no quantitative specification of thresholds. Also, these types of investigations rely on the subject's own evaluation and description of the stimulus. In contrast to these bedside examinations, experiments conducted in the laboratory employ both objective and quantitative methods.

Numerous estimates of sensitivity have been used both experimentally and clinically. The most commonly used methods of subjective sensory assessments include numerical or verbal category scales and visual analog or color scales. The subject's estimates on these scales are assumed to reflect the sensory or pain intensity in question.[1-6] These rating scales can be constructed in order to focus on different aspects of the specific sensory modality or to differentiate between different sensations, and are often specially designed for a single test situation.[7-9] The validity of subject's sensory level estimations is limited by the degree of training of the subject, the skill of the observer, and the type of experimental stimulation used.[10] Subjective sensory level estimates may be substantially distorted by psychological and social factors. In order to reduce this type of bias, threshold determinations should be used in which the subject reports any registered sensation. Different stimulation methods have been developed in order to determine the threshold most reliable.[11,12]

II. Physiology of Human Cutaneous Tactile Sensation

Microneurographic recordings of afferent axons have provided the basic knowledge of the function of human low-threshold mechanoreceptors.[13] The receptors can be classified by their receptive fields and their adaptive properties. In glabrous skin, rapidly adapting, superficial receptors (Meissner's corpuscles) correspond to well-defined fields, which constitute the terminal arborizations of a single axon, whereas the deeper located Pacinian corpuscles have larger receptive fields with poorly defined borders. Also, the Merkel cell complex and the Ruffini receptor are low-threshold slow-adapting mechanoreceptors, and the Ruffini receptor seems to have a special directional sensitivity.

III. Stimulators for Elicitation of Nonpainful Skin Sensation

A. Electrical Stimulators

Electrical current conducted through the skin has been widely used for the stimulation of cutaneous prepain and pain sensations.[14-18] However, cutaneous electrical stimulation has many physiological disadvantages. Electrical stimulation produces a volley of impulses which are quite unlike the more gradual onset, buildup, slow trail-off, and after-discharge that are produced by natural skin stimuli.[19] Electrical stimulation is an unnatural type of stimulus which activates many different types of sensory receptors simultaneously. Both tactile and vibrational sensations and a paraesthetic sensation often defined as tingling are reported by the subjects; this mixing of sensory modalities results in sensory-discriminative problems for the subject.[20]

B. Tactile Stimulators
1. Horsehair Stimulators

Quantitative tactile stimulators were initially constructed by Blix and von Frey. Blix constructed a pivoted lever system with a horsehair stylus attached to one end and a counterweight attached to the other.[21] The counterweight was used for changing the stylus force on the skin. Von Frey produced graded tactile stimuli of the skin by calibrated horsehairs of various thickness and lengths attached to a handle.[22] However, quantitative experiments using these tactile stimulators require high skills of the investigator.

2. Ballistometers
Many different and sophisticated ballistometric methods in which the skin is mechanically stimulated by different types of falling weights have been developed.[23,24] Due to lack of standardization, these instruments can generally not be used for comparative studies between different laboratories.

C. Other Types of Mechanical Stimulators
Estimation of spatial discrimination has traditionally been performed by a two-stylus apparatus in which distance between the styli can be varied. Also, the subject's recognition of different type of sandpaper has been used for estimation of tactile discriminative sensory function.

1. Vibration Stimulators
Tuning forks have been the classical instrument in clinical investigations of the sensitivity to mechanical vibration.[25,26] Specifically designed electromagnetic instruments have been widely used in laboratory settings.[27-29] Application of displacement, force, and acceleration transducers on the mechanical stimulators seem to be mandatory for standardization of stimulation parameters. Buchthal, who recorded peripheral nerve action potentials, found that displacement, velocity, and acceleration were pivotal parameters for the stimulation of tactile receptors.[30]

IV. Physiology of Human Cutaneous Thermal Sensation

The sensations of nonpainful cold and warmth are mediated by two different populations of receptors; differences have been found in the cutaneous distribution and the density.[31-33] Microneurographic investigations have shown that the two types of receptors are innervated by different fiber populations which can be identified by differences in conduction velocities. Cold is mediated by small myelinated Aδ-fibers and warmth by unmyelinated C-fibers.[34-37] LaMotte et al. found that the conduction velocity of warmth-mediating fibers was 1.5 ± 0.5 m/s corresponding to that of the C-fiber population. In contrast Aδ-fibers exhibit nerve conduction velocities above 6 m/s.[38]

A. Stimulators for Experimental Investigation of the Thermal Senses
Thermal stimulators were introduced by Goldscheider in 1884. Currently, two different types of thermal stimulators are employed. The sensation of both cold and heat can be elicited by application of hot or cold solid bodies, thermodes, on the skin surface.[39-41] The sensation of warmth can be induced by nontouching radiant heat stimulators. These are based on either focused light from electrical bulbs or radiation from high-power lasers.[42-51]

B. Thermodes
For elicitation of cold or warmth sensations different types of thermodes have been developed.[52] Solid metal bodies which have been preheated or precooled to specific temperatures can be applied to the skin surface.[53] An example of such a system is the Minnesota thermal discs, which consists of four circular 10-cm² discs made of either copper, stainless steel, glass, or polyvinyl chloride.[52] The discs are maintained at normal room temperature and when placed on the skin act as heat sinks with different thermal conductivity. The copper conducts away the most heat and feels coldest; the polyvinyl chloride surface conducts the least and feels "warm". Other heating or cooling systems have been developed based on hollow metallic elements in which hot or cold water or water vapor flows or by electrical heating of thin metal foils placed on the skin surface. Accurate thermal stimulators can be constructed using electrothermal devices (Peltier elements). These elements change temperature according to polarity of an electrical current (cooling or heating) and the capacity of heat transfer is controlled by the magnitude of electric current through the element.[39,53]

C. Nontouching Thermal Stimulation
Techniques allowing thermal stimulation of the skin without physical contact were developed by Hardy et al.[42-44] Their initial studies resulted in construction of the first reliable optical instrument, the Hardy-Wolff-Goodell radiant heat dolorimeter.[54] The mode of action was based on pain stimulation by visible and infrared radiation generated by an electric bulb, from which the light was transmitted through a lens system to the skin. Numerous investigators have employed this apparatus or modifications of it, and significant knowledge of normal and pathological pain mechanisms has since been accumulated.[55-57] The basic properties of the method were reviewed by Procacci.[17,58] The method can induce skin damage due to excessive heating of the skin and, if such damage occurs, nociceptors in both the test area and the area surrounding the stimulated tissue may be sensitized by release or production of local inflammatory mediators.[38,59-66]

D. Infrared Lasers
1. The CO_2 Laser
The far infrared light from a CO_2 laser has recently been introduced as a new source of thermal energy for ex-

perimental heat stimulation of the skin. This laser emits a spectrally narrow band of radiation in the far infrared region at 10,600 nm. CO_2 laser radiation has been extensively used for cutaneous pain stimulation, and this method allows selective thermal stimulations at both noxious and non-noxious levels to be applied to the skin without significant stimulation of other sensory modalities.[45,67–73] One major problem which has been observed by several investigators, however, is that the CO_2 laser radiation may induce superficial tissue damage, especially when the stimulus is very short (0.1 ms to 0.1 s). Energy from the CO_2 laser beam is, for practical purposes, totally absorbed in water molecules in the stratum corneum.[74] This results in a rapid temperature rise, in some instances to tissue-destructive levels.[75] From the stratum corneum the laser energy is transferred to the lower layers of epidermis and to the papillary dermis by heat conduction.

Recent experiments with visible light lasers as prepain and pain stimulator indicate that visible light, which in part is conveyed by radiation through the upper layers of the skin, is a better heat energy source for cutaneous stimulation.[70,76–78]

E. Visible Lasers
1. The Argon Laser

The argon laser was introduced into medicine in the mid 1960s. In dermatology, it was initially used for treatment of port wine stains. Willer et al. introduced this laser to cutaneous neurophysiology for elicitation of flexor reflexes.[79] In contrast to the CO_2 laser the argon laser emits visible radiation (488 and 515 nm), which partially penetrates the upper layers of the skin before it is absorbed and transformed into heat. The argon laser also possesses a range of technical characteristics, e.g., high output power (5 to 10 W) which can be gated into single pulses by simple external electronics, and a flexible fiber optic light guide system. We find that the argon laser is particularly well suited for experimental heat stimulation.[73,75] Part of the incident light from the argon laser is reflected by the surface of the skin (105 and 197 nm) and only 50 to 90% (depending on skin color) of the laser energy is absorbed and utilized for stimulation. Knowledge of the ratio between reflected and absorbed argon laser energy is of importance for standardization of the argon laser technique. An apparatus for the recording and characterization of optical parameters of the skin in order to determine the ratio between the absorbed and reflected light has been constructed.[76] When corrections for differences in surface reflection of the light are performed, argon laser stimulation has proven to be accurate and reproducible both in basic cutaneous neurophysiology and in clinical investigations.[49–51,77–85]

References

1. Wolff, B.B., Behavioral measurement of human pain, in *The Psychology of Pain,* Sternbach, R.A., Ed., Raven Press, New York, 1978, 129.
2. Grossi, E., Borghi, C., Cerchiari, E.L., Della Puppa, T., and Francucci, B., Analogue chromatic continuous scale (ACCS): a new method for pain assessment, *Clin. Exp. Rheumatol.,* 1, 337, 1983.
3. Gaston-Johansson, F., Pain assessment: differences in quality and intensity of the words pain, ache and hurt, *Pain,* 20, 69, 1984.
4. Chapman, C.R., Casey, K.L., Dubner, R., Foley, K.M., Gracely, R.H., and Reading, A.E., Pain measurement: an overview, *Pain,* 22, 1, 1985.
5. Gracely, R.H., Lota, L., Walter, D.J., and Dubner, R., A multiple random staircase method of psychophysical pain assessment, *Pain,* 32, 55, 1988.
6. Clark, W.C., Measurement of pain in humans: signal detection theory and pain, *Neurosci. Res. Progr. Bull.,* 16, 14, 1978.
7. Chapman, C.R., Measurement of pain: problems and issues, in *Advances in Pain Research and Therapy,* Vol. 1, Bonica, J.J. and Albe-Fessard, D., Eds., Raven Press, New York, 1975, 345.
8. Melzack, R., The McGill Pain Questionnaire: major properties and scoring methods, *Pain,* 1, 277, 1975.
9. Dubuisson, D. and Melzack, R., Classifications of clinical pain descriptions by multiple group discriminant analysis, *Exp. Neurol.,* 51, 480, 1976.
10. Campbell, J.A. and Lahuerta, J., Physical methods used in pain measurements: a review, *J. R. Soc. Med.,* 76, 409, 1983.
11. Svensson, P., Bjerring, P., Arendt-Nielsen, L., and Kaaber, S., Variability of argon laser-induced sensory and pain thresholds on human oral mucosa, *Anesth. Progr.,* 38, 79, 1991.
12. Arendt-Nielsen, L., Quantification of Laser Induced Pain Perception, Ph.D. thesis, University of Aalborg, Denmark, 1987.
13. Darian-Smith, I., Johnson, K.O., LaMotte, C., Shigenaga, Y., Kenins, P., and Champness, P., Warm fibers innervating palmar and digital skin of the monkey: responses to thermal stimuli, *J. Neurophysiol.,* 42, 1297, 1979.
14. Torebjörk, H.E. and Hallin, R.G., Responses in human A and C fibres to repeated electrical intradermal stimulation, *J. Neurol. Neurosurg. Psych.,* 37, 653, 1974.
15. Bromm, B. and Scharein, E., Principal component analysis of pain-related cerebral potentials to mechanical and electrical stimulation in man, *Electroenceph. Clin. Neurophysiol.,* 53, 94, 1982.
16. Notermans, S.L.H., Measurement of the pain threshold determined by electrical stimulation and its clinical application in neurological and neurosurgical patients, *Neurology,* 17, 58, 1967.
17. Procacci, P., Della Corte, M., Zoppi, M., Romano, S., Maresca, M., and Voegelin, M.R., Pain threshold measurements in man, in *Recent Advances on Pain,* Bonica, J.J., Procacci, P., and Pagni, C.A., Eds., Charles C Thomas, Springfield, IL, 1976, 105.
18. Wolff, H.G., *Headache and other Head Pain,* 2nd ed., Oxford University Press, New York, 1963.
19. Nathan, P.W., The gate-control theory of pain. A critical review, *Brain,* 99, 123, 1976.
20. Dubner, R., Sessle, B.J., and Storey, A.T., *The Neural Basis of Oral and Facial Function,* Plenum Press, New York, 1978.

21. Blix, M., Über Wirkung und Schicksal des Trichloräthyl und Trichlorbutylalkohol im Tierorganismus, *Z. Biol.*, 20, 141, 1884.
22. von Frey, M., Beiträge zur Physiologie des Schmerzsinnes. Ber ü d Verhandlungen d k Sachs Ges d Wiss z Leipzig, 46, 185, 283, 1894.
23. Carmon, A. and Dyson, J.A., New instrumentation for research on tactile sensitivity and discrimination, *Cortex*, 3, 406, 1967.
24. Benussi, V., Kinematohaptische Erscheinungen (Vorlaüfiges Mitteilung über Scheinbewegungsauffassung auf Grund Haptischer Eindrucke, *Arch. Psychol.*, 29, 385, 1913.
25. Pearson, G.H.J., Effect of age on vibratory sensibility, *Arch. Neurol. Psychiatry*, 20, 482, 1928.
26. Steiness, J.B., Vibration perception in normal subjects and in diabetics. A biothesiometric study, *Acta Med. Scand.*, 158, 315, 1957.
27. Gilmer, B.V.H., The measurement of the sensitivity of the skin to mechanical vibration, *J. Gen. Psychol.*, 13, 42, 1935.
28. Geldard, F.A., The perception of mechanical vibration. I. History of a controversy. II. The response of pressure receptors. III. The frequency function. IV. Is there a separate "vibratory sense"?, *J. Gen. Psychol.*, 22, 243, 271, 281, 291, 1940.
29. Seiler, J. and Ricker, K., Das Vibrationsempfinden. Eine apparative Schwellenbestimmung, *Z. Neurol.*, 200, 70, 1971.
30. Buchthal, F., Human nerve potentials evoked by tactile stimuli, *Acta Physiol. Scand.*, Suppl. 502, 5, 19, 1982.
31. Konietzny, F. and Hensel, H., The neural basis of the sensory quality of warmth, in *Sensory Functions of the Skin of Humans,* Kenshalo, D.R., Ed., Plenum Press, New York, 1979, 241.
32. Bessou, P. and Perl, E.R., Response of cutaneous sensory units with unmyelinated fibers to noxious stimuli, *J. Neurophysiol.*, 32, 1025, 1969.
33. Burgess, P.R. and Pearl, E.R., Cutaneous mechanoreceptors and nociceptors, in *Handbook of Sensory Physiology,* Vol. 2, Iggo, A., Ed., Springer-Verlag, Berlin, 1973, 29.
34. Beck, P.W., Handwerker, H.O., and Zimmermann, M., Nervous outflow from cat's foot during noxious radiant heat stimulation, *Brain Res.*, 67, 373, 1974.
35. Adriaensen, H., Gybels, J., Handwerker, H.O., and Van Hees, J., Response properties of thin myelinated (Aδ) fibres in human skin nerves, *J. Neurophysiol.*, 49, 111, 1983.
36. Adriaensen, H., Gybels, J., Handwerker, H.O., and Van Hees, J., Receptive properties of human Aδ fibers, in *Advances in Pain Research and Therapy,* Vol. 5, Bonica, J.J., et al., Eds., Raven Press, New York, 1983, 95.
37. Van Hees, J. and Gybels, J.M., Pain related to single afferent C fibers from human skin, *Brain Res.*, 48, 397, 1972.
38. LaMotte, R.H., Thalhammer, J.G., Torebjörk, H.E., and Robinson, C.J., Peripheral neural mechanisms of cutaneous hyperalgesia following mild injury by heat, *J. Neurosci.*, 2, 765, 1982.
39. Fruhsdorfer, H., Lindblom, U., and Schmidt, W.G., Method for quantitative estimation of thermal thresholds in patients, *J. Neur. Neurosurg. Psychiatry*, 39, 1071, 1976.
40. Hilz, M.J., Claus, D., Balk, M., and Neundorfer, B., Influence of caffeine, sweating and local hyperemisation on "Marstock" thermotesting, *Acta Neurol. Scand.*, 86, 19, 1992.
41. Rosenfalck, A., Biomedical technology in primary health care, in *Methods for Assessment of Technology in Primary Health Care,* Danish Hospital Institute, Copenhagen, 1986, 69.
42. Hardy, J.D., Wolff, H.G., and Goodell, H., Studies on pain. A new method for measuring pain threshold observations on the spatial summation of pain, *J. Clin. Invest.*, 19, 649, 1940.
43. Hardy, J.D., Influence of thermal radiation on human skin, in *Proc. Int. Photobiological Congress,* Veenmum and Zonen, Amsterdam, 1954, 205.
44. Hardy, J.D., Wolff, H.G., and Goodell, H., Methods for the study of pain thresholds, in *Pain Sensations and Reactions,* Hardy, J.D., Wolff, H.G., and Goodell, H., Eds., Williams & Wilkins, Baltimore, 1952, 52.
45. Carmon, A., Mor, J., and Goldberg, J., Application of laser to psychophysiological study of pain in man, in *Advances in Pain Research and Therapy,* Vol. 1, Bonica, J.J. and Albe-Fessard, D., Eds., Raven Press, New York, 1976, 375.
46. Mor, J. and Carmon, A., Laser emitted radiant heat for pain research, *Pain*, 3, 233, 1975.
47. Bjerring, P. and Arendt-Nielsen, L., Use of a new argon laser technique to evaluate changes in sensory and pain thresholds in human skin following topical capsaicin treatment, *Skin Pharmacol.*, 2, 162, 1989.
48. Bjerring, P., Arendt-Nielsen, L., and Søderberg, U., Argon laser induced cutaneous sensory and pain thresholds in post herpetic neuralgia. Quantitative modulation by topical capsaicin, *Acta Dermatol. Venereol.*, 70, 121, 1990.
49. Bjerring, P., Effect of antihistamines on argon laser induced cutaneous sensory and pain threshold and on histamine induced wheal and flare, *Skin Pharmacol.*, 2, 210, 1989.
50. Bjerring, P. and Arendt-Nielsen, L., Use of a new argon laser technique to evaluate changes in sensory and pain thresholds in human skin following topical capsaicin treatment, *Skin Pharmacol.*, 2, 162, 1989.
51. Bjerring, P., Arendt-Nielsen, L., and Søderberg, U., Argon laser induced cutaneous sensory and pain thresholds in post herpetic neuralgia. Quantitative modulation by topical capsaicin, *Acta Dermatol. Venereol.*, 70, 121, 1990.
52. Dyck, P.J., Karnes, J., O'Brien, P.J., and Zimmerman, I.R., Detection thresholds of cutaneous sensation in humans, in *Peripheral Neuropathy,* 2nd ed., Vol. 1, Dyck, P.J., Thomas, P.K., Lambert, E.H., and Bunge, P., Eds., W.B. Saunders, Philadelphia, 1984.
53. Arendt-Nielsen, L. and Bjerring, P., The effect of topically applied anaesthetics (Emla(R) cream) on thresholds to thermode and argon laser stimulation, *Acta Anaesthesiol. Scand.*, 33, 469, 1989.
54. Hardy, J.D., Wolff, H.G., and Goodell, H., Studies on pain. A new method for measuring pain threshold observations on the spatial summation of pain, *J. Clin. Invest.*, 19, 649, 1940.
55. Beck, P.W., Handwerker, H.O., and Zimmermann, M., Nervous outflow from cat's foot during noxious radiant heat stimulation, *Brain Res.*, 67, 373, 1974.
56. Hendler, E., Crosbie, R., and Hardy, J.D., Measurement of heating of the skin during exposure to infrared radiation, *J. Appl. Physiol.*, 12, 177, 1958.
57. Buettner, K., Effects of extreme heat and cold on human skin. I. Analysis of temperature changes caused by different kinds of heat application, *J. Appl. Physiol.*, 3, 691, 1951.
58. Procacci, P., Methods for the study of pain threshold in man, in *Advances in Pain Research and Therapy,* Vol. 3, Bonica, J.J., et al., Eds., Raven Press, New York, 1979, 781.
59. Lynn, B., Cutaneous hyperalgesia, *Br. Med. Bull.*, 33, 103, 1977.

60. LaMotte, R.H., Thalhammer, J.G., Robinson, C.J., Peripheral neural correlates of magnitude of cutaneous pain and hyperalgesia: a comparison of neural events in monkey with sensory judgments in human, *J. Neurophysiol.*, 50, 1, 1983.
61. Meyer, R.A. and Campbell, J.N., Myelinated nociceptive afferents account for the hyperalgesia that follows a burn to the hand, *Science*, 213, 1527, 1981.
62. Campbell, J.N., Raja, S.N., Meyer, R.A., and Mackinnon, S.E., Myelinated afferents signal the hyperalgesia associated with nerve injury, *Pain*, 32, 89, 1988.
63. LaMotte, R.H., Can the sensitization of nociceptors account for hyperalgesia after skin injury?, *Hum. Neurobiol.*, 3, 47, 1984.
64. Lewis, T., Experiments relating to cutaneous hyperalgesia and its spread through somatic nerves, *Clin. Sci.*, 2, 373, 1935.
65. LaMotte, R.H., Torebjörk, H.E., Robinson, C.J., and Thalhammer, J.G., Time-intensity profiles of cutaneous pain in normal and hyperalgesic skin: a comparison with C-fiber nociceptor activities in monkey and human, *J. Neurophysiol.*, 51, 1434, 1984.
66. Hilton, S.M., Bradykinin formation in human skin as a factor in heat vasodilatation, *J. Physiol.*, 157, 589, 1958.
67. Arendt-Nielsen, L. and Bjerring, P., Sensory and pain threshold characteristics to laser stimuli, *J. Neurol. Neurosurg. Psych.*, 51, 35, 1988.
68. Biehl, R., Treede, R.-D., and Bromm, B., Pain ratings of short radiant heat pulses, in *Pain Measurement in Man. Neurophysiological Correlates of Pain*, Bromm, B., Ed., Elsevier, Amsterdam, 1984, 397.
69. Bromm, B., Jahnke, M.T., and Treede, R.D., Responses of human cutaneous afferent to CO_2 laser stimuli causing pain, *Exp. Brain Res.*, 55, 158, 1984.
70. Carmon, A., Friedman, Y., Coger, R., and Kenton, B., Single trial analysis of evoked potentials to noxious thermal stimulation in man, *Pain*, 8, 21, 1980.
71. Pertovaara, A., Reinikainen, K., and Harf, R., The activation of unmyelinated or myelinated afferent fibers by brief infrared laser pulses varies with skin type, *Brain Res.*, 307, 341, 1984.
72. Meyer, R.A., Walker, R.E., Mountcastle, V.B., A laser stimulator for the study of cutaneous thermal and pain sensation, *IEEE Trans. Biomed. Eng.*, 23, 54, 1976.
73. Svensson, P., Bjerring, P., Arendt-Nielsen, L., Nielsen, J., and Kaaber, S., A comparison of four laser types for experimental pain stimulation on oral mucosa and hairy skin, *Lasers Surg. Med.*, 11, 313, 1991.
74. Mayayo, E., Trelles, M.A., Rigau, J., Sanchez, J., and Sala, P., Experimental effects of argon and CO_2 lasers on skin and vessels. Lasers in Medical Science 1989, Proc. 1st Plenary Workshop on Safety and Laser-Tissue Interaction, Suppl., 1989, 95.
75. Bjerring, P., Quantitative monitoring of acute experimental cutaneous pain and inflammatory vascular reactions, *Acta Dermatol. Venereol.*, 71 (Suppl. 161), 12, 1991.
76. Bjerring, P. and Andersen, P.H., Skin reflectance spectrophotometry, *Photodermatology*, 4, 167, 1987.
77. Diffey, B.L., Oliver, R.J., and Farr, P.M., A portable instrument for quantifying erythema induced by ultraviolet radiation, *Br. J. Dermatol.*, 111, 663, 1984.
78. Sindrup, S.H., Brøsen, K., Bjerring, P., Arendt-Nielsen, L., Larsen, U., Angelo, H.R., and Gram, L.F., Codeine increases pain thresholds to copper vapour laser stimuli in extensive but not poor metabolizers of sparteine, *Clin. Pharmacol. Ther.*, 49, 686, 1991.
79. Willer, J.C., Boureau, F., and Berny, J., Nociceptive flexion reflexes elicited by noxious laser radiant heat in man, *Pain*, 7, 15, 1979.
80. Arendt-Nielsen, L., Zachariae, R., and Bjerring, P., Quantitative evaluation of hypnotically suggested hyperaesthesia and analgesia by painful laser stimulation, *Pain*, 42, 243, 1990.
81. Arendt-Nielsen, L. and Bjerring, P., Laser induced pain for evaluation of local analgesia: a comparison of topical application (EMLA) and local injection (Lidocaine), *Anesth. Analg.*, 67, 115, 1988.
82. Arendt-Nielsen, L. and Bjerring, P., Evaluation of acetaminophen (Setamol) analgesia by argon laser induced pain related cortical responses, in *Electrophysical Kinesiology*, Proceedings ISEK, Wallinga, W., Boom, H.K.B., and deVries, J., Eds., Elsevier, Amsterdam, 1988, 195.
83. Arendt-Nielsen, L. and Bjerring, P., The effect of topically applied anaesthetics (E-mla(R) cream) on thresholds to thermode and argon laser stimulation, *Acta Anaesthesiol. Scand.*, 33, 469, 1989.
84. Arendt-Nielsen, L., Øberg, B., and Bjerring, P., Analgesic efficacy of i.m. alfentanil, *Br. J. Anaesth.*, 65, 164, 1990.
85. Arendt-Nielsen, L., Bjerring, P., and Nielsen, J., Regional variations in analgesic efficacy of Emla cream — quantitatively evaluated by argon laser stimulation, *Acta Dermatol. Venereol.*, 70, 314, 1990.

Chapter 20.2
Quantification of Cutaneous Pain

Peter Bjerring
Department of Dermatology
Marselisborg Hospital
University Hospital of Aarhus
Aarhus, Denmark

I. The Cutaneous Nociceptors — Morphological Aspects

Cutaneous pain is normally mediated by high threshold skin receptors, the nociceptors, which respond to potentially damaging high-intensity stimuli.[1] Also, stimulation of the afferent nerve fibers from these receptors along the passage to the central nervous system may elicit pain.[2] Morphologically, the receptor structures of pain-mediating nerves of the skin are free nerve endings. The nociceptors of these free nerve endings are mainly located subepidermally close to blood vessels and mast cells, but some nerve fibers penetrate the basal membrane and enter the epidermis.[3,4] These fibers have mainly been observed to terminate in association with epidermal Merkel cells, and it is not known if they also mediate nociceptive responses.

II. The Cutaneous Nociceptors — Physiological Aspects

Different types of cutaneous nociceptors can be identified based on their response to various excitatory stimuli. They may be divided into three main groups, according to their assumed natural stimuli, e.g., mechanosensitive, thermosensitive, and nociceptive nerve fibers.[5-8] However, most nociceptors respond to several excitants and are thus polymodal nociceptors. The exact biochemical structure and the molecular reactions leading to the generation of action potentials in the nociceptor have not yet been elucidated. Nociceptors without any specialized function respond to noxious mechanical and heat stimuli. Recently, it has been suggested by LaMotte et al. that selective nociceptors may exist in the skin which are specifically responsive to chemical stimuli.[9]

Nociceptors signal not only when the noxious threshold has been exceeded, but they also supply information on the stimulus intensity by means of the firing rate and the rate of increase in firing rate.[10] Also a slow, spontaneous discharge of the nociceptor may occur. Only at higher discharge frequencies or by involvement of more nociceptors is pain perceived.[11,12] During thermal stimulation, the discharge frequency of the nociceptors is linearly related to the skin temperature.[13,14] That the threshold at which the nociceptors begin to fire does not correspond with the threshold for pain perception inspired the suggestion that firing rates lower than those which elicit perception of pain may be used for transmission of other types of information, for example, on homeostasis.[11,15-17]

Microneurographic studies have been used to delineate the innervation area and type of sensory stimuli which may excite single nerve fibers.[18-20] These studies have contributed to the elucidation of some components of the structure of peripheral nerve fiber organization.

III. Pain Ratings and Descriptions

As for the nonpainful sensory modalities, a series of numerical estimates of subjective pain have been developed. Also for the painful stimulations, the quality of subjective pain level estimation is limited by the skill of the observer and the type of experimental pain stimulation used.[21]

IV. Quantitative Subjective Pain Measurement Techniques

After brief experimental thermal stimulation of the skin, three well-defined subjective perceptions can be identified: a lower level for sensation, below which nothing can be felt (the sensory threshold), a lower level for the sharp, pin-pricking pain perception (the pain threshold), and an upper level for pain tolerance (the tolerance level).[22] Determinations of the tolerance level may lead to tissue damage and will often be a matter for ethical considerations.[23] The reliability of threshold determinations seems to be strongly dependent on the stimulus used. In a comparison of different pain stimulators electrical stimulation had the best reproducibility.[24] Under normalized laboratory conditions using healthy volunteers, electrically induced pain thresholds had a coefficient of variation of less than 5% during immediate test-retest sessions, and less than 15% for intersession tests. However, electrical stimulation is a nonphysiological and "nonfamiliar" sensation to which the subjects have to become accustomed. Other types of stimulation may be regarded as more natural. Infrared CO_2 laser stimuli are perceived as light touching to sharp pinprick, eventually followed by a long-lasting, burning pain.[25-29] Argon laser stimuli are perceived a faint sensation of warmth to very strong pin pricks followed by a long-lasting, burning pain.[26,30,31] Quantification of the pain tolerance level should not be performed using brief stimuli, as these may induce tissue damage. For determination of pain tolerance test, a 2-cm diameter argon laser beam of 2 W may be used. The level of tolerance is determined as the elapsed time of laser stimulation which is needed to reach the tolerance level. Normally this ranges between 10 and 30 s.

The most reproducible measurable pain perception seems to be the threshold for the slightest sharp painful sensation which can be perceived (the pain threshold).

When the argon laser light power and the physiological and optical parameters of the skin are thoroughly controlled, the cutaneous pain threshold is a very sensitive and reproducible parameter.[32] We found the coefficients of variation to be less than 5% for the pricking pain threshold.[26] This is the summarized variance, i.e., the sum of variances of stimulation, skin optics and thermodynamics, nociceptor sensitivity and density, and finally the central processing and verbal reporting.

V. Factors that May Influence the Validity of Sensory and Pain Threshold Determinations

Determination of sensory and pain threshold depends on the registration of a binary response (presence or absence of a specific sensational quality). Binary responses have been widely used in measurement of experimental pain. These methods are based on the volunteer's ability to report whether a subjective sensation is present or not. Often, a series of short stimuli of increasing or decreasing intensities is applied. Under normal experimental conditions, the subject's ability to communicate information of sensory or pain perception after stimulation is only impaired if the pain stimulus is very strong and the perception is near the tolerance threshold.

Psychological factors may, under special conditions, modulate the pain perceived and the reliability of pain threshold has therefore been disputed. It was found to be susceptible to the influence of psychological manipulations by guided suggestions (experimenter bias) or autosuggestions (placebo response) or both.[32,33] We investigated the effect of guided suggestions during hypnosis and found that both hyperalgesia and analgesia could be induced in susceptible individuals.[34] In such situations, objective quantifications, e.g., evoked brain potentials, may be necessary as an objective correlate to the subjective perception.[34,35]

Different physiological and psychological factors which can affect the variability of pain thresholds, e.g., cooperation and awareness of the volunteers, must be taken into consideration. We found a variation in both thermal sensitivity and pain perception during the day, as part of a circadian rhythm. This has previously been described for pricking pain in men, but not in women.[36,37] Part of this circadian variation can be inhibited by naloxone.[38] Also, a sinusoidal variation of the pain threshold with a cycle duration of approximately 1 month, the circatrigintan rhythm, has been described.[22,36] In fertile women this threshold varies according to the phase of the menstrual cycle (25 days), and in young men the cycle time was determined to be 24 days. Older men and postmenopausal women had a cycle time of 30 to 32 days. Therefore, standardization measures should be taken when sequential determinations of pain thresholds or long-term comparative studies are performed.

Both spatial and temporal summation in the central nervous system may influence sensory and pain determination.[38-41] The duration of pain stimulation should be standardized. Ideally, the painful stimulus should be kept extremely short, but in the case of a laser stimulation this would require a very high energy density and a substantial risk of burning the skin surface during transmission of the stimulation energy to the deeper skin layers.

VI. Objective Quantification of Pain Perception

A. Evoked Brain Potentials

Objective methods for quantification of experimental pain perception are based on the registration of (electro)physiological or behavioral patterns. Recent investigations suggest that recordings of electrical brain responses after peripheral cutaneous laser stimulation are superior to other types of experimental pain stimulation.[28,31,34,35,42,43] Recordings of somatosensory brain potentials have also been performed after short electrical stimulations, mechanical stimulation, CO_2 laser stimulation, and focused ultrasound stimulation.[44-48]

Cortical responses evoked by noxious laser stimuli have two components, an early complex with a latency of 170 to 200 ms which seem unrelated to pain intensity and a late complex with a latency of 300 to 400 ms.[49,50] The amplitude of the late complex has been found to correlate strongly to the subjective feeling of pain.[15,25,35,50]

B. Spectral Changes of the EEG Signal

During a longer lasting or continuous painful stimulation, e.g., the cold pressor test (see below), a shift in the relative distribution of the different frequencies of the EEG as well as the amplitude occurs. Computerized analysis of EEG is required in order to visualize the pain-related alterations. Specific changes can be identified if the painful event is associated with anxiety.[51,52]

VII. Methods for Standardized Elicitation of Experimental Cutaneous Pain

Cutaneous pain can be elicited experimentally by a series of different stimulation paradigms. These may be based on either (1) fixed stimulation intensity and variation of pain duration or (2) fixed stimulation duration and variation of the pain intensity. Short experimental painful stimulations are normally repeated until a specific predetermined perceptional limit is reached or until the perception is reported to be consistent with a specific perceptional rating. A series of different methods and their limitations have been extensively reviewed.[2,53-55] In general, experimental pain stimulators can be classified according to the physical nature of the stimulus, mechanical, electrical, chemical, or thermal.

A. Mechanical Stimulation

Elicitation of pain by mechanical stimulation is always accompanied by firm touching or stabbing of the skin.[56-60] Mechanical stimulation of the skin by shear, pressure, or vibration may generate pain which in part originates from underlying structures, e.g., tendons, muscles, fascia, ligaments, and periost. This may obscure the subject's interpretation of the cutaneous nociceptive signal and make subjective ratings difficult.

Pricking of the skin's surface or inserting needles into the skin, with subsequent determination of pain threshold or ratings of suprathreshold pain sensations on a visual analog scale (VAS), seem to be adequate methods for determination of analgesia of the skin.[61,62] Juhlin and Evers reviewed the method of pin pricking for evaluation of the effect of topical analgesia (EMLA cream, an eutectic mixture of local analgesics) for minor skin surgery and for venous cannulations in children.[63,64] Other investigators have inserted thick needles or cannulas into the skin to investigate the effect of topical EMLA cream analgesia in the deeper layers of the skin and in the subcutaneous tissues. Determinations of this type have been used to estimate the utility of EMLA application for procedures such as IV cannulation or other forms of mechanical incision procedures, such as split skin grafting or minor skin surgery.[65-75] Needle insertion methods may be associated with a significant receptor sensitization, which implies a state of enhanced nociceptor responsiveness.[76] The sensitization of the nociceptors is mediated by local analgesic inflammatory mediators released in response to tissue injury.[77-79] These mediators exert their effect mainly on thin myelinated Aδ-fibers and tend to decrease perception thresholds leading to hyperalgesia and hyperpathia.[80-83] The use of experimental pain stimulation methods which induce local damage and receptor sensitization is declining due to lack of stability of baseline receptor excitability.

B. Electrical Stimulation

Electrical stimulation is a technically convenient method for elicitation of experimental pain, and the use of electric current is a method which has been widely employed for direct stimulation of nervous pathways and for cutaneous sensory and pain threshold and suprathreshold pain stimulation.[18,44,84-89] A considerable experimental bias may be anticipated as the electrical stimulation of mixed cutaneous nerves primarily activate thick, myelinated nerve fibers, which convey afferent signals of touch and temperature. Only stronger stimuli activate thin pain-mediating fibers. The resulting afferent central input is a compound signal which comprises different sensory modalities. The subjective perception is therefore usually described as a peculiar, unfamiliar sensation.[90,91]

C. Chemical Stimulation

Induction of experimental pain by application of chemicals was introduced by Lewis and has since been used extensively to elicit experimental cutaneous pain.[92-98] For an extensive review, see Reference 99. Application or injection of chemicals has also been used for elicitation of oral mucosal pain as well as for the study of pain in muscles and ligaments.[92,100] Chemical stimulation of cutaneous pain is always followed by some degree of injury to the skin resulting in sensitization and hyperalgesia.[101] This excludes repeated measurements on the same or adjacent skin areas. Although considered of reduced significance today a commonly used technique was to apply chemical substances on the exposed base of cantharidin- or suction-induced skin blisters.[96,100,102] It was difficult to obtain reproducible results because the blister formation requires either removal of the epidermis or at least detachment of epidermis from the dermis and formation of cantharidin- or suction-induced skin blisters may thus induce inflammation and pain per se. Neurogen-dependent inflammation and sensitization is also generated as nerve fibers which cross the basal membrane will be ruptured during formation of the blister. After chemical stimulation, the maximal pain sensation does not always reach its peak immediately. This could be due to an indirect action of the algesic substances, presumably by generation or activation of endogenous secondary algesic tissue- or plasma-derived pain-producing substances. Also, temporal and spatial summations in the central nervous system of afferent impulses from chemogenic sensitive C-fibers might in part account for the observed latency from a chemical pain stimulation to the perception of pain.[9,39,40]

D. Thermal Stimulation with Contact Elements

Experimental pain may also be evoked by skin contact with high- or low-temperature probes.[22,103,104] The heat transfer is accomplished by direct contact between thermodes and the skin and therefore not only nociceptors but also receptors for other sensory modalities will be stimulated.[105]

E. The Cold Pressor Test

The cold pressor test is a test mainly developed to monitor pain tolerance, i.e., the maximal pain that a subject accepts. The test evaluates both the sensory neuronal transmission systems and higher decisive functions. During the cold pressor test a longer lasting, slowly developing pain is produced by immersing the nondominant hand into ice water for a few minutes. Several modifications of the test have been made, and two principally different paradigms have been used: either immersion of the hand in ice water for as long as the subject will tolerate, and the time until the hand is withdrawn is measured, or the hand is immersed for 2 or 5 min and the actual pain is continuously rated on a VAS connected to a computer.[51,106,107]

F. Nontouching Thermal Stimulation

Both nonpainful heat sensation and heat-induced painful stimulation of the skin can be performed by radiant heat.[108,109] The first instruments delivered focused light from an electrical bulb to the skin, but more recent systems employ infrared or visible light delivered from lasers for pain stimulation (see Chapter 20.1). We found that visible light, which in part is conveyed by radiation in the skin, is the most reliable heat energy source for cutaneous stimulation.[26] However, as part of the visible laser energy is reflected off the skin surface, it is essential that the actual percentage of skin reflectance at the laser wavelength can be monitored.[26,110,111] This is most conveniently accomplished by using a visible laser as light source.

References

1. Robinson, C.J., Torebjörk, H.E., and LaMotte, R.H., Psychophysical detection and pain ratings of incremental thermal stimuli: a comparison with nociceptor responses in humans, *Brain Res.*, 274, 87, 1983.
2. Zimmermann, M., Neurophysiology of nociception, in *International Review of Physiology, Neurophysiology, II*, Vol. 10, Porter, R., Ed., University Park Press, Baltimore, 1976, 179.
3. Cauna, N., The free penicillate nerve endings of the human hairy skin, *J. Anat.*, 115, 277, 1973.
4. Cauna, N., Morphological basis of sensation in hairy skin, *Progr. Brain Res.*, 43, 35, 1976.
5. Vallbo, Å.B., Hagbarth, K.-E., Torebjörk, H.E., and Wallin, B.G., Somatosensory proprioceptive and sympathetic activity in human peripheral nerves, *Physiol. Rev.*, 59, 919, 1979.
6. Raja, S.N., Meyer, R.A., and Campbell, J.N., Peripheral mechanisms of somatic pain, *Anesthesiology*, 68, 571, 1988.
7. Sinclar, D., *Mechanisms of Cutaneous Sensation*, Oxford University Press, New York, 1981.
8. Hensel, H., Iggo, A., and Witt, I., A quantitative study of sensitive cutaneous thermoreceptor with C afferent fibres, *J. Physiol.*, 153, 113, 1960.
9. LaMotte, R.H., Simone, D.A., Baumann, T.K., Shain, C.N., and Alreja, M., Hypothesis for novel classes of chemoreceptors mediating chemogenic pain and itch, in *Proc. 5th World Congress on Pain*, Dubner, R., Gebhart, G.F., and Bond, M.R., Eds., Elsevier, Amsterdam, 1988, 529.
10. Gybels, J., Handwerker, H.O., and Van Hees, J., A comparison between the discharges of human nociceptive nerve fibres and the subject's ratings of his sensations, *J. Physiol.*, 292, 193, 1979.
11. Bromm, B., Jahnke, M.T., and Treede, R.D., Responses of human cutaneous afferent to CO_2 laser stimuli causing pain, *Exp. Brain Res.*, 55, 158, 1984.
12. Gad, J. and Goldscheider, A., Über die Summation von Hautreizen, *Z. Klin. Med.*, 20, 339, 1892.
13. Van Hees, J., Human C-fiber input during painful and nonpainful skin stimulation with radiant heat, in *Advances in Pain Research and Therapy*, Vol. 1, Bonica, J.J. and Albe-Fessard, D., Eds., Raven Press, New York, 1976, 35.
14. Zotterman, Y., Ed., *Sensory Functions of the Skin in Primates*, Pergamon Press, Oxford, 1976.
15. Kenton, B., Coger, R., Crue, B., Pinsky, J., Friedman, Y., and Carmon, A., Peripheral fiber correlates to noxious thermal stimulation in humans, *Neurosci. Lett.*, 17, 301, 1980.
16. Khayutin, V.M., Baraz, L.A., Lukoshkova, E.V., Sonina, R.S., and Chernilovskaya, P.E., Chemosensitive spinal afferents: thresholds of specific and nociceptive reflexes as compared with thresholds of excitation for receptors and axons, in *Progress in Brain Research*, Vol. 43, Iggo, A. and Ilyinski, O.B., Eds., Elsevier, Amsterdam, 1976, 293.
17. Zimmermann, M., Peripheral and central nervous mechanisms of nociception, pain, and pain therapy: facts and hypotheses, in *Advances in Pain Research and Therapy*, Vol. 3, Bonica, J.J., et al., Eds., Raven Press, New York, 1979, 3.
18. Torebjörk, H.E. and Hallin, R.G., Identification of afferent C units in intact human skin nerves, *Brain Res.*, 67, 387, 1974.
19. Baranowski, R., Lynn, B., and Pini, A., The effects of locally applied capsaicin on conduction in cutaneous nerves in four mammalian species, *Br. J. Pharmacol.*, 89, 267, 1986.
20. Lynn, B. and Carpenter, S.E., Primary afferent units from the hairy skin of the rat hind limb, *Brain Res.*, 238, 29, 1982.
21. Carlsson, A.M., Assessment of chronic pain. I. Aspects of the reliability and validity of the visual analogue scale, *Pain*, 16, 87, 1983.
22. Procacci, P., Della Corte, M., Zoppi, M., Romano, S., Maresca, M., and Voegelin, M.R., Pain threshold measurements in man, in *Recent Advances on Pain*, Bonica, J.J., Procacci, P., and Pagni, C.A., Eds., Charles C Thomas, Springfield, IL, 1976, 105.
23. Sternbach, R.A., Ethical problems in human pain research, in *Advances in Pain Research and Therapy*, Vol. 3, Bonica, J., et al., Eds., Raven Press, New York, 1979, 837.
24. Wolff, B.B., Laboratory methods of pain measurement, in *Pain Measurement and Assessment*, Melzack, R., Ed., Raven Press, New York, 1983, 7.
25. Carmon, A., Mor, J., and Goldberg, J., Application of laser to psychophysiological study of pain in man, in *Advances in Pain Research and Therapy*, Vol. 1, Bonica, J.J. and Albe-Fessard, D., Eds., Raven Press, New York, 1976, 375.
26. Arendt-Nielsen, L. and Bjerring, P., Sensory and pain threshold characteristics to laser stimuli, *J. Neurol. Neurosurg. Psych.*, 51, 35, 1988.
27. Biehl, R., Treede, R.-D., and Bromm, B., Pain ratings of shirt radiant heat pulses, in *Pain Measurement in Man. Neurophysiological Correlates of Pain*, Elsevier, Amsterdam, 1984, 397.

28. Sindrup, S.H., Brøsen, K., Bjerring, P., Arendt-Nielsen, L., Larsen, U., Angelo, H.R., and Gram, L.F., Codeine increases pain thresholds to copper vapour laser stimuli in extensive but not poor metabolizers of sparteine, *Clin. Pharmacol. Ther.*, 49, 686, 1991.
29. Carmon, A., Dotan, Y., and Sarne, Y., Correlation of subjective pain experience with cerebral evoked responses to noxious thermal stimulations, *Exp. Brain Res.*, 33, 445, 1978.
30. Willer, J.C., Boureau, F., and Berny, J., Nociceptive flexion reflexes elicited by noxious laser radiant heat in man, *Pain*, 7, 15, 1979.
31. Bjerring, P. and Arendt-Nielsen, L., Argon laser induced single cortical responses: a new method to quantify pre-pain and pain perceptions, *J. Neurol. Neurosurg. Psychiatry*, 51, 43, 1988.
32. Svensson, P., Arendt-Nielsen, L., Bjerring, P., and Kaaber, S., Vertex potentials evoked by nociceptive laser stimulation of oral mucosa: a comparison of four stimulation paradigms, *Anesthesia Pain Control Dentistry*, in press.
33. Wolff, B.B. and Horland, A.A., Effect of suggestion upon experimental pain: a validation study, *J. Abnorm. Psychol.*, 72, 402, 1967.
34. Arendt-Nielsen, L., Zachariae, R., and Bjerring, P., Quantitative evaluation of hypnotically suggested hyperaesthesia and analgesia by painful laser stimulation, *Pain*, 42, 243, 1990.
35. Arendt-Nielsen, L. and Bjerring, P., Cortical response characteristics to noxious argon laser stimulation, *Clin. Evoked Pot.*, 5, 11, 1988.
36. Procacci, P., Buzzelli, G., Passeri, I., Sassi, R., Voegelin, M.R., and Zoppi, M., Studies on the cutaneous pricking pain threshold in man. Circadian and circatrigintan changes, *Res. Clin. Stud. Headache*, 3, 260, 1972.
37. Strian, F., Lautenbacher, S., Galfe, G., and Hölzl, R., Diurnal variations in pain perception and thermal sensitivity, *Pain*, 36, 125, 1989.
38. Davis, G.C., Buchsbaum, M.S., and Bunney, W.E., Naloxone decreases diurnal variation in pain sensitivity and somatosensory evoked potentials, *Life Sci.*, 23, 1449, 1978.
39. Hardy, J.D., Wolff, H.G., and Goodell, H., Studies on pain. A new method for measuring pain threshold observations on the spatial summation of pain, *J. Clin. Invest.*, 19, 649, 1940.
40. Price, D.D., Hu, J.W., Dubner, R., and Gracely, R.H., Peripheral suppression of first pain and central summation of second pain evoked by noxious heat pulses, *Pain*, 3, 57, 1977.
41. LaMotte, R.H., Intensive and temporal determinants of thermal pain, in *Sensory Functions of the Skin of Humans*, Kenshalo, D.R., Ed., Plenum Press, New York, 1979, 327.
42. Arendt-Nielsen, L. and Bjerring, P., Evaluation of acetaminophen (Setamol) analgesia by argon laser induced pain related cortical responses, in *Electrophysical Kinesiology*, Proceedings ISEK, Wallinga, W., Boom, H.K.B., and deVries, J., Eds., Elsevier, Amsterdam, 1988, 195.
43. Arendt-Nielsen, L. and Bjerring, P., Selective averaging of argon laser induced pre-pain and pain related cortical responses, *J. Neurosci. Methods*, 24, 117, 1988.
44. Buchthal, F. and Rosenfalck, A., Evoked action potentials and conduction velocity in human sensory nerves, *Brain Res.*, 3, 1, 1966.
45. Chapman, C.R. and Benedetti, C., Nitrous oxide effects on cerebral evoked potentials: partial reversal with a narcotic antagonist, *Anesthesiology*, 51, 135, 1979.
46. Benedetti, C., Chapman, C.R., Colpitt, Y.H., and Chen, A.C., Effect of nitrous oxide concentration on event-related potentials during painful tooth stimulation, *Anesthesiology*, 56, 360, 1982.
47. Bromm, B. and Treede, R.-D., Nerve fibre discharges, cerebral potentials and sensations induced by CO_2 laser stimulation, *Hum. Neurobiol.*, 3, 33, 1984.
48. Wright, A. and Davies, I.I., The recording of brain evoked potentials resulting from intra-articular focused ultrasonic stimulation: a new experimental model for investigating joint pain in humans, *Neurosci. Lett.*, 97, 145, 1989.
49. Carmon, A., Friedman, Y., Coger, R., and Kenton, B., Single trial analysis of evoked potentials to noxious thermal stimulation in man, *Pain*, 8, 21, 1980.
50. Coger, R.W., Kenton, B., Pinsky, J.J., Crue, B.L., Carmon, A., and Friedman, Y., Somatosensory evoked potentials and noxious stimulation in patients with intractable non-cancer pain syndromes, *Psychiatry Res.*, 2, 279, 1980.
51. Stowell, A., Cerebral slow waves related to the perception of pain in man, *Brain Res. Bull.*, 2, 23, 1977.
52. Backonja, M., Howland, E.W., Wang, J., Smith, J., Salinsky, M., and Cleeland, C.S., Tonic changes in alpha power during immersion of the hand in cold water, *Electroencephalogr. Clin. Neurophysiol.*, 79(3), 192, 1991.
53. Procacci, P., Della Corte, M., Zoppi, M., Romano, S., Maresca, M., and Voegelin, M.R., Pain threshold measurements in man, in *Recent Advances on Pain*, Bonica, J.J., Procacci, P., and Pagni, C.A., Eds., Charles C Thomas, Springfield, IL, 1976, 105.
54. Hardy, J.D., The pain threshold and the nature of pain sensation, in *The Assessment of Pain in Man and Animals*, Keele, C.A. and Smith, R., Eds., Livingstone, Edinburgh, 1962, 170.
55. Keele, C.A., Addendum, in *UFAW Symposium — The Assessment of Pain in Man and Animals*, Keele, C.A. and Smith, R., Eds., Livingstone, London, 1962, 41.
56. Procacci, P., Methods for the study of pain threshold in man, in *Advances in Pain Research and Therapy*, Vol. 3, Bonica, J.J., et al., Eds., Raven Press, New York, 1979, 781.
57. Della Corte, M., Procacci, P., Bozza, G., and Buzzelli, G., A study on the cutaneous pricking pain threshold in normal man, *Arch. Physiol.*, 64, 141, 1965.
58. Stowell, H., Human evoked responses to potentially noxious tactile stimulation. I, *Act. Nerv. Super.*, 17, 1, 1975.
59. Stowell, H., Human evoked responses to potentially noxious tactile stimulation. II, *Act. Nerv. Super.*, 17, 94, 1975.
60. Wolff, B.B., The role of laboratory pain induction methods in the systematic study of human pain, *Int. J. Res. Acupunct. Electro. Ther.*, 2, 271, 1977.
61. McCafferty, D.F., Woolfson, A.D., McClelland, K.H., and Boston, V., Comparative in vivo and in vitro assessment of the percutaneous absorption of a local anaesthetic, *Br. J. Anaesth.*, 60, 64, 1988.
62. McCafferty, D.F., Woolfson, A.D., and Boston, V., In vivo assessment of percutaneous local anaesthetic preparations, *Br. J. Anaesth.*, 62, 17, 1989.
63. Juhlin, L. and Evers, H., EMLA, a new topical anaesthetic, *Adv. Dermatol.*, 5, 75, 1990.
64. Hallén, A., Ljunghall, K., and Wallin, J., Topical anaesthesia with local anaesthetic (lidocaine and prilocaine, EMLA) cream for cautery of genital warts, *Eur. J. Anaesthesiol.*, 4, 441, 1984a.
65. Hallén, B. and Uppfeldt, A., Does lidocaine-prilocaine cream permit pain free insertion of IV catheters in children?, *Anesthesiology*, 57, 340, 1982.
66. Hallén, B., Olsson, G.L., and Uppfeldt, A., Pain-free venepuncture, *Anaesthesia*, 39, 969, 1984.
67. Hallén, B., Carlsson, P., and Uppfeldt, A., Clinical study of a lignocaine-prilocaine cream to relieve the pain of venipuncture, *Br. J. Anaesth.*, 57, 326, 1985.
68. Cooper, C.M., Gerrish, S.P., Hardwick, M., and Kay, R., EMLA cream reduces the pain of venepuncture in children, *Acta Oncol.*, 26, 467, 1987.
69. Dybvik, T. and Kolflaath, J., EMLA cream used as a local anesthetic before insertion of a peripheral venous catheter (Venflon) in children, *Clin. Allergy*, 17, 307, 1987.

70. Soliman, I.E., Broadman, L.M., Hannallah, R.S., and McGill, W.A., Comparison of the analgesic effects of EMLA (eutectic mixture of local anesthetics) to intradermal lidocaine infiltration prior to venous cannulation in unpremedicated children, *Anaesthesiology*, 68, 804, 1988.
71. Ohlsén, L., Englesson, S., and Evers, H., An anaesthetic lidocaine/prilocaine cream (EMLA) for epicutaneous application tested for cutting split skin grafts, *Scand. J. Plastic Reconstr. Surg.*, 19, 201, 1985.
72. Goodacre, T.E., Sanders, R., Watts, D.A., and Stoker, M., Split skin grafting using topical local anaesthesia (EMLA): a comparison with infiltrated anaesthesia, *Br. J. Plastic Surg.*, 41, 533, 1988.
73. Juhlin, L., Evers, H., and Broberg, F., A lidocaine-prilocaine cream for superficial skin surgery and painful lesions, *Acta Dermatol. Venereol.*, 60, 544, 1980.
74. Hallén, A., Ljunghall, K., and Wallin, J., Topical anaesthesia with local anaesthetic (lidocaine and prilocaine, EMLA) cream for cautery of genital warts, *Genitourin Med.*, 63, 316, 1987.
75. Rosdahl, I., Edmar, B., Gisslén, H., Nordin, P., and Lillieborg, S., Curettage of molluscum contagiosum in children: analgesia by topical application of a lidocaine/prilocaine cream (EMLA), *Acta Dermatol. Venereol.*, 68, 149, 1988.
76. Bjerring, P. and Arendt-Nielsen, L., Depth and duration of skin analgesia to needle insertion after topical application of EMLA cream, *Br. J. Anaesth.*, 64, 173, 1990.
77. Fitzgerald, M., The spread of sensitization of polymodal nociceptors in the rabbit from nearby injury and by antidromic nerve stimulation, *J. Physiol.*, 297, 207, 1979.
78. Juan, H., The pain enhancing effect of PGI2, *Agents Actions*, 9, 204, 1979b.
79. Juan, H. and Lembeck, F., Action of peptides and other algesic agents on paravascular pain receptors of the isolated perfused rabbit ear, *Naunyn-Schmiedeberg's Arch. Pharmacol.*, 283, 151, 1974.
80. Lynn, B., Cutaneous hyperalgesia, *Br. Med. Bull.*, 33, 103, 1977.
81. LaMotte, R.H., Thalhammer, J.G., and Robinson, C.J., Peripheral neural correlates of magnitude of cutaneous pain and hyperalgesia: a comparison of neural events in monkey with sensory judgments in human, *J. Neurophysiol.*, 50, 1, 1983.
82. Meyer, R.A. and Campbell, J.N., Myelinated nociceptive afferents account for the hyperalgesia that follows a burn to the hand, *Science*, 213, 1527, 1981.
83. Campbell, J.N., Raja, S.N., Meyer, R.A., and Mackinnon, S.E., Myelinated afferents signal the hyperalgesia associated with nerve injury, *Pain*, 32, 89, 1988.
84. Hallin, R.G. and Torebjörk, H.E., Electrically induced A and C fibre responses in intact human skin nerves, *Exp. Brain Res.*, 16, 309, 1973.
85. Torebjörk, H.E. and Hallin, R.G., Recordings of impulses in unmyelinated nerve fibres in man; afferent C fibre activity, *Acta Anaesth. Scand.*, Suppl. 70, 124, 1978.
86. White, D.M. and Helme, R.D., Release of substance P from peripheral nerve terminals following electrical stimulation of the sciatic nerve, *Brain Res.*, 336, 27, 1985.
87. Gibson, R.H., Electrical stimulation of pain and touch, in *The Skin Senses*, Kenshalo, D.R., Ed., Charles C Thomas, Springfield, IL, 1968, 223.
88. Laitinen, L.V. and Eriksson, A.T., Electrical stimulation in the measurement of cutaneous sensibility, *Pain*, 22, 139, 1985.
89. Notermans, S.L.H., Measurement of the pain threshold determined by electrical stimulation and its clinical application. I. Method and factors possibly influencing the pain threshold, *Neurology*, 16, 1071, 1966.
90. Nathan, P.W., The gate-control theory of pain. A critical review, *Brain*, 99, 123, 1976.
91. Dubner, R., Bushnell, M.C., and Duncan, G.H., Behavioral and neural correlates of nociception, in *Current Topics in Pain Research and Therapy*, Yokota, T. and Dubner, R., Eds., Excerpta Medica, Amsterdam, 1983, 45.
92. Lewis, T., *Pain*, Macmillan, New York, 1942, 57.
93. Torebjörk, H.E., Afferent C units responding to mechanical, thermal, and chemical stimuli in human non-glabrous skin, *Acta Physiol. Scand.*, 92, 374, 1974.
94. Rosenthal, S.R., Histamine as the chemical mediator for cutaneous pain, *J. Invest. Dermatol.*, 69, 98, 1977.
95. LaMotte, R.H., Psychophysical and neurophysical studies of chemically induced cutaneous pain and itch. The case of the missing nociceptor, in *Progress in Brain Research*, Vol. 74, Hamann, W. and Iggo, A., Eds., Elsevier, Amsterdam, 1988, 331.
96. Becerra-Cabal, L., LaMotte, R.H., Ngeow, J., and Putterman, G.J., Chemically induced itch, pain and hyperalgesia by intraepidermal injection, *Soc. Neurosci. Abstr.*, 1063, 1983.
97. Acachi, K.I. and Ishii, Y., Vocalization response to close-arterial injection of bradykinin and other algesic agents in guinea pigs and its application to quantitative assessment of analgesic agents, *J. Pharmacol. Exp. Ther.*, 209, 117, 1978.
98. Simone, D.A., Baumann, T.K., Shain, C.N., and LaMotte, R.H., Magnitude scaling of chemogenic pain and hyperalgesia in humans, *Soc. Neurosci. Abstr.*, 13, 109, 1987.
99. Keele, C.A. and Armstrong, D., Eds., *Substances Producing Pain and Itch*, Edward Arnold, London, 1964.
100. Lawless, H., Oral chemical irritation: psychophysical properties, *Chem. Senses*, 9, 143, 1984.
101. Simone, D.A., Ngeow, J.Y.F., Putterman, G.J., and LaMotte, R.H., Hyperalgesia to heat after intradermal injection of capsaicin, *Brain Res.*, 418, 201, 1987.
102. Kiistala, U. and Mustakallio, K.K., Dermo-epidermal separation with suction. Electron microscopic and histochemical study of initial events of blistering on human skin, *J. Invest. Dermatol.*, 48, 466, 1967.
103. Fruhstorfer, H., Goldberg, J.M., Lindblom, U., and Schmidt, W.G., Temperature sensitivity and pain thresholds in patients with peripheral neuropathy, in *Sensory Functions of the Skin in Primates*, Zotterman, Y, Ed., Pergamon Press, Oxford, 1976, 507.
104. Jamal, G.A., Hansen, S., Weir, A.I., and Ballantyne, J.P., An improved automated method for the measurement of thermal thresholds. I. Normal subjects, *J. Neurol. Neurosurg. Psych.*, 48, 354, 1985.
105. Carmon, A., Mor, J., and Goldberg, J., Evoked cerebral responses to noxious thermal stimuli in humans, *Exp. Brain Res.*, 25, 103, 1976.
106. Miro, J. and Raich, R.M., Pain sensitivity range: a useful parameter to measure experimentally induced pain?, *Methods Find Exp. Clin. Pharmacol.*, 14(5), 389, 1992.
107. Moret, V., Forster, A., Laverriere, M.C., Lambert, H., Gaillard, R.C., Bourgeois, P., Haynal, A., Gemperle, M., and Buchser, E., Mechanism of analgesia induced by hypnosis and acupuncture: is there a difference?, *Pain*, 45(2), 135, 1991.
108. Hardy, J.D., Influence of thermal radiation on human skin, in *Proc. Int. Photobiological Congress*, Veenmum and Zonen, Amsterdam, 1954, 205.
109. Hardy, J.D., Wolff, H.G., and Goodell, H., Methods for the study of pain thresholds, in *Pain Sensations and Reactions*, Hardy, J.D., Wolff, H.G., and Goodell, H., Eds., Williams & Wilkins, Baltimore, 1952, 52.
110. Bjerring, P. and Andersen, P.H., Skin reflectance spectrophotometry, *Photodermatology*, 4, 167, 1987.
111. Diffey, B.L., Oliver, R.J., and Farr, P.M., A portable instrument for quantifying erythema induced by ultraviolet radiation, *Br. J. Dermatol.*, 111, 663, 1984.

Section H: Sweat Glands

21.0

Sweat Gland Distribution

21.1 Techniques for the Localization of Sweat Glands 497
 P. Dykes

Chapter 21.1
Techniques for the Localization of Sweat Glands

Peter Dykes
Department of Dermatology
University of Wales College of Medicine
Cardiff, U.K.

I. Introduction

The eccrine sweat gland is one of the major cutaneous appendages, with approximately two to four million units distributed over almost the entire body surface. The density varies greatly with body site; from 60 to 100 glands per square centimeter on the trunk and limbs to 600 to 700 glands per square centimeter on the soles and palms.

The principle function of the eccrine sweat gland is thermoregulation following physical exercise or exposure to a hot environment. Major illness or even death can occur following sweat gland failure. Another important function of sweat glands is to moisturize the skin of the palms and soles during physical activity and hence improve grip. For a review of sweat gland structure and function see Reference 1.

The eccrine sweat gland is also an excretory organ and some of the substances that it delivers to the skin surface may have important physiological functions. For example, the presence of lactate in sweat at approximately five times the plasma concentration may be significant in stratum corneum function; topical lactate-containing formulations are used extensively for the treatment of hyperkeratosis and dry skin conditions. Similarly, the urea present in sweat may have a role in skin moisturization. There have been reports of various proteins and enzymes in sweat. These include epidermal growth factor, kallikrein, and immunoglobulin A.[2,3] However, the possibility of contamination from the epidermis or stratum corneum has not been rigorously excluded in all these cases and further studies are needed in order to verify some of these findings.

Various drugs may be excreted in sweat, such as phenytoin, sulfadiazine, phenobarbital, and ethanol. Of particular interest to dermatologists are the reports that antifungal drugs such as griseofulvin and ketoconazole may reach the skin surface rapidly via sweat.[4] After a single oral dose of griseofulvin, the highest concentrations in the stratum corneum are reached within a few hours.[5] This may only be explained by secretion into sweat and rapid transport to the skin surface. As would be predicted, the blocking of sweat secretion by topical antiperspirants prevents the accumulation of griseofulvin in the stratum corneum. It has also been shown that orally administered ketoconazole is delivered to the stratum corneum in a similar manner.[6]

Clinically, abnormal sweat production is associated with a wide spectrum of diseases including several of dermatological interest. Impaired sweating or anhydrosis is a common feature of diabetes mellitus, and is related to the polyneuropathy seen in this disorder.[7] Subclinical autonomic neuropathy can be demonstrated by a decreased sweat gland responsiveness in patients with chronic bowel dysfunction.[8] A reduction in functional eccrine sweat glands is a feature of radiotherapy and may be a useful indicator of the cumulative doses received.[9] In alopecia areata a decrease in number and functional activity of sweat glands has been reported, but there did not appear to be a correlation between these changes and disease activity.[10] The sweating response to moderate thermal stress is reduced in patients with atopic eczema.[11]

Considering the importance of eccrine sweat gland function in a variety of situations it is not surprising that there have been several attempts to measure their number and functional status. These include the following:

1. Gravimetric methods based on collection of sweat in bags or pads held in close proximity to the skin.[11,12] Change in body weight has also been used as an index of gland output.
2. Direct measurement of water loss from the skin surface using devices to detect changes in relative humidity.[7]
3. Microcannulation of individual sweat ducts.[13]
4. Measurement of electrical conductivity/resistance of the skin.
5. Visualization of individual sweat droplets. This can be achieved by a variety of methods and gives information concerning the number of glands per unit skin surface area and in some circumstances the rate of sweat production.

Visualization methods are by far the most commonly used methods of assessment of sweat gland function. The main methods used are the subject of this review.

II. Object

We will detect sweat gland activity at different body sites following physical or pharmacological stimulus or in clinical conditions where sweat gland function is considered to be abnormal.

III. Methodological Principle

The methods for visualization of sweat gland function fall into two main categories. In the first category are the staining methods which use a variety of dyes or stains to detect the sweat itself. In the second category are the molding methods which detect the physical presence of liquid on the skin surface. These methods can be used after induction of sweating by vigorous exercise, thermal stimulation, or injection of sudorific agents such as phenylephrine[14] or methylcholine.[15]

A. Staining Methods

There are several methods for sweat detection which rely on the use of dyes or stains. A method based on the dye bromophenol blue was used extensively in the early 1950s by Herrmann and co-workers.[16] The method relies on filter paper impregnated with powdered bromophenol blue which is applied to the sweating skin surface for a few seconds. After removal a series of blue puncta are apparent at the site of sweat droplets. Alternatively, the bromophenol blue may be suspended in a grease and applied as a thin film to the skin surface.[17] The sweat droplet pattern is visible within a few moments and this can be photographed and the number of active glands recorded.

Injected dyes have been used to define sweat gland function.[18] A range of stains and dyes were injected intradermally at test sites and after pharmacological stimulation the sweat duct orifices were clearly marked by the stained sweat. Methylene blue was recommended as the dye of choice but the method has the drawback of permanent tattoo-like markings occurring under certain circumstances. An alternative way of staining expressed sweat was described by Juhlin and Shelley.[19] A 5% solution of o-phthalaldialdehyde in xylene is painted directly onto the skin surface and within 2 to 3 min a black reaction product is formed at the sweat duct orifice. It is thought that the interaction of o-phthalaldialdehyde with ammonia in the sweat leads to the formation of the black material.

By far the most popular staining method is that based on the iodine starch reaction. This approach has the advantage of simplicity and sensitivity without the use of chemicals which may be irritant or allergenic in nature. At its simplest a 2% solution of iodine in ethanol is painted onto the skin surface.[20] After evaporation of the solvent, sheets of ordinary paper are held against the skin surface for a few seconds. Upon removal a dark blue imprint is apparent wherever sweating is occurring. An example of the pattern produced by sweating fingers is given in Figure 1. A variation on this method is the application of a starch in castor oil solution to the iodine painted skin. Active sweat glands appear as dark blue spots on the skin surface. Sophisticated image analysis methods may be applied to photographs or camera images[21] in order to estimate the number and activity of sweat glands in this situation.

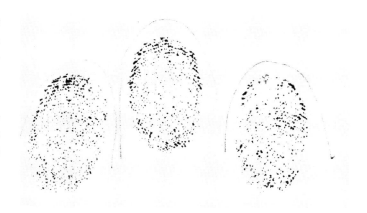

FIGURE 1 Paper imprint taken from fingers previously painted with 2% iodine in ethanol.

The simplest and most versatile iodine starch method appears to be the one-step procedure described by Sato et al.[22] In this method a preparation of iodinated starch is sprayed onto the skin surface using an atomizer. Sweat drops are visualized directly on the skin surface as dark blue or purple spots. This method has the advantage of its simplicity, the fact that it can be used at any body site over any area, and that it can be repeated at the same site after removal of the iodinated starch by simply wiping the skin surface. The safety of the operator and the patient should be considered when using this method as inhalation of iodinated starch is not desirable.

B. Molding Methods

The methods based on silicone rubber impression material or plastic solutions rely on the hydrophobicity of the applied material. After application to the skin surface any sweat droplets emerging from the eccrine pores cause a withdrawal of the hydrophobic material leading to imprints in the surface of the applied material.

Historically, the earliest attempts at molding the skin surface used solutions of polyvinyl chloride (PVC) or polyvinyl formal in a suitable solvent such as ethylene chloride.[23] The method, as modified by Harris et

FIGURE 2 Silicone rubber replica taken from the back of an actively sweating adult male.

al.,[17] uses a mixture of 5 g plastic, 10 g di-*n*-butyl phthalate (as a plasticizer), 40 g colloidal graphite, and 100 g of ethylene chloride. The colloidal graphite provides sharp contrast between the plastic material and the imprints or holes where the sweat droplets appear. After evaporation of the solvent the film of plastic is removed by applying a transparent adhesive tape to the skin surface and gently peeling the tape and plastic imprint away from the skin. If an area of tape is used that is larger than the plastic film, the tape can be used to attach the imprint to a glass slide for projection or microscopic examination. Active sweat glands appear as holes in the imprint and these can be counted without difficulty.

The most widely used method is that based on the use of silicone rubber impression material.[7,9,10,17] This has the advantage over the plastic method of materials being readily available, simpler to prepare and use, and safe to apply to any part of the body surface. The method consists of premixing the silicone rubber with the polymerizing agent and applying the mixture directly to the skin surface. Polymerization occurs within a few minutes, usually 2 to 3 min, but this can be adjusted by varying the ratio of silicone rubber to polymerizer. The imprint can then be peeled carefully from the skin surface and prepared for visualization. According to the thickness of the application two variations of the method can be achieved. If the silicone rubber solution is applied thickly the sweat droplets appear as spherical depressions in the replica surface. These can be viewed by light microscopy or by scanning electron microscopy. For scanning electron microscopy a positive image must be prepared using material such as Araldite® epoxy resin. The positive image can then be sputter coated and viewed down the scanning electron microscope. If it is assumed that the sweat droplet is spherical, measurement of the diameter will give an approximate indication of the output of the sweat gland over the time taken for the silicone rubber to polymerize.

If a thin application of the silicone rubber impression material is made, the sweat droplets appear as holes in a membrane. An example is given in Figure 2 where a replica prepared with Silaplus® silicone dental correction material (Minerva Dental Ltd., Oxford Street, Cardiff) is illustrated. The replica can be analyzed in a number of ways. The number of glands per unit skin surface area can be determined by direct counting using a low power microscope. Alternatively the replica can be mounted and projected as a magnified image onto a grid system to allow counting. Modern image analysis systems can also be used to obtain information rapidly and accurately.[10]

A variation of the molding technique is the use of a thin layer of petroleum jelly applied to the skin surface.[15] Beading of sweat droplets occurs and these are photographed under standardized conditions which enhance the contrast between the sweat and petroleum jelly. Photographic slides are produced in this method and these can be projected onto a calibrated grid system to count the number of glands per unit skin surface area.

IV. Sources of Error

The staining methods are those which are prone to the most error. This occurs principally as a result of the merging together of dots in the images generated. This can occur either as a result of sideways movement during the imprint stage or due to excessive sweating causing a running together of the droplets on the skin surface prior to imprinting. The print of the skin surface is also a mirror image of the skin and so in some circumstances it may be difficult to localize accurately the area of skin from which the print was taken. In a direct comparative study[17] the images created with bromophenol blue and iodine starch methods were considered to be less sharp than those obtained by plastic or silicone replica techniques. In addition in the same group of subjects fewer active ducts seemed to be recorded by the staining techniques and this suggests a lower sensitivity.

A possible source of error with the mold technique is the inclusion of air bubbles. This is usually a result of inexperience on the part of the operator when applying the mold material. With a little practice this can be overcome and it is the authors experience that with silicone rubber materials air bubbles are a rare problem.

V. Correlation with Other Methods

Very little information is available concerning the correlation of sweat gland activity, as measured by staining or replica techniques, with histological data. The number of active glands recorded by the silicone rubber method on a maximally sweating adult human back was reported as agreeing with histological data.[17] The number of eccrine gland recorded in the digit pads of mice by the silicone rubber method is also reported to correspond to the number of glands seen on histological sections.[24]

VI. Recommendations

In a clinical situation where information is required as to the location of areas of hyperhydrosis, the one-step iodine starch method using an atomizer is the method of choice.[22] However, in most other clinical and experimental situations the method of choice is the silicone rubber method. It is the simplest procedure and also the most accurate and gives a permanent record of sweat gland activity which can be viewed by projection techniques and by both light and scanning electron microscopy. No reagent preparation is required; the silicone rubber material and polymerizer are conveniently stored in the refrigerator. No equipment such as an atomizer is required and the method is clean to use as it is nonstaining. As silicone rubber is nontoxic, the method can be repeatedly used on the same site with no risk of an irritant or allergic reaction developing. In addition it is safe for use on delicate areas such as the face.

References

1. Sato, K., Kang, W.H., Saga, K., and Sato, K.T., Biology of sweat glands and their disorders. I. Normal sweat gland function, *J. Am. Acad. Dermatol.*, 20, 537, 1989.
2. Pesonen, K., Viinikka, L., Koskimies, A., Banks, A.R., Nicolson, M., and Perheentupa, J., Size heterogeneity of epidermal growth factor in human body fluids, *Life Sci.*, 40, 2489, 1987.
3. Okada, T., Konishi, H., Ito, M., Nagura, H., and Asai, J., Identification of immunoglobulin A in human sweat and sweat glands, *J. Invest. Dermatol.*, 90, 648, 1988.
4. Shah, V.P., Epstein, W.L., and Riegelman, S., Role of sweat in accumulation of orally administered griseofulvin in skin, *J. Clin. Invest.*, 53, 1673, 1974.
5. Epstein, W.L., Shah, V.P., and Riegelman, S., Griseofulvin levels in stratum corneum. Study after oral administration in man, *Arch. Dermatol.*, 106, 344, 1972.
6. Harris, R., Jones, H.E., and Artis, W.M., Orally administered ketoconazole: route of delivery to the human stratum corneum, *Antimicrob. Agents Chemother.*, 24, 876, 1983.
7. Kennedy, W.R., Sakuta, M., Sutherland, D., and Goetz, F.C., The sweating deficiency in diabetes mellitus: methods of quantitation and clinical correlation, *Neurology*, 34, 758, 1984.
8. Altomare, D., Pilot, M.A., Scott, M., Williams, N., Rubino, M., Ilincic, L., and Waldron, D., Detection of subclinical autonomic neuropathy in constipated patients using a sweat test, *Gut*, 33, 1539, 1992.
9. Morris, W.J., Dische, S., and Mott, G., A pilot study of a method of estimating the number of functional sweat glands in irradiated human skin, *Radiother. Oncol.*, 25, 49, 1992.
10. Elieff, D., Sundby, S., Kennedy, W., and Hordinsky, M., Decreased sweat gland number and function in patients with alopecia areata, *Br. J. Dermatol.*, 125, 130, 1991.
11. Parkinnen, M.U., Kiistala, R., and Kiistala, U., Sweating response to moderate thermal stress in atopic dermatitis, *Br. J. Dermatol.*, 126, 346, 1992.
12. Rees, J.L. and Cox, N.H., Effect of isotretinoin on eccrine gland function, *Br. J. Dermatol.*, 119, 79, 1988.
13. Schultz, I.J., Micropuncture studies of the sweat formation in cystic fibrosis patients, *J. Clin. Invest.*, 48, 1470, 1969.
14. Banjar, W.M.A., Bradshaw, E.M., and Szabadi, E., Seasonal variation in responsiveness of human eccrine sweat glands to phenylephrine, *Br. J. Clin. Pharmacol.*, 27, 276, 1989.
15. Kenney, W.L. and Fowler, S.R., Methylcholine activated eccrine sweat gland density and output as a function of age, *J. Appl. Physiol.*, 65, 1082, 1988.
16. Herrmann, F., Prose, P.F., and Sulzberger, M.B., Studies on sweating. V. Studies on quantity and distribution of thermogenic sweat delivery to the skin, *J. Invest. Dermatol.*, 18, 71, 1952.
17. Harris, D.R., Polk, B.F., and Willis, I., Evaluating sweat gland activity with imprint techniques, *J. Invest. Dermatol.*, 58, 78, 1972.
18. Hurley, H.J. and Witkowski, J., Dye clearance and eccrine sweat secretion in human skin, *J. Invest. Dermatol.*, 36, 259, 1961.
19. Juhlin, L. and Shelley, W.B., A stain for sweat pores, *Nature*, 213, 408, 1967.
20. Muller, S.A. and Kierland, R.R., The use of a modified starch iodine test for investigating local sweating responses to intradermal injection of methacoline, *J. Invest. Dermatol.*, 32, 126, 1959.

21. Sauermann, G., Hoppe, U., and Kligman, A.M., The determination of the antiperspirant activity of aluminium chlorhydrate by digital image analysis, *Int. J. Cosmet. Sci.,* 14, 32, 1992.
22. Sato, K.T., Richardson, A., Timm, D.E., and Sato, K., One step iodine starch method for direct visualisation of sweating, *Am. J. Med. Sci.,* 295, 528, 1988.
23. Thomson, M.L. and Sutarman, A., The identification and enumeration of active sweat glands in man from plastic impressions of the skin, *Trans. R. Soc. Trop. Med. Hyg.,* 47, 412, 1953.
24. Kennedy, W.R., Dakuta, M., Sutherland, D., and Goetz, F., Rodent eccrine sweat glands: a case of multiple efferent innervation, *Neuroscience,* 11, 741, 1984.

22.0

Sweat Gland Activity

22.1 Methods for the Collection of Eccrine Sweat 503
J.H. Barth

22.2 Methods for the Collection of Apocrine Sweat 507
J.H. Barth

Chapter 22.1
Methods for the Collection of Eccrine Sweat

Julian H. Barth
Department of Chemical Pathology and Immunology
General Infirmary at Leeds
Leeds, U.K.

I. Introduction

Measurements of eccrine sweating have been used to study the physiology of the eccrine gland itself and of the whole body response to environmental changes, e.g., heat acclimatization. Second, the function of the eccrine gland has been used as a diagnostic test for cystic fibrosis (CF). The concentration of sweat chloride is considered the most reliable single test in the diagnosis of CF. Although gene probes are available for the common forms of CF, there is sufficient genetic heterogeneity for the sweat test to remain the standard diagnostic tool.

The study of the physiology of the eccrine gland and its adaptation to changes in climate depend on the ability to measure the volume of fluid lost. Therefore, a spectrum of methods has evolved from sophisticated measurements of total body weight or whole body chambers to small units adhered to the skin surface with humidity-sensitive flow cells.[1,2] Moreover, the total losses of elements in sweat have been studied by complex washing procedures before and after sweat stimulation in saunas.[3] However, for studies of the metabolism of the eccrine gland where sweat composition rather than volume is appropriate, stimulation by iontophoresis of cholinergic agonists is the most convenient method.

II. Methods

A. Principle

The sudomotor processes of eccrine sweat glands are predominantly mediated by post-ganglionic sympathetic cholinergic neurones. The process of iontophoresis is the migration of small ions under the influence of an electrical current. Cholinergic agonists, e.g., pilocarpine, are transferred from the surface to the eccrine glands in the dermis.

B. Stimulation of Sweat Production

Wash and dry the skin of the flexor surface of the forearm and place two 5-cm^2 gauzes over the washed area. Moisten gauzes well with a freshly prepared solution of 1% pilocarpine nitrate. Place two gauzes saturated with saline on the extensor surface of the arm. The positive electrode of a DC power supply designed for iontophoresis is placed over the gauze with pilocarpine, and the negative electrode is placed on the saline-soaked gauze. Ensure good contact and secure the electrodes with rubber strips or similar means. If the arm is too small to secure the electrodes, as in small children, use the thigh or the interscapular area of the back of the patient.

Apply a current of 0.16 mA/cm^2 for 5 min.[4] The current will tend to increase during this time interval and should be maintained at the above setting. After 5 min remove the electrodes, clean the skin with distilled water, and dry the area.

C. Harvesting Sweat
1. Filter Paper Technique

With the use of forceps, place two prewashed Whatman No: 42 filter papers (diameter 5.5 cm) into a weighing bottle, stopper, and determine the combined weight accurately, using an analytical balance. Handle the weighing bottle with tissue or gauze to avoid direct contact with the fingers in order to avoid transfer of sweat from the operator's fingers which will increase its weight.

With the aid of forceps, place the preweighed filter paper over the skin area that was exposed to pilocarpine. Place a plastic sheet over this area and seal airtight with surgical tape. Allow the sweat to accumulate on the gauze or filter paper. This usually takes approximately 20 to 30 min but the time of sweat collection may be extended as long as necessary. In general, the appearance of droplets on the plastic sheet indicates that enough sweat has accumulated. These droplets must be included in the collection.

Remove the filter paper with forceps, place it immediately into the weighing bottle and stopper. Handle the bottle with tissues as above.

Weigh the bottle accurately (within 1 mg) to determine the weight of the filter paper and calculate the amount of sweat by difference. A minimum amount of 100 mg sweat is required for reliable quantitation. Generally, 200 to 500 µl of sweat is obtained (1 g sweat is assumed to be 1 ml).

2. Macroduct™ Sweat Collection System

The Macroduct™ (Wescor, Inc.) is the cornerstone of this system (see Figures 1 and 2). It is a small sweat-collecting unit with a shallow concave surface which is held against the previously stimulated skin. At the apex, a small aperture leads to a spirally configured plastic capillary tube which withdraws the sweat as it pools in the concavity. The duration required for sweat collection is similar to the filter paper method. The absolute volume obtained is dependent only on the method required for the sweat analysis.

D. Analysis of Sweat

The technique for elution of the sweat from the filter papers will depend on the particular analysis required. If the Macroduct™ is used, the sweat can be analyzed directly.

E. Problems with Iontophoresis

If the patient complains about discomfort, the test should be discontinued since the discomfort does not ease with time. A tickling sensation at the site of the electrode is a common finding and should be disregarded. After the test, there is usually some redness which disappears within a few hours; there may also be small grayish papules, or miliaria, these may take 2 or 3 days to fade.

Insufficient sweat may be collected using the filter paper technique above. This problem is not usually encountered by experienced operators. It may, however, be obviated by the Macroduct™ system which allows accurate analysis on very small quantities of sweat.

F. Interpretation of the Analysis of Sweat

Sodium and chloride elevations in sweat are seen even in the absence of gastrointestinal or respiratory symptoms or when pancreatic insufficiency cannot be demonstrated by other tests. In most affected infants, the test becomes positive between 3 and 5 weeks of age. Only 1 to 2% of CF patients have sweat chloride values below the value of 60 mmol/l and only 1 in 1000 has a value <50 mmol/l.[5] The exceptions are observed predominantly in CF patients who do not have pancreatic involvement.

Falsely high and low concentrations of sweat electrolytes may occur in a number of metabolic systemic and

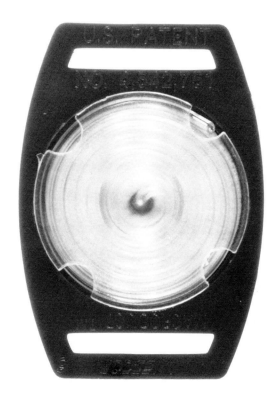

FIGURE 1 Wescor Macroduct™: illustration shows the spiral capillary tube used for the collection of pilocarpine-stimulated sweat. The sweat pools in the concave surface (not shown) which is adjacent to the skin and is drained into the tubing by capillary action.

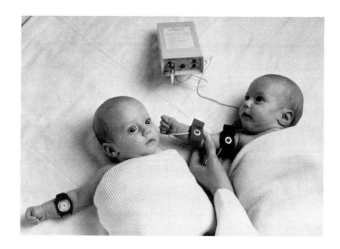

FIGURE 2 Wescor Macroduct™: illustration shows two infants. The infant on the right has the Wescor iontophoretic system in operation and the infant on the left has the collection system in operation.

cutaneous disorders as outlined in the Tables 1 to 3. Further variability may be due to physiological factors such as sweat flow rate, salt intake, and heat acclimatization.

TABLE 1 Conditions Likely to be Associated with Elevated Sweat Electrolyte Concentrations

Systemic Disorders	Cutaneous Disorders
Cystic fibrosis	Anhydrotic ectodermal dysplasia
Anorexia nervosa[6]	
Untreated adrenal insufficiency[7,8]	
Hereditary nephrogenic diabetes insipidus[9]	
Glucose-6-phosphatase deficiency[10]	
Hypothyroidism[11]	
Mucopolysaccharidoses[12]	
Malnutrition[13]	
Pupillatonia, hyporeflexia, and segmental hypohydrosis with autonomic dysfunction[14]	
Fucosidosis[15]	
Familial cholestasis[16]	
Hypogammaglobulinemia[17,18]	

TABLE 2 Conditions Likely to be Associated with Reduced Sweat Electrolyte Concentrations

Systemic Disorders	Cutaneous Disorders
Cystinosis[19]	
Hyperaldosteronism (primary or secondary, e.g., dehydration or severe stress)	
Hyperthyroidism[20]	

TABLE 3 Therapies with the Potential to Interfere with Sweat Electrolyte Concentrations

Increased Sodium and/or Chloride	Decreased Sodium and/or Chloride
All sodium salts, e.g., intravenous antibiotics	Corticosteroids
Anticholinergic drugs, e.g., premedications	Thiazide diuretics[9]

References

1. Kuno, Y., *Human Perspiration,* Charles C Thomas, Springfield, IL, 1956.
2. Graichen, H., Rascati, R., and Gonzales, R.R., Automatic dew point temperature sensor, *Am. J. Physiol.,* 52, 1658, 1982.
3. Brune, M., Magnusson, B., Persson, H., and Hallberg, L., Iron losses in sweat, *Am. J. Clin. Nutr.,* 43, 438, 1986.
4. Gibson, L.E., di Sant'Agnes, P.A., and Schwachmann, H., Procedure for the Quantitative Iontophoretic Sweat Test for Cystic Fibrosis, Cystic Fibrosis Foundation, 6000 Executive Blvd, Suite 510, Rockville, MD 20852, 1985.
5. Davis, P.B., Hubbard, V.S., and di Sant'Agnese, P.A., Low sweat electrolytes in a patient with cystic fibrosis, *Am. J. Med.,* 69, 643, 1980.
6. Beck, R., Goldberg, E., Durie, P.R., and Levison, H., Elevated sweat chloride levels in anorexia nervosa, *J. Pediatr.,* 108, 260, 1986.
7. Conn, J.W., Electrolyte composition of sweat: clinical implication as an index of adrenal function, *Arch. Intern. Med.,* 83, 416, 1949.
8. Morse, W.I., Cochrane, W.A., and Landrigan, P.L., Familial hypoparathyroidism with pernicious anaemia, steatorrhoea and adrenal insufficiency: a variant of mucoviscidosis, *N. Engl. J. Med.,* 264, 1021, 1961.
9. Lobeck, C.C., Banta, R.A., and Mangos, J.A., Study of sweat in pitressin-resistant diabetes insipidus, *J. Pediatr.,* 62, 868, 1963.
10. Harris, R.C. and Cohen, H.I., Sweat electrolytes in glycogen storage disease type 1, *Pediatrics,* 31, 1044, 1963.
11. Strickland, A.L., Sweat electrolytes in thyroid disorders, *J. Pediatr.,* 82, 284, 1973.
12. Durand, P., Bossone, C., Della Cella, G., and Liotta, A., Le mucopolisaccaridosi, *Recent Prog. Med.,* 44, 279, 1968.
13. Mace, J.W. and Scharberger, J.E., Elevated sweat chloride in a child with malnutrition, *Clin. Pediatr.,* 10, 285, 1971.
14. Esterley, N.B., Cantolino, S.J., Alter, B.P., and Brusilow, S.W., Pupillatonia, hyporeflexia and segmental hypohydrosis: autonomic dysfunction in a child, *J. Pediatr.,* 73, 852, 1968.
15. Durand, P., Borrone, C., and Della Cella, G., Fucosidosis, *J. Pediatr.,* 75, 665, 1975.
16. Lloyd-Still, J.D., Familial cholestasis with elevated sweat electrolyte concentrations, *J. Pediatr.,* 99, 580, 1981.
17. Corkey, C.W.B. and Gelfand, E.W., Hypogammglobulinaemia and antibody deficiency in patients with elevated sweat chloride concentrations, *J. Pediatr.,* 100, 420, 1982.
18. Rosario, N.A., Neto, R.S., Nitta, A., and Marinoni, L.P., Hypogammglobulinaemia and elevated sweat chloride values, *J. Pediatr.,* 102, 163, 1983.
19. Gahl, W.A., Hubbard, V.S., and Orloff, S., Decreased sweat production in cystinosis, *J. Pediatr.,* 104, 904, 1984.
20. Gibinski, K., Powierza-Kaczynska, C., Zmudzinski, J., Gec, L., and Dosiak, J., Thyroid control of sweat gland function, *Metabolism,* 21, 843, 1972.

Chapter 22.2
Methods for the Collection of Apocrine Sweat

Julian H. Barth
Department of Chemical Pathology and Immunology
General Infirmary at Leeds
Leeds, U.K.

I. Introduction

The secretions of the apocrine gland enter the hair follicle above the point of entry of the sebaceous duct. Several methods have been developed which employ either absorbent material to collect the secretions or their volatile products or direct cannulation of the apocrine duct to collect the secretions; the choice of technique will depend on the question to be solved. However, since the products of the apocrine and sebaceous glands intermingle in the infundibulum of the pilary duct, it is practically impossible to obtain pure apocrine sweat from an *in vivo* technique. All the following methods potentially suffer from contamination by either eccrine or sebaceous secretions. At present, it is probable that the only technique suitable for the study of pure apocrine sweat is the use of the isolated gland model.[1]

II. Methods

A. Visualization of Physiological Apocrine Secretions

The study of the physiological secretion of apocrine glands is difficult due to both the intermittent nature of secretion and also the effect of eccrine secretions. These latter secretions effectively dilute the apocrine secretions, spread them over the skin surface, and therefore enhance their evaporation. This use of plaster-of-Paris discs held on the axillary skin by adherent polyethylene holders has been suggested as a method of overcoming this problem.[2] Eccrine fluid is watery and is adsorbed by the discs whereas the viscid lipid apocrine secretions collect on the disc surface and can be observed after collections are made over a period of at least 3 to 4 hours. The apocrine droplets can be seen by fluorescence under ultraviolet light.

B. Harvesting Apocrine Secretions

The pool of apocrine sweat within the collecting duct can be squeezed onto the skin surface by stimulating the smooth muscle cells with adrenergic agents. Shelley and Hurley were the first to use this approach and they collected the apocrine sweat by cannulation of the apo-pilosebaceous duct with a fine capillary tube.[3]

The axillary skin should be shaved the day before the planned collection as surface manipulation is sufficient to stimulate muscular contraction of the apocrine apparatus; this results in an emptying of the reservoir of secretions. The skin is cleansed prior to the procedure with a nonionic detergent, e.g., 0.1% Triton X-100, rinsed with water, blotted dry, and finally washed with hexane. Secretion is stimulated with an intradermal injection of adrenaline 1:2000 in physiological saline. Droplets of apocrine secretion should appear almost immediately. They are visible with the naked eye but are more clearly seen with a magnifying lens. The droplets have been harvested by canulation of the ducts with individually drawn out capillaries, but sufficient fluid can be collected from the surface with a commercially available capillary (10-μl volume). Approximately 1 μl of milky fluid can be obtained from each apocrine-related orifice; however, not all the apo-pilosebaceous ducts within the visibly adrenaline-blanched area release any fluid. It should be possible to collect a total volume of 3- to 5-μL secretions.

C. Collection of Volatile Axillary Secretions

The study of axillary odor and the search for human pheromones demand a different technique from those outlined above. Cotton gauze squares 10 cm^2 which have been thoroughly cleaned and autoclaved are held in place in the axilla for periods of 6 to 9 hours.[4] The gauze pads need to be cleaned with the same procedure as will be subsequently employed to remove the absorbed material.

Individual experimental design will demand the need to standardize on such factors as axillary shaving, use of deodorants and perfumes, and frequency of washing and types of soaps employed.

References

1. Barth, J.H., Ridden, J., Philpott, M.P., Greenall, M.J., and Kealey, T., Lipogenesis by isolated human apocrine sweat glands: testosterone has no effect during long term organ maintenance, *J. Invest. Dermatol.*, 92, 333, 1989.
2. Fox, R.H., Mullan, B.J., and Thornton, C., A technique for studying apocrine gland function in man, *J. Physiol.*, 239, 75P, 1974.
3. Shelley, W.B. and Hurley, H.J., Jr., The physiology of the human axillary apocrine sweat gland, *J. Invest. Dermatol.*, 20, 285, 1953.
4. Preti, G., Cutler, W.B., Christensen, C.M., Lawley, H., Huggins, G.R., and Garcia, C.-R., Human axillary extracts: analysis of compounds from samples which influence menstrual timing, *J. Chem. Ecol.*, 13, 717, 1987.

Section I: Sebaceous Glands

23.0

Distribution and Follicular Morphology

23.1 The Follicular Biopsy 511
    O.H. Mills, Jr.

Chapter 23.1
The Follicular Biopsy

Otto H. Mills, Jr.
Hill Top Research, Inc.
University of Medicine and Dentistry of New Jersey
Robert Wood Johnson Medical School
New Jersey

I. Introduction

The follicular biopsy[1] is an extension of a technique reported in the British literature by Marks and Dawber.[2] These investigators used a cyanoacrylate polymer to obtain a thin sheet of stratum corneum cells in order to study the skin's topography under the stereomicroscope. This "skin surface biopsy" can be stained and light microscopy reveals bacteria, fungi, and other surface workings. In the course of their report they noted that "hair follicle and sweat gland openings are seen particularly well." Also, Holmes and colleagues have used cyanoacrylate polymer on the back of acne patients.[3] They noted that they could extract vellus hairs coated with a material that was composed mainly of lipid and keratin.

The follicle is central to a number of abnormalities including acne vulgaris, keratosis pilaris, ichthyotic conditions, and fungal infections. As hyperkeratinization of the sebaceous follicle is key in the pathology of acne, a noninvasive way to sample this impaction, particularly on the face, opens up many study possibilities. Two key features of the follicular biopsy is its noninvasiveness and speed of sampling.

Following is a review of some of the past work done using the follicular biopsy, a synopsis of some of the current research underway with this technique, as well as suggestions and projections for future applications of this noninvasive technique.

II. Method for Performing the Follicular Biopsy

The cyanoacrylate adhesive product we currently use in our laboratories is Loctite® (Prism™ 460 Series, Loctite Canada, Inc). One droplet (approximately 8 mg/cm²) is placed on the surface of the skin. This is spread uniformly by pressing a glass microscope slide to the skin's surface with light firm pressure. Sixty (60) seconds is allowed for polymerization. This can be extended if the temperature of the subject's skin or the environment is elevated. The slide with its confluent sheet is then removed slowly, peeling away with attention to preserving the sample with intact contents of sebaceous follicles. During this time we have not seen any instances of contact allergy or other concerns. Avoiding the eye area is, of course, important. The follicular biopsy sample can be viewed immediately under a dissecting scope using light projected at an angle. If present, sebaceous follicle impactions can be clearly seen and an assessment of size and density performed. Histological evaluation of a punch biopsy done on follicles immediately following the follicular biopsy procedure could not detect horny material in the follicular canal. Also present was a slight perifollicular lymphocytic infiltrate with minor irritation to the follicular epithelium[4] (Figures 1 and 2).

III. Aging Skin

As the skin ages, sebaceous follicles appear to develop several abnormalities. This is particularly the case with facial skin and the added effects of solar exposure. One such abnormality which appears to increase with age is trichostasis follicularis. Although this condition is seen in all decades of life it appears more prominently on the nose, cheeks, and forehead as age increases. The follicular biopsy extracts these bundles of hair and keratin easily. In 1982, Mills, Huber, McGinley, and Kligman (unpublished data) further adapted the follicular biopsy by placing the adhesive on the skin and applying sheets of polyethylene film which was able to conform to the nose and cheek areas better than the rigid surface of the glass slide.

Ten volunteer patients with dense to very dense involvement of trichostasis follicularis were selected. In general, the lesion distribution centered on the nose extending to the adjacent areas of the cheek. Seven of the ten subjects were females. No other medical problem was evident at the time of examination.

FIGURE 1 Transverse section of sebaceous follicle.

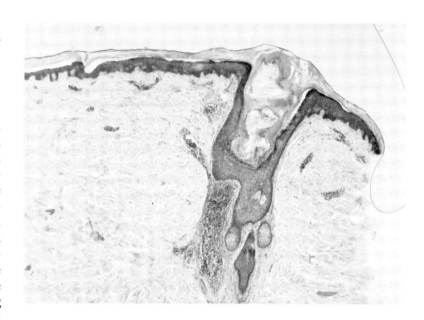

The cyanoacrylate polymer was placed on the area to be treated. Care was taken to protect the eyes. The adhesive was spread to coat the area with a visible film (approximately 8 mg/cm^2). This layer of adhesive was allowed to dry (polymerized in 60 seconds). A piece of polyethylene was placed in contact with the adhesive and held in place for 30 seconds. An additional 30 seconds was allowed for polymerization. The polyethylene film was removed from the skin as one would remove tape — peeling away from the skin. Four patients required a second procedure (2 to 4 weeks later) in order to extract a few remaining lesions.

By clinical inspection, eight of the ten patients so treated remained free of clinically evident trichostasis spinulosa for 1 year following the last procedure. Five of these subjects were followed for 1.5 years post-follicular biopsy extraction and remained 80% free of clinically observable follicular impactions.

IV. Sampling for Demodex Folliculorum

The mite *Demodex folliculorum* has been cited as playing a role in a number of dermatological conditions including rosacea and a variety of folliculitis.[5] Historically, collection of the mite from human skin has involved a number of scrapping techniques. The follicular biopsy presents an effective way to retrieve this organism from the follicular canal. In order to visualize the mites, the follicular biopsy sample should be placed under the microscope and a drop of peanut oil is added. By using a small probe it is possible to disrupt the keratinous impactions and often see mites in addition to those already visible.[1,6]

V. In Vitro Assay for Comedolytic Potential

Another possible application for the follicular biopsy is harvesting microcomedones for study in an *in vitro* comedolytic assay. As there are two topically applied agents currently used in the treatment of acne which have comedolytic activity (*all-trans* retinoic acid and salicylic acid), it may be possible to apply these as benchmarks and create a screening assay for potential comedolytic agents. With the use of microscopy and histological stains it may be possible to detect cellular changes that would suggest potential comedolytic activity. This *in vitro* approach should be explored.

VI. Evaluating Prophylactic Potential of Anti-Acne Modalities

As the follicular biopsy evacuates the follicular canal, certain numbers of the impactions reform to some degree in approximately 1 to 3 months. Thus, it may be possible to begin to evaluate the potential of agents to reduce or eliminate the reformation of these impactions. In order to do this kind of investigation it is key to well define the sample area and assure that subsequent sampling occurs in the exact area. This could be done on the face or back. The former, of course, has many more follicles while the later offers the opportunity to compare formulations.

VII. Evaluating Microbial Densities

In addition to the mite, the keratinous impactions in the sebaceous follicle, in particular microcomedones, contain aerobic and anaerobic bacteria as well as yeast-like fungi. Once retrieved by the follicular biopsy, microcomedones can be studied quantitatively for microbiology on an individual or pooled basis. The lesions are cut off the slide using a #15 scalpel under the dissecting scope.[1] They are then placed singly or pooled into test tubes of appropriate plating fluids, homogenized, and serially plated. One hundred twenty (120) individual microcomedones were homogenized and diluted in Tween 80. These yielded 410,000 ± 0.79

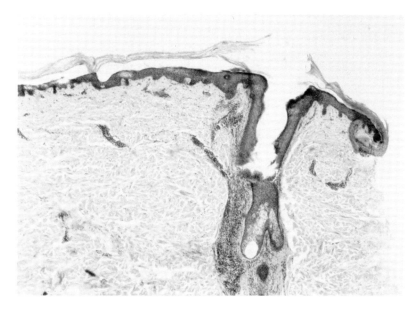

FIGURE 2 Transverse section of sebaceous follicle following a follicular biopsy.

P. acnes microorganisms per lesion whereas closed comedones (75) yielded 100,000 ± 2.09 and open (75) 67,000 ± 2.11. The respective incidence was 98, 92, and 92%.[7]

When the scrub cup technique[8] and the follicular biopsy have been compared in microbiology studies it becomes clear that one microcomedone can easily yield the equivalent or more organisms than the surface scrub's sampling.[1] The means in one study for *P. acnes* was 1.9×10^5 cm^2 vs. 6.9×10^5 per microcomedone.[1]

VIII. Lipid Analysis

Of interest to those studying acne is the free fatty acid to triglycerides ratio found in sebum as well as the total amount of sebum being excreted. As with microorganisms, the microcomedones extracted by the follicular biopsy can be studied individually or pooled together. Lipids can be extracted using hexane and incorporating methyl nervonate allows for qualitating and quantitating component lipids via thin layer chromatography.[7] When comparisons were done using the scrub cup technique the mean free fatty acid to triglyceride ratio for the scrub was 0.29 per square centimeter and 1.09 per microcomedone.[7] Quantitatively, the means were 165.7 μg/cm^2/3 h vs. 10.3 μg per microcomedone. Clearly the hydrolysis of glyceride was greater in the microcomedone than in the surface lipids. This outcome is consistent with the high density of *P. acnes* found in each lesion.

IX. Biochemical Changes

The follicular biopsy permits biochemical study of abnormal sebaceous follicle. The presence of neutrophiles in microcomedones has been confirmed using anti-neutrophile antiserum and chemical assay of lysozyme. It has been possible to estimate IGG, IGM, and the C_3 component of complement using immoprecipitation techniques.[9] These confirming findings are important in piecing together the pathology of inflammatory acne vulgaris and suggests further application of this approach.

X. Analyzing Drug Levels

Where the abnormal hyperkeratinization of the sebaceous follicle is central to the pathology of disease, it is important to know if orally or topically administered drugs are reaching the target tissue. As the follicular biopsy retrieves the microcomedone in acne, analysis for the presence of administered drug and its metabolites can be done using high-pressure liquid chromatography. A study was conducted using two concentrations of benzoyl peroxide and one of topically applied erythromycin analyzing for their presence in microcomedones and for the degree of reduction of *P. acnes*.[10]

The findings were very interesting and indicated that there was a significant correlation between *P. acnes* reduction and the active drug concentration in the microcomedone. Less anaerobe reduction occurred in subjects who showed low concentrations of benzoyl peroxide and erythromycin and high concentrations of benzoic acid or anhydroerythromycin.

These results also suggest the possibility of investigating individual responses of acne patients to active drugs. This approach may also shed light on why some patients are "resistant" to acne therapies. Drug metabolism or deactivation of the active drug may be occurring in some individuals. This remains to be investigated further.

XI. Ultraviolet Examination

Some years ago, Gunter Kahn[11] and co-workers showed that, by shining ultraviolet light on the surface of the skin, he could detect a coral red fluorescence located follicularly. This was due to *P. acnes* coporphyrins. Cornelius did additional work to define the nature of these porphyrins.[12] This approach has been used on follicular biopsy samples to evaluate the density of *P. acnes* before and after therapy.[13] It has also been used to separate keratinous impactions which show the presence and location within the follicle of the anaerobe and, therefore, allow for the separation of these for further study. Also, by using the follicular biopsy to assess porphyrin folliculitis indicating the presence of *P. acnes,* one reduces the possibility of materials blocking this process when done on the surface — topically applied.

XII. Image Analysis

The follicular biopsy sample is ideal for the application of digital image analysis. In the most widely used method to date, follicular biopsies are evaluated under polarized light which enables rapid measurement of density and size distribution of microcomedones.[14] The images are captured through a light microscope, two polarizing filters, and a high-resolution video camera. A system is then used to digitize and analyze the captured video image. In a series of studies,[4,14] there has been a high correlation between the traditional stereomicroscope scores for comedogenicity and the readout by digital image analysis (>0.9). This work represents an advance over traditional grading in that these can be central data interpretation of studies conducted at multiple investigative sites and the method allows for mathematical analysis of results and easy data storage.

XIII. Comedogenic and Comedolytic Evaluations

Not unlike other disorders of the skin, the course of acne needs to be studied by looking at both intrinsic and extrinsic factors which influence pathology. Acne-prone individuals appear to be susceptible to the effects of certain formulations causing hyperkeratinization of the follicle (comedogenics). In order to assay for comedogenic potential both animal[15] and human models[16] have been used. The follicular biopsy plays a key role in the latter assay in detecting subclinical microcomedone formation.

The original human comedolytic assay[17] used an extrinsic agent to induct comedones and then treat these lesions. As microcomedones occur naturally, we are currently developing a human comedolytic assay which will use these lesions.

XIV. Conclusion

The follicular biopsy offers a noninvasive approach for sampling the contents of the follicle. The biopsy is done quickly and the extracted material can be saved for further study. The information gained by sampling inside the follicle is important to understanding the follicular canal environment. Surface sampling techniques intended to gather information about the follicle may also include epidermal contributions.

Part of unraveling the key steps in the pathology of follicular abnormalities may well come from further study of the samples retrieved by the follicular biopsy.

References

1. Mills, O.H. and Kligman, A.M., The follicular biopsy, *Dermatologica,* 167, 57, 1983.
2. Marks, R. and Dawber, R.P.R., Skin surface biopsy: an improved technique for the examination of the horny layer, *Br. J. Dermatol.,* 84, 117, 1971.
3. Holmes, R.L., Williams, M., and Cunliffe, W.F., Pilosebaceous duct obstruction and acne, *Br. J. Dermatol.,* 87, 327, 1972.
4. Ayres, J.C., Mills, O.H., Lyssikatos, J.C., Kligman, A.M., and Groh, D.G., Assessment of a new method for determining the acnegenic potential of topically applied materials on human subjects, *Proc. 17th Annu. IFSCC Int. Congress,* Vol. 2, Yokohama, Japan, 1992, 889.
5. Ayres, S. and Ayres, S., *Demodectic* eruptions (demodicidosis) in the human. 30 years experience with 2 commonly unrecognized entities: *Pityriasis folliculorum (demodex)* and acne rosacea (*demodes* type), *Arch. Dermatol.,* 83, 816, 1961.
6. Brazeau, C., Mills, O.H., and Kligman, A.M., Does a High Population Density of *Demodex folliculorum* Contribute to "Sensitive" Skin in Adult Females? 18th World Congress of Dermatology: Progress and Perspectives, New York, June 12–18, 1992, 28A.
7. Mills, O.H., McGinley, K.J., and Kligman, A.M., The follicular biopsy in the study of acne, Program, 40th Annu. Meet., American Academy of Dermatology, 117, 1981, San Francisco.
8. Williamson, P.E. and Kligman, A.M., A new method for the quantitative investigation of cutaneous bacteria, *J. Invest. Dermatol.,* 45, 498, 1965.
9. Webster, G.B. and Kligman, A.M., A new method for the assay of inflammatory mediators in follicular casts, *J. Invest. Dermatol.,* 73, 266, 1979.
10. Wortzman, M., Scott, R., Wong, P., and Mills, O., A quantitative method for analysis of drug levels in the microcomedone, *J. Invest. Dermatol.,* 82, 413, 1984.
11. Martin, R.J., Kahn, G., Gooding, J.W., and Brown, G., Cutaneous porphyrin fluorescence as an indicator of antibiotic absorption and effectiveness, *Cutis,* 12, 758, 1973.
12. Cornelius, C.E. and Ludwig, G.D., Red fluorescence of comedones: production of porphyrins by *Corynebacterium acnes, J. Invest. Dermatol.,* 49, 368, 1967.
13. Weinstein, M., Mills, O., Berger, R., Dammers, K., and Baker, M., Follicular localization of propionibacterium acnes by ultraviolet examination, *Clin. Res.,* 36(3), 1988.
14. Groh, D.G., Mills, O.H., and Kligman, A.M., The quantitative assessment of cyanoacrylate follicular biopsies by image analysis, *J. Soc. Cosmetic Chem.,* 43, 101, 1992.

15. Kligman, A.M. and Mills, O.H., Acne cosmetica, *Arch. Dermatol.,* 106, 843, 1972.
16. Mills, O.H. and Kligman, A.M., A human model for assessing comedogenic substances, *Arch. Dermatol.,* 118, 903, 1982.
17. Mills, O.H. and Kligman, A.M., A human model for assaying comedolytic substances, *Br. J. Dermatol.,* 107, 543, 1982.

24.0

Sebum Excretion

24.1 Sebum-Absorbent Tape and Image Analysis 517
C. el-Gammal, S. el-Gammal, A. Pagnoni, and A.M. Kligman

24.2 Gravimetric Technique for Measuring Sebum Excretion Rate (SER) 523
W.J. Cunliffe and J.P. Taylor

Chapter 24.1
Sebum-Absorbent Tape and Image Analysis

Claudia el-Gammal and Stephan el-Gammal
Dermatological Department of the Ruhr-University
Bochum, Germany

Alessandra Pagnoni and Albert M. Kligman
S.K.I.N. Inc.
Conshohocken, Pennsylvania
Dermatological Department
University of Pennsylvania School of Medicine
Philadelphia, Pennsylvania

I. Summary

With the introduction of Sebutapes® a method became available which allows not only to measure sebum output as single global value of the lipid amount present on a given surface, but also to assess differences among the activity of individual sebaceous follicles. Using image analysis, parameters of sebum production can be quantified.

Sebutapes® are microporous, white adherent tapes which are applied on defatted skin; absorbed lipids become visible as transparent spots. Viewing these against a black background in reflection mode results in a black and white picture. Image analysis is based upon gray-level thresholding and filtering of the image. Prior to analysis, application of a shading correction algorithm eliminates gray-level gradients due to unequal illumination. Information about several factors can be obtained:

- About the percent area covered with sebum spots representing the overall amount of sebum produced by the follicles in this area (sebum excretion rate).
- About the number of sebum droplets. It correlates with the number of active follicles. However, high sebum production leads to confluence of the spots resulting in falsely low counts.
- About the mean and maximum size of the sebum spots. Particularly large, confluent spots are typical for patients with seborrhea and acne patients.

For proper interpretation several factors have to be considered. Due to the reservoir effect of the sebaceous follicle sebum output after defatting is highest in the beginning, reaching a constant level after a few hours. Sebum spots are subject to changes in their size and transparency depending on storage time. Crystallization processes within the droplets significantly affect the resulting values. Sebutapes® should be evaluated at a defined time after removal, preferably within 24 hours. When immediate evaluation is not possible, storage in a freezer at –30°C is advisable. Sebutape® images captured by the analysis program can be stored on discs.

For decades there have been various attempts to quantify the excretion of sebaceous glands. They all depend on recovery of lipid from the skin surface. Mostly the forehead is used as a site of high sebum production where the contribution of epidermal lipids to the overall lipid content is negligible.

Already at the turn of the last century absorbance of sebum into paper and consecutive weighing was performed. Based on these approaches Strauss and Pochi[1] developed the so-called cigarette paper method: the forehead is wiped with gauze and a stack of four cigarette papers is applied in a standard-size rectangle over a 3-hour period. The absorbed lipids are then extracted with diethyl ether and weighed after evaporation of the ether. Another, much more recently introduced material for the collection of sebum is bentonite clay, which is spread on the skin as an aqueous gel containing 0.2% carboxymethylcellulose.[2] As the clay can absorb large amounts of sebum, it is possible to leave it on the skin for long periods of time. A standardized protocol includes a 14-hour sebum depletion step followed by a 3-hour measurement step.

Since chromatographic techniques have become available, the composition of the collected and extracted lipids can be determined. Skin surface lipids originate both from the stratum corneum and the sebaceous glands; distinguishing the contributions of these two sources has been a problem at first. Today it is known that epidermal lipids consist of triglycerides, free fatty acids, and cholesterol while sebum is composed of triglycerides, wax esters, squalene, cholesterol, and cholesterol esters.[3]

The disadvantage of methods involving lipid collection into certain materials is that they are rather time consuming. To quickly obtain information about the lipids present on the skin surface, two similar instruments have been developed: the Lipometre[4] and the Sebumeter®.[5,6] Both use photometry to determine the lipid amount, with fat changing the light transmission through opalescent glass or plastic film. With these instruments usually casual surface lipid levels are measured (amount recovered from skin that has not been recently defatted or protected in any way). More accurate information is provided after standardized defatting and measurement at a defined time point several hours later.

II. Sebutapes® — Material and Means of Application

All above-mentioned techniques have in common that they yield a single global value of the lipid amount present on a given surface. Differences among the activity of individual sebaceous follicles cannot be evaluated. With the introduction of the "Sebutapes®" (CuDerm Corp., Dallas) in 1986, however, a reliable morphologic method became available, with which not only the sebum excretion rate in general but also the output of individual follicles can be monitored.[7-10]

Sebutapes® are white, open-celled, microporous, hydrophobic films coated with an adhesive layer which adheres to the skin surface. They come in sizes of 2.5 mm^2. Usually, they are applied on the forehead; however, sebum output can theoretically be measured on all parts of the body, except in areas covered with terminal hairs preventing the tape from sticking to the skin.

Before application, it is important to defat the skin in order to remove surface lipids, cell debris, or residual cosmetics. Ethanol, hexane, or Pannoclean™ (Pannoc Chemie, Herentals) are all suitable; however, it is important to stick to a certain protocol. We use the following procedure: the skin is washed with a detergent solution, Triton X 100 0.1%, rinsed with tap water, and then gently rubbed with a cotton swab soaked in hexane for 15 seconds. Most research groups agree that an application time of 1 hour is sufficient and gives the most accurate results.

III. Image Analytical Evaluation

Sebum droplets which are extruded from the follicles are absorbed by the Sebutape® and become visible as transparent spots. Viewing these against a black background results in a black and white image. For this purpose the samples are fixed on black filing cards or on glass slides which are placed on a blackboard before analysis. Information about several parameters can be obtained:

1. The percent area of the Sebutape® covered with sebum spots represents the overall amount of sebum produced by the follicles in this area. Many studies have demonstrated the reliability and reproducibility of this value and its good correlation with results gained by other methods.[7,10]
2. The number of sebum spots provides information about the number of active follicles in this area. Generally each droplet corresponds to one active follicle. However, the more sebum is produced, the more confluence of the spots occurs, resulting in falsely low follicle counts. Confluent follicle spots can be recognized by their irregular, e.g., triangular, shape, while droplets produced by a single follicle are round or elliptic. The droplet count is especially interesting when comparing the same site in a person at different time points, e.g., during a treatment phase. However, when different locations and persons with large variations in sebum excretion rates are assessed, this number is of limited value.
3. The sizes of the sebum spots are important as well. Their mean sizes are much higher in the center of the face — on the nose, chin, midforehead — than in its lateral parts such as the cheeks. In seborrhoic patients the spot sizes are very irregular, ranging from tiny spots of 30-μm diameter to large, confluent areas of more than 3 mm in diameter.

To quantify these parameters, image analysis is an indispensible tool. Evaluation is based on gray level thresholding of the image, thus telling the computer program what to recognize as "foreground" for the calculations. Figure 1 shows the image of a Sebutape® with its histogram of gray level distribution in the region of interest. The gray levels range from 0, which is black, to 255, which corresponds to white. The histogram of the displayed Sebutape® shows two peaks, the left one representing the black spots caused by the

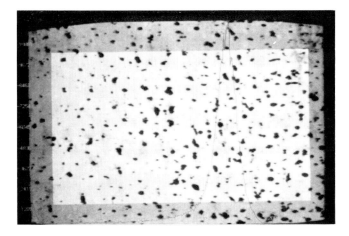

FIGURE 1 Sebutape® from the forehead of a 27-year-old woman with its histogram of the gray level distribution in the region of interest (inner light rectangle).

sebum, the right larger one representing the white or light gray background. According to the histogram curve a trigger is applied on the image; the gray level value of its upper border must be lower than the gray levels of the background. Black or dark gray sebum spots are detected. From the marked areas, the computer program calculates the total area covered, the spot number, and the mean and maximum spot size.

IV. Problems and Artifacts

A. Storage of Sebutapes®

When Sebutapes® are stored over longer periods of time at room temperature, several changes within the tapes occur which influence the results:

1. Within the first 24 hours after removal the spots gradually enlarge. This effect is especially pronounced in samples with large sebum spots and a high percent area covered. It is most probably due to spreading of the collected sebum within the tapes. Immediate evaluation may yield considerably lower values than analysis 1 day after removal. Figure 2 shows the same Sebutape® evaluated under the exact same conditions 5 min and 24 hours after removal from the skin. The increase in spot sizes lead to an increase of the total area covered from 6.77 to 8.12. The number of spots however decreased due to confluence from 165 to 148.
2. After longer periods of time (weeks) many of the spots may become whitish, probably due to crystallization processes within the sebum. Figure 3 demonstrates this phenomenon showing the same Sebutape® photographed 15 min and 30 days after removal from the skin. Spots that have turned white or light gray are difficult to distinguish from

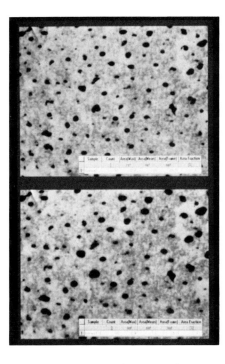

FIGURE 2 Sebutape® from the forehead of a 16-year-old boy immediately after removal from the skin (upper image) and 24 hours after removal, stored at room temperature (lower image). There is considerable enlargement of the spots from the upper to the lower image.

their surroundings and are not detected any more by gray level thresholding of the image.

In Figure 4 the influence of room temperature storage on the percent area covered is shown. A Sebutape® from the forehead was evaluated by image analysis in exactly the same area under the same conditions at different times. In the first hours there was a gradual increase in the percent area covered due to the mentioned enlargement of the spots. The values then remained stable during the following week. However, after 4 and 8 weeks, evaluation resulted in a much lower percent area covered because many of the spots had turned white.

To obtain most accurate and comparable results, Sebutapes® should be evaluated either immediately or at a certain constant time interval after removal from the skin, e.g., at 24 hours. Storing the samples in a freezer at −30°C will prevent or at least considerably delay the turning white of the sebum spots. Also, the increase in size of the spots can be kept to a minimum that way. We therefore recommend to store all Sebutapes® in a freezer when intending to evaluate them at later time points. A convenient way to record Sebutapes® is to save the images captured by the analysis program on hard or floppy discs. We use the TIFF-picture format which enables to transfer the images to other computers and programs. Image analytical evaluation can then be performed at any time independently from gaining the samples.

FIGURE 3 (A) Sebutape® from the forehead of a 28 year-old woman 15 minutes after obtaining the sample.

B. Illumination, Filtering of Images, and Interactive Analysis

Homogeneous illumination of the specimens is often a problem. In our experience, best results are obtained when Sebutapes® are placed in a white light box and illuminated from two sides by means of two fiber-optic light carriers. We use a high-resolution black and white CCD video camera mounted on a stereo microscope. To capture the images, the video signal is digitized by an IP8 frame grabber board using 255 gray levels.

To further eliminate gray level gradients due to ununiform illumination a shading correction algorithm can be applied on the image frozen by the analysis program. The image is blurred using a median filter with a large matrix and then subtracted from the original image.

In order to enhance contrast, it is very useful to apply an edge detection filter on Sebutape® images before analysis. This leads to a sharper image, and very small sebum spots can especially be recognized more readily by application of a trigger level. On filtered images, a lower gray threshold is more necessary than without filtering. Once a certain filter is chosen, one has to make sure that all evaluations are performed only after filtering of the image to assure comparability of the results.

Some image analysis programs allow to interactively manipulate the binary picture before analysis. Spots that have not been recognized can be added, artifacts can be deleted, or such spots can be cut apart, which resulted from confluence of follicles. Although working on the images in that way may improve accuracy, in our opinion such procedures should be avoided as far as possible. They not only reduce the objectivity and reproducibility of the evaluation but are also very time consuming.

V. The So-Called Reservoir Effect

When Sebutapes® are applied on the skin successively in the same area, each for 1 hour, the sebum output is highest in the beginning, reaching a constant level after a few hours. This phenomenon has been called "reservoir effect" of the sebaceous gland and has been ascribed to the fact that after degreasing the delivery of sebum from the follicular reservoir to the stratum corneum surface is increased.[2,3,11] The pilosebaceous infundibulum contains considerable amounts of sebum which have been es-

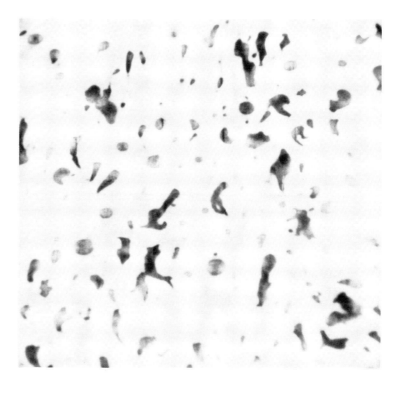

FIGURE 3 (B) The same Sebutape® 30 days after removal from the skin, stored at room temperature. Many of the sebum spots have turned light gray or white.

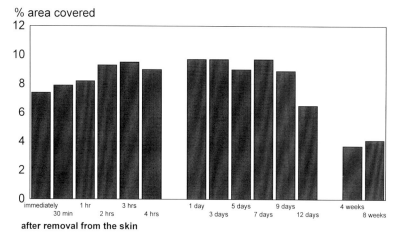

FIGURE 4 Percent area covered of a Sebutape® from the forehead evaluated by image analysis at different time points after removal.

timated at a few milligrams per square centimeter. Sebum collected in Sebutapes® during the first hours of application originates from this reservoir and does not necessarily reflect the production within the sebaceous gland. The latter is more accurately expressed by the plateau level reached after a few hours of sampling.

Figure 5 illustrates the reservoir effect showing the percent area covered of six-consecutively over 1 hour applied Sebutapes® on the nose, chin, and forehead.

VI. Applications of Sebutapes®

Sebutapes® have largely replaced and outdated older methods of measuring sebum excretion rates like the solvent extraction and gravimetric methods. Since their introduction many questions regarding the sebum output in general as well as the activity of single follicles have been investigated.

Pierard[12] disclosed four patterns of sebum excretion rates which are relatively specific for certain age groups: in infants very few, tiny spots are present, the pubertal pattern is characterized by about 100 spots/cm^2 which are small and rather uniform in size, the adult pattern shows a greater number of spots (120 to 280/cm^2) with greater variation in size, and in old age a decline in spot number, however, not in size compared to the adult pattern can be observed. Additionally the authors distinguish the so-called acne pattern which can be recognized by an unusually great number of irregularly shaped and very large spots. Correlation of Sebutapes® with cyanoacrylate surface biopsies[13] revealed that at least 50% of the sebaceous follicles are quiescent without any detectable sebum excretion and that there is no relation between the size of a follicular opening and its sebum output.

Using Sebutapes®, the hypothesis could be confirmed that sebum output underlies seasonal variations with increased delivery of sebum to the skin surface at higher ambient temperatures.[14] The number of active follicles however does not change with season.[14] In women, the ovarian cycle influences the sebum output with a peak during the week of ovulation; this effect is most pronounced in very oily persons.[15]

Blume et al.[16] measured the growth of vellus hair on the forehead, cheek, back, chest, and shoulder of men and compared it to the sebum excretion rate as assessed by Sebutapes®. They found no link between the two parameters.

Sebutapes® are unique with regard to the fact that the sebum production of individual follicles can be assessed. As opposed to other techniques they provide not only a single value but a morphological pattern of sebum production. Image analysis makes it possible to exactly quantity these patterns in a time-economic manner.

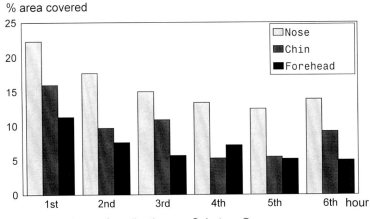

FIGURE 5 Percent area covered of consecutively applied Sebutapes® on the forehead, nose, and chin, each for 1 hour for a total time of 6 hours. The sebum excretion rate decreases in the beginning reaching a constant level after 3 hours.

References

1. Strauss, J.S. and Pochi, P.E., The quantitative gravimetric determination of sebum production, *J. Invest. Dermatol.*, 36, 293, 1961.
2. Downing, D.T., Stranieri, A.M., and Strauss, J.S., The effect of accumulated lipids on measurements of sebum secretion in human skin, *J. Invest. Dermatol.*, 79, 226, 1982.
3. Stewart, M.E., Downing, D.T., and Strauss, J.S., Sebum secretion and sebaceous lipids, *Dermatol. Clin.*, 1, 335, 1983.
4. Saint-Leger, D., Berrebi, C., Duboz, C., and Agache, P., The lipometre: an easy tool for rapid quantitation of skin surface lipids (SSL) in man, *Arch. Dermatol. Res.*, 265, 79, 1979.
5. Thune, P. and Gustavsen, T., Comparison of two techniques for quantitative measurements of skin surface lipids, *Acta Dermatol. Venereol.*, Suppl., 134, 30, 1984.
6. Dikstein, S., Zlotogorski, A., Avriel, E., Katz, M., and Harms, M., Comparison of the Sebumeter and the Lipometer, *Bioeng. Skin*, 3, 197, 1987.
7. Kligman, A.M., Miller, D.L., and McGinley, K.J., Sebutape: a device for visualizing and measuring human sebaceous secretion, *J. Soc. Cosmet. Chem.*, 37, 369, 1986.
8. Nordstrom, K.M., Schmus, H.G., McGinley, K.J., and Leyden, J.J., Measurement of sebum output using a lipid absorbent tape, *J. Invest. Dermatol.*, 87, 260, 1986.
9. Pierard, G.E., Follicle to follicle heterogeneity of sebum excretion, *Dermatologica*, 173, 61, 1986.
10. Serup, J., Formation of oiliness and sebum output — comparison of a lipid-absorbant and occlusive tape method with photometry, *Clin. Exp. Dermatol.*, 16, 258, 1991.
11. Saint-Leger, D. and Cohen, E., Practical study of qualitative and quantitative sebum excretion on the human forehead, *Br. J. Dermatol.*, 113, 551, 1985.
12. Pierard, G.E., Rate and topography of follicular sebum excretion, *Dermatologica*, 175, 280, 1987.
13. Pierard, G.E., Pierard-Franchimont, C., Le, T., and Lapiere, C., Patterns of follicular sebum excretion rate during lifetime, *Arch. Dermatol. Res.*, 279, 104, 1987.
14. Pierard-Franchimont, C., Pierard, G.E., and Kligman, A.M., Seasonal modulation of sebum excretion, *Dermatologica*, 181, 21, 1990.
15. Pierard-Franchimont, C., Pierard, G.E., and Kligman, A.M., Rhythm of sebum excretion during the menstrual cycle, *Dermatologica*, 182, 211, 1991.
16. Blume, U., Ferracin, J., Verschoore, M., Czernielewski, J.M., and Schaefer, H., Physiology of the vellus hair follicle: hair growth and sebum excretion, *Br. J. Dermatol.*, 124, 21, 1991.

Chapter 24.2
Gravimetric Technique for Measuring Sebum Excretion Rate (SER)

W.J. Cunliffe and J.P. Taylor
Leeds Foundation for Dermatological Research
Leeds General Infirmary Great George Street
Leeds, U.K.

I. Introduction

The sebaceous glands produce sebum by a holocrine process, in which the cells synthesize lipid as they move toward the center of the gland and eventually disintegrate. The lipid cell contents are then discharged via the sebaceous duct into the pilosebaceous follicle, when they are excreted to the skin surface as sebum. Autoradiographic studies have shown that it takes 2 or 3 weeks for a tritium-labeled cell to travel from the basal layer of the sebaceous gland to the sebaceous duct, and a further week or so for the sebum to reach the skin surface.[1-3] There is, in addition, a large reservoir of sebum in the pilosebaceous orifice, so that the rate of delivery of sebum to the skin surface may not accurately reflect metabolic events in the glands themselves. Nevertheless, changes in sebaceous gland size or mitotic activity of the sebaceous basal layer will eventually affect the sebum excretion rate. Normally this remains reasonably constant in a given subject and, therefore, the rates in individual subjects can be compared, and changes, as a result of experimental procedures, can be measured.

The most reliable method of assessing sebum excretion is a simple gravimetric method in which sebum is collected onto absorbent papers held in contact with an area of cleaned forehead skin.[4] This area is delineated with adhesive tape and an elastic headband is used to hold the papers in place. After a timed collection period, the lipid is either extracted with ether into a preweighed container which is again weighed after evaporation of the ether or simply weighed more directly by pre- and post-weighing the collecting papers.

II. Practical Details

The subsequent sections outline the techniques in detail.

1. Subjects should wash their hair on the evening prior to the test and apply no makeup or other topical application thereafter, but the face should be washed with soap in the normal way on the morning of the test.
2. The samples should, where possible, be collected at the same time of the day, since there may be circadian variation in both rate of sebum excretion and its composition.[5,6]
3. The room should be ventilated and should not be too hot because sweat reduces the capacity of the papers to absorb lipids. For the same reason, the subjects should not be allowed to take hot drinks during the test. Environmental temperature can have a marked effect on the apparent sebum excretion rate[7] probably by its effect on sebum viscosity,[8] but a room temperature of about 20 to 24°C is satisfactory.
4. The type of absorbent paper used is critical, because there is considerable variation in the amount of sebum absorbed by different papers, and even between different batches of paper from the same maker. The paper originally used by British workers is no longer available but a suitable alternative is the "Special velin tissue, non-fluff", obtainable from the General Papers & Box Company, Severn Road, Treforest Industrial Estate, Pontypridd CF37 5SP England (tel. no. GB 0443–841977).[9]
5. The papers must be degreased prior to use by immersing them in Analar ether for three 10-minute washes using fresh ether each time. A small, but progressive decrease in absorbency is found if the papers are immersed for more than 6 hours.[10] Thereafter, the paper must be handled with forceps, and it can be stored conveniently in ether-cleansed tin foil. Some types of tin foil treated with ether readily shed fragments of the shiny surface which stick to the papers, so it is best to use the dull surface.
6. The adhesive tape used to delineate the collection area on the forehead must be clean. Currently we use Durapore® surgical tape and prepare it as a

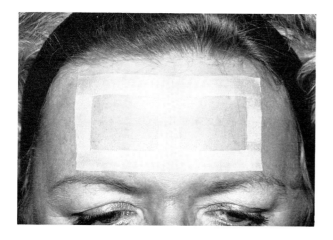

FIGURE 1 Area of collection, delineated by tape.

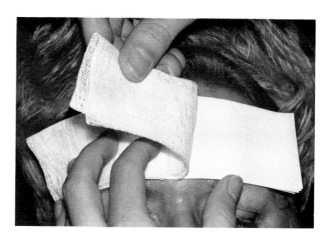

FIGURE 3 Papers covered with gauze.

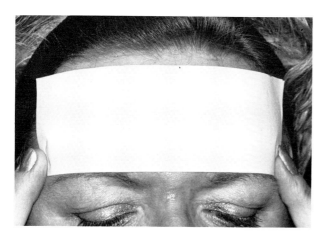

FIGURE 2 Absorbent papers on the forehead.

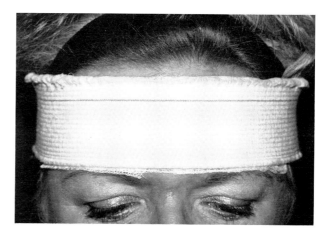

FIGURE 4 Papers and gauze held in position with an elasticated band.

template which ensures that the area of collection consists of two perfectly shaped rectangles (Figure 1). The template is made such that collections can be made simultaneously from both sides of the forehead.

7. The skin must be prepared, prior to the timed collection, by the serial application of preliminary absorbent papers until the accumulated surface lipid has been removed and only "follicular" pinpoint pattern of freshly excreted sebum is visible. Five sheets of absorbent papers, cut to the length of the forehead (Figure 2), are applied and covered with gauze (Figure 3); then a broad elastic band (50 to 60 cm long) (Figure 4) is applied and is attached at the back of head with nylon mesh (Figure 5); the sheet of paper in contact with skin is removed at 10 min intervals (three times) to remove the variable amount of surface lipid collected. Novice sebum collectors usually underestimate the time required for this preliminary procedure and the use of Sebutape® to view the "follicular patterns" (Figure 6) will often show that the papers are still heavily lipid laden. A correct pattern is shown in Figure 6. The time for the skin preparation varies for each subject, but most will require three or four removals of paper at 10-min intervals. It is now recommended that the papers are held up to the light to simply visualize the follicular pattern which is seen as small dots, each dot representing sebum emerging.

8. Once prepared as above the timed sebum collection can begin. This is done by placing ether-cleansed papers over the area marked out by the template (Figure 7) and replacing the elastic headband after lining with a single sheet of paper and a piece of gauze. The investigator should check the position of the collection papers periodically throughout the test, as many subject inadvertently move their elastic headbands.

FIGURE 5 The band is secured at the back of the head with nylon mesh.

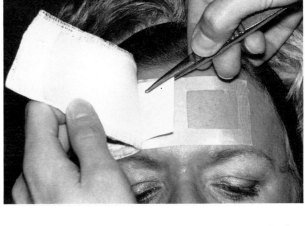

FIGURE 7 Final absorbent papers being placed on the forehead with forceps.

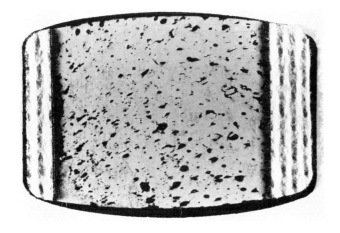

FIGURE 6 Pattern of sebum excretion using Sebutape® in a normal subject

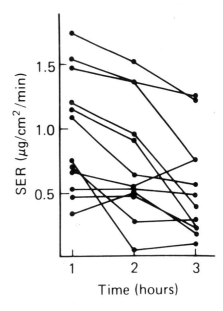

FIGURE 8 Decrease of sebum excreted during a 3-hour collection period.

9. The collection time should be standardized to about 3 hours and the time noted precisely. The hourly output of collected sebum decreases with time and it could be argued that the greater rate in the early hours is spurious and should be disregarded (Figure 8). The sebum collected during the first and subsequent 3-hour collection periods is linearly related, however, and the first collection appears to produce a valid index of sebaceous activity (Figure 9).[10] This observation emphasises that the method does not directly measure the rate of secretion from the sebaceous glands to the follicular reservoir.

10. The routine for extracting and weighing lipid is as follows: (a) the papers are transferred to beakers; sebum lipid is extracted with 3 × 25 ml analar diethyl ether and transferred into large tared flasks; the ether is removed by rotary evaporation (300mm Hg at 30°C) leaving a small volume which is transferred to preweighed 5-ml flasks which are evaporated to dryness and reweighed; (b) since the weight of lipid is very small, the flasks are handled with ether-cleaned forceps and kept in a desiccator for at least 24 hours before weighing in a microbalance accurate to 0.01 mg.

11. Good weighing technique is vital. The balance must be accurate to 10 µg (e.g., Mettler M5) and it should preferably be kept in a draft-free balance room built with a concrete floor over a main structural crossbeam of the building. Room temperature should be constant, and humidity inside the balance can be controlled by the use of silica gel. The effects of static electricity can be reduced by the use of aluminum cups in place of glass. One further development is the so-called *direct gravimetric technique* in which the papers placed

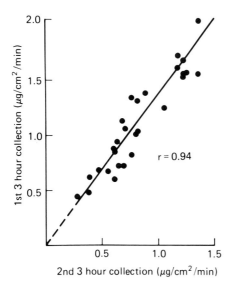

FIGURE 9 Correlation between sebum collected in two consecutive 3-hour periods.

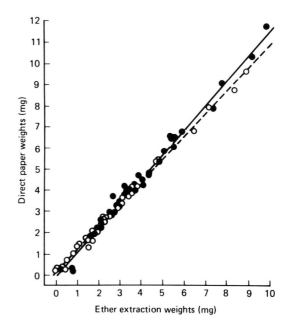

FIGURE 10 Sebum weights determined by direct vs. ether extraction techniques. •-•, results from 43 normal subjects (r = 0.991, $p < 0.001$). ○-○, results from 32 premeasured sebum standards (r = 0.997, $p < 0.001$).

on the forehead are preweighed (Figure 7). After the collection period the papers are reweighed. This technique is as accurate as the original method of Strauss and Pochi and eliminates the extraction procedure (Figure 10). The methodology and attention to detail is, therefore, similar to the original gravimetric technique, but without the extraction technique.

A. Calculation of SER

The SER is determined by the following formula:

$$\frac{\text{weight of sebum } (\mu g)}{\text{area } (cm^2) \times \text{time } (mins)}$$

Normal values for the sebum excretion rate by this method for subjects of different ages and sex have been given by Pochi and Strauss[12] but obviously each worker must obtain his own normal range. It should be noted that the experimental error of the method is proportionally larger with low sebum excretion rates, and the variance of the method should therefore be determined at each end of the range.

Although the sebum excretion rate is usually expressed in terms of the area of skin studied, it can also be related to the number of functioning glands by applying the "replica" technique[13] in which a hardened plaster of Paris (dental grade) replica of the skin surface is stained with 0.25% osmium tetroxide. This allows the number of patent follicular orifices to be counted, and it also produces a permanent record of their position in relation to other microscopic features of the skin surface. Cunliffe and Cotterill[14] have described an alternative technique in which a red dye (saturated Oil Red O) is used to stain the follicular orifices, followed by direct microphotography of the skin surface.

III. Factors Affecting the Measurement of SER

The SER is influenced by many factors which are relevant to its measurement in practice:[15]

1. Age, sex, and endocrine status. Age, sex, and endocrine status all affect SER. Some oral contraceptives inflame SER a little with the exception of antiandrogen-containing pills which can reduce SER by 30%. The effect of the menstrual cycle is probably of minor importance to clinical studies of SER.
2. Duration of collection. Duration of collection affects the apparent SER and variance of the measurements is most marked in the first and second hour.
3. Skin temperature. A 1°C change in skin temperature produces a 10% change in SER.[7]
4. Sweat gland function. Sweating reduces the amount of lipid collected by absorbent papers and a thermally neutral environment is necessary during measurement of SER. If the patient sweats obviously during the procedure the paper technique cannot be used. This event is very uncommon in the U.K.
5. Diurnal variation. SER is maximal in the mid-morning and minimal during the late evening and early hours of the morning.[5,6]

6. Site variation. Sebaceous glands are found predominantly on the scalp, face, back, and chest and the SER on the forehead is five times that on the back but when related to the number of glands at each site there is no difference.[16]
7. Disease. SER correlates with acne severity and past history of the disease.[17] Changes also occur with endocrine disease, Parkinsonism, and certain drugs.[15]

IV. Conclusion

Thus, in measuring SER, it is desirable to make collections at the same time of the day, in a room of constant ambient temperature, preferably at comparable stages of the menstrual cycle in females, to stop all relevant therapy 4 to 6 weeks before the measurement. Since the response of the sebaceous gland is slow, such assessments should be after at least 4 weeks' treatment. As results vary with details of technique (e.g., duration of collection, absorbency of papers), changes in SER are more reliable than absolute rates in comparing results from different laboratories. The method of choice will possibly depend on the requirement and local facilities. The techniques are easy to learn and a visit to the nearest "sebum" center will possibly pay dividends.

References

1. Epstein, E.H. and Epstein, W.L., New cell formation in the human sebaceous gland, *J. Invest. Dermatol.*, 46, 453, 1966.
2. Weinstein, G.D., Cell kinetics of human sebaceous glands, *J. Invest. Dermatol.*, 62, 144, 1974.
3. Plewig, G. and Christopher, E., Renewal rates of human sebaceous glands, *Acta Dermato-Venereol.*, 54, 177, 1974.
4. Strauss, J.S. and Pochi, P.E., The quantitative gravimetric determination of sebum production, *J. Invest. Dermatol.*, 36, 293, 1961.
5. Burton, J.L., Cunliffe, W.J., and Shuster, S., Circadian rhythm in sebum excretion, *Br. J. Dermatol.*, 82, 497, 1970.
6. Cotterill, J.A., Cunliffe, W.J., and Williamson, B., Variations in skin surface lipid composition and sebum excretion rate with time, *Acta Dermato-Venereol.*, 53, 271, 1973.
7. Cunliffe, W.J., Burton, J.L., and Shuster, S., The effect of local temperature variation on the sebum excretion rate, *Br. J. Dermatol.*, 83, 650, 1970.
8. Burton, J.L., The physical properties of sebum in acne vulgaris, *Clin. Sci.*, 39, 757, 1970.
9. Cunliffe, W.J., Williams, S.M., and Tan, S.G., Sebum excretion rate investigations — a new absorbent paper, *Br. J. Dermatol.*, 93, 347, 1975.
10. Cunliffe, W.J. and Shuster, S., The rate of sebum excretion in man, *Br. J. Dermatol.*, 81, 691, 1969a.
11. Lookingbill, D.P. and Cunliffe, W.J., A direct gravimetric technique for measuring sebum excretion rate, *Br. J. Dermatol.*, 114, 75, 1986.
12. Pochi, P.E. and Strauss, J.S.J., The effect of ageing on the activity of sebaceous glands in man, in *Advances in Biology of Skin*, Vol. 6, Montagna, W., Ed., Pergamom Press, New York, 1965, 121.
13. Sarkany, I. and Gaylarde, P., A method for demonstration of the distribution of sebum on the skin surface, *Br. J. Dermatol.*, 80, 744, 1968.
14. Cunliffe, W.J. and Cotterill, J.A., in *The Acnes: Clinical Features, Pathogenesis and Treatment*, W.B. Saunders, London, 1975, 293.
15. Cunliffe, W.J., *Acne*, Martin Dunitz, London, 1989.
16. Cunliffe, W.J., Perera, W.D.H., Thackray, P., et al., Pilosebaceous duct physiology, *Br. J. Dermatol.*, 95, 153, 1976.
17. Cunliffe, W.J. and Shuster, S., The pathogenesis of acne, *Lancet*, 1, 685, 1969.

Section J: Hair

25.0 Physical Properties of Hair

25.1 Hair Color 531
R.P.R. Dawber

25.2 Measurement of the Mechanical Strength of Hair 535
R.R. Wickett

Chapter 25.1
Hair Color

R.P.R. Dawber
Department of Dermatology
Churchill Hospital, Headington
Oxford, U.K.

I. Introduction

The color of the skin as perceived by the human eye is due to a complex mix of many components, melanin being a relatively minor component in very white caucasoids and a major part in black negroid individuals. One might easily (incorrectly) assume, however, that hair color[1,2] is due almost entirely to melanin pigments; in fact, this is untrue if one views hair from root to tip, many physical factors contributing.

It is also important to recognize that gray hair is not totally without pigment; pigment-free albino hair has a yellow hue, the basic color of the keratinous protein, i.e., the sulfur and iron content of hair do not contribute any "inorganic" color. In many lower animals other pigments such as porphyrins and carotenoids contribute to color.

The major factors governing hair color are (1) melanin pigments, (2) the presence or absence of weathering, (3) the shape of the fiber, and (4) the relative, greasiness of hair.

II. Melanin Pigments[1,2]

It is generally agreed that hair pigmentation is due to melanin of two types:

1. Eumelanin, the most common, giving shades from brown to black.
2. Pheomelanin, giving the yellow-blond, ginger, and red colors.

It has traditionally been stated that genetic control allows any one individual to produce either eumelanin or pheomelanin throughout life, but not both; this is probably untrue. The actual range of colors produced by melanins in human hair is limited to shades of gray, yellow, brown, red, and black. It should be noted that, whatever the color perceived by the eye, isolated melanin is brown in color and gives a dark brown solution in aqueous alkaline hydrogen peroxide.[3]

Unlike with hair dyes, the actual color of hair also depends on the sites and density of pigment deposition within the fiber and the shape of the pigment granules containing the pigment.[4,5] This aspect of color is most clearly seen by transmission and scanning electron microscopy, though the indistinct granules can easily be seen by light microscopy (Figures 1 and 2). In cutting sections to study the characteristics of hair cortical melanin granules, it is useful to use a diamond knife since glass knives are easily damaged by the coarse granules and may fail to show fine detail. Melanin granules are distributed throughout the hair cortex but in greater concentration toward the periphery; paracortex contains more granules than the less-dense orthocortex. Based on transmission electron microscopic changes and scanning electron microscopic examination of extracted pigment granules, one can generally relate the ultramicroscopic findings to particular hair colors.

The pigment granules of black to brunette hair have an oval (rice grain) shape with a more or less homogenous inner structure and sharp boundaries; the surface is finely grained with a thin surrounding membrane-like layer of osmophilic material.[4] Black hair granules are also relatively hard and have a high refractive index. Dark hair shades have greater numbers of pigment granules per unit area than pale shades. Blonde and red hair granules are more sparse, smaller, ellipsoid or spherical, and have a pitted surface.[3] Other distinguishing granular morphological changes have been specifically related to Italian brown hair, Japanese black hair, and Asian black hair.

The other characteristics contributing to the color of hair are physical phenomena.

III. Weathering

Hair is a dead structure and begins to break down superficially almost as soon as it leaves the skin — the "weathering effect".[6,7] Because dense hair transmits light evenly across its structure, the only interfaces at

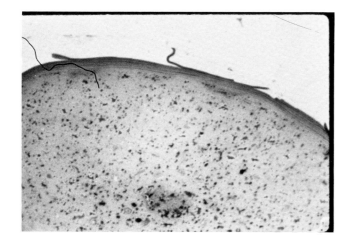

FIGURE 1 Hair fiber cross-section showing melanin granules scattered throughout the cortex (light micrograph).

which light is significantly reflected (and refracted) are the surface and from the medulla when this is present (intermittent and only in terminal hairs). Flat oval hairs reflect more surface light than round ones and thus are more likely to show a paler hue. Weathered hair has many more cortical "interfaces" for light to be "reflected" back to the eye and thus looks paler. This is the main reason for long hair appearing paler toward the tip when viewed with the eye (reflected light) but darker under routine (transmitted light) microscopy (Figure 3a and b). This optical effect is best understood by examining hairs from the hereditary hair cortical defect pili annulati.[8] The cortex contains many air spaces in the alternating abnormal bands which appear white clinically and dark on routine microscopy (Figure 4a and b). Filling the air spaces with a liquid of similar refractive index to keratin allows light to be transmitted and virtually abolishes the "optical darkening" seen microscopically (Figure 5a and b).

Melanin pigment protects against weathering to some degree. Blond scalp hair weathers more *in vivo* than dark hair; also the oxidation of melanin as the hair moves away from the scalp contributes to the weathering effects of UV — proven *in vitro* and seen *in vivo* by the decreasing pigmentation seen toward the tip in long hair, even when weathering is minimal. Also, this effect probably contributes to the greater degree of weathering seen (mostly in light-colored hair) in summer months and hot climates.

IV. Hair Fiber Shape

This only contributes in altering "shade" or hue of the hair. Circular fibers reflect less light than oval or flat ones. Therefore, the latter may appear paler, whatever the intrinsic color is, depending on the angle of reflected light perceived by the eye.

V. Hair Greasiness

Dry hair with little sebum along its surface appears paler than wet or greasy hair. After shampooing with basic shampoo containing detergent and no conditioner, when dry, hair of brown to black shades usually appears paler.

VI. Ethnic and Genetic Variations[9]

Racial hair color variations are more complex than the traditional simplistic view of the three types (1) white caucasoid — light/brown/red colors, (2) mongoloid — black, and (3) Negroid — black.

Many early studies on ethnic hair color differences are difficult to relate since they did not compare like with like, using different, only descriptive terms for hair color — general terms such as blond, flax, platinum

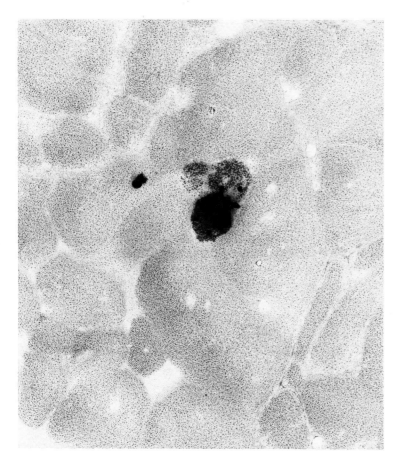

FIGURE 2 Hair fiber cross-section showing a central dark pigment granule surrounded by cortical macrofibrillary bundles (Silver methanamine stain; transmission electron micrograph × 50,000).

 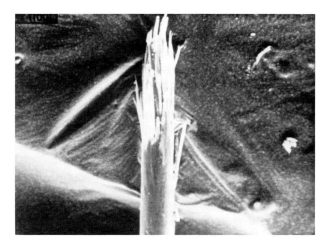

FIGURE 3 (a) Tip of a weathered, long hair. This appears dark by transmitted light microscopy. (b) Similar to (a) — scanning electron micrograph.

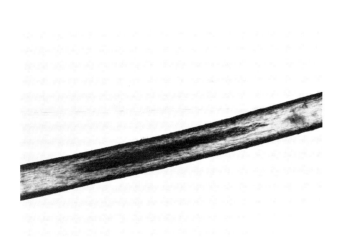 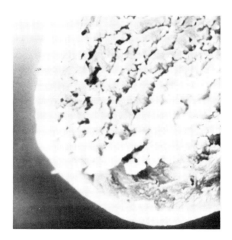

FIGURE 4 Pili annulati. (a) An abnormal band seen centrally — transmitted light micrograph. (b) Cross-section of an abnormal band showing the cortical spaces.

FIGURE 5 Pili annulati. (a) Cut through abnormal (dark) band. (b) Same as (a) but mounted in liquid of same optical density as keratin — optical "darkening" virtually abolished.

blond, golden brown, etc. may be vivid but lack measured exactness. Color matching is the more usual method now used, as in assessing hair dye, e.g., Fischer-Saller Hair Colour Scale.[1] The Munsell Notation[10] used to assess hue and color of dyes may possibly be a better method which can be semiquantitated.

Hair color is therefore far more complex than the limited shades of melanin pigment present within the fibers. In studying hair color in disease it is therefore very important to consider all the racial, congenital, pigmentary, and optical factors described above.

References

1. Rook, A. and Dawber, R.P.R., The colour of the hair, in *Diseases of the Hair and Scalp,* 2nd ed., Blackwell Scientific, Oxford, 1991, 370.
2. Zviak, C. and Dawber, R.P.R., Hair colour, in *The Science of Hair Colour,* 1st ed., Marcel Dekker, New York, 1986, 23.
3. Arnaud, J.C. and Bore, P., The isolation of melanin pigments from human hair, *J. Soc. Cosmetic Chem.,* 32, 137, 1981.
4. Orfanos, C. and Ruska, H., Die Feinstruktur des menschlichen Haares III Das Haarpigment, *Arch. Klin. Exp. Dermatol.,* 231, 279, 1968.
5. Swift, J.A., The histology of keratin fibres, in *The Chemistry of Natural Protein Fibres,* Asquith, R.S., Ed., John Wiley & Sons, London, 1977, 1.
6. Robinson, V.N.E., A study of damaged hair, *J. Soc. Cosmetic Chem.,* 27, 155, 1976.
7. Dawber, R.P.R. and Comaish, S., Scanning electron microscopy of normal and abnormal hair shafts, *Arch. Dermatol.,* 101, 316, 1970.
8. Gummer, C.L. and Dawber, R.P.R., Pili annulati: electron histochemical studies on affected hairs, *Br. J. Dermatol.,* 105, 303, 1981.
9. Wasserman, H.P., *Ethnic Pigmentation,* 1st ed., Excerpta Medica, Amsterdam, 1974.
10. Zviak, C. and Camp, M., Development of hair products, in *The Science of Hair Care,* 1st ed., Zviak, C., Ed., Marcel Dekker, New York, 1986, 326.

Chapter 25.2
Measurement of the Mechanical Strength of Hair

R. Randall Wickett
University of Cincinnati
College of Pharmacy
Cincinnati, Ohio

I. Introduction

A. Stress/Strain Curves

The mechanical behavior of hair and other keratin fibers is most conveniently and thus most frequently measured in extension. Figure 1 illustrates typical stress/strain curves for adjacent sections of the same hair obtained in either air at 40% RH or immersed in water. These curves can be characterized by three different regions. In the first region the curve is approximately linear and a slope can be determined. This is called the Hookean region and it extends to about 102% of the equilibrium length of the fiber (2% strain). Between 2 and 4% strain the curve "turns over" into the yield region. In the yield region, very little increase in force is required to increase extension. In the post-yield region, which begins at about 25% strain in the dry fiber and 28% strain in the wet fiber, the force again increases markedly with strain. For this particular hair, under the conditions tested, the slope in the post-yield region was about one fifth of that in the Hookean region of the dry fiber. There is little difference in post-yield slopes between the wet and dry sections of the fiber.

Published reports on the mechanical properties of keratin fibers date back to the work of Speakman[1-3] in the 1920s. Since that time, extensive research on hair and wool has led to an interpretation of each region of the stress/strain curves of keratin fibers in terms of changes occurring in the molecular structure. Feughelman[4] has reviewed this work very thoroughly and his review is highly recommended to anyone with a serious interest in the mechanical properties of hair. The discussion below is confined to a brief overview.

B. The Two-Phase Model of the Hookean Region

In the region of the force extension curve below 1.5% extension keratin fibers are generally considered to behave as Hookean springs. Bendit[5,6] has pointed out that curve is not truly Hookean in this region. However, the curve is approximately linear and Young's modulus of elasticity can be calculated.[4,7] Since this region as been referred to as Hookean for at least 50 years, it is likely that the terminology will persist for the foreseeable future.[4]

The mechanical properties of hair or wool in the Hookean region are well explained by the two-phase model of Feughelman.[4,8-14] This model considers the mechanical properties of the fiber to be determined by a water-impenetrable phase, C, the microfibrils, and a water permeable phase, M, the matrix. The microfibrils are primarily composed of α-helical proteins aligned parallel to the fiber axis[15] and the matrix is composed of water and high sulfur proteins that may be globular.[16] The composite may be modeled mechanically as a fixed Hookean spring in parallel with a spring and viscous dashpot in series.[4,10] The spring contributes about 1.4×10^9 Pa to the Young's modulus and is contained in the water-impenetrable microfibrils.

The main resistance to extension of the microfibrils probably comes from the hydrogen bond network in the α-helical proteins. When a keratin fiber is immersed in liquid nitrogen to prevent segment mobility, the modulus is close to that of ice as expected for a hydrogen-bonded network.[17] The matrix contributes viscous forces which decay with time causing stress relaxation. The viscosity of the matrix decreases greatly as the water content of the fiber increases.[4,11]

The two-phase model of keratin fibers accounts well for the effects of water on the mechanical properties,[2,11,18] the effect strain rate on Young's modulus,[4,19] stress relaxation behavior in the Hookean region,[13,20] and the behavior of wet, dry, and permanently set fibers in torsion.[2,9,21]

C. α-β Transformation in the Yield Region

Above extensions of about 2 to 3% the stress strain curve "turns over" into the yield region. The stress does not increase markedly until about 25% extension. The mechanical properties of a fiber extended into this region can be recovered by relaxing the fiber in water overnight if the fiber is not held too long in extension[3] and the extension is carefully confined to

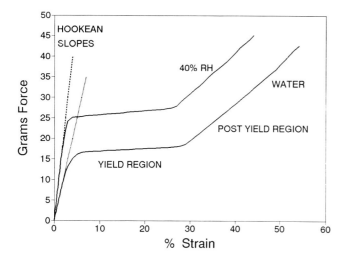

FIGURE 1 Stress/strain curves for different sections of the same hair in air at 40% RH and in water.

the yield region. As we shall see, this fact is of great practical importance in designing protocols to measure hair strength.

High-angle X-ray diffraction results have demonstrated that there is a progressive loss of α-helical content and a concomitant increase in β-sheet as a fiber is extended through the yield region.[24] By the end of the yield region about 30% of the original α-helix has been unfolded. It has also been shown[24,25] that the mechanical behavior of keratin in the yield region can be completely accounted for by application of a Burte-Halsey[26] model. The fiber is considered to contain a continuum of units which can exist in a short state, A (α-helix), or an extended state, B (β-sheet), with an energy barrier between the states. The yield region corresponds to a phase transition between state A and state B at constant stress. This first-order phase transition, producing a large length change at constant stress and temperature, is analogous to the transformation of water to steam producing a large volume change at constant temperature and pressure.

D. The Series Zone Model and the Post-Yield Region

Speakman[3] found the post-yield slope to be independent of the water content of the fiber. The increase in stiffness in the post-yield region was shown to result from a covalently bonded network involving cystine. The post-yield slope has been shown to be dependant on the disulfide content of the fibers.[27,28] Extension to about 50% strain leads to loss of all α-helical structure in the fiber[24] as judged by X-ray diffraction.

The behavior of keratin fibers in the yield and post-yield regions has been interpreted in terms of a series-zone model.[29,30] The model postulates two alternating zones, X and Y, along the microfibrils. The X zones contain the 30% of the α-helices that unfold reversibly in the yield region. The Y zones contain regions of α-helix that cannot be unfolded without the breakdown of disulfide bonds. Thus, unfolding of the Y zones in the post-yield region is irreversible.

E. Variations Among Fiber Types

Much of the work on the mechanical properties of keratin fibers has been carried out with wool. In the discussion above the term "keratin fiber" has been used to refer to either hair or wool. The mechanical behavior of the two fibers is very similar. A comparative study by Menkart et al.[31] found the elastic modulus and stress at 20% extension of hair to be somewhat higher than that of wool while Chaiken and Chemberlain[32] found the dynamic elastic modulus of hair to be nearly equivalent to that of wool. Tolgyesi et al.[33] found beard hair to behave similarly to head hair in extension but to have a slightly lower elastic modulus and stress at 20% extension. In general, head hair, beard hair, and wool may be considered similar enough in behavior that all of the conclusions about structure and mechanical properties discussed above can be considered to be equally valid for each structure.

II. Object

A. Tensile Measurements of Hair Damage

Measurements of hair tensile properties are most frequently made to assess the effects of chemical treatments on hair strength.[7] The mechanical properties of hair are greatly affected when the number of disulfide cross-links is reduced. This is especially true of wet hair in the yield and post-yield regions of the stress-strain curve.[4,7,27,28,34–37] A typical protocol is to strain an untreated hair into the yield region and measure either the force or the work of extension. The work of extension is the area under the force vs. extension curve.

Beyak et al.[38] assessed the effects of bleaching and permanent waving on the relative stress to extend a hair to 15% strain before and after treatment, the 15% index (I15), in water. They found that a 30-min bleach treatment reduced I15 by an average of 10% and a 5-min permanent wave treatment reduced I15 by about 13%. Wolfram et al.[39] reported that a 30-min bleaching treatment reduced the yield-stress by about 12% and Robbins[7] presented data showing that a commercial permanent wave caused an 18% reduction in the work to extend a hair to 20% in the wet state compared to an 11% decrease in the work to extend hair to 20% in the dry state. Tensile measurements have also been used to characterize the effect of ultraviolet radiation,[40] surfactant binding,[41] and chlorine treatments[42] on hair. These studies have all shown tensile measurements to be very useful for the study of damaging treatments that break disulfide bonds in the cortex. However, Robbins and Crawford[43] have shown that treatments that cause severe damage to hair cuticle may have little or no effect on tensile properties.

B. Bending and Torsional Measurements

A further objective of mechanical measurements on hair is to understand the processes involved in setting and permanent waving. Measurements of extensional properties are not necessarily the best way to achieve this goal. Bogaty[44] pointed out that the behavior of hair under torsional and bending strains is very important to the permanent waving process because forming a curl from straight hair involves a combination of twisting and bending deformations. He found that permanent waving decreased the torsional rigidity of hair in the wet state but actually increased it slightly at 65% RH. Wolfram and Albrecht[45] made torsional measurements on hair and concluded that the cuticle is very stiff in the dry state and may make a significant contribution to the torsional rigidity, especially for fine hairs. However, in the wet state the cuticle was found to be so plasticized as to make no contribution to mechanical behavior.

Scott and Robbins[46,47] described a practical, balanced fiber method for measuring the bending stiffness of hair. A long hair is draped over a small wire with small weights attached to each end. The bending stiffness can be calculated from the distance between the two ends. It is also possible to measure bending strength by a three-point beam deflection method and this method has been applied to measuring the stiffness of beard hairs.[48] The balanced fiber method has the disadvantage of requiring a relatively long fiber but in the author's experience it is far easier to use than three-point bending methods.

C. Chemical Relaxation Methods

The dramatic effect that breaking disulfide bonds has on the tensile properties of hair can be used to study the kinetics of the reduction reaction. If a hair is stress relaxed to a constant level of force at a constant extension in water or buffer and then the solution is switched to reducing agent, the force decays with time due to the breaking of disulfide bonds[49–52] and kinetic parameters can be determined. An alternate method is to repeatedly stretch the hair in the linear region, measuring the reduction in elastic modulus as the reaction proceeds.[53] Wortman and co-workers[54,55] have used a combination of these methods to attempt to predict permanent set based on relaxation parameters. I have recently reviewed these methods along with other work on the effects of permanent waving on the physical properties of hair.[56]

III. Methodological Principle

A. Overview

This discussion of methodology will focus on use of extensional measurements to evaluate hair strength in tension, including determination of elastic modulus, yield stress, and breaking strength. These are the most widely used means to evaluate treatment effects on hair. Robbins[7] has provided a thorough discussion of other methods for the evaluation of hair physical properties.

B. Modulus Calculation

Young's modulus of elasticity, E, (Y in some older papers) is calculated from the slope of the Hookean region of the stress/strain curve (Figure 1). E is defined as follows:

$$E = \Delta F * L / \Delta L * A \quad (1)$$

Where ΔF is the change in force induced by a change in length, ΔL, L is the equilibrium length of the fiber, and A is the cross sectional area. For example, assume that a cylindrical hair of 80 μm (8×10^{-5} M) diameter, 5×10^{-9} M^2 cross sectional area, requires 10.2 grams-force (0.1 Newtons) to extend 1%.

$$E = 100 * 0.1 N / 5 \times 10^{-9} M^2 = 2 \times 10^9 \text{ N/M}^2 \quad (2)$$

A Newton/M^2 is 1 Pa so the modulus is 2×10^9 Pa. Older papers report E or Y in dynes/cm^2. A pascal is 10 dynes/cm^2 so the modulus of this fiber is 2×10^{10} dynes/cm^2. When reporting E, the rate of extension must be specified because the slope of the Hookean region is known to vary with extension rate[7,19] due to stress relaxation in the matrix.[4] Young's modulus is typically about 1.5 to 2.0×10^9 Pa for wet hair and 3.5 to 4.5×10^9 Pa for dry hair, depending on strain rate and hair source used.

C. Tensile Testers

Stress/strain measurements on hair are usually made with commercially available tensile testers. Most measurements reported in the literature over the years have been made using one of the several varieties manufactured by the Instron® Corporation, usually a table model such as the 4201. The Instron® uses a large, very robust screw drive system to move a either a cross head containing a load cell or a large mechanical stage at a constant rate of extension. Modern Instrons® are computer controlled and a wide variety of stress/strain protocols can be programmed in using "canned" software provided with the instrument. By using a programming language such as PASCAL and drivers provided with the instrument, software can be created to collect data using virtually any protocol that the programmer imagines.

Advantages of the Instron® are its high precision and flexibility. Forces from less than 0.1 g to several kilograms can be measured, depending on the load cell selected, and extensions from less than a millimeter to more than a meter can be accurately performed. Disadvantages are its cost, $50,000 and up, depending on options, and its rather large size. Even a table model Instron® weighs a few hundred pounds and takes up a considerable amount of lab space. It is definitely not a portable instrument.

A recently available option for measuring the mechanical properties of hair is the Dia-stron Rheometer (DR). The DR is portable. The measuring jig weighs just 3 kg and the control unit is about the size of a typical personal computer box. The cost of the DR is about one fourth that of an Instron®. The curves in Figure 1 were obtained using the DR.

While completely adequate for obtaining stress/strain curves from hair, the DR has some limitations. The light weight of the DR measuring jig that makes it portable also makes it susceptible to vibrational noise, such as might arise from general activity in the lab. Stress/strain and stress relaxation protocols are entered into one of two "methods" available to the user at any one time. Each protocol allows one extension, and one compression phase with data collection after each and each extension/compression cycle can be repeated several times. This gives considerable flexibility to the operator but not the extreme flexibility provided by the Instron®. I should also point out that the software we obtained with the DR does not calculate Young's modulus correctly. Whenever using a software package provided with an instrument it is wise to check all calculations by hand.

IV. Sources of Error

A. Variability in Hair

A major source of error in measurements on hair is the inherent variability in the thickness and shape of hair. This can be compensated for to some extent by determining the dimensions of each hair and normalizing the result to cross-sectional area. Measurements of breaking strength are particularly prone to variability and even different sections of the same hair will often break at different strengths. Another source of error when evaluating treatment effects is the inherent difference in treatment response between individuals. For example, reaction rates to reducing agent may vary by a factor of ten between individuals who do not have a history of chemical treatment and may vary more if chemically treated hair is used.[50,52] Some possible solutions to the variability problem are discussed below.

B. Relative Humidity

When making dry measurements exact control of relative humidity is very important. This is well illustrated by the fact that simple room relative humidity gauges often use a horse hair to move the dial as the room RH changes. Thus, variations in RH can be a significant source of error when making dry measurements of elastic modulus or yield stress. As can be seen in Figure 1, post-yield slope and breaking strength are much less sensitive to RH.

C. Other Sources of Error

Other sources of error are slipping of the hair gripping system, damage to the hair by the gripping system, and slack in the hair resulting in an inaccurate value of equilibrium length. Slack can be easily accounted for by modern computerized methods which can recalculate the equilibrium length when significant force is first sensed. Gripping problems are discussed in detail below.

V. Recommendations

A. Reducing Variability

Use each hair as its own control! The best way to account for variability between hairs when evaluating treatment effects is to use each hair as its own control. As discussed above the mechanical properties of a keratin fiber can be recovered if the hair is not strained into the post-yield region.[1,4,7,22] A typical protocol is that used by Beyak.[38] Hairs were strained in water to 15% extension on an Instron® tensile tester and then soaked in water for 16 hours, treated, and rerun. The change in grams-force required to reach 15% extension was evaluated. Force values at 15% extension from 25 hairs ranged from 11.4 to 35.5 g in the first run. A second run without a treatment between showed an average change in force for each hair of only 0.33% of the original force with a standard deviation of 2.68%. The changes observed ranged from +5.9% to –5.7%. Bleach treatments caused an average reduction in force at 15% strain of 10 to 20% depending on treatment time. This study and many others like it[7,39–42] clearly show the value of using each hair as its own control.

When doing breaking or chemical relaxation studies it is not possible to use each hair segment as its own control. The next best approach is to cut the hairs into different sections and compare results between sections, as was done to obtain the data in Table 1, discussed below. Breaking values on different sections are not as reproducible as reruns of stress/strain curves to 20% extension on the same section of hair but still provide a large improvement over simply comparing different hairs.

When it is not possible to either use each hair as its own control or to compare different sections of the same hair, one must attempt to at least normalize for the dimensional variations in the hair as described below. The only other alternative is to run literally hundreds of samples to get statistical significance.

B. Determination of Hair Diameters

Determining the cross-sectional dimensions of a hair is a difficult problem. Not only is hair a fine fiber, it is not necessarily uniform in cross section. While Caucasian hair is generally considered elliptical in shape, significant variations from ellipticity can occur. Robbins[7] has reviewed methods for diameter determination and recommends the linear density method as the method of choice. I concur with his recommendation. To use the linear density method a hair is cut to a given length and weighed on a microbalance. The fiber density is assumed to be 1.32 g/cm^3 (Reference 7) and the cross sectional area, A, in square centimeters, is then given by

$$A = W/(1.32*L) \quad (3)$$

where W is the weight of the hair in grams and L is the length of the segment measured in centimeters. If necessary the effective diameter of the hair can then be determined from simple geometry if the hair is assumed to be circular in cross section.

C. Gripping the Hair

An annoying difficulty when making tensile measurements on hair is the problem of obtaining a good grip on the hair without causing damage at the gripping point. This is especially vexing when one wants to stress the hair all the way to breaking. The plastic "fiber grips" provided by Instron® are totally inadequate as most hairs will slip from the grips well before they break.

I have previously crimped each end of the hair into short sections of fine diameter aluminum tubing which can then be gripped by any good spring-loaded clamp, but the sections must be cut and filed carefully to avoid having any sharp edges that could cut the hair.

What seems to be a reasonable solution to the gripping problem has recently been provided by Dia-stron®. They sell 14-mm sections of 2-mm diameter metal tubing with plastic tubing inside. These sections can be crimped onto the hair to provide good "handles" for gripping. Dia-stron® also sells a crimping tool for this purpose that is convenient for crimping onto each end of a standard 3-cm hair section. We have found that an adequate crimp can also be obtained by careful use of an electrician's crimping tool available in nearly any hardware store. Even using the Dia-stron® crimping system one must be careful to get a good crimp if making a hair-breaking measurement. We usually move the sample block slightly and recrimp at least twice to insure a good crimp. Otherwise, the hair may slip before breaking.

D. Breaking Strength From an Inexpensive Device

If you want to screen a treatment for its effect on hair strength and do not have access to a tensile tester, I recommend trying a simple and inexpensive device we have developed to measure the dry-breaking strength of hair.[57] The device is based on a lever principle and is shown schematically in Figure 2. A meter stick pivots on a bolt through a hole in its center. The hair is fastened to one end and a weight on a sliding hanger is moved slowly and carefully along the other side away from the center hole. The breaking strength is Wt* (L2/L1) — Wc where Wt is the total weight of the sliding hanger and attached weight, L2 is the distance from the center hole to the slide at break, L1 is the distance from the center hole to the hair, and Wc is the weight of the clip used to hold the hair. I have made sliding weight hangers from either a mirror mount bracket or a closet door mount. No doubt other objects could be used for this purpose. Our current hanger weighs 30 g and we use an additional 150 g of balance weights attached using wire. If the hairs are mounted into the Dia-stron® crimps discussed above the grips can be made from alligator clips.

Ten hairs were cut into two sections each and the breaking strength of one section was determined on the DR and the other on the lever device. The results are shown in Table 1. The agreement between the

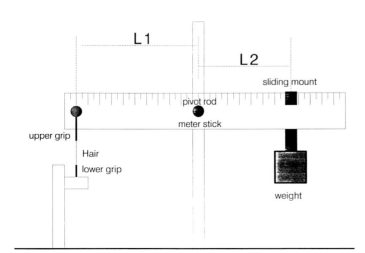

FIGURE 2 Schematic diagram of a simple device to measure hair breaking strength.

TABLE 1 Dia-stron® Hair Breaking Compared to Lever Breaking

Hair #	Dia-stron®	Lever	Difference Diast – Lever
3	183.0	193.0	+10
5	164.0	159.0	–5
1	126.0	150.0	+24
10	106.0	111.0	+5
9	105.0	97.0	–8
2	102.0	89.0	–13
8	101.5	95.0	–6.5
7	93.5	95.0	+1.5
6	79.5	65.0	–14.5
4	70.0	82.0	+12
Average	113.05	113.6	0.55

average values for breaking strength obtained from each device is remarkable. Of course, the lever device does not allow one to produce a stress/strain curve but it seems to work well as a potential screening tool considering that it took about $1/2$ hour and $12 worth of materials purchased from a local hardware store to construct.

References

1. Speakman, J.B., The gel structure of the wool fibre, *J. Text. Inst.*, 17, T457, 1926.
2. Speakman, J.B., The rigidity of wool and its changes with adsorption of water-vapor, *Trans. Farady Soc.*, 25, 92, 1929.
3. Speakman, J.B., The intracellular structure of the wool fibre, *J. Text. Inst.*, 18, T431, 1927.
4. Feughelman, M., The physical properties of alpha keratin fibers, *J. Soc. Cosmet. Chem.*, 33, 385, 1982.
5. Bendit, E.G., Properties of the matrix in keratins. II. The "Hookean" region in the stress-strain curves of keratins, *Text. Res. J.*, 48, 717, 1978.
6. Bendit, E.G., There is no Hookean region in the stress-strain curve of keratin or other viscoelastic fibers, *J. Macromol. Sci. Phys.*, B17(1), 129, 1980.
7. Robbins, C.R., *Chemical and Physical Behavior of Human Hair*, 2nd ed., Springer-Verlag, New York, chapter 8.
8. Feughelman, M., A two phase structure for keratin fibers, *Text. Res. J.*, 29, 223, 1959.
9. Mitchell, T.W. and Feughelman, M., The torsional properties of single wool fibers. I. Torque-twist relationships and torsional relaxation in wet and dry fibers, *Text. Res. J.*, 30, 662, 1960.
10. Feughelman, M., The relation between structure and the mechanical properties of keratin fibers, *Appl. Polymer Symp.*, No. 18, 757, 1971.
11. Feughelman, M. and Robinson, M.S., Some mechanical properties of wool fibers in the "Hookean" region from zero to 100% relative humidity, *Text. Res. J.*, 41, 469, 1971.
12. Feughelman, M., Keratin, in *Encyclopedia of Polymer Science and Engineering*, Vol. 8, 2nd ed., John Wiley & Sons, New York, 1987.
13. Feughelman, M. and Robinson, M.S., Stress relaxation of wool fibers in water at extensions in the Hookean region over the temperature range 0°–90°C, *Text. Res. J.*, 39, 196, 1969.
14. Feughelman, M., A note on the water impenetrable component of α-keratin fibers, *Text. Res. J.*, 59, 739, 1989.
15. Fraser, R.D.B., MacRea, T.P., and Suzuki, E., Structure of the α-keratin microfibril, *J. Mol. Biol.*, 108, 435, 1976.
16. Fraser, R.D.B., Macrae, T.P., and Rogers, G.E., Molecular organization in alpha keratin, *Nature*, 193, 1052, 1962.
17. Feughelman, M. and Robinson, M.S., The tensile behavior of wool fibers in liquid nitrogen, *Text. Res. J.*, 37, 705, 1967.
18. Breuer, M.M., The binding of small molecules to hair. I. The hydration of hair and the effect of water on the mechanical properties of hair, *J. Soc. Cosmet. Chem.*, 23, 447, 1972.
19. Sikorski, J. and Woods, H.J., The effect of rate of extension on Young's modulus of keratin fibers, *Leeds Phil. Soc.*, 5, 313, 1950.
20. Wortmann, F.J. and De Jong, S., Analysis of the humidity-time superposition for wool fibers, *Text. Res. J.*, 55, 750, 1985.
21. Feughelman, M., Microfibril/matrix relationships in the mechanical properties of keratin fibers. I. The torsional properties of "melted" and permanently set keratin fibers, *Text. Res. J.*, 48, 518, 1978.
22. Feughelman, M., A note on the recoverability of mechanical properties of wool, *J. Text. Inst.*, 59, T548, 1968.
23. Bendit, The α-β transformation in keratin, *Nature*, 179, 535, 1957.
24. Feughelman, M., Creep of wool fibres in water, *J. Text. Inst.*, 45, T630, 1954.
25. Feughelman, M. and Rigby, B.J., A two energy state model for the stress relaxation and creep of wool fibres in water, Proc. Int. Wool Textile Conference, Australia, 1955, D-62.
26. Burte, H. and Halsey, G., A new theory of non-linear viscoelasticity, *Text. Res. J.*, 17, 465, 1947.
27. Feughelman, M., The mechanical properties of permanently set and cystine reduced wool fibers at various relative humidities and the structure of wool, *Text. Res. J.*, 33, 1013, 1963.
28. Cannell, D.W. and Carothers, L.E., Permanent waving: utilization of the post-yield slope as a formulation parameter, *J. Soc. Cosmet. Chem.*, 29, 685, 1978.
29. Feughelman, M. and Haly, A.R., Structural features of keratin suggested by its mechanical properties, *Biochim. Biophys. Acta*, 32, 596, 1959.
30. Feughelman, M., The post-yield region and the structure of keratin, *Text. Res. J.*, 34, 539, 1964.
31. Menkart, J., Wolfram, L.J., and Mao, I., Caucasian hair, Negro hair and wool: similarities and differences, *J. Soc. Cosmet. Chem.*, 17, 769, 1966.
32. Chaiken, M. and Chemberlain, W.H., The propagation of longitudinal stress pulses in textile fibers, *J. Text. Inst.*, 46, T44, 1955.
33. Tolgyesi, E., Coble, D.W., Fang, F.S., and Kairinen, E.O., A comparative study of beard and scalp hair, *J. Soc. Cosmet. Chem.*, 34, 361, 1983.
34. Weigman, H.D., Rebenfield, L., and Danziger, C., The role of sulfhydryl groups in the mechanism of permanent setting of wool, III Cirtel, Paris, 1965, Sect. 2, 319.
35. Hermans, K.W., Hair keratin, reaction, penetration and swelling in mercaptan solutions, *Trans. Faraday Soc.*, 59, 1633, 1963.
36. Weigman, H.D. and Danziger, C.J., Effects of cross-links on the mechanical properties of keratin fibers, in *Applied Polymer Symposium No. 18*, Part 2, John Wiley & Sons, New York, 1971, 795.
37. Robinson, M.S. and Rigby, B.J., Thiol differences along keratin fibers: stress/strain and stress-relaxation behavior as a function of temperature and extension, *Text. Res. J.*, 55, 597, 1985.

38. Beyak, R., Meyer, C.F., and Kass, G.S., Elasticity and tensile properties of human hair. I. Single fiber test method, *J. Soc. Cosmet. Chem.,* 20, 615, 1969.
39. Wolfram, L.J., Hall, K., and Hui, I., The mechanism of hair bleaching, *J. Soc. Cosmet. Chem.,* 21, 875, 1970.
40. Beyak, R., Kass, G.S., and Meyer, C.F., Elasticity and tensile properties of human hair. II. Light radiation effects, *J. Soc. Cosmet. Chem.,* 22, 667, 1971.
41. Breuer, M.M., The interaction between surfactants and keratinous tissue, *J. Soc. Cosmet. Chem.,* 30, 41, 1979.
42. Fair, N.B. and Gupta, B.S., The chlorine-hair interaction. II. Effect of chlorination at varied pH levels on hair properties, *J. Soc. Cosmet. Chem.,* 38, 371, 1987.
43. Robbins, C.R. and Crawford, R.J., Cuticle damage and tensile properties of human hair, *J. Soc. Cosmet. Chem.,* 42, 59, 1991.
44. Bogaty, H., Torsional properties of hair in relation to permanent waving and setting, *J. Soc. Cosmet. Chem.,* 18, 575, 1967.
45. Wolfram, L.J. and Albrecht, L., Torsional behavior of human hair, *J. Soc. Cosmet. Chem.,* 36, 87, 1985.
46. Scott, G.V. and Robbins, C.R., A convenient method for measuring fiber stiffness, *Text. Res. J.,* 39, 975, 1969.
47. Scott, G.V. and Robbins, C.R., Stiffness of human hair fibers, *J. Soc. Cosmet. Chem.,* 29, 469, 1978.
48. Savenije, E.P.W. and De Vos, R., Mechanical properties of human beard hair, *Bioeng. Skin,* 2, 215, 1986.
49. Reese, C. and Eyring, H., Mechanical properties and the structure of hair, *Text. Res. J.,* 20, 743, 1950.
50. Wickett, R.R., Kinetic studies of hair reduction using a single fiber technique, *J. Soc. Cosmet. Chem.,* 34, 301, 1983.
51. Wickett, R.R. and Barman, B.G., Factors affecting the kinetics of disulfide bond reduction in hair, *J. Soc. Cosmet. Chem.,* 36, 75, 1985.
52. Wickett, R.R. and Mermelstein, R., Single fiber stress decay studies of hair reduction and depilation, *J. Soc. Cosmet. Chem.,* 37, 461, 1986.
53. Szadurski, J.S. and Erlman, G., The hair loop test — a new method of evaluating perm lotions, *Cosmet. Toiletries,* 41(12), 41, 1984.
54. Wortman, F.J. and Souren, I., Extensional properties of hair and permanent waving, *J. Soc. Cosmet. Chem.,* 38, 125, 1987.
55. Wortman, F.J. and Kure, N., Bending relaxation properties of human hair and permanent waving performance, *J. Soc. Cosmet. Chem.,* 41, 123, 1990.
56. Wickett, R.R., Disulfide bond reduction in permanent waving, *Cosmet. Toiletries,* 106(7), 37, 1991.
57. Wickett, R.R., An inexpensive device to measure hair breaking strength, manuscript in preparation.

26.0 Hair Growth

26.1 Measurement of Hair Growth 543
J.H. Barth and D.H. Rushton

26.2 Microscopy of the Hair — The Trichogram 549
U. Blume-Peytavi and C.E. Orfanos

Chapter 26.1
Measurement of Hair Growth

Julian H. Barth and D. Hugh Rushton
Department of Chemical Pathology and Immunology
Leeds General Infirmary
Leeds, U.K.
School of Pharmacy & Biomedical Sciences
University of Portsmouth
Portsmouth, U.K.

I. Introduction

Techniques for the measurement of hair growth were initially developed to satisfy the needs of the wool industry, human anthropologists, and the human cosmetics industry. Indeed, the major developments have been established for the former two rather than the latter purpose.

The methods described in this chapter derive from our interest in human hair growth and are divided into subjective and objective methods. The former are more suitable for body hair growth since both hair form and the density of hair varies considerably from one site to another. The objective methods are more suitable for the denser growth on the scalp since cosmetic changes are only apparent after there has been a reduction of approximately 20% in hair quantity.

II. Subjective Measurement

A. Scoring Systems for Scalp Hair Growth

The patterns produced by the gradual process of scalp hair loss in male pattern balding were first described in men by Hamilton[1] and in the female by Ludwig.[2] The recognition that an individual's scalp hair loss fits one of these patterning systems is essential for the diagnosis of scalp hair loss, but the staging is regrettably too insensitive for all but the most crude of measurement. An attempt has been made to scan the appearance of the scalp into a computer imaging system[3] but this is impeded by the presence of hair; more sensitivity is only afforded by shaving an area of scalp.[4]

B. Scoring Systems for Body Hair Growth

Subjective grading systems have been designed by Pedersen,[5] Beek,[6] Garn,[7] Shah,[8] Ferriman and Gallwey,[9] Lorenzo,[10] and Lunde;[11] see Table 1. All these methods are based on subdividing the body into zones; each is scored and the total is summed. However, it is the study by Ferriman and Gallwey that has been granted approval by most hirsuties investigators as it is the least complex and easiest to use. Unfortunately, as hirsutism is a subjective condition, the subjective methods of its measurement are subject to considerable observer variation (see Figure 1). It could be concluded that the difference is due to variation in the degree of hair growth in different populations of hirsute women but it is more probably due to a lack of observer conformity. It should be noted that, in the myriad of publications which have employed subjective measurement of body hair growth, there are few attempts to evaluate the precision of the method used. In our hands, the Ferriman and Gallwey scoring system has a repeatability coefficient of 3.2 (Ferriman and Gallwey units) at a mean score of 26.[12]

III. Objective Measurements

Quantitative evaluation of scalp hair requires techniques that are sensitive enough to assess fundamental variables such as hair density, fiber diameter, proportion of anagen hair (i.e., actively growing), and linear growth rate. Such information provides essential details for determining normal morphology as well as understanding changes arising from disease.

A. Presampling Considerations

Presampling factors are probably the most difficult problems to standardize. These include shampooing, combing, and other cosmetic procedures, all of which must be standardized to provide unbiased measurements. Unfortunately, ideal protocols for hair measurements are impractical since they would include 3-month shampoo-free periods for telogen counts. Therefore, subjects should continue with their normal daily routines which are used for background data.

TABLE 1 Summary of Subjective Methods for the Evaluation of Hair Growth on the Face, Trunk, and Limbs

Authors	Number of Sites Evaluated	Specific Features Evaluated
Pedersen	12	Evaluation only of presence of hair at each site but no quantitative assessment
Beek	19	Evaluation only of presence of hair at each site but no quantitative assessment
Garn	11	Evaluation of patterns of hair growth at each site and form of hair shaft present, e.g., curly or straight; no quantitative assessment
Shah	9	Quantitative assessment of terminal hairs >0.5 cm; (Q)uality (scored 1–3), (D)ensity (0–3) and (P)roportion of zone covered with hair (0–1); total score = $Q \times D \times P$
Ferriman and Gallwey	11	Quantitative score based on distribution of hair on each site
Lorenzo	7 subzones	Quantitative score based upon density and extent of involvement
Lunde	18	Quantitative assessment of length and density of terminal hairs >0.5cm

The shampooing interval prior to sampling should be kept standard.

B. Presampling Recommendations

Hair should be shampooed daily or on alternate days during the month prior to sampling, and not less than on alternate days throughout the study period. On the day of sampling specimens should be obtained, or visual imaging performed, within 3 hours of shampooing. Such actions produce minimal influences upon the derived anagen estimate, while providing excellent conditions for photographic or image analysis reproductions. It should also be pointed out that an inadequate shampooing action can have a detrimental influence upon the derived anagen estimate. Inadequate and inefficient shampooing can be found in subjects experiencing excessive hair loss, which results in poor cleansing and only partial removal of the hairs due to be shed. As excessive amounts of hair are seen at each wash, the shampooing frequency is decreased in a belief that this will reduce future loses. When this situation is found sufficient time must elapse to allow the reestablishment of shedding levels representative of the problem under investigation.

C. Sampling Criteria

The selection of area to be sampled is also of critical importance. Sites must be chosen carefully to represent the changes occurring. Problems arising from patterned presentations within the sampling distribution require careful attention to avoid biased estimates. No real problems should be encountered where the distribution of hair density is uniform, assuming sufficient hairs are obtained. Approximately 100 hairs are required to estimate variables by proportion; consequently, a sufficiently hairy area will need to be sampled.

Two sample sites providing the required total number of hairs give better estimates (smaller sampling variation) than a single site.[13] However, the surrounding area should be indistinguishable (with respect to hair density) from the sample sites (Figure 1). This arrangement allows resampling within ±5 mm of the original sites without the need to find the exact initial spot. Disturbances such as male pattern baldness often present problems because of patterned distributions in affected areas. Hair density within the vertex or frontal hair line can sometimes vary tremendously and exact resampling is required. This can be achieved by placing a small tattoo between the two original sample sites or within the center of a single site. Where tattooing is not possible, location coordinates are essential for accurate relocation at some future date.

Resampling frequencies depend upon the variables being estimated and the method of evaluation being employed. Where plucking techniques are used hair density cannot be reestimated for 6 months; with noninvasive methods (hair weights or phototrichograms)

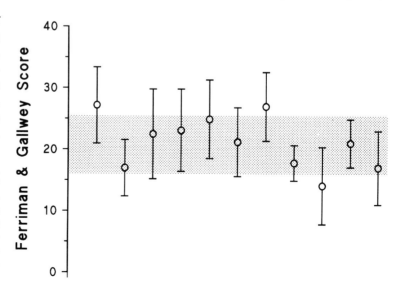

FIGURE 1 Comparison of Ferriman and Gallwey scores of hirsute women from 11 studies (mean ± SD). It is assumed that the severity of hirsuties of women attending different centers is similar and therefore this graph illustrates the variation in hirsuties grading perceived by different investigators. (Redrawn from *Diseases of the Hair and Scalp*, Rook and Dawber, Eds., Blackwell Scientific, Oxford, 1991.)

resampling can be performed within 1 month. However, estimates for the percentage of hair in the anagen growth phase need to be made 12 months apart due to seasonal variation[14] and is a requirement independent of the method of evaluation.

D. The Unit Area Trichogram

The unit area trichogram evaluates scalp hair variables with excellent reproducibility. It is a semi-invasive (plucking) technique in which all the hairs within a defined area (usually >30 mm^2) are counted and measured.[15,16] The area to be sampled should first be degreased with an acetone/isopropanol (60:40 v/v) mixture to remove surface lipids, which can cause blurring of the delineating sample line. The area to be sampled should be identified prior to epilation; a roller ball pen produces the sharpest line compared to circles drawn with fiber, felt, or ball point pens. The sample area can be quantitatively measured from an enlarged black and white photograph containing a scale bar or a computer image. The hairs should be rapidly epilated in a single smooth action in the direction of growth in order to minimize trauma to the roots.

E. Examining Hair Root Status

All epilated hairs are placed directly onto double-sided tape (15 mm wide) attached to a 25-mm microscope slide. The collected hairs are subsequently realigned so that their root ends protrude from the tape edge toward the center, thereby allowing easy visualization within the microscope (Figure 2). Each hair is classified microscopically (magnification × 40) according to its growth phase (anagen, catagen, or telogen), diameter, and length. The typical microscopical (dry-mounted) appearance of the various hair growth stages anagen, catagen, and telogen are shown in (Figure 3). Catagen hairs are classified with the telogen population for data analysis since catagen is effectively the first stage of telogen. The catagen bulb structure is fully keratinized and encased in a shrivelled-up translucent outer root sheath (Figure 3) and for these reasons we classify catagen fibers with the telogen population.

F. Problems with Anagen, Catagen, and Telogen Classifications

A possible area of confusion and contention is the reporting of "dysplastic" or "dystrophic" hairs which are usually observed in plucked samples. Apart from a few rare congenital diseases, dysplastic or dystrophic hair shafts are never seen on the scalps of men, women, or children, nor in scalp biopsy. It is our belief that these terms are in reality descriptions for traumatic features produced by the procedure of hair epilation. Classification of such roots is preferable to their exclusion from data analysis and a few simple guidelines should ensure that difficult root presentations are assigned to the correct population.

The principle feature to keep in mind is that actively growing hairs (anagen) have soft prekeratinized tissues up to approximately 2.0 mm above the apex of the dermal matrix.[13] This soft tissue bends easily and distorts upon epilation. Catagen and telogen hairs differ since they are fully keratinized and are, therefore, rigid and do not easily bend. Moreover, the diameters of catagen and telogen fibers taper toward the bulb and this zone is devoid of, or has reduced, pigmentation. The following therefore serves as a guide to assigning difficult fibers to the anagen or telogen populations. If the root end exhibits bending with or without pigmentation, we assign it to the anagen population, if not, to the telogen population. Broken hairs arising from the epilation procedure also cause difficulties with interpretation but should be assigned to the telogen population if the proximal end exhibits tapering and/or loss of pigment; otherwise, the fiber should be classified as anagen. Fortunately, the occurrence of such hairs is <5% when hair is epilated singly.

G. Determination of Hair Diameter

The shaft diameters can be measured with an eyepiece micrometer using the shafts mounted as in Figure 2. Major and minor axes should be measured and averaged, but single plane estimates are acceptable if only the anagen or telogen diameter is required. Anagen fibers should be measured 10 to 15 mm from the root end to avoid the influences of weathering. Telogen fibers

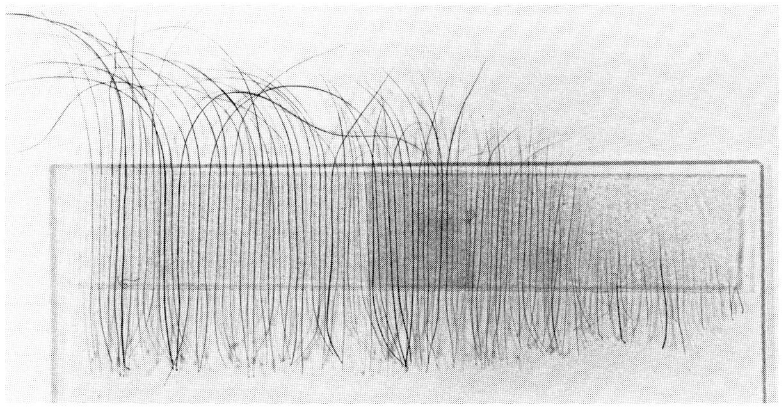

FIGURE 2 Plucked hairs mounted on double sided tape and ordered by length. The plucked hairs are initially collected onto another strip of adhesive tape and then during transfer into the illustrated format have their length measured.

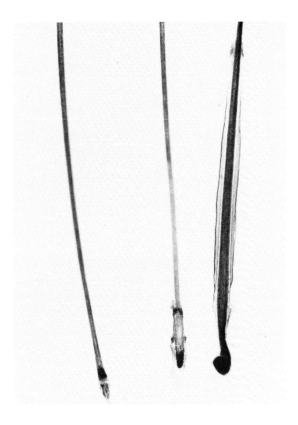

FIGURE 3 Photomicrograph of three hair shafts. The left shaft has the characteristic drumstick appearance of a telogen bulb. The middle shaft has a tapered straight bulb surrounded by a loose atrophic sack and is a catagen bulb. The shaft on the right has a densely pigmented and distorted bulb with an extravagant collar of external root sheath and is typical of an anagen bulb.

should be measured 10 mm from the last sign of tapering, which is usually 15 to 20 mm from the bulb. Since telogen hairs of equal length have correspondingly smaller diameters than their anagen counterparts,[13] studies involving hair diameter comparisons should be controlled for anagen to telogen ratios. This will ensure that any changes are true diameter effects and not simply a consequence of changes in the anagen or telogen populations. Recently, studies on linear hair growth in scalp hair has found a correlation between fiber diameter and the rate of daily growth.[17] Thicker hairs grow faster than finer hairs; as a result, studies involving linear growth rate must now specify diameter values.

H. Evaluation of Hair Shaft

A novel method for the measurement of hair growth on the limbs has been described: the Vellus Index.[18] This method is only appropriate for hair growth on the body and limbs. It utilizes hair shaved from the skin and the ratio of pigmented, medullated terminal to non-pigmented, nonmedullated vellus shafts is simply determined by microscopic examination of the shafts. It has the great advantage of not requiring samples to be shaved from a defined area. This method has been shown to separate men from women and that hirsute women have an intermediate amount of hair. It has not yet been used for monitoring anti-androgen therapy.

I. Summary of Published Methods of Objective Hair Measurement

The methods available for hair evaluation are compared in Table 2. Scalp biopsies give a similar degree of quantitative information to the unit area trichogram but suffer from being unable to resample the original site. Visual hair counting would appear to be of little value in determining the number of hairs present on the scalp, and has now been abandoned. Telogen hair counts offer no obvious information about the rate or severity of hair shedding. There are too many unknown variables which cloud the significance of total telogen counts and, without data for the total numbers of hair present or the duration of anagen, daily counts of telogen hairs are meaningless. Noninvasive methods such as the phototrichogram currently suffer from two-dimensional imaging problems;[19] however, these may be improved with the introduction of three-dimensional image analysis and then phototrichograms may become the method of choice.

The ability to estimate hair quality demands quantitative techniques capable of measuring fundamental hair components. Interrelating these variables provides additional data giving a sensitive measure of changes occurring. Although time consuming and mildly painful, the unit area trichogram offers the trained investigator a valid method for determining fiber thickness, density, and proportion actively growing, and at the present time (1993) appears the method of choice.

TABLE 2 Summary of Methods of Hair Evaluation

	Unit Area Trichogram	Phototrichogram	Visual Counting	Scalp Biopsy	Telogen Counts	Sample Weight
Sample time (h)	4–6	4–8	1–2	24–48	1–2	4–6
Number of hairs measured	50–180	90–200	50–450	5–20	20–500	115–310
Accuracy	Excellent	Good	Poor	Good	Poor	Good
Precision (%)	±5	±15	±35	±5	Unknown	Unknown
Repeat sampling	6 monthly	Monthly	Immediate	No	Daily	Bimonthly

References

1. Hamilton, J.B., Patterned loss of hair in man: types and incidence, *Ann. N.Y. Acad. Sci.*, 53, 708, 1951.
2. Ludwig, E., Classification of the types of androgenetic alopecia (common baldness) occurring in the female sex, *Br. J. Dermatol.*, 97, 247, 1977.
3. Gibbons, R.D., Fiedler-Weiss, V.C., West, D.P., and Lapin, G., Quantification of scalp hair — a computer-aided methodology, *J. Invest. Dermatol.*, 86, 782, 1986.
4. Van Neste, D., Dumortier, M., and De Coster, W., Phototrichogram analysis: technical aspects and problems in relation with automated quantitative evaluation of hair growth to computer assisted image analysis, in: *Trends in Human Hair Growth and Alopecia Research*, Van Neste, D., Lachappelle, J.M., and Antoine, J.L., Eds., Kluwer Academic, Dordrecht, The Netherlands, 1989, 155.
5. Pedersen, J., Hypertrichosis in women, *Acta Dermatol. Venereol.*, 23, 1, 1943.
6. Beek, C.H., A study on the extension and distribution of the human body hair, *Dermatologica*, 101, 317, 1950.
7. Garn, S.M., Types and distribution of the hair in man, *Ann. N.Y. Acad. Sci.*, 53, 498, 1951.
8. Shah, P.N., Human body hair — a quantitative study, *Am. J. Obstet. Gynecol.*, 73, 1255, 1957.
9. Ferriman, D. and Gallwey, J.D., Clinical assessment of body hair growth in women, *J. Clin. Endocrinol.*, 21, 1440, 1961.
10. Lorenzo, E.M., Familial study of hirsutism, *J. Clin. Endocrinol. Metab.*, 31, 556, 1970.
11. Lunde, O., A study of body hair density and distribution in normal women, *Am. J. Phys. Anthropol.*, 64, 179, 1984.
12. Barth, J.H., Cherry, C.A., Wojnarowska, F., and Dawber, R.P.R., Cyproterone acetate for severe hirsutism: results of a double-blind dose-ranging study, *Clin. Endocrinol.*, 35, 5, 1991.
13. Rushton, D.H., Chemical and Morphological Properties of Scalp Hair in Normal and Abnormal States, Ph.D. thesis, University of Wales, Cardiff, 1988.
14. Randall, V.A. and Ebling, F.J.G., Seasonal changes in human hair growth, *Br. J. Dermatol.*, 124, 146, 1991.
15. Rushton, D.H., Ramsay, I.D., James, K.C., Norris, M.J., and Gilkes, J.J.H., Biochemical and trichological characterisation of diffuse alopecia in women, *Br. J. Dermatol.*, 123, 187, 1990.
16. Rushton, D.H., Ramsay, I.D., Norris, M.J., and Gilkes, J.J.H., Natural progression of male pattern baldness in young men, *Clin. Exp. Dermatol.*, 16, 188, 1991.
17. Van Neste, D.J.J., de Brouwer, B., and Dumortier, M., Reduced linear hair growth rates of vellus and of terminal hairs produced by human balding scalp grafted onto nude mice, *Ann. N.Y. Acad. Sci.*, 642, 480, 1991.
18. Madanes, A.E. and Novotny, M., The vellus index: a new method of assessing hair growth, *Fertil. Steril.*, 48, 1064, 1987.
19. Rushton, D.H., de Brouwer, B., De Coster, W., and Van Neste, D.J.J., Comparative evaluation of scalp hair by phototrichogram and unit area trichogram analysis within the same subjects, *Acta Dermatol. Venereol.*, 73, 150, 1993.

Chapter 26.2
Microscopy of the Hair — the Trichogram

Ulrike Blume-Peytavi and Constantin E. Orfanos
Department of Dermatology
University Medical Center Steglitz
The Free University of Berlin
Berlin, Germany

I. Introduction

Dermatologists are frequently consulted for focal or diffuse effluvium, hair thinning, and changes in hair structure. In man, hair has no vital biological function but an incalculable psychological one. Hair is a major esthetic display feature of the human body, especially in social and sexual interaction. In the diagnosis of hair disease, clinical history and daily counting of shed hair by the patient are useful, but an objective and reproducible method is needed for exact quantification and characterization of hair loss.

In most rodents and other mammals, hair does not grow continuously, and the hair follicle undergoes cyclic changes of growth, regression, and inactivity periods throughout life.[1,2] In humans, however, these cyclic changes are not synchronized and seem to be regulated by complex interactions between the individual epithelial and mesenchymal components of the hair follicle.[3-6] A large variety of endogenous and exogenous factors may possibly intervene to disturb normal growth regulation processes.[4,7]

Each individual hair follicle undergoes a cyclic rhythm of

1. A growing or *anagen phase* during which the hair is produced. The hair with medulla, cortex, cuticle, and inner and outer root sheaths is formed by the highly active matrix cell population located in the hair bulb region, probably the most rapidly growing cell population of all normal tissues, reaching a maximum cell turnover rate entering mitosis every 24 hours. Different anagen stages as proanagen (anagen I to V) and metanagen (anagen VI) can be distinguished by histological examination. The normal duration of anagen in any individual scalp follicle is genetically determined and ranges from 2 to 6 years, with an approximate average anagen duration of 1000 days.[8] The growing anagen hair is normally firmly bound to the hair root and can only be plucked with some pain sensation. The anagen phase is followed by a transition or catagen phase.

2. The transition or *catagen phase* lasts only a few days (2 to 3 weeks). Catagen is characterized by a decreasing and finally stopping activity of the matrix followed by involution and keratinization of the bulb. Subsequently, the bulb moves upward toward the surface of the skin to prepare the following.

3. In the resting or *telogen phase* the hair is shed; during telogen, the follicle is located just below the orifice of the sebaceous gland. The club hair is held in its envelope by the intercellular junctions and can be pulled without any pain. The subsequent hair cycle starts spontaneously under the influence of an unknown stimulus at the end of telogen or can be induced by plucking the club hair. Normally, the telogen hair either falls out at the beginning of the new anagen phase or is retained in the follicle until metanagen is well established during the next hair cycle. In the human scalp, the telogen phase lasts between 2 and 3 months.

The duration of hair cycle phases in each individual is age dependent[9] and varies according to body region,[5,10] produced hair type (terminal, lanugo, or vellus hair),[11] and with seasonal changes[12] (see Table 1).

II. Object

Studies on the dynamics of the human follicular cycle largely depend on trichogram examinations, a microscopic evaluation of plucked hairs with subsequent quantitative measuring of the number of individual hair roots. Morphological examination of hair roots was first introduced by Van Scott et al.,[13] as an indicator of hair growth in toxicologic studies on cytostatic drugs. Initially, the term "trichogram" was used by

TABLE 1 Duration of the Anagen and the Telogen Phase in Different Regions of the Body Site[3-6,16,20,30,36-38]

Body region	Anagen	Telogen
Scalp	2–6 years	3–4 months
Eyebrows	4–8 weeks	3 months
Beard	12 months	3 months
Moustache	16 months	6 months
Axilla	Months	3 months
Pubic region	47 weeks	2 weeks
Hand	10 weeks	7 weeks
Finger	12 months	9 months
Arms	13 months	13 months
Leg	21 months	19 months

Pecoraro and co-workers[14,15] to describe several trichometric parameters, such as growth rate, thickness of the hair shaft, and telogen rate. In the following years, the trichogram technique was developed and standardized to serve as a reliable diagnostic measure in hair diseases. Today, this term is used to describe the examination of a simultaneously epilated tuft of 50 to 80 hairs and the count of various hair root types under the light microscope by a standardized technique. A major target of trichogram measurements is to count and evaluate the status of individual hair roots and to establish the anagen/telogen hair ratio.[16-19]

Trichogram measurements are based on the hair cycle and therefore serve as a standard method for quantifying hair follicles in their different growth cycle phases.[20] Thus this simple and repeatable technique is a diagnostic measure to

- Assess hair growth capacity
- Obtain an overview on the current state of growth of the hair follicles[21]
- Detect hair growth cycle disturbances[22,23]
- Classify different forms of hair loss and alopecia[18,24]
- Elucidate pathogenetic mechanisms leading to diffuse hair loss
- Detect exogenous damage or toxic effects on hair structure
- Assess the effectiveness of hair growth-promoting treatments[25]

In addition, microscopic examination of hair shaft morphology can also help to detect malformations of the hair shaft as in primary genotrichoses (e.g., monilethrix, trichorrhexis nodosa) and in secondary genotrichoses (e.g., amino acid disorders, phenylketonuria),[26] etc.

III. The Trichogram Technique

A. Methodological Principles

Materials necessary for performing a trichogram are trichogram forceps, a pair of scissors, glass slides 76 × 26 mm, cover slips, hair clips, Corbit™ balsam (I. Hecht, Kiel, Germany), preparation needles, and a bioccular microscope or a Reichert projection microscope Visopan® (Reichert & Jung, Germany) with variable objectives (4×, 10×, 40×, 63×) (See Figure 1).

The reproducibility of trichogram measurements depends on the maintenance of high standards in obtaining the hair samples. Trichogram examinations should always be performed 5 days after the last shampooing of the hair. Patients have to interrupt all local treatments including cosmetic procedures 2 to 3 weeks earlier. On the 5th day after the last hair shampooing, hairs are taken from at least two specified sites. In *diffuse effluvium* and in *androgenetic alopecia* one sample is taken 2 cm behind the frontal hair line and 2 cm from the midline; another sample is obtained from the occipital region, 2 cm lateral from the protuberans occipitalis. In circumscribed alopecia, such as *alopecia areata,* one sample is taken from the border of the lesion and the control is taken from the contralateral, obviously not affected region. In the examined area, surrounding hair is fixed with clips and a bundle of 60 to 80 hairs is grasped with a trichogram forceps whose jaws are protected with rubber. This facilitates grasping hairs of different thickness and minimizes hair traumatization (see Figure 2).

Subsequently, forceps are completely closed and hairs are then plucked by twisting and lifting the hair shafts in the direction of emergence from the scalp. To prevent dehydration of the hair root shafts, they are immediately placed with their roots on a glass slide in embedding medium (e.g., Corbit balsam). Hair shafts are cut off approximately 1 cm above the root sheath. After they have been arranged one by one with a preparation needle, they are covered with a coverslip and are left to dry for 24 hours. Plucking scalp hair may leave a small bald patch that may become slightly erythematous for a short time afterward. Performing a trichogram may cause slight discomfort caused by the removal of the whole group of hairs. Nevertheless, when correctly performed, trichogram is not painful and the plucked hairs grow back after 3 to 6 weeks. Correctly embedded hair roots can be stored for a long time.

B. Evaluation

Evaluation of trichogram is performed with a bioccular microscope or ideally with a projection microscope. The embedded hair roots are individually analyzed by microscopic examination of the mounted glass slides. First, the overall shape, the presence or absence of hair root sheaths, the external contours, and the hair shaft diameters are examined. In addition, the appearance and the size of the bulb are also important as atrophy and shrinkage reflect reduced growth activity.[27] At every stage of the hair growth cycle the hair follicle is

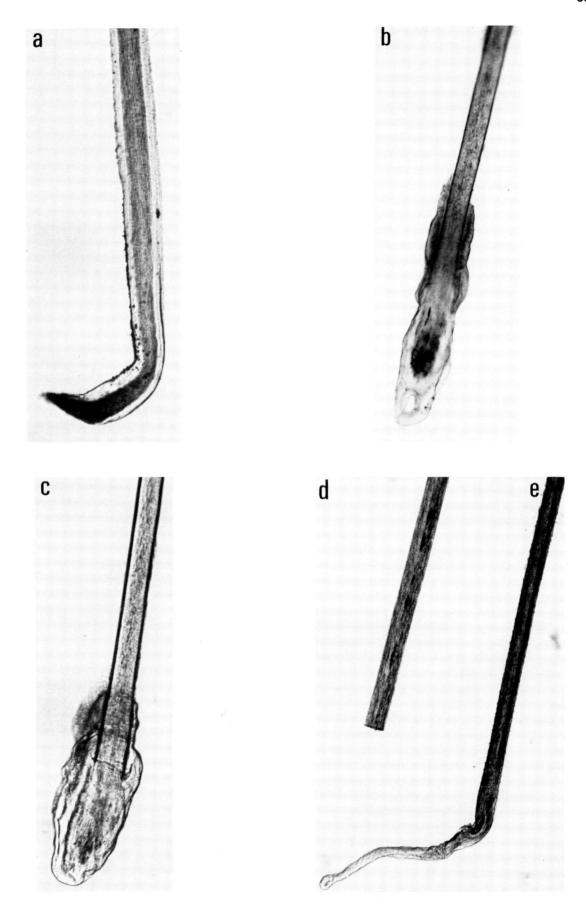

FIGURE 1 Material necessary for trichogram measurements: two trichogram forceps, glass slides, cover slips, a pair of scissors, Corbit balsam.

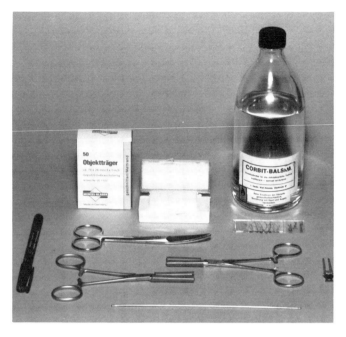

FIGURE 2 Plucking of a tuft of 50 to 80 hairs, firmly grasped with a trichogram forceps with rubber-protected jaws and pulled in the direction of emergence from the scalp.

characterized by a typical morphological structure. Based on microscopic examination of plucked hair roots, it is possible to decide on the hair growth phase of the examined hair as anagen, catagen, or telogen (Figure 3a to c) and in addition also dysplastic anagen, dystrophic, and broken-off hairs (Figure 3d and e) can be distinguished. Finally, the percentage of the different hair root types is calculated.

C. Types of Hair Roots

Anagen hair — The root has a large base with an equally large diameter throughout. A firm inner and frequently outer root sheath can be clearly distinguished. Often, a more than 20° angulation of root and shaft can be observed (Figure 3a).

Catagen hair — It generally has an equal diameter throughout or can become narrower toward the base, frequently with a missing or wrinkled root sheath; contours are not deformed (Figure 3b).

Telogen hair — The root is characteristically club shaped with smooth contours, and its sheaths are contracted around the tip (Figure 3c).

Dysplastic hairs — They are thin but obviously growing anagen hairs, with a diminished matrix diameter and plucked without root sheaths; the lower end of the hair shaft is usually wavy or bent like the handle of a walking cane (Figure 3e).

Dystrophic hairs — They are mainly described as thin, nongrowing anagen hairs, often with defective keratinization; changes are so severe that the root has broken off at the narrowest level with a tapering, pencil-like broken end tip, sometimes showing individual bundles of keratin fibrils.

Broken-off hairs — They are normally growing anagen hair shafts which break off on plucking (Figure 3d). They can be easily recognized since, in contrast to dystrophic hair, the broken ends appear smooth with a remaining diameter equal to that of the hair shaft. They may amount to up to 5 to 6% of the total number, but their prevalence is higher when hair is fragile or when the plucking technique was not adequate.

The existence of variable percentages of such "abnormal hair roots" may be of interest in various diseases where the hair can serve as a marker for an underlying systemic defect or disease (see Table 2).

IV. Criteria for Standardizing the Trichogram Procedure and Eliminating Sources of Error

The trichogram technique provides reliable results under the condition that hair samples have been obtained under a standardized procedure; al-

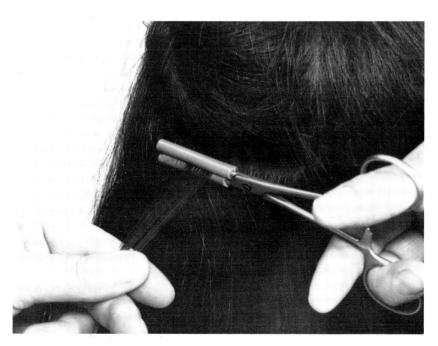

FIGURE 3 Different hair root types which can be distinguished by microscopic examination of individual hair roots. (a) anagen, (b) catagen, (c) telogen, (d) broken-off hair, (e) dysplastic anagen hair root.

TABLE 2 Normal Distribution Pattern of Hair Roots in the Trichogram of the Scalp[20]

Hair Root	%
Anagen	60–80
Dysplastic anagen	5–20[a]
Catagen	1–3
Telogen	12–15
Dystrophic	<2
Broken-off hairs	5–6

[a] *In children, adolescents, and in persons with thin hair the rate of dysplastic anagen can occasionally reach >50%.*

ready minor variations may lead to misinterpretation and false results.

- There should be a 5-day period prior to examination to avoid an artificial reduction of the telogen rate.[28]
- The same standardized scalp regions (see above) should always be chosen to permit an optimal comparison of trichogram measurement results, for example, for regular follow-up of individual patients and for clinical studies.[20]
- Correct plucking speed (sharp quick pull) and exact plucking in the direction of the emergence angle of the hairs from the scalp are very important to obtain a reliable hair root pattern.[29] Slow or hesitant traction, very rapid plucking, or the wrong pulling direction may induce distorsions and alterations of the hair shafts and cause serious damage to the hair roots.
- More than 50 hairs have to be evaluated as the error increases significantly with decreasing size of the hair sample.[20,30]
- Trichograms with >10% broken-off hairs cannot be evaluated correctly and should be repeated.[20]
- Trichogram measurements should always be performed and evaluated by the same experienced investigator to maintain optimal and comparable examination conditions. A short training period with an experienced colleague is recommended.

V. Comparison with Other Techniques

The trichogram is a semi-invasive technique with plucking of the entire hair. This makes it unsuitable for the monthly follow-up of patients, for studying the hair growth rate, and duration of the growing stage of individual hairs, and studying seasonal variations. Noninvasive microscopic techniques such as optical microscopy with image analysis,[31] phototrichogram,[32,33] or the unit area trichogram[34] should be chosen for this purpose. The trichogram technique is accurate and reliable for the diagnosis of hair diseases and is highly suitable because of its handiness.[35] Although trichogram measurements are only confined to two small scalp regions, earlier studies provided evidence that, except in alopecia areata, the trichogram technique of one site is representative for the neighboring areas.[30]

Trichogram evaluation of hair root pattern is particularly useful at the beginning of abnormal hair loss, when hair density still seems normal at clinical examination. In this case, trichogram results may permit an early diagnosis, early determination of prognosis, and an early start of treatment to prevent further continuation of effluvium. In addition, normal trichogram results may allow a favorable prognosis and help to avoid various expensive and useless treatments.

References

1. Kligman, A.M., The human hair cycle, *J. Invest. Dermatol.*, 33, 307, 1959.
2. Uno, H., Biology of hair growth, *Sem. Reprod. Endocrinol.*, 4, 131, 1986.
3. Ebling, F.J.G., The hair, in *Textbook of Dermatology*, Rook, A., Wilkinson, D.S., Ebling, F.J.G., Champion, R.H., and Burton, J.L., Eds., Blackwell Scientific, Oxford, 1986, 25.
4. Ebling, F.J.G., The biology of hair, *Dermatol. Clin.*, 5, 467, 1987.
5. Rook, A. and Dawber, R.P.R., The comparative physiology, embryology and physiology of human hair, in *Diseases of the Hair and Scalp*, Rook, A. and Dawber, R.P.R., Eds., Blackwell Scientific, Oxford, 1982, 1.
6. Sato, Y., The hair cycle and its control mechanism, in *Biology and Disease of the Hair*, Koboti, T. and Montagna, W., Eds., University Park Press, Baltimore, 1986, 3.
7. Braun-Falco, O. and Kint, A., Dynamik des normalen und pathologischen Haarwachstums, *Arch. Klin. Exp. Dermatol.*, 221, 75, 1966.
8. Orentreich, N., Scalp hair replacement in man, in *Advances in Biology of Skin*, Vol. 9, Montagna, W. and Dobson, R.L., Eds., Pergamon Press, Oxford, 1967, 99.
9. Barman, J.M., Astore, J., and Pecoraro, V., The trichogram of people over 50 years but apparently not bald, in *Advances in Biology of Skin Hair Growth*, Vol. 9, Montagna, W. and Dobson, R.L., Eds., Pergamon Press, Oxford, 1964, 211.
10. Trotter, M., The life cycles of hair in selected regions of the body, *Am. J. Phys. Anthropol.*, 7, 427, 1924.
11. Blume, U., Ferracin, J., Verschoore, M., Czernielewski, J.M., and Schaefer, H., Physiology of the vellus hair follicle: hair growth and sebum excretion, *Br. J. Dermatol.*, 124, 21, 1991.
12. Randall, V.A. and Ebling, F.J.G., Seasonal changes in human hair growth, *Br. J. Dermatol.*, 124, 146, 1991.
13. Van Scott, E.J., Reinertson, R.P.A., and Steinmüller, R., The growing hair roots of the human scalp and morphologic changes therein following amethopterin therapy, *J. Invest. Dermatol.*, 29, 197, 1957.
14. Pecoraro, V., Astore, J., Barman, J., and Araujo, C.S., The normal trichogram in the child before the age of puberty, *J. Invest. Dermatol.*, 42, 427, 1964.
15. Pecoraro, V., Astore, J., and Barman, J., The normal trichogram of the pregnant woman, in *Advances in Biology of Skin*, Vol. 9, Montagna, W. and Dobson, J.M., Eds., Pergamon Press, New York, 1967, 203.

16. Barman, J.M., Astore, J., and Pecoraro, V., The normal trichogram of the adult, *J. Invest. Dermatol.*, 44, 233, 1965.
17. Grosshans, E., Che pfer, M.P., and Maleville, J., Le trichogramme. A propos d'une méthode d'étude des cheveux, *J. Med. Strasbourg*, 378, 1972.
18. Metz, H.G. and Landes, E., Der Haarausfall und seine Untersuchung, *Fortschr. Med.*, 88, 1327, 1970.
19. Zaun, H. and Ludwig, E., Zur Definition ungewöhnlicher Haarwurzeln im Trichogramm, *Hautarzt*, 27, 606, 1976.
20. Orfanos, C.E., Androgenetic alopecia, in *Hair and Hair Diseases*, Orfanos, C.E. and Happle, R., Eds., Springer-Verlag, Berlin, 1991, 485.
21. Jost, B., Meiers, H.G., Schmidt-Elmendorff, H., and Pfaffenrath, V., Trichometrische Quantifizierung und Verlaufsbeurteilung des Hirsutismus, *Dtsch. Med. Wschr.*, 99, 2395, 1974.
22. James, K.C. and Rushton, D.H., Evaluation techniques for male pattern baldness, *J. Am. Acad. Dermatol.*, 14, 849, 1983.
23. Orfanos, C.E. and Hertel, H., Haarwachstumsstörungen bei Hyperprolaktinämie, *Z. Hautkr.*, 63, 23, 1988.
24. Sterry, W., Konrads, A., and Nase, J., Alopecie bei Schilddrüsenerkrankungen: Charakteristische Trichogramme, *Hautarzt*, 31, 308, 1980.
25. Orfanos, C.E., Meiers, H.G., Friedrich, H.C., Ludwig, E., Mahrle, G., and Zaun, H., Haarausfall, Trichogramm und hormonelle Haartherapeutika, *Dtsch. Arztbl.*, 3603, 1974.
26. Blume, U., Föhles, J., Gollnick, H., and Orfanos, C.E., Genotrichoses: clinical manifestations and diagnostic techniques, Proc. 18th World Congress of Dermatology, New York, in press.
27. Barth, J.H., Measurement of hair growth, *Clin. Exp. Dermatol.*, 11, 127, 1986.
28. Braun-Falco, O. and Fischer, C., Über den Einfluß des Haarewaschens auf das Haarwurzelmuster, *Arch. Klin. Exp. Dermatol.*, 227, 419, 1966.
29. Bassukas, I.D. and Hornstein, O.P., Effects of plucking on the anatomy of the anagen hair bulb. A light microscopic study, *Arch. Dermatol. Res.*, 281, 188, 1989.
30. Braun-Falco, O. and Heilgemeir, G.P., The trichogram. Structural and functional basis, performance and interpretation, *Sem. Dermatol.*, 1, 40, 1985.
31. Hayashi, S., Miyamoto, I., and Takeda, K., Measurement of human hair growth by optical microscopy and image analysis, *Br. J. Dermatol.*, 25, 123, 1991.
32. Fiquet, C. and Courtois, M., Une technique originale d'appréciation de la croissance et de la chute des cheveux, *Cutis*, 3, 975, 1979.
33. Van Neste, D., Dumortier, M., and De Coster, W., Phototrichogram analysis: technical aspects and problems in relation to automated quantitative evaluation of hair growth by computer-assisted image analysis, in *Trends in Human Hair Growth and Alopecia Research*, Van Neste, D., Lachapelle, J.M., and Antoine, J.L., Eds., Kluwer Academic, Dordrecht, The Netherlands, 1989, 147.
34. Rushton, H., James K.C., and Mortimer, C.H., The unit area trichogram in the assessment of androgen-dependent alopecia, *Br. J. Dermatol.*, 109, 429, 1983.
35. Meiers, H.G., Trichogramm (=Haarwurzelstatus, =Haarbild). Methode und Aussagefähigkeit, *Akt. Dermatol.*, 1, 31, 1975.
36. Astore, J., Pecoraro, V., and Pecoraro, E.G., The normal trichogram in pubic hair, *Br. J. Dermatol.*, 101, 441, 1979.
37. Saitoh, M., Uzuka, M., and Sakamoto, M., Human hair cycle, *J. Invest. Dermatol.*, 54, 65, 1970.
38. Seago, S.V. and Ebling, F.J.G., The hair cycle on the human thigh and upper arm, *Br. J. Dermatol.*, 113, 9, 1985.

Section K: Nails

27.0

Nail Thickness and Structure

27.1 Measurement of Nail Plate Thickness 557
G.B.E. Jemec

Chapter 27.1
Measurement of Nail Thickness

Gregor B.E. Jemec
Department of Dermatology
Bispebjerg Hospital
University of Copenhagen
Copenhagen, Denmark

I. Introduction

The nail is a well-defined keratin structure immediately available for physical studies, yet only few measurements have been made of the nail. Only longitudinal nail growth and nail thickness have been measured reproducibly. The measuring of nail thickness is of relevance when calculating the volume of a nail *in vivo*. Nail volume can be of interest in connection with studies of nail growth and descriptive studies of various nail diseases. Furthermore, quantitative studies of the nail may help the development of additional techniques for the measurement of the physical properties of human nails.

II. Variables

The parameter studied is simple linear distance between the external and the profound surface of the nail plate. As can be seen from the outline in Figure 1 the nailplate is of uneven thickness, being thicker toward the free edge.[1] All measurements of nail plate thickness should therefore mention where the thickness has been measured. The thickness is most appropriately expressed in fractions of millimeters. The thickness of fingernails in adults is approximately 0.3 to 0.9 mm, with the first finger's nail being the thickest, and the fifth finger's nail being the thinnest. Sex is of relevance as men appear to have thicker nails than women.

III. Methods

Two methods have been described for the noninvasive measurement of nail thickness: direct measurement with a micrometer or caliper and high-frequency ultrasound.

The direct measurement of nail thickness poses a few requirements on both nail and instrument. The nail must have a sufficient free edge of the nail in order to allow a secure and reproducible attachment of the measuring instrument. The instrument must grip the free edge, preferably exerting a standardized pressure, and measure fractions of a millimeter.[2] The most useful instrument is a spring-loaded electronic slide gauge or a micrometer. If a micrometer is used, it must have jaws that allow the investigator to grip the free edge of the nail squarely. The nail is difficult to compress, and spring loading is therefore less essential in this setting than in a setting where softer tissues are measured mechanically. The positioning of the instrument is always very important. The correct thickness of the nail plate is the shortest distance measured (see Figure 2). Some practice is required, and several measurements should be made in order for the investigator to develop a feel for the technique. In one published study the coefficient of variation for such measurements was 5.3% (SD = 2.4%). The disadvantages of the technique are increased variability if the nails are cut and the fact that the nail plate thickens as it passes over the nail bed. Most often absolute precision is however not necessary and a good relative value such as that obtained using mechanical measurements at the free edge of the nail is quite sufficient.

High-frequency ultrasound offers the advantage of being able to measure nail plate thickness over the entire nail. It is again important to make repeated measurements as any angling of the transducer will cause an overestimate of the nail thickness. The transducer should be held at a right angle to the surface of the nail. Simple machines can be used for these measurements as the nail plate has a distinctive echo profile (see Figure 3).[3] High-frequency ultrasound furthermore offers some concept of the functional structure of the nail plate, which may be of added benefit in an experimental setting.

The speed of sound within the nail is 2459 m/s, which can be used for calculations of actual distance from ultrasound examination.[3] Because the nail is not an even unilamellar structure the site of the measurement should be specified. The ultrasound measurements are influenced by nail hydration, and it is possible that

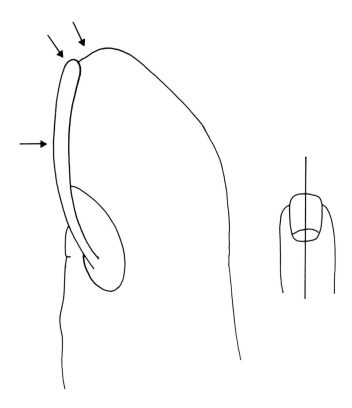

FIGURE 1 A schematic presentation of the nail plate. Note that the thickness of the nail plate increases along the nail bed.[1]

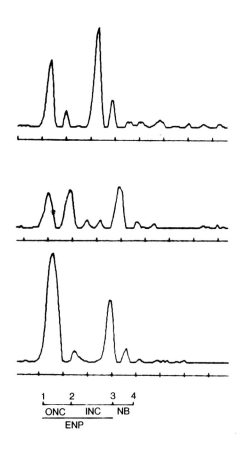

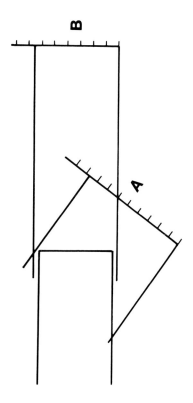

FIGURE 2 Any angling of the measuring instrument will cause overestimation of the nail thickness. This applies equally to calipers and to ultrasound.

FIGURE 3 Ultrasound profile of human thumbnails obtained with the Dermascan-A® 20 MH2 A-mode scanner (Cordex Technology, Hadskud, Denmark). ENP is the entire nail plate, ONC is the outer nail compartment, INC is the inner nail compartment, NB is the nail bed. (From Jemec, G.B.E. and Serup, J., *Arch. Dermatol.*, 125, 643, 1989. With permission.)

repeated measurements and the subsequent increased hydration of the nail may lead to a slower transmission time, and hence to an overestimate of the nail thickness.[2] Only a limited number of ultrasound measurements should be made of each nail. The coefficient of variation for ultrasound measurements of nail thickness was 4% (SD = 1.3%) in one study. The main disadvantages of using ultrasound for the measurement of nail thickness is the cost of the necessary hardware and the difficulty in transporting an ultrasound machine. The increased sensitivity of the method is however of great use in a laboratory setting.

IV. Correlation Between Methods

The two methods correlate very well and have been compared a number of times. The absolute thickness of nail plate as well as the thickness of the free edge of the nail correlate very well with the ultrasound measurements of nail plate thickness. Finlay et al. found that distal nail thickness as measured with a modified micrometer correlated well with the ultrasound measurement of the more proximal nail plate ($r = 0.79$, $p < 0.0001$).[4]

V. Recommendations

1. Position the instrument so as to avoid any angling. Positioning is easier with an A-mode scanner with a small pencil-shaped probe.
2. If you are using ultrasound make sure that the measurements are made at the same point as the nail plate is of somewhat uneven thickness.
3. Hydration may play a role when ultrasound measurements are made — do not soak the nail plate in water before measuring.

References

1. Johnson, M. and Shuster, S., Continuous formation of nail along the bed, *Br. J. Dermatol.,* 128, 277, 1993.
2. Finlay, A.Y., Western, B., and Edwards, C., Ultrasound velocity in human fingernail and effects of hydration: validation of in vivo nail thickness measurement techniques, *Br. J. Dermatol.,* 123, 365, 1990.
3. Jemec, G.B.E. and Serup, J., Ultrasound structure of the human nail plate, *Arch. Dermatol.,* 125, 643, 1989.
4. Finlay, A.Y., Moseley, H., and Duggan, T.C., Ultrasound transmission time: an in vivo guide to nail thickness, *Br. J. Dermatol.,* 117, 765, 1987.

28.0 Nail Growth

28.1 Measurement of Longitudinal Nail Growth 561
J.S. Roth and R.K. Scher

Chapter 28.1
Measurement of Longitudinal Nail Growth

Jeffrey S. Roth and Richard K. Scher
Department of Dermatology
College of Physicians and Surgeons of Columbia University
New York, New York

I. Introduction

Initial thoughts on the longitudinal measurement of nail growth turn to the simple and intuitive application of a ruler to the nail and recording of differences in length over time. While this is sound in principle, precision in nail growth measurement demands more rigorous technique and has been the subject of numerous articles over many decades. Achieving consistency, reproducibility, and accuracy of results requires that there be consensus about the definition of stable landmarks and avoiding sources of variability such as nail plate wear. In addition, nail growth has been reported to vary with season, time of day, digit, sex, and state of health,[1] making it important to establish a biologically meaningful interval over which nail growth should be measured. Though a technique may claim to allow measurement of nail growth over a 15-min period,[2] this may not be meaningful biologically.

Who might be interested in measuring nail growth? Research applications would include the response of the onychomycotic nail to novel antifungals under investigation or to quantify the response of psoriatic nails to treatment. Such applications would require relative precision to allow meaningful statistical evaluation of study data. Clinical applications might be the measurement of Beau's lines to time recent trauma to the nail matrix from disease or chemotherapy, the quantification of nail growth in a patient who complains of nails which grow poorly, or to establish efficacy of specific therapy for nail disease in the clinic setting. Clinical and research applications may impact differently on the choice of technique. Research techniques should be precise, may involve specialized equipment, need not be quick, and need not (though should) be inexpensive. Clinical methods should be rapid, inexpensive, need no specialized equipment, and can be relatively less precise. Both research and clinical applications should do no harm to the patient (an intuitive "given" which is not always adhered to).

The following will take into account the history of the treatment of this issue and a review of the major techniques which have been described.

II. Historical Overview

The development of methods for measuring nail growth spans many decades. Interest in the growth of nails initially stemmed from the notion that the health of the nail and the vigor of its growth closely parallels the general health. While this notion remains generally sound, it is now appreciated that, although the nail is not a direct reflection of the general health, it may offer clues to underlying systemic illness.

Early reports of methods of measurement and of patterns of nail growth were flawed by poor scientific rigor and anecdotal experience. Thus, Berthold[3] measured only the growth of his own fingernails (using the lunula as a fixed landmark). Similarly, Sharpley-Schafer[4] measured only his own hand. Bloch[5] criticizes the previous pioneering work of Dufour,[6] published in 1872, which measures the distance between the cuticle and a permanent mark made in the nail plate using silver nitrate, stating that the cuticle is a constant landmark only in the *well groomed.* (He offers in place of Dufour's methodology two techniques, one in which a permanent mark in the nail plate is measured against a fixed india ink tatoo on the dorsal finger and another in which the fixed point on the nail plate is measured against the knuckle of the finger bent at a fixed angle.) LeGros Clark and Buxton[7] criticize the measurement of nail growth using the proximal nail fold as a landmark. Their methodology in turn is derided as too vague in subsequent studies. Internal inconsistencies abound as well. For example, they use as their fixed landmark a point on the nail "*about 2 mm* from the margin of the lunula", yet they go on to give measurement of change in "micromillimeters" (microns). William Bennett Bean, perhaps the most indefatigable nail watcher of all time, measured his own fingernails over a 35-year period both longitudinally and by weighing his nail clippings.[8,9] His perseverance allowed him to observe the deceleration of nail growth with advancing years, from a rate of 0.123 mm/day at the age of 32 to 0.095 mm/day at age 67.

Intuitively, measuring nail growth requires no sophisticated instrumentation; nevertheless, it has been subject to sometimes faddish application of available technology as it became available. The modern age brought with it a flurry of new techniques: photography,[10,11] magnifying devices such as the "biomicrometer" of Basler[18] and the tool of LeGros Clark and Buxton,[7] the "split image rangefinder adapted to a trinocular microscope" described by Orentreich et al.,[2] and time-lapse photography.[12]

Authorities have opined on the suitability of anatomic structures for use as landmarks from which to measure the growing nail. Some favor the proximal nail fold.[1] Others prefer the lunula, though it is conceded that on some fingers the lunular margin is in fact blurred under magnification and therefore unsuitable for submillimeter measurements.[7] Furthermore, many individuals lack a visible lunula. Some favor the use of bony landmarks by X-ray or physical examination, relying on the intimate relationship between the distal phalanx and the nail plate.[10,12,14]

The results of animal studies may be difficult to extrapolate to humans, owing to fundamental differences in the shape of the nail in other species such as the rat.[17]

In summary, such a "simple" issue as how to measure the growing nail is beset with good-natured controversy. We intend to endorse no single technique but to advise a common sense approach: simplicity, reliability, ease, inexpensiveness, and, above all, avoidance of harm to the patient.

III. Measurement of Nail Growth

We will approach the measurement of nail growth by presenting techniques and points of view in each of several aspects of the problem.

A. Fixed Landmarks

The issue of finding a stable landmark as a point of reference from which to measure nail growth is important since, if this point varies between determinations, the measurements will be rendered meaningless.

Several proximal (fixed) landmarks have been proposed:

1. The cuticle. This landmark appears to be relatively stable in patients who do not manipulate or cut back the cuticle, especially if submillimeter measurements are not needed. The cuticle is obviously invalid as a point of reference if pushed back or cut.

2. The proximal nail fold. This is similar to the cuticle in its ability to be pushed back, though not as easily or permanently as the cuticle. Whether the edge of the proximal nail fold can be precisely defined for very fine measures is unclear. This has been shown to have a small interobserver error rate[13] and may be the most suitable landmark.

3. The distal interphalangeal (DIP) joint. This is among the more precise landmarks. It can be used in two ways: one, in which a mold or rigid brace is constructed so that the DIP joint is flexed at an identical angle at each determination and, second, as a radiographic landmark (used with a radiopaque marker cemented to the nail plate). The first technique requires a simple tool and may be unsuitable for submillimeter measurements. The second technique, though perhaps the most precise, involves exposure to ionizing radiation and is therefore suitable only in research settings with informed consent, if ever.

4. The lunula. Using this structure has several disadvantages. First, not everyone has a visible lunula, especially on the index, third, ring, and small fingers. Second, while the lunula appears well defined, under magnification its edge is somewhat vague and would therefore introduce unnecessary error into measurement.

5. A structure cemented to the skin of the dorsal distal phalanx. This has the advantage of being relatively stable and easy to see, but may come loose during a meaningful interval of nail growth. This technique was cited more than 50 years ago[12] and later echoed.[2]

B. Distal Landmarks

Once the issue of proximal landmarks is settled, it becomes relatively easy to choose a way of marking the growing portion of the nail. It makes no difference if the nail is etched, drilled, or if a radio-opaque band is cemented to the nail plate. The only specific requirement is that the marking be permanent over the course of measurement.

C. Technique of Measurement

Several options exist in this regard:

1. A straight ruler with at least millimeter (and preferably submillimeter) increments indicated. Such a device should be used with a loupe of 8× magnification or greater.

2. A caliper similar to those used in electrocardiography[16] which is then applied to the straight rule. This is preferable as the direct application of the straight rule to the curved nail plate may introduce inaccuracy.

3. Devices such as the "split image rangefinder"[2] have only very specialized applications, as does time-lapse photography.[12]

IV. Summary

The technique employed to measure the longitudinal growth of nails should satisfy several criteria: it should be simple, disfigure as little as possible, be safe, and be biologically relevant. The extremes of precision which many methods aim to achieve are only of value in research settings. Except in specialized situations when very short-term growth rates are measured, the rate of nail plate growth is so variable between day and night and season[15] in an individual as to make overly precise or overly frequent measurements meaningless.

No dermatologic diagnosis rests on accurate measurement of longitudinal growth of the nail. In pursuing his interest in this parameter, therefore, it behooves the physician to follow the dictum *primum non nocere,* "first, do no harm".

References

1. Dawber, R. and Baran, R., Nail growth, *Cutis,* 39, 99, 1987.
2. Orentreich, N., Markofsky, J., and Vogelman, J.H., The effect of aging on the rate of linear nail growth, *J. Invest. Dermatol.,* 73, 126, 1979.
3. Berthold, Beobachtungen uber das quantitative Verhaltniss der Nagel — und Haarbildung beim Menschen, *Mullers Arch.,* 156, 1850.
4. Sharpley-Schafer, E., Relative growth of nails on right hand and left hand respectively: on seasonal variations in rate; and on influences of nerve section upon it, *Proc. R. Soc. Edinburgh,* 51(1), 8.
5. Bloch, A.M., Etude de la croissance des ongles, *C.R. Soc. Biol.,* 58, 253, 1905.
6. Dufour, 1872.
7. LeGros Clark, W.E. and Buxton, L.H.D., Studies in nail growth, *Br. J. Dermatol. Syph.,* 50, 221, 1938.
8. Bean, W.B., A note on fingernail growth, *J. Invest. Dermatol.,* 20, 2, 1953.
9. Bean W.B., Nail growth. Thirty five years of observation, *Arch. Int. Med.,* 140, 73, 1980.
10. Babcock, M.J., Methods of measuring fingernail growth rates in nutritional studies, *J. Nutr.,* 55, 323, 1955.
11. Sibinga, M.S., Observations on growth of fingernails in health and disease, *Pediatrics,* 24, 225, 1959.
12. Morton, R., Visual assessment of nail growth, *J. Audiovis. Media Med.,* 14(1), 31, 1991.
13. Dawber, R., Fingernail growth in normal and psoriatic subjects, *Br. J. Dermatol.,* 82, 454, 1970.
14. Kandil, E., Accurate measurement of nail growth, *Int. J. Dermatol.,* 11(1), 54, 1972.
15. Scher, R.K. and Daniel, R.C., *Nails: Therapy, Diagnosis, Surgery,* W.B. Saunders, Philadelphia, 1991.
16. Hillman, R.W., Fingernail growth in the human subject. Rates and variations in 300 individuals, *Hum. Biol.,* 27, 255, 1955.
17. Godwin, K.O., An experimental study of nail growth, *J. Nutr.,* 69, 121, 1959.
18. Basler A., Growth processes in fully developed organisms, *Med. Klin.,* 33, 1664, 1937.

Section L: Clinical Evaluation and Quantification

29.0

Clinical Scoring and Grading

29.1 Standard Schemes to Assess Skin Diseases 567
R.A. Logan

29.2 Clinical Grading of Experimental Skin Reactions 575
T. Agner

Chapter 29.1
Standard Schemes to Assess Skin Diseases

Richard A. Logan
Mid Glamorgan Health Authority
Bridgen General Hospital
Wales, U.K.

I. Introduction

No other organ is as accessible to scrutiny as the skin. This allows it to be studied by a diversity of methods, as demonstrated by the breadth of topics covered in this book. Many of these are laboratory techniques and designed for research purposes. Although they provide valuable information regarding disease processes, these techniques will always remain supplementary to clinical analysis.

All potential new treatments must be assessed in comparison to alternatives, ideally by controlled clinical trials. Unfortunately, skin therapies rarely lead to the rapid and complete clearance of physical signs. More often there is a reduction in severity which must be measured and statistically analyzed to reach a valid conclusion regarding the relative effectiveness of the treatments in question.

Little attention has been paid in the literature to the important subject of quantitative clinical assessment of skin disorders.[1] This trend is perpetuated in the latest edition of the standard British dermatology reference textbook which devotes only one small paragraph (unreferenced) to the clinical grading of acne.[2] However, through medical audit, we are increasingly called upon to demonstrate that what we do as clinicians is cost effective. To do that we need data which are quantitative as well as qualitative. In the context of a busy hospital (or office) practice, the general standard of clinical notekeeping often leaves a great deal to be desired.[3] Comments such as "much better" or "fine" convey little to a doctor who did not see the patient at their last visit, and probably very little to the original author when reviewing the notes a few months later. How much better it is to be able to quantify disease numerically.

To assess skin diseases clinical assessment schemes are needed which are quantitative, reproducible, easy to learn, and preferably quick to perform. The wider adoption of standard, valid schemes for clinical assessment would facilitate comparison of the results of different trials. If simple enough, they could also be used in ordinary clinical practice away from the research setting.

II. Variables

A. Basic Features of Clinical Assessment Schemes for Skin Disease

There are three fundamental components of skin disease activity that need to be included in devising a clinical assessment scheme:

1. Extent of lesions.
2. Severity of skin symptoms.
3. Impact of disease on patient's lifestyle.

1. Extent of Skin Lesions

The method appropriate for measuring skin lesions depends of course on the disease being studied. For example, it is possible to count the papules of rosacea or acne or the number of solar keratoses. On the other hand for atopic eczema or psoriasis the area of skin involvement is an appropriate parameter.

The standard method of clinically assessing area of skin involvement is by the Rule of Nines[4] (Figure 1). The Rule of Nines does not apply accurately to children for whom modified figures according to age are available.[5,6] However, when using the Rule of Nines it has been shown that there may be significant interobserver variation in assessing the area of skin involvement with psoriasis.[1] A simpler technique for estimation of area of skin surface area is the Rule of Hand[7] whereby the area of the flat of one closed hand is taken to represent 1% of the body surface area. This is attractive but may not be very accurate. When reexamined recently using a standard nomogram for body surface area (BSA)[8] the hand area was found to represent a lower proportion of BSA (0.76% for males and 0.70% for females).[9]

Estimation of skin area involved must be accompanied by a measurement of the degree of skin changes such as erythema, scaling, thickening, etc. Such changes are usually estimated numerically, e.g., 0 = absent, 1 = mild, 2 = moderate, 3 = severe.

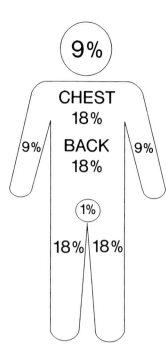

FIGURE 1 Body surface area by Rule of Nines.[4]

2. Severity of Skin Symptoms

Probably the most pervasive of symptoms in dermatology is itching. Although itching can be indirectly inferred from the extent of excoriation a simple assessment may be made by asking the patient to mark a visual analog scale ranging from 0 (no itching) to 10 cm (intolerable itching) and including this in a standard assessment form. Pruritus may also be indirectly measured by weighing the amount of cream or ointment used by the patient.

3. Impact of Disease on Patient's Lifestyle

In assessing skin disease it is now not sufficient to consider just the number of spots or the size of the lesions. This must be correlated with some assessment of the degree to which the disease is affecting the patient's lifestyle. Recently there has been a welcome (and overdue) interest in the concept of disability caused by skin disease. Various indices of disability have been devised for some skin disorders and will be discussed.

III. Methods Available

Clinical assessment schemes for skin diseases are disease-specific and rarely applicable to more than one skin disorder. The most widely used schemes for three common inflammatory skin diseases will be discussed.

A. Acne Vulgaris
1. Leeds Acne Grading Scale

I think this is the most effective and simple method of clinical assessment that has been devised for any skin disease. The Leeds Acne Grading Scale (Figure 2) is based on a large study of 435 patients with acne of varying severity.[10] The technique requires examination of the patient in a good light. Lesions are inspected visually and also palpated to detect those which may be nodular or even cystic. The acne can then be assigned a grade which ranges in practice from 0.25 to 0.75 (physiological acne) to grade 10 (severe nodulocystic acne). In recent years it has become rare to see examples of grades higher than 7. This is probably due to better management by family doctors, but also earlier referral to hospitals since the advent of 13-cis retinoic acid therapy. The patient's acne should be graded separately for (1) the face (including the neck), (2) chest, and (3) back, because acne frequently varies in severity at these sites, and also responds to treatment differently. With practice this procedure takes less than 1 min.

Training in this technique for research purposes may take up to 8 hours, as it will be necessary to be able to assess minor changes in acne grade. However, for routine clinical use reasonable confidence and competence can be achieved in about 2 hours. As well as assigning the patient's acne a grade it is useful to describe the types of lesion present, e.g., predominantly comedones, or mixed inflamed and noninflamed lesions, as this will have a bearing on the type of treatment chosen. I use this scheme routinely in my clinical practice, and find it invaluable. Acne patients are often unaware of slow improvement in their disease when they study their spots daily in the mirror — it helps to be able to show them that there has been objective improvement between their hospital visits.

A high degree of "within doctor" and "between doctor" accuracy of grading has been shown following adequate training.[4] The acne grading can be made more difficult under some circumstances. The irritant dermatitis caused by some topical therapies, and also by 13-cis retinoic acid, can make acne look worse than it is. Conversely, hair styles, cosmetics, and suntanning can mask the true extent of the disease.

2. Other Methods of Acne Grading

It is possible to count the numbers of individual types of lesion for each body site as we did in one study.[5] However, this is extremely laborious and only really essential if it were necessary to detect the effect of a treatment on the different types of lesion. It is possible to grade acne reasonably accurately using standardized photography.[5,6] The disadvantages are the need for a studio setup, and also that photography produces a two-dimensional record and may miss nodular or cystic disease which occasionally is only obvious by palpation. A combination of assessment by photograph and palpation has been described which overcomes one of these deficiencies.[7]

3. Acne Disability

There can be a disparity between the physical degree of acne and the psychological distress apparently associated with it. For example, occasional patients (usually female) are seen with what amounts to physiological acne who may be suicidal allegedly because of their spots. To assist in the full assessment of acne patients an Acne Disability Index has recently been devised based on a questionnaire survey of 100 patients with

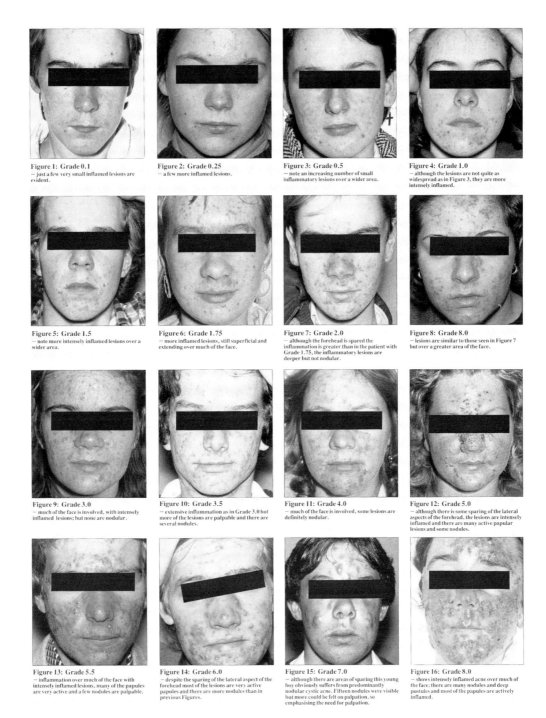

FIGURE 2 The Leeds Acne Grading Scale. (Reproduced from *Retinoids Today and Tomorrow,* 15, 1989, with permission.)

acne.[14] The same authors have refined and simplified the questionnaire to five questions on social and emotional aspects of everyday life.[15] The responses to these questions were shown to correlate significantly with the physician's assessment of acne severity. However, they also quickly identify those patients whose disability is much greater than the clinical degree of their acne. It is suggested that such patients would benefit from more aggressive therapy and supportive counseling.

B. Psoriasis
1. Psoriasis Area and Severity Index (PASI)

The PASI score[16] has been used widely in clinical trials in recent years. It is used mainly to assess differences in psoriasis activity before and after treatments which are not expected to clear the disease completely, e.g., a recent comparative study of calcipotriol and betamethasone 17-valearate ointments.[17] Trials of more powerful modalities such as photochemotherapy may use complete clearance of disease as the end point without employing the PASI system.[18,19] The PASI score is calculated as follows:

$$PASI = 0.1(E_h + I_h + D_h)A_h$$
(h = head: corresponding to 10% BSA)

$$0.3 (E_t + I_t + D_t)A_t$$
(t = trunk: corresponding to 30% BSA)

$$0.2(E_u + I_u + D_u)A_u$$
(u = upper extremity: 20% BSA)

$$0.4(E_l + I_l + D_l)A_l$$
(l = lower extremity: 40% BSA)

The area of involvement of the four main body areas (A_h, A_t, A_u, A_l) is given a numerical value ranging thus: 0 = no involvement; 1 = <10%; 2 = 10 to <30%; 3 = 30 to <50%; 4 = 50 to <70%; 5 = 70 to <90%, and 6 = 90 to 100%. The severity of three symptoms (E = erythema, I = infiltration, D = desquamation) is assessed for each area and given a numerical grading on a scale of 0 to 4: 0 = complete lack of involvement; 1 = slight involvement; 2 = moderate involvement; 3 = striking involvement, and 4 = severest possible involvement.

The PASI can vary from 0 to 72 in steps of 0.1. In practice, however, most patients attending hospital with psoriasis will have PASI scores ranging between approximately 5 and 40. A patient with a score of >10 will probably need hospital admission.[16] Patients with PASI scores greater than 40 are very rare, such that almost half of the PASI range is redundant, thereby reducing the discriminating power of the system.

The major advantages of the PASI system are its simplicity, it is relatively quick to assess, and requires no specialized equipment. It is therefore suitable for routine use in clinical research studies, although too time consuming for general outpatient (office) work.

a. Limitations of the PASI Score[20]

1. It relies greatly on the accuracy of estimation of the area of skin involvement. The inaccuracies in this have been discussed.[1] These were confirmed again in a more recent study[21] in which it was shown that there is a high interobserver error between inexperienced observers when using the Rule of Nines for estimating the area of psoriasis involvement. They generally grossly overestimate (by up to 11-fold). Such inaccuracies in area assessment would clearly produce a very unreliable PASI score. More accurate assessments were made by a tracing technique and a method of photographic image analysis. However, these two alternative techniques of measuring body area are slow (taking about 2 hours per patient) and require specialized technical equipment.[21]
2. It discriminates poorly between different forms of the disease. For example, a patient with erythrodermic psoriasis which could be considered as the most severe form of the disease may have a PASI score of 24 if the degree of scaling and induration were not very marked.[20] This could be exactly the same score as a patient such as the one in Figure 3 with more scaling and erythema.
3. Some psoriasis treatments may themselves distort the PASI score, e.g., dithranol (anthralin) may produce erythema which could create an increase in the score when, in fact, the patient was getting better.
4. Patients with localized disease, e.g., on the scalp or the hands (Figures 4 and 5), may have a low PASI score as the area of involvement is small. Such a patient may nevertheless be seriously disabled by their localized disease. For studies of localized disease, e.g., scalp psoriasis, specially adapted simplified scoring systems have been used.[22,23]

van der Kerkhof[20] suggests that, to overcome its limitations, in clinical trials the PASI score should be supplemented by an assessment of the symptoms before and after treatment. The simplest method of doing this is by a global assessment (by both patient and investigator) of the degree of efficacy of the treatment such as that used in a recent study:[24] none, slight, moderate, good, very good. Alternatively the use of a disability index may assist in this.

2. Psoriasis Disability Index (PDI)[25]

A questionnaire survey of 54 patients with psoriasis was conducted. This consisted of 28 questions covering most aspects of functional disability in the areas of daily activity, work, personal relationships, leisure, and treatment. Responses were recorded on a linear scale and correlated with the area of psoriasis involved both before and again after hospital admission for treatment of the psoriasis. Out of the initial 28 questions, 10 were found to provide the most discriminating evidence of psoriasis disability and were used to form the PDI. The PDI was found to fall in most patients after hospital admission and, as mentioned before, the total area of involvement was found not to reliably predict the degree of disability. Also, the system was able to identify those patients for whom disability was not improved by hospital admission. Such patients may require greater attention and psychological support.

The PDI has more recently been compared to the sickness impact profile (SIP).[26] The SIP is a general measure of measuring the impact of disease on a patient's functional behavior, and has been applied

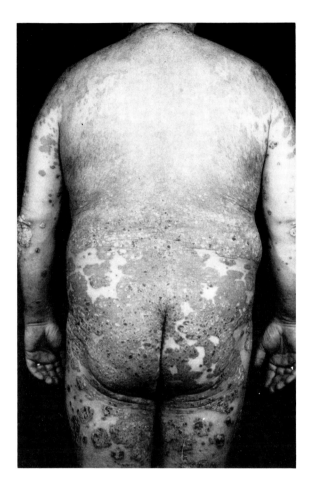

FIGURE 3 A patient with extensive but stable psoriasis. Because of induration and scaling this patient may have a PASI score equal to or greater than a patient with erythrodermic psoriasis.

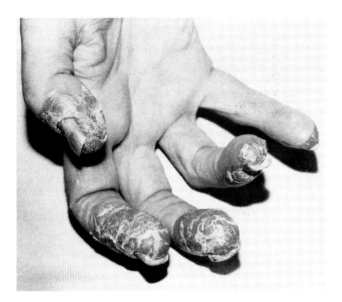

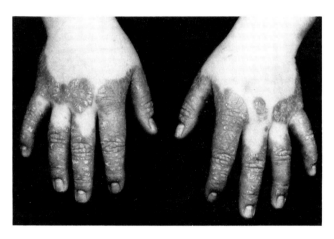

FIGURES 4 and 5 Two cases of hand psoriasis. Low PASI score, but degree of disability (physical and cosmetic) very great.

to other diseases such as angina and hypertension. It was found that the PDI and SIP correlated well together, although only the PDI correlated well with the PASI score. It was suggested that the SIP could be used to compare the impact of psoriasis on a person's lifestyle with that of other, nondermatological diseases. Such as approach may provide useful evidence in support of dermatological services.

C. Atopic Dermatitis (AD)

There have been many different systems used for the grading of AD in recent years. So far an equivalent to the PASI score has not yet been widely adopted for AD. To one who has not been closely involved in this area the main difficulties in devising a suitable scheme appear to be

1. AD is a much more labile disorder than psoriasis or acne. Within a few hours the disease can become much more (or less) active, and assessment at single time points does not necessarily give a true reflection of overall disease activity.

2. The cardinal symptom of AD is itching. The extent of the eruption is of secondary importance to the patient, and thus assessment of pruritus must be weighted positively in any scoring system.

3. Assessing the area of disease involvement is much more difficult than in psoriasis where the lesions are usually sharply defined. Atopic dermatitis can demonstrate a field effect, with no obvious boundaries. Close inspection, especially with tangential lighting, can reveal subtle papular involvement which is not obvious upon cursory examination, and would certainly not be apparent on a photograph. This problem can be overcome by concentrating disease assessment on specified body areas rather than the whole body.

4. Disease activity is different in children and adults, e.g., vesiculation and weeping is much more common in children. Thus, inclusion of assessment of these signs is vital in children, but probably not a great discriminator in adult AD.

1. Indications for AD Scoring Systems

Scoring systems are needed for two main purposes. It is unlikely that a single system will be suitable for both of these.

a. Grading AD Patients into Mild, Moderate, or Severe

After the 3rd International Symposium on Atopic Dermatitis held in Oslo in 1988 an attractively simple scheme was proposed for baseline severity grading of AD patients.[27] Briefly patients are given a score of 1 to 3 for three disease parameters:

1. Disease extent by Rule of Nines (area adjusted for infants)
2. Disease course over previous year
3. Itch severity

Adding the three scores stratifies patients into mild, moderate, or severe AD with scores of 3 to 4, 4.5 to 7.5, and 8 to 9, respectively.

b. Following the Course of AD During Drug Trials

After the same Oslo conference in 1988, Hanifin[28] published guidelines for the more-detailed grading of AD that is required during the course of clinical trials. The basic requirements are

1. Baseline recording of activity of dermatitis, with emphasis on severity (70%) over extent (30%).[29,30]
2. Recording of the stability of the disease prior to the start of the study, to avoid misinterpretation of flares of AD as being adverse events, rather than natural fluctuations of disease activity.
3. The following signs/symptoms may be recorded:
   a. Required: erythema; induration/papulation; pruritus/excoriation
   b. May be helpful in long trials: lichenification; scaling/dryness
   c. Necessary for pediatric studies: erosion/oozing/weeping
4. Numerical scoring of severity of signs/symptoms on a scale of 0 (none) to 3 (severe). For short studies under 2 weeks, it is sufficient to record activity in a single target area (usually the most severely affected), but for longer studies a general assessment is necessary.

2. Examples of Evaluation Schemes for AD

Several different schemes for evaluating AD are in current use and examples are given below.

1. The "20 area" severity assessment chart[31-33] divides the skin surface into 20 separate zones of approximately equal area (Figure 6). With some

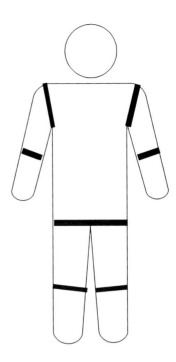

FIGURE 6 The "20 area" assessment chart for atopic dermatitis.

slight variations, a score of 0 to 3 is given for the maximum severity within each zone of (1) erythema and (2) surface damage. The area of involvement was estimated either for individual areas (1 = <33%; 2 = 34 to 66%; 3 = >67%) and multiplying the area score by the severity score for the area or by counting the number of zones in which any of the clinical features occurred. The resulting zone scores were then added to give a total score derived from both the severity of the signs and the area of involvement. Pruritus was recorded as a score of 0 to 3 (none, mild, moderate or severe) or by asking parents about the child's ability to sleep.[33] As an outside observer, the problems with this system are the lack of emphasis (weighting) on itch; the slightly varying methodology (which may have been indicated for specific reasons in each of the trials) prevents direct comparison of results; assessing the whole body surface may be unnecessary.

2. Another system[34] assessed ten severity criteria and ten body sites. The criteria were erythema, edema, vesicles, exudation, crusts, excoriations, scales, lichenification, pruritus, and loss of sleep. Each of these was given a score ranging from 0 (no lesion) to 7 (extremely severe) by choosing the most severely affected part of the body for each criterion. A topographic score of 0 to 3 was given for the extent of involvement for each of face, neck, anterior and posterior trunk, buttocks, arms, hands, legs, knees, and feet. This system has the merit of being simple to perform. It also produced results which correlated highly with those derived from a

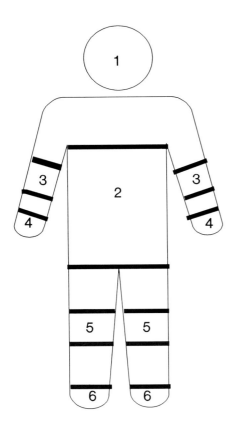

FIGURE 7 This system of atopic dermatitis assessment concentrates on the 6 body zones which are most likely to be affected.[35]

more elaborate and time-consuming system which involved detailed scoring for each zone of the body. However, it appears that some of the severity criteria could be considered redundant, e.g., crusts are a marker of exudation, loss of sleep is probably due to pruritus. Indeed, the authors suggested that the number of clinical criteria evaluated might be reduced to as little as 5 without invalidating the score. Again, pruritus does not seem to be adequately emphasized in the scoring, and there is no record of prior disease stability.

3. A recent study reduces to 6 both the clinical signs and the body areas (Figure 7) assessed.[35] By scoring the variables 0 (none) to 3 (severe) for each area a total body score (maximum 108) is derived. The extent of disease is estimated by the Rule of Nines, sleep and itch by the patient on a visual analog scale, and topical, steroid use measured by weighing tubes. There is thus, as yet, no consensus in the systems for clinical grading of atopic dermatitis although the recommendations[27,28] are available.

4. Finally an interesting and novel proposal for AD grading has recently been published.[36] The extent and severity of the AD is painted onto body schemes using a color code (green = mild; blue = moderate; red = severe). The area and severity of disease is determined by a system of point counting and the final ADASI score (Atopic Dermatitis Area and Severity Index) is calculated from a formula in which itching appears positively weighted. I confess to not understand the mathematics involved but, if this system is validated by further work, it has merits: it is quick (2 minutes per body scheme), it avoids laborious counting procedures, and it neatly combines information on the extent and severity of the disease. However, the color coding of the disease seems to be subjective. Nevertheless, often the best inventions come from the simplest ideas and the ADASI score may help to clarify the muddy waters in the area of clinical assessment of atopic dermatitis.

IV. Recommendations

In this chapter I have not attempted to produce a comprehensive catalog of all the available clinical scoring systems for skin diseases. The three diseases discussed do demonstrate some of the difficulties which have to be overcome in this field. I commend the Leeds Acne Grading Scale for its simplicity, practicality, and cheapness. The PASI score has recognized limitations which will have to be overcome through conference and modification. Guidelines are in place for the development of widely acceptable grading schemes for atopic dermatitis.[27,28]

In striving for the ideal method of clinical assessment, let us not see only the spots and forget the patient's own perception of their disease.

References

1. Marks, R., Barton, S.P., Shuttleworth, D., and Finlay, A.Y., Assessment of disease progress in psoriasis, *Arch. Dermatol.*, 125, 235, 1989.
2. Ebling, F.J.G. and Cunliffe, W.J., Disorders of the sebaceous glands, in *Textbook of Dermatology*, 5th ed., Champion, R.H., Burton, J.L., and Ebling, F.J.G., Eds., Blackwell Scientific, Oxford, 1992, 1725.
3. Shuttleworth, D., Finlay, A.Y., Rademaker, M., Shaw, C.D., and Simpson, N.B., Psoriasis consultation audit: a two centre study, *Br. J. Dermatol.*, 123, 99, 1990.
4. Wallace, A.B., The exposure treatment of burns, *Lancet*, i, 501, 1951.
5. O'Neill, J.A., Burns in children, in *Burns; A Team Approach*, Artz, C.P., Moncrief, J.A., and Pruitt, B.A., Eds., W.B. Saunders, Philadelphia, 1979, 343.
6. Kennedy, C.T.C., Reactions to mechanical and thermal injury, in *Textbook of Dermatology*, 5th ed., Champion, R.H., Burton, J.L., and Ebling, F.J.G., Eds., Blackwell Scientific, Oxford, 1992, 812.
7. Averill, R.W. and Arndt, K.A., The Rule of Hand, presented at The National Clinical Dermatology Conference, Chicago, June 25–26, 1983.
8. Sendroy, J. and Cecchini, L.P., Determination of human body surface area from height and weight, *J. Appl. Physiol.*, 7, 1, 1954.
9. Long, C.C., Finlay, A.Y., and Averill, R.W., The Rule of Hand: 4 Hand areas = 2 FTU = 1g, *Arch. Dermatol.*, 128, 1129, 1992.

10. Burke, B.M. and Cunliffe, W.J., The assessment of acne vulgaris, *Br. J. Dermatol.,* 111, 83, 1984.
11. Prendiville, J.S., Logan, R.A., and Russell-Jones, R., A comparison of dapsone against 13-cis retinoic acid in the treatment of nodular/cystic acne, *Clin. Exp. Dermatol.,* 13, 67, 1988.
12. Cook, C.H., Centner, R.L., and Michaels, S.E., An acne grading method using photographic standards, *Arch. Dermatol.,* 115, 571, 1979.
13. Allen, A.M., Clinical trials in dermatology. III. Measuring responses to treatment, *Int. J. Dermatol.,* 19, 1, 1980.
14. Motley, R.J. and Finlay, A.Y., How much disability is caused by acne?, *Clin. Exp. Dermatol.,* 14, 194, 1989.
15. Motley, R.J. and Finlay, A.Y., Practical use of a disability index in the routine management of acne, *Clin. Exp. Dermatol.,* 17, 1, 1992.
16. Fredriksson, T. and Pettersson, U., Severe psoriasis — oral therapy with a new retinoid, *Dermatologica,* 157, 238, 1978.
17. Cunliffe, W.J., Berth-Jones, J., Claudy, A., Fairiss, G., Goldin, G., Gratton, D., Henderson, C.A., Holden, C.A., Maddin, W.S., Ortonne, J.-P., and Young, M., Comparative study of calicipotriol (MC903) ointment and betamethasone 17-valearate ointment in patients with psoriasis vulgaris, *J. Am. Acad. Dermatol.,* 26, 736, 1992.
18. Tanew, A., Guggenbichler, A., Hönigsmann, H., Geiger, J.M., and Fritsch, P., Photochemotherapy for severe psoriasis without or in combination with acitretin: a randomised, double-blind comparison study, *J. Am. Acad. Dermatol.,* 25, 682, 1991.
19. Petzelbauer, P., Hönigsmann, H., Langer, K., Anegg, B., Strohal, R., Tanew, A., and Wolff, K. Cyclosporin A in combination with photochemotherapy (PUVA) in the treatment of psoriasis, *Br. J. Dermatol.,* 123, 641, 1990.
20. van de Kerkhof, P.C.M., On the limitations of the psoriasis area and severity index (PASI), *Br. J. Dermatol.,* 126, 205, 1992.
21. Ramsay, B. and Lawrence, C.M., Measurement of involved surface area in patients with psoriasis, *Br. J. Dermatol.,* 124, 565, 1991.
22. Olsen, E.A., Cram, D.L., Ellis, C.N., Hickman, J.G., Jacobson, C., Jenkins, E.E., Lasser, A.E., Lebwohl, M., Leibsohn, E., Medansky, R.S., Oestreicher, M.I., Savin, R.C., Scher, R.K., Shavin, J.S., Smith, R.D., and Day, R.M., A double-blind, vehicle-controlled study of clobetasol propionate 0.05% (Temovate) scalp application in the treatment of moderate to severe scalp psoriasis, *J. Am. Acad. Dermatol.,* 24, 443, 1991.
23. Curley, R.K., Vickers, C.F.H., Norris, T., and Glover, D.R., A comparative study of betamethasone dipropionate with salicylic acid and betamethasone valearate for the treatment of steroid-responsive dermatoses of the scalp, *J. Dermatol. Treat.,* 1, 203, 1990.
24. Timonen, P., Friend, D., Abeywickrama, K., Laburte, C., von Graffenfried, B., and Feutren, G., Efficacy of low-dose cyclosporin A in psoriasis: results of dose-finding studies, *Br. J. Dermatol.,* 122 (Suppl. 36), 33, 1990.
25. Finlay, A.Y. and Kelly, S.E., Psoriasis — an index of disability, *Clin. Exp. Dermatol.,* 12, 8, 1987.
26. Finlay, A.Y., Khan, G.K., Luscombe, D.K., and Salek, M.S., Validation of Sickness Impact Profile and Psoriasis Disability Index in psoriasis, *Br. J. Dermatol.,* 123, 751, 1990.
27. Rajka, G. and Langeland, T. Grading of the severity of atopic dermatitis, *Acta Dermatol. Venereol.,* Suppl. 144, 13, 1989.
28. Hanifin, J.M., Standardised grading of subjects for clinical research studies in atopic dermatitis: workshop report, *Acta Dermatol. Venereol.,* Suppl. 144, 28, 1989.
29. Queille-Roussel, C., Raynaud, F., and Saurat, J.H., A prospective computerised study of 500 cases of atopic dermatitis in childhood. I. Initial analysis of 250 parameters, *Acta Dermatol. Venereol.,* Suppl. 114, 87, 1985.
30. Seymour, J.L., Keswick, B.H., Hanifin, J.M., Jordan, W.P., and Milligan, M.C., Clinical effects of diaper types on the skin of normal infants and infants with atopic dermatitis, *J. Am. Acad. Dermatol.,* 17, 988, 1987.
31. Atherton, D.J., Sewell, M., Soothill, J.F., and Wells, R.S., A double blind, controlled cross-over trial of an antigen avoidance diet in atopic eczema, *Lancet,* i, 401, 1978.
32. Heddle, R.J., Soothill, J.F., Bulpitt, C.J., and Atherton, D.J., Combined oral and nasal beclomethasone dipropionate in children with atopic eczema; a randomised, controlled trial, *Br. Med. J.,* 289, 651, 1984.
33. Sheehan, M.P. and Atherton, D.J., A controlled trial of traditional Chinese medicinal plants in widespread non-exudative atopic eczema, *Br. J. Dermatol.,* 126, 179, 1992.
34. Costa, C., Rilliet, A., Nicolet, M., and Saurat, J.-H., Scoring atopic dermatitis: the simpler the better?, *Acta Dermatol. Venereol.,* 69, 41, 1989.
35. Sowden, J.M., Berth-Jones, J., Ross, J.S., Motley, R.J., Marks, R., Finlay, A.Y., Salek, M.S., Graham-Brown, R.A.C., Allen, B.R., and Camp, R.D.R., Double-blind, controlled, cross-over study of cyclosporin in adults with severe, refractory atopic dermatitis, *Lancet,* 338, 137, 1991.
36. Bahmer, F.A. and Schubert, H.-J., Quantification of the extent and severity of atopic dermatitis, *Arch Dermatol.,* 127, 1239, 1991.

# Chapter 29.2
Clinical Grading of Experimental Skin Reactions

Tove Agner
Department of Dermatology
Gentofte Hospital
Hellerup, Denmark

I. Introduction

During the last decade a number of sophisticated methods has been proposed for evaluation of patch test reactions and other experimental reactions.[1] Among the more successful and frequently used methods laser Doppler flowmetry for measurement of increase in skin blood flow, colorimetry for measurement of erythema, ultrasound A, B, or C-scan for measurement of edema formation and measurement of transepidermal water loss for evaluation of skin barrier function can be mentioned. These methods are objective and highly sensitive and therefore important devices in experimental studies. However, application of complexed measuring methods is often disturbed by heterogeneity in the clinical picture, even within patch test reactions. Each method measures one single parameter only, contributing to the complete clinical picture. Clinical (visual) evaluation of skin reactions incorporates more variables at a time, although essentially subjective. The eye and the finger of a skilled dermatologist therefore until now remain the most full-scaled instruments for evaluation of experimental skin reactions.

II. Allergic and Irritant Skin Reactions

Experimental skin reactions can be elicited either by allergens or by an irritant trauma to the skin. Application of allergens or irritants as patch tests has been a commonly used method for experimental skin studies and accordingly patch test reactions will be discussed in detail.

A. Allergic Skin Reactions

An allergic (type 4 allergy) skin reaction is an immunological process depending on the activation of specifically sensitized T-cells. The reaction requires that a preceding sensitization has taken place. If patch tests are used patches are normally left on the skin for 48 hours for allergy testing. Evaluation of the skin response is usually performed 48 and 72 hours after application, as well as after 1 week. Recording of skin reactions to patch tests as recommended by the International Contact Dermatitis Research Group is described in Table 1.

Reading of skin reactions requires some experience. The main features are erythema, infiltration, papules, and vesicles. Allergic reactions often show a delayed response with maximum at 72 hours after application or even later, i.e., for neomycin a maximum reaction after 1 week is not unusual. Testing with an allergen of relevance may cause exacerbation of skin symptoms in skin areas previously involved with contact dermatitis, and consequently further confirm the relevance of the response.

B. Irritant Skin Reactions

In experimental studies primary irritants are frequently used, but irritant reactions may also be induced by UV-light, cold, heat, or friction. A primary irritant is a substance which can damage the skin by a direct cytotoxic action without preceding sensitization.[2] The skin response to an irritant depends upon the chemical structure of the irritant. Wrinkling, papules in a follicular distribution, and bullae are generally interpreted as irritant reactions. Traditionally, allergic and irritant reactions are differentiated by the more sharp demarcation of the irritant reaction, which only rarely affects the skin outside the exposed patch test area. Irritant reactions will often be more acute, the maximum response being immediately after the patch is removed. Differentiation between irritant and allergic reactions may, however, sometimes cause difficulties, and exceptions from the above broadly outlined rules are plenty. Sodium lauryl sulfate (SLS) and anthralin are examples of irritants eliciting delayed skin reactions.[3] Benzalkonium chloride is an example of an irritant which may elicit reactions clinically very similar to allergic reactions. In a blinded study using replica technique it was possible to differentiate between epidermal changes caused by six different irritants.[4] Clinical interpretation

TABLE 1 Recording of Allergic Patch Test Reactions According to the International Contact Dermatitis Research Group

?+	Doubtful reaction; faint erythema only
+	Weak positive reaction; erythema, infiltration, possibly papules
++	Strong positive reaction; erythema, infiltration, papules, and vesicles
+++	Extreme positive reaction; intense erythema, infiltration, and coalescing vesicles

of the skin response is today the only generally available method for differentiation between allergic and irritant reactions.

III. Reproducibility of Patch Test Reactions

A great variation in skin response to irritants between individuals is a well-known fact. Within the same individual variation in skin response to identical patch tests to irritants or allergens may be considerable.[5-7] Dahl et al. found that, for simultaneous patch testing with SLS, corresponding sites on the right and the left side were scored identically in only 53% of cases.[8] In patch tests with a series of 39 substances performed on 41 patients on one or both sides of the back Gollhausen et al.[6] found that 44% of positive reactions were nonreproducible at concomitant testing, and 40% were nonreproducible at sequential testing with an interval of 1 week. Reproducibility of allergic patch test reactions to lanolin was recently studied.[9] Patients with positive patch test response to lanolin within the previous 5 years were retested, and 41% demonstrated persistance of the positive patch test.

The current skin susceptibility to allergens or irritants is influenced by a number of individual- and environment-related variables discussed in the following.

IV. Variation in Skin Susceptibility

A. Influence of Gender and Menstrual Cycle

Hand eczema is known to occur more frequently among women than among men.[10,11] Most experimental studies have found no sex relation in skin susceptibility.[12-14] The difference possibly reflects dissimilarities in exposure to environmental hazards rather than endogenous differences between the men and women.

For women, increased reactivity of the skin prior to and during the menstrual phase was reported in an experimental study.[15] Increased reactivity to SLS at day 1 in the menstrual cycle as compared to days 9 to 11 was found. The clinical implications of this finding are not yet clarified.

B. Influence of Age

Increased susceptibility to irritants in childhood[16] and increased susceptibility to SLS in young compared to elderly females[17] has been reported. Within the age range 18 to 50 years no significant influence of age on skin susceptibility should be expected.[18]

C. Influence of Test Region

Within the same individual susceptibility to exogenous irritant stimuli differs between anatomical regions. Skin susceptibility to irritants is usually ranked as leg < forearm < back < forehead.[16] Regional variation in skin susceptibility may exist, even within narrow anatomical areas, and influence the response.[19]

D. Influence of Skin Color

Blue eyes and fair skin were reported to correlate with the intensity of the inflammatory response to an irritant stimulus[20] and in Caucasians cutaneous sensitivity to UV light and to seven different chemical irritants was reported to correlate positively.[21] In a study measuring skin color by a tristimulus colorimeter, an association between fair skin and increased susceptibility to SLS was found.[18] This finding supports the hypothesis of an association between skin color and skin susceptibility. Tanning may influence the susceptibility to irritants. An increased blister formation time, indicating a decreased skin susceptibility, after exposure to ammonium hydroxide has been demonstrated in tanned skin.[22] A diminished reaction to SLS after UVB exposure was reported.[23]

E. Influence of Skin Barrier Function

Evaporimetry is a method used for measurement of transepidermal water loss (TEWL). TEWL is the passive diffusion of water through the stratum corneum, and is accepted as a practical indication of the skin barrier toward water. The use of individual baseline TEWL to predict the skin response to surfactants was suggested by Murahata et al.[24] This relationship between skin susceptibility to detergents and high baseline TEWL has later been further supported.[14,18] Baseline TEWL is important for the current skin response to a number of substances, although the correlation between baseline TEWL and susceptibility to development of irritant contact dermatitis still needs confirmation by long-term clinical studies.

F. Influence of Season

Seasonal variation in skin susceptibility is a clinical observation, adequately supported by experimental data. Significantly increased skin response to DMSO was found in the winter compared to the summer,[16] and the same variation was demonstrated for propylene glycol.[25,26] Soap testing has greatest sensitivity in the winter, when the damaging action of soap is at its peak.[27] A statistically significant seasonal variation in healthy volunteers exposed to SLS was reported.[28] A significantly lower hydration state of the skin during winter than during summer was demonstrated in the same study.

G. Influence of Pathophysiological Conditions

The significance of a history of atopic dermatitis for the development of irritant eczema is well documented.[11,29,30]

When patch tested with SLS in a quiet phase of the disease, patients with atopic dermatitis were found to react more severely than healthy controls as assessed by measurement of TEWL.[31,32]

Skin susceptibility in uninvolved skin in patients with hand eczema was originally thoroughly studied by Björnberg.[12] He found that skin reactivity to primary irritants was not increased in patients with healed eczema when tested at skin distant from the previously affected skin area.[12] van der Valk et al. found no difference in skin response to SLS in subjects with healed contact dermatitis, as compared to controls.[31] No increased skin reactivity to SLS in patients with chronic or healed eczema was found reported in a study including 39 patients with hand eczema and matched controls,[33] but patients with acute eczema showed an increased skin reactivity to SLS compared to controls.

The phenomenon of dermatitis in a localized area of the body resulting in generalized hyperirritability of the skin, including areas distant from the eczema, is well known, and patch testing should not be performed when an active eczema is present.[34] The exact etiology of hyperreactive skin remains unknown.

Apart from the presence of an acute eczema, other internal factors are also known to influence skin reactivity. Skin response may be influenced by medicaments such as prednisolone, and possibly by severe chronic illness.[35] Hardening, either due to immunological desensitization or to increase in the thickness of the stratum corneum after exposure to UVB or irritants, is also a possible modulator of skin irritation.[36]

V. Recommendations

Although clinical recording of experimental test reactions may seem simple, a skilled observer is necessary to grade the reactions correctly. Interpretation of test results calls for some imagination and knowledge about factors that influence skin susceptibility and may lead to false-positive or false-negative reactions. Age, gender, and biophysical properties of the test region, including skin color and seasonal variations, should be considered. Atopy, current eczema, drugs, and present acute or chronic illnesses should also be taken into account, all of these factors being bricks in the puzzle of skin susceptibility. When unexpected results occur, retesting may be helpful.

References

1. Berardesca, E. and Maibach, H.I., Bioengineering and the patch test, *Contact Dermatitis*, 18, 3, 1988a.
2. Kligman, A.M. and Wooding, W.M., A method for the measurement and evaluation of irritants on human skin, *J. Invest. Dermatol.*, 49, 78, 1967.
3. Bruynzeel, D.P., van Ketel, W.G., Scheper, R.J., and von Blomberg-van der Flier, B.M.E., Delayed time course of irritation by sodium lauryl sulphate: observations on treshold reactions, *Contact Dermatits*, 8, 236, 1982.
4. Agner, T. and Serup, J., Skin reaction to experimental irritants assessed by polysulfide rubber replica, *Contact Dermatitis*, 17, 205, 1987.
5. Freeman, S. and Maibach, H.I., Study of irritant contact dermatitis produced by repeat patch testing with sodium lauryl sulphate and assessed by visual methods, transepidermal water loss and laser Doppler velocimetry, *J. Am. Acad. Dermatol.*, 19, 496, 1988.
6. Gollhausen, R., Przybilla, B., and Ring, J., Reproducibility of patch tests, *J. Am. Acad. Dermatol.*, 21, 1196, 1989.
7. Agner, T. and Serup, J., Individual and instrumental variations in irritant patch-test reactions — clinical evaluation and quantification by bioengineering methods, *Clin. Exp. Dermatol.*, 15, 29, 1990.
8. Dahl, M.V., Pass, F., and Trancik, R.J., Sodium lauryl sulfate irritant patch tests. II. Variation of test responses among subjects and comparison to variations of allergic responses elicited by Toxicodendron extract, *J. Am. Acad. Dermatol.*, 11, 474, 1984.
9. Carmichael, A.J., Foulds, I.S., and Bransbury, D.S., Loss of lanolin patch-test positivity, *Br. J. Dermatol.*, 125, 573, 1991.
10. Rystedt, I., Factors influencing the occurrence of hand eczema in adults with a history of atopic dermatitis in childhood, *Contact Dermatitis*, 12, 185, 1985.
11. Meding, B. and Swanbeck, G., Epidemiology of different types of hand eczema in an industrial city, *Acta Dermatol. Venereol.*, (Stoc) 69, 227, 1989.
12. Björnberg, A., *Skin Reactions to Primary Irritants in Patients with Hand Eczema*, Isacson, Göteborg, 1968.
13. Lammintausta, K., Maibach, H.I., and Wilson, D., Irritant reactivity in males and females, *Contact Dermatitis*, 17, 276, 1987.
14. Tupker, R.A., Coenraads, P.J., Pinnagoda, J., and Nater, J.P., Baseline transepidermal water loss (TEWL) as a prediction of susceptibility to sodium lauryl sulphate, *Contact Dermatitis*, 20, 265, 1989.
15. Agner, T., Damm, P., and Skouby, S.O., Menstrual cycle and skin reactivity, *J. Am. Acad. Dermatol.*, 24, 566, 1991.
16. Frosch, P.J., *Hautirritation und empfindliche Haut*, Grosse Verlag, Berlin, 1985.
17. Cua, A.B., Wilhelm, K.P., and Maibach, H.I., Cutaneous sodium lauryl sulphate irritation potential: age and regional variability, *Contact Dermatitis*, 23, 276, 1990.
18. Agner, T., Basal transepidermal water loss, skin thickness, skin blood flow and skin colour in relation to sodium-lauryl-sulphate-induced irritation in normal skin, *Contact Dermatitis*, 25, 108, 1991.
19. Flannigan, S.A., Smith, R.E., McGovern, J.P., Intraregional variation between contact irritant patch test sites, *Contact Dermatitis*, 10, 123, 1984.

20. Björnberg, A., Löwhagen, G., and Tengberg, J., Relationship between intensities of skin test reactions to glass-fibres and chemical irritants, *Contact Dermatitis,* 5, 171, 1979.
21. Frosch, P.J. and Wissing, C., Cutaneous sensitivity to ultraviolet light and chemical irritants, *Arch. Dermatol. Res.,* 272, 269, 1982.
22. Frosch, P.J. and Kligman, A.M., Rapid blister formation in human skin with ammonium hydroxide, *Br. J. Dermatol.,* 96, 461, 1977.
23. Larmi, E., Lahti, A., and Hannuksela, M., Effect of ultraviolet B on nonimmunologic contact reactions induced by dimethyl sulfoxide, phenol and sodium lauryl sulphate, *Photodermatology,* 6, 258, 1989.
24. Murahata, R., Crove, D.M., and Roheim, J.R., The use of transepidermal water loss to measure and predict the irritation response to surfactants, *Int. J. Cosmet. Sci.,* 8, 225, 1986.
25. Warshaw, T.G. and Hermann, F., Studies on skin reactions to propylenglycol, *J. Invest. Dermatol.,* 19, 423, 1952.
26. Hannuksela, M., Pirilä, V., and Salo, O.P., Skin reactions to propylene glycol, *Contact Dermatitis,* 1, 112, 1975.
27. Frosch, P.J. and Kligman, A.M., The soap chamber test, *J. Am. Acad. Dermatol.,* 1, 35, 1979.
28. Agner, T. and Serup, J., Seasonal variation of skin resistance to irritants, *Br. J. Dermatol.,* 121, 323, 1989.
29. Rystedt, I., Work-related hand eczema in atopics, *Contact Dermatitis* 12, 164, 1985.
30. Rystedt, I., Atopic background in patients with occupational hand eczema, *Contact Dermatitis,* 12, 247, 1985.
31. van der Valk, P.G.M., Nater, J.P., and Bleumink, E., Vulnerability of the skin to surfactants in different groups of eczema patients and controls as measured by water vapour loss, *Clin. Exp. Dermatol.,* 10, 98, 1985.
32. Agner, T., Susceptibility of atopic dermatitis patients to irritant dermatitis caused by sodium lauryl sulphate, *Acta Dermatol. Venereol.,* 71, 296, 1991.
33. Agner, T., Skin susceptibility in uninvolved skin of hand eczema patients and healthy controls, *Br. J. Dermatol.,* 125, 140, 1991.
34. Roper, S. and Jones, H.E., A new look at conditioned hyperirritability, *J. Am. Acad. Dermatol.,* 7, 643, 1982.
35. Johnson, M.W., Maibach, H.I., and Salmon, S.E., Skin reactivity in patients with cancer, *N. Engl. J. Med.,* 284, 1255, 1971.
36. Elsner, P., Widmer, J., and Burg, G., Dynamics of postirritant hyposensitivity following experimental cumulative irritant dermatitis assessed by noninvasive bioengineering methods, abstract, First Congress of the European Society of Contact Dermatitis, Brussels, October 8–10, 1992.

30.0

Evaluation of Lesional Extension

30.1 Measurement of Healing and Area of Skin Pathologies 581
S. Sarin

Chapter 30.1
Measurement of Healing and Area of Skin Pathologies

Sanjeev Sarin
Department of Surgery
University College and Middlesex School of Medicine
The Middlesex Hospital
London, U.K.

I. Introduction

The accurate measurement of healing in skin defects is fundamental to assess treatment and to research into the rate of healing. Though there are numerous skin pathologies in which the area can be measured, the most common application is in ulcers of the lower limb. The literature is awash with trials comparing the effect of various topical preparations on the rate of ulcer healing (measured as changes in area), though few have considered whether rate of change of area is an appropriate measure. Measurement of the rate of change of area is logistically attractive as it can be assessed within a few weeks, thereby considerably reducing the time spent on data collection. Measurement of the rate of change of area relies on the presupposition that the rate of change in ulcer area is a good predictor of the time that the ulcer will take to heal. This assumes that the rate of healing of ulcers is uniform and, therefore, the ulcer showing greater percentage reduction in area over a few weeks of treatment will be quicker to heal. This supposition may be incorrect. Stacey et al. have shown that the individual healing curves of ulcers vary considerably, with some ulcers actually getting larger before starting to heal.[1] They were unable to show the predictive value of the rate of reduction of ulcer size during 1-month sample periods beginning when the patient first attended. They suggested that other variables were also likely to influence the rate of healing and showed that the size of the ulcer at presentation affected healing rates.[1] Bulstrode et al. suggested that the effectiveness of treatment prior to entry into a trial was also important with "clean" ulcers healing faster than those that first required cleansing.[2] In an attempt to reduce the uncontrolled variables at first presentation, Bulstrode et al. constructed healing curves with the point when healing was complete, taken as the reference. A standardized healing curve was then constructed for 7 weekly points prior to the reference point by averaging the measured areas of 30 ulcers that had healed.[2] They found that the resultant standardized curve had three phases: in the first there was either no healing or even an increase in the size of the ulcer; in the second, there was rapid healing; in the third, progressively slower healing until the skin defect finally closed. They found that the rate of healing (percent change in area) in the 5th week prior to complete healing correlated highly with the time to complete healing. However, the relevant time span during which the rate of healing can be used as a predictor of time to complete healing can only be known after the ulcer has healed because the point at which an individual patient presents within the evolution of the ulcer cannot be determined. Therefore, though this is an attractive model of wound healing, it would appear to have little clinical application in reducing the length of time taken for data collection in a prospective trial. Stacey et al. have suggested that the only alternative is to measure the time to complete healing from when the patient is first entered into the study.[1] However, some ulcers do not heal for many months; so, in practice, it would be appropriate to report the number of healed ulcers over a preset time interval. Such an approach would both restrict the length of the study and allow valid comparisons between different treatment groups.

In view of these difficulties, careful thought must be given to whether rate of change of area or time to complete healing is chosen in the design of the study. Second, it is necessary to limit the effects of other variables by accurately matching the treatment groups. Therefore, the groups should be similar in primary pathology, age and sex of patient, ulcer size, and "cleanliness" of ulcer bed at the beginning of treatment. In the case of patients with venous ulceration, the degree of calf pump dysfunction and mobility of the patient also are likely to influence the results. Such stringent matching criteria are essential if type II errors are to be avoided and the results of the study useful.

II. Methods Available

Measurement of such skin lesions is complicated by the fact that skin is not flat, and any measurement technique has to allow for all three dimensions of an irregular surface. The measurement technique should also be reproducible and accurately reflect the clinical stage of healing. It is obvious that the errors in measurement have to be considerably less than the changes being measured. Finally, ease of measurement, discomfort to the patient, and cost are important considerations.

One of the common problems encountered when measuring the area of a skin lesion is defining the point where the abnormal skin meets normal. In order to be reproducible, the "edge" of the lesion has to be consistently identifiable. This problem assumes enormous importance in those pathologies where the skin lesion gradually and indiscernibly fades away to meet normal skin. A typical example is lipodermatosclerosis, where it is usually impossible to reliably determine the edge of a lesion, even on histological section. Alternatively, there are skin lesions in which the edge is quite distinct and ulcers are an example in point. Bulstrode et al. macroscopically marked the edges of six ulcers (in pig's skin) with india ink and found on histological section that they had identified the epithelial margins to within a mean of 240 µm.[3] Certainly, such a small error is unlikely to be important in large ulcers, but will assume greater significance as the ulcer size reduces.

Measurement of a skin defect can be achieved using indirect, direct, or other methods.

A. Indirect Methods

Indirect methods rely on making measurements on a scaled model of the ulcer and two methods are commonly used. In the first, a transparent acetate sheet is laid around the limb in as close contact with the leg as possible. The margin of the ulcer is then traced with a pen.[3] Care must be taken to ensure that the area traced is accurate and reproducible. The alternative is to photograph the ulcer, with a ruler in the same plane, and then project the slide at a distance where the scale on the photographed ruler matches exactly the scale on the projection screen.[3,4] The projected ulcer is then traced onto a transparent acetate screen and measurements made. The advantage of the first method is that it is quick and inexpensive. However, it does involve placing the acetate sheet in direct contact with the limb/ulcer and may predispose to further contamination of the ulcer bed and, because the tracing is done directly on the limb, accuracy may be compromised by movement or patient discomfort. The photographing technique, on the other hand, allows the tracing of ulcer size to be undertaken at the observer's convenience, without contamination of the ulcer bed or discomfort to the patient. Disadvantages include additional time, equipment, and cost. Second, because photographs are two-dimensional, it may be difficult to accurately assess circumferential ulcers.

Once the ulcer trace has been obtained, there are a number of different methods of measuring the area. The ulcer outline can be cut out of the acetate sheet and weighed. Comparison with the weight of a known area of acetate sheet allows the ulcer area to be calculated.[5,6] Alternatively, the acetate sheet can be laid onto graph paper and the number of squares within the ulcer outline counted.[7] Both techniques are time consuming and tedious. The area can also be measured using a compensating plannimeter, though there are little data available as to the reproducibility or accuracy of this mechanical device.[4,8] Coleridge Smith and Scurr have used a digitizing pad of an area measuring device (MOP, Kontron, Germany) and found a very good correlation with the weighing technique.[9] Another method of digitizing the ulcer trace is described by Stacey et al. (Graf/Bar Mark II, Science Accessories Corp.).[1] The apparatus consists of two arrays of microphones perpendicular to each other. A stylus containing an ultrasonic pulse generator is used to outline the ulcer trace. The position of the stylus is accurately located by the microphones, the data fed into a computer, and the ulcer area derived. Digitizing the data has obvious advantages in that the area can be outlined and derived relatively quickly but is dependent on having the necessary equipment and expertise available.

B. Direct Methods

The theoretical advantage of direct methods of measurement is that there are no intermediate steps. Avoidance of multiple processes will save time and may make the measurement more accurate.

1. Sonic Digitizer

Coleridge Smith and Scurr used a sonic digitizer (Graf/Bar, Science Accessories Corp.) to directly measure 38 venous ulcers and found that this method correlated well with indirect methods of measurement.[9] A stylus with an ultrasonic impulse generator is pointed at the edge of the ulcer and the impulse generated, picked up by two microphones. The tip of the stylus is thus located to within 0.1 mm and the cartesian coordinates transmitted to a microcomputer. The procedure has to be repeated in 5- to 25-mm steps along the entire edge of the ulcer and thus a digitized trace of the ulcer outline obtained. The ulcer area can be then computed. They found that the technique was dependent on correct positioning of the stylus and that larger ulcers had to be digitized in smaller sections. Providing care was taken to ensure correct positioning, they found that the technique was accurate, did not involve any contact with the ulcer, and took approximately 1 min.

2. Stereophotogrammetry
Bulstrode et al. have described a technique of measurement of both area and volume with an accuracy (a measure of the bias of the system) and precision (a measure reflecting reproducibility of the technique) of less than 5%.[3] They used a stereo camera and then projected the polarized image onto a rotating, roughened aluminum disc, thereby producing a smooth three-dimensional image. A "floating" light spot was used to identify the x, y, and z coordinates of the ulcer edge and base which were then directly entered into a microcomputer. The entire ulcer area was then plotted with measurements taken at 2-mm intervals. In this way the area of the defect could be calculated. Second, using measurements on the normal skin adjacent to the ulcer, they derived an estimate of the surface of the limb as if the ulcer was not present. Subtracting the "true" depth coordinate from the "estimated surface" coordinate of the limb allowed them to calculate the volume of the defect present. They verified the area measurement, both on ulcers drawn on artificial legs and in the clinical setting, by comparison with direct tracing on acetate sheets and scaled photography. They found that their system was more accurate and more reproducible than either of the other two methods, particularly when the ulcer areas were small (<100 mm^2). However, in their report, direct tracing and scaled photography were very inaccurate and variable. In other hands these two techniques may be more reproducible. Additionally, the equipment is cumbersome, the technique, time consuming (taking approximately 30 min to complete all measurements), and requires extensive training.

C. Other Methods
If the skin defect is assumed to be elliptical in shape, the area can be derived by multiplying the product of the two maximal perpendicular diameters by π and dividing by 4. Stacey et al. found that the area measured by this technique was approximately the same as the area derived by more sophisticated digitizing methods and, importantly, reflected the progress of ulcer healing.[1] In fact, even the product of the two maximal perpendicular diameters has been shown to be highly correlated with the actual area of irregular ulcers, even though the ulcers were not elliptical.[1] This is the simplest technique described, needing only a ruler in the way of equipment, and has the advantages of speed and minimal discomfort for the patient. However, its disadvantages are that it does not measure the actual area, only a representation of it, and may be quite inaccurate when the defect being measured is very irregular.

III. Correlation Between Methods

Correlation between all methods of measurement of area of skin defects is high. Coleridge Smith and Scurr found that the indirect techniques of weighing the ulcer tracings or digitized area measurements correlated very well with direct measurement of the area using a sonic digitizer ($r = 0.98$ and 0.99).[9] Stacey et al. have demonstrated that the product of the two maximum perpendicular diameters correlates well with digitized measurement of ulcer area from a scaled photograph ($r = 1.033$).[1] However, neither report details the variability of the measurement technique. In both, means of three measurements were used. Bulstrode et al. suggested that, though area measurements from direct tracing and scaled photographs were well correlated, repeated measurements showed increasing variability as the size of the ulcer decreased.[3] They found that stereophotogrammetry was between five and ten times more accurate than either of the other two methods and was less variable. Thomas and Wysocki compared digitizing of scaled photographs and wound tracings, and found the acetate tracing to be more accurate.[10]

IV. Recommendations

There are many ways of measuring areas in dermatological conditions and some of the most commonly used techniques have been described. The author's own preference is for tracing the area onto a transparent acetate sheet followed by digitized analysis. In our hands, the variation between repeated measurement of the same ulcer is less than 10% using this technique, which we feel is acceptable. However, the researcher must make his or her own decision as to which technique to use. The most important point to be considered is whether change in the area of the skin defect is a useful measure. Once that is decided, cost, availability of equipment and time, and finally discomfort to the patient are further considerations. As all the techniques described are entirely observer dependent, prior to beginning the study, an assessment of the observer variation must be undertaken and reported in the results. Finally, great care must be given to study design, in order to ensure that the true results of the study are not masked by the use of inappropriate controls.

References

1. Stacey, M.C., Burnand, K.G., Layer, G.T., Pattison, M., and Browse, N.L., Measurement of the healing of venous ulcers, *Austr. N.Z. J. Surg.,* 61, 844, 1991.
2. Bulstrode, C.J.K., Goode, A.W., and Scott, P.J., Measurement and prediction of progress in delayed wound healing, *J. R. Soc. Med.,* 80, 210, 1987.
3. Bulstrode, C.J.K., Goode, A.W., and Scott, P.J., Stereophotogrammetry for measuring rates of cutaneous healing: a comparison with conventional techniques, *Clin. Sci.,* 71, 437, 1986.
4. Myers, M.B. and Cherry, G., Zinc and the healing of chronic leg ulcers, *Am. J. Surg.,* 120, 77, 1970.
5. Burnand, K.G., Clemenson, G., Morland, M., Jarret, P.E.M., and Browse, N.L., Venous lipodermatosclerosis: treatment by fibrinolytic enhancement and elastic compression, *Br. Med. J.,* 280, 7, 1980.
6. Rundle, J.S.H., Elton, R.A., Cameron, S.H., Watson, N., and Ruckley, C.V., Porcine dermis in varicose ulcers — a clinical trial, *Vasa,* 10, 246, 1981.
7. Gowland Hopkins, N.F. and Jamieson, C.W., Antibiotic concentration in the exudate of venous ulcers: the prediction of ulcer healing rate, *Br. J. Surg.,* 70, 532, 1983.
8. Ormiston, M.C., Seymour, M.T.J., Venn, G.E., Cohen, R.I., and Fox, J.A., Controlled trial of Iodosorb in chronic venous ulcers, *Br. Med. J.,* 291, 308, 1985.
9. Coleridge Smith, P.D. and Scurr, J.H., A direct method for measuring venous ulcers, *Br. J. Surg.,* 73, 320, 1986.
10. Thomas, A.C. and Wysocki, A.B., The healing wound: a comparison of three clinically useful methods of measurement, *Decubitus,* 3, 18, 1990.

Section M: Experimental Test Procedures

31.0

Standard Testing of Skin Reactivity

31.1 Irritant Patch Test Techniques 587
P.J. Frosch and B. Pilz

31.2 Allergic Patch Test Techniques 593
T. Fischer

31.3 Type I Allergy Skin Tests 607
L.J. Petersen

31.4 Ultraviolet Radiation Dosimetry 619
B.L. Diffey

31.5 Phototesting: Phototoxicity and Photoallergy 627
T. Horio

Chapter 31.1
Irritant Patch Test Techniques

Peter J. Frosch and Beate Pilz
Department of Dermatology
Städtische Kliniken Dortmund
University of Witten Herdecke
Dortmund, Germany

I. Introduction

Test techniques for the allergic skin reactivity have been fairly standardized over recent years and found general acceptance (see following chapter). In contrast the subject of *irritant patch test techniques* is very heterogenous and lacks standardization. The reasons are manifold: (1) The number of irritants is even greater than the number of allergens and ranges from caustic acids and alkalis over moderate irritants (most surfactants) to very mild ones (water, various oils, etc.). (2) The irritant response of human skin does not only depend on the irritant itself but also on a number of exogenous and endogenous factors. (3) The requirements for safety testing of agents coming into contact with the skin are different and depend on the type of product and the intended usage.

In this chapter, a survey on important techniques will be given. These are used more or less frequently in various laboratories all over the world and have proven to be useful in answering certain questions mainly regarding safety testing. The list of described techniques is not complete but has on the other hand the advantage that the authors have gathered extensive experience with these.

II. Exogenous and Endogenous Factors Determining the Irritant Response

The intensity of the cutaneous reaction depends not only on the inherent toxicity of the agent but also on the volume, concentration, mode of application, and exposure time (Table 1). Furthermore, the type of skin the irritant is applied to plays an eminent role: diseased skin with a broken stratum corneum barrier will react stronger than normal skin; regional variations in the same individual are wide with high sensitivity in the face and intertriginous areas and low sensitivity in palmoplantar locations; apart from large interindividual differences in the same race there is also weak evidence for a different suspectibility to chemical irritants of black vs. white skin. For further details the reader is referred to a recent review.[1]

III. Open Tests

Table 2 provides a summary of major tests where the irritant is applied for a rather short time (seconds until minutes) using various devices. The main goal of the investigators who developed these techniques was the determination of the individual's sensitivity to this and possibly other irritants.[2] The best known is the alkali resistance test of Burckhardt.[3] One drop of 0.5 N NaOH is applied to ventral forearm skin under a glass block for a total of 30 min with renewal of the irritant after 10 and 20 min. The alkali resistance is considered to be "normal" if erosions do not develop before the 20-min exposure ("reduced" after 10 min, "increased" after 30-min exposure). The method has been refined by Locher[4] with nitrazine yellow as an indicator for early follicular erosions. This technique is helpful in differentiating a panel of subjects in regard to its sensitivity to sodium hydroxide. The test is also strongly positive on diseased skin, e.g., eczema, when the statum corneum barrier is reduced. It is mainly a measure of the stratum corneum integrity. The test cannot reliably predict the reactivity of an individual to another irritant, especially if it is chemically very different and other pathomechanisms are involved (e.g., organic solvents).[5] Frosch and Wissing[6] found a significant correlation between the alkali resistance test and the reaction to DMSO (5-min exposure), sodium lauryl sulfate, benzalkonium chloride, and croton oil (with chamber test); there was no significant correlation with the response to kerosene and skin type regarding UVB sensitivity. The alkali resistance test has not fulfilled the expectations in predicting the risk for developing hand eczema.[7] Based on current insights into the multifactorial pathogenesis of hand eczema with atopy as a major endogenous factor, this is not surprising.

Basically, this holds true for the other open tests of Table 2. They are listed because they have proven to

TABLE 1 Exogenous and Endogenous Factors Determining the Irritant Response

Exogenous Factors
Type of irritant (chemical structure, pH)
Amount of irritant penetrating (solubility, vehicle, concentration, method and time of application)
Body site
Body temperature
Mechanical factors (pressure, friction, abrasion)
Climatic conditions (temperature, humidity, wind)

Endogenous Factors
Individual susceptibility to irritants
Primary hyperirritable ("sensitive") skin
Atopy (particularly atopic dermatitis)
Inability to develop hardening
Secondary hyperirritability (status eczematicus)
Racial factors
Age
Sensitivity to UV light

Adapted from Frosch, P.J., Cutaneous Irritation, in Textbook of Contact Dermatitis, *Rycroft, R.J.G., Menné, T., Frosch, P.J., and Benezra, C., Springer-Verlag, Berlin, 1992, 28.*

provide useful reproducible results under certain experimental conditions. The principle of these techniques can be used to study new chemicals, if nothing is known about its irritant nature in humans. *In vitro* and animal tests should precede the applications in human volunteers of cause. If a strong irritant is suspected the "drop test" according to Burckhardt and Schmidt[3] can be chosen. A more accurate exposure is provided with chambers or cups made of glass or plastic. These actually do not present open applications anymore because the irritant is applied in an excessive amount for a short time period (seconds or minutes).

The reader is referred to the detailed publications of the DMSO test and the ammonium hydroxide blistering technique.[8-10] The former is more practical because the DMSO wheal is not as injurious as a superficial blister. Furthermore, the response to DMSO can be quantified by measuring the transepidermal water loss. There is a good correlation between the whealing response and the increase of TEWL. Sensitive skin with a weak stratum corneum barrier usually shows a strong reaction to DMSO. In a panel of 44 subjects a significant correlation was found to other irritants such as sodium lauryl sulfate, benzalkonium chloride, croton oil, and kerosene.[6] More work should be done with this technique combining with modern biophysical techniques in order to delineate its value in the clinical setting and in product safety testing.

IV. Occlusive Tests

A series of occlusive tests has been developed for product safety testing, mainly cosmetics but also topical drugs applied to the skin for prolonged time periods on both normal and diseased skin. Animal tests such as that of Draize[11] have been frequently proven to be too insensitive for mild irritants. Only a strong and moderate irritant can be detected when they produce an irritant reaction after a 24-hour occlusive exposure to intact and scarified skin of albino guinea pigs. There are first positive results due to an increased sensitivity of the guinea pig to certain irritants. Dissatisfied with the results of this technique and other animal tests, Kligman and Wooding[12] were the first to standardize a repetitive patch test technique for human volunteers. The test material is applied occlusively with gauze and occlusive tape to the back of a minimum of ten persons. For a weak irritant the exposure is continued for 10 and a maximum of 20 days using various concentrations up to the undiluted form of the irritant. The ID 50 is defined as the concentration which has produced a visible irritation in 50% of the test panel. The IT 50 is the number of days of continued exposure which has caused a threshold reaction in 50% of the tested persons.

Lanman et al.[13] have developed the *21-day cumulative irritation test*. The material is exposed occlusively on 20 subjects over 21 days. The intensity of the cutaneous irritant reaction is scored on a 5-point scale. The exposure is discontinued if a strong (4+) reaction occurs. The daily recorded scores when the material is renewed are accumulated over 21 days. This test has proved to be quite sensitive for even mild irritants. A positive (irritating) control and a negative control should always be used as a reference.[14] In most countries health authorities require this test before releasing a new drug which is in contact with the skin over a prolonged time.

Many mostly unpublished studies show a good correlation between field observations where the product is used in the intended way and the 21-day cumulative irritation test. It is a rare event that a product which is completely negative in the Lanman test produces a considerable number of complaints under regular usage conditions.

The *24-hour occlusive patch test* is frequently performed as the first orientation after *in vitro* tests and animal studies suggest a low toxicity. The test is frequently performed on volunteers or on patients in dermatological clinics. However, the interpretation of a negative nonirritant response must be done with great caution. Only strong and moderate irritants can be identified in this way. Mild irritants cannot be identified with this test.

Based on the concept of Kligman and Wooding, Willis et al.[15] recently described a method with a series of irritants and a 2-day occlusive exposure with small Finn chambers. The following irritants and concentration are used:

TABLE 2 Various Methods with Open (Nonocclusive) Application of the Irritant

Test (Ref.)	Method	Interpretation of Result
"Drop test" (3)	Number of drops until visible irritation occurs (various strong irritants)	Aimed at detecting differences in reactivity; also suitable for new irritants to determine irritating potential
Alkali neutralization test (4)	n/80 NaOH Phenolphthalein	Developed to recognize subjects with increased hand eczema risk
Alkali resistance test	Erosions occurring by n/2 NaOH after 10, 20, 30 min	Primarily developed to detect individual differences; also suitable to detect damaged skin; primarily a measure of stratum corneum barrier
Acid resistance (22)	5 N HCl	Same as for alkali resistance test
Resistance to organic solvents (23)	Threshold for irritation (xylene, benzene, petrol, etc., 5–30 min)	Great interindividual and regional differences in all following test methods
"Cup test" (24)	Glass cups, KOH, and detergents	
Pain threshold (25)	Chloroform-ethanol (1:1), 6-cm^2 exposure area	
DMSO (8)	DMSO (90%, 95%, 100%), applied for 5 min, plastic block	
DMSO/TEWL (10)	DMSO test combined with measurement of transepidermal water loss	
Ammonium hydroxide (9)	Determination of minimal blistering time of ammonium hydroxide (1:1 aqueous dilution)	

Note: The goals for development were different, but most of them focused on determination of the individual's skin sensitivity to this particular irritant. Adapted from Frosch, P.J., Hautirritation und empfindliche Haut, Grosse Verlag, Berlin, 1985, 7.

Benzalkonium chloride 0.5%, 1% aq.
Sodium lauryl sulfate 1%, 2%, 5% aq.
Croton oil 0.6%, 0.8%, 1% in yellow soft paraffin
Dithranol 0.02% in yellow soft paraffin
Nonanoic acid 40%, 60%, 70%, 80% in propan-1-ol
Propylene glycol 50%, 100% aq.
Sodium hydroxide 2%, 3%, 5% aq.

They also described a detailed scoring system:

-    no visible reaction
- +/− faint, patchy erythema
- 1+ erythema with edema
- 2+ erythema with edema and papules
- 3+ erythema with edema and vesicles
- 4+ intense erythema with bullous formation

With this method Willis et al. found the following thresholds for each irritant so as to produce mild or moderate reactions (no greater than 2+) in as close to 75% of individuals tested:

Benzalkonium chloride 0.5%
Sodium lauryl sulfate 5%
Croton oil 0.8%
Dithranol 0.02%
Nonanoic acid 80%
Propylene glycol 100%
Sodium hydroxide 2%

These threshold values for the used irritants may vary even on the same test panel if tested at different times. Therefore, the absolute value is limited. However, this is a good approach to obtain more accurate data on cutaneous irritability. The objectivity can be even more increased with the modern biophysical techniques available (e.g., TEWL, BFV, etc.).

In the past it was extremely difficult to assess the irritancy of soaps and synthetic detergents. Frosch and Kligman[16] have developed the *soap chamber test* which has stood the test of time and is now considered to be a standard test for these types of irritants. The basic principal is the repetitive exposure of rather high concentrations of the soap/detergent (4 to 8%, concentration to be determined by preliminary testing) for a 5-day period. The exposure starts with 24 hours on the first day in order to breach the barrier and increase penetration. On the following days the material is only applied for 6 hours. Final readings are taken on the 7th day using a 4-point scale for erythema, scaling, and fissures. The results of the chamber test can be validated with wash tests in the antecubital space — a good correlation has been found in general. Large differences between the irritancy of soaps and detergents have been described with various commercial products. Synthetic detergent bars containing the surfactant group of isethionates or condesation products of fatty acids showed a very good skin tolerance, whereas detergent bars with a high content of sodium lauryl sulfate were very irritating. Liquid detergents with alkylether sulfates or sulfosuccinates have been also found to be very mild.[16a]

This method can be further refined by combining it with bioengineering techniques, particularly measuring the transepidermal water loss. Then the concentration of the irritants can be reduced and invisible damage can be detected at a very early stage. Thus, severe necrotic reactions can be avoided.

As in contact allergy the development of the *Duhring chamber* must be considered a major step in standardizing irritant patch test techniques. In former times patches were made in each laboratory in a different way using mainly gauze or filter paper and occlusive tapes. In most older publications no further details are given. The idea of Pirilä[17] when he developed the Finn chamber ensuring an accurate exposure of the material under occlusive conditions spread rapidly over the world. Taking this as a template we developed the larger Duhring chamber, accepting up to 0.2 ml.[17a] Nowadays the large Finn chamber of 12-mm diameter is commercially available from Hermal (Reinbek, Germany); 80 µl is a safe application volume for liquids and ointments. In the U.S. a similar chamber made from plastic is available from Hill Top Laboratories.

V. Tests on Damaged Skin

The 21-day cumulative irritation test is a very tedious and for the subject an often uncomfortable procedure. Furthermore, in a few instances it was observed that even under these extreme conditions the sensitivity of the test was too low. On the other hand false-positive reactions with folliculitis, especially when the test is performed in the warm period of the year, is another problem. These are the reasons for the development of the *chamber scarification test*.[18] The stratum corneum as the main barrier to penetrating irritants is breached by scarifying the skin with a fine needle eight times in a vertical chessboard-like pattern. The test site is the volar aspect of the forearm; the test material is applied with Duhring or large Finn chambers for 3 days with daily renewals. The scoring is performed on the 4th day on a five point scale. Irritants are classified on mean values of a minimum of five test persons: I (very low) 0 to 0.4; II (slight) 0.5 to 1.4; III (moderate) 1.5 to 2.4; IV (strong) 2.5 to 4.0.

The increase in test sensitivity by the scarification is dramatic for most irritants. If the irritant is very polar and thus poorly penetrating, differences in threshold values for normal and scarified skin can be in the range of 10 to 100. The *scarification index* results from the determination of the threshold concentration on normal and scarified skin. High values indicate that the test material can lead easily to skin irritation if the barrier is damaged. The risk is even higher if the test material is applied to intertrigineous areas where there are nearly occlusive conditions.

Using this test, 80 dermatological and cosmetic ingredients were tested. The method is also suitable for testing formulated products such as creams, lotions, and ointments. Excessive experience with this test has shown that a product with a low score in the scarification test (class I) will also show good skin tolerance if used by a large population. The combination of the 3-week cumulative irritation test on normal skin and the scarification test for 3 days on a panel of 15 subjects seems to be the optimal way of predictive testing in regard to low-level irritancy. The scarification test has also proved its value in detecting the irritant capacity of eye medicaments.[19]

VI. Test for Identification of Sensitive Skin

The reader is referred to a monograph on this subject,[2] summarized recently in the Textbook of Contact Dermatitis.[1] As already mentioned above there is still no ideal test technique which can be recommended for general usage. However, subjects with sensitive skin do exist. They probably comprise about 15% of the normal population and are frequently, but not always, atopics. They usually react to irritants applied in an open way such as sodium hydroxide, DMSO, or ammonium hydroxide more vigorously than other "tough subjects". Subjects with a low minimal blistering time of ammonium hydroxide have been shown to react stronger to SLS and to two soaps in comparison to subjects with a high minimal blistering time. The same holds true for another panel of 30 subjects who showed a correlation between the minimal blistering time to ammonium hydroxide and to 21-day cumulative irritation test with propylene carbonate, Triton X 400, decanol, and lactic acid. Therefore, for product safty testing a panel of highly sensitive subjects can be selected using these techniques.

VII. The Stinging Phenomenon

Some individuals will develop a delayed type of stinging induced by sunscreens or other cosmetics. The phenomenon occurs usually only on the face and is aggravated by a hot and humid climate. This is a real and reproducible phenomenon. For identification of susceptible subjects the "stinging test" was developed;[20] 10% lactic acid is applied at room temperature to the nasolabial fold and the reaction is scored on a 4-point scale after 10 seconds, 2.5 min, 5 min, and 8 min. The sensitivity is increased by profuse sweating in a sauna. A number of ingredients of cosmetics and topical drugs has been identified to cause delayed type of stinging (sunscreens such as Escalol 506 or Givtan F, the insect repellent diethyltoluamide, tretinoin, and others).

The stinging potential does not always correlate with the irritation potential and probably reflects a direct nerve stimulation. New studies of Lammintausta et al.[21] have shown that the skin of "stingers" may also react more intensely to an irritant such as sodium lauryl sulfate. This was not visible clinically but has been recognized by the use of measuring cutaneous blood flow with the laser Doppler technique.

VIII. Test for the Evaluation of Skin Barrier Creams

As another example of an irritant patch test technique providing useful information, we refer to our recent publication on the repetitive irritation test in humans for the evaluation of the efficacy of skin barrier creams. Sodium lauryl sulfate is applied for 30 min consecutively ten times over a period of 2 weeks to the backs of human volunteers. The barrier cream is applied 30 min before the irritant. The irritant response is scored on a clinical scale and quantified by measuring the transepidermal water loss, blood flow, conductance for skin hydration, and erythema by chromametry. Large differences between formulated commercial products have been found. While some were extremely effective in suppressing the irritation due to SLS completely, others failed or even aggravated the irritant response. Further irritants such as sodium hydroxide, lactic acid, and organic solvents like toluene have been tested — the results will be published in the near future.

References

1. Frosch, P.J., Cutaneous Irritation, in *Textbook of Contact Dermatitis*, Rycroft, R.J.G., Menné, T., Frosch, P.J., and Benezra, C., Springer-Verlag, Berlin, 1992, 28.
2. Frosch, P.J., *Hautirritation und empfindliche Haut*, Grosse Verlag, Berlin, 1985, 7.
3. Burckhardt, W. and Schmidt, R., Die Epicutanprobe durch wiederholte Benetzung, *Hautarzt*, 15, 555, 1964.
4. Locher, G., Permeabilitätsprüfung der Haut Ekzemkranker und Hautgesunder für der neuen Indikator Nitrazingelb "Geigy", Modifizierung der Alkalieresistenzprobe, pH Verlauf in der Tiefe des Stratum corneum, *Dermatologica*, 124, 159, 1962.
5. Björnberg, A., Skin reactions to primary irritants in men and women, *Acta Dermatoven*, 55, 191, 1975.
6. Frosch, P.J. and Wissing, Ch., Cutaneous sensitivity to ultraviolet light and chemical irritants, *Arch. Dermatol. Res.*, 272, 269, 1982.
7. Björnberg, A., Skin Reactions to Primary Irritants in Patients with Hand Eczema. An Investigation with Matched Controls, Thesis, Gothenburg, Sweden, 1968.
8. Frosch, P.J., Duncan, S., and Kligman, S.M., Cutaneous biometrics. I. The response of human skin to dimethyl sulfoxide, *Br. J. Dermatol.*, 102, 263, 1989.
9. Frosch, P.J. and Kligman, A.M., Rapid blister formation in human skin with ammonium hydroxide, *Br. J. Dermatol.*, 96, 461, 1977.
10. Frosch, P.J., Methoden zur Charakterisierung der Hautempfindlichkeit — Ammoniak-MBZ und DMSO-Test, *Hautarzt*, 32 (Suppl. V), 449, 1981.
11. Draize, J.H., Woodard, G., and Calvery, H.O., Methods for the study of irritation and toxicity of substances applied topically to the skin and mucous membranes, *J. Pharmacol. Exp. Ther.*, 82, 377, 1944.
12. Kligman, A.M. and Wooding, W.M., A method for the measurement and evaluation of irritation human skin, *J. Invest. Dermatol.*, 49, 78, 1967.
13. Lanman, B.M., Elvers, W.B., and Howard, C.J., The role of human patch testing in a product development program, Proc. Joint Confer. on Cosmetic Science, The Toilet Goods Association, Washington DC, 1968, 135.
14. Steinberg, M., Akers, W.A., Weeks, M.H., Mc Creesh, A., and Maibach, H.I., A comparison of test techniques based on rabbit and human skin response to irritants with recommendations regarding the evaluation of mildly to moderately irritating compounds, in *Animal Models in Dermatology*, Churchill Livingstone, Edinburgh, 1975, 1.
15. Willis, C.M., Stephens, C.J.M., and Wilkinson, J.D., Experimentally-induced irritant contact dermatitis, *Contact Dermatitis*, 18, 20, 1988.
16. Frosch, P.J. and Kligman, A.M., The soap chamber test: a new method for assessing the irritancy of soaps, *J. Am. Acad. Dermatol.*, 1, 35, 1979.
16a. Kästner, W. and Frosch, P.J., Hautirritation verschiedener anioischer Tenside im Duhring-Kammer-Test am Menschen im Vergleich zu tierexperimentellen Modellen, *Fette, Seifen, Anstrichmittel*, 83, 33, 1981.
17. Pirilä, V., Chamber test versus patch test for epicutaneous testing, *Contact Dermatitis*, 1, 48, 1975.
17a. Frosch, P.J. and Kligman, A.M., The Duhring-chamber: an improved technique for epicutaneous testing of irritant and allergic reactions, *Contact Dermatitis*, 5, 73, 1979.
18. Frosch, P.J. and Kligman, A.M., The chamber-scarification test for irritancy, *Contact Dermatitis*, 2, 314, 1976.
19. Frosch, P.J., Weickel, R., Schmitt, Th., and Krastel, H., Nebenwirkungen von ophthalmologischen Externa, *Zeitschr. Hautkr.*, 63, 126, 1988.
20. Frosch, P.J. and Kligman, A.M., A method for appraising the stinging capacity of topically applied substances, *J. Soc. Cosmet. Chem.*, 28, 197, 1977.
21. Lammintausta, K., Maibach, H.I., and Wilson, D., Mechanisms of subjective (sensory) irritation, *Dermatosen*, 36, 45, 1988.
22. Oberriether, P., Untersuchungen über die Säureneutralisation und Säureresistenz der Haut anhand neuer Methoden, *Dermatologica*, 108, 279, 1954.
23. Leder, M., Die Benzintoleranz der Haut, *Dermatologica*, 88, 316, 1943.
24. Blohm, S.G., A new technique for epicutaneous testing, *Acta Dermatol. Venereol.*, 40, 46, 1960.
25. Klaschka, F., Mengel, G., and Nörenberg, M., Quantitative und qualitative Hornschicht-Diagnostik, *Arch. Derm. Forsch.*, 244, 69, 1972.

Chapter 31.2
Allergic Patch Test Techniques

Torkel Fischer
Occupational Dermatology
National Institute of Occupational Health
Solna, Sweden

I. Introduction

Over the last hundred years, patch testing has developed from the brief exposure to raw materials or application of fabric soaked with solutions of various substances to the present, more sophisticated methods. Individual intolerance to wood and other natural products was recognized in antiquity. A method to test idiosyncratic reactions was recognized during the early 19th century using a strip of blotting paper. In 1895, Jadassohn introduced the first device to demonstrate contact sensitivity. This was the beginning of modern patch testing.[1-4]

The present patch test systems include a device and test material.[5] The Al test™ was the standard method of applying antigens for several years. It consists of several filter paper discs of cellulose attached to a strip of plastic-coated aluminum foil.[6] The now most widely used method to apply patch tests employs aluminum or polypropylene plastic chambers and cups. The patches are fixed with tape to the skin.[7-10]

Allergens, incorporated in a petrolatum vehicle, are applied to the filter paper discs or in cups.[11-17] The test strip is fixed with tape to the upper part of the back and left in place for 48 hours.[18]

A future developmental line, which improves and simplifies diagnostic testing is the ready-to-apply TRUE Test™ (KABI-Pharmacia, Hillerød, Denmark) where allergens are incorporated in a thin, hydrophilic film layer which is coated on polyester patches.[5,19,20]

Uncommon methods include mucosal tests and methods for testing highly volatile substances such as freons.[21-23]

II. Objective

The apparently simple method of patch testing holds many difficulties and pitfalls, both technical and interpretive. The ideal patch test should indicate contact sensitization and produce no false-positive and no false-negative reactions. The allergens should be chemically defined and possess high purity. An optimal amount of allergen per square area should be applied to produce a reaction of high quality, easy to evaluate in both weakly and strongly sensitized individuals. The test should produce a minimum of irritant reactions, and the risk of sensitizing the patient actively must be kept low. The test should be easy to apply, have good stability, be acceptable to patients, and not unnecessarily limit their activities.

III. The Test Apparatus

A. The Tape

Most tapes in current use possess adequate adherence, and keep the patches in close approximation with the skin. Excessive exercise and sweating and high humidity together with oily skin during the test period may decrease the adhesiveness of the tape and remove the corners and even the whole test strip.

Old types of adhesive tapes consist of a colored cotton fabric covered with a sticky mass composed of crepe rubber, rosin, zinc oxide, lanolin, antioxidants, and preservative. Using this material, severe tape allergy was common and usually due to the rosin. The antioxidants, preservatives, and lanolin could also be responsible for the sensitization.[24-26] Some modern tapes have polyvinyl chloride backing.[27] Additives to this plastic may cause allergic reactions. Rosin derivatives are present as adhesive constituents in many modern tapes, causing allergic reactions occasionally.[28,29] Such rosin derivatives are not always identified with an ordinary colophony test.

Modern tapes like Scanpor® (Norgesplaster, Kristiansand, Norway) are made from finely meshed cellulose fiber material and an acrylate adhesive. Allergic reactions can occur due to uncured acrylate or to formaldehyde, which may be present in low concentration.[30-32] As with all tapes, mildly irritant, acneiform, and miliarial reactions occur with some frequency.[33]

Individuals with a tendency to dermographism may present with transient urticarial reactions after tape removal. Depigmentation from patch test tape is rare.[5]

B. The Patch

The test device should keep the test material in close contact with the skin in a small, sharply delineated area. With accurate application technique, most available chambers fulfill these criteria. Even with the correct technique, however, small amounts of test material may diffuse a millimeter or so around the patch.

Today, most patch tests are performed with aluminum or plastic chambers with a diameter of 10 mm (Finn Chambers) and a volume of 25 µl.[7] These chambers give test reactions of good accuracy and use a smaller test area than Al test or large plastic chambers like van der Bend chambers.[35–37] With Finn chambers 60 to 80 patches can be applied at the same time depending on the size of the individual's back. A filter paper must be added to the Finn chambers for liquid allergen preparations.

Sensitization to aluminum is a rare complication when testing with Al chambers. It presents as multiple ring-shaped test reactions. It is most commonly seen in individuals who have been desensitized previously with aluminum-containing allergen extracts.[38–41]

Metal salts like mercury, cobalt, and nickel which have a higher positive redox potential interact with aluminum.

Aluminum may also interact with organic substances such as methylene blue, probably forming a double salt of low solubility and fixing strongly to the stratum corneum.[42] Aluminum also increases the rate of acrylate polymerization, probably through a catalytic reaction.[43] Mixing allergens in petrolatum, will minimize such interactions and they have probably minimal clinical significance. Plastic chambers and plastic coated Finn chambers will eliminate this effect.

There is a significant variation in skin absorption of allergen with different patch test systems.[44]

C. The Vehicle

The vehicles for patch test allergens must be inert, stable, nonirritant, and protect the allergen against environmental influences.[45–52] Lanolin and olive oil were first tried, but soon abandoned because of their sensitizing properties. Yellow petrolatum was then tried. Semisolid, pure white petrolatum is presently the standard vehicle.

Petrolatum has advantages and disadvantages and is suitable for many allergens. It has been regularly questioned for its efficacy.[53,54] It gives good occlusion, has good own stability, and keeps most allergens stable. It is highly lipophilic, while most allergens are hydrophilic, thus retaining the release. This occurs most likely with metal salts.[17,55–57]

Rarely there have been reported cases of petrolatum allergy caused by phenanthrene derivatives.[58–60] "Softisan", a hydrophilic lanolin substitute, was launched a couple of years ago as a vehicle, but has been little tested so far.[61]

Modern vehicles are hydrophilic gels such as hydroxypropyl cellulose and polyvinylpyrrolidone (PVP), which can preserve the allergen in a dried, ready to apply form as it is used in the TRUE Test™.[5,19,62]

Water is the optimal vehicle for some allergens, e.g., formaldehyde and Cl$^+$Me isothiazolinon. Some investigators claim that the aqueous vehicle will give more reliable results also with nickel.[63]

For special allergens solvents like ethanol, acetone, methyl ethyl ketone, and ethyl ether are recommended. Stability is low for most of these solutions and they are difficult to handle and preserve.[64,65]

In research projects, vehicles containing salicylic acid, detergents, dimethyl sulfoxide, and alkalis may be applied to increase the penetration. Such vehicles increase as a rule the irritant properties of the test material and may promote subliminal irritancy of the allergen.[66]

D. The Antigens

Of more than 6 million chemicals identified, about 3000 are known to have contact-sensitizing properties.[65] New allergens are identified each year in predictive tests and in patients with contact dermatitis. Contact allergens are simple chemical substances, with molecular weights rarely greater than 1000 Da and in general less than 500 Da. Most allergens used in patch testing are well-defined chemical compounds. Furthermore, a large number of less well-defined biologic materials such as natural resins, fragrances, and lanolin may also sensitize.[67–70] Attempts have been made to describe the chemical basis of sensitization, so far with limited success.

1. The National and International Standard Series of Allergens

A small number of substances account for the majority of contact allergies. National and international groups of patch test specialists recommend that allergen series, including mixes of allergens, be used for routine testing. The series are revised yearly or biyearly on the basis of changes in the sensitivity of the population and new scientific information. Such series comprise 20 to 25 test substances containing 40 to 50 different allergens. The present European standard contains 23 items, 7 of them mixes representing all together 45 allergens.

Allergens, which present more than 1% positive test reactions in multiple studies, performed in several countries and, where the knowledge about that particular allergy gives a significant contribution to the understanding of the patient's disease, are candidates to be introduced in the international standard series.

The new allergen must not be detected with other materials in the tray.

If the frequency of positive reactions to a particular allergen in patients routinely tested for contact allergy is less than 0.5%, it is a candidate for replacement from the standard series.

2. Other Standardized Allergens and Antigentic Materials
a. Chemicals

In addition to the standard series, many other allergens can be obtained as standardized test materials finely ground in petrolatum or dissolved in distilled water.

More than 500 such antigens are available from different manufacturers standardized as to concentration and vehicle.

The clinician should utilize commercially standardized test material whenever possible. Well-reputed manufacturers are Chemotechnique Diagnostics AB, Ringugnsgatan 7, S-216 16 Malmö, Sweden and Hermal Kurt Herrman AG, P.O. Box 1228, D-2057 Reinbeck, Hamburg, Germany.[71,72]

Their catalogs contain good information not only on test materials but also on the occurrence of allergens, cross-reactivity, and other valuable test information.

Catalogs and textbooks on contact dermatitis and patch testing show slight differences in recommendations on concentrations and vehicles.

Patch test material may also be prepared from raw materials obtained from chemical supply companies or from manufacturers of products suspected to cause contact allergy. Recommendations about test material are given in the literature.[63–65,73–83]

b. Natural Compounds

Natural compounds may contain several different allergenic compounds which are more or less well characterized. The concentration of antigen differs from batch to batch depending on origin and storage.

Balsam of Peru, lanolin, and colophony are examples of natural mixes. Balsam of Peru contains several sensitizing fragrances and will identify approximately 50% of fragrance allergy detected by the fragrance mix. Conversely, the fragrance mix with seven pure perfume chemicals and one (oak moss) natural component included will indicate approximately 70 to 80% of fragrance sensitivity detected by Balsam of Peru.

With turpentine and colophony much of the antigenicity seems to be due to peroxides rather than pure terpenes and resin acids.[69,84]

The responsible allergen in lanolin is unknown and follows the acidic fraction. Most of it can be extracted with methanol and has been claimed to be a diol.[67,68]

3. Testing with Nonstandard Allergens and Antigenic Materials

Testing with nonstandard materials should be done with great care. Complications such as scarring, necrosis, keloids, pigmentation, and depigmentation may appear when a too high concentration is used.

Ask for detailed product information. There are often only a few ingredients of interest in the product, which may cause allergy while the remainder are substances known to be innocent. If the suspected allergen is available as ready prepared test material from manufacturer, patch test with such a material. Impurities and contaminants may be the cause of the dermatitis. Therefore, test also with a sample of the product suspected to cause dermatitis.

Solid product such as textiles, rubber, plants, wood, and paper can usually be applied as wafer-thin, smooth sheets or as finely dispensed particles in petrolatum. As a rule plants and woods should be tested as extracts. The material is extracted in water, ethanol, acetone, or ether. The filtered extract is evaporated, weighed, and redituled or mixed with petrolatum to appropriate concentration.

Cosmetics and "leave-on" skin care products can usually be tested as is. Testing with soap, shampoos, and detergents will seldom give valuable information since, for testing purposes, they must be diluted in water to 0.1 to 1%, and then the concentration of a causative agent, e.g., a perfume, is too low to elicit reactions. For testing base make-up, moisturizers, sunscreens, and lipsticks, the material should be diluted in petrolatum 1 to 10%.

To rule out irritant reactions, a sufficient number of controls should be used. Often the structure and known properties of a chemical may indicate the possibility and risk of irritation.

IV. Preparation of Patch Test Material

Standardized test material is necessary to obtain accurate, reliable, and reproducible patch test results. Test results are impossible to compare between different investigators and clinics if different materials, concentrations, vehicles, and methods are used.

The basis of this standardization is an attempt to balance sensitivity and specificity of the biologic assay — the patch test for a given antigen.

A. Raw Materials — Purity, Stability

Selection of chemicals for patch testing is difficult and demanding. Metal salts are almost without exception marginal irritants. Salts with low irritant potential seldom present the antigen well. It is also important to choose the right chemical. Cinnamic alcohol has itself low or no sensitizing potential while its metabolite cinnamic aldehyde is a potent allergen.[85]

Purity of the test material is of the utmost importance. Chemicals of best quality should be used.[86] The patch test allergens sold by large manufacturers are according to their product catalogs, chemically defined, and pure. The physician should request the manufacturer's data on chemical analyses. Some allergies previously reported were in fact caused by impurities or degradation products.[87,88]

Stability depends largely on the allergen and varies considerably. Metal salts, for example, are stable. Contrary to what might be expected, formaldehyde in aqueous solution is also stable. *p*-Phenylenediamine, colorless when fresh, rapidly degrades, which is indicated by blue-black discoloration. Its antigenic properties, however, do not change significantly. In the case of *p*-phenylenediamine, the base gives a higher proportion of positive reactions. Aging of fragrances decreases their antigenicity. Complex antigens may lose antigenicity with increased purity.

Photoallergens stored in sunlight deteriorate rapidly. This is why such chemicals used for patch testing should be kept in darkness. It is a good general rule that allergens should be kept in a refrigerator to minimize degradation.

B. The Pharmaceutical Preparation

The present method of preparing test material from solids involves micronizing and mixing the allergen with the vehicle, a process in which it is difficult to obtain sufficiently fine particle size. Most test materials at present use have a particle size of 2 to 200 μm.[13,16] Petrolatum keeps the particles in closed compartments against the skin where they dissolve slowly and form small areas of high allergen concentration. Follicular openings are especially sensitive to irritant reactions, as many substances readily enter the epidermis in these areas. Thus, large particles of allergen with irritant properties may cause patchy irritant reactions when in contact with follicular areas.[13,15] Large particles also cause inhomogeneities in the amount of allergen applied.

Lipid materials may be dissolved, melted, or solubilized into the petrolatum phase. Hydrophilic and liquid materials are difficult to incorporate in petrolatum. Fragrance mix is therefore sometimes prepared with the detergent sorbitan sesquioleate.[71] This may solve a pharmaceutical problem, but at the same time change the irritant and sensitizing properties of the material.[89]

Another example is Balsam of Peru, a liquid test material which when included in petrolatum forms small instable droplets both inside and outside the petrolatum.[13]

In the TRUE Test™, the allergens are evenly incorporated in hydrophilic gels.

To overcome the problem to incorporate a gas in the dry TRUE Test™ vehicle, a pre-allergen, N-hydroxymethylsuccinimide was used.[90]

By using buffer solutions for acid and alkaline products, the test concentration can be raised.[91] Addition of detergents and penetration promotors such as DMSO, urea, lactic acid, and sorbitan sesquioleate increases the patch test response. In the future such additives and also pH adjusters and antioxidants to the test material will probably be accepted even if it means a more complex test material.[92,93] Additives may reduce the sensitizing properties of test materials.[94,95]

C. The Allergen Dose — Bioavailability

In the future, skin permeation and skin deposition studies will become an integral part of the optimization of patch test materials.

On patch testing, an optimal dose of a hapten should reach the immunoreactive epidermis. There are at least eight parameters which influence the dose of hapten penetrating into the skin:

- Chemical used
- Allergen dose
- Vehicle
- Patch test system and tape
- Exposure time
- Environmental factors — temperature, humidity
- Individual skin factors such as hydration and oiliness
- Individual habits of dressing, movement, work, hobbies, and sleep

The penetration capacity depends upon the salts used; for example, there is a large difference between the penetration of nickel achieved by nickel sulfate and nickel chloride.[17,18,96] The higher penetration of nickel from the chloride is probably explained by the partitioning between skin and vehicle of the salts. Chromium salts demonstrate the same type of difference.[97-99]

Most important is the concentration of the allergen. A change in concentration, from 0.1 to 1.0%, for example, may elicit irritant reactions with many allergens, and a reduction in concentration, as in the case of nickel sulfate from 5 to 2.5%, may result in false-negative reactions in a small but significant number of nickel-sensitive patients.[100-102]

All too often, test concentrations recommended are based upon testing a small number of people and are suboptimal. One should seek the original references for the data.

The concentration of an allergen is normally given as a percentage. One catalog presents molality together with percentage (weight/weight).[72] In TRUE Test dose is given in milligrams per square centimeter. In research projects, doses of allergen are best compared when presented as moles/cm² skin area.[102]

D. The Clinical Standardization

Extremes of pH, polarity, and solvent properties increase the risk of irritation, and suggest a test concentration of 0.01 to 1%. To minimize the risk of irritant reactions, one should start with an open test and, if this is negative, continue with occlusive patch testing, then starting with the lowest concentration. To find a nonirritant concentration most of "supposedly innocent" chemicals can be tested at three concentrations, either aqueous or in petrolatum, 0.1, 1.0, and 10%. A positive reaction should not be accepted as allergic unless at least 20 controls are negative. This number of controls is a minimum.

To fine tune the dose for patch testing, a serial dilution technique is used. Common allergens should be optimized in this way. For most allergens the sensitivities of allergic individuals present a geometric distribution, with a standard deviation of +3 and minus a third of a mean.[100,102]

For some allergens, especially metal salts, irritant reactions and weak allergic reactions will appear together within a certain dosage range. The choice of test dose for such allergens is a balance between demands of sensitivity vs. specificity. The dose should be high enough to detect weak sensitivities, but allow only an insignificant number of irritant reactions. One must also take care not to sensitize. Few allergens sensitize when tested below their threshold of irritancy.[103]

The optimal dose for an allergen is related to the test method, as the bioavailability of different allergens differs between methods. An example is formaldehyde, where a concentration of 2% as recommended in the A1 test causes irritation with Finn chamber method where 1% is optimal. For most allergens, the equivalent TRUE Test dose is one third of that used with chamber methods.

E. Mixes

Patch testing with single allergens gives the most reliable information. The use of mixes of closely related allergens saves time and skin space.[104] In preparing a mix, it is a challenge to choose a vehicle and optimize allergens that possess different chemical reactivity, solubility, and stability, as was shown in the standardization of caine mix for the TRUE Test™.[105–110] This involves problems of concentration, interference, stability, and formulation. The major problem with mixes is that they often cause irritant reactions because of a high total concentration of allergen. To avoid this, the allergens are often incorporated in suboptimal amounts, resulting in false-negative reactions in weak allergies as is the case with mercaptomix. An interesting example of interference is the thiuram mix, which contains three thiuram disulfides and one monosulfide.

The three disulfides change to an equilibrium of six different thiuram disulfides, whereas the monosulfide remains stable. The standard European series includes seven mixes containing two to eight substances.[71,72] The fragrance mix in petrolatum is borderline irritant although most of the ingredients of this mix are incorporated in suboptimal concentrations. The only compound included in a concentration close to irritancy is cinnamic aldehyde. TRUE Test incorporates the fragrance mix in betacyclodextrin and is thus less irritating than the mix in petrolatum.

When one obtains a doubtful or weak positive reaction to a mix, it is advisable to repeat the test using the separate ingredients.

F. The Packing

The packing should protect the test material from the environmental influences of air, humidity, light, UV radiation, microbes, and contamination. The polypropylene syringes that are currently used have a high resistance to chemicals and protect the test substances against air, humidity, microbial, and other contamination.[111] Heat, light, and ultraviolet (UV) do penetrate the syringes to some extent, requiring that the syringes be kept in a cool, dark place for storage.[112] At room temperature, mercury in petrolatum sediments loses its homogeneity. This also may happen with nickel and chromium salts at increased temperature.[113]

G. Ready-To-Use Test Systems

The TRUE Test™ represents a new generation of patch tests that solves many of the problems indicated above. It is an easy and time-saving ready-to-use test method. The allergens are incorporated in hydrophilic gels, preferably cellulose derivatives or polyvinyl pyrrolidone. The allergen in gel preparation is printed on an impermeable backing or polyester and dried to a thin film. The coated sheet is cut into 9 × 9 mm squares — test patches — that are mounted on a tape, covered with protective sheet, and packed in an airtight and light-impermeable envelope.[5,19,20,102]

Cellulose derivatives have low sensitizing potential according to guinea pig maximization (GPM) tests, adhere adequately to the plastic backing, and are compatible with most allergens. The surface distribution and the amount of allergen can be dosed with high accuracy. When the test patch contacts the skin, perspired water hydrates the film to a gel. The gel — covered with plastic — will produce optimal contact with the skin and good skin permeation opportunities resulting in high allergen bioavailability.

The test has been subject to a comprehensive control program. Each allergen is standardized with serial dilution technique on patients sensitive to the allergen. The intensity of the test reactions is evaluated against tests with the Finn chamber method. For most allergens, equivalent reactions will be obtained for

TRUE Test™ at an application of 20 to 40% of the amount of allergen normally applied in Finn chambers. Analytic methods were defined for the different allergens. The amount of allergen on the test patches has been controlled by qualitative, quantitative analysis and *in vivo* tests in patients with material kept at different temperatures for different time intervals. *In vitro* experiments have determined the permeation profiles of several allergens through skin and different membranes. There was good agreement between the clinical and the diffusion chamber experiments. Multicenter studies have verified good concordance between TRUE Test™ and chamber methods.[114–117]

Other ready-to-use test methods are the Epiquick, no longer marketed, and the RPT systems which basically use the chamber methods with petrolatum material.[118,119]

V. The Test Technique

The amount of test material applied to the patch is critical to achieve reliable results. The dose of allergen should cover the test area completely, but without extrusion. Finn chambers have a volume of 25 µl; the application of 15 to 18 µl of test material gives the best results.[14]

Aqueous solutions and other volatile test materials must be applied immediately before application to prevent evaporation of vehicle which reduces skin contact. Application of one drop from the bottle tip to the filter paper disc is sufficient.

Exact dose of liquid test preparations are obtained via a digital pipette with disposable plastic tips.

Allergens frequently causing strong, cross-, or concomitant reactions should be positioned at distance from each other on the test strip.

A. The Test Application

Start preparing the test strip by applying test material to each disk. A 5-mm ribbon of petrolatum-based antigen equals 15 to 20 µl.

Remember that the order of the antigens will be left to right reversed when they are applied to the patient's back.

Patch test materials are standardized for application to the lateral aspects of the upper back. If, for some reason, this is not possible, the outer aspects of the upper arms may be used. Do not apply patches to the midline.

As many as 50 to 60 tests can be applied at a time, using the Finn Chamber technique.

The tape strips should be applied from below with firm pressure to remove air bubbles and provide uniform adhesion. Each test site should then be pressed lightly to assure contact with the skin.

The most frequent cause of loosening is extrusion of excessive test material; the next most frequent cause is the presence of extremely sebaceous skin.

The back may be cleaned with plain water, no soap, and then let dry before application when it is wet, dirty, or oily.

Ethanol washing may be used to degrease the test area, but must evaporate completely prior to application. Otherwise, irritant reactions from both tape and patches may appear.

The skin must be normal and hairless. Patches applied directly over a heavy coat of hair will decrease the skin contact and result in false-negative reactions. Hair should be removed by shaving or clipping without soap, shaving foam, or cream. The combination of clipping, petrolatum, and tape sometimes contributes to irritation and interferes with the reading of the test.

B. The Application Time

At least with metal salts only small amounts enter the epidermis from a patch test. An influx equilibrium from the test material probably occurs within hours after the test application. In the case of chromium such an equilibrium occurs by 5 hours.[99]

Some individuals present positive reactions after less than 1 hour of application of nickel salts. Nevertheless, there is an increased frequency of positive patch test reactions with prolonged exposure time for at least 48 hours. Sensitized patients who have been free of contact dermatitis for a long period of time need this continuous, long exposure time to react. The reason for this is probably a sluggish immune response, slow coupling, or propagation of the antigen. Therefore, 48 hours exposure before removal of the test is the general recommendation, although 24 hours exposure and even shorter times have been suggested. The 48-hour patch test will probably remain our standard procedure to avoid significant loss of positive reactions.[18,120–124]

C. The Marking of the Test Area

The marking of the test area must remain recognizable on the skin for at least 72 hours and at the same time not discolor clothing or sensitize the patient. If test strips with constant distances between the discs are used, only two marks per strip are needed.

Various water-resistant skin markers have been recommended. An aqueous solution of methylrosanilin 1% with silver nitrate 20% in ethanol (70%), dispensed with a felt-tipped marking pen, is a popular marker. Application of transparent tape over the ink markings is another method to reduce contamination of clothing.[125] Another method to avoid discoloration of the patients' clothes is to use an inexpensive fluorescent ink pen with sodium fluorescein.[126] To visualize the ink, a conveniently sized flashlight with a black light bulb can be used. It illuminates the ink surrounding the test site to a bright yellow-green when the room is darkened. Reapplication should be done at each return visit. Skin markers may cause irritant skin reactions.[127,128] A third method is to rip off the edge of the patch test tape by a serrated template and leaving a small strip of tape as a marker for orientation. Then the test strips must be removed at the testing office and not by the patients.

D. The Patient Instruction About Patch Testing

Testing for contact allergy should include patient information both orally and with an instruction sheet about contact dermatitis, the aim of the test, and when to remove it. The handout should also instruct the patient to avoid excessive exercise and sweating, not to wet the test area, and to avoid scratching and rubbing even if there is itching. The patient should not expose the test area to sunlight or sunlamps.[5]

VI. The Patch Test Reaction

A. The Evaluation Time

At the time of removal of the test, irritant reactions from occlusion and tape are most pronounced. Preliminary inspection should wait until these effects have subsided, usually within an hour.

Optimally, readings should be done at 48, 72, and 96 hours and 1 week after test application. A single reading at 48 hours will lose approximately 30% of the positive reactions, and at the same time often show irritant reactions.

An acceptable compromise for only a single reading is evaluation time at 72 to 96 hours after test application. Then only 5 to 10% of the positive reactions are missed. Two readings at days 3 and 5 are strongly recommended.[129–133]

Neomycin and *p*-phenylenediamine often show maximum reactions on days 4 to 7 after application. This may occur with other allergens as well. Such a reactivity can often be traced back to a weak erythema on day 3.

Patients must be instructed to report any late reactions immediately.

B. Reading Parameters

The following parameters can be evaluated clinically from a patch test reaction: erythema, infiltration (edema), fine structure, surface distribution, and the area involved. These variables have distinct qualities in discrimination of the test reaction.[5,134]

Erythema is an intensity parameter but does not discriminate between allergic and irritant reactions.

Infiltration is essentially an intensity parameter, but should be present in allergic reactions. Reactions without infiltration are almost always irritant; however, infiltration itself is no proof of allergy.

The fine structure is the most important parameter discriminating between allergic and irritant reactions. Allergic reactions present the following structural panorama in order of increasing strength: discrete papules, distinct papules, papulovesicles, and coalescing vesicles. Homogenous reactions of such types are usually allergic, but irritant reactions do occur that are impossible to discriminate from allergy. On rare occasions, allergic reactions present only erythema and infiltration. Irritant reactions present many different types of fine structure: faint homogenous erythema without infiltration fine wrinkling (silk paper structure), erythema and papules in follicular distribution, petechiae, pustules, bullae, and necrosis.[135–142] Petechial reactions are rarely allergic. Patch testing using chambers with aqueous vehicles may result in ring-shaped allergic and irritant reactions.[41,143,144] If a reaction presents a mixture of allergic and irritant fine structure components, a retesting is recommended. The background may be alterations of the test area such as scars, acne, miliaria, and folliculitis.

In general, allergic reactions homogeneously cover the test area, while many irritant reactions are irregular, patchy, ring shaped, or follicular in appearance. Exceptions are weak allergic reactions, which may present with a patchy or follicular appearance, which can be best shown in the most dilute steps of a serial dilution test.[97,100–102]

The surface involvement of the reaction is a variable of quality rather than of discrimination. Extrusion of the reaction beyond the area of the patch is less common in irritant than in allergic reactions. However, strong irritant reactions may also present this characteristic.

Patch test reactions should be recorded according to principles, recommended by the International Contact Dermatitis Research Group.[6,134]

C. Histopathology of the Patch Test Reaction

Light and electronic microscopy have been of minimal help to distinguish between allergic and irritant patch test reactions. In the guinea pig, the proportion of basophilic cells in the allergic reaction is higher than in an irritant reaction. Human experiments have not verified this difference.[145,146]

The monoclonal antibody techniques used to subgroup lymphoid cells have been of limited value. No distinct differences between the cell infiltrate of allergic and irritant reactions has been reported in humans.[147–150] Research on cytokines in the suction blister fluid from allergic patch test reactions as compared to irritant appears promising.[151–153]

D. Objective Evaluation Methods

Several instruments and principles have been used to obtain an objective evaluation method.[154–156]

Measurements of reflectance at different wavelengths offer a grading of reaction intensity.[155,157] Infrared radiation using thermography gives an image of allergic and irritant reactions.[158] Skin conductivity and impedance measurements have gained new interest.[159] Evaluation of water evaporation, carbon dioxide diffusion, or transcutaneous oxygen resistance give some differential diagnostic information.[160–163] To determine skin edema with caliper is a simple method.[164] Epidermal water content may be measured with microwave resonance.[165] Blood flow measurements with laser Doppler technique have been used and a new scanning technique is promising.[166-169] Ultrasound imaging techniques are useful.[170,171] Most of these methods however only assess special features and grade the intensity of the test reactions.

Irritant and allergic reactions have different time courses, and this is the most common basis for their clinical discrimination.

VII. The Patch Test Interpretation

A positive test response may be interpreted as either allergic or nonallergic. This evaluation may be true or false. A negative patch test response may mean either that no allergy exists or that the patient is sensitized but the patch test has failed.

A. Sensitivity and Specificity

Sensitivity is defined as the ability of a test to detect contact allergy. Specificity means the capability of the test to discriminate between true allergic and nonallergic test reactions. Thus, low sensitivity indicates a risk of false-negative test reactions and low specificity the risk of false-positive. Taken together, sensitivity and specificity describe the accuracy of the test; this integer of the variables is an expression of the quality of the test, which considers both false-positive and false-negative test reactions.[102]

1. Certifying the Patch Test Reaction

The accuracy of ? or + ? reactions is low; just one or a few percent have an allergic background. The accuracy of + reactions depends on the allergen and the patch test material. For metal allergens in petrolatum, it is about 20%; for other allergens, it is between 20 and 50%. The accuracy of + + reactions is 80 to 90% and of + + + reactions 95 to 100%. Positive patch tests in general have a reproducibility of 60 to 80% on retest.[172,173] If the result of a test is important to the patient and the reaction is weak or questionable, retest after 4 to 6 weeks with the same test material, with another test technique, or with serial dilution tests.[101-103] A retest with mixes should normally be performed with separate ingredients in an increased concentration. Intracutaneous testing, especially with metals, may be of value to confirm dubious test reactions.[174,175]

2. False-Positive Reactions

False-positive reactions are frequent. The most common background of false-positive reactions is irritation, caused by a too high concentration of allergen, inhomogenous preparation, or inappropriate vehicle for the test material. Another reason may be a low threshold of irritancy in atopic or sundamaged skin, exaggerated local sensitivity in the skin due to a close by strong patch test reaction, an adjacent folliculitis, or to sensitivity to the material in the tape or patches. Multiple allergic reactions should raise the suspicion of spillover effects from other nearby reactions, or the presence of irritable skin, "the angry back" or "excited skin syndrome".

3. False-Negative Reactions

The frequency of false-negative reactions is uncertain.[176–180] There may be several technical reasons for a false-negative response, most often insufficient penetration of allergen due to insufficient amount of allergen applied, improper vehicle which does not release the allergen, or inappropriate test technique. The reaction may also be suppressed. UV radiation and grenz rays have both local and systemic suppressive effect on patch test reactions, but there seems to be no seasonal influence of sunlight on patch tests.[181–185] Systemic and topical corticosteroids are well-known suppressors.[186,187] Antihistamines are regarded as inert, but cinnarizine seems to influence patch test reactions.[187,188] Sodium chromoglycate also inhibits allergic patch test reactions, as does systemic and topical cyclosporine.[189–191] Unexpected factors like dietary zinc may contribute to the intensity of the allergic response.[192]

B. Relevance

Relevance describes the relation between the test reaction and the patient's disease. A relevant patch test reaction means that the reaction explains at least in part the patient's disease. The evaluation of relevance is the most difficult and intricate part of the patch test procedure and is highly dependent on the skill of the physician. Such a determination is based on the history and an understanding of the sources of the antigen. From a practical point of view, unknown or hidden relevance is the same thing as negative relevance. A physician experienced in patch testing can predict by history 80% of nickel sensitivity, 50% of rubber, colophony, and fragrance sensitivity, but only 10 to 20% of reactions to other allergens. A complimentary history after the test shows that about 80% of positive patch test reactions are relevant and gives important information to the patient. Remaining positive reactions are of unknown relevance.[193–197]

VIII. Adverse Reactions from Patch Testing

Patch testing to antigens at standard concentration and in standard vehicles is a safe procedure. In a small percentage of persons, there is aggravation of the dermatitis and increased itching, sometimes for a week or so. In 1 or 2 of 1000 people undergoing patch testing, there is a general flare-up of the dermatitis that may last for one to several weeks. General toxic reactions and induction of allergy with patch testing rarely occur. Formaldehyde has been suspected as a cause in a few cases of anaphylaxis.[198]

A patch test reaction normally persists 1 to 2 weeks. Test reaction may remain longer than 1 month, and persistence up to 8 months has been reported. Hypopigmented, scarred, keloid, and pseudolymphomatous reactions are rare late sequelae.[199–201]

A Koebner reaction may occur in patients with psoriasis at the site of a positive response.

Active sensitization is an infrequent complication of patch testing. When a patch test site flares up later than day 10 to 14 after application, active sensitization may have occurred. Such reactions are common with DNCB (dinitrochlorobenzene) and "natural" primin, which possess high sensitizing potency. Synthetic primin is easier to standardize and has a relatively low risk. In the standard series, cobalt and p-phenylenediamine involve the highest risk for sensitization.[202] Cl$^+$Me-isothiazolinone, recently included in the European standard series, has a significant risk to sensitize in 0.03% aqueous solution; however, with the 0.01% test used as standard, the risk is insignificant.[203–205] Unfortunately, this concentration will diagnose only 50% of the contact sensitivity.

IX. Correlation With Other Methods

A. Open Tests

In open testing a product or allergen, dissolved in water or some solvent, is applied on the volar forearm and allowed to spread. No occlusion is used. An open test is recommended as the first step when testing poorly defined or unknown substances or products. The test site should be checked at regular intervals during the first 30 to 60 min after application, and after 3 to 4 days.

A single open test application will give allergic reactions only in strongly sensitized individuals and is used essentially to indicate strong irritants.

Repeated open application test (ROAT) is used to mimic the natural situation of contacts with allergenic materials.[206] ROAT may be used both with pure allergens and formulated products such as a cosmetics, shampoos, and topical medicament. Commercial products as well as special test substances are used. They are applied twice daily for 7 days or until reaction. The application is performed on the outer aspect of the upper arm or on the scapular area of the back on a test area of at least 5 × 5 cm, preferably 10 × 10 cm. The smaller area needs approximately 0.1 ml and the larger area 0.5 ml test material. A positive response, erythema, and eczema will usually start on day 2 to 4. These tests are increasingly used to evaluate the clinical significance of allergy to an ingredient in a formulated product previously found reactive by ordinary patch testing.

Reactions to ROAT are weaker than ordinary patch test reactions.

B. *In Vitro* Tests

In vitro tests include different lymphocyte and mediator tests. They have still too low accuracy for patient testing, but can be used on group basis as a scientific proof of type IV sensitization.

X. Recommendations

Contact dermatitis accounts for about 10% of dermatological patients, and in 30 to 40% of these patients contact allergy is responsible for at least part of the skin disease. The patch test constitutes the dermatologist's most important tool in reaching a correct diagnosis and at present the only practical method to verify contact sensitization. Accurate patch testing shows several positive reactions unexpected by history.[73–81,207,208] The patch test is a model of the disease being investigated, because the organ tested is the same as the one affected with the disease, and the same mechanism for production of the disease is used. Therefore, patch testing is the only recommended method for clinical use. In research, however, accurate *in vitro* methods are under development which can be a valuable tool to verify contact sensitization.

References

1. Foussereau, J., History of epicutaneous testing: the blotting paper and other methods, *Contact Dermatitis*, 11, 219, 1984.
2. Staedeler, Über die eigentümlichen Bestandteile der Anacardium Fruchte, *Ann. Chem. Pharm.*, 63, 117, 1847.
3. Sulzberger, M.B., The patch test — who should and should not use it and why, *Contact Dermatitis*, 1, 117, 1975.
4. Jadassohn, J., Zur Kenntnis der medikamentöse Dermatosen, *Verhandlungen der Deutschen Dermatologischen Gesellschaft. Fünfter Kongress, Graz*, 1895, Vienna, W. Braunmüller, 1896, 106.
5. Fischer, T. and Maibach, H.I., Improved, but not perfect, patch testing, *Am. J. Contact Dermatitis*, 1, 73, 1990.
6. Fregert, S., *Manual of Contact Dermatitis*, 2nd ed., Munksgaard, Copenhagen, 1981.

7. Pirilä, V., Chamber test versus patch test for epicutaneous testing, *Contact Dermatitis,* 1, 48, 1975.
8. Frosch, P.J. and Kligman, A.M., The Duhring chamber. An improved technique for testing of irritant and allergic reactions, *Contact Dermatitis,* 5, 73, 1979.
9. Dooms-Goossens, A., Van der Bend Chamber Test, *Nieuwsbrief Contactdermatologie,* December 1984.
10. Quisno, R.A. and Doyle, R.L., A new occlusive patch test system with a plastic chamber, *J. Soc. Cosmet. Chem.,* 34, 13, 1983.
11. Fischer, T. and Maibach, H.I., The Finn chamber patch test technique, *Contact Dermatitis,* 11, 137, 1984.
12. Fischer, T. and Rystedt, I., False positive, follicular and irritant patch test reactions to metal salts, *Contact Dermatitis,* 12, 93, 1985.
13. Fischer, T. and Maibach, H.I., Patch test allergens in petrolatum: a reappraisal, *Contact Dermatitis,* 11, 224, 1984.
14. Fischer, T. and Maibach, H.I., The amount of nickel applied with a standard patch test, *Contact Dermatitis,* 11, 285, 1984.
15. Vanneste, D., Martin, P., and Lachapelle, J.M., Comparative study of the density of particles in suspensions for patch testing, *Contact Dermatitis,* 6, 197, 1980.
16. Puschmann, M., Galenische Qualität von Testsubstansen, *Z. Hautkr.,* 62, 1710, 1987.
17. Kalimo, K., Lammintausta, K., Mäki, J., Teuho, J., and Jansén, C., Nickel penetration in allergic individuals: bioavailability versus X-ray microanalysis dection, *Contact Dermatitis,* 12, 255, 1985.
18. Kalimo, K. and Lammintausta, K., 24 and 48 h allergen exposure in patch testing. Comparative study with 11 common contact allergens and $NiCl_2$, *Contact Dermatitis,* 10, 25, 1984.
19. Fischer, T. and Maibach, H.I., Easier patch testing; TRUE Test™, *J. Am. Acad. Dermatol.,* 20, 447, 1989.
20. Fischer, T. and Maibach, H.I., The thin layer rapid use epicutaneous test (TRUE-test), a new patch test method with high accuracy, *Br. J. Dermatol.,* 112, 63, 1985.
21. van Loon, L.A.J., van Elsas, P.W., van Joost, T.H., and Davidson, C.L., Contact stomatitis and dermatitis to nickel and palladium, *Contact Dermatitis,* 11, 294, 1984.
22. Axell, T., Spiechowiez, E., Glantz, P.-O., Andersson, G., and Lassron, Å., A new method for intraoral patch testing, *Contact Dermatitis,* 15, 58, 1986.
23. van Ketel, W.G., Allergic contact dermatitis from propellants in deodorant sprays in combination with allergy to ethyl chloride, *Contact Dermatitis,* 2, 115, 1976.
24. Calnan, C.D., Diethyldithiocarbamate in adhesive tape, *Contact Dermatitis,* 4, 61, 1978.
25. Cronin, E. and Calnan, C.D., Allergy to hydroabietic alcohol in adhesive tape, *Contact Dermatitis,* 4, 57, 1978.
26. Rasmussen, J.E. and Fisher, A.A., Allergic contact dermatitis to a salicylic acid plaster, *Contact Dermatitis,* 2, 237, 1976.
27. Fregert, S., Trulson, L., and Zimerson, E., Contact allergic reactions to diphenylthiourea and phenylisothiocyanate in PVC adhesive tape, *Contact Dermatitis,* 8, 38, 1982.
28. Sjöborg, S. and Fregert, S., Allergic contact dermatitis from a colophony derivative in a tape skin closure, *Contact Dermatitis,* 10, 114, 1984.
29. Hausen, B.M., Jensen, S., and Mohnert, J., Contact allergy to colophony. IV. The sensitizing potency of commercial products. An investigation of French and American modified colophony derivatives, *Contact Dermatitis,* 20, 133, 1989.
30. Jordan, W.P., Cross-sensitization pattern in acrylate allergies, *Contact Dermatitis,* 1, 13, 1975.
31. Waegemaekers, T.H. and van der Walle, H.B., The sensitizing potential of 2-ethylhexyl acrylate in guinea pig, *Contact Dermatitis,* 9, 372, 1983.
32. Andersen, K.E., Hjorth, N., Bundgaard, H., and Johansen, M., Formaldehyde in a hypoallergenic non-woven textile acrylate tape, *Contact Dermatitis,* 9, 228, 1983.
33. Marks, J.G. and Rainey, M.A., Cutaneous reactions to surgical preparations and dressings, *Contact Dermatitis,* 10, 1, 1984.
34. Mathias, C.G.T., Dermographism and patch test reactions, *Contact Dermatitis,* 10, 110, 1984.
35. Cronin, E., Comparison of Al-test and Finn chamber, *Contact Dermatitis,* 4, 301, 1978.
36. Hammershøy, O., Contradictory results following patch testing with Finn Chambers, *Contact Dermatitis,* 6, 216, 1980.
37. Peltonen, L., Comparison of Al-test and Finn chamber test, *Contact Dermatitis,* 7, 192, 1981.
38. Clemmensen, O. and Knudsen, H.E., Contact sensitivity to aluminium in a patient hyposensitized with aluminium precipitated grass pollen, *Contact Dermatitis,* 6, 305, 1980.
39. Fischer, T. and Rystedt, I., A case of contact sensitivity to aluminium, *Contact Dermatitis,* 8, 343, 1982.
40. Cox, N.H., Moss, C., and Forsyth, A., Allergy to non-toxoid constituents of vaccines and implications for patch testing, *Contact Dermatitis,* 18, 143, 1988.
41. Veien, N.K., Hattel, T., Justesen, O., and Nørholm, A., Aluminium allergy, *Contact Dermatitis,* 15, 295, 1986.
42. Fischer, T. and Maibach, H.I., The Finn chamber methylene blue ring, *Contact Dermatitis,* 12, 12, 1985.
43. Björkner, B. and Niklasson, B., Influence of the vehicle on the elicitation of contact allergic reactions to acrylic compounds in guinea pig, *Contact Dermatitis,* 11, 268, 1984.
44. Kim, H.O., Wester, R.C., McMaster, J.A., Bucks, D.A.W., and Maibach, H.I., Skin absorption from patch test systems, *Contact Dermatitis,* 17, 178, 1987.
45. Atkinson, J.C. and Rodi, S.B., Effect of vehicles and elicitation concentration in contact dermatitis testing. XI. Statistical analysis of data, *Contact Dermatitis,* 2, 330, 1976.
46. Cohen, H.A., The role of carrier in sensitivity to chromium and cobalt, *Arch. Dermatol.,* 112, 37, 1976.
47. van Ketel, W.G., Patch testing with nickel sulphate in DMSO, *Contact Dermatitis,* 4, 167, 1978.
48. Marzulli, F.N. and Maibach, H.I., Effects of vehicles and elicitation concentration in contact dermatitis testing. I. Experimental contact sensitization in humans, *Contact Dermatitis,* 2, 325, 1976.
49. Marzulli, F.N. and Maibach, H.I., Further studies of the effects of vehicles and the elicitation concentration in experimental contact sensitization testing in humans, *Contact Dermatitis,* 6, 131, 1980.
50. Samsoen, M., Stampf, J.L., Lelievre, G., et al., Patch testing with hexavalent chromium salts in different vehicles and with nickel and cobalt in petrolatum, *Dermatosen,* 30, 181, 1982.
51. Skog, E. and Wahlberg, J.E., Patch testing with potassium dichromate in different vehicles, *Arch. Dermatol.,* 99, 697, 1969.
52. Suhonen, R., Photoepicutaneous testing. Influence of the vehicle, occlusion time and concentration of the test substances on the results, *Contact Dermatitis,* 2, 218, 1976.
53. van Ketel, W.G., Petrolatum again: an adequate vehicle in cases of metal allergy?, *Contact Dermatitis,* 5, 192, 1979.
54. Wahlberg, J.E., The vehicle role of petrolatum, *Acta Dermatol. Venereol.,* 51, 129, 1971.
55. Depuis, G. and Benezra, C., *Allergic Contact Dermatitis to Single Chemicals: A Molecular Approach,* Marcel Dekker, New York, 1982.

56. Fischer, T. and Maibach, H.I., Recovery of nickel sulphate from a standard patch test, *Contact Dermatitis,* 11, 134, 1984.
57. Mendelow, A.Y., Forsyth, A., Florence, A.T., and Baille, A.J., Patch testing for nickel allergy. The influence of the vehicle on the response rate to topical nickel sulphate, *Contact Dermatitis,* 13, 29, 1985.
58. Dooms-Goossens, A. and Degraff, H., Contact allergy to petrolatums. I. Sensitizing capacity of different brands of yellow and white petrolatums, *Contact Dermatitis,* 9, 175, 1983.
59. Dooms-Goossens, A. and Degraff, H., Contact allergy to petrolatums. II. Attempts to identify the nature of the allergens, *Contact Dermatitis,* 9, 247, 1983.
60. Dooms-Goossens, A. and Dooms, M., Contact allergy to petrolatums. III. Allergenicity prediction and pharmacopoeial requirements, *Contact Dermatitis,* 9, 352, 1983.
61. Väänänen, A. and Hannuksela, M., Softisan — a new vehicle for patch testing, *Contact Dermatitis,* 14, 215, 1986.
62. Fullerton, A., Menné, T., and Hoelgaard, A., Patch testing with nickel chloride in a hydrogel, *Contact Dermatitis,* 20, 17, 1989.
63. Fisher, A.A., *Contact Dermatitis,* 3rd ed., Lea & Febiger, Philadelphia, 1986.
64. Fossereau, J., Benezra, C., and Maibach, H.I., *Occupational Contact Dermatitis, Clinical and Chemical Aspects,* Munksgaard, Copenhagen, 1982.
65. de Grooth, A.C., *Patch Testing. Test Concentrations and Vehicles for 2800 Allergens,* Elsevier, Amsterdam, 1986.
66. Wahlberg, J.E., Patch testing, in *Textbook of Contact Dermatitis,* Rycroft, R.J.G., Menné, T., Frosch, P.J., and Benezra, C., Eds., Springer-Verlag, Berlin, 1992, 239.
67. Clark, E.W., Blondeel, A., Cronin, E., Oleffe, J.A., and Wilkinson, D.S., Lanolin of reduced sensitizing potential, *Contact Dermatitis,* 7, 80, 1981.
68. Fregert, S., Dahlquist, I., and Trulsson, L., An attempt to isolate and identify allergens in lanolin, *Contact Dermatitis,* 10, 16, 1984.
69. Karlberg, A.-T. and Lidén, C., Comparison of colophony patch test preparations, *Contact Dermatitis,* 18, 158, 1988.
70. Matthies, C., Dooms-Goossens, A., Lachapelle, J.-M., Lahti, A., Menné, T., White, I.R., and Wilkinson, J., Patch testing with fractionated balsam of Peru, *Contact Dermatitis,* 19, 384, 1988.
71. *The Trolab Guide for Patch Testing,* Hermal, Kurt Herrmann, Reinbek, Germany, 1993.
72. Patch Test Allergens, Product catalogue, Chemotechnique diagnostis AB, Malmö, Sweden, 1992.
73. Hjorth, N., Routine patch tests, *Trans. St. John's Hosp. Dermatol. Soc.,* 49, 99, 1963.
74. Lepine, E.M., Results of routine office patch testing, *Contact Dermatitis,* 2, 89, 1976.
75. Gailhofer, G. and Ludvan, M., Zur Änderungen der Allergenspectrums bei Kontaktekzem in der Jahren 1975–1984, *Dermatosen,* 35, 12, 1987.
76. Magnusson, B., Blohm, S.-G., Fregert, S., Hjorth, N., Høvding, G., Pirilä, V., and Skog, E., Routine patch testing. III. Frequency of contact allergy at six Scandinavian clinics, *Acta Dermatol. Venereol.,* 46, 396, 1966.
77. Young, E. and Honwing, R.H., Patch test results with standard allergens over a decade, *Contact Dermatitis,* 17, 64, 1987.
78. Mitchell, J.C., Adams, R.M., Glendenning, W.E., Fisher, A., Kanof, N., Larsen, W., Maibach, H.I., Rudner, E.J., Schnorr, W., Stores, F., and Taylor, J., Results of patch tests with substances abandoned, *Contact Dermatitis,* 8, 336, 1982.
79. North American Contact Dermatitis Group, The frequency of contact sensitivity in North America 1972–74, *Contact Dermatitis,* 1, 277, 1975.
80. Pevny, I., Brennenstuhl, M., and Raxinskas, G., Patch testing in children. II. Results and case reports, *Contact Dermatitis,* 11, 302, 1984.
81. Goh, C.L., Epidemiology of contact dermatitis in Singapore, *Int. J. Dermatol.,* 27, 308, 1988.
82. Adams, R.M., *Occupational Skin Disease,* Grune & Stratton, Philadelphia, 1990.
83. Cronin, E., *Contact Dermatitis,* Churchill Livingstone, New York, 1980.
84. Karlberg, A.-T., Air oxidation increases the allergenic potential of of tall oil rosin. Colophony allergens also identified in tall oil rosin, *Am. J. Contact Dermatitis,* 2, 43, 1991.
85. Weibel, H., Kontakteksem. Biofarmaceutiske aspekter, *Pharm. AS Danmarks Farm. Højskole,* Thesis 1988.
86. Fregert, S., Publication of allergens, *Contact Dermatitis,* 12, 123, 1985.
87. Bruze, M., Fregert, S., and Gruvberger, B., Occurrence of para-aminobenzoic acid and benzocaine as contaminants in sunscreen agents of para-aminobenzoic acid type, *Photodermatology,* 1, 277, 1984.
88. Fregert, S., Björkner, B., Bruze, M., Dahlquist, I., Gruvberger, B., Persson, K., Trulsson, L., and Zimerson, E., *Yrkesdermatologi,* Studentlitteratur, 1990, 161.
89. Enders, F., Przybilla, B., and Ring, J., Patch testing with fragrance mix and its constituents: discrepancies largely due to the presence or absence of sorbitan sesquioleate, *Contact Dermatitis,* 24, 238, 1991.
90. Andersen, K.E., Svensson, L., Fischer, T., and Gunnarsson, Y., Experience with n-hydroxymethylsuccinimid (HMS) in TRUE test™ patches for the diagnosis of formaldehyde contact allergy, *Contact Dermatitis,* 23, 291, 1990.
91. Bruze, M., Use of buffer solutions for patch testing, *Contact Dermatitis,* 10, 267, 1984.
92. van Ketel, W.G., Patch testing with nickel sulphate in DMSO, *Contact Dermatitis,* 4, 167, 1978.
93. Wahlberg, J.E. and Skog, E., The effect of urea and lactic acid on the percutaneous absorption of hydrocortisone, *Acta Derm.,* 53, 207, 1973.
94. Söderberg, T.A., Elmros, T., Gref, R., and Hallmans, G., Inhibitory effect of zinc oxide on contact allergy due to colophony, *Contact Dermatitis,* 23, 346, 1990.
95. Fischer, T., On 8-hydroxyquinoline zinc oxide incompatibility, *Dermatologica,* 149, 129, 1974.
96. Wahlberg, J.E., Nickel chloride or nickel sulfate? Irritancy from patch test preparations as assessed by laser Doppler flowmetry, *Dermatol. Clin.,* 8, 41, 1990.
97. Wahlberg, J.E., Thresholds of sensitivity in metal contact allergy, *Berufsdermatosen,* 21, 22, 1973.
98. Gammelgaard, B., Fullerton, A., Avnstorp, C., and Menné, T., Permeation of chromium salts through human skin in vitro, *Contact Dermatitis,* 27, 302, 1992.
99. Lidén, S. and Lundberg, E., Penetration of chromium in human skin in vivo, *J. Invest. Dermatol.,* 72, 42 1979.
100. Marzulli, F.N. and Maibach, H.I., The use of graded concentrations in studying skin sensitizers: experimental contact sensitization in man, *Cosmet. Toxicol.,* 12, 219, 1974.
101. Fischer, T. and Rystedt, I., Cobalt allergy in hard metal workers, *Contact Dermatitis,* 9, 115, 1983.
102. Fischer, T. and Maibach, H.I., The science of patch test standardization, in *Immunology and Allergy Clinics of North America,* Vol. 9. Maibach, H.I., Ed., W.B. Saunders, Philadelphia, 1989, 417.
103. Bruze, M., Dahlquist, I., Fregert, S., Gruvberger, B., and Persson, K., Contact allergy to the active ingredients of Kathon CG, *Contact Dermatitis,* 16, 183, 1987.

104. Mitchell, J.C., Patch testing with mixes. Note on mercaptobenzothiazole mix, *Contact Dermatitis*, 7, 98, 1981.
105. Lynde, C.W., Mitchell, J.C., Adams, R.M., Maibach, H.I., Schorr, W.J., Storrs, F.J., and Taylor, J., Patch testing with mercaptobenzothiazole and mercapto-mixes, *Contact Dermatitis*, 8, 273, 1982.
106. van Ketel, W.G., Thiuram mix, *Contact Dermatitis*, 2, 232, 1976.
107. Menné, T. and Hjorth, N., Routine patch testing with paraben esters, *Contact Dermatitis*, 19, 189, 1988.
108. Beck, M.H. and Holden, A., Benzocaine — an unsatisfactory indicator of topical local anaesthetic sensitization for the U.K., *Br. J. Dermatol.*, 118, 91, 1988.
109. Kreilgård, B., Hansen, J., and Fischer, T., Optimization of the TRUE Test caine mix, *Contact Dermatitis*, 21, 23, 1989.
110. Lammintausta, K. and Kalimo, K., Sensitivity to rubber. Study with rubber mixes and individual rubber chemicals, *Dermatosen*, 33, 204, 1985.
111. Hjorth, N. and Trolle-Lassen, C., Quick and easy method for application of patch tests and storage of test substances, *Acta Derm. Venereol.*, 43, 324, 1963.
112. Bruze, M. and Fregert, S., Studies on purity and stability of photopatch test substances, *Contact Dermatitis*, 9, 33, 1983.
113. Goh, C.L. and Kwok, S.F., The influence of temperature on the concentration homogeneity of patch test materials, *Contact Dermatitis*, 15, 231, 1986.
114. Ruhnek-Forsbeck, M., Fischer, T., Meding, B., Pettersson, L., Stenberg, B., Strand, A., Sundberg, K., Svensson, L., Wahlberg, J.E., Widström, L., Wrangsjö, K., and Billberg, K., Comparative multicenter study with True Test™ and Finn Chamber patch test methods in eight Swedish hospitals, *Acta Derm. Venereol.*, 68, 123, 1988.
115. Lachapelle, J.M., Bruynzeel, D.P., Ducombs, G., Hannuksela, M., Ring, J., White, I.R., Wilkinson, J., Fischer, T., and Billberg, K., European multicenter study of the TRUE Test™, *Contact Dermatitis*, 19, 91, 1988.
116. Stenberg, B., Billberg, K., Fischer, T., Nordin, L., Pettersson, L., Ruhnek-Forsbeck, M., Sundberg, K., Swanbeck, G., Svensson, L., Wahlberg, J.E., and Wrangsjö, K., Swedish multicenter study with TRUE Test™, panel 2, in *Current Topics in Contact Dermatitis*, Frosch, P.J., Dooms-Goossens, A., Lachapelle, J.M., Rycroft, R.J., and Scheper, R.J., Eds., Springer-Verlag, Berlin, 1989, 518
117. Wilkinson, J.D., Bruynzeel, D.P., Ducombs, G., Frosch, P.J., Gunnarson, Y., Hannuksela, M., Lachapelle, J.-M., Ring, J., Shaw, S., and White, I.R., European multicenter study of the TRUE Test™, Panel 2, *Contact Dermatitis*, 22, 218, 1990.
118. Hornstein, M., Anvendung eines neuen Fertigpflasters in der Praxis (Epiquick), *Z. Hautkr.*, 62, 1719, 1987.
119. Ayala, F., Deledda, S., Francalancid, S., Lisi, P., Sertoli, A., and Valsecci, R., Inchiesta Policentrica GIRDCA sul Rapid Patch Test (RPT), *Boll. Dermatol. Allerg. Prof.*, 1, 33, 1992.
120. Jordan, W.P., 24-, 48-, and 48/48 hour patch tests, *Contact Dermatitis*, 6, 151, 1980.
121. Menné, T., Nickel allergy — reliability of patch test. Evaluated in female twins, *Dermatosen*, 29, 156, 1981.
122. Bruze, M., Patch testing with nickel sulphate under occlusion for 5 hours, *Acta Derm. Venereol.*, 68, 361, 1988.
123. Rudzki, E., Zakrzewski, Z., Prokopczyk, G., and Koslowska, A., Patch tests with potassium dichromate removed after 24 and 48 hours, *Contact Dermatitis*, 2, 309, 1976.
124. Skog, E. and Forsbeck, M., Comparison between 24- and 48-hour exposure time in patch testing, *Contact Dermatitis*, 4, 362, 1978.
125. McKee, S. and Burrows, D., Marking of patch tests, *Contact Dermatitis*, 9, 40, 1983.
126. Jordan, W.P., Fluorescent marking ink for patch test site identification, *Contact Dermatitis Newslett.*, 10, 229, 1971.
127. Björnberg, A., Toxic reactions to a patch test skin marker containing fuchsin-silver nitrate, *Contact Dermatitis*, 3, 101, 1977.
128. Dooms-Goossens, A., Toxic reactions to skin markers, *Contact Dermatitis*, 3, 280, 1977.
129. Paramsothy, Y., Collins, M., and Smith, A.G., Contact dermatitis in patients with leg ulcers. The prevalence of late positive reactions and evidence against systemic ampliative allergy, *Contact Dermatitis*, 18, 30, 1988.
130. Macfarlane, A.W., Curley, R.K., Graham, R.M., Lewis-Jones, M.S., and King, C.M., Delayed patch test reactions at days 7 and 9, *Contact Dermatitis*, 20, 127, 1989.
131. Mathias, C.G.T. and Maibach, H.I., When to read a patch test?, *Int. J. Dermatol.*, 18, 127, 1979.
132. Mitchell, J.C., Day 7 (D7) patch test reading — valuable or not?, *Contact Dermatitis*, 4, 139, 1978.
133. Rietschel, R.L., Adams, R.M., Maibach, H.I., Storrs, F.J., and Rosenthal, L.E., The case for patch test readings beyond day 2, *J. Am. Acad. Dermatol.*, 18, 42, 1988.
134. Wilkinson, D.S., Fregert, S., Magnusson, B., et al., Terminology of contact dermatitis, *Acta Derm. Venereol.*, 50, 287, 1970.
135. Hannuksela, M., Pirilä, V., and Salo, O.P., Skin reactions to propyleneglycol, *Contact Dermatitis*, 1, 112, 1975.
136. Trancik, R.J. and Maibach, H.I., Propyleneglycol: irritation or sensitization?, *Contact Dermatitis*, 8, 185, 1982.
137. Romaguera, C. and Grimalt, F., PPPP syndrome, *Contact Dermatitis*, 3, 102, 1977.
138. Schmidt, H., Schultz-Larsen, F., Ölholm-Larsen, P., and Søgaard, H., Petechial reaction following patch testing with cobalt, *Contact Dermatitis*, 6, 91, 1980.
139. Fischer, T. and Rystedt, I., Patch testing with sodium tungstate, *Contact Dermatitis*, 9, 69, 1983.
140. Wahlberg, J.E. and Maibach, H.I., Sterile cutaneous pustules: a manifestation of primary irritancy? Identification of contact pustulogens, *J. Invest. Dermatol.*, 76, 381, 1981.
141. Bruynzeel, D.P., van den Hoogenband, H.M., and Koedijk, F., Purpuric vasculitis-like eruption in a patient sensitive to balsam of Peru, *Contact Dermatitis*, 11, 207, 1984.
142. Roed-Petersen, J., Clemmensen, O.J., Menné, T., and Larsen, E., Purpuric contact dermatitis from black rubber chemicals, *Contact Dermatitis*, 18, 166, 1988.
143. Fyad, A., Masmoudi, M.L., Lachapelle, J.-M., The "edge effect" with patch test materials, *Contact Dermatitis*, 16, 147, 1987.
144. Lachapelle, J.-M., Tennstedt, D., Fyad, A., Masmoudi, M.L., and Nouaigui, H., Ring-shaped positive allergic patch test reactions to allergens in liquid vehicles, *Contact Dermatitis*, 18, 234, 1988.
145. Skoog, M.-L. and Groth, O., Cellular infiltrate in experimental allergic and toxic contact dermatitis. Effects of systematically administered corticosteroids, *Int. Arch. Allergy Appl. Immunol.*, 65, 168, 1981.
146. Kanerva, L., Ranki, A., and Lauharanta, J., Lymphocytes and Langerhans cells in patch tests, *Contact Dermatitis*, 11, 150, 1984.
147. Scheynius, A., Fischer, T., Forsum, U., and Klareskog, L., Phenotypic characterization in situ of the cells in allergic and irritant contact dermatitis in man, *Clin. Exp. Immunol.*, 55, 81, 1984.
148. Scheynius, A. and Fischer, T., Phenotypic differentiation between allergic and irritant patch test reactions in man, *Contact Dermatitis*, 14, 297, 1986.

149. Gawkrodger, D.J., Carr, M.M., McVittie, E., et al., Keratinocyte expression of MHC class II antigens in allergic sensitization and challenge reactions and in irritant contact dermatitis, *J. Invest. Dermatol.,* 88, 11, 1987.
150. Avnstorp, C., Ralfkiaer, E., Jørgensen, J., and Lange-Wantzin, G., Sequential immunophenotypic study of lymphoid infiltrate in allergic and irritant reactions, *Contact Dermatitis,* 16, 239, 1987.
151. Thestrup-Pedersen, K., Grönhöj-Larsen, C., and Rönnevig, J., The immunology of contact dermatitis. A review with special reference to the pathophysiology of eczema, *Contact Dermatitis,* 80, 81, 1989.
152. Von Blomberg, B.M.E., Recent Advances in Contact Allergy Research, Thesis, *Free University Press,* Amsterdam, 1989.
153. Belsito, D., The mechanism of allergic contact dermatitis, in *Color Text of Contact Dermatitis,* Larsen, W.G., Adams, R.M., and Maibach, H.I., Eds., W.B. Saunders, Philadelphia, 1992, 1.
154. Berardesca, E. and Maibach, H.I., Review article. Bioengineering and the patch test, *Contact Dermatitis,* 18, 3, 1988.
155. Agner, T., Basal transepidermal water loss, skin thickness, skin blood flow and skin colour in relation to sodium-lauryl-sulphate-induced irritation in normal skin, *Contact Dermatitis,* 25, 108, 1991.
156. van Neste, D. and de Brouwer, B., Monitoring of skin response to sodium lauryl sulphate: clinical scores versus bioengineering methods, *Contact Dermatitis,* 27, 151, 1992.
157. Mendelow, A.Y., Forsyth, A., Feather, J.W., Baille, A.J., and Florence, A.T., Skin reflectance measurements of patch test responses, *Contact Dermatitis,* 15, 73, 1986.
158. Raskin, M.M. and Zies, P.M., Role of thermography in allergic contact dermatitis, *Contact Dermatitis,* 3, 206, 1977.
159. Emtestam, L. and Ollmar, S., Electrical impedance index in human skin: measurement after occlusion in 5 anatomical regions and in mild irritant dermatitis, *Contact Dermatitis,* 28, 104, 1993.
160. Serup, J. and Staberg, B., Differentiation of allergic and irritant reactions by transepidermal water loss, *Contact Dermatitis,* 16, 129, 1987.
161. Malten, K.E. and Thiele, F.A.J., Evaluation of skin damage II: Water loss and carbon dioxide release measurements related to skin resistance measurements, *Br. J. Dermatol.,* 89, 565, 1973.
162. Prens, E.P., van Joost, T.H., and Szeketee, J., Quantification of patch test reactions by transcutaneous PO_2 measurement, *Contact Dermatitis,* 16, 142, 1987.
163. Pinnagoda, J., Tupker, R.A., Agner, T., and Serup, J., Guidelines for transepidermal water loss (TEWL) measurements. A Report from the Standardization Group of the European Society of Contact Dermatitis, *Contact Dermatitis,* 22, 164, 1990.
164. Wahlberg, J.E., Assessment of skin irritancy: measurement of skin fold thickness, *Contact Dermatitis,* 9, 12, 1983.
165. Jacques, S.L., Maibach, H.I., and Suskind, C., Water content in stratum corneum measured by a focused microwave probe: normal and psoriatic, *Bioeng. Skin,* 3, 118, 1981.
166. Li, Q., Aoyama, K., and Matsushita, T., Evaluation of contact allergy to chemicals using laser Doppler flowmetry (LDF) technique, *Contact Dermatitis,* 26, 27, 1992.
167. Gawkrodger, D.J., McDonagh, A.J.G., and Wright, A.L., Quantification of allergic and irritant patch test reactions using laser-Doppler flowmetry and erythema index, *Contact Dermatitis,* 24, 172, 1991.
168. Staberg, B. and Serup, J., Allergic and irritant skin reactions evaluated by laser Doppler flowmetry, *Contact Dermatitis,* 18, 40, 1988.
169. Nilsson, G., Otto, U., and Wahlberg, J.E., Assessment of skin irritancy in man by laser Doppler flowmetry, *Contact Dermatitis,* 8, 401, 1982.
170. Serup, J. and Staberg, B., Ultrasound for assessment of allergic and irritant patch test reactions, *Contact Dermatitis,* 17, 80, 1987.
171. Seidenari, S. and di Nardo, A., Cutaneous reactivity to allergens at 24-h increases from the antercubital fossa to the wrist: an echographic evaluation by means of a new image analysis system, *Contact Dermatitis,* 26, 171, 1992.
172. Lachapelle, J.-M., A left versus right side comparative study of Epiquick® patch test results in 100 consecutive patients, *Contact Dermatitis,* 20, 51, 1989.
173. Gollhausen, R., Przybilla, B., and Ring, J., Reproducibility of patch test results: comparison of True test™ and Finn Chamber test, in *Current Topics in Contact Dermatitis,* Frosch, P.J., Dooms-Goossens, A., Lachapelle, J.M., Rycroft, R.J., and Scheper, R.J., Eds., Springer-Verlag, Berlin, New York, 1989, pp. 518, 524.
174. Ruzicka, T., Gerstmeier, M., Przybilla, B., and Ring, J., Allergy to local anesthetics. Comparison of patch test with prick and intradermal test results, *J. Am. Acad. Dermatol.,* 16, 1202, 1987.
175. Möller, H., Intradermal testing in doubtful cases of contact allergy to metals, *Contact Dermatitis,* 20, 120, 1989.
176. Podmore, P., Burrows, D., and Bingham, E.A., Prediction of patch test results, *Contact Dermatitis,* 11, 283, 1984.
177. Cronin, E., Patch testing with nickel, *Contact Dermatitis,* 1, 56, 1975.
178. Dooms-Goossens, A., Naret, C., Chrispeels, M.T., and Degreef, H., Is 5% nickel sulphate patch test concentration adequate?, *Contact Dermatitis,* 6, 232, 1980.
179. Epstein, E., Contact dermatitis to neomycine with false negative patch tests: allergy established by intradermal and usage tests, *Contact Dermatitis,* 6, 219, 1980.
180. Epstein, E., Contact Dermatitis to 5-fluorouracil with false negative patch tests, *Contact Dermatitis,* 6, 220, 1980.
181. Sjöwall, P. and Christensen, O.B., Local and systemic effect of ultraviolet irradiation (UV B and UV A) on human allergic contact dermatitis, *Arch. Dermatol. Res.,* 66, 290, 1986.
182. Bruze, M., Seasonal influence on routine patch test results, *Contact Dermatitis,* 14, 184, 1986.
183. Dooms-Goossens, A., Lesaffre, E., Heidbuchel, M., Dooms, M., and Degreef, UV sunlight and patch test reactions in humans, *Contact Dermatitis,* 19, 36, 1988.
184. Lindelöf, B., Liden, S., and Lagerholm, B., The effect of grenz rays on the expression of allergic contact dermatitis in man, *Scand. J. Immunol.,* 21, 463, 1985.
185. Ek, L., Lindelöf, B., and Liden, S., The duration of Grenz ray-induced suppression of allergic contact dermatitis and its correlation with the density of Langerhans cells in human epidermis, *Clin. Exp. Dermatol.,* 14, 206, 1989.
186. Sukanto, H., Nater, J. P., and Bleumink, E., Influence of topically applied corticosteroids on patch test reactions, *Contact Dermatitis,* 7, 180, 1981.
187. Feuerman, E. and Levy, A., A study of the effect of prednisone and an antihistamine on patch test reactions, *Br. J. Dermatol.,* 86, 68, 1972.
188. Lembo, G., Presti, M.L., Balato, N., Ayala, F., and Santoianni, P., Influence of cinnarizine on patch test reactions, *Contact Dermatitis,* 13, 341, 1985.
189. Meffert, H., Wischnewski, G. G., and Gunther, W., Disodium cromoglycate inhibits allergic patch test reactions, *Contact Dermatitis,* 12, 18, 1985.

190. Biren, C.A., Barr, R.J., Ganderup, G.S., Lemus, L.L., and McCullough, J.L., Topical cyclosporine: effects on allergic contact dermatitis in guinea pigs, *Contact Dermatitis,* 20, 10, 1989.
191. Aldridge, R.D., Sewell, H.F., King, G., et al. Topical cyclosporin A in nickel contact hypersensitivity: results of a preliminary clinical and immunohistochemical investigation, *Clin. Exp. Immunol.,* 66, 582, 1986.
192. Warner, R.D., Dorn, C.R., Blakeslee, J.R., Gerken, D.F., Gordon, J.C., and Angrick, E.J., Zinc effects on nickel dermatitis in the guinea pig, *Contact Dermatitis,* 19, 98, 1988.
193. Agrup, G., Dahlquist, I., Fregert, S., et al., Value of history and testing in suspected contact dermatitis, *Arch. Dermatol.,* 101, 212, 1970.
194. Cronin, E., Clinical prediction of patch test results, *Trans. St. John's Dermatol.,* 58, 153, 1972.
195. Kiefer, M., Nickel sensitivity. Relationship between history and patch test reaction, *Contact Dermatitis,* 5, 398, 1979.
196. Rystedt, I., Evaluation and relevance of isolated test reactions to cobalt, *Contact Dermatitis,* 5, 122, 1979.
197. Rystedt, I. and Fischer, T., Relationship between cobalt and nickel sensitization in hard metal workers, *Contact Dermatitis,* 9, 195, 1983.
198. Orlandini, A., Viotti, G., and Magno, L., Anaphylactoid reaction induced by patch testing with formaldehyde in an asthmatic, *Contact Dermatitis,* 19, 383, 1988.
199. Björkner, B.E., Contact allergy and depigmentation from Alstroemeria, *Contact Dermatitis,* 8, 178, 1982.
200. Monti, M., Berti, E., Caviccini, S., and Sala, F., Unusual cutaneous reaction after gold chloride patch test, *Contact Dermatitis,* 9, 150, 1983.
201. Calnan, C., Keloid formation after patch test, *Contact Dermatitis,* 7, 279, 1981.
202. Agrup, G., Sensitization induced by patch testing, *Br. J. Dermatol.,* 80, 631, 1968.
203. de Groot, A.C., Bos, J.D., Jagtman, B.A., Bruynzeel, D.P., van Joost, T., and Weyland, J.W., Contact allergy to perservatives. II, *Contact Dermatitis,* 15, 218, 1986.
204. Weaver, J.E., Cardin, C.W., and Maibach, H.I., Dose response assessments of Kathon biocid. I. Diagnostic use and diagnostic threshold patch testing with sensitized humans, *Contact Dermatitis,* 12, 141, 1985.
205. Björkner, B., Bruze, M., Dahlquist, I., Freger, S., Gruvberger, B., and Persson, K., Contact allergy to the perservative Kathon® CG, *Contact Dermatitis,* 14, 85, 1986.
206. Hannuksela, M. and Salo, H., The repeated open application test (ROAT), *Contact Dermatitis,* 14, 221, 1986.
207. Hammershöy, O., Standard patch test results in 3,225 consecutive Danish patients from 1973 to 1977, *Contact Dermatitis,* 6, 263, 1980.
208. Hirano, S. and Yoshikawa, K., Patch testing with European and American standard allergens in Japanese patients, *Contact Dermatitis,* 8, 48, 1982.

Chapter 31.3
Type I Allergy Skin Tests

Lars Jelstrup Petersen
Department of Dermatology
Laboratory of Immunology and Biochemistry
Bispebjerg Hospital
University of Copenhagen
Copenhagen, Denmark

I. Introduction

Skin testing has been utilized for decades in the evaluation of atopic diseases. The simplicity, rapidity of performance and interpretation, low cost, and sensitivity of skin tests may explain the frequent use of this technique in identification of allergic diseases.[1] Besides the application of skin tests in routine diagnostic work, skin tests can be used in experimental immunopharmacological investigations,[2,3] standardization of allergen preparations,[4-9] and evaluation of changes in skin sensitivity during specific immunotherapy.[10-12] The use of skin testing in the latter issues requires thorough insight in skin testing methodology and interpretation, and will not be discussed further.

Skin tests represent a most widely used diagnostic tool in allergy. However, the lack of standardization of equipment, allergen extracts, and technical performance have sabotaged a uniform standard on the use of skin tests. Today, several approved techniques are applied, and many different commercial sources of antigen extracts are available. There is an increasing concern on the use of skin tests, and a number of recommendations on the use of skin tests have recently been published.[6,13-17] The following text goes into detail on these issues.

II. Pathophysiology of Skin Tests

The knowledge of the mechanisms involved in IgE-mediated hypersensitivity has increased considerably during the recent years.[18] Also, the scenario of mediators released and generated in immediate hypersensitivity is increasingly discovered.[19-22]

In short, the basic event taking place in the generation of a type I allergic reaction is contact between a specific antigen with IgE antibodies fixed on the surface of mast cells and basophils. The IgE-Fc receptors aggregate and activate the cell to release its vasoactive mediators. In order to elicit an IgE-mediated hypersensitivity reaction, the subjects have to be sensitized, i.e., the subject must have produced specific IgE antibodies which fix in great amounts on the surface of mast cells and basophils. When first sensitized, a cross-linking of allergen with two IgE antibodies on the cell causes degranulation of these cells. The early phase of the allergic type I reaction is mainly caused by degranulation of mast cells, which release of a number of preformed and newly generated vasoactive substances, including histamine, leukotriene C_4, prostaglandin D_2, platelet activating factor, proteases, and chemotactic factors.[19,20,22] A schematic representation of the mediators and the role of mast cell mediators in immediate allergic reactions is shown in Figure 1. In contrast to mucosa mast cells and in some cases circulating basophils, skin mast cells can also release histamine upon non-IgE stimulations, e.g., lectins, calcium ionophores, neuropeptides, compound 48/80, and complement fragments. Also, various cytokines and drugs like opiates may induce mast cell degranulation. Sensory neuropeptides localized in cutaneous neurons, including substance P, seem to be involved in some parts of the early reaction as well. The clinical picture, the wheal and flare response, is caused by leakage of capillaries leading to plasma extravasation and vasodilatation of the microvasculature. This reaction, which resolves within 1 hour, is inconsistently followed by a late-phase reaction developing about 6 hours after skin testing. The late-phase response is foremost characterized by infiltration of mononuclear cells, neutrophils, basophils, and eosinophils. The biochemical events taking place in the late phase reaction are very complex.[23,24] Clinically, the late phase reaction is an ill-defined edematous reaction that resolves within 24 hours.

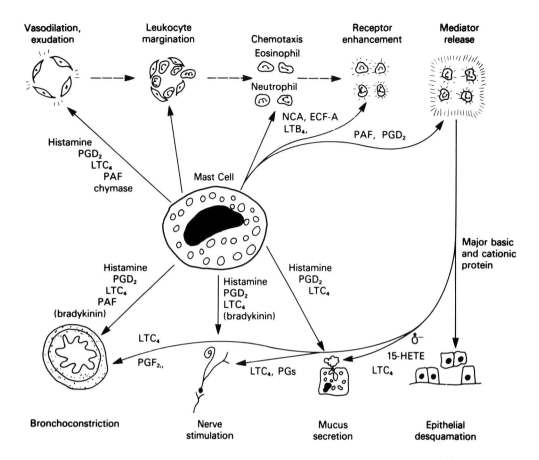

FIGURE 1 A schematic representation of the mediators released in an allergic type I reaction.

III. General Considerations of Skin Testing

It must always be remembered that skin tests cannot replace a detailed clinical history and a thorough physical examination. The information obtained from the history and the examination may serve as a basis for choice of allergens and interpretation of the skin responses. A list of general rules of allergy testing is shown in Table 1.

A. Precautions

Before any skin tests are initiated, some precautions must be taken into consideration. The items are listed in Table 2. No skin test should be performed without a physician immediately available, emergency equipment should be at hand, the skin test should not take place if allergic symptoms (predominately asthma) or dermatitis are present,[13,25] allergens must be controlled for concentration, potency, and stability, a positive and a negative control must be included, the patient must be evaluated for dermographism, the medication and schedule of medication must be recorded in the patient's file, and the skin responses must be measured at the proper time.

1. Adverse Reactions

The hazard of skin tests seems rather insignificant.[26,27] Since 1945 six fatalities have been reported as a consequence of skin testing, none of these with prick test.[26] In skin prick testing, development of large, local inflammatory responses and mild systemic reactions may occur, but life-threating reactions are extremely rare. In a very large survey, no severe allergic reactions (anaphylaxis, asthma) were seen with skin prick tests.[27] The experience of nonallergic reactions (syncope, near syncope, or malaise) was considerably smaller than that observed with venipuncture. However, in intradermal testing, the importance of the security precautions must be emphasized, since intradermal testing is about 10,000 times more sensitive than skin prick testing.

It is evident that skin testing should not spread disease. Therefore, to avoid viral or bacterial contamination, all material used for skin testing should be noncontaminated and disposable needles and syringes should be used.

B. Performance

Skin tests are normally performed on the volar surface of the forearm or on the back. If only a small number

TABLE 1 General Rules for Allergy Tests

1. Be sure that simpler, less expensive, and safer techniques have received an adequate trial
2. Ask yourself: *Are the tests really necessary?*
3. Always *believe* the history; no matter how outlandish it may seem, accept it for a fact until cautious testing proves or disproves it; consider the patient's safety above all else
4. Avoid innumerable and casually selected allergens; know *why* each test will help management
5. Limit the number of tests done at any one time
6. Test highly suspected allergens separately
7. Record important allergic sensitivities in a highly visible, specific place in the patient's record
8. Establish procedures to make certain that you or your office personnel do not inadvertently test the patient for factors to which he or she is anaphylactically sensitive
9. Never perform any type of test without having emergency equipment immediately at hand
10. Perform challenge tests only when the benefits makes the risk worthwhile; document the decision to perform a test with written consent

Note: It must be emphasized that these rules express a rather conservative attitude, and they are recommended mainly for intracutaneous testing. Due to the frequent use of skin prick tests with highly standardized allergen extracts, some of the items are less relevant.

From Mansmann, H.C., Allergy tests in clinical diagnosis, in Allergic Diseases of Infancy, Childhood and Adolescence, *Bierman, C.W. and Pearlman, D.S., Eds., W.B. Saunders, Philadelphia, 1980, chap. 21. Reprinted with permission.*

TABLE 2 General Skin Test Precautions

1. Skin test should never be performed unless a physician is available immediately to treat constitutional reactions
2. Have emergency equipment at hand: rubber tourniquet, airways, laryngoscope, hand resuscitation bag, epinephrine, vasopressor agent, theophylline, injectable antihistamine, and oxygen must be available and an intravenous set with intravenous fluids should be checked routinely
3. Do not test if the patient has moderate to severe symptoms: it is preferable not to test if *any* asthmatic symptoms are present
4. Double check known and suspected list by history
5. Be certain that the test concentrations are known and appropriate
6. Determine and record medications patient is taking and time of last dose

Note: For comments, please refer to Table 1.

From Mansmann, H.C., Allergy tests in clinical diagnosis, in Allergic Diseases of Infancy, Childhood and Adolescence, *Bierman, C.W. and Pearlman, D.S., Eds., W.B. Saunders, Philadelphia, 1980, chap. 21. Reprinted with permission.*

of tests are performed, the forearm is most convenient.[13] It is not necessary to clean the test area with alcohol; if alcohol is used, the skin must be totally dry before skin testing.

Before application of the allergens, the positions of the extracts are marked with an appropriate code. To avoid interference between test sites, test sites should be separated by 5 cm in intradermal skin testing and by 3 cm in prick tests.[7,13] Some studies have suggested an even larger space between skin tests.[28,29] Avoid skin testing within 3 cm from the anticubital fossa and 5 cm from the wrist.

Precision of skin test methods varies with technique and the person performing the test.[30–34] Therefore, it is highly recommended to assure the precision of the technique in the hands of the persons doing skin tests, both in daily routine as well as in experimental studies.

C. Factors Influencing Skin Testing
1. Skin Testing and Drug Treatment

Several drugs can interfere with skin tests by modulating the wheal and flare response, thereby making the interpretation of the tests difficult. The wheal and flare reaction obtained by skin testing may be significantly suppressed by a number of antiallergic drugs.[2]

a. Antihistamines

In general, H_1-antihistamines should be discontinued at least 3 days before skin testing.[35–37] This recommendation include conventional antihistamines such as chlorpheniramine, promethazine, cyproheptadine, and the recent nonsedating antihistamines such as terfenadine, loratadine, cetirizine, and acrivastine. Clemastine and hydroxine may suppress the wheal reaction for up to 10 days.[36,37] The duration of the antihistaminergic effect of astemizole may last more than 2 months. Since the duration of the wheal-suppressive effects of each drug may vary considerably among subjects, a positive histamine control may suggest that allergen testing is adequate, since the reaction to histamine and allergen seems comparable.[35]

The H_2-antihistamines ranitidine and cimetidine exert little or no effect on the wheal response.[38,39]

b. Theophylline
The effect of theophylline on the wheal reaction is insignificant, and it is of no practical importance.[40–42]

c. β-Adrenergic Agents
Inhaled β_2-agonists in doses for use in asthma therapy seem not to suppress the wheal size.[41–43] Orally administered and parenteral given adrenergics diminish the wheal response,[44,45] and should be avoided for 72 hours. β-blocking drugs significantly increase skin reactivity.[13]

d. Corticosteroids
Systemically administered glucocorticosteroids up to 30 mg of prednisolone per day for 1 week and chronic low dose (less than 10 mg prednisolone per day) have no effect on skin reactivity.[46,47] However, a recent study showed that low-dose corticosteroid suppressed the wheal size to the mast cell secretagogue codeine, but did not affect histamine-induced wheal formation.[48] Topical application of steroid formulations potently suppresses skin responses markedly.[49]

e. Various Drugs and Pharmacological Modalities
Tricyclic antidepressants may interfere with skin testing and should therefore be avoided for 2 weeks. Phototherapy may also interfere with skin reactivity. A single, low-dose of UVA increases skin reactivity, whereas psoralene plus UVA decreases the response.[50]

f. Specific Immunotherapy
Specific immunotherapy significantly decreases skin reactivity to the allergen used for immunotherapy as shown by various allergenic materials.[51,52] Skin reactivity to other allergens is not affected. The decrease in skin reactivity is evident within the first week of immunotherapy.[11]

2. Individual Factors Affecting Skin Testing
The size of the wheal and flare reaction is related to a number of individual factors as mentioned below, but depends also on the degree of sensitization of mast cells and the skin sensitivity to the mediators released.[53–55]

a. Anatomical Region
It is generally believed that skin reactions to allergen varies within different body regions.[56] The mid and upper back are more reactive than the lower back.[57] The back is more reactive than the forearm for some substances,[58,59] but not for all reaction producers.[59] Proximal locations of the forearm are the most reactive parts;[53] the ulnar side of the arm is more reactive than the radial.[53,56] Skin reactivity may differ slightly from the right to the left arm.[60] Some authors have found no differences in wheal size in different locations on the arm.[61]

b. Age
Skin reactivity to histamine is moderate in infants. It is possible to perform skin testing in infants after 3 months of age.[62] Since the sensitivity to allergens parallels the reactivity to histamine and mast cell releasers, a comparison with the positive reference is recommended. The wheal size increase in young adults, and gradually declines during aging.[63–65]

c. Circadian Variations
Skin reactions show diurnal variations with maximal skin reactivity during the late evening and the night.[66-68] However, skin reactivity seems constant during normal working hours.[69,70]

d. Other Factors
There seems to be no difference in skin reactivity between men and women, but skin reactivity may change during the menstruation cycle.[67] Wheal sizes are greater in pigmented skin.[71] The skin sensitivity to pollen increases during the pollen season, possibly due to increased IgE production.[72,73]

D. Allergen Extracts
Today, a large number of allergen extracts are available. In order to ensure an adequate diagnostic and scientific program in skin testing, they must be available in standardized and stable preparations with known composition, potency, and stability.[4,6,13,74] The lack of standardization and validation of commercial antigen extracts may lead to considerable variability in skin tests.[75,76] Batch-to-batch variation must be minimized since the potency may vary by up to a 100-fold between batches.[5,77]

The potency of the allergen extracts determines the sensitivity and specificity of skin tests. Allergen extracts with low potency do not diagnose patients with low sensitivity. The potency of allergen extracts used to diagnose specific allergy in patients attending a university allergy clinic may differ from a random sample of subjects in the community.

In order to ensure quality and stability of allergen extracts, a number of requirements must be fulfilled.[14] The requirements are listed in Table 3. Compared to the natural occurring allergen, the extract must contain all the potential allergenic molecules, the ratio between the different allergenic substances must be maintained, and the potency level should be similar. The concentration must be known and characterized according to either Nordic Biological Units (BU/ml), the U.S. Allergy Units (AU/ml), or the Index of Reactivity (IR).[78] Potency should also be given in International Units (IU/ml). The extract must not be contaminated with other allergens. Irritating or toxic substances should be avoided. Meanwhile, native insect venoms

TABLE 3 Requirements for Allergenic Extracts

1. It must contain all potential allergenic molecules
2. The ratio between the different allergenic molecules must be relevant
3. The allergens in the extract must be of native character
4. The potency level in allergen extracts must be relevant
5. The concentration and content of major allergens must be known and characterized according to acknowledged standards
6. The allergen must not be contaminated with other allergens
7. Irritating and toxic substances in the extracts should be avoided if possible or kept at a minimum
8. The stability of the extracts must be documented within accepted limits
9. Batch-to-batch variation must be minimized

contain both allergens together with a number of vasoactive substances. Some allergen extracts are contaminated with histamine.[79] For a short review on manufacturer-specific units used to label skin tests extracts, and units prescribed by authorities, refer to Reference 9.

It is important that the potency of the extracts and dilutions are kept constant and bacterial contamination is avoided during storage. The potency must be kept constant by use of freeze dried extracts. It is recommended that all allergens should be stored at 4 to 8° C to delay loss of potency. However, a recent report showed a 20% deterioration of *Alternaria* extract after 1 year of storage at 4°C;[80] 50% glycerol should be added to stabilize the diluted extracts. However, 50% glycerol should not be added for intradermal skin testing since it produces pain and inflammation at the test site.[81] Intradermal test solutions can be diluted in 0.03% human serum albumin preserved with 0.4% phenol.[82]

The quality of some allergens, primarily some food allergens, is not well characterized.[83–86] Therefore, skin testing with fresh fruit, fish, meat, nut, etc. can be performed with a modified skin prick procedure. At first the allergy pricker is introduced into the allergenic substance, then into the skin. However, some allergens, e.g., flour, need extraction.

E. Positive and Negative Controls

It is necessary to include a positive and a negative control in each skin test. To ensure normal reactivity of the skin, the skin test must include a positive control and to exclude patients with dermographism a negative control must be performed.

1. Positive Controls

Positive controls can include histamine or mast cell secretagogues such as compound 48/80, codeine, or morphine.[15,87–89] Rabbit and murine anti-human IgE have been used experientially to mimic the allergen-induced skin response in nonsensitized individuals.[90] Exogenous histamine elicits a skin response which is a measure of skin sensitivity, whereas mast cell secretagogues also take into account local differences in mast cell density and the mediator-releasing properties of these cells. The use of histamine is considered appropriate as a positive control in most countries. For skin prick test, histamine dihydrochloride 1 mg/ml (5.43 mM) has been used, but recent studies have indicated that histamine 10 mg/ml is preferable.[6,91] In intradermal testing, histamine 0.01 to 0.1 mg/ml or codeine 0.1 to 1.0 mg/ml can be used.[13,92]

2. Negative Controls

To evaluate unspecific reactions, tests should be performed with a negative control. Normally, the negative control is the diluent. For skin prick test that means 50% glycerol in isotonic saline; for intradermal test it means the diluent with the concentration of human serum albumin or phenol used with the allergen extracts.

F. Measurement of Skin Reactions
1. General Aspects

Skin challenge with allergens in atopic subjects elicits an immediate wheal-and-flare reaction, which should be inspected and recorded after 15 min.[13,16,93,94] The kinetics of the skin responses may vary between the different positive controls and allergens, but, since the wheal-and-flare reaction slowly fade, a fixed interval from the skin challenge to the recording of skin responses is adequate. In some patients, skin reactions develop slowly and reactions must therefore be reinspected 25 to 30 min after skin testing. The largest reaction is recorded. Some subjects develop an allergic late phase reaction. Investigating skin reactivity to allergens, the early wheal-and-flare reaction is usually determined. From a clinical point of view, the late-phase reaction may be interesting since the late-phase reaction is considered to be a measure of IgE sensitivity.[95,96] From a practical point of view, this reaction is not easily recorded. There is no proper method for measuring this phenomenon in the skin, and it usually develops several hours after allergen exposition.

2. Tracing of Skin Responses

The contours of the wheal-and-flare reactions should be outlined with a fine tip pen, and transferred to the patient's medical record by means of hypoallergic micropore tape.

3. Wheal or Flare Size?

In prick tests, the wheal size is normally used. In intradermal testing both wheal size and flare size are recorded.[93,97] Some investigators have found the flare size to be most reproducible,[8,57] whereas others have found the wheal size to be less variable.[98,99] Meanwhile, in food allergy, for example, the size of the flare reaction is a poor predictor of clinical allergy.[99,100] There may be disparity between wheal-and-flare reaction in skin tests. Using skin prick test in food-allergic children with atopic dermatitis, Sampson[85] found positive wheal responses were present 8 to 38% of the time with no evidence of flare, depending on the allergen extracts used.

4. Qualitative and Quantitative Measurement of Skin Responses

The size of the skin reactions should be measured as the mean of the longest and the midpoint orthogonal diameters, or the sum of these diameters. For a simple evaluation of IgE sensitization, the size of the allergen-induced wheal has been evaluated in relation to a fixed cutoff value or have been compared with that of the reference wheal as described later in this text. In several cases, a more quantitative approach is essential, for example, for the determination of biological activity of allergen extracts, quantitative skin testing, monitoring the effect of specific immunotherapy, immunopharmacological interventions, and methodological skin test procedures.

Quantitative assessment of the outlined skin reactions can be performed using the sum of orthogonal diameters, mean diameter, cutting/weighting, computer digitizing, and scanning. Comparing the different methods, excellent agreements have been found.[101-103] For small wheal sizes the coefficient of variation on wheal size determined by diameters is substantial. Problems with pen thickness and ink diffusion in skin and on adhesive tape have major impact on the measurement, especially with small wheal sizes.

Other methods, such as laser Doppler flowmetry and ultrasound have been applied for measurement of wheal-and-flare reactions.[104,105] The advantage of objective methods for assessment of skin reactions compared with conventional tracings in the diagnostics and scientific use of skin tests needs to be substantiated.

IV. Methods of Skin Testing

Two skin test methods are frequently used, the intradermal test technique and skin prick test. The previously used scratch technique is not recommended for either clinical or experimental use due to its traumatic nature, resulting in many false-positive reactions.[106]

A. Intradermal Test

The intradermal method was originally described by Mantoux. The allergen or mast cell secretagogue is injected intradermally, where the reaction producer stimulates skin mast cells to release their inflammatory mediators. The intradermal method is 10.000 to 30.000 more sensitive than skin prick test, and should be performed in low-sensitivity patients, and with allergen extracts of low potency.[8,13,107] If intradermal testing is not proceeded by skin prick tests, the allergen exposure may be too high. This may account for the increased risk of adverse reactions with intradermal skin tests compared with skin prick tests.[26]

1. Technical Performance

A sterile, small-volume syringe with a built-in fine needle, e.g., a 26- or a 27-gauge needle, is recommended.[13] The syringe should contain 100 μl of the test substance, and all air bubbles should be expelled. Space each injection by 5 cm to avoid interaction between test sites. The needle is introduced into the skin at a 45° angle to the skin surface, the bevel of the syringe pointing downward. The point of the needle is gently inserted in the skin in a forward lifting motion so as to pick up the skin with the tip of the needle. When the tip of the needle is in the skin, the motion is shifted to a forward, downward pressure, with the barrel lowered to minimize the angle to the skin surface. The needle is positioned in the skin with the bevel entirely within the skin. The test solution in volume from 10 to 50 μl is then introduced into the skin, and a small bleb, 3 mm in diameter, is generated. If no bleb is formed, leakage occurring from the bleb or air is splashed into the skin and a new test must be performed an appropriate distance from the first one. Avoid injection of allergen extracts directly into the dermal vasculature bed.

2. Reproducibility

To assure reproducibility, single intradermal injections are sufficient if the test is performed appropriately.[8] Coefficient of variation on the wheal and flare size is about 5 to 20%.[98,108] The use of very precise microinjection syringes offers no advantage on reproducibility compared to disposable syringes. The design and size of the needle may have some impact on the patient compliance.[94]

3. Cutoff Limits

Intradermal skin tests are very sensitive. False-positive reactions may occur, especially if the allergen extracts contain vasoactive components or if pH, osmolarity, or preservatives are not controlled.[109] To discriminate between allergic and nonallergic subjects, the skin reaction must exceed a certain value. Using the grading presented in Table 4, a 2+ reaction is most used as a criterion for a positive skin test.

B. Skin Prick Test

In skin prick testing, the allergen is introduced into the dermis via a small canal from the surface of the epidermis to the middle dermis. Not until recent

TABLE 4 Grading of Skin Responses with Intradermal Tests

Grade	Erythema (mm)	Wheal (mm)
0	<5	<5
+/−	5–10	5–10
1+	11–20	5–10
2+	21–30	5–10
3+	31–40	10–15 or with pseudopods
4+	>40	>15 or with many pseudopods

Note: A 2+ reaction is considered a positive reaction.

According to Bousquet, J., In vivo methods for study of allergy: Skin tests, techniques, and interpretation, in Allergy, Principles and Practice, *Middleton, E., Jr., Reed, C.E., Ellis, E.F., Adkinson, N.F., Jr., and Yunginger, J.W., Eds., C.V. Mosby, St. Louis, 1988, chap. 19.*

years, the skin prick test was performed with a fine disposable needle or an ordinary blood lancet.[110] In 1979, Østerballe and Weeke introduced a lancet with a 1-mm tip and shoulders to prevent deep penetration.[101] Since then other devices have been presented.[30,32,58,97,111]

1. Technique

Using the method described by Pepys, the tip of the needle is introduced into the superficial skin layers through a drop of allergen extract at an angle 60 to 70° to the skin surface. A small canal is then formed, through which the allergen solution penetrates. This technique has been modified after the introduction of specially made SPT needles. Perpendicular to the skin, the lancet should be pressed, not jabbed, through a droplet of allergenic extract.[6,30,32,97] The pressure and duration of the pressure is very important. The lancet is calmly removed again through the droplet. Use a new lancet for each skin test solution. Some advocate immediate blotting of the allergen after prick testing, whereas others leave the allergen in place until tracing. The droplets are then removed with a tissue. Ensure that the allergen extracts do not mix when removed. If bleeding occurs, repeat the skin test in an appropriate distance from the first puncture.

2. Reproducibility

Appropriately performed, the reproducibility of skin prick tests is adequate.[8,60,112,113] However, the coefficient of variation may range from 10 to more than 50%,[60,98,108] depending on a number of factors, mainly the size of the wheal reaction. Recent studies have shown that the coefficient of variation, sensitivity, and specificity may vary considerably among different prick test devices.[32,114–116] This particular subject receives much attention.[117] It is well known that the reproducibility of the skin prick test technique varies considerably among subjects performing the tests. Therefore, it is recommended that the coefficient of variation on duplicate tests in a number of subjects for each investigator is recorded and evaluated. A coefficient of variation less than 40% based on area and 20% based on mean diameters is recommended.[6,7,118] Among skilled technicians, this variation can be reduced with 50%.[34] In general, skin prick test is less reproducible than intradermal tests.

It is highly recommended that skin prick tests are performed in duplicate since false-negative results occur in clearly allergic subjects.[6] In an epidemiological study, it was found that the prevalence of sensitization decreased from 49 to 34% when the limit for sensitization was changed from one to two positive skin prick tests.[119] The risk of false-negative results is high in patients with low skin sensitivity. For scientific purposes, quadruplicate tests are recommended.[6,7]

3. Evaluation of Results

Most studies on skin prick tests have measured the wheal size, not the flare reaction.[30,58,89] Generally, the wheal reaction to the negative control is imperceptible, but it may amount to 2.5 mm even without otherwise demonstrable dermographics. A wheal >3 mm is considered positive,[6] but in some studies a cutoff limit of 2 mm in diameter is used.[120] A small wheal size may be positive if the negative control is zero, but it depends on the potency of the allergen preparation and the technique used. The difference between 2- and 3-mm wheal diameters denotes a tenfold difference in skin sensitivity. A grading system calculating the percentage of the allergen wheal size compared with the positive control may be used. However, allergen potency is not taken into account, and the choice of positive control is of major importance.

V. Interpretation of Skin Tests

A positive skin test does not necessarily mean that the disease is allergic, since a number of nonallergic patients demonstrate positive skin tests.[121,122] For example, positive skin tests are present in over 80% of patients with atopic dermatitis,[123] but most of these positive tests have no clinical relevance.[124] Likewise, a negative skin test does not rule out an allergic disease. We know that the difference in sensitivity between the most- and the least-sensitive patients is about 10^4.[6] Also, in the very early onset of allergy, skin and *in vitro* tests might show negative results on a specific allergen which elicits symptoms upon provocation in the target organ.[125] The term latent allergy[126,127] must also be kept in mind. Some subjects become sensitized to allergens in the environment without elicitation of clinical allergy in response to exposure to such antigens in daily life. Some may develop clinical allergy and some may not. Also, reactive IgE can be found years after disappearance of clinical allergy, e.g., allergy to milk in children.

A. Prevalence of Positive Skin Tests

The prevalence of positive skin tests in asymptomatic subjects has been thoroughly investigated. In large studies, the frequency of positive skin tests in the general population ranges from 10 to 40%.[128-132] In children and adolescents, the prevalence is 20 to 50%;[59,133-137] in young and middle-aged adults the frequency is 30 to 55%.[134,138,139] In aging subjects the frequency decreases to values about 25%.[140] The frequency of positive skin tests in the general population or selected groups of subjects depends on a number of factors, e.g., geographical region, criteria for selection of the target population, age of the subjects, choice of skin test method and its application, the quality, concentration, and type of allergen extracts, the number of allergens tested, value of cutoff limit, and individual variations as described previously.[141,142]

VI. Comparisons With *In Vitro* Diagnostic Procedures

The riddle in all diagnostic work is the lack of an answer book. In this context, allergic diseases exert no exception. All diagnostic tests are indicative, but no one is a true marker of allergy.

In allergic diseases, the clinical history is of utmost importance. Sometimes the history is very informative, e.g., in animal dander allergy and when the symptoms are related to well-defined pollen seasons. However, in many cases, further examinations are needed. Skin tests together with *in vitro* tests and organ provocations are used to confirm our suspicion of the existence of a specific IgE-mediated allergic disease.

The definition of an allergic disease is the development of allergic symptoms when the patient is exposed to a specific allergen. Consequently, allergen provocation in the target organ, e.g., bronchial, conjunctival, nasal, or food challenge, has been regarded as a very useful test. Even though the allergen exposure during the test (high dose — short duration) is different from exposure in daily life, it is difficult to believe that nonreacting patients suffer from an allergic disease to the allergen in question. In many allergy diagnostic comparative trials as mentioned below, only provocation-positive patients have been included.

Radio-allergoabsorbent test (RAST) and basophil histamine release (HR) test are commonly used methods in the diagnosis of allergic diseases.[143,144] In order to compare the diagnostic sensitivity and specificity of skin test against *in vitro* diagnostic tests, the performance of the tests has to be standardized. The results of *in vitro* allergy tests also rely on the allergen material used and the sensitivity and specificity of the method. For example, the use of fresh egg and milk improved the sensitivity of skin test and basophil HR in patients with double-blind, placebo-controlled food challenged documented food hypersensitivity compared with commercial extracts.[145] Comparisons with different RAST systems may show conflicting results due to various reasons.[146,147]

Generally, the concordance between skin test and *in vitro* tests is good, but it certainly depends on the type and quality of allergens used.[82,148-156] In general, there is a good correlation between strong positive skin tests and strong *in vitro* diagnostic tests; likewise, a negative skin test is often seen with a negative radio-absorbent test or HR test. For example, skin prick test results are excellent negative indicators of immediate food hypersensitivity provided that the food allergens are of appropriate quality.[124] Like most diagnostic procedures, the problems arise with borderline results.

Acknowledgement

H.-J. Malling, M.D., Ph.D., is thanked for comments to the manuscript.

References

1. Skin testing and radioallergosorbent testing (RAST) for diagnosis of specific allergens responsible for IgE-mediated diseases. Position statement, *J. Allergy Clin. Immunol.*, 72, 515, 1983.
2. Pipkorn, U., Pharmacological influence of anti-allergic medication on in vivo allergen testing, *Allergy*, 43, 81, 1988.
3. Van Neste, D., Ghys, L., Antoine, J.L., and Rihoux, J.P. Pharmacological modulation by cetirizine and atropine of the histamine- and methacholine-induced wheals and flares in human skin, *Skin Pharmacol.*, 2, 93, 1989.
4. Aas, K., Backman, A., Belin, L., and Weeke, B., Standardization of allergen extracts with appropriate methods. The combined use of skin prick testing and radio-allergosorbent test, *Allergy*, 33, 130, 1978.
5. Björkstén, F., Haahtela, T., Backman, A., and Suoniemi, I., Assay of the biological activity of allergen skin test preparations, *J. Allergy Clin. Immunol.*, 73, 324, 1984.
6. Dreborg, S., *The Skin Prick Test. Methodological Studies and Clinical Applications*, Thesis, Linköping University Medical Dissertations no. 239, 1987.
7. Malling, H.-J., Proposed guidelines for quantitative skin prick test procedures to determine the biological activity of allergenic extracts using parallel line assay, *Allergy*, 42, 391, 1987.
8. Turkeltaub, P.C., Rastogi, S.C., Baer, H., Anderson, M.C., and Norman, P.S., A standard quantitative skin-test assay of allergen potency and stability: studies on the allergen dose response curve and effect on wheal, erythema, and patient selection on assay results, *J. Allergy Clin. Immunol.*, 70, 343, 1982.
9. Dreborg, S., Standardization of allergenic preparations by *in vitro* and *in vivo* methods, *Allergy*, 48, 63, 1993.
10. Van Metre, T.E., Adkinson, N.F., Jr., Kagey-Sobotka, A., Marsh, D.G., Norman, P.S., and Rosenberg, G.L., How should we use skin testing to quantify IgE sensitivity?, *J. Allergy Clin. Immunol.*, 86, 583, 1990.
11. Bousquet, J., Calvayrac, P., Guerin, B., Hejjaoui, A., Dhivert, H., Hewitt, B., and Michel, F.B., Immunotherapy with a standardized *Dermatophagoides pteronyssinus* extract. I. In vivo and in vitro parameters after a short course of treatment, *J. Allergy Clin. Immunol.*, 76, 734, 1985.

12. Basomba, A., Evaluation of changes in skin sensitivity by means of skin tests, *Allergy,* 48, 71, 1993.
13. Leonard Bernstein, I., Proceedings of the task force on guidelines for standardizing old and new technologies used for the diagnosis and treatment of allergic diseases, *J. Allergy Clin. Immunol.,* 82, 487, 1988.
14. Dreborg, S., Skin tests used in type I allergy testing. Position paper, EAACI subcommittee on skin tests, *Allergy,* Suppl. 10, 1, 1989.
15. Bousquet, J., In vivo methods for study of allergy: skin tests, techniques, and interpretation, in *Allergy, Principles and Practice,* Middleton, E., Jr., Reed, C.E., Ellis, E.F., Adkinson, N.F., Jr., and Yunginger, J.W., Eds., C.V. Mosby, St. Louis, 1988, 419.
16. Report on skin test standardizations, The committee on skin test standardization of the Netherlands Society of Allergy, *Clin. Allergy,* 18, 305, 1988.
17. Dreborg, S. and Frew, A., Allergen standardization and skin tests, Position paper, EAACI subcommittee on allergen standardization and skin tests, *Allergy,* Supp. 14, 49, 1993.
18. Ishizaka, T., Mechanisms of IgE-mediated hypersensitivity, in *Allergy, Principles and Practice,* Middleton, E., Jr., Reed, C.E., Ellis, E.F., Adkinson, N.F., Jr., and Yunginger, J.W., Eds., C.V. Mosby, St. Louis, 1988, 71.
19. Holgate, S.T., Robinson, C., and Church, M.K., Mediators of immediate hypersensitivity, in *Allergy, Principles and Practice,* Middleton, E., Jr., Reed, C.E., Ellis, E.F., Adkinson, N.F., Jr., and Yunginger, J.W., Eds., C.V. Mosby, St. Louis, 1988, 135.
20. Burrall, B.A., Payan, P.G., and Goeztl, E.J., Arachidonic acid-derived mediators of hypersensitivity and inflammation, in *Allergy, Principles and Practice,* Middleton, E., Jr., Reed, C.E., Ellis, E.F., Adkinson, N.F., Jr., and Yunginger, J.W., Eds., C.V. Mosby, St. Louis, 1988, 164.
21. Dale, M.M. and Foreman, J.C., *Textbook of Immunopharmacology,* Blackwell Scientific, Oxford, 1989.
22. Benyon, R.C. and Church, M.K., Mast cells in the skin, in *Handbook of Atopic Eczema,* Ruzicka, T., Ring, J., and Przybilla, B., Eds., Springer-Verlag, Berlin, 1991, 173.
23. Lemanske, R.F., Jr. and Kaliner, M.A., Late-phase allergic reactions, in *Allergy, Principles and Practice,* Middleton, E., Jr., Reed, C.E., Ellis, E.F., Adkinson, N.F., Jr., and Yunginger, J.W., Eds., C.V. Mosby, St. Louis, 1988, 224.
24. Slifman, N.R., Adolphson, C.R., and Gleich, G.J., Eosinophils: biochemical and cellular aspects, in *Allergy, Principles and Practice,* Middleton, E., Jr., Reed, C.E., Ellis, E.F., Adkinson, N.F., Jr., and Yunginger, J.W., Eds., C.V. Mosby, St. Louis, 1988, 179.
25. Uehara, M., Reduced histamine reaction in atopic dermatitis, *Arch. Dermatol.,* 118, 244, 1982.
26. Lockey, R.F., Benedict, L.M., Turkeltaub, P.C., and Bukantz, S.C., Fatalities from immunotherapy (IT) and skin testing (ST), *J. Allergy Clin. Immunol.,* 79, 660, 1987.
27. Turkeltaub, P.C. and Gergen, P.J., The risk of adverse reactions from percutaneous prick-puncture allergen skin testing, venipuncture, and body measurements: data from the second National Health and Nutrition Examination Survey 1976–80 (NHANES II), *J. Allergy Clin. Immunol.,* 84, 886, 1989.
28. Terho, E.O., Husman, K., Kivekäs, J., Riihimäki, H., Histamine control affects the wheal produced by the adjacent diluent control in skin prick tests, *Allergy,* 44, 30, 1989.
29. Koller, D.Y., Pirker, C., Jarisch, R., and Gotz, M., Influence of the histamine control on skin reactivity in skin testing, *Allergy,* 47, 58, 1992.
30. Basomba, A., Sastra, A., Peláez, A., Romar, A., Campos, A., and García-Villmanzo, A., Standardization of the prick test. A comparative study of three methods, *Allergy,* 40, 395, 1985.
31. Brown, W.G., Halonen, M.J., Kalterborn, W.T., and Barbee, R.A., The relationship of respiratory allergy, skin test reactivity, and serum IgE in a community based sample, *J. Allergy Clin. Immunol.,* 63, 328, 1979.
32. Holgersson, M., Stråhlenheim, G., and Dreborg, S., The precision of skin prick test with Phazet™, the Østerballe needle and the bifurcated needle, *Allergy,* Suppl. 4, 64, 1985.
33. Dreborg, S., Nilsson, G., and Zetterström, O., The precision of intracutaneous skin test with timothy pollen allergen preparation using two different techniques, *Ann. Allergy,* 58, 33, 1987.
34. Dreborg, S., Basomba, A., Belin, L., Durham, S., Einarsson, R., Eriksson, N.E., Frostad, A.B., Grimmer, Ø., Halvorsen, R., Holgersson, M., Kay, A.B., Nilsson, G., Malling, H.-J., Sjögren, I., Weeke, B., Våla, I.-J., and Zetterström, O., Biological equilibration of allergenic preparations: methodological aspects and reproducibility, *Clin. Allergy,* 17, 537, 1987.
35. Galant, S., Zippin, C., Bullock, J., and Crisp, J., Allergy skin tests. I. Antihistamine inhibition, *Ann. Allergy,* 30, 53, 1972.
36. Cook, T.J., MacQueen, D.M., Wittig, H.J., Thornby, J.I., Lantos, R.L., and Virtue, C.M., Degree and duration of skin test suppression and side effects with antihistamines. A double blind controlled study with five antihistamines, *J. Allergy Clin. Immunol.,* 51, 71, 1973.
37. Long, W.F., Taylor, R.J., Wagner, C.J., Leavengood, D.C., and Nelson, H.S., Skin test suppression by antihistamines and the development of subsensitivity, *J. Allergy Clin. Immunol.,* 76, 113, 1985.
38. Harvey, R.P. and Schocket, A.L., The effect of H_1 and H_2 blockade on cutaneous histamine response in man, *J. Allergy Clin. Immunol.,* 65, 136, 1980.
39. Hägermark, Ö., Strandberg, K., and Grönneberg, R., Effects of histamine receptor antagonists on histamine-induced responses in human skin, *Acta Derm.,* 59, 297, 1979.
40. Spector, S.L., Effect on a selective $beta_2$ adrenergic agonist and theophylline on skin test reactivity and cardiovascular parameters, *J. Allergy Clin. Immunol.,* 64, 23, 1979.
41. Chipps, B.E., Sobotka, A.K., Saunders, J.P., Teets, K.C., Norman, P.S., and Lichtenstein, L.M., Effect of theophylline and terbutaline on immediate skin tests, *J. Allergy Clin. Immunol.,* 65, 61, 1980.
42. Abramowitz, P.W., Perez, M.M., Johnson, C.E., McLean, J.A., Effect of theophylline, terbutaline, and their combination on the immediate hypersensitivity skin-test reaction, *J. Allergy Clin. Immunol.,* 66, 123, 1980.
43. Imbeau, S.A., Harruff, R., Hirscher, and M., Reed, C.E., Terbutaline's effects on the allergy skin test, *J. Allergy Clin. Immunol.,* 62, 193, 1978.
44. Grönneberg, R., Hägermark, Ö., and Strandberg, K., Effect in man of oral terbutaline on cutaneous reactions induced by allergen and cold stimulation, *Allergy,* 35, 143, 1980.
45. Kram, J.A., Bourne, H.R., Maibach, H.I., and Melmon, K.L., Cutaneous immediate hypersensitivity in man: effects of systemically administered adrenergic drugs, *J. Allergy Clin. Immunol.,* 56, 387, 1975.
46. Galant, S.P., Bullock, J., Wong, D., and Maibach, H.I., The inhibitory effect of antiallergic drugs on allergen and histamine induced wheal and flare responses, *J. Allergy Clin. Immunol.,* 51, 11, 1973.

47. Slott, R.I., and Zweiman, B., A controlled study of the effect of corticosteroids on immediate skin test reactivity, *J. Allergy Clin. Immunol.*, 54, 229, 1974.
48. Olson, R., Karpink, M.H., Shelanski, S., Atkins, P.C., and Zweiman, B., Skin reactivity to codeine and histamine during prolonged corticosteroid therapy, *J. Allergy Clin. Immunol.*, 86, 153, 1990.
49. Pipkorn, U., Hammarlund, A., and Enerbäck, L., Prolonged treatment with topical glucocorticoids results in an inhibition of the allergen-induced weal-and-flare response and a reduction in skin mast cell numbers and histamine content, *Clin. Exp. Allergy*, 19, 19, 1989.
50. Yen, A., Gigli, I., and Barrett, K.E., Modulation of human cutaneous mast cell responsiveness by a single, low-dose, PUVA treatment, *J. Allergy Clin. Immunol.*, 88, 395, 1991.
51. Berg, T., Nordvall, S.L., and Lanner, Å., Clinical studies of a purified timothy pollen extract. Desentization therapy with a purified timothy pollen preparation compared to a crude timothy pollen extract. I. Results of tests in vivo, *Int. Arch. Allergy Appl. Immunol.*, 63, 266, 1980.
52. Graft, D.F., Schuberth, K.C., Kagey-Sobotka, A., Kwiterovich, K.A., Niv, Y., Lichtenstein, L.M., and Valentine, M.D., The development of negative skin tests in children treated with venom immunotherapy, *J. Allergy Clin. Immunol.*, 73, 61, 1984.
53. Swain, H.H. and Becker, E.L., Quantitative studies in skin testing. V. The whealing reactions of histamine and ragweed pollen extract, *J. Allergy*, 23, 441, 1952.
54. Barbee, R.A., Brown, W.G., Kaltenborn, W., and Halonen, M., Allergen skin-test reactivity in a community population sample: correlation with age, histamine skin reactions, and total serum immunoglobulin E, *J. Allergy Clin. Immunol.*, 68, 15, 1981.
55. Stuckey, M.S., Witt, C.S., Schmitt, L.H., Warlow, R., Lattimore, M., and Dawkins, R.L., Histamine sensitivity influences reactivity to allergens, *J. Allergy Clin. Immunol.*, 75, 373, 1985.
56. Bowman, K.L., Pertinent factors influencing comparative skin tests on the arm, *J. Allergy*, 7, 39, 1935.
57. Galant, S.P. and Maibach, H.I., Reproducibility of allergy epicutaneous test techniques, *J. Allergy Clin. Immunol.*, 51, 245, 1973.
58. Kjellman, N.I.M., Dreborg, S., and Fälth-Magnusson, K., Allergy screening including a comparison of prick test results with allergen-coated lancets (Phazet) and liquid extracts, *Allergy*, 43, 277, 1988.
59. Haahtela, T.M.K., The prevalence of allergic conditions and immediate skin test reactions among Finnish adolescents, *Clin. Allergy*, 9, 53, 1979.
60. Taudorf, E., Malling, H.-J., Laursen, L.C., Lanner, Å, and Weeke, B., Reproducibility of histamine skin prick test. Inter- and intravariation using histamine dihydrochloride 1, 5, and 10 mg/ml, *Allergy*, 40, 344, 1985.
61. Clark, C.W., Mitchell, J., Nunn, A.J., and Pepys, J., Reproducibility of prick skin tests to five common allergens, *Clin. Allergy*, 12, 1, 1982.
62. Ménardo, J.L., Bousquet, J., Rodière, M., Astruc, J., and Michel, F.-B., Skin test reactivity in infancy, *J. Allergy Clin. Immunol.*, 75, 646, 1985.
63. Schwarzenbach, H.R., Nakagawa, T., Conroy, M.C., and de Weck, A.L., Skin reactivity, basophil degranulation and IgE levels in ageing, *Clin Allergy*, 12, 465, 1982.
64. Skassa-Brociek, W., Mandersfield, J.-C., Michel, F.-B., and Bousquet, J., Skin test reactivity to histamine from infancy to old age, *J. Allergy Clin. Immunol.*, 80, 711, 1987.
65. Niemeijer, N.R. and de Monchy, J.G., Age-dependency of sensitization to aero-allergens in asthmatics, *Allergy*, 47, 431, 1992.
66. Lee, R.E., Smolensky, M.H., Leach, C.S., and McGovern, J.P. Circadian rhythms in the cutaneous reactivity to histamine and selected antigens, including phase relationship to urinary cortisol excretion, *Ann. Allergy*, 38, 231, 1977.
67. McGovern, J.P., Smolensky, M.H., and Reinberg, A., Circadian and ciramensual rhythmicity in cutaneous reactivity to histamine and allergenic extracts, in *Chronobiology in Allergy and Immunology*, McGovern, J.P., Smolensky, M.H., and Reinberg, A., Eds., Charles C Thomas, Springfield, IL, 1977, 79.
68. Reinberg, A., Sidi, E., and Ghata, J., Circadian reactivity rhythms of human skin to histamine or allergen and the adrenal cycle, *J. Allergy*, 36, 273, 1965.
69. Vichyanond, P. and Nelson, H.S., Circadian variation of skin reactivity and allergy skin tests, *J. Allergy Clin. Immunol.*, 83, 1101, 1989.
70. Paquet, F., Boulet, L.-P., Bédard, G., Tremblay, G., and Cormier, Y., Influence of time of administration on allergic skin prick tests response, *Ann. Allergy*, 67, 163, 1991.
71. Van Niekerk, C.H. and Prinsloo, A.E.M., Effect of skin pigmentation on the response to intradermal histamine, *Int. Arch. Allergy Appl. Immunol.*, 76, 73, 1985.
72. Connell, J.T., Quantitative intranasal pollen challenges. III. The priming effect in allergic rhinitis, *J. Allergy*, 43, 33, 1969.
73. Haahtela, T. and Jokela, H., Influence of the pollen season on immediate skin test reactivity to common allergens, *Allergy*, 35, 15, 1980.
74. Norman, P.S., Why standardized extracts? (editorial), *J. Allergy Clin. Immunol.*, 77, 405, 1986.
75. Imber, W.E., Allergic skin testing: a clinical investigation, *J. Allergy Clin. Immunol.*, 60, 47, 1977.
76. Dirksen, A., Malling, H.-J., Mosbech, H., Søborg, M., and Biering, I., HEP versus PNU standardization of allergen extracts in skin prick testing. A comparative randomized in vivo study, *Allergy*, 40, 620, 1985.
77. Baer, H., Godfrey, H., Maloney, C.J., Norman, P.S., and Lichtenstein, L.M., The potency and antigen E content of commercially prepared ragweed extracts, *J. Allergy*, 45, 347, 1970.
78. Schaeffer, M., Sisk, C., and Brede, H.D., *Regulatory Control and Standardization of Allergenic Extracts*, Arbeiten aus dem Paul-Ehrlich-Institut, Fourth International Paul-Ehrlich-Seminar, October 16–17, 1985, Gustav Fisher Verlag, Stuttgart, 1987.
79. Williams, P.B., Nolte, H., Dolen, W.K., Koepke, J.W., and Selner, J.C., The histamine content of allergen extracts, *J. Allergy Clin. Immunol.*, 89, 738, 1992.
80. Horst, M., Hejjaoui, A., Horst, V., Michel, F.B., and Bousquet, J., Double-blind, placebo-controlled, rush immunotherapy with a standardized *Alternaria* extract, *J. Allergy Clin. Immunol.*, 85, 460, 1990.
81. Nelson, H.S., Effect of preservatives and conditions of storage on the potency of allergenic extracts, *J. Allergy Clin. Immunol.*, 67, 64, 1981.
82. Norman, P.S., Lichtenstein, L.M., and Ishizaka, K., Diagnostic tests in ragweed hay fever. A comparison of direct skin tests, IgE antibody measurements, and basophil histamine release, *J. Allergy Clin. Immunol.*, 52, 210, 1973.
83. Dreborg, S. and Fouchard, T. Allergy to apple, carrot and potato in children with birch pollen allergy, *Allergy*, 38, 167, 1983.
84. Halmepuro, L., Vuontela, K., Kalimo, K., and Björkstén, F., Cross-reactivity of IgE antibodies with allergens in birch pollen, fruits and vegetables, *Int. Arch. Allergy Appl. Immunol.*, 74, 235, 1984.

85. Sampson, H.A., Comparative study of commercial food antigen extracts for the diagnosis of food hypersensitivity, *J. Allergy Clin. Immunol.*, 82, 718, 1988.
86. Bernhisel-Broardbent, J., Strauss, D., and Sampson, H.A., Fish hypersensitivity. II. Clinical relevance of altered fish allergenicity caused by various preparation methods, *J. Allergy Clin. Immunol.*, 90, 622, 1992.
87. Voorhorst, R. and Nikkels, A.H., Atopic skin test reevaluated. IV. The use of compound 48/80 in routine skin testing, *Ann. Allergy*, 38, 255, 1977.
88. Casale, T.B., Bowman, S., and Kaliner, M., Induction of cutaneous mast cell degranulation by opiates and endogenous peptides: evidence of opiate and nonopiate receptor participation, *J. Allergy Clin. Immunol.*, 73, 755, 1984.
89. Dreborg, S., Holgersson, M., Nilsson, G., and Zetterström, O., Dose response relationship of allergen, histamine, and histamine releasers in skin prick test and precision of the method, *Allergy*, 42, 117, 1987.
90. Weiss, M.E., Trent, P., Fisher, R., Norman, P.S., Waterbury, W.E., and Adkinson, N.F., Jr., Rabbit F(ab')2 anti-human IgE is a universal skin test reagent in the evaluation of skin mast cell degranulation *in vivo*, *J. Allergy Clin. Immunol.*, 83, 1040, 1989.
91. Malling, H.-J., Skin prick testing and the use of histamine references, *Allergy*, 39, 596, 1984.
92. Nelson, H.S., Diagnostic procedures in allergy. I. Allergy skin testing, *Ann. Allergy*, 51, 411, 1983.
93. Voorhorst, R., Perfection of skin testing techniques, *Allergy*, 35, 247, 1980.
94. Voorhorst, R. and van Krieken, H., Atopic skin test re-evaluated. I. Perfection of skin testing technique, *Ann. Allergy*, 31, 137, 1973.
95. Solley, G.O., Gleich, G.J., Jordon, R.E., and Schroeter, A.L., The late phase of the immediate wheal and flare reactions. Its dependence upon IgE antibodies, *J. Clin. Invest.*, 58, 408, 1976.
96. Pienkowski, M.M., Norman, P.S., and Lichtenstein, L.M., Suppression of late-phase skin reactions by immunotherapy with ragweed extract, *J. Allergy Clin. Immunol.*, 76, 729, 1985.
97. Malling, H.-J., Andersen, C.E., Boas, M.-B., Holgersen, F., Munch, E.P., and Weeke, B., The allergy pricker. Qualitative aspects of skin prick testing with a precision needle, *Allergy*, 37, 563, 1982.
98. Voorhorst, R. and van Krieken, H., Atopic skin test re-evaluated. II. Variability in results of skin testing done in octuplicate, *Ann. Allergy*, 31, 195, 1973.
99. May, C.D., Objective clinical and laboratory studies of immediate hypersensitivity reactions to foods in asthmatic children, *J. Allergy Clin. Immunol.*, 58, 500, 1976.
100. Sampson, H.A., Role of immediate food hypersensitivity in the pathogenesis of atopic dermatitis, *J. Allergy Clin. Immunol.*, 71, 473, 1983.
101. Østerballe, O. and Weeke, B., A new lancet for skin prick testing, *Allergy*, 34, 209, 1979.
102. Poulsen, L.K., Liisberg, C., Bindslev-Jensen, C., and Malling, H.J., Precise area determination of skin prick tests. Validation of a scanning device and software for a personal computer, *Clin. Exp. Allergy*, 23, 61, 1993.
103. Ownby, D.R., Computerized measurement of allergen-induced skin reactions, *J. Allergy Clin. Immunol.*, 69, 536, 1982.
104. Serup, J., Diameter, thickness, area, and volume of skin-prick histamine wheals, *Allergy*, 39, 359, 1984.
105. Serup, J. and Staberg, B., Quantification of wheal reactions with laser Doppler flowmetry, *Allergy*, 40, 233, 1985.
106. Indrajana, T., Spieksma, F.T.M., and Voorhorst, R., Comparative study of the intracutaneous, scratch and prick tests in allergy, *Ann. Allergy*, 29, 639, 1971.
107. Belin, L.G. and Norman, P.S., Diagnostic tests in the skin and serum of workers sensitized to *Bacillus subtilis* enzymes, *Allergy*, 7, 55, 1977.
108. Malling, H.-J., Diagnosis and immunotherapy of mould allergy. II. Reproducibility and relationship between skin sensitivity estimated by end point titration and histamine equivalent reaction using skin prick test and intradermal test, *Allergy*, 40, 354, 1985.
109. Chipps, B.E., Talamo, R.C., Mellits, E.D., and Valentine, M.D., Immediate (IgE-mediated) skin testing in the diagnosis of allergic diseases, *Ann. Allergy*, 41, 211, 1978.
110. Pepys, J., Skin testing, *Br. J. Hosp. Med.*, 14, 412, 1975.
111. Brown, H.M., Su, S., and Thantrey, N., Prick testing for allergens standardized by using a precision needle, *Clin. Allergy*, 11, 95, 1981.
112. Aas, K., Some variables in skin prick testing, *Allergy*, 35, 250, 1980.
113. Malling, H.-J., Reproducibility of skin sensitivity using a quantitative skin prick test, *Allergy*, 40, 400, 1985.
114. Adinoff, A.D., Rosloniec, D.M., McCall, L.L., and Nelson, H.S., A comparison of six epicutaneous devices in the performance of immediate hypersensitivity skin testing, *J. Allergy Clin. Immunol.*, 84, 168, 1989.
115. Demoly, P., Bousquet, J., Manderscheid, J.C., Dreborg, S., Dhivert, H., and Michel, F.-B., Precision of skin prick and puncture tests with nine methods, *J. Allergy Clin. Immunol.*, 88, 758, 1991.
116. Engler, D.B., DeJarnatt, A.C., Sim, T.C., Lee, J.L., and Grant, J.A., Comparison of the sensitivity and precision of four skin test devices, *J. Allergy Clin. Immunol.*, 90, 985, 1992.
117. Bousquet, J. and Michel, F.-B., Precision of prick and puncture tests, editorial, *J. Allergy Clin. Immunol.*, 90, 870, 1992.
118. Malling, H.J., Quantitative skin prick testing. Dose-response of histamine- and allergen-induced wheal reactions, *Allergy*, 42, 196, 1987.
119. Haahtela, T. and Jaakomaeki, I., Relationship of allergen, specific IgE antibodies and allergic disorders in unselected adolescents, *Allergy*, 36, 251, 1981.
120. Hattevig, G., Kjellman, B., Johansson, S.G.O., and Björkstén, B., Clinical symptoms and IgE responses to common food proteins in atopic and healthy children, *Clin. Allergy*, 14, 551, 1984.
121. Barbee, R.A., Lebowitz, M.D., Thompson, H.C., and Burrows, B., Immediate skin-test reactivity in a general population sample, *Ann. Intern. Med.*, 84, 129, 1976.
122. Hagy, G.W. and Settipane, G.A., Risk factors for developing asthma and allergic rhinitis. A 7-year follow-up study of college student, *J. Allergy Clin. Immunol.*, 58, 330, 1976.
123. Rajka, G., Prurigo Besnier (atopic dermatitis) with special reference to the role of allergic factors. II. The evaluation of the results of skin reactions, *Acta Derm.*, 41, 1, 1961.
124. Sampson, H.A. and Albergo, R., Comparison of results of skin tests, RAST, and double-blind, placebo-controlled food challenges in children with atopic dermatitis, *J. Allergy Clin. Immunol.*, 74, 26, 1984.
125. Huggins, K.G. and Brostoff, J., Local production of specific IgE antibodies in allergic-rhinitis patients with negative skin tests, *Lancet*, 2, 148, 1975.
126. Juhlin-Dannfeldt, C., About the occurrence of various forms of pollen allergy in Sweden, *Acta Med. Scand.*, 26, 563, 1948.
127. Horak, F., Manifestation of allergic rhinitis in latent-sensitized patients. A prospective study, *Acta Otorhinolanryngol.*, 242, 239, 1985.

128. Lindblad, J.H. and Farr, R.S., The incidence of positive intradermal reactions and the demonstration of skin sensitizing antibody to extracts of ragweed and dust in humans without history of asthma or rhinitis, *J. Allergy*, 32, 392, 1961.
129. Gergen, P.J., Turkeltaub, P.C., and Kovar, M.G., The prevalence of allergic skin test reactivity to eight common aeroallergens in the U.S. population. Results from the second National Health and Nutrition Examination Survey, *J. Allergy Clin. Immunol.*, 80, 669, 1987.
130. Andersson, H.R., The epidemiological and allergic features of asthma in the New Guinea Highlands, *Clin. Allergy*, 4, 171, 1974.
131. Woolcock, A.J., Colman, M.H., and Jones, M.W., Atopy and bronchial reactivity in Australian and Melanesian populations, *Clin. Allergy*, 8, 155, 1978.
132. Backer, V., Ulrik, C.S., Hansen, K.K., Laursen, E.M., Dirksen, A., and Bach-Mortensen, N., Atopy and bronchial responsiveness in a random population sample of 527 children and adolescents, *Ann. Allergy*, 69, 116, 1992.
133. Van Asperen, P.P., Kemp, A.S., and Mellis, C.M., Skin test reactivity and clinical allergen sensitivity in infancy, *J. Allergy Clin. Immunol.*, 73, 381, 1984.
134. Barbee, R.A., Kaltenborn, W., Lebowitz, M.D., and Burrows, B., Longitudinal changes in allergen skin test reactivity in a community population sample, *J. Allergy Clin. Immunol.*, 79, 16, 1987.
135. Corbo, G.M., Foresi, A., Morandini, S., Valente, S., Mattoli, S., and Ciappi, G., Probit analysis applied to the allergen dose-response curve: a method for epidemiologic surveys, *J. Allergy Clin. Immunol.*, 81, 41, 1988.
136. Godfrey, R.C. and Griffiths, M., The prevalence of immediate positive skin tests to *Dermatophagoides pteronyssinus* and grass pollen in schoolchildren, *Clin. Allergy*, 6, 79, 1976.
137. Haahtela, T., Björkstén, F., Heiskala, M., and Suoniemi, I., Skin prick test reactivity to common allergens in Finnish adolescents, *Allergy*, 35, 425, 1980.
138. Grow, M.H. and Herman, N.B., Intracutaneous tests in normal individuals, *J. Allergy*, 7, 108, 1936.
139. Johnsen, N.N. and Mygind, N., Incidence of latent and clinical respiratory tract allergy in medical students, *Ugeskr Læger*, 140, 596, 1978.
140. D'Souza, M. and Davies, R., The distribution of allergic disorders and atopy in the community and their relationship to total levels of serum IgE antibody, *Am. Rev. Respir. Dis.*, 115, 211, 1977.
141. Pastorello, E., Skin tests for diagnosis of IgE-mediated allergy, *Allergy*, 48, 57, 1993.
142. Haahtela, H., Skin tests used for epidemiologic studies, *Allergy*, 48, 76, 1993.
143. Homburger, H.A. and Katzman, J.A., Methods in laboratory immunology. Principles and interpretation of laboratory tests for allergy, in *Allergy, Principles and Practice*, Middleton, E., Jr., Reed, C.E., Ellis, E.F., Adkinson, N.F., Jr., and Yunginger, J.W., Eds., C.V. Mosby, St. Louis, 1988, 402.
144. Nolte, H., Update: clinical aspects of basophil histamine release, *Immunol. Allergy Practice*, XIV, 255, 1992.
145. Norgaard, A., Skovm P.S., and Bindslev-Jensen, C., Egg and milk allergy in adults: comparison between fresh foods and commercial allergen extracts in skin prick test and histamine release from basophils, *Clin. Exp. Allergy*, 22, 940, 1992.
146. Jeep, S., Kirchhof, E., O'Conner, A., and Kunkel, G., Comparison of the Phadebas RAST with the Pharmacia CAP system for insect venom, *Allergy*, 47, 212, 1992.
147. Williams, P.B., Dolen, W.K., Koepke, J.W., and Selner, J.C., Comparison of skin testing and three in vitro assays for specific IgE in the clinical evaluation of immediate hypersensitivity, *Ann. Allergy*, 68, 35, 1992.
148. Berg, T.L. and Johansson, S.G., Allergy diagnosis with the radioallergoabsorbent test: a comparison with the results of skin and provocation test in an unselected group of children with asthma and hay fever, *J. Allergy Clin. Immunol.*, 54, 209, 1974.
149. Bryant, D.H., Burns, M.W., and Lazarus, L., The correlation between skin tests, bronchial provocation tests and the serum level of IgE specific for common allergens in patients with asthma, *Clin. Allergy*, 5, 145, 1975.
150. Eriksson, N.E. and Ahlstedt, S., Diagnosis of reaginic allergy with house dust, animal dander and pollen allergens in adult patients. V. A comparison between the enzyme-linked immunosorbent assay (ELISA), provocation tests, skin tests and RAST, *Int. Arch. Allergy Appl. Immunol.*, 54, 88, 1977.
151. Griese, M., Kusenbach, G., and Reinhardt, D., Histamine release test in comparison to standard tests in diagnosis of childhood allergic asthma, *Ann. Allergy*, 65, 46, 1990.
152. Nolte, H., Strom, K., and Schiøtz, P.O., Diagnostic value of a glass fibre based histamine analysis for allergy testing in children, *Allergy*, 45, 213, 1990.
153. van der Zee, J.S., de Grott, H., van Swieten, P., Jansen, H.M., and Alberse, R.C., Discrepancies between the skin test and IgE antibody assays: study of histamine release, complement activation in vitro, and occurrence of allergen-specific IgG, *J. Allergy Clin. Immunol.*, 82, 270, 1988.
154. Kerrebijn, K.F., Degenhart, H.J., and Hammers, A., Relation between skin test, inhalation test, and histamine release from leukocytes and IgE in house-dust mite allergy, *Arch. Dis. Child.*, 51, 252, 1976.
155. Petersson, G., Dreborg, S., and Ingestad, R., Clinical history, skin prick test and RAST in the diagnosis of birch and timothy pollinosis, *Allergy*, 41, 398, 1986.
156. Østergaard, P.A., Ebbesen, F., Nolte, H., and Skov, P.S., Basophil histamine release in the diagnosis of house dust mite and dander allergy of asthmatic children. Comparison between prick test, RAST, basophil histamine release and bronchial provocation, *Allergy*, 45, 1, 1990.

Chapter 31.4
Ultraviolet Radiation Dosimetry

B. L. Diffey
Regional Medical Physics Department
Dryburn Hospital
Durham, U.K.

I. Introduction

Dosimetry is the science of radiation measurement. No persuasion should be needed as to the importance of dosimetry and in dermatology there are two principal reasons why ultraviolet radiation (UVR) should be measured:[1]

1. To allow consistent radiation exposure of patients over many months and years within a local department.
2. To allow the results of irradiations made in different departments to be published and compared.

It is important to distinguish between these two objectives. The first requires *precision,* or reproducibility. The radiometer is used as a monitor to give a reference measurement and so it needs to be stable. *Accuracy,* i.e., absolute calibration against some accepted standard, is not essential. The second objective requires both precision and accuracy. Here the radiometer must not only be stable from one day to the next, but also the display (in, say, milliwatts per square centimeter) must be traceable to absolute standards. While electrooptical technology has improved over the years, resulting in the availability of versatile and precise ultraviolet radiometric equipment, these improvements have not been accompanied by improved accuracy. Indeed the accuracy of administered radiation doses in psoralen photochemotherapy leaves much to be desired *(vide infra).*

II. Radiometric Terms and Units

In clinical and photobiological UVR dosimetry it is customary to use the terminology of radiometry rather than that of photometry, since photometry is based on visible light measurements that simulate the human eye's photopic response curve and, strictly speaking, a source that emits only UVR has a zero intensity in photometric terms.

The common radiometric terminology is listed in Table 1. Terms relating to a beam of radiation passing through space are the "radiant energy" and "radiant flux". Terms relating to a source of radiation are the "radiation intensity" and the "radiance". The term "irradiance", which is the most commonly used term in photobiology, relates to the object (e.g., patient) struck by the radiation. The radiometric quantities in Table 1 may also be expressed in terms of wavelength by adding the prefix "spectral".

The time integral of the irradiance is strictly termed the "radiant exposure", but is sometimes expressed as "exposure dose", or even more loosely as "dose". The term "dose" in photobiology is analogous to the term "exposure" in radiobiology and not to "absorbed dose". As yet the problems of estimating the energy absorbed by critical targets in the skin remain unsolved.

A. Radiometric Calculations

The most frequent radiometric calculation is to determine the time for which a patient, who is prescribed a certain dose (in J/cm^2), should be exposed when the radiometer indicates an irradiance in mW/cm^2. The relationship between these three quantities (time, dose, and irradiance) is simply

$$\text{Exposure time (minutes)} = \frac{1000 \times \text{prescribed dose} \left(\text{J/cm}^2\right)}{60 \times \text{measured irradiance} \left(\text{mW/cm}^2\right)} \quad (1)$$

B. Units of Biologically Effective Ultraviolet Radiation

In addition to the radiometric quantities given above, derived quantities of effective irradiance and dose related to a specific photobiological action spectrum are often used in photomedicine. The effective irradiance is obtained by weighting the spectral irradiance of the radiation at wavelength λnm by the effectiveness of

TABLE 1 Radiometric Terms and Units

Term	Unit	Symbol
Wavelength	nm	λ
Radiant energy	J	Q
Radiant flux	W	φ
Radiant intensity	Wsr^{-1}	I
Radiance	Wcm^{-2} sr^{-1}	L
Irradiance	Wcm^{-2}	E
Radiant exposure	Jcm^{-2}	H

radiation of this wavelength to cause a particular photobiological effect (e.g., minimal erythema) and summing over all wavelengths present in the source spectrum. This can be expressed mathematically as

$$\Sigma E(\lambda) \cdot S(\lambda) \cdot \Delta\lambda \qquad W/m^2 \qquad (2)$$

E(λ) is the spectral irradiance in W/m²/nm at wavelength λnm and Δλ is the wavelength interval used in the summation. S(λ) is a measure of the effectiveness of radiation of wavelength λnm relative to some reference wavelength in producing a particular biological endpoint. As it is a ratio, S(λ) has no units. The effective irradiance is equivalent to a hypothetical irradiance of monochromatic radiation having a wavelength at which S(λ) is equal to unity. The time integral of effective irradiance is the effective radiant exposure (also called the *effective dose*).

A unit of effective dose commonly used in photodermatology is the *minimal erythema dose* (MED). One MED has been defined as the lowest radiant exposure of UVR that is sufficient to produce erythema with sharp margins 8 to 24 hours after exposure.[2] Another end point often used is a just-perceptible reddening of exposed skin. The dose of UVR necessary to produce this *minimal perceptible erythema* is sometimes also referred to as an MED.[3] Furthermore, in unacclimatized white skin there is a four- to fivefold range in MED for exposure to UVB radiation.[4] When the term MED is used as a unit of exposure dose, however, a representative value is chosen for sun-sensitive individuals. If, in Equation 2, S(λ) is chosen to be the reference action spectrum for ultraviolet erythema in human skin[5] and a value of 200 J/m² at wavelengths for which S(λ) is equal to unity (i.e., λ ≤ 298 nm) is assumed for the MED,[6] the dose (expressed in MEDs) received after an exposure period of *t* seconds is

$$t\Sigma E(\lambda) \cdot S(\lambda) \cdot \Delta\lambda / 200 \qquad (3)$$

Notwithstanding the difficulties of interpreting accurately the magnitude of such an imprecise unit as the MED, it has the advantage over radiometric units of its relationship to the biological consequences of the exposure.

III. Detection of Ultraviolet Radiation

Techniques for the measurement of UVR fall into three classes: physical, chemical, and biological. In general physical devices measure power, while chemical and biological systems measure energy.

The use of chemical methods, which measure the chemical change produced by the radiation, is called actinometry. These techniques usually form the basis of personal ultraviolet dosimeters.

Biological techniques of measurement are generally limited to the use of viruses and microorganisms. Human skin is often used as a UVR dosimeter in phototherapy in an indirect fashion; treatment times are determined by exposing small areas of the patient's skin to increasing exposures from a UV lamp and noting that exposure which produces a given degree of erythema.

A. Physical Ultraviolet Radiation Detectors

A physical UVR detector consists of an element which absorbs the radiation and a means of measuring the resulting change in some property of the element. There are two basic physical types: thermal and photon.

1. Thermal Detectors

Thermal detectors respond to heat or power and have a broad spectral response with near-uniform sensitivity from the ultraviolet to the infrared. The absorption of radiation increases the temperature of the element and this change can be detected in a variety of ways.

In the thermopile the temperature rise of the element causes a small voltage to be generated at the junction of two dissimilar metals. The heat-sensitive element in a pyroelectric detector consists of a slab of ferroelectric material which produces a change in current proportional to the rate of change of temperature of the surface of the slab, which in turn is proportional to the rate of change of irradiance. A bolometer works by sensing the change in resistance of the absorbing element. In the Golay cell the temperature rise is sensed by the expansion of an enclosed gas which presses against a flexible mirror altering its focal length. Finally, photoacoustic detectors use a microphone to detect the small fluctuations in pressure which occur when the heat generated by the radiation absorption is coupled into the gas contained in an acoustic resonator.

2. Photon Detectors

Photon detectors operate on the principle of the liberation of electrons by the absorption of a single quantum of radiation, and consequently tend to have a nonlinear spectral response. Examples of photon detectors include photoemissive types (vacuum phototube, gas filled phototube, and photomultiplier tube); junction photodetectors (e.g., Si, GaAsP, GaP photodiodes) which may be operated with a "zero bias" (sometimes called "photovoltaic mode") or a "reverse bias" (sometimes called "photoconductive mode"); and photoconductors (e.g., CdS, CdSe, PbS, InAs).

More detailed descriptions of physical optical radiation detectors can be found elsewhere.[7-9]

IV. Spectroradiometry

It is common practice to talk loosely of *UVA lamps* or *UVB lamps*. However such a label does not characterize adequately UV lamps, since nearly all UV lamps will emit both UVA and UVB, and even UVC, visible light and infrared radiation. The only correct way to specify the nature of the emitted radiation is by reference to the spectral power distribution. This is a graph (or table) which indicates the radiated power as a function of wavelength. The data are obtained by a technique known as spectroradiometry. Figure 1 shows the spectral power distribution of UVR emitted by a medium pressure mercury arc lamp (commonly called a "hot-quartz" lamp in the U.S.). This type of lamp has been used for many years for the phototherapy of skin diseases in units such as the Alpine Sunlamp (Hanovia Ltd, Slough, U.K.). In Figure 1a, the relative power emission is plotted on a linear scale. Although lamp spectra are commonly plotted using a linear scale, this representation may not be the most appropriate. Since the erythemal sensitivity of normal skin varies by four orders of magnitude over the spectral region of 200 to 400 nm, it is helpful to be able to discern components of the spectrum which may be of small amplitude in physical terms but nonetheless important in photobiological terms. Figure 1b shows the spectrum from the same lamp with the relative power plotted on a logarithmic scale. We can see now the characteristic wavelengths present in all mercury lamps are superimposed upon a low-level continuous distribution of radiation. The exact shape of the continuum, particularly at wavelengths of less than 300 nm, depends on factors including the lamp envelope material and the vapor pressure of the mercury.

A. Components of a Spectroradiometer

The three basic requirements of a spectrometer system are (1) the input optics, designed to conduct the radiation from the source into (2) the monochromator, which usually incorporates a diffraction grating as the wavelength dispersion element, and (3) an optical radiation detector.

1. Input Optics

The spectral transmission characteristics of monochromators depend upon the angular distribution and

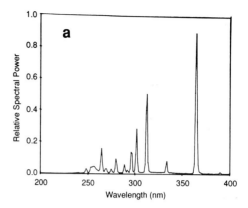

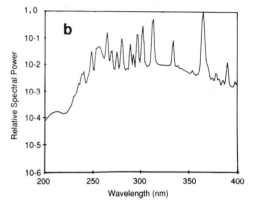

FIGURE 1 The spectral power distribution of UVR from a medium pressure mercery arc lamp (a) linear scale on ordinate and (b) logarithmic scale on ordinate.

polarization of the incident radiation as well as the position of the beam on the entrance slit. For measurement of spectral irradiance, particularly from extended sources such as linear arrays of fluorescent lamps or daylight, direct irradiation of the entrance slit should be avoided. There are two types of input optics available to ensure that the radiation from different source configurations is depolarized and follows the same optical path through the system: the integrating sphere or the diffuser. Both these types of input optics produce a cosine-weighted response, since the radiance of the source as measured through the entrance aperture varies as the cosine of the angle of incidence.

2. Monochromator

A blazed ruled diffraction grating is normally preferred to a prism as the dispersion element in the monochromator used in a spectroradiometer, mainly because of better stray radiation characteristics.

High performance spectroradiometers, used for determining low UV spectral irradiances in the presence of high irradiances at longer wavelengths, demand extremely low stray radiation levels. Such systems may incorporate a double monochromator, i.e., two single ruled grating monochromators in tandem, or laser holographically produced concave diffraction gratings can be used in a single monochromator.

3. Detector

Photomultiplier tubes, incorporating a photocathode with an appropriate spectral response, are normally the detectors of choice in spectroradiometers. However, if radiation intensity is not a problem, solid-state photodiodes may be used, since they require simpler and cheaper electronic circuitry.

B. Calibration

It is important that spectroradiometers are calibrated over the wavelength range of interest using standard lamps. A tungsten filament lamp operating at a color temperature of about 3000 K can be used as a standard lamp for the spectral interval 250 to 2500 nm, although workers concerned solely with the ultraviolet region (200 to 400 nm) may prefer to use a deuterium lamp.

C. Sources of Error in Spectroradiometry

Accurate spectroradiometry, even where only relative spectral power distributions are used, requires careful attention to detail. Factors which can affect accuracy include wavelength calibration, bandwidth, stray radiation, polarization, angular dependence, linearity, and calibration sources.[10]

V. Narrow Band Radiometry

Although spectroradiometry is the fundamental way to characterize the radiant emission from a light source, radiation output is normally measured by techniques of narrow band radiometry. Narrow band radiometers generally combine a detector (such as a vacuum phototube or a solid-state photodiode) with a wavelength-selective device (such as a color glass filter or interference filter) and suitable input optics [such as a quartz hemispherical diffuser or polytetrafluoroethylene (PTFE) window].

A. The Problem of Spectral Sensitivity

Naive users of narrow band radiometers often gain the impression from commercial literature that instruments are readily available to measure UVA, UVB, or UVC. In order to meet the criterion for a UVB radiometer, say, the sensor should have a uniform spectral response from 280 to 315 nm (the UVB waveband) with zero response outside this interval. In other words, the electrical output from the sensor should depend only on the total power within the UVB waveband received by the sensor and not on how the power is distributed with respect to wavelength. In practice no such sensor exists with this ideal spectral response (neither does one exist that measures UVA or UVC correctly for that matter). All radiometers that combine a photodetector with an optical filter have a nonuniform spectral sensitivity within their normal spectral band. This problem is discussed more fully in the next section using the calibration of dosimeters used in PUVA therapy as an example.

B. Calibration of Ultraviolet A Dosimeters Used in Psoralen Photochemotherapy (PUVA)

PUVA therapy — the combination of oral psoralens and long-wave ultraviolet radiation (UVA) — is widely used in the treatment of a variety of skin diseases, particularly psoriasis.[11] Most dermatology departments in the U.K. have PUVA units incorporating UVA fluorescent lamps.[12] The spectral emission (i.e., a plot of the intensity of radiation at each wavelength) of these lamps is remarkably similar no matter which make of PUVA unit is used. This spectrum is shown in Figure 2.

Shortly after the introduction of PUVA therapy in the early 1970s, the protagonists of the treatment stressed that "... *careful attention to dosimetry is essential and ... the dosimetry system is a key to both the effectiveness and safety of PUVA*". These authors[13] recommended that UVA exposure of the patient be expressed in radiometric units; the intensity (or irradiance) of the radiation beam is measured in milliwatts per square centimeter (mW/cm^2) and the prescribed dose (or radiant exposure) is expressed in Joules per square centimeter (J/cm^2). It is important to realize that all sensors that are used to measure UVA irradiance show a wavelength-dependent response (Figure 3) and so produce an electrical signal that depends not only on the irradiance from the optical source but also on its spectral emission.[14]

Despite these early recommendations subsequent intercomparisons of UVA dosimeters carried out in the U.K.,[15] Belgium,[16] and France[17] have shown wide variations in accuracy, with threefold differences in sensitivity at the extreme ends of the range. It is likely that inconsistencies in UVA dosimetry are a major factor in the tenfold range of doses prescribed at the start of treatment found in the U.K.[12]

So what do users do about calibration? Many dermatology centers with UVA dosimeters rely upon the initial calibration provided by the manufacturer. Yet it is not clear from the literature provided by some suppliers of UVA dosimeters used in PUVA therapy exactly *how* they calibrate their dosimeters — so what the meter reading in mW/cm^2 actually means is unclear. Other manufacturers calibrate their dosimeters with narrow band radiation around the peak spectral sensitivity. For the dosimeter with the spectral sensitivity shown in Figure 3 this would be at 360 nm. This means that the calibrated dosimeter would record the correct irradiance from a source emitting monochro-

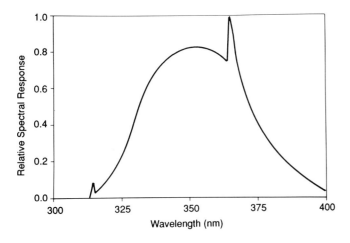

FIGURE 2 The spectral power distribution of UVA fluorescent lamps used in PUVA therapy. Lamps which have this spectrum include Philips TL85/100W/09, Philips TL100W/09 R-UVA, Philips UVA 100W-P, Sylvania F85/100W-PUVA, and Thorn UV 75/85W.

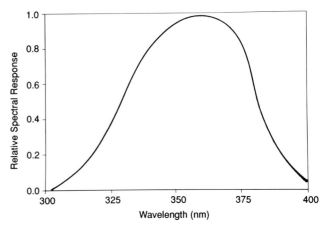

FIGURE 3 The spectral sensitivity of a typical UVA dosimeter. Dosimeters with similar sensitivity include[9] the IL1352 phototherapy radiometer (International Light Inc, Newburyport), PUVA meter, and UV meter (H Waldmann GmbH & Co., Schwenningen, Germany), Blak-Ray J221 and UVX-36 (Ultra-Violet Products, Inc., San Gabriel), and Uvichek (Rank Hilger, Margate, U.K.).

matic radiation at 360 nm but when irradiated with any other spectral power distribution, such as that shown in Figure 2, would indicate an irradiance that was different from the true UVA irradiance. Expressed mathematically, the true UVA irradiance is

$$\int_{315}^{400} E(\lambda) d\lambda \quad mW/cm^2 \qquad (4)$$

whereas the meter would indicate

$$\int_{0}^{\infty} E(\lambda) S(\lambda) d\lambda \quad mW/cm^2 \qquad (5)$$

$E(\lambda)$ is the spectral irradiance in mW/cm²/nm and $S(\lambda)$ is the relative spectral sensitivity normalized to unity at the wavelength (in this case 360 nm) where the dosimeter is most sensitive. Clearly these two quantities are numerically different unless $E(\lambda)$ is zero at all wavelengths other than 360 nm.

More enlightened users might ask national standards laboratories, such as the National Physical Laboratory in the U.K., to calibrate their UVA dosimeter. Unfortunately, a similar problem can exist. When UVA dosimeters are sent to standards laboratories for calibration, an optically filtered medium pressure mercury arc lamp is often used as the source. The spectral power distribution of this source is confined to a narrow band of radiation centered at 365 nm (one of the characteristic lines present in the mercury spectrum) and is very different from that shown in Figure 2. The consequence is that dosimeters calibrated in this way will underestimate the true UVA irradiance from UVA fluorescent lamps by about 25%.

If the purpose of UVA dosimetry is to measure the UVA irradiance to which patients are exposed and to record the UVA doses that patients receive, the logical approach is to calibrate dosimeters using lamps with the same spectral power distribution of radiation as those used for treatment.

The technique used in the author's laboratory is to measure the spectral irradiance at 20 cm from a bank of eight UVA fluorescent lamps (Philips TL20W/09N) using a spectroradiometer. It is important that users of spectroradiometers have their own standard lamps (either deuterium or tungsten) regularly calibrated by standards laboratories so that these can be used to provide an absolute spectral sensitivity calibration of the spectroradiometer. By summing the spectral irradiance across the UVA waveband (315 to 400 nm) the absolute UVA irradiance can be derived. Calibration of a UVA dosimeter simply involves removing the spectroradiometer, placing the entrance aperture of the sensor at the same point as the input optics of the spectroradiometer, and adjusting the meter display so that it reads the UVA irradiance determined spectroradiometrically.

VI. Broad Band Radiometry

Broad band radiometry uses a detector which responds equally to all wavelengths of optical radiation. The most common detector used is the thermopile and this is especially useful in measuring the irradiance from an irradiation monochromator used in the investigation of skin photosensitivity.[18] Until a few years ago, commercial thermopiles were hand-made, expensive, and fragile devices. A major advance came with the production of multiple junction thermopiles based on thin-film technology. These devices are rugged and much less expensive and typified by the Dexter range of thermopiles (Dexter Research Center, Michigan) which have found a role in dermatological photobiology.[19,20]

A. Calibration

Thermopiles measure absolute radiant power and calibration can only be achieved satisfactorily by national standards laboratories, such as the National Physical Laboratory in the U.K.[21,22]

VII. Radiometer Stability

It should be remembered that the sensitivity of all radiometers will change with time; frequent exposure to high-intensity sources of light will accelerate this change. For this reason it is always a sound policy to acquire two radiometers, preferably of the same type, one of which has a calibration traceable to a national standards laboratory. This radiometer should be reserved solely for intercomparisons with the other radiometer(s) used for routine purposes. A measurement of the same source is made with each radiometer and a ratio calculated. It is the stability of this ratio over a period of months and years which indicates long-term stability and good precision.

VIII. Personal Ultraviolet Radiation Dosimetry

We have seen that UVR is generally measured with thermal or photon detectors, often used in conjunction with optical filters. A different yet complementary approach is the use of various photosensitive films as UVR dosimeters. The principle is to relate the degree of deterioration of the films, usually in terms of changes in their optical properties, to the incident UVR dose. The principal advantages of the film dosimeter are that it provides a simple means of integrating UVR exposure continuously and that it allows numerous sites, inaccessible to bulky and expensive instrumentation, to be compared simultaneously. Personal ultraviolet dosimetry can be useful in establishing the effect of photoprotective agents in the treatment of photosensitivity.[23]

A. Requirements of Personal UV Dosimeters

Ultraviolet dosimeters designed for personal use should have the following characteristics:

1. The physical or chemical change produced in the dosimeter (e.g., increase in optical absorbance) should, ideally, increase linearly with UV dose. If not, the dose response curve should at least be monotonic, that is, any given dosimeter response is effected by only one radiation dose.
2. The dosimeter should exhibit photoaddition; each wavelength acts independently and the effect of polychromatic radiation is the sum of the effects of all wavelengths involved.
3. The dosimeter response should depend only on dose and be independent of dose rate.
4. The spectral sensitivity of the dosimeter should, ideally, match the action spectrum of the photobiological effect being monitored.
5. The dosimeter response should be independent of temperature and humidity; it should exhibit no "dark effect" (continuing response when radiation exposure terminated); it should be stable on long-term storage.
6. The dosimeter should be easy to handle and not impose restrictions on the activities of the wearers.
7. The dosimeter should not require laborious processing and should be easy to convert the physical or chemical response to a measure of ultraviolet exposure dose.
8. The cost per dosimeter should be low so that large-scale monitoring is feasible.

B. Types of Personal UV Dosimeters
1. Polysulfone Film

Perhaps the most commonly used material for studies of personal UV dosimetry has been the thermoplastic, polysulfone, which was first suggested as a possible dosimeter for UVR by Davis et al.[24] Since then, the use of polysulfone film as a personal UV dosimeter has been exploited for monitoring both environmental and artificial UVR.[25] The basis of the method is that, when film is exposed to UVR at wavelengths less than 330 nm, its UV absorption increased. The increase in absorbance measured at a wavelength of 330 nm increases with UV dose. In practice the film (40 μm thick) is mounted in cardboard photographic holders and normally worn on the lapel site.

2. Plastic Films Incorporating Photosensitizing Drugs

In field studies of drug-induced photosensitivity the possibility of using a dosimeter which incorporates the relevant drug is an attractive proposition. To this end, several drugs which are known to have photosensitizing effects in humans have been incorporated as the chromophore in a polyvinyl chloride (PVC) film. Photoactive drugs which have been used in this way include phenothiazine,[26] 8-methoxypsoralen,[27] nalidixic acid,[28] and benoxaprofen.[29]

3. Diazo Systems

Diazo systems, which are based on diazonium compounds, are one of the oldest photochemical non-silver processes. The two fundamental properties of the diazo type process which make it suitable for use as a UV dosimeter are

1. the ability to be decomposed by ultraviolet radiation
2. the ability of the undecomposed diazonium compound to couple with a color former to produce a stable image

Diazonium compounds are sensitive principally to the UVA and blue regions of the spectrum. Their spectral sensitivity, together with the simplicity, economy, and convenience of the diazo system, have led to their use as film badge dosimeters for UVA and blue radiation.[30,31]

4. Photosensitive Papers

One drawback of the film dosimeters described above is that they require laboratory equipment to facilitate readout. An alternative approach is to use a system whereby the photochemical process initiates a color change so that visual comparisons with stable printed color standards enable the user to obtain a reasonably accurate and continuously readable integrated measure of his exposure to UVR. An example of a dosimeter based on this principle has been described by Zweig and Henderson.[32] This dosimeter is a polycarbonate film matrix incorporating a chromophore which converts to a red photoproduct following exposure to UVR of wavelengths less than 350 nm. The depth of red color developed depends solely on the radiant exposure.

Another type of photodosimeter is based on the reversible color change of photochromic aziridine formulations.[33] The colorless aziridine undergoes isomerization following ultraviolet exposure to form the blue-colored azomethine slide.

Photosensitive papers form the basis of dosimeters designed for consumer use while sunbathing.[34]

5. Thermoluminescent Materials

Several thermoluminescent (TL) materials have been investigated as possible UV dosimeters. Many materials (e.g., LiF:Mg; $CaSO_4$: Tm; CaF_2: natural) require pre-irradiation with high doses of gamma radiation and partial annealing before showing sensitivity to UVR (so-called "transferred thermoluminescence"), whereas other materials (e.g., MgO; Al_2O_3:Si; CaF_2:Dy) have proved to be directly sensitive to UVR. It is probably true to say that TL materials have yet to find an established role as dosimeters for UVR.

6. Polycarbonate Plastic

On exposure to UVR the transparent plastic, CR-39 (allyl diglycol carbonate), alters its optical properties. These changes are the basis of its use as UVR dosimeter.[35] After exposure the plastic is etched in 6 N KOH at 80°C for 3 hours, rinsed, and allowed to dry. The degree of UV dose-dependent front surface damage, visible as opacification, is quantified by measuring the transmission at 700 nm. A novel use of this material has been to construct CR-39 contact lenses which can be used for measuring ultraviolet exposure to the front surface of the eye.[36]

References

1. Mackenzie, L.A., UV radiometry in dermatology, *Photodermatology*, 2, 86, 1985.
2. Willis, I. and Kligman, A.M., Aminobenzoic acid and its esters, *Arch. Dermatol.*, 102, 405, 1970.
3. Epstein, J.H., Polymorphous light eruptions: wavelength dependency and energy studies, *Arch. Dermatol.*, 85, 82, 1962.
4. Diffey, B.L. and Farr, P.M., The normal range in diagnostic phototesting, *Br. J. Dermatol.*, 120, 517, 1989.
5. McKinlay, A.F. and Diffey, B.L., A reference action spectrum for ultraviolet induced erythema in human skin, *CIE J.*, 6, 17, 1987.
6. Urbach, F., Man and ultraviolet radiation, in *Human Exposure to Ultraviolet Radiation: Risks and Regulations*, Passchier, W.F. and Bosnjakovic, B.F.M., Eds., Excerpta Medica, Amsterdam, 1987, 3.
7. Rabek, J.F., *Experimental Methods in Photochemistry and Photophysics*, John Wiley & Sons, New York, 1982.
8. Dereniak, E.L. and Crowe, D.G., *Optical Radiation Detectors*, John Wiley & Sons, New York, 1984.
9. Wilson, A.D., Optical radiation detectors, in *Radiation Measurement in Photobiology*, Diffey, B.L., Ed., Academic Press, London, 1989, chap. 2.
10. Moore, J.R., Sources of error in spectroradiometry, *Lighting Res. Tech.*, 12, 213, 1980.
11. Green, C., Diffey, B.L., and Hawk, J.L.M., Ultraviolet radiation in the treatment of skin disease, *Phys. Med. Biol.*, 37, 1, 1992.
12. Farr, P.M. and Diffey, B.L., PUVA treatment of psoriasis in the United Kingdom, *Br. J. Dermatol.*, 124, 365, 1991.
13. Wolff, K., Gschnait, F., Honigsmann, H., et al., Phototesting and dosimetry for photochemotherapy, *Br. J. Dermatol.*, 96, 1, 1977.
14. Stobbart, D. and Diffey, B.L., A comparison of some commercially available UVA meters used in photochemotherapy, *Clin. Phys. Physiol. Meas.*, 1, 267, 1980.
15. Diffey, B.L., Challoner, A.V.J., and Key, P.J., A survey of the ultraviolet radiation emissions of photochemotherapy units, *Br. J. Dermatol.*, 102, 301, 1980.
16. Diffey, B.L. and Roelandts, R., Status of ultraviolet A dosimetry in methoxalen plus ultraviolet A therapy, *J. Am. Acad. Dermatol.*, 15, 1209, 1986.
17. Roelandts, R., Diffey, B.L., and Bocquet, J.L., Une dosimetrie UVA precise en photobiologie et un photodermatologie est-elle une illusion?, *Ann. Dermatol. Venereol.*, 115, 1261, 1988.
18. Magnus, I.A., *Dermatological Photobiology*, Blackwell Scientific, Oxford, 1976, chap. 9.
19. Diffey, B.L., Farr, P.M., and Ive, F.A., The establishment and clinical value of a dermatological photobiology service in a district general hospital, *Br. J. Dermatol.*, 110, 187, 1984.

20. Mountford, P.J. and Davies, V.J., Ultraviolet radiometry of clinical sources with a miniature multijunction thermopile, *Clin. Phys. Physiol. Meas.,* 8, 325, 1987.
21. Gillham, E.J., Radiometric Standards and Measurements, Notes on Applied Science No. 23, HMSO, London, 1961.
22. Goodman, T.M., Calibration of light sources and detectors, in *Radiation Measurement in Photobiology,* Diffey, B.L., Ed., Academic Press, London, 1989, chap. 3.
23. Corbett, M.F., Hawk, J.L.M., Herxheimer, A., and Magnus, I.A., Controlled therapeutic trials in polymorphic light eruption, *Br. J. Dermatol.,* 82, 107, 1982.
24. Davis, A., Deane, G.H.W., and Diffey, B.L., Possible dosimeter for ultraviolet radiation, *Nature,* 261, 169, 1976.
25. Diffey, B.L., Ultraviolet radiation dosimetry with polysulphone film, in *Radiation Measurement in Photobiology,* Diffey, B.L., Ed., Academic Press, London, 1989, chap. 7.
26. Diffey, B.L., Davis, A., Johnson, M., and Harrington, T.R., A dosimeter for long wave ultraviolet radiation, *Br. J. Dermatol.,* 97, 127, 1977.
27. Diffey, B.L. and Davis, A., A new dosimeter for the measurement of natural ultraviolet radiation in the study of photodermatoses and drug photosensitivity, *Phys. Med. Biol.,* 23, 318, 1978.
28. Tate, T.J., Diffey, B.L., and Davis, A., An ultraviolet radiation dosimeter based on the photosensitising drug nalidixic acid, *Photochem. Photobiol.,* 31, 27, 1980.
29. Diffey, B.L., Oliver, I., and Davis, A., A personal dosimeter for quantifying the biologically-effective sunlight exposure of patients receiving benoxaprofen, *Phys. Med. Biol.,* 27, 1507, 1982.
30. Jackson, S.A., A film badge dosimeter for UVA radiation, *J. Biomed. Eng.,* 2, 63, 1980.
31. Moseley, H., Robertson, J., and O'Donoghue, J., The suitability of diazochrome KBL film for UV dosimetry, *Phys. Med. Biol.,* 29, 679, 1984.
32. Zweig, A. and Henderson, W.A., Jr., A photochemical mid-ultraviolet dosimeter for practical use as a sunburn dosimeter, *Photochem. Photobiol.,* 24, 543, 1976.
33. Fanselow, D.L., Pathak, M.A., Crone, M.A., Ersfield, D.A., Raber, P.B., Trancik, R.J., and Dahl, M.V., Reusable ultraviolet monitors: design, characteristics and efficacy, *J. Am. Acad. Dermatol.,* 9, 714, 1983.
34. Moseley, H., Mackie, R.M., and Ferguson, J., The suitability of suncheck patches and Tanscan cards for monitoring the sunburning effectiveness of sunlight, *Br. J. Dermatol.,* in press.
35. Wong, C.F., Fleming, R., and Carter, S.J., A new dosimeter for ultraviolet B radiation, *Photochem. Photobiol.,* 50, 611, 1989.
36. Sydenham, M.M., Wong, C.F., Hirst, L.W., and Collins, M.J., Ocular UVB dosimetry made possible for the first time using a CR-39 contact lens.

Chapter 31.5
Phototesting: Phototoxicity and Photoallergy

Takeshi Horio
Department of Dermatology
Kansai Medical University
Osaka, Japan

I. Phototesting

In some patients it is possible to make a confident diagnosis of photosensitivity based on the distribution pattern and history of cutaneous changes. However, subsequent phototesting is required to make an accurate decision. Phototesting is usually employed to make a miniature of skin lesions in patients with suspected photosensitivity disease by means of irradiation with artificial light sources. The existence of photosensitivity, precise diagnosis, action spectrum, or etiologic factors can be confirmed when the characteristic skin changes are reproduced in phototesting. This type of *in vivo* testing is useful for several photosensitivity diseases listed in Table 1. Light is exposed to the patient's skin with or without an application of chemicals which are suspected to cause the photosensitivity state.

II. Light Source

The selection of a light source is the most essential step in phototesting. The sun is the reasonable light source since clinical photosensitivity reaction in patients is induced by natural sunlight. However, light energy from the sun is unpredictable and hard to control. It is impossible to obtain the irradiance at any intensity and at any time desired.

A variety of artificial light sources are commercially available.[1] However, it is imperative that the light source for phototesting has adequate irradiance in the action spectrum of the disease being examined. Otherwise, phototesting may yield a false-negative result. The emission spectrum should be in long-wave ultraviolet (320 < UVA < 400nm), mid-wave ultraviolet light (290 < UVB < 320nm), and or visible light (400- to 800-nm) range. Infrared radiation seems to be irresponsible to photosensitivity reaction. The light source for *in vivo* testing should not emit short-wave ultraviolet light (UVC < 290nm), which is not included in the natural sunlight, and may induce a false-positive reaction. Furthermore, the light source is desirable to be easily and economically available for practice use.

Fluorescent tubes are useful and convenient light sources for UVA and UVB irradiation. So-called "black light" and "sunlamp" produce irradiation primarily in the UVA and in the UVB range, respectively. These fluorescent tubes have a large field size enough to test a number of materials at once in photopatch testing. A slide projector can be used as visible light source to examine especially solar urticaria.

The "solar simulator" equipped with xenon lamp emits a broad waveband with a mix of UVB, UVA, and visible light similar to that occurring in natural sunlight. However, this light source is expensive and has a small field size.

III. Phototoxic and Photoallergic Reactions

The most common cause of photosensitivity diseases is drug or chemical substance. Drug photosensitivity reactions are cutaneous responses to the combined action of a chemical and a physical agent. The activating light must be absorbed by the drug to initiate the photosensitizing reaction, since photobiologic responses are predicated upon photochemical reactions. Thus, drugs which do not absorb light energy do not induce a photosensitivity reaction. Drug-induced photosensitivity can be divided into phototoxic and photoallergic reaction based on the mechanisms involved, whether nonimmunological or immunological.

The mechanism of action of drug-induced phototoxic reaction is not uniform depending on the responsible chemicals. Phototoxic reactions can occur theoretically in 100% of the population if sufficient doses of the drug are administered and appropriate wavelengths (action spectrum) of light are irradiated. Experimentally, it can develop without an incubation period after first exposure to the causative drugs. Clinically, however, there exists an apparent refractory period, because the photosensitizing chemicals and the required wavelengths are only rarely present at the start of drug administration in the proper amounts for a reaction to occur.

Drug-induced photoallergy usually involves a delayed hypersensitivity response. It is especially well-established that photoallergic contact dermatitis develops through a T cell-mediated type IV immunologic reaction.[2,3] Experimental photoallergic contact dermatitis can be induced in guinea pigs and mice using a modification of animal model of contact hypersensitivity. Therefore, the cutaneous change is clinically eczematous eruption which is identical with allergic contact dermatitis. However, the

TABLE 1 Indication of Phototesting

Disease	Action Spectrum
Polymorphous light eruption	UVA and/or UVB
Photocontact dermatitis	UVA
Drug-induced photosensitivity	UVA
Chronic actinic dermatitis	UVA, UVB (visible light)
Solar urticaria	UVA, UVB, and/or visible light
Hydroa vacciniforme	UVA
Lupus erythematosus	UVB
Xeroderma pigmentosum	UVB (UVA)
Cockayne's syndrome	UVB

difference in clinical features is not always clear cut between a phototoxic and photoallergic reaction. Action spectrum studies cannot differentiate two reactions.

A single drug may cause phototoxic as well as photoallergic reactions. Photoallergic reaction is usually induced with less doses of the drug and light than phototoxic reaction. Like an allergic drug eruption or contact dermatitis there is an incubation period at least of 7 days in man.

Some patients with drug-induced photoallergic dermatitis may retain a persistent reactivity to sunlight that continues long after the exposure to the causative photosensitizing compound has ceased. These patients are called "persistent light reactors".[4]

IV. Phototesting without Chemicals

A. UVA Irradiation

The UVA-sensitive photodermatoses include drug-induced photosensitivity (phototoxic and photoallergic), photocontact dermatitis, chronic actinic dermatitis, certain cases of polymorphous light eruption, hydroa vacciniforme, and some cases of solar urticaria (Table 1). There is no standardized procedure for UVA phototesting. The irradiation dose to reproduce the clinical lesion depends on the severity of photosensitivity in the patients, and therefore it varies not only among diseases but also among patients even of an identical disease.

In our photodermatology unit, UVA at doses of 1.5, 3.0, 6.0, and 9.0 J/cm^2 is exposed to four areas (2×2 cm each) on normal appearing skin (usually on the back) using fluorescent black light tubes. Exposed areas are read 24 and 48 hours after irradiation. Only the patients with possible solar urticaria are evaluated for wheal formation until 30 min after exposure. In normal subjects, the UVA exposure under this condition does not produce any reaction except immediate pigment darkening which disappears within a few hours after irradiation. Therefore, an erythematous reaction can be estimated as an abnormal photosensitive state.

Patients with chronic actinic dermatitis are most reactive to UVA irradiation,[6] while most cases with polymorphous light eruption are least susceptible. When no reaction is produced with this phototesting in patients who are strongly suspected to have photosensitivity disease, the UVA exposure is repeated two or three times at 24- or 48-hour intervals at the same area.

Eczematous or dermatitic changes appear in the UVA-exposed skin of the patients with drug- or chemical-induced photosensitivity, chronic actinic dermatitis, and polymorphous light eruption. In hydroa vacciniforme, edematous erythema appears with vesicle or bullae, which may develop into necrosis, when sufficient dose of UVA is exposed (Figure 1).

B. UVB Irradiation

The UVB-sensitive photodermatoses are less common than the UVA-reactive diseases. The UVB radiation easily produces sunburn reaction even in normal persons. Therefore, it is not always easy to interpret the reaction induced by UVB irradiation because the photosensitive patients can develop pathological changes mixed with a physiological reaction. The reactions should be evaluated quantitatively and qualitatively.

There are two procedures of UVB phototesting: the minimal erythema dose (MED) and the delayed erythema dose (DED) tests.[15]

The MED test is described in detail elsewhere in this text. In patients with xeroderma pigmentosum or Cockayne's syndrome, the MED to induce sunburn reaction is lowered. The cutaneous reaction is indistinguishable from that of normal subjects macroscopically and microscopically. However, the time course of UVB erythema is characteristic in these photosensitive genodermatoses in comparison with normal sunburn reaction. The peak of reaction is delayed reaching at 48 to 72 hours after irradiation, and the erythema persists longer extending to 5 to 7 days.

Patients with other photodermatoses such as chronic actinic dermatitis often react to much lower UVB dose than MED in normal subjects. However, the UVB-induced reaction is not a simple sunburn but an eczematous change in these patients.

It may be necessary to deliver larger amounts of UVB (DED) for reproduction of skin eruption in certain photosensitivity diseases, in which MED is normal, such as polymorphous light eruption, some patients with drug-induced photosensitivity, and the systemic as well as discoid lupus erythematosus. There is no established procedure for this test. The cutaneous changes of other photoaggravated dermatoses may be also induced with high doses of UVB radiation.

A single or divided exposure of 6 to 8 MEDs on nonlesional skin is often used for DED test. Sunburn erythema appears at 10 to 24 hours after the last irradiation. As it is subsiding, a second, erythematous, or papular reaction may become visible at 3 to 10 days after the irradiation. It is evaluated as positive when the reaction is identical with the clinical eruption (Figure 2). The positive responses may persist for 4 to 14 days. In most instances, repeated exposures of divided dose at 24- to 48-hour intervals are more useful than single exposure of the same

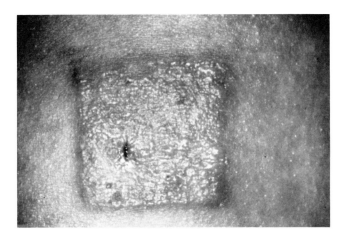

FIGURE 1 Phototesting in hydroa vacciniforme. An edematous erythema with vesicles was induced with UVA radiation.

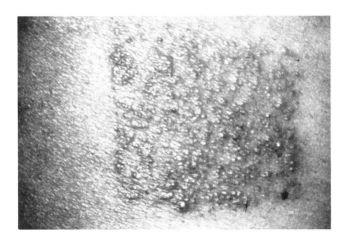

FIGURE 2 Phototesting in polymorphous light eruption. 2 MEDs of UVB were exposed three times at a 48-hour interval to the same area. Papular reaction appeared 3 days after the last irradiation.

UVB dose. Using this technique, Epstein demonstrated a positive response in 90% of the patients with polymorphous light eruption.[7]

It may be necessary to irradiate previously diseased skin in some patients to demonstrate the positive phototest results.

C. Visible Light Irradiation

The phototesting with visible light source is of diagnostic value especially in solar urticaria whose action spectrum lies in this range. A slide projector is easily available and valuable light source for provocative phototesting. The exposure dose varies depending on the severity of the disease and on the light source used.

We routinely use a slide projector equipped with a 500-W bulb, since a larger bulb emits too much infrared radiation and may heat the exposed skin. A wheal formation can be produced in the patients with solar urticaria immediately after the exposure for 1 to 10 min at a distance of 20 cm from the projector lens (Figure 3).

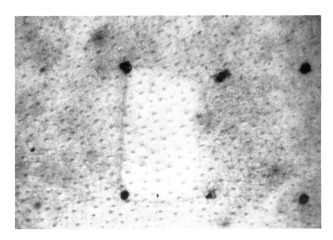

FIGURE 3 Urticarial wheal induced in a patient with solar urticaria by exposure to a slide projector.

Precise action spectra of solar urticaria can be obtained by using cutoff glass filters in combination with a slide projector. In selected patients, not only action spectra but also inhibition and/or augmentation spectra can be found.[8,9] In these cases, pre- or post-irradiation with wavelengths longer than action spectrum may inhibit or augment the urticarial response.

In a few patients with chronic actinic dermatitis, visible light radiation may evoke an eczematous change at 24 to 48 hours after exposure. These patients are also photosensitive to UVA range.

V. Photopatch Testing

A. Principle and Indication

The principle of photopatch testing is exactly the same as that of patch testing in allergic contact dermatitis. It is performed in order to reproduce the photoallergic contact dermatitis in miniature by topical application of the offending photosensitizer and subsequent exposure by activating ultraviolet light. Photopatch testing may be valuable in some patients who are photoallergically sensitized by systematically administered drugs. It cannot be applied for diagnosis or to prove the cause of phototoxic dermatitis, because phototoxic reaction is not specific to the photosensitive patients, but can occur in all normal persons.

B. Method

Photopatch testing is performed with the following procedure, although a certain modification may be made.

1. Application of chemical substances in duplicate patches
2. Covering of patches with light opaque material
3. Evaluation of uncovered patches 48 hours later
4. Irradiation of one set of patches with UVA
5. Covering of both sets
6. Evaluation after additional 24, 48, and 72 hours

C. Test Site

Any site of the whole body skin can reveal positive photopatch test result. However, the test area must be large enough, since the chemicals to be examined are applied in duplicate. The patches are placed symmetrically on nonlesional skin of the interscapular or lumbar regions of the back.

D. Materials

Suspected photoallergenic chemicals must be selected to be examined. The chemical substances listed in Table 2 are often used as the photopatch test series. Selection should be modified by information from the patient's history.

Most substances are usually incorporated into petrolatum in 1% concentration. However, photosensitizing chemicals have the ability to induce a phototoxic as well as photoallergic reaction. Therefore, test materials must be used in nonirritating and non-phototoxic concentrations. For example, chlorpromazine is a strongly phototoxic substance, and also frequently induces photoallergic reaction. It is commonly used in 0.1 to 0.01% concentration. In contrast, the phototoxicity of halogenated salicylanilides is low in *in vivo* tests. However, higher concentrations can inadvertently photosensitize patients after repeated photopatch testing. Furthermore, Harber observed a "broadening of the base" phenomenon in several patients with 1% tetrachlorosalicylanilide (TCSA).[1] These patients were initially photosensitive only to TCSA but developed photosensitivity to bithionol on the fourth photopatch testing.

E. Light Shielding

In the severely photosensitive patients, even trace amounts of light exposure to patched sites may yield false-positive test results. Epstein designated such a reaction as "masked photopatch test", since it is a positive photopatch test due to unintentional exposure to light.[10] Willis and Kligman were able to elicit positive photopatch test reaction to halogenated salicylanilide with light exposure through cotton and woolen cloth, adhesive tape, and bond paper.[11] To avoid false-positive reaction, the patches must be completely covered with light opaque material. For this purpose Finn chamber or ALTest is recommended. During the irradiation of one set of patches, another set for patch testing must be carefully shielded.

F. Irradiation

The action spectrum of photoallergic contact dermatitis is primarily in the UVA range. Therefore, the most important instrument for photopatch testing is an artificial light source with sufficient spectral irradiance in the UVA range. A bank of multiple black light lamps is convenient equipment for this purpose because of large field size, unnecessity of filters, and easy availability. Test sites are irradiated with 1 to 5J/cm^2 or with 50% of threshold dose of UVA.

Willis and Kligman irradiated test areas immediately after application of photosensitizers.[11] This came from

TABLE 2 Compounds for Routine Photopatch Testing

Substance	%	Vehicle
Tetrachlorosalicylanilide	0.1–1.0	Petrolatum Ethanol
Tribromosalicylanilide	1.0	Petrolatum Ethanol
Dibromosalicylanilide	1.0	Petrolatum Ethanol
Bithionol	1.0	Petrolatum
Hexachlorophene	1.0	Petrolatum
Trichlorocarbanilide	1.0	Petrolatum
Irgasam	1.0	Petrolatum
Fentichlor	1.0	Petrolatum
Chlorpromazine	0.01–0.1	Petrolatum Ethanol
Perphenazine	0.01–0.1	Petrolatum Ethanol
Promethazine	1.0	Petrolatum
Levomepromazine	1.0	Petrolatum
Thioridazine	1.0	Petrolatum
Diphenhydramine	1.0	Petrolatum
Musk ambrette	1.0–5.0	Petrolatum
6-Methylcoumarin	1.0–5.0	Petrolatum
Para-aminobenzoic acid	5.0	Ethanol
Ketoprofen	1.0–3.0	Petrolatum
Suprofen	1.0	Petrolatum
Piroxicam	1.0	Petrolatum
Dibucaine	0.1	Water
Psoralens	0.001	Ethanol

their thesis that the role of light energy is simply to transform the photosensitizer into a more potent contact sensitizer and therefore that it is not necessary to allow time for penetration to occur prior to irradiation.[12] Usually, however, test sites are irradiated 24 to 48 hours after application, since the immediate irradiation may yield a false-negative result.

G. Interpretation

Results of photopatch testing are evaluated comparing with another set of patch testing. When the irradiated patch shows a positive reaction and the nonirradiated patch shows a negative one to the identical test material, the result is interpreted as the presence of photoallergic contact dermatitis (Figure 4). If both sites show equally positive reactions, plain contact dermatitis is present because the light exposure does not play any role on the reaction. In a few cases, both irradiated and nonirradiated sites show positive results, but the former reaction is more pronounced than the latter. Such a result is interpreted as a coexistence of allergic and photoallergic contact dermatitis to the same chemical. If there is no reaction at either site, there is no contact or photocontact sensitization to the substances tested.

VI. Drug Phototesting

Photopatch testing may be of no value in the diagnosis of photosensitivity reaction induced by systematically administered drugs. In principle, the causative agent should be administered through the same route as in

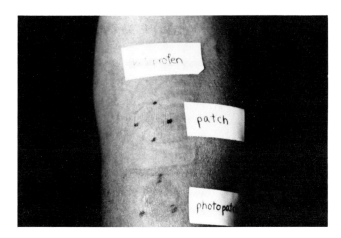

FIGURE 4 Positive photopatch testing to ketoprofen. Patch testing yielded a negative result.

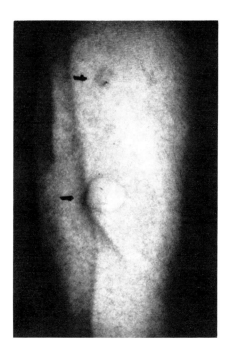

FIGURE 5 A wheal formation in a patient with solar urticaria at an injection site of *in vitro*-irradiated patient's serum.

the clinical use to reproduce the reaction. Unlike the provocative test in ordinary drug eruption, the drug phototesting can be safely performed because the hypersensitivity reaction appears only on the skin localized in the exposed area.

Drug phototesting is performed according to the following steps in our photodermatology clinic.

1. Discontinuation of drug intake for 1 week
2. Determination of MED (especially to UVA)
3. Readministration of drug for 1 day
4. Reexamination of MED
5. Comparison of MED before and after drug readministration

When the MED is significantly reduced after readministration, the drug can be estimated as the responsible photosensitizer. The above-mentioned procedure must be modified depending on the sensitizing drugs, degree of patient's photosensitivity, or type of eruption.

VII. Serum Phototesting

Solar urticaria can be passively transferred to normal persons by means of an intradermal injection of patient's serum and subsequent exposure to the activating wavelengths.[13] This is a modification of the Prausnitz-Küstner technique used in immediate hypersensitivity reaction. Therefore, the passive transfer test is a useful procedure to examine whether the disease process is allergic in nature. However, this test is not performed at the present time, since it can transfer not only allergic reaction but also viral diseases.

Some patients with solar urticaria develop a wheal at the site of injection of their own serum, which was previously exposed to light *in vitro*.[14,15] This indicates that the wheal-forming factor is a substance produced by light energy in the patient's serum (Figure 5).

References

1. Harber, L.C. and Bickers, D.R., *Photosensitivity Diseases*, B.C. Decker, Philadelphia, 1989.
2. Takigawa, M. and Miyachi, Y., Mechanisms of contact photosensitivity in mice. I. T cell regulation of contact photosensitivity to tetrachlorosalicylanilide under the genetic restrictions of the major histocompatibility complex, *J. Invest. Dermatol.*, 79, 108, 1982.
3. Maguire, H.C. and Kaidbey, K., Experimental photoallergic contact dermatitis: a mouse model, *J. Invest. Dermatol.*, 79, 147, 1982.
4. Wilkinson, D.S., Patch test reactions to certain halogenated salicylanilides, *Br. J. Dermatol.*, 74, 302, 1962.
5. Jillson, O.F. and Curwen, W.L., Phototoxicity, photoallergy and photoskin tests, *Arch. Dermatol.*, 80, 678, 1959.
6. Norris, P.G. and Hawk, J.L.M., Chronic actinic dermatitis: a unifying concept, *Arch. Dermatol.*, 126, 376, 1990.
7. Epstein, J.H., Polymorphous light eruption, *Ann. Allergy*, 24, 397, 1966.
8. Hasei, K. and Ichihashi, M., Solar urticaria: determinations of action and inhibition spectra, *Arch. Dermatol.*, 118, 346, 1982.
9. Horio, T. and Fujigaki, K., Augmentation spectrum in solar urticaria, *J. Am. Acad. Dermatol.*, 18, 1189, 1988.
10. Epstein, S., "Masked" photopatch tests, *J. Invest. Dermatol.*, 41, 369, 1963.
11. Willis, I. and Kligman, A.M., Photocontact allergic reactions: elicitation by low doses of long ultraviolet rays, *Arch. Dermatol.*, 100, 535, 1969.
12. Willis, I. and Kligman, A.M., The mechanism of photoallergic contact dermatitis, *J. Invest. Dermatol.*, 51, 378, 1968.
13. Horio, T., Solar urticaria — sun, skin and serum, *Photodermatology*, 4, 115, 1987.
14. Horio, T. and Minami, K., Solar urticaria: photoallergen in a patient's serum, *Arch. Dermatol.*, 113, 157, 1977.
15. Horio, T., Photoallergic urticaria induced by visible light: additional cases and further studies, *Arch. Dermatol.*, 114, 1761, 1978.

32.0

Special Experimental Techniques

32.1 Skin Chamber Techniques 633
B. Zweiman

32.2 Skin Microdialysis 641
L.J. Petersen

Chapter 32.1
Skin Chamber Techniques

Burton Zweiman
Division of Allergy and Immunology
University of Pennsylvania School of Medicine
Philadelphia, Pennsylvania

I. Introduction

Investigators have long been interested in exploring the *in vivo* inflammatory responses in human skin for a number of reasons. (1) Important information about normal sequential biologic events such as host responses against microbial invaders and antigens can be obtained more feasibly than in other areas of the body such as the lower airways. (2) The clinical investigator would be interested in assessing pathogenic events in a variety of skin disorders. (3) The clinical pharmacologist would value a reliable way of determining the effects of pharmacologic and other modulatory agents on both the humoral and cellular manifestations in experimentally induced inflammatory reactions. A side benefit might be an approach to measure delivery of the therapeutic agent to the skin without having to extract it from sizable portions of tissue. The intended end result of these objectives could be an enhanced overall knowledge of human inflammatory responses.

Earlier approaches to skin inflammation involved assessment of the gross and histologic responses to agents applied to or injected within the skin. These approaches will be described elsewhere in the book. Although much valuable information can be obtained, as reviewed,[1] one cannot obtain samples of the interstitial fluid from the dermis to determine levels of inflammatory mediators, etc. Also, it is very difficult to carry out sequential studies at the same skin site.

Therefore, attempts have been made to obtain sequential samples found in the skin by relatively noninvasive methods, that is, designs which can obtain the desired information without excessively damaging the skin.[2] The major limitation to studies of dermal inflammation has been the relatively impermeable barrier posed by the epidermis, particularly the stratum corneum. Therefore, a number of approaches to removing the epidermis have been made (Table 1). Earlier attempts to remove the epidermis involved scraping with a scalpel blade, abrasion with a grinding cylinder, or tape stripping.[3-5] A number of important observations were made about leukocyte exudative responses on cover glasses or skin chambers appended to these sites, these techniques were limited by the irregular depth of the abrasion and induction of some point bleeding. Some local discomfort and a persistent scar could occur. Therefore, other attempts at less traumatic approaches have been made. Some of these involving the induction of skin blisters will now be described.

A. Objective

The objective of this discussion is to describe (1) the several ways in which skin blisters have been induced and their use; (2) the design and use of skin collection chambers; and (3) the application of these approaches in one type of experimentally induced inflammatory reaction (IgE-mediated) and certain skin disorders.

II. Methodologic Principle

A. Skin Blisters
1. Induction

It has been recognized that induction of a skin blister with its floor at the dermal-epidermal junction would provide a valuable although quite small amount of interstitial fluid for sampling. In early studies, vesicants such as cantharides were applied to the skin surface. This was replaced by suction techniques[2,6] in which blisters could be induced within 2 hours by negative pressures of about 200 mm Hg or 0.3 kg/cm² exerted through devices with openings of various diameters (up to about 15 mm). Some investigators, including our group, have found that blisters can be induced in a shorter time, 60 to 90 min, by a combination of relatively gentle heat and suction. The method currently used in our laboratory is

1. A circular plexiglass chamber apparatus (constructed especially for us) consists of an outer chamber (70-mm diameter) with two flanges (15 mm) perforated by slots for insertion of a binding strap. This outer chamber also has two outlets into which metal correcting tubes are inserted with a tight seal. Sealed within the outer chamber

TABLE 1 Some Applications of the Skin Chamber Approach

1. Sequential assessment of cellular and humoral events in inflammation
2. Comparison of exudative cell responses with those in the underlying dermis
3. Investigation of pathogenic processes in skin diseases
4. Study of modulatory effects of systemic therapies on skin inflammation
5. Assessment of drug delivery to inflammatory skin reactions

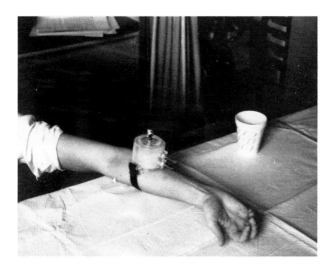

FIGURE 1 Heat/suction chamber apparatus applied firmly to the forearm of a volunteer subject with a binding strap. The outlet at the top of the inner chamber is connected to a tubing from the vacuum source. The outlets in the outer chamber are connected to entrance and exit tubing attached to a heater water pump.

is an inner chamber (45-mm diameter) in which a 10-mm diameter circular opening has been drilled in the bottom. Into the top of the inner chamber is drilled an 8-mm diameter outlet into which a metal connecting tube is tightly sealed. This chamber apparatus is applied firmly to the skin of the forearm using Velcro-tipped binding straps (Figure 1).

2. The two outlets in the outer chamber are connected by appropriate tubing to the outlet and inlet ports of a thermostat-controlled (55°C) heated water pump (Haake, Saddlebrook, NJ) so that this heated water circulates in the outer chamber. The outlet in the top of the inner chamber is then connected to a vacuum line. When negative pressure is conveyed to the inner chamber, suction is exerted on the opening in the bottom of the inner chamber (which is now tightly appended to the skin surface) while the heated water in the surrounding outer chamber heats the local skin surface. The transparent nature of the plexiglass allows continuous inspection to determine when adequate blisters are formed, seen protruding into the inner chamber through the opening in the bottom. The blister induction apparatus is then removed. We can simultaneously induce two blisters on the forearm of average-sized adults with little or no discomfort. Other investigators using devices employing suction only can induce at least three blisters on the forearm depending on the diameter of the blister.[2,6]

3. In our lab, we can aspirate about 25 μl from individual blisters. Such small fluid samples can be analyzed for levels of inflammatory mediators if sufficiently sensitive assay systems are available. For example, histamine levels have been compared in the fluid of blisters raised over involved and uninvolved skin in certain skin disorders like urticaria.[8] However, there may be a sizable nonspecific release of some mediators into these fluids due to the antecedent trauma resulting from the blister induction. For this reason, some investigators wait 24 hours after blister induction before using the site, to allow for this traumatically induced mediator release to be less of a factor. This approach requires protecting the blister sites from inadvertent external trauma with overlying shields. However, in our experience, it is not unusual for blisters to collapse at least partially by 24 hours.

2. Skin Chambers

Use of the skin blisters themselves is limited by (1) the factors described above; (2) inability to obtain samples repeatedly from the same blister; and (3) inability to challenge the site of the blister repeatedly with agents like antigens. Therefore, several investigative groups have modified the skin blister technique to use collection chambers.[2,6,7,9] The technique used currently by us is

1. The blisters are inspected carefully after induction, not using any in which there is gross evidence of local bleeding.
2. The blister surface is cleansed with 70% ethanol and the contents are aspirated and stored at -70°C for future use when appropriate.
3. The blisters are unroofed aseptically with small forceps and scissors. The blister roof is set aside for use in epidermal cell studies, if appropriate.
4. The blister base is washed thoroughly at least twice with sterile phosphate-buffered saline (PBS) or other isotonic solution as indicated.
5. Circular collection chambers are appended to each blister base by firm cross-hatched application of nonirritating tape (e.g., Durapore, 3M Company,

Minneapolis) so that a tight seal between the chambers and underlying skin is achieved without causing excessive pressure on underlying structures. Several variations in technique have been used by us and other investigators.

a. We have used a variety of containers, all made out of various types of rigid plastic material An example is shown in Figure 2. Each chamber has two rubber-sealed ports. One is used for introduction of fluids into the chamber and their subsequent removal with a tuberculin syringe and 27-gauge needle; a separate 27-gauge needle is inserted through the top port just into the interior of the chamber to allow air to evacuate. We generally inject 0.3 or 0.5 ml of fluid into each chamber, (depending on the protocol) looking carefully for any sign of leakage at the chamber base.

b. We do not use an adhesive to adhere these chambers to the skin and have not encountered major problems during challenges as long as 9 hours if the subjects keep their arms relatively quiet. Other have used adhesives like cyanoacrylates "Skin Bond®" or "Bond Fast®", reportedly allowing firm adherence of the chambers to the skin for as long as 48 hours.[2,9] Dr. Charlesworth reports that occasionally prominent irritant reactions occur under metal skin chamber adhered with Skin Bond® for longer than 8 hours (personal communication). This can be reduced by placing a plastic film barrier such as Saran® (Dow Chemical, U.S.) wrap between the metal chamber and the skin.

c. At the end of a challenge time period the chamber contents are aspirated into a tuberculin syringe with a 27-gauge needle inserted into the port (see above). The volume is measured and the fluid centrifuged (1000 rpm for 10 min at 4°C) for separation and subsequent study of humoral and cellular components. In cases where the site is to be rechallenged, we find it very important to thoroughly wash the site with PBS or like solution to remove any residual challenge agent, mediators, etc. left on the blister base. Some investigators do this wash with the collection chamber in place. We prefer to remove the chamber and wash the blister base thoroughly before applying a fresh collection chamber. The latter action is a precaution against leaving any material on the inner surface of the first collection chamber.

d. In this manner, sequential challenges with the same or different solutions can be carried out for up to 48 hours (see above). To reduce the rate of fibrin formation at the blister base, we have found it helpful to add heparin (final

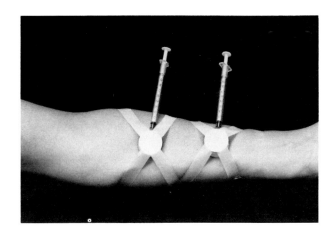

FIGURE 2 A collection skin chamber of another shape appended to the skin blister base with straps of nonirritating tape. A tuberculin syringe with needle attached is inserted in one of the two ports.

concentration 10 units/ml) to all solutions placed in the chambers. In previous quality control studies, we had found that this concentration of heparin did not cause gross bleeding in the site or affect levels of mediators released into the overlying skin chambers.

e. Depending on the requirements of the individual study, the cell-free chamber fluids are divided into aliquots, stored at –70°C for further analysis. The cells are resuspended in PBS enumerated with an automated counter (Coulter). Aliquots are cytocentrifuged (Shandon Cytocentrifuge) in replicate for differential counts, immunocytochemistry, etc. or functional studies are performed. Since the total cell yield varies considerably depending on the condition of challenge and the donor, it may be necessary to pool cells obtained from replicate challenge sites.

f. At the end of the skin chamber incubation, we frequently also assess the population of inflammatory cells exuding into the blister base at that time period.[10,11] Sterile well-cleaned circular cover glasses, 10-mm diameter, are appended with gentle pressure to the blister bases for 15 to 30 minutes (depending on the study). They are then removed carefully with sterile forceps, air dried, stained with Wright's stain (Figure 3), or stored at –20°C in a wrapped container for future fixation and other types of study. We have found no impressively greater adherence of cells to cover glasses previously coated with human serum albumin before sterilization. Indeed, relatively few cells are found adhering to a second cover glass appended to the site following removal of the first cover glass after 30-min incubation.

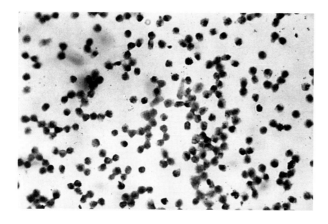

FIGURE 3 Imprint of exuding leukocytes adhered to a cover glass appended for 30 min to the blister base of a skin chamber site following a 5-hour pollen antigen challenge in a sensitive subject (Wright's stain, 600× magnification). Generally, over 95% of the exuding cells are granulocytes.

3. Biopsy of the Blister Base

It has sometimes been important to assess the histologic patterns in the dermis of the blister bases following completion of the skin chamber challenge at that site. We have found that 3-mm diameter punch biopsies at the center of the blister bases are remarkably well tolerated when performed after local infiltration with 2% lidocaine solution in a circular pattern around the periphery of the blister base. Such biopsied tissue is washed in PBS and either processed immediately or stored in appropriate fixative or carrying medium, depending on the nature of the study.[11,12] Meticulous care should be taken in positioning and sectioning these biopsies since they are quite small and have no epidermal layer that would help in positioning.

4. Associated Blood Studies

It is sometimes desirable to compare patterns/events in the blood obtained during the course of the skin chamber challenge.[13] Whenever feasible, we obtain the blood specimens from veins not in the region of the skin chamber (unless the latter is desired). An infusion set needle is inserted in a vein prior to initiation of the skin chamber challenge, kept patent with a heparin solution in low concentration, and capped for future removal of blood specimens if desired. The first 3.0 ml of subsequently aspirated blood is discarded to eliminate any diluting or other effects of heparin in the tubing.

III. Sources of Error

We and other investigators have found the skin chamber approach to be a valuable, relatively noninvasive probe of *in vivo* human inflammatory events in health and disease. However, there are some limitations and caveats which should be mentioned:

1. Despite attempts at making the blister induction as gentle as feasible, this process does involve some trauma leading to a variable inflammatory response by itself. This may impact more on comparisons of the inflammatory cell exudation at the control site as compared to that at the site challenged with antigen or other agonist.[14] We have generally found relatively little in the way of nonspecific release of mast cell mediators like histamine and tryptase at control sites provided that the blister bases are irrigated thoroughly before skin chambers are appended. However, granulocyte contents may be released nonspecifically at times.

2. The skin chambers capture only those components which diffuse from the underlying dermis. This may be affected by (1) molecular weight and charge of the compound; (2) diffusion of the released agents in "other directions" after release (e.g., absorption into local venules and lymphatics); (3) chemotactic/chemokinetic properties of different inflammatory cell types. For example, we find relatively few eosinophils exuding into the overlying chamber fluid even when they are relatively prominent in the underlying dermis. (4) Barriers such as fibrin may act as impediments. Even with heparin in the solutions, we find that fibrin deposition is sometimes prominent in blister bases which have been exposed overnight only to the diluent solution (negative control challenges). Although much of the fibrin can be removed it still may interfere with passage of mediators and cells from the underlying dermis. (5) Also, there is inevitable dilution of any humoral agent diffusing from the underlying dermis by the volume of diluent placed in the skin chamber at the outset. Thus, the price one pays for injecting larger volume of diluent into the skin chamber (to have more fluid for replicate analyses at the end) are lower levels of the mediators assayed.

3. The diffusion of different agonists (placed initially in the skin chamber) into the underlying dermis also may vary considerably. Although the epidermal barrier has been removed, the delivery from the skin chamber to the dermis may not be as rapid as by intradermal injection, particularly for larger molecular weight compounds. For example, we had found that antigen skin chamber challenge of sensitive subjects led to a more prolonged release of histamine than the very prominent but relatively short-lived histamine release induced by the nonimmunologic mast cell activator codeine.[15] We wondered whether these differences could be due to the much lower molecular weight of codeine than the antigens employed. The former would then diffuse away from the site rapidly while the antigen solution both entered and left

the site more slowly. However, when sites were continuously challenged hourly with fresh antigen and codeine solutions and subsequent removal of the fluid for analysis, the patterns of histamine release were the same as described above, that is, there was an initial peak of histamine release, followed by a plateau of lower level histamine release at the sites repeatedly challenged with antigen; in contrast, repeated codeine challenge induced an equally prominent initial peak of histamine release, a small release in the 2nd hour, and then no more release than at the buffer diluent challenge sites over the next 4 hours (Figure 4). These findings suggest that the different patterns of histamine release were due to differing biologic effects of the two agonists placed in the skin chambers and not a difference in their retention in the dermis.[16] Absorption of agonists into the underlying dermis could also be affected by a fibrin layer formed at the blister base after prolonged incubation, as noted above. This is particularly true in the case in sites of IgE-mediated allergic skin reactions where we have found increased fibrin formation.[17]

4. It is conceivable that some inflammatory mediators diffusing from the underlying reaction site into the skin chamber fluid over a period of hours may be metabolized to some degree by enzymes released locally by accumulating inflammatory cells such as histaminase, acetylhydrolyses, etc. Our early quality control studies indicated that exogenous histamine placed into skin chambers over blister bases was not degraded over a period of at least 1 hour. However, that environment is not quite the same as an acute inflammatory site containing many more granulocytes. We have found that much of the leukotriene B4 released into skin chambers overlying IgE-mediated reactions was omega transformed.[18] This is additional evidence that the levels of particular mediators in these skin chamber fluids are underestimates of the concentrations released in the underlying reaction site.

5. Because there is increased local vascular permeability in most inflammatory reactions secondary to the release of vasoactive mediators, it is not surprising that there is a gradually increasing concentration of serum proteins in the chamber fluids after Ag challenge periods of at least 2 to 3 hours (experience in our lab and that of Dr. R. Gronneberg, Stockholm). Indeed, such plasma proteins may be the substrate for formation of the increased kallikrein levels we have found released at allergic reaction sites.[19] However, the elevated serum levels of certain neutrophil and eosinophil components may indicate a contribution of "serum protein leak" as well as release in the underlying dermis to the increased levels of proteins like major basic protein we have observed in the overlying skin chamber fluids.[11] Thus, it is very important to get an estimate of such serum protein leaks by concomitant measurement of the chamber fluid and serum levels of other serum proteins (e.g., albumin, IgG) which one would not expect to be released within the dermis. The presence of serum proteins in the chamber fluid may affect functional measurement of certain enzymes (e.g., chymase) which are readily inhibited by certain serum proteins. Indeed, it was only because our colleague, Dr. Allen Kaplan, utilized a technique which assayed activated Hageman Factor and kallikrein bound to their serum inhibitor (C1 — esterase inhibitor) that we could detect elevated levels of these pro-inflammatory compounds in skin chambers overlying sites of allergic reactions.[19] Serum proteins present in the chamber could also act as a chemotactic agent, likely due to C5a, although such effects are generally not prominent until the serum comprises at least 50% of the total chamber fluid volume.[5,6] Some investigators report that the leukocytes recovered from serum

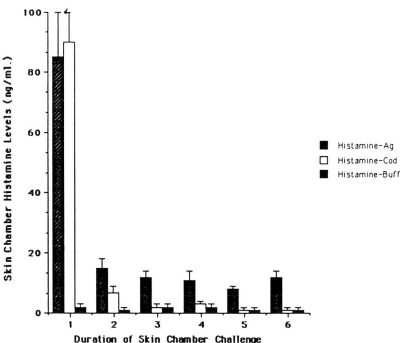

FIGURE 4 A graphic illustration of typical temporal patterns of histamine release into skin chambers overlying sites of hourly blister base challenge with pollen antigen (Ag), codeine (Cod), and buffer diluent (Buff) in a very sensitive subject. Fluids were removed hourly for histamine analysis and were replaced with fresh solutions of the like substance.

containing chambers were deactivated for *in vitro* chemotaxis as compared to autologous blood leukocytes.[20] However, we have found similar *in vitro* responses of chamber fluid and autologous neutrophils for fMLP and activated serum but not to PAF.[21]

6. As in many *in vivo* approaches, there is some variability in sequential skin chamber responses at replicate sites in the same individual. Scheza and Forsgren[6] found about 19% variability in leukocyte exudation if single sites were challenged sequentially with autologus serum; this variability was reduced to 14% of the mean of duplicate chambers studied each time point was used in the comparisons. We have generally found <10% variation in histamine release in duplicate chambers challenged with antigen in sensitive subjects at the same time.

7. There is the possibility that a prominent reaction occurring at one skin chamber site may affect the levels of mediators seen in an adjacent site if the two are placed close to one another. We have not observed this using two sites/forearms as long as the sites reasonably spaced with the control challenge sites were always distal to the site containing active agonist. However, Charlesworth reports "contamination" of adjacent site fluids by reactions induced by high antigen concentrations (such as 1000 PNU/ml of pollen extracts [Charlesworth, E. N., personal communication]). He also has observed less histamine release and cellular exudation in the distal of two similar antigen challenge sites in the same arm.

8. In an attempt to reduce the early effects of blister induction on mediator release, some investigators induce the blister 24 hours before unroofing and application of the collection chamber, as described above. However, this may lead to greater leukocyte exudation at the control site than when the chambers are placed soon after blister induction.

IV. Correlation with Other Methods

It is difficult to compare the skin chamber approaches just described with other *in vivo* methods because there is no closely paralleled technique. However, we and others have compared the extent of humoral and cellular responses in skin chamber overlying inflammatory sites with other *in vivo* and *in vitro* parameters. For example, we have found that the total amount of histamine and the number of leukocytes exuding into skin chambers overlying allergic challenge sites each correlated with the size of the gross late phase reaction seen 6 hours after intradermal injection of the same antigen in the same subject.[22] There was also a correlation of the degree of skin chamber kallikrein formation with gross late reactivity.

In contrast, we have found that the amount of histamine released in skin chambers overlying antigen-induced IgE-mediated reactions does not correlate precisely with the immediate wheal and flare response to intradermal antigen. This variance may be due to factors such as mast cell and/or basophil "releasability", end organ response to histamine, and even the gender of the subject.[23,24]

Scheza amd Forsgren found no correlation between *in vivo* mobilization of leukocytes into chambers filled with autologous serum and the *in vitro* chemotaxis of blood leukocytes from the same subject to autologous blood.[20] We have also been impressed that the degree of granulocyte exudation and granule protein release into the skin chamber does not always parallel the extent of neutrophil and eosinophil accumulation in biopsies of the underlying dermis. The latter may be due to variable degrees of leukocyte activation in such reaction sites.[25]

V. Recommendations

To summarize, some of the favorable aspects of the skin chamber model are

1. Relatively noninvasive and only modestly traumatic with little in the way of discomfort and local sequella. The blister base will heal rapidly leaving a pigmented area for varying length of time (generally not more than 1 year). Although scarring is not obvious since the local dermis is preserved, one should avoid studying individuals known to form keloids.
2. At least semiquantitative with fairly good reproducibility.
3. Allows measurement of multiple mediators in the same fluid (depending on the sensitivity of the assays) and sequential assessment over a period up to 48 hours.
4. Allows measurement of levels of therapeutic agents given orally in studies of drug delivery to the skin. Can also investigate local effects of topically applied drugs.

On the other hand, there are some limitations to keep in mind.

1. Modest inflammatory effects secondary to the blister induction.
2. Chamber fluid levels of mediators are indirect measures of those in the underlying dermis and are likely underestimates.
3. Effects of local enzyme action and exuding serum proteins on some of the mediators released into the chamber fluid and then awaiting aspiration.

With these factors in mind, one should standardize the skin chamber approach as much as feasible, at least within the individual laboratory. Then, I feel that it would be valuable probe for study of

1. Certain skin diseases, e.g., chronic urticaria.[26]
2. Therapeutic modulators, e.g., corticosteroid,[27,28] antihistamines.[2,29]
3. Molecular and cellular mechanism underlying the pathogenesis of inflammatory reactions in health and disease.

The use of skin chambers in a number of investigative approaches is listed in Table 1. Hopefully, with technical modifications leading to improvement, the skin chamber model will be of increasing value to basic and clinical investigators.

Acknowledgment

My great appreciation is extended to my laboratory colleagues who helped in the development and use of this skin chamber approach over the years. Particular gratitude goes to Mrs. Carolyn von Allmen whose careful and thoughtful approaches have contributed to valuable modifications and productive research.

References

1. Zweiman, B., Mediators of allergic inflammation in the skin, *Clin. Allergy*, 18, 419, 1988.
2. Michel, L. and Dubertret, L., Skin models for analysis of IgE dependent reactions in vivo, in *Advances in Allergology and Clinical Immunology*, Godard, Ph., Bousquet, J., and Michel, F.B., Eds., Parthenon Press, Carnforth, 1992, 505.
3. Rebuck, J.W. and Crowley, J.H., A method of studying leukocyte functioning in vivo, *Ann. N.Y. Acad. Sci.*, 59, 757, 1955.
4. Senn, H., Holland, J.F., and Bannerjee, T., Kinetic and comparative studies of localized leukocyte mobilization in normal man, *J. Lab. Clin. Med.*, 74, 742, 1969.
5. Goldberg, B.S., Wetton, W.R., Kohler, P.F., Harris, M.B., and Humbert, J.R., Transcutaneous leukocyte migration in vivo:
6. Scheza, A. and Forsgren, A., A skin chamber technique for leukocyte migration studies: description and reproducibility, *Acta Path. Microb. Immunol. Scand.*, C93, 25, 1985.
7. Dunksy, E.H. and Zweiman, B., The direct demonstration of histamine release in the skin using a skin chamber technique, *J. Allergy Clin. Immunol.*, 62, 167, 1978.
8. Kaplan, A.P., Horakova, A., and Katz, S., Assessment of tissue fluid histamine levels in patients with urticaria, *J. Allergy Clin. Immunol.*, 61, 350, 1978.
9. Charlesworth, E.N., Heard, A.F., Soter, N.A., Kagey-Sobotka, A., Norman, P.S., and Lichtenstein, L.M., Cutaneous late phase response to allergen. Mediator release and inflammatory cell infiltration, *J. Clin. Invest.*, 83, 1529, 1989.
10. Ting, S., Zweiman, B., Lavker, R.M., and Dunsky, E.H., Histamine suppression of in vivo eosinophil accumulating and histamine release in human allergic reactions, *J. Allergy Clin. Immunol.*, 61, 65, 1981.
11. Zweiman, B., Atkins, P.C., von Allmen, C., and Gleich, G.J., Release of eosinophil granule proteins during IgE-mediated allergic skin reactions, *J. Allergy Clin. Immunol.*, 87, 984, 1991.
12. Zweiman, B., Lavker, R.M., Presti, C., and Atkins, P.C., Comparisons of inflammatory responses in IgE-mediated and codeine-induced skin reactions, *J. Allergy Clin. Immunol.*, 91, 963, 1993.
13. Taylor, M., Zweiman, B., Moskovitz, A., von Allmen, C., and Atkins, P.C., Platelet activating factor and leukotriene B4 induced release of lactoferrin from blood neutrophils of atopic and non-atopic individuals, *J. Allergy. Clin. Immunol.*, 86, 740, 1990.
14. Bedard, P.M., Zweiman, B., and Atkins, P.C., Quantitation by myeloperoxidase assay of neutrophil accumulation at the site of in vivo allergic reactions, *J. Allergy Clin. Immunol.*, 3, 94, 1983.
15. Shalit, M., Schwartz, L.B., von Allmen, C., Atkins, P.C., Lavker, R.M., and Zweiman, B., Release of histamine and tryptase during continuous and interrupted cutaneous challenge with allergen in humans, *J. Allergy Clin. Immunol.*, 86, 117, 1990.
16. Atkins, P.C., von Allmen, C., Moskovitz, A., Valenzano, M., and Zweiman, B., Fibrin formation during ongoing cutaneous allergic reactions. Comparison of responses to antigen and codeine, *J. Allergy Clin. Immunol.*, 91, 956, 1993.
17. Atkins, P.C., Kaplan, A.P., von Allmen, C., Moskovitz, A., and Zweiman, B., Activation of the coagulation pathway during ongoing allergic cutaneous reactions in humans, *J. Allergy Clin. Immunol.*, 89, 552, 1992.
18. Shalit, M., Valone, F.H., Atkins, P.C., Ratnoff, W.D., Goetzl, E.J., and Zweiman, B., Late appearance of phospholipid platelet activating factor and leukotriene B4 in human skin after repeated antigen challenge, *J. Allergy Clin. Immunol.*, 83, 691, 1989.
19. Atkins, P.C., Miragliotta, G., Talbot, S.F., Zweiman, B., and Kaplan, A.P., Activation of plasma Hageman factor and kallikrein in ongoing allergic reaction in the skin, *J. Immunol.*, 139, 2744, 1987.
20. Scheza, A. and Forsgren, A., Functional properties of polymorphonuclear leukocytes accumulated in a skin chamber, *Acta Path. Microb. Immunol. Scand.*, C43, 31, 1985.
21. Fleekop, P.D., Atkins, P.C., von Allmen, C., Valenzano, M., Shalit, M., and Zweiman, B., Cellular inflammatory response in human allergic skin reactions, *J. Allergy Clin. Immunol.*, 80, 140, 1987.
22. Bedard, P.M., Zweiman, B., and Atkins, P.C., Antigen induced local mediator release and cellular inflammatory responses in atopic subjects, *J. Allergy Clin. Immunol.*, 71, 394, 1983.
23. Atkins, P.C., von Allmen, C., Valenzano, M., Olson, R., Shalit, M., and Zweiman, B., Determinants of in vivo histamine release in cutaneous allergic reactions in humans, *J. Allergy Clin. Immunol.*, 86, 371, 1990.
24. Atkins, P.C., von Allmen, C., Valenzano, M., and Zweiman, B., The effects of gender upon allergen induced histamine release in ongoing allergic cutaneous reactions, *J. Allergy Clin. Immunol.*, 91, 1031, 1993.
25. Zweiman, B., Kucich, U., Shalit, M., von Allmen, C., Moskovitz, A., Weinbaum, G., and Atkins, P.C., Release of lactoferrin and elastase in human allergic skin reactions, *J. Immunol.*, 144, 3953, 1990.
26. Bedard, P.M., Brunet, C., Pelletier, G., and Hebert, J., Increased compound 48/80 induced local histamine release from nonlesional skin of patients with chronic urticaria, *J. Allergy Clin. Immunol.*, 78, 1121, 1986.
27. Atkins, P.C., Schwartz, L.B., Adkinson, N.F., von Allmen, C., Valenzano, M., and Zweiman, B., In vivo antigen in-

duced cutaneous mediator release: simultaneous comparisons of histamine, tryptase and prostaglandin D2 release and the effect of oral corticosteroid administration, *J. Allergy Clin. Immunol.,* 86, 360, 1990.
28. Charlesworth, E.N., Sobotka, A., Schleimer, R.P., Norman, P.S., and Lichtenstein, L.M., Prednisone inhibits the appearance of inflammatory mediators and the influx of eosinophils and basophils associated with the cutaneous lab phase response to allergen, *J. Immunol.,* 146, 671, 1991.
29. Charlesworth, E.N., Kagey-Sobotka, A., Norman, P.S., and Lichtenstein, L.M., Effects of cetirizine on mast cell mediator release and cellular traffic during the cutaneous late phase response, *J. Allergy Clin. Immunol.,* 83, 905, 1989.

Chapter 32.2
Skin Microdialysis

Lars Jelstrup Petersen
Department of Dermatology
Laboratory of Immunology and Biochemistry
Bispebjerg Hospital
University of Copenhagen
Copenhagen, Denmark

I. Introduction

Measurements pertinent to the extracellular environment can be informative in many situations. This liquid compartment is the environment where the traffic of compounds and exchange of chemical solutes take place, and the extracellular liquid is the site where drugs are acting on cell surface-bound receptors. Meanwhile, such examinations in skin have been restricted by lack of appropriate methods. A number of techniques, including skin perfusion methods,[1] skin window techniques,[2–4] and local venous drainage,[5,6] have been presented as methods to monitor inflammatory skin reactions and compartmental pharmacokinetics. None of these techniques, however, are able to give a quantitative and kinetic picture of mediator and drug trafficking in the extracellular environment. Also, no present methods allow simultaneous extracellular sampling and measurement of clinical observations.

Microdialysis is a new bioanalytical sampling technique, which opens up the possibility to sample substances within the extracellular water space in various organs. Microdialysis has been used for more than 20 years in experimental brain research in animals. Originally described by Bito et al.,[7] the method has been continuously improved.[8,9] During the last years microdialysis has become practical for the routine measurement of neurochemicals *in vivo*. The technique has been rapidly adopted by many neuroscientists, and the method is becoming the preferred technique in this experimental field. The growth of papers on microdialysis is shown in Figure 1.

The microdialysis technique possess several features that suggest a possible role for microdialysis outside the central nervous system in animals. The microdialysis system is a closed liquid system that protects the surrounding tissue from damage, contamination, and irritation. The microdialysis probe remove chemical substances from the extracellular environment without removing liquid. Microdialysis can be performed within the intact target tissues, samples can be collected continuously, and it is possible to recover and/or introduce atraumatically endogenous and exogenous solutes in the tissue. In addition, it is possible to perform quantitative measurements.

The microdialysis method has only lately been used in peripheral tissues, mainly in subcutaneous adipose tissue[10–15] and muscle,[16–18] but also in myocardium,[19,20] eye,[21] liver,[22] and oral mucosa.[23] The interest in peripheral microdialysis is escalating.[24] For an overview on microdialysis applications, refer to recent publications.[15,25–29]

Application of the microdialysis technique in human skin is only recently described.[30–34] The aim of this paper is to give a short introduction to microdialysis and its application in skin.

II. Principle of Microdialysis

A hollow dialysis fiber permeable to water and small solutes is introduced in a tissue, connected to a microinjection pump, and perfused constantly with a physiological fluid. The liquid equilibrates with the extracellular fluid compartment since substances on both sides of the dialysis membrane diffuse according to their diffusion gradients. This means that solutes in the tissue diffuse into the perfusate and substances in the perfusate diffuse the opposite direction, i.e., from the fiber to the tissue. It is the concentration gradient only that determines the direction of the substance flux.

In principle, the microdialysis method is a simple technique. Meanwhile, as described later in this text, issues like choice of dialysis membrane, construction of the probe, perfusion setup, interactions between the dialysis probe, and the tissue with respect to exchange of solutes make interpretation of microdialysis experiments complex.[35–40]

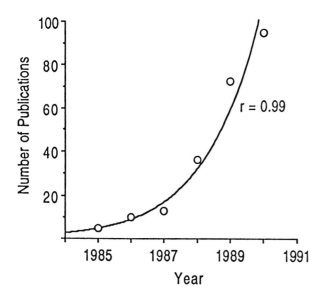

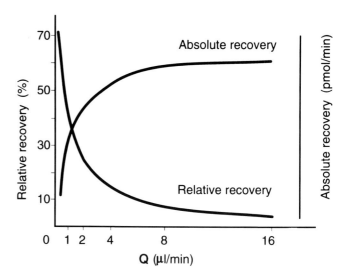

FIGURE 1 The increase in the number of papers published during the last 6 years is fit almost perfectly by an exponential curve. (From *Microdialysis in the Neurosciences,* Robinson, T. E. and Justice, J. B., Jr., Eds., Elsevier, Amsterdam, 1991. With permission.)

FIGURE 2 The relative and absolute recovery of a substance as a function of flow rate. For further details refer to the text. (From Application Note No. 1, CMA/Microdialysis, Sweden. With permission.)

A. Recovery

An important issue of microdialysis is recovery. The dialysing property of a microdialysis probe is expressed as its recovery of a particular substance. Two different terms of recoveries are used: relative recovery and absolute recovery. Relative recovery is defined as the concentration of a solute in the dialysate in percent of the concentration of that solute in the surrounding medium. The constantly perfused probe restricts establishment of 100% diffusion equilibrium across the dialysis membrane. Therefore, relative recovery is below 100%. The absolute recovery is defined as the amount (mole or weight) of substance recovered per time unit.

1. Factors Influencing Recovery

A number of parameters influences recovery. Among those, the perfusion velocity is of utmost importance. The impact on varying perfusion velocities on relative and absolute recovery is illustrated in Figure 2. It appears that relative recovery decreases and absolute recovery increases with increasing perfusion velocity. At high perfusion velocities, relative recovery is very modest and absolute recovery reaches a level with no increase despite increasing perfusion velocity. Recovery is also influenced by other parameters, for example, the ambient temperature,[41] and the dialysis membrane material,[42] as shown in Table 1. However, since the relative recovery is independent of the concentration in the extracellular environment, the dialysate reflects ongoing reactions in the tissue. In some situations, the absolute recovery may decline during an experiment due to drainage of substances by the dialysis probe, thereby, an artificial concentration gradient is created in the tissue which may have profound impact on the interpretation of the results and may disturb tissue functions. The issue of recovery is closely connected with attempts to perform quantitative microdialysis experiments, which is discussed later.

III. Microdialysis Instrumentation

In order to perform microdialysis studies, we need several tools, including a microdialysis probe, a perfusion medium, a microdialysis perfusion pump, a sampling unit, and an analytical technique. Most microdialysis instruments are made especially for microdialysis purposes, but also the analytical technique must be adjusted to meet requirements that follow the small sampling volume obtained by microdialysis.

A. Dialysis Probes

A number of dialysis probes have been presented for use in animal brains. Most probes can be used in

TABLE 1 Some Variables with Impact on Recovery of Solutes by Microdialysis Probes

Probe characteristics	Molecular weight cut off the membrane
	Membrane material
	Membrane surface area
	Wall thickness
	Inner diameter
Perfusion characteristics	Perfusion velocity
	Perfusate composition
Substance characteristics	Molecular weight
	Physical configuration
	Chemical properties

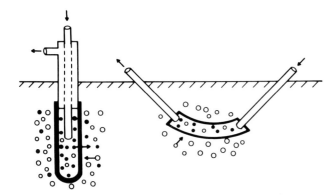

FIGURE 3 Two different probes used for microdialysis experiments. The left probe is a double-lumen concentric probe, and the right probe is a single-lumen serial probe. The microdialysis allows sampling of substances from the extracellular environment (open circles) as well as local delivery of substances (closed circles). (From Arner, P. and Bolinder, J., *J. Intern. Med.*, 230, 381, 1991. With permission.)

human skin provided that the guide cannula permits intracutaneous insertion. The probes differ mainly in the construction of the connection of the afferent and the efferent tubings with the dialysis membrane and dialysis membrane material used. The dialysis membrane can be arranged in a serial or a parallel arrangement as shown in Figure 3. The concentric probe is primarily used in brain studies, since this probe can be positioned very accurately in different neuronal compartments. The serial probe has a large surface area prone for diffusion which preferable in peripheral tissues. Both types of dialysis probes have been used in human skin.[30,32]

The dialysis membrane allows diffusion of solutes with small molecular weights. Most membranes do not allow substances greater than 20,000 Da molecular weight to pass the membrane. It is, however, possible to made probes with greater molecular weight cutoff values, but these probes are infrequently used. It must be emphasized that the nominal molecular weight cutoff value is of minor value in practice since the recovery of solutes declines most rapidly with increasing molecular weight. In fact, the recovery of solutes with molecular weight greater than 5000 Da is almost negligible with a 20,000-Da probe.[43] Even though the cutoff value may be identical among dialysis membranes, the recovery of solutes may show considerable variations.[44] Also the ability of different probes to detect rapid changes in the surrounding medium may show substantial differences.

B. Choice of Perfusate

The objective of the perfusion fluid (perfusate) is to act as a vehicle for substances to be collected from or delivered to the tissue. The perfusate should not interfere with the biochemical events taking place in the tissue surrounding the probe. Therefore, the perfusate should be identical with the extracellular water in the target tissue. Meanwhile, the lack of knowledge of the specific composition of the interstitial ionic and solute composition in various organs obstruct the construction of a perfusate identical with the extracellular water phase.

1. Perfusate Composition

Over the years, a variety of different fluids have been used in microdialysis studies.[35] These solutions vary widely with respect to ionic composition, osmolarity, and pH. Only recently has focus been put on the composition of the perfusion fluid. Inappropriate composition of the perfusate has shown to have a major impact in the central nervous system as well as in peripheral tissues.

a. Impact on Dialysis in the Central Nervous System

The blood-brain barrier prevents free diffusion of various substances, and very effective dialysis systems may discontinue the normal homeostatic mechanisms. For example, perfusate with low or absent calcium content can drain the tissue surrounding the dialysis probe,[45] as shown in Figure 4. Incorrect ionic composition may interfere with the release and metabolism of extracellular solutes.[46] Anisoosmotic fluids may also disturb the extracellular environment.[16,47] Even with apparently physiological perfusate, microdialysis may interfere with normal homeostasis, as indicated by increased sensitivity to ischemic insult in microdialysed neurons compared to nondialysed neurons.[48]

b. Impact on Dialysis in Peripheral Tissues

In peripheral tissues, drainage of substances can also take place. Lönnroth et al.[10] showed that a glucose-free perfusate caused minor drainage of extracellular glucose in subcutaneous adipose tissue. This drainage of glucose was, however, shown to have major impact on the local metabolic activity since drainage of glucose inhibited local lactate production in subcutaneous adipose tissue.[49] Whether similar findings can be found in human skin are not clarified. We found no detectable drainage of glucose during prolonged perfusion without glucose added to the perfusate.[32]

c. Sampling Considerations

Since the protein content in the extracellular environment is low, there is no need to add protein. In a few situations, perfusate with protein is recommended for sampling reasons. Some neuropeptides are very "sticky" and 0.2 to 0.5% albumin may be added to prevent adhesion to membrane material, tubes, and vials.

C. Perfusion Pumps

The perfusion pump used in microdialysis studies must be available to deliver a very small volume of liquid with a constant flush. In most peripheral

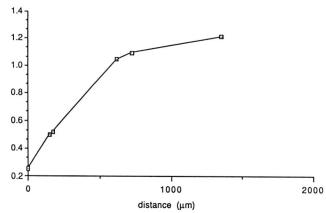

FIGURE 4 An example of the impact of dialysis on the interstitial milieu in the vicinity of the probe. Perfusion of the probe with calcium-free perfusate was started 1 hour following implantation into the cortex. The interstitial Ca^{2+} concentration was measured 30 min after the perfusion start by a Ca^{2+}-sensitive microelectrode at various distances from the perfused dialysis probe. (From Benveniste, H. and Hansen, A. J., Practical aspects of using microdialysis for determination of brain interstitial concentrations, in *Microdialysis in the Neurosciences*, Robinson, T. E. and Justice, J. B., Jr., Eds., Elsevier, Amsterdam, 1991, chap. 4. With permission.)

microdialysis applications, the probe is perfused at a rate of 1 to 5 μl/min. CMA/Microdialysis (Sweden) has made different perfusion pumps especially for microdialysis applications. The CMA/100 is a three-syringe pump which is able to deliver from 1 nl/min to 1 ml/min with syringes from 10 μl to 10 ml. Recently, a dual syringe perfusion pump with the possibility to use two different perfusion rates separately has been presented (CMA/102).

D. Sampling Unit

The dialysate leaving the efferent tube must be sampled for subsequent analysis. Depending on the setup and the substance studied, one can use small single-use vials or special-made fraction collectors. In several situations, the use of closed circuit between the efferent dialysis tube and the sampling system is required since a number of substances are degraded by air exposure. One may also prefer an automatized setup. For such purposes CMA/Microdialysis produces several fraction collectors dedicated for microdialysis experiments. These models can be delivered with single to triple probe sampling, with and without cooling systems, and some have the possibility for on-line injection into HPLC systems.

E. Analytical Techniques

This text will not discuss analytical aspects in detail since analysis of microdialysis samples does not differ from other analyses except for a few items. The analytical technique must be adjusted to handle the small samples, usually in the range of 2 to 30 μl. Meanwhile, since most microdialysis probes have molecular weight cutoff levels far below protein levels, enzymatic degradation of the substances is avoided; microdialysis is both a sampling technique and a separation technique. Therefore, no purification or separation steps are required for most substances before analysis.

In the neurosciences, microdialysis with on-line HPLC is used in several applications. Meanwhile, this setup is restricted to trained users of both techniques since defects in either the microdialysis technique or the analytical system might ravage the entire experiment.

IV. Practical Aspects of Skin Microdialysis

The practical performance of skin microdialysis certainly depends on the type of dialysis probe used. For example, to avoid bubbles in the probe, the commercially available concentric probe must be perfused prior to insertion. During this preinsertion perfusion stage the probe must be gently shivered to drive out bubbles. Verification of no bubbles can be performed under a stereomicroscope. Our serial 200-μm inner diameter fibers do not need to be perfused prior to insertion since the fibers do not easily trap bubbles. Also, the ultrafiltration capacity of this fiber is very low and the mechanical strength is high whereby we can perfuse the fiber for a short period with high-velocity perfusion to drive out air without damaging the membrane.

A. Implantation Procedure
1. Local Anesthetics

Since insertion of a cannula perpendicular to and through the skin surface is rather painful, it is pertinent to reduce the discomfort in some way. In our skin microdialysis studies we pretreat the skin area with an anesthetic cream containing lidocaine and prilocaine (EMLA, Astra, Sweden).[32-34] Other authors infiltrate the skin with lidocaine.[30,31] In each situation, control measurements must be performed to confirm that the nerve blockade does not interfere with the measurements.

2. Insertion

All dialysis probes are fragile and cannot be inserted into the skin without a guide. Our home-made serial fibers[32] are inserted into the skin through a 0.6-mm outer diameter (23-gauge) guide cannula which penetrates the most superficial parts of the dermis, runs in parallel with the skin surface in 20 mm, and then leaves the skin again. Then the 216-μm thick fiber is led through the guide cannula and the guide is removed.

Implantation of commercially available CMA/Microdialysis probes in skin has been done by Anderson.[30,31] A small bleb is formed with lidocaine, the 1.6-mm thick guide cannula penetrates the bleb and is led in parallel with the skin surface about 3 cm from the inlet point, the probe is inserted, and the guide is removed.

3. Dermal Localization of Probes
The physical size of the probe and the guide cannula determines the intradermal localization. The localization can be visualized by ultrasound scanning. Studies have confirmed intracutaneous localization of more than 30 serial fibers scanned 3 times along the probe. Mean depth is about 0.6 mm (Petersen et al., unpublished observations). Ultrasound scannings on concentric probes have shown that mean depth was 1.22 mm.[31]

4. Tissue Damage
Implantation of a microdialysis probe in a tissue elicits acute and chronic inflammatory reactions. Immediately after implantation, the level of endogenous substances is usually high. Following an interval of 30 to 120 min a steady state is reached. The rapid fall in the concentration of most substances in the perfusate after insertion is probably due to an initial lesion of the tissue and establishment of a new steady state of the substances proximate to the probe. Meanwhile, it should be kept in mind that the time-dependent decline in recovery may be due to establishment of a new gradient across the dialysis membrane since the phenomenon can also be seen *in vitro*.[37,45]

There is also a period of acute disturbed tissue function. In the brain of animals, one can see decreased cerebral blood flow, increased glucose metabolism, altered neurotransmitter release, etc. after implantation. The duration of this reaction varies from 30 min to several hours, depending on the technical performance, size of probe, anatomical localization, target substance, etc.

In the skin, implantation of a microdialysis probe causes an acute inflammatory reaction easily seen as redness of the skin. This hyperemia lasts about 60 to 120 min as evidenced by laser Doppler flowmetry.[31,33] The duration of the hyperemia may vary, depending on whether the flow is measured on the top of the probe or in remote sites. Also, the insertion of a dialysis probe causes histamine release which gradually declines and reaches a steady-state level after approximately 40 min.[31]

The tissue damage following chronically implanted dialysis probes has been studied mainly in animals. Histologically, 1 to 2 days after implantation only minimal tissue changes were found, whereas connective tissue formation may be seen several days after insertion.[50,51] The changes were still present 60 days after implantation. The tissue reactions depends on the membrane material used. An interesting finding was that neuronal damage may occur in remote sites.[51] No histologically visible damage was seen in human adipose tissue 7 days after implantation of small serial fibers.[29]

B. Handling of Perfusate
When filling the syringes with perfusate, air bubbles must be avoided. They may stick inside the membrane and decrease the dialysing surface, thereby limiting the surface area prone for diffusion. Air bubbles may also totally block the perfusion. Air bubbles can be avoided by using slightly heated fluids, or by using degassed perfusate (by vacuum or chemically degassed by helium). The perfusate must be free of any particles that may block the probe.

V. Interpretation of Microdialysis Results

Data obtained from microdialysis experiments can be presented as dialysate concentrations or attempts can be performed to estimate true extracellular concentrations of the solutes.

A. Dialysate Concentrations
Since recovery is concentration independent, dialysate concentrations can be used to compare measurements in different anatomical locations and experiments on different populations of subjects, e.g., patients with normal and inflamed skin. Also, dialysate concentrations can be utilized to follow the release and elimination pattern of various transmitters and mediators. Differences in dialysate concentrations among subjects or after stimulation in the same individuals may arise from differences in extracellular tissue concentrations but it may also be influenced by variations or changes in *in vivo* recovery. Therefore, a number of *in vivo* calibration techniques have been presented to estimate *in vivo* recovery and thereby the possibility to calculate absolute tissue concentrations.

B. Calibration Techniques
Several methods have been developed in an attempt to estimate extracellular concentrations of various solutes. Zetterström et al.[52] used the recovery obtained *in vitro* to estimate *in vivo* concentrations. Measurements *in vivo* were then transformed into extracellular concentrations to calculate the tissue concentration simply by dividing the dialysate concentration with the known *in vitro* recovery. Meanwhile, an abundance of reports have clearly demonstrated that recovery *in vitro* differs from recovery *in vivo*, mainly due to altered diffusion characteristics in the tissue.[35,53] Consequently, a number of methods for quantitative microdialysis have been presented which take into account this difference. This subject is, however, very complex, and it is still under debate. Most of the methods are performed on the assumption that the solute in question is in steady-state concentration during the calibration

procedure or rely on complicated mathematical calculations. Many of these methods are recently reviewed.[25] It is beyond the scope of this short text to go into detail on these methods. Examples of such quantitative methods are the varying perfusion method,[54] the slow perfusion method,[55] the concentration difference method,[10] retrodialysis,[56,57] and various reference methods as mentioned below. Simulations *in vitro* and comparative studies *in vivo* have confirmed the advantage of calibration methods for estimating of extracellular concentrations of free unbound drug concentrations.[54,55] Estimation of extracellular concentrations with various methods seems to give similar results.[55] A few of these methods will be discussed shortly.

1. The Difference Method

The concept of the difference method presented by Lönnroth et al.[10] is based on the fact that the direction of flux of various diffusible substances is directed by the concentration gradient only. That means that perfusing a probe with or without low concentrations of the substance in question results in a positive net flux into the dialysate. On the contrary, upon perfusing the probe with a concentration which is higher than in the surrounding medium, a negative net flux occurs. By varying the perfusate concentration over a large range of concentrations, it is possible to calculate the equilibrium concentration where no net flux over the dialysis membrane takes place. Accordingly, this corresponds to the tissue concentration. The concept is graphically presented in Figure 5A. We have used this method in human skin to quantitate skin glucose levels.[32] An example is shown in Figure 5B.

2. Retrodialysis

The concept of retrodialysis is based on the same idea as used with the difference method: the direction of substances flux is bidirectional and it is only the concentration gradient that determines the flux across the dialysis membrane. Retrodialysis is mostly used with exogenous substances. The method estimates the loss of substance in question to estimate *in vivo* recovery. Provided that the concentration of the target substance is negligible or absent from the tissue (which must always be established) retrodialysis can be used to estimate *in vivo* recovery of the substance. This procedure is mainly used in pharmacokinetic studies, e.g., investigating drugs like zidovudine.[56,58]

3. Reference Methods

The use of most calibration methods assumes that the substance in question is in steady state during the calibration procedure. In many cases this may not be true. The use of internal or external reference substances to indicate recovery or changes in recovery of microdialysis probes is therefore an interesting approach.

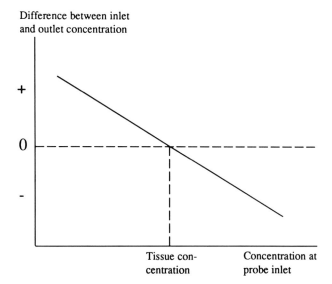

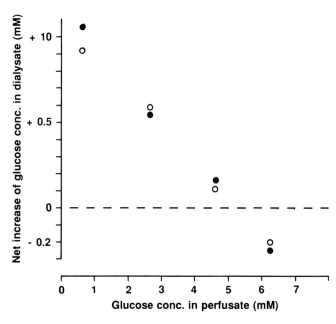

FIGURE 5 (A) The idea of the difference method. The flux of substances is based on the concentration difference only. Therefore, by varying the concentration in the perfusate it is possible by linear regression to calculate absolute tissue concentrations. Please refer to details in the text. (Reprinted with permission from Microdialysis, User's Guide, 4th ed., CMA/Microdialysis, Sweden.) (B) An example of the use of the difference method in human skin. In this example two probes (open and closed circles, respectively) were perfused with four different glucose concentrations, and the net increase/decrease was calculated. Glucose rapidly equilibrates in the total water phase. The estimated skin glucose concentration was 5.2 and 5.3 mM, respectively. Plasma glucose was 5.3 mM. (From Petersen, L.J., Kristensen, J.K., and Bülow, J., *J. Invest. Dermatol.*, 99, 357, 1992. With permission.)

In order to use a reference substance it is critical that the reference substance acts like the substance under study, i.e., the *in vivo* diffusion characteristics should be similar for the reference and the target substance. Also, the reference must be metabolically inert in order to not disturb the extracellular milieu

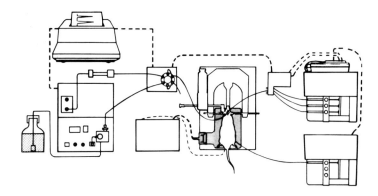

FIGURE 6 An experimental system for anesthetized animals. The microperfusion pump with a system controller is connected with a syringe selector to the probe. The dialysate is collected via an on-line injector and led into a HPLC. The temperature is kept constant by the temperature controller and the second microperfusion pump allows systemic drug injection and/or continuous microdialysis of blood. (From Application Note No. 6, CMA/Microdialysis, Sweden. With permission.)

eicosanoids,[65,66] oxygen-free radicals,[67] growth factors,[68] and various cytokines,[69,70] including interleukin-1 by microdialysis, may imply an expanding use of microdialysis in clinical and experimental dermatologic research. Full automated animal studies can be set up too as shown in Figure 6.

Microdialysis also opens up the possibility to monitor compartmental pharmacokinetics.[71–77]

VII. Conclusion

The use of microdialysis in peripheral tissues is scarcely described. Only recently have microdialysis studies been carried out in human skin. Meanwhile, we consider the microdialysis method to possess several advantages compared to other techniques. For the first time, it is possible to assess *in situ* mediator levels in intact skin and to monitor drug transport and metabolism following percutaneous and parenteral drug delivery. It is possible to quantify the concentration of various substances, and the continuous perfusion allows us to follow the kinetics of miscellaneous mediators more closely, e.g., histamine elimination following mast cell degranulation by the neuropeptide substance P[34] or other stimuli as shown in Figure 7. With the skin microdialysis technique, it is possible to monitor clinical observations such as wheal, flare, and itch reactions together with fluid sampling.

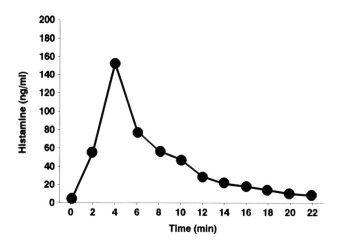

FIGURE 7 An example of histamine release in human skin by intracutaneous injected murine codeine. At time zero, anti-IgE was injected and dialysates were collected in 2-min intervals. Peak histamine was found in the second sample (2 to 4 min). In a monoexponential plot, the dialysate histamine elimination curve was linear (r = –0.99); dialysate histamine half life was 4.5 min.

studied. It is preferable that the reference and the target substance can be analyzed simultaneously. Examples are the use of caffeine for recovery of theophylline.[59]

It must be emphasized that most reference substances are used in animal research, and are not usable or validated in humans. Examples of such reference standards are antipyrine,[60] ^{3}H-mannitol,[61] and ^{14}C-labeled lactate.[57]

VI. Skin Microdialysis — Recent Studies and Future Applications

The use of the microdialysis technique in skin is only recently described. The skin microdialysis technique has been applied for studies on percutaneous absorption of ethanol,[30] quantification of glucose levels,[32] and measurement of histamine levels in normal and challenged skin sites.[31,33,34]

The ability to recover substances like histamine,[31,33,34] bradykinin,[23] serotonin,[62] various neuropeptides,[63,64]

References

1. Søndergaard, J. and Greawes, M.W., Direct recovery of histamine from cutaneous anaphylaxis in man, *Acta Derm.*, 51, 98, 1971.
2. Dunsky, E.H. and Zweiman, B., The direct demonstration of histamine release in allergic reactions in the skin using a skin chamber technique, *J. Allergy Clin. Immunol.*, 62, 127, 1978.
3. Ting, S., Dunsky, E.H., Lavker, R.M., and Zweiman, B., Patterns of mast cell alterations and in vivo mediator release in human allergic skin reactions, *J. Allergy Clin. Immunol.*, 66, 417, 1980.
4. Talbot, S.F., Atkins, P.C., Valenzano, M., and Zweiman, B., Correlations of in vivo mediator release with late cutaneous allergic responses in humans. Kinetics of histamine release, *J. Allergy Clin. Immunol.*, 74, 819, 1984.
5. Heavy, D.J., Ind, P.W., Miyatake, A., Brown, M.J., MacDermot, J., and Dollery, C.T., Histamine release locally after intradermal antigen challenge in man, *Br. J. Clin. Pharmacol.*, 18, 915, 1984.

6. McBride, P., Jacobs, R., Bradley, D., and Kaliner, M., Use of plasma histamine levels to monitor cutaneous mast cell degranulation, *J. Allergy Clin. Immunol.*, 83, 374, 1989.
7. Bito, L., Davson, H., Levin, E., Murray, M., and Snider, N., The concentrations of free amino acids and other electrolytes in cerebrospinal fluid, in vivo dialysate of brain, and blood plasma of the dog, *J. Neurochem.*, 13, 1057, 1966.
8. Delgado, J.M., DeFeudis, F.V., Roth, R.H., Ryugo, D.K., and Mitruka, B.M., Dialytrode for long-term intracerebral perfusion in awake monkeys, *Arch. Int. Pharmacodyn. Ther.*, 198, 9, 1972.
9. Ungerstedt, U. and Pycock, C., Functional correlates of dopamine neurotransmission, *Bull. Schweiz. Akad. Med. Wiss.*, 30, 44, 1974.
10. Lönnroth, P., Jansson, P.-A., and Smith, U., A microdialysis method allowing characterization of intercellular water space in humans, *Am. J. Physiol. (Endocrinol. Metab.)*, 253, E228, 1987.
11. Arner, P., Bolinder, J., Eliasson, A., Lundin, A., and Ungerstedt, U., Microdialysis of adipose tissue and blood for in vivo lipolysis studies, *Am. J. Physiol. (Endocrinol. Metab.)*, 255, E737, 1988.
12. Jansson, P.-A., Fowelin, J., Smith, U., and Lönnroth, P., Characterization by microdialysis of intercellular glucose level in subcutaneous tissue in humans, *Am. J. Physiol. (Endocrinol. Metab.)*, 255, E218, 1988.
13. Bolinder, J., Hagström, E., Ungerstedt, U., and Arner, P., Microdialysis of subcutaneous adipose tissue *in vivo* for continuous glucose monitoring in man, *Scand. J. Clin. Lab. Invest.*, 49, 465, 1989.
14. Hagström, E., Arner, P., Ungerstedt, U., and Bolinder, J., Subcutaneous adipose tissue: a source of lactate production after glucose ingestion in humans, *Am. J. Physiol. (Endocrinol. Metab.)*, 258, E888, 1990.
15. Jansson, P.-A., Microdialysis of Human Subcutaneous Tissue. A New Technique to Study Fat Cell Metabolism In Situ, Dissertation, University of Göteborg, Göteborg, Sweden, 1991.
16. Lehmann, A., Effects of microdialysis-perfusion with anisoosmotic media on extracellular amino acids in the rat hippocampus and skeletal muscle, *J. Neurochem.*, 53, 525, 1988.
17. de Boer, J., Postema, F., Plijter-Groendijk, H., and Korf, J., Continuous monitoring of extracellular lactate concentration by microdialysis for the study of rat muscle metabolism in vivo, *Pflügers Arch.*, 419, 1, 1991.
18. Grønlund, B., Astrup, A., Bie, P., and Christensen, N.J., Noradrenaline release in skeletal muscle and in adipose tissue studied by microdialysis, *Clin. Sci.*, 80, 595, 1991.
19. Van Wylen, D.G., Willis, J., Sodhi, J., Weiss, R.J., Lasley, R.D., and Mentzer, R.M., Cardiac microdialysis to estimate interstitial adenosine and coronary blood flow, *Am. J. Physiol. (Heart Circ. Physiol.)*, 27, H1642, 1990.
20. Dorheim, T.A., Wang, T., Mentzer, R., and Van Wylen, D.G.L., Interstitial purine metabolites during regional myocardial ischemia, *J. Surg. Res.*, 48, 491, 1990.
21. Ben-Nun, J., Cooper, R.L., Cringle, S.J., and Constable, I.J., Ocular dialysis. A new technique for in vivo intraocular pharmacokinetic measurements, *Arch. Ophthalmol.*, 106, 254, 1988.
22. Scott, D.O., Sorensen, L.R., and Lunte, C.E., *In vivo* microdialysis sampling coupled to liquid chromatography for the study of acetaminophen metabolism, *J. Chromatogr.*, 506, 461, 1990.
23. Hargreaves, K.M. and Costello, A., Glucocorticoids suppress levels of immunoreactive bradykinin in inflamed tissue as evaluated by microdialysis probes, *Clin. Pharmacol. Ther.*, 48, 168, 1990.
24. Editorial, Microdialysis, *Lancet*, 330, 1326, 1992.
25. Robinson, T.E. and Justice, J.B., Jr., Eds., *Microdialysis in the Neurosciences*, Elsevier, Amsterdam, 1991.
26. Hamberger, A., Jacobson, I., Nyström, B., and Sandberg, M., Microdialysis sampling of the neuronal environment in basic and clinical research, *J. Intern. Med.*, 230, 375, 1991.
27. Ungerstedt, U., Microdialysis-principles and applications for studies in animals and man, *J. Intern. Med.*, 230, 365, 1991.
28. Arner, P. and Bolinder, J., Microdialysis of adipose tissue, *J. Intern. Med.*, 230, 381, 1991.
29. Lönnroth, P. and Smith, U., Microdialysis — a novel technique for clinical investigations, *J. Intern. Med.*, 227, 295, 1990.
30. Anderson, C., Anderson, T., and Molander, M., Ethanol absorption across human skin measured by in vivo microdialysis technique, *Acta Derm. Venereol.*, 71, 389, 1991.
31. Anderson, C., Anderson, T., and Andersson, R.G.G., In vivo microdialysis estimation of histamine in human skin, *Skin Pharmacol.*, 5, 177, 1992.
32. Petersen, L.J., Kristensen, J.K., and Bülow, J., Microdialysis of the interstitial water space in human skin in vivo: quantitative measurement of cutaneous glucose concentrations, *J. Invest. Dermatol.*, 99, 357, 1992.
33. Petersen, L.J., Bindslev-Jensen, C., Søndergaard, J., and Skov, P.S., Histamine release in immediate-type hypersensitivity reactions in intact human skin measured by microdialysis. A preliminary study, *Allergy*, 47, 635, 1992.
34. Petersen, L.J., Poulseu, L.K., Søndergaard, J., and Skov, P.S., Measurement of substance P-induced histamine release in intact human skin by a microdialysis technique, *J. Allergy Clin. Immunol.*, in press.
35. Benveniste, H. and Hüttemeier, P.C., Microdialysis — theory and application, *Prog. Neurobiol.*, 35, 195, 1990.
36. Lindefors, N., Amberg, G., and Ungerstedt, U., Intracerebral microdialysis. I. Experimental studies of diffusion kinetics, *J. Pharmacol. Methods*, 22, 141, 1989.
37. Amberg, G. and Lindefors, N., Intracerebral microdialysis. II. Mathematical studies on diffusion kinetics, *J. Pharmacol. Methods*, 22, 157, 1989.
38. Morrison, P.F., Bungay, P.M., Hsiao, J.K., Ball, B.A., Mefford, I.N., and Dedrick, R.L., Quantitative microdialysis: analysis of transients and application to pharmacokinetics in brain, *J. Neurochem.*, 57, 103, 1991.
39. Bungay, P.M., Morrison, P.F., and Dedrick, R.L., Steady-state theory for quantitative microdialysis of solutes and water *in vivo* and *in vitro*, *Life Sci.*, 46, 105, 1990.
40. Dykstra, K.H., Hsiao, J.K., Morrison, P.F., Bungay, P.M., Mefford, I.N., Scully, M.M., and Dedrick, R.L., Quantitative examination of tissue concentration profiles associated with microdialysis, *J. Neurochem.*, 58, 931, 1992.
41. Parry, T.J., Carter, T.L., and McElligott, J.G., Physical and chemical considerations in the in vitro calibration of microdialysis probes for biogenic amine neurotransmitters and metabolites, *J. Neurosci. Methods*, 32, 175, 1990.
42. Tao, R. and Hjorth, S., Differences in the in vitro and in vivo 5-hydroxytryptamine extraction performance among three common microdialysis membranes, *J. Neurochem.*, 59, 1778, 1992.
43. Application note No. 1, Principles of recovery, CMA/Microdialysis, Sweden.
44. Maidenment, N.T. and Evans, C.J., Measurement of extracellular neuropeptides in the brain: microdialysis linked to solid-phase radioimmunoassays with sub-femtomole limits of detection, in *Microdialysis in the Neurosciences*, Robinson, T.E. and Justice, J.B., Jr., Eds., Elsevier, Amsterdam, 1991, chap. 12.

45. Benveniste, H., Hansen, A.J., and Ottosen, N.S., Determination of brain interstitial concentrations by microdialysis, *J. Neurochem.*, 52, 1741, 1989.
46. Osborne, P.G., O'Connor, W.T., and Ungerstedt, U., Effect of varying the ionic concentration of a microdialysis perfusate on basal striatal dopamine levels in awake rats, *J. Neurochem.*, 56, 452, 1991.
47. Solís, J.M., Herranz, A.S., Herreras, O., Lerma, J., and Martin del Rio, R., Does taurine act as an osmoregulatory substance in the rat brain?, *Neurosci. Lett.*, 91, 53, 1988.
48. Phebus, L.A., Mincy, R.E., and Clemens, J.A., Microdialysis perfusion increases the sensitivity of rat striatal neurons to ischemic insult, *Brain Res.*, 578, 339, 1992.
49. Jansson, P.-A., Smith, U., and Lönnroth, P., Evidence for lactate production by human adipose tissue in vivo, *Diabetologia*, 33, 253, 1990.
50. Benveniste, H. and Diemer, N.H., Cellular reactions to implantation of a microdialysis tube in the rat hippocampus, *Acta Neuropathol.*, 74, 234, 1988.
51. Shuaib, A., Xu, K., Crain, B., Sirén, A.-L., Feuerstein, G., Hallenbeck, J., and Davis, J.N., Assessment of damage from implantation of microdialysis probes in the rat hippocampus with silver degeneration staining, *Neurosci. Lett.*, 112, 149, 1990.
52. Zetterström, T., Vernet, L., Ungerstedt, U., Tossman, U., Jonzon, B., and Fredholm, B.B., Purine levels in the intact rat brain. Studies with an implanted perfused hollow fibre, *Neurosci. Lett.*, 29, 111, 1982.
53. Alexander, G.M., Grothusen, J.R., and Schwartzman, R.J., Flow dependent changes in the effective surface area of microdialysis probes, *Life Sci.*, 43, 595, 1988.
54. Ståhle, L., Segervärd, S., and Ungerstedt, U., A comparison between three methods for estimation of extracellular concentrations of exogenous and endogenous compounds by microdialysis, *J. Pharmacol. Methods*, 25, 41, 1991.
55. Smith, A.D., Olson, R.J., and Justice, J.B., Jr., Quantitative microdialysis of dopamine in the striatum: effect of circadian variation, *J. Neurosci. Methods*, 44, 33, 1992.
56. Wang, Y.-F., Wong, S.L., and Sawchuk, R.J., Comparison of *in vitro* and *in vivo* calibration of microdialysis probes using retrodialysis, *Curr. Sep.*, 10, 88, 1991.
57. Scheller, D. and Kolb, J., The internal reference technique in microdialysis: a practical approach to monitoring dialysis efficacy and to calculating tissue concentration from dialysate samples, *J. Neurosci. Methods*, 40, 31, 1991.
58. Wong, S.L., Wang, Y., and Sawchuk, R.J., Analysis of zidovudine distribution to specific regions in rabbit brain using microdialysis, *Pharm. Res.*, 9, 332, 1992.
59. Larsson, C.I., The use of an "internal standard" for control of the recovery in microdialysis, *Life Sci.*, 49, PL73, 1991.
60. Yorkel, R.A., Allen, D.D., Burgio, D.E., and McNamara, P.J., Antipyrine as a dialyzable reference to correct differences in efficiency among and within sampling devices during in vivo microdialysis, *J. Pharmacol. Toxicol. Methods*, 27, 135, 1992.
61. Fink-Jensen, A., Judge, M.E., and Hansen, A.J., Can *in vivo* recovery be determined by mannitol?, *Curr. Sep.*, 10, 82, 1991.
62. Wester, P., Dietrich, W.D., Prado, R., Watson, B.D., and Globus, M.Y., Serotonin release into plasma during common carotid artery thrombosis in rats, *Stroke*, 23, 870, 1992.
63. Kendrick, K.M., Microdialysis measurement of in vivo neuropeptide release, *J. Neurosci. Methods*, 34, 35, 1990.
64. Brodin, E., Linderoth, B., Gazelius, B., and Ungerstedt, U., In vivo release of substance P in cat dorsal horn studied with microdialysis, *Neurosci. Lett.*, 76, 357, 1987.
65. Yergey, J.A. and Heyes, M.P., Brain eicosanoid formation following acute penetration injury as studied by in vivo microdialysis, *J. Cereb. Blood Flow Metab.*, 10, 143, 1990.
66. Patel, P.M., Drummond, J.C., Mitchell, M.D., Yaksh, T.L., and Cole, D.J., Eicosanoid production in the caudate nucleus and dorsal hippocampus after forebrain ischemia: a microdialysis study, *J. Cereb. Blood Flow Metab.*, 12, 88, 1992.
67. Zini, I., Tomasi, A., Grimaldi, R., Vannini, V., and Agnati, L.F., Detection of free radicals during brain ischemia and reperfusion by spin trapping and microdialysis, *Neurosci. Lett.*, 138, 279, 1992.
68. Yamaguchi, F., Itano, T., Mizobuchi, M., Miyamoto, O., Janjua, N.A., Matsui, H., Tokuda, M., Ohmoto, T., Hosokawa, K., and Hatase, O., Insulin-like growth factor I (IGF-I) distribution in the tissue and extracellular compartment in different regions in rat brain, *Brain Res.*, 533, 344, 1990.
69. Woodroofe, M.N., Sarna, G.S., Wadhwa, M., Hayes, G.M., Loughlin, A.J., Tinker, A., and Cizner, M.L., Detection of interleukin-1 and interleukin-6 in adult rat brain, following mechanical injury, by in vivo microdialysis: evidence of a role in microglia in cytokine production, *J. Neuroimmunol.*, 33, 227, 1991.
70. Yan, H.Q., Banos, M.A., Herregodts, P., Hooghe, R., and Hooghe-Peters, E.L., Expression of interleukin (IL)-1 beta, IL-6 and their respective receptors in the normal rat brain and after injury, *Eur. J. Immunol.*, 22, 2963, 1992.
71. Herrera, A.M., Scott, D.O., and Lunte, C.E., Microdialysis sampling for determination of plasma protein binding of drugs, *Pharm. Res.*, 7, 1077, 1990.
72. Lönnroth, P., Carlsten, J., Johnson, L., and Smith, U., Measurements by microdialysis of free concentrations of propranolol, *J. Chromatogr.*, 568, 419, 1991.
73. Scott, D.O., Sorenson, L.R., Steele, K.L., Puckett, D.L., and Lunte, C.E., In vivo microdialysis sampling for pharmacokinetic investigations, *Pharm. Res.*, 8, 389, 1991.
74. Caprioli, R.M. and Lin, S.-N., On-line analysis of penicillin blood levels in the live rat by combined microdialysis/fast-atom bombardment mass spectrometry, *Proc. Natl. Acad. Sci. U.S.A.*, 87, 240, 1990.
75. Saisho, Y. and Umeda, T., Continuous monitoring of unbound flomoxef levels in rat blood using microdialysis and its new pharmacokinetic analysis, *Chem. Pharm. Bull.*, 39, 808, 1991.
76. Ault, J.M., Lunte, C.E., Meltzer, N.M., and Riley, C.M., Microdialysis sampling for the investigation of dermal drug transport, *Pharm. Res.*, 9, 1256, 1992.
77. Ståhle, L., Pharmacokinetic estimations from microdialysis data, *Eur. J. Clin. Pharmacol.*, 43, 289, 1992.

Section N: Legal and Ethical Aspects of Noninvasive Techniques

33.0

Regulatory Aspects

33.1 FDA and EEC Regulations Related to Skin: Documentation
 and Measuring Devices 653
 E.K. Seidenschnur

Chapter 33.1

FDA and EEC Regulations Related to Skin: Documentation and Measuring Devices

Edel K. Seidenschnur
Department of Regulatory Affairs
Leo Pharmaceutical Products
Ballerup, Denmark

I. Introduction

Needless to repeat the words stated by Prof. A. M. Kligman[1] 5 years ago that, with the noninvasive instrumental measuring techniques, scientific measurements have joined the clinical art of observation in dermatology. What appeared to be obvious possibilities at that time has become even more so since then.

To which extent, however, these new methods will actually be used in the practical development of skin product as performed by industrial companies or remain experimental possibilities for determination of parameters related to the skin and its function may depend on the usability of the testing methods within the scope of regulatory testing for efficacy and safety applied to skin products before their approval or release for marketing.

A. Functional Application of Noninvasive Methods

1. Description of Normal Skin

With the new noninvasive testing methods, it has become possible to provide multifaceted descriptions of skin conditions on the normal healthy subject, young, and aged. However, the term "normal" as the absence of clinically recognizable states of disease may become more nuanced when expressed in instrumental parameters. As an example, Froch and Wissing[2] confirmed in 1982 by clinical observation of reactions the correlation between light sensitivity and susceptibility to irritants in studies comparing DMSO responses in normal individuals with their minimal erythema dose, when exposed to the sunburn spectrum of a high-pressure mercury lamp. On this basis, skin typing on test responses was suggested as a replacement for skin typing based on complexion and sunburn history. With subsequent quantification of the wheaking response to skin application of DMSO by measurement of transepidermal water loss, electrical conductance, and ultrasound skin thickness,[3] a quick and simple procedure was available for skin sensitivity typing. However, although relevant also for the normal skin, it is likely that such testing may remain within the dermatological clinic and will find its field of application in diagnostics.

No doubt the refined technique is also changing the basic ideas about the nature and function of skin. Thus, the almost impermeable barrier known as human skin has now altered itself by means of testing methodology and become a semipermeable barrier allowing the absorption of compounds into the epidermis. This is illustrated by investigations using a battery of noninvasive methods showing that the lipid phase of an oil-in-water emulsion is absorbed into epidermis within the hours following application, providing a lasting improvement of skin hydration.

The more nuanced pattern obtained by refined techniques may provide possibilities for developing a more substantial basis for the diversification already existing in the perception of vaguely defined conditions. As an example, the dermatologist has an impression of dry skin based on clinical observations and composed of visual signs being beyond normal. However, the subjective feeling of "dry skin" seems to be something else, as suggested by results of a study[4] showing that only 8% of a number of healthy volunteers perceiving their skin as being dry exhibit skin dryness according to the dermatologist's definition. Attempts made in this study to correlate the clinical evaluation and subjective impression with objective measures comprised determination of scaling by densitometry and skin hydration by electrical capacitance, but these measures did not suffice to establish a nonobjective, precise definition of dry skin. Perhaps testing for transepidermal water loss would show results differentiating with the subjective perception also within conditions of dryness considered normal.

2. Diagnostics in Skin Diseases

Diagnostics is probably so far the area that has attracted the major interest for the very simple reason that dermatology deals with abnormalities referring the normal condition to the less interesting baseline position. It has been possible to some extent to establish diagnostic patterns of parameters, making it possible to designate skin diseases by functional parameters and not only by description of visual signs as usually done clinically.

At the present time there is sufficient experience with bioengineering techniques in the measurement of parameters in contact dermatitis and atopy to suggest testing programs for diagnoses in these diseases.[5] Attempts to replace direct visual evaluation and photography of skin color and surface contours as basic elements of the clinical diagnosis in dermatitis have led to instrumental recording of color in a way simulating the color perception of the human eye and to preparation of precise casts of surface contours by polysulfide rubber replica. The use of skin casting does not only provide a lasting picture in three dimensions, it also allows studies of details not possible by two-dimensional photographs both by direct observation in three dimensions and by allowing quantified measurement of the contours.

To judge the moisture balance of the skin which is often disturbed in dermatitis, although not appearing as a visible sign, determination of transepidermal water loss expressing the diffusional water loss through the skin has proved generally reliable and more accurately related to the functional condition of the skin as a water barrier than direct conductance or capacity measurement of the skin surface state of hydration, which has been the initial approach to an estimate of the moisture factor, and which still serves as nuanced differentiation between different types of dermatitis, as it has found its place also in the characterization of different types of eczema.[6]

Also in establishing diagnostic parameters of inflammation, the direct approach of observing skin temperature as reflecting inflammatory processes has been abandoned and replaced by measurement of the vasodilatation by laser Doppler flowmetry and by determination of the edema by ultrasound measurement of skin thickness. However, thermographic monitoring of skin surface temperature gradients contributes to the distinction between irritant reactions and allergic reactions in dermatitis.[5]

Furthermore, noninvasive techniques may also lead to better understanding of certain diseases, in addition to being diagnostic tools. As an example, studies of skin surface contours in patients with atopic dermatitis led to the hypothesis that skin roughness observed and subjectively expressed feeling of dry skin associated with a defective water barrier function observed directly as low skin hydration and as increased transepidermal water loss was due to changes in stratum corneum lipids.[7]

3. Observation of Changes

The last but not least interesting area of application of noninvasive measuring techniques is the provision of new possibilities of observation of changes occurring in the skin as a consequence of some kind of treatment by means of substances or products applied to the skin, or by physical exposure.

In dermatology, clinical determination of an effect is no longer limited to a gross picture of normalization or degree of improvement. With the specific testing methods it is now possible to observe more precisely what the mechanism of the change actually is. It is thereby possible to obtain a differentiated picture of effects of a treatment. This possibility opens new doors in the screening of substances in the attempt to relate a compound to a potential use. It also provides a possibility of separating useful effects on disease-related parameters from effects that could constitute the reason for possible unwanted effects.

By means of determination of electrical conductance and capacitance, it has been possible to measure directly the effect of moisturizers on the hydration of normal skin. It has also been demonstrated that such treatment has a measurable effect on the size of scaling flakes of dry skin leading to an invisible desquamation without discomfort.[8]

Systematic studies by similar techniques on the hydration effect of traditional bath treatment of skin[9] showed short-lasting and very minor effect on hydration by exposure to water bath, but an improved hydration state of epidermis lasting some hours by addition of even small amounts of lipids, a measurement which makes it possible to compare and distinguish effects of various treatments previously left to subjective preference.

As another example of systematic efficacy testing using instrumental methods, urea creams of different concentrations (3 and 10%) as hydrating agent were tested in a group of healthy volunteers.[10] Chrometric determinations of skin color, electrical conductance, and capacity measurements of the hydration of various skin layers, evaporimetric determination of transepidermal water loss as an expression of the skin's water barrier function, as well as evaluation of scaling and epidermal desquamation by microscopy and light densitometry on tape harvested components of stratum corneum were employed for evaluation of hydration effect and compared to the clinical evaluation of the hydration state of the skin. Efficacy parameters correlated well with blind, clinical evaluation and showed improved hydration with both strengths. With the high concentration, improvement of the epidermal barrier function was noted which may indicate a more lasting effect than the direct hydration ability, and which might be of value in dermatology.

B. Methods Expanding the Scope of Skin Testing

Being given the designation "noninvasive methods", the whole category has probably been named from what is a minor characteristic which, although true enough, may be a feature of less importance for the impact it may have on research related to skin.

Evidently not much of the information that may be gathered from these new instrumental measurements could be derived from invasive methods in the sense that this means taking out biopsies from the skin and/or inserting instruments. Even in the field most justified for biopsies, namely the observation of pathological changes in the skin layers, it seems that the noninvasive ultrasound technique may add complementary information, in particular with respect to extension and depth of lesions,[11] but also by its ability to be repetitive in time and extension which provides the possibility of showing a dynamic development. However, specific identification and typing of pathological changes still require biopsied, stained samples for microscopy.

Furthermore, it seems that the new methods have their origin in clinical determinations by the attempt to replace directly sensed observation of physical conditions, such as color, thickness, and scaling with objective measures of exactly the same thing. In some cases, such as, for example, with respect to color, it has been possible, with slight modifications from the first direct simulation of the clinical observation,[12] to develop a quantitative technique simulating the visual perception and at the same time providing a precise, standardized measure capable of monitoring subtle nuances.[13] Other parameters have escaped the direct connection between perceived ideas and physical measurement.

Lack of correlation between electrically measured skin hydration and the perception of dry skin[4] may exclude these methods from the task of identifying persons with dry skin within a normal population. This does not, however, prevent the same methods from being used in diagnostics on the basis of proved correlation with state of disease as it has been seen in the different types of eczema.[6] Neither does it prevent use in the relative quantification of an effect of change accomplished by a treatment aiming at improvement of the hydration state.[9]

Probably some of the most important quantifications are those describing the ability of the skin to maintain a dynamic function, such as, for example, the measurement of transepidermal water loss expressing the function of the skin as a water-retaining barrier, not only because it has proved to reflect the hydration state of the skin, but also because this parameter reflects disturbance of the function as present in various states of disease or provoked by irritants in testing procedures.[5]

Also when the abnormality or reaction itself is clearly observed clinically, the overall gross picture may be more precise and refined when supplemented by quantitative determination of single parameters. A skin inflammation is characterized by redness, heat, and abnormal skin thickness, which can be measured directly by chromameter, contact thermometer, and ultrasound, and the vasodilatation being an essential feature of the inflammation can be measured by laser Doppler flowmetry as an increased blood flow.[5]

Clearly the value of the noninvasive testing methods lies in their ability to provide accurate and reproducible data on parameters solely or partially providing a pattern describing the skin in its normal or diseased state and capable of monitoring changes expressing effects of treatment applied to the skin.

II. Legislation Related to Skin Products

A. Requirements for Research in Human Beings

Being a methodology to be used within the framework of laws and guidelines regulating this primary target activity, noninvasive testing methods find their legal position under the conditions and requirements prevailing for clinical investigations in patients and healthy subjects, and having as their objective the protection of the human being.

However, the fact that much scientific work beyond definition of diagnoses and establishment of methodology relates to the possible achievement of an improvement of prevailing characteristics, the noninvasive techniques are qualified to find a place in the search for treatment of skin diseases and possibly also treatment of skin being normal according to the dermatologist's definition, but nevertheless needing improvement from the individual's point of view.

On this background it is worth looking at legislation related to drugs and cosmetics in order to see the methodologies in this perspective, but also to speculate on a possible impact on legislation from new methodology which may shed new light on the way of thinking underlying existing legislation.

B. Legislation on Drugs, Cosmetics, and Devices

1. Legal Definitions in the U.S. and the EEC

In the U.S., basic legislation dates back to June 1938, at which time the Federal Food, Drug, and Cosmetic Act became effective.[14]

According to stated definitions,[15] "*drug* is intended to affect the structure or any function of the body of man;" "a *cosmetic* is intended to be rubbed, poured, sprinkled, or sprayed on, introduced into, or otherwise applied to the human body or any part thereof for cleansing, beautifying promoting attractiveness, or altering the appearance;" "a *device* is for use in the diagnosis, cure, mitigation, treatment or prevention of disease in man."

Within the European Economic Community, drugs or medicines, cosmetics, and devices are contained in separate directives, being at various stages of establishment ranging from the Medicines Directive from 1965[16] with later supplements and amendments up to the Directive on Medical Devices,[17] which is expected to be approved in 1993, with the Directive on Cosmetics in between dating from 1976[18] in its first version and currently undergoing modifications with a significant amendment pending at present.[19]

According to the directives and proposals, "a *medicinal product* is a substance for treating or preventing disease, or for making a medical diagnosis, or for restoring, correcting or modifying physiological functions;" "a *cosmetic product* is intended for application to various parts of the human body to clean, perfume or protect them, for maintaining good condition, change appearance, or correct body odours;" "a *medical device* is for diagnosis, prevention, monitoring, treatment or alleviation of disease, injury or handicap."

A directive issued in the EEC in not a law by itself since it exists supranationally directed toward member states and not their citizens. To become effective it will need to be implemented into the national laws prevailing in the member states. Usually a directive will contain an estimate for the time within which it is intended to come into force, but delays of different lengths in the member states are not unusual. Referring to the three defined areas, the situation is that the Medicines Directive,[17] with its later appended directives providing detailed requirements,[20,21] has been fully implemented in all member states since the beginning of the 1980s.

The Cosmetics Directive works primarily by means of annexes containing positive and negative listings, respectively, on substances to be used or not used in cosmetic products. Such procedure provides an almost instant implementation in member states by publication of decrees.

The Directive of Medical Devices was presented as a final draft in 1992. With its approval a time schedule for its implementation is expected to appear. However, since its principles by now have been known for some years, implementation is expected to occur shortly after leaving time enough for administrative purposes only.

2. Differentiation between Drugs and Cosmetics

Definitions within the three areas of drugs, cosmetics, and devices in the U.S. and the EEC are based on the intended purpose of use. However, within the area of interest for skin testing, this identifying purpose seems to be identical only with respect to devices.

Within the EEC, drugs are defined by the desired therapeutic goal, whereas the U.S. definition merely contains the means by which a drug may work. This difference leaves the possibility that a substance may be a drug in the U.S., without having the therapeutic purpose it needs to have according to an EEC definition. The EEC definition classifies by idea which indirectly brings about the requirement of efficacy to be proved. The U.S. definition classifies a drug by its pharmacological properties, irrespective of their application in health or disease.

As an example showing this difference, a chewing gum containing nicotine to be used to help people stop smoking was classified as a drug in the U.S. In Europe, however, the first application submitted for registration as a medicine was not accepted, with reference to the fact that smoking is no disease. Registrations in the EEC countries were obtained only after receiving acceptance by the EEC Commission's Committee on Proprietary Medicine, who decided that the product could be regarded in a medical context.

This difference in definition of drugs leaves different areas for cosmetics as a category. Although definitions of cosmetics are similar by nature in the U.S. and the EEC, it is of significance that the U.S. classification as cosmetics necessitates as a prerequisite that a substance is not regarded as a drug. This difference is illustrated by reference to the fact that UV filters used to protect against sunburn are considered as cosmetics in the EEC[22] but clearly ruled as drugs in the U.S.[23]

C. Requirements for Drugs and Cosmetics
1. Efficacy for Drugs, not for Cosmetics

The fundamental thinking underlying the definitions of drugs and cosmetics is the fact or the assumption that a drug has an effect. In Europe the definition is not intentional in that respect, but substantial in the respect that the intended effect is related to a disease. The broadness of the definition in Europe is narrowed down by subsequent articles in the directives and their annexes stating clearly that a disease to be treated has to be well recognized as such, and by giving detailed requirements on how to demonstrate efficacy.

In the U.S., however, a substance can be defined as a drug solely by a pharmacological effect or even by what can be regarded as a physical effect as being the case with UV filters. Legal drug classification does not necessarily mean that a substance is for therapeutic use. However, the requirement of efficacy is still there when it comes to actual use, since a product will be approved in relation to its use, and the claimed effect has to be documented. If the claimed effect is therapeutic the drug becomes a medicine.

For cosmetics, definitions do not envisage any effect. Realizing that any effect will categorize a substance as a drug may have the consequence that substances with an effect that would be considered desirable for a cosmetic according to the consumers viewpoint may be avoided by the manufacturer, or ineffective low concentrations of an ingredient employed in order to avoid the drug classification and its consequences with regard to the documentation requirements such classification would have in the U.S. As an example in directly illustrating this, an application has been filed on tretinoin to obtain approval as a medicine in the U.S. for treatment of the visible signs of aging caused by excessive exposure to sunlight.[24] This is probably the clinical condition under which it has been possible to present a statistically valid demonstration of efficacy as required for a medicine. It may be speculated, however, whether potential users may come from groups being exposed to weathering to such an extent that their skin may be considered in a state of disease demanding therapy or whether the need to be served in reality may be of cosmetic nature caused by the wish to make improvement toward the ideal of natural healthy skin.

To fulfill the need for antiwrinkle remedies in cosmetics, substances are also used which have not been documented with regard to effect.[25] Such products are being sold as cosmetics with a labeling from which it can be understood that they possess an effect, although their marketing as cosmetics is only possible because there is none.

In Europe the legal gap between the definition of medicine as aiming at the treatment of diseases and the definition of a cosmetic by purpose and implying that it has no effect has been realized. An attempt to fill it out is being made under the revision of the Cosmetic Directive proposed in 1992.[19] Although the definition of cosmetics still unchanged focuses on the application of cosmetics and their intended purpose, a new article contains a demand for documentation of a claimed effect which means that in the future it is likely that a beneficial effect can be contained within the scope of cosmetics in the EEC.

2. Safety Requirement for All Products

Not surprising, the fundamental prerequisite of safety to human beings is an essential element of legislation on drugs, cosmetics, and devices both in the U.S. and within the EEC. Historically, legislation on medicine has originated from the basic understanding that medicine shall not be harmful in its normal use although safety in medicine is being something relative as expressed by the benefit-risk concept. This is, however, not the case in cosmetics where a more narrow margin applies to safety which in this field really means innocuous as stated clearly in Article 2 of the EEC Directive on Cosmetics.[18] Such requirement is less directly stated in the U.S. legislation but managed by the definition as either being different from drugs by having no activity or being a drug and as such further defined as either a substance known to be effective and safe or a new drug for which efficacy and safety is to be demonstrated.

In the U.S., legislation on devices is designed as parallel to that of drugs in the sense that accompanying labeling shall describe the article, give precise directions for use, and that proper use according to instructions given shall be safe to human beings, but requirements subjecting devices to preclearance appear only when falling within the legal definition of a drug.

The proposal for an EEC Directive on Devices classifies devices into four different classes, by a distinction reflecting primarily different levels of safety requirements necessitating varying degrees of control.

In neither parts of the world, instruments for noninvasive measurements employed for diagnoses and control in dermatology will necessitate a preclearance procedure, although claimed abilities and proper use will have to be documented under the legal responsibility of the manufacturer. However, the methods and the results provided by the instruments can be expected to play a significant role within the scope of legislation pertaining to drugs and cosmetics.

III. Regulations and Administrative Procedures

A. Cosmetics in Regulatory Practice
1. Regulatory Procedures

For cosmetics the EEC directive is directly pointing out substances not to be used, and within certain categories listing substances allowed to be used or to be used for specified purposes in concentrations up to a maximum limit. Decisions are taken on the basis of safety evaluation, taking into account all information available. For new substances safety testing is typically presented for approval by manufacturers interested in marketing substances to be used in cosmetics. No specific testing requirements are prevailing as rules or guidelines under the EEC directives, but standards exist primarily established by the industry itself and also by international organizations as, for example, the OECD.

The FDA regulations state manufacturers' duty to substantiate safety of their products, but do not state exactly how cosmetic products shall be substantiated with respect to safety. Instead of control, consumer awareness is trusted as a source of information on potential safety problems. The fact that the FDA receives relatively few complaints about adverse reactions from consumers relating to cosmetic products[26] may suggest that the safety control of cosmetic products exercised by the industry itself is done to the fulfillment of the law.

2. Standard Testing

A testing battery employed by the industry in Europe typically includes *in vitro* testing of chemical and physical properties, *in vivo* animal studies to test tolerability to skin and mucoid surfaces, and to some extent experimental studies in human volunteers with limited exposure. In use safety testing simulating normal conditions of use is usually limited to groups of volunteers involving 50 to 100 subjects.[27] The intended effect of being of technological nature or functional for a cosmetic product seems to be beyond the interest so far of the European listings of substances. The new proposal in the revised Cosmetic Directive[19] demanding documentation for claimed effects seems to be meant as a possibility rather than a requirement since it does not seem to interfere with procedures for listings of substances to be allowed or not to be used.

Also in the U.S., safety testing of cosmetics and substances intended for cosmetic use is following the industry's own standards which as a basis will include testing for potentials of irritation of skin and mucoid surfaces as well as eye irritation, tests on the possible contact and photosensitization, as well as acute toxicity by the oral and percutaneous routes.[28] Testing is usually performed on animals but experimental and controlled usage tests in human volunteers also form part of a standard testing battery.

3. Cosmetic Drug Products

An interesting group of products are those identified by U.S. law as drugs but used by consumers as cosmetics and having to comply with rules for both, which means a full testing program to demonstrate efficacy and safety as for a medicine and also demonstration of innocuousness by intended and unintended human exposure to substantiate cosmetic safety. Using a new antiperspirant as an example, such a testing program has been published.[29]

B. Drug Product Approvals
1. Regulatory Instruments

Approval procedures for new drugs have been established for decades by law in European countries as well as the U.S.[16,20,30] Application procedures, requirements for documentation, and the handling of the material are regulated to a highly developed level by detailed rules, procedural precedence, and also by guidelines which are continuously undergoing revision in order to reflect a current level of efficacy and safety testing. No doubt, the most strongly regulated field is the safety documentation which reflects the fact that responsibility in decision making is felt most burdensome and therefore demanding a high degree of certainty in this field with regard to the underlying documentation.

2. Testing Principles

For a dermal product being a drug there is major emphasis on a possible potential for systemic toxicity which is studied along the lines for systemic drugs in case the substance is absorbed through skin or in case there is a possibility of other systemic exposure such as unintended inhalation of a product to be used in the form of a spray. In case the absorption of a dermal drug is low or negligible, toxicity by dermal application becomes the most important study. In drug regulations it seems that study conditions prevailing for systemic drugs are taken over which means that two species and dosing periods of 6 to 12 months apply for products to be used for a long time or repeatedly. Usually, open application is used, and observations comprise clinical evaluation of dermal effects, but focus is made on pathology and blood and urine chemistry.

To some extent the testing battery usually employed in cosmetics to reveal any possible dermal reactions such as irritation, sensitization, and phototoxicity are employed for drug substances as well. Certainly to the extent a substance is a drug by legal definition, but intended for widespread use in cosmetic or toilet products, it is important that these sensitive methods are employed to guarantee that only an innocuous substance will be selected for use. However in medicine, such testing does not serve the purpose of selecting possible candidates for use, since such effects may well be considered acceptable depending on the therapeutic benefit of the substance. Testing results will, however, provide the information necessary to establish proper labeling and user instructions to accompany a drug product.

C. Testing Procedures Applied to Skin Products
1. Safety Testing

1. The standard testing of *skin irritation potential* is the Draize Test which exists in numerous modification applied to rabbits' intact and abraded skin. By principle it is a patch test with 24 hours of application and with visual evaluation of skin reaction.[26] Variations of patch testing on rabbits involving repetitive application are used as well. The Draize testing is used for substances as well as products and can be used with varying concentrations of testing material to obtain sufficient sensitivity.

2. Single-exposure Draize testing for regulatory purpose is usually performed in rabbits and not primarily in humans because rabbit skin is more sensitive. Human studies using patch testing usually comprise repeated exposure to have sufficient sensitivity. The *21-day accumulative irritancy test* seems to be considered a standard testing regimen.[28] It employs repeated daily appli-

cations to a group of 8 to 25 volunteers under occlusive patches with daily readings of results graded on a 4-step visual scale.

3. A special variation of the Draize test is *the eye irritation test* which has, as its main function in evaluation of skin products, to predict a possible irritation from unintended exposure to the eye[28] but which is also employed as a measure for possible irritancy to mucoid surfaces.[27]

4. Second in the listing of safety topics to be tested for a skin product is the *contact sensitization potential*. No doubt, the maximization test used both on animals and in human beings stands as the golden standard since it is considered sensitive enough to detect weak or potential allergens, which means that substances showing no allergic response in this test are considered with confidence. On the other hand, it has been pointed out that the test sometimes is difficult to interpret correctly since the method has identified allergens among substances previously used in the market for many years without problems. Although maximum allergy is intended, it is important that concentrations used are low enough to avoid a possible irritant effect which cannot be distinguished from the allergic reaction in the visual reading of the result.

5. A test for predicting *human contact sensitization potential* of a substance or a product, taking into account the variability among human beings in being prone to allergy, is the repeat insult patch test (RIPT) which calls for 50 to 200 subjects comprising both sexes and intended to include persons with allergies or sensitive skin. The use of occlusive patches makes the test very sensitive but it also increases the risk of interference of result observed with a possible irritancy of a test substance.[28]

6. The so-called *photosensitivity testing* may be performed as a 24-hour patch test with subsequent UV-irradiation before reading of a result expressing an acute phototoxicity, or as a photoallergenicity testing with induction by repeated application under patches followed by UV-irradiation after each exposure and challenge at a later date. Reading may reveal not only redness of skin, but also hyperpigmentation.[28]

7. Among less frequently used safety testing, although testing to show very common effects, is the *acnegenicity test* which is done more often now for fine cosmetics to be used on the face in order to substantiate a positive claim of its negative effect. The rabbit ear assay on comedogenicity with repeated exposure to produce enlarged pores and hyperkeratosis is the standard test throughout many years whereas the pustologenicity testing using occlusive patch application to the back of human beings prone to acne is more recent. The latter is thought to correlate better with actually occurring acnegenicity in human beings although empirical data to show the possible correlation are still missing.[26]

8. *The stinging test* is performed merely to evaluate a possible reaction of stinging and burning sensitization in subjects with light complexions and sensitive skin who can be identified as stingers by their reactions to application of lactic acid. This testing is not really a safety testing but serves to provide a basis for information on labeling.

9. As probably the most important safety testing of a new skin product is the *controlled usage test*, which is done simulating intended use over a prolonged period usually at least 4 weeks in a relatively large number of subjects often ranging from 40 to 200, and involving frequent clinical observations of skin reactions as well as reporting of the subject's own observations. As a modification, it can be done by patch testing which has advantages with respect to compliance and controlled exposure.[28]

A parallel can be drawn between the controlled usage testing a cosmetics and the therapeutic trial on a dermatological product as having in common the objective of clinical observations related to safety. It should, however, be taken into account that reactions observed may have entirely different consequences, since adverse reactions are usually considered unacceptable for a cosmetic, but observed as a fact for a medicine as long as they do not outbalance the therapeutic benefit.

2. Efficacy Testing

The major difference so far between the two main groups of skin products is the testing of efficacy, which is supposed to be absent for a cosmetic by legal definitions, but which is subjected to a very stringent documentation requirement with regard to drugs.

Efficacy testing in dermatology is usually based upon visual examination of clinical signs of a disease or its absence. Because of the fact that results obtained in this way are highly variable, much effort has been done to reduce the uncertainly associated with the use of such a nonobjective method. First, attempts are done to reduce the impact of subjectivity in the clinical evaluation by the use of graded scales and by differentiated reading of the various skin parameters which constitute the clinical score. Second, clinical therapeutic trials are to the possible extent done as double-blind clinical studies comparing with positive control and/or with placebo. Multi-center design, blind analysis of data, and application of suitable statistical methods serve to ensure that the results arising expressed as differences between averages of groups have sufficient validity.

As a consequence of these testing conditions clinical efficacy studies have become very extensive, and the time necessary for planning large multi-center studies across national borders and the time needed for subsequent analysis of data have made therapeutic clinical studies extremely expensive. To the extent new measuring methods can contribute to provide more accurate data and thereby eliminate some of the problems being a basic reason for the necessity of extensive clinical studies, they would have an important resource-saving effect.

D. Development of Regulatory Methods
1. Function of Guidelines

Usually guidelines for testing originate from a substantial basis of knowledge which guarantees generation of valid results as the outcome when following proper guidelines. This is essentially the intrinsic value of a guideline and the justification for their existence as a regulatory instrument.

On this background it seems self-explanatory that guidelines usually seem to reflect the past when appearing. Very few examples of guidelines exist for testing of efficacy or safety of drugs, which have been created on the basis of entirely new methodologies on which insufficient experience has been available to constitute a basis for estimation of validity of testing results. The testing procedures in the EEC guideline on mutagenicity were criticized for that particular reason when introduced for regulatory testing on new drugs in Europe. Beyond any doubt, experience helping to evaluate validity and predictability of these testing procedures has by now accumulated as a consequence of the ruling.

Realizing that official guidelines usually do not have the function of promoting the development of new testing procedures it may be interesting to analyze by which means the introduction of new methodologies in the regulatory field actually can take place.

2. Introduction of New Methods

The intersurface growth layer for that development is probably the communication between regulatory bodies and scientists from the industry during the development of new drugs. Although this interaction usually has as its primary goal to define the elements of documentation in a specific New Drug Application it also provides a mutual forum for the evaluation of usability of new methods for testing purposes.

In the U.S. this communication is structured into pre-IND and pre-NDA meetings at which occasions the FDA staff is prepared to give advice and review development plans presented to them.

In Europe access to communication with regulatory bodies is less institutionalized and by far more diversified compared to the U.S. structure of interaction. This difference is directly arising from the difference in staffing of regulatory bodies. The FDA is primarily doing their evaluation of the documentation by their own staff of full-time employed assessors who are available also for pre-application discussions. In Europe medical evaluation is to a great extent done by outside experts working on a consultant basis for the regulatory bodies either as permanent members of advisory boards, as ad hoc advisers providing high level expertise, or as rapporteurs being responsible for in-depth studying of an application's medical documentation.

Although documentation contained in NDAs are confidential and thereby not available to other applicants through communication with regulatory bodies, knowledge derived from studying such documentation will inevitably contribute to a personal perception of the state of the art and as such be communicated in general terms as principles to subsequent applicants. The creation by predecessors of models becoming standard for the next applicant provides a highly efficient tool for development of new methodologies within a legal framework which from watching the surface may appear bureaucratic. Advice given by the FDA and their viewpoints communicated in public are associated with a high degree of certainty and provides a reliable basis for companies' development work.

In Europe, no voice speaks so far for the entire area, and even within national borderlines it may be difficult to obtain advice from regulatory bodies reaching beyond the scope of personal views. The fact that sometimes no recipe more operational than general guidelines can be given for the development of a particular new drug is often found frustrating by companies. Also the differences existing among individual countries within the EEC in spite of 20 years attempts to harmonize makes it difficult to design a documentation program for a new drug with the aim of a cost-effective development toward the necessary marketing authorizations.

On the other hand the European situation provides a freedom of choice on how to document efficacy as long as procedures and methodologies can be defended as eligible and demonstrated as valid. To facilitate design of a development program under these circumstances, advice is often taken from leading experts who are asked to suggest models and procedures representing the front lines of science in the field as being the best guarantee of a successful approval procedure later on.

Thus, irrespective of the pathways, valuable new methodologies soon find their ways into the regulatory systems as far as drugs are concerned.

IV. Use of Noninvasive Skin Testing Methods

A. Efficacy Testing on Drug Products
1. Objective Measures to Replace Clinical Observations

Obviously, it is most desirable to replace simple visual observations with objective measurements. An example of this being attempted in regulatory testing is the recent Interim Guidance from the FDA on bioequivalence testing of topical corticosteroids by the vasoconstrictor assay.[31] As stated in this text it is no longer acceptable to measure pharmacological effects solely by a human observer, given that sophisticated methods capable of detecting physical and chemical changes are available. On this background, the use of a chromameter to determine blanching is encouraged. However, it is worth noticing that chromametric measurement is not suggested as a replacement. It is intended to be used concurrently with visual scores. This cautious attitude has a legal background as seen in the light of the regulatory function of a bioequivalence test as a demonstration of a similar potential of therapeutic effect as the sole basis for approval at least with respect to efficacy documentation.

The situation would probably be different if application of chromametry to determine erythema is done for purposes independent of past history, since colorimetric measurements by instruments have been shown to be precise and accurate. Adapted by the CIE system of color coordinators to take into account the nonlinear color perception of the human eye, commercially available colorimeters have been demonstrated to correlate well with clinical scoring.[13] In addition to being an instrumental measurement of what is visible to the naked eye, colorimetric measurement of erythema has also been demonstrated to correlate well with changes in the underlying blood flow as measured by laser Doppler flowmetry and being the reason for the change in skin surface color. This has been demonstrated with color changes provoked by irritants in patch testing on volunteers.[13] Previously, spectrophotometric measurements have also been used to assess the erythema induced by ultraviolet irradiation.[12]

Besides the principle of measurement of an effect in a way attempting to quantify the way it has been visible to the eye, an alternative approach is to determine the effect itself which would be the logical thing to do if a past history of clinical evaluation was non-existing. Measurement by laser Doppler technique of changes in cutaneous blood flow as caused by a vasodilatation is an example of such attempt.[13,32] Considering that erythema caused by vasodilatation is an important parameter both in efficacy studies and also nearly in all the standard testing applied to skin products such as testing for primary irritancy, sensitization, and photosensitivity, it is envisaged that such measurements will have a role to play in the future safety testing of skin products.

2. Precise Quantifiable Methods to Replace Clinical Scores

The measurements of cutaneous blood flow as an evaluation of reactions to irritants applied under patches[13] demonstrated that values differentiated well within the range of low clinical scores but insufficiently at high clinical scores. This will have the practical consequence when used in testing for primary irritancy that concentration shall be held at a low level not to overload the measuring capacity. The broad distribution of values within low clinical scores also suggests that the measuring of blood flow provides a much more nuanced measure of a change in vascularization compared to the visual perception of the erythema. The ability of the method to distinguish subtle nuances has also been demonstrated by its ability to show interindividual variation in human volunteers exposed to a standard irritant patch testing.[33] This ability to show individual differences which in the same study is shown also by the ultrasound measurement of skin thickness and by determination of transepidermal water loss as an expression of the barrier function of the skin may at the first glance look as an inconvenient variability.

However, it also provides possibilities for refining a testing technique by identification of subjects being similar in those parameters, and thereby assuming similarity in testing sensitivity. Narrowing down subject variability in the standard safety testing on skin products by selection of uniformly responding groups of volunteers may provide a tool for a more precise evaluation of safety parameters. It may also provide a basis for proper information to users about the fact that subjects with particular sensitive skin may react to a product.

It seems interesting to speculate whether instrumental tests of the above-mentioned colorimetric tests for erythema and ultrasound testing of skin thickness can be used in efficacy testing to replace visual evaluation. In such studies patients will serve as their own control from entry to end of treatment and will therefore provide relative numerical data on the relevant parameters to demonstrate a possible progress from disease to healthy skin, and as such create a correlation to the graded scales usually employed. Large-scale therapeutic studies may thus be the tool to provide necessary information about correlation between clinical diagnoses and instrumental methods for measuring the parameters reflecting the clinical manifestation.

3. High Degree of Standardization Necessary

Evidently testing on human beings and animals of parameters easily fluctuating with changes of the biological system such as erythema and skin temperature are subjected to be influenced by the surrounding room in which climatic factors temperature, humidity, sunlight, and air convection will have an impact on the results measured and in some instances also on the parameters to be tested. This will easily be the case with respect to skin temperature and transepidermal water loss. In that sense the instrumental methods may provide a new set of problems for the user to whom standardization of testing conditions to the extent necessary may become an obstacle.[34] Indeed, full validity of data is prevailing only to the extent that methods can be used independently from the time and place for their generation.

It may seem possible to solve this problem with regard to the animal safety testing which is always performed in laboratories equipped for the purpose and with high awareness of the importance of good laboratory practice. It may also very well be possible to implement the necessary standardization in testing facilities doing experimental skin testing for safety evaluation in volunteers, since such testing also takes place in special facilities and under the proper awareness of conditions as being important for the results.

Obviously the instrumental quantitative measurement will also be of interest to therapeutic trials carried out for efficacy testing of new drugs. With intraobserver variations of 10 to 30% and interobserver variability of 20 to 40% which has been reported,[35] it seems attractive to replace subjective clinical judgment of skin parameters with nonsubjective methods. Also, the fact that instruments for these new methods are becoming readily available on a commercial basis makes it believable that their introduction into hospital wards and dermatological clinics will take place within a short period of time.

However, it is important to realize that more precise data will be obtained only if the instrumental methods are used under observation of the standardized conditions with respect to subjects and surroundings necessary to obtain reliable results. This may be a critical issue in the clinical practice which is organized primarily to handle efficiently the treatment of patients with respect to their therapy and nursing and less may be to maintain conditions for instrumental measurements.

In order to derive the full benefit of the new techniques in therapeutic trials which should be a significant reduction in the sample size needed to obtain data of statistical validity, it is important to emphasize validation of the whole testing setup in clinical centers selected for participation in clinical trials.

4. Techniques Providing New Possibilities

Of particular interest in clinical testing is of course those instrumental methods monitoring disease parameters or diagnoses which cannot be recorded visibly and where alternative methods may be invasive. Thus, in-depth imaging of skin structure by ultrasound scanning now provides measures not only for diagnosis but also for continuous measurement for possible changes due to therapeutic treatment. This technique also has the practical advantage that the in-depth structure of the skin is not sensitive to subjects' conditions and that surroundings have no impact. This means that conditional variability of data is small and independent of time and space. However, this technique provides results which can be interpreted with respect to its meaning as related to clinical diagnosis which will require special training at least for its diagnostic use.[5] This disadvantage, however, is to some extent counterbalanced by the fact that, once established in relation to the individual patient's diagnosis with suitable reference to a normal baseline condition, changes of the picture can be continuously recorded and analyzed by computer imaging technique under conditions to ensure blinded, unbiased interpretation as required according to Good Clinical Practice requirements prevailing both in the U.S. and Europe. Not surprisingly it has been reported that the FDA is encouraging the use of techniques providing quantifiable imaging in disease detection and monitoring.[35]

B. Substantiate Claimed Effects of Cosmetics

The possibilities of observing effects with the new methods for measurement on parameters which have not previously been accessible for precise determinations will probably be of particular interest with respect to cosmetic products which from the consumers' viewpoint are expected to exert desirable effects on the skin, but which have generally been left in a gray zone with regard to documentation in that respect. No doubt, the fact that creams and lotions are sold in enormous quantities throughout decades indicate that they do fulfill those expectations, if not to the desired effect then at least to a satisfactory extent. It is not necessarily a reflection solely of advertising.

Recently, studies on series of cosmetic preparations involving measurement of skin hydration by measuring changes in electrical conductance in superficial skin layers and capacitance in epidermis, respectively, as well as assessment of scaling by desquamation taping and by light transmission have demonstrated significantly improved hydration and skin function.[36,37]

Although demonstration of beneficial effects of cosmetic products may be of little significance from a legal and regulatory viewpoint for the time being, this may well change, in light of the statement inserted in the proposed amendment of the EEC Directive on Cosmetics.[19] According to the amended proposal claimed effects shall be documented and this may express a consumer demand seeking its satisfaction on legal grounds although it should be possible to get such reasonable demand fulfilled through market mechanisms in view of the assumption that proper documentation of usability properties should fall within the interest of the manufacturers as well.

C. Minimize Use of Animals in Safety Testing

Another aspect which may work to promote the introduction of new technologies in testing of skin products is the fact that animal testing is considered undesirable and in some areas even unacceptable.

In the testing of drugs there is a public understanding that the use of animals is part of the price to pay in order to obtain new valuable therapeutics safe to human beings. However, attempts are ongoing within the regulatory area to assure that animals are used only when absolutely necessary. Use of animals is controlled in many countries by rigid procedures, and research work aiming at finding alternative testing methods not employing mammalian species is encouraged.

However, when talking about cosmetics it seems to be a prevailing public viewpoint that animal testing is unacceptable for that purpose. This is made clear by the proposed amendment to the EEC Directive on Cosmetics in which it is suggested that animal testing of substances for the sole purpose of cosmetics should be prohibited.[19]

Looking at the safety testing usually applied to skin products it is evident that most skin testing principles can be applied to both animals and human beings. Furthermore, the experimental testing on human beings usually involve limited skin areas thereby limiting the possible risk associated with the testing. Introducing the new noninvasive instrumental methods with their ability to determine in a precise and finely nuanced way the various parameters on even more limited skin areas it seems realistic to think that the numbers of animals used in the general safety testing of skin products can be reduced considerably. Seen in that light the use of animals may remain mainly in the field of basic systemic toxicity testing where no alternatives have appeared so far, and to those special fields where animal strains having special properties are fundamental to valuable testing models.

V. Dermatology and Cosmetics — Areas on the Move

A. Borderlines Become Fluent

1. Cosmetic Problems in Dermatology

An example has already been mentioned showing how dermatologists' definition of a clinical manifestation of a skin problem may differ significantly from individuals' perception, namely the study on skin dryness in healthy volunteers in which it was attempted to establish correlations between instrumental parameters on hydration and visual observation of symptoms of dry skin. Apparently many healthy subjects find skin dryness a problem. This may reflect the current idea that healthy and normal more and more means ideal and optimal, a concept which leaves no room for the natural changes associated with age and living conditions. What is observed by the individual as deviating from ideal may be looked upon as a problem for which a remedy is sought. How does the dermatologist react to this?

As long as he is unable to observe a problem via his clinical examination he is likely to declare it nonexisting, but, at the very moment possibilities to diagnose a condition as different from the baseline value identified as "normal", he will also be able to suggest remedies that can normalize the condition, provided such remedies exist.

The existence of remedies is another factor determining what is therapy demanding condition and what is referred to as "cosmetic problems".

All through history the most pertinent visible sign of aging, wrinkled skin, has been the inevitable fate which could only be subjected to camouflage cosmetics. Certainly few reports on clinical studies have been received with a broader public interest than Prof. Kligman's demonstration of an antiwrinkle effect of retinoic acid containing products available in most countries for treatment of acne.

Prescriptions of the product for treatment of visible signs of aging can be expected to appear as documentation accumulates that can lead to approval of the indication. The patients will be there waiting.

2. Treatment in Cosmetics

The cosmetic industry is no doubt well aware of the consumer's need for remedies that can help what they see as problems, and that can improve the appearance of the skin and prevent it from aging. However, they have traditionally been left to use words and pictures in advertising media as the way they could meet the customer's need, and to provide promises instead of documented solutions to the problems. Their understanding of the customer's essential needs can be observed clearly from their advertising in glossy magazines by which the message about anti-aging cream, hydration emulsions, and wrinkle treatment is given.

B. Consumer Demand for Efficacy of Cosmetics

1. Conflicting with Definition of Cosmetics

The consumer has certainly a need to be fulfilled, although perhaps a subjective one from a dermatologist's point of view. It may be a medical need or a cosmetic one which to the individual means skin care, because such distinction is of minor interest to the individual who does not really care whether a product is a drug or a cosmetic according to legal definitions. What matters is the same for both categories, namely, that the product is expected to work according to claims and it that shall be safe.

Knowing the scientific abilities of dermatologists and industrial laboratories, the consumer is no longer prepared to accept advertising promises. The strict separation of dermatological skin products and cosmetics and the existing dilemmas with respect to documentation of efficacy is no doubt a consequence of the legal definitions of drugs and cosmetics and the regulatory requirements prevailing for drugs. Cosmetics are assumed to have no effects; if substances have effects they are no longer cosmetics, and the drug categorization is associated with requirements of efficacy documentation which have allowed the penetration of relatively few new skin products over the last 2 decades because the burden of proof has been a serious problem given the simple clinical observation as the main tool for evaluation of efficacy.

2. Evidence Possible by New Techniques

With the new instrumental technique it has become possible to show effects of skin products when applied to the human skin in its healthy or diseased state, and demonstrate how products work on functional parameters. It is possible to target products to affect narrowly certain parameters and thus design products to suit a well-defined purpose.

Such possibilities seem to be interesting for all parties involved and should not only be reserved for safety testing of skin products and for efficacy trials of drugs. Used to document the actual effect of cosmetic products exerted on the normal skin, the new methods may provide to the cosmetic industry the possibility of demonstrating the actual ability of their product.

For the consumer it will be a long-awaited opportunity to receive clear knowledge about the product they buy and to select products as a basis of evidence and not on the basis of promises.

VI. Impact of New Methods on Regulations

Being useful for diagnostic purposes in diseases and capable of observing changes taking effect in the normal skin, bioengineering techniques provide measures of observation of changes that can be used to evaluate the effect of intended treatment with respect to desirable outcome, as well as to show effects that may provide reason to expect unwanted reactions of an intended treatment. This can be used in the goal-oriented development of therapeutic remedies for dermatology, and it can be used for products to serve the much-desired improvement of the normal skin enjoying the attention of the cosmetic industry.

Thus, the development of the new instrumental testing methods constitutes one basic prerequisite to allow science to enter not only dermatology but also to pass the borderline and enter into the area of cosmetic efficacy. Another important prerequisite for such expansion has to do with the existing legal structures categorizing cosmetics as having no effect with the consequence that category may be changed to drug if a product actually has an effect.

As an important first step in the direction of a possible future change, the proposed amendment of the EEC Cosmetic Directive, the requirement for documentation of claimed effect, can be seen as an opening. Although this new statement is unlikely to provide much change from the beginning, it has significant importance by the fact that it introduces the viewpoint that a cosmetic product may have an effect.

The clear legal definition in the U.S. of cosmetics as having no effect and the classification as drugs of everything having a recognized effect is not surprising considering that these definitions date back to 1938 and have remained basically unchanged since then. Possibilities of observing effects are now changing and, with the methodologies now available, such a definition of cosmetic products seems to be an oversimplification warranting a revision.

References

1. Kligman, A.M., in *Cutaneous Investigation in Health and Disease. Noninvasive Methods and Instrumentation*, Leveque, J.-L., Ed., New York, 1989.
2. Froch, P.J. and Wissing, C., Cutaneous sensitivity to ultraviolet light and chemical irritants, *Arch. Dermatol. Res.*, 272, 269, 1982.
3. Agner, T. and Serup, J., Quantification of the DMSO-response — a test for assessment of the skin, *Clin. Exp. Dermatol.*, 14, 214, 1989.
4. Jemec, G.B.E. and Serup, J., Scaling, dry skin and gender, *Acta Derm. Venereol.*, Suppl. 177, 26, 1992.
5. Serup, J., Characterization of contact dermatitis and atopy using bioengineering techniques. A Survey, *Acta Derm. Venereol.*, Suppl. 177, 14, 1992.

6. Agner, T. and Serup, J., Comparison of two electrical methods for measurement of skin hydration. An experimental study on irritant patch test reactions, *Bioeng. Skin,* 4, 263, 1988.
7. Linde, Y.W., Dry skin in atopic dermatitis, *Acta Derm. Venereol.,* Suppl. 177, 9, 1992.
8. Serup, J., Winther, A., and Blichmann, C., A simple method for the study of scale pattern and effects of a moisturizer — qualitative and quantitative evaluation by D-Squame® Tape compared with parameters of epidermal hydration, *Clin. Exp. Dermatol.,* 14, 277, 1989.
9. Stender, I.M., Blichmann, C., and Serup, J., Effects of oil and water baths on the hydration state of the epidermis, *Clin. Exp. Dermatol.,* 15, 206, 1990.
10. Serup, J., A double-blind comparison of two creams containing urea as the active ingredient, *Acta Derm. Venereol.,* Suppl. 177, 34, 1992.
11. Serup, J., Ten years' experience with high-frequency ultrasound examination of the skin: development and refinement of technique and equipment, in *Ultrasound in Dermatology,* Altmeyer, P., el-Gammal, S., and Hoffman, K., Eds., Springer-Verlag, Berlin, 1992.
12. Diffey, B.L., Oliver, R.J., and Farr, P.M., A portable instrument for quantifying erythema induced by ultraviolet radiation, *Br. J. Dermatol.,* 111, 663, 1984.
13. Serup, J. and Agner, T., Colorimetric quantification of erythema — a comparison of two colorimeters (Lange Micro Color and Minolta Chroma Meter CR-200) with a clinical scoring scheme and laser-Doppler flowmetry, *Clin. Exp. Dermatol.,* 15, 267, 1990.
14. Food Drug and Cosmetic Act, U.S. Code Title 21, Section 301.
15. Food Drug and Cosmetic Act, U.S. Code Title 21, Section 321.
16. Council Directive of January 26, 1965 (65/65/EEC) on the Approximation of Provisions Laid Down by Law Regulation or Administrative Action Relating to (Proprietary) Medicinal Products, Off. J. Eur. Com., 9.2.65; No. 369/65; revised edition in the Rules Governing Medicinal Products in the European Community. Vol. 1, September 1991 (ISBN 92–826–3166–4).
17. Proposal for a Council Directive Concerning Medical Devices (91/C237/03) COM (91) 287 final — SYN353, Official Journal of the European Communities, 12.9.91; No. C237/3.
18. Council Directive of July 27, 1976 (76/768/EEC) about the Approximation of the Laws on Cosmetics in the Member States, Official Journal of the European Communities, 27.9.76; No. L262/169.
19. Commission; Amended Proposal for Council Directive on the Sixth Amendment of Council Directive 76/768/EEC about the Approximation of the Laws on Cosmetics in the Member States, (92/C249/04) COM (92)364 final — SYN307, Official Journal of the European Communities, 26.9.92; No. 249/5.
20. Council Directive of May 20, 1975 (75/319/EEC) on the approximation of provisions laid down by law, regulation or administrative action relating to proprietary medicinal products, Off. J. Eur. Com., 9.6.75; No. L147, revised edition in the Rules Governing Medicinal Products in the European Community, Vol. 1, September 1991 (ISBN 92–826–3166–4).
21. Council Directive of May 20, 1975 (75/318/EEC) on the approximation of the laws of Member States relating to analytical, pharmacotoxicological and clinical standards and protocols in respect of the testing of proprietary medicinal products, Off. J. Eur. Com., 9.6.75; No. L147, revised edition in the Rules Governing Medicinal Products in the European Community, Vol. 1, September 1991 (ISBN 92–826–3166–4).
22. Janousek, A., *Regulatory Aspects of Sunscreens in Europe, Sunscreens, Development, Evaluation and Regulatory Aspects,* Lowe, N.J. and Schaath, N.A., Eds., Cosmetic Science and Technology Series, Vol. 10, 137.
23. Murphy, E.G., *Regulatory Aspects of Sunscreens in the United States,* Lowe, N.J. and Schaath, N.A., Eds., Cosmetic Science and Technology Series, Vol. 10.
24. SCRIP, No. 1711/12, April 22nd–24th 1992, 25.
25. Ziolkowsky, B., Die Behandlung lichtgeschädigter Haut sowie deres Schutz vor vorzeitigen Alterung durch specielle Wirkstoffe, *Kosmetikjahrbuch,* 1990, 382.
26. Bronaugh, R.L. and Maibach, H.I., Cosmetic Safety in Cosmetic Safety, A Primer for Cosmetic Industry, Whittam, J.H., Ed., Cosmetic Science and Technology Series, Vol. 5, 11.
27. Matthies, W., Experimentelle Methoden zur Ermittlung der Haut- und Schleimhautverträglichkeit von Rohstoffen und Fertigerzeugnissen, *Kosmetikjahrbuch,* 1990, 376.
28. Kaufmann, P.J. and Rappaport, M.J., Skin Care Products, A Primer for Cosmetic Industry, Whittam, J.H., Ed., Cosmetic Science and Technology Series, Vol. 5, 179.
29. Morton, J.J.P. and Palazzolo, M.J., Antiperspirants, A Primer for Cosmetic Industry, Whitman, J.H., Ed., Cosmetic Science and Technology Series, Vol. 5, 221.
30. Food Drug and Cosmetic Act, U.S. Code Title 21, Section 355.
31. FDA, Division of Bioequivalence: Interim Guidance; Topical Corticosteroids: In Vivo Bioequivalence and In Vitro Release Methods, July 1, 1992.
32. Staberg, B., et al., Patch test responses evaluated by cutaneous blood flow measurements, *Arch. Dermatol.,* 120, 741, 1984.
33. Agner, T. and Serup, J., Individual and instrumental variations in irritant patch-test reactions — clinical evaluation and quantification by bioengineering methods, *Clin. Exper. Dermatol.,* 15, 29, 1990.
34. Pinnagoda, J., et al., Guidelines for transepidermal water loss (TEWL) measurement, *Contact Dermatitis,* 22, 164, 1990.
35. Conklin, J.J., et al., Image processing techniques applied to drug and medical device development, in *New Drug Approval Process,* 2nd ed., Guarino, R.A., Ed., 1993, 145.
36. Blichmann, C.W., Serup, J., and Winther, A., Effects of a single application of a moisturizer, *Acta Derm. Venereol.,* 69, 327, 1989.
37. Serup, J., A three-hour test for rapid comparison of effects of moisturizers and active constituents (urea), *Acta Derm. Venereol.,* Suppl. 117, 29, 1992.

34.0

Ethical Aspects

34.1 Ethical Considerations 667
     P. Sohl and G.B.E. Jemec

Chapter 34.1
Ethical Considerations

Patricia Sohl and Gregor B.E. Jemec
Department of Dermatology
Bispebjerg Hospital
University of Copenhagen
Copenhagen, Denmark

Ethics is that branch of philosophy that deals with values relating to human conduct. Ethics is concerned with concepts such as good or bad. Although some people may think science to be strictly confined to the realm of rational deduction, the concepts of good and bad are ubiquitous elements of human life, and are therefore also present in the pursuit of knowledge through science.

The good and the bad have preoccupied Man through the ages, and many schools of thought exist offering systems to distinguish good from bad choices. The two principal lines of thought differ in the focus of their attention. The line of deontological ethics is focused on universal rules of conduct, while the utilitarian line focuses on the results of conduct. In reality, such a strict separation of general obligations and specific results is rarely possible, and ethical considerations therefore often contain both deontological and utilitarian elements most often taking the form of utilitarianism within the confines of general obligations.

Ethics plays a role in our assessment of scientific matters as well, and the aim of this chapter is to give some structure to such assessments. The very nature of clinical science obliges the researcher to pay extra attention to ethics, because clinical science is by definition carried out with other human beings, i.e., involves human subjects. The key element of the ethical considerations is the fact that human subjects are by definition self-determining persons. Self-determination in turn means that normal ethical considerations in terms of rights and obligations have to be respected as in any other interpersonal relationship. In the conduct of a clinical trial the investigator therefore has to take into account several levels of ethical considerations and obligations, which all stem from the basic respect for individual self-determination and autonomy.

A derivative principle of this rule is that human beings must themselves agree to what they will or will not participate in. In society in general this is, for example, reflected by common legal requirements for veracity in commerce, the foundations of all commercial law. In interhuman relations it is also important to realize that moral principles often have a substantial nonrational value to as human beings, and especially so in a clinical setting. Not only is the patient in a position of increased vulnerability, but the physician is in a relative state of authority, and an additional level of professional ethical considerations is therefore added. Professions traditionally have specific areas of expertise and a monopoly in the exercise thereof; two classic examples of professions are clergy and medicine. Other members of society are dependent on this expert knowledge, and professions are therefore generally granted high levels of autonomy within society. Professional ethics confirm basic social moral principles of interhuman activities and, in cases where professions have the power to injure or constrain the expression of self-determination of others, the ethics of the profession is often expressed as a code (e.g., the Hippocratic Oath) and act as a guarantee to society that these powers will only be used to serve the best interests of fellow human beings. Physicians are granted clinical autonomy by society, i.e., the freedom to chose what is best for the patient, but this is only granted on the condition that the patient's best interests are in fact served. The special knowledge of the professions therefore imposes extra ethical obligations on them.

From an ethical point of view the noninvasive methods pose no real problems. By the very nature of the techniques they pose only nominal or no real risk to the human subject. In addition they are not connected with unpleasant or fearful procedures, and therefore also avoid the more relative risk of predominantly mental discomfort typically associated with, for example, venepuncture. The benign nature of the techniques however does not make all ethical considerations superfluous, as the technique chosen forms but one part of the study and the ethics concerns themselves with the entire process. It is of little redeeming value to use an ethically advantageous method in a generally unethical study.

The need for protection of individual self-determination and autonomy became apparent following the atrocities of World War II. A number of initiatives have been made by both physicians and society to promote this, and for medical research these efforts have been outlined in greater detail in the 1975 Helsinki Declaration II.

The key elements of the declaration are

1. Scientific design
2. A risk-benefit evaluation
3. Informed consent
4. Procedures to implement these goals

Scientific design is the theme of many books, and is further described in Chapter 00. In ethical terms scientific design is important because a good design is a prerequisite for any risk-benefit evaluation and hence for any informed consent. A good scientific design will allow the researcher to answer a relevant question with a minimum risk to his or her fellow human beings. Literature review, sample size, and statistical analysis are all necessary in order to evaluate risk, to minimize exposure, and to avoid superfluous reexposure of human subjects. Bad science in its most extreme exposes human subjects to unnecessary risk of no benefit to anyone.

Risk and benefit are core elements of any ethical analysis, and a very advantageous point for noninvasive methods. A classic of thinking about risk and benefit is to think of toxicity (risk) and efficacy (benefit) in drug studies. The measurement technique chosen is however very rarely the only element involved in an estimate of risk. Fear of disease, substances tested, prognosis of disease, and additional exposure to allergens or irritants all are elements of the risk evaluation.

In order for society and individuals to accept the inevitable risks associated with clinical trials, an element of motivation is necessary. Motivation can be derived from many different levels of abstraction, i.e., physical, individual, or societal motivation. It is important to notice here that risk and benefit are not at necessarily opposite ends of a spectrum for the individual human subject. For example, the epiluminiscence evaluation of an obvious malignant melanoma has little benefit for the individual patient, but great value for society as an aid in the development of a practically viable screening method.

Informed consent is often the only point of ethical contention in the implementation of clinical studies, which is why this point will receive special attention.

Informed consent is a procedure, *not* a finished product, i.e., not merely a signed consent form. The procedure is intended to provide the basis for a fully informed choice by a self-determining autonomous person, and all requirements are intended to safeguard the autonomy of the individual. From this it is also clear that human subjects can withdraw their consent at any point during the study.

Informed consent must be able to satisfy certain requirements. It must be

1. Informative
2. Understandable
3. Free
4. Competent

Informative — The information must contain all necessary data to make it possible for human subjects to come to a decision. All side effects with a potentially fatal outcome must be listed. It is important to realize that the obligation to inform a subject of a potential side effect of a trial is independent of the chance of this side effect occurring. The decision to accept or reject a given risk is strictly personal and cannot be generalized. For similar reasons, all minor potential side effects should be listed. The investigator cannot decide on behalf of another human being what is or is not sufficiently unpleasant to warrant nonparticipation in a study.

In practical terms it is necessary for the investigator writing an information sheet to know the reasons for writing and the reader as well. A good grasp of the reasons, i.e., promoting autonomy, and of the reader may require the help of another person not involved in research and not involved in the trial at hand. Most human subjects need risk to be explained, and explaining it to someone before writing it may help you identify areas of importance in your explanation. Not all human subjects are academics with a special interest in the particular study.

Understandable — Structuring the information in an appropriate way is usually a difficult task for most investigators. The writing must be clear, simple and brief, accurate and complete, polite, and human. This is of course self-evident, but is more difficult to achieve in reality than many investigators would care to admit. There are many sources of inspiration for the structuring of the text, e.g., instructions for the use of household goods, the tabloid press, and advertising. The text should be simple, short, and specific. Use short words, not long words. Use short sentences, not long sentences. Simple words are better than complicated foreign words. For most academics the exercise of constructing an information sheet is difficult simply because of their training, and several rewrites are usually necessary to construct a good information sheet. Remember that at all times attention must be paid to the contents, so as not to reduce the informative value of the writing as it is simplified.

Free — Absence of coercion is a prerequisite for valid consent; this is the case in contract law as well as informed consent. Unintended coercion can take place in many ways. Time is one factor and human subjects must have sufficient time to consider their participation in a drug trial. No fixed rules for this time can be given, but 24 hours will suffice in most cases. In the clinical environment there are other possibilities for unintended coercion, which physicians are not always fully aware of. Participation in a trial must not in any way compromise future treatment. This may seem obvious, but is often a latent fear in patients. It is therefore useful if the attending physician is *not* the person to do the informed consent procedure. He or she may however freely answer any questions, and should at all times confirm that, no matter which decision the patient comes to regarding the trial, this will have no influence whatsoever on their treatment and care. Whenever a potential state of dependency exists between the human subjects and the investigators, e.g., employees or students, similar care must be taken to ensure that consent is given freely.

Competent — Finally, consent must be competent to be valid. The consent of a 2-year old is invalid. The person must understand the consequences of the consent. In some areas of research this element causes genuine problems, e.g., pediatric, obstetric, psychiatric research. In these cases substituted consent can be used. Substituted consent is consent given by a family member or guardian on behalf of the incompetent human subject. Substituted consent is an inferior solution, different from true informed consent for obvious reasons. The substituted consent attempts to secure individual autonomy through a paternalistic procedure in which a guardian is appointed. The guardian can estimate risk independently and can promote benefit, but cannot by the very nature of the relationship promote autonomy which is the core of informed consent.

As can be seen from the preceding paragraphs the informed consent process requires the mastering of several steps to function properly. It does however offer the investigator the benefits not only of ethical behavior, access to journals which will publish the results, trouble-free interaction with the ethical committee, etc., but also more tangible benefits for the trial at hand. The informed consent enables you to motivate the human subjects and obliges them additionally to participate in the study, because they fully understand what they have agreed to.

Procedure — The ethical committee or the institutional review board structures the procedures required by the Helsinki Declaration II. The actual organization of these committees may differ from country to country, and there is no evidence for linking one particular type of organization or procedure with performance. In addition to any specific national rules and regulations which may exist in this area, it is mainly a question of understanding the role of the ethical committee for the individual investigator. The role of the ethical committee is to evaluate a planned study in terms of its ethical aspects, and thereby promote the interests of society and protect the autonomy of individuals in general. The committees must therefore contain experts as well as nonexpert members in a balanced way, and protocols submitted to the committee should always have a resume which is written so that it can be understood by the lay members of the committee. Most ethical committees limit themselves to drawing up requirements for trials or studies to comply with the Helsinki declaration, but readers should be aware that some committees have gone on to monitor the ethical aspects of ongoing trials.

Ethical considerations are necessary in all clinical trials. The guiding principle is the promotion of self-determination and autonomy for the human subjects involved. This is done through the appropriate design of relevant studies, which takes into account the risks as well as the benefits for not only the individual but society as well.

Investigators should always check the national professional regulations and rules before submitting protocols. Wait for approval from the ethical committee before starting the study. It is mandatory that studies are conducted only following fully informed consent. The process of informed consent requires the investigator to communicate the risks and benefits of a proposed study to the human subjects, who then freely and without any coercion give their consent. This can only be achieved without conflict if sufficient attention is paid to the clarity and contents of the information given. Investigators should furthermore remember that the ethical process is not as much an obstacle to research as an additional effort to ensure the quality of science.

References

1. Silberman, W.M., *Human Experimentation. A Guided Step into the Unknown,* Oxford University Press, Oxford, 1975.
2. Brody, H., *Ethical Decisions in Medicine,* Little, Brown, Boston, 1981.
3. Katz, J., Ed., *Experimentation with Human Beings,* Russel Sage, New York, 1972.
4. Gowers, E., *The Complete Plain Words,* Penguin Books, London, 1977.

35.0

Clinical Aspects

35.1 Good Clinical Practice 671
      E.F. Hvidberg

Chapter 35.1
Good Clinical Practice

Eigill F. Hvidberg
Department of Clinical Pharmacology
University Hospital of Copenhagen
Copenhagen, Denmark

I. Introduction

This chapter intends to explain the general, regulatory aspects of clinical trials, in particular requirements according to Good Clinical Practice. Specific requirements related to methods in skin research are described in Chapter 33.1.

The concept of Good Clinical Practice (GCP) is — in a strict regulatory sense — a set of requirements for the performance of clinical trials on drugs* if such trials are initiated to generate data for the inclusion in submissions to drug authorities in order to obtain a marketing license (i.e., registration). Essentially, however, GCP is a compilation of quality standards for clinical (drug) research based on common sense principles that any scientist should follow. The principles laid down in the GCP apply, therefore, to all biomedical research in humans, and GCP will consequently serve as a guidance for any kind of scientific experimentation in man.

II. What Good Clinical Practice Means

The term GCP was coined in the U.S., although never officially used. The term does not capture its basic idea but has now been established almost as an institution in line with Good Laboratory Practice (GLP) and Good Manufacturing Practice (GMP). It includes, of course, only requirements for clinical experimentation (of drugs) and has nothing to do with the doctor's handling of his patients in general although the coherence is obvious. The term should rather have been "Good Clinical Research Practice" (U.K.)[1] or "Good Clinical Practice" (Nordic),[2] and in the European Community (EC) it is denoted "Good Clinical Practice for Trials on Medicinal Products"[13] and similarly for WHO,[4] but the term GCP is now almost universally used and cannot easily be changed. However, confusion may still arise among clinicians who are not used to drug regulatory terminology, and explanations should be readily available.[5,6] Conceptually, GCP contains several elements (Figure 1), but as an entity it can be defined in various ways. The most simple definition is presented in the EC-GCP document[3] "A standard by which clinical trials are designed, implemented, and reported so that there is public assurance that the data are credible, and the rights, integrity, and confidentiality of subjects are protected." Another — though similar — explanation is offered by WHO:[4] "Good Clinical Practice is a standard for clinical studies which encompasses the design, conduct, monitoring, termination, audit, analyses, reporting and documentation of the studies and which ensures that the studies are scientifically and ethically sound and the clinical properties of the diagnostic/therapeutic/prophylactic product under investigation are properly documented." Although officially confined to drug regulation these definitions clearly indicate why the GCP concept is introduced also in this book.

The history of GCP is not very long, starting in U.S. 2 to 3 decades ago,[7] partly as a consequence of revealed fraud or scientific misconduct. During the 1980s, several national or regional GCPs were issued (e.g., France, Nordic countries, Italy, Germany), but also the pharmaceutical industry and its organizations were quite active.[1] In the EC, CPMP (the EC drug registration committee) worked out the previously mentioned GCP Note for Guidance[3] to be used in the 12 member states. It has been in force since July 1991. At present, a GCP guideline is being drafted by WHO[4] and GCP is also a target for ICH (International Conference of Harmonisation) activities (see later). All these initiatives have had a great impact on clinical drug research and indirectly they are bound to influence all clinical investigations, *in vivo* research of the skin being no exception.

III. The Principles of Good Clinical Practice

There are few differences between the various GCP documents as regards the stated principles and standards.

---

* "Drug" is in the context of this chapter used equivalent with "medicinal product" (EC), "pharmaceutical product" (WHO), and "therapeutic drug".

FIGURE 1 Good Clinical Practice — basic elements.

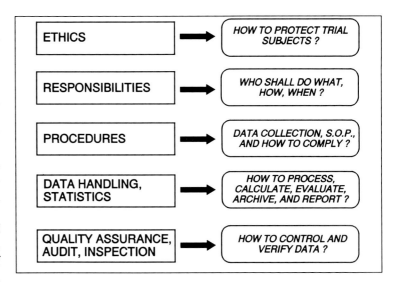

Practical and regulatory requirements may be at variance and these can be of great importance in the context of drug regulation related to clinical trials. In the following, such details will partly be left out. Several paragraphs are quoted from the EC document or the WHO draft guideline (although not always the entire text or by word-for-word quotations). In essence, the principles of GCP are presented in a condensed form:

(1) *Prerequisites for conducting a clinical trial* include a justification for the trial, thorough benefit/risk considerations, and knowledge of basic ethical principles, as they are expressed in the Declaration of Helsinki and the International Ethical Guidelines for Biomedical Research Involving Human Subjects from CIOMS (Council for the International Organization of Medical Sciences).[8] Investigators must have appropriate expertise, qualifications, and competence to undertake the proposed study. Preclinical studies to provide sufficient evidence of the potential safety of the drug(s) and methods involved are mandatory. Finally, it is necessary to be aware of and follow existing national regulations for clinical trials.

(2) *A protocol* must be drawn up, agreed upon, and only changed if written agreement is obtained from all parties. Such changes should be appended as amendments. The protocol must be strictly followed and it must be scientifically and ethically appraised by suitably constituted review bodies (see below), independent of investigator and sponsor. Most guidelines contain a model list of items to be contained in a protocol and it is strongly advised to consult such lists prior to the construction of the protocol.

(3) *The protection of the trial subjects* includes that the personal integrity and welfare of the subjects (as defined in the Declaration of Helsinki) is the ultimate responsibility of the investigator, who must also take into consideration the scientific validity of the trial for which all involved are responsible. An *ethics committee* (or similar review board) must be consulted and, for example, preview the protocol. Such a committee should have documented policies, procedures, and rules for its composition as a basis for its work. The authority under which the committee is established should be known and most guidelines further instructions for the ethics committees. Subjects must not be entered into the trial until the relevant ethics committee(s) has issued its favorable opinion on the procedures and documentation in writing. The ethics committee must be informed of all subsequent protocol amendments and of any serious adverse events occurring during the trial likely to affect the safety of the subjects or the conduct of the trial. The ethics committee should be asked for its opinion if a reevaluation of the ethical aspects of the trial appears to be called for.

Information on the details of the trial should be given in both oral and written form to the participants of the trial whenever possible. No subject should be forced to participate in the trial. *Informed consent* from the trial subject is, therefore, obligatory. Subjects, their relatives, guardians or, if necessary, legal representatives must be given ample opportunity to inquire about details of the trial. The information must make clear that the trial is a research procedure, participation is voluntary, and that refusal to participate or withdraw from the trial at any stage is without prejudice to the subject's care and welfare. Subjects must be allowed sufficient time to decide whether or not they wish to participate. The subject must be made aware and consent that personal information may be scrutinized during audit/inspection by competent authorities and properly authorized persons/sponsor, and that participation and personal information in the trial will be treated as confidential and will not be publicly available. The subject must have access to information about insurances and other procedures for compensation and treatment should he/she be injured/disabled by participating in the trial. The subject should know the circumstances under which the investigator or the sponsor might terminate the subject's participation in the study. Consent must be documented either by the subject's dated signature or by the signature of an independent witness who records the subject's consent. In either case, the signature confirms that the consent is based on information which has been given, and that the subject has freely chosen to participate without prejudice to legal and ethical rights while reserving the right to withdraw at his/her own initiative from the study at any time, without having to give any reason. In a nontherapeutic study, i.e., when there is no direct clinical benefit to the subject, consent must always be given by the subject and his/her signature obtained.

(4) *The investigator* is responsible for adequate and safe medical care of those subjects who participate for the duration of the trial and the investigator must ensure that appropriate medical care is maintained after the trial for a period that is dependent upon the nature of the disease and the trial and the interventions made.

Details are available from the different guidelines for the requirements concerning the qualifications of the investigator, his/her selection of trial subjects, the compliance with the protocol, his/her responsibilities for information to and obtaining consent from trial subject, and the trial site facilities and staff. The investigator undertakes to ensure that the observations and findings are recorded correctly and completely in the CRFs and signed by the appropriate person after delegation according to the protocol.

It is clear that the investigator takes on board a load of administrative responsibilities by carrying out a clinical trial and, to a certain degree, he/she must exercise a "bookkeeper" mentality. A summary of the investigator's general responsibilities according to GCP is given at the end of this chapter.

(5) *The sponsor* may be a pharmaceutical company, but may also be an investigator or an independent institution or organization that initiates, funds, organizes, and oversees the conduct of a trial. The sponsor is responsible for the overall adequacy and reliability of the data and information that are presented to the investigator before the start of the clinical trial or that become available during the trial, as well as responsible for the pharmaceutical product(s) involved. The sponsor, investigator, or both are responsible as stipulated in the national regulations for the necessary contacts with the drug regulatory authority and independent ethics committee, such as notification or submission of the trial protocol, reporting adverse events and submitting reports on the trial. In clinical trials in which the investigator is a sponsor, he/she is responsible for the corresponding functions, including monitoring. The sponsor should set up a system of quality assurance (see later), and establish written detail standard operating procedures (SOP) to comply with GCP.

There are several additional requirements for the sponsor in relation to the selection and information of investigator, compliance with national regulatory supply of the investigational drug, etc.

(6) *The monitor* is the principal communication link between the sponsor and the investigator. The monitor is appointed by the sponsor and must be appropriately trained and fully aware of all aspects of the drug under investigation and the requirements of the protocol, its annexes, and amendments. The monitor should have adequate medical, pharmaceutical, and/or scientific qualifications and should always be available at any time to the investigator for consultation or reporting of adverse events. The monitor should follow a predetermined written set of SOPs. The main responsibility of the monitor is to oversee progress of the trial and to ensure that this is conducted and reported in accordance with the protocol. Any unwarranted deviation from the protocol or any transgression of the principles embodies in GCP should be reported promptly by the monitor both to the sponsor and the interested ethics committee. The responsibilities and work of the monitor are detailed in the guideline documents.

(7) It is required that the *safety* must be carefully monitored; thus, any adverse event must be recorded. The trial protocol should clearly state method(s) by which adverse events will be monitored. It should also describe how this information is to be handled and analyzed by the investigator and sponsor and their responsibilities to report to each other and to the regulatory authority(s). The sponsor should provide adverse event reporting forms. National regulations may require the sponsor and/or the investigator to report certain types of adverse events/reactions (e.g., serious, previously unknown, etc.) to the regulatory authority and ethics committee. If required, all such reports should be accompanied by an assessment of causality and possible impact on the trial and on future use of the product. The investigator has to report adverse events to the sponsor immediately and to the regulatory authority and the ethics committee in accordance with national regulations. Normally, adverse events associated with the use of the product must be reported to the regulatory authority within specified time limits. Reports on adverse events submitted by the investigator to the drug regulatory authority should contain both subject and trial identification data (i.e., unique code number assigned to each subject in the trial). When reporting adverse events to the sponsor, the investigator should not include the names of individual subjects, personal identification numbers, or addresses. The unique code number assigned to the trial subject should be used in the report and the investigator should retain the code. The name of the investigator reporting the adverse events should be stated. After the trial has ended, all recorded adverse events should be listed, evaluated, and discussed in the final report. During the conduct of the trial, the sponsor has to report adverse events/reactions to the drug regulatory authority according to national regulations. The sponsor is responsible for reporting adverse events/reactions with the trial product to the local health authority as required by national regulations and to the other investigators involved in clinical trials of the same product. The sponsor should also report as soon as possible to the investigator as well as internationally and nationally to drug regulatory authorities any trial with the same product that has been stopped anywhere in the world due to

action taken by any regulatory authorities or any other withdrawals from the market for safety reasons.

(8) Requirements for *record keeping and handling of data* are stated to record, transfer, and where necessary convert efficiently and without error the information gathered on the trial subject into data which can be used in the report. All steps involved in data management should be documented in order to allow for a step-by-step retrospective assessment of data quality and study performance (audit paper trail concept). Documentation is facilitated by the use of check lists and forms giving details of action taken, dates, and the individuals responsible. A basic aspect of the integrity of data is the safeguarding of "blinding" with regard to treatment assignment. It starts with the randomization of patients into treatment groups. It is maintained through all steps of data processing up to the moment when the decision to break the code is formally taken. In the event of electronic data handling, confidentiality of the database must be secured by safety procedures such as passwords and written assurances from all staff involved. Provision must be made for the satisfactory maintenance and back-up procedures of the database. Details are pointed out for computer entry safeguards, CRF corrections, and record keeping and archiving.

(9) Concerning *statistics and calculations* the use of qualified biostatistical expertise is necessary before and throughout the entire trial procedure, commencing with the design of the protocol and CRFs and ending with completion of the final report and/or publication of results. Where and by whom the statistical work should be carried out is agreed upon between the sponsor and the investigator and recorded in the protocol. The scientific integrity of a clinical trial and the credibility of the data produced depend first on the design of the trial. In the case of comparative trials, the protocol should, therefore, describe the following: an *a priori* rationale for the targeted difference between treatments which the trial is being designed to detect, and the power to detect that difference, taking into account clinical and scientific information and professional judgment on the clinical significance of statistical differences, and measures taken to avoid bias, particularly methods of randomization, when relevant, and selection of patients.

In the case of randomization of subjects, the procedure must be documented. Where a sealed code for each individual treatment has been supplied in a blinded, randomized study, it should be kept both at the site of the investigation and with the sponsor. In the case of a blinded trial, the protocol must state the conditions under which the code is allowed to be broken and by whom. A system is also required to enable access to the information on treatment schedule of one trial subject at a time. If the code is broken, it must be justified and documented in the CRF. The type(s) of statistical analyses to be used must be specified in the protocol, and any other subsequent deviations from this plan should be described and justified in the final report of the trial. The planning of the analysis and its subsequent execution must be carried out or confirmed by an identified, appropriately qualified, and experienced statistician. The possibility and circumstances of interim analyses must also be specified in the protocol. The investigator and monitor must ensure that the data are of the highest quality possible at the point of collection and the statistician must ensure the integrity of the data during processing. The results of analyses should be presented in such a manner as to facilitate interpretation of their clinical importance, e.g. by estimates of the magnitude of the treatment effect/difference and confidence intervals, rather than sole reliance on significance testing. An account must be made of missing, unused, or spurious data excluded during statistical analyses. All such exclusions must be documented so that they can be reviewed if necessary.

(10) *Drug accountability* is a joint responsibility of investigator, sponsor, and monitor. It means to keep track of supplies and storage, dispensing, and return of investigational drug(s), any of these being adequately and properly documented in accordance with regulatory requirements and (where applicable) in the protocol. The investigator must not supply the investigational drug(s) to any person not targeted to receive it and assure that it is used only in accordance with the protocol. Correct labeling and packaging is the responsibility of the sponsor.

(11) The sponsor is responsible for the implementation of a system of *quality assurance* in order to ensure that all sites, data, and documents are available for verification. All observations and findings should be verifiable in order to ensure the credibility of data and to assure that the conclusions presented are derived correctly from the raw data. Verification processes must, therefore, be specified and justified. Statistically controlled sampling may be an acceptable method of data verification in a trial. Quality control must be applied to each stage of data handling to ensure that all data are reliable and have been processed correctly. The sponsor's audit should be conducted by persons/facilities independent of those carrying out the trial. Any or all of the recommendations, requests, or documents addressed in the guidelines or in national regulations may be subject to, and must be available for, an audit through the sponsor or a nominated independent organization and/or competent authorities (inspection). Sponsor and investigational sites, facilities, and laboratories and all data (including patient files) and documentation must be available for institutional and independent sponsor audit as well as for inspection by competent authorities.

(12) Because a *multicenter trial* is conducted simultaneously by several investigators at different sites following the same protocol, some special administrative arrangements are normally needed. Ideally, the trial should begin and end simultaneously at all sites. A number of aspects are rendered more complex and lists of these are provided in several guidelines. A multicenter trial, therefore, may require a special administrative system, the scale of which will depend on the number of trial sites involved, study end points, and present knowledge of the investigational drug.

(13) *The role of authorities* is to provide the legal framework for clinical trials. The aim should be twofold: to protect the safety and rights of the subjects participating in a trial and to allow only trials which may lead to conclusive data. Regulatory authorities should have a mandate to revise or terminate trials. The system must allow for on-site inspection of the quality of the data obtained, with due concern to subject confidentiality.

Finally, all guidelines include a *glossary* with explanations/definitions of the terms used in the particular document. It is utterly important to make oneself acquainted with the terminology, etc., in order to understand the text correctly. It should, however, be noted that various GCP guidelines to a certain extent make use of slightly different definitions, which may cause some confusion.

IV. GCP and "Normal Medical Practice"

It is important to realize that the standards set out in the (regulatory) guidelines of GCP must strike the right balance between optimal clinical research and careful clinical work.[5] Trial regulations must not interfere unduly with the patient/doctor relationship and not give the impression that considerations regarding research nor bureaucratic rules supersede the respect for the individual and his/her treatment and care. The latter should rather influence both planning and conducting trials. This is, however, easy to state in general terms but may be difficult to observe in practice. There are several conflicts hidden in this problem, which a few examples will show. Placebo-controlled trials may put the patient in a situation that is not in the best interest of the individual, which is basically required of the doctor to take care of after. However, if the drug (or procedure) under investigation is not known to be superior to placebo treatment (the reason for the trial), this principle will not be jeopardized. Another example is how much the increased general consideration given to patients' participating in a trial may bias the outcome. Finally, the special interest of the patient (or the investigational treatment) may influence the patient/doctor relationship.

It is the responsibility of the investigator thoroughly to have considered such problems before for the individual trial is commenced.

V. GCP and Techniques Applied

It should be recognized that, in dealing with techniques that aim to characterize different pathological conditions and/or disease stages, or changes therein due to various interventions, it is of utmost importance to establish a connection between the outcome, based on the method applied on one hand, and the disease (i.e., the diagnosis) or the changes in the clinical situation (i.e., therapeutic results) on the other hand. In general, techniques must, as for all other instruments, be validated also in this respect, provided this reasoning is applicable to the research project in question (this may not be the case for all physiological studies). Expert knowledge is, accordingly, a *sine qua non* to avoid that very accurate and sophisticated measurements become meaningless in a clinical context. This superior problem may be seen as parallel to the question of "surrogate endpoints"[9] or it is, indeed, a part of it. Related to this spectrum of problems is the question of standardizing the apparatus involved, calibration to surrounding conditions, and other important factors as recently discussed by Serup.[10]

How could these issues be derived from GCP? Neither the EC document nor other GCPs, such as the Nordic or the WHO guidelines, go into details about such requirements. However, it is clear that prerequisites for meaningful and relevant results, generated by any technique used in biomedical research, are that such techniques are standardized (in the broadest sense) and that the results should be unambiguous, relevant, and coherent with the clinical condition and changes of it. As far as clinical drug development is concerned, GCP rules will, more or less indirectly, cover such requirements, which consequently will impact any noninvasive technique used in skin research.

VI. Legal Aspects and Non-Drug Biomedical Research

The legal status of GCP requirements varies among countries. This is not only due to differences in legal systems but also because GCP guidelines often are recommendations and, therefore, not legally binding. The U.S. has the most elaborate system, fully legislated, regulated, and controlled.[7] In the EC, some of the GCP requirements have been incorporated in a Directive (92/507/EEC) thereby increasing their legal status.[11] However, only if incorporated in the national legislation can the GCP requirements be enforced. This is now being done in most of the EC member states but the process takes time and implementation is difficult, also for financial and other resource-related reasons. Potential trialists are strongly advised to obtain information from the national authorities, as noncompliance with local regulations may cause serious troubles for the involved parties. As previously stated GCP recommendations are — strictly speaking — created to serve drug regulatory purposes. The requirements are basically made to secure that the data on new drugs are dependable so that a correct evaluation of safety and effectiveness can be done, and that the trial subjects are not harmed. However, the principles behind these aims are not different from those governing all experimental research in humans. GCP requirements are only a part of trial regulation (concerning drugs). In most countries, a notification (or an application) must be sent to the drug regulatory authorities before a trial can be initiated. Relevant documents (e.g., the protocol) must be included. Furthermore, regulation includes the rights of the authorities to intervene, the demands for reporting adverse reactions, submission of trial reports, and many other details. How these requirements apply to trials where an investigational drug is not involved varies, but the opinion of an ethics committee should, in principle, always be sought for projects within the area of biomedical research in humans.

In some countries (e.g., France, Denmark), legislation on biomedical research is in force and in the U.S. the rights of the human subjects are protected by similar laws. In many other countries, however, this responsibility has partly been taken over by medical professional bodies. In the EC, a directive on clinical trial regulation is, currently, being prepared by the Commission.

The requirements for trial audit and inspection of trial sites and data by the authorities were included in GCP in order to protect the society from decisions made on a false basis caused by scientific misconduct or fraud in clinical trials. A whole spectrum of irregularities is known to exist, reaching from inadvertent wrong registration of an observation to full-blown fraud by fabricating data or inventing patients. Misconduct (of any kind) is not only confined to investigators; sponsors may also be involved. The requirements for control and quality assurance should limit such problems, and the prophylactic value of these demands may be considered high.

Several cases of scientific or professional misconduct and of fraud have been revealed during the last few years and the problem has become more visible. Actions are now being taken in many countries to combat irregularities and fraudulent behavior in clinical trials, and the responsibilities of the parties directly involved in trials, as well as the authorities and the medical community at large, are discussed. In the U.K., for example, initiatives have been taken by the pharmaceutical industry in collaboration with the General Medical Council (GMC). Many aspects of fraud and malpractice are reviewed in a report from the Association of the British Pharmaceutical Industry (ABPI).[12]

GCP covers — as previously stated — officially only drug-related trials, and the requirements for quality assurance are limited to such trials. However, other biomedical research projects may just as well have problems with scientific and/or professional misconduct. The GCP demands for audit, etc. should accordingly be extended to include all trials.

VII. Insurance, Liability, Financing

In a clinical trial, the participating subjects may suffer injuries and satisfactory insurance should, therefore, be arranged to cover the trial subjects, as well as the sponsor and the investigator(s). The basic rule is that no trial should commence without the insurance problems having been settled. However, the legal framework and other conditions are complicated and vary among countries. A clear overview of the insurance and liability problems are, therefore, difficult to establish, and there seems to exist some disagreement on these points between those involved. In trials without a sponsoring organization, company, or institution, the investigator may not be aware of the liabilities involved, and he/she is urged to consider the problems and attend to relevant legislation and expertise. In sponsored trials, the sponsor has usually an insurance coverage of clinical trials.

All financial arrangements related to a clinical trial should be fully transparent and a budget made. Information should be available about the sources of economical support (funding from private/public foundations or funds, sponsoring companies, academic institutions, etc.). The distribution of expenditures must be clearly apparent, especially who is paid for what and how much. Objections to this are rarely advanced because financing of projects is considered confidential matters. However, the mere possibility of bias creating conditions should be enough to keep everything in the open.

VIII. Harmonization of GCP

The most interesting exercises in harmonization of GCPs have been done in Europe. In the EC, the professional, scientific, and ethical basis were agreed upon without major problems and the resulting "Note for Guidance" has been in force since 1991.[3] However, harmonizing the regulatory procedures was (and will become) more difficult, the main problem easily turning out to be how a unified implementation throughout the community can be established.

Another example is the Nordic Good Clinical Trial Practice (GCTP) worked out by a committee under the Nordic Council of Medicines in 1989.[2] The harmonization process was more or less parallel to the work in the EC and, although with a different setup, it is quite close to the EC guideline in principles and scope. It should be mentioned that the Australian GCP is based on the Nordic document. In the EFTA member states outside the Nordic countries, the GCP requirements do not differ essentially from the EC document. Some of the countries from the former East-Block countries have taken steps toward harmonization of GCP requirements and, for example Hungary, are quite far in this direction.

On a more global level, two major initiatives for harmonizing GCP are now appearing, one being the so-called "International Conference on Harmonisation" (ICH),[13,14] the other a WHO initiative. ICH is a trilateral cooperation between the drug authorities and pharmaceutical manufacturers' organizations in three regions, namely the EC, U.S., and Japan. ICH has depicted GCP as one of the many objects for global harmonization of drug registration requirements.[15] Although the existing GCPs from the three regions are different at various points there are also many similarities.

The GCP initiative taken by WHO is to issue a guideline that can be adopted worldwide.[4] The basic idea is not to harmonize between nations or regions as such, but to support that global standards are established for clinical drug testing. It is also meant to assist those countries that have a less-developed drug regulatory system, and is not intended to replace or compete with existing guidelines or regulations. It will probably be quite close to the existing major GCPs (EC, U.S., Nordic, Japanese). The previous given description of GCP requirements is partly based on the WHO draft. Finally, it should be mentioned that numerous national and international workshops, courses, and other meetings on GCP are held all over the world these years. Many problems are here discussed among the people that are responsible for the practical execution of clinical drug trials. Such interactions at a practical level result in a broader understanding of common difficulties paving the way for harmonization and the subsequent implementation.

IX. The Superior Objective

The superior purpose of the establishment of GCP and clinical trial regulation should be emphasized. The most simple — and possibly most naive — approach is that both the patient, and in the long run also the society, should benefit from such actions. Regulation is not an objective in itself, but meant as a support and a guidance to improve health care standards individually and collectively, also by the advancement of science, prompted by such investigations. These endpoints may sometimes be clouded in the process of inventing new and elaborated regulatory means to fulfill that goal, but trial regulation should not put too much constraint on the imaginative scientific researcher. The investigator must still have the key position in clinical research, but must take upon himself a wider spectrum of responsibilities.

X. Summary of Investigator's Responsibilities According to GCP

The investigator

- Must be aware that GCP guidelines are produced to facilitate correct trial data in every way.
- Must be well aware of the scientific and clinical aspects of the trial, but also know the regulatory aspects, including the responsibilities involved.
- Is morally bound to follow the principles laid down in the GCP guidelines, because they presuppose that the investigator is both professionally qualified and has the necessary time and facilities at his/her disposal.
- Must have a thorough understanding of the ethical principles for biomedical research in humans and know the procedures for obtaining informed consent from trial subjects.
- Must accept to participate in a team work (with sponsor, monitor, authorities, etc.), in mutual trust and confidence.
- Must tolerate supervision and audit/inspection.
- Must accept much paper work, book keeping, reporting, etc.
- Must prepare, agree on, and sign the protocol (in drug trials at least) making him/her legally responsible for his/her part of the trial, implying thorough familiarity with the investigational product, as well as with the methods and techniques of the trial.
- Commits himself to collect, record, and report all data properly and conduct the study in accordance with the protocol.
- Is responsible for the handling, archiving, and reporting the generated data, for the randomization and for the statistical handling of the project.
- Is responsible for the accountability of the investigational drug, for adverse event reporting, and for preparation of the final report.
- Has the full responsibility for the trial subjects. This extends to guarantee the confidentiality of all personal information, as well as for the medical care of the subjects during and after the trial.

References

1. *Guidelines on Good Clinical Research Practice,* ABPI, London, 1988.
2. Good Clinical Trial Practice. Nordic Guidelines, NLN Publication No. 28, Nordiska Läkemedelsnämden, Nordic Council on Medicines, Uppsala, 1989.
3. Good Clinical Practice for trials on medicinal products in the European Community, III/3976/88-EN Final, in *The Rules Governing Medicinal Products in the European Community,* Vol. 3 Commission of the European Communities, London, 1990.
4. WHO Guidelines for Good Clinical Practice (GCP) for Trials on Pharmaceutical Products, Division of Drug Management & Policies, World Health Organization, Geneva, Draft, September 1992.
5. Hvidberg, E.F., Good clinical practice: a way to better drugs, *Br. Med. J.,* 299, 580, 1989.
6. Gerlis, L., Good clinical practice in clinical research, *Lancet,* I, 1008, 1989.
7. Kessler, D.A., The regulation of investigational drugs, *N. Engl. J. Med.,* 320, 281, 1989.
8. International Ethical Guidelines for Biomedical Research Involving Human Subjects, CIOMS, Geneva, 1993.
9. Boissel, J.-P., Collet, J.-P., Moleur, P., and Haugh, M., Surrogate endpoints: a basis for a rational approach, *Eur. J. Clin. Pharmacol.,* 43, 235, 1992.
10. Serup, J., Bioengineering and the skin: from standard error to standard operating procedure, *Acta Derm. Venereol. (Stoclch),* Suppl. 185, 5, 1994.
11. Allen, M.E. and Vandenburg, M.J., Good clinical practice: rules, regulations and their impact on the investigator, *Br. J. Clin. Pharmacol.,* 33, 463, 1992.
12. Fraud and Malpractice in the Context of Clinical Research, A report prepared by a working party of the Medical Committee of the Association of the British Pharmaceutical Industry, London, 1992.
13. *Proc. 1st Int. Conf. on Harmonisation, Brussels 1991,* D'Arcy and Harron, Eds., The Queen's University of Belfast, 1992.
14. Hvidberg, E.F., Regulatory implications of Good Clinical Practice. Towards harmonisation, *Drugs,* 45 (2), 171, 1993.
15. Maurice, N.P., Case studies, in *Proc. 1st Int. Conf. on Harmonisation, Brussels 1991,* D'Arcy and Harron, Eds., The Queens University of Belfast, 1992.

Index

Index

A

A-mode scanning, 239–240, 269
   sensitivity of, 291
   in skin thickness measurement, 263, 289–292
   sources of error in, 291
Abscesses, 281–287
Acanthosis, 189
   epidermal echoes in, 242
   pattern of, 52
   reduction of, 276
   ultrasound examination of, 248
Accountability, 674
Accuracy, 33
   definition of, 18, 33–34
   verification bias and, 38–40
Acid-base balance, 197
Acne
   detection of, 514
   grading systems for, 568
   inflammatory lesions of, 248
   microorganisms of, 512–513
   pilosebaceous keratin in, 121
Acne Disability Index, 568–569
Acne estivalis, 29
Acne vulgaris
   bacteria in, 212
   biochemical changes in, 513
   clinical assessment schemes for, 568–569
   follicle in, 511
   prophylactic potential of agents for, 512
Acnegenicity test, 659
Acrivastine, 609
Acrosclerosis skin-phalanx distance, 290
Actinic aging, 63, 325
Actinic elastosis, 276
Actinic keratoses, 27
Adhesive-coated disks, 10–11, 154
Adhesive materials, 149–151
Adhesive plaster stripping, 433
Adhesive tape stripping, 150
   in allergic patch test, 593
   in assessing barrier function, 183
   effect on skin hydration, 168
   in sebum excretion rate measurement, 523–524
   technique of, 154
Advertising claims, 6–7
Aerobic corneforms, 207
Age
   skin hydration and, 168
   skin integument differences with, 23–24
   in skin mechanical properties, 332–333
   skin reactivity and, 576, 610
Aging
   actinic, 63, 325
   cosmetics for, 663
   evaluation of, 11–12
   image analysis of skin surface changes in, 91–95
   intrinsic, 323–325
   sebaceous follicles in, 511–512
   skin effects of, 332–333
   skin elasticity and, 351
   skin mechanical properties and, 339
   skin slackness and, 346, 347
   skin softness and, 351
   skin thickness changes with, 291
   UV-induced, 27
Agonists, 636–637
Air bearing table, 99
Air conditioning, 28–29
Air convections, 177
Air plethysmography, 439, 453–455
Air turbulence, 182, 183
Al test, 593
Alkali resistance test, 587
Alkylether sulfates, 589
Allergens
   bioavailability of, 596–597
   erythema due to, 392
   extracts of, 610–611
   mixes of, 597
   national and international standard series of, 594–595
   nonstandard, 595
   potency of, 610–611
Allergic disease, 614
Allergic patch test, 247, 263–264, 593–601
Allergic reaction
   A-mode scan in reading, 290
   clinical grading of, 575
   echographic quantification of, 263–264
   late phase, 607
   photosensitive, 627–628
   reliability of, 4
   type I, 220, 607–614
   ultrasound examination of, 390
Allergy
   general rules for testing, 609
   plant, 29
Alopecia areata, 550
   dermatoglyphics in, 63
   exclamation point hairs in, 52
Ambulatory venous pressure, 440, 444–447
Ammonium hydroxide, 588, 590
Amplitude display, 232
Anagen hair, 545, 549, 552
Analgesia, topical, 491
Analgesic, 467–468
ANAT3D, 234
Androgenetic alopecia, 550
Animal testing, 663
Anionic injury, 4
Anisoosmotic fluids, 643
Anisotropy, 341–344
Ankle blood pressure, 449, 450, 452
ANOVA blood flow model, 204
Anthralin, irritation response to, 392, 575–576

Anti-acne modalities, 512
Anti-aging advertisements, 6–7
Antifungal drugs, 497
Antigenic materials, 595
Antigens, 594–595
Antihistamines
    affecting skin test reaction, 609
    assessment of effect, 382
Anti-inflammatory drugs, 467–468
Antiwrinkle remedies, 657, 663
Apocrine secretions, 507
Apocrine sweat glands, 507
Appearance, deceiving, 4
A priori probability, 35
Aqueous vehicle, 594
Araldinte epoxy resin, 499
Argon lasers, 485
Argyria, 62
Arm-foot pressure differential, 439
Artefacts, skin surface replica, 93–95
Arterial inflow measurement, 453–455
Arterial insufficiency, 416–417
Arterial ischemia, 251–252
Arterial occlusion
    in diabetes mellitus, 369
    transcutaneous oxygen tension in, 189
Arterial occlusive disease, 189, 191–193
Arterial-venous shunts, newborn, 200
Arteriography, 455
Arteriolar vasomotion, 411
Arteriovenous anastomoses, 399, 449
Arthritis, 468
ASCII format, 423
Aspartate aminotransferase (AST) test, 33, 35–40
Atopic dermatitis
    bacteria in, 220
    clinical assessment schemes for, 571–573
    during drug trials, 572
    grading systems for, 572
    hydration measurements in, 168
    severity criteria of, 572–573
    transcutaneous oxygen tension in, 190
    vessel changes in, 65
Atopic Dermatitis Area and Severity Index, 573
Atraumatic local labeling, 430
Atropy, 291
Auspitz's sign, 52
Autofocus principle, 98
Autoregressive modeling, 415
Average roughness, 84–85
Axial resolution, 233
Axillary secretions, volatile, 507
Axillary skin, 507
Azelaic acid cream, 395

B

B-mode scanning, 239, 240, 245, 269
    for deep venous trunks, 439
    equipment in, 257–259
    in vivo structure analysis and, 257–267
    of normal skin, 240
    of skin lesion, 303
    tissue characterization in, 259–267
    venous leg ulcer, 252
B-scan systems, 269–270
Background threshold, 425

Back-scattered echo signals, 126
Backscattered light, 425
Bacteria, 29, 207–215
Ball rebound tests, 359
Ballistic measurements, 363–364
Ballistometer, 359–360, 484
Ballistometry, 347, 359–364
Balsam of Peru, 595, 596
Barrier function
    with aging, 23
    assessment of, 182–183
    evaluation in clinical conditions, 173–174
    regional efficiency variations in, 24
Basal-cell carcinoma
    echopoor, 271–272
    inversion-recovery-sequence of, 310
    nests of, 312–313
    NMR examination of, 306, 309
    relaxation times of, 309
    superficial, 309
    ultrasound examination of, 249–251
Basal cell tumor nests, 313
Base excess, 197
Basophil histamine release test, 614
Bayes' theorem, 34, 35, 38
Bayesian inference, 34
Beagley-Gibson Grading System, 64
Bean, William Bennet, 561–562
Beau's lines, 561
Bending behavior, hair fiber, 537
Benefits, risk and, 668
Benign skin conditions, 63–65
Benoxaprofen, 624
Benzalkonium chloride, 587, 588
Benzoyl peroxide, 513
Beta-adrenergic agents, 610
Betacarotine, 375
Betamethasone 17-valearate ointments, 569–570
Biased prevalence, 34
Bifocals, 50
Bilirubin, 375
Bioavailability, 201
Bioengineering, 3–7, 9–14, 17–21
Bioengineering study, 17–19
Biologic elasticity (BE), 11–12
Biological zero, 406, 426
Biomechanical behavior, 341–342
Biomechanics, 25, 325–326
Biomedical research, non-drug, 675–676
Biometrological practice, good, 10
Biopsy
    of blister base, 636
    follicular, 511–514
    invasive punch techniques of, 121
    with low-power skin surface examination, 54
    skin surface, 63, 73, 121–123
    thickness of, 272–273
Biopsy needle, ultrasound-guided, 288
Birfurcation points, vasomotion at, 411
Bisphosphonate therapy, 468
Black light, 627
Blistering time, 590
Blisters, 491, 633–639
Blood-brain barrier, 643
Blood clearance, 479
Blood flow, see also Cutaneous blood flow
    autoregulation of, 192

calculating rate of, 432–433
dynamic, 367–369
measurement of, 449
oscillatory, 413–414
temporal variations in, 411–412
Blood flux, 399–403, 413
Blood pressure, systolic, 449–450
Blood studies, 636
Body hair growth, 543
Body mass index, 291
Body region, skin integument differences with, 23–24
Body surface area (BSA), 567
Body temperature, 458–459
Body weight, 245
Bond Fast, 635
Breaking strength, hair, 538, 539–540
Breast, 282, 283
Breathing motion, 301
Breslow thickness, 241
Brightness display, 232–233
Brinnel hardness, 359
Broken-off hair roots, 552
Bromophenol blue staining, 498
Brown-black globules, asymmetric, 70
Buffer capacity, 224
Buffer solutions, 596
Bürger's model, 322
Burte-Halsey model, 536

C

C-fibers, 491
C-mode scanning, 239, 269
in skin thickness measurement, 291
of skin tumor, 274
Cadmium Telluride (Chloride) detector, 429
Cafe au lait spots, 62–63
Calcipotriol, 569–570
Calcitonin, 468
Calcium inophores, 607
Calibration, 10
Calibration methods
errors in, 18
in optical profilometry, 98–100
Calipers
for nail growth measurement, 562
for skin thickness measurement, 293–296
Callus, powdered human stratum corneum from, 201–202
Cancer, skin, sun-induced, 27
Cannula insertion, 644–645
Capacitance, 165–170
Capacitance instrument, 165–167, 169
Capacitor, 166
CapiFlow, 367
Capillaries
in atopic skin, 65
blood flow velocity in, 412
in disease states, 65–67
leakage of, 607
partial carbon dioxide pressure of, 199
ringlike arrangement of, 65
torturous loops of, 368
Capillaroscopy, 367–369
Capillary blood-cell velocity (CBV), 367–369
Capillary loops, 399
Capillary microscopy, 57
Capillary network, 399

Carbon dioxide
elimination of, 185
exchange of, 197
lasers, 484–485
partial pressure of, 197–200
Carbon replication, 74–75
Catagen hair, 545, 549, 552
CBF, see Cutaneous blood flow
CDSS, 94
Cellulitis, 285
Central nervous system, dialysis impact on, 643
Cerebral blood flow measurement, 192–193
Cetirizine, 609
Chamber scarification test, 590
Chamber technique, 143
Chapping, 28–29
Charged-coupled device (CCD), 235
Chemical, patch testing, 595–596
Chemical allergens, 595
Chemical pain stimulation, 491
Chemical relaxation studies, hair, 537, 538
Chemical shift imaging sequence, 137
Chemotactic factors, 607
Chi-square test, 45
Children, skin irritation in, 24
Chromameter, 387–395
Chromametry, 654, 661
Chromates, 29
Chromatographic techniques, 518
Chromium salts, 596
Chromophore content calculation, 374–375
Chromophores, 377, 378
CIE system, 661
Cigarette paper method, 517
Cimetidine, 609
Cinematography, time-lapse, 63
Circadian rhythm, skin reactivity and, 610
Clark-type electrode, 185–186
Classification algorithms, 232
Cleanliness-healing relationship, 581–582
Cleansers, 4
Clemastine, 609
Climate, 27
Clinical assessment, 567–573
Clinical definitions, 17
Clinical documentation, 235
Clinical practice, good, 671–677
Clinical reference point, 17
Clinical score system, 661
Clinical trial, 672–675, 677, 683
Coagulase-negative staphylococcus, 207
Coefficient of variation, 45
Cohesograph, 147
Cold pressor test, 492
Cold stimulators, 484
Cold stress test, 470
Colinergic agonists, 503
Collagen bundles, 331
Collagen fibers, 335
Colloidal clearance, 475, 476, 478
Colophony, 595
Color, hair, 531–534
Color-flow imaging, 439
Color-flow scanning, 438
Color-order systems, 385, 387
Color perfusion image, 409
Color space values, 386

Color systems, CIE, 385–388, 390
Colorimeters, 377, 387–390, 661
Colorimetric assessment
   in pH measurement, 224
   of Sebutape, 12
Colorimetry
   accuracy of, 5
   applications of, 390–395
   CIE, 385–395
   in skin reaction grading, 575
   tristimulus measurements in, 392
Comedogenic evaluation, 514
Comedogenicity, 514
Comedolytic agents, 512
Comedolytic evaluation, 514
Comedolytic potential, 512
Comedone extraction, 213
Comedone extractor, 212
Comet-tail artifact, 280–281
Commercial connection, 6–7
Commercial law, 667
Competence, informed consent, 669
Complement fragments, 607
Compounds 48/80, 607
Compression period, 360
Computed tomographic scanning, 302
Computed tomography, X-ray radiation with, 305
Computer
   in chromatometry, 394
   in magnetic resonance techniques, 301
   in spectrophotometry, 374
   in stylus method, 84
   in ultrasound, 258
Computer imaging, 235–236, 301
Conditional probabilities, 34
Conductance, 161–165
Conduction, 460–464
Conductometry, high-frequency, 163
Confocal microscopy, 7, 138
Connective tissue bands, 127
Connective tissue bundles, 272
Connective tissue disease
   lymph flow in, 478–479
   mixed, capillaroscopic study in, 369
   skin mechanical properties and, 333
   torsional measurement in, 327
   ultrasonography of, 248–249
Contact allergens, 594
Contact allergy, 29
Contact dermatitis
   A-mode scanning in, 290
   allergic, 29, 629–630
   photoallergic, 627–628
   study design for, 18–19
Contact plates, 207–209, 213, 219
Contact sensitization potential, 659
Contactants, skin, 29
Continuous data, t-test for, 44
Continuous respiratory monitoring, 199
Continuous-wave Doppler venous scanning, 437–439
Controlled usage test, 659
Convection, 460
Copper constantan, 460, 461
Core body temperature, 458, 465–466
Corium, 271–272, 274–276, 312
Corneocytes, 143–147, 156
Corneometer, 13, 165, 166

Corneometer CM420, 163
Corpora aliena, 65
Corticosteroid atrophy, 249
Corticosteroids, 610
   bioequivalence testing of, 661
   blanching effect of, 392–393
   evaluation of topical activity, 264
   screening for anti-inflammatory potency of, 5
   skin atrophy with, 7
Cosmetics
   approval of, 658
   chemicals in, 203
   definition of, 656, 664
   delayed stinging reaction to, 590–591
   dermatology and, 663–664
   versus drugs, 656
   efficacy of, 664
   efficacy on stratum corneum compliance, 326
   formulations of, 80
   legislation on, 655–656
   regulatory procedures for, 657
   requirements for, 656–657
   safety of, 657
   skin pH and, 225
   substantial claimed effects of, 662–663
   testing human skin microcirculation stimulation by, 203
   testing of, 595, 658–660
   treatment in, 663–664
Cosmetics Directive, 656
Cost-benefit analysis, 40
Cost-effectiveness analysis, 40
CR-39 contact lenses, 625
Creams, pH and, 224
Creep, 335, 342
   deformation, 321–322, 324
   elevation and, 346–347
Critical value, 36
Cross relaxation, 306
Croton oil, 587, 588
Crow's feet, 63, 93
Crusting, 162
Cubital vein, 240
Cultures218–219
Cumulative irritation test, 21-day, 588, 590
Cutaneous blood flow, 399–400
   factors influencing, 400–401
   laser Doppler measurement of, 399–403, 405–409
   measurement of, 3
   oscillatory, 413–414
   periodic fluctuations in, 411–417
   postocclusive reactive hyperemia and, 21
   rate of, 401–403, 429–435
   temperature and, 424
   transcutaneous oxygen tension and, 188
   variables affecting, 20
Cutaneous lymphomas, 190
Cutaneous microcirculation, 367–369
Cutaneous microphotographs, Beagley-Gibson Grading System for, 64
Cutaneous morphologic signs, 51–52
Cutaneous pain quantification, 489–492
Cutaneous respiration185–193
Cutaneous scar, 282
Cutaneous tactile sensation, 483
Cutaneous thermal sensation, 484–485
Cutaneous vessel, disease-related changes, 65–67
Cuticle, 562

Cutometer, 11, 329, 330–331, 335–339
Cyanacrylate polymer, 511, 512
Cyanacrylate surface biopsy, 521
Cyanoacrylate glue, 121, 212, 213, 635
Cyanocrylate cements, 150
Cyclic blood flow oscillations, 411
Cycloid arch coefficient E, 94
Cystic fibrosis, 503

D

D-Squame discs, 149–151
D-Squame image analysis, 154–156
D-Squame tape, 146
Damaged skin, irritation tests on, 590
Dansyl chloride, 7, 144–145, 154
Data, 10, 18, 44, 45
Decision analysis, 40
Deep venous disease, 453, 454
Deep-venous thrombosis, 439, 448
Definitive reference test, 35
Deformation, 321, 354
   delayed viscoelastic, 323
   inhomogeneous, 343
   parameters of, 337, 338
   relaxation time of, 324–325
   time-dependent, 335
Delayed erythema dose (DED) tests, 628
Delta 10 dermatoscope, 59, 60
Demodex follicularum, 512
Deoxyhemoglobin, cutaneous, 375
Depigmenting agents, bleaching effect of, 395
Depot clearance, 476
Dermaflex, 329–333
Dermal echoes, 243
Dermal fluid, 351
Dermal loss factor, 356–357
Dermal Torque Meter, 327–328
Dermaphot, 59, 60
Dermascan C, 243, 244, 257–258, 269–270, 274–275
Dermaspectrometer, 380
Dermatitis
   atopic, 65
   contact, 18–19, 29, 290, 627–630
   skin reaction and, 577
   sub-clinical atopic, 5
   visual alterations in, 52
Dermatocosmetics,165–170, 339, see also Cosmetics
Dermatoglyphics, 63, 64, 122, 127
Dermatological examination, gross, 54
Dermatology
   A-mode scanning in, 290–291
   bioengineering in, 3–7
   cosmetic problems in, 663
   cosmetics and, 663–664
   digital imaging in, 229–236
   FDA and EEC regulations in, 653–664
   magnetic resonance techniques in, 299–300
   present status of replication techniques in, 80
   reflectance spectrophotometry in, 375
   ultrasound applications in, 247–252
Dermatophytes, 217
Dermatoscope, 57, 58–60
Dermatoscopy, 57
   archiving and follow-up in, 70–71
   of benign conditions, 63–65
   of cutaneous vessel changes in disease, 65–67

   infrared photography and, 60–62
   instrumentation in, 58–60
   of malignant conditions, 67–70
   object of, 57–58
   surface microscopy and, 63
   ultraviolet photography and, 62–63
   validation of, 70
Dermatosis
   conductance and, 162
   inflammatory, 53–54
   low-humidity occupational, 28
Dermis
   acoustic density of, 242
   characterization of, NMR imaging in, 134–135
   echogenicity of by site, sex, and age, 262
   NMR imaging of, 133
   organization of, 89
   papillary, 329
   reticular, 329
   ultrasound velocity in, 241
   water behavior in, 135–136
Dermo-hypodermis junction, 138
Dermometrology, 9–14
Desquamation, 143–147, 153–154
Desquamation index, 156
Detergent scrub technique, 210–213
Detergents
   irritancy of, 183, 589
   skin pH and, 224
   testing of, 595
Devices, 4
DHA, see Dihydroxyacetone
Dia-stron, 539, 540
Diabetes
   differentiation of, 190
   foot lesions in, 451
   screening for venous insufficiency in, 447
Diabetes mellitus, capillaroscopy for, 367, 369
Diagnostic tests, 33–41
Dialysate concentration, 645
Dialysis membrane, 643
Dialysis probe, 642–645
Dialysis sampling unit, 644
Diazo systems, 625
Diazonium compounds, 625
Dichotomous data, 45
Difference method, 646
Diffuse effluvium, 550
Digital image processing, 231
Digital imaging, 229–236
Digitized area measurements, 583
Digitizing pad, 582
Dihydroxyacetone (DHA), 146–147
Discoid lupus erythematosus
   papillary capillary destruction in, 66–67
   transcutaneous oxygen tension in, 190
Displacement calibration, 357
Distal interphalangeal (DIP) joint, 562
Distensibility, 330, 333
DMSO, 587, 588, 653
DNCB reaction, 601
Doppler color scanning, 438
Doppler effect, 422
Doppler phenomenon, 399
Doppler scanners, continuous-wave high-frequency, 279
Doppler shift, 233, 406
Doppler ultrasonography, 239

of skin lesions, 279
of skin tumors, 280
of subcutaneous tissue abnormalities, 280–287
in venous disease, 437–440
Dosimeters, personal UV, 624–625
Dosimetry, ultraviolet radiation, 619–625
Dot pitch, 230
Draize Test, 658–659
Drop test, 588
Drugs
  affecting skin test reaction, 609–610
  approval of, 658
  versus cosmetics, 656
  efficacy for, 656–657
  efficacy testing of, 661–662
  legislation on, 655–656
  phototesting for, 630–631
  risk and benefits of, 668
Dry skin
  D-Squame and image analysis of, 153–156
  in elderly, 23
  evaluation of, 10–11
  quantification of, 153–154
  signs of, 19–20
  treatment of, 497
Dry solid surface replication, carbon and metal/carbon, 74-75
Dryness, 354
Duhring chamber, 590
Duplex scanning, 437–440
Durapore surgical tape, 523–524
Dynamic spring rate (DSR), 356
Dysplastic hair roots, 552
Dysplastic nevus
  epiluminescence microscopic scoring protocol for, 70
  ultrasound examination of, 249
Dystrophic hair roots, 552

E

EC-GCP document, 671
Eccrine gland duct, 128, 129
Eccrine sweat collection, 503–505
Eccrine sweat glands, 497–500
Echographic characteristics, 262–263
Echographic image, 257–259
Echography, 244, 266
Echolucent areas, 127
Echolucent band (ELB), 276
Echopoor areas (ELA), 276
Echorich areas (ERA), 129
Ectatic vessels, 66–67
Ectodermal dysplasia, 63–64
Eczema, 64, 577
  cutaneous vessel changes in, 66
  gender differences in, 576
  photoallergic, 627–628
  staphylococcal bacteria in, 208
Edema
  echo disturbance with, 290
  formation of, 575
  lymph flow in, 478–479
  skin mechanical properties and, 333
  ultrasound examination of, 243, 390
Effective dose, 620
Efficacy
  consumer demand for, 664
  toxicity and, 668
Efficacy testing
  of drugs, 656–657, 661–662
  for skin products, 659–660
Ehlers-Danlos syndrome, 333, 339
Elastase, 351
Elastic deformation, 323
Elastic fibers, 331
Elastic force-displacement, 361
Elastic modulus, 325
Elasticity
  aging and, 332, 333, 351
  energy of, 360–361
  formula for, 330
  of hair, 537
  measurement of, 329, 331
  parameters of, 320–321
  ratio, 337
  reduction in, 11
  retinoic acid effects on, 326
  sun exposure and, 332
Elasticity meter, 335–339
Elastin, 331
Elastin fibers, 335
ELA, see Echopoor areas
ELB, see Echolucent band
Elderly, see also Age; Aging
  dry skin in, 23
  skin acoustic properties of, 261–262
  ultrasound structure of skin of, 243
Electric current, 353
Electrical conductance, 159–164
Electrical stimulation, 491
Electrical stimulators, 483
Electrical zero, 426
Electrochemical carbon dioxide electrode, 197–200
Electrodes, 166
Electrodynamometer, gas-bearing, 353–357
Electroencephalographic signal, 490–491
Electrolytes, 504–505
Electrometric procedures, 224–225
Electronic fluorimeter, 145–146
Elevation, 346–347
Elongation tests, 320
Elongation-to-initial sample length ratio, 320
Emissivity, skin, 464–465
EMLA cream analgesia, 491
Endocrine disease, 291
Endocrine factors, 332
Entry echo, 271
Environmental factors
  influencing skin, 27–29
  skin hydration and, 167
  in transepidermal water loss, 174–175
Eosinophil accumulation, 638
Epidermal cell population, 143
Epidermal-dermal junction, neoplastic disorders of, 51
Epidermal desquamation, 143, 434
Epidermal echoes, 242
Epidermal growth factor, 497
Epidermal pigmentary disorders, 62–63
Epidermis, 329
  alterations in, 52
  capacitance measurement of, 165–170
  error in thickness measurement of, 138
  fine structures of, 129–130
  function of, 153

high-resolution ultrasound of, 125–130
hydration profile reconstruction for, 136–138
in vivo NMR examination of, 133–139
NMR imaging in, 134–135
pathological, 134
substructures, visualization of, 128
ultrasound structure of, 242–243
ultrasound velocity in, 241
visualization of, 128–129
water behavior in, 135–136
Epiluminescence microscopy, 59, 67–70, 272
ERA, see Echorich areas
Erythema
with allergic patch testing, 599
determination of, 661
increased skin temperature with, 468
due to irritants and contact allergens, 392
measurement of, 377–383, 389
methods of measuring, 575
scale thickness and, 382–383
UV-induced, 389
dose-response curves of, 393–395
measurement of, 390–391
UVB and UVC, 381–382
variations of, 380
well-demarcated, 53
Erythrasma, 62
Erythrocyte number, 400
Erythromycin, 513
Ethanol
excreted in sweat, 497
washing, 598
Ethics, 667–669
Ethics committee, 672
Ethylene glycol monomethyl ether (EGME), 144
Eumelanin, 531
European Concensus Document on Chronic Critical Leg Ischaemia, 452
European Economic Community
Directive on Cosmetics, 663, 664
guidelines of, 660
legal regulations on skin products, 655–656
European Society of Contact Dermatitis, standardization group of, 20
Evaporimeter, 4, 18–19, 174–175, 180
accuracy, reproducibility and sensitivity of, 180–181
calibration and accuracy of, 177
effectiveness of, 183
environmental influences on, 175–176
external and environmental factors affecting, 181–182
measuring probe of, 179
sensors of, 19
variability of, 176–177
variables related to, 175–177
Evaporimeter EP1, 174
Evaporimetry, 3, 4
Evoked brain potentials, 490
Examination room temperature, 466
Exclamation point hairs, 52
Exercise
heat generation in, 459
testing for venous insufficiency, 443–445
Exfoliative cytology, 149
Exolift, 212
Experimental conditions, 10
Explanatory mistake, 18
Eye irritation test, 659

F

Facial skin
dryness of, 326
transcutaneous oxygen tension in, 187
False negatives, 34
False positives, 34
Fast Fourier transform, 414–415
Fat necrosis, 286
Femoral venous obstruction, 439
Fibrillin network abnormality, 327
Filter paper, water-saturated, 160
Filter paper technique, 503–504
Filtering, 85
Filters, digital imaging, 231–232
Financing, 676
Finn Chamber, 590
Finn Chamber technique, 598
Flaking, see Desquamation; Scaling
Flow-motion, 411–412
Flowmeter guidelines, 20
Fluorescence, 154
Fluorescence comparator, 145
Fluorescence comparator technique, 146
Fluorescence microlymphangiography, 475
Fluorescence microscopic techniques, 146
Fluorescence photographic photometric technique, 145
Fluorescent dyes, 144–145
Fluorescent ink pen, 598–599
Fluorescent tubes, 627
Flusher-blushers, 4–5
Flux, 406
Flux motion, 408–409, 413
Flux responses, 408
Follicle, 511
changes with hair growth cycle, 550–552
cycle dynamics of, 549–553
plugging of, 52
sampling methods for, 212–213
Follicular biopsy, 511–514
Follicular cast, 121–123
Fontana technique, 121
Food and Drug Administration (FDA)
Interim Guidance from, 661
regulatory procedures of, 657–658, 660
Food hypersensitivity testing, 614
Foot skin lesions, 450–452
Force
versus indentation, 362
vector, 354
Force-displacement, 353
Force relaxation, 323
Force relaxation time, 324–325
Forearm skin
B-mode scanning of, 245
echographic characteristics of, 262–263
high-resolution ultrasonography of, 126–127
regional differences in, 263
transcutaneous oxygen tension rate in, 188
Forehead, pH of, 223, 224
Foreign body
metallic, 280, 285, 314
small white opacification at site of, 52
Formaldehyde, 29
Fourier analysis, cutaneous furrows, 111–112
Fourier transform, two-dimensional, 107, 108, 110
Fourier transformation, 86
Fractional removal rate, 476

Fragrance stability, 596
Free fatty acids, 513
Free induction decay (FID), 300
Frequency-response analysis, 99–100
Frequential analysis, 111–112
Frosch-Kligman 5-day soap chamber test, 4
Fungal infections
    follicle in, 511
    mapping of, 217–220
    tropical climate-related, 29
    ultraviolet photography of, 62
Furrows, 91–95, 111–114, 116

G

Gadolinium, 301
Galileo thermoscope, 457
Gamma camera studies, 480–481
Gas-bearing dynanometer, 4
Gas-bearing electrodynamometer, 353–354
    application of, 354
    data reduction and, 357
    description of, 355–357
Gas exchange measurement, 193
Gaussian distribution functions, 37
Gaussian reference distribution, 38
Gel layer thickness, 244
Gender
    skin hydration and, 168
    skin integument differences with, 23
    skin mechanical properties and, 339
    skin reactivity and, 576
Generalizability, 41
Glucocorticosteroids, 610
Glucose tolerance test, 190
Glycosaminoglycans, 332
Gold sputtering, 80
Gold standard, 35
Good Clinical Practice (GCP), 19, 671–677
Good Laboratory Practice (GLP), 19
Good Manufacturing Practice (GMP), 19, 671
Gorlin's syndrome, 80
Grafts, monitoring survival of, 299–300
Gravimetric technique, 523–527
Gravimetric water measurement, 160–161
Ground substance, 351
Guard ring, 346

H

Hair, 127
    α-β transformation in yield region of, 535–536
    bending and torsional measurements of, 537
    breaking strength of, 539–540
    breaking values of, 538
    chemical relaxation methods for, 537
    color of, 531–534
    cross-section of, 532
    cyclic changes in, 549
    diameter of, 539, 545–546
    dysplastic or dystrophic, 545
    ethnic and genetic variations of, 532–534
    fiber shape of, 532
    greasiness of, 532
    gripping system for, 538, 539
    long, weathered fiber, 533
    mechanical strength of, 535–540
    microscopy of, 549–553
    reducing variability of, 538–539
    relative humidity and, 538
    shedding rate of, 546
    stress/strain curve for, 535
    tensile measurement of damage to, 536–537
    tensile testers for, 537–538
    thinning of, 549
    variability of, 538
    Young's modulus of elasticity of, 537
Hair growth
    body, 543
    follicular changes with, 550–552
    measurement of, 543–546
    phases of, 549
    presampling of, 543–544
    sampling criteria for, 544–545
    scalp, 543
Hair roots, 552, 553, 545
Hair shaft, 52, 546
Hand, Rule of, 567
Hand-forearm immersion test, 183
Hand wash products, 211
Hapten, 596
Hardness measurement, 349–352
Hardness measures, 359
Harvesting methods, 149–151
Healing
    measurement of, 118–119
    measuring in area of skin pathologies, 581–583
    theoretical and experimental evolution of, 117–118
Heat
    conduction in cutaneous blood flow rate measurement, 434
    stimulation of, 485
    stimulators of, 484
    transfer methods of, 460, 465–466
Heel, 135–138
Helium uptake, 434
Helsinki Declaration II, 668, 669
Hemangioma, 287
Hemocytometer, 143
Hemoglobin, 377–380
Henderson-Hasselbalch equation, 198
Herpes, 190
Herschel, John, 457, 458
Herschel, Sir William, 457, 458
High-frequency skin conductance, 160–161
Hippocratic Oath, 667
Histamine, 607
    cutaneous blood flow and, 402–403
    flare reaction, 382
    recovery of, 647
    release of, 647
    in skin chambers, 638
    skin distensibility and, 333
    wheals and, 266, 390
Histamine-induced hyperemia, 190–191
Histologic diagnosis, ultrasonic, 234
Histological sections, 270
Histology
    versus nuclear magnetic resonance, 138
    versus ultrasonography, 241
Histometry, 130
Histopathology, patch test reaction, 599–600
Hommel Tester, 83, 84
Hookean region, two-phase model of, 535
Horny cell sampling, 154
Horny layer

measuring water content of, 165
staining of, 144
water content of, 6
Horsehair stimulators, 483–484
Hough transform, 112–113, 114
HSL clearance, 476
H_2-antihistamines, 609
Human contact sensitization potential, 659
Human research requirements, 655
Humidity
  effect on skin, 28–29
  hair effects of, 538
  skin conductance and, 161
  transepidermal water loss and, 176, 177, 182
Hyaluronidase, 351
Hydration, see also Water
  depth of detection of, 167
  electrically measured, 655
  factors influencing, 167–168
  humidity and, 167
  measurements of, 166–170
  profile reconstruction, 136–138
  seasonal influences on, 167
  skin sites for testing, 167
  in ultrasound nail measurement, 559
Hydrochloric acid test, 266
Hydrocortisone, 203
Hydrophilic gels, 594
Hydroxine, 609
Hydroxypropyl cellulose, 594
Hygrometer, 160, 161
Hyperbilirubinemia, neonatal, 375
Hyperemia, 186
  insufficient, 187
  postocclusive reactive, 21
  reactive, 192
  testing of, 190–191
Hyperhydrosis, 500
Hyperkeratinization, 513
Hyperkeratosis, 189, 497
Hyperparakeratosis, 125, 127–129
Hyperpigmentation agents, 395
Hyperproliferative skin disease, 121
Hyperrreactive skin, 577
Hypersensitivity reaction, IgE-mediated, 607
Hypertension, chronic venous, 453–455
Hypertrophic scar, 190
Hypoechoic streaks, 281
Hypoplasia, appendage, 63–64
Hysteresis, 333, 335

I

ICC, see Intracorneal cohesion
ICH, see International Conference on Harmonisation
Ichthyosis, 66
Ichthyotic conditions, 511
IgE-Fc receptors, 607
Image analysis, 272
  automation in, 95
  automation of, 92
  correlation with stylus method, 87
  D-Square, 154–155
  of furrows and wrinkles, 89–95
  illumination in, 154–155
  indications for, 259–260
  lighting angle choice in, 92–93
  measurement parameters in, 90–92
  methodology of, 90–92
  objectiveness and reproducibility of, 155
  problems and artifacts of, 519–520
  procedures in, 155
  recommendations for, 155–156
  with Sebutapes, 518–520
  software for, 259
  ultrasound, 243
Image analyzer, 90
Image capturing procedure, 425
Image-processing methods, 257
Image segmentation, 260, 267
Image subtraction mode, 423
Imaging, 229–230
  advances in, 7
  nonvisible-light, 232–234
  visible-light, 234–236
Immunoglobulin A, 497
Immunoglobulin E-mediated hypersensitivity, 607
Immunological desensitization, 577
Immunosuppression, local UVB-induced, 28
Immunotherapy, 610
Impacting masses, 359, 360
Impedance, 159–166, 271
Impression/replica bacteria sampling methods, 207–209
Indentation, 360–362
Indentometry, 347, 349–352
Independent data, 44
Industrial activities, 29
Infections, tropical climate-related, 29
Inflammation
  diagnostic parameters of, 654
  increased skin temperature with, 467–468
  neurogen-dependent, 491
  ultrasound image analysis of, 243
Inflammatory mediators, 637
Inflammatory response, 263, 633–639
Inflammatory skin disease, 247–248, 260, 275–276
Information sheet, 668
Informed consent, 668–669, 672–673
Infrared lasers, 484–485
Infrared photography, 60–62
Infrared radiation, 464
Infrared thermography, 458
Infundibular follicular tissue, 122
Injection trauma, 479–480
Instron, 537–538
Instruments
  definitions and validation of, 18
  dermatoscopic, 58–60
Insurance, 676
Integument, 23–25, 78
Intermittent claudication, 409
International Conference on Harmonisation (ICH), 671, 677
International Society for Bioengineering of the Skin (ISBS), 3
Interstitial protein clearance, 476
Intracorneal cohesion (ICC), 147
Intradermal test, 612
Intraindividual variance, 45
Intravital dyes, 475
Invasive tumor mass, 273, 275
Inversion-recovery sequence, 307–308, 310, 314–315
Investigator responsibilities, 673, 677
In vitro diagnostic procedures, 614
Iodine starch method, 498
Iontophoresis, 503, 504
Iotasol, 475
Iotralan, 475

Irritancy test, 21-day accumulative, 658
Irritancy testing, 173
Irritant patch test, 587–591
Irritant patch test reactions
 A-mode scan in reading, 290
 determining intensity of, 259–260
 exogenous and endogenous factors determining, 587
 inflammatory component of, 264–266
Irritant reactions, 247
 clinical grading of, 575–576
 evaluation of, 264–266
 ultrasound examination of, 390
Irritants, 264–266
 erythema due to, 392
 thresholds of, 589
Irritation, skin
 age and, 24
 of dermatocosmetics, 168
 racial differences in, 25
 sex differences in, 23
 sodium-lauryl-sulfate, 264–265
 subclinical, 265
Ischemia, 408–409
Isotope clearance, 479–481

J

Joint infections, 468
Junctional nevus, 68

K

Kallikrein, 497
Kaposi sarcoma, 250, 252
Kelvin, Lord, 470
Keratin
 follicular, 511
 nail, 557
 of pilosebaceous duct, 121–123
Keratin fibers, 536
Keratinization, 153
Keratinocytes, 83
Keratinous impactions, 513–514
Keratosis, 273, 511
Kerosene reaction, 588
Koebner reaction, 601
^{85}Krypton washout, 434

L

Laboratory facilities, 18, 244
Laboratory room conditions, 21
Labscan 6000, 387
Lactate-containing formulations, 497
Lactic acid testing, 591
Lamé's coefficient, 321
Lange equipment, 389
Lange Micro Color instrument, 388
Langer lines, 329
Langerhans cells, 28
Langer's lines, 341–344
Lanolin, 594, 595
Larmour frequency, 300, 301
Laser beam, 426
Laser diodes, 107
Laser Doppler devices, 399–403
Laser Doppler flow (LDF), 381–382
Laser Doppler flowmeter probe, 443–444
Laser Doppler flowmetry, 3, 399, 405, 421
 application of, 193
 calibration of, 406
 in cutaneous blood flow measurement, 401–403
 in efficacy testing, 661
 erythema index and, 381–382
 guidelines for, 20
 laser Doppler perfusion imaging and, 427
 new technology and, 409
 problems in clinical practice, 406–408
 recommendations for, 428
 sensitivity of, 390
 in skin reaction grading, 575
 theory of, 405–406
 vasomotion detection by, 412–414
 in vasomotion monitoring, 408–409
Laser Doppler imaging, 421–428, 455
Laser Doppler technique, 190, 400–403
Laser Doppler velocimetry, 4, 5, 202–204
Laser focusing system, 108
Laser profilometry, 86–87, 97–104
Lasers, 402–403, 484–485
Lateral resolution, 233
Law
 commercial, 667
 Good Clinical Practice and, 675–676
 on skin products, 655–657
Leakage phenomenon, 414
Lectins, 607
Leeds acne grading scale, 568, 569
Leg
 assessment of dryness of, 155
 high-resolution ultrasonography of, 126–127
Leg ulcer
 contour evolution of, 118
 geometric parameters of, 117
 pH in, 301
 strain gauge plethysmography in, 451–452
 theoretical evolution of, 118
Leukocyte mobilization, 638
Leukotriene C_4, 607
Levarometry, 345–347
Liability, 676
Lichen ruber planus, 65, 66
Lichen sclerosis et atrophicus (LSEA), 64–65
Lichen sclerosus, 52
Lifestyle, 568
Light effects, 177
Light-emitting diodes (LEDs), 379
Light microscopy
 skin replication for, 73–80
 in skin surface biopsy, 121, 122
Light reactors, persistent, 628
Light reflection rheography, 443–448
Light sources, 627
Lighting angle, 92–93
Likelihood, 35
Likelihood ratio, 35
Limit of detection, 18
Limit of quantification, 18
Linear variable differential transformer (LVDT), 345, 350
Linearity, 18
Lipids
 acne and, 513
 dissolution of, 596
 extraction from sebum, 525
 follicular, 511
 racial differences in corneum content, 24

skin surface removal of, 162
stratum corneum water content and, 25
Lipoma, 286–287
Lipometer Sebumeter, 12
Lipometre, 518
Lipophilic yeasts, 217–220
Liposclerosis
  arteriography with, 455
  effects on microcirculation testing, 448
Liquid crystal, 462–464
Liver biopsy, 33
Load, 338
Load-extension curve, 343
Load-extension response, 341–342
Longitudinal relaxation, 306
Longitudinal relaxation time, see T1 relaxation time
Loratadine, 609
Lunula, 562
Lupus erythematosus, 66
Lymph clearance rates, 480
Lymph flow, 475–480
Lymph node
  benign fat-infiltrated, 280
  uptake of, 476
Lymph transport, 476, 478
Lymphangiography, 475
Lymphatic clearance, 476
Lymphatic system function, 475, 476
Lymphatic vessels, 475
Lymphedema, 252
Lymphography, indirect, 475
Lymphoma, 251
Lymphoscintigraphy, 475–476, 480–481

M

M-mode scanning, 239
Macroduct sweat collection system, 504
Macrophotography, high-resolution ultraviolet, 6
Macrorelief surface quantification, 114–118
Magnetic resonance coils, 306–307
Magnetic resonance imaging, 234, 299, see also Nuclear magnetic resonance
  versus nuclear magnetic resonance, 138
  for skin, 276
  for skin disease, 313–314
  in skin evaluation, 7
Magnetic resonance microscopy, 306–307, 313–314
Magnetic resonance principle, 306
Magnetic resonance spectroscopy, 299, 300–302, 306–307
Magnetic resonance surface coils, 299–300
Magnetic resonance techniques
  increasing importance of, 305–306
  recommendations for, 302–303
  types of, 299
  uses in dermatology, 299–300
Magnification, 49–54
Male pattern baldness, 544
Malignancy, 67–70
Marfan's syndrome, 327
Marginal probability, 33
Masked photopatch testing, 630
Mast cell degranulation, 607
Measurement guidelines, 19–21
Measuring standards, 18
Mechanical detector, 109
Mechanical properties, see also Biomechanics
  aging effects on, 339
  anatomical site differences in, 339
  biological and environmental variables in, 331
  Cutometer in measurement of, 335–339
  gender differences in, 339
  Langer's lines and, 341–342
  of normal skin, 332–333
  in pathological conditions, 333
  quantitative measurement of, 353–357
  skin aging and, 323–325
  standardized measurement of, 331–332
  suction chamber measurement of, 329–333
Mechanical stimulation, 491
Mechanical stimulators, 483–484
Mechanical testing, 320
Mechanoreceptors, low-threshold, 483
Medical applications, 326–327
Medical practice, normal, 675
Medicinal product, 656
Meissner's corpuscles, 483
Melanin granules, 61
Melanin pigment
  evaluation of, 375
  in hair, 531, 532, 534
  measurement of indices, 377–383
Melanocytic nevus, 69–70
Melanoma
  differentiating benign and malignant, 139
  epiluminescence microscopy appearance of, 70
  malignant
    amelanotic, 311, 313, 314
    color Doppler examination of, 284
    diagnosis of, 305
    metastasis from, 287
    NMR examination of, 306, 310–311
    relaxation time determination in, 314
    staging of, 275
    ultrasonography of, 241, 249–250, 274, 284
  MRI for, 234
Melanoma cell, 63
Melasma, 375
Menstrual cycle, 576
Merkel cell complex, 483
Metabolic acidosis, 197
Metabolic alkalosis, 197
Metabolism, 299
Metal/carbon replication, 74–75
Metal salts, 595–596
Metastasis, 284, 287
Methodology, committee on standardization of, 3
8-Methoxypsoralen, 624
Methyl nicotinate, 203
Methylene blue stain, 498
Methylnicotinate-induced vasodilatation, 25
Methylrosanilin, 598
Microangiopathy
  air plethysmographic evaluation of, 453–455
  high and low perfusion, 408
Microbiology, skin, 207–215
Microcirculation
  dynamic capillaroscopy of disorders, 367–369
  layers of, 407–408
  vasomotion in, 412
Microcomedones, 213, 512–514
Microdialysis, 641–647
Microrelief study, 118–119
Microrelief surface quantification, 109–114
Microscan System, 60, 63
Microscopy, see also Dermatoscopy

epiluminescence, 59, 67–70
in evaluating fluorescent skin samples, 146
of hair, 549–553
intravital, 412
light, 73–80, 121
magnetic resonance, 306–307
noninvasive, 57–71
scanning electron, 73–80, 121
skin replication for, 73–80
skin surface, 63
Microtopography, two-dimensional, 5
Microvasculature, cutaneous, 399
Microwave thermography, 458
Miliaria, 29
Miminal melanogenic dose (MMD), 394
Minimal erythema dose (MED), 391, 394, 620, 628, 631
Minimal perceptible erythema, 620
Minoxidil, topical, 204
Modena group, 247–248
Moisture balance, 654
Moisture content, 23
Moisturization, 24
Moisturizers, 159, 163, 168–169
Molding methods, 498–500
Monochromator, 621–622
Morphea, 272
  active plaques of, 327
  surface patterns of, 64–65
  transcutaneous oxygen tension in, 190
  ultrasonography of, 275–276
Morphological features, 49
Motion
  artifacts of, 138
    in laser Doppler imaging, 425–426
    in photoplethysmography, 444–445
  differential equation of, 362–363
Mottle pigmentation, 53–54
Mucopolysaccharide, 351
Multicenter trial, 675
Multipoint measuring system, 428
Multiwavelength systems, 407
Munsell color-order system, 385, 387
Muscle fascia, 243
Muscle-fiber bundles, 273
Mutagenicity, 660
Myelinated nerve fibers, 491

N

Nail
  growth, 561–563
  thickness of, 557–559
  ultrasound thickness measurement of, 252
Nail fold
  capillary microscopy of, 57
  proximal, 562
Nail plate, 291, 558
Nalidixic acid, 624
Natural compounds, 595
Necrotizing fasciitis, 192
Needle insertion pain stimulation, 491
Neomycin
  allergic reaction to, 575
  reaction time of, 599
Neonate
  arterial-venous shunts in, 200
  regional skin variations in, 24

transcutaneous carbon dioxide partial pressure in, 199
ultrasound structure of skin of, 243
Neoplasm
  pigmented, 53–54
  ultrasonography of, 249–251
Neurological dysfunction, 469
Neuropeptides, 607
Neutrophil accumulation, 638
Nevocellular nevus, 310–311
Nevus
  dysplastic, 70, 249
  echo phenomena of, 273
  ultrasound examination of, 249, 280
Nevus flammeus, 67
N(H) values, 134, 135–136
Nickel, 596
Nickel sulfate sensitivity, 263–265
Nines, Rule of, 567, 570
NMR, see Nuclear magnetic resonance
Nociceptors, 489–490
Nodular prurigo, 248
Noise level, 426
Noninvasive instruments, 5–6
Noninvasive methods
  development of, 305
  ethical considerations in, 667–669
  expanding scope of, 655
  functional application of, 653–654
  pitfalls and errors of, 18
  use of in skin testing, 661–663
Noninvasive oil immersion skin examination, 49–54
Noninvasive technology, 9–13
Noninvasiveness, 3
Nontouching thermal stimulation, 484
Nonvisible-light imaging, 232–234
Nordic Good Clinical Trial Practice (GCTP), 677
Nova DPM 9003, 165
Novatherm liquid crystal detector system, 463
Nuclear magnetic resonance
  correlation with other methods, 138
  in vivo high-resolution equipment, 133
  methodological principle of, 133–138
  object of, 133
  phenomenon of, 299
  received signal intensity in, 134
  recommendations for, 138–139
  in skin disorder examination, 305–315
  in skin examination, 299–303
  sources of error in, 138
Nuclear magnetic resonance-microscopy, 272
Nuclear spin, 300
Nutritive blood flow, 408, 417

O

Occlusive arterial disease, 450–452
Occlusive irritant patch test, 588–590
Odds ratios, 35
Oil-glass interface, 50–51
Olive oil, 594
Optical detector, 109
Optical focus profilometer, 108
Optical profilometer, 98–104
Optical sensor, 99
Oral thermistor thermometer, 462
Organic solvents, 591
Orthokeratosis, 129

Orthostatic change, 188
Osteoporosis, 291
Oxygen consumption, 188–191
Oxygen inhalation, 191
Oxygen tension, transcutaneous, 185–193
Oxyhemoglobins
    cutaneous, 375
    light absorption of, 378–379

P

P-value, 43–44
Pacemakers, 302
Pads, 208–209, 212
Paget's disease, 468
Pain, 489–492
Pain-mediating fibers, 491
Paired data, 44
Palms
    B-scan of, 129
    high-resolution ultrasonography of, 127
    transcutaneous oxygen tension in, 187
Papillary dermis, 128
Papillomatosis, 220
Parameter calculations, 101
Paraocular photography (POP), 63
Parapsoriasis en plaque, 64
Parositosis, 63
Partial volume effect, 138
Patch, 594
Patch tests
    allergic, 575, 593–601
    in efficacy testing, 661
    indications for, 266
    in vitro, 601
    irritant, 264–266, 587–591
    open, 601
    patient instruction about, 599
    reactions, 190–191
        A-mode scan in reading, 290
        reproducibility of, 576
        ultrasound examination of, 263–264, 390
    reading of, 4
    UV radiation and reactivity to, 28
Pedical flaps, 190
Pemphigoid, 190
Pendulum, 360–364
Percutaneous absorption, 25, 201
Percutaneous water loss, insensible, 160
Performance error, 18
Perfusate, 643, 645
Perfusion, temporal changes in, 425–426
Perfusion pumps, 643–644
Peripheral ischemia, 416–417
Peripheral tissues
    microdialysis in, 647
    perfusate in, 643
Peripheral vascular disease
    carbon dioxide partial pressure in, 200
    laser Doppler imaging in, 421
Perspex, 350
Petrolatum, 594, 596
pH, 197
    on cheek, 223
    intercellular, 301
    measurement in skin surface, 223–225
pH meter, 198, 223–224
Pharmaceutical preparation, 596

Phenanthrene derivatives, 594
Phenobarbital excretion, 497
Phenomenon measurement, 18
Phenothiazine, 624
p-Phenylenediamine, 599
Phenytoin excretion, 497
Pheomelanin, 531
Pheromones, 507
Phlebography, 440
Phosphomonoester concentration, 300
Phosphomonoester-to-phosphodiester ratio, 300–301
Photoacoustic detectors, 620
Photoallergen stability, 596
Photoallergic reactions, 627–628
Photochemotherapy, 390–391, 569–570
Photodamaged skin, 53–54
Photodermatoses, 628–629
Photodetectors with semiconductors (PSD), 108–109
Photoelectric reflectance meters, 377
Photography
    dermatoscopic, 57, 59
    versus digital imaging, 235, 236
    infrared, 60–62
    for measuring ulcer, 582
    in nail growth measurement, 562
    paraocular, 63
    resolution of, 235
    ultraviolet, 62–63
    value of, 5–6
Photometry, 518
Photon detectors, 620–621
Photopatch testing, 629–630
Photoplethysmography, 402, 405, 443–448
Photopulse plethysmography, 204
Photosensitive papers, 625
Photosensitivity, 624, 627–628, 659
Photosensitizing drug plastic films, 624
Phototesting, 627–631
Phototherapy, 375
Phototoxic reactions, 627–628
Phototrichogram, 544, 546, 553
Physical capacity, 291
Physiology, racial differences in, 24
Pierard's method, 346
Pigment network alterations, 54
Pigmentation
    fading, DHA-induced, 146–147
    hair, 531–534
    intensity of, 380
    irritation response and, 392
    spectrophotometric characterization of, 373–375
    susceptibility to irritants and, 25
    UV-induced, 393–395
Pigmented cutaneous lesions, 69–70
Pigments, 53–54
Pili annulati, 533
Pilocarpine, 503
Pilosebaceous follicle bacteria, 209, 212
Pilosebaceous infundibulum, 520–521
Pilosebaceous keratin, 121–123
Pilo-sebaceous units, 7
Pityriasis, 219–220
Pityriasis alba, 162
Pityrosporum folliculitis, 220
Pityrosporum ovale, 217–219
Pixels, 230, 259
    elimination of, 114, 115
    size of, 133, 138–139

Plant allergies, 29
Plastic film barriers, 635
Plastic impression technique, 75
Plastic surgery
    laser Doppler imaging in, 427
    transcutaneous oxygen tension in, 190
Platelet activating factor, 607
Plethysmography
    air, 439, 453–455
    strain gauge, 449–452
    venous occlusion, 406
Pocket Doppler instrument, 437
Pocket Doppler scanning, 437, 438
Poikiloderma, 66–67
Poisson's ratio, 320–321
Polycarbonate plastic, 625
Polysulfone film, 624
Polyvinyl chloride molding solutions, 499
Polyvinylpyrrolidone (PVP), 594
Popliteal vein patency, 439
Porokerastosis, superficial actinic, 80
Porphyria cutania tarda, 62
Porphyrin folliculitis, 514
Port-wine stains, 68
Positive predictive value, 35
Posterior probability, 35
Postthrombotic syndrome, 415–416
Posture
    Langer's lines and, 344
    physiological vasoreactions to change, 191–192
    transcutaneous oxygen tension and, 187
Post-yield slope, 536
Powdered human stratum corneum (PHSC), 201–202
Power, statistical, 43–45
Preatrophy, 53
Precision, 18
Prediction rule, 40, 41
Predictive irritancy testing, 173
Predictive value, 34, 35–36
Prednisolone
    skin reaction to, 577
    thermal imaging in evaluation of, 468
    wheal formation and, 610
Pressure-sensitive adhesive discs, 149, 150–151
Pressure-sensitive adhesives, 149
Prevalence, observed, 33
Primin reaction, 601
Prior odds, 35
Prior probability, 35
Prism grating technique, three-stage, 412
Probe-to-skin distance, 258
Profile filter, 85
Profilometric study equipment, 107
Profilometry, 6
    correlation with other methods, 94
    development of, 89
    disadvantages of, 107
    improved apparatus for, 107–109
    laser, 97–104
    mechanical, 94
    methodology of, 89–93
    objective of, 89
    recommendations for, 94–95
    sources of error in, 93–94
Pro-inflammatory compounds, 637–638
Prony spectral line estimation (PSLE), 414, 415
Prophylaxis, 4–5
Propionibacteria, 207, 212, 214

Propionibacterium acnes, 207, 212
Proportional full-thickness strain, 329
Propylene carbonate cumulative irritation test, 590
Prostaglandin E_2, 607
Prostaglandin E_2-induced hyperemia, 190–191
Proteases, 607
Protein
    clearance of, 476–481
    plasma, extravascular circulation of, 479
    serum levels and inflammatory mediators, 639
Proteoglycans, 335
Protocol, 672
Proton density, 134–137
Proton resonance imaging, see Magnetic resonance imaging
Pseudocolor scale, 259
Psoralen photochemotherapy (PUVA), 622–623
Psoriasis
    clinical assessment schemes for, 4, 569–571
    cutaneous vessel changes in, 65
    epidermal echoes in, 242
    of hand, 571
    hydration measurements in, 168
    phosphomonoester-to-phosphodiester ratio in, 300–301
    replication techniques in documenting, 80
    skin color in, 380
    skin lymph flow in, 476
    skin mechanical properties and, 333
    transcutaneous oxygen tension in, 190
    transepidermal water loss in, 174
    treatment of, 570
    ultrasound examination of, 248, 266–267
    whorls in, 63
Psoriasis area and severity index (PASI), 44, 569–571
Psoriasis Disability Index (PDI), 570–571
Psoriasis pustulosa, 65
Psoriasis vulgaris, 52
Psoriatic corneocytes, 122
Psoriatic plaques, 125, 128
    erythema and melanin indices in, 382–383
    high-resolution ultrasonography of, 127
    MR image of, 136
PTB parameter specimen, 99, 100
Pulsatile flux waves, 408–409
Pulse sequence, 301
Punch biopsy, 121, 241

Q

Quality assurance, 674
Quality control, 9–10
Quantimet, 90
Quantitative measurement, 235–236; see also specific techniques
Quantitative structure-activity relationships (QSAR), 204

R

Race
    hair color variations with, 532–534
    skin integument differences with, 24–25
Radiation, thermal, 460, 464–466
Radiative heat transfer, 465–466
Radioactive isotopes, 434
Radio-allergoabsorbent test (RAST), 614
Radiodermatitis, chronic, 66–67
Radiofrequency coil, 133
Radiography, 242
Radiolabeled tracers, 477, 479

Radiometer, 619, 622
Radiometric calculations, 619
Radiometry, 619–620, 622–624
Range, 18
Ranitidine, 609
Ratio measurements, 155
Raynaud's disease
    capillaroscopy for, 367–369
    skin flow pulsatility in, 405
Raynaud's phenomenon, 468, 469
Real time scanning, 239
Receiver operating characteristic (ROC) curves, 34, 36–41
Reference calibration method, 646–647
Refilling time, 445–447
Reflectance chromameters, 390–395
Reflectance spectrophotometers, 377
    configuration of, 385–386
    in melanin and erythema index measurement, 379–383
Reflectance spectrophotometry, 373–375
Reflection errors, 426
Reflux testing, 437–438
Regression methods, 40
Regulations, 653–664
Regulatory methods, 660
Regulatory requirements, 671–673
Relative elastic recovery (RER), 11
Relative humidity
    hair effects of, 538
    skin hydration and, 167, 170
    transepidermal water loss and, 174, 177, 182
Relative perfusion units, 407
Relaxation
    intensity of, 323
    process of, 306
Relaxation time, 299, 301
    determination of, 314–315
    force, 324–325
    longitudinal (T1), 307–308
        in basal-cell carcinoma, 309
        comparison of, 314
        for malignant melanoma, 311–313
        of skin layers, 137
    transversal (T2), 308
Reliability, diagnostic, 40, 41
Repeatability, 18
Repeated open application test (ROAT), 601
Replication, 73–80
Representative population, 33–34
Reproducibility, 18
Reservoir effect, 520–521
Resolution
    definition of, 270
    in magnetic resonance imaging, 301
Respiratory acidosis, 197
Respiratory alkalosis, 197
Restitution, coefficient of, 359, 360–361
Restitution period, 360
Retinoic acid
    screening assay for, 512
    topical, 326
Retinoids, 339
Retinopathy, diabetic, 369
Retrodialysis, 646
Reversal of conditioning, 34
Rheological model, 322
Risk, 668
Rosacea, inflammatory, 4–5
Roughness
    parameters of
        average, 84–85
        extension to three dimensions, 109–110
        in stylus method, 87–88
        standard, 99, 109
Routine diagnostic findings, 270–271
Ruffini receptor, 483
Ruggedness, 18

S

Safety
    for drugs and cosmetics, 657
    testing for
        animals in, 663
        for skin products, 658–659
        standardization of, 662
Salicylic acid screening assay, 512
Sample size calculation, 43–45
Sapheno-popliteal junction, 438
Sarcoma, soft-tissue, 287–288
Saturation-recovery sequence, 308, 311
Scabies, 52
Scale thickness, 151, 156
Scaling
    conductance and, 162
    D-Squame and image analysis of, 153–156
    pattern of, 154
    thickness of and erythema, 382–383
Scalp hair
    distribution patterns of, 553
    growth of, 543–546
Scaly skin, 168
Scanning electron microscopy
    of pilosebaceous keratin, 123
    skin replication for, 73–80
    in skin surface biopsy, 121–122
Scanpor, 593
Scarification index, 590
Schade's measuring system, 349
Scientific design, 668
Scientific experiment, 43
Scientific misconduct, 676
Scintillation detector system, 480–481
Scleroderma, 339
    A-mode skin thickness measurement in, 290
    progression and regression of, 276
    progressive systemic, 275–276
    skin mechanical properties and, 333
    skin thickness measurement in, 327
    systemic, 190
    ultrasound examination of, 248–249, 251, 275–276
Sclerosis, progressive systemic, 369
Sclerotic skin lesions, 190
Scraping methods, 150
Scrub cup technique, 513
Seasonal variation
    skin effects of, 27–29
    skin hydration and, 167
    in skin reaction, 576–577
    in transepidermal water loss, 177
Sebaceous cyst, breast, 282
Sebaceous follicle, 511–514
Sebaceous gland
    reservoir effect of, 520–521
    sebum production by, 523–527
Seborrheic dermatitis, 220
Seborrheic eczema, 64

Seborrheic keratosis, 273
Sebum
  excretion rate of
    gravimetric measurement of, 523–527
    patterns of, 521
  hair color and, 532
  high production of, 517
  image analysis of, 517–521
  interactive analysis of, 520
  lipids in, 513
  output at skin surface, 12–13
Sebum-trapping tape, 5
Sebutape, 12, 13, 517–521
Segmentation algorithms, 232
Selection bias, 34
Sellotape stripping, 121, 209, 212
Sensation, 483–485
Sensibility assessment, 483–485
Sensitivity
  definition of, 18
  limitations of, 35–36
  observed, 34
  tests for identifying, 590
  verification bias and, 38–40
Sensitization, active, 601
Sensory assessment, 483
Series zone model, 536
Serum phototesting, 631
ServoMed, 4, 174, 175
Shadowing method, 90–91, 94
Shampoo testing, 595
Shear mode, 354
Shear modulus, 321
Shear viscosity, 360–361
Shore scleroscope, 359
Sickle cell disease, 417
Sickness impact profile (SIP), 570–571
Signal to noise ratio, 17
Significance level, 43–44
Silaplus silicon dental correction material, 499
SILFLO resin, 89–90, 115–116
Silicone elastomer replication, 75–79
Silicone rubber plastics, 73
Silicone rubber replicas, 97–98, see also Skin surface replica
  for sweat gland evaluation, 498–500
  in wound quantification, 115
  wrinkle quantification with, 114–115
Silver nitrate, 144
Sine bar, 99
Skicon, 4, 165
Skicon-100, 166
Skin, see also Cutaneous; Dermis; Epidermis; Stratum corneum
  abnormalities of, 280
  acoustic density of, 241, 242
  aging
    mechanical behavior and, 323–325
    stiffness and, 333
  atopic, vessel changes in, 65
  atrophic, 7, 291
  bacteria sampling from, 207–215
  bacterial habitats of, 207
  bacterial sampling methods for, 207–215
  ballistometry of, 359–364
  barrier function of
    evaluation of, 173–174
    reactivity and, 576
  bioengineering of, 3–7
  biomechanics of, 25, 341–342
  blood flow measurement of, 202–204
  body temperature and, 458–459
  cancer of, 299; see also Skin tumors
  capacitance of, 10, 11
  color
    colorimetric assessment of, 385–395
    diurnal variations in, 383
    measurement of, 393
    melanin indices and, 377–383
    reactivity and, 576
    UV sensitivity and, 394–395
  conductance and capacitance of, 160
  conductance measurement of, 161–162
  dermatoglyphic pattern of, 51
  description of normal, 653
  digital imaging of, 229–236
  diurnal echogenicity variation in, 246–247
  dry, see also Xerosis
    D-Squame and image analysis of, 153–156
    evaluation of, 10–11
  echographic characteristics of volar forearm, 262–263
  electrical conductance and impedance in, 159–164
  electrode application to, 199
  emissivity of, 464–465
  entry echo of, 127, 128
  extensibility and Langer's lines, 342–343
  flow perfusion during sleep, 406–407
  functions of, 353
  gross examination of, 54
  humidity effect on, 28–29
  hydration state of, 159–164
  impedance of, 160, 165–166
  inside of, 97
  integument, 23–25
    sex differences in, 23
  in vivo study of architecture of, 125
  irritation potential testing of, 658–659
  laser Doppler imaging of, 421–428
  lymph flow in, 476
  mapping fungi of, 217–220
  measurement of excretion from, 89
  mechanical parameters of, 320–322
  mechanical properties of
    changes in, 11–12
    equipment and determination of, 329–331
    medical applications of, 326–327
    quantitative measurement of, 353–357, 331–333, 335–339
  microdialysis of, 641–647
  micropathy caused by chronic venous hypertension, 453-455
  microstructure of, 101
  microvasculature of, 399
  neoplasm, ultrasound examination of, 249–251
  noninvasive assessment of penetration, 201–205
  noninvasive oil immersion examination of, 49–54
  normal ultrasound anatomy of, 279–280
  normal variations in, 260–262
  nuclear magnetic resonance examination of, 299–303
  observation of changes in, 654
  occlusion of, 183
  outside of, 97
  pigments and color of, 373–375
  Pityrosporum ovale distribution in, 219–220
  profile in stylus method, 84–85
  radiative heat transfer from, 465–466
  recorded mechanical parameters of, 12
  reflectance variations in, 388–389

reflectivity of, 261–262
regional texture differences in, 58
replication, for light and scanning electron microscopy, 73–80
respiration of, 185–193
scraping of, 150
seasonal variation effects on, 27–29
sebum output as surface of, 12–13
senile cutaneous vessel changes in, 66–67
sensation stimulators, 483–484
sensibility of, 483–485
sensitive, tests for, 590
slackness of, 345–347
softness of, 349–352
temperature effect on, 28
temperature of, 457–470
texture of in benign conditions, 63
thermal patterns of, 466–467
thickness of
    A-mode measurement of, 289–292
    caliper measurement of, 293–296
    in histamine wheals, 390
    versus mean density of, 246
    in scleroderma, 327
torque relaxation, 320, 321
tropical climate-related infections of, 29
twistometry measurement of elasticity, 319–328
ultrasound examination of, 279–280
    high-frequency, 239–252
    normal image in, 271–272
ultrasound structure of, 242–243
ultrasound velocity in, 241
UV effects on, 27–28
UV sensitivity of, 389
variation in susceptibility of, 576–577
water sorption-desorption test of, 162, 163
Skin annulus deformation, 321, 322
Skin barrier cream tests, 591
Skin-blanching assay, 392–393
Skin blisters, 634–639
Skin Bond, 635
Skin chamber incubation, 635
Skin chamber techniques, 633–639
Skin diseases
    connective tissue, 248–249
    diagnostics in, 653–654
    high-frequency sonography of, 269, 272–276
    impact on lifestyle, 568
    measuring healing in, 581–583
    NMR examination of, 305–315
    severity of symptoms of, 568
    standard assessment schemes for, 567–573
    variables in, 567–568
Skin flaps
    carbon dioxide partial pressure in viability of, 200
    monitoring survival of, 299–300
Skin flux methods, 454–455
Skin layers, 134–138, see also Dermis; Epidermis; Stratum corneum
Skin lesions
    benign, 280
    edge of, 582
    extent of, 567
    false-color renditions of, 61–62
    malignant, 280
    measuring methods for, 582–583
    severity of, 568
    strain gauge plethysmography in feet and toes, 450–452

    transcutaneous oxygen tension and, 187
Skin lines, 63
Skin markers, 598–599
Skin microtopography, 89–95
Skin prick test, 612–613
Skin products, 655–660
Skin reactions, 575–577, see also Allergic reactions; Irritation, skin
Skin rheological model, 322
Skin structure, 97–104, 210, 329
Skin surface
    acidic nature of, 223
    biopsy of
        cyanoacrylate cements in, 150
        examples of, 122
        method of, 121–123
        standard fluorescent, 145
    hydration measurements of, 166–170
    hydration of, 165
    mechanical stimuli to, 319
    microscopy techniques for, 50
    normal replicas of, 76
    parameters of, 101–104
    pH measurement in, 223–225
    quantification of, 109–118, 272
    stratified, 101–104
    stylus contour measurement of, 83–88
    temperature and transepidermal water loss, 177
    textural analysis of, 111–112
    three-dimensional evaluation of, 107–119
    water absorption by, 164
    zones of, 126–127
Skin surface coils, 299–302
Skin Surface Hygrometer, 160, 161
Skin surface replica, see also Silicone rubber replicas
    artefacts of, 93–95
    casting of, 85–86
    durability of, 95
    image analysis of furrows and wrinkles, 89–95
    for laser profilometry, 97–98
    materials used for, 84
    profilometry applied to, 89–90
    reproducibility of, 94
    three-dimensional analysis methods for, 91
Skin testing, 607–614, 655, 657–663
Skin tumors
    A-mode scanning of, 290–291
    B-mode scanning for, 260
    3D reconstruction of, 274–275
    diagnostic techniques for, 315
    dimensions of, 125
    echo phenomena of, 273
    epidermis changes with, 313
    MRI evaluation of, 299
    NMR examination of, 306
    overestimation of depth of, 273
    prognostic classification of, 305
    thickness of, 272–273
    ultrasound examination of, 233, 272–275, 280
    weighted sonometric measurement of, 272
Skinfold
    inelastic, 331
    thickness measurement of
        versus A-mode scanning, 291
        gender differences in, 294–295
        sites of, 293–294
        typical values of, 295
    thickness of, 187

Skinfold caliper, 242, 293–296
Slackness, measurement of, 345–347
Soap chamber test, 589
Soaps
  mildness of, 4
  skin pH and, 224, 225
  testing of, 595
Sodium hydroxide testing, 591
Sodium lauryl sulfate
  irritant reaction to, 23, 264–265, 575–576, 587–588
  in skin barrier cream testing, 591
Softisan, 594
Softness measurement, 349–352
Software
  image analysis, 259
  for laser Doppler imaging, 421, 423
Solar radiation, 27–28
Solar simulator, 627
Soles, 127, 129
Solvent vehicle, 594
Sonic digitizer, 582–583
Spacing device, 244
Spatial distortions, 138
Specificity, 34–36, 38–40
Spectral methods, 414
Spectral sensitivity, 622
Spectrophotometer, 373–374
  configuration of, 385–386
  evaluation of, 389
  in melanin and erythema index measurement, 379–383
  in skin color assessment, 385
Spectrophotometric techniques
  for skin pigment and color, 373–375
  upgrading of, 7
Spectrophotometry, 4–5, 661
Spectroradiometer, 621–622
Spectroradiometry, 621–622
Spin echo image
  T1, 309
  three-dimensional, 315
  two-dimensional, 134
Spin-echo sequence, 307
Spin-grid relaxation, 306
Spin-spin interaction, 306
Spitz nevus, 69–70
Split image rangefinder, 563
Squame collection, 143
Squamometry, 11
Squamous cell carcinoma
  MR image of, 136
  ultrasound examination of, 249
Squamous inflammatory dermatitis, 63
Staining methods, 144–147, 498–500
Standard deviation, 45
Standard operating procedure, 19
Standardization, 9–10
  need for, 13–14
  in safety testing, 662
Staphylococcus, coagulase-negative, 210
Staphylococcus aureus, 208
Statistical analysis
  concepts and methods of, 33–35
  of sensitivity, specificity, and predictive value, 33–41
  of skin surface contour measurement, 85
  of surface quantification, 110–111
Statistical test, power of, 43–44, 45
Stefan-Boltzmann's constant, 464
Stereological parameters, 91–92

Stereophotogrammetry, 583
Sterile bag technique, 211
Sticky slides, 150
Stiffness, 354
Stimulators, 483–484
Stinging phenomenon, 590–591
Stinging test, 659
Stow-Severinghaus electrode, 198
Strain-versus-time curves, 337
Strategic error, 18
Stratum corneum, 329
  barrier function of, 182–183
  echopoor, 128
  function of, 149, 159
  harvesting techniques for, 149–151
  high-resolution ultrasound of, 125
  hydration of, 23, 153, 326
  hydration profile reconstruction for, 136–138
  loss of, 143, see also Desquamation
  mean histological thickness of, 130
  measuring turnover of, 153–154
  moisturization of, 24
  MR imaging of, 313
  powdered human, 201–202
  quantification of, 10–11
  racial differences in, 24
  renewal time of, 146
  skin biomechanics and, 325–326
  sonometric measurements of, 129–130
  stiffness of, 354
  thickness of, 24, 127
  ultrasound velocity in, 241
  visual alterations in, 52
  water content of
    lipids and, 25
    measurement, 159–164
Stratum corneum/stratum Malpighii interface, 129
Stratum disjunctum cells, 79
Stratum Malpighii, 128–129
Streptococcus, group A beta-hemolytic, 207
Stress-relaxation, 321
Stress-strain curves, 337–338, 356
  hair, 535
  of hair sections, 536
Stress-strain deformation coefficients, 356–357
Stress-strain modulus, 353, 354
Stress-strain properties, 335
Stress testing, 469
Stretchability, 354
Student's *t*-test, 44
Study design, monoinstrumental and multi-instrumental, 17–19
Stylus instruments, 100–101
Stylus method, 83–88
Stylus profilometer, 108
Stylus sensor, 107
Subcutaneous fat, 127
  echogenicity of, 281
  ultrasound structure of, 243, 252
  ultrasound velocity in, 241
Subcutaneous fatty tissue
  NRM of, 312
  ultrasonography of, 272
Subcutaneous hematoma, 280
Subcutaneous space, 243, 329
Subcutaneous tissue
  abnormalities of, 280–288
  abscess in, 286
  hypoechoic mass in, 287–288

trauma to, 280
ultrasound anatomy of, 279–280
^{133}Xe labeling of, 430–431
Subject-related error, 18
Subpapillary plexus
  tortuous, 66
  vascular, 53, 128
Substance P, 607
Suction chamber method, 329–333, 335–339, 343–344
Sudeck's algodystrophy, 470
Sudomotor processes, 503
Sulfadiazine excretion, 497
Sulfosuccinates, 589
Sun exposure
  erythema and pigmentation with, 393–395
  skin mechanical properties and, 332–333
  skin moisture and, 24
  viscoelastic properties and, 339
Sunburn, 27
Sunburn erythema, 628–629
Sunscreen
  delayed stinging reaction to, 590–591
  evaluation of, 375
Surface anisotropy, 110–111
Surface microscopy, 63
Surface texture measurement, 83–88
Surgery, 427, see also Plastic surgery
Surrogate endpoints, 675
Swabbing methods, 209–210, 213
Sweat
  abnormal production of, 497
  analysis of, 504
  apocrine, 507
  collection methods for, 503–505, 507
  droplets of, 64
  drugs excreted in, 497
  harvesting of, 503–504
  interpreting analysis of, 504–505
  production stimulation of, 503
  proteins and enzymes in, 497
Sweat glands
  activity in clinical conditions, 174
  localization techniques for, 497–500
Sweating
  function of, 28
  transepidermal water loss and, 177
  ^{133}Xe loss through, 433
Swept-gain curve, 258–259
Synchronicity, 413
Systolic blood pressure, 449–450

T

T1 relaxation time, 135, 306–309, 314–315
T2 relaxation time, 134–135, 306
  determination of, 308, 315
  for malignant melanoma, 311, 313
Tactile sensation, 483–485
Tactile stimulators, 483–484
Tape stripping, 161–162
Technical error, 18
Techniques, comparison of, 9–14
Technological error, 18
Telangiectases, 53
Telangiectasia, 66–67
Teledermatology, 235
Telogen hair, 552, 545–546, 549
Temperature
  ambient, 458–459
    effects on skin, 28
    in examination room, 466
  carbon dioxide partial pressure and, 198
  electrochemical electrode and, 199
  skin, 402, 467–469
  of TEWL measuring probe, 181–182
  thermal imaging of, 457–470
  transepidermal water loss and, 176, 177, 181
Tensile strength, hair, 536–537
Tensile testers, 537–538
Tension arches, 272
Tension tests, 343
Terfenadine, 609
Test area preconditioning, 12
Testing precondition errors, 18
Tetrachlorsalicylanilide (TCSA), 144
Tetralogy of Fallot, 367–368
Tewameter, 179–184
TEWL, see Transepidermal water loss
Textural analysis, 111–112
Theophylline, 610
Thermal clearance principle, 462
Thermal clearance probe, 463
Thermal detectors, 620
Thermal imaging
  applications of, 466–470
  contact, 458
  historical background of, 457–458
  non-contact, 457
  systems for measuring radiative heat transfer, 466
  techniques of, 459–464
  thermal radiation principles and, 464–466
Thermal injury, 249
Thermal pain stimulation, 492
Thermal radiation principles, 464–466
Thermal sensation, physiology of, 484–485
Thermal skin patterns, 466–467
Thermal stimulators, 484
Thermistors, 461
Thermocouples, 457, 460–461
Thermodes, 484
Thermogram, 457
Thermography
  infrared, 466–467, 469–470
  versus laser Doppler imaging, 427
  methods of, 402
Thermoluminescent materials, 625
Thermometer, 457
Thermometry, 457, 460–464
Thermopile, 620, 624
Thermoregulation
  capillaries in, 399
  skin in, 28, 458–459
Thermoregulatory flow, 408
Three-dimensional parameters, 102–104
Three-dimensional reconstructions, 274–275
Three-dimensional scanning, 240
Three-dimensional topographical measurement, 118–119
TIFF format, 423
Time imaging magnetic resonance systems, 306
Tissue
  damaged in microdialysis, 645
  differentiation by NMR, 314–315
  flux measurement in, 406
  hypoxia of, 417
  MRS localization of, 301–302
  perfusion, rhythmic variations in, 425–426

sound reflectance in, 233
ultrasound characterization of, 259–267
Toe blood pressure, 449–452
Toes, strain gauge plethysmography for, 450–452
Toluene testing, 591
Topical medicament allergy, 29
Torque, 320–323
Torsional behavior, dynamic, 323–324
Torsional equipment, 319–320
Torsional testing
  equipment for, 319–320
  of hair fiber, 537
  mechanical, 320–322
  skin aging and, 323–325
  stratum corneum mechanical behavior and, 325–326
  validity and interpretation of, 322–323
Touchless acoustic stylus, 85
Toxicity, 668
Traction forces, 331
Transcutaneous carbon dioxide partial pressure, 197–200
Transcutaneous oxygen tension, 186–193
Transducer
  resolution of, 128–129
  in stylus method, 84
Transepidermal water loss
  barrier function and, 173–174
  complexity of, 10–11
  in elderly, 23
  environmental variables in, 174–175, 177, 181–182
  external factors influencing, 181–182
  individual-related variables in, 177
  instrument-related variables in, 175–177
  instruments for measuring, 179–181
  in lesional skin, 162
  measurement of, 3, 4, 19, 173–177, 291
    applications of, 182–183
    comparison of methods, 179–184
    guidelines for, 20
    instruments for, 179–181
    measuring probe for, 179–180
  in neonates, 24
  in psoriasis, 174
  racial differences in, 25
  reactivity and, 576
  slow rate of, 165
  whealing response and, 588
Transferred thermoluminescence, 625
Transmission electron microscopy, 74–75
Transversal relaxation time, 308
Trapezoidal approximation, 37–38
Traumatic scars, 249
Traverse unit, 84
Triamcinolone acetonide, 7
Triangulation system, 108–109
Trichogram, 549–553
Trichostasis follicularis, 511
Tricyclic antidepressants, 610
Triglycerides, 513
Tropical climate, 29
True negatives, 34
True positives, 34
TRUE Test, 594, 596–598
Tuberculin reactions, positive, 190
Tumors, see also Skin tumors
  Breslow's thickness, 273
  contours of, 3
  lateral and axial expansion of, 305

relaxation times of, 299
T1 relaxation time of, 314
ultrasound classification of, 249–250
Turgor, 6
Turpentine, 595
Twisting strain, 319
Twistometer, 11, 327–328
Twistometry, 319–328
Two-dimensional ultrasound, see B-mode scanning

U

Ulcers
  changes at edges of, 189
  chronic, cutaneous vessel changes in, 67
  chronic diabetic, 451
  determining boundary of, 116
  effects on microcirculation testing, 448
  geometric parameters of, 117
  healing of, 581–582
  leg, 251–252, 301, 450, 451–452
  measurement methods for, 582–583
  replica scanning of, 116
  theoretical and contour evolution of, 118
  venous, 189–190, 581–582
Ulnar loops, 63
Ultrasonic digitizer, 582–583
Ultrasonography, 4
  A-mode, 232, 269, 289–292
  B-mode, 232–233, 269, 289
    in vivo structure analysis and, 257–267
    of skin lesion, 303
  biological variables in, 245–246
  C-mode, 233, 289
  comet-tail artifact in, 280–281
  cutaneous, 232–234
  dermatology applications of, 247–252
  Doppler, 239, 279–287, 437–440
  equipment, laboratory facility, and examiner in, 244–245
  in experimental animals, 252
  high-frequency, 239–252, 269–276, 557–559
  high-resolution
    of human epidermis, 125–130
    physical principles and techniques of, 240–241
  versus histology, 241
  image analysis of, 243
  introduction of, 239
  of irritant patch test, 264–266
  methodological principles of, 289–290
  50-MHz, 126–127
  100-MHz, 126, 127–128
  modes of, 239
  versus MRI, 302
  versus nuclear magnetic resonance, 138
  parameters of, 258–259
  positioning for, 244–245
  real-time, 280
  resolution of, 240–241, 270
  in skin examination, 279–288
  of skin thickness, 324, 661
  versus skinfold caliper and radiography, 242
  technical developments in, 289
  tissue characterization in, 259–267
  torque measurement in, 324
  uses of, 3, 233–234
  for venous obstruction, 438–439, 440
  of wheals, 266–267

Ultrasound beam, 289
Ultrasound equipment, 257–259
Ultrasound phenomena, 271
Ultrasound scanner, 239, 257–259
Ultrasound wave velocity, 241, 289–290
Ultraviolet A dosimeters, 622–623
Ultraviolet B erythema, 27
Ultraviolet L flux, 28
Ultraviolet lamps, 621
Ultraviolet light, 373
Ultraviolet photography, 62–63
Ultraviolet radiation
   aging and, 27
   A wavelength, 27–28, 621–623, 627–628
   B wavelength, 27, 28, 381–382, 621, 627–629
   biological effective units of, 619–620
   C wavelength, 381–382, 621
   detection of, 620–621
   dose-response curves of erythema and pigmentation with, 393–395
   erythema and, 389
   exposure time to, 619
   for follicular function, 513–514
   measuring pigmentation induced by, 393
   reasons for measuring, 619
   skin mechanical properties and, 332–333
   sunscreen evaluation with, 375
Ultraviolet radiation detectors, 620–621
Ultraviolet radiation dosimetry, personal, 624–625
Ultraviolet radiation-induced erythema, 390–391
Uniaxial tension tests, 343
Unit area trichogram, 545, 546
Urticaria
   cholinergic, 174
   increased skin temperature with, 468
   solar, 629, 631
   wheals of, 247–248

V

Validation, 18
Validity, 33–34, 40
Varicose veins, 453, 454
Vascular permeability, 637–638
Vascular surgery, 427
Vascular system
   in skin atrophy, 52–53
   sympathetic control of, 467
   ulceration, 251–252
Vasoactive agents, 4–5, 607
Vasoconstrictor assay, 392–393, 661
Vasomotion, 408–409, 411–417
Vasomotor waves, 408–409
Vasospastic reaction, 470
Vasospasticity assessment, 469
Vellus Index, 546
Velocimetry, 399
Velocity, 360–361
Velvet pads, 212
Venoarteriolar response, 408
Venography
   ascending, 438
   ultrasound-guided, 280
Venous disease, 437–440
Venous incompetence, 440
Venous insufficiency
   chronic
      causes of, 437
      vasomotion in, 415–416
   photoplethysmography and light-reflection rheography for, 443–448
   ultrasound examination of, 251–252
Venous leg ulcer, 251–252
Venous obstruction, 438–440
Venous occlusion plethysmography, 402, 406, 434, 453
Venous pattern mapping, 61
Venous pressure, 415–416
Venous reflux tests, 437–438
Venous ulceration
   cleanliness and healing of, 581–582
   transcutaneous oxygen tension in, 189–190
Venules, 408
Verification bias, 38–40
Verrucous hypertrophic lichen planus growth, 302
Vibration stimulators, 484
Vibrational sensation, 483, 484
Video image analysis, 154–155
Video imaging, 514
Villiform projections, 76
Viscoelastic deformation, delayed, 323
Viscoelastic materials, 353
Viscoelastic properties, 339
Viscoelastic ratio (VER), 11
Viscoelastic structure, 331
Viscoelasticity, 325, 335
Viscosity, 353
   of dermal fluid, 351
   parameters of, 321–322
   shear, 360–361
Visible lasers, 485
Visible light, 629
Visible-light imaging, 234–236
Vitiligo, 375
Vitiligo lesions, 62–63
Volume flux measurement, 406, 407–408
Volume of distribution, 480
Voxel reconstruction, 308

W

Washing bacteria sampling methods, 210–211
Washout model, 431–432
Water, see also Hydration
   behavior in epidermis and dermis, 135–136
   as coupling medium, 127
   evaporation of, 179
   gravimetric determination of, 160–161
   in horny layer pliability, 7
   measurement of in stratum corneum, 159–164
   in skin elasticity, 331
Water barrier, 19
   defective function of, 654
   deficient, 162
Water sorption-desorption test, 162, 163
Water vapor
   boundary layer for, 174
   mantel, 19
   passing skin, 173–177, see also Transepidermal water loss
Waveform pattern analysis, 414
Wavelength, 619–620
Weathering, 531–532
Wheal-and-flare reaction, 607, 611–612
Wheals
   antihistamines and, 609